Conversion Factors (continued)

Volume	$1\ \text{cm}^3 = 10^{-6}\ \text{m}^3 = 3.53 \times 10^{-5}\ \text{ft}^3 = 6.10 \times 10^{-2}\ \text{in}^3$
	$1\ \text{m}^3 = 10^6\ \text{cm}^3 = 10^3\ \text{L} = 35.3\ \text{ft}^3 = 6.10 \times 10^4\ \text{in}^3 = 264\ \text{gal}$
	$1\ \text{L} = 10^3\ \text{cm}^3 = 10^{-3}\ \text{m}^3 = 1.056\ \text{qt} = 0.264\ \text{gal}$
	$1\ \text{in}^3 = 5.79 \times 10^{-4}\ \text{ft}^3 = 16.4\ \text{cm}^3 = 1.64 \times 10^{-5}\ \text{m}^3$
	$1\ \text{ft}^3 = 1728\ \text{in}^3 = 7.48\ \text{gal} = 0.0283\ \text{m}^3 = 28.3\ \text{L}$
	$1\ \text{qt} = 2\ \text{pt} = 946.5\ \text{cm}^3 = 0.946\ \text{L}$
	$1\ \text{gal} = 4\ \text{qt} = 231\ \text{in}^3 = 3.785\ \text{L} = 0.134\ \text{ft}^3$
Time	$1\ \text{h} = 60\ \text{min} = 3600\ \text{s} = 4.167 \times 10^{-2}\ \text{d}$
	$1\ \text{d} = 24\ \text{h} = 1440\ \text{min} = 8.64 \times 10^4\ \text{s}$
	$1\ \text{y} = 365\ \text{d} = 8.77 \times 10^3\ \text{h} = 5.26 \times 10^5\ \text{min} = 3.16 \times 10^7\ \text{s}$
Angle	$360^\circ = 2\pi\ \text{rad}$
	$180^\circ = \pi\ \text{rad}$ $1\ \text{rad} = 57.3^\circ$
	$90^\circ = \pi/2\ \text{rad}$
	$60^\circ = \pi/3\ \text{rad}$ $1^\circ = 0.0175\ \text{rad}$
	$45^\circ = \pi/4\ \text{rad}$
	$30^\circ = \pi/6\ \text{rad}$
Speed	$1\ \text{m/s} = 3.6\ \text{km/h} = 3.28\ \text{ft/s} = 2.24\ \text{mi/h}$
	$1\ \text{km/h} = 0.278\ \text{m/s} = 0.621\ \text{mi/h} = 0.911\ \text{ft/s}$
	$1\ \text{ft/s} = 0.682\ \text{mi/h} = 0.305\ \text{m/s} = 1.10\ \text{km/h}$
	$1\ \text{mi/h} = 1.467\ \text{ft/s} = 1.609\ \text{km/h} = 0.447\ \text{m/s}$
	$60\ \text{mi/h} = 88\ \text{ft/s}$
Force	$1\ \text{newton} = 10^5\ \text{dyn} = 0.225\ \text{lb}$
	$1\ \text{dyne} = 10^{-5}\ \text{N} = 2.25 \times 10^{-6}\ \text{lb}$
	$1\ \text{lb} = 4.45 \times 10^5\ \text{dyn} = 4.45\ \text{N}$
	Equivalent weight of 1-kg mass $= 2.2\ \text{lb} = 9.8\ \text{N}$
Pressure	$1\ \text{Pa}\ (\text{N/m}^2) = 1.45 \times 10^{-4}\ \text{lb/in}^2 = 7.5 \times 10^{-3}\ \text{torr (mm Hg)} = 10\ \text{dyn/cm}^2$
	$1\ \text{torr (mm Hg)} = 133\ \text{Pa}\ (\text{N/m}^2) = 0.02\ \text{lb/in}^2 = 1333\ \text{dyn/cm}^2$
	$1\ \text{atm} = 14.7\ \text{lb/in}^2 = 1.013 \times 10^5\ \text{N/m}^2 = 1.013 \times 10^6\ \text{dyn/cm}^2$
	$= 30\ \text{in. Hg} = 76\ \text{cm Hg}$
	$1\ \text{bar} = 10^5\ \text{N/m}^2 = 10^6\ \text{dyn/cm}^2$
	$1\ \text{millibar} = 10^2\ \text{N/m}^2 = 10^3\ \text{dyn/cm}^2$
Energy	$1\ \text{J} = 10^7\ \text{ergs} = 0.738\ \text{ft-lb} = 0.239\ \text{cal} = 9.48 \times 10^{-4}\ \text{Btu} = 2.78 \times 10^{-7}\ \text{kWh}$
	$1\ \text{kcal} = 4186\ \text{J} = 3.968\ \text{Btu}$
	$1\ \text{Btu} = 1055\ \text{J} = 778\ \text{ft-lb} = 0.252\ \text{kcal}$
	$1\ \text{cal} = 4.186\ \text{J} = 3.97 \times 10^{-3}\ \text{Btu} = 3.09\ \text{ft-lb}$
	$1\ \text{ft-lb} = 2.69 \times 10^6\ \text{J} = 1.29 \times 10^{-3}\ \text{Btu}$
	$1\ \text{eV} = 1.60 \times 10^{-19}\ \text{J}$
Power	$1\ \text{W} = 0.738\ \text{ft-lb/s} = 1.34 \times 10^{-3}\ \text{hp} = 3.41\ \text{Btu/h}$
	$1\ \text{ft-lb/s} = 1.36\ \text{W} = 1.82 \times 10^{-3}\ \text{hp}$
	$1\ \text{hp} = 550\ \text{ft-lb/s} = 745.7\ \text{W} = 2545\ \text{Btu/h}$

Technical College Physics

Third Edition

Technical College Physics

Third Edition

Jerry D. Wilson

Chair, Division of Biological and Physical Sciences
Lander College, Greenwood, South Carolina

Saunders College Publishing

A Harcourt Brace Jovanovich College Publisher

Fort Worth Philadelphia San Diego New York Orlando Austin San Antonio
Toronto Montreal London Sydney Tokyo

Text Typeface: Caledonia
Compositor: Bi-Comp, Inc.
Acquisitions Editor: John Vondeling
Developmental Editor: Ellen Newman
Managing Editor: Carol Field
Senior Project Manager: Marc Sherman
Manager of Art and Design: Carol Bleistine
Art Director: Christine Schueler
Art and Design Coordinator: Caroline McGowan
Text Designer and Layout Artist: Rebecca Lemna
Cover Designer: Lawrence R. Didona
Text Artwork: J & R Technical Services
Production Manager: Charlene Squibb
Director of EDP: Tim Frelick
Marketing Manager: Marjorie Waldron

Cover: A carbon dioxide laser cutting steel. © Chuck O'Rear/Westlight

Printed in the United States of America

TECHNICAL COLLEGE PHYSICS, 3rd edition

ISBN 0-03-073898-9

Library of Congress Catalog Card Number: 91-052811

234 071 98765432

THIS BOOK IS PRINTED ON **ACID-FREE, RECYCLED** PAPER

Preface

The third edition of *Technical College Physics*, like the previous editions, presents the basic principles of introductory physics, along with many technical applications. Not only does the student gain a foundation in physical principles, but also gleans insight into how these principles are used—the basics plus applications.

Changes in the Third Edition

A number of changes and improvements have been made in preparing the third edition of this text. Many of these changes are in response to comments and suggestions offered by users of the second edition and reviewers of the manuscript. The following represent the major changes in the third edition:

1. New two-color design, which is used to highlight important concepts throughout the text. Color is also used in the figures to improve clarity and understanding.
2. Up-to-date information about current topics, such as MagLev, the train of the future.
3. Some sections have been rewritten and reorganized to improve clarity according to reviewers' suggestions.
4. Problems have been revised and keyed to specific sections of the text. They are graded in difficulty for the instructor's convenience in assigning problems.
5. Additional emphasis on SI system units in Chapter 1.
6. All problems in the text and answers in the Instructor's Resource Manual have been checked for accuracy.

The topic coverage is still of sufficient scope to allow the instructor some choice of course content in a three-quarter or two-semester sequence. Also, there are special interest topics. For example, in the Electricity and Magnetism section, you will find such topics as household circuits, wire gages, electrical safety, induction and dielectric heating, and magnetic levitation (the MagLev train); in the Optics section, nonreflecting lenses, fiber optics, color, and LCDs; in the Modern Physics section, xerography and electrostatic copiers, nuclear waste and proliferation, and solar energy technology. Scan the Table of Contents for yourself.

Some special pedagogical features of the text are:

- A section on problem solving
- Solved example problems in each section
- Numerous illustrations and photographs of principles and technical applications
- Use of both SI and British systems of units, but with comprehensive coverage and explanation of the SI system in the first chapter
- Summary and definitions of important terms and summary of important formulas at the end of each chapter

- End-of-chapter Questions and Problems (with level of difficulty indicated) listed by section
- Special Features and Chapter Supplements, including

  The Meter
  Automobile Efficiency
  Anti-Lock Brakes
  Radar and the Doppler Effect
  Heat Pump Cooling and Heating
  Superconductivity (updated)
  Personal Safety and Electrical Effects
  Lasers in the Supermarket

Nor is the historical background of physics neglected. There are Special Features on:

Isaac Newton
Galileo Galilei
Marie and Pierre Curie

Did Galileo really drop things from the Tower of Pisa? See

Galileo and the Leaning Tower of Pisa

Ancillaries

The following ancillaries are available with this text:

Instructor's Resource Manual with Test Bank This manual contains solutions to all problems as well as answers to all questions in the text. The Test Bank includes 25 questions and 2 prepared quizzes for each chapter; questions are varied and feature completion, multiple choice, and mathematical problems where applicable. The format of the Test Bank allows instructors to duplicate pages for distribution to students. In addition, 100 transparency masters containing figures from the text are included for the instructors use in the classroom. The manual has been reviewed for accuracy.

ExaMaster™ Computerized Test Bank Available for the IBM computers, the test bank contains approximately 1000 open-ended and multiple-choice problems, representing every chapter of the text. The test bank allows instructors to customize tests by rearranging, editing, or adding questions. The software program solves all problems and prints the answers on a separate grading key.

Student Study Guide The Study Guide contains chapter summary, sample problems with solutions, and worked-out solutions to selected problems from the text.

The preparation of a text requires a great deal of assistance. I greatly appreciate the photographs of technical applications and products supplied by many companies. The text was improved by the helpful comments and suggestions of Professor Baher Hanna, The Michael J. Owens Technical College; Professor Lawrence Josbeno, Corning Community College; Professor Fred B. Otto, Maine Maritime Academy; and Professor Michael W. Wolf, Embry-Riddle Aeronautical University. I appreciate the careful accuracy review of all the problems in the text and in the Instructor's Resource Manual by Roger L. Hanke, North Central Technical College. Thanks are also extended to the editorial and production staff at Saunders College Publishing, in particular Ellen Newman and Marc Sherman.

Jerry D. Wilson
Lander College
November 1991

Contents

* Indicates Optional Section

Highway distance shown in British and metric units.

Chapter 1

Measurement and Systems of Units

When you can measure what you are speaking about and express it in numbers, you know something about it; but when you cannot measure it, when you cannot express it in numbers, your knowledge is of meager and unsatisfactory kind.

Lord Kelvin

1.1 What We Measure

How tall are you? How much do you weigh? What time is it? The answers to these and many other such questions require that measurements be made. Rarely does a day go by in which we do not make or use measurements. Many times we are not aware that we are doing this. For example, when you tell someone what time it is, you are stating a measurement. Your clock is the measurement instrument.

Let's think of some things we commonly measure. With a little thought, you might say length. (I am 5 feet 8 inches tall, or I live about three miles or five kilometers from school.) Some of us frequently weigh ourselves. (I weigh 160 pounds or 712 newtons.) And then there is time. (Class periods are 50 minutes long.)

When buying gas for your car, you may buy 10 gallons of gasoline. So we should add volume or capacity to our list of commonly measured items. Summarizing, we have

Commonly Measured Items

Length

Weight (mass)*

Time

Volume or capacity

There are other things we measure, but let's keep the list simple for our discussion. It should be noted that volume really involves length measurements. Recall that the volume (V) of a rectangular box is its length (l) times its width (w) times its height (h) or, in an equation, $V = l \times w \times h$. However, width and height are lengths too. So volume is a combination of lengths.

In science and technology, things are described as simply as possible. This is done in terms of basic **fundamental properties,** which include length, mass, and time. These properties describe the concepts of *space, matter,* and *time.*

> **Length describes an object's size and specifies its position in space.**

> **Mass is the quantity of matter an object contains.**

> **Time is an involved concept and is sometimes characterized as the continuous, forward flow of events. Time "flows" forward, never backward. Two events define an interval of time.**

> **(There is another fundamental property that is associated with electricity, the electric charge. More will be said about this property in Chapter 19.)**

The vast majority of the things we observe in nature can be measured or described in terms of these four fundamental properties and their various combinations.

Mass and Weight

Before discussing how we make measurements, let's first distinguish between mass and weight, since there is an important difference.

> **Mass is the quantity of matter an object contains.**

> **Weight is the force of gravitational attraction on an object by some celestial body, most commonly Earth.**

Of course, mass and weight are related—the greater the mass of an object, the greater its weight. The relationship between weight (w) and mass (m) is expressed by the equation

$$w = mg \qquad \textbf{(Eq. 1.1)}$$

where g is called the acceleration due to gravity and is taken to be constant near the surface of the Earth. (More will be said about this later.)

Mass is the fundamental property, since in general the mass of an object does not change. However, an object's weight can change due to variations in the value of g. For example, an object will have the same mass or quantity of matter on Earth and on the moon; but the object will weigh only about $\frac{1}{6}$ as much on the moon as on Earth (Fig. 1.1). This is because the value of g on the moon (g_m) is $\frac{1}{6}$ the value of g on Earth ($g_m = \frac{1}{6}g$).

Notice from Figure 1.1 that when we "mass" an object with a balance, we are comparing it with an object of known mass. In a balanced condition, the weights of the objects are also equal, i.e., $w_1 = m_1g = m_2g = w_2$. As can be seen, the g's cancel and $m_1 = m_2$, whether on Earth or on the moon. However, the weights as determined by spring scales are different for the different cases. If you weigh 180 lb on Earth, you would weigh only 30 lb on the moon, since $g_m = \frac{1}{6}g$, but your mass would be the same.

Density

The mass or quantity of matter that an object contains depends not only on its size, but also on how compact the matter in the object is. For example, a block of iron has more mass than a block of ice of equal size or volume. For a block of ice to have equal mass, it would have to be several times the size of the iron block.

The compactness of matter in a substance is expressed in terms of its **mass density** (ρ), which is the mass (m) per volume (V).

$$\rho = \frac{m}{V} \qquad \textbf{(Eq. 1.2)}$$

* The distinction between mass and weight will be made shortly.

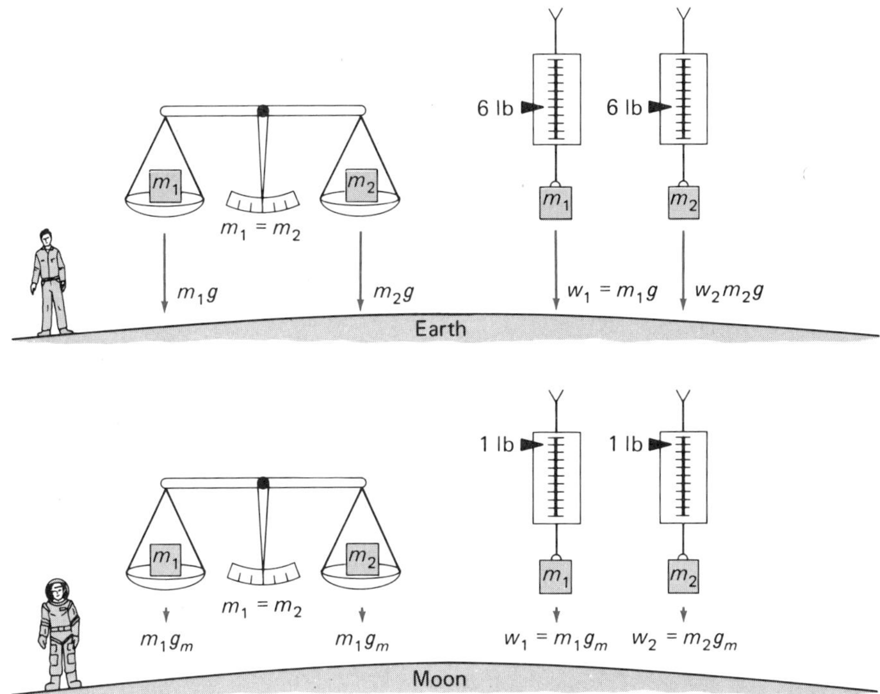

Figure 1.1
Mass is a fundamental property and weight is not. Two objects with the same mass have the same quantity of matter, on either the Earth or the moon. However, their weights are less on the moon because the gravitational attraction there is less, about 1/6 that on Earth.

Density is not a fundamental property, but a combination of these properties (mass and length), and it is a **derived quantity.**

The density of iron is about $8\frac{3}{4}$ times that of ice, or iron is $8\frac{3}{4}$ times denser than ice. This means that for a given volume, iron has $8\frac{3}{4}$ times more matter than ice. (How much larger would a block of ice have to be if it had a mass equal to that of a block of iron?)

A weight density is often used in engineering applications. The **weight density** (D) of an object is its weight per volume:

$$D = \frac{w}{V} \qquad \text{(Eq. 1.3)}$$

Since $w = mg$, we have $D = w/V = mg/V = (m/V)g = \rho g$, or the weight density is just the mass density ρ times g.

We often say that density is the mass (or weight) per *unit* volume. We'll come back to this once we have defined some units.

1.2 Units and Systems of Units

Now that we know what we measure, how do we do it? It's really a matter of choice. For example, a table has a certain length no matter how we choose to describe it. One person might measure the table in feet, another in yards, and still another in meters. Certainly the length of the table doesn't change, only the choice of units used to describe it.

The measurement *unit* is the key word. If a particular unit becomes popular and/or officially accepted, it becomes a standard unit. **A standard unit has a fixed and reproducible value for the purpose of taking accurate measurements.** For example, the foot and the meter are standard units. To measure something in feet, we compare it to a standard foot ruler or yardstick. To measure something in meters, we compare it to a meterstick standard (Fig. 1.2).

But who establishes or chooses a standard unit? Traditionally, it has been the head of state

Figure 1.2
A measurement is a quantitative comparison of a fundamental property to a standard. The standard units of measurement may be different, but the length of the table is the same in any case.

or government. Early standards were referenced to parts of the human body. (What could be more convenient?) Some units of the British, or English, system, which is the customary system of units in the United States, originated from anatomical references. For example, the inch was referenced to the thumb (Fig. 1.3), which of course varied from person to person.

In the 1300's King Edward II of England decreed the official inch to be equal to three barleycorns taken from the middle of the ear and laid end to end (which was not a great improvement). He also decreed the foot to be equal to 12 three-barleycorn inches. (Perhaps this was the length of his royal foot.) Another English monarch, King Henry I, established the yard as the distance from the tip of his nose to the end of the middle finger of his outstretched arm.

Later, material standards were made. King Henry VII had an iron bar made that was to be used as the yard standard (the first yardstick?). Today, most governments have agencies that maintain and establish material standards for common and scientific measurements. In the United States, this is the responsibility of the National Institute of Standards and Technology (formerly the National Bureau of Standards) of the Department of Commerce.

Systems of Units

There are two major systems of units in the world today—the British and metric systems. As you are no doubt aware, the United States is in a unit transition period. There is a great deal of discussion on the pros and cons of conversion from the British system to the metric system. Regardless of the controversy, metric units are coming into increasingly common use. Hence, it is important for a person in a technical field to be familiar with metric units.

The metric system will be emphasized in the text so you can become more familiar with its units. However, the British system will not be completely ignored, since you will still commonly use these units in many everyday measurements. By learning the metric system, you will be "ahead of the game" when an official conversion takes place. Many think that this will be relatively soon, even though it will involve an enormous cost for retooling in order to change standards. But, as may be seen from Figure 1.4, we are an island in a metric world, and international trade and commerce exert a great pressure for conversion.

The next section will show how easy it is to convert from one system to the other, once you have learned the metric units.

The Metric System

The need for a more uniform and convenient system of units led to the development of the metric system, which is now used in most countries around the world. Let's take a look at the length, mass, volume, and time units in the metric system, along with their British counterparts.

Length

The metric standard of length is the **meter** (see Special Feature 1.1). The meter, abbreviated m, is a little longer than a yard—3.37 inches longer (Fig. 1.5).

With a length standard selected, the next job is to define submultiple and multiple units. For

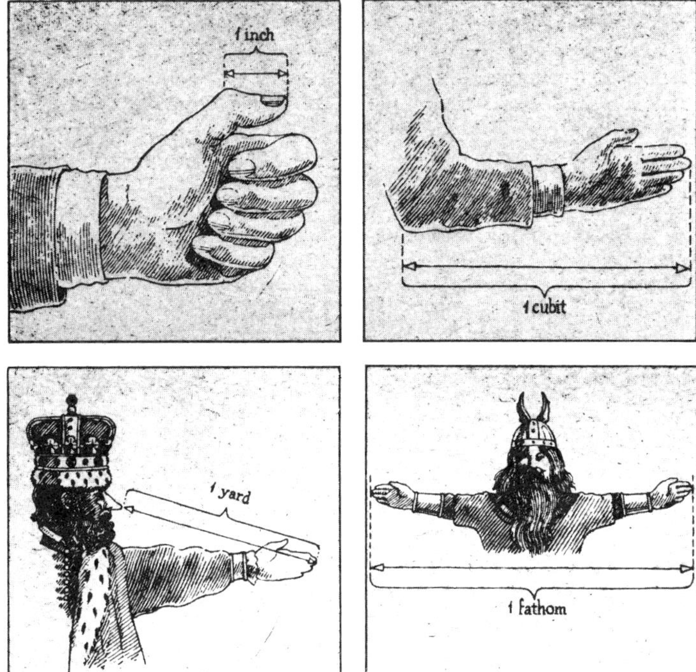

Figure 1.3
Anatomical units. Many units were originally defined in reference to parts of the human body. Variations from person to person gave rise to the need for standard units. (Courtesy Hagley Museum, Wilmington, Delaware.)

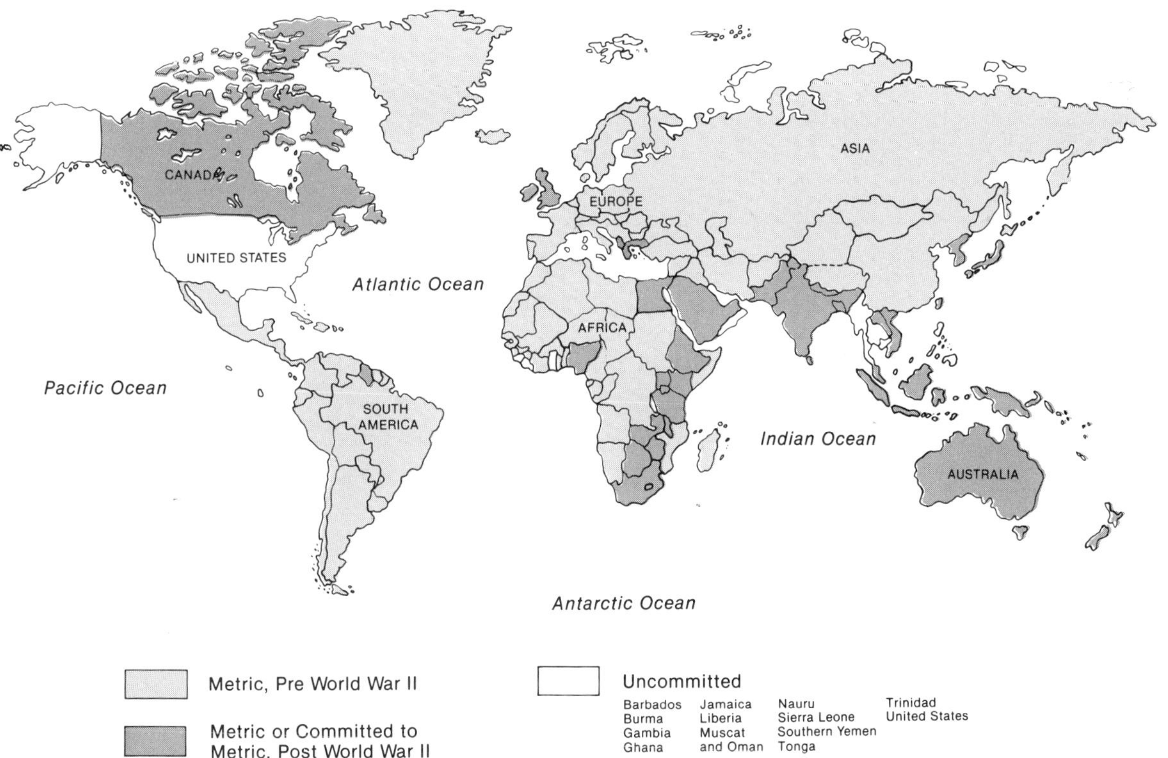

Figure 1.4
Islands in a metric world. The map shows the few countries that are uncommitted to, or have not officially adopted, the metric system. (Courtesy U.S. Dept. of Commerce.)

SPECIAL FEATURE 1.1

The Meter

In 1790, in the midst of the French Revolution, the National Assembly of France requested that the French Academy of Sciences "deduce an invariable standard for all the measures and all the weights." The commission appointed by the Academy proposed a system that was simple and scientific. The name *metre*, which we spell meter, was assigned to the unit of length. This name was derived from the Greek word *metron*, meaning "to measure."

The length of the meter was defined as one ten-millionth of the distance along a meridian from the North Pole to the Equator (Fig. 1). A portion of a meridian running near Dunkirk in France and Barcelona in Spain was surveyed and the length of a meter determined. Based on this result, a 1-meter bar of platinum was constructed. This bar became the "Meter of the Archives," from which copies were made.

The use of metric weights and measures was legalized in the United States in 1866, and since 1893 the yard has been defined in terms of the meter. Metal bar meter lengths are used for common measurement reference standards, but these lengths are affected by temperature variations. In 1960, the meter was defined in terms of the wavelength of light. In 1983, a new definition was adopted that references the meter to the distance light travels in a vacuum (see Table 1.3).

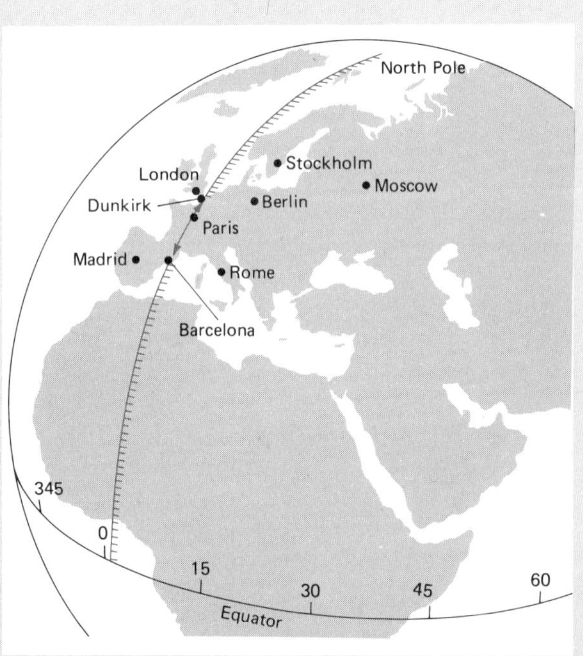

Figure 1
Definition of the meter. The meter was originally defined as one ten-millionth of the distance from the North Pole to the Equator along a meridian that ran through France.

example, in the British system, 1 ft = 12 in. and 3 ft = 1 yd.* The metric system is a decimal, or base-10, system. That is, larger and smaller units are obtained by multiplying or dividing standard units by factors of ten.

A list of the metric prefixes used to indicate these factors is given in Table 1.1. However, only three prefixes are usually needed to describe everyday measurements of length (see Fig. 1.6):

milli—0.001 (1/1000)
1 millimeter (mm) = 0.001 meter
or 1 m = 1000 mm

centi—0.01 (1/100)
1 centimeter (cm) = 0.01 meter
or 1 m = 100 cm

kilo—1000 (pronounced kil-oh-meter)
1 kilometer (km) = 1000 m
or 1 m = 0.001 km

Let's take a closer look at the decimal base of the metric system, which is one of its greatest advantages. You are already familiar with a similar decimal system—our money system. The

* These relationships are called *equivalence statements*. The two sides of an *equation* must be equal in both magnitude (numerically) and units. Although an equal sign is frequently used in equivalence statements, e.g., 1 yd = 36 in., this really means "one yard is equivalent to 36 inches," or "the same length as."

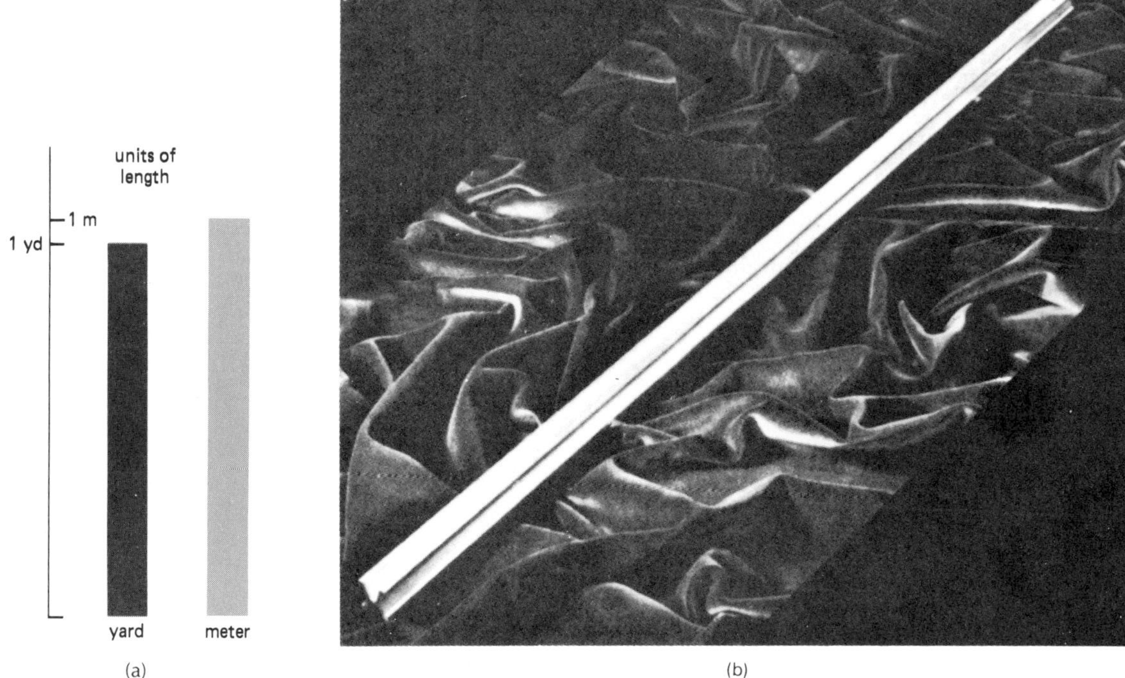

Figure 1.5
The meter and the yard. (a) The meter is slightly longer than a yard (3.37 inches longer). (b) The prototype meter bar that is the United States copy of the Standard Meter (Prototype Meter No. 27). The bar was sent from France in 1890. (Courtesy U.S. Dept. of Commerce.)

Table 1.1
Metric Prefixes

Prefix (Abbreviation)	Pronunciation*	Value	Meaning
exa (E)	ex'a (*a* as in *a*bout)	10^{18}	One quintrillion times
peta (P)	as in *peta*l	10^{15}	One quadrillion times
tera (T)	as in *terra*ce	10^{12}	One trillion times
giga (G)	jig'a (*a* as in *a*bout)	10^{9}	One billion times
mega (M)	as in *mega*phone	10^{6}	One million times
kilo (k)	as in *kilo*watt	10^{3}	One thousand times
hecto (h)	heck toe	10^{2}	One hundred times
deka (da)	deck'a (*a* as in *a*bout)	10	Ten times
deci (d)	as in *deci*mal	10^{-1}	One tenth of
centi (c)	as in *senti*ment	10^{-2}	One hundredth of
milli (m)	as in *mili*tary	10^{-3}	One thousandth of
micro (μ)	as in *micro*phone	10^{-6}	One millionth of
nano (n)	nan'oh (*an* as in *an*t)	10^{-9}	One billionth of
pico (p)	peek'oh	10^{-12}	One trillionth of
femto (f)	fem'toe (*fem* as in *fem*inine)	10^{-15}	One quadrillionth of
atto (a)	as in an*at*omy	10^{-18}	One quintrillionth of

* Source: Metric Guide for Educational Materials, American National Metric Council.

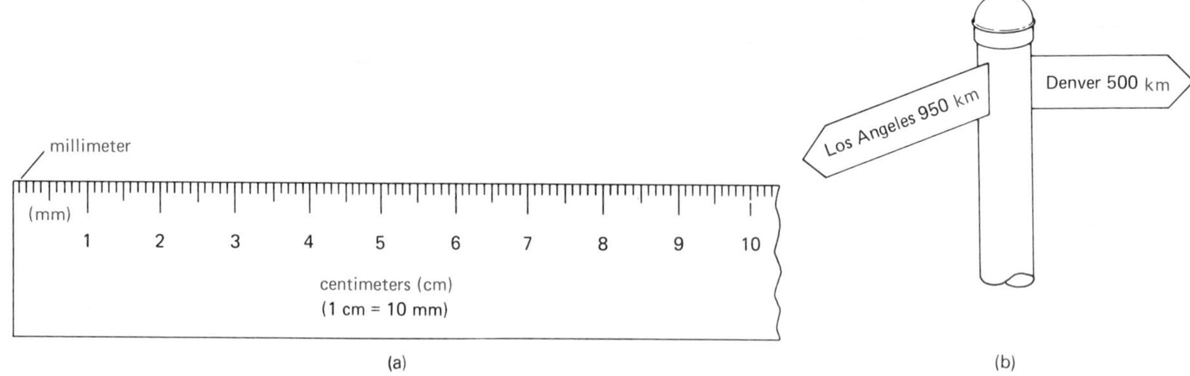

Figure 1.6
Metric prefixes. Only three metric prefixes are needed to describe most everyday measurements: (a) millimeter (mm) and centimeter (cm), and (b) kilometer (km).

dollar is divided into cents, with 100 cents (pennies) making one dollar. If a dollar is compared to a meter, then a cent or penny is comparable to a centimeter. In fact, we could call a penny a "centidollar." For example,

150 cm = 1.50 m and 150 cents = $1.50

We can carry this analogy a step further. You may have heard how property taxes (or school-bond levies) are assessed in mils. A mil is $\frac{1}{10}$ of a cent, and there are 1000 mils in a dollar. Hence a mil, or "millidollar," is analogous to a millimeter.*

Notice how much easier it is to convert from one unit to another in the metric decimal system than in the British system. (The British system is a duodecimal, or base-12, system with 12 inches in one standard foot.) For example, 118 cm can be directly determined to be 1.18 m. Can you tell as quickly how many feet there are in 118 inches?

Mass and Volume

In the metric system, the mass standard was originally related to length. The quantity of water in a particular metric volume was originally used to define the standard metric mass unit. A container 10 cm on a side has a volume of 10 cm × 10 cm × 10 cm = 1000 cm^3 (cubic centimeter, sometimes abbreviated cc). See Figure 1.7. The mass of the quantity of water that filled the container (1000 cm^3) was defined to be 1 **kilogram** (kg).†

Since the metric prefix "kilo-" means 1000, it follows that 1 kg = 1000 grams (g), and one cubic centimeter of water has a mass of 1 g. The gram unit is often divided into milligrams (1 g = 1000 mg or 1 mg = 0.001 g), which is a convenient unit for small quantities. For larger quantities, a **metric ton** (sometimes written *tonne*) is defined to be 1000 kg.

Density is the mass per *unit* volume. In the metric system standard units, this would be kg/m^3. That is, for a substance with a uniform density, each cubic meter (m^3) would contain a certain number of kilograms (kg). For example, the density of iron is $\rho_{iron} = 7900\ kg/m^3$, or each cubic meter (unit volume) contains 7900 kg of iron. On a smaller scale, $\rho_{iron} = 7.9\ g/cm^3$, or each cubic centimeter contains 7.9 g of iron. Notice that the density of water is $\rho_{water} = 1.0\ g/cm^3$ in these units. Why?

The unit of mass in the British system is operationally defined in terms of weight or force. As

* The mil is sometimes used in British length measurement. Here, 1 mil = $\frac{1}{1000}$ (0.001) inch.

† At standard atmospheric pressure and at the temperature of the maximum density of water (4°C). See Chapter 17.

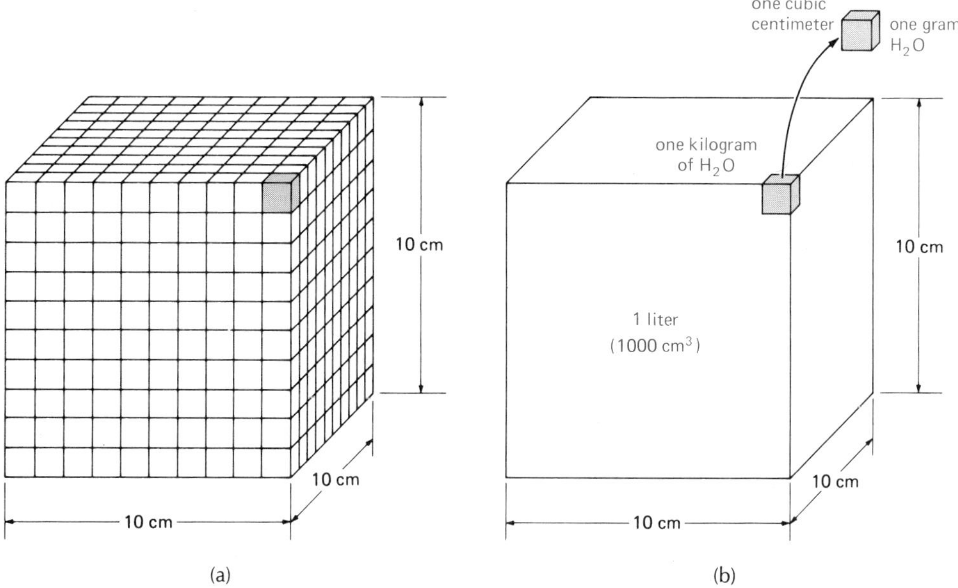

Figure 1.7
Mass units are related to volume in the metric system. (a) A cube 10 cm on a side has a volume of 1000 cm^3. (b) The amount of water that fills this volume has a mass of 1 kilogram, and 1 cm^3 of water has a mass of 1 gram. (The volume 1000 cm^3 is defined to be a liter.)

a result, the British system is said to be a gravitational system. The standard unit of force or weight is the pound (lb). The standard unit of mass, the *slug*, is then defined by the pound through Equation 1.1, $w = mg$, using a standard value of g.* A mass of one slug on Earth has a weight of

$$w = mg = (1 \text{ slug})(32 \text{ ft/s}^2) = 32 \text{ lb}$$

where the value of g in the British system is 32 ft/s^2. It is possible that you may never have heard of the standard British unit of mass since it is rarely used. Matter is generally described in terms of weight (pounds) in the British system. Notice how the value of g is a combination of standard units. Such combinations are referred to as **derived units.**

* It is unfortunate that the official abbreviation for the gram (g) is the same as the commonly used symbol for the acceleration due to gravity (g). Notice that the g for the acceleration due to gravity is italic, since it is in general a variable, whereas the g for gram is not. When working with both quantities, gram is sometimes abbreviated gm for distinction.

The relationship between the kilogram and the pound is that 1 kg mass has an equivalent weight of 2.2 lb on the surface of the Earth.

This relationship is illustrated in Figure 1.8, along with the metal prototype kilogram standard. The unit of force in the metric system is the newton (N). One kilogram has a *weight* force of 9.8 N since $w = mg = (1 \text{ kg})(9.8 \text{ m/s}^2) = 9.8$ N, where $g = 9.8 \text{ m/s}^2$ in the metric system. We will say more about this later.

The metric unit of volume or capacity is the volume used in defining the kilogram. **A volume of 1000 cm^3 is defined to be one liter.** The abbreviation for the liter is either a small "el" (l) or a capital L. The latter is generally preferred in the United States so as to avoid confusion with the numeral one. A liter is a bit larger than a U.S. quart (Fig. 1.9). We commonly write

$$1 \text{ L} = 1.056 \text{ qt}$$

A common smaller unit of volume is the milliliter (mL). From the previous discussion, we know that

$$1 \text{ L} = 1000 \text{ cc} = 1000 \text{ mL}$$

and

$$1 \text{ cc} = 1 \text{ mL}$$

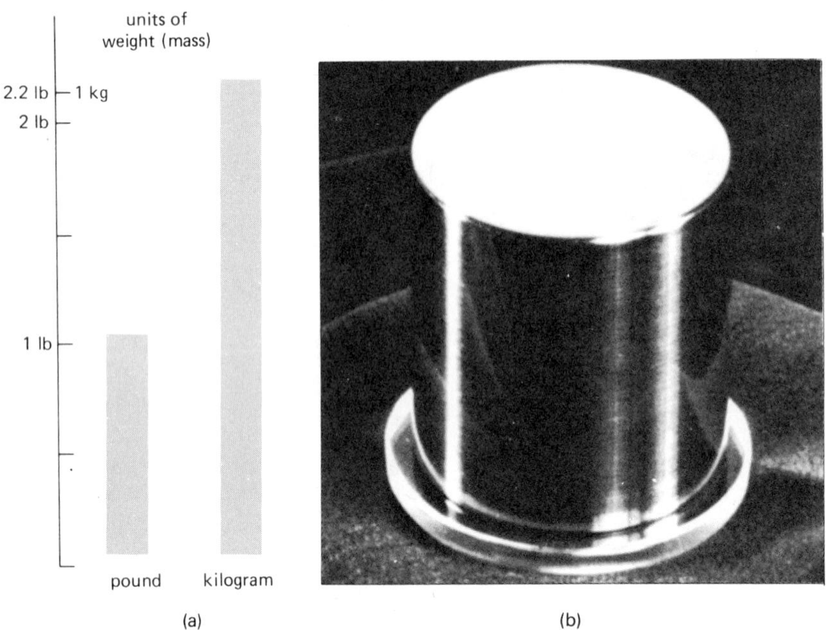

Figure 1.8
The kilogram and the pound. (a) The metric kilogram unit of mass has an equivalent weight of 2.2 lb. (b) The prototype kilogram standard cylinder (Prototype No. 20) is the national standard mass for the United States. (Courtesy U.S. Dept. of Commerce.)

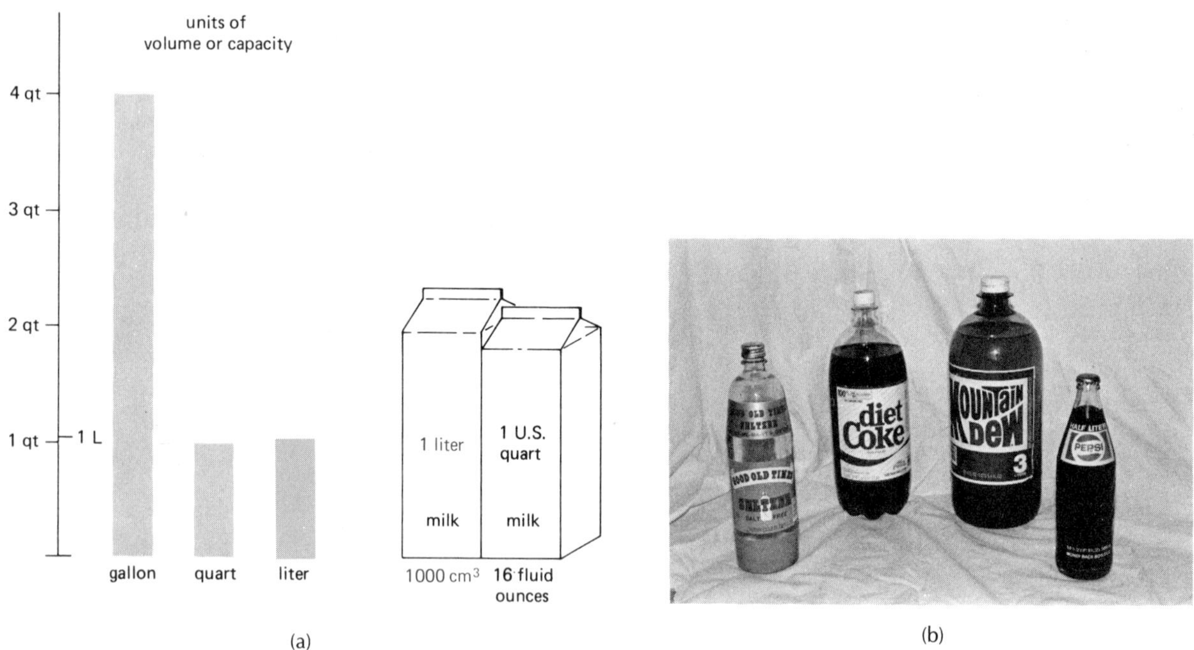

Figure 1.9
The liter and the quart. (a) The liter is slightly larger than the quart (1 L = 1.056 qt). (b) One, two, three, and one-half liters. The liter is in common use in the United States.

Table 1.2
Standard Units in Measurement Systems

Fundamental Property	British (English) fps (foot-pound-second)	Metric mks or SI (meter-kilogram-second)
length	foot (ft)	meter (m)
mass	slug*	kilogram (kg)
time	second (s)	second (s)

* The slug was originally defined in terms of the pound force (weight). A force of one pound (lb) acting on a mass of one slug produces an acceleration of one foot per second squared.

Time

The standard unit of time may give you some relief. This is the **second** (s) in both the British and metric systems.* Two events define an interval of time, similar to two marks or positions defining a length interval. (Notice how we often say, for example, that class periods are 50 minutes *long* or 50 minutes in *length*.)

Prior to 1956, the apparent daily motion of the Sun was used to define the second. A solar day is the time that elapses between two successive crossings of the same meridian by the Sun. A second was defined as 1/86,400 of an *apparent* solar day. (One day = 24 hours = 1440 minutes = 86,400 seconds.)

However, the Earth's motions vary slightly, and not all solar days are the same length. To remedy this, an average, or mean, solar day was computed from the apparent solar days during a one-year period, and this average was used as the length of a solar day.

Even so, the mean solar day is not exactly the same in each yearly period, owing to minor variations in the Earth's motions plus a steady slowing of the Earth's rate of rotation due to tidal friction. An atomic standard is now used for a more precise time reference, as was done for the length standard (see Table 1.3).

A summation of the standard units in the metric and British systems is given in Table 1.2. A subsystem of metric units is sometimes used when small quantities are involved. This is the cgs (*c*entimeter-*g*ram-*s*econd) system.

* The units of electricity are the same in both systems too. The unit of electric charge is the coulomb, and the unit of electric current is the ampere (see Chapters 19 and 21).

SPECIAL FEATURE 1.2

The SI System

The SI system is the *modernized* version of the metric system, established by international agreement. SI stands for the International System of Units (from the French, *Le Système International d'Unités*). It was initiated by a request from the International Union of Pure and Applied Physics and various other scientific organizations for the adoption of a practical international system of units—that is, an international "language" by which everyone could communicate in the same unit quantities.

After a great deal of work in revision and simplification, the International General Conference on Weights and Measures (which resulted from the 1875 Treaty of the Meter) in 1960 adopted six base units—the meter, kilogram, second, ampere, kelvin, and candela (corresponding to the properties of length, mass, time, electric current, temperature, and luminous (light) intensity, respectively). In 1971, a seventh base unit for the amount of substance, the mole, was added (see Table 1.3).

The seven properties and base units listed in Table 1.3 are thought to be the smallest number needed for a full description of everything that is observed and measured. Combinations of base units form derived units. You will note that the table also lists two supplementary units, for two- and three-dimensional angular measurements. It is a matter of disagreement whether these geometric units are base units or derived units. Even so, according to our National Institute of Standards and Technology (formerly the National Bureau of Standards), this system "provides a logical and interconnected framework for all measurements in science, industry, and commerce."

The SI System

You may have heard people refer to the SI system and wondered how this differs from the metric system. The SI is a specialized and modernized version of the metric system that designates the use of specific units for various properties. (See Special Feature 1.2.)

All You Will Need to Know About Metric

(For Your Everyday Life)

Metric is based on Decimal system

The metric system is simple to learn. For use in your everyday life you will need to know only ten units. You will also need to get used to a few new temperatures. Of course, there are other units which most persons will not need to learn. There are even some metric units with which you are already familiar: those for time and electricity are the same as you use now.

BASIC UNITS

METER: a little longer than a yard (about 1.1 yards)
LITER: a little larger than a quart (about 1.06 quarts)
GRAM: a little more than the weight of a paper clip

(comparative sizes are shown)

1 METER

1 YARD

25 DEGREES FAHRENHEIT

COMMON PREFIXES
(to be used with basic units)

milli: one-thousandth (0.001)
centi: one-hundredth (0.01)
kilo: one-thousand times (1000)

For example:
1000 millimeters = 1 meter
100 centimeters = 1 meter
1000 meters = 1 kilometer

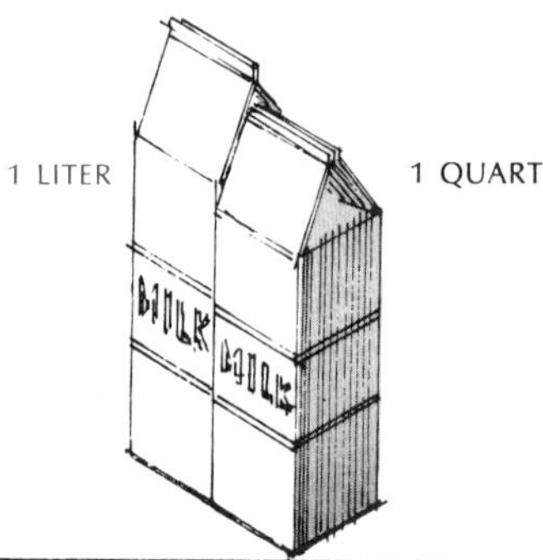

OTHER COMMONLY USED UNITS

millimeter:	0.001 meter	diameter of paper clip wire
centimeter:	0.01 meter	a little more than the width of a paper clip (about 0.4 inch)
kilometer:	1000 meters	somewhat further than ½ mile (about 0.6 mile)
kilogram:	1000 grams	a little more than 2 pounds (about 2.2 pounds)
milliliter:	0.001 liter	five of them make a teaspoon

25 DEGREES CELSIUS

OTHER USEFUL UNITS

hectare: about 2½ acres
metric ton: about one ton

1 POUND

WEATHER UNITS: **FOR TEMPERATURE** degrees Celsius

FOR PRESSURE kilopascals are used
100 kilopascals = 29.5 inches of Hg (14.5 psi)

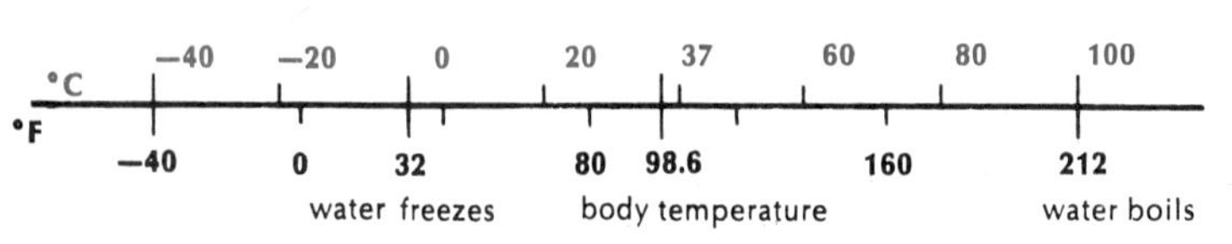

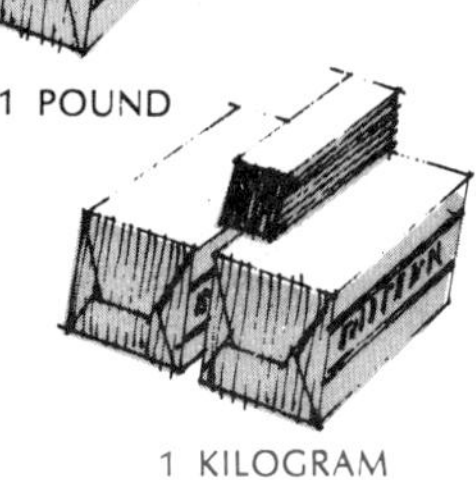

1 KILOGRAM

Figure 1.10
Common British-metric comparisons. (Courtesy U.S. Dept. of Commerce.)

As may be seen from Table 1.3, the SI system consists of seven base units and two supplementary units. The most common measurements involve length, mass, time, electric current, and temperature. Various units will be considered later in the text.

As the United States converts to the metric system, we will need to become familiar with some new units in our everyday activities. To help people "think metric," the National Institute of Standards and Technology has published a flyer, *All You Will Need to Know About Metric (for Your Everyday Life)*, as shown in Figure 1.10.

1.3 Conversion Factors

Suppose you are asked how many feet are in five yards. "Fifteen" would no doubt be your prompt reply. Then, how many feet are there in 48 inches? Four, of course. These are examples of converting from one unit to another. Such common conversions are done frequently and often can be carried out in one's head. But suppose you are asked how many meters there are in 5 yards, or how many centimeters there are in 48 inches. How can you make these conversions?

They can be made basically in the same way conversions were made in the previous examples—by using conversion factors. In this case, however, you are being asked to convert units from one system to another rather than converting units within the same system.

Let's examine what was actually done in the examples of conversions within the British system. In the case of converting yards to feet, we know from experience that there are three feet in one yard, i.e., 3 ft = 1 yd.* By dividing both sides of this equation by 1 yd, we have

$$\frac{3\text{ ft}}{1\text{ yd}} = \frac{1\text{ yd}}{1\text{ yd}} \quad \text{or} \quad \frac{3\text{ ft}}{1\text{ yd}} = 1$$

since the ratio of 1 yd to 1 yd is one. Hence, we have the ratio, or conversion factor, of 3 ft per 1 yd, or 3 ft/yd. (The 1 in the denominator of a ratio is commonly omitted for convenience.) What you did mathematically in your head in converting 5 yd to feet was

$$5\,\cancel{\text{yd}}\left(\frac{3\text{ ft}}{1\,\cancel{\text{yd}}}\right) = 15\text{ ft}$$

That is, you multiplied by a conversion factor. Notice the yd units "cancel" in the equation as though they were numbers.

Another important point is that the sides of an equation are equal not only in magnitude but also in units. When "cleaned up," the previous equation states

$$5 \times 3\text{ ft} = 15\text{ ft}$$

The equality sign indicates that the magnitudes (numbers) *and* units on both sides of the equation are equal. (An exception is made in expressions such as 3 ft = 1 yd for demonstration purposes. Technically, we should say a length of 3 ft is *equivalent* to a length of 1 yd. Conversion factors are used in ratio form rather than as equations.)

Similarly, you convert 48 inches to feet, using the conversion factor 1 ft/12 in., by the operation

$$48\,\cancel{\text{in.}}\left(\frac{1\text{ ft}}{12\,\cancel{\text{in.}}}\right) = 4\text{ ft}$$

If you want to go the other way, i.e., convert feet to inches, simply invert the conversion factor, 12 in./1 ft or 12 in./ft, and

$$4\,\cancel{\text{ft}}\left(\frac{12\text{ in.}}{\cancel{\text{ft}}}\right) = 48\text{ in.}$$

Now we return to the conversion of yards to meters and inches to centimeters. To perform these operations, we simply need to know the appropriate conversion factors. Common conversion factors are listed in Appendix 1 and inside the front cover. From the table, 1 ft =

(*Text continues on p. 16*)

* Keep in mind that this is an equivalence statement. Technically, as an equation it is not dimensionally correct, as will be pointed out shortly. Remember, the quantities of an equation must be equal not only in magnitude (numerically), but also in units.

Table 1.3

The Modernized Metric System

Name of Unit (Abbreviation)	Property Measured	Definition	Further Information
Seven Base Units			
meter (m) 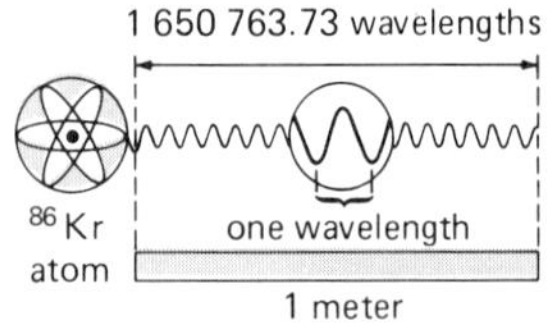	length	Before 1983 the meter was defined as 1 650 763.73 wavelengths in vacuum of the orange-red line of the spectrum of krypton-86. In October 1983, the General Conference on Weights and Measures meeting in Paris adopted a new definition for the meter. The new method defines the meter as the distance traveled by light in vacuum during 1/299 792 458 of a second. Looking at it another way, the new definition establishes the speed of light as 299 792 459 meters per second. Using time, our most accurate measurement, to define length is considered to be a real breakthrough.	The SI unit of area is the **square meter** (m^2). The SI unit of volume is the **cubic meter** (m^3). The liter (0.001 cubic meter), although not an SI unit, is commonly used to measure fluid volume.
1 650 763.73 wavelengths; ^{86}Kr atom; one wavelength; 1 meter. An interferometer is used to measure length by means of light waves			
kilogram (kg) 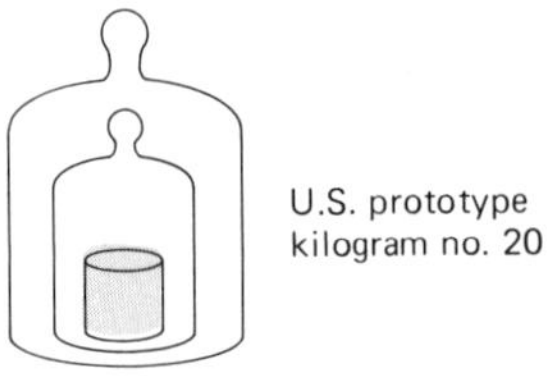	mass	The standard for the unit of mass, the kilogram, is a cylinder of platinum-iridium alloy kept by the international Bureau of Weights and Measures in Paris. A duplicate in the custody of the National Institute of Standards and Technology serves as the mass standard for the United States. This is the only base unit still defined by an artifact.	The SI unit of force is the **newton** (N). One newton is the force that, when applied to a 1-kg mass, will give the mass an acceleration of 1 (meter per second) per second. $1\ N = 1\ kg \cdot m/s^2$ 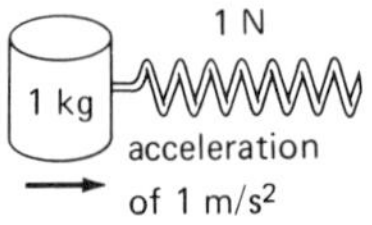 (1 N; 1 kg; acceleration of 1 m/s²) The SI unit for pressure is the **pascal** (Pa). $1\ Pa = 1\ N/m^2$ The SI unit for work and energy of any kind is the **joule** (J). $1\ J = 1\ N \cdot m$ The SI unit for power of any kind is the **watt** (W). $1\ W = 1\ J/s$
U.S. prototype kilogram no. 20			
second (s)	time	The second is defined as the duration of 9 192 631 770 cycles of the radiation associated with a specified transition of the cesium-133 atom. It is realized by tuning an oscillator to the resonance frequency of cesium-133 atoms as they pass through a system of magnets and a resonant cavity into a detector.	The number of periods, or cycles, per second is called *frequency*. The SI unit for frequency is the **hertz** (Hz). One hertz equals one cycle per second. The SI unit for speed is the **meter per second** (m/s). The SI unit for acceleration is the **(meter per second) per second** (m/s^2).
cavity; transition region; oscillating field; cesium source; detector; deflection magnet; deflection magnet; oscillator; NBS atomic time-scale system. Schematic diagram of an atomic beam spectrometer or "clock." Only those atoms whose magnetic moments are "flipped" in the transition region reach the detector. When 9 192 631 770 oscillations have occurred, the clock indicates one second has passed.			
ampere (A) 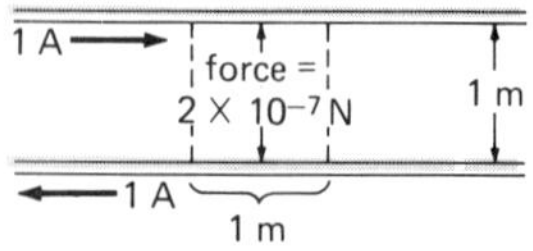	electric current	The ampere is defined as that current which, if maintained in each of two long parallel wires separated by one meter in free space, would produce a force between the two wires (due to their magnetic fields) of 2×10^{-7} newton for each meter of length.	The SI unit of voltage is the **volt** (V). $1\ V = 1\ W/A$ The SI unit of electric resistance is the **ohm** (Ω). $1\ \Omega = 1\ V/A$
1 A; force = 2×10^{-7} N; 1 m; 1 A; 1 m			

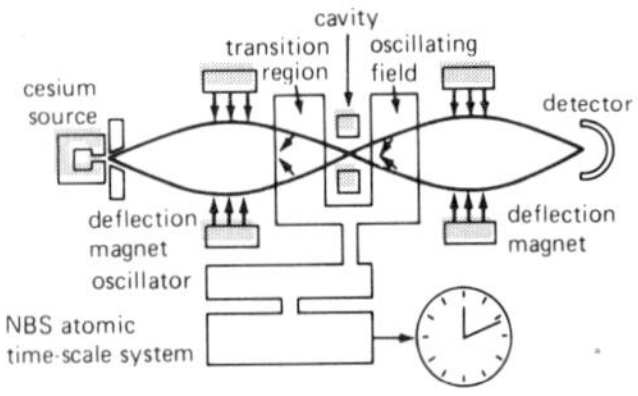

Name of Unit (Abbreviation)	Property Measured	Definition	Further Information
kelvin (K)	temperature	The kelvin is defined as 1/273.16 of the thermodynamic temperature of the triple point of water. The temperature 0 K is called "absolute zero."	On the commonly used Celsius temperature scale, water freezes at about 0°C and boils at about 100°C. The °C is defined as an interval of 1 K, and the Celsius temperature 0°C is defined as 273.15 K. 1.8 Fahrenheit degrees are equal to 1.0°C or 1.0 K; the Fahrenheit scale uses 32°F as a temperature corresponding to 0°C.
mole (mol)	amount of substance	The mole is the amount of substance of a system that contains as many elementary entities as there are atoms in 0.012 kg of carbon-12.	When the mole is used, the elementary entities must be specified and may be atoms, molecules, ions, electrons, other particles, or specified groups of such particles. The SI unit of concentration (of amount of substance) is the **mole per cubic meter** (mol/m³).
candela (cd)	luminous intensity	The candela is defined as the luminous intensity of 1/600 000 of a square meter of a blackbody at the temperature of freezing platinum (2045 K).	The SI unit of light flux is the **lumen** (lm). A source having an intensity of one candela in all directions radiates a light flux of 4π lumens.

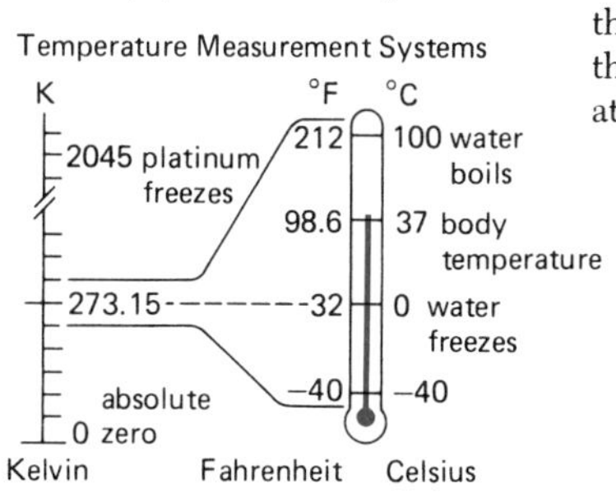

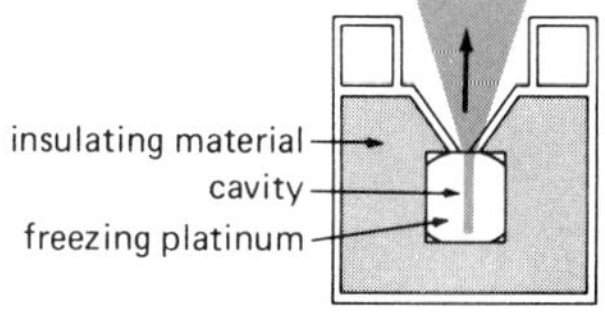

Two Supplementary Units

Name of Unit (Abbreviation)	Property Measured	Definition
radian (rad)	plane angle	The radian is the plane angle with its vertex at the center of a circle that is subtended by an arc equal in length to the radius.
steradian (sr)	solid angle	The steradian is the solid angle with its vertex at the center of a sphere that is subtended by an area of the spherical surface equal to that of a square with sides equal in length to the radius.

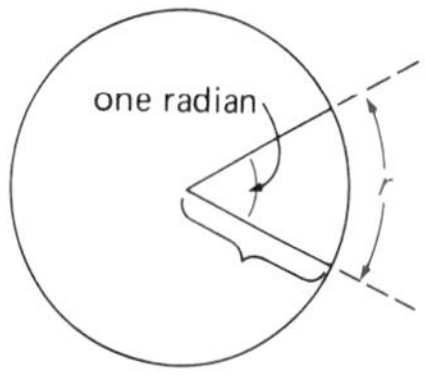

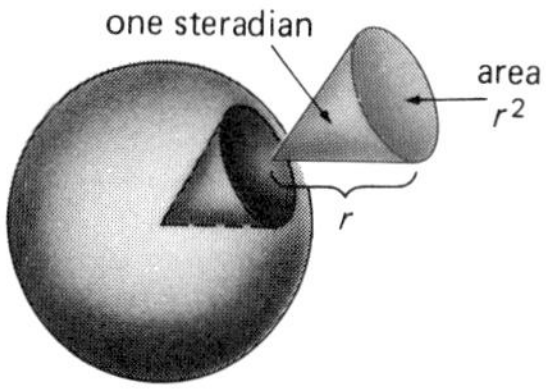

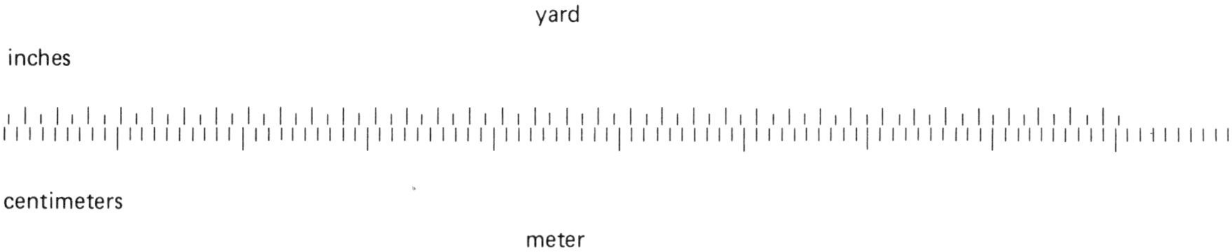

0.3048 m, and 1 yd = 3 ft = 0.914 m, or 0.914 m/yd, and

$$5\ \text{yd}\left(\frac{0.914\ \text{m}}{\text{yd}}\right) = 4.57\ \text{m}$$

When a conversion factor is in the "equation" form, e.g., 1 yd = 0.914 m, it may not be evident whether you should multiply or divide by the non-unity quantity. However, in ratio form, you can simply let the units do the "thinking" for you by using the ratio in the proper form so that the units cancel. For example, suppose you try using the form 1 yd/0.914 m and set up the conversion operation in the form

$$5\ \text{yd}\left(\frac{1\ \text{yd}}{0.914\ \text{m}}\right) = ?$$

You can immediately see that the units do not cancel properly. You would end up with yd^2/m, which doesn't make sense. The units are "telling" you to invert the conversion factor. Equivalently, you may have found in the table of conversion factors that 1 m = 1.094 yd. In this case, you form the ratio 1 m/1.094 yd, and

$$5\ \text{yd}\left(\frac{1\ \text{m}}{1.094\ \text{yd}}\right) = 4.57\ \text{m}$$

The key to the conversion process is to use a conversion factor in a form such that the unit or units that are being converted cancel, and you are left with the desired unit(s). The rest is simple mathematics. Can you now convert 48 inches to centimeters?

EXAMPLE 1.1 Convert 48.0 inches to centimeters.

Solution: From the conversion tables, 1 in. = 2.54 cm, and

$$48.0\ \text{in.}\left(\frac{2.54\ \text{cm}}{\text{in.}}\right) = 122\ \text{cm}$$

EXAMPLE 1.2 A contractor receives a request to build a room addition with floor measurements of 3.00 m × 4.00 m. What is the floor area in square feet?

Solution: Converting the meter measurements to feet (1 m = 3.28 ft from table),

$$3.00\ \text{m}\left(\frac{3.28\ \text{ft}}{1\ \text{m}}\right) = 9.84\ \text{ft}$$

$$4.00\ \text{m}\left(\frac{3.28\ \text{ft}}{1\ \text{m}}\right) = 13.1\ \text{ft}$$

Then

$$\text{Area} = 9.84\ \text{ft} \times 13.1\ \text{ft} = 129\ \text{ft}^2$$

Alternative Solution: Converting the area in square meters to square feet,

$$12.0\ \text{m}^2\left(\frac{3.28\ \text{ft}}{\text{m}}\right)^2 =$$

$$12.0\ \text{m}^2\left(\frac{10.76\ \text{ft}^2}{\text{m}^2}\right) = 129\ \text{ft}^2$$

(Here, the 1 was left out of the denominator of the ratio for convenience.)

On occasion, a conversion table may not list the direct conversion factor you need, or a table may not be available. In the latter case you can probably make the conversion using commonly known intermediate factors. For example, how many seconds are in five days? We all know the number of hours in a day, the number of minutes in an hour, and the number of seconds in a minute, so we can perform a multiple conversion:

$$5\ \text{days}\left(\frac{24\ \text{hr}}{1\ \text{day}}\right)\left(\frac{60\ \text{min}}{1\ \text{h}}\right)\left(\frac{60\ \text{s}}{1\ \text{min}}\right) = 432{,}000\ \text{s}$$

Had the conversion factor 1 day = 86,400 s been known, the conversion could have been done directly: 5 days (86,400 s/day) = 432,000 s.

Similarly, suppose you want to convert 2 ft to meters, but you don't know and can't find the direct conversion factor. A good "cross-over" conversion factor between the British and metric systems is 1 in. = 2.54 cm. Knowing this, you can do the conversion using commonly known factors, e.g.,

$$2\ \text{ft}\left(\frac{12\ \text{in.}}{1\ \text{ft}}\right)\left(\frac{2.54\ \text{cm}}{1\ \text{in.}}\right)\left(\frac{1\ \text{m}}{100\ \text{cm}}\right) = 0.61\ \text{m}$$

EXAMPLE 1.3 A machinist wishing to replace a damaged shaft measures its length and diameter to be 1 ft, $4\frac{1}{8}$ in. and $\frac{3}{4}$ in., respectively. However, the catalog of a foreign manufacturer lists the dimensions of shafts in centimeters. What size shaft should be ordered?

Solution: Given that the shaft length l = 1 ft, $4\frac{1}{8}$ in. = $16\frac{1}{8}$ in. = 16.125 in., and the diameter $d = \frac{3}{4}$ in. = 0.75 in., convert these lengths to centimeters.

$$l = 16.125 \text{ in.} \left(\frac{2.54 \text{ cm}}{\text{in.}}\right) = 41.0 \text{ cm}$$

and

$$d = 0.75 \text{ in.} \left(\frac{2.54 \text{ cm}}{\text{in.}}\right) = 1.9 \text{ cm}$$

Metric units are increasingly being used in domestic and commercial applications. For example, soft drinks are now packaged in two-liter bottles. See Figure 1.9. (Is this more or less than one-half gallon?) Also, several American automobiles are being manufactured with metric components. British (sometimes called English) wrench sets do not fit the nuts and bolts of these and imported foreign cars, as we sometimes discover painfully when trying to make repairs. Metric wrench sets and combined British/metric sets are now readily available (Fig. 1.11).

Table 1.4
Common Automotive Wrench Sizes

English	Metric Equivalent	Metric	English Equivalent
$\frac{3}{8}''$ (0.3750 in.)	9.525 mm	9 mm	0.3543 in.
$\frac{7}{16}''$ (0.4375 in.)	11.113 mm	10 mm	0.3937 in.
$\frac{1}{2}''$ (0.5000 in.)	12.700 mm	11 mm	0.4331 in.
$\frac{9}{16}''$ (0.5625 in.)	14.287 mm	12 mm	0.4724 in.
$\frac{5}{8}''$ (0.6250 in.)	15.875 mm	13 mm	0.5118 in.
$\frac{11}{16}''$ (0.6875 in.)	17.462 mm	14 mm	0.5512 in.
$\frac{3}{4}''$ (0.7500 in.)	19.050 mm	15 mm	0.5906 in.
$\frac{13}{16}''$ (0.8125 in.)	20.637 mm	16 mm	0.6299 in.
$\frac{7}{8}''$ (0.8750 in.)	22.225 mm	17 mm	0.6693 in.
$1''$ (1.0000 in.)	25.400 mm	19 mm	0.7480 in.
$1\frac{1}{16}''$ (1.0625 in.)	26.988 mm	21 mm	0.8268 in.
$1\frac{1}{8}''$ (1.1250 in.)	28.575 mm	22 mm	0.8661 in.
$1\frac{1}{4}''$ (1.2500 in.)	31.750 mm	24 mm	0.9449 in.
		27 mm	1.0630 in.

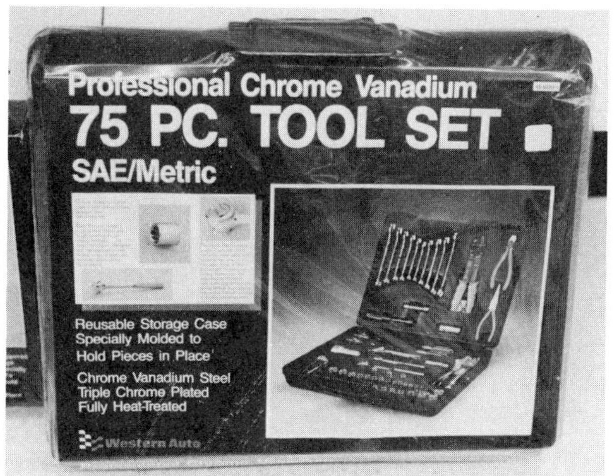

Figure 1.11
Combination tool set. Wrenches may be purchased in combination sets of SAE (Society of Automotive Engineers) or English sizes and metric sizes. Individual sets are also available.

Unfortunately, there is no convenient correspondence between the fixed wrench sizes in the different systems, as Table 1.4 shows. In attempting to loosen a $\frac{1}{2}$-in. nut with a metric wrench set, one quickly finds that a 12-mm wrench won't fit and that a 13-mm wrench is slightly too large. When one tries to loosen the nut with a 13-mm wrench, the wrench will slip and most likely will round the edges of the nut.

Some socket sets now have sockets cut for similar metric and British sizes, e.g., $\frac{1}{2}$ in. and 13 mm, or 1 in. and 25 mm. (Alternate cuts are made to fit the slightly different sizes.) However, the general solution is to have both English and metric wrench sets available when doing automotive repairs, or hope that the job can be done with an adjustable wrench.

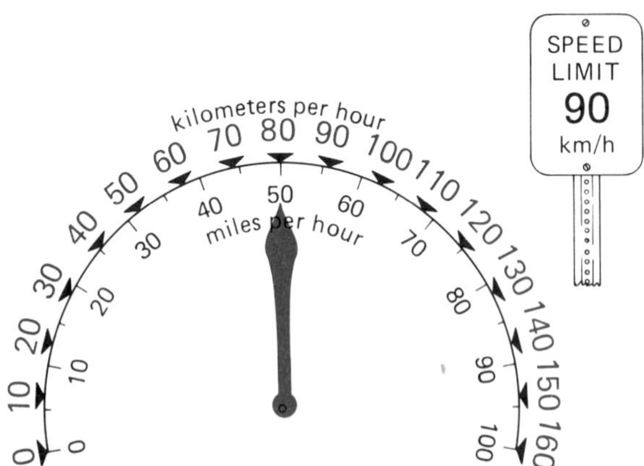

Figure 1.12
Getting used to metric. On conversion to the metric system, our highway signs would be in km/h. The 55 mi/h speed limit would probably be replaced by a 90 km/h (55.9 mi/h) speed limit.

You may have noticed that newer cars in the United States have both miles per hour (mi/h or mph) and kilometers per hour (km/h or kph) on the speedometers. As illustrated in Figure 1.12, the 55 mi/h speed limit would probably be posted as 90 km/h (55.9 mi/h). Notice that 100 km/h is close to 65 mi/h (62 mi/h).

When we convert completely to the metric system, road signs will show not only the speed limits in km/h, but also the distances in kilometers. For instance, a road sign would show a town to be 16 km away rather than 10 mi. Some signs already show both. (See the chapter introductory photo.)

EXAMPLE 1.4 A set of road signs shows a town to be 3 mi further ahead and the speed limit to be 40 mi/h. If the signs were replaced with metric signs, what would the new signs read?

Solution: Using the conversion factor between mile and kilometer, 1.61 km/mi or 0.62 mi/km,

$$3 \text{ mi} \left(\frac{1.61 \text{ km}}{\text{mi}}\right) \simeq 5 \text{ km}$$

and

$$40 \frac{\text{mi}}{\text{h}} \left(\frac{1 \text{ km}}{0.62 \text{ mi}}\right) \simeq 65 \text{ km/h}$$

Note how either form of the conversion factor can be easily used by observing unit cancellation.

EXAMPLE 1.5 A clipboard measuring 0.200 m × 0.225 m has an area of 0.0450 m^2. What is the area of the board in square centimeters (cm^2)?

Solution: This is a simple conversion factor from m^2 to cm^2. It is sometimes thought that the conversion factor would be 100 cm^2/m^2, but this is incorrect, as the following calculation shows.

$$0.0450 \text{ m}^2 \left(\frac{100 \text{ cm}}{\text{m}}\right)^2 = 0.0450 \text{ m}^2 \left(\frac{10{,}000 \text{ cm}^2}{\text{m}^2}\right)$$
$$= 450 \text{ cm}^2$$

Note that there are 10,000, or 10^4, cm^2 per m^2.

Of course, there's another way to arrive at the solution. The dimensions in meters could be converted to centimeters and the area then calculated. Without writing the conversion factors explicitly, it should be easy to see that 0.200 m and 0.225 m are equivalent lengths of 20.0 cm and 22.5 cm, respectively. Then,

$$\text{Area} = (20.0 \text{ cm}) \times (22.5 \text{ cm}) = 450 \text{ cm}^2$$

1.4 Problem Solving

In science and technology, physical principles are applied to various situations. This leads to problem solving, which is perhaps one of the most difficult aspects of physics for students and scientists alike. In short, the essence of physics is investigating nature, applying principles, and solving problems. You will find problem exercises at the end of each chapter in the text. The purpose of these problems is to give you practice in the applications of the physical principles you have studied.

How does one go about solving problems? There is no set way. Different people may approach a problem differently and achieve the same results. However, the basic approach is to first analyze and understand the problem; then the appropriate physical principles and equations may be applied to its solution.

In general, you will experience two types of problems. These might be termed (1) "plug-in" problems, and (2) "second-step" problems. The plug-in problems are generally easier and are straightforward. A set of data is given, and by "plugging" it into an equation and turning the mathematical "crank," you obtain the answer. You are no doubt familiar with this type of problem.

Second-step problems are a bit more difficult. In this case, there is a "second thought" process in analyzing and solving a problem and a second application of part of the given data. More intricate problems may require a third-step or even a fourth-step application.

In any case, the main emphasis should be in first *analyzing and understanding the problem* before any mathematical operations are carried out. A suggested procedure in solving problems is as follows:

Step 1. Read the problem and write down what is given. That is, list the given quantities along with their units. Make certain you extract all of the information. In some cases, certain pieces of information will not be given explicitly, as illustrated in Example 1.7. If appropriate, make a sketch or drawing of the situation to promote "visual understanding."

Step 2. Inspect the listed quantities to see if they are all in the same system of units. If not, apply the appropriate conversion factor(s). This step is done to avoid mixed units in computations.

Step 3. Determine what is wanted. This is very important. You can't solve a problem if you don't know what you are looking for. Express this in the appropriate mathematical symbol, e.g., length x.

Step 4. Think about the solution of the problem, that is, how you get what is wanted from the data given. Ask yourself, What is the principle(s) involved? What is the appropriate equation(s) to use?

Step 5. Perform the calculations and express your result with the appropriate number of significant figures and *units*. It is a good practice to mentally inspect your result for proper "order of magnitude." If you get a ridiculously large or small value, you have probably made a mathematical error or are applying the wrong principle. (See Special Feature 1.3 on significant figures.)

Here are some simple applications of the procedures:

EXAMPLE 1.6 A rectangular piece of metal has a length of 1.20 m, a width of 0.75 m, and a thickness of 20 cm. What is its volume?

Solution:

Step 1. Given: l = 1.20 m, w = 0.75 m, and t = 20 cm.

Step 2. t = 20 cm (1 m/100 cm) = 0.20 m

Step 3. Wanted: volume V.

Step 4. Volume is given by $V = l \times w \times t$

Step 5.
$$V = l \times w \times t = (1.20 \text{ m}) \times (0.75 \text{ m}) \times (0.20 \text{ m}) = 0.18 \text{ m}^3$$

This is a "plug-in" problem.

EXAMPLE 1.7 A cylindrical container has an inside diameter of 10 cm and a height of 20 cm. If the container is filled with water, what is the mass of this amount of water? (Here a sketch is helpful.)

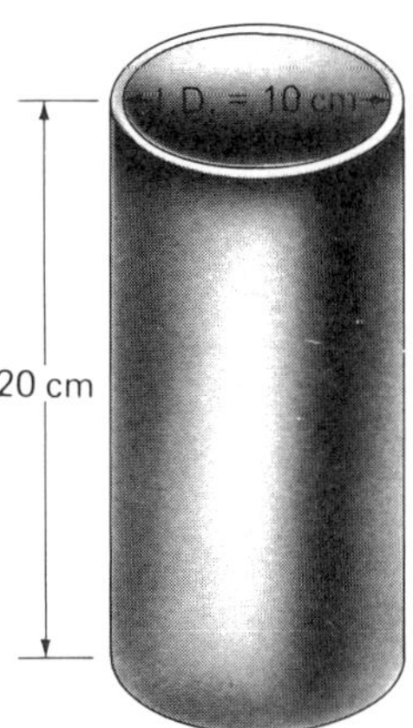

Solution:

Step 1. Given: d = 10 cm or radius r = 5.0 cm, and h = 20 cm.

Step 2. Units are okay.

Step 3. Wanted: mass m.

Step 4. As you may be thinking, the volume of the container can be computed from the given dimensions, and because the mass of a volume of water is wanted, the problem involves density ($\rho = m/V$). However, the density of water is not given. In this case, you are expected to look up the density of water in a reference table, e.g., Table 14.1, or you may remember that it is 1.0 g/cm^3. Then we can add, "Given (indirectly): $\rho = 1.0$ g/cm^3."

First find the volume of the container, $V = Ah = (\pi r^2)h$, and then compute the mass from the density expression,

$$\rho = m/V \qquad \text{or} \qquad m = \rho V$$

Step 5.

$$V = Ah = (\pi r^2)h$$
$$= \pi(5.0 \text{ cm})^2(20 \text{ cm}) = 1.6 \times 10^3 \text{ cm}^3$$

where scientific notation is used to express the answer with two significant figures. (If you are not familiar with the scientific powers-of-ten notation, this will be reviewed in the next chapter.)

$$m = \rho V = (1.0 \text{ g/cm}^3)(1.6 \times 10^3 \text{ cm}^3)$$
$$= 1.6 \times 10^3 \text{ g}$$

or 1.6 kg.

Notice how the units are carried along in the equation. This allows dimensional analysis. The sides of a physical equation must be equal not only in magnitude (numbers), but also in units. The two cm^3 units in the last equation cancel and we have grams (g) = grams (g).

It is often convenient and desirable to carry along the units in computations. However, as the combinations of units become more complex, the inclusion of units may make the equation a bit "messy," or the unit cancellation for a specially named combination of units may not be evident. At this point, keep in mind that if the units of the given quantities are in the same system, then the unit of the answer will also be in that system.

EXAMPLE 1.8 An automobile traveling at 30 m/s comes uniformly to a stop in 5.0 s. How far in kilometers did the auto travel in this time?

Solution:

Step 1. Given: initial speed $v_o = 30$ m/s and $t = 5.0$ s. Also, you are expected to recognize an implicit piece of information—the final speed is zero, $v_f = 0$.

Step 2. Units are in same system. You will change the answer to km after calculations.

Step 3. Wanted: the distance traveled d.

Step 4. It will be learned later that the average speed $\bar{v}$ and distance may be calculated from the formulas below.

Step 5.

$$\bar{v} = \frac{v_f + v_o}{2} = \frac{0 + 30 \text{ m/s}}{2} = 15 \text{ m/s}$$

Then,

$$d = \bar{v}t = (15 \text{ m/s})(5.0 \text{ s}) = 75 \text{ m}$$

or

$$d = 0.75 \text{ km}$$

Wait! Take a look at the answer in km. Three-quarters of a kilometer is on the order of $\frac{1}{2}$ mi (1 km is equivalent to 0.62 mi), and this would be a long distance for stopping a car. Checking the conversion explicitly,

$$75 \text{ m } (1 \text{ km}/1000 \text{ m}) = 0.075 \text{ km}$$

That's better.

In some instances in problem solving, there may be more than one unknown or quantity involved in a solution. Usually, different conditions will give different equations containing the unknown quantities.

EXAMPLE 1.9 Without stating a problem, suppose you have

Given: $A = 9$ m, $a = 0.2$, and $b = 0.4$.

Wanted: x_1 and x_2 (two unknowns)

Equations: $A - x_1 - x_2 = 0$ (Equation 1)

$ax_1 - bx_2 = 0$ (Equation 2)

Note that there are two equations and two unknowns. Basically, you need the same number of equations as there are unknowns. With two equations and two unknowns, three equations and three unknowns, and so on, a problem is essentially solved. The numerical values of the unknowns can be found with a little mathematical manipulation.

Can you find the values of x_1 and x_2 in this two-unknown example? Putting the given values into the equations, we have

$$9 \text{ m} - x_1 - x_2 = 0 \qquad \text{(Equation 1)}$$

$$(0.2)x_1 - (0.4)x_2 = 0 \qquad \text{(Equation 2)}$$

Then, one might proceed like this: rewriting Equation 2,

$$x_1 = \left(\frac{0.4}{0.2}\right) x_2 = 2x_2 \qquad \text{(Equation 2)}$$

Substituting this x_1 into Equation 1,

$$9 \text{ m} - 2x_2 - x_2 = 0 \qquad \text{or} \qquad x_2 = \frac{9 \text{ m}}{3} = 3 \text{ m}$$

Then, substituting this value of x_2 into Equation 2,

$$x_1 = 2x_2 = (2)(3 \text{ m}) = 6 \text{ m}$$

and the problem is completely solved.

SPECIAL FEATURE 1.3

Calculations and Significant Figures

After doing a calculation, you may wonder how many figures or digits you should retain in your answer. This is particularly true when using a calculator, which may supply you with an eight- or ten-digit result. For example, the following mathematical operation done on a calculator gives

$$\frac{2.7}{1.74} = 1.551724138$$

The calculator displays a ten-digit answer for the division of two numbers with two and three digits, respectively.

The number of digits reported in a measured quantity is an indication of the fineness of the instrument scale. These digits are also called significant figures, and they are determined by how many digits can be read on the scale of a measurement instrument. The last figure of a reported, or data, quantity is doubtful; it is an estimate of the fraction of the smallest division on a measurement scale. For example, a measured value of 2.4 cm indicates that the smallest division on the measurement scale was a centimeter (.4 is the estimated fraction of the division). A measured value of 2.45 cm indicates that the smallest scale division was a millimeter. Why? (You'll get more instruction on this in the laboratory.)

More significant figures, i.e., a better reading, cannot be gained by mathematical calculations. The rule is as follows:

> In the multiplication or division of two or more numerical measurements, the number of significant figures in the final answer can be no greater than the number of significant figures in the measurement with the least number of significant figures.

This means that the result of a mathematical operation should be rounded off to the corresponding least number of significant figures. In the preceding calculation, this would be two significant figures, so that

$$\frac{2.7}{1.74} = 1.6$$

To determine the number of significant figures in a quantity, read the numbers from left to right and count all the digits, beginning with the first nonzero digit. For example, 2.05 cm and 0.0205 cm both have three significant figures, and 1.5 m and 0.15 m both have two significant figures. In numbers such as 100 km (without a decimal point), the zeros may or may not be significant. Suppose there were two significant figures; then, using scientific notation, 1.0×10^2 km would be an appropriate way to express this.

With regard to **rounding off a number,** use the following general method.

> **Treat the excess digits to the right of the last significant figure as a decimal fraction. If the fraction is less than 0.5, leave the last significant figure unchanged. If the fraction is greater than or equal to 0.5, increase the last significant figure by one.**

(In this method, five digits—0, 1, 2, 3, 4—are rounded down, and five digits—5, 6, 7, 8, 9—are rounded up.)

When you are working problems, the number of significant figures used in a calculation can affect your answer. This is particularly true for problems

(continued)

SPECIAL FEATURE 1.3

Calculations and Significant Figures (*continued*)

with more than one part or calculation. For example,

A car travels a distance of 553 ft. (a) How far does it travel in meters? (b) If it takes 20 s to travel this distance, what is the average speed of the car? (Average speed is given by the distance divided by time, $\bar{v} = d/t$.)

Solution

(a) To change the distance in feet to meters, we multiply by a conversion factor:

$$d = 553 \text{ ft} \left(\frac{0.3048 \text{ m}}{\text{ft}}\right)$$

$$= 168.5544 \text{ m (calculator result)}$$

$$= 169 \text{ m (rounded to 3 places)}$$

(b) Dividing this distance by the time,

$$\bar{v} = \frac{d}{t} = \frac{169 \text{ m}}{20 \text{ s}}$$

$$= 8.45 \text{ m/s (calculator result)}$$

$$= 8.5 \text{ m/s (rounded to 2 places)}$$

But suppose you were just asked to find the average speed. This can be done in one step with the conversion factor included:

$$\bar{v} = \frac{d}{t} = \frac{553 \text{ ft (0.3048 m/ft)}}{20 \text{ s}}$$

$$= 8.42772 \text{ m/s (calculator result)}$$

$$= 8.4 \text{ m/s (rounded to 2 places)}$$

As can be seen, the doubtful figures carried through in the calculations and the rounding off give slightly different answers. Both are technically "correct" for the problem solution. To minimize rounding errors, *only the final answer should be rounded.* However, it is often instructive and necessary to work a problem in steps, so slightly different answers in some instances would be expected. This may also occur when a problem is worked by different methods or approaches.

In general, problems in this text will use two or three significant figures. In some cases, more significant figures may be assumed and answers reported with more figures for clarity. For example, suppose it is given that the floor dimensions of a room are 3.2 m by 4.8 m, and the floor area is wanted. Then, 3.2 m × 4.8 m (= 15.36 m^2) = 15.4 m^2. Technically, the answer should be rounded to 15 m^2. Why?

Significant figures and rounding procedures are important in approximating uncertainty, but the prime consideration is learning how to solve problems.

Important Terms

fundamental properties the basic physical properties of measurement; these are length, mass, time (and electric charge), which describe the concepts of space, matter, time (and electricity), respectively

length a dimension of space that describes an object's size and locates its position relative to some reference

mass the quantity of matter an object (or system) contains

time the forward flow of events; two events define an interval of time

standard unit a fixed and reproducible value for the purpose of taking accurate measurements

weight the gravitational attraction on a body by a celestial object, most commonly Earth

density

mass the mass per unit volume of an object (or system)

weight the weight per unit volume of an object (or system)

Important Formulas

weight: $w = mg$

density:

(mass) $\rho = m/V$

(weight) $D = w/V$

Since $w = mg$, then $D = w/V = mg/V = (m/V)g = \rho g$

or $D = \rho g$

Questions

What We Measure

1. Explain why length, mass, and time are "fundamental" properties. Suggest some possible substitutes.
2. Why is weight not a fundamental property?
3. What would your weight be on the surface of the moon?
4. If the mass density of iron is $8\frac{3}{4}$ times that of ice, how many times greater is the weight density of iron than that of ice?
5. A quantity (mass) of gas in a container is transferred to another container with twice the volume. Is the density of the gas affected? If so, how? (The temperature of the gas is the same in both containers.)

Measurement Units

6. Some anatomical units of measurement are shown in Figure 1.3. Give the equivalent lengths in metric units.
7. A dime (10¢) would be given what name in the "centidollar" system? (Use a metric prefix.)
8. Can a nickel (5¢) or a quarter (25¢) be given a name with a metric prefix in the "centidollar" system? Explain.
9. Give the value of the metric prefixes for the following units: (a) megaton, (b) microsecond, (c) milligram, (d) nanosecond.
10. Express the following in terms of metric prefixes: (a) one thousandth of a second, (b) one millionth of a meter, (c) one billionth of a second.
11. The national debt is currently on the order of 2 trillion dollars. Express this in the appropriate metric-prefix dollar unit.
12. The volume of automobile cylinders is sometimes given in cubic inches. How is this given in the metric system?
13. Since the British system is a gravitational system using weight, which is not a fundamental property, does this make the system invalid? Explain.

Conversion Factors

14. Even though we commonly write conversion factors in equation form, e.g., 1 in. = 2.54 cm, is this technically correct? Explain.
15. Explain how the conversion factors 1.61 km/mi and 0.62 mi/km are related.
16. The program for a basketball game lists the height of a person playing a team's center position as 155 cm. Would you suspect a misprint? Explain.
17. Explain whether each of the following is likely or unlikely.
    (a) One student is 0.20 m taller than another student.
    (b) A dog's tail is 45 cm long.
    (c) The height of a room's ceiling is 2 m.
    (d) The normal speed on an interstate highway is 60 kph.
    (e) The floor area of a room in a house is 14 m^2.

Problems

Levels of difficulty are indicated by asterisks for your convenience. (No asterisk for least difficult to one or two asterisks for more difficult.)

1.1–1.2 What We Measure and Units

1. What is the numerical value of the acceleration due to gravity on the surface of the moon? (g = 9.8 m/s^2 = 32 ft/s^2 on Earth.)
2. How much would an astronaut weighing 180 lb on Earth weigh on the moon?
3. An astronaut on the moon picks up a rock sample and records its weight as 3.5 lb, measured on a scale in the moon lander. What would be the measured weight of the rock when it is brought back to Earth? Does its mass change too? Explain.
4. What are (a) the surface area and (b) the volume of a sphere with a diameter of 0.20 m? (*Hint:* see Appendix 5.)

*5. An open cylindrical container has a height of 8.0 cm and a diameter of 6.0 cm (inside measurements). What are (a) the inside surface area of the cylinder and (b) its volume in mL? (*Hint:* see Appendix 5.)

*6. A paper cone with a base diameter of 6.0 in. and a depth of 10 in. is used to serve popcorn in a movie theater. (a) What is the volume of the container? (b) Is the inside surface area of the cone numerically larger than the volume? Justify your answer. (*Hint:* see Appendix 5.)

*7. A cube 0.25 m on a side has a mass of 3.0 kg. What is the density of the material making up the cube?

*8. A solid sphere of radius 8.0 cm has a mass of 1000 g. What is the density of the material making up the sphere? (*Hint:* see Appendix 5.)

*9. (a) What is the average mass density of Earth? (b) Why do we say *average* density, and does weight density make any sense in this case? Explain. (*Hint:* see Appendix 2.)

*10. Compare the densities of Earth and the moon. Which is greater and by what factor? (*Hint:* see Appendix 2.)

*11. (a) What is your approximate average weight density? Explain your approximation. (b)

How could you determine this by actual measurements? (*Hint:* think of submerging yourself in water.)

*12. A metal cube 6.0 in. on a side has a weight of 4.0 lb. What is the weight density of the metal in lb/ft^3?

*13. The weight density of water is 62.4 lb/ft^3. What would be the weight of the water in a filled cylindrical bucket 1.0 ft tall with a radius of 6.0 in.?

*14. Mercury has a density of 13.6 g/cm^3. What would be the volume of 68 g of mercury?

*15. An aquarium tank has inside dimensions of 1.5 m by 0.80 m by 0.60 m. If it is filled with seawater ($\rho = 1020$ kg/m^3), what mass of water would it contain?

**16. A rectangular water tank is used to quench hot pieces of metal. If the tank has a square base 3.0 ft on a side, and contains 1 ton of water (2000 lb) when filled, what is the depth of water in the tank? ($D_{H_2O} = 62.4$ lb/ft^3.)

**17. A conical pile of sand near a construction site has a base circumference of 31.4 ft and a height of 4.5 ft. If the sand has a density of 75 lb/ft^3, how many tons of sand are in the pile?

1.3 Conversion Factors

18. Perform the following conversions:
    (a) 60 mm to inches
    (b) 16.0 m to feet
    (c) 16.0 cm to inches
    (d) 20 km to miles

19. Perform the following conversions:
    (a) 20 mi to kilometers
    (b) 10.0 ft to centimeters
    (c) 30 in. to meters
    (d) 500 yd to kilometers

20. Just for fun, how would the following popular expressions be expressed in the metric system?
    (a) I wouldn't touch that with a ten-foot pole.
    (b) The cowboy wore a ten-gallon hat.
    (c) Give him an inch and he'll take a mile.
    (d) This thing must weigh a ton.

*21. Bottled water and soft-drink beverages are now commonly purchased in 1-L, 2-L, and 3-L bottles. Are these larger or smaller than their nearest British equivalents, and how much larger or smaller?

*22. Which is longer, a 50.0-yard dash or a 50.0-meter dash, and by how much?

*23. What is the density of water in kg/m^3? (*Note:* it is *not* 1 kg/m^3.)

*24. Which is greater and by how much: a metric ton or a customary ton (2000 lb)?

(a)

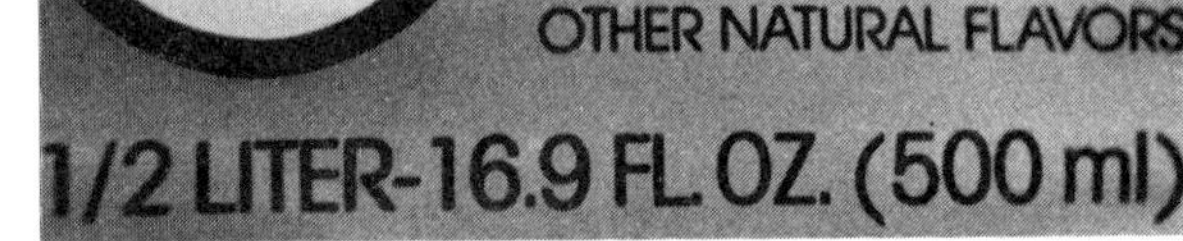

(b)

Figure 1.13
See Problem 27.

*25. At a metric weight-watchers' club, how would each of the following weight losses be reported? (a) 22 lb, (b) 11 lb, (c) 2.2 lb, (d) 1.1 lb, (e) 0.55 lb.

*26. How many cubic centimeters (cm^3) are there in one cubic meter?

*27. Most commercial products now list British and metric units (Fig. 1.13). Notice that the metric units come first on the soda cans. From the figure, determine conversion factors for (a) ounces (weight), pounds, and grams, and (b) liters, (fluid) ounces, and pints. Note that the ounce (oz) unit has two meanings in the British system—weight and fluid (volume).

*28. Give your height and mass in metric units. How about your mass in the British unit?

*29. A homemaker ordinarily plans a meal with $\frac{1}{4}$ lb of meat per serving. Twelve people will be at a

Figure 1.14
See Problem 30.

Figure 1.15
See Problem 37.

meal on Sunday. However, the butcher shop went metric. About how many kilos of meat should be purchased?

*30. In metric countries, gasoline is sold in liters. Some stations in this country also sell gas in liters. Help the attendant in Figure 1.14 by telling him how many gallons the motorist is asking for.

*31. A machinist measures the sides of a rectangular work piece to be 6.00 in. × 10.0 in. What is the area of the piece in cm^2?

*32. The area of a wall is found to be 100 ft^2. How many square yards is this?

*33. A person in a foreign country measures the dimensions of a floor to be 8.00 m × 10.0 m. He orders tile from an American supplier who sells it in ft^2. How many square feet should the supplier send?

*34. What is the cubic-inch equivalent of a 2.50-L engine?

*35. A draftsman draws a diagram of a piece of machinery to a scale of 1.0 in. : 1 ft (or 1.0 in. to the foot). If the actual length of the piece is 54 inches, what is its length to scale on the diagram?

*36. A contractor builds a house from a blueprint with a scale of 0.25 in. : 1 ft ($\frac{1}{4}$ in. to the foot). On the blueprint the floor plan of a room is 2.5 in. × 3.0 in. What is the actual area of the room floor?

**37. Noah of Biblical fame was instructed to build an ark 300 cubits long, 50 cubits wide, and 30 cubits high (Fig. 1.15). What were the dimensions and volume (assume rectangular) of the ark in the metric system? (See Question 6.)

**38. If two runners ran the 100-yard dash and the 100-meter dash in the same time of 15 seconds, which would be faster and by how much?

Figure 1.16
See Problem 39.

**39. (a) For the two soda sizes in Figure 1.16, which is the better buy? (b) What are the price per mL and price per fl. oz for each?

1.4 Problem Solving

No specific problems are given here, but there will be plenty of applications in the following chapters.

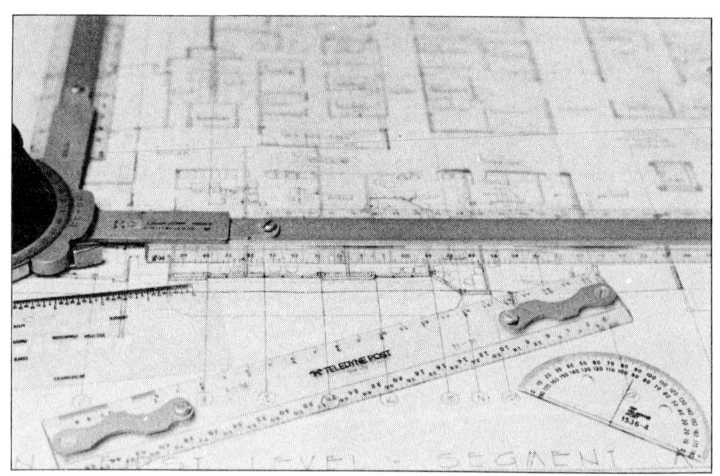

Chapter 2

Technical Mathematics

A mechanic ordinarily can perform a particular operation with any one of several tools, but one tool is usually most applicable. Hence, the greater the variety of tools mechanics have at their disposal, the better prepared they are to carry out their trade.

For a person in a technical field, mathematics provides some of the basic tools of the trade. One might think of different areas of mathematics as being analogous to different tools; certain mathematical tools are more applicable than others to particular jobs or situations. The greater knowledge a person has of a variety of mathematics, the better prepared he or she is to analyze and understand physical situations.

In this chapter, some of the tools of technical mathematics will be presented. It is assumed that the student has a basic working knowledge of algebra. Here attention will be given to scientific notation (or powers of ten), elementary trigonometry, and vectors.

2.1 Scientific Notation (Powers of Ten)

Many physical applications and descriptions involve very large and very small numbers. For example, the average distance from Earth to the Sun is 93,000,000 miles, and the mass of an electron is 0.000000000000000000000000000911 grams. To express and perform mathematical operations with such large numbers and small decimal fractions, it is convenient to use what is known as **scientific notation** (also called **powers of ten**). This is a simple notation that uses the decimal base of our numbering system.

Expressing Large Numbers

Consider the number 72,000. This can be written in terms of multiples of 10 as follows:

$$
\begin{aligned}
72{,}000 &= 7200 \times 10 &&= 7200 \times 10^1 \\
&= 720 \times 100 &&= 720 \times 10 \times 10 \\
& &&= 720 \times 10^2 \\
&= 72 \times 1000 &&= 72 \times 10 \times 10 \times 10 \\
& &&= 72 \times 10^3 \\
&= 7.2 \times 10{,}000 &&= 7.2 \times 10 \times 10 \times 10 \times 10 \\
& &&= 7.2 \times 10^4
\end{aligned}
$$

where the multiples of ten have been expressed as powers of ten. That is, $100 = 10 \times 10 = 10^2$, $1000 = 10 \times 10 \times 10 = 10^3$, and so on.

Notice that the exponent is equal to the number of places the decimal point has been shifted to the left:

$$\underset{4\,3\,2\,1}{7\,2\,0\,0\,0.} = 7.2 \times 10^4$$

The exponent superscript of a 10 is called the *power* of ten. For example, 10^3 is ten raised to the third power. We commonly refer to quantities raised to the second and third powers as being squared and cubed, respectively. A quantity raised to the first power, e.g., 10^1, is just the quantity itself: $10^1 = 10$. By definition, a quantity raised to the zero power is equal to one, e.g., $10^0 = 1$.

Then, expressing the average distance from Earth to the Sun, 93,000,000 miles, in powers-of-ten notation, we have

$$\begin{aligned} 93{,}000{,}000 \text{ mi} &= 93 \times 10^6 \text{ mi} \\ &\quad \text{(decimal shifted 6 places)} \\ &= 9.3 \times 10^7 \text{ mi} \\ &\quad \text{(decimal shifted 7 places)} \end{aligned}$$

This means that the numerical prefix of 93 or 9.3 is multiplied by 10 six or seven times, respectively. It is customary to express a number in scientific notation with one digit to the left of the decimal point, e.g., 9.3×10^7. However, 93×10^6 is the same number, and this form may be more convenient for mathematical operations in some instances. Hence, we may state the following general rule:

To express large numbers in scientific notation, the exponent or power of ten is equal to the number of places the decimal point is shifted to the left.

EXAMPLE 2.1 Expressing large numbers in scientific notation (significant figures assumed):

$$80000 = 8.0 \times 10^4$$

$$5280000 = 5.28 \times 10^6$$

$$270.6 = 2.706 \times 10^2$$

$$100000 = 1.0 \times 10^5 = 10^5$$

(In the last case, a numerical prefix of 1 is sometimes omitted as being understood.)

Expressing Small Numbers

Very small numbers in the form of decimal fractions are expressed in scientific notation in a similar manner. Decimal fractions indicate fractions of multiples of ten; for example,

$$0.048 = \frac{48}{1000} = \frac{48}{100 \times 10} = \frac{48}{10 \times 10 \times 10}$$

or

$$= \frac{48}{10^3} = 48 \times 10^{-3}$$

where the reciprocal relation $1/10^x = 10^{-x}$ is used. Notice that the negative exponent is equal to the number of places the decimal point has been shifted to the right:

$$0\,.\underset{1\,2\,3}{0\,4\,8} = 48 \times 10^{-3}$$

or, in standard form with one number or digit to the left of the decimal point,

$$0.048 = 4.8 \times 10^{-2}$$

Hence,

To express a decimal fraction in scientific notation, the negative exponent is equal to the number of places the decimal point is shifted to the right.

EXAMPLE 2.2 Expressing small numbers (decimal fractions) in scientific notation:

$$0.000037 = 3.7 \times 10^{-5}$$

$$0.0168 = 1.68 \times 10^{-2} = 16.8 \times 10^{-3}$$

$$0.0002 = 2 \times 10^{-4}$$

$$0.000001 = 1 \times 10^{-6} = 10^{-6}$$

(A numerical prefix of 1 is sometimes omitted as being understood.)

Changing the Power of Ten

It should be noted that a number can be expressed in different forms in powers-of-ten notation. Although it is customary to have one digit to the left of the decimal point in the numerical prefix, it is often convenient to have larger or smaller prefixes for mathematical operations. This is accomplished by shifting the decimal point, which changes the power of ten. The effect of this operation is as follows:

The exponent is decreased by one for every place the decimal point is shifted to the right.

The exponent is increased by one for every place the decimal point is shifted to the left.

For example,

$$6.4 \times 10^2 = 64 \times 10^1 = 0.64 \times 10^3$$

$$2.8 \times 10^{-3} = 28 \times 10^{-4} = 0.28 \times 10^{-2}$$

Multiplication and Division

One of the major advantages of scientific notation is the convenience of mathematical operations, particularly multiplication and division. Long multiplication or division with large and small numbers is prone to error. This is particularly true when many zeros are involved, because it is difficult to keep track of the zeros. Even on a calculator you may miss a zero.

However, there is no such problem when using scientific notation. Consider multiplying 800,000 times 120,000,000 or, in scientific notation, 8.0×10^5 times 1.2×10^8. Then, simply,

$$(8.0 \times 10^5)(1.2 \times 10^8) = 9.6 \times 10^{13}$$

For multiplication of numbers in scientific notation, the numerical prefixes are multiplied and the exponents of the powers of ten are added.

Division is just as simple. Consider dividing 120,000,000 by 600,000, or

$$(1.2 \times 10^8)/(6.0 \times 10^5).$$

For the division of numbers in scientific notation, the numerical prefixes are divided and the exponents of the powers of ten are subtracted (the exponent in the denominator is subtracted from the exponent in the numerator).

Then,

$$\frac{1.2 \times 10^8}{6.0 \times 10^5} = \frac{12 \times 10^7}{6.0 \times 10^5} = \frac{12}{6.0} \times 10^{(7-5)}$$
$$= 2.0 \times 10^2$$

where it was convenient to shift the decimal in the numerator before dividing. Notice that subtracting the exponents is equivalent to bringing the power of ten in the denominator to the numerator, which requires a sign change of the exponent, and adding the exponents.

EXAMPLE 2.3 Multiplication and division with scientific notation:

$$(0.04 \times 10^{-3})(2 \times 10^8) = (4 \times 10^{-5})(2 \times 10^8)$$
$$= 8 \times 10^3$$

$$(3 \times 10^{-7})(2 \times 10^{-3}) = 6 \times 10^{-10}$$

$$(6.0 \times 10^4)^2 = 36 \times 10^8 = 3.6 \times 10^9$$

$$(2.0 \times 10^{-3})^3 = 8.0 \times 10^{-9}$$

$$(8 \times 10^6)/(4 \times 10^{-2}) = 2 \times 10^{6-(-2)} = 2 \times 10^8$$

$$(1.0 \times 10^3)(6.0 \times 10^{-12})/(4.0 \times 10^5) =$$
$$\frac{(1.0 \times 6.0)}{4.0} \times 10^{(3-12-5)} = 1.5 \times 10^{-14}$$

Square Root

In taking the square root of a number in powers-of-ten notation, it is convenient for the exponent to be an even number.

As can be seen from the third operation in the previous example, when a number is squared the exponent is doubled. Since taking the square root is the reverse process, the exponent is halved or divided by 2 when taking the square root and, hence, should be even. The square root of the numerical prefix is taken in the ordinary manner.

EXAMPLE 2.4 Taking the square roots of numbers in powers-of-ten notation:

$$\sqrt{4 \times 10^8} = (4 \times 10^8)^{1/2} = 2 \times 10^4$$

The ½ power, i.e., $(\ )^{1/2}$, indicates the square root, as does the symbol $\sqrt{\ }$. As stated previously, the square root of the power of ten is taken by dividing by 2 or, equivalently, multiplying by ½.

$$(9 \times 10^{-4})^{1/2} = 3 \times 10^{-2}$$

$$(3.6 \times 10^{13})^{1/2} = (36 \times 10^{12})^{1/2} = 6.0 \times 10^6$$

$$(40 \times 10^{-7})^{1/2} = (4.0 \times 10^{-6})^{1/2} = 2.0 \times 10^{-3}$$

If the exponent is not even, as in the last two examples, the decimal is shifted to make it so.

Cube roots follow the same principle. When a number in scientific notation is cubed, the exponent is added three times or multiplied by three. Hence, in taking a cube root, it is convenient for the exponent to be divisible by three. The symbols here are $\sqrt[3]{\ }$ and $(\ )^{1/3}$.

Addition and Subtraction

Addition and subtraction of numbers expressed in scientific notation require that the exponents of the numbers be equal.

This requirement ensures that the decimal points are in column alignment, as is necessary when a column of numbers is added and subtracted. The powers of ten are made equal by shifting the decimal points of one or more of the numbers. Subtraction is simply a form of addition—the addition of negative quantities.

EXAMPLE 2.5 Addition and subtraction of numbers in powers-of-ten notation:

$$(2.4 \times 10^3) + (7.1 \times 10^3) = 9.5 \times 10^3$$

$$(9 \times 10^{-3}) - (0.3 \times 10^{-2}) =$$

$$(9 \times 10^{-3}) - (3 \times 10^{-3}) = 6 \times 10^{-3}$$

2.2 Angular Measure and Trigonometry

Angular measurement is commonplace in technical applications. Angles are commonly measured with a protractor, as shown in the introductory photograph of this chapter. Angles are associated with circular arcs. We say a particular angle *subtends* a particular portion of a circle, or arc length s, as illustrated in Figure 2.1.

Angles are commonly measured in degrees and, as is well known, there are 360° in a complete circle or revolution. Degrees are divided into smaller units of minutes and seconds: 1° = 60 minutes and 1 minute = 60 seconds. These second units have no relationship to the time unit of second.

In many instances, it is convenient to express an angle in terms of the corresponding arc length rather than in degrees. This is done in terms of radian measure. An angle in radians is equal to the ratio of the subtended arc length and the radius of the circle,

$$\theta = \frac{s}{r} \quad (\theta \text{ in radians}) \qquad \textbf{(Eq. 2.1)}$$

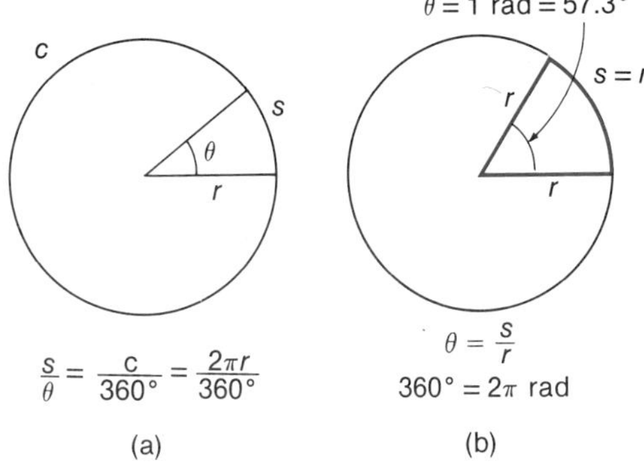

Figure 2.1
Angular measure. (a) An angle θ is commonly measured in degrees, with 360° in a complete revolution. The arc length s subtended by an angle θ is related to the angle and radius by the ratio shown in the figure. (b) Angles are also expressed in radian (rad) measure. An angle θ in radians is defined as the ratio of the arc length to the radius, $\theta = s/r$, and hence $\theta = 1$ rad when $s = r$. Then, $360° = 2\pi$ rad and $s = r\theta$.

and, if $s = r$, then $\theta = r/r = 1$ rad. Thus,

one radian (abbreviated rad) is the angle that subtends an arc length equal to that of the radius of the circle (Fig. 2.1).

To relate radian measure to degrees, we note that if the arc length is equal to the circumference c of the circle, $s = c = 2\pi r$. Then, by Equation 2.1,

$$\theta = \frac{s}{r} = \frac{2\pi r}{r} \qquad \textbf{(Eq. 2.2)}$$

$$= 2\pi \text{ rad} \quad \text{(for complete circle)}$$

Since there are 360° in a complete circle, we have

$$360° = 2\pi \text{ rad}$$

or

$$1 \text{ rad} = \frac{360°}{2\pi} = 57.3°$$

With 2π rad in a complete circle (360°), it follows that

$$180° = \pi \text{ rad}$$

$$90° = \pi/2 \text{ rad}$$

$$60° = \pi/3 \text{ rad}$$

$$45° = \pi/4 \text{ rad}$$

$$30° = \pi/6 \text{ rad}$$

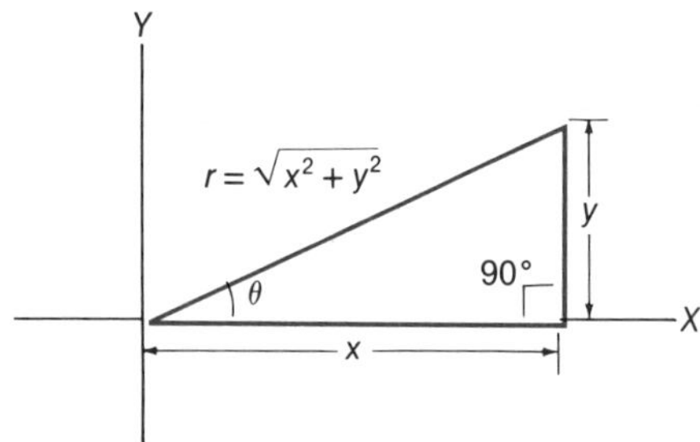

Figure 2.2
A right triangle. One of the interior angles of the triangle is equal to 90°. The sides or legs of the triangle (x and y) are related to the hypotenuse by Pythagoras' theorem.

and any of these relationships may be used as a conversion factor for degrees and radians.

From Equation 2.1 it can be seen that an angle expressed in radian measure is a pure number with no physical units, since it is the ratio of two lengths: $\theta = s/r$. An arc length of a circle is then given by the product of the circle radius and the angle in radians that the arc subtends, i.e. (from Eq. 2.1),

$$s = r\theta$$

EXAMPLE 2.6 What is the arc length that subtends an angle of 45° on a circle with a radius of 2.00 cm?

Solution: Since $r = 2.00$ cm and $\theta = 45° = \pi/4$ rad,

$$s = r\theta = (2.00 \text{ cm})(\pi/4) = \pi/2 = 1.57 \text{ cm}$$

Trigonometry

Trigonometry is the branch of mathematics that deals with the properties of triangles. Angular measure and triangles go hand in hand, since triangles involve angles. Of particular interest in many applications is the right triangle, which is a triangle with one interior angle equal to 90°, or a right angle (Fig. 2.2).

When the lengths of two sides of a right triangle are known, the length of the third side can be calculated from Pythagoras' theorem, which states that the sum of the squares of the sides or legs of a right triangle is equal to the square of the hypotenuse, i.e.,

$$r^2 = x^2 + y^2 \qquad \text{or} \qquad r = \sqrt{x^2 + y^2} \qquad \textbf{(Eq. 2.3)}$$

However, suppose you measure one of the interior non-right angles and the length of one side of a triangle. Can you calculate the lengths of the other two sides? Certainly not from Pythagoras' theorem, since it has no provision for angular measure. This is where trigonometry comes in.

Consider the triangles with the common 45° angle shown in Figure 2.3. The ratio of y/x is the same for all triangles: $1/1 = 1$, $2/2 = 1$, and $3/3 = 1$. In fact, the ratio of y/x is the same for any 45° right triangle. The ratio of y/x is given the special name of tangent (tan) in trigonometry, so we say

$$\tan 45° = 1$$

This idea can be generalized for any angle θ, and

$$\tan \theta = y/x$$

Rather than saying "the tangent of θ is y over x," it is sometimes convenient to say it is the ratio of "the side opposite (the angle θ)" and the "side adjacent (the angle θ)," i.e.,

$$\tan \theta = \frac{y}{x} = \frac{\text{side opposite}}{\text{side adjacent}} \qquad \textbf{(Eq. 2.4)}$$

This is particularly true when the sides of the triangle are not horizontal or vertical, as for triangle 3 in Figure 2.3.

The tangent is equal to the slope ($\Delta y/\Delta x$) of the hypotenuse. In figuring the slope of a roof or the grade of a hill (Fig. 2.4), we often say it is the "rise" (Δy, side opposite) over the "run" (Δx, side adjacent). The slope or tangent is commonly expressed as a percent, e.g., a 25% slope for an angle $\theta = 14°$ ($\tan 14° = 0.25$).

The ratios of y/x, i.e., the tangents, for various angles of right triangles have been tabulated in trigonometric tables, as found in Appendix 3. The angles in "trig" tables are usually listed in both degrees and radians. If you have a calculator with trig functions, you will note that it has a switch that can be set to either degrees or radians.

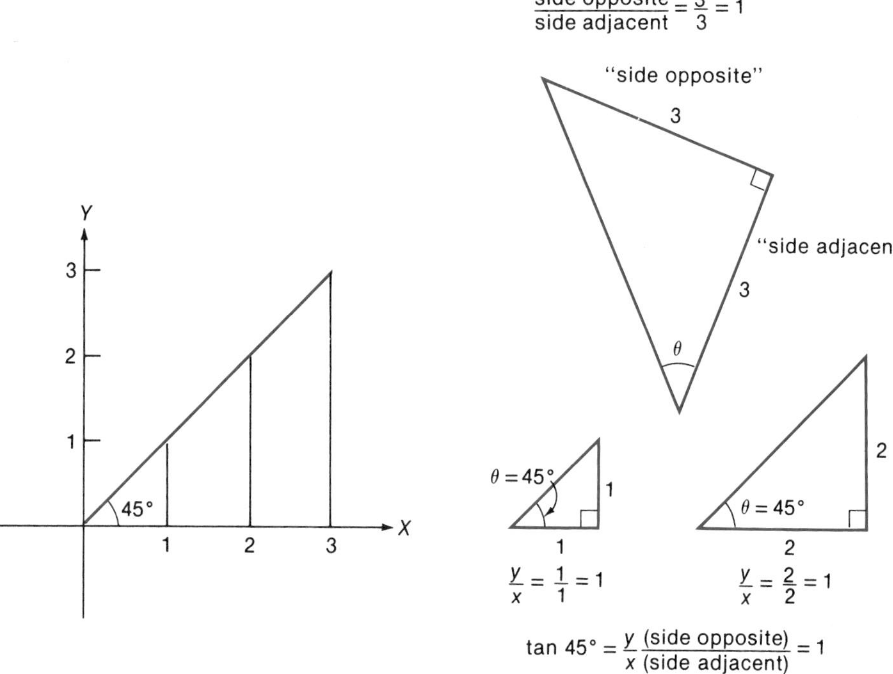

Figure 2.3

Basic trigonometry. The ratio of the sides of any 45° right triangle is equal to one. The ratio of y/x of a right triangle is called the tangent (tan) of the angle θ. Hence, tan 45° = 1. When the triangle is not oriented with the angle measured from the x-axis, it is convenient to say tan θ = side opposite/side adjacent. For the large triangle to the right, tan θ = 3/3 = 1, and θ = 45°.

Similarly, two other trigonometric functions, the sine (sin) and cosine (cos), are defined in terms of the ratios of the other combinations of the sides of a right triangle:

$$\sin\theta = \frac{y}{r} = \frac{\text{side opposite}}{\text{hypotenuse}}$$

$$\cos\theta = \frac{x}{r} = \frac{\text{side adjacent}}{\text{hypotenuse}} \qquad \textbf{(Eq. 2.5)}$$

These ratios are also constant for a particular angle θ, and the tabulated values are also found in the trig tables.

Now, back to the original question: Knowing one of the interior non-right angles and the length of one side of a right triangle, can you find the lengths of the other sides? With a knowledge of basic trigonometry, the answer is yes.

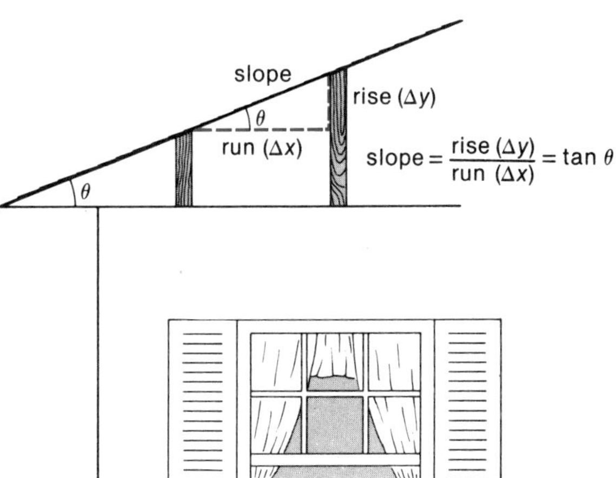

Figure 2.4

Slope. The slope of a roof or grade of a hill is equal to tan $\theta = y/x$, or the "rise over the run."

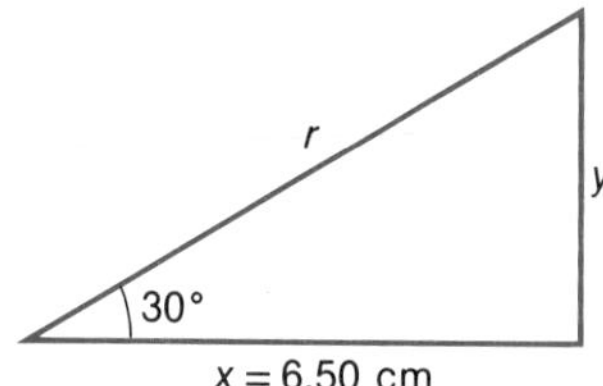

Figure 2.5
See Example 2.7.

EXAMPLE 2.7 A 30° right triangle has one leg (say x) equal to 6.50 cm. What are the lengths of the other two sides (Fig. 2.5)?

Solution: Since $\tan\theta = y/x$, then $y = x\tan\theta$, and

$$y = x\tan 30° = (6.50\text{ cm})(0.577) = 3.75\text{ cm}$$

where the value of tan 30° was obtained from the trig tables. The Pythagorean theorem could now be used to calculate the length of the hypotenuse of the triangle, but also

$$\sin\theta = \frac{y}{r} \quad \text{or} \quad r = \frac{y}{\sin\theta}$$

and

$$r = \frac{y}{\sin 30°} = \frac{(3.75\text{ cm})}{(0.500)} = 7.50\text{ cm}$$

Suppose the length of r were known initially instead of the length of x, and you wanted to find the length of x. In this case, the cosine function, which involves the ratio of x/r, would be used. Since $\cos\theta = x/r$, then $x = r\cos\theta$, and

$$x = r\cos 30° = (7.50\text{ cm})(0.866) = 6.50\text{ cm}$$

You will notice that in the tables the values of the trigonometric functions for the limit angles of 0° and 90° are given as

$$\cos 0° = 1 \qquad \cos 90° = 0$$

$$\sin 0° = 0 \qquad \sin 90° = 1$$

$$\tan 0° = 0 \qquad \tan 90° = \infty^*$$

* The symbol ∞ means infinity or an infinitely large number. Technically, the tangent approaches infinity as the angle approaches 90°. In terms of $\tan 90° = y/x$, as x approaches zero ($x \to 0$), $\tan\theta \to \infty$.

Table 2.1
Trig Functions for Some Angles

Angle				
Degree	***Radian***	**Cosine**	**Sine**	**Tangent**
0°	0 rad	1.00	0	0
30°	$\pi/6$ rad	0.866	0.500	(0.577)
45°	$\pi/4$ rad	0.707	0.707	1.00
60°	$\pi/3$ rad	0.500	0.866	(1.732)
90°	$\pi/2$ rad	0	1.00	∞*

* Approaches infinity.

Also notice that the sine of a particular angle θ is equal to the *co*sine of the *co*mplementary angle (the other interior non-right angle, i.e., $90° - \theta$). That is, $\sin\theta = \cos(90° - \theta)$ and $\cos\theta = \sin(90° - \theta)$. For example,

$$\sin 30° = \cos 60° = 0.500$$

and
$$\cos 30° = \sin 60° = 0.866$$

The cosine and sine of 45° are equal. Since $x = y$ in this instance, as we have seen, it follows that

$$\cos 45° = \sin 45° = 0.707 = \left(\frac{1}{\sqrt{2}}\right)$$

These functions are summarized in Table 2.1.

Inverse trigonometric functions can be used to find the values of angles in the cases in which the lengths of two sides of a triangle or their ratio is known. Also commonly called "arc" functions, the inverse functions are written arctangent (arctan), arccosine (arccos), and arcsine (arcsin).

As an example of an inverse function, consider $\theta = \arctan(0.268)$, which is read "θ is the angle whose tangent is 0.268." Finding the value of 0.268 in the tangent column in the trig tables and reading "backward," or from the tangent value to the angle, one sees that $\theta = 15°$ (or 0.262 rad).†

The inverse functions are also denoted by $\tan^{-1}$, $\cos^{-1}$, and $\sin^{-1}$. (This is the form usually found on calculators.) For example, the arctangent in the previous example can be written

† When finding an inverse function on a calculator, it is important to note whether it is set on the DEG (degree) or RAD (radian) mode.

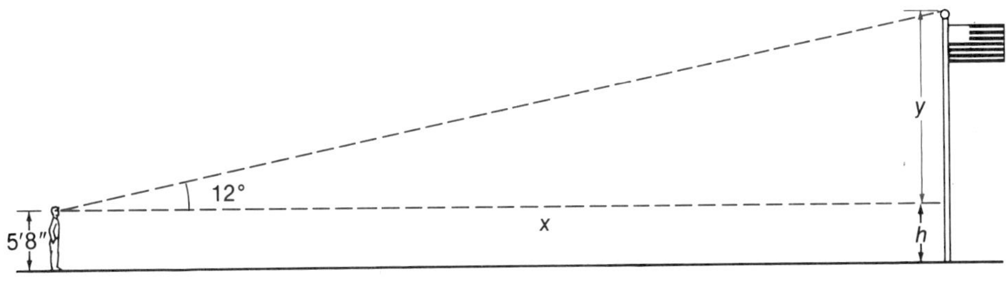

Figure 2.6
See Example 2.8.

$\theta = \tan^{-1}(0.268)$. The negative exponent *does not* mean the reciprocal as in the algebraic case of $10^{-1} = \frac{1}{10}$.

EXAMPLE 2.8 A student 5 ft, 8 in. tall stands 100 ft from a flag pole. Sighting on the top of the flag pole, the student measures the angle between the line of sight and the horizontal to be 12°. What is the height of the flag pole?

Solution: Using the triangle shown in Figure 2.6, with $x = 100$ ft, $\theta = 12°$, and $\tan\theta = y/x$ or $y = x \tan\theta$,

$$y = x \tan 12° = (100 \text{ ft})(0.213) = 21.3 \text{ ft}$$

Then, assuming the student's eye level to be about 4 inches below the top of the head, the height of the sighting level is $h = 5$ ft, 4 in. $= 5.3$ ft. The height of the flag pole from the ground is then $y + h = 21.3$ ft $+ 5.3 = 26.6$ ft.

On a Cartesian x-y graph, the standard angle is referenced to the positive (+) x-axis. However, a particular triangle may not be in the first quadrant; for example, if the legs of the triangle are $-x$ and $+y$, the triangle lies in the second quadrant (Fig. 2.7). In such a case, the standard angle is greater than 90°, and you may have noticed that the trig tables only list angles to 90°. But we are really interested in an interior angle of the triangle, and the interior angle measured from the $-x$-axis is equal to $180° - \theta = 180° - 110° = 70°$ in this case.

The cosine of an angle in the second quadrant is negative because

$$\frac{\text{Adj}}{\text{Hyp}} = \frac{-x}{r} = -\left(\frac{x}{r}\right) = -\cos\theta$$

(θ is measured from the $-x$-axis)

and

$$-\cos 70° = -0.342$$

The sine in the second quadrant is positive. Why?

The cosine and sine of a standard angle greater than 90° (and less than 180°) are also given by the identities $\cos\theta = -\cos(180° - \theta)$ and $\sin\theta = \sin(180° - \theta)$. (See Appendix 3.) However, it is usually more convenient to use a graphical approach, as was done in the previous examples, to determine the sign of a particular function in a given quadrant. For example, the sine of an angle in the third quadrant (measured from the $-x$-axis) is negative, i.e., $-y/r = -(y/r) = -\sin\theta$.

A convenient device to help you remember the signs of the trig functions in the various quadrants is the phrase *A*ll-*S*tate *T*echnical *C*ollege or some similar saying. The letters indicate which trig functions are *positive* in a particular quadrant (other functions are negative): first quadrant, *A*ll functions are positive; second quadrant, the *S*ine is positive; third quadrant, the *T*angent is positive; and fourth quadrant, the *C*osine is positive.

If you use a calculator to find the value of a trig function, the sign will be indicated when the standard reference angle is entered, e.g., $\cos 110° = -0.342$ and $\sin 230° = -0.766$.

The trigonometry of non-right triangles is discussed in the Chapter Supplement.

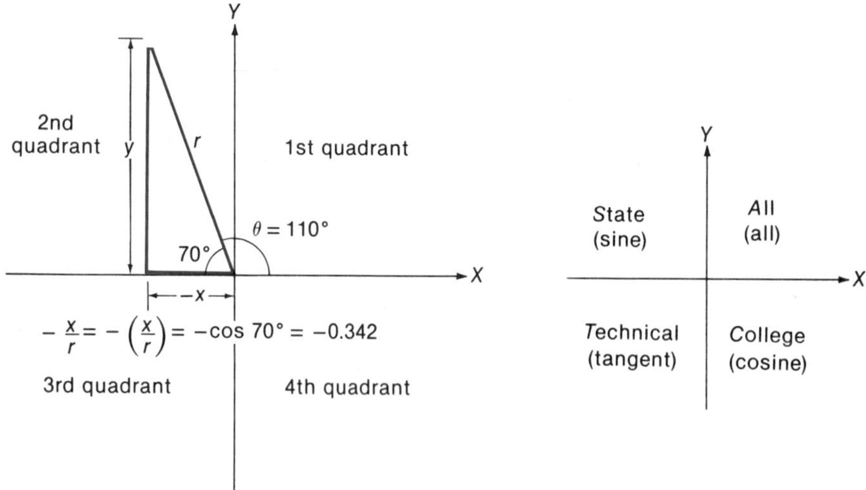

Figure 2.7
Angles greater than 90°. Trigonometric functions for angles greater than 90° are given by identity relationships. However, it is usually more convenient to measure the angle from the x-axis in that quadrant, e.g., $-x/r = -\cos 70° = \cos 110°$. A convenient memory device to know which functions are positive in the various quadrants is shown to the right.

2.3 Vectors

Physical quantities are classified as being either scalars or vectors.

A scalar quantity has magnitude only (along with units). For example, temperature measurement is a scalar quantity, having only magnitude or numerical value, e.g., 68°. Another example is speed. The speed of an automobile may be reported as 25 mph.

A vector quantity has magnitude *and* direction. For example, displacement is a vector. The magnitude of a displacement is the straight-line distance between two points, e.g., the three blocks illustrated in Figure 2.8. If the automobile travels this distance going east, then its vector displacement is three blocks *eastward*—a magnitude *plus* direction. Similarly, in expressing the velocity (a vector) of the automobile, we might say that it traveled 25 mph eastward. The velocity vector has magnitude (25 mph) and direction (eastward).

Vectors are represented graphically by arrows (Fig. 2.9). The length of the arrow is proportional to the magnitude of the vector quantity, and the arrowhead indicates its direction. In textbooks, vectors are commonly represented by boldface letters (e.g., **A**) and the scalar (magnitude) proportion of the vector is represented by italic letters (e.g., A). To indicate a vector when writing on a blackboard or paper, a common notation is $\vec{A}$.

The magnitude and direction of a vector are indicated explicitly by $\mathbf{A} = A\mathbf{a}$, where **a** is a unit vector. A unit vector has a magnitude of one (unity) and points in the direction of **A**. Hence, the notation $\mathbf{A} = A\mathbf{a}$ is read "The vector **A** has a magnitude of A in the direction of **a**." Two vec-

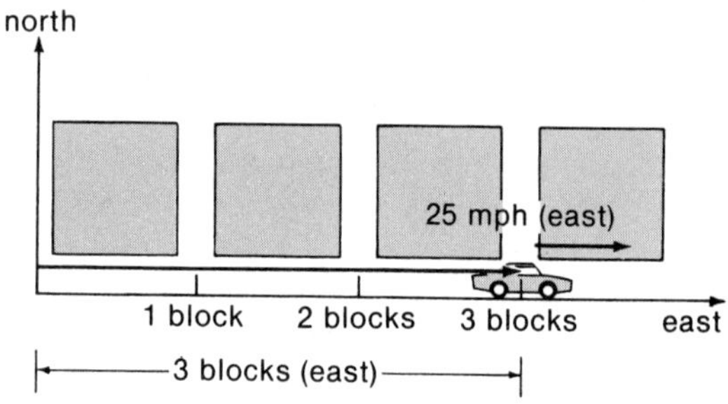

Figure 2.8
Displacement is a vector quantity. The displacement of the automobile is 3 blocks (magnitude) eastward (direction). Similarly, velocity is a vector quantity. The car is traveling 25 mi/h (magnitude) eastward (direction).

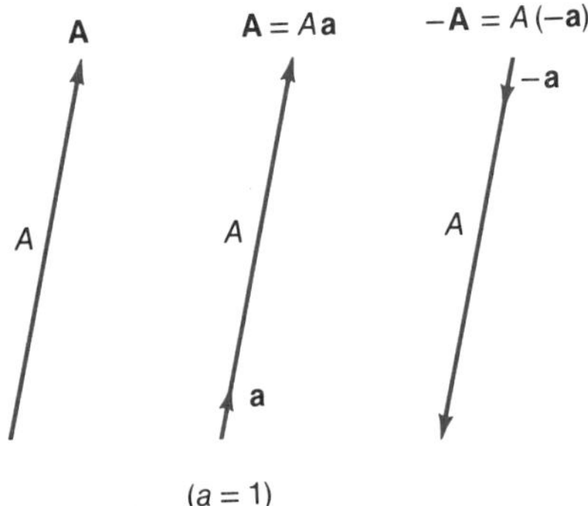

Figure 2.9
Vector representation. Vectors may be represented by arrows; the length of an arrow is proportional to the magnitude, and the arrowhead indicates the vector direction. A unit vector **a** with a magnitude of one gives direction. A vector −**A** is equal in magnitude to **A**, but is in the opposite direction (−**a**).

tors that have the same magnitude and direction are said to be *equal* (Fig. 2.9).

The negative vector of a given vector **A** is defined as a vector having the same magnitude as **A** and a direction opposite to that of **A**. The negative vector of **A** is written −**A** (Fig. 2.9). When written explicitly, $-\mathbf{A} = -A\mathbf{a} = A(-\mathbf{a})$, where the negative goes with the unit vector, since A is simply a magnitude or number without direction. The unit vector −**a** is in the opposite direction of **a**. The vectors **A** and −**A** are commonly referred to as being equal and opposite.

Vector Addition (and Subtraction)

Since vectors have both magnitude and direction, the common scalar addition does not in general apply to the addition of vectors. The direction of a vector is an important consideration in vector addition. Descriptions of several methods of vector addition follow.

Parallelogram Method To add two vectors **A** and **B** by the parallelogram method, a parallelogram is constructed with **A** and **B** as two sides, as shown in Figure 2.10. The diagonal vector **R** is then the vector sum or the resultant of **A** and **B**, i.e.,

$$\mathbf{A} + \mathbf{B} = \mathbf{R} \qquad \text{(Eq. 2.6)}$$

The resultant **R** is described by its magnitude and direction, with the direction usually being specified by an angle θ relative to **A** or **B** (usually the lower vector). These properties can be obtained by graphical analysis. If the vector sides of the parallelogram are drawn to scale, e.g., on graph paper, the length of the resultant can be measured and its magnitude determined according to the scale of the drawing. The angular direction of the resultant can be measured with a protractor.

Physically, the resultant vector quantity is equivalent to the combined effect of the added vector quantities. For example, suppose the vectors $\mathbf{F}_A$ and $\mathbf{F}_B$ represent forces being applied by pulls on ropes tied to an object (Fig. 2.11). The same effect would be accomplished by pulling on a single rope in the direction of $\mathbf{F}_R$ with a force of that magnitude.

Triangle Method The triangle method may be thought of as a simplification of the parallelogram method. Notice in Figure 2.12 that the side of the parallelogram opposite **B** is the same

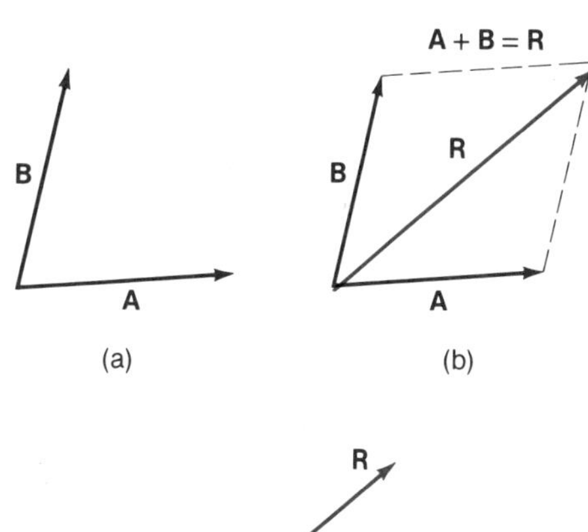

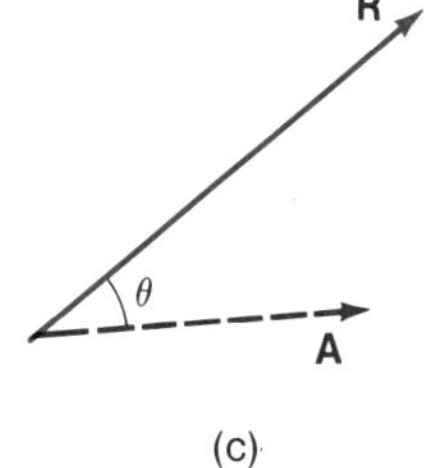

Figure 2.10
Vector addition—parallelogram method. (a) The vector sum of **A** + **B** is given by completing the parallelogram. (b) The vector sum or resultant is the diagonal vector **R**. (c) The direction of **R** is specified by the angle θ relative to one of the vectors.

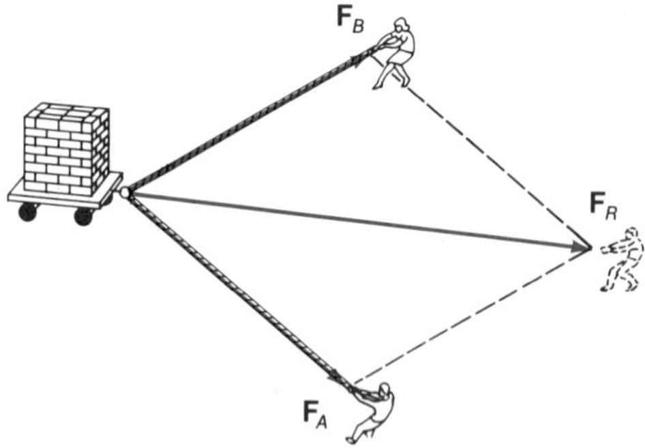

Figure 2.11
Vector physical equivalents. The resultant force vector $\mathbf{F}_R$ is physically equivalent to the combined effect of $\mathbf{F}_A$ and $\mathbf{F}_B$.

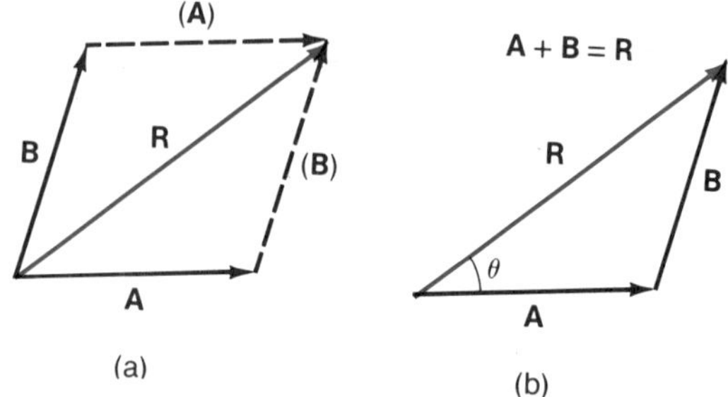

Figure 2.12
Vector addition—triangle method. (a) In the parallelogram method, the side of the parallelogram opposite **B** is the same as **B**. (b) The resultant **R** is found by shifting **B** to form a triangle, i.e., by placing **B** so the "tip" of **A** is connected to the "tail" of **B**. The resultant **R** is the third side of the triangle.

length or magnitude as **B**. By indicating the direction with an arrowhead, this opposite side is equivalent to **B** (Fig. 2.12).

> **In general, for analysis purposes, a vector may be translated or moved as long as its length (magnitude) and orientation (direction) are not changed.**

Hence, the **B** vector is placed so the "tip" of **A** is connected to the "tail" of **B**, and the completed triangle gives the result **R** as the third side of the triangle. The properties of **R** can again be determined graphically.

Also, by use of the mathematical relationships of triangles, the properties can be determined analytically. This is particularly easy when the triangle is a right triangle. Otherwise, the laws of sines and cosines must be used. See the Chapter Supplement.

EXAMPLE 2.9 A man walks three blocks east and then four blocks north. What is his resultant displacement (magnitude of straight-line distance and direction) from his starting point?

Solution: The individual displacements the man walked are vector quantities. Letting east and north correspond to the x and y directions, respectively, on a graph, the displacement vectors can be written $\mathbf{A} = A\mathbf{a} = 3\mathbf{x}$ and $\mathbf{B} = B\mathbf{b} = 4\mathbf{y}$, where **x** and **y** are unit vectors in the x and y directions. The vectors are plotted in Figure 2.13 using the triangle "tip-to-tail" method.

The resultant displacement **R** can be determined from the graph by measuring its length (magnitude) and its directional angle. However, the properties of **R** can also be conveniently determined by analytical methods. Since the triangle in this case is a right triangle, the magnitude of **R** is given by Pythagoras' theorem:

$$R = \sqrt{x^2 + y^2} = \sqrt{(3)^2 + (4)^2} = 5 \text{ blocks}$$

To specify the angular direction of **R** relative to **A**, by trigonometry

$$\tan \theta = \frac{y}{x} = \frac{4}{3} = 1.33$$

and

$$\theta = \tan^{-1} (1.33) = 53°$$

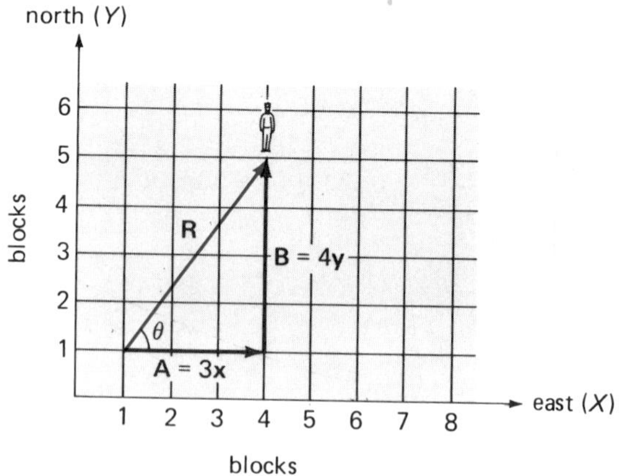

Figure 2.13
See Example 2.9.

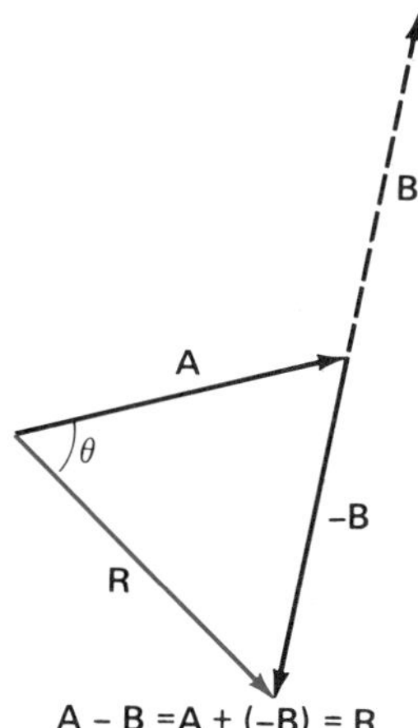

Figure 2.14
Vector subtraction. Vector subtraction is a special case of vector addition. The resultant of **A** − **B** is obtained by adding −**B** to **A**, that is, **A** + (−**B**). The vector −**B** is in the opposite direction of **B**.

Vector subtraction is simply a special case of vector addition. The subtraction of a vector is defined as the addition of the corresponding negative vector. That is, the vector **R** = **A** − **B**, representing the vector difference between **A** and **B**, is obtained by adding the negative vector −**B** to **A** (Fig. 2.14), as can be seen from

$$\mathbf{A} - \mathbf{B} = \mathbf{A} + (-\mathbf{B}) \qquad \textbf{(Eq. 2.7)}$$

Polygon Method The polygon method of vector addition is an extension of the triangle method for the addition of several vectors, e.g., **A** + **B** + **C**. As illustrated in Figure 2.15, the vectors are placed tip-to-tail and the resultant **R** of the three vectors is formed by drawing a vector from the tail of the initial vector to the tip of the last vector, which completes a polygon. From the second diagram of the figure, it can be seen that this is equivalent to adding **A** + **B** by the triangle method, then adding the resultant of **A** + **B** to **C** by another triangle.

This method can be used for any number of vectors and is convenient for graphical analysis. However, analytical analysis can become quite involved. The analytical analysis of the sum of several vectors is more convenient by the component method.

Resolving a Vector and the Component Method

In Example 2.9, two vectors along the x and y axes were added to obtain the resultant **R**. Essentially, **R** was made up of two component vectors. In general, we might think of any vector in the x-y plane as being composed of two component vectors along the x and y axes. Hence, we can resolve the vector into rectangular compo-

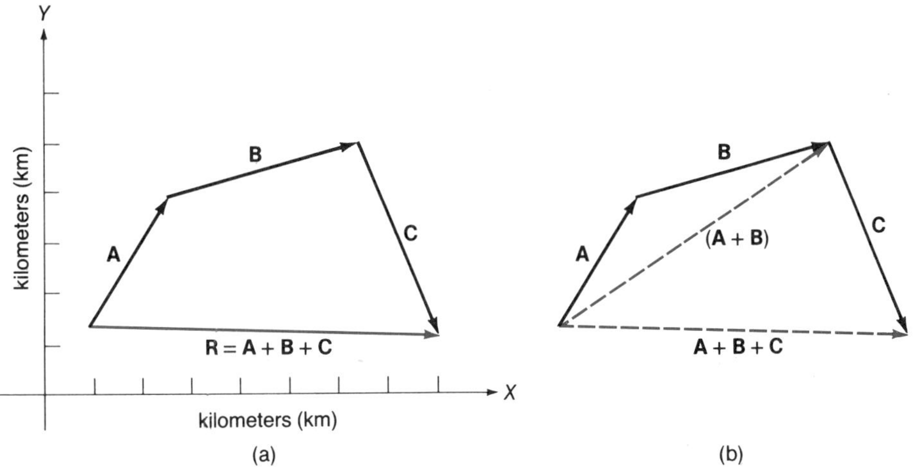

Figure 2.15
Vector addition—polygon method. (a) The sum of three or more vectors is given by the side of the polygon formed by placing the vectors "tip to tail." (b) The polygon method is an extension of the triangle method. Note that the resultant of **A** + **B** when added to **C** is the resultant **A** + **B** + **C**.

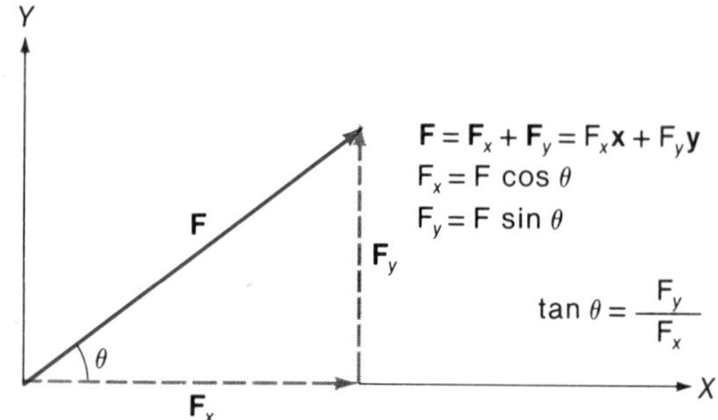

Figure 2.16
Vector components. The vector **F** can be resolved into rectangular components $\mathbf{F}_x$ and $\mathbf{F}_y$, where $\mathbf{F}_x + \mathbf{F}_y = \mathbf{F}$.

nents. For example, the vector **F** in Figure 2.16 is the sum (or is composed) of $\mathbf{F}_x$ and $\mathbf{F}_y$, i.e., $\mathbf{F} = \mathbf{F}_x + \mathbf{F}_y$. In resolving a vector into rectangular components, we essentially "break down" the vector into x and y components—sort of the inverse of addition.

Given the magnitude of **F** and its directional angle θ relative to the x-axis, by trigonometry the magnitudes of the rectangular components are given by

$$F_x = F \cos \theta$$
$$F_y = F \sin \theta \qquad \textbf{(Eq. 2.8)}$$

Hence, we may write

$$\mathbf{F} = \mathbf{F}_x + \mathbf{F}_y = (F \cos \theta)\mathbf{x} + (F \sin \theta)\mathbf{y}$$

Vector addition by the component method makes use of resolving vectors into rectangular components. The procedure is as follows:

1. Resolve the vectors to be added into their x and y components.
2. Add all the x components together vectorially and all of the y components together vectorially to obtain the resultants of the x and y components, respectively.
3. Add the resultants of the rectangular components vectorially to obtain the resultant sum of the initial vectors.

Expressing this mathematically, suppose we wish to find the resultant of $\mathbf{F}_1 + \mathbf{F}_2 + \mathbf{F}_3$ (= **F**). Then, resolving each vector into rectangular components and adding vectorially (Steps 1 and 2),

$$\mathbf{F}_x = \mathbf{F}_{x1} + \mathbf{F}_{x2} + \mathbf{F}_{x3}$$
$$= (F_1 \cos \theta_1)\mathbf{x} + (F_2 \cos \theta_2)\mathbf{x} + (F_3 \cos \theta_3)\mathbf{x}$$
$$\mathbf{F}_y = \mathbf{F}_{y1} + \mathbf{F}_{y2} + \mathbf{F}_{y3}$$
$$= (F_1 \sin \theta_1)\mathbf{y} + (F_2 \sin \theta_2)\mathbf{y} + (F_3 \sin \theta_3)\mathbf{y}$$

and (Step 3)

$$\mathbf{F} = \mathbf{F}_x + \mathbf{F}_y = F_x\mathbf{x} + F_y\mathbf{y}$$

Since $\mathbf{F}_x$ and $\mathbf{F}_y$ are rectangular components, they form a right triangle, and the magnitude of **F** and its angular direction relative to the x-axis are given by

$$F = \sqrt{F_x^2 + F_y^2}$$
$$\theta = \tan^{-1}\left(\frac{F_y}{F_x}\right) \qquad \textbf{(Eq. 2.9)}$$

Note that a vector can be represented either in rectangular form with unit vectors or in magnitude-angle form.

EXAMPLE 2.10 Find the resultant of the vectors shown in Figure 2.17 by the component method.

Solution: The vectors represent forces with magnitudes given in newtons (N). We will say more about forces in a later chapter. The focus here is the vector addition.

Step 1. Resolve the vectors into x and y components. (See Fig. 2.17.)

Step 2. Vectorially add all the x components together and all the y components together.

$$\mathbf{F}_x = \mathbf{F}_{x1} + \mathbf{F}_{x2} + \mathbf{F}_{x3}$$
$$= (F_1 \cos 30°) - (F_2 \cos 45°) + 0$$
$$= (4.00 \text{ N})(0.866) - (6.00 \text{ N})(0.707)$$
$$= 3.46 - 4.24 = -0.78 \text{ N}$$

where the directions of the components are indicated by plus and minus signs; for example, the x component ($\mathbf{F}_{x2}$) of $\mathbf{F}_2$ is in the $-x$ direction, so $\mathbf{F}_{x2} = -(F_2 \cos 45°)$. Another way of looking at this is that the cosine of 45° in the second quadrant is negative as discussed in the previous section. Since $\mathbf{F}_3$ is along the negative y-axis, it has no x component, i.e., $\mathbf{F}_{x3} = 0$.

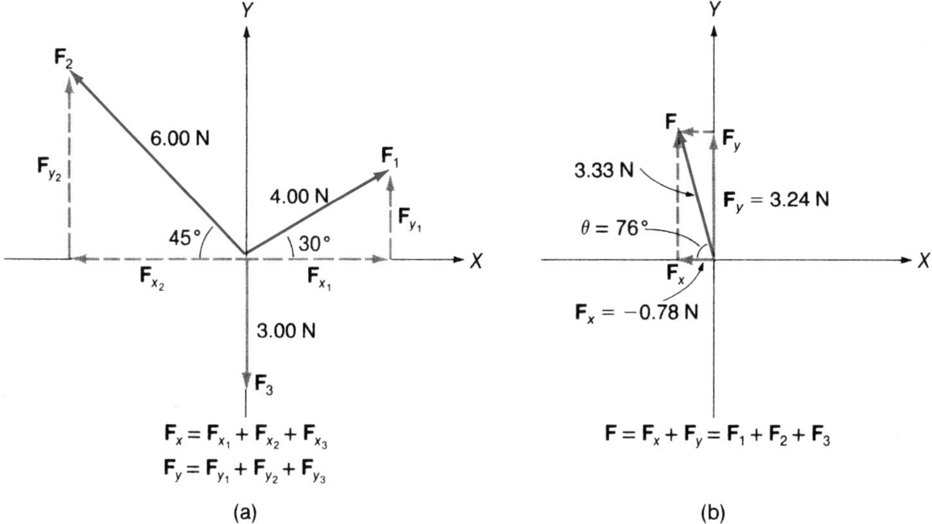

Figure 2.17
Vector addition—component method. (a) Several vectors may be added by resolving the vectors into rectangular components and summing the x and y components. (b) The resultant of the summed components is the resultant of the original vectors. See Example 2.10.

Similarly, for the y components,

$$\begin{aligned}\mathbf{F}_y &= \mathbf{F}_{y1} + \mathbf{F}_{y2} + \mathbf{F}_{y3} \\ &= (F_1 \sin 30°) + (F_2 \sin 45°) - F_3 \\ &= (4.00\text{ N})(0.500) + (6.00\text{ N})(0.707) \\ &\qquad - 3.00\text{ N} \\ &= 2.00 + 4.24 - 3.00 = 3.24\text{ N}\end{aligned}$$

Step 3.

$$\mathbf{F} = \mathbf{F}_x + \mathbf{F}_y = -0.78\text{ N }\mathbf{x} + 3.24\text{ N }\mathbf{y}$$

or, in magnitude-angle form,

$$\mathbf{F} = \sqrt{F_x^2 + F_y^2} = \sqrt{(0.78)^2 + (3.24)^2} = 3.33\text{ N}$$

and

$$\theta = \tan^{-1}\left|\frac{F_y}{F_x}\right| = \tan^{-1}\left(\frac{3.24}{0.78}\right) = 76°$$

where the angle is measured relative to the $-x$-axis. This means that the resultant **F** is in the second quadrant since $\mathbf{F}_x$ is in that direction, as illustrated in (b) in Figure 2.17.

Learning to resolve vectors into components is *very important*. You will be doing this over and over again in the following chapters, not only to find a vector sum by the component method, but also to find the effects of a particular vector quantity in a component direction. Many quantities are represented as vectors—for example, velocity, acceleration, force, and electric and magnetic fields.

A vector may be resolved into any number of components. This is similar to the reverse process of adding a number of vectors by the polygon method. All of the added vectors could be considered to be components of the resultant. However, in general, vectors are resolved into only two (x and y) or three (x, y, and z) rectangular components for analysis.

In some instances, a vector is resolved into components that do not correspond to the standard horizontal and vertical x–y axes, as the following example shows. However, the components are still rectangular, i.e., 90° apart.

EXAMPLE 2.11 A block of mass m rests on an inclined plane as shown in Figure 2.18. Resolve the weight force ($w = mg$) of the block into two components, one parallel to the plane and one perpendicular to the surface of the plane.

Solution: The weight force is vertically downward, and it has components parallel and perpendicular to the plane as shown in Figure 2.18.

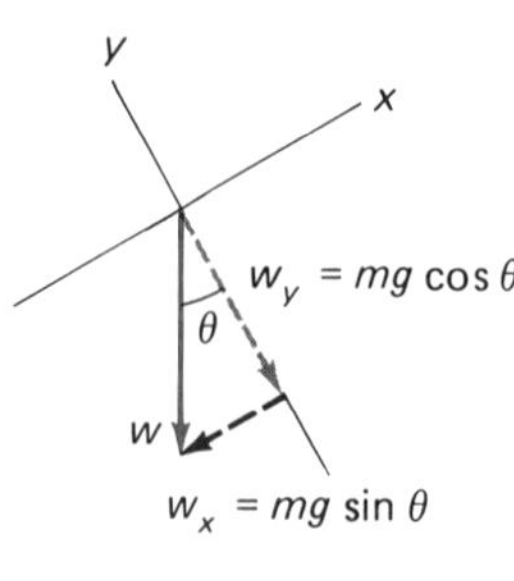

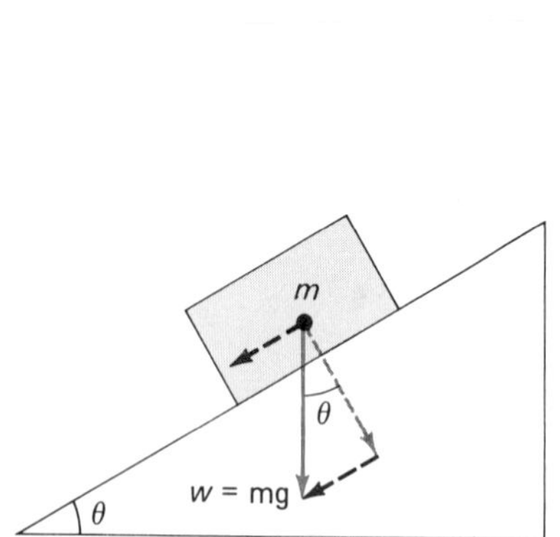

Figure 2.18
Vector resolution. The resolution of the weight force of a block on an inclined plane into components parallel and perpendicular to the surface of the plane. See Example 2.11.

The parallel component is drawn twice for illustration (shifted from the vector triangle to the center of the block). The angle θ in the vector triangle is the same as the plane's angle of incline by similar triangles. The components are rectangular, as may be seen by considering the x–y axes to be oriented along and perpendicular to the surface of the plane.

Then, using the appropriate trig relationships, e.g., $\sin\theta$ = side opposite/hypotenuse, we have (component parallel to the surface of the plane)

$$w_x = w\sin\theta = mg\sin\theta$$

and (component perpendicular to the surface of the plane)

$$w_y = w\cos\theta = mg\cos\theta$$

Important Terms

scientific (powers of ten) notation a notation that allows very large numbers and very small decimal fractions to be expressed conveniently in a "shorthand" form using powers of ten (see general rules below)

radian (rad) a unit of angular measure; one radian is the angle that subtends an arc length equal to that of the radius of the circle

scalar a quantity that has magnitude only

vector a quantity that has both magnitude *and* direction

General Rules for Scientific (Powers of Ten) Notation Operations

1. Expressing number:
   The exponent or power of ten is equal to
   (a) the number of places the decimal point is shifted to the left (positive exponent) for large numbers,
   (b) the number of places the decimal point is shifted to the right (negative exponent) for decimal fractions.
2. Shifting decimal:
   The exponent or power of ten is
   (a) decreased by one for every place the decimal point is shifted to the right,
   (b) increased by one for every place the decimal point is shifted to the left.
3. Multiplication: exponents are added
   Division: exponents are subtracted
   Addition and subtraction: exponents must be equal and remain the same
   Taking root:
   Square root—exponent should be even
   Cube root—exponent should be divisible by three

Important Formulas

degrees and radians:	$360° = 2\pi$ rad
or	1 rad = 57.3°
arc length of a circle:	$s = r\theta$
Pythagoras' theorem:	$r^2 = x^2 + y^2$
or	$r = \sqrt{x^2 + y^2}$

trigonometric functions:

$$\tan\theta = \frac{y}{x} \quad \left(\frac{\text{side opposite}}{\text{side adjacent}}\right)$$

$$\sin\theta = \frac{y}{r} \quad \left(\frac{\text{side opposite}}{\text{hypotenuse}}\right)$$

$$\cos\theta = \frac{x}{r} \quad \left(\frac{\text{side adjacent}}{\text{hypotenuse}}\right)$$

vector addition:

Triangle method (vectors forming right triangle):

$$\mathbf{A} + \mathbf{B} = \mathbf{R}$$

$$R = \sqrt{A^2 + B^2}$$

$$\theta = \tan^{-1}\left(\frac{B}{A}\right)$$

Component method—magnitudes of x and y components of **F**:

$$F_x = F\cos\theta$$

$$F_y = F\sin\theta$$

Summation of components:

$$\mathbf{F}_x = \mathbf{F}_{x_1} + \mathbf{F}_{x_2} + \mathbf{F}_{x3}$$

$$= (F_1\cos\theta_1)\mathbf{x} + (F_2\cos\theta_2)\mathbf{x} + (F_3\cos\theta_3)\mathbf{x}$$

$$\mathbf{F}_y = \mathbf{F}_{y1} + \mathbf{F}_{y2} + \mathbf{F}_{y3}$$

$$= (F_1\sin\theta_1)\mathbf{y} + (F_2\sin\theta_2)\mathbf{y} + (F_3\sin\theta_3)\mathbf{y}$$

and

$$\mathbf{F} = \mathbf{F}_x + \mathbf{F}_y = F_x\mathbf{x} + F_y\mathbf{y} \qquad \text{(rectangular form)}$$

Since $\mathbf{F}_x$ and $\mathbf{F}_y$ are rectangular components, they form a right triangle, and the magnitude of **F** and its angular direction relative to the x-axis are given by

$$F = \sqrt{F_x^2 + F_y^2}$$

$$\theta = \tan^{-1}\left(\frac{F_y}{F_x}\right) \qquad \text{(magnitude-angle form)}$$

Questions

Scientific Notation (Powers of Ten)

1. In expressing numbers in scientific notation, how is the numerical prefix usually expressed?
2. What is the value of 3.0×10^0?
3. Give the value of each of the following.
   (a) 10^{-3} (b) 10^{-2} (c) 10^3 (d) 10^6 (e) 10^9
4. What metric prefixes would describe the powers of ten given in Exercise 3?
5. If the decimal of the prefix in a scientific notation is (a) shifted two places to the left or (b) shifted three places to the right, how is the power of ten affected?

Angular Measure and Trigonometry

6. What is radian measure and how is it defined?
7. What are the physical units of the radian?
8. How is the arc length of a circle related to the angle the arc subtends?
9. What is the difference between a right triangle and an oblique triangle? What is the sum of the interior angles of each?
10. How is the sine of an angle related to the cosine of the complementary angle in a right triangle?
11. Show that $\tan\theta = \sin\theta/\cos\theta$.
12. What are inverse trigonometric functions and how are their values determined?

Vectors

13. Distinguish between a scalar and a vector, and give an example of each.
14. What is a unit vector?
15. Are scalars and vectors added by the same rules of addition? Explain.
16. What is meant by resolving a vector into components?
17. Into how many components may a vector be resolved? Explain.

Problems

Levels of difficulty are indicated by asterisks for your convenience.

2.1 Scientific Notation (Powers of Ten)

1. Express the following numbers in scientific notation with one digit to the left of the decimal.
   (a) 5280 (b) 0.0105 (c) 10,000 (d) 0.0001 (e) 8,500,000 (f) 0.00000092
2. Convert the following numbers in scientific notation to regular number notation.
   (a) 2.1×10^{-2} (b) 1.0×10^4 (c) 31.4×10^{-3} (d) 0.0673×10^6 (e) 1.67×10^{-5} (f) 7.856×10^2
3. Express the following numbers with two digits to the left of the decimal.
   (a) 1.0×10^3 (b) 0.32×10^{-2} (c) 0.045×10^5 (d) 569×10^{-6}

*4. Perform the following operations:
(a) $(2.0 \times 10^2)(6.0 \times 10^3)$
(b) $(3.5 \times 10^4)(2.0 \times 10^{-3})$
(c) $(5.0 \times 10^{-4})(3.2 \times 10^{-2})$
(d) $\dfrac{(1.0 \times 10^{-5})(8.6 \times 10^{12})}{2.0 \times 10^4}$

*5. Perform the following operations:
(a) $6.6 \times 10^5/1.1 \times 10^3$
(b) $9.0 \times 10^6/3.0 \times 10^{-3}$
(c) $8.0 \times 10^{-2}/4.0 \times 10^{-6}$
(d) $\dfrac{1.0 \times 10^{-3}}{(2.5 \times 10^{-12})(2.0 \times 10^6)}$

*6. Perform the following operations by long multiplication and division, then convert the numbers to scientific notation and perform the operations (assume two significant figures):
(a) (4500)(2,000,000)
(b) (30,000)(0.00015)
(c) 6000/0.0012
(d) 0.060/15,000
(e) 0.00081/0.027

*7. Compute the following:
(a) $(4.0 \times 10^6)^{1/2}$
(b) $(9.0 \times 10^{12})^{1/2}$
(c) $(1.6 \times 10^5)^{1/2}$
(d) $(1.0 \times 10^{-3})^{1/2}$
(e) $(25{,}000{,}000)^{1/2}$

*8. Compute the following:
(a) $(8.0 \times 10^{12})^{1/3}$
(b) $(1.0 \times 10^{15})^{1/3}$
(c) $(2.7 \times 10^{-8})^{1/3}$
(d) $(1.25 \times 10^8)^{1/3}$

*9. Compute the following sums and differences (assume any number of significant figures):
(a) $(4.32 \times 10^4) + (2.5 \times 10^3)$
(b) $(9.2 \times 10^7) + (6.0 \times 10^8)$
(c) $(1.2 \times 10^{-5}) + (3.5 \times 10^{-4})$
(d) $(6.6 \times 10^8) - (1.1 \times 10^8)$
(e) $(4.0 \times 10^5) - (2.0 \times 10^4)$
(f) $(9.5 \times 10^{-2}) - (4.0 \times 10^{-3})$

2.2 Angular Measure and Trigonometry

10. Express the following angles in radians (in terms of π).
(a) 90° (d) 270°
(b) 15° (e) 300°
(c) 120° (f) 540°

11. Express the following angles in degrees.
(a) $\pi/4$ rad (d) $\pi/24$ rad
(b) $5\pi/6$ rad (e) 1.1 rad
(c) $3\pi/4$ rad (f) 2.4 rad

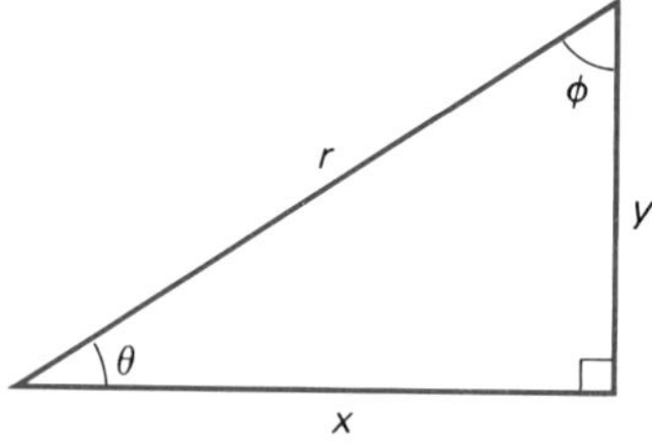

Figure 2.19
See Problems 14–22.

*12. A circle has a radius of 2.0 m. What are the arc lengths subtended by the following angles?
(a) $\pi/2$ rad (c) π rad
(b) 30° (d) 45°

*13. An angle of 60° on a circle subtends an arc length of 4.0 cm. What is the diameter of the circle?

(Problems 14–16: two sides of a right triangle are known.)

*14. If the right triangle in Figure 2.19 has sides x = 4.0 m and y = 3.0 m, what are (a) the length of side r; (b) $\sin\theta$, $\cos\theta$, and $\tan\theta$; and (c) $\sin\phi$, $\cos\phi$, and $\tan\phi$? Note any relationships among the θ and ϕ trig functions. (d) What are θ and ϕ in degrees and in radians?

*15. Let x = 3.0 m and r = 6.0 m in Figure 2.19. What are (a) the length of side y; and (b) $\sin\theta$, $\cos\theta$, and $\tan\theta$? (c) What are θ and ϕ in degrees and radians?

*16. Let y = 2.5 m and r = 7.0 m in Figure 2.19. What are (a) the length of side x; and (b) $\sin\theta$, $\cos\theta$, and $\tan\theta$? (c) What are θ and ϕ in degrees and radians?

(Problems 17–22: one side and one angle of a right triangle are known. Use trig relationships whenever possible in solutions.)

*17. Let x = 4.0 m and θ = 30° in Figure 2.19. What are the lengths of sides y and r?

*18. Let y = 6.0 m and θ = 45° in Figure 2.19. What are the lengths of the sides x and r?

*19. Let r = 10 m and θ = 60° in Figure 2.19. What are the lengths of x and y?

**20. In Figure 2.19, let y = 8.00 m and ϕ = 25°. What are the lengths of sides x and r? (Use trig functions of ϕ.)

**21. In Figure 2.19, let x = 6.00 m and ϕ = 30°. What are the lengths of y and r? (Use trig functions of ϕ.)

**22. In Figure 2.19, let r = 9.00 m and ϕ = 50°. What are the lengths of sides x and y? (Use trig functions of ϕ.)

*23. Given a vector with a magnitude of 10 N at an angle of 45°, express the vector in terms of rectangular components with unit vectors.

*24. Resolve the following vectors into rectangular components.
(a) F = 15.0 N at an angle of 60° relative to the $-x$-axis (second quadrant), or θ = 150°.
(b) F = 12.0 N at an angle of 30° relative to the $-x$-axis (third quadrant), or θ = 210°.
(c) F = 6.0 N at an angle of 50° relative to the x-axis (fourth quadrant), or θ = 310°.
(d) F = 9.0 N at an angle of θ = 270°.

*25. You have a piece of sheet metal in the form of a right triangle. The hypotenuse and one side are 20.0 cm and 16.4 cm, respectively. (a) What is the interior angle between these two sides? (b) What is the length of the other leg of the triangle? (Use trigonometry.)

*26. A contractor builds a roof with a rise of 3.6 in. to a run of 1.0 ft. What are (a) the slope of the roof expressed as a percent and (b) the angle of the roof relative to the horizontal?

*27. A road is banked with a 7.0 percent grade. (a) What is the rise for a run of 100 ft? (b) What is the angle of the grade (relative to the horizontal)?

*28. A student standing 35 meters from a building measures the angle between the level ground and a meterstick pointed at the top of the building to be 26°. (a) How tall is the building? (b) Assuming an average height of 3.4 m per floor or story, how many stories does the building have?

*29. Referring to Figure 2.20(a), if side A of the triangle is 12 m long and θ = 45°, what are the lengths of sides B and C?

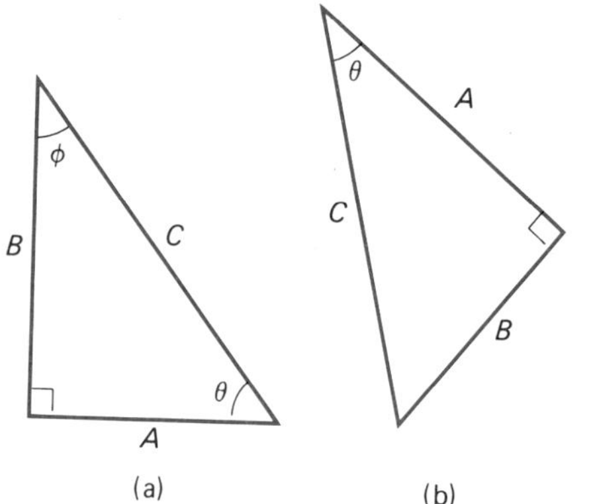

Figure 2.20
See Problems 29–34.

*30. If side B of the triangle in Figure 2.20(a) is 45 cm long and θ = 30°, what are the lengths of sides A and C?

**31. If side C of the triangle in Figure 2.20(a) is 15.0 m long and ϕ = 15°, what are the lengths of sides A and B?

**32. Referring to Figure 2.20(b), if the lengths of sides A and C of the triangle are 9.0 m and 15 m in length, respectively, what are (a) the length of side B and (b) the angle θ?

**33. If side B of the triangle in Figure 2.20(b) is 12.0 cm long and θ = 30°, what are the lengths of sides A and C?

**34. If side C of the triangle in Figure 2.20(b) is 2.0 m long and θ = 37°, what are the lengths of sides A and B?

2.3 Vectors

35. What is the vector sum of (a) $\mathbf{A} = 3\mathbf{x}$ and $\mathbf{B} = 5\mathbf{x}$ and (b) $\mathbf{A} = 4\mathbf{x}$ and $\mathbf{B} = -7\mathbf{x}$? (Express vectors in meters.)

36. Using a protractor and graph paper, plot $\mathbf{A}$ = 10.0 m, at an angle of 30° relative to the $+x$-axis, and $\mathbf{B}$ = 12.0 m, at an angle of 70° relative to the $+x$-axis, to scale and determine the resultant (magnitude and direction) by the parallelogram method.

37. What is the resultant (magnitude and direction) of the two force vectors $\mathbf{F}_1$ = 5.0 N $\mathbf{x}$ and $\mathbf{F}_2$ = 5.0 N $\mathbf{y}$? (Use the triangle method.)

*38. Applying the triangle method twice, graphically find the resultant of $\mathbf{F}_1$ = 6.0 N $\mathbf{x}$, $\mathbf{F}_2$ = 3.5 N $\mathbf{y}$, and $\mathbf{F}_3$ = 2.5 N at an angle of 45° relative to the $-x$-axis (second quadrant).

*39. Find the resultant of the vectors in Problem 36 using the component method.

*40. Find the resultant of the vectors in Problem 38 using the component method.

*41. Find the resultant of the following vectors: $\mathbf{d}_1$ = 4.5 m $\mathbf{x}$ − 1.2 m $\mathbf{y}$, $\mathbf{d}_2$ = 1.7 m $\mathbf{x}$ + 1.9 m $\mathbf{y}$, $\mathbf{d}_3$ = −3.4 m $\mathbf{x}$, and $\mathbf{d}_4$ = −1.0 m $\mathbf{x}$ − 4.6 m $\mathbf{y}$ in (a) unit vector form and (b) magnitude-angle form.

*42. A jogger jogs 2 blocks west, 3 blocks south, and 2 blocks further west. What is the jogger's displacement from the starting point?

**43. Find the resultant of the vectors shown in Figure 2.21(a) on p. 44. Express in both unit-vector and magnitude-angle forms.

**44. Find the resultant of the vectors shown in Figure 2.21(b). Express in both unit-vector and magnitude-angle forms.

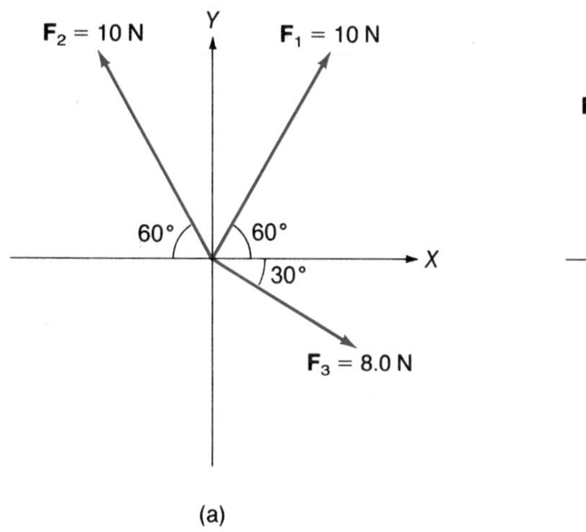

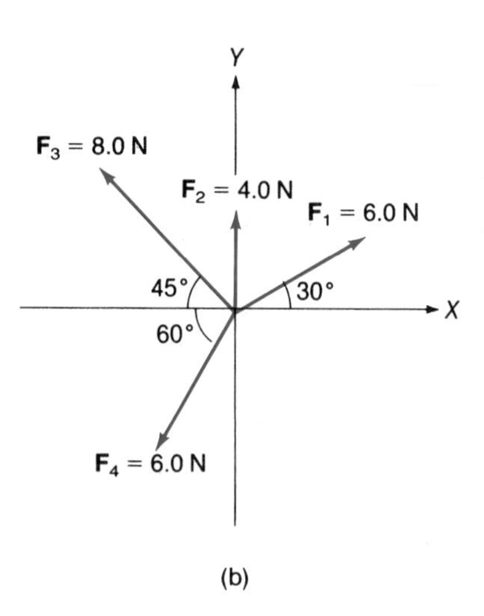

Figure 2.21
See Problems 43 and 44.

45. Determine the resultant of the following vectors: (a) **A = 6.0 N **x**, **B** = 8.0 N at an angle of 30° relative to the $+x$-axis, and **C** = 10 N at an angle of 60° relative to the $-x$-axis (second quadrant); (b) **A** = 8.0 N at an angle of 30° relative to the $+x$-axis, **B** = 4.0 N **y**, and **C** = 15.0 N at an angle of 10° relative to the $-y$-axis (third quadrant).

Chapter Supplement

Laws of Sines and Cosines

This chapter was concerned with the trigonometry of right triangles. Many triangles do not contain right angles. Such triangles are called *oblique* triangles. A general example is shown in Figure 2S.1, where triangle ABC has angles α, β, and γ, with opposite sides a, b, and c, respectively. Relationships between the angles and sides of oblique triangles are given by the law of sines and the law of cosines.

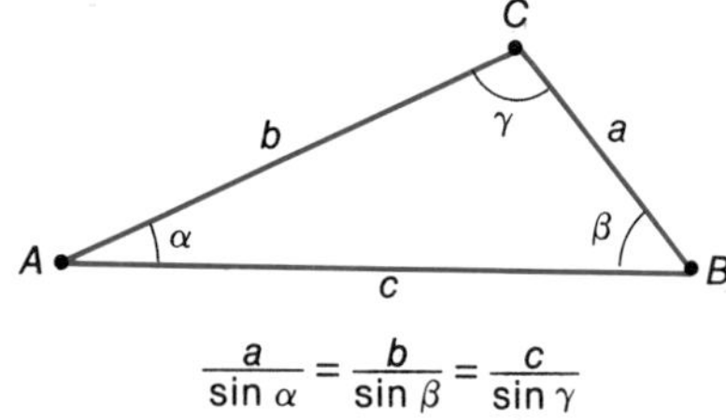

Figure 2.S1
Geometry for the laws of sines and cosines. See text for description.

The law of sines is as follows:

$$\frac{a}{\sin\alpha} = \frac{b}{\sin\beta} = \frac{c}{\sin\gamma} \qquad \textbf{(Eq. 2S.1)}$$

Three relationships are implied:

$$\frac{a}{\sin\alpha} = \frac{b}{\sin\beta}$$

$$\frac{a}{\sin\alpha} = \frac{c}{\sin\gamma}$$

$$\frac{b}{\sin\beta} = \frac{c}{\sin\gamma}$$

EXAMPLE 2S.1 An oblique triangle has side a = 46 cm, α = 63°, and γ = 71°. What are the lengths of the other sides of the triangle?

Solution: Since $\alpha + \beta + \gamma = 180°$ for any plane triangle,

$$\beta = 180° - \alpha - \gamma = 180° - 63° - 71° = 46°$$

Then, using the individual relationships of the law of sines,

$$b = \frac{a \sin \beta}{\sin \alpha} = \frac{(46 \text{ cm})(\sin 46°)}{\sin 63°}$$

$$= \frac{(46 \text{ cm})(0.719)}{(0.891)} = 37 \text{ cm}$$

$$c = \frac{a \sin \gamma}{\sin \alpha} = \frac{(46 \text{ cm})(\sin 71°)}{\sin 63°}$$

$$= \frac{(46 \text{ cm})(0.946)}{(0.891)} = 49 \text{ cm}$$

The law of cosines is expressed in the following three general forms:

$$a^2 = b^2 + c^2 - 2bc \cos \alpha \quad \text{(a)}$$

$$b^2 = a^2 + c^2 - 2ac \cos \beta \quad \text{(b)} \quad \textbf{(Eq. 2S.2)}$$

$$c^2 = a^2 + b^2 - 2ab \cos \gamma \quad \text{(c)}$$

The law of cosines is used when the somewhat simpler law of sines is not applicable, as illustrated in the following example.

EXAMPLE 2S.2 An oblique triangular flower bed has two sides with lengths $a = 4.5$ m and $b = 3.5$ m with the angle $\gamma = 110°$. What are the values of the other side and other angles of the flower bed? (See Fig. 2S.2.)

Solution: Notice that with the given data the law of sines is not applicable. Applying the appropriate form of the law of cosines to find the length of c (Eq. 2S.2c),

$$c^2 = a^2 + b^2 - 2ab \cos \gamma$$

$$= (4.5 \text{ m})^2 + (3.5 \text{ m})^2 - 2(4.5 \text{ m})(3.5 \text{ m}) \cos 110°$$

Since cos 110° exceeds the values of the angles listed in most trig tables, we use the trigonometric identity (see Appendix 3),

$$\cos \theta = -\cos(180° - \theta)$$

and

$$\cos 110° = -\cos(180° - 110°) = -\cos 70°$$

Hence,

$$c^2 = (4.5 \text{ m})^2 + (3.5 \text{ m})^2 + 2(4.5 \text{ m})(3.5 \text{ m})(\cos 70°)$$

$$= 43 \text{ m}^2$$

and

$$c = (43 \text{ m}^2)^{1/2} = 6.6 \text{ m}$$

The law of sines could now be applied to find α or β, but continuing with the law of cosines and solving for β using Equation 2S.2b (or, alternatively, for α using Eq. 2S.2a),

$$\cos \beta = \frac{a^2 + c^2 - b^2}{2ac}$$

$$= \frac{(4.5 \text{ m})^2 + (6.6 \text{ m})^2 - (3.5 \text{ m})^2}{2(4.5 \text{ m})(6.6 \text{ m})} = 0.87$$

and

$$\beta = \cos^{-1}(0.87) = 30°$$

Then,

$$\alpha = 180° - \beta - \gamma = 180° - 30° - 110° = 40°$$

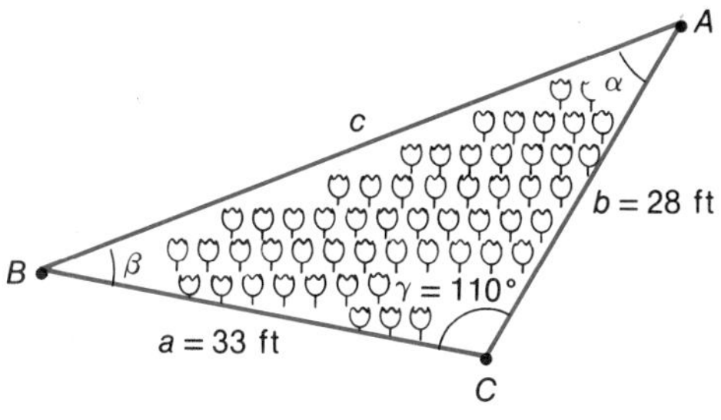

Figure 2.S2
See Example 2S.2.

Problems

1. An oblique triangle has side $b = 10.0$ m, $\beta = 40°$, and $\alpha = 60°$. What are the lengths of the other sides of the triangle?
2. An oblique triangle has sides $b = 6.0$ cm and $c = 21.0$ cm with $\alpha = 50°$. What are the values of the other side and other angles?
3. Find the resultant of the vectors in Problem 36 using the triangle method with the laws of sines and cosines.

The Piscataqua River bridge between Portsmouth, New Hampshire, and Kittery, Maine. (Courtesy Bethlehem Steel Corporation.)

Chapter 3

Statics and Equilibrium

Mechanics, which in general deals with the study of motion, is one of the oldest branches of physics. Objects in motion are commonplace—a person jogging, the moving parts of a machine, a thrown ball, and so on. Even on the submicroscopic level, the atoms making up matter are in incessant motion.

An important part of physics is the description and analysis of motion, and several of the following chapters will be devoted to this. Another important branch of mechanics is statics, which is the study of objects *not* in motion. For example, an important consideration in designing a bridge, as shown in the introductory photograph of this chapter, is that the component parts and the bridge as a whole do not move.

Knowing why objects do not move provides a better understanding of the mechanics of motion. Hence, before considering the description and analysis of the motion of objects, it is instructive to consider the static situations of objects at rest. In this chapter, the physical conditions required for an object to be in static equilibrium will be investigated and applied.

3.1 Particle Statics

Real objects can rotate as they travel. For example, a baseball usually spins when it is thrown. To avoid discussing rotational motion in our initial study of static conditions, it is convenient to introduce the concept of a particle.

> **A *particle* implies an amount of matter that is assumed to occupy a single point in space. As a result, a particle is an object without dimensions, so rotational motion is not a consideration.**

Actually, there is no such thing as a particle, i.e., a "point" particle as just defined, in nature. However, the particle concept is quite useful because in many instances real objects and certain points in systems can be treated as though they were particles. For example, in Figure 3.1, the ball can be treated as a particle. The pulls or forces on the ball act on a common point represented by a particle. Forces acting on or through a common point are called **concurrent forces.**

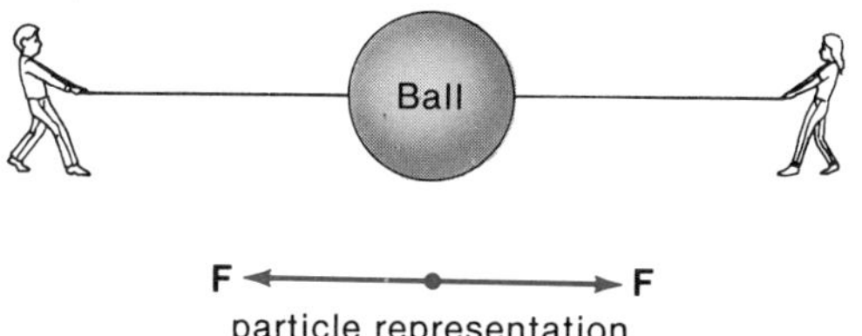

Figure 3.1
Particle representation. In some instances, it is convenient and possible to consider real objects as particles. A "point" particle has no dimensions, so rotational motion is not a consideration. Forces acting through a common point are said to be concurrent.

A *force* is required to cause an object at rest to be set in motion.

For example, we may push on an object or pull on a rope attached to the object. A force represents the action of one body on another and can be symbolized by a vector, as illustrated in Figure 3.1. The units of force in the SI and British systems are the newton (N) and the pound (lb), respectively.

In the previous chapter, several methods of finding vector sums, or resultants, were presented. These methods of vector addition apply in determining the resultant of two or more forces acting on a particle. In the examples of finding resultant vectors in Chapter 2, the resultants were nonzero.

However, it is possible for the resultant to be zero. In the case of concurrent forces acting on a particle, a zero resultant means that the net effect of the given forces is zero, or the forces "cancel" each other. The simplest case is two equal and opposite concurrent forces (Fig. 3.2). Even when several forces act on a particle, if their resultant is zero the net motional effect is as though no force acted at all. Since a particle at rest will not be moved in the absence of a net force, it follows that

When the resultant force acting on a particle at rest is zero, the particle is in static equilibrium.

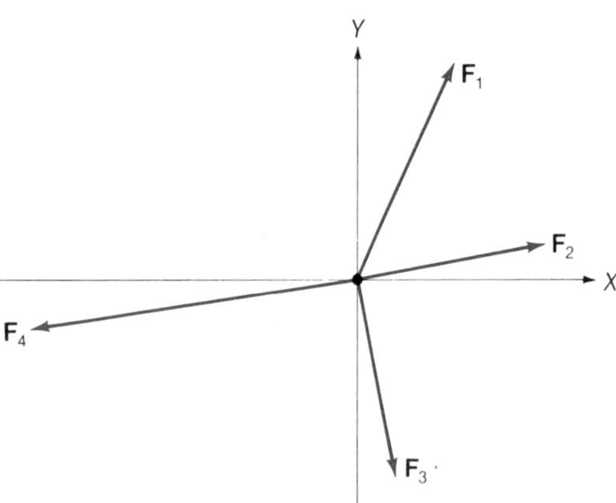

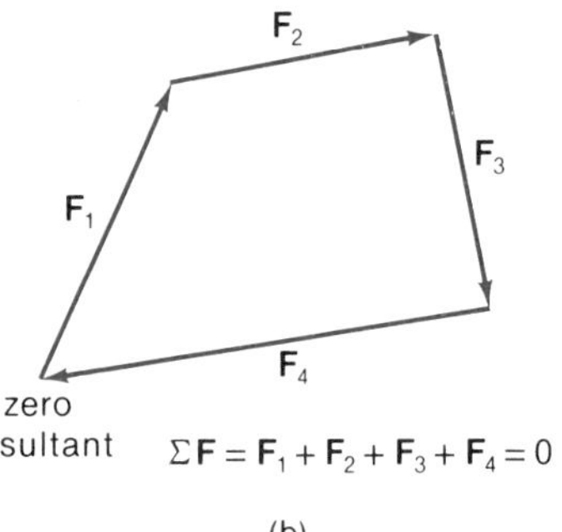

Figure 3.2
Zero resultants. (a) The resultant of two equal and opposite forces is zero, $\Sigma \mathbf{F} = 0$. The forces "cancel" each other and their net effect is zero. (b) The resultant of several concurrent forces may be zero, as illustrated by the polygon "tip-to-tail" method of vector addition.

This condition may be represented in vector notation as

$$\mathbf{F}_1 + \mathbf{F}_2 + \mathbf{F}_3 + \cdots$$

or

$$\boxed{\Sigma\,\mathbf{F} = 0} \qquad \textbf{(Eq. 3.1)}$$

where the Greek letter sigma, Σ, means "the sum of," and the equation is read "The sum of the forces is zero." Suppose there were three forces. Written explicitly, $\Sigma\,\mathbf{F} = \mathbf{F}_1 + \mathbf{F}_2 + \mathbf{F}_3 = 0$.*

For coplanar forces (vectors in the same plane), it follows that an equivalent representation of the condition for static equilibrium of a particle is that the vector sums of the rectangular components are zero, i.e.,

$$\Sigma\,\mathbf{F}_x = 0 \quad \text{and} \quad \Sigma\,\mathbf{F}_y = 0 \qquad \textbf{(Eq. 3.2)}$$

This can be seen from the component method of vector addition (Chapter 2),

$$\Sigma\,\mathbf{F} = \Sigma\,\mathbf{F}_x + \Sigma\,\mathbf{F}_y = 0$$

which requires the conditions of Equation 3.2. (Why?)

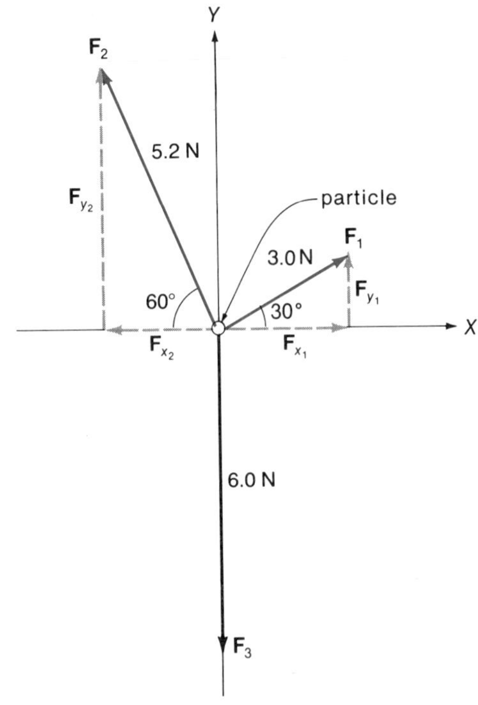

Figure 3.3
See Example 3.1.

EXAMPLE 3.1 Is the particle acted upon by the coplanar forces shown in Figure 3.3 in static equilibrium?

Solution: Resolving the vectors into rectangular components and summing (where signs indicate directions):

$$\begin{aligned}\Sigma\,\mathbf{F}_x &= F_1 \cos 30^\circ - F_2 \cos 60^\circ + F_{3_x}\\ &= (3.0\text{ N})(0.866) - (5.2\text{ N})(0.50) + 0\\ &= 2.6\text{ N} - 2.6\text{ N} = 0\end{aligned}$$

$$\begin{aligned}\Sigma\,\mathbf{F}_y &= F_1 \sin 30^\circ - F_2 \sin 60^\circ + F_{3_y}\\ &= (3.0\text{ N})(0.50) + (5.2\text{ N})(0.866) - 6.0\text{ N}\\ &= 1.5\text{ N} + 4.5\text{ N} - 6.0\text{ N} = 0\end{aligned}$$

Hence, the particle is in static equilibrium.

EXAMPLE 3.2 Given the two forces acting on a particle as shown in Figure 3.4(a), what are the magnitude and direction of a third force necessary for the particle to be in static equilibrium?

Solution: Resolving the vectors into components and summing:

$$\begin{aligned}\Sigma\,\mathbf{F}_x &= F_1 \cos 45^\circ - F_2 \cos 65^\circ\\ &= (12.0\text{ N})(0.707) - (15.0\text{ N})(0.4226)\\ &= 8.48\text{ N} - 6.34\text{ N} = 2.14\text{ N}\end{aligned}$$

$$\begin{aligned}\Sigma\,\mathbf{F}_y &= -F_1 \sin 45^\circ + F_2 \sin 65^\circ\\ &= -(12.0\text{ N})(0.707) + (15.0\text{ N})(0.906)\\ &= -8.48\text{ N} + 13.59\text{ N} = 5.11\text{ N}\end{aligned}$$

The resultant is then

$$\mathbf{F} = \mathbf{F}_x + \mathbf{F}_y = 2.14\,\mathbf{x} + 5.11\,\mathbf{y}$$

* In more detailed notation this would be written $\Sigma_{i=1}^{3}\,\mathbf{F}_i = 0$. The summation is from $i = 1$ to 3.

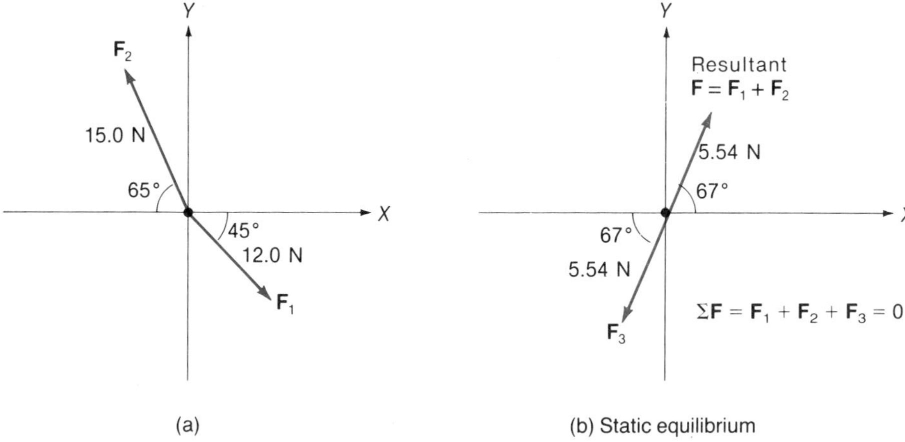

Figure 3.4
See Example 3.2.

which has a magnitude and direction of

$$F = \sqrt{F_x^2 + F_y^2} = \sqrt{(2.14)^2 + (5.11)^2} = 5.54 \text{ N}$$

$$\theta = \tan^{-1}\left(\frac{F_y}{F_x}\right) = \tan^{-1}\left(\frac{5.11}{2.14}\right)$$

$$= \tan^{-1}(2.39) = 67°$$

Then, to have static equilibrium, there must be a third force, $\mathbf{F}_3$, acting on the particle that is equal and opposite to this resultant; that is,

$$\mathbf{F}_3 = -\mathbf{F} = -2.14\ \mathbf{x} - 5.11\ \mathbf{y}$$

which lies in the third quadrant 67° from the $-x$-axis, as illustrated in Figure 3.4(b).

The condition for particle static equilibrium can easily be extended to three dimensions. The overall condition is still $\Sigma\,\mathbf{F} = 0$, and the summation of the components in each dimension or direction (x, y, and z) must be zero, i.e.,

$$\Sigma\,\mathbf{F}_x = 0 \quad \Sigma\,\mathbf{F}_y = 0 \quad \Sigma\,\mathbf{F}_z = 0$$

However, our present discussion will be limited to two-dimensional, or coplanar, forces.

3.2 Free-Body Diagrams

When we analyze a physical problem or situation, it is often convenient to make a sketch of the physical conditions of the problem. In engineering mechanics, such a sketch is commonly called a **space diagram.** For example, a space diagram of a 15.0-lb sign suspended by means of a cord is shown in Figure 3.5.

Suppose we want to know the magnitude of the tension force in the cord. This must not exceed the tensile strength of the cord; if it did, the cord would break when the sign was hung. With a cord of sufficient strength, the sign will hang in static equilibrium. (Imagine that you are holding the end of the cord instead of its being attached to a hook. To support the sign, you would have to pull on the cord with a force equal to the tension force.)

This is one of the many actual situations that can be reduced to a problem concerning the static equilibrium of a particle. An equivalent force situation is that of a 15.0-lb particle suspended by a cord (mass neglected).

A drawing that shows an object as a particle (a dot) and the concurrent forces acting on it (as emerging arrows) is called a *free-body diagram* (Fig. 3.5).

We use the component method to determine T_1 and T_2, under the conditions for particle static equilibrium,

$$\Sigma\,\mathbf{F}_x = T_{x_1} - T_{x_2} = 0 \qquad \text{or} \qquad T_{x_1} = T_{x_2}$$

where signs are used to indicate direction. Hence, the magnitudes of the x components are equal, which should be evident.* This gives no

* Otherwise there would be motion in the x direction. Also, since $T_1 = T_2$, then $T_{y1} = T_{y2}$.

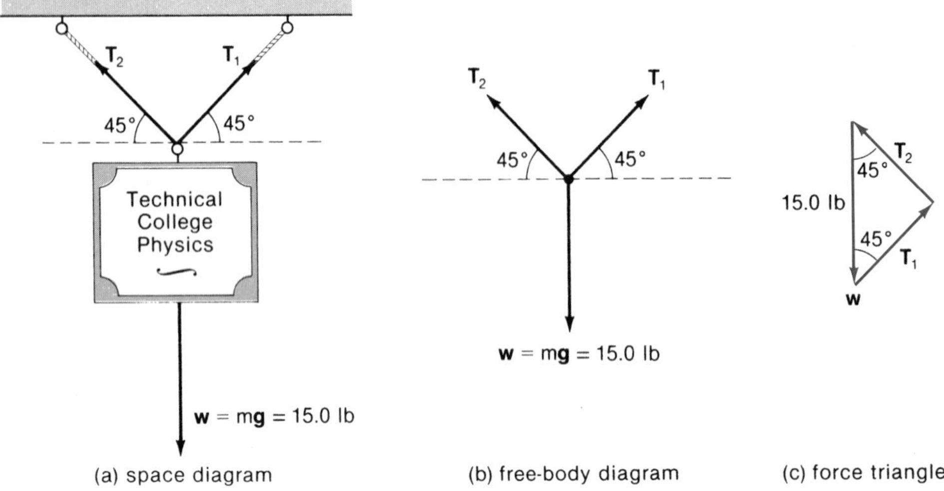

Figure 3.5
Diagrams. (a) A sketch of the physical conditions of a situation is called a space diagram. (b) A drawing showing the concurrent forces acting through a point is called a free-body diagram. (c) A force triangle showing the resultant is zero, and the object is in static equilibrium.

information about the magnitude of the tension force, but looking at the y components,

$$\Sigma \mathbf{F}_y = T_{y1} + T_{y2} - w = 2T_{y1} - w$$
$$= 2(T_1 \sin 45°) - 15.0 \text{ lb} = 0$$

or

$$2T_1(0.707) = 15.0 \text{ lb}$$

and

$$T_1 = \frac{(15.0 \text{ lb})}{2(0.707)} = 10.6 \text{ lb} = T_2$$

Hence, hanging the sign with a cord having a tensile strength of 10 lb would not be safe.

The triangle method is also illustrated in Figure 3.5. From the force triangle (a right triangle in this case), it can be seen that

$$T_1 = (15.0 \text{ lb}) \cos 45°$$
$$= (15.0 \text{ lb})(0.707)$$
$$= 10.6 \text{ lb} = T_2$$

In this particular example, the triangle method is somewhat simpler because the force triangle is a right triangle. Were the force triangle an oblique triangle, the solution could be obtained using the law of sines (see the Chapter 2 supplement).

Summarizing the **steps used in solving static particle problems:**

1. **Sketch a space diagram of the problem.**
2. **Isolate a particle representation.**
3. **Draw a free-body diagram of the particle.**
4. **Sum the forces applying the condition for particle static equilibrium ($\Sigma \mathbf{F} = 0$, or usually in component form $\Sigma \mathbf{F}_x = 0$ and $\Sigma \mathbf{F}_y = 0$) by using the triangle or component method, and solve for the unknown(s).**

EXAMPLE 3.3 A 25.0-N object is suspended by cords as shown in Figure 3.6. What are the tensions in the cords? (Neglect cord masses.)

Solution: The free-body diagram of the static condition, taking the point of connection as a particle, is shown in the figure. The weight force of the object is vertically downward. Summing the x and y components, we have

$$\Sigma \mathbf{F}_x = T_1 - T_2 \cos 60° = T_1 - T_2(0.500) = 0$$

$$\Sigma \mathbf{F}_y = T_2 \sin 60° - 25.0 \text{ N}$$
$$= T_2(0.866) - 25.0 \text{ N} = 0$$

From the second equation,

$$T_2 = \frac{(25.0 \text{ N})}{0.866} = 28.9 \text{ N}$$

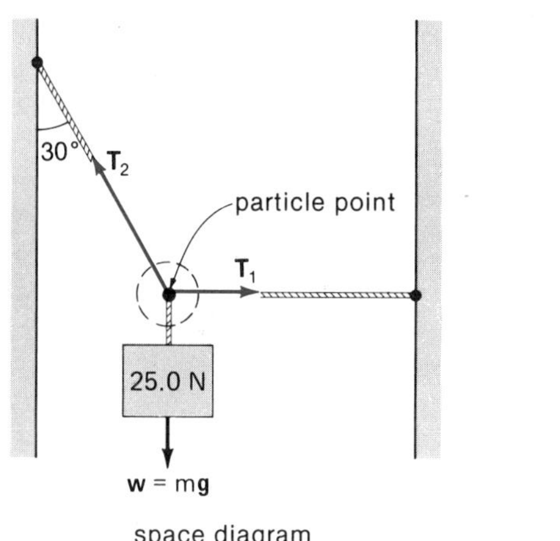

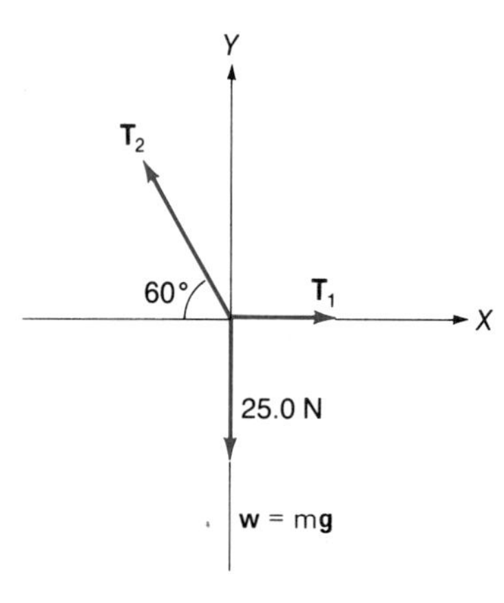

Figure 3.6
See Example 3.3.

and from the first equation,

$$T_1 = T_2(0.500) = (28.9\ \text{N})(0.500) = 14.5\ \text{N}$$

Note that the magnitudes of the tensions are not the same in this case, just as in the following example, where the suspension angles are different.

EXAMPLE 3.4 A picture is suspended by cords as shown in Figure 3.7. If the cord has a tensile strength of 10.0 lb, what is the maximum picture weight that can be suspended?

Solution: The free-body diagram of the particle representation of the point of connection is shown in the figure. Summing the x and y com-

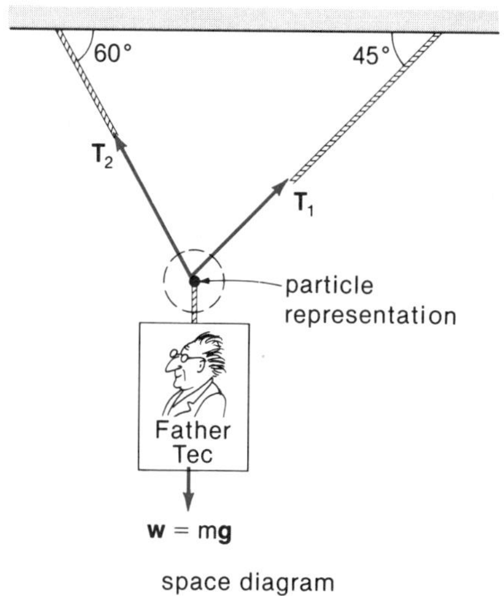

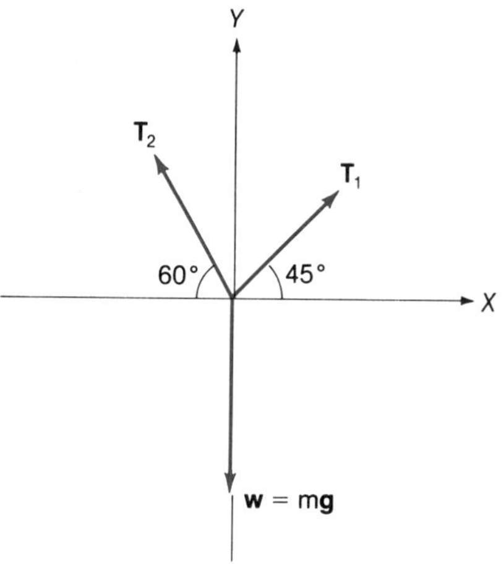

Figure 3.7
See Example 3.4.

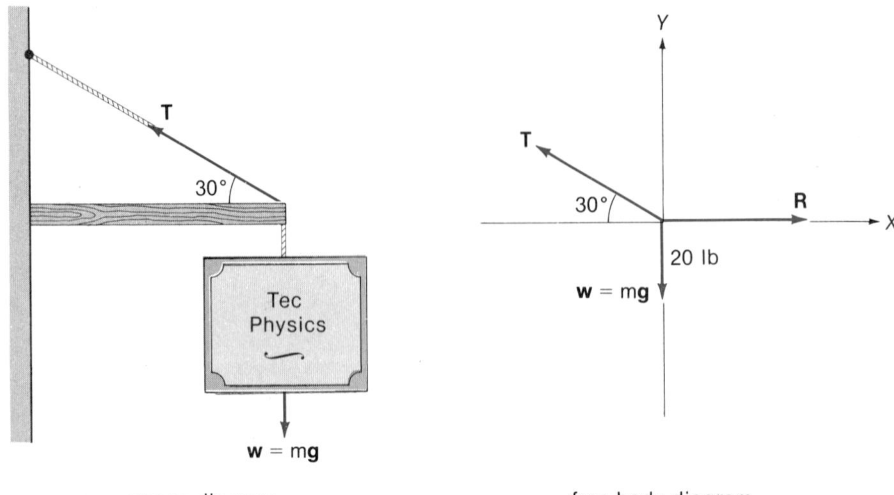

Figure 3.8
See Example 3.5.

ponents and applying the conditions of static equilibrium,

$$\Sigma\, \mathbf{F}_x = T_1 \cos 45^\circ - T_2 \cos 60^\circ$$
$$= T_1(0.707) - T_2(0.500) = 0$$
$$\Sigma\, \mathbf{F}_y = T_1 \sin 45^\circ + T_2 \sin 60^\circ - w$$
$$= T_1(0.707) + T_2(0.866) - w = 0$$

From the first equation,

$$T_2 = \frac{(0.707)T_1}{(0.500)} = (1.41)T_1$$

and T_2 is greater than T_1, so the cord with tension T_2 would break first. This would occur when $T_2 = 10.0$ lb, which is the given tensile strength of the cord. From the second equation, using $T_1 = T_2/1.41$,

$$T_1(0.707) + T_2(0.866) - w =$$
$$\frac{T_2(0.707)}{1.41} + T_2(0.866) - w = 0$$

or

$$T_2\left(\frac{0.707}{1.41} + 0.866\right) = w$$

and

$$T_2(1.37) = w$$

Then, with $T_2 = 10.0$ lb, $w = 13.7$ lb. Hence, the maximum weight that can be suspended is 13.7 lb (actually a little less if the cord breaks when the tension is exactly 10.0 lb).

EXAMPLE 3.5 A 20.0-lb sign hangs by means of a beam and rope, as illustrated in Figure 3.8. What are the magnitudes of the tension T in the rope and the reaction force of the beam?

Solution: The free-body diagram for the particle representation is shown in the figure. The reaction force R of the beam is in the x direction, i.e., the beam "pushes" on the rope in that direction. (Think what you'd do if you were to replace the beam.)

Then, summing the forces,

$$\Sigma\, \mathbf{F}_x = R - T_x = R - T \cos 30^\circ = 0$$
$$\Sigma\, \mathbf{F}_y = T_y - w = T \sin 30^\circ - 20.0 \text{ lb} = 0$$

From the second equation,

$$T = \frac{(20.0 \text{ lb})}{\sin 30^\circ} = \frac{(20.0 \text{ lb})}{(0.500)} = 40.0 \text{ lb}$$

Then, from the first equation,

$$R = T \cos 30^\circ = (40.0 \text{ lb})(0.866) = 34.6 \text{ lb}$$

Another common situation of static equilibrium is that of an object resting on a surface. In

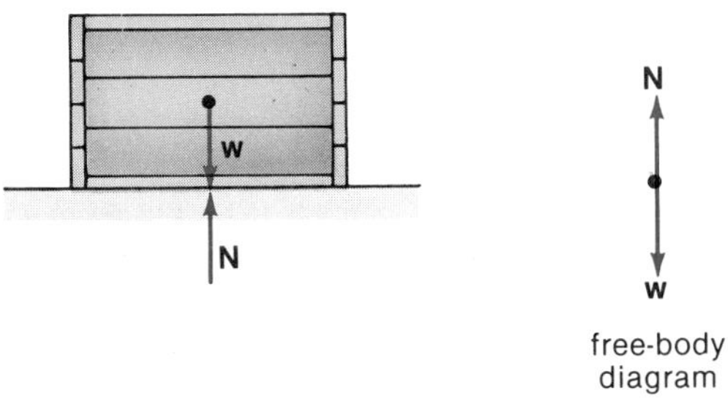

Figure 3.9
Normal force. The vector **N** is the normal reaction force of the surface on the crate due to its weight **w**. The normal force is perpendicular to the surface.

many instances, the object can be treated as a particle. Consider a crate sitting on a horizontal surface, as illustrated in Figure 3.9. We know the crate experiences or has a downward weight force w. But since the crate is at rest, or in static equilibrium, there must be another force to satisfy the $\Sigma \mathbf{F} = 0$ condition.

Where and what is this force? Evidently, it must act upward on the crate. We call this force the normal force **N**. It is the reaction force of the surface *on* the crate due to its weight (see free-body diagram of particle representation in Fig. 3.9). If you were to hold the crate stationary, you would have to exert an upward force equal and opposite to its weight.

Also, in real situations when a horizontal force is applied to an object on a surface and the object is in static equilibrium, then there is an opposing force of friction f as shown in Figure 3.10. In this case, the tension force T exerted by the string is balanced by the frictional force so that $\Sigma \mathbf{F}_x = 0$. The forces on the suspended mass must also be balanced, since it too is in static equilibrium.

EXAMPLE 3.6 A 10-lb block on a horizontal surface is attached to a suspended weight of 3 lb (Fig. 3.10). What is the magnitude of the frictional force required for static equilibrium?

Solution: Each object is isolated for particle representation and, drawing the free-body diagrams as shown in the figure, we see that $T = w_2 = 3$ lb. The tension in the string is transmitted undiminished to the block. From the other free-body diagram,

$$f = T = 3 \text{ lb}$$

Note that we say that the tension in a string connecting two stationary objects is *transmitted undiminished.* That is, the magnitude of the tension acting on each object is the same. If this were not the case, the string would not be in static equilibrium.

The normal force is always normal, or perpendicular, to the plane of the surfaces in contact. In the case of an object on a horizontal surface, the normal force is vertically upward. However, for an inclined surface, as in the case of a ramp or an inclined plane, the normal force is at an angle θ to the vertical, and θ is equal to the angle of the incline (Fig. 3.11). Note that the normal force is normal, or perpendicular, to the surface. Since we are free to select the orientation of the reference axes, it is customary to take the x-axis along the inclined plane. The y-axis is then normal to the plane.

Looking at the free-body diagram of the particle representation with the weight resolved into x and y components, we see that the normal

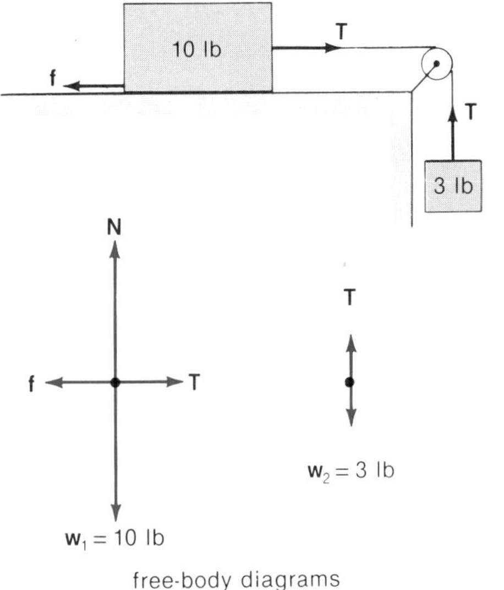

Figure 3.10
See Example 3.6.

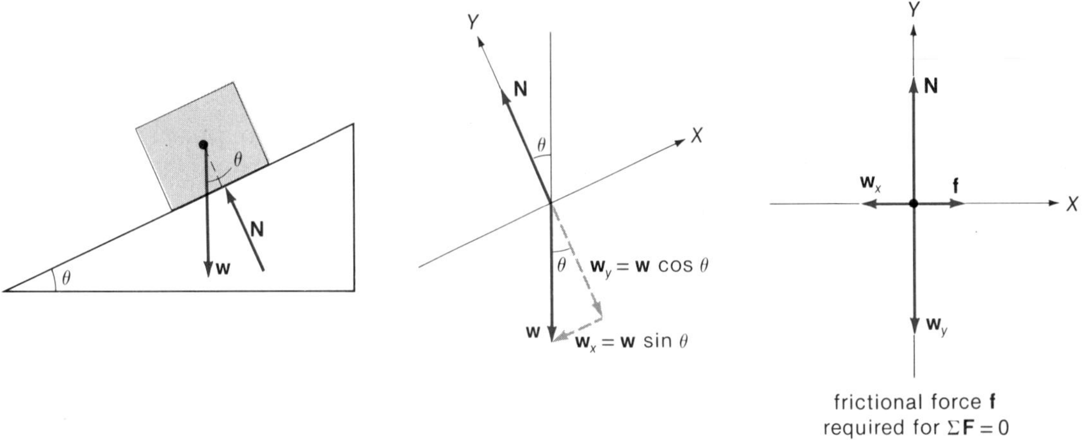

Figure 3.11
A block on an inclined plane in static equilibrium. The normal force **N** is perpendicular to the surface of the plane. This is balanced by the normal component of the weight force $\mathbf{w}_y$. The component of the weight force down the plane, $\mathbf{w}_x$, is balanced by the frictional force **f**.

force **N** is balanced by the normal weight component $\mathbf{w}_y$. But what balances the weight component $\mathbf{w}_x$ down the plane to satisfy the static equilibrium condition $\Sigma \mathbf{F}_x = 0$?

There must be a force acting up the plane opposing $\mathbf{w}_x$, and there is the ever-present force of friction **f**. The frictional force always opposes motion, and in this case the force of static friction prevents the object from moving down the plane. More will be said about the force of friction in Chapter 7, when it is considered in detail. For the present discussion, we shall treat friction simply as a force that in some situations helps satisfy the $\Sigma \mathbf{F} = 0$ condition.

EXAMPLE 3.7 If the weight of the block on the inclined plane in Figure 3.11 is 500 N and the angle of incline is 30°, what is the magnitude of the force of friction for static equilibrium?

Solution: The magnitude of the force of friction can be obtained directly from the $\Sigma \mathbf{F}_x = 0$ condition. With $w = 500$ N and $\theta = 30°$, we have

$$\Sigma \mathbf{F}_x = f - w_x = f - w \sin 30°$$
$$= f - (500 \text{ N})(0.500) = 0$$

and

$$f = 250 \text{ N}$$

It is often convenient and instructive to neglect frictional forces. This is done by considering hypothetical frictionless surfaces. Such idealizations are commonplace in the study of physics. They permit a simplification of the complexity of nature so that the basic concepts can be studied more easily. Once these are analyzed and understood, real situations can be dealt with more easily.

It should be evident that without the frictional forces in Figures 3.10 and 3.11 there would be unbalanced forces, or no static equilibrium, and the blocks would move. For static equilibrium, even with frictionless surfaces, the sum of the forces must be zero.

EXAMPLE 3.8 Two blocks are in static equilibrium on two frictionless planes, as illustrated in Figure 3.12. If the weight of w_1 is 4.0 N, what is the weight of w_2?

Solution: A free-body diagram for each block is shown in the figure. From the diagram for w_1, we have

$$\Sigma \mathbf{F}_x = T - w_{1_x} = T - w_1 \sin 60° = 0$$

and

$$T = w_1 \sin 60° = (4.0 \text{ N})(0.866) = 3.5 \text{ N}$$

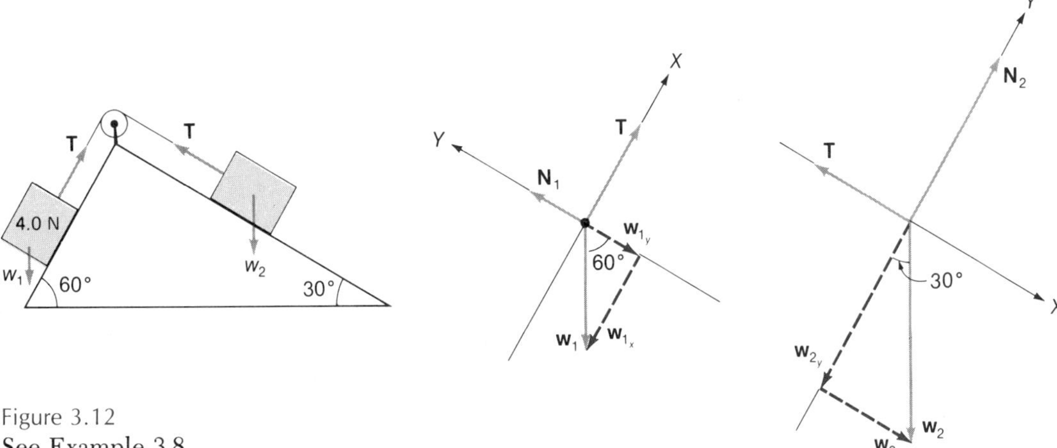

Figure 3.12
See Example 3.8.

From the diagram for w_2,

$$\Sigma\, \mathbf{F}_x = w_{2_x} - T = w_2 \sin 30° - T = 0$$

and

$$w_2 = \frac{T}{\sin 30°} = \frac{3.5\ \text{N}}{0.500} = 7.0\ \text{N}$$

since T is transmitted undiminished in the cord. Summations for the y components were not made because they were not necessary for the solution of the problem.

EXAMPLE 3.9 A block weighing 300 N resting on a frictionless surface, as illustrated in Figure 3.13, is in static equilibrium. What is the weight of w_2?

Solution: Drawing the free-body diagrams and summing the forces with $w_1 = 300$ N,

$$\Sigma\, \mathbf{F}_x = T - w_{x_1} = T - w_1 \sin 45° = 0$$

and

$$T = w_1 \sin 45° = (300\ \text{N})(0.707) = 212\ \text{N}$$

And from the free-body diagram for w_2,

$$T - w_2 = 0$$

and

$$w_2 = T = 212\ \text{N}$$

3.3 Rigid Body Statics

A **rigid body** is a solid object in which all of the particles making up the object maintain fixed distances from each other. As seen in the last section, a stationary rigid body does not move when the resultant of the concurrent forces acting on the body is zero, i.e., $\Sigma\, \mathbf{F} = 0$ (Fig. 3.14). This is the condition for *static translational (nonrotational) equilibrium* for a particle, or for a rigid body that may be represented as a particle.

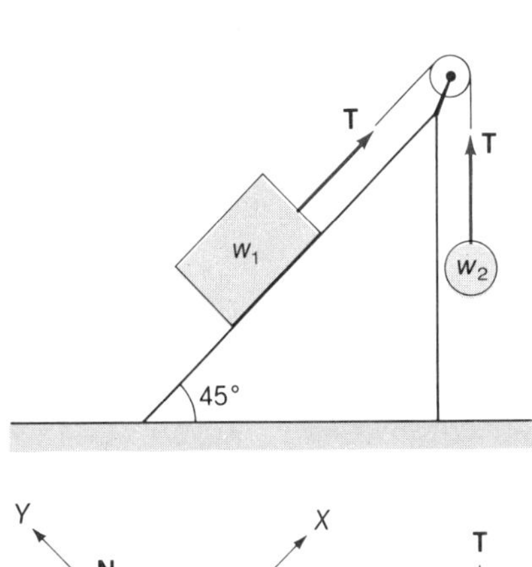

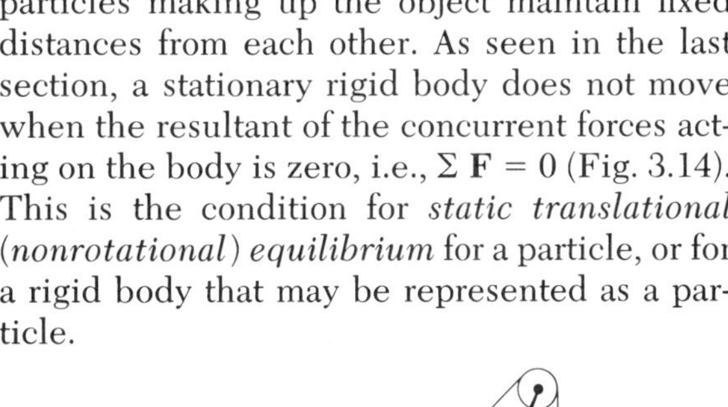

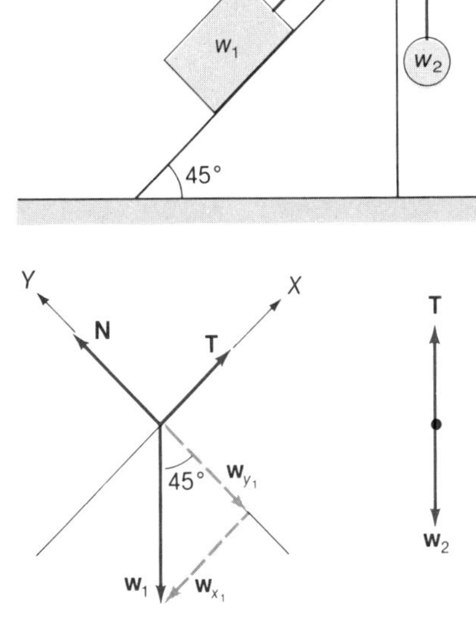

Figure 3.13
See Example 3.9.

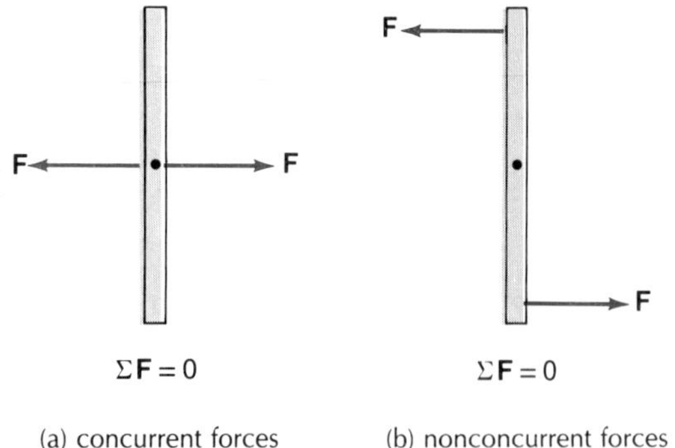

Figure 3.14
Equal and opposite forces. (a) The sum of the concurrent forces is zero, which satisfies the condition for translational static equilibrium. (b) The sum of the nonconcurrent forces is also zero, but the object will rotate.

However, suppose the forces acting on a body are nonconcurrent (not acting through a common point), as illustrated in the second diagram of Figure 3.14. The condition $\Sigma \mathbf{F} = 0$ is still satisfied, since the forces are equal and opposite and $\Sigma \mathbf{F} = F - F = 0$.

But it should be apparent that the object is not in static equilibrium, since the applied forces would cause it to rotate counterclockwise. Evidently there must be another condition in addition to $\Sigma \mathbf{F} = 0$ for a rigid body acted on by nonconcurrent forces to be in static equilibrium.

This condition, which ensures *static rotational equilibrium,* is expressed in terms of torque or moment of force. A **torque** (τ) tends to produce rotational motion and is defined as the moment arm $r_\perp$ times the applied force F,

$$\text{Torque} = \text{moment arm} \times \text{force}$$

$$\tau = r_\perp F \qquad \textbf{(Eq. 3.3)}$$

where the moment arm $r_\perp$ is the perpendicular distance from the axis of rotation to the line of action of the force (Fig. 3.15). The line of action of a force is a line along the direction of the force. Since torque is a distance times a force, the units of torque are meter-newton (m-N) and foot-pound (ft-lb) in the metric and British systems, respectively.

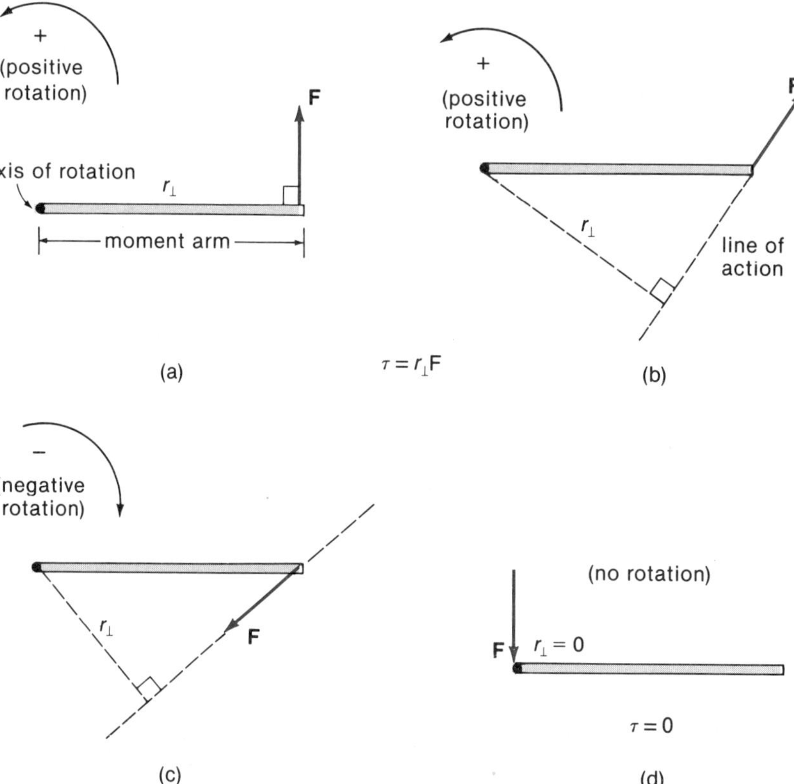

Figure 3.15
Torque. Torque magnitude is the product of the moment arm $r_\perp$ and the applied force F, here acting on a pivoted rod. The moment arm is the perpendicular distance from the axis of rotation to the line of action of the force. The torques in (a) and (b) produce positive or counterclockwise rotations. In (c) the rotation would be negative or clockwise, and in (d) the torque is zero since the moment arm is zero.

Rotational motion takes place about an axis of rotation. In the case of the nonconcurrent forces in Figure 3.14, the axis of rotation would be through a point in the body between the lines of action of the two forces. In many cases, as in Figure 3.15, a rigid body is constrained to rotate about a fixed pivot point that defines an axis of rotation. Notice that if a force acts through the axis of rotation, the torque is zero since $r_\perp = 0$.

A torque tends to produce a rotation, which can be in either a clockwise or a counterclockwise sense. Similarly to linear motions, the directions of these two rotational motions are indicated by the sign convention of positive (+) and negative (−) rotations. The convention is arbitrary. However, since it is customary to measure angular displacement in the counterclockwise sense (from the $+x$-axis), this rotational direction is commonly taken as positive and the clockwise sense as negative.

Following this convention for our discussion of static equilibrium, we will consider a torque produced by a force that tends to produce a counterclockwise rotation to be positive, and a torque produced by a force that tends to produce a clockwise rotation to be negative (see Fig. 3.15).

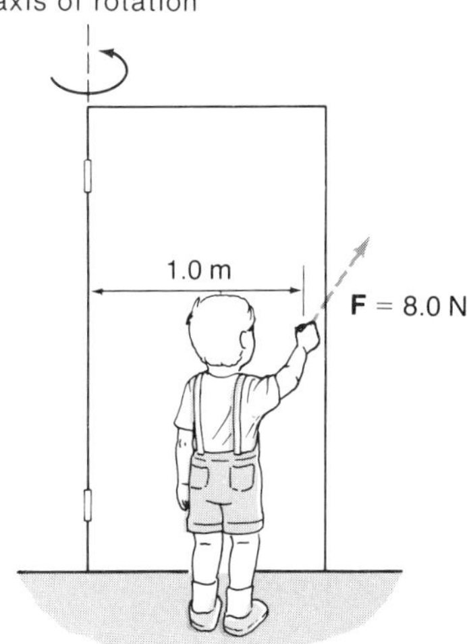

Figure 3.16
A commonly applied torque. See Example 3.10.

EXAMPLE 3.10 A person pushes perpendicularly on a door with a force of 8.0 N at a distance of 1.0 m from the hinges (axis of rotation; see Fig. 3.16). What is the applied torque?

Solution: With $F = 8.0$ N and $r_\perp = 1.0$ m,

$$\tau = r_\perp F = (1.0 \text{ m})(8.0 \text{ N}) = +8.0 \text{ m-N}$$

The torque is positive by our convention, since it tends to produce a counterclockwise rotation (as viewed from above).

Now, back to the conditions for the static equilibrium of a rigid body. The condition $\Sigma\, \mathbf{F} = 0$ is necessary for static translational (nonrotational) equilibrium. The condition for **static rotational equilibrium** is similarly expressed:

> When the resultant torque acting on a nonrotating rigid body is zero, the body is in static rotational equilibrium.

Or, in vector notation,

$$\boxed{\Sigma\, \boldsymbol{\tau} = 0} \qquad \textbf{(Eq. 3.4)}$$

Hence, the *two* conditions for the static equilibrium of a rigid body are

$\Sigma\, \mathbf{F} = 0$	(translational equilibrium)
$\Sigma\, \boldsymbol{\tau} = 0$	(rotational equilibrium)

(Eq. 3.5)

In addition to the previously given four steps in solving particle static equilibrium problems (where the "particle" locations for a rigid body are the points of force applications), we now have the additional steps for the $\Sigma\, \boldsymbol{\tau} = 0$ **(rotational equilibrium) condition:**

5. **Choose a convenient pivot point or axis of rotation of the system.**
6. **Sum the torques about the pivot point or axis of rotation, applying the condition for rotational equilibrium ($\Sigma\, \boldsymbol{\tau} = 0$).**

   The equations from the $\Sigma\, \mathbf{F} = 0$ and $\Sigma\, \boldsymbol{\tau} = 0$ conditions give a set of n equations with n unknowns from which the complete solution to the problem can be obtained.

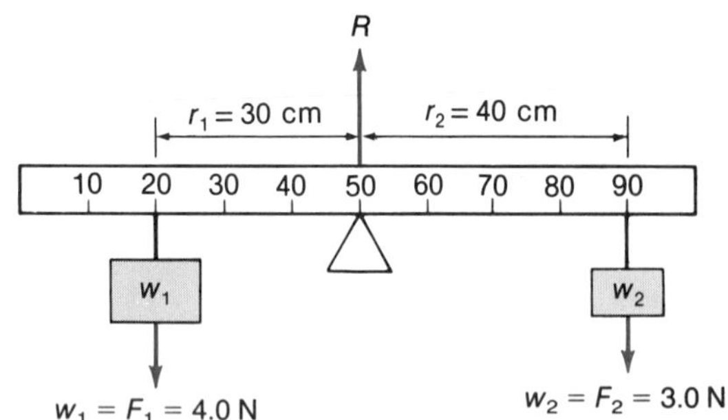

Figure 3.17
See Example 3.11.

These steps are illustrated in the following examples.*

EXAMPLE 3.11 A meterstick with suspended weights is balanced on a pivot, as shown in Figure 3.17. Show that the conditions for static equilibrium are satisfied. (Neglect the weight of the meterstick.)

Solution: From the figure, $F_1 = 4.0$ N with $r_1 = 30$ cm $= 0.30$ m, and $F_2 = 3.0$ N with $r_2 = 40$ cm $= 0.40$ m.

Since the stick cannot move linearly, the condition $\Sigma\,\mathbf{F} = 0$ is satisfied automatically. The reaction force R on the pivot on the meterstick must equal the weight forces, and

$$\Sigma\,\mathbf{F} = R - F_1 - F_2 = R - 4.0\text{ N} - 3.0\text{ N} = 0$$

and

$$R = 7.0\text{ N}$$

Then the stick will not rotate about the pivot point if $\Sigma\,\boldsymbol{\tau} = 0$, and

$$\begin{aligned}\Sigma\,\boldsymbol{\tau} &= r_1F_1 - r_2F_2 = 0\\ &= (0.30\text{ m})(4.0\text{ N}) - (0.40\text{ m})(3.0\text{ N}) = 0\\ &= 1.2\text{ m-N} - 1.2\text{ m-N} = 0\end{aligned}$$

Hence, the conditions for static equilibrium are satisfied.

* Although these steps may appear complicated, they are quite simple and mechanical. You do the same thing over and over, viz., sum the forces in particular directions ($\Sigma\,\mathbf{F} = 0$) and sum the torques about a chosen axis ($\Sigma\,\boldsymbol{\tau} = 0$). The rest is simply the mathematics of solving for what is wanted.

It should be noted that when a system is in static equilibrium, the sum of the torques is zero about *any* point or axis of rotation. In solving problems, a point or axis is usually chosen that conveniently eliminates one or more of the torques, since the forces acting through this point or axis have zero torques ($r = 0$). For example, in Example 3.11 the pivot point at the center of the meterstick eliminates the torque for the reaction force **R**.

Suppose the pivot point were chosen to be at the zero end of the meterstick. In this case,

$$\Sigma\,\boldsymbol{\tau} = -r_1F_1 + r_2R - r_3F_2 = 0$$

where the respective r's are measured from the end of the meterstick (not the same as in the figure), and

$$\begin{aligned}\Sigma\,\boldsymbol{\tau} &= -(0.20\text{ m})(4.0\text{ N}) + (0.50\text{ m})(7.0\text{ N})\\ &\quad - (0.90\text{ m})(3.0\text{ N}) = 0\\ &= -0.80\text{ m-N} + 3.5\text{ m-N} - 2.7\text{ m-N} = 0\end{aligned}$$

and the condition for rotational static equilibrium is satisfied about this zero-end pivot point as it is about any point.

EXAMPLE 3.12 Find the reaction force **R** of the hinge for a horizontal crane boom assembly in static equilibrium as illustrated in Figure 3.18. (Assume any number of significant figures.)

Solution: The vertical forces are the 400-lb load and the weight of the boom, 150 lb, which may be considered to be concentrated at the center (of gravity) of the boom. (More will be said about the center of gravity in the next section.) The tension force **T** and the reaction force **R** both have vertical and horizontal components.

Then, applying the first condition for static equilibrium to the force components,

$$\Sigma\,\mathbf{F}_x = R_x - T\cos 30^\circ = 0$$

$$\Sigma\,\mathbf{F}_y = R_y + T\sin 30^\circ - 150\text{ lb} - 400\text{ lb} = 0$$

Since there are three unknowns, R_x, R_y, and T, and only two equations, the components of **R** cannot be found from these equations alone. However, there is the other condition for static equilibrium, $\Sigma\,\boldsymbol{\tau} = 0$.

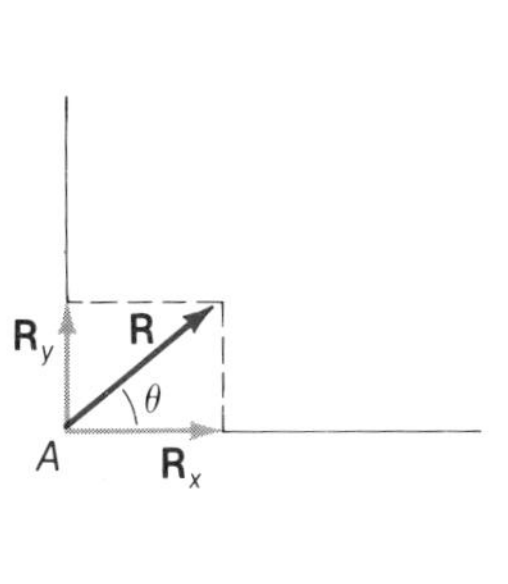

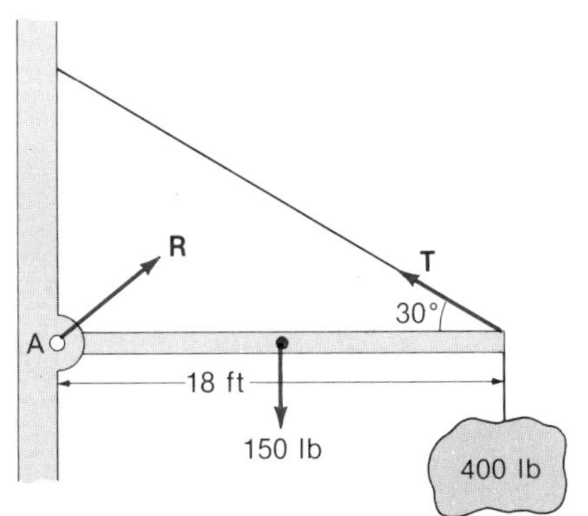

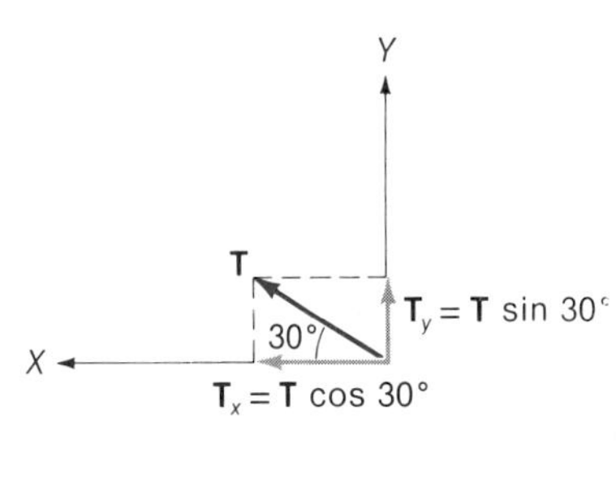

Figure 3.18
See Example 3.12.

Summing the torques about an axis through point A,

$$\Sigma \boldsymbol{\tau} = \Sigma rF$$

$$= (18 \text{ ft})T_y - (9.0 \text{ ft})(150 \text{ lb}) - (18 \text{ ft})(400 \text{ lb}) = 0$$

$$= (18 \text{ ft})(T \sin 30°) - 1350 \text{ ft-lb} - 7200 \text{ ft-lb} = 0$$

$$= (18 \text{ ft})T(0.50) - 8550 \text{ ft-lb} = 0$$

and

$$T = \frac{(8550 \text{ ft-lb})}{(0.50(18 \text{ ft})} = 950 \text{ lb}$$

Notice that only T_y contributes to the torques, since T_x acts through the axis of rotation. By choosing to sum the torques about an axis through A, we could ignore the torque due to R. Why?

Using the value of T in the first two equations,

$$R_x = T \cos 30° = (950 \text{ lb})(0.866)$$

$$= 823 \text{ lb}$$

$$R_y = 550 \text{ lb} - T \sin 30° = 550 \text{ lb} - (950 \text{ lb})(0.50)$$

$$= 75 \text{ lb}$$

Then,

$$R = (R_x^2 + R_y^2)^{1/2} = [(823)^2 + (75)^2]^{1/2} = 826 \text{ lb}$$

and

$$\theta = \tan^{-1}\left(\frac{R_y}{R_x}\right) = \tan^{-1}\left(\frac{75}{823}\right) = 5.2°$$

3.4 Center of Gravity and Equilibrium

As has been stated, a rigid body is a solid object in which all of the particles making up the object maintain a fixed distance from each other. Each particle in a rigid body is acted on by gravity and has weight (Fig. 3.19). The total weight of the body is the sum of the weights of the individual particles, $w = Mg = (\Sigma\, m_i)g$, where the total mass M is equal to the sum of the masses of the individual particles, $\Sigma\, m_i$.

If an object is pivoted at some arbitrary point, the weights of the individual particles will produce torques, both clockwise and counterclockwise, about an axis through the pivot point. In general, the sums of the clockwise and counterclockwise torques are not equal ($\Sigma\, \boldsymbol{\tau} \neq 0$), and the object will tend to rotate. However, regardless of the shape and size of the body, there is one point about which the particle torques sum to zero. This point is called the **center of gravity.**

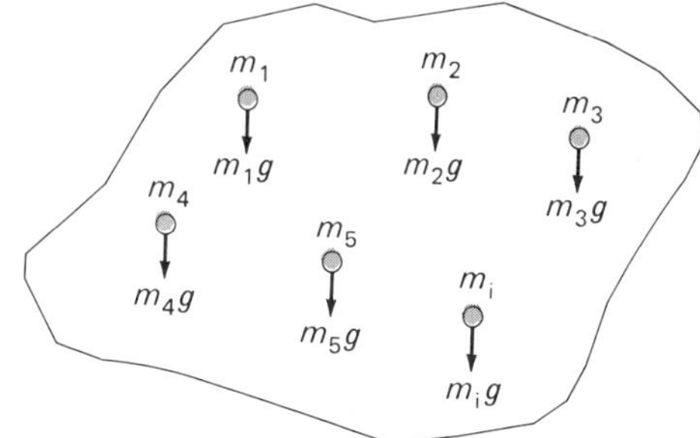

Figure 3.19
Rigid body. A rigid body is made up of particles that maintain fixed distances from each other. The total weight of the body is the sum of the weight of the individual particles, $w = Mg = (\Sigma m_i)g$.

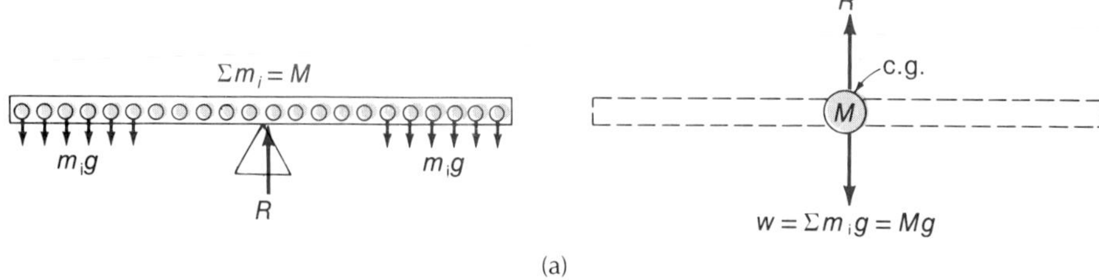

(a)

(b)

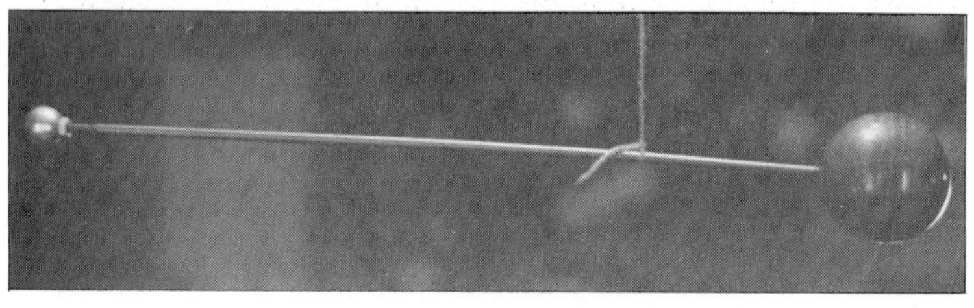

(c)

Figure 3.20
Center of gravity. (a) A uniform rod. With the pivot point at the center of the rod, the sum of the torques due to the individual particles on each side of the center point is zero. This locates the center of gravity, where all of the weight or mass may be considered to be concentrated. (b) A student balances a meterstick at its center of gravity. (c) Nonuniform object. For two balls of different weights on a rod, the center of gravity is nearer the heavier ball. Why?

Consider, for example, the rod shown in Figure 3.20. If the mass (particles) is uniformly distributed and the rod is pivoted at its center, then the sum of the torques due to the weight forces of the particles is zero. The center of gravity of the rod is then at its midpoint.

The rod is in static equilibrium, so the condition $\Sigma \mathbf{F} = 0$ must hold. Hence,

$$\Sigma \mathbf{F} = R - \Sigma(m_i)g = 0$$

and

$$R = \Sigma(m_i)g = Mg = w$$

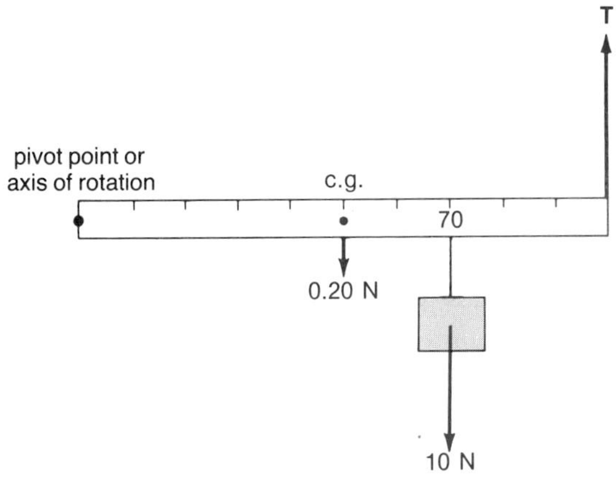

Figure 3.21
See Example 3.13.

where w is the total weight of the rod. The force **R** supports the total weight of the rod, and an equivalent force situation is as though the total weight (or mass) were concentrated at the center of gravity (Fig. 3.20). Hence,

> **The center of gravity is the point at which the entire weight (or mass) of an object may be considered to be concentrated.**

Another way of looking at it is that the center of gravity is simply the average location of the weight distribution of a body. (The center of gravity or weight coincides with the **center of mass** in a uniform gravitational field.)

This representation is helpful in computing the torque due to the weight of the rigid body. For a long structure like a rod, the center of gravity is located at the "balance" point of the structure and can be found experimentally by balancing the structure on a pivot point, even if its mass is not uniformly distributed.

EXAMPLE 3.13 A uniform meterstick weighing 0.20 N and pivoted at one end has a weight of 10 N suspended at the 70-cm position (Fig. 3.21). What magnitude of the force T applied to the end of the meterstick is required to maintain the stick in a horizontal position?

Solution: Considering the weight of the meterstick to be concentrated at its center of gravity (c.g.) at 0.50 m, from the equilibrium condition $\Sigma\, \boldsymbol{\tau} = 0$,

$$\Sigma\, \boldsymbol{\tau} = \Sigma(rF) = (1.0 \text{ m})T - (0.50 \text{ m})(0.20 \text{ N}) - (0.70 \text{ m})(10 \text{ N}) = 0$$

and

$$T = \frac{(0.10 \text{ m-N} + 7.0 \text{ m-N})}{(1.0 \text{ m})} = 7.1 \text{ N}$$

Notice that the condition $\Sigma\, \mathbf{F} = 0$ is not satisfied for the three forces shown in the figure. Why? Remember the upward reaction force at the pivot point. Can you determine its magnitude?

The preceding example shows how to balance a uniform body that has its center of gravity at its midpoint. However, if the mass is not uniformly distributed, the center of gravity of the system is not in general at its midpoint. In some instances, the condition that $\Sigma\, \boldsymbol{\tau} = 0$ for static rotational equilibrium can be used to locate the horizontal position of the center of gravity of a nonuniform structure, as the following example shows.

EXAMPLE 3.14 A rigger wishes to lift a loaded platform, as illustrated in Figure 3.22. Over

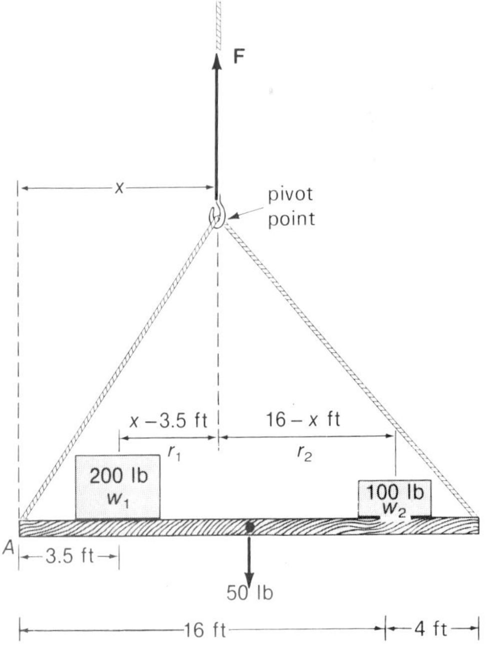

Figure 3.22
A rigging problem. See Example 3.14. (Vector arrows not to scale.)

what horizontal position should the rigging hook be attached to the support line? (Neglect the weight of the support line.)

Solution: In the figure the loads are shown at the locations on the platform in reference to point A, with the load or weight of the uniform platform (20 ft in length) at its midpoint (c.g.). To prevent the platform from tipping or rotating, the rigging hook would have to be attached directly above the center of gravity of the system. Assuming the c.g. of the system to be above a point a distance x from A and applying the condition $\Sigma\, \boldsymbol{\tau} = 0$ about this point, we have (with the moment arms shown in the figure and $r_p = 10\text{ ft} - x$ for the moment arm of the weight of the platform, assumed to be at its midpoint)

$$\begin{aligned}\Sigma\, \boldsymbol{\tau} &= \Sigma(rF)\\ &= r_1 w_1 - r_p w_p - r_2 w_2\\ &= (x - 3.5\text{ ft})(200\text{ lb}) - (10\text{ ft} - x)(50\text{ lb})\\ &\quad - (16\text{ ft} - x)(100\text{ lb}) = 0\end{aligned}$$

or

$$(x - 3.5\text{ ft})(200\text{ lb}) = (10\text{ ft} - x)(50\text{ lb}) + (16\text{ ft} - x)(100\text{ lb})$$

and solving for x,

$$(350\text{ lb})x = 2800\text{ ft-lb}$$

so

$$x = \frac{(2800\text{ ft-lb})}{(350\text{ lb})} = 8.0\text{ ft}$$

Thus, the c.g. is located 8.0 ft from point A. The condition for static translational equilibrium gives the minimum force the rig would have to apply to lift the load:

$$\Sigma\, \mathbf{F} = 0 = F - w_1 - w_p - w_2$$

or

$$F - 200\text{ lb} - 50\text{ lb} - 100\text{ lb} = 0$$

and

$$F = 350\text{ lb}$$

This is the upward force F for static equilibrium; hence, a force greater than F would be required to lift the loaded platform.

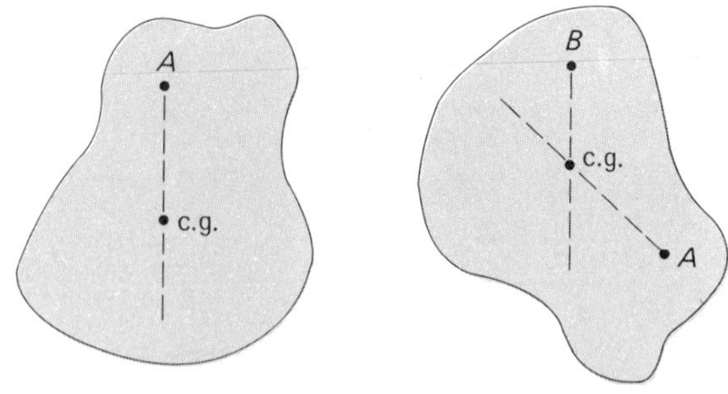

Figure 3.23
Center of gravity of an irregularly shaped, flat object. The center of gravity lies directly below the point of suspension, so its location may be determined by the intersection of vertical lines from two or more suspensions.

As the preceding example illustrates, the location of the center of gravity of an object or system depends on its shape and/or its mass distribution. For symmetrically or regularly shaped objects with uniform mass distributions, such as spheres, rods, and cylinders, the center of gravity is at the center of symmetry of the object. For some objects, such as a ring or coat hanger, the center of gravity lies outside the body. Why?

The center of gravity of an irregularly shaped (and nonuniform) object can be determined experimentally. Consider a flat sheet of material as illustrated in Figure 3.23. When pivoted about an arbitrary point, the object will not in general be in rotational equilibrium, and it will rotate and oscillate until the center of gravity comes to rest vertically below the pivot point. The torque due to the concentrated weight at the c.g. is then zero, since the line of action of the weight force passes through the pivot point (moment arm is zero).

If the object is then pivoted at another point, the same thing occurs, and the center of gravity is located at the intersection of the two lines, as shown in the figure. If the object were three-dimensional, the intersection of three lines from three pivot points would locate the c.g.

Equilibrium

An object is in static equilibrium when the two conditions $\Sigma \mathbf{F} = 0$ and $\Sigma \boldsymbol{\tau} = 0$ are satisfied. Equilibrium is further classified as being *stable, unstable,* or *neutral.* The terms essentially describe the behavior of objects when they are disturbed and thus moved slightly from their equilibrium positions.

An object in **stable equilibrium** will tend to return to its equilibrium position when slightly disturbed. For example, the cone in the first diagram in Figure 3.24 is in stable equilibrium. If the cone is disturbed slightly by tilting it on edge, a restoring torque acts to return it to its equilibrium position. When an object in stable equilibrium is disturbed, the center of gravity is raised. The behavior of the center of gravity is analogous to that of a particle in a bowl.

When an object is slightly disturbed and a displacing torque causes it to topple or fall over, the object is said to be in **unstable equilibrium.** As shown in Figure 3.24, a cone on its point is in unstable equilibrium. A slight disturbance from equilibrium causes the center of gravity to be lowered in this case. The center of gravity may be likened to a particle initially sitting on the top of an inverted bowl.

The transition between stable and unstable equilibrium is directly related to the relative position of the center of gravity. **As long as the c.g. is vertically over and inside the base original support of the object, there will be a restoring torque that tends to return the object to its equilibrium position.** However, if the center of gravity is not above or is outside the object's base of support, the object is in unstable equilibrium. In this case, a displacing torque will cause a rotation about the point or line of contact and the object will topple. This is illustrated in Figure 3.25. (Notice that the center of gravity lies outside the body.)

When an object is slightly disturbed and its center of gravity is neither raised nor lowered, it

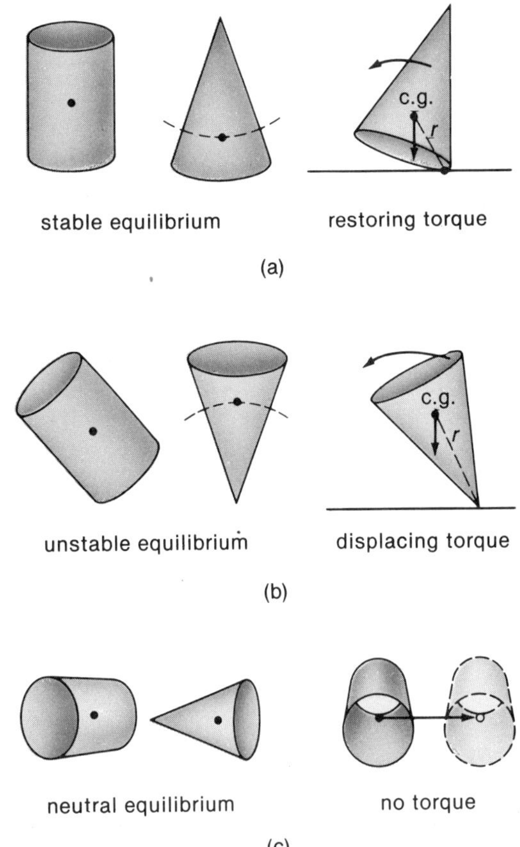

Figure 3.24
Types of equilibrium. (a) When an object in stable equilibrium is slightly disturbed, the center of gravity (c.g.) is raised and a restoring torque acts to return the object to its equilibrium position. (b) In unstable equilibrium, a disturbance lowers the c.g. and gives rise to a displacing torque that causes the object to topple. (c) In neutral equilibrium, the c.g. is neither raised nor lowered when the object is disturbed.

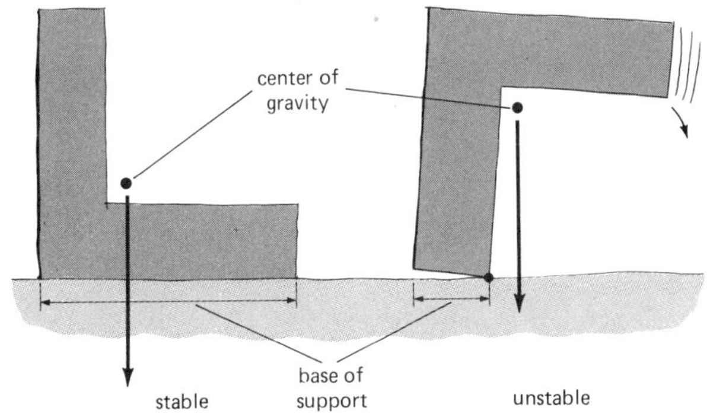

Figure 3.25
Stability. A body is in stable equilibrium if its center of gravity lies vertically above and inside its base of support.

Figure 3.26
Low to the ground. Racing cars have low centers of gravity and wide bases for greater stability.

is in **neutral equilibrium**—for example, when a cylinder lying on its side is displaced in the direction of the surface (Fig. 3.24).

The c.g. and stability are important design considerations in many applications. It should be evident that objects with wide bases and low centers of gravity are more stable and less likely to tip over than objects with small bases and high centers of gravity. An example of the application of these principles is the design of racing cars, which are very low to the ground (Fig. 3.26).

Important Terms

particle an amount of matter that is assumed to occupy a single point in space

(particle) static equilibrium the condition when the resultant force acting on a particle at rest is zero

rigid body a solid object in which all the particles making up the object maintain fixed distances from each other

torque the product of moment arm and force; a torque tends to produce rotational motion

rigid body statics:

translational (nonrotational) equilibrium the sum of the forces acting on a rigid body at rest is zero

rotational equilibrium the sum of the torques acting on a nonrotating rigid body is zero

center of gravity the average location of the weight distribution of a body, or the point at which the entire weight (or mass) of an object may be considered to be concentrated; the center of gravity coincides with the **center of mass** in a uniform gravitational field

stable equilibrium an object in stable equilibrium when slightly displaced tends to return to its equilibrium position due to a restoring torque; as long as the c.g. is vertically over and inside the base of support of an object, there will be a restoring torque that tends to return the object to its equilibrium position

unstable equilibrium the principle that an object in unstable equilibrium topples when slightly displaced due to a displacing torque

Important Formulas

condition for particle static equilibrium:	$\Sigma \mathbf{F} = 0$
or	$\Sigma \mathbf{F}_x = 0$ $\Sigma \mathbf{F}_y = 0$ (in two dimensions)
torque:	$\tau = r_{\perp}F$ (moment arm times force)
conditions for rigid body static equilibrium:	$\Sigma \mathbf{F} = 0$ $\Sigma \boldsymbol{\tau} = 0$

steps (1–4) for solving particle static and translational (nonrotational) equilibrium problems:

1. Sketch a space diagram of the problem.
2. Isolate a particle representation.
3. Draw a free-body diagram of the particle.
4. Sum the forces applying the condition for particle static equilibrium ($\Sigma \mathbf{F} = 0$, or usually in component form $\Sigma \mathbf{F}_x = 0$ and $\Sigma \mathbf{F}_y = 0$) by using the triangle or component method, and solve for the unknown(s).

steps (5–6) for solving rotational equilibrium problems:

5. Choose a convenient pivot point or axis of rotation of the system.
6. Sum the torques about the pivot point or axis of rotation, applying the condition for rotational equilibrium ($\Sigma \boldsymbol{\tau} = 0$).

Apply all steps for rigid body statics problems. The equations from the $\Sigma \mathbf{F} = 0$ and $\Sigma \boldsymbol{\tau} = 0$ conditions give a set of n equations with n unknowns from which the complete solution to the problem can be obtained.

Questions

Particle Statics

1. Do "point" particles really exist? Explain.
2. How does the particle representation of an object avoid the consideration of rotational motion?
3. What are concurrent forces?
4. A particle is acted upon by two forces. Describe the situations (a) when the particle could be in static equilibrium and (b) when the particle would not be in static equilibrium.

Free-Body Diagrams

5. Distinguish between space diagrams and free-body diagrams.
6. Summarize the steps used in solving particle static problems.
7. What is a normal force?
8. For an object on an inclined plane, what is the direction of the normal force?

Rigid Body Statics

9. How could water be made a rigid body?
10. Which of the following are rigid bodies? (a) a screwdriver, (b) Jell-O, (c) a raw egg, (d) a hard-boiled egg.
11. Referring to Figure 3.14(b), could rotational static equilibrium be achieved with (a) one additional applied force? (b) two additional applied forces? Explain.
12. In Figure 3.27(a), when the free end of the hinged board is held horizontally and released, it falls downward. As shown in (b), to keep the board horizontal a hammer is hung on the end! Explain.

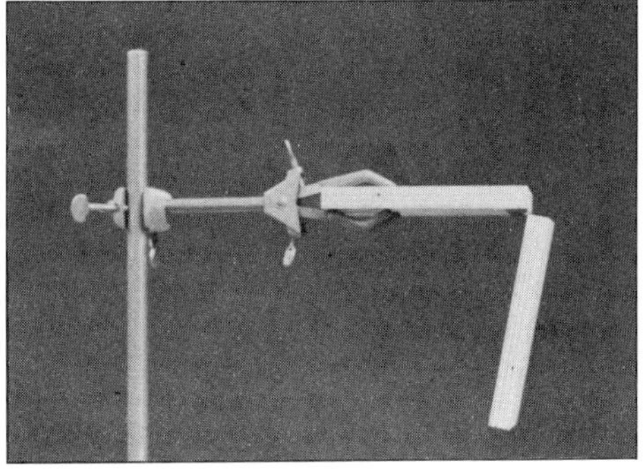

(a)

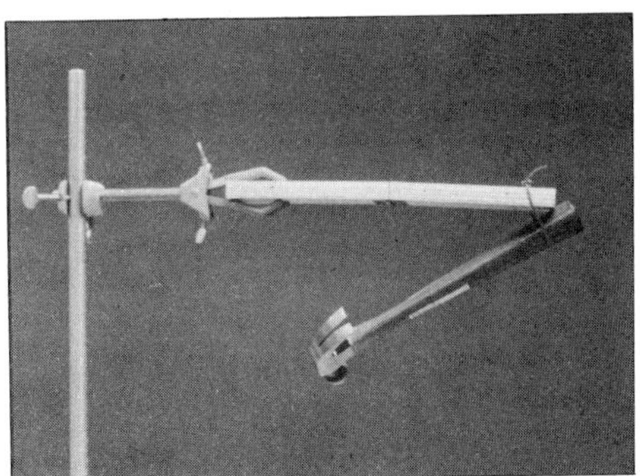

(b)

Figure 3.27
A suspended weight keeps the board from falling? See Question 12.

Center of Gravity and Equilibrium

13. Is it possible for a rigid body to have a center of mass and not a center of gravity? Explain.
14. Where are the centers of gravity of (a) a donut and (b) a boomerang located? How could you determine the locations of their centers of gravity?
15. Is it possible for a person's center of gravity to lie outside of his or her body? Explain.
16. What is the most stable position you can assume?
17. A common child's toy with a rounded, weighted bottom always rights itself when knocked over. Explain why this toy can't be knocked down to stay.
18. How does standing with your legs apart—for example, in a moving bus—increase your stability? Is stability increased in all directions? (Better hang on.)
19. Does it make a difference how the trailer of a tractor-trailer rig is loaded? (Consider both sideways and vertical load distributions for a truck that will go around some banked curves on a trip.)
20. Does putting "big wheels" on a pickup truck make it more stable (Fig. 3.28)? Explain. (*Hint:* what is the purpose of the bar behind the cab in the photo?)
21. What is the effect of balancing an auto tire on its center of gravity, and what is the purpose?
22. The Leaning Tower of Pisa leans because of the settling of the ground under one side of its base (see Fig. 7.6). (a) If this continues, when will the tower fall? (b) Corrective actions have been taken.* What else might be done to keep the tower from toppling?
23. How many different positions of stable and unstable equilibrium are there for a cube? (Consider that resting on each different side is a different position.)
24. Explain the balancing acts shown in Figure 3.29.

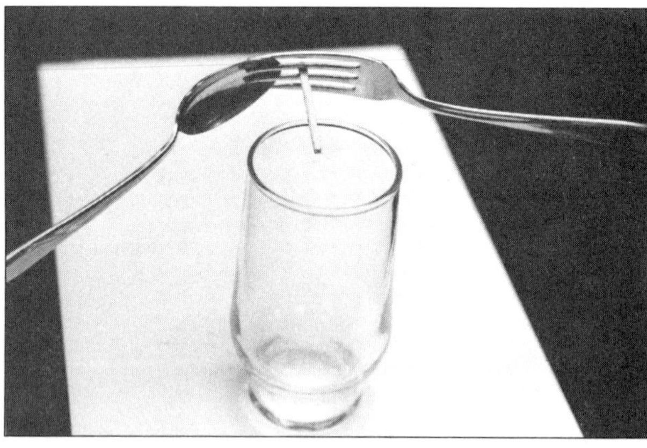

(a)

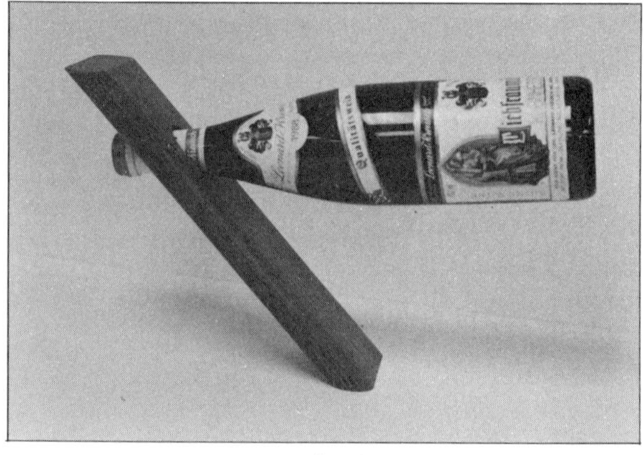

(b)

Figure 3.29
Balancing acts? See Question 24.

Figure 3.28
Stable situation? See Question 20.

* The Leaning Tower (56.2 m high) was begun in 1174 and completed in the 14th century. The settling began in the early stages of construction, and the third, fifth, and top (eighth) stories were straightened somewhat to compensate for the lean. The foundation is only as wide as the circumference of the tower. The foundation has been injected with cement, but the tower is still threatened with collapse. The top of the tower leans about 5 m from the vertical, and at the time of this writing is closed for repairs.

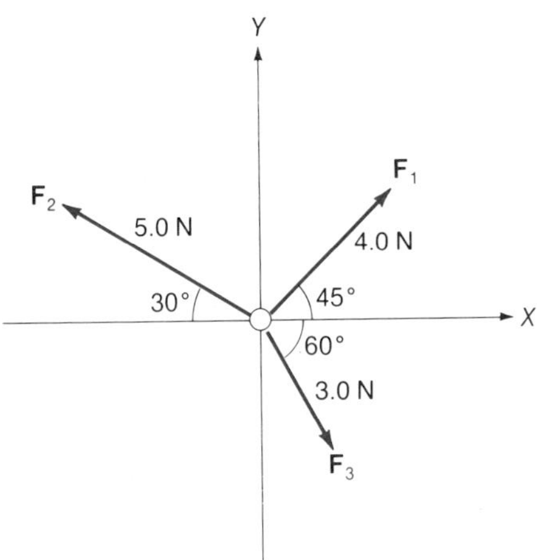

Figure 3.30
See Problem 5.

Problems

Levels of difficulty are indicated by asterisks for your convenience.

3.1–3.2 Particle Statics and Free-Body Diagrams

1. Three concurrent forces, $\mathbf{F}_1 = 10\mathbf{x} - 8\mathbf{y}$, $\mathbf{F}_2 = -4\mathbf{x} - 2\mathbf{y}$, and $\mathbf{F}_3 = -6\mathbf{x} + 10\mathbf{y}$ (magnitudes in N), act on a particle at the origin. Is the particle in static equilibrium?
2. Two concurrent forces, $\mathbf{F}_1 = 7\mathbf{x} - 5\mathbf{y}$ and $\mathbf{F}_2 = -3\mathbf{x} - 2\mathbf{y}$ (magnitudes in lb), act on a particle at the origin. What additional third force would put the particle in static equilibrium?
3. Three concurrent forces, $\mathbf{F}_1 = 6\mathbf{x} + 3\mathbf{y}$, $\mathbf{F}_2 = -4\mathbf{x} + 5\mathbf{y}$, and $\mathbf{F}_3 = -2\mathbf{x} - 4\mathbf{y}$ (magnitudes in N), act on a particle at the origin. Is the particle in static equilibrium? If not, what additional force will make it so?
4. Three concurrent forces, $\mathbf{F}_1 = 4\mathbf{x} - 2\mathbf{y} + 3\mathbf{z}$, $\mathbf{F}_2 = 3\mathbf{x} + 5\mathbf{y} - 7\mathbf{z}$, and $\mathbf{F}_3 = -7\mathbf{x} - 3\mathbf{y} + 4\mathbf{z}$ (magnitudes in N), act on a particle at the origin. Is the particle in static equilibrium? If not, what additional force will make it so?

*5. Is the particle in Figure 3.30 in static equilibrium? If not, what additional force will make it so?

*6. Is the particle in Figure 3.31 in static equilibrium? If not, what additional force will put it in equilibrium?

*7. What is the minimum tensile strength of the cables shown in Figure 3.32 that could support the 100-lb weight?

*8. What would be the tension T in the horizontal cable shown in Figure 3.33 if the reaction force were along the strut? (Neglect the weight of the strut.)

*9. What is the tension in each of the cables in Figure 3.34?

*10. A 100-N block is at rest on a 45° inclined plane. What are (a) the normal force on the block and (b) the frictional force?

*11. A 150-lb crate is at rest on a 20° inclined plane. What are (a) the normal force on the crate and (b) the frictional force?

*12. A block is kept at rest on a 37° inclined plane by a frictional force of 50.0 N along the plane. What is the normal force on the block?

*13. If $w_1 = 9.50$ N in Figure 3.12, what would be the weight of w_2 if the blocks were in static equilibrium?

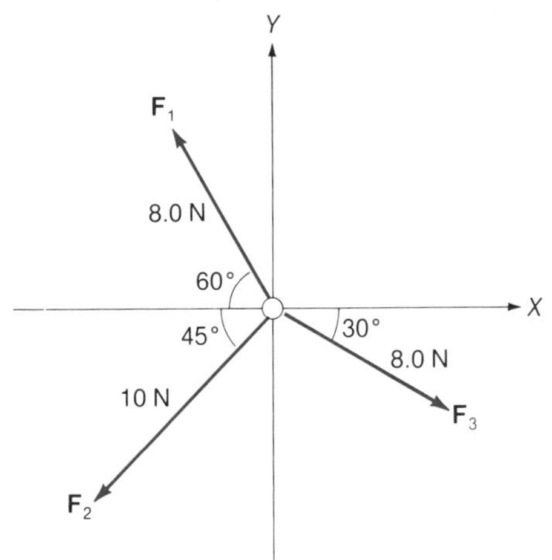

Figure 3.31
See Problem 6.

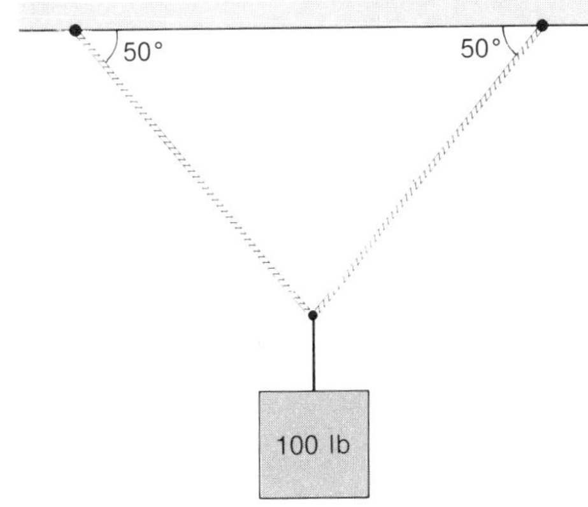

Figure 3.32
See Problem 7.

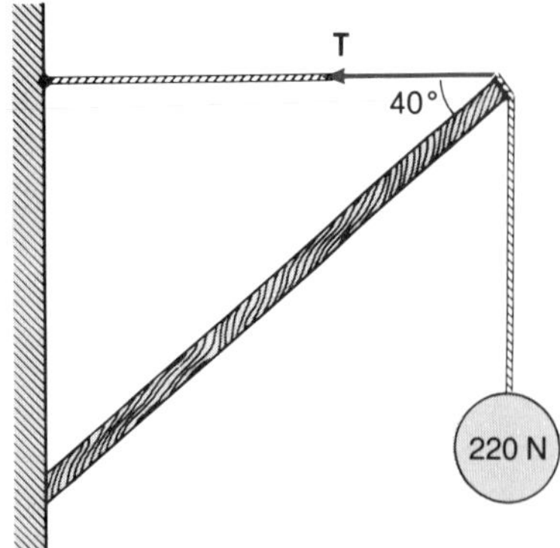

Figure 3.33
See Problems 8 and 19.

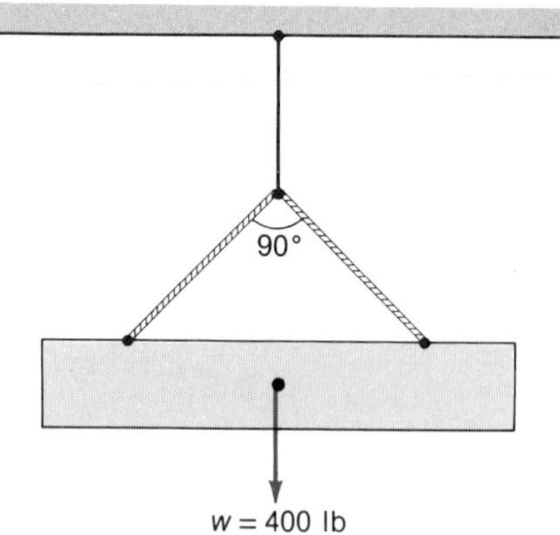

Figure 3.34
See Problem 9.

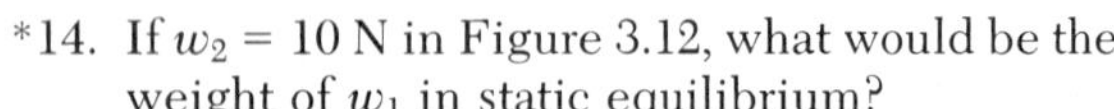

*14. If w_2 = 10 N in Figure 3.12, what would be the weight of w_1 in static equilibrium?

*15. If w_1 = 5.0 lb on the frictionless surface in Figure 3.13, what would be the weight of w_2 if the system were in static equilibrium?

*16. In a situation similar to that in Figure 3.13, the angle of incline of the frictionless surface is 30° and w_2 = 10 N. What is the weight of w_1 for static equilibrium?

**17. What is the magnitude of the normal force on the block in Figure 3.35?

**18. Suppose the angles made by the cables in Figure 3.31 were 45° and 30°. What would be the tensions in the cables?

**19. Suppose the length of the strut in Figure 3.33 were 4.0 m and the distance between the base of the strut at the wall and the point where the cable is attached to the wall were 2.0 m. What would be the tension in the horizontal cable if the reaction force were along the strut? Neglect the weight of the strut. (The angle is not 40° in this case.)

**20. A block weighing 2.00 tons is supported by a crane (Fig. 3.36). What are the magnitudes of the thrust force of the boom and the tension in the boom support cable?

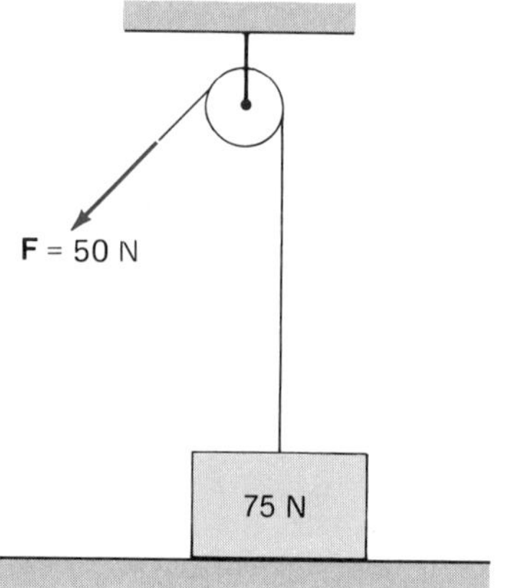

Figure 3.35
See Problem 17.

3.3 Rigid Body Statics

21. Taking the origin of a Cartesian coordinate system as the axis of rotation, compute the moment arm for the following forces applied at a position of (2.0 m, 0) if (a) the force is perpendicular to the x-axis, (b) the force is at an angle of 30° relative to the $+x$-axis, (c) the force is at an angle of 120° relative to the $+x$-axis, and (d) the force is along the x-axis.

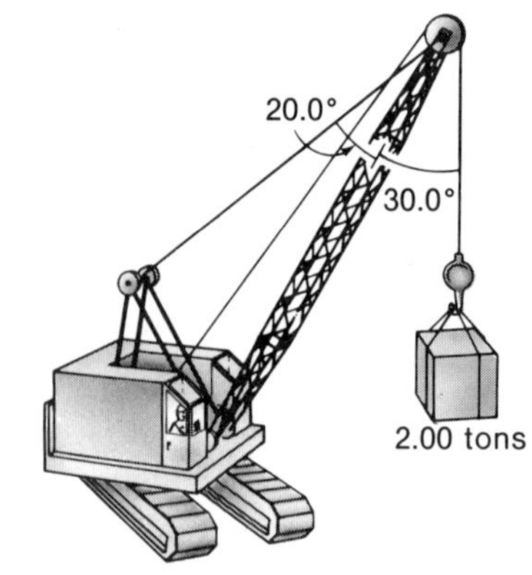

Figure 3.36
See Problem 20.

22. A force of 5.0 lb acts on a rigid body with its line of action 18 in. from the axis of rotation. What is the magnitude of the torque?
23. What is the torque for the situation in Figure 3.15(c) if the length of the pivoted rod is 4.0 ft and the 10-lb force is applied at an angle of 30° relative to the rod?
24. What is the torque for the situation in Figure 3.15(b) if the length of the pivoted rod is 1.2 m and the 50-N force is applied at an angle of 120° relative to the rod?
*25. A meterstick is pivoted at its center (cf. Fig. 3.17) and has a weight of 24.0 N suspended from the left end of the stick. (a) At what position would a 40.0-N weight have to be suspended for the system to be in static equilibrium? (b) What is the reaction force of the support on the stick?
*26. A yardstick pivoted at its center (cf. Fig. 3.17) has a 0.50-lb weight suspended at the 8.0-in. position. (a) At what position would a 1.0-lb weight have to be suspended for the system to be in static equilibrium? (b) What is the reaction force of the support on the stick?
*27. A pivoted meterstick as in Figure 3.17 has a 20-N weight suspended at the 10-cm position. Could a suspended 15-N weight be used to put the system in static equilibrium? If so, what is the position of suspension?
*28. Show that the system shown in Figure 3.17 is in static equilibrium with the axis of rotation taken at the 70-cm position of the meterstick.
**29. Show that the system shown in Figure 3.18 is in static equilibrium with the axis of rotation taken through the midpoint of the boom.
**30. A truck on a 100-ft bridge is 30.0 ft from one end (Fig. 3.37). If the truck weighs 20.0 tons and the bridge weighs 10,000 lb, what are the upward forces exerted by the support piers at each end? (*Hint:* take the axis of rotation through one support pier.)

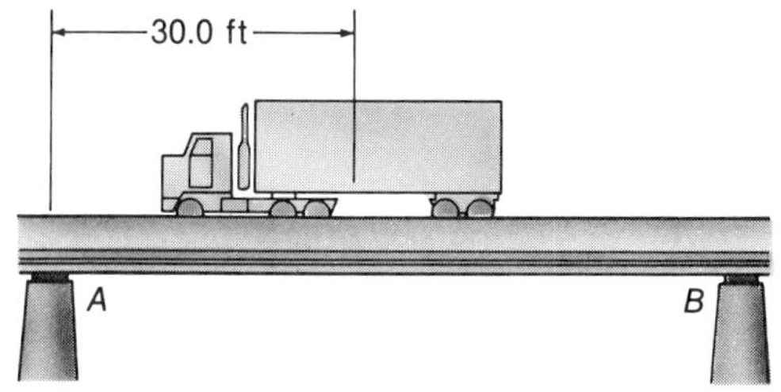

Figure 3.37
See Problems 30 and 31.

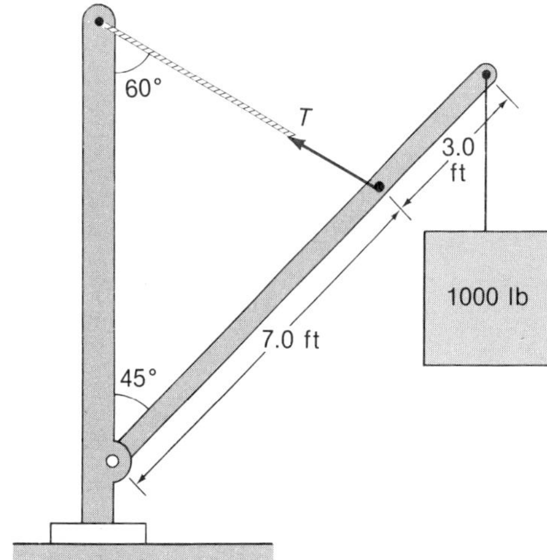

Figure 3.38
See Problems 33 and 34.

**31. Suppose the truck in Figure 3.37 were (a) at the center of the bridge and (b) 90.0 ft from pier *A*. What would be the upward forces exerted by the support piers in each of these cases?
**32. Suppose for the horizontal boom assembly in Figure 3.18 the 400-lb weight were 16 ft from the boom pivot. What would be the tension in the support cable?
**33. What is the tension in the cable for the boom and cable shown in Figure 3.38? Neglect the weight of the boom. (*Hint:* for the tension torque, use the tension force component perpendicular to the boom.)
**34. What is the tension in the cable for the boom and cable in Figure 3.38 if the weight of the uniform boom (500 lb) is considered?

3.4 Center of Gravity and Equilibrium

*35. Suppose the mass of the meterstick shown in Figure 3.21 were not uniformly distributed and an applied force of $T = 7.15$ N were required to have the system in static equilibrium. What would be the location of the center of gravity of the stick?
*36. A painter's scaffold 10.0 ft long is supported by ropes attached to each end. The uniform scaffold weighs 100 lb and the painter, who weighs 180 lb, stands 4.0 ft from one end. What are the tensions in the ropes supporting the scaffold?

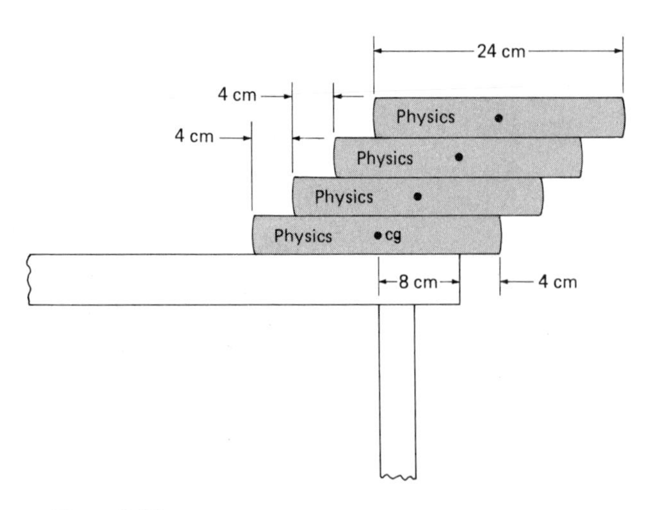

Figure 3.39
See Problem 38.

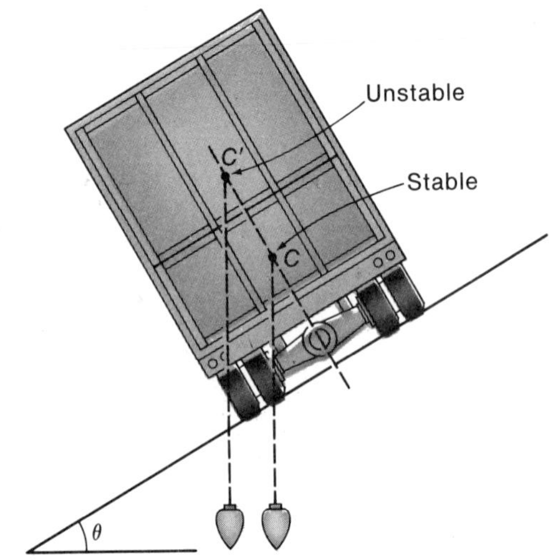

Figure 3.40
See Problem 39.

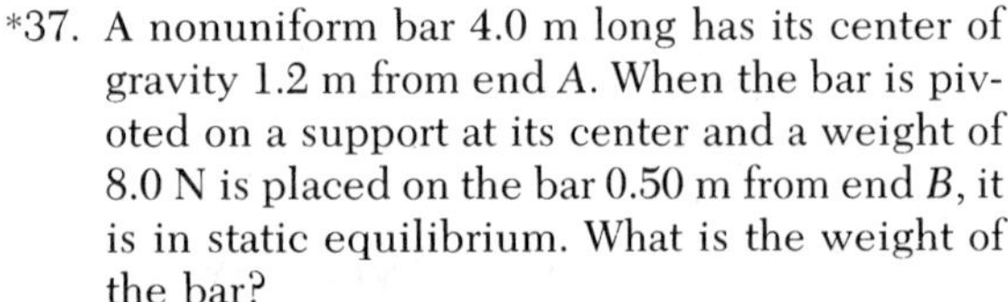

*37. A nonuniform bar 4.0 m long has its center of gravity 1.2 m from end *A*. When the bar is pivoted on a support at its center and a weight of 8.0 N is placed on the bar 0.50 m from end *B*, it is in static equilibrium. What is the weight of the bar?

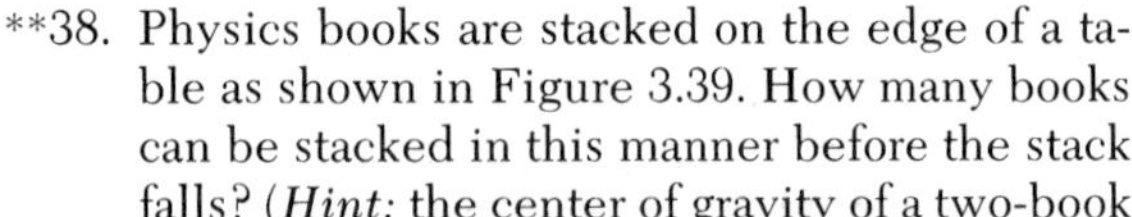

**38. Physics books are stacked on the edge of a table as shown in Figure 3.39. How many books can be stacked in this manner before the stack falls? (*Hint:* the center of gravity of a two-book "system" is midway between the centers of gravity of the two books.)

**39. The principle of stable and unstable equilibrium is illustrated in Figure 3.40. If the truck is loaded so that its center of gravity is midway of its width and 10 ft above the road surface, what will be the angle of incline at which it will just be in unstable equilibrium if the truck has an outer rear wheelbase of 12 ft?

Chapter 4

Motion: Description and Analysis

In the preceding chapter, we learned the conditions under which a stationary object or particle remains at rest—in static equilibrium. It follows that when an object or particle is not in static equilibrium, it must be in motion. The physical description and analysis of motion are an important part of physics. In addition to common, everyday examples of motion—people walking, cars traveling, etc.—there are many technical applications.

It has been implied in previous chapters that motion involves the fundamental properties of length and time and that forces produce motion. However, in this chapter, only **kinematics**—the study of motion without regard to what produces the motion—will be considered. The study of force and motion—dynamics—will be the subject of later chapters.

4.1 Motion, Speed, and Velocity

When an object is undergoing a continuous change of position, we say that the object is moving or in **motion.** In many cases it is desirable to quantitatively describe motion.

Average Speed

One method of describing the motion of an object is by its **average speed** $\bar{v}$, which is defined as

$$\bar{v} = \frac{\Delta d}{\Delta t} \qquad \textbf{(Eq. 4.1)}$$

or

$$\text{average speed} = \frac{\text{scalar distance}}{\text{time to travel the distance}}$$

which is often written simply as

$$v = \frac{d}{t} \qquad \textbf{(Eq. 4.1)}$$

where the **distance** d is the *actual* path length traveled by the object and t is the time interval taken to travel the distance. A bar over a symbol ($\bar{v}$) indicates that it is an average value, in this case over some time interval. The delta (Δ)

means "change in," but is often omitted since it is understood we are talking about intervals. For example, instead of writing the time interval as $\Delta t = t - t_o$, it is customary to write Δt simply as t, where it is understood that the initial time t_o is taken as $t_o = 0$.

The actual path length or distance traveled by an object may be in a straight line or along a curved path. Distance is a scalar quantity, as in time. Hence, the average speed is a scalar.

As can be seen from Equation 4.1, average speed has units of length per time. In standard SI units, they are m/s, and in the British system, ft/s. Commonly used nonstandard units are km/h (kph) and mi/h (mph). The conversion factors are

$$1 \text{ m/s} = 3.6 \text{ km/h} = 2.2 \text{ mi/h}$$

$$1 \text{ ft/s} = 0.68 \text{ mi/h} = 1.1 \text{ km/h}$$

[Notice that there is another 2.2 conversion factor in going from metric to British units, i.e., 1 m/s = 2.2 mi/h (rounded off to two places). Recall that 1 kg = 2.2 lb or, more correctly, "is equivalent to."] A convenient conversion factor from mi/h to ft/s is

$$60 \text{ mi/h} = 88 \text{ ft/s} \quad (= 96 \text{ km/h})$$

$$30 \text{ mi/h} = 44 \text{ ft/s} \quad (= 48 \text{ km/h})$$

and

$$15 \text{ mi/h} = 22 \text{ ft/s} \quad (= 24 \text{ km/h})$$

EXAMPLE 4.1 A roller coaster makes a run on a 1.5-mi track in 3.0 min. What is its average speed in (a) mi/h and (b) ft/s?

Solution: (a) Since $d = 1.5$ mi and $t = 3.0 \text{ min} \times 1 \text{ h}/60 \text{ min} = 0.050 \text{ h}$,

$$\bar{v} = \frac{d}{t} = \frac{1.5 \text{ mi}}{0.050 \text{ h}} = 30 \text{ mi/h}$$

(b) As can be seen from the previous conversion factors, 30 mi/h = 44 ft/s.

EXAMPLE 4.2 A hand-operated forklift in a factory (Fig. 4.1) moves a load 30 m in 2.0 min. What is the average speed of the forklift in (a) m/s and (b) km/h?

Figure 4.1
A hand-operated forklift used to move loads in a factory. See Example 4.2. (Courtesy General Electric Co.)

Solution: (a) With $d = 30$ m and $t = 2.0 \text{ min} \times 60 \text{ s/min} = 120 \text{ s}$,

the average speed is

$$\bar{v} = \frac{d}{t} = \frac{30 \text{ m}}{120 \text{ s}} = 0.25 \text{ m/s}$$

(b) Converting to km/h,

$$\frac{1}{4} \text{ m/s} \left(\frac{3.6 \text{ km/h}}{1 \text{ m/s}}\right) = 0.90 \text{ km/h}$$

Notice that by expressing 0.25 as a fraction ($\frac{1}{4}$), the calculation can be done quickly "in your head." This is sometimes convenient when you don't have a calculator handy.

Average Velocity

For straight-line motion, the change in position of an object can be described by its displacement. **Displacement,** a vector, is the straight-line distance between two points plus a direction (Fig. 4.2). The average velocity is defined as

$$\bar{\mathbf{v}} = \frac{\Delta \mathbf{s}}{\Delta t} \frac{\text{(vector displacement)}}{\text{(time for displacement)}} \qquad \textbf{(Eq. 4.2)}$$

This is often written simply as

$$\bar{\mathbf{v}} = \frac{\mathbf{s}}{t} \qquad \textbf{(Eq. 4.2)}$$

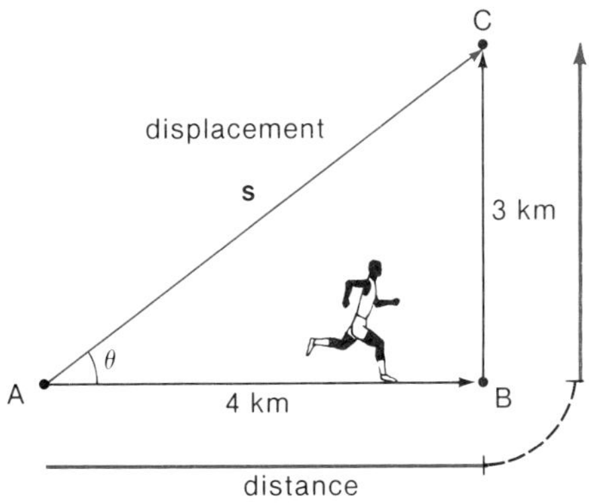

Figure 4.2
Distance and displacement. Distance is the actual path length traveled and is a scalar quantity. Displacement is a vector quantity and is the straight-line distance between two points plus a direction. See Example 4.3.

Like average speed, the average velocity describes a time rate of change of position *but* in a specific direction. For example, if the motion is in the x or y direction, the average velocity could be written $\bar{\mathbf{v}} = \mathbf{x}/t$ or $\bar{\mathbf{v}} = \mathbf{y}/t$, respectively.

It should be evident that the units for average velocity are the same as those for average speed (m/s or ft/s). Also, for straight-line motion *in one direction,* the magnitude of the average velocity is equal to the average speed $\bar{v} = d/t = s/t$, since the magnitude of the displacement is equal to the distance in this case.

EXAMPLE 4.3 A city jogger jogs 4.0 km eastward along a street in 25 min, then turns north and jogs 3.0 km in 20 min (Fig. 4.2). What are the jogger's average speed and average velocity?

Solution: The total distance (actual path length) is

$$\begin{aligned} d &= d_1 + d_2 \\ &= 4.0 \text{ km} + 3.0 \text{ km} = 7.0 \text{ km} \end{aligned}$$

and the total time is

$$\begin{aligned} t &= t_1 + t_2 \\ &= 25 \text{ min} + 20 \text{ min} = 45 \text{ min} = 0.75 \text{ h} \end{aligned}$$

The average speed is then

$$\bar{v} = \frac{d}{t} = \frac{7.0 \text{ km}}{0.75 \text{ h}} = 9.3 \text{ km/h}$$

[The time could have been left in minutes, in which case the average speed would be in units of km/min. However, the time was converted to hours to obtain the more commonly used unit of km/h (and 1 km/h = 0.62 mi/h).]

The magnitude of the jogger's vector displacement is the straight-line distance from A to C. This is 5.0 km, as may be determined from Pythagoras' theorem or by observing that the displacement is the hypotenuse of a 3-4-5 triangle. Hence, the magnitude of the average velocity is

$$\bar{v} = \frac{s}{t} = \frac{5.0 \text{ km}}{0.75 \text{ h}} = 6.7 \text{ km/h}$$

The direction of the displacement and the average velocity can be specified by the angle θ (see Fig. 4.2). Using trigonometry,

$$\theta = \tan^{-1}\left(\frac{y}{x}\right) = \tan^{-1}\left(\frac{3.0 \text{ km}}{4.0 \text{ km}}\right) = 37°$$

or the direction of the average velocity is 37° north of east.

EXAMPLE 4.4 It is possible to get from one side of a city to the other (A to B in Fig. 4.3) by two routes: a bypass highway and a direct cross-town route.

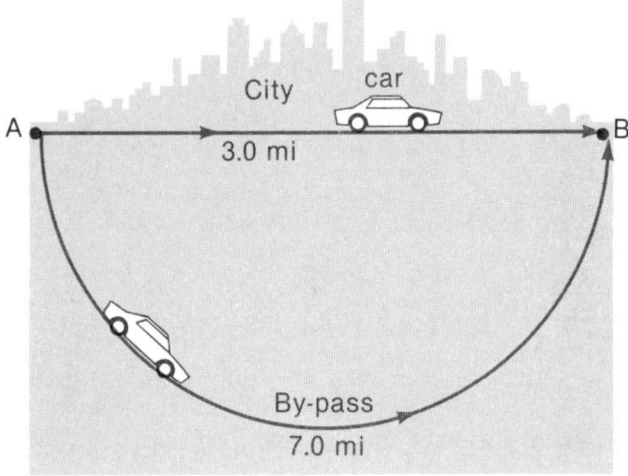

Figure 4.3
Illustration for Example 4.4.

(a) If it takes a car 8.5 min to travel the 3.0-mi direct route through the city, what are its average speed and average velocity?

(b) If another car takes 8.5 min to travel the 7.0-mi bypass route, what is its average speed?

(c) If the first car returns to its original starting point by the direct route in another 10 min, what are its average speed and average velocity for the total trip?

Solution: (a) By the direct route, $d = 3.0$ mi and $t = 8.5$ min (1 h/60 min) = 0.14 h. Thus, the average speed is

$$\bar{v} = \frac{d}{t} = \frac{3.0 \text{ mi}}{0.14 \text{ h}} = 21 \text{ mi/h}$$

For this straight-line motion in one direction, the distance is equal to the magnitude of the displacement, $d = s$, and the average speed is equal to the magnitude of the average velocity. Thus, the average velocity is $\bar{\mathbf{v}} = 21$ mi/h $\mathbf{x}$, where the motion is taken to be in the x direction.

(b) For the bypass route, $d = 7.0$ mi and $t = 8.5$ min = 0.14 h, and the average speed is

$$\bar{v} = \frac{d}{t} = \frac{7.0 \text{ mi}}{0.14 \text{ h}} = 50 \text{ mi/h}$$

Notice that the average velocity has no meaning for the curved bypass route, since the car does not travel in a single vectorial direction.

(c) The time for the return direct trip is $t_2 = 10$ min (1 h/60 min) = 0.17 h. For the total trip from A to B and B to A, the distance is

$$d = d_1 + d_2 = 3.0 \text{ mi} + 3.0 \text{ mi} = 6.0 \text{ mi}$$

and

$$t = t_1 + t_2 = 0.14 \text{ h} + 0.17 \text{ h} = 0.31 \text{ h}$$

The average speed is then

$$\bar{v} = \frac{d}{t} = \frac{6.0 \text{ mi}}{0.31 \text{ h}} = 19 \text{ mi/h}$$

The average velocity for the total trip is zero, since the total displacement is zero, i.e.,

$$\mathbf{s} = \mathbf{s}_1 + \mathbf{s}_2 = 3 \text{ mi} - 3 \text{ mi} = 0.$$

Uniform Speed and Velocity

In general, the average speed and average velocity do not really tell us anything about the actual motion of a car or some other object *between* two points, since they are averages over a given time interval. During this time, the car probably varied its speed, perhaps even stopping (at a traffic light).

In the special case of an object traveling the same distance each time interval, the object travels with a constant or uniform speed.

For example, as illustrated in Figure 4.4(a), the car travels 15 meters each second. Hence, it has a uniform speed of $\bar{v} = 15$ m/s (or 54 km/h or about 33 mi/h).

In the case of an object moving with a uniform speed (constant magnitude) along a straight-line path in one direction (constant direction), its average velocity is constant.

If the magnitude of the displacement x is plotted *versus* time t for uniform motion, a straight line is obtained, as shown in Figure 4.4. The slope of the line $\Delta x/\Delta t$ is constant for any Δx and corresponding Δt intervals. Since $\bar{v} = \Delta x/\Delta t$, the average speed is equal to the slope of the line. Notice that $\tan \theta = \Delta x/\Delta t$, so $\bar{v} = \tan \theta$.

If the motion is nonuniform, the curve is not a straight line, as illustrated in the second graph (part b) of Figure 4.4. The slope of the curve, and hence the average speed, varies along the curve.

Instantaneous Speed and Velocity

To describe motion in more and more detail, we can continually shorten the measured time interval Δt, which will shorten the corresponding distance interval Δx. Letting Δt approach zero ($\Delta t \rightarrow 0$) so that Δt is an "instant," we obtain the **instantaneous speed.** As the name implies, this is the speed of an object at a particular time or instant at a particular point x. The speed and velocity are no longer averages, so the bar over the symbols is no longer applicable.

We may then write the **instantaneous velocity** as $\mathbf{v} = v\,\mathbf{x}$, where v is the instantaneous speed in the x direction. For example, the speedometer of an automobile gives an almost instantaneous speed. This reading plus the di-

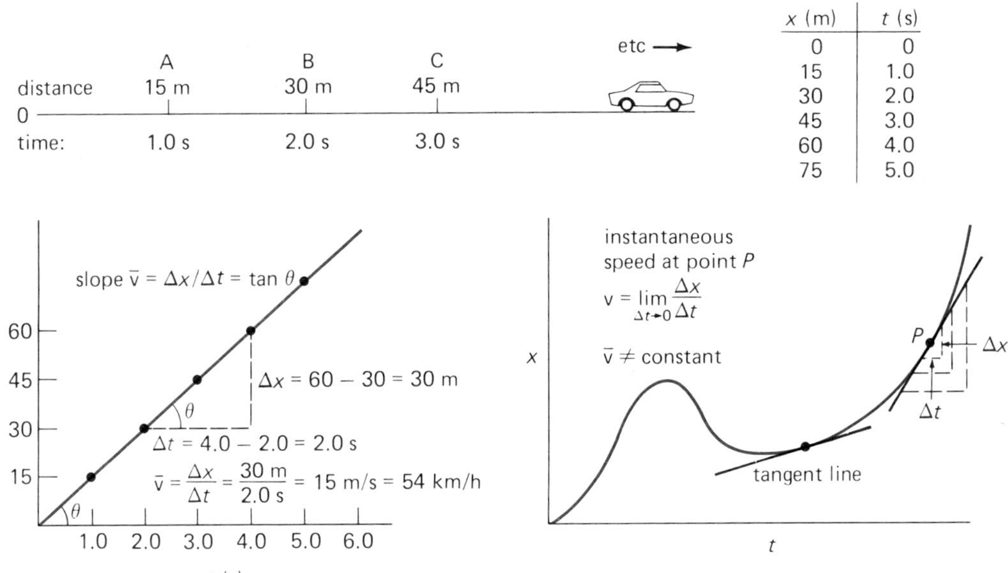

Figure 4.4
Uniform and nonuniform motions. (a) For uniform motion, a graph of distance versus time (x versus t) is a straight line. The slope of the line is equal to the average speed. Since the motion is uniform, the average speed is equal to the instantaneous speed, $\bar{v} = v$. (b) For nonuniform motion, a graph of x versus t is not a straight line. The slope of the curve, and hence the average speed, varies along the curve. The instantaneous speed at a point is equal to the slope of a line "tangent" to the curve at that point. This is obtained by letting the time interval Δt approach zero.

rection of the automobile at that instant is its instantaneous velocity.

In the case of uniform motion where the average speed is constant [as in Fig. 4.4(a)], the instantaneous speed is constant and equal to the average speed, i.e., $v = \bar{v}$. This corresponds to driving an automobile at a constant speed as indicated on the speedometer. If the highway is straight, then the instantaneous velocity is also constant (constant direction).

For nonuniform motion, the instantaneous speed at a particular point is equal to the slope of a line "tangent" to the curve at that point. As illustrated in Figure 4.4(b), this line represents the slope or tangent of the Δx-Δt triangle as Δt approaches zero.

4.2 Acceleration

In most real situations, the velocity of an object in motion is not constant but varies with time, i.e., the object speeds up or slows down. This variation in velocity with time is described by acceleration. **Acceleration** is the time rate of change of velocity. The **average acceleration** is defined as

$$\bar{\mathbf{a}} = \frac{\Delta \mathbf{v}}{\Delta t} = \frac{\mathbf{v}_2 - \mathbf{v}_1}{t_2 - t_1} \quad \textbf{(Eq. 4.3)}$$

$$= \frac{\text{(change in velocity)}}{\text{(time to make change)}}$$

where $\mathbf{v}_1$ and $\mathbf{v}_2$ are the instantaneous velocities at t_1 and t_2, respectively.

Since velocity is a vector, the average acceleration is also a vector. Note that an acceleration results from a change in velocity (a vector), which may be a change in the velocity magnitude *and/or* direction (Fig. 4.5). Similar to instantaneous velocity, by letting Δt approach zero, we obtain an **instantaneous acceleration,** which is the rate of change of velocity at a particular instant of time.

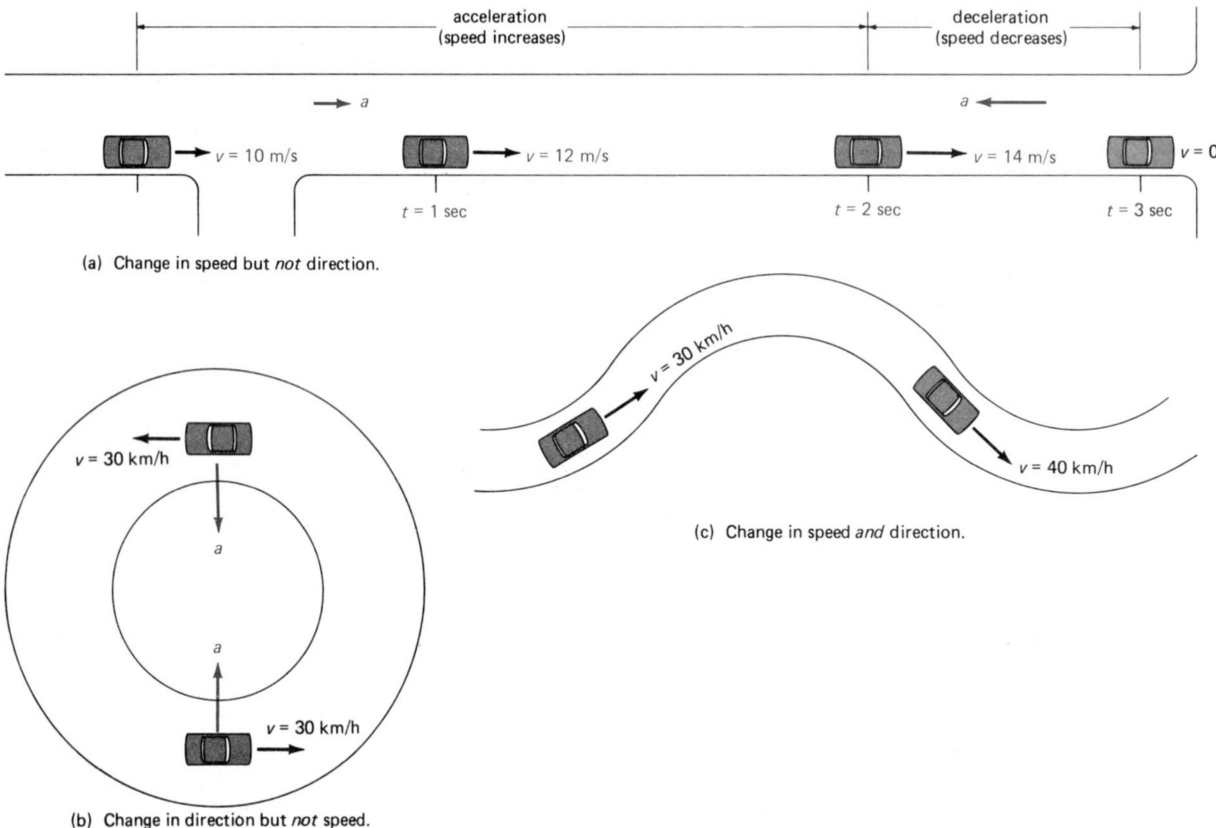

Figure 4.5
Acceleration = time rate of change of velocity (velocity/time) due to a change in speed and/or direction. [The centripetal acceleration in (b) is discussed in a later chapter.]

The units of acceleration can be derived from Equation 4.3. For example, in the SI system, the standard units of velocity are m/s, so $a = v/t = (\text{m/s})/\text{s} = \text{m/s}\cdot\text{s} = \text{m/s}^2$ (read "meter per second squared"). Similarly in the British system, the units of acceleration are ft/s^2.

To help grasp the idea of acceleration, think about the effect of the gas pedal, or "accelerator," of an automobile during travel along a straight road. When you push down on the accelerator, the car speeds up, or the velocity changes (increases), as indicated on the speedometer. In this case, the acceleration (vector) is in the same direction as the motion. When you let up on the accelerator, the car slows down and the velocity changes (decreases). The acceleration in this case is sometimes referred to as a *deceleration,* and its direction is opposite to that of velocity.

The description of accelerated motion can be complex and require advanced mathematics. For our purposes, we shall restrict our discussion to constant, or uniform, accelerations. For a constant acceleration, the velocity changes uniformly with time. An example of straight-line motion with constant acceleration is shown in Figure 4.6. Since **a** is constant, the average acceleration is equal to the instantaneous acceleration $\bar{\mathbf{a}} = \mathbf{a}$, and we may write Equation 4.3 as

$$a = \frac{v_2 - v_1}{t_2 - t_1} \qquad \text{(constant acceleration)}$$

(The boldface vector notation will be omitted from now on for simplicity, with the understanding that the vector directions for straight-line motions will be indicated by plus and minus signs.)

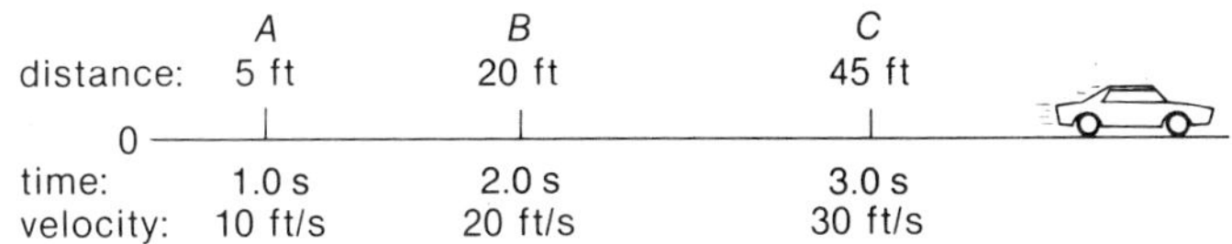

v (ft/s)	t (s)
0	0
10	1.0
20	2.0
30	3.0
40	4.0
50	5.0

x (ft)	t (s)
0	0
5	1.0
20	2.0
45	3.0
80	4.0
125	5.0

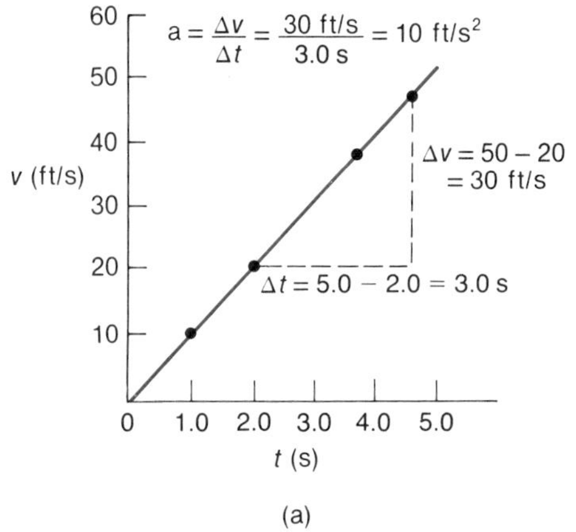

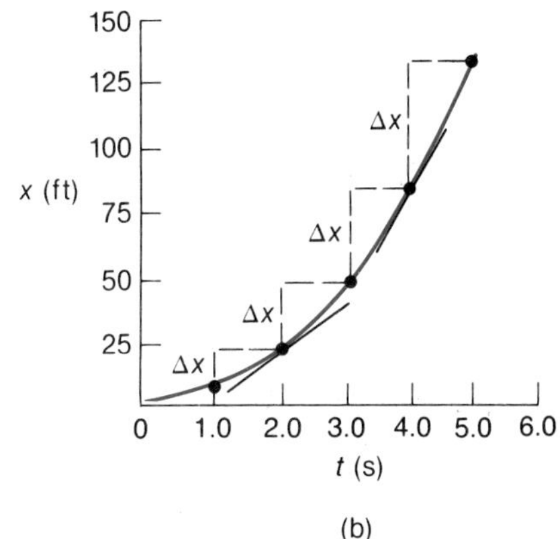

Figure 4.6

Constant acceleration. (a) For constant acceleration, the velocity increases uniformly with time, and the v versus t graph is a straight line. The slope of the line is equal to the magnitude of the acceleration. (b) The x versus t graph in this case is a curved line, since the distance x traveled for each unit time interval increases with time.

It is customary to take the initial time as $t_1 = t_o = 0$, with a corresponding initial or "original" velocity of $v_1 = v_o$. Also, t_2 is taken as an arbitrary final time t, with a corresponding final velocity $v_2 = v_f$. The equation then becomes

$$a = \frac{v_f - v_o}{t} \qquad \textbf{(Eq. 4.4)}$$

or

$$v_f = v_o + at$$

(final velocity = initial velocity + change in velocity).

As shown in Figure 4.6 for constant acceleration, the graph of this equation is a straight line with a slope of $a = \Delta v/\Delta t$. The initial velocity v_o is the intercept of the v-axis. For the case in Figure 4.6, the intercept is at the origin, $v_o = 0$, and the car was initially at rest. The x versus t graph in this case is a curved line, since the distance traveled for each time interval increases with time.

When the velocity varies uniformly with time, i.e., the acceleration is constant, the average velocity is given by

$$\bar{v} = \frac{v_f + v_o}{2} \qquad \textbf{(Eq. 4.5)}$$

(constant a)

For example, suppose the speeds of an object undergoing uniform acceleration motion are 10 m/s, 20 m/s, 30 m/s, 40 m/s, and 50 m/s at corresponding equal time intervals. How would you find the magnitude of the average velocity? Add the quantities and divide by 5? This would be correct, and an average speed of $\bar{v} = 30$ m/s would be obtained. But since the speed increases uniformly, the average speed is also given by the average of the initial and final values, $\bar{v} = (v_f + v_o)/2 = (50 + 10)/2 = 30$ m/s.

The average velocity can also be obtained by using Equation 4.2,

$$\bar{v} = \frac{x}{t} \quad \text{or} \quad x = \bar{v}t \qquad \textbf{(Eq. 4.6)}$$

where the displacement is taken to be in the x direction and the direction of the displacement is indicated by sign.

Hence, given the initial velocity v_o of an object and its (constant) acceleration a, one can calculate the final velocity v_f from Equation 4.4, the average velocity $\bar{v}$ from Equation 4.5, and finally the position of the object x from Equation 4.6. This requires three mathematical steps. However, if this is done algebraically for a general case, all three steps will be expressed in a single equation. Starting with Equation 4.6 and substituting Equation 4.5 into it,

$$x = \bar{v}t = \left(\frac{v_f + v_o}{2}\right) t$$

But, by Equation 4.4, $v_f = v_o + at$, and substituting this expression for v_f, we have

$$x = \left(\frac{v_o + at + v_o}{2}\right) t$$

or

$$x = v_o t + \tfrac{1}{2}at^2 \qquad \textbf{(Eq. 4.7)}$$

Hence, given v_o and a, the position x at any time t can be determined from one equation rather than having to use three equations.

EXAMPLE 4.5 An automobile traveling initially at 30 mph accelerates uniformly at a rate of 10 ft/s^2 for 5.0 s. How far does the automobile travel during this time?

Solution: It is given that $v_o = 30$ mph $= 44$ ft/s, $a = 10$ ft/s^2, and $t = 5.0$ s. (Notice that 30 mph was changed to 44 ft/s. All quantities must be expressed in consistent units before making calculations. Standard units are ordinarily used.) Then, by Equation 4.4,

$$v_f = v_o + at$$
$$= 44 \text{ ft/s} + (10 \text{ ft/s}^2)(5.0 \text{ s}) = 94 \text{ ft/s}$$

By Equation 4.5,

$$\bar{v} = \frac{v_f + v_o}{2} = \frac{94 \text{ ft/s} + 44 \text{ ft/s}}{2} = 69 \text{ ft/s}$$

And, by Equation 4.6,

$$x = \bar{v}t = (69 \text{ ft/s})(5.0 \text{ s}) = 345 \text{ ft}$$

(additional significant figures assumed). However, Equation 4.7 could have been used directly:

$$x = v_o t + \tfrac{1}{2}at^2$$
$$= (44 \text{ ft/s})(5.0 \text{ s}) + \tfrac{1}{2}(10 \text{ ft/s}^2)(5.0 \text{ s})^2$$
$$= 345 \text{ ft}$$

and the calculation is much more convenient by this equation.

Another convenient combination of Equations 4.4, 4.5, and 4.6 is obtained by the elimination of t. Using Equation 4.4 in the form $t = (v_f - v_o)/a$,

$$x = \bar{v}t = \left(\frac{v_f + v_o}{2}\right)\left(\frac{v_f - v_o}{a}\right) = \frac{v_f^2 - v_o^2}{2a}$$

or

$$v_f^2 = v_o^2 + 2ax \qquad \textbf{(Eq. 4.8)}$$

The equations used to describe straight-line motion with constant acceleration are summarized in Table 4.1.

EXAMPLE 4.6 A worker sitting on a scaffold 30 m above the ground accidentally drops a wrench. If the wrench falls with a uniform acceleration of 9.8 m/s^2, (a) how fast will it be traveling when it strikes the ground? (b) What is the time of fall?

Table 4.1
Summary of Equations Used to Describe Motion

$x = \bar{v}t$	(Eq. 4.6)	
$\bar{v} = \dfrac{v_f + v_o}{2}$	(Eq. 4.5)	Constant acceleration
$v_f = v_o + at$	(Eq. 4.4)	Constant acceleration
$x = v_o t + \frac{1}{2}at^2$	(Eq. 4.7)	Constant acceleration
$v_f^2 = v_o^2 + 2ax$	(Eq. 4.8)	Constant acceleration

Solution: (a) It is given that $x = 30$ m, $a = 9.8$ m/s^2, and $v_o = 0$. The latter is implied: since the wrench is dropped, it is initially at rest at $t = 0$. Then,

$$v_f^2 = v_o^2 + 2ax = 0 + 2(9.8 \text{ m/s}^2)(30 \text{ m}) = 588 \text{ m}^2/\text{s}^2$$

and

$$v_f = \sqrt{588 \text{ m}^2/\text{s}^2} = 24 \text{ m/s}$$

(b) The time of fall can be found from either Equation 4.4 or Equation 4.7. Using Equation 4.7 with $v_o = 0$,

$$x = \tfrac{1}{2}at^2 \quad \text{or} \quad t = \sqrt{\frac{2x}{a}} = \sqrt{\frac{2(30 \text{ m})}{9.8 \text{ m/s}^2}} = 2.5 \text{ s}$$

EXAMPLE 4.7 A car traveling 60 mph slows down at a uniform rate to 15 mph in 10 s. (a) What is the car's acceleration? (b) How far does the car travel in the first 5 s?

Solution: (a) It is given that $v_o = 60$ mph $= 88$ ft/s, $v_f = 15$ mph $= 22$ ft/s, and $t = 10$ s. Then, using Equation 4.4,

$$a = \frac{v_f - v_o}{t} = \frac{22 \text{ ft/s} - 88 \text{ ft/s}}{10 \text{ s}} = -6.6 \text{ ft/s}^2$$

The minus sign indicates that the acceleration is in the direction opposite the motion, and in this case is a deceleration. (Note that a negative acceleration does not always indicate a deceleration, or a slowing down. If an object were moving in the negative direction (e.g., $-x$ direction), a negative acceleration would speed it up.)

(b) Using Equation 4.7 with $t = 5.0$ s and $a = -6.6$ ft/s^2,

$$\begin{aligned} x &= v_o t + \tfrac{1}{2}at^2 \\ &= (88 \text{ ft/s})(5.0 \text{ s}) - \tfrac{1}{2}(6.6 \text{ ft/s}^2)(5.0 \text{ s})^2 \\ &= 3.6 \times 10^2 \text{ ft} \end{aligned}$$

Note that the equation must be "told" the direction of the acceleration by the sign convention. (Why was the answer written in scientific notation?)

4.3 Free Fall

The situation in Example 4.6 was an example of an object in free fall, i.e., falling freely (neglecting air resistance) under the influence of gravity (Fig. 4.7).* You may have wondered why a value of 9.8 m/s^2 was chosen for the magnitude of the acceleration in the example. This is actually the value of the acceleration due to gravity in SI units. Because this acceleration enters into so many physical applications, it is given the special symbol designation g, and its value in the various systems of units is

$$g = 9.8 \text{ m/s}^2 = 32 \text{ ft/s}^2$$

The direction of g is vertically downward toward the Earth.

The acceleration due to gravity is essentially constant near the Earth's surface. As such, the previous motion equations for constant acceleration apply to free fall. Although x was used as the vertical dimension in Example 4.6, it is customary to designate the vertical dimension as y and the horizontal dimension as x. Also, upward is taken to be the positive y direction and downward to be the negative y direction. With these

* We commonly think of "falling" as traveling downward. Technically, however, a freely falling body is any object moving freely under the influence of gravity, regardless of its initial conditions of motion. This includes objects projected upward or downward, since they are moving freely once released. Such projectile motion will be considered generally in Chapter 5.

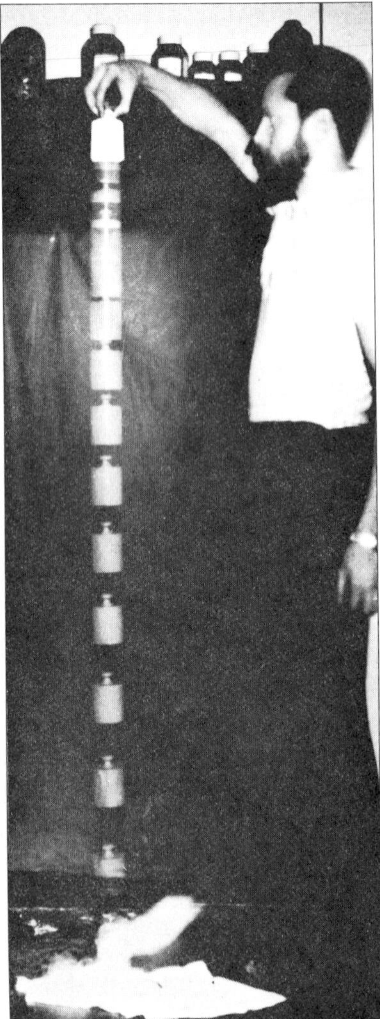

Figure 4.7
An object in free fall (air resistance neglected). A falling object has a constant acceleration of 9.80 m/s^2 due to gravity. This flash exposure was made at a rate of 30 flashes per second. Notice how the distance traveled each time interval increases with time. (Courtesy of R. Stevenson and R. B. Moore.)

conventions we may write the equations of motion in the y direction for an object with an acceleration of $a_y = -g$ as

$$
\begin{aligned}
y &= \bar{v}_y t && \text{(a)} \\
\bar{v}_y &= \frac{v_f + v_o}{2} && \text{(b)} \\
v_y &= v_{y_o} - gt && \text{(c)} \\
y &= v_{y_o} t - \tfrac{1}{2} g t^2 && \text{(d)} \\
v_y^2 &= v_{y_o}^2 - 2gy && \text{(e)}
\end{aligned}
\qquad \textbf{(Eq. 4.9)}
$$

where v_y corresponds to v_f in the y dimension and the downward negative direction of g has been explicitly indicated by a minus sign.

EXAMPLE 4.8 An object is dropped from rest. What is its velocity 3.0 s after being released?

Solution: Since the object is initially at rest, $v_{y_o} = 0$. Also, since the object is falling, it is understood that it has an acceleration of $g = 32$ ft/s^2. (If m/s^2 had been chosen as the units of g, the problem would be worked in the SI system. British units will be used here so that an easy comparison can be made in mi/h.)

Using Equation 4.9c with $t = 3.0$ s,

$$v_y = v_{y_o} - gt = 0 - (32 \text{ ft/s}^2)(3.0 \text{ s}) = -96 \text{ ft/s}$$

The minus sign indicates that the velocity is downward in keeping with the sign convention.

Notice that an object in free fall falls rather rapidly. After falling 3.0 s, the object is moving with a speed in excess of 60 mph (88 ft/s), actually about 65 mph.

EXAMPLE 4.9 A person at the edge of a vertical cliff drops a stone over the edge. (a) How long does it take the stone to fall 200 m? (b) What is the stone's velocity on falling this distance?

Solution: (a) It is given that $y_o = 0$ and $y = -200$ m (minus because displacement is downward). Then, Equation 4.9d with $v_{y_o} = 0$ is $y = -\frac{1}{2}gt^2$ and

$$t = \sqrt{\frac{2y}{-g}} = \sqrt{\frac{2(-200 \text{ m})}{-9.8 \text{ m/s}^2}} = 6.4 \text{ s}$$

(b) With $t = 6.4$ s,

$$v_y = v_{y_o} - gt = 0 - (9.8 \text{ m/s}^2)(6.4 \text{ s}) = -63 \text{ m/s}$$

(Why minus?)

It should be noted that the minus signs can be avoided by considering the downward direction to be positive. This is appropriate when all of the vector quantities are in the same direction. Another way of looking at this is that for falling objects, the acceleration due to gravity increases the velocity and g is a positive acceleration. In this case, $+g$ is used in the equations. [In the next chapter on projectiles, when an object is thrown or projected upward, g must be taken as a "slowing" (negative) acceleration, or deceleration.]

To illustrate this, the downward direction will be taken as positive in the following example.

EXAMPLE 4.10 In a manufacturing process, blocks of material on a conveyor belt moving 0.80 ft/s drop onto another conveyor belt 2.5 ft below, as illustrated in Figure 4.8.

(a) If the process requires that a block drop off the upper belt just as the preceding block hits the lower belt, how far apart should the blocks be spaced on the upper belt?

(b) If the blocks are to be 9 in. apart on the lower belt, what is the required speed of the belt?

Solution: It is given that $v_{y_o} = 0$, $y = 2.5$ ft, and $v_{x_1} = 0.80$ ft/s.

(a) The time it takes for a block to fall from the upper belt to the lower one is found by Equation 4.9d,

$$y = v_{y_o}t + \tfrac{1}{2}gt^2 = 0 + \tfrac{1}{2}gt^2$$

and

$$t = \sqrt{\frac{2y}{g}} = \sqrt{\frac{2(2.5 \text{ ft/s})}{32 \text{ ft/s}^2}} = 0.40 \text{ s}$$

Then, as one block drops, the next block should be behind it at a distance

$$x = v_{x_1}t = (0.80 \text{ ft/s})(0.40 \text{ s}) = 0.32 \text{ ft}$$

So the blocks should be spaced a distance of 0.32 ft (12 in./ft) = 3.8 in. apart.

(b) After landing on the lower belt, a block must move 9 inches, or 0.75 ft, in the 0.40 s it takes for the next block to fall. Then

$$v_{x_2} = \frac{x}{t} = \frac{0.75 \text{ ft}}{0.40 \text{ s}} = 1.9 \text{ ft/s}$$

Important Terms

motion the continuous change of position of an object

distance the actual path length traveled by an object (a scalar quantity)

speed

- **average** the distance traveled divided by the time to travel the distance ($\bar{v} = d/t$)
- **instantaneous** the speed at a given time or instant ($\Delta t \to 0$)

displacement the straight-line distance between two points plus a direction (a vector quantity)

velocity

- **average** the displacement traveled divided by the time to travel the displacement ($\bar{\mathbf{v}} = \mathbf{s}/t$)
- **instantaneous** a vector with the instantaneous speed as the magnitude and with the direction of motion at that instant

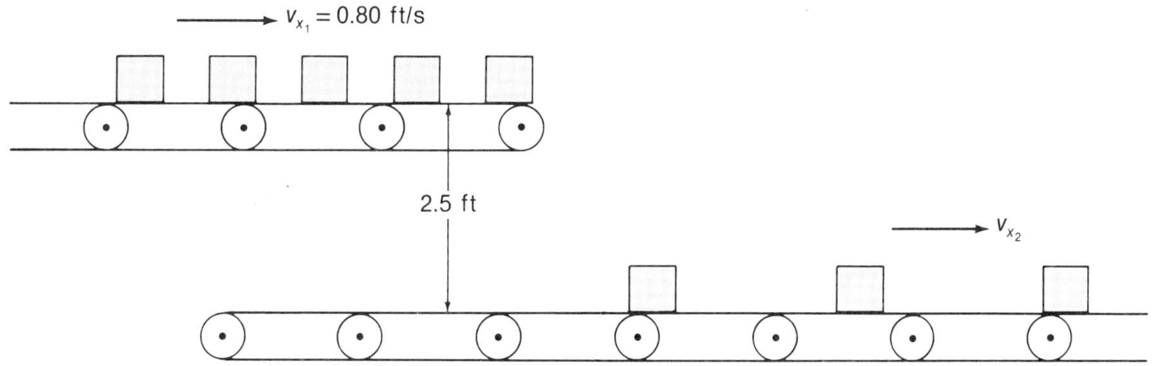

Figure 4.8

Object in motion on a conveyor belt. See Example 4.10.

acceleration the time rate of change of velocity

free fall an object in motion under the influence of (acceleration due to) gravity with air resistance neglected

Important Formulas

average speed: $\bar{v} = \dfrac{d}{t}$

average velocity: $\bar{\mathbf{v}} = \dfrac{\mathbf{s}}{t}$

acceleration (average): $\bar{\mathbf{a}} = \dfrac{\Delta \mathbf{v}}{\Delta t} = \dfrac{\mathbf{v}_2 - \mathbf{v}_1}{t_2 - t_1}$

equations describing straight-line motion: $x = \bar{v}t$

$$\left.\begin{aligned} \bar{v} &= \frac{v_f + v_o}{2} \\ v_f &= v_o + at \\ x &= v_o t + \tfrac{1}{2}at^2 \\ v_f^2 &= v_o^2 + 2ax \end{aligned}\right\} \text{Constant acceleration}$$

acceleration due to gravity: $g = 9.8\ \text{m/s}^2 = 32\ \text{ft/s}^2$

free fall in y direction (downward negative direction, g expressed explicitly):

$$y = \bar{v}_y t$$

$$\bar{v}_y = \frac{v_f + v_o}{2}$$

$$v_y = v_{y_o} - gt$$

$$y = v_{y_o} t - \tfrac{1}{2}gt^2$$

$$v_y^2 = v_{y_o}^2 - 2gy$$

Questions

Motion, Speed, and Velocity

1. Is the distance between two points always greater than the magnitude of the displacement between the points? Can it be less?
2. Is it possible to be in the same place and to have moved?
3. Does average speed tell what an object in motion actually did over the measured time interval?
4. What is the difference between average speed and average velocity?
5. After taking an auto trip, why is average velocity not used to express how fast the trip was made?
6. How is uniform speed related to the velocity for straight-line motion?
7. How may the average speed be determined from a graph of x versus t for an object in uniform motion?
8. Can the average speed be determined from a graph of x versus t for an object in nonuniform motion? Explain.
9. Suppose you went somewhere and returned to your starting point in 30 minutes. What would be your average velocity for the trip?

Acceleration

10. A towmotor in a factory moves along a straight aisle at a uniform speed of 5 ft/s. What is the towmotor's acceleration?
11. What are the units of acceleration?
12. A motorist in a moving car applies the brakes. Does the car accelerate? Explain. Discuss what could be called an "accelerator" and a "decelerator" on a car—for example, the steering wheel and the gears in "gearing down."
13. What is a "deceleration" in terms of the initial and final velocities?
14. What is the graph of v versus t for an object with uniform acceleration, and how can the magnitude of the acceleration be determined from the graph? Can the initial velocity be determined from the graph?
15. Sketch a graph of v versus t for an object with an initial velocity v_o and a uniform acceleration.
16. Sketch a graph of x versus t for an object initially at rest and moving with a uniform acceleration.
17. Generally describe the motions illustrated in the graphs in Figure 4.9 and interpret the axis intercepts of the graphs.

Free Fall

18. How does the velocity of an object in free fall change?
19. Give the velocities of a freely falling object dropped from rest for (a) $t = 1$ s and (b) $t = 3$ s.
20. If the air resistance were taken into account for a falling object, how would this affect its acceleration?
21. An object is thrown vertically upward. (a) What is its velocity at the maximum height? (b) How does the acceleration due to gravity affect its velocity on the upward and downward parts of the path?
22. The acceleration due to gravity on the moon is 1/6 that on Earth. Does this mean that an object dropped on the moon will strike the surface in 1/6 the time taken by an object dropped from an equivalent height on Earth? (There is no problem with air resistance on the moon. Why?)

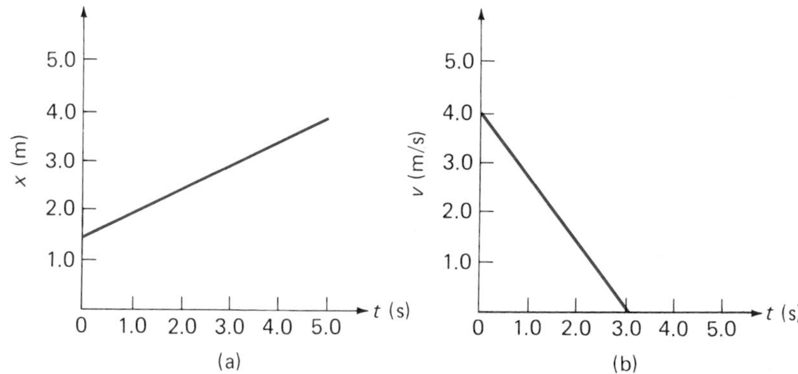

Figure 4.9
See Question 17. (a) See Problems 12 and 23. (b) See Problem 24.

Problems

Levels of difficulty are indicated by asterisks for your convenience.

4.1 Motion, Speed, and Velocity

1. A jogger jogs 4 blocks west, 4 blocks south, and 4 blocks east. What are the jogging distance and the displacement?
2. A student drives a distance of 2.5 mi to school in 15 min. What is the average speed?

*3. Suppose the student in Problem 2 makes the return trip home by the same route in 25 min. What is the average speed for the total trip?

*4. A football player catches the kickoff in a practice session on his own goal line. He runs the direct length of the field for a touchdown in 20 s. What is the player's average velocity?

*5. Two cars, A and B, travel the same 100-km distance when driven by students going home on spring break. Car A makes the trip in 2 hours and car B in $2\frac{1}{2}$ hours. (a) Which car has the greater average speed? (b) What is the average speed of each car?

*6. An automobile travels 50 km in 30 min, and a train travels 90 km in 1 h. (a) Which has the greater average speed? (b) What is the average speed of each?

*7. A racing car travels around a circular track 1.0 mi in diameter in a time of 94 s. What is the average speed in (a) mi/h and (b) ft/s?

*8. How long would it take a racing car to go around a circular track 2.0 km in diameter traveling at an average speed of 150 km/h?

*9. A runner runs around a circular track 300 ft in diameter in a time of 42.8 s. What are the runner's (a) average speed and (b) average velocity?

*10. From the graph in Figure 4.9(a), determine (a) the average velocity and (b) the instantaneous velocity.

*11. The x versus t graphs of the motion of two cars are shown in Figure 4.10. What are (a) the average speed and (b) the instantaneous speed of each car?

*12. A person walks with a constant speed of 0.50 m/s. How far does the person walk in (a) 1 s, (b) 20 s, and (c) 1.0 min?

*13. An airplane flies eastward with a constant speed of 300 km/h for 2.0 h, and then southward at 250 km/h for another 2.0 h. What is the location of the plane relative to its initial position?

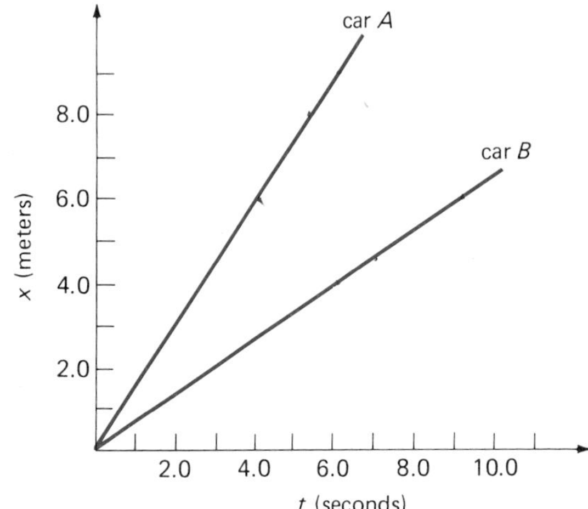

Figure 4.10
See Problem 11.

**14. According to the theory of continental drift, the continents were once part of a giant single continent (called Pangaea), which broke and drifted apart. Measurements of sea-floor spreading along midoceanic ridges show the rate of continental drift to be on the order of 4 cm/year. Assuming this rate to have been constant in the past, and taking the interlocking shapes of South America and Africa to be 5000 miles apart today, approximately how long ago did Pangaea break up?

**15. A factory worker moves a fork lift 25 m north along a straight aisle in 18 s, then right 55 m in 1.0 min, and then left 20 m in 15 s. What are (a) the average speed and (b) the average velocity?

**16. A motorboat on a still lake travels 500 yd in a direction 37° east of north in 23 s and then 400 yd south in 18 s. What are the boat's (a) average speed and (b) average velocity in both ft/s and mi/h?

**17. An airplane flies on a heading directly north at a speed of 160 km/h. At the same time a wind with a speed of 30.0 km/h blows eastward. (a) What is the effective velocity of the plane? (*Hint:* draw a vector diagram.) (b) Is the plane accelerating? Explain.

4.2 Acceleration

18. An automobile starting from rest achieves a velocity of 45 m/s in 9.0 s along a straight road. What is the car's average acceleration?

19. A motorist uniformly accelerates her car from 30 mi/h to 60 mi/h in 10 s along a straight segment of interstate highway. What is the acceleration of the car (and motorist)?

20. An object with an initial velocity of 12 m/s accelerates uniformly at a rate of 8.0 m/s^2 for 6.0 s. What is the final velocity of the object?

21. A sports car accelerates uniformly at 8.0 ft/s^2 to a speed of 60 mi/h in 4.0 s. What was the initial velocity of the car?

*22. A conveyor belt is speeded up at a constant rate from 0.50 ft/s to 1.5 ft/s in 4.0 s. (a) What is the average velocity of an object on the belt for this period? (b) How far does it travel during this time? [Work (b) by two methods.]

*23. Determine the acceleration of the motion illustrated in Figure 4.9(a).

*24. (a) Find the acceleration for the situation shown in Figure 4.9(b). (b) What are the initial and final speeds?

*25. Graphs of v versus t for two cars are shown in Figure 4.11. What are the acceleration and initial speed of each car?

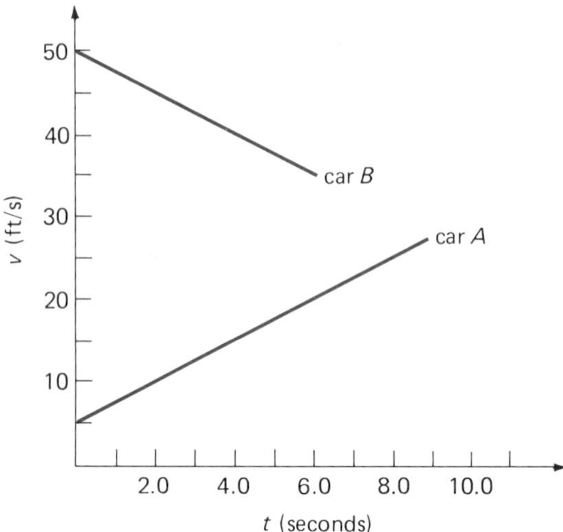

Figure 4.11
See Problem 25.

*26. A jet plane on takeoff reaches a speed of 250 km/h uniformly in 5.00 s. What are (a) the acceleration and (b) the distance traveled in 5.00 s? (c) What is the jet's speed at the end of 4.00 s?

*27. A train starting from rest accelerates uniformly at a rate of 6.0 ft/s^2 on a straight track. How far does the train travel *during* the fourth second?

*28. An automobile speeds up from 15.0 mi/h to 45.0 mi/h along a straight road at a rate of 10.0 ft/s^2. How far does the car travel during this time?

*29. A monorail train traveling at 75 km/h slows down uniformly to 15 km/h in 10 s. How far does the train travel during this time?

*30. A racing car traveling at 120 mi/h ejects a drag chute and comes uniformly to a stop in 5.00 s. (a) What is the car's acceleration? (b) How far did the car travel after ejecting the chute?

**31. A graph of displacement versus time is shown in Figure 4.12 for an object in straight-line motion.
(a) What is the average velocity for each region?
(b) Is the motion uniform or nonuniform in these regions?
(c) What is the instantaneous velocity at point E?
(d) Comment on the average acceleration in each region.

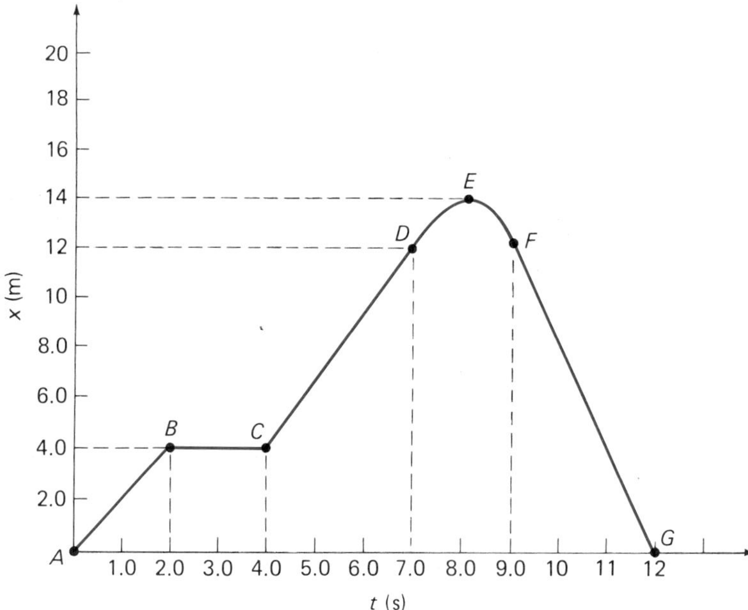

Figure 4.12
See Problem 31.

4.3 Free Fall (Assume additional significant figures if needed.)

32. (a) What is the speed of a dropped object in free fall at the end of 5.0 s? (b) How far does the object fall during this time?
33. How far does a dropped object fall during the time interval between $t = 3.0$ s and $t = 4.0$ s?
*34. A construction worker accidentally drops a wrench from the sixth story of a building 60 ft above the ground. With what speed will the wrench strike the ground? (Is this a good reason for wearing hard hats?)
*35. In a metal shot process, it is required that globules of molten metal fall for 1.5 s before hitting a container of water. From how far above the water surface should the metal be dropped?
*36. A stone is dropped into a well and a splash is heard 2.50 s later. Neglecting the time for the sound to reach the top of the well, how many meters is the water level below the top of the well?
*37. A ball is thrown downward with a speed of 4.0 m/s. (a) What is its velocity in 2.0 s? (b) How far does it travel in 2.0 s?
*38. It is desired to have a pile driver strike a pile with a speed of 40 mi/h. Assuming the driver to be in free fall, from how far above the pile should it be released?
*39. An object is dropped from a landing craft on the surface of the moon. (a) What is the object's velocity at the end of 3.0 s? (b) How far did it fall during this time?
**40. In a manufacturing process as in Figure 4.8, the upper conveyor belt moves with a speed of 1.2 ft/s and the vertical distance between the belts is 3.0 ft. (a) How far should the blocks be spaced on the upper belt so that one block drops off the belt just as the preceding block lands on the lower belt? (b) If the blocks are to be spaced 1.0 ft apart on the lower belt, what is the required speed of the belt?
**41. A ball is thrown downward with a speed of 4.0 ft/s from a height of 80 ft. How long does it take for the ball to strike the ground?

Chapter 5

Motion in a Plane

In the last chapter, the study of motion was generally confined to motion in one dimension. This is the simplest case. The next step is to consider motion in two dimensions, or in a plane.

There are several important cases of motion in a plane. In this chapter, we shall consider two of them—projectile motion and uniform circular motion. There are many technical applications in which objects are projected or move in a circle. The analysis of projectile motion is particularly easy since it involves the analysis of linear motion in each of two directions. Hence, the equations from the preceding chapter apply. In the case of uniform circular motion, some new concepts will be introduced.

5.1 Components of Motion

In Chapter 4, motion was generally limited to one-dimensional, or straight-line, situations, with the motion taken along the x- or y-axis. However, since the orientation of a set of axes is arbitrary, straight-line motion may also be described in two dimensions.

For example, suppose an object moves with a constant velocity along a path at some angle to the x-axis (Fig. 5.1). The constant velocity vector may be resolved into x and y components, v_x and v_y, where $\mathbf{v} = v_x\mathbf{x} + v_y\mathbf{y}$. The magnitudes of the displacement components in the x and y directions are then expressed by the equations

$$x = v_x t$$
$$y = v_y t \qquad \textbf{(Eq. 5.1)}$$

In this manner, we can resolve the motion into rectangular components, analyze the motion in each component direction, and then put the components "back together" for the net effect—much the same as we did for the component method of vector addition. This is illustrated in the following example.

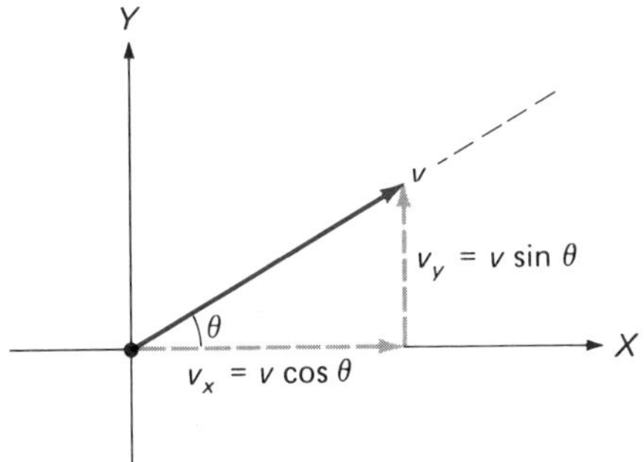

Figure 5.1
Components of motion. The vector quantities of motion may be resolved into rectangular components and the motion analyzed in the respective directions.

EXAMPLE 5.1 An airplane flies on a heading 30° north of east with a speed of 150 km/h (refer to Fig. 5.1). What is its coordinate position, and how far will it travel in 2.0 h?

Solution: Given: $v = 150$ km/h, $\theta = 30°$, and $t = 2.0$ h. Then, resolving the velocity into x and y components,

$$v_x = v \cos\theta = (150 \text{ km/h}) \cos 30° = 130 \text{ km/h}$$

and

$$v_y = v \sin\theta = (150 \text{ km/h}) \sin 30° = 75 \text{ km/h}$$

The component distances traveled in 2.0 h are then

$$x = v_x t = (130 \text{ km/h})(2.0 \text{ h}) = 260 \text{ km}$$

and

$$y = v_y t = (75 \text{ km/h})(2.0 \text{ h}) = 150 \text{ km}$$

and we have

$$(x, y) = (260 \text{ km}, 150 \text{ km})$$

The actual path length of the motion is given by

$$d = \sqrt{x^2 + y^2}$$
$$= \sqrt{(260 \text{ km/h})^2 + (150 \text{ km/h})^2} = 300 \text{ km}$$

and, as we know, at an angle of

$$\theta = \tan^{-1}\left(\frac{y}{x}\right) = \tan^{-1}\left(\frac{150 \text{ km}}{260 \text{ km}}\right) = 30°$$

north of east, or relative to the x-axis.

You may be quick to point out that this is like "cracking a nut with a sledgehammer." Why not just multiply the speed by the time and get the distance [$d = vt = (150 \text{ km/h})(2.0 \text{ h}) = 300$ km] along the path at an angle of 30°?

Indeed, this would be much simpler in the case of Example 5.1. The purpose of the example was to show you how the components of motion can be used. For straight-line motion, there is not much advantage. But, for projectile motion, which generally involves *curved* motion in a plane, there is a great advantage. By resolving the motion into components, we can analyze curved motion in terms of two straight-line component motions, as will be seen.

5.2 Projectile Motion

Projectile motion refers to the motion of an object that has been projected without having the further capability to propel itself. Examples of projectiles are a ball that has been thrown, an arrow that has been shot from a bow, and a rocket after it has burnt all of its fuel. Although this motion is in general in two dimensions, it can be analyzed by means of the equations in Chapter 4. It is instructive to consider vertical projections first, then horizontal projections, and finally projections at any angle. Air resistance will be neglected throughout the discussion.

Vertical Projections

Vertical projections are in one dimension and involve objects that have been projected vertically either upward or downward ($+y$ and $-y$ by convention). Once projected or released, the objects move freely under the influence of gravity—i.e., are in free fall. An object dropped from rest may be considered a vertical projection with an initial velocity of zero ($v_{y_o} = 0$).

Suppose instead of being dropped from rest, an object were given an initial downward velocity of 1.5 m/s, i.e., $v_{y_o} = -1.5$ m/s. What would be the object's velocity and how far would it have traveled in 3.0 s? The answers are obtained as follows:

$$v_y = v_{y_o} - gt = -1.5 \text{ m/s} - (9.8 \text{ m/s}^2)(3.0 \text{ s})$$
$$= -31 \text{ m/s}$$

$$y = v_{y_o}t - \tfrac{1}{2}gt^2 = (-1.5 \text{ m/s})(3.0 \text{ s}) - \tfrac{1}{2}(9.8 \text{ m/s}^2)(3.0 \text{ s})^2$$
$$= -49 \text{ m}$$

The minus signs indicate that v_y and y are in the downward direction with $y = 0$ taken at the starting point.

Notice that the direction of the initial velocity was specifically designated as downward by a minus sign, as was done in the last chapter for g. This "tells" the equation that the velocity and acceleration are both in the downward direction. Also, as pointed out in Chapter 4, the minus signs may be avoided by considering the downward direction to be positive, since all the vector quantities are in the same direction (or g is positive, since it increases the velocity).

However, when an object is projected vertically upward, the initial velocity and the acceleration due to gravity are in opposite directions, and this must be indicated by the normal sign convention (considering g to be a "slowing," negative acceleration in this case).

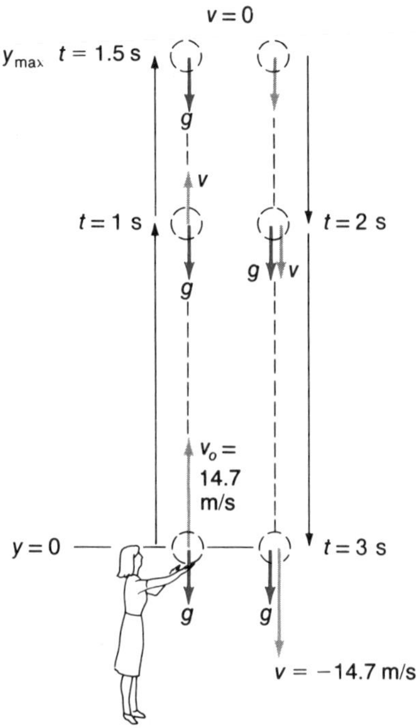

Figure 5.2
Vertical projection. See Example 5.2 (Downward motion displayed for clarity.)

EXAMPLE 5.2 A person throws a ball vertically upward with an initial velocity of 14.7 m/s. (a) How high above the point of release does the ball go? (b) How long after being thrown does it take for the ball to return to its starting point?

Solution: (a) It is given that $v_{y_o} = 14.7$ m/s. To find out how high the ball goes, we consider only the upward part of the motion (Fig. 5.2). At its highest point, the ball is instantaneously at rest, hence $v_y = 0$ at this point. Then, using Equation 4.9e,

$$v_y^2 = 0 = v_{y_o}^2 - 2gy$$

and

$$y = \frac{v_{y_o}^2}{2g} = \frac{(14.7\text{ m/s})^2}{2(9.80\text{ m/s}^2)} = 11.0\text{ m}$$

where $g = 9.80$ m/s² to three significant figures.

(b) It takes the same amount of time for a vertically projected object to fall from its highest point as it takes for it to travel upward to this point. The time it takes for the ball to travel upward, t_u, is given by Equation 4.9c,

$$v_y = v_{y_o} - gt_u = 0$$

and

$$t_u = \frac{v_{y_o}}{g} = \frac{(14.7\text{ m/s})}{(9.80\text{ m/s}^2)} = 1.50\text{ s}$$

Then, the total time is twice this upward time,

$$t = 2t_u = 2(1.50\text{ s}) = 3.00\text{ s}$$

To prove that it takes 1.50 s for the ball to fall back to its starting point from its highest point, the downward part of the ball's motion is the same as dropping a ball from a height of 11.0 m. Then, by Equation 4.9d with $v_{y_o} = 0$, the time for the ball to travel downward, t_d, is

$$-y = v_{y_o}t - \tfrac{1}{2}gt^2 = 0 - \tfrac{1}{2}gt^2$$

(minus y because distance traveled is downward)

and

$$t_d = \sqrt{\frac{2y}{g}} = \sqrt{\frac{2(11.0\text{ m})}{9.80\text{ m/s}^2}} = 1.50\text{ s}$$

While proving things, we can also show that the ball returns to its starting point with the same speed it was given initially. Again, looking at the falling part of the ball's motion, we have

$$v_y = v_{y_o} - gt_d = 0 - (9.80\text{ m/s}^2)(1.50\text{ s})$$
$$= -14.7\text{ m/s}$$

Notice how the sign convention (+ and −) is used to designate the directions of the vector quantities. The consistent use of a sign convention will be reflected in the results. For example, note that in the previous example, the velocity of the ball when it returns to its starting point is $v_y = -14.7$ m/s, which indicates that the velocity is downward. Here is another example.

EXAMPLE 5.3 What is the height of the ball in the preceding example at $t = 2.00$ s?

Solution: With $v_{y_o} = 14.7$ m/s and $t = 2.00$ s, then

$$y = v_{y_o}t - \tfrac{1}{2}gt^2 = (14.7 \text{ m/s})(2.00 \text{ s}) - \tfrac{1}{2}(9.80 \text{ m/s}^2)(2.00 \text{ s})^2$$

$$= 29.4 \text{ m} - 19.6 \text{ m} = 9.8 \text{ m}$$

and the ball is 9.8 m above its starting point (positive y, upward).

The height of the ball is given directly by the equation. Were you to split the motion into up and down parts, you would have to realize that this is the height of the ball after it has fallen for 0.50 s from its maximum height.

To further illustrate the use of the sign convention, let's compute the height of the ball at $t = 3.00$ s. We know this is the time it takes for the ball to return to its starting point ($y = 0$, Fig. 5.2). Then, with $t = 3.00$ s,

$$y = v_{y_o}t - \tfrac{1}{2}gt^2 = (14.7 \text{ m/s})(3.00 \text{ s}) - \tfrac{1}{2}(9.80 \text{ m/s}^2)(3.00 \text{ s})^2$$

$$= 44.1 \text{ m} - 44.1 \text{ m} = 0$$

What would be the implication for a time greater than 3 seconds?

Horizontal Projections

When projected horizontally, an object is given an initial horizontal velocity (v_{x_o} in the x direction). After being released or projected, there is no acceleration in the horizontal direction, so the projectile travels in the x direction with a constant velocity (v_{x_o}). The distance it travels in the x direction with time is given by $x = v_{x_o}t$.

However, upon being projected horizontally, an object has a downward acceleration due to gravity ($a_y = g$). As a result, the object is in free fall in the y direction while it travels uniformly in the x direction and describes a curved path, as illustrated in Figure 5.3. The fact that a horizontally projected object is in free fall while traveling horizontally can be demonstrated by dropping another object at the same time. They both strike the ground at the same time; hence, their vertical motions must have been identical.

The projected object travels the same amount of time in the x direction as it does in the y direction.

Time is the connecting element between the two components of projectile motion.

To analyze this motion, we consider the different parts, or the x and y components, of motion.

EXAMPLE 5.4 If in Figure 5.3 the ball is thrown from a height of 30 m and given an initial horizontal velocity of 8.0 m/s, how far from the building will it land?

Solution: It is given that $y = 30$ m, $v_{y_o} = 0$, and $v_{x_o} = 8.0$ m/s. Considering the vertical motion of the ball, the time it takes to fall 30 m in the $-y$ direction is obtained by

$$-y = v_{y_o}t - \tfrac{1}{2}gt^2 = 0 - \tfrac{1}{2}gt^2$$

and

$$t = \sqrt{\frac{2y}{g}} = \sqrt{\frac{2(30 \text{ m})}{9.8 \text{ m/s}^2}} = 2.5 \text{ s}$$

During this time the object travels in the x direction a distance

$$x = v_{x_o}t = (8.0 \text{ m/s})(2.5 \text{ s}) = 20 \text{ m}$$

and hence lands 20 m from the building.

EXAMPLE 5.5 If the buildings in Figure 5.3 were 25 m apart, what would be the required initial horizontal velocity of the projected ball so that it would hit the other ball just as they both struck the ground? (Assume the dropped ball is close to the building.)

Solution: It is given that $x = 25$ m. From Example 5.4 we know the balls strike the ground at $t = 2.5$ s after being simultaneously released. For the projected ball to travel horizontally

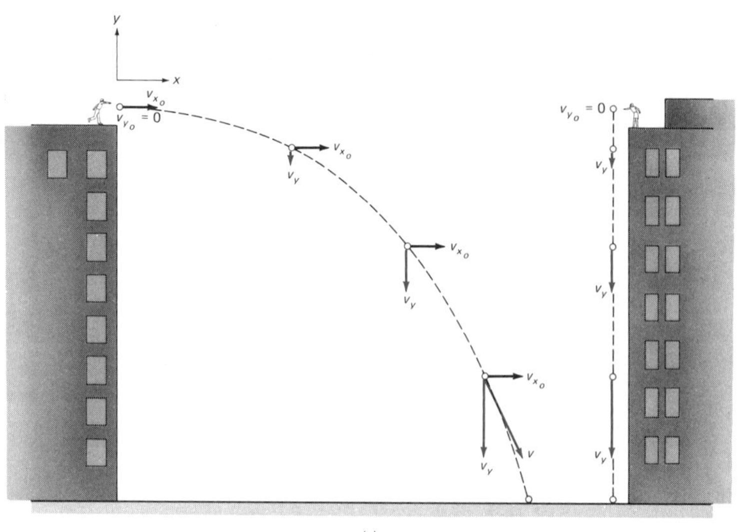

(a)

(b)

Figure 5.3
Horizontal projection. When an object is projected horizontally, it travels in the horizontal direction with a constant velocity v_{x_o} (as indicated by the constant horizontal vector lengths in the drawing). While traveling horizontally, the object is also falling under the influence of gravity $a_y = g$. The combined motions produce a curved path. Note that the downward motion of the projected ball is the same as that of a dropped ball. (Courtesy of the Educational Development Center, Newton, Mass.)

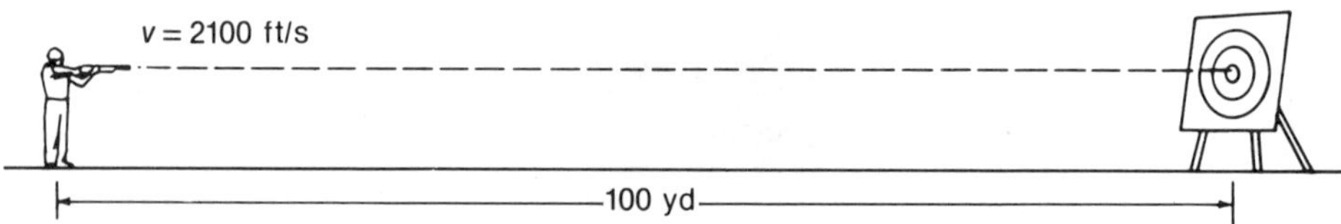

Figure 5.4
Another horizontal projection. If the bullet is aimed directly at the bull's-eye at the same height, the bullet will strike the target below the center of the bull's-eye. See Example 5.6.

25 m in this time requires an initial horizontal velocity of

$$x = v_{x_o}t$$

or

$$v_{x_o} = \frac{x}{t} = \frac{25\ \text{m}}{2.5\ \text{s}} = 10\ \text{m/s}$$

EXAMPLE 5.6 A hunter sighting in his rifle aims directly and horizontally at the center of the bull's-eye of a target 100 yd away (Fig. 5.4). If the muzzle velocity of the bullet is 2100 ft/s, how far below the center of the bull's-eye does the bullet strike the target?

Solution: It is given that $v_{x_o} = 2100$ ft/s and $x = 100$ yd $= 300$ ft. The time it takes the bullet to reach the target is found by

$$x = v_{x_o}t$$

or

$$t = \frac{x}{v_{x_o}} = \frac{(300\ \text{ft})}{(2100\ \text{ft})} = 0.143\ \text{s}$$

Then, during this time, the bullet falls a distance of

$$y = v_{y_o}t - \tfrac{1}{2}gt^2$$
$$= 0 - \tfrac{1}{2}(32\ \text{ft/s}^2)(0.143\ \text{s})^2 = -0.33\ \text{ft}$$

or

$$-0.33\ \text{ft}\ (12\ \text{in./ft}) = -4.0\ \text{in.}$$

below the center of the bull's-eye.

Projections at an Angle

In general, projectile motion is a projection at some angle θ relative to the horizontal. In this case, the object describes a curved path in two dimensions. A curve of this shape is called a *parabola* (Fig. 5.5).

(a)

(b)

Figure 5.5
Projections at angles. Objects projected at an angle relative to the horizontal describe a curved path in two dimensions called a parabola. (a) Courtesy of R. Stevenson and R. B. Moore. (b) Courtesy of R. Serway and J. Faughn.

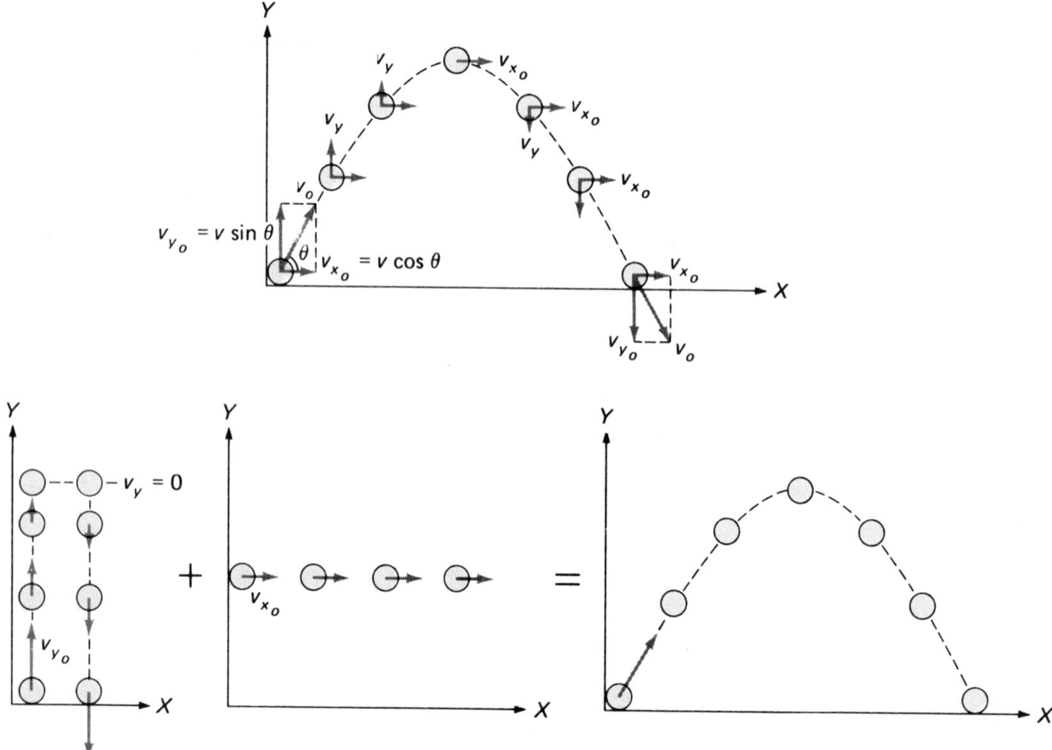

Figure 5.6
Projection at an angle. The motion is a combination of vertical and horizontal motions. The object describes a parabolic path.

The initial velocity v_o can be resolved into x and y components, and

$$v_{x_o} = v_o \cos \theta \qquad \text{(Eq. 5.2)}$$
$$v_{y_o} = v_o \sin \theta$$

As in the case of horizontal projection, there is no acceleration in the x direction. The distance the object travels in the x direction is called the **range** (R), and

$$R = x = v_{x_o}t = (v_o \cos \theta)t \qquad \text{(Eq. 5.3)}$$

While the object is traveling horizontally, it travels upward and downward vertically (Fig. 5.6). This component of the motion is analyzed as in the vertical projection case considered previously. Again, the time t is the connecting element between the two component motions.

EXAMPLE 5.7 Suppose the projectile in Figure 5.6 were given an initial velocity of 14 m/s at an angle of 60° relative to the horizontal. What would be (a) the maximum height and (b) the range of the projectile?

Solution: With $v_o = 14$ m/s and $\theta = 60°$, the x and y velocity components are

$$v_{x_o} = v_o \cos \theta = v_o \cos 60° = (14 \text{ m/s})(0.500) = 7.0 \text{ m/s}$$

$$v_{y_o} = v_o \sin \theta = v_o \sin 60° = (14 \text{ m/s})(0.866) = 12 \text{ m/s}$$

(a) Considering the vertical component of motion, the time it takes for the projectile to reach its maximum height (y_{max}), where $v_y = 0$, is given by

$$v_y = v_{y_o} - gt = 0$$

and the upward travel time t_u is

$$t_u = \frac{v_{y_o}}{g} = \frac{12 \text{ m/s}}{9.8 \text{ m/s}^2} = 1.2 \text{ s}$$

Then,

$$y_{\max} = v_{y_o}t_u - \tfrac{1}{2}gt_u^2 = (12 \text{ m/s})(1.2 \text{ s}) - \tfrac{1}{2}(9.8 \text{ m/s}^2)(1.2 \text{ s})^2$$
$$= 7.3 \text{ m}$$

Notice that $v_y^2 = v_{y_o}^2 - 2gy$ could also have been used to find $y_{\max}$.

(b) The total time for both upward and downward vertical motions is $t = 2t_u = 2(1.2 \text{ s}) = 2.4$ s. During this time, the projectile travels a horizontal distance of

$$R = x = v_{x_o}t = (7.0 \text{ m/s})(2.4 \text{ s}) = 17 \text{ m}$$

EXAMPLE 5.8 An artillery shell with a muzzle velocity of 4.0×10^2 ft/s is fired at an angle of 35° to the horizontal (Fig. 5.7). If the shell explodes 10 s after being fired, what is its location?

Solution: It is given that $v_o = 4.0 \times 10^2$ ft/s, $\theta = 35°$, and $t = 10$ s. Then,

$$v_{x_o} = v_o \cos 35° = (4.0 \times 10^2 \text{ ft/s})(0.819)$$
$$= 3.3 \times 10^2 \text{ ft/s}$$

$$v_{y_o} = v_o \sin 35° = (4.0 \times 10^2 \text{ ft/s})(0.574)$$
$$= 2.3 \times 10^2 \text{ ft/s}$$

and

$$x = v_{x_o}t$$
$$= (3.3 \times 10^2 \text{ ft/s})(10 \text{ s}) = 3.3 \times 10^3 \text{ ft}$$

$$y = v_{y_o}t - \tfrac{1}{2}gt^2$$
$$= (2.3 \times 10^2 \text{ ft/s})(10 \text{ s}) - \tfrac{1}{2}(32 \text{ ft/s}^2)(10 \text{ s})^2$$
$$= (2.3 \times 10^3 \text{ ft}) - (1.6 \times 10^3 \text{ ft})$$
$$= 0.7 \times 10^3 \text{ ft} = 700 \text{ ft}$$

Hence, $(x, y) = (3300 \text{ ft}, 700 \text{ ft})$; that is, the shell explodes 3300 ft from the artillery piece and 700 ft above the muzzle (why?).

Maximum Range

As can be seen from Equation 5.3, the greater the magnitude of the initial velocity, the greater the range. But what angle of projection gives the maximum range for a given initial speed? This is an important consideration in many applications, such as golfing, other sports, and even fire fighting when it is desired to get the maximum range of a stream of water from a hose. The range of a projectile is given by

$$R = x = v_{x_o}t = (v_o \cos \theta)t$$

The time of flight of a projectile is twice the time it takes to reach its highest point and, from $v_y = v_{y_o} - gt_u = 0$,

$$t = 2t_u = \frac{2v_{y_o}}{g} = \frac{2(v_o \sin \theta)}{g}$$

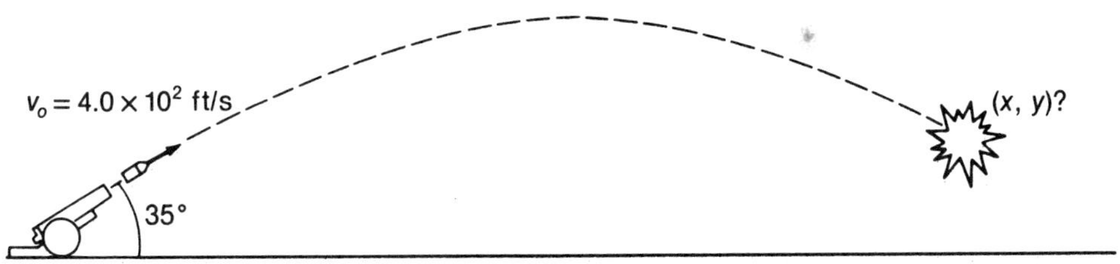

Figure 5.7

An artillery shell exploding in midair. See Example 5.8.

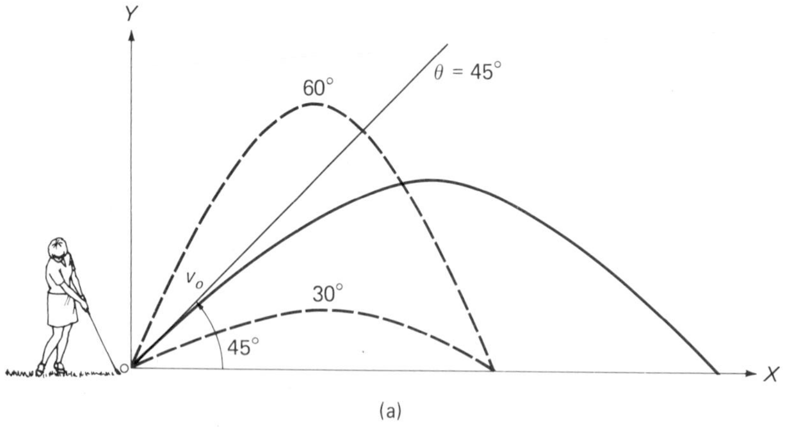

(a)

(b)

Figure 5.8

Maximum range. The maximum range for a given projection speed occurs ideally for a projection angle of 45°. An ideal situation neglects air resistance. This is an important consideration in many applications, such as (a) golf and (b) shot put. [(b) Courtesy of Harry Snavely, Ohio University.]

Substituting this into the previous equation,

$$R = (v_o \cos \theta)t = \frac{(v_o \cos \theta)2(v_o \sin \theta)}{g}$$

$$= \frac{v_o^2(2 \sin \theta \cos \theta)}{g}$$

Using the trigonometric identity

$$\sin 2\theta = 2 \sin \theta \cos \theta$$

(see Appendix 3), this equation can be written as

$$R = \frac{v_o^2 \sin 2\theta}{g} \qquad \textbf{(Eq. 5.4)}$$

Hence, R is a maximum when $\sin 2\theta = 1$, since this is the maximum value of the sine function (varies from 0 to 1), and

$$R_{max} = \frac{v_o^2}{g} \qquad \textbf{(Eq. 5.5)}$$

Since this occurs for $\sin 2\theta = 1$, and $\sin 90° = 1$, we have

$$2\theta = 90°$$

or

$$\theta = 45°$$

Thus, to get the maximum range, a projectile should *ideally* be projected at an angle of 45° (Fig. 5.8). Note that the range is the same for angles equidistant above and below the 45° angle, as illustrated in the figure.

This, however, is an ideal situation with air resistance neglected. When a ball or object is thrown or hit hard, air resistance comes into effect. In this case, the trajectories resemble those shown in Fig. 5.9. Air resistance, or friction, reduces the speed of the projectile, particularly in the horizontal direction. This causes the angle for maximum range to be less than 45°.

Other factors affect the range, such as spin on a golf ball. Reverse spin produces lift, and the angle for maximum range may be considerably less than 45°. Elevation also makes a difference. When an object is projected from a location above its landing position, the maximum range angle is less than 45°.

5.3 Uniform Circular Motion and Centripetal Acceleration

Suppose an automobile is traveling around a circular track with a uniform, or constant, speed. Is the automobile accelerating? On first thought, you might say no, since the speed is constant. But remember that acceleration is a change in *velocity*, which has both magnitude (speed) and *direction*. In going around the track, the velocity vector of the car is continually changing direction, so there is an acceleration (Fig. 5.10). Recall that a vector quantity may change as a result of a change in magnitude and/*or* direction.

The acceleration necessary for uniform circular motion is called the **centripetal ("center-seeking") acceleration.** It turns out that this acceleration is directed toward the center of the circle. The direction of the velocity is tangential to the circle (at a right angle to a radius at any point on the circular path) and is called the **tangential velocity.**

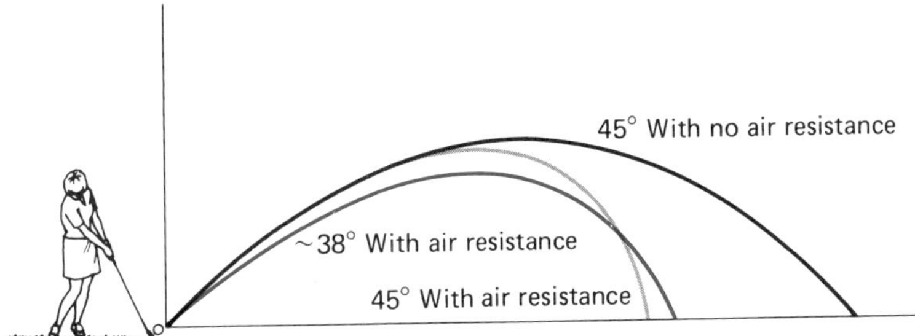

Figure 5.9
Range and air resistance. Air resistance reduces the speed of a projectile, particularly in the forward direction, which causes the angle for the maximum range to be less than 45°.

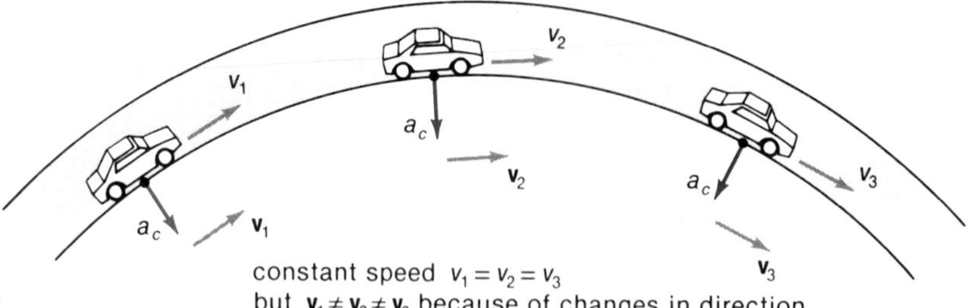

Figure 5.10

Uniform circular motion and centripetal acceleration. An object in uniform circular motion has a constant speed. However, since the velocity vector changes direction, there must be an acceleration. This is the centripetal or "center-seeking" acceleration, a_c, toward the center of the circle.

Consider a particle in uniform motion as illustrated in Figure 5.11. Forming the vector triangle of two arbitrary velocity vectors, we see that there is a change in velocity $\Delta\mathbf{v}$, where $\mathbf{v}_1 + \Delta\mathbf{v} = \mathbf{v}_2$ or $\Delta\mathbf{v} = \mathbf{v}_2 - \mathbf{v}_1$. As the time interval becomes smaller and smaller, in the limit $\Delta t \rightarrow 0$, the direction of $\Delta\mathbf{v}$ is toward the center of the circle. Hence, the instantaneous (centripetal) acceleration producing the change in velocity must also be in that direction ($\mathbf{a} = \Delta\mathbf{v}/\Delta t$).

As Δt approaches zero, the pie-shaped section of the circle bounded by the two radii (r) and arc length (s) approaches a triangle similar to that of the velocity vector triangle (Fig. 5.11). By similar triangles, we have

$$\frac{\Delta v}{v} = \frac{s}{r}$$

or

$$\Delta v = \frac{vs}{r} = \frac{v^2\,\Delta t}{r}$$

where $s = v\,\Delta t$. Then, the centripetal acceleration $\mathbf{a}_c = \Delta\mathbf{v}/\Delta t$ has a magnitude of

$$a_c = \frac{\Delta v}{\Delta t} = \frac{(v^2\,\Delta t/r)}{\Delta t} = \frac{v^2}{r}$$

and

$$\boxed{a_c = \frac{v^2}{r}} \qquad \textbf{(Eq. 5.6)}$$

with a direction toward the center of the circle.

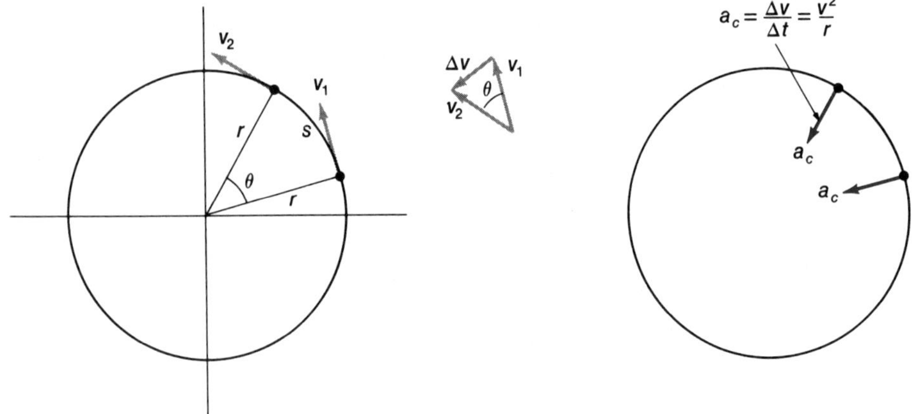

Figure 5.11

The geometry for analyzing centripetal acceleration. See text for description.

EXAMPLE 5.9 A racing car travels around a circular track that has a radius of 1000 ft with a constant speed of 60 mi/h. What is the car's centripetal acceleration?

Solution: It is given that $r = 1000 \text{ ft} = 10^3 \text{ ft}$ and $v = 60 \text{ mi/h} = 88 \text{ ft/s}$. Then,

$$a_c = \frac{v^2}{r} = \frac{(88 \text{ ft/s})^2}{10^3 \text{ ft}} = 7.7 \text{ ft/s}^2$$

toward the center of the track.

The centripetal acceleration for the car's uniform circular motion is supplied by friction on the car's tires.

An object in a circular orbit, such as the moon (approximately circular orbit), has a centripetal acceleration. In effect, such an object is continually "falling" toward the center of the circular orbit (Fig. 5.12).

As the moon or other object in circular motion travels a distance $v\Delta t$ in the direction of the tangential velocity in a time interval Δt, it is also being accelerated toward the center of the circle and travels a distance $\frac{1}{2}a_c(\Delta t^2)$. In the limit ($\Delta t \rightarrow 0$), the resultant of these motions is a circular path.

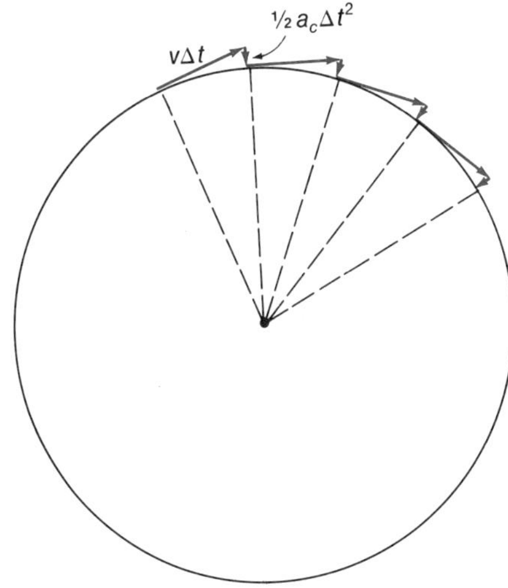

Figure 5.12

An object in circular orbit. An object traveling in a circular orbit is continually "falling" toward the center of the orbit. As it travels a tangential distance, $v\Delta t$, it also travels a distance $\frac{1}{2}a_c(\Delta t)^2$ toward the center of the circle. In the limit $\Delta t \rightarrow 0$, the combination of these motions is a circular path.

If it were not for the centripetal acceleration, the object would not describe a circular path. Should the centripetal acceleration go to zero, the object would "fly off" in a straight line in the direction of the instantaneous tangential velocity. What do you think supplies the centripetal acceleration of the moon?

EXAMPLE 5.10 A communications satellite in a circular orbit at an altitude of 500 km (310 mi) makes one complete revolution in 95 minutes. What is the centripetal acceleration of the satellite?

Solution: The altitude of the satellite is $h = 500 \text{ km} = 0.5 \times 10^6 \text{ m}$, but the radius of the circular orbit is

$$\begin{aligned} r &= R_e + h \\ &= 6.4 \times 10^6 \text{ m} + 0.5 \times 10^6 \text{ m} \\ &= 6.9 \times 10^6 \text{ m} \end{aligned}$$

where R_e is the radius of the Earth (obtained from Appendix 2). The period, or time to make one revolution, is

$$\begin{aligned} t &= 95 \text{ min} \times 60 \text{ s/min} \\ &= 5.7 \times 10^3 \text{ s} \end{aligned}$$

The distance traveled in this time is the circumference (c) of the circular orbit, $d = c = 2\pi r$, and the orbital speed is

$$\begin{aligned} v = \frac{d}{t} = \frac{2\pi r}{t} &= \frac{2\pi(6.9 \times 10^6 \text{ m})}{5.7 \times 10^3 \text{ s}} \\ &= 7.6 \times 10^3 \text{ m/s} \end{aligned}$$

Then, the centripetal acceleration of the satellite toward the Earth is

$$a_c = \frac{v^2}{r} = \frac{(7.6 \times 10^3 \text{ m/s})^2}{6.9 \times 10^6 \text{ m}} = 8.4 \text{ m/s}^2$$

Important Terms

components of motion the components of vector quantities in rectangular or x and y directions

projectile a projected object without the further capability of propelling itself

range the horizontal (x) distance traveled by a projectile

centripetal acceleration the "center-seeking" acceleration of an object in uniform circular motion

Important Formulas

velocity components:

$$v_{x_o} = v_o \cos\theta$$
$$v_{y_o} = v_o \sin\theta$$

projectile displacement components:

$$x = v_{x_o}t = (v_o \cos\theta)t$$
$$y = v_{y_o}t - \tfrac{1}{2}gt^2 = (v_o \sin\theta)t - \tfrac{1}{2}gt^2$$

projectile velocity components:

$$v_x = v_{x_o} = v_o \cos\theta$$
$$v_y = v_{y_o} - gt = v_o \sin\theta - gt$$
$$v_y^2 = v_{y_o}^2 - 2gy = v_o^2 \sin^2\theta - 2gy$$

projectile range (launch and landing positions at same elevation):

$$R = v_{x_o}t = v_o(\cos\theta)t = \frac{v_o^2 \sin 2\theta}{g}$$
$$R_{max} = \frac{v_o^2}{g} \quad (\text{at } \theta = 45°)$$

centripetal acceleration:

$$a_c = \frac{v^2}{r}$$

Questions

Components of Motion

1. What are the advantages of considering components of motion?
2. Suppose for motion in a plane there is a constant acceleration. What equations would describe the magnitudes of the displacement components (x, y)?
3. Can curved motion in a plane be analyzed in terms of straight-line motion? Explain.
4. An object has a velocity **v** in the y-z plane at an angle θ relative to the horizontal y-axis. What are the magnitudes of the vector components in this case?
5. In three dimensions, how many components of motion are there? Would one angle suffice?

Projectile Motion

6. Is there an acceleration involved in projectile motion? Explain.
7. How is free fall associated with projectile motion?
8. When a marksman sights his rifle at a target 100 yd away, is the rifle barrel pointed directly at the bull's-eye? Explain.
9. How may the range of a projectile be increased?
10. How would air resistance affect the range of a projectile?
11. If the opposite directions of the initial, upward velocity of a projectile and the acceleration due to gravity were not distinguished and the equation for the y displacement were written $y = v_{y_o}t + \frac{1}{2}gt^2$, what is physically implied by the equation?
12. Several years ago, a pilot of a jet plane fired the plane's cannon at an imaginary target, then put the plane into an evasive dive. Rather embarrassingly, the pilot shot himself down. How did this happen?
13. Is it possible for a vertical projection and a projection at an angle to have the same maximum height? Explain.
14. Suppose you and another student are playing catch with a ball. He throws the ball at an angle of 35° and you catch it, while standing still, at the height at which it was thrown. Could you throw the ball back with the same initial speed but at another angle so that he could catch it as you did? If so, at what angle? Which throw would have the greater maximum height?

Centripetal Acceleration

15. How can an object traveling with a constant speed be accelerating?
16. Is the moon "falling" toward the Earth? Explain.
17. How is the centripetal acceleration of a particle in circular motion affected (a) if the speed is doubled and (b) both the speed and radius are doubled?
18. If the radius of a particle in circular motion is doubled, how must the tangential speed be changed to keep the centripetal acceleration the same?
19. Do all of the particles in a rotating rigid body have the same centripetal acceleration?
20. Suppose the speed of an object in circular motion (constant radius) increases. What is the *total* acceleration (magnitude and direction) of the object? (*Hint:* draw a vector diagram.)

Problems

Levels of difficulty are indicated by asterisks for your convenience.

5.1 Components of Motion

1. A particle has a velocity of 5.0 m/s at an angle of 210°. What are the rectangular components of the velocity?
2. The velocity components of an object are 16 m/s **x** and 9.0 m/s **y**. What are the magnitude of the velocity and its angular direction?

*3. An object moves in the x-y plane with a constant velocity of $\mathbf{v} = 2.0\ \text{m/s}\ \mathbf{x} + 1.5\ \text{m/s}\ \mathbf{y}$. What is the position (x, y) of the object at $t = 4.0$ s?

*4. Using Equations 5.1, show that the path of an object moving with a constant velocity in the x-y plane is a straight line of the form $y = mx + b$.

*5. An airplane flies directly northeast with a constant speed of 200 km/h. What are the coordinates (x, y) of the plane after 1.5 hours?

*6. A ball rolls along the diagonal of a 3.0 m × 4.0 m table with a constant speed of 0.25 m/s. What is its position (x, y) after 6.0 seconds?

**7. A straight section of river is 300 ft wide and has a current with a speed of 4.0 ft/s. A motorboat with a speed of 15 ft/s crosses the river. (a) If the boat is headed directly toward the opposite bank, how far will it travel across *and* down the river in 10 s? (b) How long will it take the boat to get to the opposite bank, and how far downstream will the boat land from the point directly opposite its starting point? (Assume a constant boat speed at all times—*not* a good way to land.)

5.2 Projectile Motion (Neglect air resistance.)

8. A ball is thrown vertically upward with a speed of 9.8 m/s. (a) How long does it take the ball to reach its maximum height? (b) How long does it take for the ball to return to its starting point, and what is its velocity at that point?

9. A basketball player throws a ball vertically upward with a speed of 32 ft/s. (a) How high does the ball go (above its starting point)? (b) How much time elapses between the time the ball is thrown and the time the player catches the returning ball?

*10. A boy on a bridge 30 m above a river throws a stone vertically downward with a speed of 8.0 m/s. (a) What is the speed of the stone when it strikes the water? (b) How much time elapses before the stone hits the water?

*11. A painter on the ground tosses a brush vertically upward with a speed of 35 ft/s toward his partner on a scaffold 30 ft above the ground. (a) If the brush is released 5.0 ft above the ground, will the partner on the scaffold be able to catch the brush at the height of the scaffold? (b) If not, with what minimum initial velocity should the brush be thrown?

*12. An arrow is shot vertically upward with a speed of 25 m/s. (a) How high does the arrow go? (b) What is its height above the release point at $t = 4.0$ s?

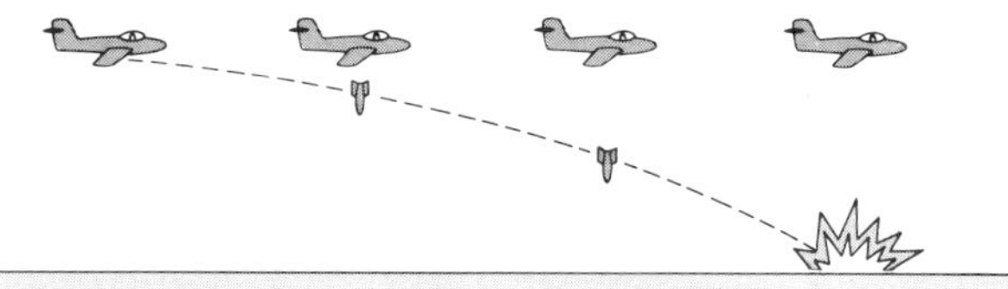

Figure 5.13
See Problem 17.

*13. An object projected vertically upward strikes the ground 4.00 s later. What is the object's initial velocity?

*14. A woman runs with a velocity of 8.0 ft/s off a horizontal diving board. If the water level in the pool is 12 ft below the diving board, how far from a point directly below the edge of the board will she hit the water?

*15. A ball is thrown horizontally with a speed of 20 m/s from the top of a building 50 m above the ground. (a) How long does it take for the ball to strike the ground? (b) How far from the building does it strike the ground?

*16. A stunt driver drives a car horizontally off a cliff, and the car lands in a lake 45 m from the base of the cliff 3.0 s later. (a) What was the initial velocity of the car? (b) How high was the cliff?

*17. A bomb is dropped from a plane flying with a horizontal velocity of 120 mi/h at an altitude of 1000 ft (Fig. 5.13). (a) How far from the point that is directly below the plane at the time of release will the bomb land? (b) Where will the plane be relative to the explosion?

*18. A baseball player hits a ball at an angle of 30° relative to the horizontal with a speed of 60 mi/h. How far from home plate will the ball travel? (Neglect the initial height of the ball above the ground.)

*19. The nozzle of a fire hose is positioned on the ground at an angle of 30° to the horizontal. If the speed of the water coming from the nozzle is 50 m/s, would the stream of water be effective in fighting a fire near the top of a 35-m building?

*20. A golfer drives a ball with a velocity of 120 ft/s at an angle of 45°. (a) What is the maximum height of the ball above the level fairway? (b) What is the length of the drive?

*21. Which will go farther, a golf ball hit at an angle of 40° or a golf ball hit at an angle of 55° with the same initial speed? (Justify your answer mathematically.)

Figure 5.14
See Problem 26.

*22. In fighting a grass fire, fire fighters direct a stream of water from a hose on the fire 140 ft away. If the nozzle speed of the water leaving the hose is 67 ft/s, what is the angle of projection?

*23. If the maximum range of an artillery shell is 1.0 mi, what is its muzzle velocity?

*24. What is the range of the stream of water in Problem 19?

**25. An artillery shell with a muzzle velocity of 55.0 m/s is fired at an angle of 60° to the horizontal. If the shell explodes 9.00 seconds later, what is the range of the shell and how far above a level field does it explode?

**26. The coyote is trying to catch the roadrunner again, this time wearing a pair of Acme jet-powered skates (Fig. 5.14). If the roadrunner makes a sudden turn at the edge of the cliff and the coyote doesn't, what is the range of the coyote from the base of the 500-ft cliff (a) if his skates drop off as he goes off the cliff with a horizontal speed of 30 mi/h? (b) What is the range if his skates stay on and in operation when he is in "flight" so as to give him a constant horizontal acceleration of 5.0 ft/s^2? (Assume he doesn't hit another cliff wall while in flight.)

5.3 Centripetal Acceleration

*27. A particle in uniform circular motion travels with a speed of 2.0 ft/s in an orbit with a radius of 6.0 in. What is the centripetal acceleration required for this motion?

*28. What is the centripetal acceleration of a ball being swung in a horizontal circle with a diameter of 3.0 m and with a uniform speed of 0.90 m/s?

*29. An object in uniform circular motion has a centripetal acceleration of 0.090 m/s^2. If the radius of its circular path is 1.5 m, what is the speed of the object?

*30. The uniform speed of a car going around a circular track is 60 mi/h. If the centripetal acceleration of the car is 25.8 ft/s^2, what is the diameter of the track?

*31. The friction on the tires of a racing car can supply a maximum centripetal acceleration of 5.0 m/s^2. (a) If the car races on a circular track with a diameter of 1.0 km, what is the maximum safe speed the car could travel in km/h? (b) What is the maximum safe speed in mi/h? (c) What would happen if the car exceeded this speed?

*32. What is the centripetal acceleration of the moon? (*Hint:* $r = 3.8 \times 10^5$ km and $T = 29.5$ days. Recall $v = d/t$.)

**33. Assuming the Earth travels in a circular orbit about the Sun, what would be its centripetal acceleration? (The average distance from the Earth to the Sun is 1.5×10^8 km.)

**34. A satellite in circular orbit about the Earth at an altitude of 800 km (500 mi) makes one revolution in about 100 minutes. What is the centripetal acceleration of the satellite?

**35. A person on Earth travels in a circular path because of the daily rotation of our planet. (a) If you were at the Equator, what would be your necessary centripetal acceleration? (b) Find your centripetal acceleration at your own latitude.

Chapter 6

Newton's Laws of Motion

The kinematic description of motion in the previous chapters was without regard to what causes the motion. The dynamic description of motion takes into account the cause of motion, namely forces. We commonly say that a force is required to produce motion or a change in motion. This is basically true, but also critical to the description of motion is how a force is related to the resulting motion—the cause and effect relationship, so to speak. An acceleration is the evidence of the action we identify as force.

One of the many achievements of the great scientist Isaac Newton was to summarize the relationships between force and motion in three general statements, which have become known as "laws" of motion. See Special Feature 6.1.

6.1 Newton's First Law of Motion: The Law of Inertia

According to Aristotle's theory of motion, which prevailed for some 1500 years after his death, a body required a force to keep it in motion. That is, the normal state of a body was one of rest, with the exception of celestial bodies, which were naturally in motion. Aristotle observed that moving objects tended to slow down and come to rest (due to friction, we now know), so this conclusion seemed logical to him.

The groundwork for the first law was laid by Galileo (Special Feature 6.2). Rather than rely solely on intuition and general observations, Galileo tested his ideas with experiments. He observed the motions of a ball on inclined planes (Fig. 6.1).

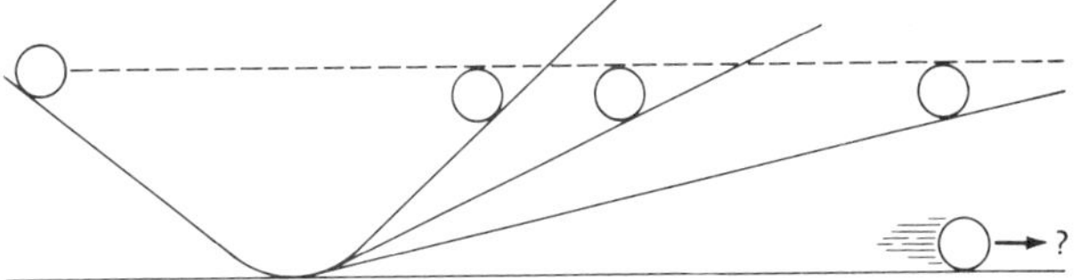

Figure 6.1
An illustration of Galileo's experiment. A ball would roll farther, the smoother the surface. In an ideal case of no friction, the ball would continue (slide) indefinitely, since there would be no force to alter its motion.

SPECIAL FEATURE 6.1

Isaac Newton

On Christmas Day, 1642, Isaac Newton was born in Woolsthorpe in Lincolnshire, England. His father had died three months earlier. When he was 14 years old, his mother was widowed for a second time and brought Isaac home from school to help run the family farm. He proved to be a lackadaisical farmer, being occupied more with mathematics than with farm chores.

At the age of 18 he entered Trinity College at Cambridge, and he received his degree four years later, in 1665. Later that year, the spread of the Great Plague caused the university to close. Newton returned home, and during the next 18 months he conceived most of the ideas for his famous discoveries in science and mathematics. Chief among these were the development of calculus mathematics and studies on light and color, motion, and gravitation. As Newton later described this period,

> I was in the prime of my age for invention, and minded Mathematics and Philosophy [science] more than at any time since.

After the plague had passed, he returned to Cambridge and at the age of 26 was appointed professor of mathematics. Newton's early published works were in the field of optics. He developed a new type of telescope—a reflecting telescope that used a mirror rather than a lens to collect light (see Chapter 29). His most notable book, *Principia Mathematica Philosophiae Naturalis* (Mathematical Principles of Natural Philosophy*), or "Principia" for short, was published in 1687 (Fig. 1). The cost of the book was borne by a contemporary, Edmund Halley, who predicted the return of a famous comet that bears his name. In the "Principia" Newton set forth his theories on gravity, tides, and motion.

A bachelor, Newton lived very austerely and is reported to have been the classic absent-minded professor, being so absorbed in his work that he forgot meals and other day-to-day activities. In later life Newton was appointed master of the mint, and he moved to London in 1701. Queen Anne knighted him in 1705 in recognition of his numerous accomplishments. Early in 1727 he was taken seriously ill, and he died on March 20 of that year.

An insight into the character of this great scientist is given by one of his statements:

> If I have been able to see farther than some, it is because I have stood on the shoulders of giants.

One of those giants was Galileo. (See Special Feature 6.2.)

* Physics was once called natural philosophy.

(a)

(b)

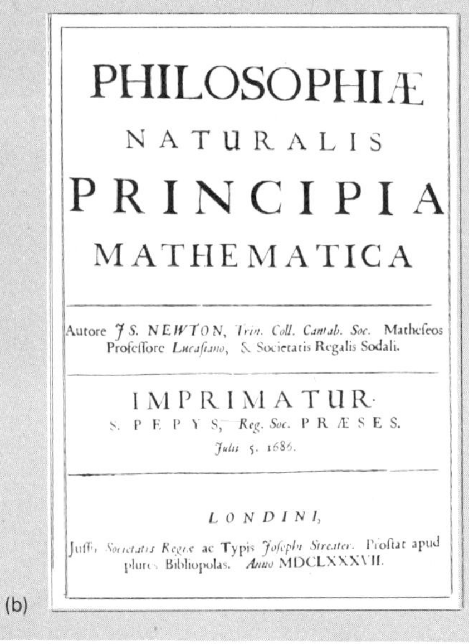

PHILOSOPHIÆ
NATURALIS
PRINCIPIA
MATHEMATICA

Autore *J. S.* NEWTON, *Trin. Coll. Cantab. Soc.* Matheseos Professore *Lucasiano*, & Societatis Regalis Sodali.

IMPRIMATUR.
S. PEPYS, *Reg. Soc.* PRÆSES.
Julii 5. 1686.

LONDINI,

Jussu *Societatis Regiæ* ac Typis *Josephi Streater*. Prostat apud plures Bibliopolas. *Anno* MDCLXXXVII.

Figure 1
(a) **Sir Isaac Newton** (1672–1727), one of the greatest physicists of all time. (b) The title page of Newton's *Principia.* Notice that the name of the famous diarist Samuel Pepys appears near the middle of the page. Pepys was president of the Royal Society, which sponsored the book. Can you read the date in Roman numerals at the bottom? (Courtesy of G. Holton, F. Rutherford, and F. Watson.)

SPECIAL FEATURE 6.2

Galileo Galilei

Now known universally by his first name, Galileo (Fig. 1) was born in Pisa on February 15, 1564. His father was a musician, and Galileo was the eldest of seven children. In 1574 the family moved to Florence, and Galileo was sent to school, where he studied Latin, Greek, and mathematics as well as topics in physics, astronomy, and what is now collectively called the humanities. At the age of 17 he returned to Pisa to study medicine. But this was quickly supplanted by mathematics, a subject in which he excelled. At the age of 25 he received an appointment as professor of mathematics at the University of Pisa, and other prestigious appointments followed.

Throughout his lifetime, Galileo engaged in a variety of scientific pursuits. These included studies of time, motion, floating bodies, and the nature of heat, as well as the construction of telescopes and microscopes (see Chapter 29). Although he is perhaps best known from the popular stories of his alleged experiments of dropping objects from the Leaning Tower of Pisa, which he probably did not do (see Special Feature 7.1), it was the telescope that played a very critical part in his life. With his telescopes he observed the features of the moon, sunspots, and various planets, including the phases of Venus.

Figure 1
Galileo Galilei (1564–1642), the mathematician, astronomer, and physicist who made many contributions to science, including the description of motion. (Courtesy G. Holton, F. Rutherford, and F. Watson.)

Galileo lived in an era in which the Aristotelian view of an Earth-centered universe was being challenged. His observations supported the Copernican Sun-centered theory of the solar system, in which the Earth moved. However, Roman Catholic Church dogma at the time held that the Earth was the stationary center of things, as supported by various Biblical references (Joshua 10:12–13, Psalms 19:4–6 and 104:5). To believe and publicly state differently was considered heresy. (A former monk, Giordano Bruno, was burned at the stake in 1600 for holding such views.)

Galileo's thought is expressed in one of his popular quotations: "The Bible shows the way to go to heaven, not the way the heavens go." As early as 1615, he was admonished by the Church about his questionable views. The publication in 1632 of his book, *The Dialogue of the Two Great World Systems*, left little doubt. He was summoned to Rome by the Inquisition, put on trial, and forced to recant his "heretical" ideas:

> I, Galileo, son of the late Vincenzo Galilei, Florentine, aged seventy years . . . have been pronounced by the Holy Office to be vehemently suspected of heresy, that is to say, of having held and believed that the Sun is the center of the world and immovable and that the Earth is not the center and moves. . . . This vehement suspicion justly conceived against me, with sincere heart and unfeigned faith I abjure, curse, and detest the aforesaid errors and heresies . . . and I swear that in future I will never again say or assert, verbally or in writing, anything that might furnish occasion for a similar suspicion. . . .*

Contrary to popular belief, he did not end his statement with *Eppur si muove* ("but it still moves"). At least there is no record of this, and it is doubtful that Galileo would have been so foolhardy.

Afterward, Galileo returned to Florence to work on less controversial topics, such as projectile motion. The remainder of his life was spent under house arrest. (He was finally "rehabilitated" by Pope John Paul II in 1984.) Totally blind the last five years of his life, he died on January 8, 1642, the same year another great scientist, Isaac Newton, was born.

* de Santillana, G., *The Crime of Galileo*, The University of Chicago Press, Chicago, 1955.

When a ball was released and allowed to roll down an incline, it would roll up an adjoining incline to about the same height (a little less, because of friction), being stopped by the retarding forces of gravity and friction. When the angle of incline of the adjoining plane was made less steep, the ball would roll to approximately the same height in each case, *but it rolled farther in the horizontal direction.* Then, when the ball was allowed to roll onto a horizontal plane, it rolled a considerable distance before coming to rest.

Galileo took great pains to make the surfaces of the plane and the rolling ball as smooth as possible to reduce friction, and he found that the smoother the surface, the greater the horizontal distance the ball would roll. So the question arose, How far would the ball travel if friction could be removed completely and the plane made infinitely long? Galileo reasoned that in this ideal case the ball would continue to travel in straight-line, uniform motion forever, since there would be nothing (no force) to alter or change its motion.

Contrary to Aristotle's ideas, Galileo concluded that material bodies exhibited the behavior or property of maintaining a state of motion. Similarly, if a ball were at rest, it would remain so, unless something caused it to move. Galileo called this property *inertia,* and we say,

> **Inertia is the property of matter that describes its resistance to changes in motion.**

That is, if an object is at rest, it seems to "want to" remain at rest. If an object is in motion, it seems to "want to" remain in motion.

Newton summarized these results in his **first law of motion,** which is also called the law of inertia:

> *Everybody preserves its state of rest, or of uniform motion in a right [straight] line unless it is compelled to change that state by forces impressed thereon.**

Or, in more modern language:

> **An object remains (a) at rest or (b) in motion with a constant velocity unless acted upon by an unbalanced force.**

It should be evident that the first condition (a) is simply that for static translational equilibrium of a particle object (Chapter 3). An object initially at rest would not be expected to move, unless acted upon by an unbalanced or net force. In other words, if $\Sigma \mathbf{F} = 0$, a stationary object remains at rest. In our discussion of force and motion, therefore, it will be understood that **a force that produces motion or a change in motion is a net or unbalanced force (nonzero resultant).**

The second part (b) of Newton's first law is implied from Galileo's experiment. If we could set an object in motion in free space, where there is no air (resistance) and all other forces are negligible, it would move with a uniform velocity until acted upon by a force.

> **A force is commonly defined as a quantity capable of producing motion or a change in motion.**

The capability of this action depends on whether or not the force is unbalanced. Thus, an unbalanced force is evidenced by what it *does* to the motion of an object. For example, when an object starts to move (a change in motion), we know a force is acting. Similarly, when a moving object has a change in velocity (magnitude and/or direction), we know a force is acting. Therefore,

> **A change in motion or the presence of a nonzero acceleration is evidence of a net force.**

Newton eventually made the idea of inertia quantitative by relating inertia to *mass.* Originally he thought of mass as a "quantity of matter," but then he effectively redefined it as a measure of inertia. Today we know that the mass of a given quantity of matter is not totally independent of conditions. (The mass of an object depends on its speed and even on its temperature, but these changes are commonly too small to measure.)

Quantity of matter is still commercially acceptable as a definition of mass, but physicists use the definition

> **Mass is a measure of inertia.**

* *Principia Mathematica Philosophiae Naturalis,* from Magie, W. F., *A Source Book in Physics,* Harvard University Press, Cambridge, Mass., 1963.

That is, mass, the fundamental property of matter, is a measure of an object's inertia—the greater an object's mass, the greater its inertia. As will be learned, Newton incorporated this idea in his second law of motion.

There are many common examples and practical applications of the concept of inertia. For example, it seems that there is something resisting motion when we push a stalled automobile, and it is easier to get a small sports car moving than a more massive van. Also, when a moving automobile is suddenly braked to a stop, loose objects will, and the passengers may, continue in motion according to Newton's first law. (The force of friction on the seat of one's pants is not sufficient to stop the motion.) It is hoped that the inertia of motion of the passengers would be overcome by forces exerted by seat belts and shoulder straps rather than by the forces exerted by the dashboard or windshield with which they may come into contact.

Practical examples of inertia include the use of a heavy (massive) ball in building demolition (Fig. 6.2) and flywheels. The latter is a case of rotational inertia. Flywheels are used to keep machinery running smoothly during periods when power is not being transmitted; for example, the flywheel of an automobile keeps the engine running smoothly between power strokes of the pistons.

Figure 6.2

An application of inertia. The inertia of a heavy wrecking ball is used in demolishing a building. The ball is seen above the stream of water being used to keep down dust. (Courtesy Crane and Excavator Division, FMC Company.)

Question: To tighten a hammer head on its handle, a person might bring the hammer downward and strike the butt of the handle sharply on a hard surface (Fig. 6.3). What does this accomplish?

Answer: As the hammer is brought downward, the handle and head are in motion. When the handle butt strikes the surface, it stops suddenly because of the large contact force of the surface on the handle. But the massive head continues in motion until it is stopped by the tapered handle. This tightens the hammer head on the handle.

6.2 Newton's Second Law of Motion: Cause and Effect

Newton's first law states that an object remains at rest or in motion with a constant velocity until acted upon by a force. In the absence of an unbalanced force, then, the acceleration of an object is zero, since there is no change in velocity. This should lead you, like Newton, to the conclusion that a force acting on an object produces an acceleration.

However, Newton recognized that inertia or mass also plays a part. For a given force F, the greater the mass of an object the less its acceleration a, or change in motion. Expressing this as a proportion,*

$$a \propto \frac{F}{m}$$

* The symbol $\propto$ means "proportional to." The expression $a \propto F$ means that a is *directly* proportional to F and they change in the same proportion. For example, if F is doubled, a is doubled. The expression $a \propto 1/m$ means that m is *indirectly* or *inversely* proportional to a and they change in the same inverse proportion. For example, if m is doubled, a is halved. Putting the expressions together, we have $a \propto F/m$. A proportion gives only relative changes. An equation can be used to calculate the exact values.

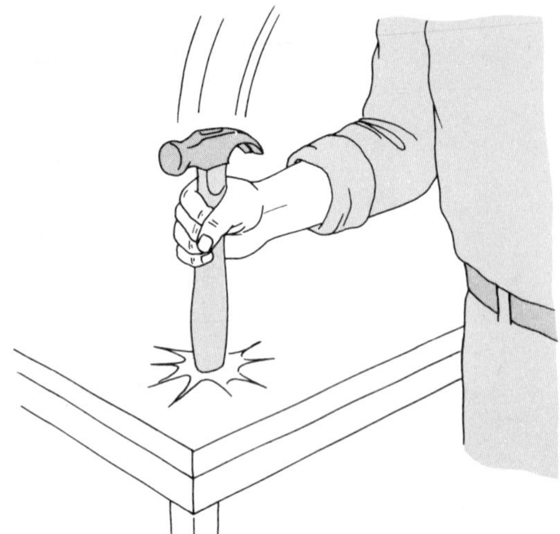

Figure 6.3
Inertia "in action" is used to tighten a loose hammer head.

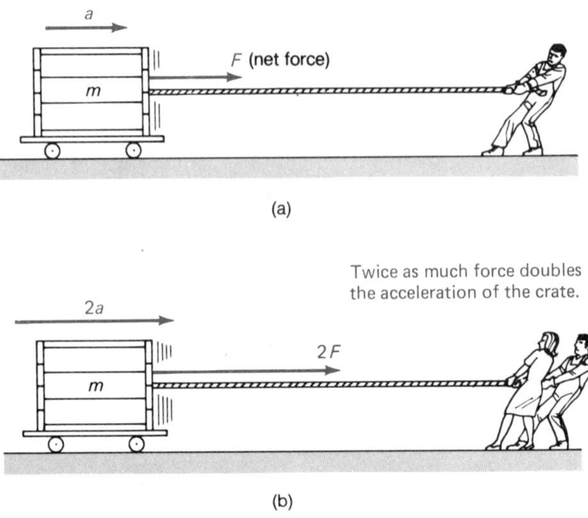

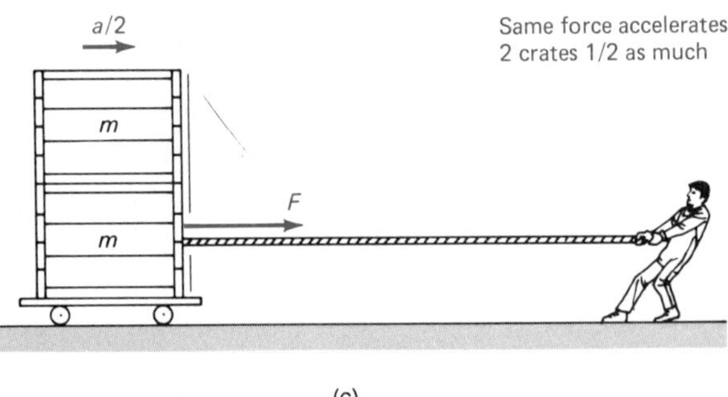

Figure 6.4
Relationships between acceleration, force, and mass are expressed by Newton's second law of motion.

Thus, the acceleration of an object depends *both* on the net force and on the mass of the object. This relationship is illustrated in Figure 6.4. If the force acting on a mass is doubled, the acceleration is doubled. If the mass is doubled and the original force remains the same, the acceleration is one half.

Newton's **second law of motion** is commonly expressed in equation form as

$$\boxed{F = ma} \qquad \textbf{(Eq. 6.1)}$$

Force = mass × acceleration

where m is the mass of the object or system that experiences a change in motion as a result of the applied force. A system may contain more than one object, and m is the *total* mass of the moving system. It should be noted that if more than one force is acting on an object or system, then F is the net or unbalanced (resultant) force.* Also, if the applied force is an average force $\bar{F}$, then $\bar{F} = m\bar{a}$.

The units of force are the newton (N) in the SI system (What could be more appropriate?) and the pound (lb) in the British system. The standard unit equivalents of these units may be seen in Equation 6.1.†

$$F = ma$$

$$\text{SI:} \quad \text{N (newton)} = (\text{kg})(\text{m/s}^2)$$

$$\text{British:} \quad \text{lb (pound)} = (\text{slug})(\text{ft/s}^2)$$

For example, in the SI system we see that 1 N is the force that produces an acceleration of 1 m/s^2 when it acts on a mass of 1 kg.

* In vector notation we would write $\mathbf{F} = m\mathbf{a}$, indicating that the resulting acceleration is in the direction of the net force.

† In the cgs metric system, the unit of force is the dyne (dyn), and dyn = (g)(cm/s^2), i.e., $F = ma$.

SPECIAL FEATURE 6.3

An Accelerometer

A carpenter's air-bubble level can be used as an accelerometer or a force meter in detecting an acceleration or a force by observing the motion of the bubble (Fig. 1). If a force toward the left is applied to a level at rest, as shown in the figure, which way will the bubble move?

Many people say that the bubble would move to the right, but actually it would move to the left, in the direction of the acceleration and the force. The incorrect answer arises from the fact that we are used to observing the bubble rather than the liquid. The correct answer is explained by Newton's first law. Because of inertia, the liquid resists the motion and "piles up" toward the rear of the level. This forces the bubble in the other direction—the direction of the acceleration. Think of pushing a stationary pan of water. What happens to the liquid?

What do you think the bubble would do (a) if you pushed the level with a constant velocity and (b) while the level came to rest after you stopped pushing? Get a level and investigate the bubble's behavior for yourself.

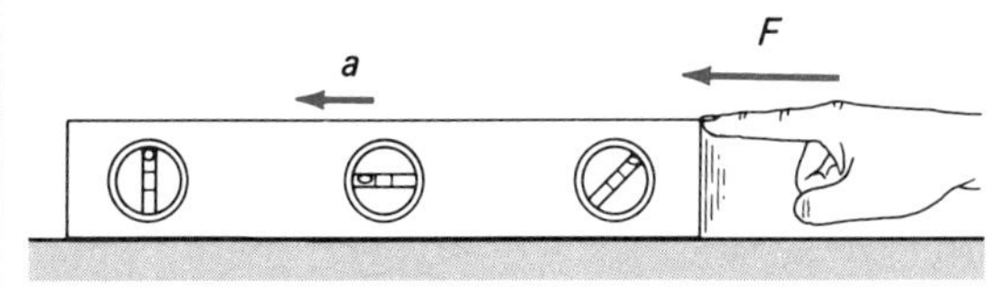

Figure 1
An accelerometer. The motion of the bubble indicates the direction of the acceleration of the level.

Also, we see from Equation 6.1 that if the unbalanced force on an object is zero, then its acceleration is zero, and it remains at rest or in motion with a constant velocity (no acceleration), which is what Newton's first law tells us.

EXAMPLE 6.1 A force of 20 N acts on an object of 4.0-kg mass that is initially at rest. (a) What is the resulting acceleration of the object and (b) how far does the object travel in 6.0 s?

Solution: (a) It is given that $F = 20$ N and $m = 4.0$ kg. Then,

$$a = \frac{F}{m} = \frac{20 \text{ N}}{4.0 \text{ kg}} = 5.0 \text{ m/s}^2$$

(b) With the object initially at rest, $v_o = 0$. We know that $a = 5.0$ m/s^2 and $t = 6.0$ s, so the distance traveled can be found using the kinematic (motion) equation

$$x = v_o t + \tfrac{1}{2}at^2 = 0 + \tfrac{1}{2}(5.0 \text{ m/s}^2)(6.0 \text{ s})^2 = 90 \text{ m}$$

Notice how the kinematic equation of Chapter 4 can be applied to the dynamic situation once the acceleration resulting from the force is known.

Newton's second law relates an applied force directly to an object's acceleration—a sort of "cause and effect" relationship: the force is the "cause" and the acceleration the "effect." The acceleration of an object is always in the direction of the applied net force. If the force is applied in the direction of the object's motion, it will increase the object's velocity (speed). If the force is applied in the direction opposite to the motion, a decrease in velocity (speed) will result (deceleration). When applied at an angle to the direction of the object's motion, a force will deflect or change the direction of the object's motion. This is also an acceleration or change in velocity (certainly a change in direction *and* possibly a change in magnitude as well).

In the case of a liquid mass, inertia may be used to indicate the direction of an acceleration or force. See Special Feature 6.3.

6.3 Applications of Newton's Second Law

Mass and Weight

Newton's second law allows a quick distinction between mass and weight. The mass (m) of a body is the quantity of matter it contains or a measure of inertia, i.e., a fundamental property. The weight of a body is the gravitational force acting on it. Usually this is the gravitational attraction of the Earth, and the force can be easily demonstrated. When we drop an object, it falls (accelerates) toward the Earth.

We commonly write the formula for weight as $w = mg$, where g is the acceleration due to gravity and has a relative constant value near the surface of the Earth of 9.8 m/s^2 (or 32 ft/s^2). Notice that this is a special form of Newton's second law,

$$F = ma$$

or

$$w = mg$$

where $a = g$. The acceleration due to gravity is given the special symbol g because it is so common.*

In the SI system, the weight force, like all forces, is expressed in newtons (N). However, a body's "weight" is often expressed in kilograms ("kilos") or in mass units (Fig. 6.5). Recall that 1 kg of mass has an *equivalent* weight of 2.2 lb or, by the above formula,

$$w = mg = (1.0\text{ kg})(9.8\text{ m/s}^2) = 9.8\text{ N}$$

Keep in mind that the weight *force* in the SI system is mg (newtons).

The unit of mass in the British system is the slug. The use of this not-so-common unit is usually indirectly avoided, as the following example shows.

Figure 6.5
Weight and mass. Weight is force, expressed in units of newtons or pounds. In the SI system, a body's "weight" is commonly expressed in kilograms ("kilos") or in units of mass.

EXAMPLE 6.2 A block weighing 16 lb is initially at rest on a horizontal frictionless surface. A constant force gives the block an acceleration of 12 ft/s^2. What is the magnitude of the applied force if the force is (a) parallel to the surface and (b) at an angle of 30° to the surface (Fig. 6.6)?

Solution: (a) It is given that $w = 16$ lb and $a = 12$ ft/s^2. The mass of the block is given by $m = w/g$ (recall $w = mg$), where g is the acceleration due to gravity and $g = 32$ ft/s^2 in the British system.

Hence, the mass of the block is $m = w/g = (16\text{ lb})/(32\text{ ft/s}^2) = \frac{1}{2}$ slug. This may be used directly in Newton's law, or we may write

$$F = ma = \left(\frac{w}{g}\right)a = \left(\frac{16\text{ lb}}{32\text{ ft/s}^2}\right)(12\text{ ft/s}^2) = 6.0\text{ lb}$$

In this manner, the slug unit is not expressed explicitly.

(b) When the force is applied at an angle to the horizontal, only the horizontal component of the force produces an acceleration in that direction. In this case, $F\cos\theta = ma$, and

$$F = \frac{ma}{\cos\theta} = \frac{\left(\frac{w}{g}\right)a}{\cos 30°} = \frac{(16\text{ lb}/32\text{ ft/s}^2)(12\text{ ft/s}^2)}{0.866} = 6.9\text{ lb}$$

* This should not be confused with the accepted abbreviation for the gram (g), which unfortunately has the same symbol. The nonstandard abbreviation gm is sometimes used for gram for distinction.

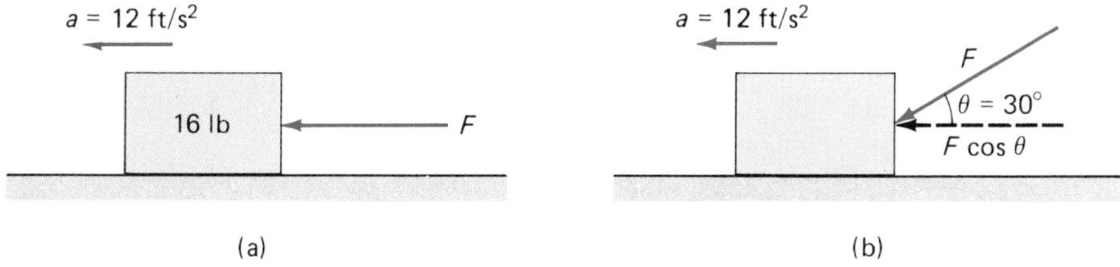

Figure 6.6
Force and acceleration. See Example 6.2.

It was pointed out in the previous section that the m in $F = ma$ is the total mass of the moving system. This is indeed the case when the system is considered as a whole. However, Newton's second law applies to a total system as well as to any part of it. That is, we can isolate parts of a system and apply the second law, as the following examples show.

EXAMPLE 6.3 Two masses m_1 and m_2 are connected by a rope that runs over a frictionless pulley as shown in Figure 6.7. Neglecting the masses of the rope and the pulley, (a) what is the acceleration of the suspended masses? (b) What is the magnitude of the tension force?

Solution: (a) Considering the system as a whole, with the tension in the rope being transmitted undiminished, the T's cancel. (A necessary condition for this is that the mass of the rope be zero or negligible. Otherwise the tension forces would have to be different to supply a net force to accelerate the mass of the rope.) Overall, then, the net force is just the difference of the weight forces, and applying Newton's second law,

$$F = F_2 - F_1$$

$$= w_2 - w_1 = m_2 g - m_1 g = ma$$

$$= (m_2 + m_1)a$$

Notice that the mass m in Newton's second law is the *total* moving mass of the system—in this case, $m = m_2 + m_1$. Solving for a,

$$a = \frac{m_2 g - m_1 g}{m_2 + m_1} = \frac{(m_2 - m_1)g}{m_2 + m_1}$$

$$= \frac{(10 \text{ kg} - 5.0 \text{ kg})(9.8 \text{ m/s}^2)}{10 \text{ kg} + 5.0 \text{ kg}}$$

$$= 3.3 \text{ m/s}^2$$

Since $m_2 > m_1$, we know that m_2 descends and m_1 rises. Notice that the single fixed pulley is simply a direction changer. The situation could have been drawn horizontally with the same result.

Newton's law applies to the total system or any part of a system. Alternatively, we could isolate and sum the forces on each mass, which by Newton's law gives two equations:

$$m_2 g - T = m_2 a$$

and

$$T - m_1 g = m_1 a$$

Eliminating T from the equations and solving for the acceleration a yields the previous equation.

(b) Knowing a, either of the equations may be used to solve for T. Choosing the second one,

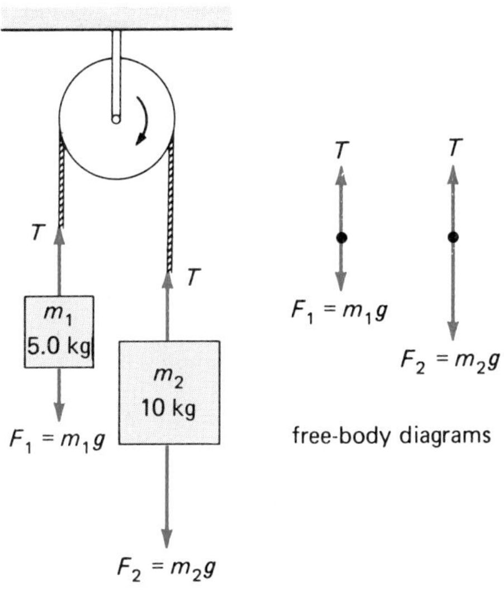

Figure 6.7
An Atwood machine. See Example 6.3.

$$T = m_1a + m_1g = m_1(a + g)$$
$$= (5.0 \text{ kg})(3.3 \text{ m/s}^2 + 9.8 \text{ m/s}^2)$$
$$= 6.6 \text{ N}$$

(*Note:* this pulley arrangement is called an Atwood machine and was originally used to measure the acceleration due to gravity g by determining a from length-time measurements.)

Let's try another problem with tension—this time with more than one string so that we have more than one tension.

EXAMPLE 6.4 Three blocks connected by light strings rest on a frictionless surface (Fig. 6.8). If a horizontal force of 9.0 N is applied to the right-end block, (a) what is the acceleration of the system? (b) What are the tensions in the strings?

Solution: Light strings imply negligible masses, so the tensions may be drawn as shown. The tensions in the strings aren't the same, as will be shown shortly.

(a) First, let's find the acceleration of the system as a whole:

$$F = ma = (m_1 + m_2 + m_3)a$$

and

$$a = \frac{F}{(m_1 + m_2 + m_3)} = \frac{(9.0 \text{ N})}{(1.0 \text{ kg} + 3.0 \text{ kg} + 2.0 \text{ kg})}$$
$$= 1.5 \text{ m/s}^2$$

(b) Then, isolating the masses as shown in the free-body diagrams, we have

$$T_1 = m_1a$$
$$T_2 - T_1 = m_2a$$
$$F - T_2 = m_3a$$

Notice from the middle equation that the tensions in the different strings cannot have the same magnitude. If they did, the net force on m_2 would be zero and it would not move. Since m_2 accelerates to the right, T_2 is greater than T_1, and their magnitudes are given by:

$$T_1 = m_1a = (1.0 \text{ kg})(1.5 \text{ m/s}^2) = 1.5 \text{ N}$$
$$T_2 = m_2a + T_1 = (3.0 \text{ kg})(1.5 \text{ m/s}^2) + 1.5 \text{ N}$$
$$= 6.0 \text{ N}$$

Here's an application of Newton's law to an inclined plane problem.

EXAMPLE 6.5 A 6.0-kg block on a frictionless inclined plane is attached to a suspended 2.5-kg mass as shown in Figure 6.9. Assuming the pulley to be frictionless and neglecting the masses of the string and the pulley, what is the acceleration of the system?

Solution: The direction of the acceleration is assumed to be up the plane, as shown in the figure. The component of m_1's weight force down the plane is $F_1 = m_1g \sin 30°$. Applying Newton's law, we have

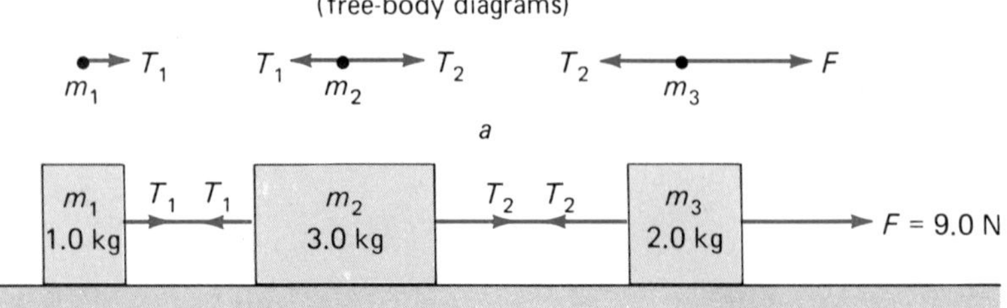

Figure 6.8
Tension forces. See Example 6.4.

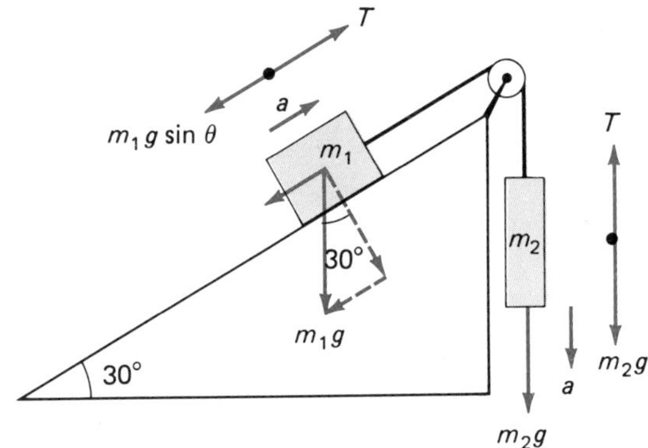

Figure 6.9
Forces and motion. See Example 6.5. (Vector arrows not to scale.)

$$F = F_2 - F_1 = m_2 g - m_1 g \sin 30° = ma = (m_1 + m_2)a$$

and

$$a = \frac{m_2 g - m_1 g \sin 30°}{m_1 + m_2}$$

$$= \frac{(2.5\text{ kg})(9.8\text{ m/s}^2) - (6.0\text{ kg})(9.8\text{ m/s}^2)(0.50)}{6.0\text{ kg} + 2.5\text{ kg}}$$

$$= \frac{24.5 - 29.4}{8.5} = -0.58\text{ m/s}^2$$

Does the negative result mean that the system is decelerating? No, it simply means that the acceleration was chosen to be in the wrong direction, and that the block moves down the plane.

It should be noted that $F = ma$ is a general expression and can be applied to any number of physical forces. Also, a force does not have to be constant as in the preceding examples. A common force that varies in magnitude is that of a spring. As one compresses or stretches a linear coil spring, progressively more force is needed.

The force F exerted *by* a spring when it is compressed or stretched a distance x is given by a relationship known as **Hooke's law,** after Robert Hooke, an English physicist and a contemporary of Newton:

$$F = -kx \qquad \textbf{(Eq. 6.2)}$$

where k is the spring constant with SI units of N/m.

The spring constant characterizes the "strength" of a spring—the greater the spring constant, the stronger the spring, or the more force is required to compress or stretch the spring a given distance. The minus sign in Equation 6.2 indicates that the displacement and the spring force are in opposite directions (Fig. 6.10). The equation in full notation is $F = -k\Delta x = -k(x - x_o)$, or $F = -kx$ with the initial position taken as zero ($x_o = 0$).

Applying Newton's second law, $F = ma = -kx$, and $a = -(k/m)x$. Note that the acceleration is not constant, but varies with position. As a result, we cannot use this acceleration with our usual kinematic equations, since they were derived for a constant acceleration. The mathematics of a varying force is beyond the scope of this text.

However, we can use Hooke's law in static situations.

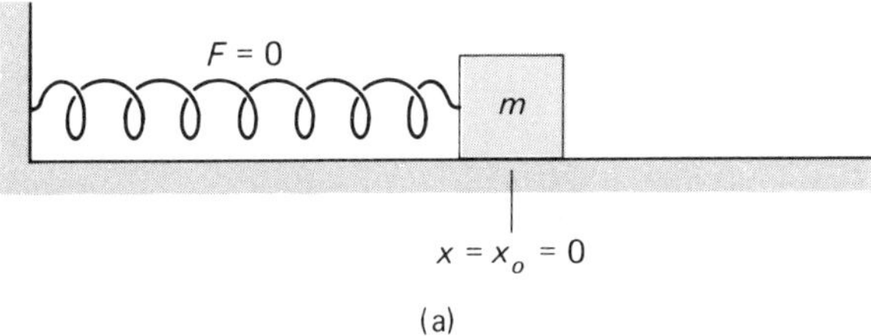

(a)

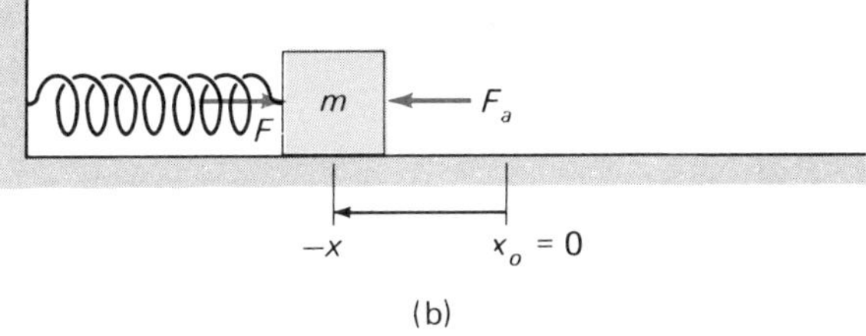

(b)

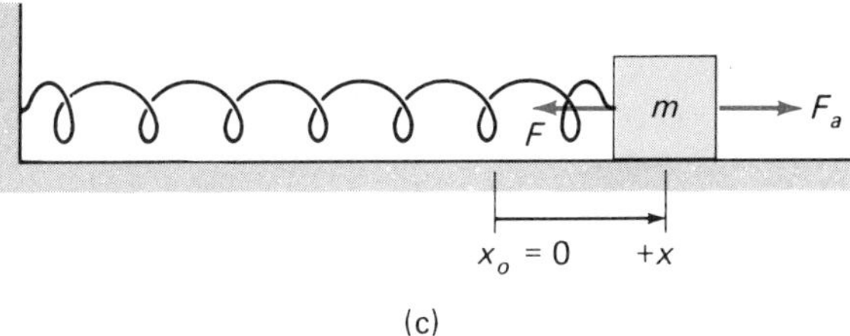

(c)

Figure 6.10
Spring force. The spring force exerted on the block by the spring is equal to $F = -k(x - x_o) = -kx$ (with $x_o = 0$). The minus sign indicates that the displacement and the spring force are in opposite directions. In equilibrium, the applied force F_a is equal and opposite to the spring force F.

EXAMPLE 6.6 A 0.50-kg mass suspended from a spring stretches the spring 10 cm. If an additional 1.0 kg is suspended, what is the total stretched distance of the spring?

Solution: Note that the spring constant is not given. With $m_1 = 0.50$ kg and $y_1 = 0.10$ m (representing the vertical direction as y), we have the magnitude of the upward spring force equal to the weight of m_1 in static equilibrium, i.e.,

$$ky_1 = m_1g$$

or

$$k = \frac{m_1g}{y_1} = \frac{(0.50 \text{ kg})(9.8 \text{ m/s}^2)}{(0.10 \text{ m})} = 49 \text{ N/m}$$

(where the directional minus sign in Hooke's law is neglected). Hence, the spring constant can be experimentally determined.

Then, knowing k, with both masses suspended,

$$y = \frac{mg}{k} = \frac{(m_1 + m_2)g}{k} = \frac{(0.50 \text{ kg} + 1.0 \text{ kg})(9.8 \text{ m/s}^2)}{(49 \text{ N/m})} = 0.30 \text{ m } (= 30 \text{ cm})$$

(Could you have guessed this answer from the initial data, knowing that the force and displacement for a spring are linearly proportional?)

The acceleration in Newton's second law need not always be a linear acceleration. For example, in the case of uniform circular motion with a radial centripetal, or "center-seeking," acceleration, $a_c = v^2/r$ (Chapter 5), we have

$$F = ma_c = \frac{mv^2}{r} \quad \textbf{(Eq. 6.3)}$$

F now represents a centripetal force that supplies the centripetal acceleration necessary for circular motion.

In the case of swinging a ball on a rope around one's head (Fig. 6.11), the centripetal force acting on the ball is supplied by the tension in the rope, which arises from the person pulling on the rope. If you know the mass, speed, and radius of the ball's circular orbit, the magnitude of the centripetal force can be calculated from Equation 6.3.

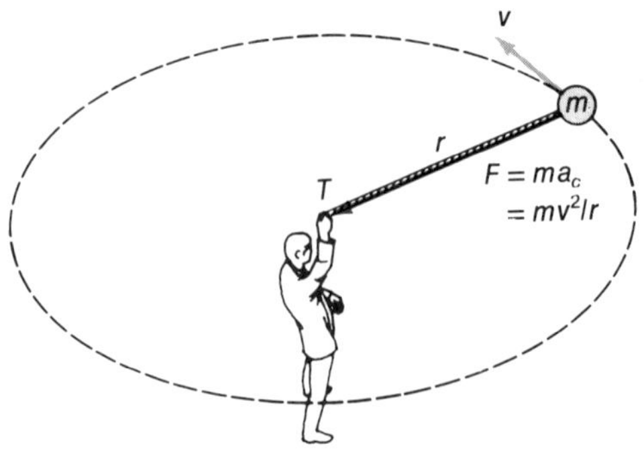

Figure 6.11
Centripetal force. The centripetal acceleration of an object in circular motion is supplied by a centripetal force. In this case, the force is supplied by the person pulling on the rope.

Of course, the horizontal rope illustrated in Figure 6.11 is an ideal, impossible situation. The rope must slant downward at some angle to the horizontal plane. Apply Newton's second law to the vertical forces acting on the ball and you'll see why.

EXAMPLE 6.7 A 1000-kg car traveling with a constant speed of 90 km/h goes around a circular curve with a radius of curvature of 75 m. What is the magnitude of the centripetal force acting on the car, and what supplies it?

Solution: With $m = 10^3$ kg,

$$v = 90 \text{ km/h} \left(\frac{0.278 \text{ m/s}}{\text{km/h}}\right) = 25 \text{ m/s}$$

and $r = 75$ m, using Equation 6.3 we have

$$F = \frac{mv^2}{r} = \frac{(10^3 \text{ kg})(25 \text{ m/s})^2}{75 \text{ m}} = 8.3 \times 10^3 \text{ N}$$

As you might guess, the centripetal force is supplied by friction between the road and the car tires. What would happen if the car hit a wet or icy spot on the road and the friction and centripetal force were reduced?

6.4 Newton's Third Law of Motion: Action and Reaction

Although we commonly talk of single forces, Newton recognized that it is impossible to have an individual force. Rather, there is a mutual interaction, and forces always occur in pairs. An example given by Newton was that if you press on a stone with your finger, the finger is also pressed upon by the stone. That is, if one object exerts a force on a second object, then the second object exerts a force on the first. This is like saying that you can't touch something without being touched.

Newton termed these forces *action* and *reaction*, and **Newton's third law** is commonly expressed as follows:

> **For every action, there is an equal and opposite reaction.**

Or, alternatively,

> **For every force, there is an equal and opposite force.**

In symbol form,

$$F_{\text{action}} = -F_{\text{reaction}}$$

where the negative sign indicates the opposite direction. Which force is the action or reaction is arbitrary and depends on how you look at the situation—it's a relative interaction.

The third law may seem contradictory to the second law. (If you have equal and opposite forces, how can there be an acceleration?) However, the second law is concerned with force(s) acting *on a particular body* and the body's resulting acceleration. In applying the second law, we look at only the forces acting on a given body. The force pair of the third law acts on *different bodies* (Fig. 6.12).

Let's take a look at some examples of the third law action-reaction force pair. When you are holding something quite heavy, you supply an upward force (action) *on the object*. After a short time, you may become painfully aware of the reaction force the heavy object exerts *on you* [Fig. 6.13(a); particularly if the cooler is full].

In Figure 6.13(a), there are two sets of force pairs. There is the upward action force *on the cooler handle* by the person, and the downward reaction force *on the person's hand* by the cooler. Also, there is a downward action force *on the cooler* due to gravity (its weight). Although it is not obvious, there is an upward reaction force *on the Earth*. Notice that there are equal and opposite forces acting *on* the held cooler, so no net force acts on it, and it is stationary (Newton's first law).

But suppose the cooler is dropped [Fig. 6.13(b)]. Now there is a net force on it, and it falls, or accelerates downward (Newton's second law). The third law action-reaction force

Figure 6.12
An example of Newton's third law. For every force there is an equal and opposite force. The forces act on *different bodies*.

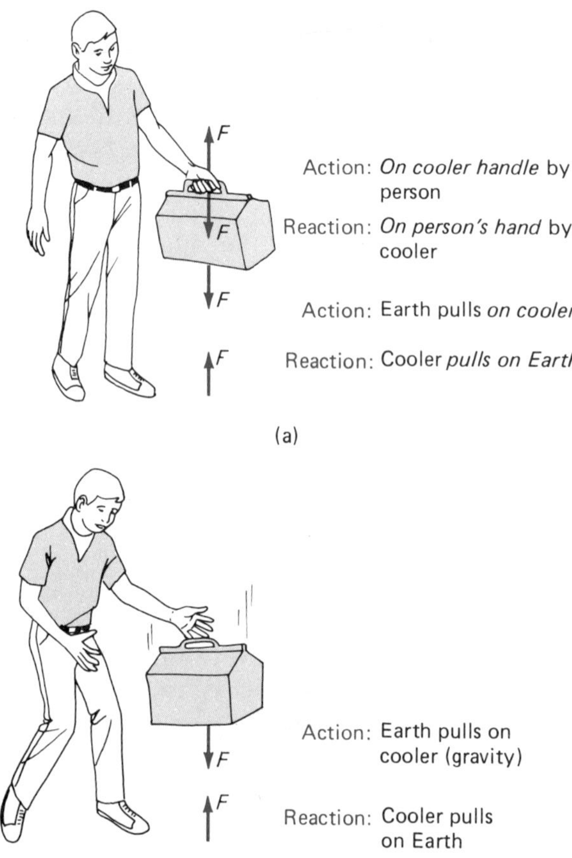

Figure 6.13
Newton's third law force pairs. (a) There are two force pairs associated with the cooler when it is held stationary. Notice that the net force on the cooler is zero. (b) When the cooler is dropped and falling, there is still a force pair, but the net force acting on the cooler is not zero, since it accelerates toward the Earth.

pair between the cooler and the Earth is still there (as it always is, since we can't turn off gravity). The Earth is so massive (6×10^{24} kg), however, that its reaction motion (acceleration) is negligible.

Another example of the action-reaction forces of Newton's third law is jet propulsion. Exhaust gases from burned fuel are accelerated out the back of a rocket or jet engine, and the rocket or aircraft is accelerated forward by the reactive force. Similarly, astronauts use hand rockets to maneuver themselves on "space walks" (Fig. 6.14).

(a)

(b)

Figure 6.14
Newton's third law in action. Rockets operate on action-reaction forces of the rocket and exhaust gases. (a) An Apollo/Saturn V rocket blasting off. (b) An astronaut on a space walk using a hand-held, self-maneuvering unit, or hand rocket. (Courtesy NASA).

Question: When a rocket blasts off, what causes it to lift off the launching pad? Is it the exhaust gases "pushing against" the pad?

Answer: The launching pad is just there as a launching place. It is actually the reactive force of the gases pushing on the rocket that causes the rocket to accelerate upward. If this were not the case, there would be no space travel, since there is nothing to "push against" in space.

Sometimes a not-so-obvious reaction force can be better understood by substituting another force in its place. For example, the reaction force of a wall on a car produces the same effect as the force applied by another car (Fig. 6.15). As another example, consider a 10-kg mass suspended by a rope fastened to a wall, as shown in Figure 6.16. A spring scale is used to measure the force. This would be the weight of the mass, $w = mg = (10\ \text{kg})(9.8\ \text{m/s}^2) = 98\ \text{N}$. The force is transmitted through the rope and the scale to the wall.

To illustrate that the wall "pulls" on the rope with a reaction force of 98 N, suppose you unfastened the rope and held the mass stationary (like the wall). You would have to pull with a force of 98 N, so the wall must be doing the same.

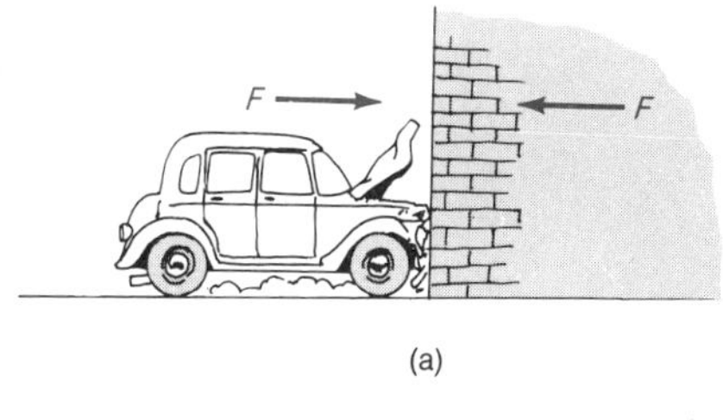

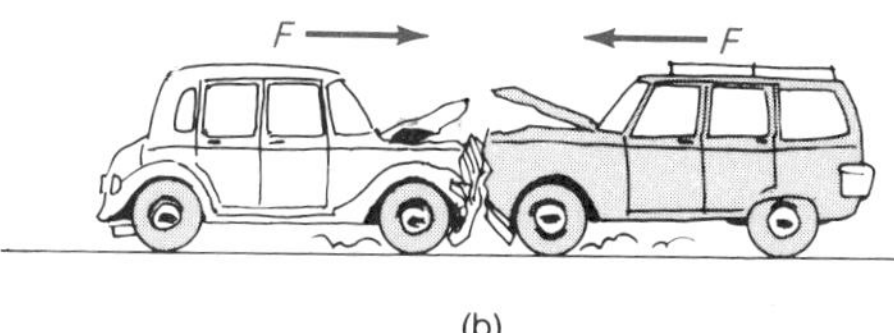

Figure 6.15
Undesirable action-reaction pairs. The reaction force of the wall on the car (a) is recognized when a substitution is made (b).

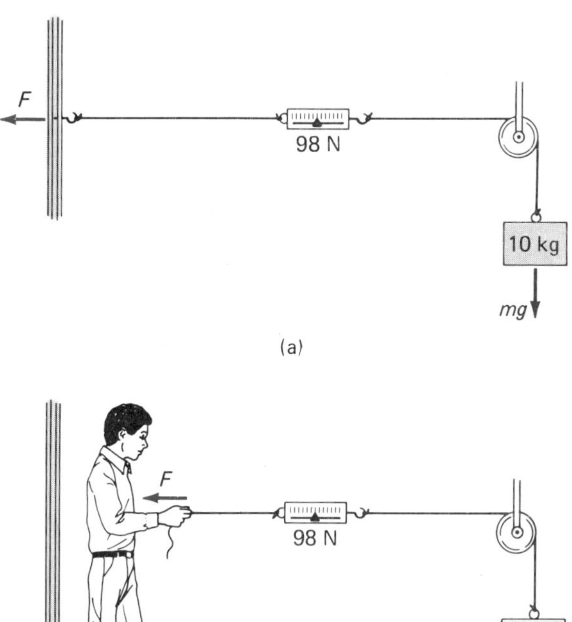

Figure 6.16
Newton's third law one more time. (a) The weight of the mass exerts a force of 98 N on the wall, and the wall "pulls" back with an equal force. (b) This can be easily seen by replacing the wall with yourself or another person.

Important Terms

inertia the property of matter that describes its resistance to changes in motion; mass is a measure of inertia

Newton's first law the law that an object remains at rest or in motion with a constant velocity unless acted upon by an unbalanced force

Newton's second law the law that the acceleration of an object is directly proportional to the applied force and inversely proportional to its mass ($a \propto F/m$)

newton (N) the unit of force in the SI system

Newton's third law the law that for every force (action) there is an equal and opposite force (reaction)

Important Formulas

Newton's second law: $F = ma$

Hooke's law: $F = -kx$

centripetal force: $F = ma_c = \dfrac{mv^2}{r}$

Questions

Newton's First Law

1. How did Galileo's and Aristotle's ideas on motion differ?
2. Can you actually isolate or find the inertia of a body?
3. Why does a load in the back of a truck that is not properly packed shift when the truck moves? Which way does it shift?
4. An old parlor trick involves suddenly pulling a tablecloth from underneath a setting of plates and glasses. Rather than falling to the floor and breaking, they remain on the table (Fig. 6.17, which also shows a modern version that is a bit hard on the tableware). Explain the "magic" of this trick.
5. Suppose you are standing on an icy surface (very little friction) and a heavy object and a light object of the same size and shape are also on the ice. How can you tell which object is heavier without picking them up?
6. A force acts on an object for a short time. Neglecting friction, what happens after this?
7. When a balanced force acts on an object, what is the result?
8. What are flywheels, and what is the principle of their operation?

Newton's Second Law and Applications

9. The amount of money ($) one receives when returning returnable bottles to a store is proportional to the number (n) of bottles returned. (a) Write a symbol relationship for this proportionality. (b) Does the relationship tell you how much money you would actually get for a certain number of bottles? (c) Write the equation relationship that would apply if you received 5¢ for each bottle returned. (d) Does the equation allow you to calculate the amount of money you would receive for a certain number of bottles, say $n = 10$?

(a)

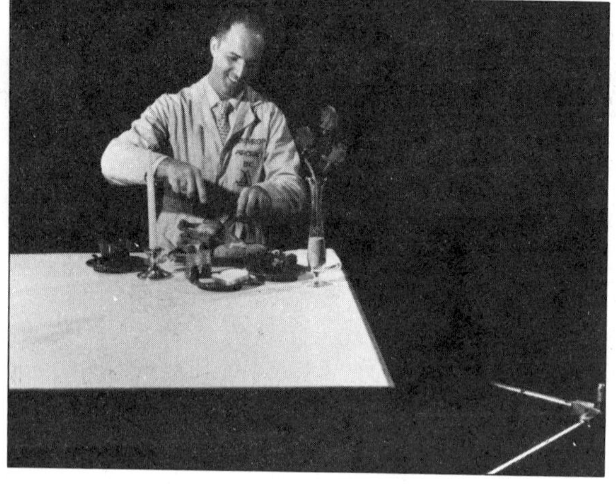

(b)

Figure 6.17
Inertia in action. See Question 4. [(b) Courtesy of J. Williams, F. Trinklein, and H. Metcalfe.]

MECHANICS

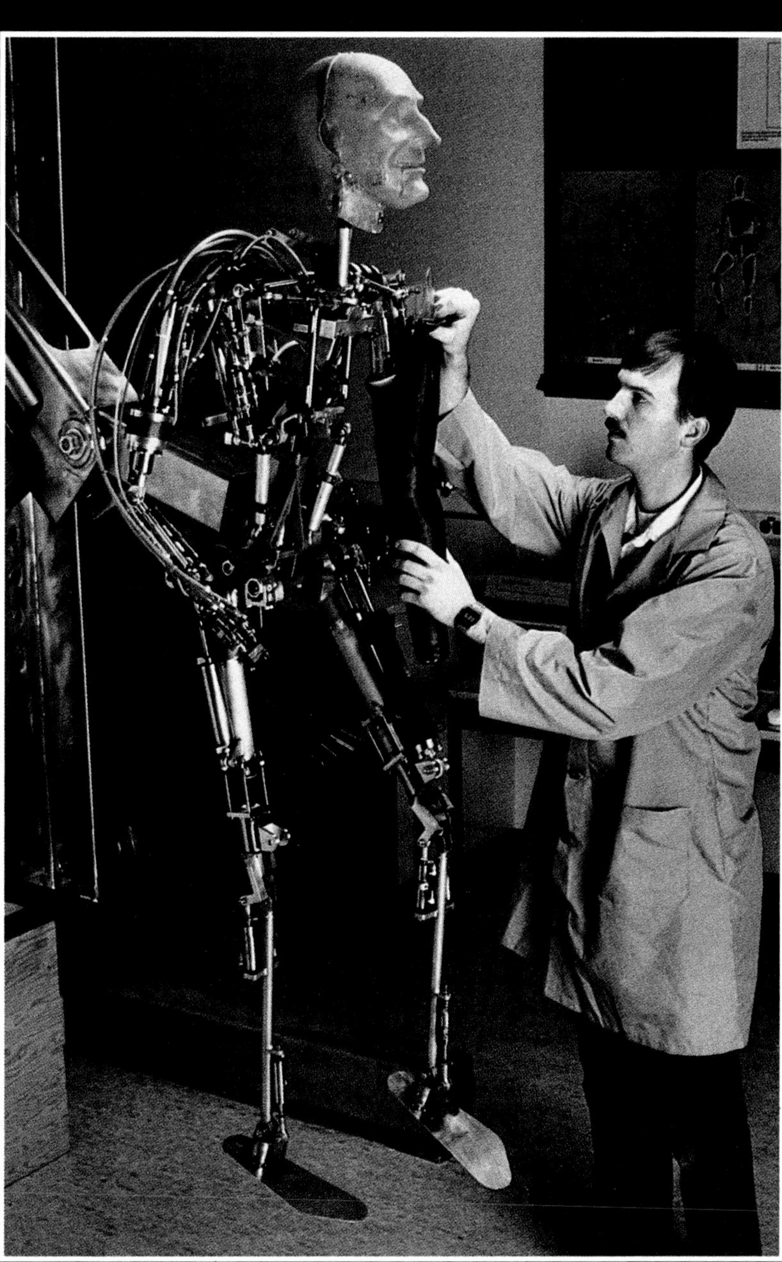

Researchers have developed a robotic mannequin so human-like it even perspires. The robot will be used to help evaluate protective clothing in a variety of hazardous environments. (Courtesy of Department of Energy)

Static, but unstable, equilibrium. A large balanced rock at the Garden of the Gods in Colorado Springs, Colorado. (Photo by David Serway)

Work and energy. The energy stored temporarily in the bent pole is a form of potential energy. What other types of energy are involved? [© Harold E. Edgerton. Courtesy of Palm Press, Inc.]

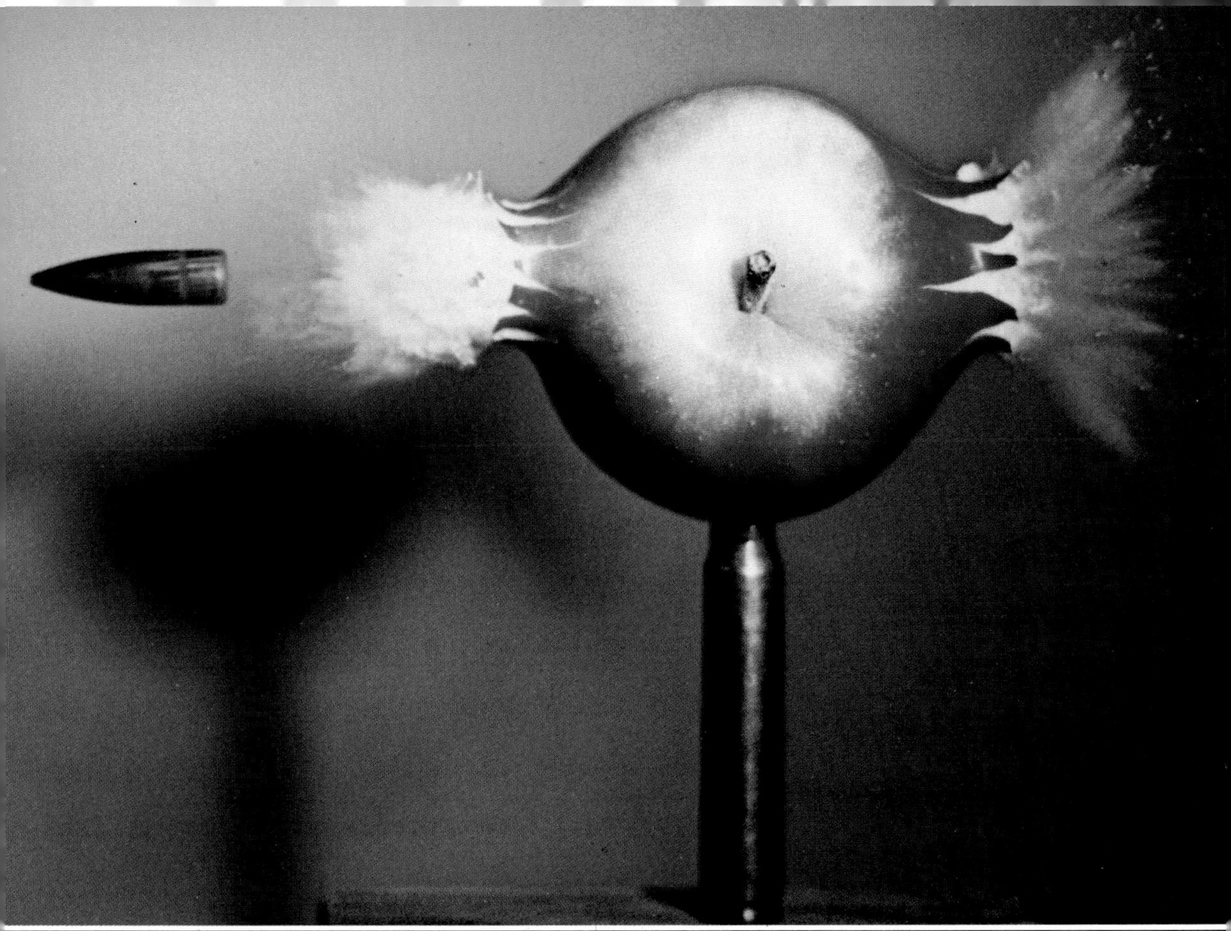

Impact and collision. An apple being pierced by a 30-caliber bullet traveling at a supersonic speed of 900 m/s. This collision was photographed with a microflash stroboscope using an exposure time of 0.33 microseconds. Shortly after the photo was taken, the apple disintegrated completely. [Shooting the Apple, 1964. © Harold E. Edgerton. Courtesy Palm Press, Inc.]

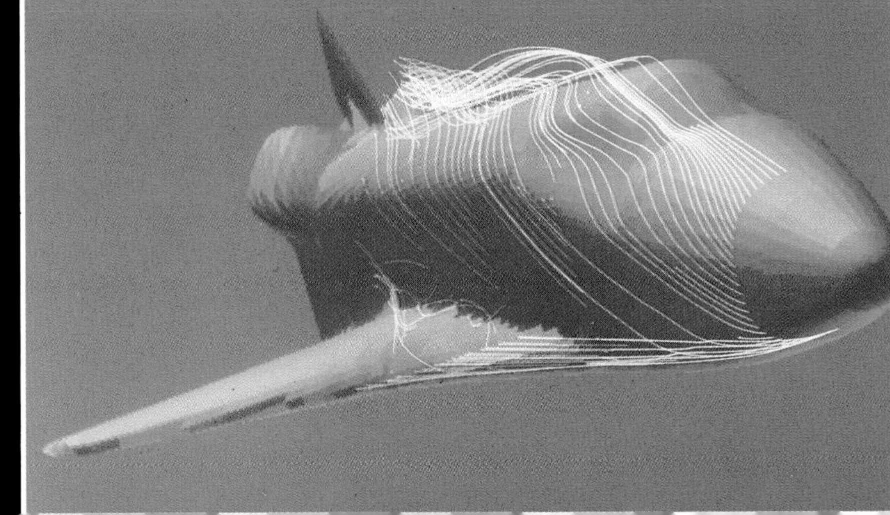

Fluid flow. A computer simulation of the pattern of air flow around a space shuttle. (Courtesy NASA)

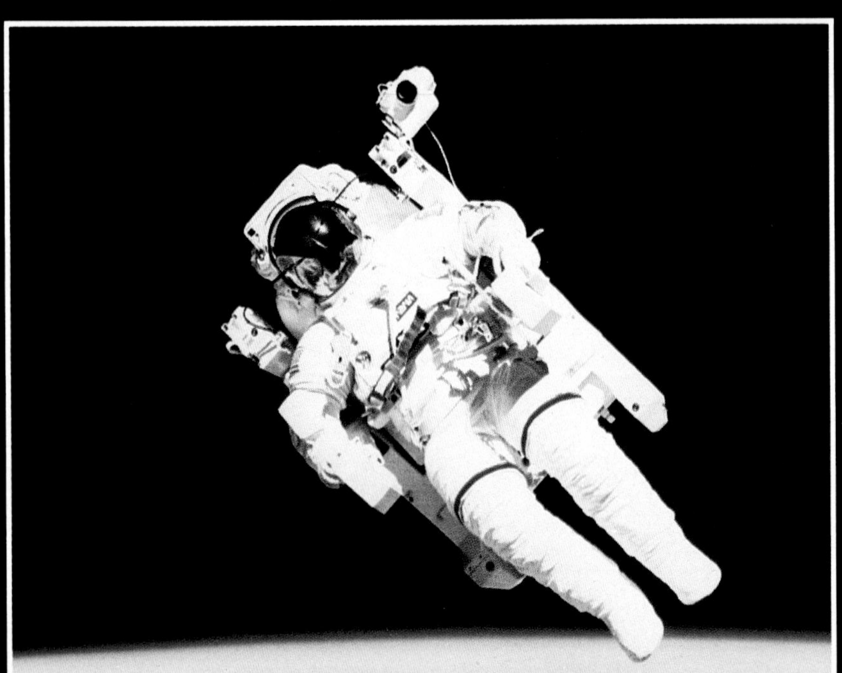

Gravity acting on the astronaut on a space walk supplies the necessary centripetal force to keep him in orbit. The astronaut is testing a nitrogen-propelled hand maneuvering unit (rocket). (Courtesy NASA)

10. Student study time is proportional to extracurricular activities. Write and explain the symbol relationship for this proportionality.
11. Newton's first law of motion may be derived from his second law of motion. Explain this statement.
12. How does the inertia of a body affect its acceleration when a force is applied?
13. For a given force, what would happen if the object on which the force acts lost $\frac{3}{4}$ of its mass?
14. In the equation $F = ma$, explain why m is the total mass of the moving system.
15. How would the mass of the rope in the Atwood machine (Fig. 6.7) be treated if it were not neglected?
16. Which is greater: 1 N, 1 lb, or 1 dyne?
17. An air-bubble level is pushed along a table surface. Where is the bubble or in which direction does it move (a) when the level is moved with a constant velocity, and (b) when the applied force is removed and the level comes to a stop?
18. (a) A child sitting in a stationary car holds a helium balloon by a string. What happens to the balloon when the car starts to move forward? (b) How could you build an accelerometer from a mass and two springs?
19. An air-bubble accelerometer sits radially on a rotating turntable. In which direction is the bubble displaced?
20. What does the minus sign in Hooke's law imply? Could there ever be a plus sign? Explain.

Newton's Third Law

21. In a tug-of-war, two teams pull on the rope with equal and opposite forces, and then one team exerts a greater force. Analyze these situations (forces) in terms of (a) Newton's second law and (b) Newton's third law.
22. What would happen if a fire fighter did not securely hold a hose emitting water under high pressure?
23. Identify the action and reaction forces of Newton's third law for each of the following cases.
    (a) A person pushing on a compressed spring.
    (b) A swimmer changing directions in starting another lap at the end of a pool.
    (c) A person jumping onto a bank from an untied canoe. (Explain the motions of the person and of the canoe.)
24. When a person pushes on a wall, the wall pushes on the person by Newton's third law. Suppose the person put a block of wood between his or her hand and the wall. Analyze the forces acting on the block of wood. Why doesn't the block move?
25. What causes a rotary lawn sprinkler to rotate?

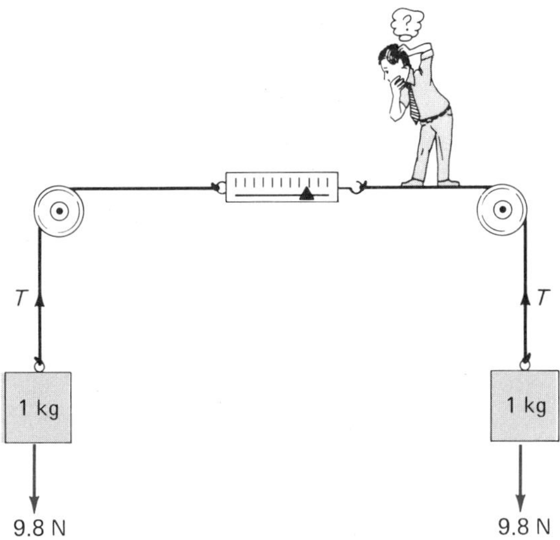

Figure 6.18
How much do they weigh? See Question 26.

26. Two masses are attached to a spring scale, as shown in Figure 6.18. (a) If each mass is 1 kg, what force in newtons would the scale read? (b) We say that the force is transmitted undiminished by the string. What would happen if this were not the case, i.e., if the tension were different in different parts of the string?
27. In approaching (falling toward) the moon, astronauts in a spaceship slow down their spacecraft by firing retrorockets in order to go into orbit and not crash into the moon. How does firing a rocket slow down the spacecraft?
28. A person places a bathroom scale (not the digital type) in the center of the floor and stands on the scale with his arms at his sides. Keeping his arms *rigid* and quickly raising his arms over his head, he notices that the scale reading changes as he brings his arms upward. Similarly, when he brings his arms quickly back to his side, the scale reading changes. What is the change in each case and why? (Try this experiment yourself.)

Problems

Levels of difficulty are indicated by asterisks for your convenience.

6.2–6.3 Newton's Second Law and Applications

1. An unbalanced force of 50 N acts on a 10-kg mass. What is the acceleration?
2. A net force of 20 lb acts on a 10-lb object. What is the resulting acceleration?

3. A force F acts on an object of mass m. If the mass is tripled, how is the acceleration affected?
4. A force F acts on an object of mass m. If the force is doubled and the mass is decreased to $\frac{1}{4}m$, what is the acceleration of the system?
5. A 64-lb object receives an acceleration of 2.0 ft/s^2. What is the magnitude of the force acting on the object?

*6. A 4.0-kg object moving with a constant velocity of 3.0 m/s is acted upon by a force of 10 N in the direction of motion for 2.0 s. What is the velocity of the object at the end of this time?

*7. A 3200-lb automobile has an initial speed of 30 mi/h. It is accelerated uniformly for 4.0 s to 60 mi/h. What is the magnitude of the net force acting on the auto?

*8. An object weighing 49 N at rest on a frictionless surface is acted upon by a horizontal force of 20 N for 6.0 s. How far does the object move in this time?

*9. A force of 25 N acts on a 5.0-kg object through a distance of 2.0 m. If the object is initially moving with a velocity of 10 m/s, what is the final velocity (a) if the force is in the direction of motion and (b) if the force is in the opposite direction of the motion?

*10. An unbalanced force of 4.0 N gives an object an acceleration of 0.25 m/s^2. How large a force would give the object an acceleration of 1.5 m/s^2?

*11. A net force of 20 lb gives an object an acceleration of 4.0 ft/s^2. What acceleration would be given the object if it were acted upon by a 28-lb force? (*Hint:* use a ratio.)

*12. How large a force is needed to accelerate a 1-metric-ton car from rest to a speed of 80 km/h in 10 s?

*13. A force $F = 0.50 \text{ N } \mathbf{x} - 1.5 \text{ N } \mathbf{y}$ acts on a 2.0-kg mass initially at rest at the origin. What are the coordinates of the mass at $t = 3.0$ s?

*14. With a horizontal rope, a 3200-lb car is towed uniformly from rest to a speed of 30 mi/h in 10 s. Could this be accomplished with a rope having a tensile strength of 400 lb? Justify your answer.

*15. If the suspended masses of an ideal Atwood machine are 2.0 lb and 3.0 lb, what is the acceleration of the masses when they are released from rest? See Figure 6.7.

*16. The suspended masses of an ideal Atwood machine are 100 g and 150 g. If the 150-g mass is released from rest 1.0 m above the floor (the 100-g mass rests on the floor), how long will it take for the 150-g mass to hit the floor? See Figure 6.7.

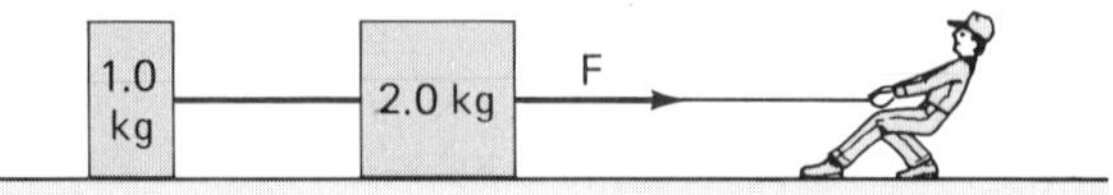

Figure 6.19
See Problems 19–22.

*17. What would be the acceleration of the masses in Figure 6.9 if (a) $m_2 = 6.0$ kg and $m_1 = 3.0$ kg, (b) $m_2 = 6.0$ kg and $m_1 = 5.0$ kg, and (c) $m_2 = 4.0$ kg and $m_1 = 4.0$ kg? (Neglect friction.)

*18. What must m_1 be in Figure 6.9 if it is to slide down the frictionless plane, with $m_2 = 1.5$ kg?

*19. Two connected blocks are on a frictionless surface as illustrated in Figure 6.19. What force F would be required to set the blocks in motion?

*20. If the force F in Figure 6.19 is 18 N, (a) what is the acceleration of the system? (Neglect friction.) (b) What is the magnitude of the tension in the connecting light cord?

*21. If the force F in Figure 6.19 is 20 N and is applied at an angle of 45° to the horizontal, (a) what is the acceleration of the system? (b) What is the magnitude of the tension in the connecting light cord?

*22. If each block in Figure 6.19 experiences a force of friction 5.0 N when in motion, what is the acceleration of the system if F is 50 N?

*23. Neglecting friction and the mass of the pulley and string in Figure 6.20, (a) what is the acceleration of the masses? (b) What is the magnitude of the tension in the connecting light cord?

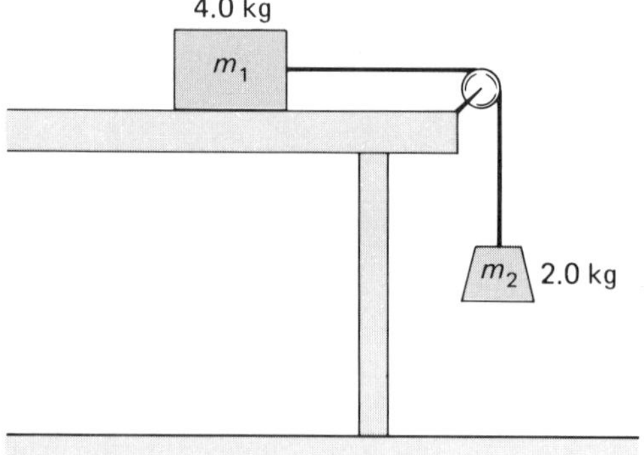

Figure 6.20
See Problems 23–25 and 38–39.

*24. If there is a force of friction between the block and the table in Figure 6.20 of 5.0 N when in motion, what is the acceleration of the system?

*25. Suppose there were another mass, m_3, suspended from another pulley at the opposite end of the table in Figure 6.20. Neglecting friction, what would be the acceleration of the system if (a) $m_3 = 2.0$ kg? (b) $m_3 = 3.0$ kg? (c) $m_3 = 1.0$ kg? Also, what would be the magnitudes of the tensions in the connecting light cords in this case?

*26. How much mass would have to be suspended from a spring with a spring constant of 75 N/m to stretch it 6.4 cm?

*27. A spring is compressed 4.0 cm by an applied force of 12 N. A force of 18 N is then applied in the opposite direction. How far is the spring extended?

*28. A 1.0-kg mass suspended on a spring stretches it a distance of 10 cm. (a) How much more mass would have to be added to stretch the spring a total of 14 cm? (b) What would be the case if only 0.25 kg were suspended from the spring?

*29. A 0.25-kg puck on an air table is attached to a 0.45-m length of string that is fixed at the other end. If the puck is swung in a circle with a speed of 3.0 m/s, what is the tension in the string?

**30. A 2.0-kg mass travels with a speed of 8.0 m/s in the x direction. If a force of 10 N acts on the mass for 2.0 s, what is the final velocity of the mass if the force is (a) in the $+x$ direction and (b) in the $-x$ direction? (c) What is the total distance the mass travels during this time in each case?

**31. A 10-g bullet traveling at a speed of 300 m/s is fired into a tree that is 0.50 m thick. (a) If the bullet comes to rest after traveling 0.30 m, what was the average force exerted on the bullet? (b) If the bullet emerges from the other side of the tree with a speed of 50 m/s, what was the average force exerted on the bullet?

**32. A wooden box weighing 245 N is pushed across the floor by a workman who exerts a horizontal force of 100 N. If the force of friction between the box and the floor is 25 N, what is the acceleration of the box? (Friction always opposes motion.)

**33. (a) Suppose the workman pushed the wooden box in Problem 32 from rest with a force of 150 N. How long would it take him to push the box 7.0 m? (Frictional force remains the same.) (b) If, while pushing the moving box, the workman decides to move it with a constant velocity, what force should he apply?

**34. A horizontal force of 4.0 N accelerates an 8.0-kg block from rest on a frictionless surface for one second. At this time, a 2.0-kg block is placed on top of the first block while the constant force is still applied. How far from the original starting point are the blocks at the end of a total of 3.0 s?

**35. A box with a mass of 5.0 kg sits in the middle of the bed of a pickup truck moving at a speed of 90 km/h. The maximum force of friction between the box and the truck bed is 15.0 N. If the truck were braked uniformly to a stop in 100 m, would the box move? If so, which way and why?

**36. An Atwood machine (Fig. 6.7) is used to measure the acceleration due to gravity, with suspended masses of 0.50 kg and 0.55 kg. When released from rest, the masses move 75 cm in 2.0 s. Compute the experimental value of g. Compare this to the accepted value of 9.8 m/s^2 and explain possible sources of error.

**37. In Example 6.5, suppose there could initially be a maximum force of friction between the block and the plane of up to 5.0 N. How would this affect the situation? (*Hint:* cf. Chapter 7.)

**38. Suppose you are told that there is a force of friction between the block and the table in Figure 6.20 of 25 N when the system is in motion. Do you believe this? Justify your answer. What would be the implication?

**39. With a mass $m_3 = 8.0$ kg suspended from another pulley at the opposite end of the table in Figure 6.20 and a force of friction between the block and the table of 20 N when the block is in motion, what would be the acceleration of the system?

**40. The force of gravity supplies the centripetal force that keeps the moon in its nearly circular orbit about the Earth. (a) What is the force of gravity acting on the moon due to the Earth? (b) What is the force of gravity acting on the Earth due to the moon? (The period of the moon is 29.5 days.)

**41. A person with a 0.25-kg ball on a light string 0.75 m in length swings it in a circle about her head with a uniform speed of 6.4 m/s. If the string makes an angle of 10° with the horizontal, what is the tension in the string?

Chapter Supplement

Centrifugal or Centripetal Force?

You have probably heard the term "centrifugal force" or "centrifugal acceleration." As passengers in a fast-moving car rounding a sharp curve or on a rotating ride in an amusement park, we "feel" a force "pushing" us outward, or away from the center of curvature. This is the so-called centrifugal or "center-fleeing" force.

Some people avoid using the term "centrifugal force" because it is a false or pseudo-force that doesn't exist according to Newton's laws. You may argue that it is real enough for you. The problem is a matter of definition and distinction between frames of reference—in other words, how one looks at the situation.

To illustrate this, consider a car rounding a corner after having traveled along a straight roadway with a constant velocity (Fig. 6S.1). An outside, stationary observer would describe the situation as follows: The car and its occupants are moving along with a constant velocity. The driver turns the wheels, and the friction force between the tires and the road supplies the necessary centripetal force for the car to round the corner.

The passenger, who is not wearing a seat belt, continues to move in a relatively straight line in accordance with Newton's first law. (The friction on the seat of his pants is not great enough to supply the necessary centripetal force.) However, the car turns "in front of" the passenger, and the door pushes up against him. This supplies the centripetal force needed for him to accelerate inwardly and around the corner with the car. Notice that in this description the passenger has no outward force acting on him.

Now, imagine yourself to be the passenger in the car. Your immediate frame of reference is the car, and the situation may be described like this: When moving along the straight roadway with a constant velocity, you are at rest with respect to your reference frame (the car). As the car rounds the corner, you notice yourself moving with respect to the car—outwardly, toward the door. From your knowledge of Newton's laws, you think that this requires a force to be acting on you, which could conveniently be called a centrifugal, or center-fleeing, force. Finally, as a result of this "force" you push up against the door and it pushes back, stopping your outward motion. You are now in equilibrium, with your outward centrifugal force being balanced by the inward door force acting on you.

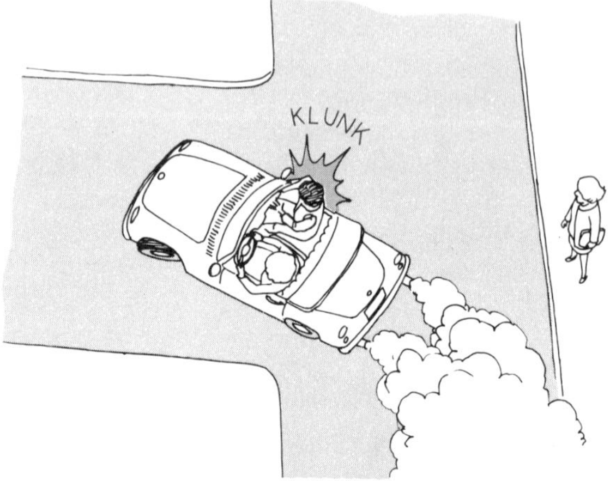

Figure 6S.1

Centrifugal force is a pseudo-force. A person in a car rounding a curve experiences what is believed to be an outward force, but to a stationary observer, the passenger in the car is moving in accordance with Newton's first law of motion and there is no "center-fleeing" force.

Centrifugal force, then, is something that is "invented" to make $F = ma$ work in an accelerating reference frame (the car changing direction or velocity in the example). An inertial, or nonaccelerating, reference frame is one for which Newton's first law (the law of inertia) holds; that is, referenced to this frame an isolated object would be stationary or moving with a constant velocity. In such a reference frame, forces described by Newton's second law are real forces, i.e., the interactions of a body with its environment.

If a reference frame is accelerating, it is said to be noninertial. Here, an observer is moving along with the accelerating frame—for example, a body in circular or rotational motion. A noninertial observer must introduce pseudo-forces to make Newton's second law work in his or her accelerating reference frame. These forces, such as the centrifugal force, *appear* to be real in the noninertial frame. However, the pseudo-forces *do not exist* when the motion is observed from an inertial frame. Hence, only to an observer in an accelerating, or noninertial, reference frame does the term "centrifugal force" have any significance.

Upon examination, an observer in the accelerating frame would find the outward centrifugal force to have the same magnitude as the centripetal force in an inertial frame, $F = mv^2/r$. However, such fictitious forces are not exerted by another body, so there is no third-law reaction force. Notice in the car example that, although the centrifugal and centripetal forces acting on the passenger when the passenger is against the door are "equal and opposite," they are *not* the force pair of Newton's third law, since they act on the same object.

So, in actuality, objects fly outward because of a *lack* of centripetal force (the real one). Centripetal force, or the lack of it, is used in various practical applications. For example, in a washing machine's spin cycle, water is separated from the clothes (Fig. 6S.2). The tub of the washer rotates rapidly, but the centripetal force exerted on the water by the clothes is not great enough to make the water travel in a circle with the spinning (accelerating) tub. The water flies off, leaving the clothes less wet.

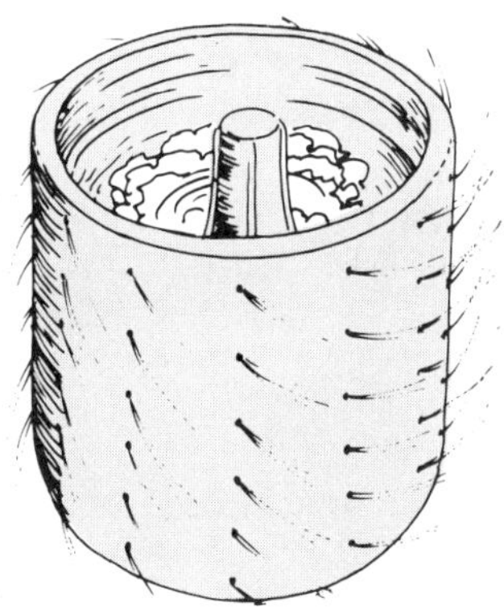

Figure 6S.2

An application of centripetal force. In a washing machine, water is separated from clothes by a spinning action. The (centripetal) force exerted on the water by the clothes is not enough to make the water travel in a circle with the clothes, and the water flies off.

A less desirable case of lack of centripetal force is when the rear wheel of an automobile spins in mud. Have you ever been pushing a stuck car and gotten a bit muddy? Stay away from behind the spinning wheel. The adhesion of the mud to the tire (centripetal force) is not great enough to hold the mud on the tire, so it comes off tangentially to the tire's circular motion.

The Centrifuge

We build machines called *centrifuges* to separate materials of different densities by spinning action. The spinning drum in a washing machine used to separate water from clothes, as described above, is a centrifuge. Here, the clothes are constrained to move in a circle by the reaction force of the drum, and the water flies off. Other applications of centrifuges include separating blood cells from plasma, and cream from milk in dairy separators (Fig. 6S.3). Ultracentrifuges with speeds of the order of 500,000 rpm (revolutions per minute) are capable of concentrating viruses in solutions.

In a container in a centrifuge, such as a test tube, the denser or heavier materials migrate farther toward the outer end of the spinning horizontal container. (The centrifuge container holder is pivoted so the container is horizontal when the centrifuge spins rapidly.)

For example, when blood samples are sufficiently centrifuged, the red cells are bottommost in the tube, with the lighter white cells on top at the bottom of the liquid plasma. This has to do with the resistive force on the cells in solution, which is the centripetal force; it slows the lighter cells more quickly than the heavier cells. (See the effect of air resistance on falling objects in the next chapter.) Here again, the outward migration is due to a lack of centripetal force. To an inertial observer, a centrifuge should really be called a "centripuge."

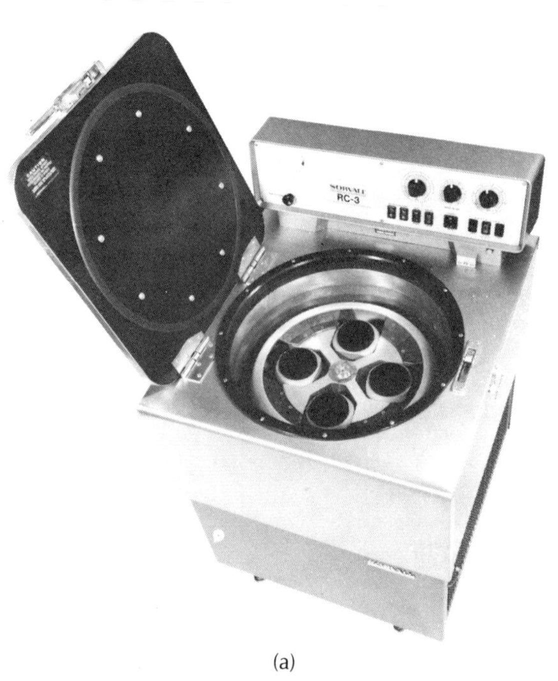

(a)

(b)

(c)

Figure 6S.3

Centrifuge applications. (a) A general-purpose refrigerated centrifuge capable of separating blood plasma and cells. (Reprinted by permission of E. I. du Pont de Nemours and Company.) (b) A dairy centrifuge used to separate milk and cream. (c) Centrifuges used to purify fuel and lubricating oils on board a seagoing vessel. (Courtesy DeSaval Separator Company.)

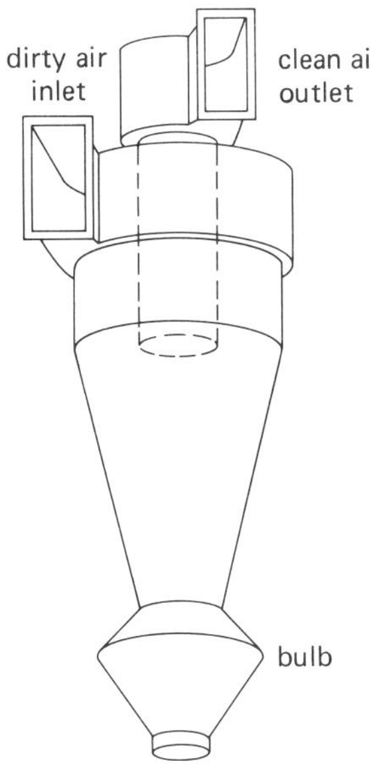

Figure 6S.4

A cyclone dust separator. Dust-laden air enters the separator, and the rapid rotation inside the cyclone provides the force to separate the material from the gas stream. The bulb at the bottom is a vortex breaker, which allows the removal of the collected dust and the return of clean air up through the center of the cyclone to the clean air outlet. (Courtesy United McGill Corporation.)

Centrifuge action is also used in "cyclone" dust separators (Fig. 6S.4). Such separators have widespread use in cement and ceramic operations, grain handling and processing, and woodworking shops. Due to a lack of centripetal acceleration, the particles are "thrown" outward and trapped.

Centrifugal Governor

A mechanical device known as a centrifugal governor can be used to control the speed of an engine (Fig. 6S.5). The control rod moves vertically as the speed of the engine rotation increases and decreases. Coupled to the engine's fuel supply, the movement of the control rod causes the engine to maintain a relatively steady speed regardless of the load. Consider the forces acting on one of the balls (Fig. 6S.5). From the free-body diagram of the ball,

$$F_y = mg \qquad \text{and} \qquad F_x = F_c$$

or

$$F_y = F\cos\theta = mg$$

$$F_x = F\sin\theta = \frac{mv^2}{r}$$

Dividing one equation by the other to eliminate F, we have

$$\cos\theta/\sin\theta = mg/(mv^2/r) = \frac{gr}{v^2}$$

It is usually more convenient to express the speed of the ball in terms of its frequency of revolution, f, or the number of revolutions per second (rps). The frequency is the reciprocal of the period of revolution T, the number of seconds per revolution; that is,

$$f(\text{rev/sec}) = \frac{1}{T(\text{sec/rev})} = 1/(2\pi r/v) = v/2\pi r$$

where T is expressed in terms of the speed and radius of the circular path ($d = 2\pi r = vT$). Hence, with $v = 2\pi rf$,

$$\frac{\cos\theta}{\sin\theta} = \frac{gr}{(2\pi rf)^2} = \frac{g}{4\pi^2 f^2 r}$$

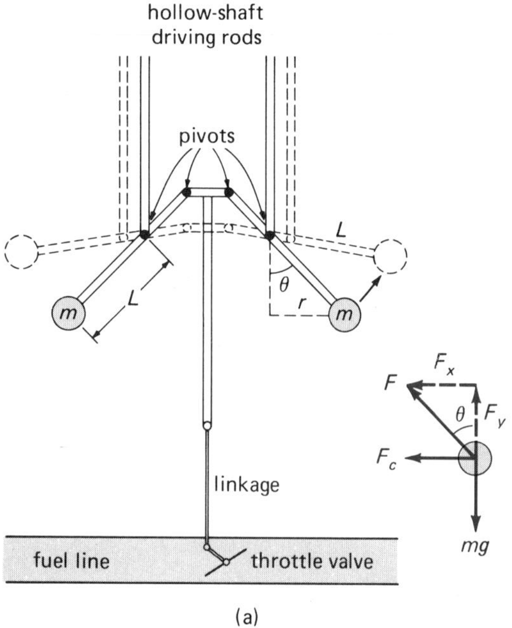

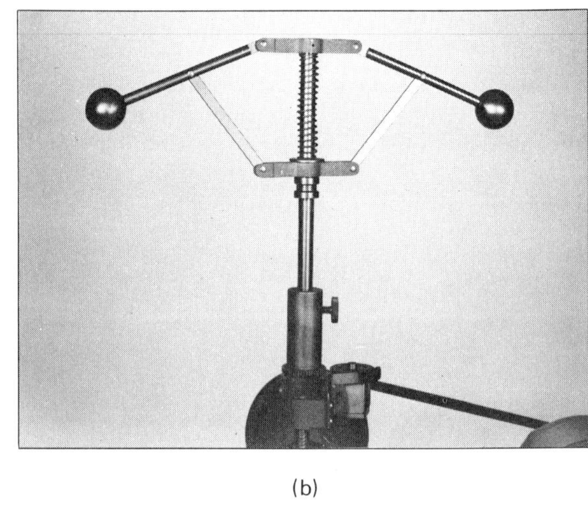

Figure 6S.5
The centrifugal governor. (a) The control rod moves vertically as the speed of the engine and the balls increases and decreases. Coupled to the engine's fuel supply, the movement of the control rod causes the engine to maintain a steady speed. (b) A laboratory model of the centrifugal governor. (Courtesy Central Scientific Co.)

But, from Figure 6S.5, $r = L \sin \theta$, where L is the length of the support rod, so

$$\frac{\cos \theta}{\sin \theta} = \frac{g}{4\pi^2 f^2 L \sin \theta}$$

and

$$\cos \theta = \frac{g}{4\pi^2 f^2 L} \qquad \textbf{(Eq. 6S.1)}$$

Hence, the greater the speed of the engine (larger f), the smaller the $\cos \theta$, and the greater the angle θ. The lowering of the control rod reduces the fuel supply to the engine so as to slow it down.

Similarly, the lower the speed of the engine (smaller f), the greater the $\cos \theta$ and the smaller θ, in which case the control rod speeds up the engine. This action maintains the engine at a relatively steady speed. As can be seen from Equation 6S.1, the critical factors in designing a centrifugal governor are the speed of the engine (f) and the length of the support rod (L).

EXAMPLE 6S.1 The arm or support rod of a centrifugal governor is at an angle of 30° when the frequency is 100 rpm (cf. Fig. 6S.5). What is the support arm angle if the frequency is increased to 101 rpm?

Solution: Forming a ratio of Equation 6S.1 for the two conditions,

$$\frac{\cos \theta_2}{\cos \theta_1} = \frac{f_1^2}{f_2^2} = \left(\frac{f_1}{f_2}\right)^2$$

and

$$\cos \theta_2 = \left(\frac{f_1}{f_2}\right)^2 \cos \theta_1$$

$$\cos \theta_2 = \left(\frac{100 \text{ rpm}}{101 \text{ rpm}}\right)^2 \cos 30^\circ$$

$$= (0.990)^2(0.866) = 0.849$$

Then,

$$\theta_2 = \cos^{-1}(0.849) = 31.9^\circ$$

Notice that it was not necessary to convert the frequencies to standard units (rps or s^{-1}) since a ratio was used.

Problems

6S.1 A centrifugal governor having a support rod 12 in. in length rotates with a frequency of 60 rpm. What is the angle between the support rod and the vertical shaft?

6S.2 The support rod of a centrifugal governor is at an angle of 35° when the frequency is 150 rpm. If the frequency decreases to 140 rpm, what is the angle of the support rod?

Chapter 7

Forces: Gravity and Friction

As seen in the previous chapter, Newton's laws relate force (the cause) to motion (the effect). In particular, Newton's second law mathematically relates the force to acceleration, $F = ma$. Furthermore, force can be related to motional quantities; for example, for linear motion, $F = ma = m(v_f - v_o)/t$, and for uniform circular motion with centripetal acceleration, $F = ma_c = mv^2/r$.

Such relationships express force in terms of motional effects. But can forces be described otherwise? The answer is yes, in some instances. Some forces can be expressed in terms of the parameters of the physical situations. Two illustrative and important examples considered in this chapter are the gravitational force and the force of friction.

7.1 Newton's Law of Gravitation

The force due to gravity is familiar to us all. When an object falls to Earth, we say it is due to gravity. Near the surface of the Earth we measure the acceleration due to gravity to be $g = 9.8$ m/s^2 = 32 ft/s^2, and $F = ma = mg$.

However, gravity is a mutual interaction between two or more masses, since by Newton's third law, if there is a force on one mass there must be an equal and opposite force on another mass or object. An expression for the gravitational interaction between two masses was formulated by Newton himself. His concern at the time was the motion of the moon in its nearly circular orbit about the Earth. As you know, and Newton knew, circular motion requires a centripetal acceleration. But what force acting on the moon causes it to centripetally accelerate toward the Earth?

Allegedly, Newton's insight was sparked by observing an apple falling to the ground (Fig. 7.1). If gravity attracted an apple toward the Earth, perhaps the same force attracted the moon toward the Earth. Using his knowledge of centripetal acceleration and planetary motion and a great deal of insight, Newton formulated

Figure 7.1
Newton and gravitation. Legend has it that Newton's insight came to him from the observation of an apple falling to Earth. Perhaps the same force that caused the apple to be accelerated also causes the moon to be centripetally accelerated toward the Earth.

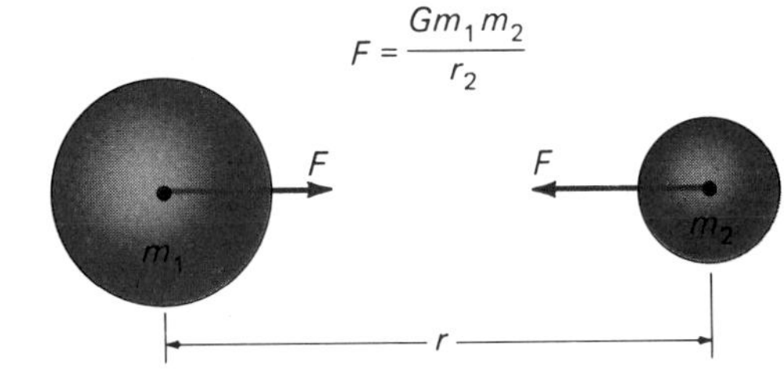

Figure 7.2
Newton's law of gravitation. The magnitude of the mutual gravitational force acting between two uniform spherical masses or two point masses is directly proportional to the product of the masses (m_1m_2) and inversely proportional to the square of the distance between the point particles or the centers of the uniform spherical masses. The gravitational force is always attractive, or toward, the other mass. By Newton's third law of motion, the forces are equal and opposite.

an expression for the attractive gravitational force that acts between the Earth and the moon, the Sun and the planets, and indeed, between any two masses in the universe.

Newton's law of gravitation expresses the force of interaction between two particles or point masses, m_1 and m_2:

$$\boxed{F = \frac{Gm_1m_2}{r^2}} \qquad \textbf{(Eq. 7.1)}$$

where r is the distance between two point masses and G is a constant of proportionality. The direction of the gravitational force acting on a mass is said to be attractive (as opposed to repulsive), or acting toward the other mass.

This means that every mass particle is attracted toward every other particle. For real objects, the net force is the vector sum of the individual particle forces, which is often difficult to compute. However, for two uniform spherical masses, the separation distance may be taken from their centers as though the total mass of a sphere were concentrated there (Fig. 7.2).

The constant G is called the *universal* gravitational constant since it is believed that the law of gravitation applies to all masses everywhere. It was not until about 70 years after Newton's death that instrumentation sensitive enough to measure this constant was developed. The value of the constant is

$$G = 6.67 \times 10^{-11} \text{ N-m}^2/\text{kg}^2$$

This universal gravitational constant is sometimes called "big" G to distinguish it from "little" g, the acceleration due to gravity.

EXAMPLE 7.1 What is the force due to gravity of the Earth on a 70.0-kg person on the Earth's surface?

Solution: The Earth is not exactly spherical, nor does it have a uniform mass distribution. However, to a good approximation we can assume these conditions and consider all of the Earth's mass to be concentrated at its center (Fig. 7.3). On a relative basis, the mass of the person may be considered to be a particle.

Then, $m_1 = m = 70.0$ kg, and from Appendix 2 we have $m_2 = M_e = 6.0 \times 10^{24}$ kg and $r = R_e = 6.4 \times 10^6$ m. Using Equation 7.1,

$$F = \frac{Gm_1m_2}{r^2} = \frac{GmM_e}{R_e^2} =$$

$$\frac{(6.67 \times 10^{-11} \text{ N-m}^2/\text{kg}^2)(70.0 \text{ kg})(6.0 \times 10^{24} \text{ kg})}{(6.4 \times 10^6 \text{ m})^2}$$

$$= 6.8 \times 10^2 \text{ N}$$

As you know, this force on a person is his or her

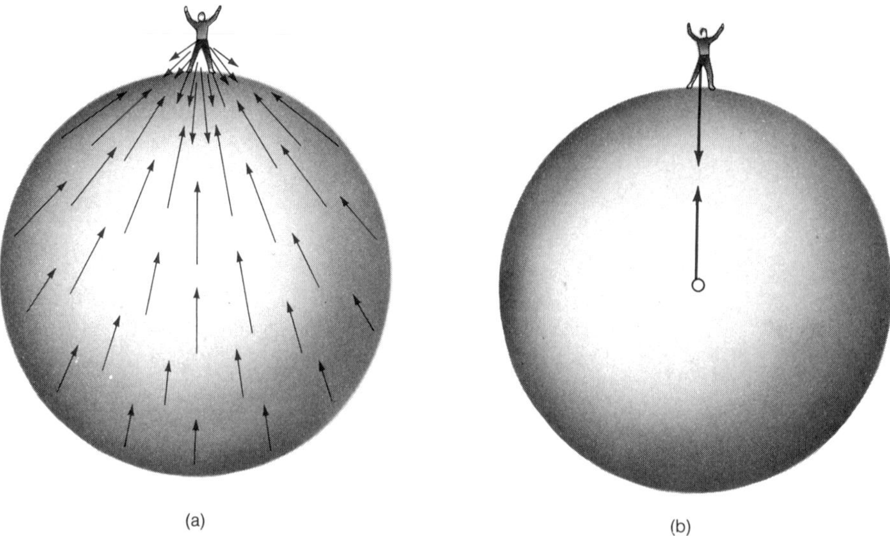

Figure 7.3
Gravity. (a) Mutual gravitation attraction exists between any two particles. (b) Approximating that the Earth is spherical with a uniform mass distribution, the mass of the Earth can be considered to be concentrated at its center. An object's weight is then the gravitational force of attraction between this and its "point" mass. The distance between the masses is the radius of the Earth.

weight. Alternatively, we could have written

$$F = w = mg = (70.0\ \text{kg})(9.80\ \text{m/s}^2) = 686\ \text{N}$$

which is quite close to our other answer. [In the British system, the person would weigh (70.0 kg)(2.2 lb/kg) = 154 lb.]

While we're on the subject of weight, let's investigate the common spring scales (Fig. 7.4). As we learned in Chapter 1, balances are used to measure mass (or weight) by balancing the gravitational force of a known mass with that of an unknown mass. A spring scale, on the other hand, uses the extension properties of a spring.

As we learned in Chapter 6, the spring force is given by $F = -kx$ (Hooke's law), where the minus sign is used to indicate that the force and the displacement are in opposite directions. For a measurement with a simple spring scale, an object's weight is balanced by the spring force and a linear scale calibrates the extension x in force (weight) or mass units, since $\Sigma\ \mathbf{F} = 0$, and $mg = kx$. Dial-type spring scales are also common.

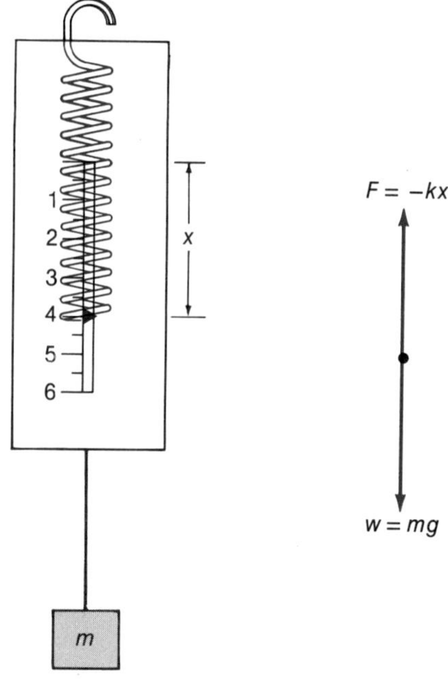

Figure 7.4
A spring scale. The weight of an object is obtained by balancing the weight force against the spring force, $F = -kx = w$. The length scale of the spring may be calibrated in weight or mass units.

Question: Why do you not experience an attraction (gravitational) toward a textbook or another person?

Answer: The gravitational force is too small for you to detect. Consider a couple of 100-kg (220-lb) football players standing 1 m apart. Then, by Newton's law with the appropriate numbers,

$$F = \frac{Gm_1m_2}{r^2} \approx 10^{-7}\ \text{N (or 0.0000001 N)}$$

A force on the order of 10^{-7} N is very small (much less than the weight of a flea). It would be even smaller between you and a textbook ($m \approx 1$ kg).

Because G is so small, to have an appreciable force requires that one or both of the interacting masses be quite large and separated by an appropriate distance.

EXAMPLE 7.2 What are the gravitational forces acting between the Earth and the moon?

Solution: From Appendix 2 we find that the mass of the Earth is $M_e = 6.0 \times 10^{24}$ kg, the mass of the moon is $M_m = 7.4 \times 10^{22}$ kg, and the distance between the Earth and the moon is $r =$ 240,000 mi $= 3.8 \times 10^8$ m. The gravitational force on the moon due to the Earth is

$$F_{me} = \frac{GM_mM_e}{r^2} =$$

$$\frac{(6.67 \times 10^{-11}\ \text{N-m}^2/\text{kg}^2)(7.4 \times 10^{22}\ \text{kg})(6.0 \times 10^{24}\ \text{kg})}{(3.8 \times 10^8\ \text{m})^2}$$

$$= 2.1 \times 10^{20}\ \text{N}$$

toward the Earth. By Newton's third law or the same calculation, the gravitational force on the Earth due to the moon is $F_{em} = 2.1 \times 10^{20}$ N toward the moon.

The gravitational attraction of the moon is a major factor causing tides on Earth. It is easy to understand how the moon gravitationally attracts the ocean water toward it to give a tidal bulge on the side of the Earth nearest the moon. However, as the Earth rotates beneath the bulge, this would seem to give only one high tide and one low tide (the water depression on the opposite side of the Earth). As you probably know, there are two high tides and two low tides daily. So what causes the other bulge for the second high tide?

Newton showed that the two high tides per day are caused by the differences in the gravitational pulls on the opposite sides of the Earth (Fig. 7.5). Keep in mind that the gravitational

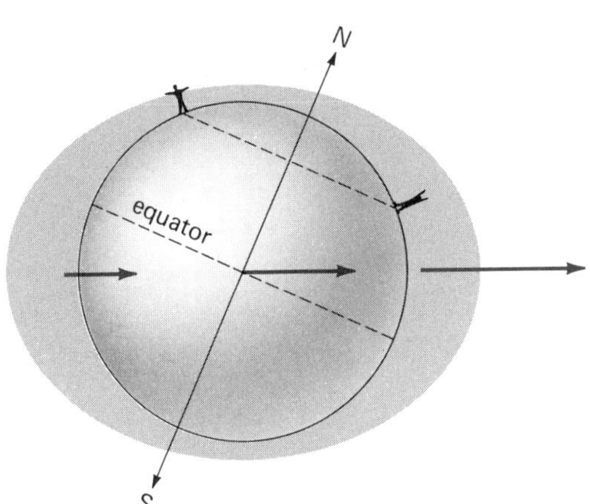

Figure 7.5

Tides. There are two tidal bulges, or two high tides per day because of the difference in the gravitational attraction of the moon at different distances. Notice that the high tides need not be equally "high" and the bulges can be above or below the equator because of the inclination of the moon's orbit.

pull of the moon acts on the Earth itself, as well as on the water on both sides of the Earth, with the attraction getting weaker the greater the distance from the moon.

The water nearest the moon has the greatest attraction, and this pull forms one tidal bulge. The Earth itself is attracted toward the moon with more force than the water on the side opposite the moon. As a result, the Earth is effectively pulled away from the water on the distant side, which gives a second and opposite tidal bulge. So, as the Earth rotates, two high tides "travel" around the Earth daily. The intervening water depressions are the two low tides.

Question: The gravitational pull on the Earth by the massive Sun is about 200 times greater than that by the moon. Why doesn't this produce mammoth tides?

Answer: Although the Sun's gravitational attraction is greater than the moon's, the *difference* between the Sun's pull on opposite sides of the Earth is appreciably less. This is because the Earth's diameter is a relatively small fraction of the distance from the Sun as compared to the moon. The Sun's gravitational pull on each side of the Earth is relatively large, but the difference is relatively small.

To illustrate this idea, consider the fractional difference between two large numbers like 20,000 and 20,002, which is relatively small compared with the fractional difference between two small numbers like 20 and 22.

7.2 A Closer Look at g

We have talked about the acceleration due to gravity quite a bit. Let's take a closer look at g with the help of Newton's law of gravitation.

The gravitational force on an object on Earth is its weight force, $F = mg$. But this must also be the same as the force given by Newton's law of gravitation for the special case of two interacting masses—m, the mass of an object, and M_e, the mass of the Earth. Hence, we may equate the two equations

$$F = mg = \frac{GmM_e}{R_e^2} \qquad \textbf{(Eq. 7.2)}$$

where the separation distance is the radius of the Earth, R_e. Cancelling the m's in the equation,

$$g = \frac{GM_e}{R_e^2} \qquad \textbf{(Eq. 7.3)}$$

Hence, we can see why the acceleration due to gravity near the Earth's surface can be considered to be constant for most practical purposes—G, M_e, and R_e are constants.

However, the acceleration due to gravity and an object's weight does decrease with appreciable altitude (Fig. 7.6). For a distance h above the Earth's surface, the separation distance in Newton's law is $r = R_e + h$, and an object's weight is given by

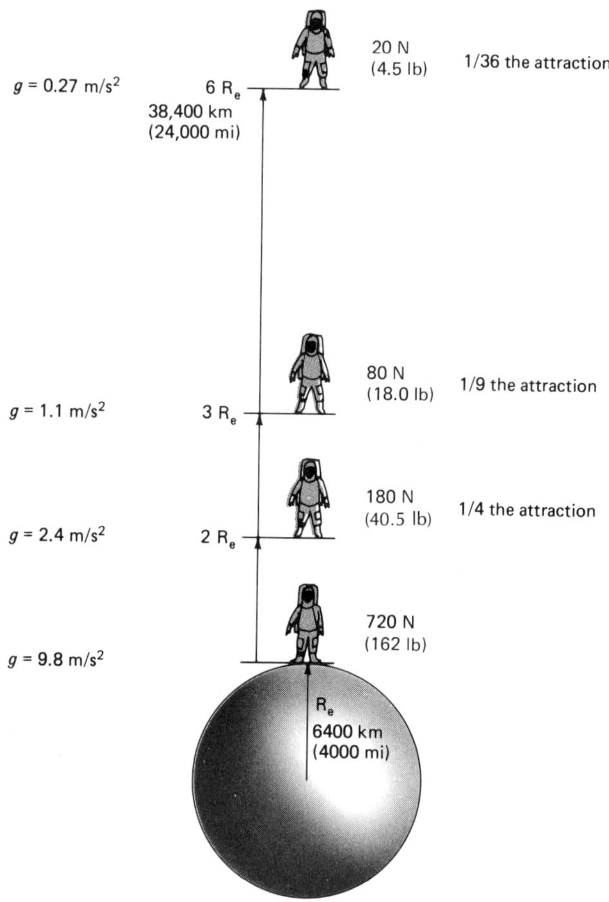

Figure 7.6
Weight and gravity. The force of gravity and g decrease as $1/r^2$ decreases, which results in small values, but an object has weight at any height or altitude. (Surface weight equivalent values are approximate for illustration.)

$$w = mg(h) = \frac{GmM_e}{(R_e + h)^2} \quad \textbf{(Eq. 7.4)}$$

where $g(h)$ indicates that g is a function of height.

Notice that Newton's law of gravitation is an "inverse square" law. That is, the force falls off (decreases) as the reciprocal of the square of the distance between the masses. However, for relatively short distances above the Earth's surface, the acceleration due to gravity may be considered to be constant for practical purposes. At an altitude of 1.6 km (1 mi), the acceleration due to gravity, and an object's weight, is only 0.05 percent less than that at the Earth's surface. At an altitude of 160 km (100 mi), the weight is just 5 percent less than the value on Earth. But keep in mind that no matter how high we go, gravity still acts, and an object still has weight.

Notice from Equation 7.3 that the acceleration due to gravity of an object falling near the Earth's surface is independent of the mass of the object. That is, the mass of the object does not appear in the equation. Hence, neglecting air resistance, all objects fall at the same rate regardless of their masses. It might be thought that heavy objects would fall faster than lighter objects. After all, if one object weighs twice as much as another, or the gravitational force is twice as great, why shouldn't the heavier object fall twice as fast? Aristotle thought that this was the case in his theory of motion, but Galileo showed, or at least believed, otherwise. See Special Feature 7.1.

Galileo gave no reason why objects in free fall have equal accelerations, but Newton's second law does. Recall that the acceleration of an object depends not only on the applied force, but *also* on its mass ($a = F/m$). If one object has twice the mass of another object, then it also has twice the inertia. Hence, it needs twice as much force to accelerate it at an equal rate. This result of Newton's law is illustrated in Figure 7.7.

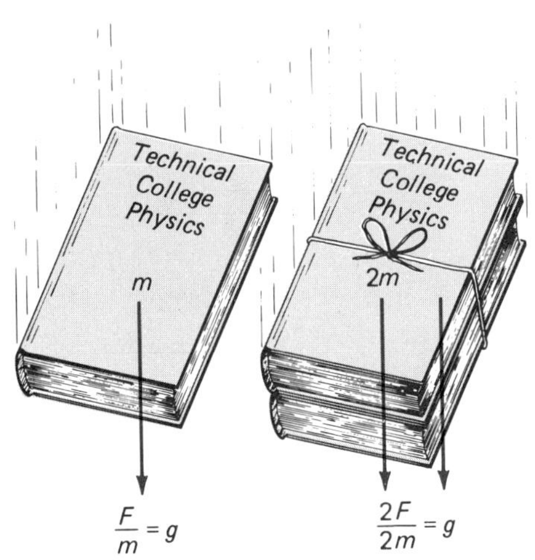

Figure 7.7
Force and inertia. A heavier object (mass $2m$) in free fall has twice the (gravitational) force acting on it as a lighter object (mass m); but having twice as much mass and inertia, it falls with the same acceleration.

Mass of the Earth

If one wishes to calculate the gravitational force between the Earth and some distant object, e.g., an artificial satellite or the moon as in Example 7.2, the mass of the Earth must be known. Determining the mass of the Earth may appear to be a sizable task. How was it done? Certainly the Earth wasn't put on a scale and weighed ($m = w/g$).

Fortunately the result of Newton's law of gravitation in the form of Equation 7.3 comes to our aid, and the mass of the Earth can be determined indirectly.

EXAMPLE 7.3 Compute the mass of the Earth.

Solution: As we use Equation 7.3, all the quantities in the equation are measurable except M_e, i.e., $g = 9.8$ m/s^2, $G = 6.67 \times 10^{-11}$ N-m^2/kg^2, and $R_e = 6400$ km $= 6.4 \times 10^6$ m. Then,

$$M_e = \frac{gR_e^2}{G} = \frac{(9.8 \text{ m/s}^2)(6.4 \times 10^6 \text{ m})^2}{6.67 \times 10^{-11} \text{ N-m}^2/\text{kg}}$$
$$= 6.0 \times 10^{24} \text{ kg}$$

Orbital Speeds of Earth Satellites

Satellites can be put into circular orbits around the Earth by means of rockets (see Chapter 9). The orbital speeds required for such orbits may be found by using Newton's law of gravitation, since this force supplies the necessary centripetal force, i.e.,

$$F_g = F_c$$

SPECIAL FEATURE 7.1

Galileo and the Leaning Tower of Pisa

There is a popular and well-known story that Galileo performed experiments with falling bodies by dropping objects from the Leaning Tower of Pisa (Fig. 7.1). However, the authenticity of this study is doubtful. The original question as addressed by Aristotle was, Why do bodies fall to the ground?

According to the Aristotelian view, the bodies were seeking their natural place at the center of the Earth, which itself was the center of the universe. How this occurred supposedly depended on the "earthiness" of a body; that is, the heavier the body, the faster it would fall in seeking its natural place. There is little doubt that Galileo questioned this view. In an early writing from about 1590 when Galileo was living in Pisa (see Special Feature 6.2), he stated:

> How ridiculous is this opinion of Aristotle is clearer than light. Who ever would believe, for example, that . . . if two stones were flung at the same moment from a *high tower*, one stone twice the size of the other, . . . that when the smaller was half-way down the larger had already reached the ground?*

And, from a later passage concerning the falling of wood and lead objects,

> . . . but a little later the motion of the lead is so accelerated that it leaves the wood behind, and if they are let go from a *high tower*, precedes it by a long space; and I have often made this test.

Here there were probably some air resistance considerations.

But in a 1638 work we find:

> Aristotle says that 'an iron ball of one hundred pounds falling from a height of one hundred cubits reaches the ground before a one-pound ball has fallen a single cubit.' I say that they arrive at the same time.

The first account of a Tower of Pisa experiment came a dozen years after Galileo's death and was written by Vincenzo Viviani, his last pupil and first biographer. Viviani relates that the falling bodies

> . . . all moved at the same speed; demonstrating this with repeated experiments from the height of the Campanile (Tower) of Pisa in the presence of the other teachers and philosophers, and the whole assembly of students. . . .

Yet there is no independent record of this from the time, and it is not mentioned in Galileo's writings. Did Galileo relate this to Viviani in his declining years, or did Viviani use his imagination in describing his teacher's experiments from a "high tower"? In debating this, many lose sight of the point. Whether Galileo dropped objects from the Tower of Pisa in myth or in fact, the acceleration due to gravity is independent of an object's mass.

* This and following quotations are from Cooper, L., *Aristotle, Galileo, and The Tower of Pisa*, Cornell University Press, Ithaca, N.Y., 1935.

or

$$\frac{GmM_e}{r^2} = \frac{mv^2}{r}$$

where m and M_e are the masses of the satellite and the Earth, respectively, and r is the radius of the orbit measured from the center of the Earth. Solving for v, we have

$$v = \sqrt{\frac{GM_e}{r}} = \sqrt{\frac{GM_e}{(R_e + h)}} \qquad \textbf{(Eq. 7.5)}$$

where $r = R_e + h$, or the radius of the Earth plus the height or altitude of the satellite's orbit above the Earth.

EXAMPLE 7.4 A satellite is in stable circular orbit at an altitude of 500 km (310 mi) above the Earth's surface. (a) What is the orbital tangential speed of the satellite? (b) What is its period of revolution?

Solution: (a) It is given that h = 500 km = 5.0×10^5 m. Using Equation 7.5 with the known values of G, M_e, and R_e,

(a)

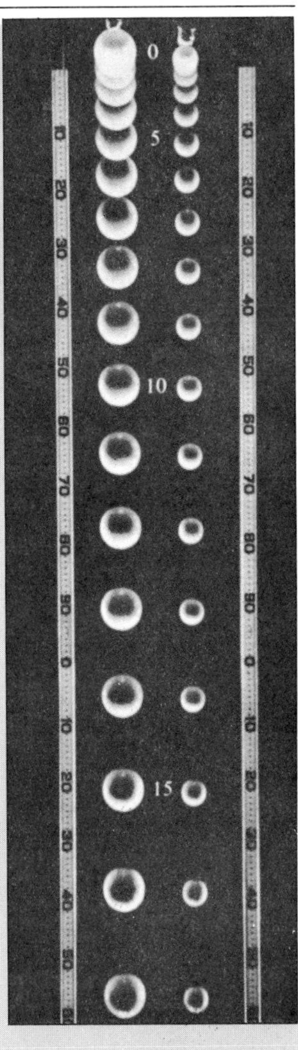

(b)

Figure 1
Free fall. Objects in free fall have the same acceleration. (a) Legend has it that Galileo demonstrated this by dropping objects of different mass or weight from the top of the Leaning Tower of Pisa. (b) A photograph of a baseball and a golf ball released simultaneously showing that they fall together or have the same acceleration.

$$v = \sqrt{GM_e/(R_e + h)}$$

$$= \left[\frac{(6.67 \times 10^{-11}\ \text{N-m}^2/\text{kg}^2)(6.0 \times 10^{24}\ \text{kg})}{(6.4 \times 10^6\ \text{m} + 0.5 \times 10^6\ \text{m})}\right]^{1/2}$$

$$= 7.6 \times 10^3\ \text{m/s}$$

or

$$7.6 \times 10^3\ \text{m/s}\left(\frac{2.24\ \text{mi/h}}{1\ \text{m/s}}\right) = 17{,}000\ \text{mi/h}$$

This speed is less than that needed for an orbit at the Earth's surface, since the required gravitational or centripetal force is less at this altitude.

(b) The period of revolution T is the time it takes for the satellite to make one revolution about the Earth. Using the equation $d = v/t$ or $t = d/v$, the distance the satellite travels in one revolution is equal to the length of the circumference of its circular path, $d = 2\pi r$, and

$$T = \frac{d}{v} = \frac{2\pi r}{v} = \frac{2\pi(6.9 \times 10^6\ \text{m})}{7.6 \times 10^3\ \text{m/s}}$$

$$= 5.7 \times 10^3\ \text{s}\ (= 95\ \text{min})$$

EXAMPLE 7.5 A **synchronous satellite** has a period of revolution that matches the period of the Earth's rotation. The revolving satellite is then "stationary" over one location on the Earth. What is the altitude of such a satellite?

Solution: The period of revolution for a synchronous satellite must be the same as the period of the Earth's rotation, $T = 24\text{ h} = 8.64 \times 10^4$ s. The period of revolution (see Example 7.4) is given by

$$T = \frac{2\pi r}{v}$$

and the orbital speed by Equation 7.5:

$$v = \sqrt{GM_e/r}$$

Combining these equations,

$$T = 2\pi r/(GM_e/r)^{1/2} = 2\pi r(r/GM_e)^{1/2}$$

Squaring both sides of the equation,

$$T^2 = \frac{4\pi^2 r^3}{GM_e}$$

and

$$r^3 = GM_e T^2/4\pi^2 =$$

$$\frac{(6.67 \times 10^{-11}\text{ N-m}^2/\text{kg}^2)(6.0 \times 10^{24}\text{ kg})(8.64 \times 10^4\text{ s})^2}{4\pi^2}$$

$$= 76 \times 10^{21}\text{ m}^3$$

Then,

$$r = (76 \times 10^{21}\text{ m}^3)^{1/3} = 4.2 \times 10^7\text{ m}$$
$$= 4.2 \times 10^4\text{ km}$$

or

$$r = 4.2 \times 10^4\text{ km}\left(\frac{0.62\text{ mi}}{\text{km}}\right) = 2.6 \times 10^4\text{ mi}$$

Since $r = R_e + h$,

$$h = r - R_e = 26{,}000\text{ mi} - 4000\text{ mi} = 22{,}000\text{ mi}$$

Hence, a synchronous satellite is in orbit approximately 22,000 mi above the Earth's surface.

7.3 Apparent Weightlessness

The terms *weightlessness* and *zero gravity* have become common as a result of the space program. Let's investigate the meaning of these terms.

Weightlessness, or the condition of "zero gravity," generally refers to the situation when astronauts "float" in space, apparently because they have no weight (Fig. 7.8). This would be the case if $g = 0$. But a better term for this condition is *apparent weightlessness,* because gravity does act on astronauts in space, and therefore they do have weight. (Recall the definition of weight.)

In the case of an astronaut in a spacecraft in circular orbit about the Earth, we know that gravity provides the necessary centripetal force for the circular motion. The astronaut therefore has weight. The "weightless" feeling arises because the upward reaction force normally provided by the floor or a chair is missing. This is because the floor or the spacecraft is "falling" toward the Earth just as fast as the astronaut. In the astronaut's frame of reference, he "floats" relative to the spacecraft.

To help understand this situation, consider the case of a person standing on a scale in an elevator (Fig. 7.9). The reaction force R of the scale on the person is actually the weight measurement indicated on the scale. In a stationary elevator ($a = 0$), $R = mg = w$, and R is equal to the true weight of the individual. However, suppose the elevator is descending with an acceleration a, and $a < g$. From the free-body diagram and Newton's second law, we have

$$mg - R = ma$$

and the *apparent* weight w' is

$$w' = R = m(g - a) \qquad \textbf{(Eq. 7.6)}$$

(where the downward direction is taken as positive in this instance). Since the person is accelerating, R must be less than mg. Hence, the scale indicates that the person weighs less than she actually does.

Now, suppose the elevator is in free fall, i.e., $a = g$. Then, from Equation 7.6,

$$w' = R = m(g - a) = m(g - g) = 0$$

and the reaction force is zero. What this tells us is that the scale is falling just as fast as the person and, hence, the person is not "pushing down" on the scale. One might call this a "weightless" condition since the scale reads zero.

Figure 7.8
Apparent weightlessness. In an orbiting spacecraft, astronauts "float" in space, apparently weightless. However, gravity supplies the necessary centripetal force to keep them in orbit along with the spacecraft, so they do have weight.

However, gravity is acting on the person, and by definition she must have weight. It is certainly not a case of zero gravity. The person is only *apparently* weightless. A similar condition applies to the astronaut in the spacecraft. If the person in the elevator were to raise herself off the scale by touching the walls of the elevator, she would "float" in the elevator as she and it both fell downward. Similarly, an astronaut "floats" in a spacecraft as both he and it "fall" toward the Earth in a circular orbit (Fig. 7.8).

A relatively new term used to describe the apparent weightless conditions experienced by astronauts orbiting the Earth is *microgravity.*

7.4 Friction: Causes and Types

In previous chapters, we have for the most part ignored friction by conveniently considering ideal frictionless surfaces. However, there is no such convenience in real applications, and an understanding of friction is of prime importance for persons in technical fields.

In some instances, friction is desirable and even promoted. For example, we spread sand on icy walkways and roads to increase the traction between the surface and our shoes and automobile tires. In other instances, such as between the contacting surfaces of moving machine parts, it is desirable to keep friction to a minimum to reduce energy losses and wear. In such cases, we commonly use lubrication to reduce friction (see the Chapter Supplement).

Friction refers to the ever-present resistance to relative motion between contacting material surfaces or within a medium. Resistive frictional forces occur between all types of media—solids, liquids, and gases. In fluids (liq-

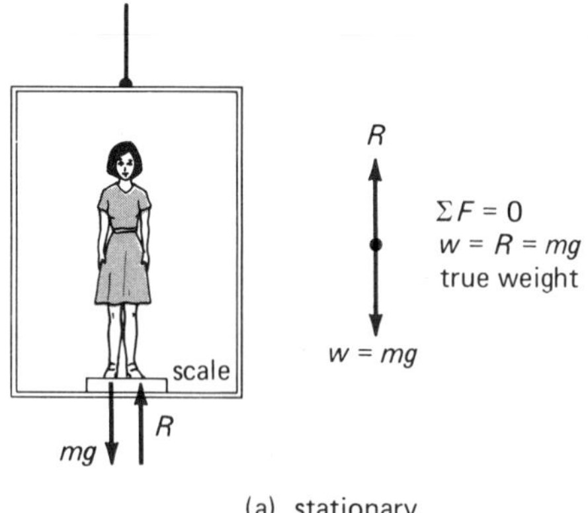

(a) stationary

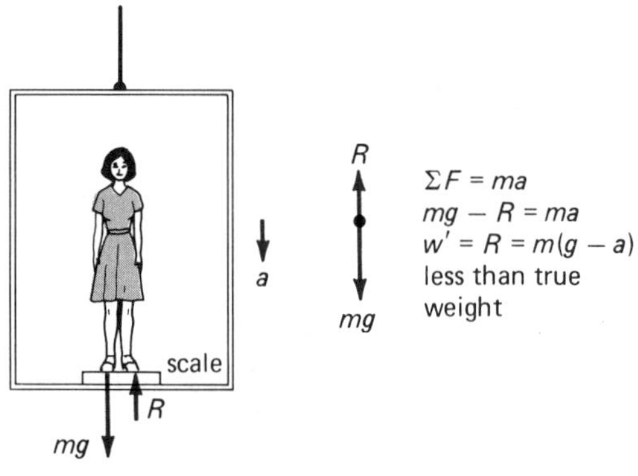

(b) descending

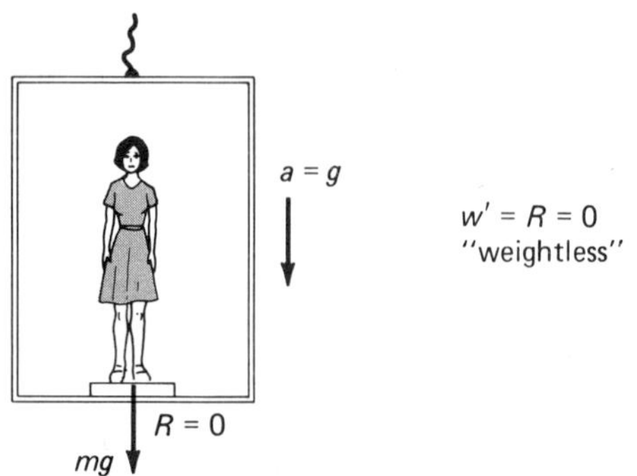

(c) descending with $a = g$

Figure 7.9

Apparent weightlessness in an elevator. (a) In a stationary elevator, a person on a scale reads true weight. The weight reading on the scale is the reaction force R of the scale. (b) If the elevator is descending with an acceleration $a < g$, the reaction force is less than the true weight. (c) If the elevator were in free fall, $a = g$, the reaction force of the scale (and the indicated weight) would be zero, since the scale would be falling as fast as the person.

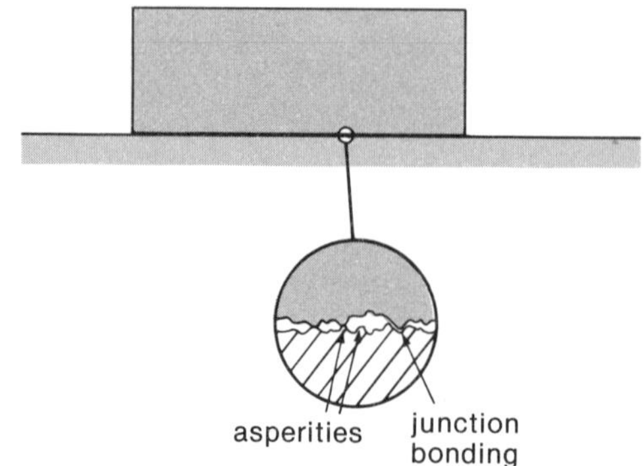

Figure 7.10

Friction due to junction bonding. When two surfaces, particularly metal surfaces, are pressed together, high-pressure spots develop at contacting asperities (high spots) and cause local welding or bonding between the surfaces. These weld spots resist motion and must be sheared before relative sliding motion can occur.

uids and gases), internal resistance is expressed in terms of viscosity, which is a measure of the resistance to fluid flow (see the Chapter Supplement and Chapter 15). However, our discussion will be concerned chiefly with friction between solid surfaces, which encompasses a majority of technical applications.

All solid surfaces are microscopically rough, no matter how smooth they may appear or feel. Early investigators thought friction to be primarily due to the mechanical interlocking of surface irregularities called *asperities* (microscopic high spots). However, modern research suggests that most friction between contacting surfaces of ordinary solids (particularly metals) is due to local adhesion between the surfaces.

When two surfaces are in contact, pressure spots between the contacting asperities cause local welding or bonding of the surfaces. These welded spots must then be broken or sheared for sliding motion to occur. The bonding is confined to a few small patches or junctions where the asperities of one material make contact with the asperities of the other material (Fig. 7.10). Studies show this bonding to be the main consideration—up to 80 percent for metal surfaces.

Another frictional contribution comes from the so-called "plowing" effect of the asperities of a hard material digging into a softer material when relative motion takes place.

The welding-shearing theory offers the most satisfactory physical explanation for metallic friction. However, the mechanism of friction varies depending on the composition of the contacting materials. For example, it is believed that molecular attraction accounts for most of the friction between elastic materials and plastics.

It is convenient to consider various types of friction that are descriptive of particular situations. These include static friction and kinetic or sliding friction, which will be dealt with shortly, along with air (resistance) friction. Before discussing them, let's look at the "laws" of friction.

Classical Laws of Friction

Early investigators of friction formulated the so-called classical laws of friction, which generally describe the frictional force between solid surfaces. According to these laws, the frictional force acting on an object is:

1. Always in a direction opposite to the motion of the object or the force attempting to produce the motion,
2. Directly proportional to the load,
3. Independent of the surface area, and
4. Independent of the sliding speed.

The first law is valid for all frictional situations.

The second law is valid over a wide range of conditions. However, there are a number of notable exceptions. The **load** is the applied force perpendicular to the contacting surfaces that presses them together. The load of an object on a horizontal surface is equal to the weight of the object (Fig. 7.11).

But, for an object on an inclined plane, the load is the perpendicular component of the weight force on the surface. For this reason, the load is expressed in terms of the normal reaction force N of the surface acting *on* the object, which is equal in magnitude to the load in any case. The normal force and the friction force act on the object. In symbol notation, the second law may be written

$$f \propto N$$

where f is the frictional force. This law is not applicable for extremely large loads.

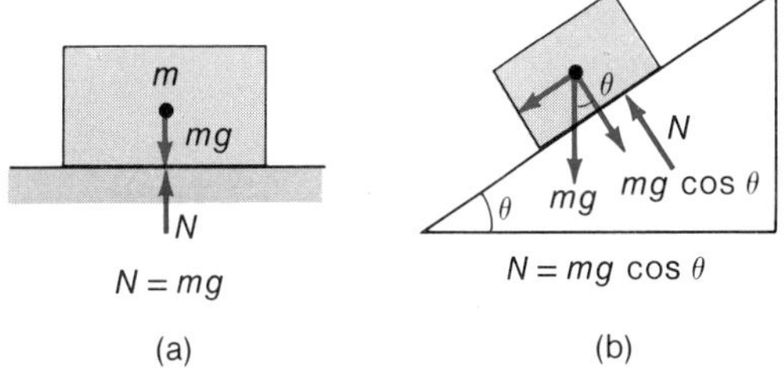

Figure 7.11
Load and normal force. The load is the applied force perpendicular to the contacting surface that presses the objects together. (a) For an object on a horizontal surface, the load is equal to the weight of the object, mg. However, the magnitude of the load is commonly expressed in terms of the normal force N, and $N = mg$ in magnitude. (b) For an object on an inclined plane, the normal force is equal to the magnitude of a component of the object's weight, $N = mg \cos \theta$.

The third law implies that the force of friction between a surface and an object, e.g., a brick, is the same regardless of on which side (area) the brick rests. The load is the same in any case, and the physical basis of the law has to do with pressure or force (load) per area. A smaller area has fewer asperities than a larger area, but the force per asperity area is greater and the welding or bonding is greater for fewer asperities. The third law is generally valid for metal-to-metal surfaces, but less valid for other surfaces, e.g., wood. It does not apply to plastic surfaces.

The fourth law is severely qualified and is only approximately valid for metal surfaces in *slow* relative motion. It does not apply to the frictional force between the surfaces of many other materials—in particular, plastics.

Even with these restrictions, considering the tools available to early investigators, the classical laws of friction give a relatively good description of friction. After all, many materials were not available at the time, e.g., synthetic plastics. The laws do serve as general guides to friction, particularly between dry metal machine surfaces.

7.5 Coefficients of Friction

Since the force of friction is in general proportional to the load or, equivalently, to the magnitude of the normal force, $f \propto N$, we may write in equation form

$$f = \mu N \qquad \textbf{(Eq. 7.7)}$$

where the Greek letter μ (mu) is a dimensionless constant called the coefficient of friction. The coefficient of friction is essentially a property of the contacting surfaces.

Coefficient of Static Friction

If a force F is applied to an object resting on a surface and the object does not move, by Newton's laws there must be an opposing force that prevents the object from moving, namely, the force of static friction (Fig. 7.12). The force of static friction f_s must be equal and opposite to the applied force. (If f_s were larger than F, the object would move in the opposite direction from the applied force.) The force of static friction exists only in response to an applied force.

If the applied force is increased and the object still doesn't move, the force of static friction must also increase. The object will not move until the applied force slightly exceeds the *maximum* force of static friction. This maximum force of static friction is expressed as

$$\boxed{f_s = \mu_s N} \qquad \textbf{(Eq. 7.8)}$$

where μ_s is the coefficient of static friction.

Hence, by measuring the applied force required to just move the object, which is essentially $F = f_s = \mu_s N$, one can experimentally determine the coefficient of static friction when the magnitude of the load or normal force is known: $\mu_s = f_s/N = F/N$. The coefficients of static friction for various contacting material surfaces are listed in Table 7.1.

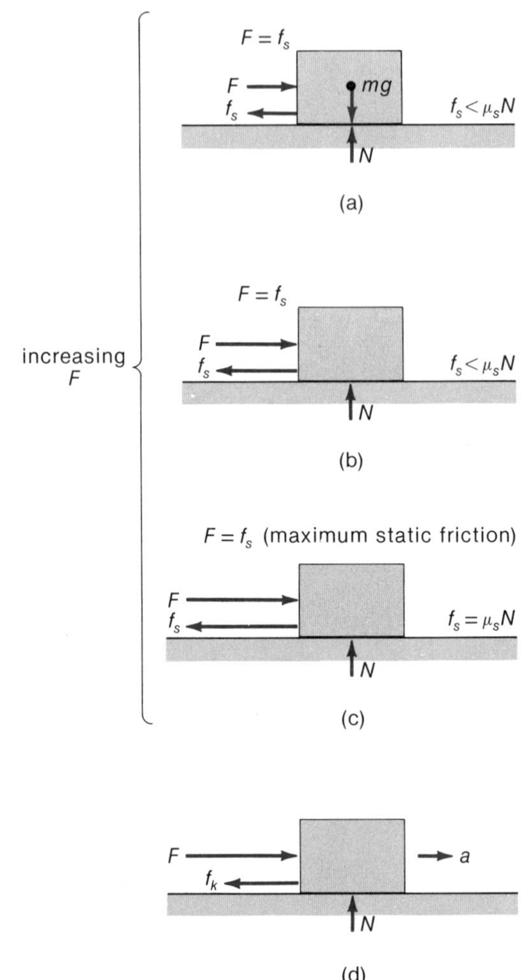

Figure 7.12
Static and kinetic friction. (a) and (b) As the applied force is increased, the force of static friction increases until a maximum value is reached, $f_s = \mu_s N$ (diagram c). (d) When $F > f_s = \mu_s N$, the block is set into motion and the force of kinetic friction, $f_k = \mu_k N$, acts on the block. By Newton's second law, $F - f_k = ma$. If the block moves with a uniform speed ($a = 0$), then $F = f_k$.

EXAMPLE 7.6 A group of workers attempts to move a 300-lb wooden crate resting on a wooden floor. If the coefficient of static friction between the wooden surfaces is 0.58 (Table 7.1), what is the applied force necessary to move the crate for cases (a) and (b) shown in Figure 7.13? (Neglect the friction of the pulley.)

Solution: (a) The load is the weight of the crate, $w = 300$ lb $= N$, and the crate will move when the applied horizontal force slightly exceeds the maximum force of static friction, i.e., just before the crate slides:

$$F = f_s = \mu_s N = (0.58)(300 \text{ lb}) = 174 \text{ lb}$$

(b) When the applied force is at an angle (15°) to the horizontal, the horizontal component of the applied force is equal to the maximum force of static friction just before the crate slides, i.e.,

$$F(\cos 15°) = f_s = \mu_s N$$

Table 7.1
Typical Coefficients of Friction*

Materials	Static Friction, μ_s Dry	Static Friction, μ_s Lubricated	Sliding Friction, μ_k Dry	Sliding Friction, μ_k Lubricated
Steel on babbit†	0.42–0.70	0.08–0.17	0.35	0.08–0.14
Steel on cast iron	0.40	0.18	0.23	0.13
Aluminum on aluminum	1.05	0.30	1.40	—
Glass on glass	1.94	0.35	0.40	0.09
Wood on wood	0.58	—	0.40	0.07–0.16
Wood on steel	0.50	—	0.30	—
Waxed wood on snow	—	0.05	—	0.03
Teflon on Teflon	0.04	—	0.04	—
Teflon on steel	0.04	—	0.04	—
Rubber on concrete (dry)	1.20	—	0.85	—
Rubber on concrete (wet)	0.80	—	0.60	—

* All values are approximate, and lubricated values depend on type of lubricant.

† An alloy of tin, copper, and antimony (or similar alloy) developed by Isaac Babbit (1799–1862), an American metallurgist, and commonly referred to as an antifriction metal. Softer metal particles of the alloy are worn down to a level slightly below the harder metal particles, providing a series of channels for a supply of lubricant.

But in this case, the load on the surface is reduced by the vertical component of the applied force, and the magnitude of the normal reaction force in this case is

$$N = w - F(\sin 15°)$$

where w is the weight of the crate. Hence,

$$F(\cos 15°) = \mu_s N = \mu_s[w - F(\sin 15°)]$$

Solving for F,

$$F(\cos 15° + \mu_s \sin 15°) = \mu_s w$$

$$F = \frac{\mu_s w}{(\cos 15° + \mu_s \sin 15°)}$$

$$= \frac{(0.58)(300 \text{ lb})}{[0.97 + (0.58)(0.26)]} = 155 \text{ lb}$$

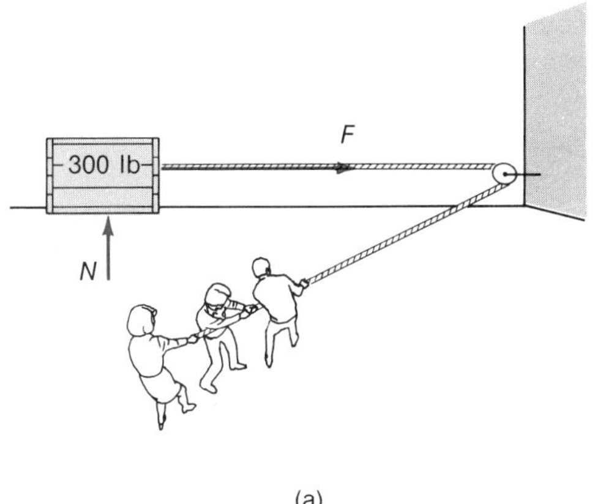

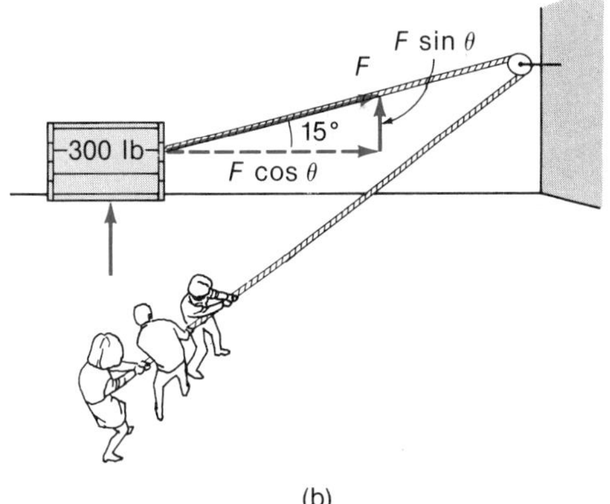

Figure 7.13
Working against friction. See Example 7.6.

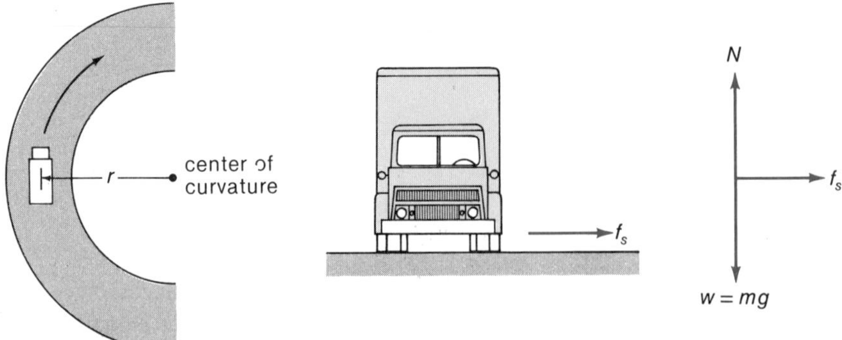

Figure 7.14
Friction supplies the centripetal force for a truck to round a curve. See Example 7.7.

EXAMPLE 7.7 In some cases friction is desirable. For example, friction is necessary for a car to negotiate a level curve, since the friction between the tires and the road supplies the centripetal force. What is the maximum speed a truck can travel around a level circular curve on a concrete road with a radius of curvature of 62 m without slipping (Fig. 7.14)?

Solution: It is given that $r = 62$ m, and assuming the pavement to be dry, from Table 7.1, $\mu_s = 1.2$. Equating the centripetal force to the maximum static frictional force acting toward the center of the curve (sum of the frictional forces on the four tires),

$$F_c = f_s$$

$$\frac{mv^2}{r} = \mu_s N = \mu_s mg$$

where the magnitude of the normal force is equal to the weight of the truck. Solving for v,

$$\begin{aligned} v &= \sqrt{\mu_s gr} \\ &= \sqrt{(1.2)(9.8 \text{ m/s}^2)(62 \text{ m})} \\ &= 27 \text{ m/s } (= 97 \text{ km/h} = 60 \text{ mi/h}) \end{aligned}$$

Hence, if the truck exceeds 60 mi/h, it will not be able to negotiate the curve and will skid outward, since friction cannot supply the necessary centripetal force.

Coefficient of Kinetic (Sliding) Friction

When the applied force on an object exceeds the maximum static force of friction, the object moves, and the motion is resisted by the force of sliding or kinetic friction f_k. This frictional force is also proportional to the load:

$$\boxed{f_k = \mu_k N} \qquad \textbf{(Eq. 7.9)}$$

where μ_k is the coefficient of kinetic or sliding friction. As can be seen from Table 7.1, μ_k is generally less than μ_s. Thus, in most cases, it takes more applied force to get an object moving than it does to keep it moving.

For relatively slow speeds, the force of kinetic friction (and the coefficient of kinetic friction) is generally independent of the speed. However, as the speed increases, the frictional force between the contacting surfaces decreases. For convenience in our discussion, we shall assume sliding friction to be independent of speed.

EXAMPLE 7.8 A 10-kg wooden crate slides down a steel ramp with an incline of 20° under the influence of gravity (Fig. 7.15). What is the acceleration of the crate?

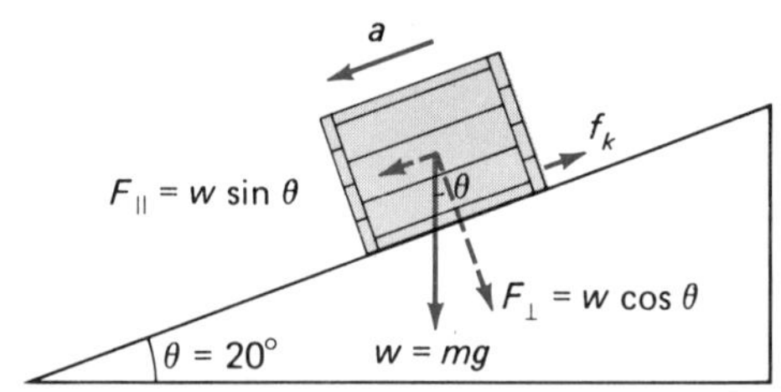

Figure 7.15
Kinetic friction opposes a crate sliding down a ramp. See Example 7.8.

Solution: From Table 7.1, $\mu_k = 0.30$. The weight of the crate is

$$w = mg = (10 \text{ kg})(9.8 \text{ m/s}^2) = 98 \text{ newtons (N)}$$

The component of the weight force acting down the plane is then

$$F_{\parallel} = w \sin 20° = (98 \text{ N})(0.34) = 33 \text{ N}$$

The load is equal to the component of the weight force perpendicular to the ramp surface, which is equal to the normal force:

$$F_{\perp} = N = w \cos 20° = (98 \text{ N})(0.94) = 92 \text{ N}$$

The force of sliding friction opposing the motion is then

$$f_k = \mu_k N = (0.30)(92 \text{ N}) = 28 \text{ N}$$

By Newton's second law,

$$F_{\parallel} - f_k = ma$$

and the acceleration is

$$a = \frac{(F_{\parallel} - f_k)}{m} = \frac{(33 \text{ N} - 28 \text{ N})}{10 \text{ kg}} = 0.50 \text{ m/s}^2$$

The coefficient of kinetic friction can be determined experimentally by adjusting an inclined surface so an object slides down the incline with a slow uniform speed. In this case, with no acceleration, the component of the weight force of the object down the plane is equal in magnitude to the frictional force up the plane, i.e.,

$$F_{\parallel} = f_k$$

or

$$w \sin \theta = \mu_k w \cos \theta$$

and

$$\mu_k = \frac{\sin \theta}{\cos \theta} = \tan \theta$$

The coefficient of kinetic friction is then equal to the tangent of the angle of the incline when the object slides down the plane with a uniform velocity.

Rolling Friction

This is another important type of friction that is often overlooked. Rolling friction occurs when one surface rotates and does not slip or slide at the point or area of contact with the other surface. This friction is primarily due to the deformation of materials. The load of a rolling object may deform a surface. At the depression region, the displaced material builds up in front of the rolling object, which must climb this "mound" of material to continue in motion. Thus, for very hard objects and surfaces, rolling friction may be quite small.

It was early discovered that it is much easier to move a load by putting it on rollers (or wheels) than by sliding it. This implies that rolling friction is much less than sliding friction. Ball bearings and roller bearings are used in machinery to take advantage of this property.

Rolling is an important consideration in terms of distance and control in bringing an automobile to a stop. Many newer cars and trucks have "anti-lock" brakes for rolling, rather than sliding, stops. See Special Feature 7.2.

7.6 Air Resistance

Although generally neglected in our previous studies of motion, air resistance, or air friction, is an important consideration in many cases. Cars, for instance, are streamlined so as to reduce this frictional "drag" and improve fuel economy. For some falling objects, the effects of air resistance are quite noticeable. For example, a feather "floats," or falls slowly, because of the upward air resistance. A common lecture demonstration is illustrated in Figure 7.16. With air in the tube, the heavier coin always beats the feather to the bottom. However, if enough air is evacuated from the tube, giving a partial vacuum, the air resistance is negligible and the feather and coin fall at the same rate.

A similar experiment with a hammer and a feather was performed on the moon in 1971 by astronaut David Scott. The moon has no atmosphere, and hence no air resistance. When released simultaneously, the hammer and feather fell at the same rate and hit the lunar surface at the same time. Of course, they fell quite a bit more slowly than they would in a vacuum chamber on Earth, since the acceleration due to gravity on the moon is one-sixth that on Earth.

The air resistance on a falling body depends on its (1) shape, (2) size (exposed area), and (3)

SPECIAL FEATURE 7.2

Anti-Lock Brakes

When a person driving a car experiences a situation requiring an emergency stop, it is an instinctive reflex to jam on the brakes and lock the wheels. The car then slides, often with a loss of control. With the car sliding, the force of kinetic (sliding) friction acts on the wheels.

To prevent sliding on wet or icy roads, people quickly learn that it is better to pump the brakes so that the car rolls, rather than slides, to a stop. This affects the stopping distance as well as the amount of control.

On this same principle, some newer cars are equipped with anti-lock braking systems (ABS). Electronic and hydraulic electric brake components of such a system supplement the main braking system, helping to prevent wheel lock-up and sliding. The ABS does not replace the main system; it is not activated by normal braking. But on hard braking, the ABS senses and alters the brake line pressure, allowing the wheels to continue to roll during braking. Basically, a control computer does the brake "pumping" for you, even though you are pushing hard on the pedal.

To get an idea of the difference in the stopping distances between rolling and sliding stops, let's take a case where the rolling friction is negligible. Static friction then acts on a rolling wheel, as opposed to the kinetic friction that occurs when the wheel is sliding. (The work of slowing the car is done by friction between the brake pads and disks or drums, but this happens inside the system. The *external* force of static friction determines the rolling condition of the wheels.)

Considering a constant force, in coming to a stop ($v_f = 0$) we have

$$v_f^2 = 0 = v_o^2 - 2ax$$

where the acceleration is negative. (Why?) Then the stopping distance is

$$x = \frac{v_o^2}{2a}$$

By Newton's second law,

$$\Sigma F = f = \mu N = \mu mg = ma$$

and the stopping acceleration has a magnitude of

$$a = \mu g$$

Hence, we have

$$x = \frac{v_o^2}{2\mu g}$$

Recall that the coefficient of kinetic (sliding) friction is generally less than the coefficient of static friction ($\mu_k < \mu_s$). For example, for rubber on wet concrete, $\mu_k = 0.60$ and $\mu_s = 0.80$. Considering this, what does the preceding equation tell you about sliding versus rolling to a stop?

speed.* The amount of air a body "catches" depends on its shape. For example, a sky diver with an unopened parachute falls quite rapidly. But when the chute opens, the shape and size of the falling "body" change, so the air resistance is increased and the descent is slowed, which makes skydiving a pleasure, or at least repeatable.

The effect of shape can be observed by moving your hand through water. With the palm of the hand in the direction of motion, there is more resistance than if you have the tips of the fingers in the direction of motion. Automobiles are now streamlined in shape to reduce air resistance and improve fuel consumption.

Air resistance is also velocity- or speed-dependent. That is, the greater the speed of a falling object, the greater the air resistance. This is because the faster an object falls, the greater the number of air molecules it hits or collides with per second. This gives a greater total resistive impulse force, or air resistance. In many cases, the air resistance is proportional to the square of the speed.

So, as an object falls, it gains speed—accelerates due to gravity—and the retarding force of air resistance increases with speed (Fig. 7.17).

* The air density is also a factor, but we will assume this to be relatively constant near the Earth's surface.

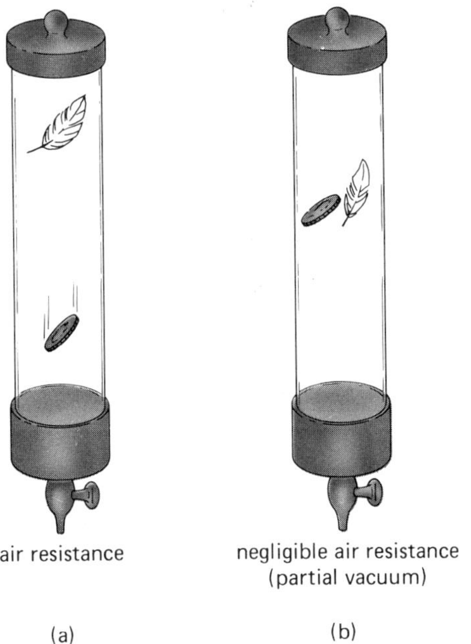

Figure 7.16
Air resistance. (a) When falling in air, a feather's motion is retarded more than a coin. (b) If most of the air is removed from the tube, the coin and feather fall together.

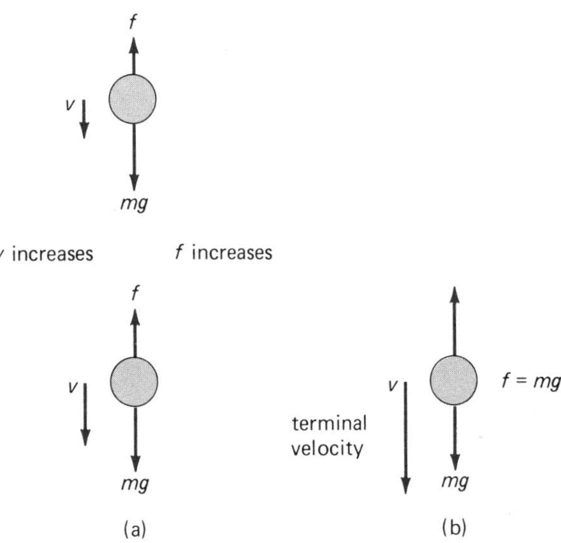

Figure 7.17
Air resistance is velocity-dependent. (a) As the velocity of a falling object increases, the force of air resistance f increases. (b) At a certain velocity, called the terminal velocity, the force of air resistance is equal to the weight of the object. There is then no longer a net acceleration and the object falls with a constant speed.

This continues until the force of air resistance equals the weight force of the falling object. The net force is then zero (downward weight force balanced by the upward force of air resistance), and the object no longer accelerates, but falls with a constant speed. We then say that the object has reached **terminal velocity.** For a sky diver with an unopened parachute, the terminal velocity is about 200 km/h (125 mi/h).

Sky divers are aware of the shape-dependence of air resistance. In falling they use a "spread-eagle" position to increase the air resistance and prolong the fall (Fig. 7.18). When the parachute is opened, the fall is slowed by the additional resistive force to a terminal velocity of about 40 km/h (25 mi/h). See Special Feature 7.3 for the effect of air resistance on falling objects of different weights.

Figure 7.18
Air resistance in action. Sky divers assume a "spread-eagle" position to increase the air resistance and prolong the time of fall.

Important Terms

universal gravitational constant (G) the constant of proportionality in Newton's law of gravitation, which is believed to have the same value throughout the universe, $G = 6.67 \times 10^{-11}$ N-m²/kg²

weight the gravitational force on an object by a celestial body, usually Earth

apparent weightlessness a condition in which a body and its reference frame are accelerating ("falling") at the same rate, and weight, or the force due to gravity, is not perceived; sometimes incorrectly re-

SPECIAL FEATURE 7.3

Falling Objects and Terminal Velocity

The effect of air resistance on falling objects is to retard their motion. The magnitude of the air resistance on an object depends on the object's shape, size, and speed. Ordinarily, heavy objects of similar shape and size dropped short distances appear to accelerate at the same rate and strike the ground at the same time. This is the essence of Galileo's alleged Leaning Tower of Pisa experiment (see Special Feature 7.1).

However, suppose the objects were dropped from a very high altitude so that speed and air resistance were appreciable factors. For example, if two metal balls of the same size and shape but made of different materials, one much more massive (heavier) than the other, were dropped from a high-altitude balloon (Fig. 1), which would strike the ground first?

Here, air resistance would be a factor, and the heavier ball would hit the ground first. As the falling balls gained speed and the air resistance increased, the weight of the lighter ball would be the first to be balanced by the force of air resistance. The heavier ball would continue to accelerate downward until it reached its greater terminal velocity. So the heavier ball would be ahead of the lighter ball and would be falling faster, thereby reaching the ground first.

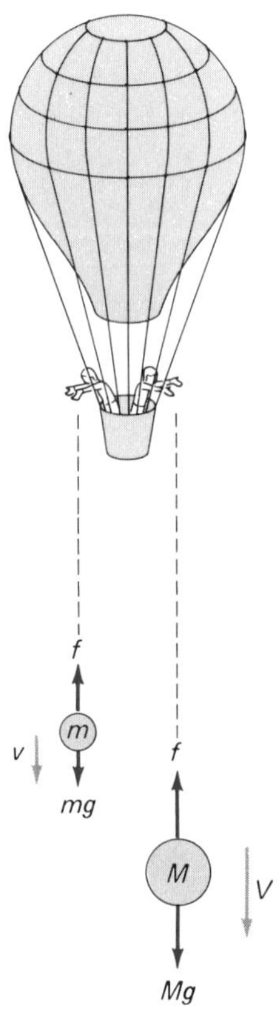

Figure 1
Air resistance and terminal velocity. Because of air resistance, a lighter (less massive) ball will reach its slower terminal velocity before a simultaneously released heavier ball of identical shape and size. As a result, the heavier ball (shown as larger for visual purposes) will strike the ground first.

---

ferred to as "zero gravity" [true weightlessness would require true zero gravity ($g = 0$)]

friction the resistance to relative motion between contacting material surfaces or within a medium

coefficient of friction the dimensionless quantity that relates the frictional force (f) to the load or normal force (N), that is, $\mu = f/N$

terminal velocity the velocity of a falling object at which the magnitude of the force of air (resistance) friction equals the magnitude of the weight (gravitational) force

Important Formulas

Newton's law of gravitation:

$$F = \frac{Gm_1m_2}{r^2}$$

$$(G = 6.67 \times 10^{-11}\ \text{N-m}^2/\text{kg}^2)$$

acceleration due to gravity at the Earth's surface:

$$g = \frac{GM_e}{R_e^2}$$

weight as a function of an object's altitude above the Earth:

$$w = mg(h) = \frac{GmM_e}{(R_e + h)^2}$$

frictional force:

$$f = \mu N$$

Questions

Newton's Law of Gravitation

1. G is a *universal* gravitational constant. Cite another universal constant. (*Hint:* consider the ratio of the circumference and the diameter of a circle.)
2. What are the value and units of G in the British system?

3. What is the magnitude of the force due to gravity between the Earth and the moon in pounds? (Cf. Example 7.2.)
4. The law of gravitation is an inverse-square law. Suppose it varied as (a) $1/r$ or (b) $1/r^3$. How would this affect weight on the surface of the Earth?
5. The Sun is more massive than the Earth (about a million times more). Why doesn't the gravitational attraction of the Sun pull us off the Earth?
6. Explain how Cavendish determined the value of G by measuring the force between two ordinary-size objects.
7. What are the units of a spring constant in SI units? How might the spring constant of a spring be determined?
8. Being an ultra-conscientious consumer, would you rather buy items by weight or by mass at the Dead Sea Supermarket? Why? How about the Top-of-Mt. Everest Supermarket?
9. If the law of gravitation were somehow repealed, what would happen to the solar system?
10. Would you expect to weigh more at full moon or at new moon? Why are these effects not detected?

A Closer Look at g

11. Does g decrease linearly with altitude? Explain.
12. If the Earth shrank to one-half its present size but kept its mass constant, what effect would this have on the acceleration due to gravity at the Earth's surface?
13. Suppose a hole were drilled through the center of the Earth to the other side. If you dropped a stone down the hole, what would happen? (Hint: remember that an object is gravitationally attracted toward the center of the Earth.)
14. An astronaut on a "moon walk" easily picks up a lot of equipment that would be too heavy for him on Earth (backpacks on the order of 300 lb, Fig. 7.19). Explain how this is possible.
15. Is it correct to speak about the weight of a planet? Explain.
16. Is the acceleration due to gravity the same at the "surfaces" (some are gaseous) of all the planets?

Apparent Weightlessness

17. How far from Earth would you have to go to be truly weightless?
18. Suppose you are standing on a chair and jump off. Are you weightless until you hit the ground? Explain.
19. If you were in an elevator standing on a scale and the elevator accelerated upward, what would the scale read and why? Describe the sensations you feel in an elevator when it (a) suddenly accelerates upward, (b) suddenly accelerates downward, and (c) moves with a constant speed.

Figure 7.19
Moon walk. See Question 14.

20. Suppose a spacecraft orbiting the Earth had an elevator. Describe the effects if an astronaut used the elevator to go up and down. Would an elevator be a good idea, and necessary?
21. Consider only the Earth and the moon. Is it possible for an astronaut on a moon trip to have zero gravity? Explain.
22. The plans for a new super space shuttle suggest that a handball or racquetball court be included so the astronauts can get exercise. Describe such a game with the space shuttle in orbit about the Earth.

Friction

23. Does asperity-welding theory apply to plastic materials? If not, why not?
24. Are the classical laws of friction completely valid? Explain.
25. Is the normal force always equal to an object's weight? Explain.
26. Explain how the frictional force between metal surfaces being independent of surface area is consistent with the asperity-welding theory.

27. When does the condition $f_s = \mu_s N$ hold?
28. Which is generally smaller, μ_s or μ_k, and what does this imply?
29. Referring to Table 7.1, explain why the coefficient of kinetic friction for aluminum on aluminum is so much greater than that for steel on steel. What would you expect for steel on aluminum?
30. How may (a) the coefficient of static friction and (b) the coefficient of kinetic friction be determined experimentally?
31. What is rolling friction and what causes it?

Air Resistance

32. Two sky divers jump at the same time from an airplane, after having agreed to open their parachutes at a particular altitude. One diver falls with his arms and legs pulled in (fetal position) and the other in a spread-eagle position. Which diver opens his parachute first?
33. Explain what effect an updraft of air would have on the terminal velocity of a falling object.
34. Snowflakes fall more slowly than sleet. Why?
35. Is it possible for a heavy object and a lighter object to have the same terminal velocity? Explain.
36. Suppose you live in the future, and moon vacations are common. Someone asks you to join a parachute club being formed for jumping on the moon. Would you sign up?

Problems

Levels of difficulty are indicated by asterisks for your convenience.

7.1 Newton's Law of Gravitation

1. A 4.0-kg mass and a 6.0-kg mass are separated by a distance of 3.0 m. What is the gravitational force on each mass?

*2. What is the gravitational attraction on you due to a mountain with a mass of 100 million metric tons 1 km away? (Consider yourself and the mountain to be particles.)

*3. A 60-kg student holds her 1-kg physics book at a distance of 0.30 m. How strongly is she (gravitationally) attracted to her book? (Consider the student and the book to be particles.)

*4. How massive would an object have to be for a person weighing 539 N ($\simeq$ 120 lb) to experience a gravitational attraction of 4.45 N (1.0 lb) when standing 10.0 m from the object? (Consider a uniform spherical object and the person as a particle.) Comment on the magnitude of this mass relative to that of Earth.

*5. The mutual force of gravitational attraction between two identical spherical masses is 2.03×10^{-5} N when their centers are separated by a distance of 3.30 m. What is the mass of the spheres?

*6. A spring has a force constant of 250 N/m. If a 5.0-kg mass is suspended on the spring, how far would it be stretched?

*7. A force of 10 N stretches a particular spring 0.025 m. If a 1.5-kg mass is suspended on the spring, how far is the spring stretched?

*8. What is the magnitude of the gravitational force between the Sun ($M_s = 2.0 \times 10^{30}$ kg) and the Earth ($M_e = 6.0 \times 10^{24}$ kg)? (Consider the Earth to be 93 million miles or 1.5×10^8 km from the Sun.) Compare this to the force between the Earth and the moon.

**9. Two uniform metal spheres with masses of 100 kg and 150 kg have radii of 6.0 cm and 8.0 cm, respectively. (a) If their surfaces are separated by a distance of 0.86 m, what is the gravitational force of attraction between the spheres? (b) What is the maximum possible force of attraction between the spheres?

7.2 A Closer Look at g

10. What is the acceleration due to gravity at an altitude of $4R_e$ (four Earth radii) above the Earth's surface?

11. If the Earth shrank to one half of its diameter and lost one half of its mass, how would the acceleration due to gravity be affected?

12. Using Equation 7.3, show that $g = 9.8$ m/s^2 at the Earth's surface.

*13. An astronaut weighs 160 lb on Earth. (a) What is the astronaut's weight on a spacecraft orbiting the Earth at an altitude of 500 mi? (*Hint:* use a ratio.) (b) At what altitude would the astronaut's weight be one half of his weight on Earth? (Take the Earth's radius to be 4000 mi.)

*14. A spacecraft is in circular orbit about the Earth at an altitude of 1000 km. An astronaut inside the spacecraft weighs 165 lb on Earth. Even though he may experience apparent weightlessness, what is his actual weight at this altitude?

*15. What would be the initial acceleration of (a) a 2.2-kg object and (b) a 22-kg object dropped from 1.6×10^3 km above the Earth's surface?

*16. A satellite is in a synchronous circular orbit about the Earth. Its weight at this altitude is what percent of its weight on Earth?

*17. A satellite is in circular orbit about the Earth at an altitude of 1000 km. What are (a) the orbital speed of the satellite and (b) its period of revolution?

*18. At what altitude above the Earth would a satellite in circular orbit have an orbital speed of 21,600 km/h?

*19. The moon has a diameter of 2160 miles and a mass $\frac{1}{81}$ that of Earth. Show that an object's weight on the moon is approximately $\frac{1}{6}$ of the object's weight on Earth.

*20. The planet Saturn is about 95 times more massive than Earth, but the acceleration due to gravity on the surface of Saturn is about the same as g on Earth. What does this tell you about the size of Saturn? Approximately how many times greater is the radius of Saturn than that of Earth?

**21. Halley's comet on its most recent visit had a closest approach to the Sun (perihelion) of about 8.8×10^7 km. What was its approximate tangential speed at this point?

**22. Compute the mass of the Sun.

7.3 Apparent Weightlessness

*23. A 160-lb person in an elevator stands on a scale. What is the weight indicated by the scale if the elevator descends at a constant velocity of 4.0 ft/s?

*24. A person weighing 128 lb stands on a scale in a "jet" elevator. What is the person's weight, read on the scale, if the elevator (a) moves upward with an acceleration of 6.0 ft/s^2 and (b) moves downward with an acceleration of 10 ft/s^2?

*25. A 70-kg person stands on a scale in an elevator. What would be the apparent change in the mass and weight of the person if the elevator (a) moved upward with an acceleration of 2.8 m/s^2 and (b) moved downward with an acceleration of 10 m/s^2? Describe what would happen in part (b).

**26. At what rate would an elevator have to accelerate upward for a person in the elevator to have an apparent weight twice that of his or her true weight?

**27. A person with a weight of 637 N stands on a scale in an elevator with a downward acceleration. If the scale reads 585 N, what is the elevator's acceleration?

7.4–7.5 Friction

28. A 10-kg block sits on a level surface, and a horizontal force of 50 N is applied to the block. (a) If the coefficient of static friction between the block and the surface is 0.75, does the block move? (b) If not, how great a force would be required to move the block?

29. An 8.0-lb machine piece is at rest on a level surface. If the coefficient of static friction between the piece and the surface is 0.50, what applied horizontal force would be required to move the piece?

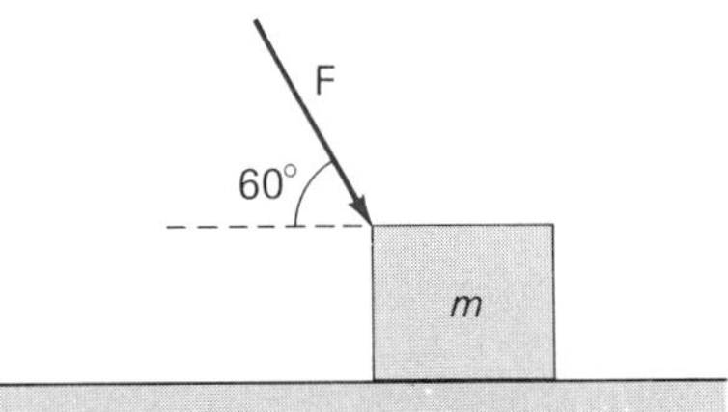

Figure 7.20
See Problem 32.

30. It is found that a horizontal force slightly in excess of 3.6 lb is required to move a 4.0-lb concrete block resting on a level wooden surface. What is the coefficient of static friction for the concrete and wooden surfaces?

*31. A force at an angle of 30° above the horizontal is applied to a 100-lb wooden crate on a wooden floor by means of a rope. What is the magnitude of the applied force necessary to move the crate? (See Table 7.1.)

*32. A force of 40 N applied to a 2.0-kg block at an angle of 60° to the horizontal moves the block along a level surface (see Fig. 7.20). If the coefficient of kinetic friction between the block and the surface is 0.27, what is the acceleration of the block?

*33. The coefficients of static and kinetic friction between a 16-lb concrete block and a level surface are 0.80 and 0.50, respectively. A horizontal force applied to the block just sets it in motion. If the force is maintained, what is the acceleration of the block?

*34. Refer to Figure 6.20 in the last chapter. (a) What is the minimum coefficient of static friction between the mass and the table that will keep the system stationary? (b) If the masses are set into motion and the coefficient of kinetic friction between the mass and the table is 0.40, what is the acceleration of the system? (Neglect the friction and mass of the pulley in both cases.)

*35. A large, wooden 250-kg packing crate is placed on a steel ramp with an incline of 37° by a loading rig. (a) Will the crate slide down the ramp? (See Table 7.1.) How much force must workers apply parallel to the ramp surface (b) to get the crate to move or to keep it from moving, and (c) to keep it moving at a constant velocity?

*36. A block weighing 44 N sits on an adjustable inclined plane. If the coefficient of static friction between the block and the plane is 0.65, above what angle of incline will the block start to move down the plane?

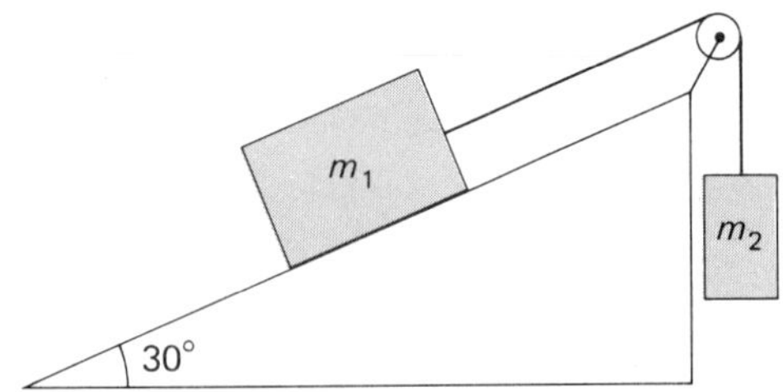

Figure 7.21
See Problems 37 and 39.

*37. Refer to Fig. 7.21. (a) If $m_1 = 20$ kg and $\mu_s = 0.85$ between the block and the inclined surface, what is the minimum mass m_2 that will cause m_1 to start to move up the plane? (Neglect the friction and mass of the pulley.) (b) If $m_2 = 1.5$ kg and $\mu_s = 0.45$ between the block and the inclined surface, what is the minimum mass for m_1 that will allow it to move down the plane?

*38. A wooden crate slides down a steel ramp that has an incline of 16° with a uniform speed of 1.0 m/s. What is the coefficient of kinetic friction between the crate and the ramp surface?

*39. Refer to Figure 7.21. If $m_1 = 15$ kg and $m_2 = 4.0$ kg, and the coefficient of kinetic friction between the block and inclined surface is 0.20, what is the acceleration of the system when the block slides on the plane? (Neglect the friction and mass of the pulley.)

*40. A 2.0-kg block sits at the top of a 30° inclined plane with a length of 3.0 m. If $\mu_k = 0.25$ between the block and the plane surface, how long will it take the block to slide down the plane when released?

*41. A skier on waxed wooden skis coasts freely from rest down a jumping ramp with a 45° incline. If it takes the skier 4.0 s to go down the ramp, what is her speed at the bottom of the ramp?

*42. What is the maximum safe speed a car can travel around a level circular curve on a dry concrete road with a radius of curvature of 150 ft?

*43. On a wet racetrack, the maximum safe speed to avoid sliding on a level circular curve with a radius of curvature of 50 m is 60 km/h. What is the coefficient of friction between the car's tires and the track surface?

*44. A car traveling at a constant speed of 90 km/h on a wet concrete road starts into a circular curve with a radius of curvature of 75 m. Will the car negotiate the curve? If not, what will happen?

**45. The proper safe speed for highway curves is an important consideration. The coefficient of friction is variable, depending on weather and highway conditions. In the design of a curved roadway, safety is promoted by banking or inclining the roadway (Fig. 7.22). The horizontal component of the normal force supplies the required centripetal force, and friction is not considered. (a) Show that, in the absence of friction, the banking angle for a given speed v is given by $\tan \theta = v^2/gr$, where r is the radius of curvature. (b) How would the frictional force enter into these considerations? (c) What is the proper banking angle for a speed of 30 mi/h on a curve with a radius of 200 ft in the absence of friction?

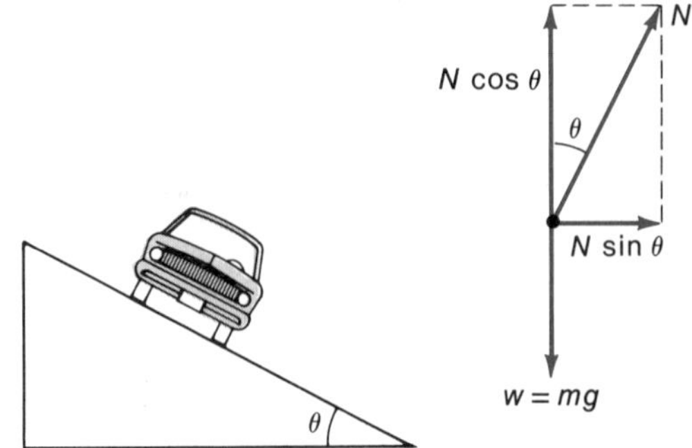

Figure 7.22
See Problem 45.

**46. An amusement park ride called the Rotor is shown in Figure 7.23. A person on the ride stands against the wall as the Rotor gradually increases its speed. After a certain critical speed is reached, the floor drops out, but the riders do not fall! (a) Show that the critical tangential speed is given by $v = (gr/\mu_s)^{1/2}$, where r is the radius of the Rotor and μ_s the coefficient of friction between a person and the Rotor wall. (b) Is the critical speed the same for an adult and a small child with the same coefficient of friction? (c) What effect might the clothes a person is wearing have on the critical speed? (d) If the radius of the Rotor is 8.0 m and the minimum safe coefficient of friction for all riders is taken to be 1.1, above what critical speed, in mi/h, is the floor dropped?

**47. Consider the air resistance on a falling object to be given by $f = \mu_a v$, where μ_a is the "coefficient of air resistance" and v the speed of the object. (a) Derive an expression for the terminal velocity v_t. (b) On what does μ_a depend? Show that it has units of kg/s. (c) What is the terminal velocity for a falling 70-kg sky diver if $\mu_a = 14$ kg/s? Express your answer in km/h and mi/h.

(a)

(b)

Figure 7.23
A Rotor ride. (a) Rotating below a certain critical speed. (b) Above the critical speed, the floor drops out, but riders do not fall. See Problem 46.

Chapter Supplement

Lubrication

In previous chapters, we generally considered frictionless surfaces for convenience. In reality, friction can be reduced but never eliminated.

To maintain equipment at maximum operating efficiency and to prolong the useful life of machine components, lubricants are used. That is, we usually oil or grease the components. Any substance used to reduce friction may be classified as a lubricant. Applications vary widely, but as a general rule:

> **Effective lubrication depends on the proper application of the proper amount of proper lubricant at the proper time.**

The selection of the proper lubricant depends on the particular situation. Some of the considerations in selecting a lubricant include

1. Proper fluidity (viscosity)
2. Chemical stability
3. Adhesiveness to surfaces
4. Sealing properties
5. Cost

A variety of substances are used as lubricants. Although the term *lubricant* usually brings to mind liquid oils and semiliquid greases, solids and gases are used as lubricants as well. There are many types of oils: mineral (petroleum) oils, animal oils, vegetable oils, and synthetic oils, e.g., silicone oils. Let's concentrate on oils, in particular automotive motor oils.

Viscosity is a fundamental aspect of liquid lubrication. As noted previously, viscosity is a measure of the resistance to fluid flow—the greater the viscosity, the slower the fluid flow, or the less the fluidity. Some liquids flow more readily than others at a particular temperature, and the same liquid will flow more readily when heated, i.e., the viscosity generally decreases with increasing temperature. Also, the flow of a liquid is less when the liquid is subjected to greater load pressures. Hence, the comparison and determination of viscosity are made under specific conditions. In general, viscosity determinations are made at given temperatures and at standard atmospheric pressure.

Figure 7S.1
A Seybolt viscometer used for the testing of petroleum products. A temperature bath maintains the test sample at constant temperature, and the motor drives a stirrer for oil circulation. (Courtesy Humbolt Mfg. Co.)

Viscosity measurements are made by means of viscometers. A common, standard such instrument for oils is the Seybolt universal viscometer (Fig. 7S.1). In this instrument, the test oil is heated to a given temperature by means of a water bath. The oil is then allowed to flow through an outlet into a receiving flask. The time, in seconds, for 60 cm^3 of oil to flow into the receiving flask is measured and reported as Seybolt Universal Seconds (SUS). Oils with SUS times of less than 32 seconds are not measured by this method.

The SUS measurement is a relative measure of viscosity, since the greater the viscosity of the oil, the more time it takes for the 60 cm^3 of oil to flow from the viscometer.

SAE Numbers

To simplify the recommendations for automotive motor oils, the Society of Automotive Engineers (SAE) established a numbering system based on a range of SUS units. The SAE viscosity numbers and the corresponding SUS ranges are shown in Table 7S.1. Thick oils, which flow slowly, have high numbers. Thin oils, which flow freely, have low numbers. In general, the higher the SAE number, the greater the viscosity. Each number represents a range of viscosity at a given temperature. From the table, the range for lower-viscosity oils is in SUS at 0°F, and for higher-viscosity oils the range is in SUS at 210°F.

Table 7S.1
SAE Viscosity Numbers of Oils

SAE Number	Viscosity Range,			
	0°F min	*0°F max*	*210°F min*	*210°F max*
Crankcase Oils				
5W	—	4000		
10W	6000	<12,000		
20W	12,000	48,000		
20			48	<58
30			58	<70
40			70	<85
50			85	110
Transmission and Axle Lubricants				
75	—	15,000		
80	15,000	100,000		
90			75	<120
140			120	<200
250			200	—

* SUS times below 32 seconds are not listed.

Because of seasonal temperature variations, it is important to use an oil of the proper SAE number or "weight" in one's automobile. In summer a "heavy" oil, e.g., SAE 30, should be used, and in winter a "light" oil, e.g., SAE 10W, should be used. The "W" denotes oils primarily for winter service—below 32°F (0°C).

Additives

Certain additives impart desired properties to lubricants. For example, sulfur is used in metal-cutting oils to prevent welding between metal parts. In general, the purposes of lubricant additives include

1. Retardation of deterioration of the lubricant
2. Improvement of viscosity ranges
3. Protection of lubricated surfaces against contaminants
4. Prevention of foaming

Motor oils are exposed to high-temperature operating conditions, which are conducive to chemical oxidation of the lubricating oil. Not all of the oxidation products are washed away by the circulating oil, and some tend to accumulate as a gummy conglomerate on engine parts.

To help disperse such residue, detergents are added to oils. These detergents are basically soaps that are soluble in oil. The detergent action is much the same as in dishwashing and laundering, where detergents are used to wet and disperse particles that water washing would not ordinarily remove. In engines, oil filters are used to filter out the dispersed oxidation products from the motor oil.

Zinc and phosphorus compound additives are used in oils as antioxidation agents and corrosion inhibitors. Corrosion in this instance refers to the etching of metal parts that may be caused by acids formed in oil oxidation processes.

An important, relatively new group of lubricant additives function as viscosity improvers. Adequate lubrication protection is required over a wide temperature range. The viscosity difference is relatively great between many minimum and maximum operating temperatures, e.g., the automobile oil temperature at a cold-weather start and the engine operating temperature in the piston region. Most engine wear occurs during warm-up.

To provide adequate lubricant flow over a wide temperature range, polymer additives are used, such as polyacrylates and polyisobutylenes. These polymeric viscosity improvers have long-chain, coiled molecules. With an increase in temperature, the molecules become uncoiled and intertwine with each other. In this manner, the normal decrease in viscosity is counteracted. With the reverse action on cooling, the oil maintains a relatively constant viscosity over a temperature range.

Polymeric additives are now commonly used in automatic transmission fluid oils and motor oils for automobiles. The viscosity improvers

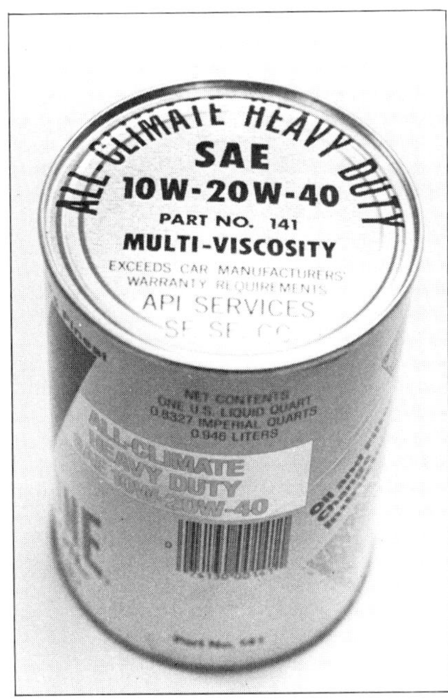

Figure 7S.2
Multigrade motor oil. Polymeric additives modify the viscosity of a motor oil so it can be used over several SAE ranges. The multi-viscosity SAE 10W–20W–40 fulfills the operating range requirements that would normally require several oils of different viscosities.

can modify the viscosity of motor oils so that the range between the high and low temperature limits spans the specifications of two or more SAE viscosity ranges.

Such modified oils are referred to as multigrade motor oils. For example, if such a multigrade oil meets the minimum viscosity requirement of SAE 10W (0°F) and the maximum viscosity requirement of SAE 40 (210°F), then it fulfills the operating range requirements that would normally require five oils. This is indicated by the multigrade designation of SAE 10W-20W-40, or SAE 10W-40 for short (Fig. 7S.2). This multigrade oil can be used in place of SAE 10W, SAE 20W, SAE 20, SAE 30, and SAE 40 oils, which must be changed seasonally.

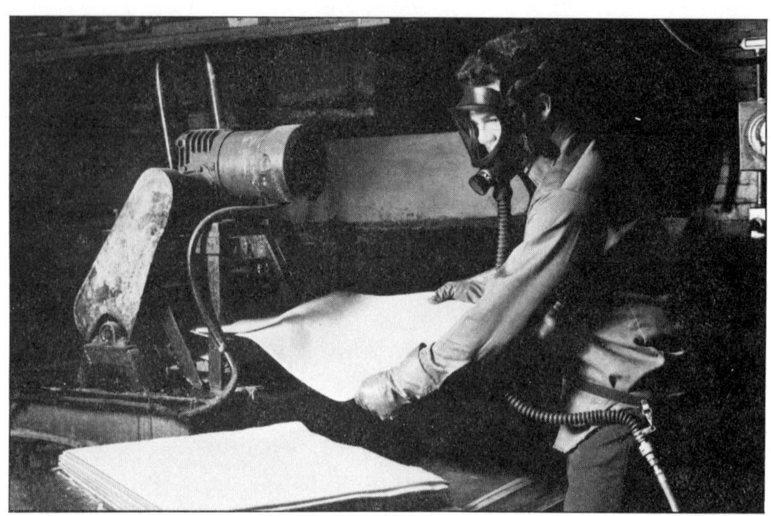

(Courtesy INCO Safety Products Company.)

Chapter 8

Work, Energy, and Power

When work is done, energy is expended and action occurs. And, as might be expected, forces are necessary in performing tasks or doing work. Less than 100 years ago, the majority of the force needed to do work was supplied by humans and animals, in particular the horse. However, in our modern society machines have all but replaced the horse. Power tools driven by various energy sources aid us in doing the work previously done by hand. Today, a large part of the labor force is "white-collar" workers, who supply more mental work than manual labor.

In technical fields, the measurement of how much work is done in a particular situation is often an important consideration. For example, an engineer must know the work capacity of a machine, its energy requirements, and how fast it can perform work, i.e., the power output of the machine. These basic concepts will be defined and explained as you "work" on this chapter.

8.1 Work

People tend to measure the work they do by how tired they become or by the amount of time spent in performing a task. However, two persons may be equally tired after doing different amounts of work. This concept of work is not applicable to machinery, which does not become "tired." Hence, a more standard technical definition of work is required.

Technically, work is done on an object by an applied force, and

Work is generally defined as the product of the magnitude of the applied force *F* (or a component thereof) and the *parallel* displacement (distance) *d* through which the force acts, i.e.,

$$\boxed{W = Fd} \qquad \textbf{(Eq. 8.1)}$$

$$\text{(work)} = \text{(force)} \times \text{(distance)}$$

This technical definition of work requires motion, since if an applied force acting on an object acts through a distance, the object is moved. Notice that the source of the force is not specified. It may be mechanical, human, gravitational, or in some other form. If the force is not constant,

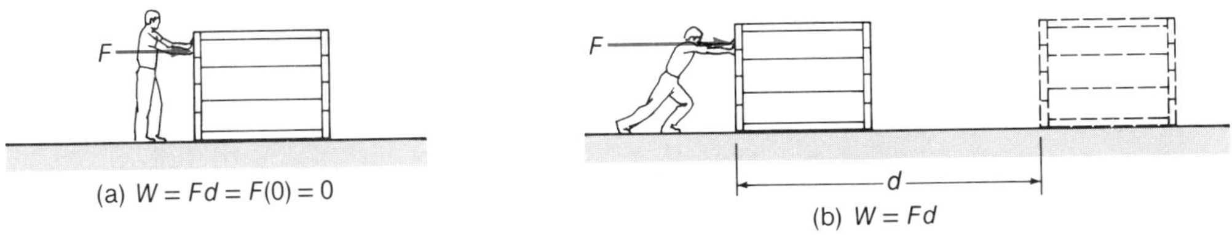

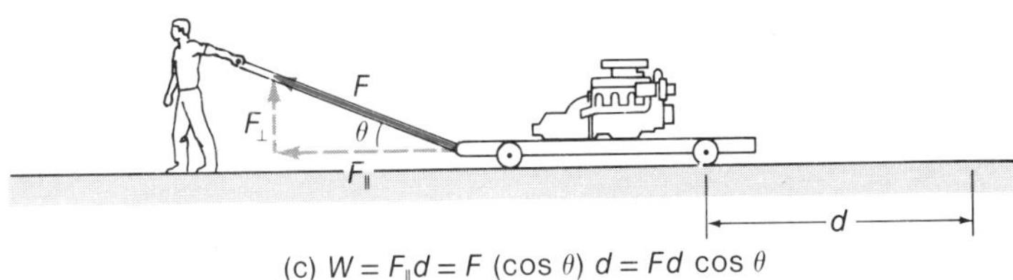

Figure 8.1
Work is done when a force acts through a distance. (a) A force is applied to the crate, but it doesn't move; hence, no work is done. (b) The applied force F acts through or moves the crate a parallel distance d, and the work done is $W = Fd$. (c) In moving the load through a distance d, only the component of the force parallel to the displacement, $F_\parallel = F \cos \theta$, does work.

then we consider the work done by the average force, as will be discussed shortly.

To see how this definition of work is applied to different situations, consider the cases illustrated in Figure 8.1. In the first case, the crate is not moving, and no work is done. Although there is an applied force, it does not act through a distance, i.e., $d = 0$ and $W = Fd = 0$.

In the second case, an applied force acts through a distance and moves the crate, so the work is $W = Fd$. Notice that the force and the distance through which the force acts are parallel.

In the third case, the applied force is not parallel to the displacement, but there is a *component* of the force that is parallel to the directed distance or displacement. As can be seen in the figure, this parallel component is $F_\parallel = F \cos \theta$. Hence, the work is

$$W = F_\parallel d = F(\cos \theta)d$$

or

$$W = Fd \cos \theta \qquad \textbf{(Eq. 8.2)}$$

The vertical component of the force $F_\perp$ does no work, since there is no displacement of the load in this direction.

Units of Work

The units of work are readily obtained from the defining equation, $Fd = W$.

The unit of work in the SI system by definition is the newton-meter (N-m):

$$\underset{\text{(newton)}}{F} \times \underset{\text{(meter)}}{d} = \underset{\text{(N-m)}}{W}$$

However, the derived unit called the **joule (J)** (pronounced "jewel") in honor of the English scientist James Prescott Joule, 1818–1889, is commonly used in place of the newton-meter, i.e.,

$$1 \text{ joule} = 1 \text{ newton-meter}$$

$$1 \text{ J} = 1 \text{ N-m}$$

In the British fps (foot-pound-second) system, we have

$$\underset{\text{(lb)}}{F} \times \underset{\text{(ft)}}{d} = \underset{\text{(ft-lb)}}{W}$$

Hence, the British unit of work is the foot-pound (ft-lb). Although the standard form of the equation gives lb-ft, the unit is expressed in the reverse order as ft-lb. A force of one pound acting through a distance of one foot does one foot-pound of work.

See Table 8.1 for a comparison of units.* These are also the units of energy, as will be seen shortly when the relationship between work and energy is described.

EXAMPLE 8.1 A worker applies a horizontal force of 75 lb to move a crate a distance of 6.0 ft along a level floor (case b, Fig. 8.1). How much work is done?

Solution: With $F = 75$ lb and $d = 6.0$ ft,

$$W = Fd = (75 \text{ lb})(6.0 \text{ ft}) = 450 \text{ ft-lb}$$

EXAMPLE 8.2 A load on a skid is moved on a lift truck by a worker who pulls on the lift handle at an angle of 30° to the horizontal with a force of 150 N (case c, Fig. 8.1). If the skid is moved a distance of 10 m, how much work is done?

Solution: The component of force parallel to the displacement is $F_{\parallel} = F \cos \theta$. Then,

$$W = F_{\parallel}d = F(\cos 30°)d$$
$$= (150 \text{ N})(0.866)(10 \text{ m}) = 1.3 \times 10^3 \text{ N-m}$$

Work is a scalar quantity (magnitude only). However, because of the vector directions of the force and displacement, work may be either positive or negative. When the force and displacement are in the same direction, the work is positive; when they are in opposite directions, the work is negative.

Table 8.1
Units of Work (and Energy)

1 J = 0.738 ft-lb = 10^7 erg
1 ft-lb = 1.36 J = 1.36×10^7 erg
1 erg = 10^{-7} J = 7.38×10^{-8} ft-lb

* In the cgs system, the unit of work is the dyne-centimeter: F (dyne) × d (centimeter) = W (dyn-cm)

Here again, a derived unit, the erg, is equivalent to a dyn-cm: 1 erg = 1 dyn-cm

As seen from the table, the erg is a relatively small unit.

Negative Work

In the previous examples, the work was positive since the force and the displacement were in the same direction. When the displacement of an object is in the opposite direction to a particular force, the work is negative, and we say that work is done *against* the force. An example of this is the work done against the force of friction. Suppose that the crate in Example 8.1 were moved with a constant velocity by the applied force. In this case, with the acceleration being zero the net force must be zero, so $F = f =$ 75 lb, where the frictional force f is equal (and opposite) to the applied force F (f is not shown in Fig. 8.1). The applied force does $W = Fd =$ 450 ft-lb of work. The frictional force is also "applied" to the load, and the work done by the frictional force is $W_f = -fd = -450$ ft-lb. The minus sign indicates that the frictional force is in the opposite direction to the displacement. Hence, we say the applied force does work against the force of friction.

The total or net work (W_T) in this case is zero, since the net force on the crate is zero.

$$W_T = W - W_f = Fd - fd = (F - f)d = 0$$

since $F = f$ in magnitude for a constant velocity.

EXAMPLE 8.3 How much work is required to lift a 5.0-kg box a vertical distance of 1.0 m (Fig. 8.2)?

Solution: To lift the box, a vertical upward force must be applied equal to the box's weight (actually slightly greater; why?). With $m = 5.0$ kg, the weight of the box is

$$w = mg = (5.0 \text{ kg})(9.8 \text{ m/s}^2) = 49 \text{ N}$$

Then, in lifting the box a distance of 1.0 m, the work done is

$$W = Fd = (49 \text{ N})(1.0 \text{ m}) = 49 \text{ J}$$

This is the same amount of (negative) work done *against* the gravitational weight force since w is downward and in the opposite direction to the displacement.

However, there is an important difference between this case and the frictional case. The

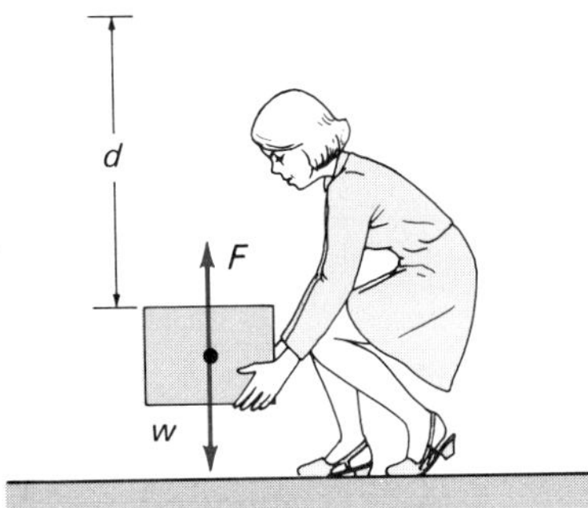

Figure 8.2
Work is done against gravity in lifting an object. See Example 8.3.

work done in raising the box is "stored," so to speak, as will be learned in the next section.

Work Done by Variable Forces

In the previous discussion, a constant applied force was used. However, in many applications, the applied force may be changing. In most such instances, advanced mathematics is required to calculate the work done by variable forces. But, if the average force ($\overline{F}$) can be determined, then the (average) work done by the variable force is

$$W = \overline{F}d \qquad \textbf{(Eq. 8.3)}$$

An example of a variable force that is relatively easy to analyze in terms of the average force is the spring force. Recall from Chapter 6 (Eq. 6.2) that the spring force is given by

$$F = -kx$$

where k is the spring constant and expresses the "stiffness" of the spring, and x is the distance the spring has been compressed or extended from its equilibrium position x_o, i.e., $\Delta x = x - x_o = x$, with $x_o = 0$. The negative sign indicates that the spring force is in the direction opposite to the displacement.

As can be seen from the equation, the spring force is a variable force—the greater the displacement, the greater the force. Since the force is a linear function of displacement with a constant k, the average force $\overline{F}$ can be calculated in the same way as the average velocity $\bar{v}$ was calculated in Chapter 4 where v was a linear function of time with a constant acceleration. This is shown graphically in Figure 8.3 for the applied force, which is opposite to the spring force. This avoids using the minus sign. (Why?) In the limit, the two forces are equal in magnitude.

The average force is then

$$\overline{F} = \frac{F_f + F_o}{2} = \frac{kx + 0}{2} = \tfrac{1}{2}kx$$

where $F_o = 0$ at $x_o = 0$. This is the average force applied to the spring in compressing or extending it (F and x in the same direction).

Then, the work done in compressing or extending the spring by an applied force is

$$W = \overline{F}x = \tfrac{1}{2}kx^2 \qquad \textbf{(Eq. 8.4)}$$

This is the work done by the applied force *against* the spring force.

Notice from Figure 8.3 that the work done is equal to the area under the F versus x curve, in this case a triangle. The area of a triangle is $A = \frac{1}{2}$(the altitude) $\times$ (the base), or $A = \frac{1}{2}ab$. From the figure, the altitude of the force triangle is $a = F = kx$ and the base is $b = x$. Hence, $W = \frac{1}{2}ab = \frac{1}{2}kx^2$, which is the area of the triangle or the area under the curve.

EXAMPLE 8.4 A heavy-coil spring is extended 0.010 m when a mass of 5.0 kg is suspended from it. How much work is required to compress the spring 0.15 m from its equilibrium position?

Solution: The extension condition allows computation of the spring constant. In equilibrium, the magnitude of the spring force is equal to the suspended weight, and

$$k = \frac{F}{x} = \frac{mg}{x} = \frac{(5.0\text{ kg})(9.8\text{ m/s}^2)}{0.010\text{ m}}$$
$$= 4.9 \times 10^3\text{ N/m}$$

Then, the work required to compress the spring a distance of $x = 0.15$ m is

$$W = \tfrac{1}{2}kx^2 = \tfrac{1}{2}(4.9 \times 10^3\text{ N/m})(0.15\text{ m})^2$$
$$= 55\text{ N-m} = 55\text{ J}$$

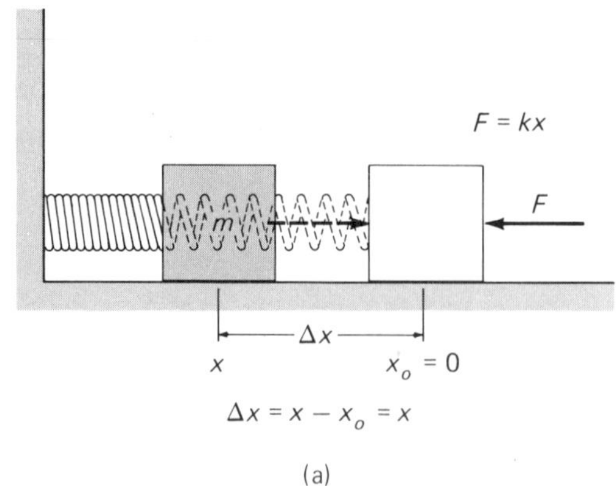

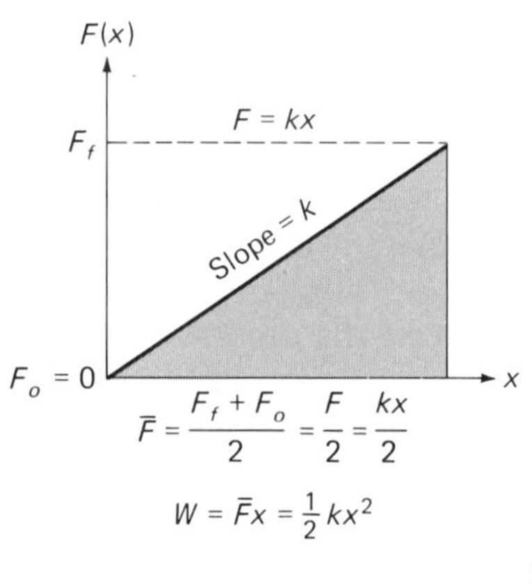

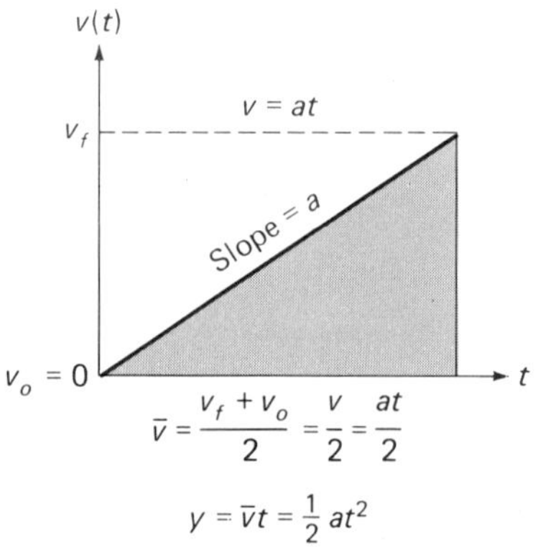

Figure 8.3
Variable spring force. (a) The force needed to compress (or extend) a spring is a variable force, $F = kx$ ($x_o = 0$). The force exerted by the spring is in the opposite direction to the displacement. (b) The work done against the spring force is $W = \frac{1}{2}kx^2$. The average force $\bar{F}$ is graphically analogous to the average speed $\bar{v}$ of an object with a constant acceleration or a uniformly varying velocity. Notice that the work is equal to the area under the curve or the area of the shaded triangle.

8.2 Energy

Energy (Greek *energeia*; *en* = in, *ergon* = work) is closely associated with work. In fact, energy is sometimes defined as the ability to do work. That is, **when a body possesses energy, it has the capability to do work.** There are many forms of energy, e.g., gravitational, chemical, electrical, and nuclear. However, in this initial discussion we will consider energy primarily in the mechanical sense.

Kinetic Energy

Kinetic energy is the energy of motion. An object in motion possesses kinetic energy and hence has the ability to do work.

Work is required to set an object into motion or to change its velocity (an acceleration), and this action requires a force acting through a distance. Suppose a constant, unbalanced force acts on an object. From our study of force and motion, we know that the instantaneous velocity of the object is given by Equation 4.8,

$$v_f^2 = v_o^2 + 2ax$$

The work done by an applied force ($F = ma$ in terms of Newton's second law) acting through a distance x can then be expressed

$$W = Fx = max = \frac{m(v_f^2 - v_o^2)}{2} = \tfrac{1}{2}mv_f^2 - \tfrac{1}{2}mv_o^2 \qquad \textbf{(Eq. 8.5)}$$

The resulting product of one-half times the mass times the square of the instantaneous velocity is defined to be the instantaneous motional or kinetic energy (KE) of the object (Fig. 8.4):

$$\boxed{KE = \tfrac{1}{2}mv^2} \qquad \textbf{(Eq. 8.6)}$$

Hence, Equation 8.5 tells us that the work done by an unbalanced force acting on an object is equal to the change in the kinetic energy of the object:

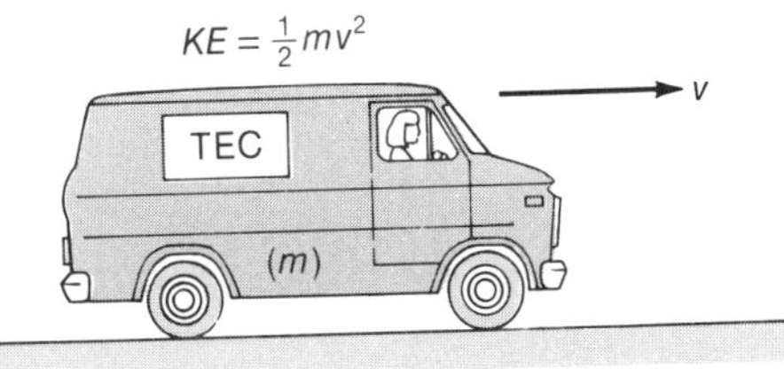

Figure 8.4
Kinetic energy is the energy of motion. The kinetic energy of a moving object is equal to one-half the product of its mass and the square of its velocity, $KE = \frac{1}{2}mv^2$.

$$W = \tfrac{1}{2}mv_f^2 - \tfrac{1}{2}mv_o^2$$
$$W = KE_f - KE_o = \Delta KE \qquad \textbf{(Eq. 8.7)}$$

This expression is called the **work-energy theorem,** and it relates work to energy. Notice that the units of energy are the same as those of work.

Also note that if the change in kinetic energy (ΔKE) is positive, then the work is positive and the force acts to speed up the object. If the change in kinetic energy is negative, then the work is negative and the force acts to slow the motion (force and displacement in opposite directions).

EXAMPLE 8.5 A 3.0-kg object moving initially with a velocity of 2.0 m/s is acted upon by a force, and the velocity is increased to 6.0 m/s. (a) What is the final kinetic energy of the object? (b) How much work was done by the applied force? (c) How much work is required to bring the object to a halt?

Solution: It is given that m = 3.0 kg, v_o = 2.0 m/s, and v_f = 6.0 m/s.

(a) The final kinetic energy of the object is

$$KE_f = \tfrac{1}{2}mv_f^2 = \tfrac{1}{2}(3.0 \text{ kg})(6.0 \text{ m/s})^2 = 54 \text{ J}$$

(b) The work done by the applied force is equal to the change in the kinetic energy of the object (work-energy theorem):

$$W = \Delta KE = KE_f - KE_o = \tfrac{1}{2}mv_f^2 - \tfrac{1}{2}mv_o^2$$
$$= 54 \text{ J} - \tfrac{1}{2}(3.0 \text{ kg})(2.0 \text{ m/s})^2$$
$$= 54 \text{ J} - 6.0 \text{ J} = 48 \text{ J}$$

Notice that the change in kinetic energy involves calculating the kinetic energy for each velocity and subtracting; it is the *difference in the squares of the velocities*, $\frac{1}{2}m(v_f^2 - v_o^2)$. This is not the same as taking the difference in the velocities and then *incorrectly* computing the change in kinetic energy with the change in velocities since $(v_f - v_o)^2 \neq v_f^2 - v_o^2$.

Also notice that we do not have to consider whether the force is constant or variable. In general, only the instantaneous velocities and kinetic energies are needed.

(c) The work required to bring the object to rest cannot be calculated from an applied braking force, since this is not given. But, considering the initial velocity to be v_o = 6.0 m/s and the final velocity to be zero, v_f = 0, we have by the work-energy theorem

$$W = KE_f - KE_o = -KE_o$$
$$= -\tfrac{1}{2}mv_o^2$$
$$= -\tfrac{1}{2}(3.0 \text{ kg})(6.0 \text{ m/s})^2 = -54 \text{ J}$$

The work is negative because energy is lost. (The braking force and the displacement of the object are in opposite directions.)

Another way of looking at this situation is that the moving object has 54 J of energy that can do work. For example, suppose a moving car runs into the fender of another car and comes to a stop. The kinetic energy of the moving car goes into doing work on (denting) the fender of the other car. This fender-bender case would not be considered an example of energy doing "useful" work.

Potential Energy

Another form of energy that lends itself to the idea that energy is "stored" work is potential energy. **Potential energy is the energy of an object due to its position.** That is, potential energy is associated with the location of an object, since work is required to move an object from one position to another. There are various forms of potential energy. We will consider two common types here—gravitational potential energy and the potential energy of a spring.

Gravitational Potential Energy

As was shown in Example 8.3, work is required to lift an object. Such a change in position

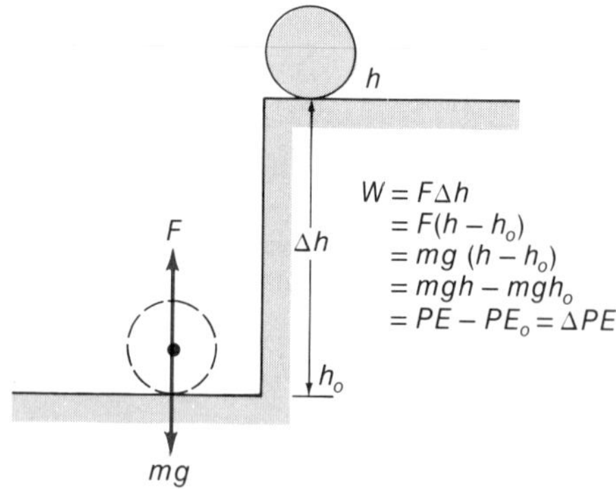

Figure 8.5
Gravitational potential energy. The work done in lifting an object is equal to the change in the potential energy of the object. Potential energy is the energy of position. At a height h, the object has more potential energy than at height h_o.

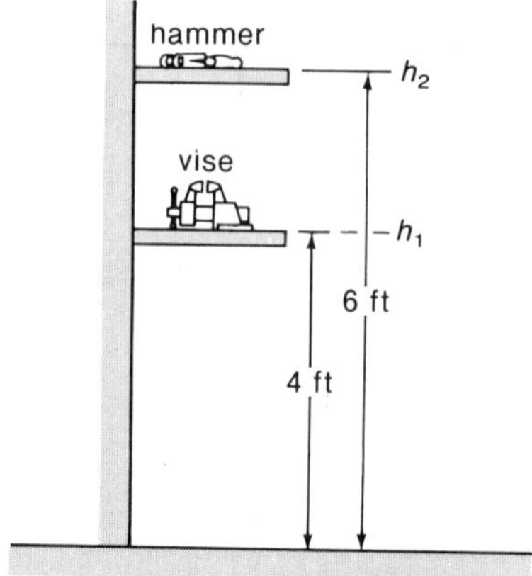

Figure 8.6
Potential energy is the energy of position. The hammer and vise have different potential energies relative to the floor. See Example 8.6.

(height) changes the potential energy (PE) of an object. The work is done against the force of gravity, $F = mg$. Thus, for the general case of lifting an object a distance Δh (Fig. 8.5),

$$W = F\Delta h = mg(h - h_o) = mgh - mgh_o$$

$$= PE - PE_o = \Delta PE \quad \textbf{(Eq. 8.8)}$$

If $h_o = 0$, as it is commonly taken to be, then

$$\boxed{PE = mgh = wh} \quad \textbf{(Eq. 8.9)}$$

where $w = mg$ is the weight of the object.

EXAMPLE 8.6 A 10-lb vise and a 3.0-lb hammer rest on shelves that are 4.0 ft and 6.0 ft high, respectively (Fig. 8.6). (a) What is the potential energy of each object relative to the floor? (b) How much work is required to lift the vise to the upper shelf, and what is its potential energy there?

Solution: (a) Using Equation 8.9 for the vise,

$$PE_v = m_v g h_1 = w_v h_1 = (10 \text{ lb})(4.0 \text{ ft}) = 40 \text{ ft-lb}$$

and for the hammer,

$$PE_h = w_h h_2 = (3.0 \text{ lb})(6.0 \text{ ft}) = 18 \text{ ft-lb}$$

(b) The work required in lifting the vise to the upper shelf, or through a distance $\Delta h = h_2 - h_1 = 6.0 \text{ ft} - 4.0 \text{ ft} = 2.0 \text{ ft}$, is

$$W = F\Delta h = w_v \Delta h = (10 \text{ lb})(2.0 \text{ ft}) = 20 \text{ ft-lb}$$

Notice that this is just the change in the potential energy of the vise,

$$W = w_v \Delta h = w_v(h_2 - h_1) = PE_2 - PE_1 = \Delta PE$$

so the potential energy of the vise is increased by 20 ft-lb. The vise's potential energy on the top shelf relative to the floor ($h_o = 0$) is then

$$40 \text{ ft-lb} + 20 \text{ ft-lb} = 60 \text{ ft-lb}$$

or, by direct calculation,

$$PE_v = w_v h_2$$
$$= (10 \text{ lb})(6.0 \text{ ft})$$
$$= 60 \text{ ft-lb}$$

Zero Reference

It should be noted that the position or reference point for zero potential energy, $PE = 0$ or $h = 0$, is arbitrary. That is, we may measure h from any point we like. For example, in Figure 8.7, *relative* to ground level at a height of $2h$, the mass has a potential energy of $PE_2 = mg(2h) = 2mgh$.

If the mass is at the bottom of a hole a distance $-h$ below the reference point, its potential energy is $PE_1 = mg(-h) = -mgh$, relative to ground level. The difference in the potential energies of the mass at the two positions is then

$$\Delta PE = PE_2 - PE_1 = 2mgh - (-mgh) = 3mgh$$

The same result without the negative potential energy can be obtained by measuring the height relative to the bottom of the hole. From

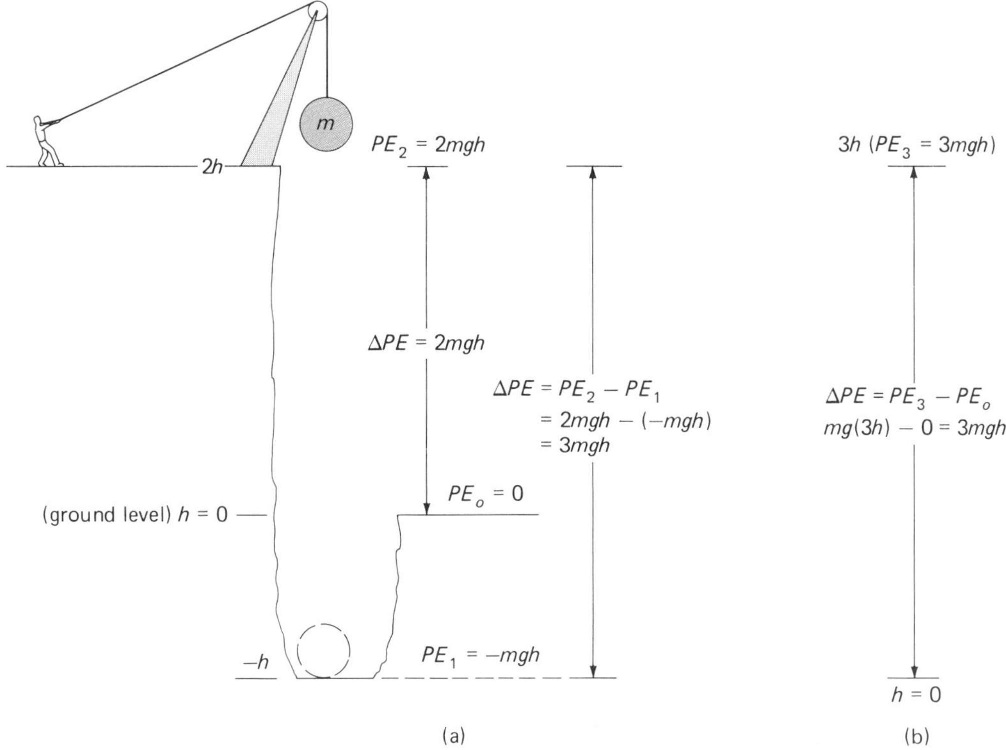

Figure 8.7
The reference point for zero potential energy is arbitrary. (a) In reference to ground level, the mass at the top of the cliff ($2h$) has a potential energy $PE_2 = 2mgh$. At the bottom of the hole ($-h$), the potential energy of the mass is $PE_1 = -mgh$. The difference in the potential energies of the two positions is $\Delta PE = 3mgh$. (b) If the bottom of the hole is taken as $h = 0$ (zero potential energy), the mass at the top of the cliff has a potential energy of $PE_3 = 3mgh$, but the difference in potential energy is the same $\Delta PE = 3mgh$.

this reference point, the height of the top of the building is $3h$, and the difference in the potential energies is

$$\Delta PE = mg\Delta h = mg(3h - 0) = 3mgh$$

Hence, the zero position for calculating ΔPE is arbitrary.

An object has potential energy at a particular location relative to another height (and kinetic energy when traveling at a particular speed). However, it is the *change* in energy that is important when something is being done or occurs.

Potential Energy of a Spring

The potential energy of an extended or compressed spring is the "stored" work done in extending or compressing the spring. As given in the previous section, the work done in compressing or extending a spring is given by Equation 8.4,

$$W = \tfrac{1}{2}kx^2$$

where x is the distance the spring has been compressed or extended from its equilibrium position x_o, i.e., $\Delta x = x - x_o = x$, with $x_o = 0$.

Since this work is "stored" in a spring as potential energy, the potential energy of a spring is given by the same expression:

$$PE = \tfrac{1}{2}kx^2 \qquad \textbf{(Eq. 8.10)}$$

EXAMPLE 8.7 A spring with a spring constant of 800 N/m is compressed 0.20 m. What is the potential energy of the compressed spring?

Solution: Using Equation 8.10,

$$PE = \tfrac{1}{2}kx^2 = \tfrac{1}{2}(800\ \text{N/m})(0.20\ \text{m})^2 = 16\ \text{J}$$

SPECIAL FEATURE 8.1

Rest Energy

A relatively new energy category is rest energy. Every object has an inherent amount of energy by virtue of having mass. This rest energy has nothing to do with an object's kinetic energy or potential energy of position. It is associated with an object's mass when at rest, or its **rest mass.**

Early in the 1900's, Albert Einstein formulated a relationship between an object's rest mass, m_o, and its rest energy, E_o,

$$E_o = m_oc^2 \qquad \textbf{(Eq. 8.11)}$$

where c is the speed of light (3×10^8 m/s). In his theory of relativity, Einstein showed that the mass of an object increases with increasing speed, so that its mass m may be greater than m_o. However, this mass increase is appreciable only at speeds approaching the speed of light (186,000 mi/s), and for ordinary situations we write $m = m_o$.

Einstein's equation has been experimentally verified many times, and we now say that *mass is a form of energy*. This intrinsic mass energy is enormous. For example, if we could convert 1 kg of (rest) mass completely to energy, the energy release would be

$$E_o = m_oc^2 = (1 \text{ kg})(3 \times 10^8 \text{ m/s})^2 = 9 \times 10^{16} \text{ J}$$

This is equivalent to the energy content of about 15 million barrels of crude oil (about one day's consumption in the entire United States).

If we could easily convert rest mass energy into useful work, our energy resources would be unlimited. However, matter is not freely converted into energy, but in some circumstances, a significant amount of rest energy can be converted into energy of other forms. This occurs chiefly in nuclear reactions, which will be discussed in Chapter 29. Such conversion of matter into energy occurs in the Sun and stars, providing their energy source, and in the awesome energy release of nuclear weapons.

In the Sun, 400 million tons of matter are converted to energy each second. Even at this rate, the Sun could survive for more than 10^{10} years. On a smaller, terrestrial scale are nuclear weapons, which are rated in megatons. A megaton of energy is the equivalent energy of 1 million tons of exploded TNT. This is 4.2×10^{15} J, so

$$1 \text{ megaton (of energy)} = 4.2 \times 10^{15} \text{ J}$$

Using the previous calculation of the equivalent energy of 1 kg of rest mass, we see that the amount of matter converted to energy in a 1-megaton nuclear blast is

$$(4.2 \times 10^{15} \text{ J/megaton})(1 \text{ kg}/9 \times 10^{16} \text{ J}) = 0.047 \text{ kg/megaton}$$

or 47 grams.

Some mass-energy conversion also takes place in ordinary chemical reactions, such as the burning of a match. A certain amount of matter is converted to energy; however, the mass loss is so tiny that it is impossible to detect.

Another type of energy is discussed in Special Feature 8.1. Here, the "stored" energy is not always readily released.

8.3 The Conservation of Energy

When we say something is conserved, we mean that it has the same constant value at any time. For example, if a class enrollment is conserved, no students drop the class, and the enrollment (and hopefully the attendance) is constant throughout the quarter or semester.

Energy is continually being converted from one form to another. It is often said that the total energy of the universe is conserved. This means that if we add up all the energy in the various forms in the universe, we would expect to get the same (constant) value at any time. Since the universe is the largest system we can imagine, the total energy must be in it somewhere in some form, and hence constant. In fact, **the total energy of any physical situation is always conserved.** It is there somewhere in some form. (Since mass may be converted to energy, and vice versa, in certain instances, we sometimes refer to the conservation of mass-energy.)

However, this is not always a very practical conservation law, since some forms of energy are difficult to measure or determine. Of more importance in most practical applications is the *conservation of mechanical energy.* The total mechanical energy E of a system is defined as the sum of the kinetic and potential energies:

$$E = KE + PE$$
$$E = \frac{1}{2}mv^2 + PE \qquad \textbf{(Eq. 8.12)}$$

where the potential energy may be in various forms—for example, gravitational or spring.

When the total mechanical energy of a system is conserved, it is constant at any time. That is, the kinetic and potential energies always add up to the same value. The values of the kinetic and potential energies may change, but their sum does not. Then, E at a time t must equal E_o at an initial time t_o, i.e.,

$$E = E_o$$

or

$$KE + PE = KE_o + PE_o$$

Then

$$KE - KE_o = -(PE - PE_o)$$

and

$$\Delta KE = -(\Delta PE)$$

The minus sign indicates that an increase in the kinetic energy ΔKE of the system is equal to the decrease in potential energy $-(\Delta PE)$ and vice versa. Such a system is called a *conservative system.*

A water analogy of a conservative system is illustrated in Figure 8.8. The total amount of water in the beakers, say 500 cc (cm^3), is analogous to the total mechanical energy. The water may be poured back and forth between the "kinetic energy" beaker and the "potential energy" beaker. It may be all in one beaker or the other, or it may be divided between the two, but the sum of the amounts in each beaker always is 500 cc, which is constant. Hence, the total water is "conserved."

Conservative mechanical systems are ideal situations. Due to ever-present frictional losses, all mechanical systems are nonconservative to some degree. In the water analogy, this corresponds to spilling or losing some water from the beakers. The total mechanical energy of a nonconservative system is not conserved, since the sum of KE and PE is no longer equal to the original constant value.

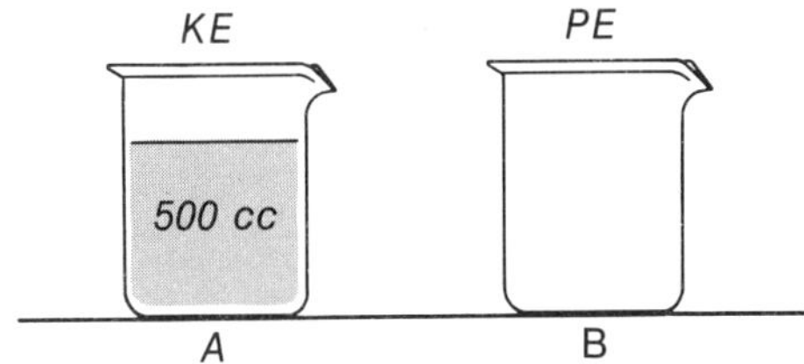

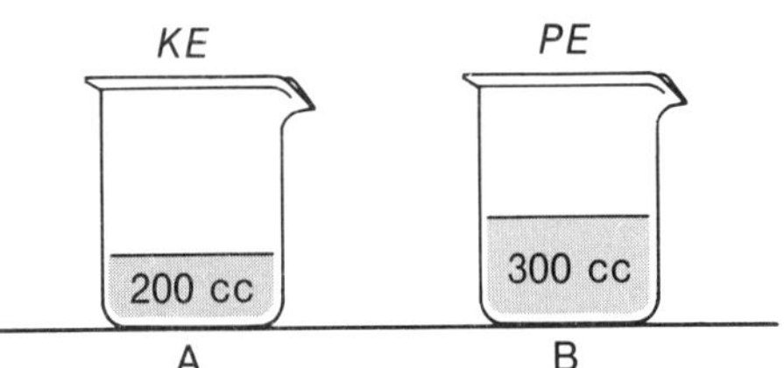

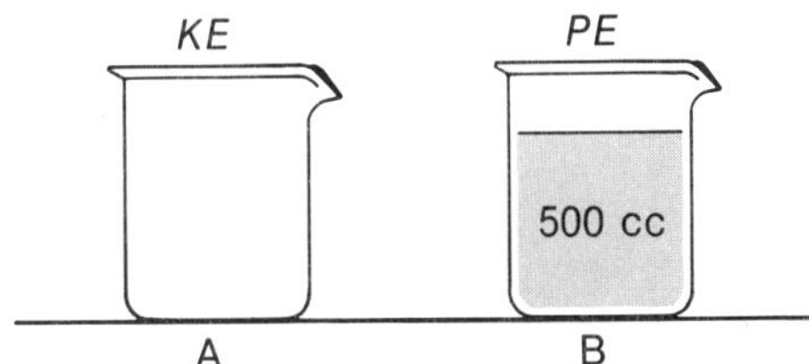

Figure 8.8
A water analogy of a conservative system. The total amount of water is 500 cc (cm^3) and is in either beaker A or beaker B or both, i.e., A + B = 500 cc. The total amount of water is the same, or constant, in all instances, analogous to the total mechanical energy of a conservative system. If some of the water were spilled from the beakers, the beaker system would no longer be conservative.

Basically, the lost mechanical energy is no longer available to do useful work. However, the *total* energy is still conserved, as this takes into account the lost energy (Q), and we may write

$$E_t = KE + PE + Q \qquad \textbf{(Eq. 8.13)}$$

In real mechanical situations, if the energy losses are small enough to be considered negligible, we can approximate a conservative sys-

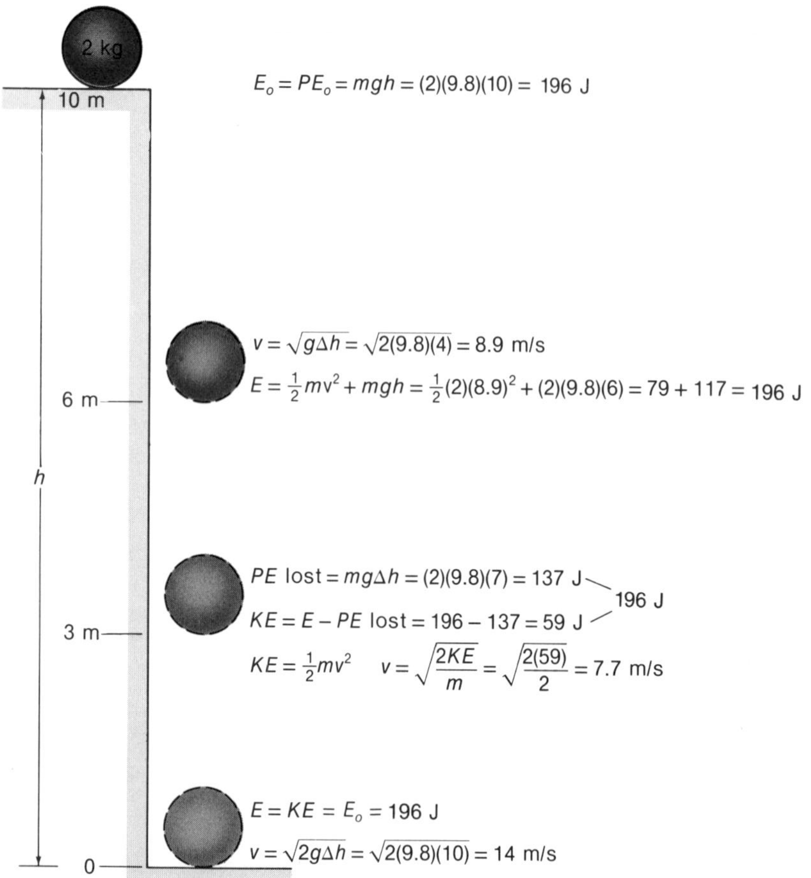

Figure 8.9
A conservative system. Neglecting air resistance, the total mechanical energy of the falling ball is constant. During the fall, it loses potential energy and gains kinetic energy. When the ball strikes the ground, the total mechanical energy is no longer conserved.

tem. Suppose, for example, that a 2.0-kg ball is dropped from a height of 10 m (Fig. 8.9). Neglecting air resistance, we see the system is conservative while the ball is falling.

Upon release ($v_o = 0$), the total mechanical energy is all potential energy, $E_o = mgh_o$. Then, during its fall, the ball loses potential energy and gains kinetic energy, and

$$E_o = E$$

or

$$mgh_o = \tfrac{1}{2}mv^2 + mgh$$

where h is the height of the ball at any time during its fall.

The velocity of the ball may be found at any height from Equation 8.12,

$$\tfrac{1}{2}mv^2 = mg(h_o - h) = mg\Delta h$$

and

$$v = \sqrt{2g\Delta h} \qquad \textbf{(Eq. 8.14)}$$

where Δh is the distance the ball falls. The term $mg\Delta h$ is just the potential energy ΔPE lost by the ball. Notice how this is used to find the kinetic energy of the ball at the 3-m height in the figure. When the ball strikes the ground, the mechanical energy is no longer conserved.

EXAMPLE 8.8 A mechanical pile driver with a driver weight of 480 lb and a vertical length of 20 ft is used to drive a small pile into the ground (Fig. 8.10). (a) How much energy is delivered to the pile on the initial strike? (b) With what velocity does the driver strike the pile? (Neglect friction.)

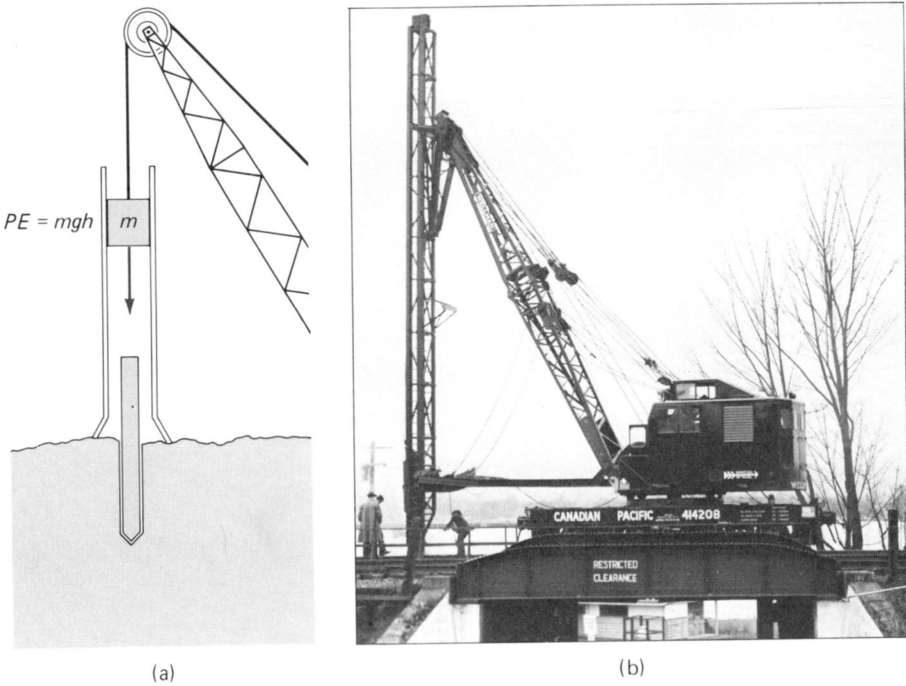

Figure 8.10
The pile driver. A mechanical pile driver uses potential energy to drive a piling. See Example 8.8 (b). (Courtesy of the Ohio Locomotive Crane Co.)

Solution: (a) With the driver at the top position (h_o = 20 ft), it has a total energy of

$$E_o = PE_o = mgh_o = wh_o = (480\ \text{lb})(20\ \text{ft}) = 9600\ \text{ft-lb}$$

The energy is converted to kinetic energy as the weight falls and is delivered to the pile when the driver strikes the pile.

(b) Just before the driving weight strikes the pile, it has a velocity of

$$v = \sqrt{2g\Delta h} = \sqrt{2(32\ \text{ft/s}^2)(20\ \text{ft})} = 36\ \text{ft/s}$$

in the downward direction. Notice that the striking velocity does not depend on the mass or weight of the driver, but its energy does.

As the pile is driven into the ground, the height of fall of the driving weight increases, and the striking velocity increases with each successive strike. The energy of the driver goes into the work of driving the pile into the ground. Once the driver strikes the pile, the mechanical energy of the system is no longer conserved. Why?

An example of the continuous conversion of kinetic energy and potential energy is an ideal simple pendulum (Fig. 8.11). At the pendulum bob's maximum arc of swing, or maximum

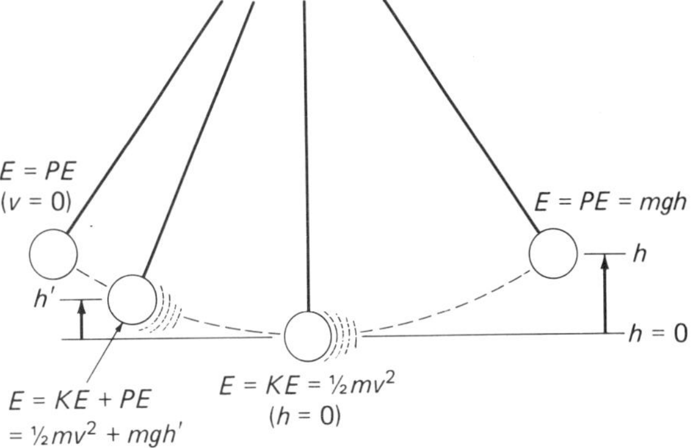

Figure 8.11
A simple pendulum. As the pendulum swings, there is a continuous conversion of mechanical energy between kinetic energy and potential energy. At the top of the swing (v = 0), all the energy is potential energy. At the bottom of the swing (h = 0), all the energy is kinetic energy.

height, the bob stops momentarily ($v = 0$) and all of the mechanical energy is potential energy. At the bottom of the swing ($h = 0$), all of the mechanical energy of the pendulum is kinetic energy. In between, the total mechanical energy is made up of both kinetic and potential energies (in the absence of any energy loss).

EXAMPLE 8.9 A simple pendulum, as in Figure 8.11, with a mass of 0.50 kg is released from a height of 0.25 m above its equilibrium position (lowest position, $h = 0$). (a) What is the magnitude of its velocity at the bottom of the swing? (b) What is the magnitude of its velocity when the pendulum bob is at a height of 0.10 m? (c) Will the pendulum swing to a height greater than 0.25 m at the opposite side of the swing?

Solution: (a) Upon release ($v_o = 0$, $KE_o = 0$), all the mechanical energy is potential energy, and with $h = 0.25$ m,

$$\begin{aligned} E_o = PE &= mgh \\ &= (0.50 \text{ kg})(9.8 \text{ m/s}^2)(0.25 \text{ m}) = 1.2 \text{ J} \end{aligned}$$

At the bottom of the swing, all the mechanical energy is kinetic energy ($h = 0$, $PE = 0$), and

$$E = KE = E_o = 1.2 \text{ J}$$

and

$$v = \sqrt{\frac{2E}{m}} = \sqrt{\frac{2(1.2 \text{ J})}{0.50 \text{ kg}}} = 2.2 \text{ m/s}$$

(b) At a height $h' = 0.10$ m, by the conservation of mechanical energy

$$E = E_o$$

or

$$\tfrac{1}{2}mv^2 + mgh' = mgh$$

and

$$\begin{aligned} v &= \sqrt{2g(h - h')} \\ &= [2(9.8 \text{ m/s}^2)(0.25 \text{ m} - 0.10 \text{ m})]^{1/2} \\ &= 1.7 \text{ m/s} \end{aligned}$$

This is also the magnitude of the velocity at a height of 0.10 m on the opposite side of the swing. Why?

(c) No, the pendulum will not swing to a height greater than 0.25 m. This would violate the conservation of mechanical energy. In practice, the maximum height of the pendulum's swing would gradually decrease owing to energy losses from the friction of the support and air resistance.

In some instances, it is convenient to calculate the energy loss to friction, or to some other reason, from the conservation of *total* energy. Suppose the initial total energy is completely mechanical energy and that some of this is lost to friction in a mechanical process. Then

$$\text{Initial total energy} = \text{final total energy}$$

$$KE_o + PE_o = KE + PE + Q$$

or

$$E_o = E + Q$$

where Q is the lost mechanical energy (cf. Eq. 8.13), and

$$Q = E_o - E = -\Delta E \qquad \textbf{(Eq. 8.15)}$$

Hence, Q is equal to the decrease in the mechanical energy of the system.

EXAMPLE 8.10 A spring with a spring constant of 160 N/m is attached to a mass of 0.50 kg resting on a rough horizontal surface. The spring is compressed a distance of 20 cm and then released. As the mass passes through the equilibrium position ($x_o = 0$), it has a speed of 2.6 m/s. How much energy was lost to the friction between the mass and the surface?

Solution: It is given that $k = 160$ N/m, $m = 0.50$ kg, $x = 20$ cm $= 0.20$ m, and $v = 2.6$ m/s. When the spring is compressed, the total mechanical energy of the system is all potential energy:

$$E_o = PE = \tfrac{1}{2}kx^2 = \tfrac{1}{2}(160 \text{ N/m})(0.20)^2 = 3.2 \text{ J}$$

As the mass passes through the equilibrium position, $x_o = 0$, its total mechanical energy is equal to its kinetic energy since the potential energy is zero ($PE = \tfrac{1}{2}kx_o^2 = 0$), and

$$E = \tfrac{1}{2}mv^2 = \tfrac{1}{2}(0.50 \text{ kg})(2.6 \text{ m/s})^2 = 1.7 \text{ J}$$

The energy lost to friction is then

$$\begin{aligned} Q &= -\Delta E = E_o - E \\ &= 3.2 \text{ J} - 1.7 \text{ J} = 1.5 \text{ J} \end{aligned}$$

8.4 **Power**

The amount of work done, or the amount of energy expended, in performing a task is of prime importance. However, the time it takes to do a job or the rate at which work is done is equally important in many technical applications. For example, a particular machine or person may be able to do a job in 10 min, while another machine or person may be able to do the same job in 5 min. Obviously, time is an important factor in work output.

The time factor is taken into account in the expression for power. **Power is the time rate of doing work (or expending energy).** Expressed mathematically,

$$\boxed{P = \frac{W}{t} = \frac{Fd}{t}} \qquad \textbf{(Eq. 8.16)}$$

$$\text{Power} = \frac{\text{work}}{\text{time}}$$

This is actually the average power done over the time interval t. Hence, if one machine can do a certain amount of work in half the time of another machine, the first machine has twice as much power as the second. Computations of the motor power are shown in Figure 8.12.

Units of Power

In the British system, the unit of power is the ft-lb/s:

$$\frac{W\ (\text{lb-ft})}{t\ (\text{s})} = P\ (\text{ft-lb/s})$$

However, a common unit used to express power is the **horsepower** (hp), and

$$1\ \text{hp} = 550\ \text{ft-lb/s}$$

This unit was originated by James Watt (1736–1819), the Scottish engineer who developed an improved steam engine. The value of the horsepower was based on his experiments with strong dray horses and is about 50 percent greater than the work an average horse can do in a working day. The horsepower is the common unit used to rate motors and engines. A metric horsepower, which is slightly less than the British horsepower, is used in some countries.

In the metric system, the unit of power is the joule/second (J/s = N-m/s):

$$\frac{W\ (\text{J})}{t\ (\text{s})} = P\ (\text{J/s})$$

However, the derived unit **watt** in honor of James Watt, is commonly used:

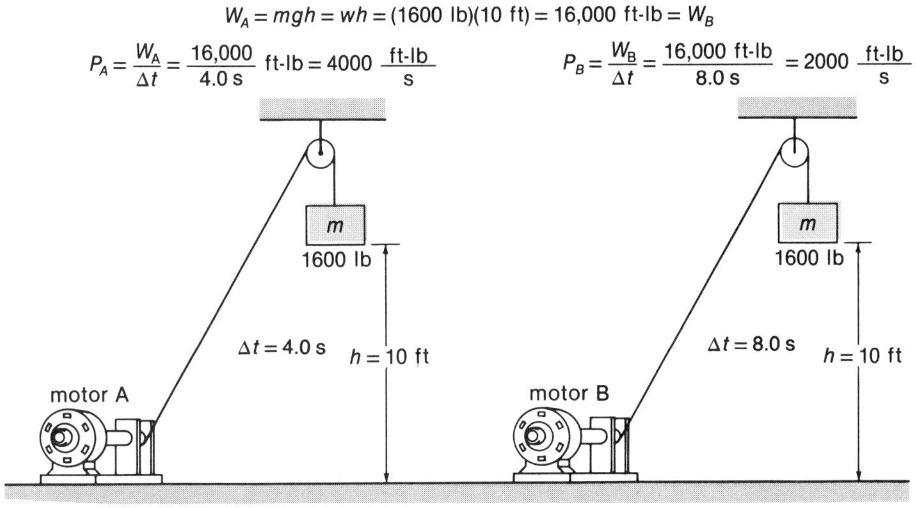

$$W_A = mgh = wh = (1600\ \text{lb})(10\ \text{ft}) = 16{,}000\ \text{ft-lb} = W_B$$

$$P_A = \frac{W_A}{\Delta t} = \frac{16{,}000}{4.0\ \text{s}}\ \text{ft-lb} = 4000\ \frac{\text{ft-lb}}{\text{s}} \qquad P_B = \frac{W_B}{\Delta t} = \frac{16{,}000\ \text{ft-lb}}{8.0\ \text{s}} = 2000\ \frac{\text{ft-lb}}{\text{s}}$$

Figure 8.12

Power is the time rate of doing work. Both motors do the same amount of work in lifting equal loads to the same height. However, motor A delivers more power than motor B, since it does the work in less time. In this case, the power output of motor A is twice that of motor B in doing the same work in one-half the time.

Table 8.2
Units of Power

$$1 \text{ hp} = 550 \text{ ft-lb/s} = 33{,}000 \text{ ft-lb/min}$$
$$= 746 \text{ W} = 7.46 \times 10^7 \text{ erg/s}$$
$$1 \text{ W} = 10^7 \text{ erg/s} = 0.74 \text{ ft-lb/s} = 1.34 \times 10^{-3} \text{ hp}$$
$$1 \text{ erg/s} = 10^{-7} \text{ W} = 7.4 \times 10^{-8} \text{ ft-lb/s}$$
$$= 1.34 \times 10^{-10} \text{ hp}$$

$$1 \text{ watt} = 1 \text{ joule/second}$$

$$1 \text{ W} = 1 \text{ J/s}$$

A comparison of the units of power in the various systems is given in Table 8.2.*

In the SI system, large values of power are expressed in kilowatts (kW). You have no doubt heard this term used in the electrical sense. We pay the electric company for the number of kilowatt-hours (kWh) we use. The kilowatt-hour is a unit of energy or work rather than power, as can be seen from Equation 8.16:

$$P = \frac{W}{t}$$

and

$$W = P \cdot t$$

Hence, we pay the electric "power" company for electrical energy that is used to do work in the units of power × time, or kWh.

EXAMPLE 8.11 An elevator with an empty weight of 800 lb has a load capacity of 2000 lb. If the design of the elevator is such that it can be raised in a 120-ft shaft in 30.0 s, what is the minimum horsepower rating of the lift motor, neglecting frictional losses?

Solution: Assuming that the elevator is loaded to capacity, the weight being lifted is F = 2000 lb + 800 lb = 2800 lb. Then, with the elevator traveling a distance d = 120 ft in a time t = 30.0 s, the power supplied by the lift motor is

$$P = W/t = Fd/t = (2800 \text{ lb})(120 \text{ ft})/(30.0 \text{ s})$$
$$= 11{,}200 \text{ ft-lb/s}$$

and

$$11{,}200 \text{ ft-lb/s } (1 \text{ hp}/550 \text{ ft-lb/s}) = 20.4 \text{ hp}$$

Would you design or install an elevator with a lift motor having the minimum required horsepower?

* In the cgs system, the unit of power is the erg/second (dyne-cm/s):

$$W \text{ (erg)}/t \text{ (s)} = P \text{ (erg/s)}$$

There is no derived unit for power in the cgs system.

The distance and time in the preceding example imply a speed, and power can be easily expressed in terms of the speed of the action. From Equation 8.16, we have for a constant force

$$P = \frac{W}{t} = \frac{Fd}{t} = F\bar{v}$$

or

$$P = F\bar{v} \qquad \textbf{(Eq. 8.17)}$$

where $d/t = \bar{v}$ is the average speed.

EXAMPLE 8.12 A motorized windlass pulls a load along a surface with a constant speed of 0.30 m/s (Fig. 8.13). If the windlass maintains a tension of 2.6×10^4 N in the line, (a) how much power is supplied by the motor and (b) how much energy is expended each second?

Solution: (a) With $F = 2.6 \times 10^4$ N and $\bar{v} = v = 0.30$ m/s,

$$P = F\bar{v} = (2.6 \times 10^4 \text{ N})(0.30 \text{ m/s})$$
$$= 7.8 \times 10^3 \text{ W}$$
$$= 7.8 \text{ kW}$$

(b) With $P = 7.8 \times 10^3 \text{ W} = 7.8 \times 10^3$ J/s, there are 7.8×10^3 J of energy expended each second, i.e.,

$$W \text{ (energy)} = P \cdot t$$
$$= (7.8 \times 10^3 \text{ J/s})(1 \text{ s}) \doteq 7.8 \times 10^3 \text{ J}$$

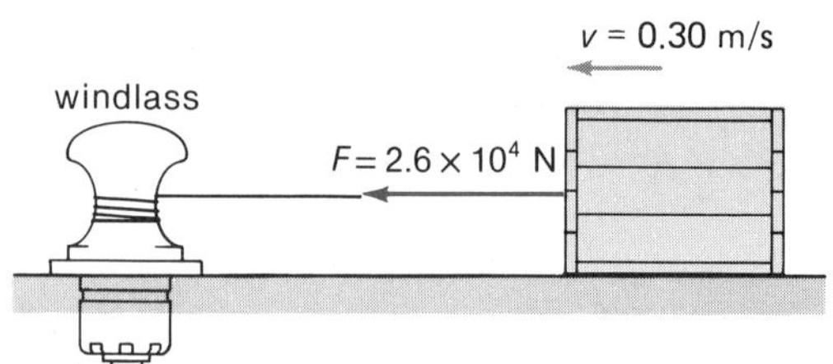

Figure 8.13
A windlass pulling a load along a surface. See Example 8.12.

EXAMPLE 8.13 An electric pump in a well delivers water to the surface at a rate of 80 gal/min from a depth of 50 ft. Assuming no power losses, how much power is supplied by the pump motor? (A gallon of water weighs 8.3 lb.)

Solution: The weight of the water lifted each second is

$$\begin{aligned} w/t &= 80 \text{ gal/min } (8.3 \text{ lb/gal})(1 \text{ min}/60 \text{ s}) \\ &= 11 \text{ lb/s} \end{aligned}$$

The power required to lift the water a distance of $d = 50$ ft at this rate is then

$$\begin{aligned} P = W/t = (F/t)d = (w/t)d &= (11 \text{ lb/s})(50 \text{ ft}) \\ &= 550 \text{ ft-lb/s} \end{aligned}$$

Hence, the pump motor must be at least a "one-horse" motor, since 550 ft-lb/s = 1 hp.

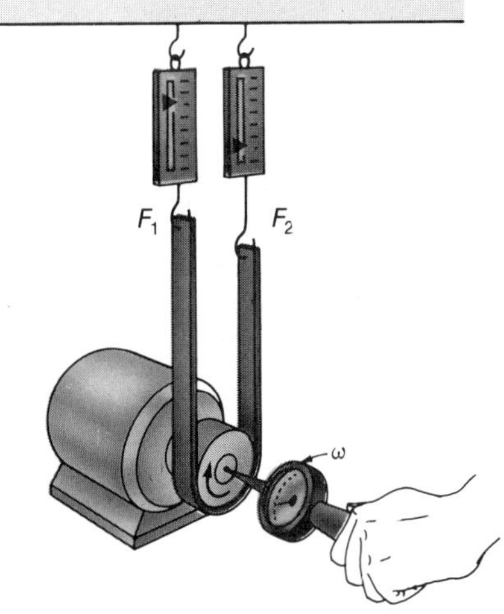

Figure 8.14
A Prony brake. Using the difference between the forces of friction and the angular speed measured by a tachometer, the brake power of a motor can be computed. See text for description.

Power Measurement

It is often necessary to determine the power input and/or the power output of a machine, e.g., a motor or engine. One method of doing this is by the use of a dynamometer. A dynamometer is a device that measures force.

In one common type of dynamometer, the force is measured by braking action. One of the first such dynamometers was developed by the Frenchman Gaspard de Prony (1755–1839) and is called a Prony brake. The principle of the Prony brake is illustrated in Figure 8.14. The shaft of the test motor is coupled directly to a revolving drum with a tension friction belt. As the motor turns, the frictional force on the belt is measured.

In Figure 8.14, the measurement is made by the spring balances, and $f = F_2 - F_1$. In one rotation, the force acts through a distance equal to the circumference ($2\pi r$) of the drum, and the work per revolution is

$$W = fd = (F_2 - F_1)(2\pi r)$$

where r is the radius of the drum. The power is then given by multiplying the work by the angular speed ω of the motor, which is usually given in revolutions/min (rpm), i.e.,

$$\begin{aligned} P &= (F_2 - F_1)(2\pi r)\omega \\ \text{Power} &= (\text{force})(\text{distance/rev})(\text{rev/min}) \\ &= \frac{\text{force} \times \text{distance}}{\text{time}} \\ &= \text{work/time} \end{aligned}$$

which is dimensionally correct, as the analysis shows. Expressed in "brake" horsepower (bhp),

$$P \text{ (bhp)} = \frac{(F_2 - F_1)(2\pi r)\omega}{33{,}000} \qquad \textbf{(Eq. 8.18)}$$

where the force is in pounds, the radius of the drum is in feet, and the angular speed is in rpm. The factor 1/33,000 converts the ft-lb/min to horsepower through the conversion factor 1 hp = 33,000 ft-lb/min.

EXAMPLE 8.14 A small motor is tested on a Prony brake with a drum radius of 6.0 in. If the spring balance readings are 35.0 lb and 32.0 lb, respectively, and the angular speed of the motor shaft is 1750 rpm, as determined by a tachometer, what is the brake horsepower of the motor?

Solution: With $r = 6.0$ in. $= 0.50$ ft, $F_2 = 35.0$ lb, $F_1 = 32.0$ lb, and $\omega = 1750$ rpm,

(a)

(b)

Figure 8.15
Motor testing. (a) A hydraulic motor being tested for maximum stall torque under operating conditions. A dynamometer is connected to a roller chain that passes over a sprocket on the output shaft of the hydraulic motor. When peak force has been exerted by the motor, a relief valve opens. (b) Design testing an electrical motor for horsepower, torque, speed, and electrical characteristics. To effect a stalling force, the motor is connected to drive a generator to which an electrical load is applied, thus retarding the motor. The dynamometers measure the resultant horsepower and torque. (Courtesy W. C. Dillon & Company, Inc.)

$$P = \frac{(F_2 - F_1)(2\pi r)\omega}{33{,}000}$$

$$= \frac{(35.0\text{ lb} - 32.0\text{ lb})2\pi(0.50\text{ ft})(1750\text{ rpm})}{33{,}000}$$

$$= 0.50\text{ hp (bhp)}$$

Other brake dynamometers make use of hydraulic and electromagnetic brakes. Another type of dynamometer uses an electric generator. The test motor or engine is used to turn the generator, and the power output is determined from the generated current (I) and voltage (V), $P = IV$ (see Chapter 21).

The amount of force, or torque, necessary to stall a motor is another commonly measured quantity. Various arrangements and mechanical dynamometers are used (Fig. 8.15).

Efficiency

The output of work or power of a machine is never equal to its input. In every mechanical machine, friction causes some of the energy or power to be lost, regardless of how well the machine is lubricated. Other energy losses may be due to cooling, such as in the cooling system of an automobile engine.

The energy or power losses of a machine are important considerations. They are expressed in terms of the machine's **efficiency** (Eff), which is the ratio of a machine's work output (or power output) and the work input (or power input):

$$(\%)\text{ Efficiency} = \frac{\text{work output}}{\text{work input}}\ (\times\ 100\%)$$

(Eq. 8.19)

$$= \frac{\text{power output}}{\text{power input}}\ (\times\ 100\%)$$

SPECIAL FEATURE 8.2

Automobile Efficiency*

Automobiles powered by gasoline engines are known to be very inefficient machines. Even under ideal conditions, less than 15 percent of the available energy in the fuel is used to power the vehicle. This situation is much worse in stop-and-go driving in the city.

Many mechanisms contribute to the energy losses in a typical automobile. About two thirds of the energy available from the fuel is lost in the engine. Part of this energy ends up in the atmosphere via the exhaust system, and part is used in the engine's cooling system.

About 10 percent of the available energy is lost in the automobile's drive-train mechanism. This loss occurs through friction in the transmission, drive shaft, wheel and axle bearings, and differential. Friction in other moving parts accounts for about 6 percent of the energy loss. Approximately 4 percent of the available energy is used to operate fuel and oil pumps and such accessories as power steering, air conditioning, power brakes, and electrical components.

Finally, about 14 percent of the available energy is used to propel the automobile ("useful" work). This energy is used mainly to overcome road friction and air resistance. These energy losses are summarized in Table 8.3.

Table 8.3
Energy Losses in a Typical Automobile

Mechanism	Power Loss (%)
Exhaust (heat)	33
Cooling system	33
Drive train	10
Internal friction	6
Accessories	4
Propulsion of vehicle ("useful" work)	14

* From Serway, R. A., *Physics for Scientists and Engineers*, 3rd ed., Saunders College Publishing, Philadelphia, 1990.

or

$$(\%)\ \text{Eff} = \frac{W_{out}}{W_{in}} (\times\ 100\%) = \frac{P_{out}}{P_{in}} (\times\ 100\%)$$

Efficiency is commonly expressed as a percentage rather than as a decimal fraction. Since there are always energy losses due to friction, a machine's efficiency is always less than 1.0 (100 percent).

Suppose, for example, that a machine has an efficiency of 60 percent (or 0.60). This means that 60 percent of the power input goes into doing useful work, or that the machine's power output is 60 percent of its power input. The other 40 percent of the power input is lost to friction or some other cause.

The typical efficiencies of some machines are listed in Table 8.4. You may be surprised by the relatively low efficiency of the automobile. See Special Feature 8.2 to find out where all the energy is lost.

Table 8.4
Typical Efficiencies of Some Machines

Machine	Eff (%)
Compressor	85
Electric motor	Up to 95
Automobile	<15
Human muscle*	20–25
Steam locomotive	5–10

* Technically not a machine, but used to perform work.

EXAMPLE 8.15 A small gasoline engine has a power input of 4.6 hp and a measured brake horsepower of 3.5 hp. (a) What is the engine's efficiency? (b) How much energy is lost in 10 min of operation?

Solution: (a) With $P_{in} = 4.6$ hp and $P_{out} = 3.5$ hp,

$$\% \text{ Eff} = \frac{P_{out}}{P_{in}} \times 100\% = \frac{3.5 \text{ hp}}{4.6 \text{ hp}} \times 100\% = 76\%$$

(b) With a 76% efficiency, 100% − 76% = 24% (or 0.24) of the power input is lost, or

$$\text{Power lost} = (0.24)P_{in} = (0.24)(4.6 \text{ hp}) = 1.1 \text{ hp}$$

or, directly,

$$\text{Power lost} = P_{in} - P_{out}$$
$$= 4.6 \text{ hp} - 3.5 \text{ hp} = 1.1 \text{ hp}$$

Converting to ft-lb/s,

$$1.1 \text{ hp } (550 \text{ ft-lb/s}/1 \text{ hp}) = 605 \text{ ft-lb/s}$$

Hence, 605 ft-lb of energy are lost each second, and in 10 min = 600 s,

$$W = P \cdot t = (605 \text{ ft-lb/s})(600 \text{ s})$$
$$= 3.63 \times 10^5 \text{ ft-lb}$$

Important Terms

work the product of the magnitude of the applied force and the parallel displacement through which the force acts

energy the ability to do work; that is, when a body possesses energy, it has the capability to do work

kinetic energy the energy of motion

potential energy the energy of position

rest energy the inherent energy of an object by virtue of having mass

conservation of total energy the principle that the total energy (in any form) is conserved (constant)

conservation of mechanical energy the principle that the total mechanical energy ($KE + PE$) of a conservative system is conserved

power the time rate of doing work (or expending energy)

efficiency the ratio of a machine's work (or power) output and the work (or power) input, which is usually expressed as a percentage

Important Formulas

work: $W = Fd$
$(W = F\|d = Fd \cos \theta)$

work done in compressing (or extending) a spring: $W = \frac{1}{2}kx^2$

kinetic energy: $KE = \frac{1}{2}mv^2$

work-energy theorem: $W = \Delta KE = KE_f - KE_o = \frac{1}{2}mv_f^2 - \frac{1}{2}mv_o^2$

potential energy:
(gravitational) $PE = mgh = wh$
(spring) $PE = \frac{1}{2}kx^2$

rest energy: $E = m_o c^2$

total mechanical energy: $E = \frac{1}{2}mv^2 + PE$

total energy: $E = \frac{1}{2}mv^2 + PE + Q$

power: $P = \dfrac{W}{t} = \dfrac{Fd}{t} = F\bar{v}$

efficiency: $(\%) \text{ Eff} = \dfrac{W_{out}}{W_{in}} (\times 100\%) = \dfrac{P_{out}}{P_{in}} (\times 100\%)$

Questions

Work

1. Is work related to how tired one becomes? Explain.
2. A worker holds a roof section of a house frame stationary so as to prevent it from falling while a carpenter secures it. How much work does the worker do?
3. While moving a load of boxes on a skid on a horizontal floor, a worker pushes down on the top of the boxes to stabilize the load. Does this worker do work? Explain.
4. When work is done by an applied force in moving an object, work is usually done *against* friction. What is the situation if the works done aren't equal? Could the work done against friction ever be greater than the work done by the applied force?
5. In applying a force at an angle to the horizontal in moving a load on a level surface, the perpendicular component of the force does no work. Considering friction, does the perpendicular force component affect the work done against friction? (Note: the force may be applied at a positive or negative angle to the horizontal.)

Energy

6. A person sits in a moving car. Does the person have kinetic energy? Explain.
7. What does a negative change in potential energy imply?
8. If the speed of an object is (a) doubled and (b) tripled, what is the change in its kinetic energy?
9. How does the speed of an automobile affect its stopping distance? (Consider a constant braking or stopping force.)
10. (a) A weight lifter holds a set of weights over his head. Is he doing work? Has he done work? Ex-

Figure 8.16
Work in lifting. See Question 10.

plain. See Figure 8.16. (b) In Figure 8.16, the short weight lifter and the tall weight lifter hold identical weights. Who does more work? (c) How could both do the same work?

11. You and another student are late to class and race to the classroom by different routes. (a) Who does more work against gravity? (b) If you beat your classmate to class, who had the greater power output?
12. Why are water towers often placed on high hills or elevations?
13. Explain the work and energy considerations in lifting an object and letting it fall. What happens to the energy when it hits the ground?
14. A ball lies on the first floor of a house. A person on that floor says it has zero potential energy. A person on the second floor says the ball has −20 J of potential energy. A person in the basement says it has +20 J of potential energy. Can they all be right? Explain.
15. A spring with a spring constant of 200 N/m is stretched 0.10 m and then stretched another 0.15 m. How much energy is stored in the spring?
16. The law of conservation of mass says that matter cannot be created or destroyed. Is this true in light of nuclear processes?
17. Show that the equation $E = m_o c^2$ is dimensionally correct, i.e., gives the proper energy units.

Conservation of Energy

18. Distinguish between total energy and total mechanical energy.
19. A pendulum released from rest swings inward and its string comes in contact with a rod, as shown in Figure 8.17. Is this a trick photograph, or does the bob really swing to the same height, as shown in the photo? Explain.

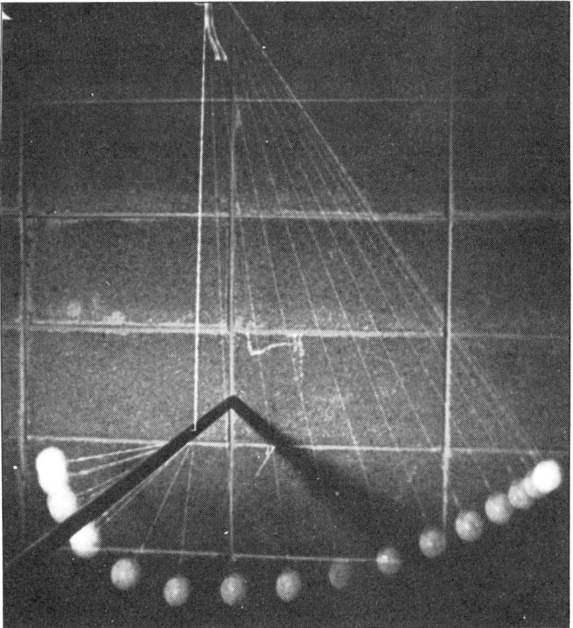

Figure 8.17
Pendulum and peg. See Question 19.

20. A pendulum is taken to the moon, and its bob is released from the same height above the ground as it was on Earth. Would there be any difference(s) in the movement of the bob? (Consider energy, speed, and maximum heights of swing.)
21. A person on a trampoline can go higher with each bounce. Explain in terms of energy considerations how this is possible. Is there a maximum height to which the person can go? Explain.
22. A baseball player slides into home plate. What happens to his kinetic energy?
23. Explain the work and energy considerations in a car rolling down an incline and coming to a stop on a horizontal surface.
24. It is sometimes said that energy cannot be created or destroyed. Is this a conservation law? Explain.

Power

25. Is the horsepower rating of a motor its power input or its power output?
26. Explain the difference in the capabilities of a 2.0-hp motor and a 4.0-hp motor.
27. What are (a) a dynamometer and (b) a Prony brake?
28. Show why efficiency is equal to the ratios of both the output and input power and work.
29. Can efficiency ever be 100 percent or greater? Explain.

Problems

Levels of difficulty are indicated by asterisks for your convenience.

8.1 Work

1. A horizontal force of 50 N moves an object 10 m on a level surface. How much work is done by the force?
2. How much work is required to lift a 4.0-kg concrete block to a height of 2.0 m?
3. A 4800-lb truck is lifted 10 ft by a hydraulic garage lift. How much work is done by the lift?

*4. A worker pushes a crate up a 10° ramp with a 35-ft-long base by applying a force of 80 lb parallel to the ramp surface. How much work does the worker do?

*5. By applying a constant force parallel to a level surface, a person does 2.0×10^3 J of work in moving a crate 5.0 m. What is the applied force?

*6. A sled is towed along a nearly frictionless, level surface by a rope at an angle θ relative to the horizontal. If the rope is pulled with a force of 300 N, how much work is done in moving the sled a distance of 15 m if (a) $\theta = 0°$, and (b) $\theta = 37°$?

*7. A person pushes a lawn mower along a level lawn with a force of 200 N at an angle of 30° to the horizontal. How much work is done in pushing the mower 25 m?

*8. An 8.0-kg block is pushed up a frictionless ramp with a 20° incline. How much work is done if the length of the ramp is 8.0 m?

*9. A 10-kg block is moved along a level surface at a constant velocity through a distance of 4.0 m. (a) If the coefficient of kinetic friction between the block and the surface is 0.50, how much work is done against friction? (b) What is the magnitude of the applied force? (c) What is the total work?

*10. How much work is done in compressing a spring 0.20 m if the spring has a spring constant of 120 N/m?

*11. When a 4.0-kg mass is suspended from a spring, the spring extends 10 cm. How much work is required to stretch the spring another 30 cm by pulling downward on the mass?

**12. A 54-lb trunk is pushed along a level floor by an applied downward force of 65 lb at an angle of 45° to the horizontal. If the coefficient of sliding friction between the trunk and the floor is 0.46, how much work is done by the applied force if the trunk slides at a constant velocity through a distance of 4.0 ft?

**13. A graph of F versus x for a force capable of variable and constant applications is shown in Figure 8.18. How much work is done by the force in this graphical application?

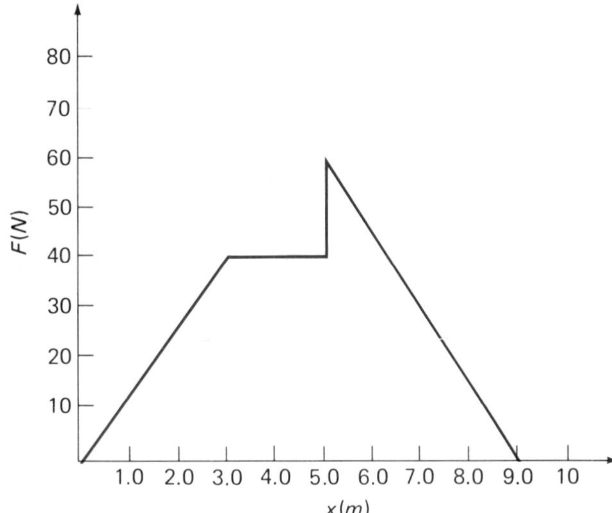

Figure 8.18
See Problem 13.

8.2 Energy

14. A 0.20-kg ball travels with a speed of 3.0 m/s. What is the kinetic energy of the ball?

15. A car with a mass of one metric ton travels at a speed of 90 km/h. How much work is required to bring the car to rest?

*16. A 2400-lb electric cart accelerates from rest at a rate of 6.0 ft/s^2. What is the kinetic energy of the cart when it has accelerated for 4.0 s?

*17. How much work is required to bring a 1000-kg racing car traveling at a speed of 60 km/h to rest?

*18. A 3200-lb car traveling at 30 mi/h speeds up to 60 mi/h. How much useful work was supplied by the engine?

*19. In a stock room, a 5.0-kg machine part originally on a shelf 0.60 m above the floor is moved to a shelf 2.0 m above the floor. (a) What is the change in potential energy? (b) What is the potential energy of the part on the top shelf relative to the floor?

*20. A spring with a spring constant of 2.0 lb/in. is compressed 8.0 in. from its equilibrium position. How much potential energy is stored in the spring?

*21. An extended spring has a potential energy of 135 J. If the spring constant is 3.0×10^3 N/m, how far is the spring extended for its equilibrium position?

*22. How much rest energy does your physics book contain? (Estimate its mass.) If this energy could be released, give an idea of its capability.

*23. Assume that 1.0×10^{-11} kg of a 1.0-kg mass is converted to energy. (a) Could this mass be detected? (b) How much energy would be released? Would this amount of energy be detectable?

**24. How many times greater would the stopping distance be for a car going 60 mi/h than for one going 30 mi/h? (Consider the braking or stopping force to be constant and the same in both cases.)

**25. A weight lifter holds a set of weights 6.0 in. off the floor. When she lifts and presses the weights, their potential energy is increased 12-fold. How high relative to the floor are the weights lifted?

8.3 The Conservation of Energy

*26. A ball is dropped from a height of 9.0 m. What is the speed of the ball just before it strikes the ground? (Use energy considerations and neglect air resistance.)

*27. An 8.00-lb block is dropped from a height of 75.0 ft. (a) What are the potential and kinetic energies of the block at $t = 1.0$ s and $t = 2.0$ s? (b) What is the speed of the block just before it hits the ground?

*28. An object is dropped from rest at a height of 200 m. At what height during its fall will its kinetic energy be twice its potential energy?

*29. In Example 8.8, suppose the pile driver drives the pile 2.0 in. every time it strikes the pile. (a) When the driver is ready for the fourth strike, what is the potential energy of the driver relative to the pile? (b) With what speed does the driver strike the pile on the fifth strike? (Neglect friction.)

*30. An ideal pendulum with a 0.20-kg bob has a length of 0.50 m. At its maximum arc, the cord of the pendulum makes an angle of 30° relative to the vertical. (a) What is the total mechanical energy of the pendulum? (b) What is the maximum speed of the pendulum? (c) Do you need to know the mass of the pendulum to find its maximum speed? Explain.

*31. A 48-lb crate is held stationary at the top of a ramp 2.0 ft high. The crate is released and after sliding down the ramp has a speed of 6.0 ft/s at the bottom. How much energy was lost to friction?

**32. A 25-g bullet with a muzzle velocity of 300 m/s is fired from a rifle into a board 4.00 cm thick. (a) In a hard board, the bullet penetrates 3.00 cm and comes to a stop. What was the average force exerted on the bullet by the board? (b) For a soft board, the bullet goes through the board and emerges with a speed of 50.0 m/s. What was the average force exerted on the bullet by the board in this case?

**33. A 1.0-kg block is released from rest at the top of a curved frictionless track as shown in Figure 8.19. (a) What are the speeds of the block at points A and B? (b) If the block goes on a level surface at point C with a coefficient of kinetic friction of 0.50, how far from point C will the block come to rest? (c) Describe the work done in each section of the track.

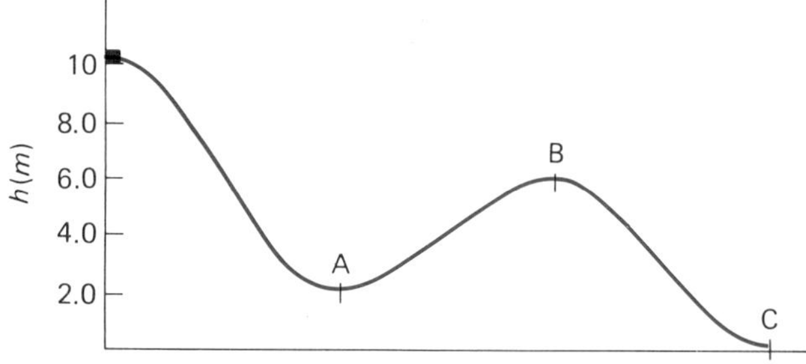

Figure 8.19
See Problem 33.

8.4 Power

34. A 180-lb mountain climber climbs a 4000-ft mountain in 3.5 h. (a) What is the average work done by the mountain climber? (b) What is the climber's average power?
35. A motor hoist lifts a 100-kg object 15 m in 4.0 s. How much power was supplied by the hoist?
36. A horse pulling a plow exerts an average force of 150 lb in plowing a furrow 50 ft long in 2.0 min. What is the average power output of the horse in horsepower?
37. Which is more efficient, machine A that has a work output of 500 J for a work input of 700 J, or machine B that does 900 J of useful work with an input of 1400 J?

*38. A constant force of 30 N acts on an object moving initially with a speed of 4.0 m/s and increases the object's speed to 6.0 m/s. How much power was supplied by the force?

*39. A motor is tested by a Prony brake with a 16-in.-diameter drum. If a horsepower of 1.0 bhp is measured with a motor shaft speed of 2100 rpm, what is the net force between the spring balances?

*40. A 12-hp motor is 75 percent efficient. How much energy is lost each second?

*41. A 100-hp motor raises a 2080-lb elevator 200 ft in 10.0 s. (a) What is the efficiency of the elevator system? (b) How much energy is lost during the 10.0 s?

*42. In Problem 18, suppose the car had an efficiency of 15 percent. How much energy is lost during the time the car accelerates?

**43. A 5.0-kg block is dropped from a height of 19.6 m. How much power is supplied by gravity in bringing the block to the ground?

**44. A pump delivers water to a water tower 25 ft high at a rate of 100 ft^3/min. What is the power, in horsepower, supplied by the pump motor in raising the water? (Water weighs 62.4 lb/ft^3.)

**45. A friend tells you that he has just invented a 5.0-kW engine that has been shown to be able to lift a 40-kg object a distance of 10 m in 0.75 s. He asks you to invest in the manufacture of this engine. (a) Would you do so with these data? (b) Suppose you checked this amazing engine out and the actual time to lift the mass were 0.80 s. Would this affect your decision?

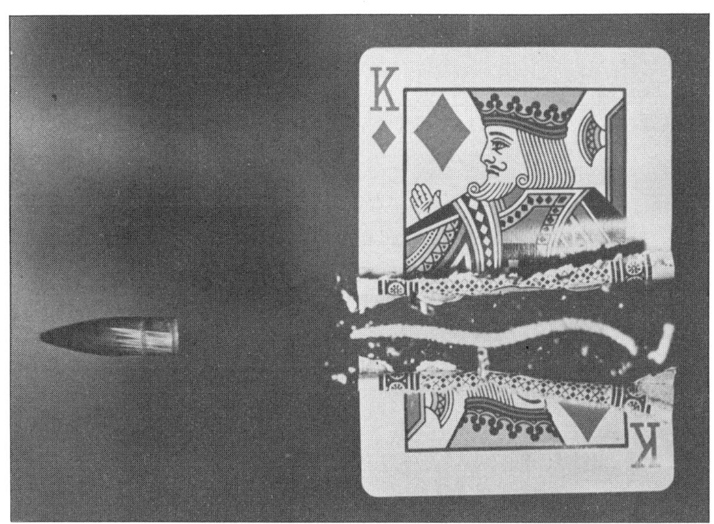

(Courtesy Dr. Harold E. Edgerton, Massachusetts Institute of Technology.)

Chapter 9

Momentum

Momentum is a commonly used term. A football team "on the move" is said to have or to have gained momentum. From this it is implied that momentum involves motion. In another instance, a running football player (with a lot of momentum) may knock down another, bigger player running toward him, and continue on toward the goal line. Here, in such a collision, the masses or inertias of the players must also be a consideration.

Such common usage of the important physical quantity of momentum gives insight into its meaning. In this chapter we will consider the technical definition of momentum, which Newton called a "quantity of motion." Knowledge of its application and conservation greatly expands our understanding of dynamical situations.

9.1 Impulse and Momentum

Let's take a look at force and motion from another point of view. When a force is applied to an object, this is often through a contact or collision of objects—for example, hitting a ball with a bat or "shooting" a pool ball with a cue stick (Fig. 9.1). The resulting motion or change in motion of an object, such as a hit ball, depends not only on the applied force, but also on the time of the contact or application of force. This may be expressed in terms of a quantity called **impulse:**

$$\text{Impulse} = \overline{\mathbf{F}}\Delta t \qquad \textbf{(Eq. 9.1)}$$

where $\overline{\mathbf{F}}$ is the average force that is applied during the contact time Δt.

The force involved in an impulse is usually not a steady force, but varies with time. For example, when a bat hits a ball, the force on the ball increases rapidly from an initial zero value, and the ball is deformed (Fig. 9.2). The force decreases as the ball recoils, and the ball returns to its original shape on leaving the bat. (This is an example of elasticity. All solid materials, even steel, are elastic to some degree.)

The force changes with time in such cases, and the impulse force may be difficult to deter-

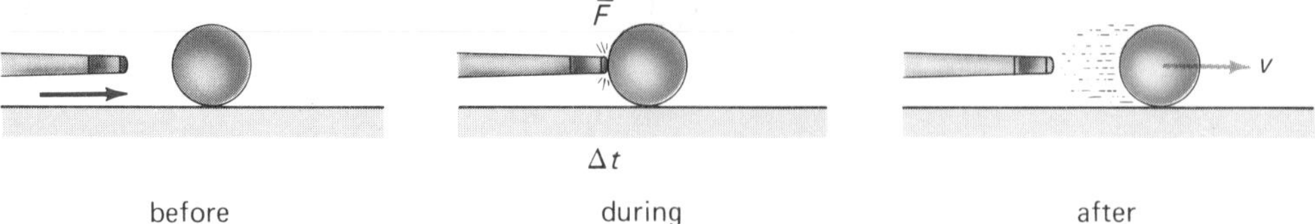

Figure 9.1
An example of impulse. The impulse occurs during the contact time Δt, and impulse = $\bar{F}\,\Delta t$.

mine. In the impulse testing of materials, drop testing machines measure force versus time (as shown in the graph in Fig. 9.2) by electromechanical means. The impulse may be approximated by using the average force. Notice from Figure 9.2 that the impulse ($\bar{\mathbf{F}}\Delta t$) is the area

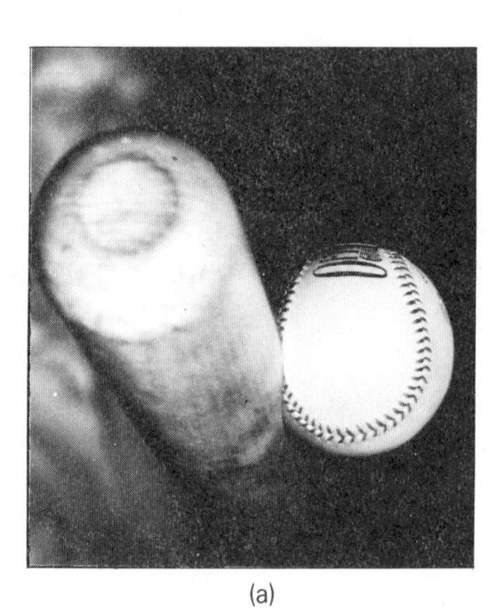

(a)

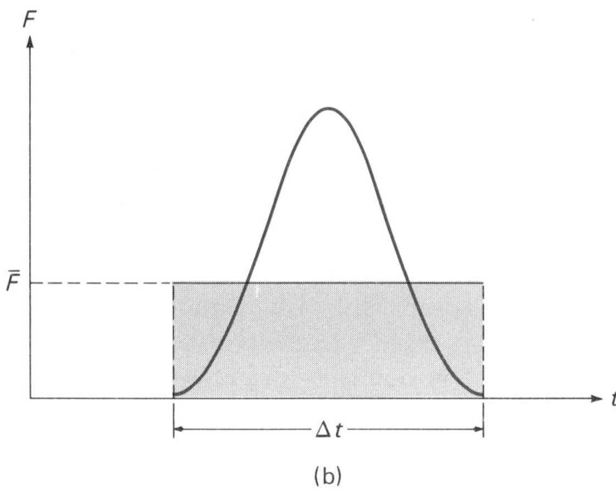

Figure 9.2
Collision impulse. (a) Objects deform in contact collisions and (b) the force varies with time.
(Courtesy of A. and J. Turk.)

under the average force line. The impulse is related to the change in motion or velocity of an object that receives the impulse. This can be seen directly from Newton's second law, $\bar{\mathbf{F}} = m\bar{\mathbf{a}}$, since

$$\bar{\mathbf{F}} = m\bar{\mathbf{a}} = m\,\frac{\Delta \mathbf{v}}{\Delta t} = \frac{m(\mathbf{v}_f - \mathbf{v}_o)}{\Delta t}$$

and

$$\bar{\mathbf{F}}\Delta t = m\mathbf{v}_f - m\mathbf{v}_o \qquad \textbf{(Eq. 9.2)}$$

The quantity $m\mathbf{v}$ is called (linear) momentum, which Newton called a "quantity of motion." The **linear momentum p** of an object then is simply the product of the mass m of an object and its instantaneous velocity **v**,

$$\boxed{\mathbf{p} = m\mathbf{v}} \qquad \textbf{(Eq. 9.3)}$$

momentum = mass × velocity

Then, in terms of momentum, the impulse (Eq. 9.2) is

$$\boxed{\bar{\mathbf{F}}\Delta t = \mathbf{p}_f - \mathbf{p}_o = \Delta \mathbf{p}} \qquad \textbf{(Eq. 9.4)}$$

Hence, **the impulse is equal to the change in momentum.***

For the case in Figure 9.1, the initial velocity of the pool ball is zero, $\mathbf{v}_o = 0$, so $\mathbf{p}_o = 0$, and $\bar{\mathbf{F}}\Delta t = \mathbf{p}$. (The momentum vector is in the same direction as the velocity.) However, suppose a batter hits a pitched ball. To find the impulse of the collision, one needs to look at the momentum before and after collision, as illustrated in Figure 9.3. Let's assume that the ball leaves the bat with the same speed as it had coming in, i.e., $v_f = v_o$. Then, in this case,

$$\begin{aligned}\bar{\mathbf{F}}\Delta t &= \Delta \mathbf{p} = \mathbf{p}_f - \mathbf{p}_o = m\mathbf{v}_f - m\mathbf{v}_o \\ &= mv_f - m(-v_o) \\ &= 2m\mathbf{v}\end{aligned}$$

* Newton originally expressed his second law in terms of momentum, i.e., $\mathbf{F} = \Delta \mathbf{p}/\Delta t$, or force is equal to the time rate of change of momentum.

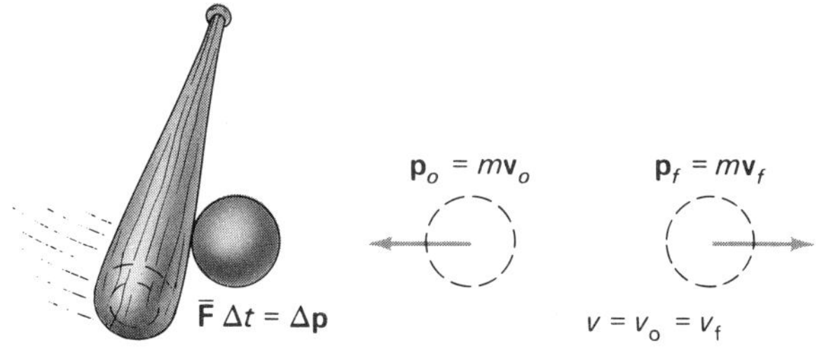

Figure 9.3
Impulse and change in momentum. The impulse is equal to the change in the momentum of the object. Here, the ball leaves the bat with the same speed as it had before being hit. The impulse is equal to $2mv$ in the direction of the final velocity.

where the minus sign is used for v_o to indicate the direction of the velocity ($\mathbf{v}_f$ and $\mathbf{v}_o$ are in opposite directions).

Notice that the unit of momentum, $\mathbf{p} = m\mathbf{v}$, is kg-m/s or, in terms of impulse, N-s.

EXAMPLE 9.1 A softball with a mass of 0.20 kg is pitched with a speed of 30 m/s. When hit by the batter, the ball leaves the bat with a speed of 40 m/s in the opposite direction. What is the magnitude of the impulse of the collision?

Solution: With $m = 0.20$ kg, $v_o = 30$ m/s, and $v_f = 40$ m/s,

$$\begin{aligned}\bar{\mathbf{F}}\Delta t = \Delta\mathbf{p} &= m[v_f - (-v_o)] = m(v_f + v_o)\\ &= (0.20\text{ kg})(40\text{ m/s} + 30\text{ m/s})\\ &= 14\text{ kg-m/s (or N-s)}\end{aligned}$$

EXAMPLE 9.2 An automobile traveling at 30 mi/h is braked to a sudden stop in 3.0 s. (a) What is the average force exerted on a 128-lb passenger by a shoulder strap? (b) What is the average force exerted on a passenger without a shoulder strap who is stopped by a padded dashboard in 0.50 s?

Solution: It is given that $v_o = 30$ mi/h $= 44$ ft/s and $v_f = 0$.

(a) With $\Delta t = 3.0$ s,

$$\begin{aligned}\bar{\mathbf{F}} &= \frac{\Delta\mathbf{p}}{\Delta t} = \frac{m(v_f - v_o)}{\Delta t} = \left(\frac{w}{g}\right)\frac{(v_f - v_o)}{\Delta t}\\ &= \left(\frac{128\text{ lb}}{32\text{ ft/s}^2}\right)\frac{(0 - 44\text{ ft/s})}{3.0\text{ s}}\\ &= -59\text{ lb}\end{aligned}$$

The minus sign indicates that the force exerted by the strap on the person was in the opposite direction to the initial velocity, which would be needed in order to bring the person to a stop.

(b) With $\Delta t = 0.50$ s,

$$\begin{aligned}\bar{\mathbf{F}} &= \frac{m(v_f - v_o)}{\Delta t}\\ &= \left(\frac{128\text{ lb}}{32\text{ ft/s}^2}\right)\frac{(0 - 44\text{ ft/s})}{0.50\text{ s}}\\ &= -350\text{ lb}\end{aligned}$$

Does this give you a good reason for wearing shoulder straps?

In bringing something to a stop, the change in momentum is a fixed value, mv_o, and is equal to the magnitude of the impulse. However, the impulse is equal to the product of the force and the time, and these quantities may be manipulated to affect the individual magnitudes. A well-known example is in catching a ball.

Suppose someone throws you a hard ball with a velocity v and you catch it with your arms rigidly extended. As we all know, the ball "stings" your hands. The change in the magnitude of the momentum is just mv, and

$$\mathbf{\bar{F}}\Delta t = mv$$

where the small Δt indicates that you stopped the ball in a short time. This makes the impulse force large, which stings your hands (Fig. 9.4).

After a couple of stinging catches, you get smart and learn to move your hands backward while catching the ball; that is, you learn to manipulate the impulse (Fig. 9.4). Assuming the ball is thrown with the same velocity, the change in momentum is the same for each catch. However, when you move your hands in the direction of the motion of the ball, the contact time is increased, which lessens the impulse force and the sting; in symbol form,

$$\bar{\mathbf{F}}\mathbf{\Delta}t = mv$$

Another example of manipulating the contact time to control the impulse force is in jumping

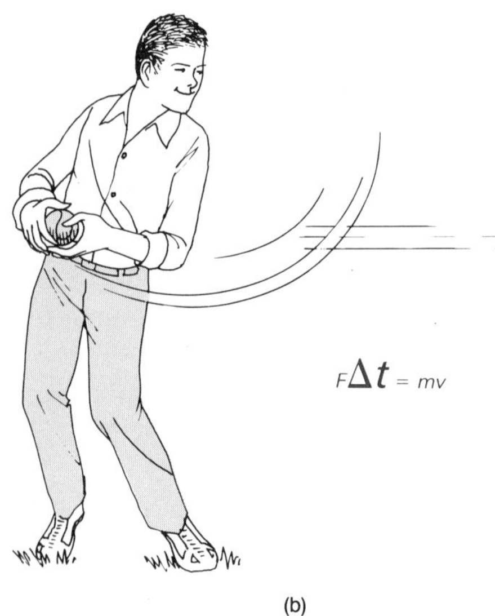

Figure 9.4
Impulse in action. In stopping (catching) a ball, the change in momentum is mv, which is equal to the impulse. (a) If the contact time is small, F is large and the ball "stings" the hands. (b) If the contact time is increased by moving the hands along with the ball, the force is smaller, and there is little or no sting.

from a high place on to a solid surface or floor. If you were to land stiff-legged, you would stop suddenly (small Δt), and the large impulse force might hurt your knees, legs, or spine. You quickly learn to bend your knees when landing so as to increase the contact time and reduce the force.

Impulse and momentum also are involved in sports. For example, in baseball or golf when it is desired to hit the ball a long distance, the player "follows through" with the swing so as to increase the contact time. As a result, the ball receives a larger impulse and a larger change in momentum, or a greater velocity.

Contact time plays an important role in automobile safety, as already shown in Example 9.2. Another application is the automobile air bag. See Special Feature 9.1.

9.2 Conservation of Linear Momentum

Recall that Newton's first law tells us that an object remains at rest or in motion with a constant velocity until acted upon by an unbalanced force. If the velocity is constant (including zero), then the momentum of an object is also constant (assuming its mass is also constant, which is usually the case). Hence, we have a condition for the conservation of momentum—no unbalanced applied force.

This condition can be seen directly from Equation 9.4, which is actually a statement of Newton's second law in momentum form. If the force on an object is zero, $\overline{\mathbf{F}} = 0$, then

$$\overline{\mathbf{F}}\Delta t = 0 = \Delta\mathbf{p} = \mathbf{p}_f - \mathbf{p}_o$$

and

$$\mathbf{p}_f = \mathbf{p}_o \qquad \textbf{(Eq. 9.5)}$$

This tells us that the momentum $\mathbf{p}_f$ at any time t is the same as the momentum at an initial time t_o, or that

In the absence of an unbalanced force, the linear momentum of an object is conserved.

The conservation of momentum also applies to a system of objects or particles, and is quite important in analyzing the motion of a system. Since momentum is a vector quantity, the total momentum **P** of a system is the vector sum of the individual momenta of the particles of the system, i.e.,

$$\mathbf{P} = \mathbf{p}_1 + \mathbf{p}_2 + \mathbf{p}_3 \ldots$$
$$= m_1\mathbf{v}_1 + m_2\mathbf{v}_2 + m_3\mathbf{v}_3 \ldots$$

Applying the previous condition of a zero unbalanced force acting on the system, we have the similar result that the total linear momentum of the system is conserved, i.e.,

$$\mathbf{P}_f = \mathbf{P}_o \qquad \textbf{(Eq. 9.6)}$$

SPECIAL FEATURE 9.1

The Automobile Air Bag

Air bags are now being installed in new automobiles as a safety feature. This had been delayed for some time for economic and manufacturing reasons. Air bags in cars will prevent many injuries in accidents, particularly for people in the front seat who now do not "buckle up for safety" with seat belts and shoulder straps. The air bag inflates automatically on the hard impact of an automobile so as to prevent the driver (or passenger) from hitting the steering wheel, the dashboard, and/or the windshield in accordance with Newton's first law (Fig. 1).

In terms of impulse, the air bag increases the stopping contact time, thereby reducing the impact force and preventing injury.

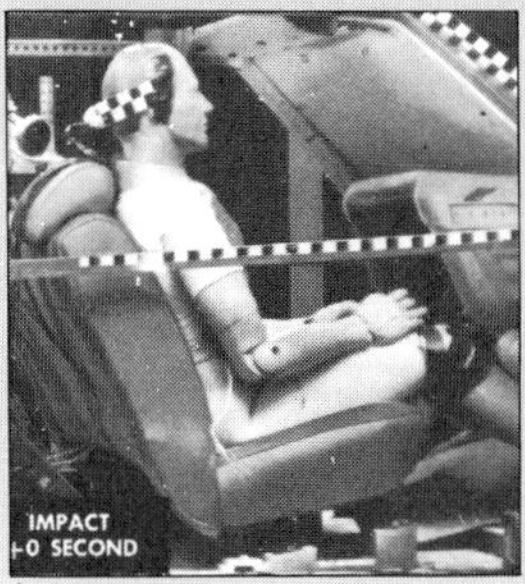

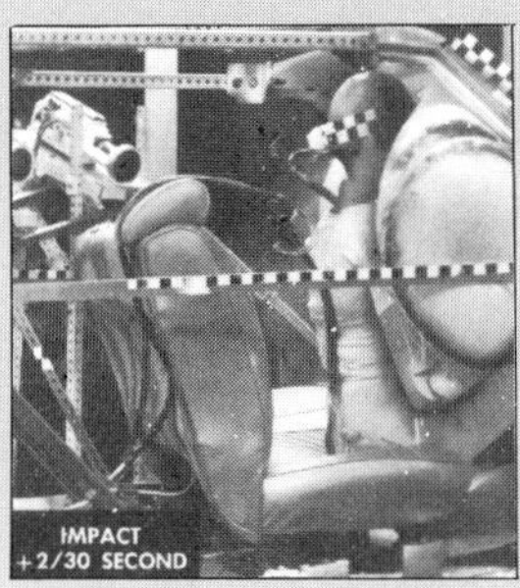

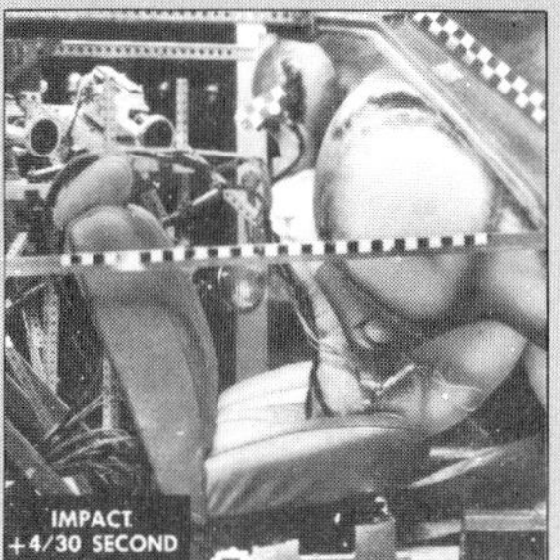

Figure 1
Automobile air bag. The air bag increases the collision contact time in stopping a person, thereby reducing the impulse force and possible injury.
(Courtesy P. Highsmith.)

or

In the absence of an unbalanced, *external* force, the total linear momentum of a system is conserved.

An external force is distinguished from an internal force that acts within the system. For example, suppose that a system is a quantity of gas and that two gas molecules collide with each other, as illustrated in Figure 9.5. During the collision, each molecule exerts an impulsive force on the other. According to Newton's third law, these forces are equal and opposite. As a

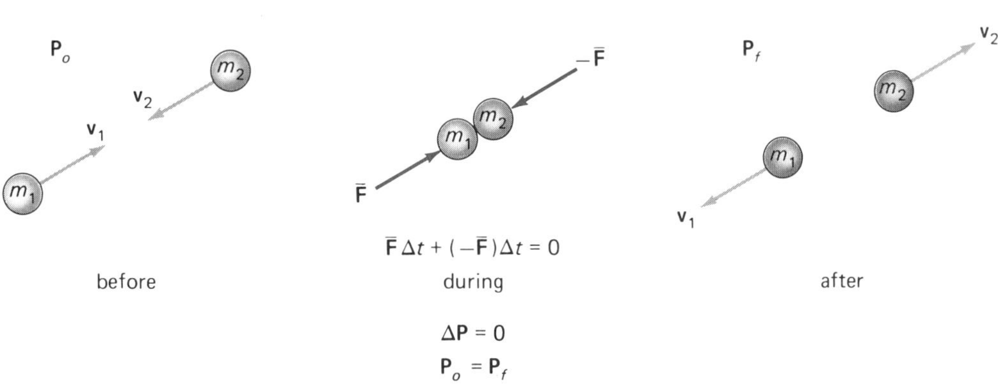

Figure 9.5
Impulse due to internal forces of a system is zero. Internal forces occur in equal and opposite pairs (Newton's third law); hence, the total impulse is zero and the total momentum is conserved.

result, the internal impulses cancel vectorially, $\bar{F}\Delta t + (-\bar{F})\Delta t = 0 = \Delta P$, and momentum is conserved. Hence, internal forces do not affect the total momentum of the system.

This does not mean that the individual momenta of the particles do not change. Momentum is usually transferred from one particle to another during collision. But the *total* linear momentum of a system is the same before and after collision, provided no external force is acting on the system. In the case of two particles,

$$\mathbf{P}_o = \mathbf{P}_f$$

$$\mathbf{p}_{1_o} + \mathbf{p}_{2_o} = \mathbf{p}_{1_f} + \mathbf{p}_{2_f} \qquad \text{(Eq. 9.7)}$$

or

$$m_1\mathbf{v}_{1_o} + m_2\mathbf{v}_{2_o} = m_1\mathbf{v}_{1_f} + m_2\mathbf{v}_{2_f}$$

If the motion of the particles is expressed in two or three dimensions, then the components of momentum are conserved in each direction. For example, if $\mathbf{P} = \mathbf{P}_x + \mathbf{P}_y$ = a constant (or conserved), then $\mathbf{P}_x$ and $\mathbf{P}_y$ must be conserved. More will be said about this in the next section, on collisions.

EXAMPLE 9.3 A man and a child, with masses of 70 kg and 30 kg, respectively, stand in the middle of an ice rink (Fig. 9.6). They "push off" each other and the man moves off with a speed of 2.0 m/s. What is the velocity of the child? (Neglect friction.)

Solution: It is given that m_m = 70 kg, m_c = 30 kg, and v_m = 2.0 m/s, where the subscripts indicate *m*an and *c*hild. When standing together, the total linear momentum of the "system" of the skaters is zero, as the initial velocities are zero. Since the "push-off" forces are *internal* forces that cancel, and there are no external forces, the total momentum is conserved. Hence,

Total momentum before = total momentum after

$$0 = \mathbf{P}_f$$

or

$$0 = m_c v_c - m_m v_m$$

and

$$m_c v_c = m_m v_m$$

Then,

$$v_c = \frac{m_m v_m}{m_c} = \frac{(70 \text{ kg})(2.0 \text{ m/s})}{30 \text{ kg}} = 4.7 \text{ m/s}$$

in the direction opposite to the man's velocity.

Figure 9.6
An example of conservation of momentum. See Example 10.3.

EXAMPLE 9.4 Two identical railroad cars roll toward each other, as shown in Figure 9.7, and become coupled during collision. What is the velocity of the coupled cars after collision? (Assume momentum is conserved.)

Solution: Let the mass of the cars be m, and from the figure, v_{o_1} = 20 km/h and v_{o_2} = 10 km/h. Then, applying the conservation of momentum with v_f as the final velocity of the coupled cars (assumed positive or in the direction of v_{o_1}),

Total momentum before = total momentum after

$$mv_{o_1} - mv_{o_2} = (m + m)v_f$$

Cancelling the m's,

$$v_{o_1} - v_{o_2} = 2v_f$$

$$20 \text{ km/h} - 10 \text{ km/h} = 2v_f$$

and

$$v_f = \frac{10 \text{ km/h}}{2} = 5.0 \text{ km/h}$$

Since v_f is positive, the velocity is in the assumed direction. What do you think would be the result if the cars approached each other with equal speeds?

9.3 Collisions

Things are always bumping into or colliding with each other—sometimes purposefully, like billiard balls, and sometimes not so purpose-

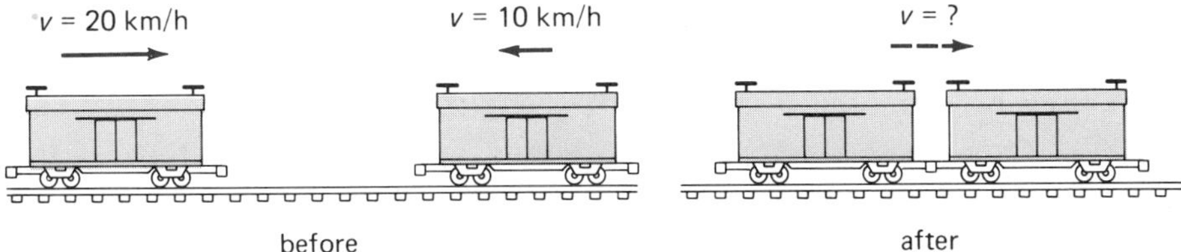

Figure 9.7
Railroad car collision. See Example 9.4.

fully, like automobiles. We ordinarily think of objects coming into contact during collisions with contact forces involved. However, this is not always the case. Action-at-a-distance forces can be involved. For example, a meteor may "collide" with the Earth (gravity acting) and miss it completely. Also, a great deal of our understanding of submicroscopic phenomena comes from collision scattering.

Our basic understanding of collisions comes from classical principles. It is convenient to consider contact collisions between spherical particles or balls for simplicity. In any interaction, momentum is involved. Therefore, in a broad sense, a **collision** is any interaction in which momentum is exchanged or transferred. In the absence of an external force, the total linear momentum is conserved in a collision.

The conservation of mechanical energy may also be a consideration. However, this depends on the elasticity of the objects—that is, whether they recover from being deformed after collision. If an object is permanently deformed, e.g., dented, the mechanical energy is not conserved since some of the energy of the colliding objects goes into doing work. In analyzing contact collisions of spherical balls, it is convenient to consider only horizontal collisions so that gravitational potential energy is not a factor and only kinetic energy is involved.

Collisions are generally classified as being elastic or inelastic, depending on whether or not the kinetic energy is conserved.

> **In an elastic collision, the kinetic energy is conserved.**

That is, the kinetic energy is the same before and after collision. The kinetic energy changes during collision, but if the objects are perfectly elastic, any energy used in deformation is restored after maximum deformation during collision.

This is an ideal case, but many hard objects, such as billiard balls, steel balls, bowling balls, and marbles, have nearly elastic collisions. We find in actuality that only atoms and subatomic particles have truly elastic collisions.

Then, **for elastic collisions the total linear momentum and the total kinetic energy are conserved:**

$$\boxed{\begin{aligned} \mathbf{P}_o &= \mathbf{P}_f \\ KE_o &= KE_f \end{aligned}} \qquad \textbf{(Eq. 9.8)}$$

(conditions for an elastic collision)

Kinetic energy is not conserved in an inelastic collision. Here, energy is lost in doing work, as in permanently deforming a ball. If two objects stick together on collision, the collision is said to be completely inelastic. This does not mean that all of the kinetic energy is lost ($v_f \neq 0$), but only an amount of energy consistent with the conservation of momentum. **Momentum is conserved in both elastic and inelastic collisions.**

Let's consider a one-dimensional, head-on, elastic collision between two balls, one initially at rest for convenience (Fig. 9.8). Assuming that both balls move in the positive direction after collision, by the conservation of linear momentum we have

$$\underset{\text{(before)}}{m_1 v_{1_o}} = \underset{\text{(after)}}{m_1 v_{1_f} + m_2 v_{2_f}} \qquad \textbf{(Eq. 9.9)}$$

From the conservation of kinetic energy,

$$\tfrac{1}{2} m_1 v_{1_o}^2 = \tfrac{1}{2} m_1 v_{1_f}^2 + \tfrac{1}{2} m_2 v_{2_f}^2 \qquad \textbf{(Eq. 9.10)}$$

Rearranging these equations,

$$m_1(v_{1_o} - v_{1_f}) = m_2 v_{2_f}$$
$$m_1(v_{1_o}^2 - v_{1_f}^2) = m_2 v_{2_f}^2$$

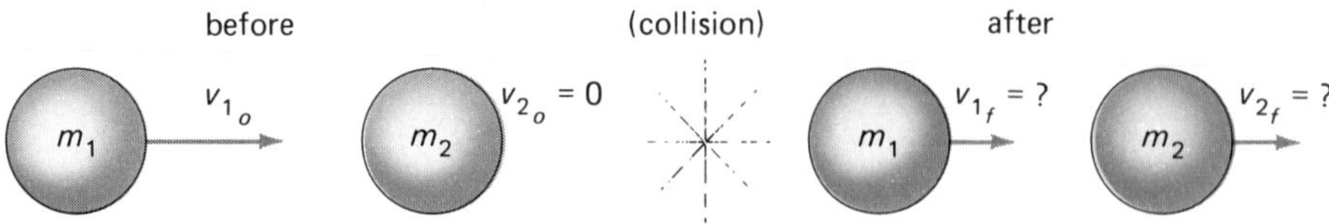

Figure 9.8
Collision analysis. General conditions in considering an elastic collision of two balls, one initially stationary. See text for description.

Then, since $v_{1_o}^2 - v_{1_f}^2 = (v_{1_o} - v_{1_f})(v_{1_o} + v_{1_f})$, dividing one equation by the other yields

$$v_{1_o} + v_{1_f} = v_{2_f} \qquad \textbf{(Eq. 9.11)}$$

Using this equation to eliminate one of the velocities in Equation 9.9 (or Eq. 9.10), we may write the final velocities in terms of the initial velocity:

$$v_{1_f} = \left(\frac{m_1 - m_2}{m_1 + m_2}\right) v_{1_o}$$

$$v_{2_f} = \left(\frac{2m_1}{m_1 + m_2}\right) v_{1_o} \qquad \textbf{(Eq. 9.12)}$$

Thus we see that the final velocities depend on the relative masses of the balls. The balls will not always move off after collision as was assumed in Figure 9.8. Given the masses of the balls, Equations 9.12 tell us what actually happens in such an *elastic* collision.

For example, let's consider three cases.

(a) $m_1 = m_2$. From Equations 9.12,

$$v_{1_f} = 0 \qquad \text{and} \qquad v_{2_f} = v_{1_o}$$

If the masses of the balls are equal, there is a *complete* transfer or exchange of momentum. This can be shown to be the case when m_2 is not initially at rest (Fig. 9.9).

(b) $m_1 > m_2$; for example, $m_1 = 2m_2$. From Equations 9.12,

$$v_{1_f} = \tfrac{1}{3}v_{1_o} \qquad \text{and} \qquad v_{2_f} = \tfrac{4}{3}v_{1_o}$$

Here, if a more massive ball strikes a less massive one, the incoming ball is slowed down and the light, stationary ball is "knocked" away with a speed greater than the initial speed of the incoming ball (Fig. 9.9).

Don't be fooled by the relative magnitudes of the velocities of the balls. m_2 is less massive, and momentum is still conserved. In this specific case, two thirds of the initial momentum is transferred to the stationary ball during collision, as can be easily shown.

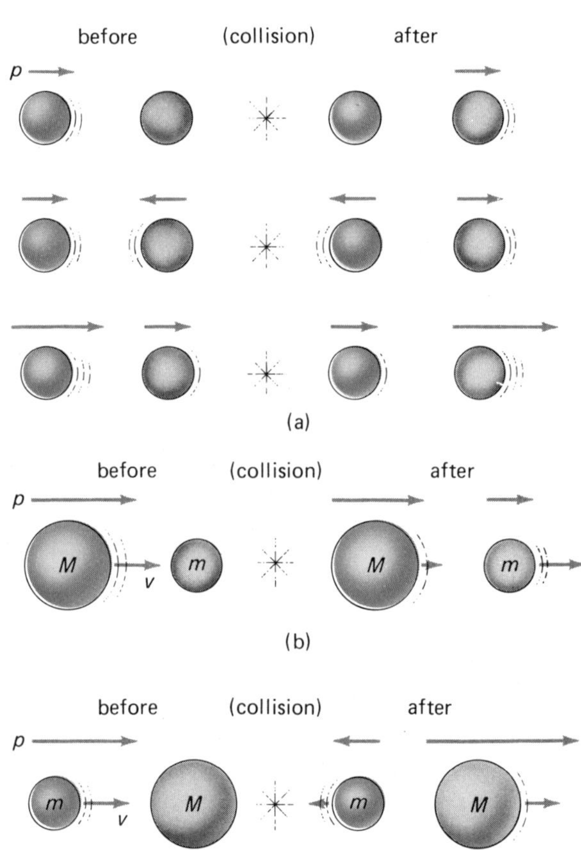

Figure 9.9
Elastic collisions. Case (a): $m_1 = m_2$. If the masses are equal, there is a complete transfer of momentum as indicated by the arrows. Case (b): $m_1 > m_2$. When a more massive ball strikes a less massive ball, the incoming ball is slowed down and the light ball speeds away with a speed greater than that of the initial speed of the incoming ball. Case (c): $m_1 < m_2$. The incoming, less massive ball rebounds in the opposite direction and the more massive ball moves in the other direction.

(c) $m_1 < m_2$; for example, $m_1 = m_2/2$ or $2m_1 = m_2$. From Equations 9.12,

$$v_{1_f} = -\tfrac{1}{3}v_{1_o} \qquad \text{and} \qquad v_{2_f} = \tfrac{2}{3}v_{1_o}$$

If a light ball strikes a more massive ball, the light ball rebounds in the opposite direction (as indicated by the minus sign), and the more massive ball moves off in the other direction (Fig. 9.9).

How about the momentum in this case? The magnitude of the initial momentum is $p_o = m_1v_{1_o}$, and the magnitude of the momentum of the more massive ball is $p_{2_f} = m_2v_{2_f} = 2m_1(\tfrac{2}{3}v_{1_o}) = \tfrac{4}{3}m_1v_{1_o}$—a violation of the conservation of momentum? No, momentum is still *vectorially* conserved, since

$$\mathbf{p}_{1_f} + \mathbf{p}_{2_f} = m_1(-\tfrac{1}{3}v_{1_o}) + \tfrac{4}{3}m_1v_{1_o} = m_1v_{1_o} = \mathbf{p}_o$$

The momentum transferred to the more massive ball is just equal to the (vector) *change* in the momentum of the light ball.

EXAMPLE 9.5 A 0.20-kg ball with a velocity of 3.0 m/s along the $+x$-axis collides elastically with a stationary 0.60-kg ball. (a) What are the final velocities of the balls? (b) Show that the kinetic energy is conserved.

Solution: It is given that $v_{1_o} = 3.0$ m/s, $m_1 = 0.20$ kg, and $m_2 = 0.60$ kg.

(a) From Equations 9.12,

$$v_{1_f} = \left(\frac{m_1 - m_2}{m_1 + m_2}\right) v_{1_o} = \left(\frac{0.20 \text{ kg} - 0.60 \text{ kg}}{0.80 \text{ kg}}\right) (3.0 \text{ m/s}) = -1.5 \text{ m/s}$$

$$v_{2_f} = \left(\frac{2m_1}{m_1 + m_2}\right) v_{1_o} = \left[\frac{2(0.20 \text{ kg})}{0.80 \text{ kg}}\right] (3.0 \text{ m/s}) = 1.5 \text{ m/s}$$

So, the balls move away from each other after collision with equal and opposite velocities.

(b) The kinetic energy is conserved if

$$KE_{1_o} = KE_{1_f} + KE_{2_f}$$

Then,

$$KE_{1_o} = \tfrac{1}{2}m_1v_{1_o}^2 = \tfrac{1}{2}(0.20 \text{ kg})(3.0 \text{ m/s})^2 = 0.90 \text{ J}$$

and

$$KE_{1_f} + KE_{2_f} = \tfrac{1}{2}m_1v_{1_f}^2 + \tfrac{1}{2}m_2v_{2_f}^2 = \tfrac{1}{2}(0.20 \text{ kg})(-1.5 \text{ m/s})^2 + \tfrac{1}{2}(0.60 \text{ kg})(1.5 \text{ m/s})^2 = 0.225 \text{ J} + 0.675 \text{ J} = 0.90 \text{ J}$$

and the kinetic energy is conserved, which of course was an initial condition for the *elastic* collisions described by Equations 9.12.

A popular novelty item is shown in Figure 9.10. When one ball swings in, one swings out; when two balls swing in, two swing out; and so on. You can now understand why this happens. During the collisions, momentum is conserved in the horizontal direction. Since the balls are of equal mass, ideally (elastically) a complete transfer of momentum occurs.

When one ball swings in, it stops as it transfers all of its momentum to the other ball [case (a)]. The momentum is transferred in collisions down the line of balls as shown in the figure, and the final ball swings out. Because the balls in the row are in contact, you do not "see" the collisions, but they occur as though the balls were separated (Fig. 9.10).

When two balls swing in, the first to make contact with a stationary ball transfers its momentum and comes to a stop. The second incoming ball then hits the stopped ball and the momentum transfer is repeated down the row—a "double shot" of momentum, so to speak—and two balls swing out at the opposite end.

EXAMPLE 9.6 If two balls swing in with a velocity v in the apparatus shown in Figure 9.10, why doesn't one ball swing out with a velocity of $2v$?

Solution: It is given that $v_o = v$ and $v_f = 2v$. In such a case, the momentum would be conserved:

$$mv + mv = m(2v)$$

or

$$\underset{\text{(in)}}{(m + m)v} = \underset{\text{(out)}}{2mv}$$

However, how about the kinetic energy for the (nearly) elastic collisions? The initial and final

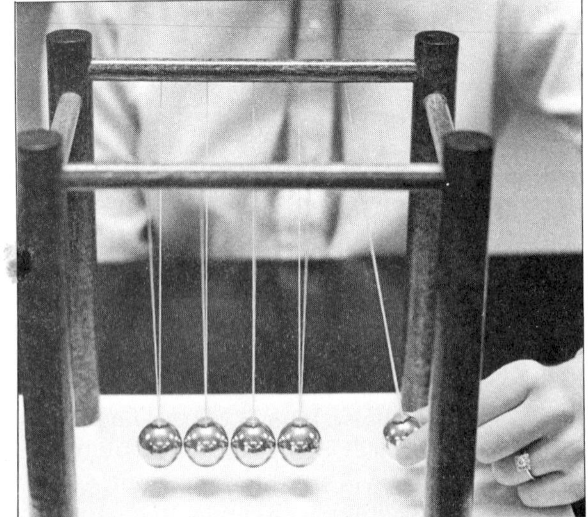

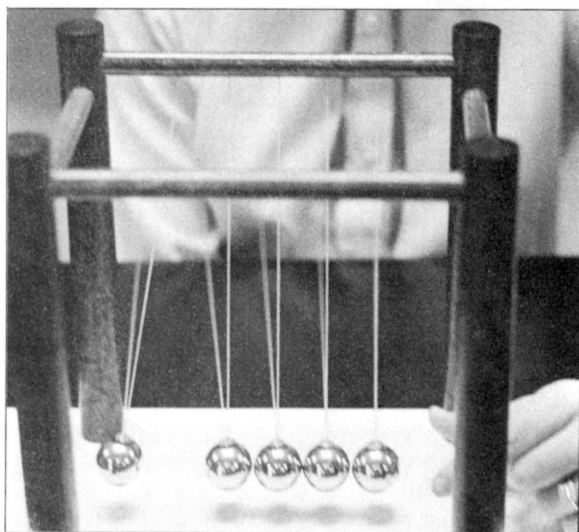

(a)

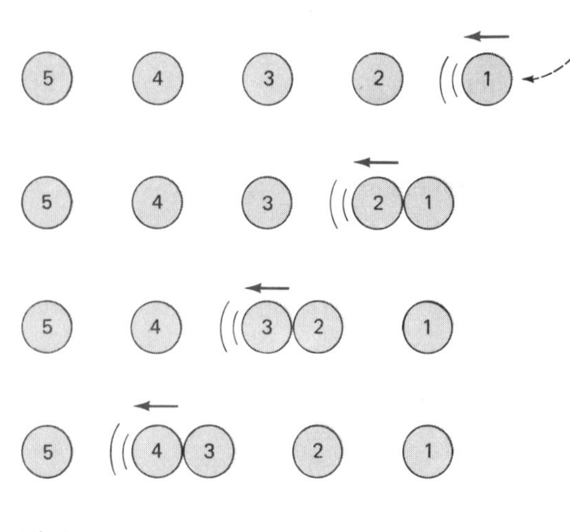

(b)

Figure 9.10

A case of momentum transfer in elastic collisions. Momentum is transferred down the line of balls, and when one ball swings in, one swings out. When two balls swing in, two balls swing out; and so on. The line drawing (b) illustrates the collision transfer process when a single ball swings in.

kinetic energies are

$$KE_o = \tfrac{1}{2}(m + m)v^2 = mv^2$$

and

$$KE_f = \tfrac{1}{2}m(2v)^2 = 2mv^2$$

Hence, $KE_o \neq KE_f$, and this violates the conservation of energy—in fact, energy would be created with $KE_f > KE_o$.

Not all collisions are head-on. But even in glancing collisions the total linear momentum is conserved. Here, the components of momentum show that the momentum is conserved as illustrated in Figure 9.11.

Initially, all the momentum is in the x direction, and the momentum in the y direction is zero. After collision, the same is true. The y components of the momenta of the two balls cancel, and the x components add up to equal the original momentum. The angles of the momenta of the balls and the directions in which they move are given by $\theta = \tan^{-1}(p_y/p_x)$ for each ball.

9.4 Rockets and Jet Propulsion

In Chapter 6, rockets were considered in terms of the action and reaction forces of Newton's third law. Rockets and jet engines can also be analyzed in terms of momentum. To give an idea how this is done, let's first consider the firing of an idealized cannon as illustrated in Figure 9.12.

When the powder charge explodes (an internal force), the cannonball accelerates down the barrel. The total momentum of the system at any instant is still zero. This is easier to see after the cannonball leaves the barrel. Then, the momentum of the cannonball moving with a constant muzzle velocity is equal and opposite to the momentum of the cannon. The speed of the

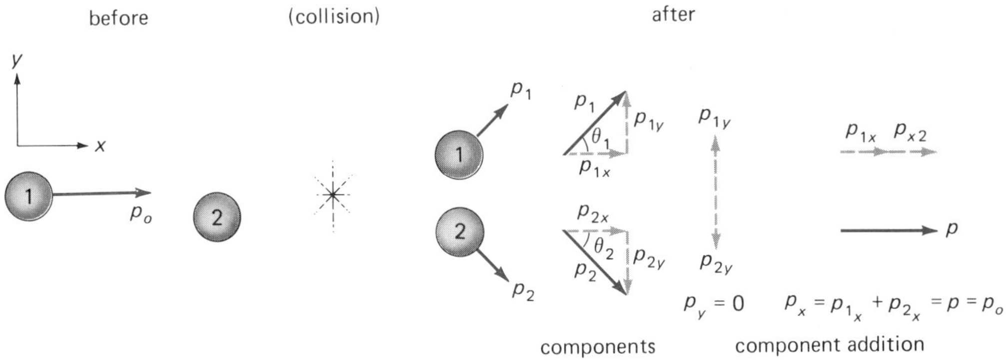

Figure 9.11
A glancing collision. The total momentum is conserved before and after collision in the component (x and y) directions.

cannonball is much greater than that of the cannon. Why?

In a similar manner, the conservation of momentum can be applied to a rocket (Fig. 9.13). Assuming the rocket blasts off from rest, the total momentum of the exhaust gases (assuming they don't strike anything) at *any instant* is equal and opposite to the momentum of the rocket.

This is difficult to analyze, since one body, the rocket, is continually losing mass, and the other body, the expelled gases, is continually gaining mass. A simple example is given in Special Feature 9.2.

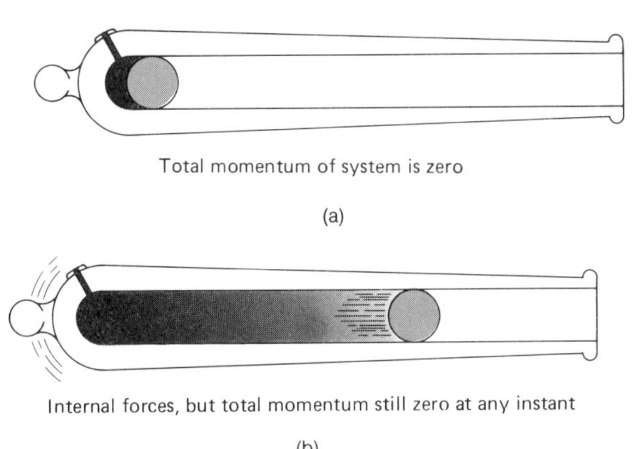

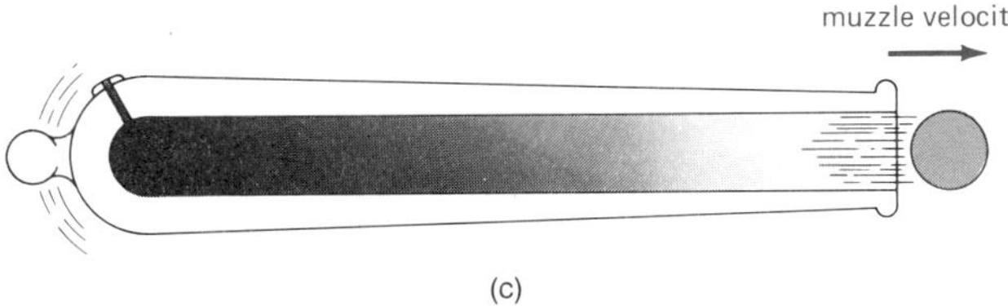

Figure 9.12
Conservation of momentum. The total momentum of the system is zero in each idealized case.

Jet aircraft engines make use of the rocket principle. In jet engines, air is taken into the engine, accelerated toward the rear, and discharged rearward, thereby developing a forward thrust. The intake air is compressed by a compressor, mixed with fuel, and fired in the combustion chamber of the engine. The exhaust gases are used to turn a turbine, which drives the compressor. The turbine also drives a propeller or fan in the case of turboprop or turbofan engines, respectively. The exhaust gases pass through propelling nozzles to produce a high-speed stream of expelled gases and hence a forward thrust on the engine and aircraft.

Figures 9.14 and 9.15 show a jet engine and a special application of a turbine-powered device. This "aerial platform," which uses a small turbofan engine, can fly for 30 minutes at speeds up to 60 mi/h. It takes off vertically; accelerates rapidly forward, backward, and sideways; hovers; and rotates on its axis (perhaps it is the transportation of the future?).

Reverse Thrust

Momentum plays an important role in stopping or braking jet aircraft. If you have flown in a jet aircraft, you no doubt recall the roar when the pilot "revved up" the engines and the braking action after the plane touched down on landing. How does applying power to the engines cause a plane to slow down?

SPECIAL FEATURE 9.2

The Rocket*

A rocket, or spaceship, consists of two primary parts: (1) the payload—the rocket hull, astronauts, instruments, and so on that we wish to propel somewhere—and (2) the fuel and anything else that gets ejected in propelling the rocket.

To simplify the situation, let's consider a rather primitive type of spacecraft, of mass 1001 kg, far out in space where no forces are acting on the craft (Fig. 1). Let m_2 be the mass of the rocketship *at any time* and m_1 be the mass of a 1-kg "fuel pellet."

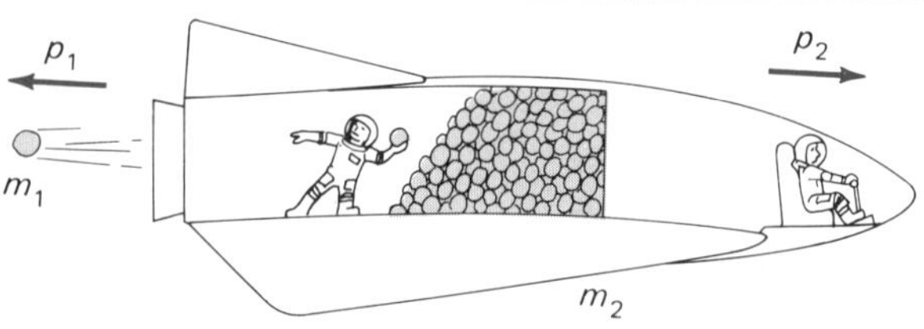

Figure 1
An "impulse" rocket propelled by the ejection of "fuel pellets." See text for description.

Suppose an astronaut throws one of the fuel pellets out the back of the rocket with a velocity of 20 m/s. We take the position where the pellet was thrown out as our zero position (see figure). Then, by the conservation of momentum,

$$m_1 v_1 = -m_2 v_2$$

or $$(1 \text{ kg})(20 \text{ m/s}) = -(1000 \text{ kg})v_2$$

and $$v_2 = -0.02 \text{ m/s}$$

where the minus sign indicates that the change in the velocity of the rocket is in the opposite direction to that of the fuel pellet, or to the right in the figure.

Hence, the rocket has changed its velocity by a small amount relative to the position in space where the fuel pellet was thrown out, and the rocket has 1 kg less mass.

The astronaut then throws out another 1-kg chunk of fuel with the same velocity, and

$$m_1 v_1 = -m_2 v_2$$

or $$(1 \text{ kg})(20 \text{ m/s}) = -(999 \text{ kg})v_2$$

and $$v_2 = -0.02002 \text{ m/s}$$

The change in the velocity of the rocket this time is slightly greater because the rocket lost mass in the first throw. Relative to the original zero position, the change in the rocket's velocity is now $-0.02 + (-0.02002) = -0.04002$ m/s.

If the astronaut continued to throw out fuel pellets, the rocket's velocity would change slightly each time and it would continue to lose mass, which would give the rocket a slightly greater velocity change each time. For example, by the time the rocket had a mass of 401 kg, the next-thrown fuel pellet would give the rocket a change in velocity of

$$m_1 v_1 = -m_2 v_2$$

$$(1 \text{ kg})(20 \text{ m/s}) = -(400 \text{ kg})v_2$$

and $$v_2 = -0.05 \text{ m/s}$$

The change in velocity for this throw is $2\frac{1}{2}$ times greater than for the first fuel pellet.

If the astronaut continued to throw out fuel pellets, the rocket would continue to lose mass, and the change in its velocity would increase each time. If the fuel pellets were thrown out often enough, the change in velocity would appear smooth, which the astronaut would interpret as an increasing acceleration.

For a real rocket during a "burn," the exhaust gas molecules are ejected out the rear of the rocket at speeds of several thousand meters per second. The average velocity of the exhaust gases and the amount of material ejected are relatively constant. Therefore, the rocket thrust and the force on the rocket are constant. The increasing acceleration is due to the rocket's decreasing mass.

Of course, the masses of the exhaust gas molecules are much less than that of the 1-kg fuel pellets of our simplified rocketship, but there are a lot of them. To show the effect of high exhaust velocities, suppose the astronaut had thrown out the first fuel pellet with a speed of 20,000 m/s. Then,

$$m_1 v_1 = -m_2 v_2$$

$$(1 \text{ kg})(20{,}000 \text{ m/s}) = -(1000 \text{ kg})v_2$$

and $$v_2 = -20 \text{ m/s}$$

A large velocity change indeed!

To further reduce the in-flight mass of a rocketship, multistage rockets are used, and the burnt-out stages are jettisoned.

* Adapted from Highsmith, P. E., and A. S. Howard, *Adventures in Physics*, W. B. Saunders Co., Philadelphia, 1972.

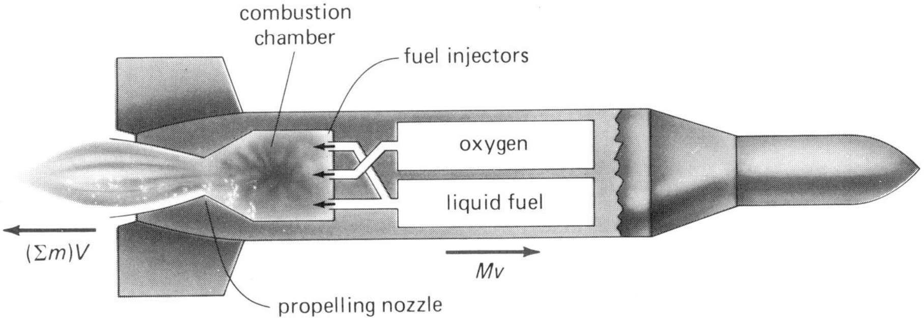

Figure 9.13

The rocket engine. The rocket engine is a jet engine but does not use atmospheric air in its combustion system. It carries oxygen and fuel, thus enabling it to operate outside the Earth's atmosphere. If originally zero, the total momentum of the system is zero at any instant, and the magnitude of the momentum of the exhaust gases $(\Sigma m)V$ is equal in magnitude to the momentum of the rocket, Mv.

Figure 9.14

Jet propulsion. An aircraft jet engine. (Courtesy Rolls-Royce, Ltd.)

Figure 9.15
Jet propulsion. A special application "aerial platform" propelled by a turbofan engine. (Courtesy Williams International.)

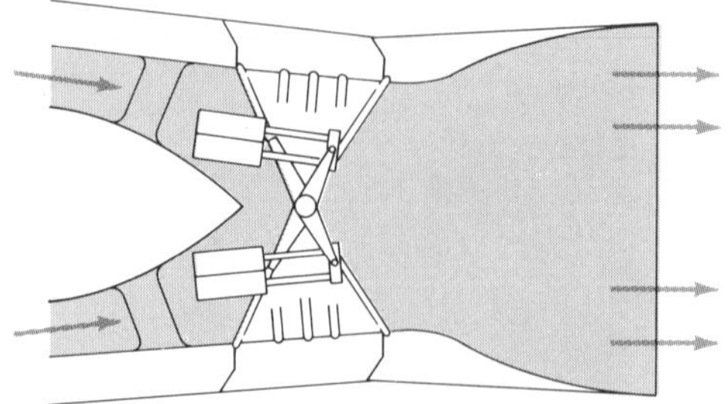

clamshell doors in forward thrust position

(a)

(b)

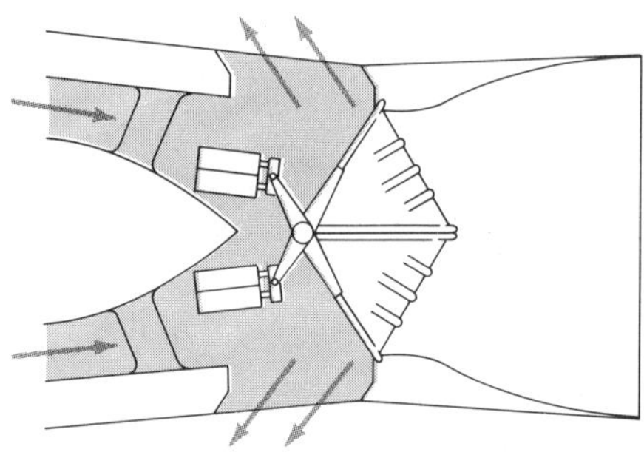

clamshell doors in reverse thrust position

(c)

Figure 9.16
Reverse thrust. (a) To produce a forward thrust, high-speed gases are expelled rearward. (b) and (c) To produce a reverse thrust for braking, blocker doors are inserted and cascade vanes direct the exhaust gases forward. This particular type of thrust reverser assembly uses "clamshell" doors. (Courtesy Rolls-Royce, Ltd.)

This is an example of applying reverse thrust, and one method is to use clamshell doors that deflect the exhaust gases from the jet engines in a forward direction (Fig. 9.16). With the doors in an in-flight open position (a), the exhaust gases go out the back of the engine, and the plane receives momentum in the forward direction, or a forward thrust. With the clamshell doors closed (c), the exhaust gases are deflected in the forward direction, and there is a (vector) component of the momentum in that direction.

By the conservation of momentum, the engine and the airplane receive a momentum in the reverse direction, which gives an impulse or reverse thrust in that direction, thus helping to slow and stop the plane. (The airplane brakes would never do it alone on normal runways.)

The deflector doors may also be exterior to the engine and form part of the engine casing. Smaller commercial jet aircraft use reverse thrust to "back away" from loading docks. Larger planes are pushed by tow motors. (Why?) The next time you are at a large airport, watch to see if you can see the deflector doors when a small plane backs out.

Important Terms

linear momentum the product of mass and velocity

impulse the product of the applied force and time of application; it is equal to the change in linear momentum

conservation of linear momentum the principle that, in the absence of an unbalanced, external force, the linear momentum of a system is conserved

collision any interaction in which momentum is exchanged or transferred

elastic collision a collision in which the total linear momentum and total kinetic energy are conserved

inelastic collision a collision in which the total kinetic energy is not conserved; the total linear momentum is conserved in an inelastic collision

rocket thrust the forward force due to a change in momentum in a rocket system

reverse thrust the rearward force due to a change in momentum in a rocket system

Important Formulas

impulse: $\overline{\mathbf{F}}\Delta t$

linear momentum: $\mathbf{p} = m\mathbf{v}$

impulse and momentum relationship: $\overline{\mathbf{F}}\Delta t = \Delta\mathbf{p} = \mathbf{p}_f - \mathbf{p}_o$

conservation of linear momentum: If $\overline{\mathbf{F}} = 0$, then $\mathbf{P}_f = \mathbf{P}_o$

final velocities in a two-body, *elastic* collision ($v_{2_o} = 0$):

$$v_{1_f} = \left(\frac{m_1 - m_2}{m_1 + m_2}\right) v_{1_o}$$

$$v_{2_f} = \left(\frac{2m_1}{m_1 + m_2}\right) v_{1_o}$$

Questions

Impulse and Momentum

1. Pole vaulters and gymnasts use padded mats for landings. How do these mats help?
2. A boxer quickly learns to move his head backward when he sees he is going to receive a jab to the head. What does this head motion accomplish?
3. A golfer using a nine-iron to chip onto the green or a wedge to get out of a sand trap uses a short "chopping" swing. Why?
4. Explain how impulsive forces are used in material testing.
5. Fragile items are packed securely in soft materials such as Styrofoam before shipping. Explain how this prevents breakage should the package be hit or dropped.
6. Most new automobiles are equipped with bumpers that collapse on a large impact. What is the purpose of such bumpers?
7. Guardrails along roadsides collapse (bend and crumple) when a car runs into them. Wouldn't it be better to install stronger guardrails so they wouldn't have to be replaced so often?
8. Why do baseball players and golfers follow through with swings for long drives?

Conservation of Linear Momentum

9. In the collision of balls of equal masses, the momentum is conserved. We might also say that the velocity is conserved. Why is this? If the balls were not of equal mass, would the velocity be conserved? Would the momentum be conserved?
10. Suppose you are standing in the middle of a frozen lake and the ice is perfectly frictionless. How can you get to the shore?
11. A person standing on the Earth jumps vertically upward. According to the conservation of momentum, what happens to the Earth? (You can move the Earth!) Can we observe this effect on the Earth? Explain.
12. A rubber ball is thrown horizontally against a wall, and the ball rebounds in the opposite direction with essentially the same speed it had just before hitting the wall. Is momentum conserved? Explain.
13. In testing explosive charges in space, an astronaut leaves a bomb there that is later detonated by remote control. The bomb explodes into many fragments that fly off in various directions. What is the total momentum of the bomb fragments?
14. A cannonball is fired horizontally from a cannon with a constant muzzle velocity. Is the horizontal momentum conserved thereafter? Is the vertical momentum conserved? (*Hint:* think in terms of components of momentum.)

Collisions

15. Show that the kinetic energy of an object of mass m can be written in terms of its momentum as $KE = p^2/2m$.
16. A ball of mass m and velocity v collides with a stationary ball of unknown mass. What is the mass of the stationary ball if after collision the incoming ball is stationary and the other ball has a velocity?

Figure 9.17
Unwanted collision. See Question 18.

17. Could Equations 9.12 be used for an inelastic collision? Explain.
18. Two Volkswagen "bugs" collide in a completely inelastic collision, as illustrated in Figure 9.17. What is the velocity of the cars immediately after collision?
19. Two cars of the same model and year (same mass) approach an intersection with the same speed. One is traveling north and the other east. At the intersection, one runs a stop sign and they collide in a completely inelastic collision. In what direction do the cars move after collision? (*Hint:* use a vector diagram.)
20. In the suspended ball set with five balls (Fig. 9.10), if three balls are allowed to swing in toward two stationary balls, explain why three balls swing out. Why not one ball with a velocity of $3v$?
21. Is momentum conserved at all times in the action of the suspended ball set shown in Figure 9.10? Explain.

Rockets and Jet Propulsion

22. How does the multistage feature improve the efficiency of a rocket?
23. When a balloon is blown up and released, it flies around in a zig-zag fashion. What causes this?
24. When a rifle or gun is fired, there is generally more recoil or "kick" the greater the caliber (the more massive the bullet) of the gun. Why is this? Would you want to fire a rifle that was only a few times heavier than the bullet? Explain.
25. A bazooka or rocket launcher used as a weapon against tanks is essentially a tube open at both ends that a soldier holds on the shoulder when firing a heavy shell. There is very little recoil on the person firing the rocket launcher. Why is this? Also, it is very important that another person loading the launcher not stand behind it when it is fired. Why is this?
26. The Army uses large-caliber "recoilless" rifles. Explain how a rifle might be made recoilless. (Actually, the rifles are not completely recoilless, but the recoil is greatly reduced.)
27. An astronaut approaching the moon must fire retrorockets to slow the craft down so it will not crash (fall) into the moon. The exhaust gases of the retrorockets are expelled toward the moon. Explain how the firing of these rockets slows down the approach.
28. Analyze the maneuverability of an astronaut using a hand rocket in space in terms of momentum. See Figure 6.14.
29. Suppose a person in a sailboat is becalmed on a lake and has a large battery-operated fan on board. Getting a bright idea, the person aims the fan at the sail in order to blow the boat ashore (Fig. 9.18). However, he finds that the boat goes nowhere. Getting an even brighter idea (and applying physics principles he learned in college), he takes the sail down and turns the fan around. The boat then moves forward. Explain the motion (or lack of it) in each case.

Figure 9.18
Jet propulsion? See Question 29.

Problems

Levels of difficulty are indicated by asterisks for your convenience.

9.1 Impulse and Momentum

1. A constant force of 12 N is applied to an object for 0.10 s. (a) What is the impulse on the object? (b) What is its change in momentum?
2. If the 2.0-kg object in Problem 1 had an initial speed of 0.50 m/s, what would be its speed after the impulse was applied? (Assume all vector quantities to be in the same direction.)
3. A 90-kg football player runs with a velocity of 5.0 m/s down the field. What is his linear momentum?
4. With what speed would a 0.010-kg bullet have to travel to have the same momentum as the football player in Problem 3?

*5. How fast must a 1600-lb car be traveling to have the same momentum as a 2-ton truck traveling at 30 mi/h?

*6. During a snowball fight, a 0.15-kg snowball traveling with a horizontal velocity of 10 m/s hits a student in the back of the head. (a) What is the impulse of the collision? (b) If the contact time of the collision is 0.10 s, what is the average force on the student's head?

*7. A 1000-kg automobile traveling at 25 m/s is brought to a stop in 5.0 s. What is the magnitude of the average braking force? (Work this problem in two ways.)

*8. A baseball player hits a 0.10-kg ball with a horizontal impulse of 1.0 N-s. (a) What is the change of momentum of the ball? (b) If the incoming speed of the ball is 30 m/s, what is its outgoing speed?

*9. A rubber ball is thrown horizontally against a wall and the ball rebounds in the opposite direction with the same speed it had just before hitting the wall. If the ball has a mass of 0.10 kg and an incoming speed of 0.50 m/s, what is the change in momentum?

*10. Suppose the ball in Problem 9 is incident on the wall at an angle of 45°. Would there be any conservation of momentum? Show that the change of momentum would be equal to $\Delta p = \sqrt{2}p$ in the direction normal (perpendicular) to the wall, where $p_f = p_o = p$.

*11. A 3200-lb automobile traveling at 60 mi/h along a straight, level road has its speed reduced to 30 mi/h in 4.0 s. What is the magnitude of the average stopping force? Work the problem in two ways: (a) using linear momentum and (b) by finding the average acceleration.

**12. The muzzle velocity of a 0.010-kg bullet is 400 m/s. If the length of the rifle barrel is 1.0 m, (a) what is the magnitude of the average force on the bullet while in the barrel? (b) What is the impulse on the bullet?

**13. A 0.25-kg ball with an initial speed of 10 m/s hits a wall head-on and rebounds in the opposite direction with a speed of 8.0 m/s. (a) What is the impulse of the collision? (b) If the collision takes place in 0.20 s, what is the average force exerted by the wall on the ball? (c) Was the collision elastic?

9.2 Conservation of Linear Momentum

*14. A hunter weighing 750 N jumps from a stationary 30-kg canoe to shore with a horizontal speed of 1.5 m/s (Fig. 9.19). With what speed does the canoe initially move away from the shore? (Neglect resistance.)

*15. A 160-lb man and his 80-lb daughter stand together on skates. If they push apart and the father receives a velocity with a magnitude of 0.50 m/s, what is the velocity of the daughter? (Neglect friction.)

*16. A single railroad car with a velocity v rolls toward two coupled railroad cars moving with a

Figure 9.19
See Problem 14.

velocity of $v/2$ in the opposite direction. If the track is level and all the cars have equal masses, what happens after collision?

*17. Two masses on a horizontal frictionless surface with a spring between them are pushed together, and a string is attached to hold them together (Fig. 9.20). If the string is burned and the $3M$ mass moves to the right with a speed of 2.0 m/s, what is the velocity of the other mass? What kind of force does the spring apply to the system?

**18. An 80-kg astronaut is stranded at rest 10 m from his spaceship in free space. In order to get back to the ship, he throws a 2.0-kg piece of equipment with a speed of 0.50 m/s directly away from the spaceship. How long will it take the astronaut to reach the ship?

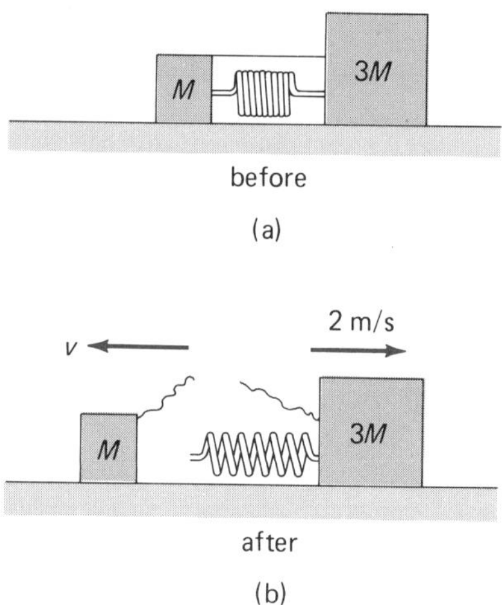

Figure 9.20
See Problem 17.

**19. A 5.0-kg block of wood at rest on a horizontal frictionless surface is hit by a 0.010-kg bullet traveling with a speed of 300 m/s. The bullet passes through the block and emerges with a speed of 200 m/s. (a) What velocity is imparted to the block? (b) How much work did the bullet do in boring through the block? (c) If the block is 10 cm thick, how long did it take for the bullet to go through the block?

9.3 Collisions

*20. A 2.0-kg object with a velocity of 3.0 m/s has a head-on, elastic collision with a stationary 2.0-kg object. What are the velocities of the objects after collision?

*21. A 10-kg ball traveling at 4.0 m/s collides elastically in a head-on collision with a 2.0-kg ball. What are (a) the velocities and (b) the total momentum of the balls after collision?

*22. In a head-on, elastic collision, a 1.0-kg sphere with a velocity of 2.6 m/s hits a sphere with a mass of 3.0 kg. What are the velocities of the spheres after collision?

*23. Two balls with masses of 0.20 kg and 0.80 kg travel toward each other with speeds of 9.0 m/s and 3.0 m/s, respectively. If the balls have a head-on, *inelastic* collision and the 0.20-kg ball recoils with a speed of 7.0 m/s, what is the velocity of the other ball?

*24. A 1000-kg railroad car initially at rest rolls down a grade through a vertical distance of 4.5 m. At the bottom of the incline on a level track, the car collides and couples with an identical

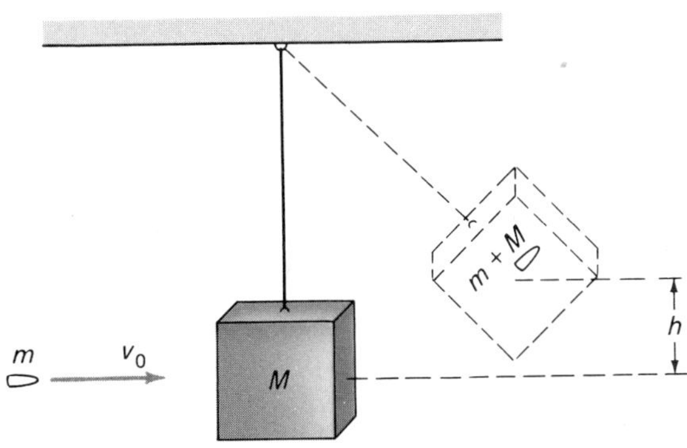

Figure 9.21
See Problem 32.

car at rest. What is the velocity of the cars immediately after collision?

*25. A loaded freight car with a total mass of 1800 kg traveling with a speed of 10 m/s collides with an identical, stationary, empty car. If after the collision the speed of the coupled cars is 9.0 m/s, what was the mass of the load?

*26. Show that the collision in Figure 9.7 is inelastic.

*27. Show that for objects of equal mass having a head-on, elastic collision it is not possible for $v_{1_f} = v_{2_f}$. (Assume $v_{2_o} = 0$.)

*28. Show that for the elastic collision of two objects initially in motion the condition on the velocities is given by $v_{1_o} - v_{2_o} = v_{2_f} - v_{1_f}$, where all motions are considered to be in the positive direction.

*29. In the case of an elastic collision of two balls with $m_1 < m_2$, specifically $m_1 = m_2/2$, show that the *change* in momentum of m_1 is equal to $\frac{4}{3}m_1 v_{1_o}$.

*30. A railroad car with a speed of 20 km/h rolls on a level track in the same direction as another car of the same mass that has a speed of 10 km/h. The faster car overtakes the slower one, and they couple together on collision. (a) What is the velocity of the cars after collision? (b) Show that the kinetic energy is not conserved in the collision.

*31. For a glancing collision, as illustrated in Figure 9.11, m_1 approaches m_2 with a speed of 2.0 m/s. (a) If the masses are equal, what are the speeds of the balls after collision? ($\theta = 45°$)

**32. The arrangement shown in Figure 9.21 is called a ballistic pendulum and is used to find the speed of projectiles. Show that the initial speed of a projectile is given by

$$v_o = \left(\frac{m + M}{m}\right)(2gh)^{1/2}$$

(*Hint:* during collision, linear momentum is conserved. Afterward, mechanical energy is conserved.)

**33. In an elastic, head-on collision with a stationary target, a particle recoils with a speed one-half that of its incident speed. (a) What is the ratio m_1/m_2 of the particle masses? (b) What is the speed of the target particle after collision in terms of the initial speed of the incident particle?

**34. An elastic, three-body collision is illustrated in Figure 9.22. (a) What is the momentum in the x direction after collision? (b) What is the momentum in the y direction?

**35. If the incoming ball in Figure 9.22 has an initial speed of 1.0 m/s, what are the final speeds of the struck balls if all of the balls have equal masses, $v_{1_f} = 0.50$ m/s, and $p_x = p_y$ for each struck ball?

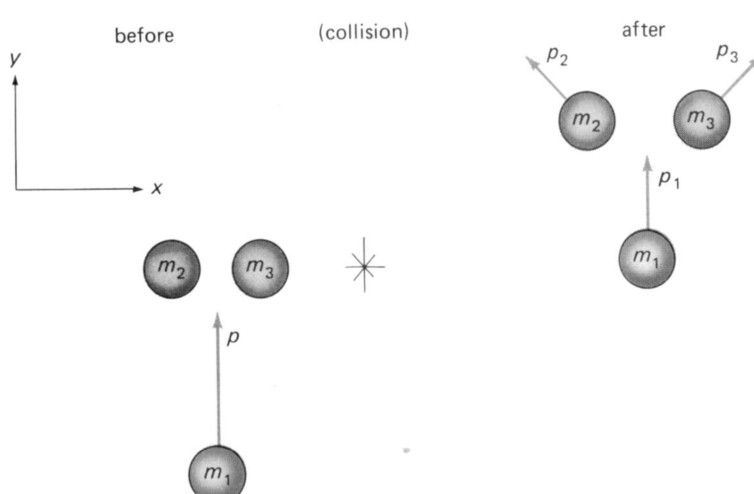

Figure 9.22
Three-body collision. See Problem 34.

Chapter 10

Rotational Motion and Dynamics

In the previous discussions on motion, it was convenient to use a particle representation for objects—i.e., a center-of-mass particle—so as to eliminate rotational considerations. When an object moves without rotation, it is said to be in *pure* translational motion. In pure translational motion, an object moves as a whole with all of its particles having the same instantaneous velocity (no rotation).

However, real objects are also subject to rotations, and rotating objects are found in many technical applications, such as a workpiece on a lathe. In *pure* rotational motion, all the particles of an object move in circles with the same instantaneous angular velocity about a fixed axis (no translation). In general, an object or body may have a combination of translational and rotational motions. Here, the center of mass of the body may be in translational motion with rotation about an axis through this point, such as a thrown, spiraling football.

The description and analysis of the rotational motion of a body are an easy extension of the circular motion of a particle, since solid objects can be considered to be made up of particles. An object in which all the particles maintain fixed distances relative to each other is called a **rigid body.** When a rigid body rotates, each particle of the body undergoes circular motion about an *axis of rotation.*

Since a rotating rigid body is made up of particles, its overall rotational motion can be analyzed in terms of a summation of the motions of the individual particles, as will be shown in this chapter.

10.1 Description of Rotational Motion

Angular Displacement

Recall from Chapter 2 that angular measurement θ is made in degrees or radians. (Revolutions are also sometimes used.) For rotational

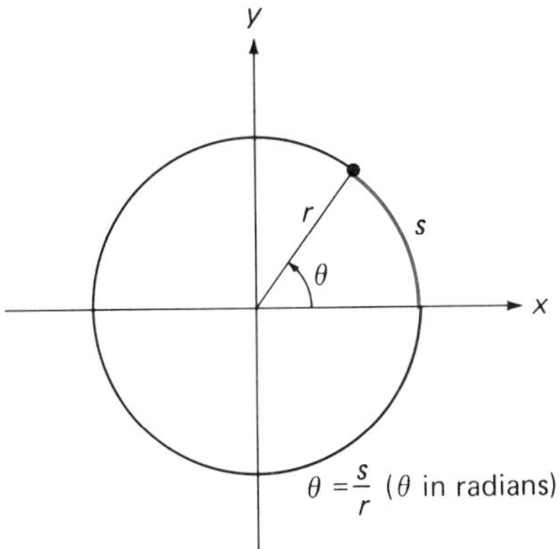

Figure 10.1
Angular distance. When $s = r$, then $\theta = s/r = r/r =$ 1 rad. For a complete revolution of 360°, $s = 2\pi r$ and $\theta = s/r = 2\pi r/r = 2\pi$ rad, or 2π rad = 360°.

Table 10.1
Radian and Degree Relationships (1 rad = 57.3°)

2π rad = 360°
π rad = 180°
$\frac{\pi}{2}$ rad = 90°
$\frac{\pi}{3}$ rad = 60°
$\frac{\pi}{4}$ rad = 45°
$\frac{\pi}{6}$ rad = 30°
$\frac{\pi}{12}$ rad = 15°

motion, radian (rad) measure is the preferred unit. If a particle rotates in a circle of radius r and moves along an arc length s (Fig. 10.1), the magnitude of the angular displacement, or angular distance, is given by

$$\theta = \frac{s}{r} \quad (\theta \text{ in radians}) \qquad \textbf{(Eq. 10.1)}$$

or

$$s = r\theta$$

Hence, when $s = r$, then $\theta = r/r = 1$ rad (= 57.3°). For a complete revolution of 360°, the arc length is the circumference of the circle, $s = 2\pi r$, and in radian measure (for one revolution, or 360°),

$$\theta = s/r = 2\pi r/r = 2\pi \text{ rad}$$

or

$$2\pi \text{ rad} = 360°$$

which is a convenient conversion factor for radians and degrees (see Table 10.1).

Notice that θ in radians is simply a ratio of two lengths, and hence has no units. The "unit" of rad is carried along to indicate how the angle is measured, but the radian is not a unit in the ordinary sense.

The angular distance through which a particle rotates is simply the magnitude of the angular displacement as given by Equation 10.1. When given a direction of rotation, we have angular displacement (analogous to linear distance and displacement). The positive (+) direction for rotations is generally taken to be counterclockwise, since the angle is usually measured in this directional sense (from the $+x$-axis). Clockwise and counterclockwise are neither unique directions nor the actual directions of the angular displacement vector. However, for simplicity, we will use these directional senses to describe rotational directions.

Angular Velocity

Analogous to the linear velocity, which tells how fast (and in what direction) a particle or object moves linearly, the angular velocity tells how fast a particle or object rotates. For example, if the disk or wheel in Figure 10.2 rotates with an angular velocity (ω) of 10π rad/s, then it makes five revolutions each second,

$$10\pi \text{ rad/s } (1 \text{ rev}/2\pi \text{ rad}) = 5 \text{ rev/s (rps)}$$

The angular velocity is sometimes referred to as the *angular frequency* since it tells the number of rotations or cycles per second.

As the name implies, angular velocity is a vector quantity.* However, the actual vector direction is not important in most cases, so only the directional sense or effect of the angular ve-

* The direction of the angular displacement and velocity may be found by a right-hand rule. If the fingers of the right hand are curled in circular direction of the motion of a particle or object, the thumb points in the direction of the vectors.

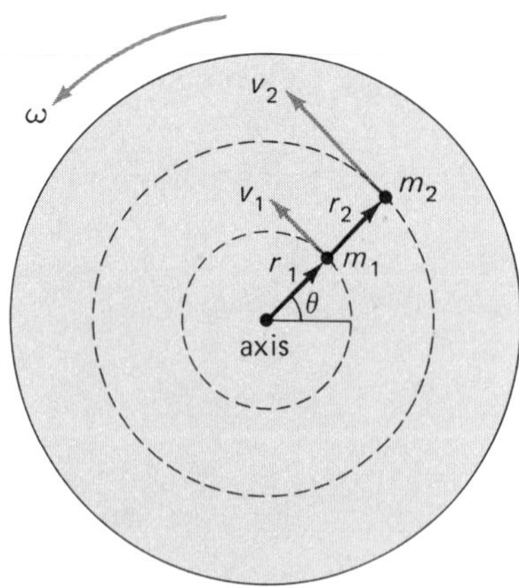

Figure 10.2
Angular velocity. The angular velocity ω is the same for all particles of the disk, but the tangential velocities are different at different radii, i.e., $v_1 = r_1\omega$ and $v_2 = r_2\omega$. The actual direction of the angular velocity vector is out of the plane of the paper.

locity will be used when necessary—counterclockwise positive, clockwise negative. The boldface notation for vector quantities will generally be omitted for convenience and only the rotational effects considered rather than vector directions.

Analogous to the case of linear velocity ($\bar{v} = x/t$), the **average angular velocity** is defined as

$$\bar{\omega} = \frac{\theta}{t} \qquad \textbf{(Eq. 10.2)}$$

with units of rad/s. At any instant, there is an instantaneous angular velocity ω, and $\bar{\omega} = \omega$ for a constant angular velocity.

The magnitude of the angular velocity is related to that of the tangential velocity, as may be readily shown. Since $s = r\theta$, dividing both sides of the equation by time, $s/t = r\theta/t$, we may write

$$v = r\omega \qquad \textbf{(Eq. 10.3)}$$

Note in Figure 10.2 that the angular velocity for both (and all) particles in the disk is the *same*. However, the tangential velocities of the particles are different because of different radii, $v_1 = r_1\omega$ and $v_2 = r_2\omega$. This is apparent from the orbital paths. Since both particles make one revolution in the same time, m_2 travels a longer distance and hence must have a greater speed.

EXAMPLE 10.1 A particle in a disk rotating with a uniform angular speed of 2.0 rps is 0.20 m from the axis of rotation. What are (a) the tangential speed of the particle and (b) the angle through which it rotates in 0.50 s?

Solution: It is given that $r = 0.20$ m and $\omega = 2.0$ rev/s (2π rad/rev) $= 4\pi$ rad/s.

(a) $v = r\omega = (0.20 \text{ m})(4\pi \text{ rad/s}) = 0.80\pi$ m/s (recall that the rad is not a unit in the ordinary sense). It is convenient to leave answers in the π symbol form unless an explicit number is required.

(b) With $t = 0.50$ s,

$$\theta = \omega t = (4\pi \text{ rad/s})(0.50 \text{ s}) = 2\pi \text{ rad}$$
(or 360°—one revolution)

Angular Acceleration

When a rotating object speeds up or slows down, an angular acceleration (or deceleration) is involved. As in the linear case, the angular acceleration (α) tells how much the angular velocity changes each second. The **average angular acceleration** is given by

$$\bar{\alpha} = \frac{\omega_f - \omega_o}{t}$$

or

$$\omega_f = \omega_o + \alpha t \qquad \textbf{(Eq. 10.4)}$$

where in the second equation $\bar{\alpha} = \alpha$; i.e., the angular acceleration is taken to be uniform or constant, as was done for the linear case in most applications. Similarly, if the angular velocity and angular acceleration have the same sign, or are in the same rotational direction, then an object speeds up (angular speed increases). If the angular velocity and angular acceleration have opposite signs, then an object slows down—an angular deceleration.

EXAMPLE 10.2 A disk rotating at an angular speed of 10 rad/s is slowed down by a uniform angular acceleration to a speed of 4.0 rad/s in 3.0 s. What is the angular acceleration?

Solution: It is given that $\omega_o = 10$ rad/s, $\omega_f = 4.0$ rad/s, and $t = 3.0$ s.

Then,

$$\alpha = \frac{\omega_f - \omega_o}{t} = \frac{4.0 \text{ rad/s} - 10 \text{ rad/s}}{3.0 \text{ s}}$$

$$= -2.0 \text{ rad/s}^2$$

The minus sign indicates that the angular acceleration is in the direction opposite that of the motion or angular velocity—that is, the disk is slowing down or decelerating. Note the unit of angular acceleration is rad/s^2 (analogous to the linear acceleration unit of m/s^2).

The magnitude of the angular acceleration is related to the magnitude of the tangential acceleration. Since $v = r\omega$, dividing both sides of the equation by time, $v/t = r\omega/t$, we have

$$a_t = r\alpha \qquad \textbf{(Eq. 10.5)}$$

Note in Figure 10.3 that when a rotating disk is given an angular acceleration α, all the particles of the disk receive the *same* angular acceleration. However, the tangential acceleration of a particular particle depends on its distance or radius from the axis of rotation, $a_{t_1} = r_1\alpha$ and $a_{t_2} = r_2\alpha$. Also note that the tangential acceleration is perpendicular to the centripetal acceleration of a particle (supplied by interparticle forces). The net acceleration is the vector sum of these, $\mathbf{a} = \mathbf{a}_t + \mathbf{a}_c$.

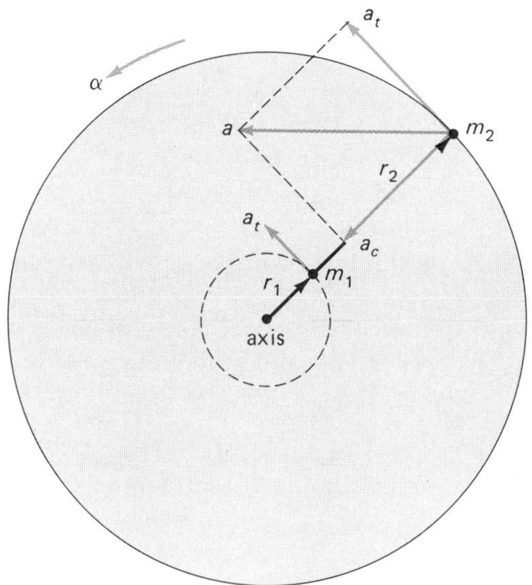

Figure 10.3

Angular acceleration. The angular acceleration is the same for all particles of the disk, but the tangential accelerations are different at different radii, since $a_t = r\alpha$. The centripetal acceleration of a particle is $a_c = v^2/r = r^2$, and the net acceleration is the vector sum, $\mathbf{a} = \mathbf{a}_t + \mathbf{a}_c$.

Table 10.2
Linear and Angular Analogies

	Linear	Angular		
	x	θ	$(s = r\theta)$	
	v	ω	$(v = r\omega)$	
	a	α	$(a = r\alpha)$	
				(Eq. 10.7)
	$x = \bar{v}t$	$\theta = \bar{\omega}t$		(a)
constant acceleration	$\bar{v} = \frac{v_f + v_o}{2}$	$\bar{\omega} = \frac{\omega_f + \omega_o}{2}$		(b)
	$v_f = v_o + at$	$\omega_f = \omega_o + \alpha t$		(c)
	$x = v_o t + \frac{1}{2}at^2$	$\theta = \omega_o t + \frac{1}{2}\alpha t^2$		(d)
	$v_f^2 = v_o^2 + 2ax$	$\omega_f^2 = \omega_o^2 + 2\alpha\theta$		(e)

At any instant, the centripetal acceleration of a particle is given by $a_c = v^2/r$. This may also be written in terms of the instantaneous angular velocity ($v = r\omega$):

$$a_c = \frac{v^2}{r} = r\omega^2 \qquad \textbf{(Eq. 10.6)}$$

Analogies Between Linear and Angular Motions

It should be evident from the previous discussion that there is a one-to-one analogy between linear and angular quantities. Rather than repeat the mathematical operations for the rotational cases as was done for linear quantities in Chapter 4, Table 10.2 simply summarizes the analogy. The equations are the same except for the changed symbols and different meanings for the physical quantities.

EXAMPLE 10.3 A circular disk initially at rest experiences a uniform angular acceleration of 0.25 rad/s^2 through two revolutions. (a) What is the angular speed at the end of two revolutions? (b) How long did it take to make the two revolutions?

Solution: It is given that $\omega_o = 0$, $\alpha = 0.25$ rad/s^2, and $\theta = 2$ rev (2π rad/rev) $= 4\pi$ rad.

(a) Using Equation 10.7(e),

$$\omega_f^2 = \omega_o^2 + 2\alpha\theta = 0 + 2(0.25 \text{ rad/s}^2)(4\pi \text{ rad})$$
$$= 2\pi \text{ rad}^2/\text{s}^2$$

and $\omega_f = \sqrt{2\pi \text{ rad}^2/\text{s}^2} = 2.5$ rad/s

(b) With $\omega_f = 2.5$ rad/s, from Equation 10.7(c),

$$t = \frac{\omega_f - \omega_o}{\alpha} = \frac{2.5 \text{ rad/s} - 0}{0.25 \text{ rad/s}^2} = 10 \text{ s}$$

10.2 Torque and Moment of Inertia

Analogous to an unbalanced force producing linear motion or the linear acceleration of a particle ($F = ma$), **an unbalanced torque produces an angular acceleration.** Recall from the discussion of statics in Chapter 3 that a torque is defined as a moment arm times a force (Fig. 10.4),

$$\boxed{\tau = r_\perp F} \qquad \textbf{(Eq. 10.8)}$$

where $r_\perp$ is the moment arm (sometimes called lever arm), or the perpendicular distance from the axis of rotation to the line of action of the force. Notice in Figure 10.4 that, if the line of action of the force is perpendicular to the radius r of the circle described by the particle in rotational motion, then $r = r_\perp$.

For the general case of the line of action of the force not being perpendicular to the radius of the circle of motion (diagram b), $r_\perp = r \sin\theta$, where θ is the angle between the line of action of the force and the radial direction. Hence, in general,

$$\tau = r_\perp F = (r \sin\theta)F$$

$$\boxed{\tau = rF \sin\theta} \qquad \textbf{(Eq. 10.9)}$$

Notice that this is equivalent to multiplying the radius r times the component of the force perpendicular to the radial direction, $F_\perp = F \sin\theta$, i.e., $\tau = r_\perp F = rF_\perp$. In either case, Equation 10.9 gives the general expression for the magnitude of a torque. Torque has units of m-N and lb-ft in the SI and British systems, respectively.

In Chapter 3 in the case of static equilibrium, the torques were always balanced, $\Sigma\tau = 0$, and there was no rotational motion. Hence, a net or unbalanced torque is required to rotationally accelerate a particle or object. For a net torque acting on a particle of mass m, the angular acceleration is related to the torque by

$$\tau = rF = r(ma) = rm(r\alpha) = (mr^2)\alpha$$

or

$$\tau = (mr^2)\alpha \qquad \textbf{(Eq. 10.10)}$$

where it is assumed that $r = r_\perp$, and the relationship between angular and tangential accelerations, $a = r\alpha$, was used.

As stated previously, a rigid body can be considered to be made up of many individual particles (Fig. 10.5). To give each individual mass an angular acceleration requires that a torque act on each particle, i.e., $\tau_1 = (m_1r_1^2)\alpha$, $\tau_2 = (m_2r_2^2)\alpha$, and so on; or, in general, $\tau_i = (m_ir_i^2)\alpha$. The angular acceleration is the same for all particles, since the rigid body rotates as a whole with an angular acceleration.

The total torque τ that angularly accelerates an object as a whole is then the sum of the individual particle torques,

$$\tau = (\tau_1 + \tau_2 + \tau_3 + \cdots \tau_N)$$
$$= (m_1r_1^2 + m_2r_2^2 + m_3r_3^2 + \cdots m_Nr_N^2)\alpha$$

or

$$\tau = \Sigma\tau_i = (\Sigma m_ir_i^2)\alpha \qquad \textbf{(Eq. 10.11)}$$

where N is the total number of particles in the rigid body.

The quantity $(\Sigma m_ir_i^2)$ is constant for a rigid body (why?) and is called the **moment of inertia** I of the body, i.e.,

$$\boxed{I = \Sigma m_ir_i^2} \qquad \textbf{(Eq. 10.12)}$$

Hence, we may write for a rigid body

$$\boxed{\tau = I\alpha} \qquad \textbf{(Eq. 10.13)}$$

which is the rotational form of Newton's second law for a rigid body (analogous to $F = ma$, where m is the total mass of the body). **The moment of inertia I is the rotational analog of mass and is a measure of the rotational inertia of a body.**

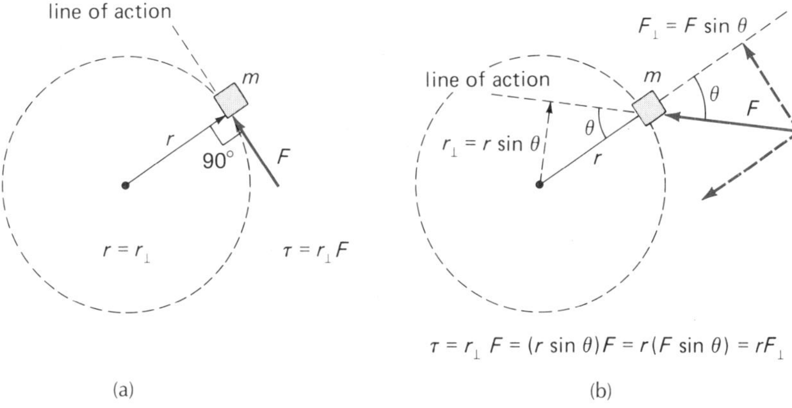

(a) (b)

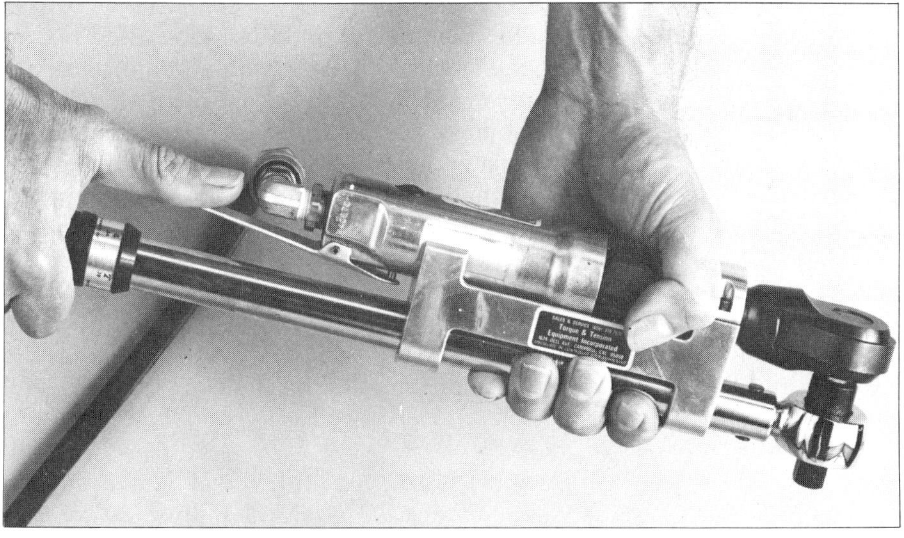

(c)

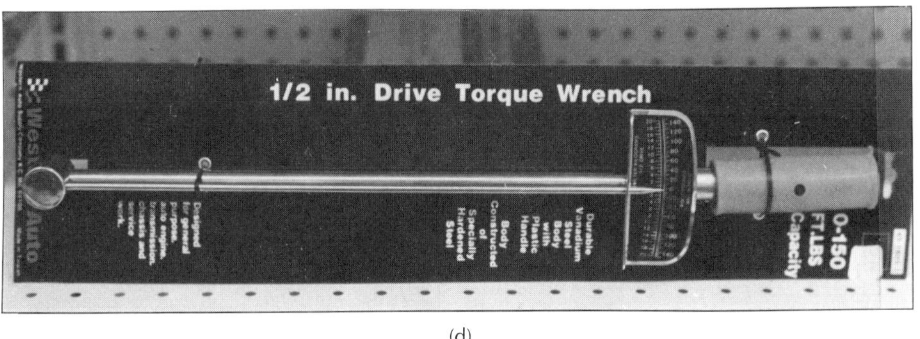

(d)

Figure 10.4

Torque. The magnitude of a torque is given by the product of the moment arm $r_\perp$ and the applied force F, i.e., $\tau = r_\perp F$. The moment arm is the perpendicular distance from the axis of rotation to the line of action of the force. (a) If the force is perpendicular to the radius of the circular path of a particle, then $r_\perp = r$. (b) For the general case, $r_\perp = r \sin\theta$, and $\tau = r_\perp F = rF \sin\theta$, where θ is the angle between the radial and force directions. Note also that $F_\perp = F \sin\theta$, and $\tau = r_\perp F = rF_\perp$. (c) A manual torque wrench. The scale indicates the magnitude of the applied torque. (d) An air-powered torque wrench. The air pressure controls the applied torque. (Courtesy of Torque and Tension Equipment Company.)

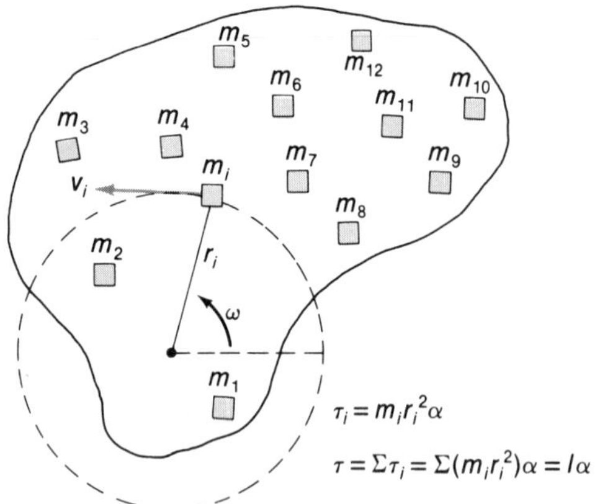

Figure 10.5
Rigid body. A rigid body can be considered to be made up of many individual particles. The total torque acting on the body is the sum of the individual particle torques.

There is also a rotational analog of Newton's first law: An object will remain at rest or in motion with a constant angular velocity until acted upon by an external torque. In other words, an object remains in rotational equilibrium unless it is acted upon by a net external torque. Is there a rotational analog of Newton's third law? That is, for every torque is there an equal and opposite torque? You should be able to convince yourself that there is.

EXAMPLE 10.4 What is the moment of inertia for a wheel as illustrated in Figure 10.6? Consider the mass of the axle and spokes to be negligible.

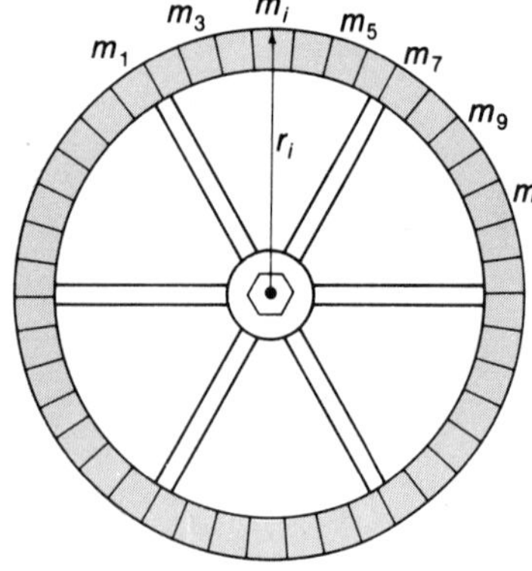

Figure 10.6
Moment of inertia. The moment of inertia of a rigid body is found by summing the individual $m_i r^2$. See Example 10.4.

Solution: Assuming the wheel to be a hoop made up of individual particles,

$$I = \Sigma m_i r_i^2 = (m_1 R^2 + m_2 R^2 + \cdots m_N R^2)$$
$$= (m_1 + m_2 + m_3 + \cdots m_N)R^2 = MR^2$$

where the sum of the masses of the individual particles is equal to the total mass M of the wheel. Hence, the rotational form of Newton's second law for a hoop-shaped rigid body can be written $\tau = I\alpha = (MR^2)\alpha$.

The moments of inertia of most rigid bodies are not so easily calculated and require the use of advanced mathematics. Table 10.3 lists the moments of inertia for several regularly shaped objects with uniform mass distribution. Basically, the moment of inertia of a body depends on how its mass is distributed around the axis of rotation. As can be seen from Table 10.3, the moment of inertia of a solid wheel (or solid disk) is one-half that of a spoked wheel (thin hoop with spokes of negligible mass), in which the mass is concentrated on the rim of the wheel.

EXAMPLE 10.5 A solid wheel and a spoked wheel of equal mass and radius (20 kg and 0.50 m) are set into rotation by descending masses of 4.0 kg (Fig. 10.7). What is the angular acceleration of each object?

Solution: From Table 10.3, the moments of inertia of the solid wheel and spoked wheel (mass of spokes considered to be negligible) are

$$\text{(solid wheel)} \quad I = \tfrac{1}{2}MR^2 = \tfrac{1}{2}(20 \text{ kg})(0.50 \text{ m})^2$$
$$= 2.5 \text{ kg-m}^2$$
$$\text{(spoked wheel)} \quad I = MR^2 = 2(\tfrac{1}{2}MR^2)$$
$$= 2(2.5) = 5.0 \text{ kg-m}^2$$

Then, using Equation 10.13, $\tau = I\alpha$, we have for the general case

$$\alpha = \frac{\tau}{I} = \frac{RT}{I} \tag{1}$$

where T is the tension in the cord. But, looking at the isolated, descending mass (the dashed

Table 10.3
Moments of Inertia of Some Common Uniform Objects

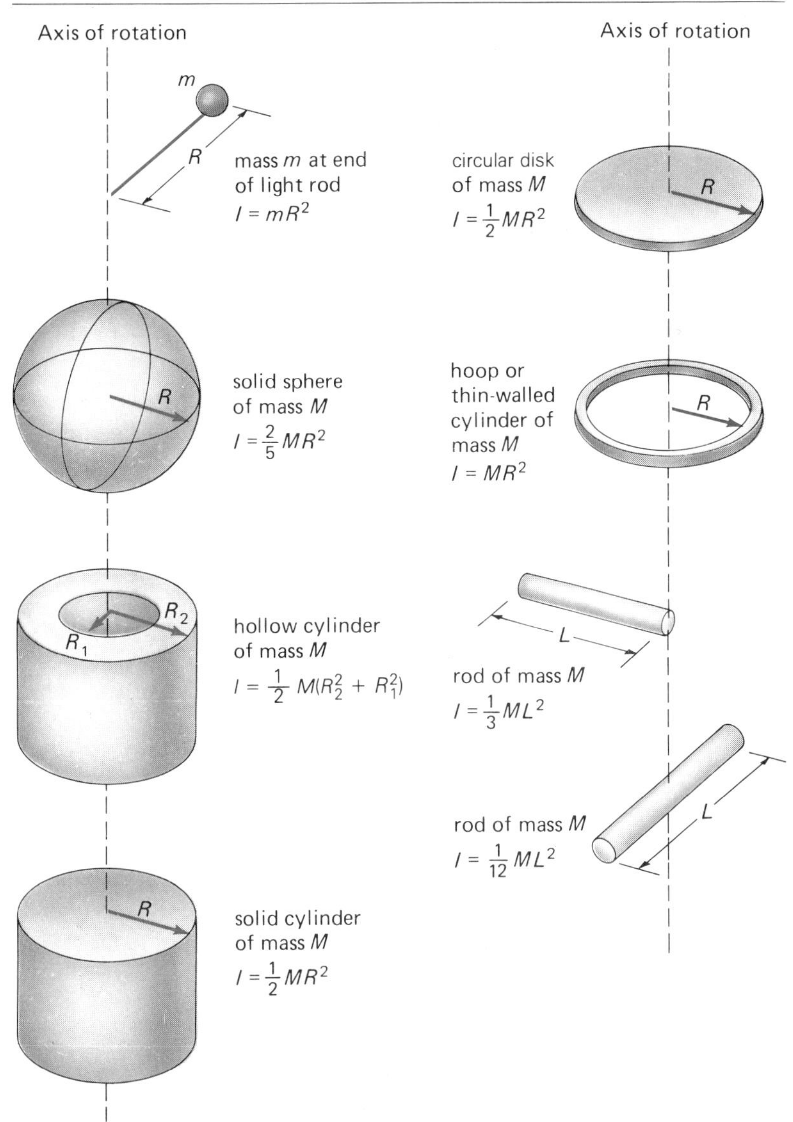

box in the figure), we have, from Newton's second law,

$$mg - T = ma \tag{2}$$

The linear acceleration of a descending mass and the angular acceleration of a pulley are related by $a = R\alpha$, so

$$T = mg - mR\alpha \tag{3}$$

Using Equation 3 to eliminate T from Equation 1 gives

$$\alpha = \frac{R(mg - mR\alpha)}{I}$$

and

$$\alpha = \frac{mgR}{I + mR^2} \tag{4}$$

Then, putting in the different values of I,

(solid wheel)

$$\alpha = \frac{mgR}{I + mR^2} = \frac{(4.0 \text{ kg})(9.8 \text{ m/s}^2)(0.50 \text{ m})}{2.5 \text{ kg-m}^2 + (4.0 \text{ kg})(0.50 \text{ m})^2}$$

$$= 5.6 \text{ rad/s}^2$$

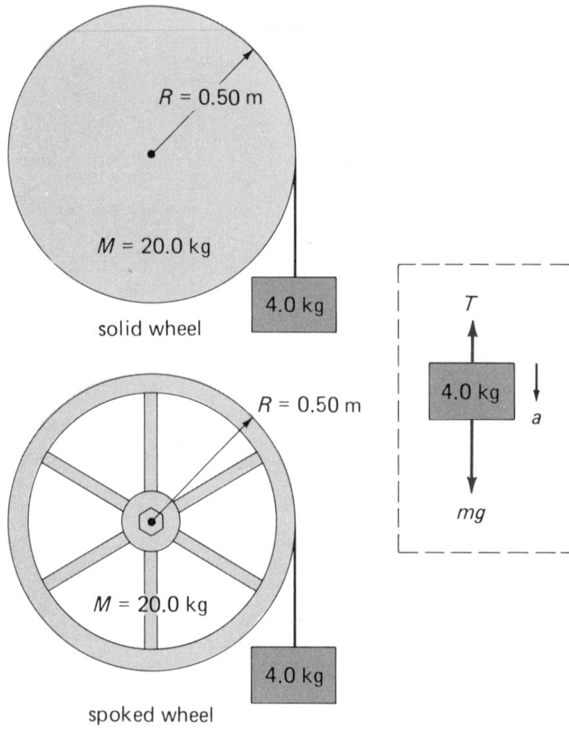

Figure 10.7
Rotational inertia. The angular accelerations of the wheels are different because of different moments of inertia. See Example 10.5.

(spoked wheel)

$$\alpha = \frac{mgR}{I + mR^2} = \frac{(4.0\text{ kg})(9.8\text{ m/s}^2)(0.50\text{ m})}{5.0\text{ kg-m}^2 + (4.0\text{ kg})(0.50\text{ m})^2}$$

$$= 3.3\text{ rad/s}^2$$

As the preceding example shows, the farther the mass is distributed from the axis of rotation, the greater the moment of inertia of a body (due to the r^2 dependence of I). A spoked wheel with its mass concentrated on its rim has twice the moment of inertia of a solid wheel of equal mass and radius. Since the spoked wheel has a greater moment of inertia, it has a greater resistance to a change in motion.

As a result of this property, flywheels, which make use of rotational inertia to keep rotary machinery running smoothly, are designed with the mass concentrated near the rim (Fig. 10.8).

10.3 Rotational Work, Power, and Kinetic Energy

The expressions for rotational work, power, and kinetic energy are again analogous to those for

(a)

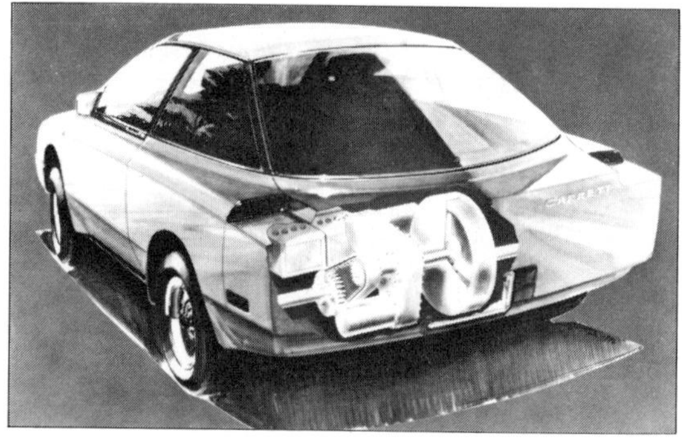

(b)

Figure 10.8
Flywheels. A flywheel has most of its mass concentrated near the rim to give a greater moment of inertia. (a) The flywheel of an old piece of machinery. (Courtesy A. and J. Turk.) (b) The flywheel at the right of the cutaway will provide better acceleration in this design of a new test car employing advanced electric vehicle technology. (Courtesy of AiResearch Manufacturing Company.)

linear motion. The equations have the same form, but different meanings for the quantities.

Work

When a force acts on a rigid body to set it into motion about an axis of rotation, the force acts through a circular arc of length s. Hence,

$$W = Fs = Fr\theta = \tau\theta$$

and

$$\boxed{W = \tau\theta} \qquad \textbf{(Eq. 10.14)}$$

where $s = r\theta$. The rotational work is thus the product of the torque and the angular displacement θ.

Power

The rotational power is the rotational work per unit of time. Hence,

$$\boxed{P = \frac{W}{t} = \frac{\tau\theta}{t} = \tau\bar{\omega}} \qquad \textbf{(Eq. 10.15)}$$

where $\bar{\omega}$ is the average angular speed.

Rotational power and its transmission are important considerations in driving rotary machines, which is a topic in Chapter 11.

EXAMPLE 10.6 A tangential force applied to a stationary flywheel with a mass of 40 kg and a radius of 0.50 m angularly accelerates the wheel uniformly through one revolution in a time of 2.0 s. How much (a) work and (b) power were done by the force in this time?

Solution: It is given that $\omega_o = 0$, $M = 40$ kg, $R = 0.50$ m, $\theta = 2\pi$ rad (one revolution), and $t = 2.0$ s.

(a) Since F is not given, the torque cannot be computed from $\tau = rF$, but $\tau = I\alpha$, where $I = MR^2$ in general for a flywheel with its mass concentrated on the rim. From Equation 10.7(d),

$$\theta = \omega_o t + \tfrac{1}{2}\alpha t^2 = 0 + \tfrac{1}{2}\alpha t^2$$

and
$$\alpha = \frac{2\theta}{t^2} = \frac{2(2\pi\ \text{rad})}{(2.0\ \text{s})^2} = \pi\ \text{rad/s}^2$$

so

$$\tau = I\alpha = MR^2\alpha = (40\ \text{kg})(0.50\ \text{m})^2(\pi\ \text{rad/s}^2)$$
$$= 10\pi\ \text{m-N}$$

Then,

$$W = \tau\theta = (10\pi\ \text{m-N})(2\pi\ \text{rad}) = 20\pi^2\ \text{J} = 200\ \text{J}$$

Note that it is convenient to keep π in symbol form in calculations.

(b) With $t = 2.0$ s,

$$P = \frac{W}{t} = \frac{200\ \text{J}}{2.0\ \text{s}} = 100\ \text{W}$$

Kinetic Energy

When a rigid body rotates, the kinetic energy of each particle of the body may be expressed in terms of the tangential velocity. Using the summation process to obtain the total **rotational kinetic energy** of the body, we have

$$KE = \Sigma\tfrac{1}{2}m_i v_i^2 = \Sigma\tfrac{1}{2}m_i(r_i\omega)^2$$
$$= \tfrac{1}{2}(\Sigma m_i r_i^2)\omega^2 = \tfrac{1}{2}I\omega^2$$

and

$$\boxed{KE = \tfrac{1}{2}I\omega^2} \qquad \textbf{(Eq. 10.16)}$$

where $v_i = r_i\omega$. The ω has no i subscript since all of the particles of a rigid body have the same angular velocity.

There is also a **work-energy theorem** for the rotational case:

$$W = \tau\theta = I\alpha\theta = I[\tfrac{1}{2}(\omega_f^2 - \omega_o^2)]$$
$$= \tfrac{1}{2}I\omega_f^2 - \tfrac{1}{2}I\omega_o^2 \qquad \textbf{(Eq. 10.17)}$$
$$= \Delta KE$$

where $\tau = I\alpha$ and $\omega_f^2 = \omega_o^2 + 2\alpha\theta$. Hence, the angular work is equal to the change in the rotational kinetic energy.

In some instances, an object may be in linear motion as well as rotational motion. A common example of this is rolling motion. A rolling object (uniform and symmetric) is obviously rotating, and it is also moving linearly (Fig. 10.9). That is, its center of mass or center of gravity is translating with a speed v. If the object rolls

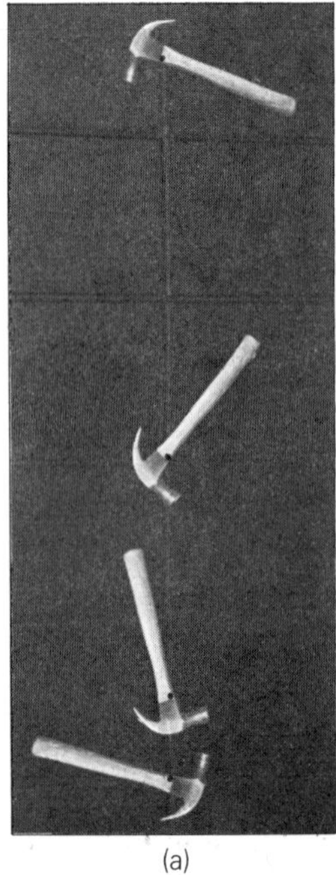

(a)

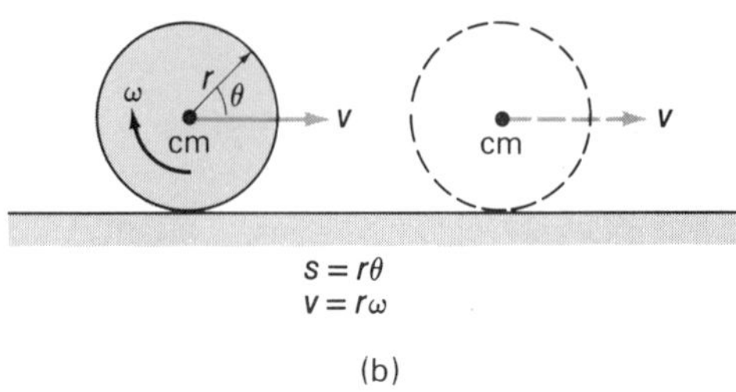

(b)

Figure 10.9
Rigid body motion. The general motion of a rigid body involves translational (linear) and rotational motions. (a) A falling hammer. Notice that the center of mass (as indicated by the dot) falls in a straight line while at the same time the hammer rotates about an axis through this point. (b) Rolling is a combination of translational and rotational motions. As an object rolls, it rotates as it moves linearly (translates). The condition for rolling without slipping is $s = r\theta$ or $v = r\omega$.

without slipping, then the linear distance s it rolls is the same as the circumference length s, and $s = r\theta$. An equivalent condition for rolling without slipping is $v = r\omega$.

Then, **a rolling object has both translational and rotational kinetic energies:**

$$\boxed{KE = \tfrac{1}{2}Mv^2 + \tfrac{1}{2}I\omega^2} \qquad \textbf{(Eq. 10.18)}$$

where M is the total mass of the object, v is its translational speed, and I is the moment of inertia about the axis of symmetry and through the center of mass of the object.

Both the translational and rotational kinetic energies of an object or a system must be taken into account when working problems using the conservation of energy, as the following examples show.

EXAMPLE 10.7 A ball initially at rest rolls without slipping down an incline of height h (Fig. 10.10). What is the linear speed of the ball when it rolls onto the horizontal surface?

Solution: By the conservation of mechanical energy, the initial potential energy of the ball at the top of the plane is equal to its kinetic energy at the bottom. However, the ball has both rotational and translational kinetic energies. Then,

$$\Delta KE = \Delta PE$$

and

$$\tfrac{1}{2}Mv^2 + \tfrac{1}{2}I\omega^2 = Mgh \qquad \textbf{(Eq. 10.19)}$$

where v is the velocity of the center of mass and I is the moment of inertia about an axis through the center of mass.

Since the ball rolls without slipping, $v = R\omega$, and the moment of inertia of the ball is $I = \tfrac{2}{5}MR^2$ (Table 10.3), so the equation becomes

$$\tfrac{1}{2}Mv^2 + \tfrac{1}{2}(\tfrac{2}{5}MR^2)(v/R)^2 = Mgh$$

and

$$\tfrac{1}{2}Mv^2 + \tfrac{1}{5}Mv^2 = \tfrac{7}{10}Mv^2 = Mgh$$

Hence,

$$v = \sqrt{\tfrac{10}{7}gh}$$

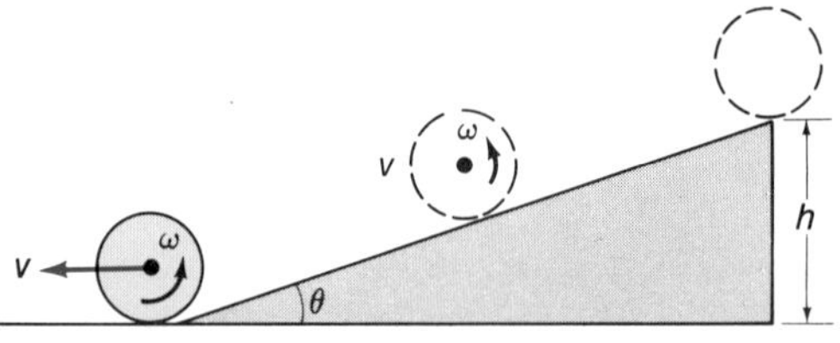

Figure 10.10
A rolling object has both translational and rotational kinetic energies. See Example 10.7.

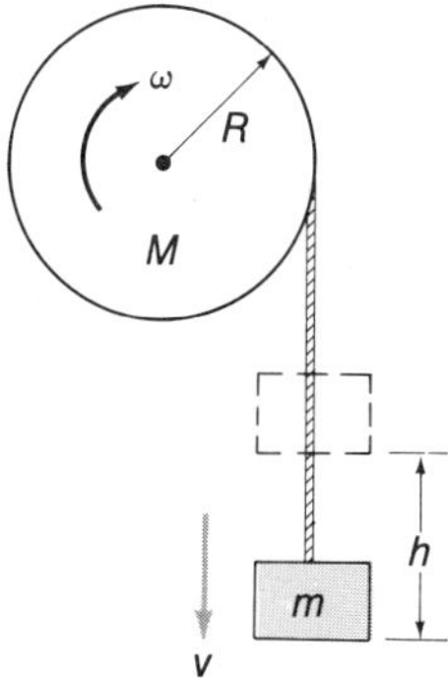

Figure 10.11
Conservation of energy. The potential energy lost by the descending mass goes into the rotational kinetic energy of the disk and the linear kinetic energy of the mass. See Example 10.8.

EXAMPLE 10.8 A mass suspended by a string wrapped around a pivoted disk is released from rest (Fig. 10.11). What is the angular speed of the disk after the suspended mass has fallen a distance h? (Neglect friction.)

Solution: In falling a distance h, the mass loses potential energy $PE = mgh$. This goes into the rotational kinetic energy of the disk *and* the translational kinetic energy of the descending mass. Hence, by the conservation of mechanical energy for the system,

$$\Delta KE = \Delta PE$$

$$\tfrac{1}{2}mv^2 + \tfrac{1}{2}I\omega^2 = mgh$$

The tangential speed of the rim of the disk is the same as that of the descending mass, $v = R\omega$, and from Table 10.3 for a disk, $I = \frac{1}{2}MR^2$. Hence,

$$\tfrac{1}{2}m(R\omega)^2 + \tfrac{1}{2}(\tfrac{1}{2}MR^2)\omega^2 = mgh$$

and

$$\tfrac{1}{2}mR^2\omega^2 + \tfrac{1}{4}MR^2\omega^2 = mgh$$

Solving for ω,

$$\omega^2 = \frac{mgh}{(\frac{2}{4}m + \frac{1}{4}M)R^2}$$

and

$$\omega = \sqrt{\frac{4mgh}{(2m + M)R^2}}$$

10.4 Angular Momentum

Just as linear momentum, $p = mv$, is an important linear dynamic quantity, so is its rotational counterpart: angular momentum. The angular momentum ℓ of a particle of mass m in rotational motion is given by

$$\ell = rp = rmv = (mr^2)\omega \quad \textbf{(Eq. 10.20)}$$

where the relationship $v = r\omega$ was used.

The **total angular momentum** L of a rigid body is given by a summation over all of the particles making up the body, as was previously done for torque, i.e.,

$$L = \Sigma\ell_i = (\Sigma m_i r_i^2)\omega = I\omega$$

and

$$\boxed{L = I\omega} \quad \textbf{(Eq. 10.21)}$$

Angular momentum is related to torque in a fashion analogous to the relationship of linear momentum and force ($F = \Delta p/\Delta t$). This may be shown as follows:

$$\tau = I\alpha = I\left(\frac{\omega_f - \omega_o}{\Delta t}\right) = \frac{I\omega_f - I\omega_o}{\Delta t}$$

$$= \frac{L_f - L_o}{\Delta t} = \frac{\Delta L}{\Delta t}$$

and

$$\boxed{\tau = \frac{\Delta L}{\Delta t}} \quad \textbf{(Eq. 10.22)}$$

where the kinematic equation $\omega_f = \omega_o + \alpha t$ was used. Hence, the torque is equal to the time rate of change of angular momentum. Notice in Equation 10.22 that the units of angular momentum are m-N-s (or kg-m^2/s) and lb-ft-s, respectively, in the SI and British systems.

Conservation of Angular Momentum

Equation 10.22 demonstrates an important concept in rotational dynamics—the **conservation of angular momentum:**

> **In the absence of an unbalanced external torque acting on a body, the angular momentum is conserved.**

As can be seen from Equation 10.22, if $\tau = 0$, then $\Delta L = L_f - L_o = 0$, $L_f = L_o$, and

$$I\omega_o = I\omega_f$$

or

$$L = I\omega = \text{constant} \qquad \textbf{(Eq. 10.23)}$$

The concept of the conservation of angular momentum has many important and interesting applications. Suppose a helicopter has only one set of rotor blades. When the helicopter is sitting on the ground with the engine off, its angular momentum is zero. The helicopter engine supplies the (internal) torque to start and maintain the rotor blades in motion. If the helicopter lifted off the ground, the helicopter body would rotate in the opposite direction to that of the rotor blades to conserve angular momentum (equal angular momentum vectors in opposite directions), which is a bit undesirable.

As a result, large helicopters are equipped with two sets of rotor blades that rotate in opposite directions (Fig. 10.12). Smaller helicopters with a single set of overhead rotor blades have an "anti-torque" tail rotor that maintains the stability of the helicopter body.

You will see a small helicopter in Figure 10.12 without a tail rotor. This is a new design which may become quite common. Low-pressure air is circulated through the tail boom to provide anti-torque and directional control. Air released through slots on the side of the boom captures the downwash from the main rotor, creating "lift" laterally or sideways along the boom, similar to lift due to airflow over the top of an airplane wing (Chapter 15). In this case, the "lift" is to the side, and provides 60 percent of the anti-torque force needed for hovering. Other air is vented through a movable thruster at the end of the boom for additional anti-torque force and directional control.

This new system greatly reduces the hazards associated with tail rotor systems. On the ground, there is danger to personnel from the rotating blades, and in flight many mishaps come from tail rotor strikes. Not only are there safety advantages in the new design, the tail boom is a hollow graphite composite which gives nearly a 20 percent weight savings over the equivalent, conventional aluminum tail assembly.

A common lecture demonstration of conservation of angular momentum is illustrated in Figure 10.13. A person on a rotating stool or turntable, holding weights with his or her arms extended, is set into rotation with an angular speed ω_o. When the person draws the weights inward, he or she spins faster as though acted upon by some invisible torque. Of course, there is no invisible torque. The action is a result of conservation of angular momentum.

(a)

(c)

Figure 10.12
Conservation of angular momentum. (a) Large helicopters have two counter-rotating rotors. With only one rotor, the body of the helicopter would rotate to conserve angular momentum. (Courtesy Boeing Vertol Company.) (b) Small helicopters have an anti-torque tail rotor to prevent the "copter" from rotating. (c) A new design with no tail rotor. What supplies the anti-torque? See text for description. (Courtesy McDonnel Douglas Helicopter Company.)

When the weights are drawn inward, the mo-

(a)

(b)

Figure 10.13
Conservation of angular momentum. (a) When the person's arms are extended, there is a large moment of inertia and the rotation (angular speed) of the stool and person is slow. (b) When the arms and weights are brought inward, the moment of inertia decreases and the angular speed increases to conserve angular momentum.

ment of inertia of the rotating system is decreased ($I = \Sigma m_i r_i^2$, and the r's of the weights are decreased). Since the rotating system is isolated and there are no external torques (neglecting frictional torque on the bearings of the turntable), the angular momentum is conserved, and $L_1 = L_o$, or $I_1\omega_1 = I_o\omega_o$. If $I_1 < I_o$ (due to drawing the weights inward), then conservation of angular momentum requires $\omega_1 > \omega_o$, and the rate of rotation increases.

Let's show this by considering only the masses of the weights. Then,

$$I_1\omega_1 = I_o\omega_o$$

and $$(mr_1^2 + mr_1^2)\omega_1 = (mr_o^2 + mr_o^2)\omega_o$$

or $$\omega_1 = \left(\frac{r_o}{r_1}\right)^2 \omega_o$$

Since $r_o > r_1$, then $\omega_1 > \omega_o$. If the weights are extended, I increases and the rotational speed decreases.

Some other examples of the conservation of angular momentum are discussed in Special Feature 10.1.

When the angular momentum of a spinning object is conserved, the angular momentum is constant in *direction* as well as magnitude, since angular momentum is a vector quantity. The direction of the angular momentum is the same as the angular velocity and is given by a right-hand rule, as mentioned earlier in a footnote.

The directional aspect of the conservation of angular momentum is used in various applications. For example, a football is passed with a spiral or spin [Fig. 10.14(a)]. Since the angular momentum is relatively constant, the football does not wobble. Similarly, a rifle barrel has "rifling" or spiral grooves so as to give the bullet a spin and directional stability.

Also, a rapidly spinning top remains upright and does not wobble and fall until its angular momentum is changed by frictional torque [Fig. 10.14(b)]. When the top slows and tilts, there is a gravitational torque as a result of the center of gravity's not being directly over the point of support. The top does not topple, as would an object in unstable equilibrium, but rotates or precesses about the vertical axis. This is a directional change in the angular momentum as a result of the torque.

Such spinning objects with constant directional properties are said to have *gyroscopic*

Conservation of Angular Momentum in Action

When angular momentum is conserved, $I\omega$ is constant. I and ω can change, but the product is still constant. In particular, the angular speed of an object can be changed by changing I through the rearrangement of the mass distribution of the object. If this is done within or *internal* to the system, the angular momentum is still conserved (no *external* torques). This is what a diver does in a tuck (smaller I) to rotate faster during a dive [Fig. 1(a)]. To reduce the rotation, the diver may "pike" (greater I), then "layout" (even greater I) to break the water cleanly. Gymnasts use similar techniques for twisting somersaults—for example, in dismounts from the horizontal bar and parallel rings, in which landing on the feet is essential.

Cats make use of a similar technique to land on their feet when dropped from a sufficient height. While falling, a cat changes its moment of inertia by instinctively reorienting its legs and tail. By the proper twisting reorientation it is possible to rotate the head one way and the feet the other, so that the feet are down when the cat lands.

The conservation of angular momentum is also used by ice skaters in doing spins [Fig. 1(b)] and by ballet dancers in doing pirouettes. Sweeping motions of the arms and legs start the body in rotational motion. When pivoting on the tip of the skate blade or on the tip of the toe, the skater or dancer is virtually free of external torques. The arms are then "tucked in" or raised vertically overhead to reduce the moments of inertia of the body, and the skater or dancer spins rapidly to conserve angular momentum.

Should you lack these talents, you can still get into the act (of conservation of angular momentum). All you need is a rotating stool or turntable and some weights, as shown in Figure 10.13. With your feet off the floor, you'll need someone (an external torque) to get you started. Try to get yourself rotating when isolated on the stool (or standing on a turntable). You'll learn to appreciate Newton's rotational first law. Don't start rotating too fast at first. When you bring in the weights toward your body (reducing I), you'll spin quite fast. You don't want to get dizzy, go into unstable equilibrium, and fall off the stool.

(a)

(b)

Figure 1
Conservation of angular momentum. (a) A diver spins by "tucking" and decreasing the moment of inertia. By the conservation of angular momentum, this causes the angular speed to increase. (b) An ice skater uses the same principle in doing spins.

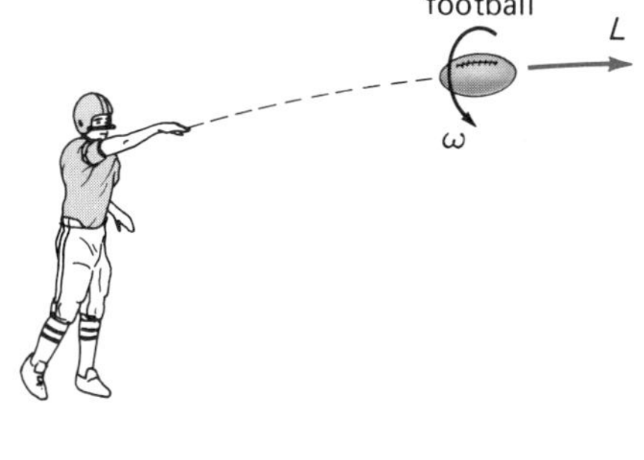

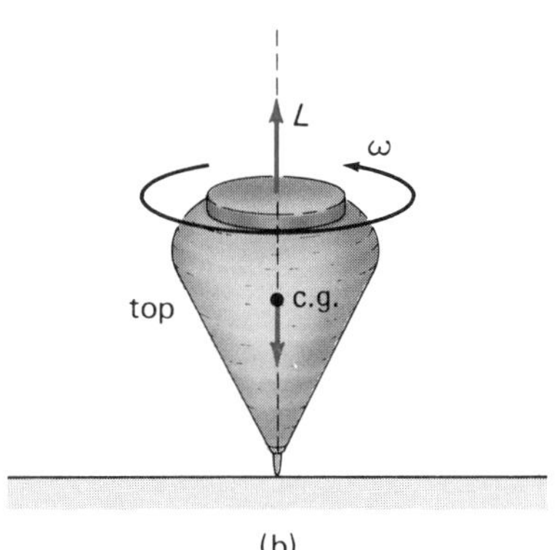

Figure 10.14

The conservation of angular momentum gives directional stability. (a) The angular momentum of a spiraling football is approximately conserved in the forward direction. (b) A rapidly spinning "sleeping" top remains upright. Notice that the center of gravity is on the axis of rotation and hence there is no gravitational torque.

action. Gyroscopic devices are used to indicate direction (gyrocompasses), to steer ships and airplanes (gyropilots, or autopilots), and to provide stabilization (gyrostabilizers).

A basic gyroscope is shown in Figure 10.15. It consists of a rotating wheel universally mounted on gimbals or rings so that the wheel is free to turn about any axis. For a perfectly balanced and friction-free gyroscope, the angular momentum is conserved, and the spin axis (or angular momentum vector) maintains a fixed direction in space—even if the gyroscope base is moved (Fig. 10.15).

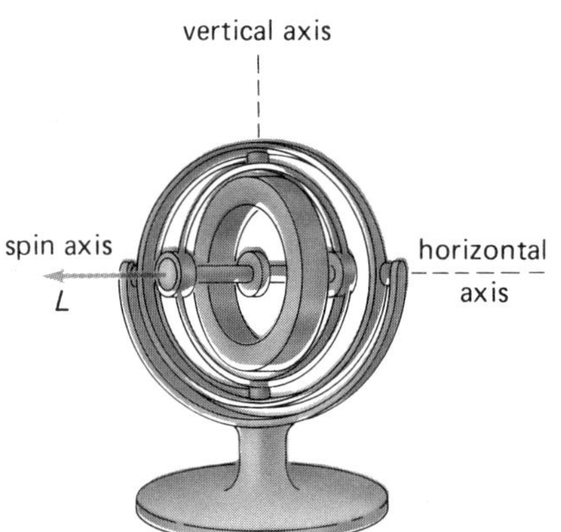

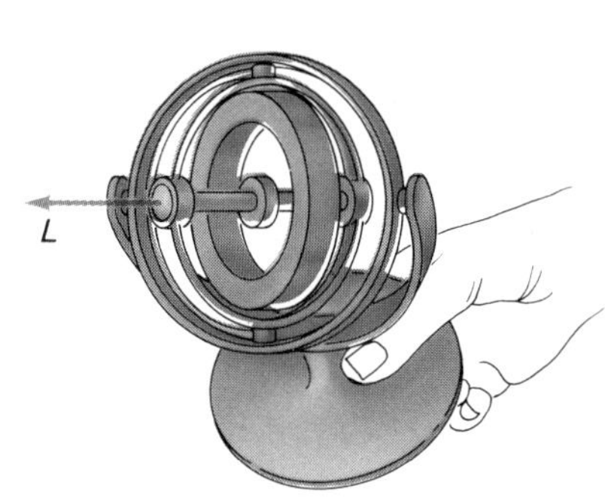

Figure 10.15

A basic gyroscope. A rotating wheel is universally mounted on gimbals or rings so the wheel is free to turn about any axis. The angular momentum is conserved and the spin axis or angular momentum vector maintains a fixed direction in space, even if the gyroscope is moved.

Important Terms

rigid body an object in which all the particles maintain fixed distances relative to each other

angular displacement the directed angle through which an object in rotation moves, measured in degrees or radians

angular velocity the time rate of change of angular displacement

angular acceleration the time rate of change of angular velocity

torque the product of the moment arm and applied force; an unbalanced torque produces an angular acceleration or angular motion

moment of inertia the summation of the products of the masses and distances squared of the particles of a body about an axis of rotation, and a measure of rotation inertia

angular momentum the product of the moment of inertia of a body and its angular velocity around an axis of rotation

conservation of angular momentum the principle that, in the absence of an unbalanced external torque acting on a body, the angular momentum is conserved

Important Formulas

		(linear analogy)
angular displacement: (in radians)	$\theta = \frac{s}{r}$ (or $s = r\theta$)	
	2π rad = 360°	(linear analogy)
average angular velocity:	$\bar{\omega} = \frac{\theta}{t}$ or $\theta = \bar{\omega}t$	$\left(\bar{v} = \frac{x}{t}\right)$
velocity relationship:	$v = r\omega$	
average angular acceleration:	$\bar{\alpha} = \frac{\omega_f - \omega_o}{t}$	
or, with $\bar{\alpha} = \alpha$:	$\omega_f = \omega_o + \alpha t$	$(v_f = v_o + at)$
kinematic equations: (constant acceleration)	$\bar{\omega} = \frac{\omega_f + \omega_o}{2}$	$\left(\bar{v} = \frac{v_f + v_o}{2}\right)$
	$\theta = \omega_o t + \frac{1}{2}\alpha t^2$	$(x = v_o t + \frac{1}{2}at^2)$
	$\omega_f^2 = \omega_o^2 + 2\alpha\theta$	$(v_f^2 = v_o^2 + 2ax)$
acceleration relationship:	$a_t = r\alpha$	
centripetal acceleration:	$a_c = \frac{v^2}{r} = r\omega^2$	
moment of inertia:	$I = \sum_i m_i r_i^2$	
torque:	$\tau = r_\perp F = rF \sin\theta$	
	$\tau = I\alpha$	$(F = ma)$
work:	$W = \tau\theta$	$(W = Fx)$
power:	$P = \frac{W}{t} = \frac{\tau\theta}{t} = \tau\bar{\omega}$	$\left(W = \frac{Fx}{t} = F\bar{v}\right)$
kinetic energy:	$KE = \frac{1}{2}I\omega^2$	$(KE = \frac{1}{2}mv^2)$
for rolling object:	$KE = \frac{1}{2}mv^2 + \frac{1}{2}I\omega^2$	
angular momentum:	$L = I\omega$	$(p = mv)$
and	$\tau = \frac{\Delta L}{\Delta t}$	$\left(F = \frac{\Delta p}{\Delta t}\right)$

Questions

Description of Rotational Motion

1. How are radians and degrees related? Is the radian actually a unit? Explain.
2. Consider a circular plate turning on a lathe. Are the angular and tangential speeds the same at any point on the plate? Explain.

3. What are the angular speeds of (a) the second hand, (b) the minute hand, and (c) the hour hand of a clock?
4. Considering the relative motion of a phonograph needle in the grooves of a record, does the needle have a greater angular speed near the outer portion of the record or near the center of the record? How about tangential speed?
5. A linear acceleration can change the magnitude and/or the direction of linear velocity. Does an angular acceleration have the same effect on a particle in circular motion? Describe the total effect if the particle remains in the same circular orbit.
6. A person was given a speeding ticket because a patrolman's radar showed the car to be exceeding the 55 mi/h speed limit. At the hearing, three witnesses in the car testified that the speedometer read exactly 55 mi/h. Yet the judge imposed a fine when he learned that new, oversized tires had recently been installed on the car. Was this a fair ruling? Explain. Would the new tires have any effect on the odometer (mileage meter) reading?

Torque and Moment of Inertia

7. (a) What is the magnitude of a torque if r and F are parallel? (b) When is the torque for a given r and F a maximum?
8. When pedaling a bicycle, when do you apply the maximum and minimum torques? (Assume the downward force of your foot is constant.) What is the value of the minimum torque?
9. A stubborn screw won't come loose when you try to remove it with a screwdriver. If you had two other screwdrivers available, one with the same diameter metal shaft and a bigger handle and one with the same diameter but a longer metal shaft and the same size handle, which one would you use to give the screw another try and why?
10. A circus tightrope walker often carries a long pole to help keep his or her balance. Explain the effect of the pole in terms of rotational dynamics.
11. Compare the rotational inertia of a baseball bat about axes through (a) the length of the bat, (b) the handle end, and (c) the opposite (larger) end.
12. Explain the difference in the rotational inertias of the cases shown in Figure 10.16. Which is easier to rotate, or takes less torque? (*Hint:* explain the facial expressions.)

Rotational Work, Power, and Kinetic Energy

13. An object rotates with a uniform angular velocity. (a) How much net work is done per revolution? (b) How much work was done on the object originally?
14. Give the units of rotational work, power, and kinetic energy.
15. The angular speed of a rotating rigid body is doubled. How much more work was done on the object?
16. Explain how $s = r\theta$ is a condition for rolling without slipping.
17. How is the rotational power related to torque?
18. A solid wooden ball and a hollow metal ball both have the same mass and radius. (a) If the balls spin at the same angular speed about an axis through their centers, which has the greater kinetic energy, and how much greater is it? (b) If the balls were released simultaneously from rest at the top of an inclined plane, which would reach the bottom first? (Justify your answer.)

(a)

(b)

Figure 10.16
Same weight, but different effect. See Question 12.

Angular Momentum

19. Define angular momentum for (a) a particle and (b) a rigid body.
20. Is the angular momentum of each particle of a rotating rigid body the same? Explain.
21. It is difficult to balance oneself on a stationary bicycle, yet it is quite easy when the bike is in motion. What's the difference?
22. Circus performers ride bicycles across tightwires. Some use long poles and others do not. How is stability enhanced in each case?
23. Suppose you are standing on a large turntable near the outside edge and you start to walk

around the outer circumference of the table. How would you move relative to someone watching you who is not on the turntable? (Consider the turntable bearings to be frictionless.)

24. Unlike helicopters, single-engine aircraft have only one propeller (or rotor blade). What keeps the body of the airplane from rotating? (*Hint:* think in terms of torques on airfoils.)

Problems

Levels of difficulty are indicated by asterisks for your convenience.

10.1 Description of Rotational Motion

1. How many radians are there in (a) 270°, (b) 100°, and (c) 30°?
2. How many degrees are there in (a) 6π rad, (b) $\pi/5$ rad, and (c) $\pi/24$ rad?
3. A pie is cut into eight equal wedge-shaped pieces. What is the wedge angle of the pieces in radians?

*4. (a) What are the magnitudes of the angular displacements of the second and minute hands of a clock in a time of 300 s? (b) If the second hand is 10 cm long, what is the tangential speed of the tip of the second hand?

*5. The pitch of the threads on a bolt is $\frac{1}{16}$ in. (This is the distance between adjacent threads, and the lateral distance the bolt would move in one revolution when being turned into a threaded hole.) How far would a nut travel on the bolt when rotated through 34.5 radians?

*6. An automobile travels halfway around a circular track in 30 s. What is its average angular speed?

*7. A motor shaft has a speed of 120 rpm. What is its angular frequency in rad/s?

*8. A lathe rotates with an angular speed of 6π rad/s. If the speed is decreased to 2π rad/s in 1.5 s, what is the average angular acceleration?

*9. The speed of a lathe is increased from 60 rpm to 120 rpm in 4.0 s. What is the average angular acceleration?

*10. A potter's wheel rotating initially with an angular speed of $\pi/2$ rad/s is accelerated uniformly at a rate of 0.50 rad/s^2 for 3.0 s. What is the angular speed of the wheel at the end of this time?

*11. A rotating piece of machinery with an initial angular speed of 1.6 rad/s accelerates uniformly at a rate of 0.80 rad/s^2 for 5.0 s. How many revolutions does the piece make during this time?

*12. A disk jockey speeds up a turntable uniformly from 33.3 rpm to 45 rpm. In doing so, the turntable goes through 4.0 revolutions. What is the angular acceleration of the table?

**13. (a) What are the average angular speeds of the Earth's rotation and revolution? (Assume a circular orbit.) (b) What is the tangential speed of the Earth at the equator due to its rotation? How about at your location? (c) What is the tangential speed of the Earth due to its revolution?

**14. A particle in a circular orbit with a radius of 10 cm and a speed of 0.50 rad/s is accelerated uniformly in 0.60 s to a speed of 1.5 rad/s while maintaining the same orbit. What is the total acceleration of the particle (a) before the angular acceleration is initially applied and (b) at the end of the 5.0-s interval (angular acceleration still applied)?

10.2 Torque and Moment of Inertia

*15. A mechanic applies a force of 50 N to the end of a 20-cm wrench at an angle of 30° relative to its length. (a) What is the magnitude of the applied torque? (b) What maximum torque would be applied for the given force?

*16. A torque of 20 lb-ft is applied to a pivoted rod by a force of 15 lb at an angle of 37° relative to the length of the rod. (a) What is the moment arm? (b) What is the length of the rod?

*17. A 0.10-kg particle constrained to move in a circle is set into motion by a force of 4.0 N acting perpendicularly to the radius of rotation. If the particle is 0.20 m from the axis of rotation, what is its angular acceleration?

*18. An unbalanced torque of 200 m-N gives an angular acceleration of 8.0 rad/s^2 to the rotor of a motor. What is the moment of inertia of the rotor?

*19. An unbalanced force of 25 lb is applied tangentially to the rim of an 8.0-lb wheel in the form of a uniform disk. If the radius of the wheel is 1.5 ft, what is the angular acceleration?

*20. The flywheel of a motor is accelerated from rest to a speed of 1500 rpm in 5.0 s. If the flywheel has a moment of inertia of 40 kg-m^2, what is the unbalanced torque on the wheel?

*21. A pulley in the form of a uniform disk has a radius of 6.0 in. and a weight of 4.8 lb. If a constant force of 3.0 lb is applied tangentially to the rim of the pulley, how many rotations will the pulley make in 4.0 s, starting from rest?

*22. A rotating rigid body can be treated as a rotating particle by using the concept of the *radius of gyration*. The radius of gyration k is the radial distance from the axis of rotation to a point at which the total mass of an object could be

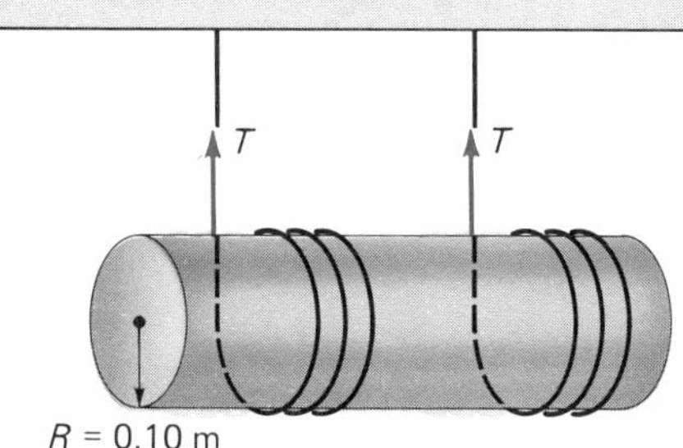

Figure 10.17
See Problem 25.

considered concentrated without changing the moment of inertia, and $I = Mk^2$. Hence, for a hoop with $I = MR^2 = Mk^2$, so $k = R$, or the hoop could be represented by a particle of mass M at a distance R from the axis of rotation. What are the radii of gyration for (a) a solid cylinder and (b) a solid sphere?

*23. The rotor of a large motor has a radius of gyration of 0.50 m and a mass of 30 kg. If an applied torque of 200 m-N produces an angular acceleration of 20 rad/s^2, what is the frictional torque on the rotor? (See Problem 22.)

*24. A rod may be pivoted at its center or at one end. (a) Which condition would give the greater moment of inertia? (b) How many times greater would it be?

**25. A solid cylinder with a radius of 0.10 m is suspended by two strings wrapped around it (Fig. 10.17). If the cylinder is released from rest, (a) what is its angular acceleration as it falls? (b) What is the angular speed when it has made four complete revolutions? (*Hint:* to find the string tension T, apply Newton's second law considering the center of mass of the cylinder, and $a = R\alpha$.)

**26. Show for a combination of two rigid bodies with the same axis of rotation that the total moment of inertia of the combination is given by $I = I_1 + I_2$.

10.3 Rotational Work, Power, and Kinetic Energy

*27. A constant torque of 50 m-N acts on a particle moving in a circle of radius 30 cm through an arc length of 75 cm. How much work is done?

*28. An applied unbalanced torque of 25 lb-ft brings a rotating object to rest after 4.0 revolutions. How much work was done?

*29. A flywheel with an initial angular speed of 15 rad/s is accelerated at a rate of 5.0 rad/s^2 to an angular speed of 20 rad/s. If the magnitude of the applied torque is 8.0 m-N, how much work was done?

*30. A flywheel in the form of a spoked wheel with a mass of 6.0 kg and a radius of 0.80 m rotates with an angular speed of 20 rad/s. What is the kinetic energy of the wheel? (Neglect the mass of the spokes.)

*31. What is the kinetic energy of the rotating Earth?

*32. The rotor of a motor rotating initially with an angular speed of 50 rad/s is slowed to 30 rad/s while making 10 revolutions. If the rotor has a moment of inertia of 12 kg-m^2, (a) how much work is done in slowing the rotor, and (b) what is the average applied torque?

*33. A solid cylinder initially at rest is set rotating about an axis along the axis of symmetry of the cylinder by an applied torque of 35 m-N that acts through 5.0 revolutions. If the cylinder has a mass of 2.5 kg and a radius of 0.40 m, (a) what is the kinetic energy of the cylinder? (b) What is its angular speed?

*34. A hoop rolls down an inclined plane from an initial height of 4.0 m. What is the translational speed of the hoop at the bottom of the incline?

*35. A solid sphere and a solid cylinder of equal radii and masses of 40 kg and 32 kg, respectively, roll together down an inclined plane from rest. Which reaches the bottom of the incline first?

*36. A torque of 30 lb-ft gives a rigid body an average angular speed of 4.0 rad/s. How much power is supplied to the body?

*37. A constant torque of 75 m-N acts on a wheel initially at rest and gives the wheel an angular speed of 40 rad/s. What is the average power delivered to the wheel?

*38. How much torque is developed by a 1.5-hp motor that has a shaft speed of 1750 rpm?

**39. A solid ball rolls down an incline into a "loop-the-loop" of radius R as shown in Figure 10.18. (a) What is the minimum speed the ball must have at the highest point of the loop so

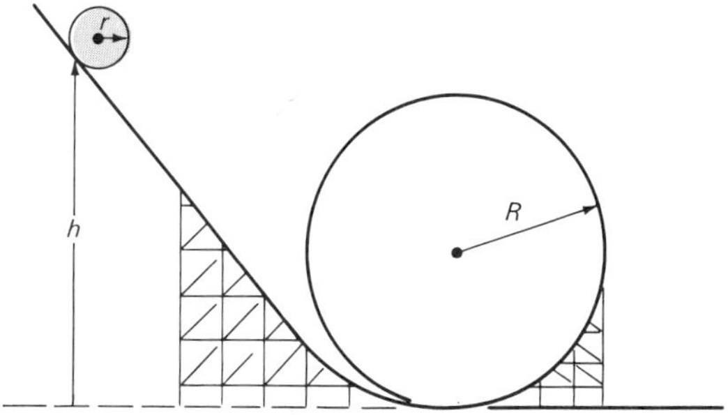

Figure 10.18
See Problem 39.

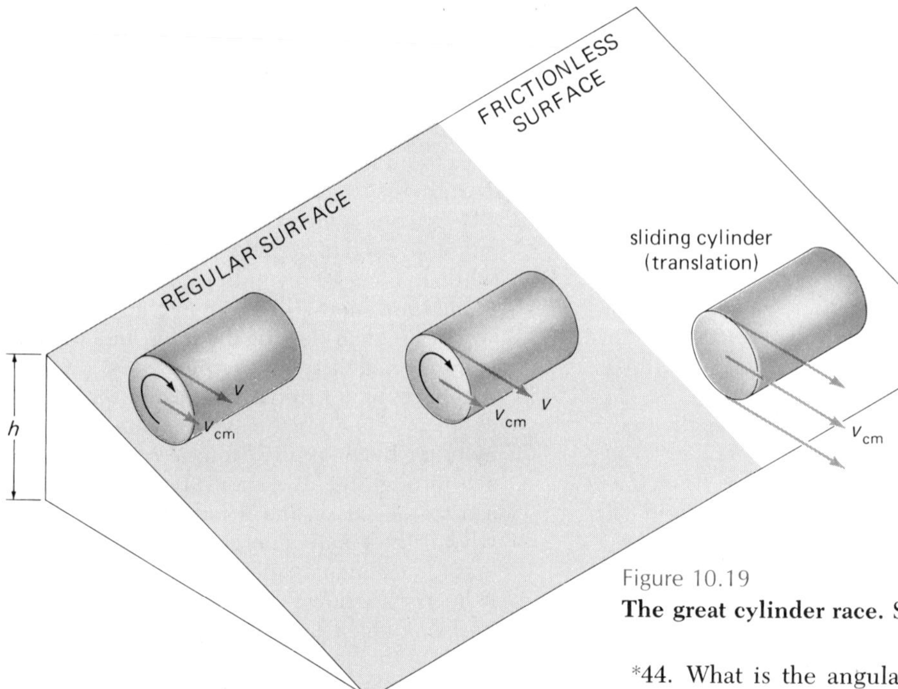

Figure 10.19
The great cylinder race. See Problem 40.

that it stays on the loop? (b) At what vertical height on the incline must the ball be released so that it will just have the required minimum speed at the top of the loop? (Height h is expressed in terms of the radius of the loop.)

**40. In the race of cylinders down an inclined plane as shown in Figure 10.19, (a) show that the sliding solid cylinder will always beat the rolling solid and hollow cylinders when all are released simultaneously from rest at the top of the inclined plane. (b) Show that the solid rolling cylinder will always beat the hollow cylinder, regardless of mass and radius. (Consider for simplicity a thin-walled cylinder.)

**41. A string is wrapped around a disk pulley with a mass of 0.50 kg and a radius of 0.25 m, and a mass of 2.0 kg is suspended on the string. If the suspended mass is released from rest, what is the kinetic energy of the pulley when the mass has descended a distance of 1.5 m?

10.4 Angular Momentum

*42. What is the angular momentum of a 0.20-kg particle traveling in a circular orbit with a radius of 40 cm at a speed of 12 rad/s?

*43. A hollow cylinder with a mass of 5.0 kg and inside and outside diameters of 50 cm and 56 cm, respectively, rotates with an angular speed of 1.4 rad/s about an axis along the axis of symmetry of the cylinder. What is the angular momentum of the cylinder?

*44. What is the angular momentum of the Earth due to its daily rotation?

*45. The angular momentum of a 70-kg turbine wheel increases from 80 m-N-s to 100 m-N-s in 5.0 s. What is the average unbalanced torque that produced this change?

*46. Suppose the Earth expanded to twice its present diameter. How would this affect its rotational speed? (Assume uniform distribution of the same mass in each case.)

*47. A skater has a moment of inertia of 100 kg-m^2 when her arms are outstretched, and a moment of inertia of 75 kg-m^2 when her arms are raised over her head. If she starts into a spin with an angular speed of 2.0 rps while her arms are outstretched, what is her angular speed when she raises her arms over her head?

**48. In a frictional clutch mechanism, a freely rotating member consists of a 1.5-kg circular disk with a radius of 20 cm and rotates with an angular speed of 65 rad/s. The engaging member is also a circular disk with a mass of 1.1 kg and a radius of 15 cm and is initially at rest. What is the angular speed of the disks when they are engaged? (*Hint:* use the result of Problem 26.)

**49. A potter's wheel with a moment of inertia of 20 kg-m^2 rotates freely with an angular speed of 10 rad/s. If a 4.0-kg piece of clay with a radius of gyration of 50 cm is dropped on the axis of rotation of the wheel, what is the resulting angular speed of the wheel and the clay? (*Hint:* use the result of Problem 26 and see Problem 22.)

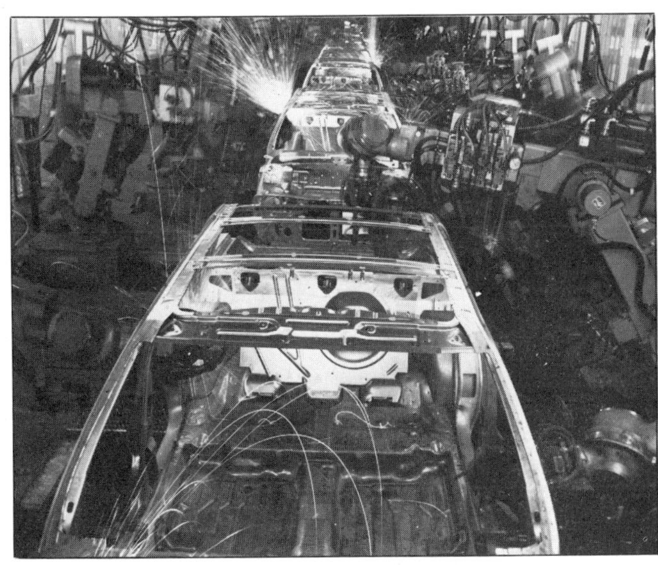

Robot welders at a Chrysler assembly plant. (Courtesy Chrysler Motors Corp.)

Chapter 11

Machines

We use machines every day, particularly in technical applications. Ordinarily, we think of machines as complex mechanical devices, such as a lathe, a gasoline engine, a typewriter, or those shown in the introductory photograph of this chapter. However, simple devices such as hand tools—pliers and hammers—are also machines. In fact, all complex machines are made up of simple machines.

What do machines do for us? Typical answers are: they do work for us, and they make doing work easier. For example, when a jack is used to lift a car, work is done and the task is made easier, since a person could not lift the car without the aid of a machine such as a jack.

But we must analyze the situation carefully. Making things *easier* tends to imply that we get something for nothing, which is not the case. Machines can and do give us a mechanical "advantage" but not without sacrificing something else, as will be learned in this chapter.

11.1 Mechanical Advantage and Efficiency

A machine is a device that multiplies force and/or changes the direction of a force. For example, a claw hammer is a machine that can be used to apply a large force in pulling out a nail. A relatively small pull or force on the handle produces a much greater force at the claws, and in a different direction.

Machines multiply force, but *work* is the critical quantity in analyzing the capabilities of machines. Recall that work is the product of the force F and the parallel distance d through which the force acts:

$$W = Fd$$

By the conservation of energy, the energy or work output of a machine must equal the energy or work input, i.e.,

$$\text{Work input} = \text{work output}$$

$$W_i = W_o + W_f \qquad \textbf{(Eq. 11.1)}$$

or

$$F_i d_i = F_o d_o + W_f$$

where W_f is the work done against, or lost to, friction or some other cause. This relationship is

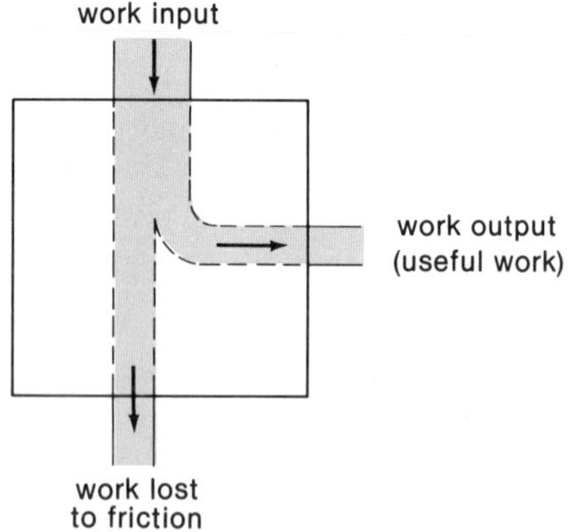

Figure 11.1
Schematic representation of a machine. By the conservation of energy, the work input must be equal to the work output plus the work lost to friction or some other cause. The greater the amount of work lost to friction, the less the useful work output.

illustrated schematically in Figure 11.1. Friction exists in every practical situation, so it must be taken into account. The quantity $W_o = F_o d_o = F_i d_i - W_f$, is the *useful* work output of a machine.

From Equation 11.1, it can be seen that the useful work output is less than the work input:

$$W_o < W_i$$

or

$$F_o d_o < F_i d_i$$

Then, if a machine multiplies force, F_o is greater than F_i, i.e., $F_o > F_i$, and it should be evident that the input distance d_i must be greater than the output distance d_o, or $d_i > d_o$.

What this means is that

> **Force multiplication is accomplished at the expense of a reduction in distance.**

For example, in pulling a nail with a claw hammer (Fig. 11.2), the greater output force acts through a shorter distance than does the smaller input force.

Hence, a machine makes work "easier" by multiplying force, but the smaller input force must move through a greater distance. More work must be put into a machine than the machine puts out (taking friction into account), so a machine does not give you something for nothing. Only in an ideal frictionless case could you "break even."

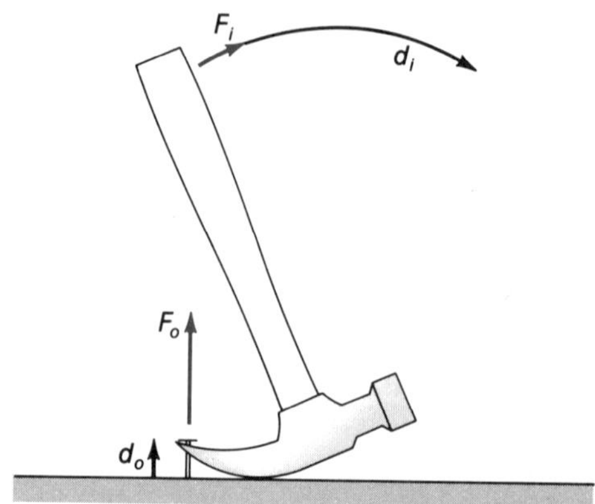

Figure 11.2
Force multiplication. The output force F_o is greater than the input force F_i, hence, there is a force multiplication. However, this is accomplished at the expense of a distance reduction, $d_i > d_o$.

Even so, machines give a definite advantage in doing work through force multiplication. This is expressed in terms of mechanical advantage, which is the factor or magnitude of the force multiplication. The **actual mechanical advantage (AMA)** of a machine is given by the ratio of the output force F_o and the input force F_i:

$$\boxed{\text{AMA} = \frac{F_o}{F_i}} \qquad \textbf{(Eq. 11.2)}$$

Suppose, for example, that the input force of a machine is 10 lb and the output force is 20 lb. Then the machine has an actual mechanical advantage of

$$\text{AMA} = \frac{F_o}{F_i} = \frac{20 \text{ lb}}{10 \text{ lb}} = 2$$

That is, force is multiplied by the machine by a factor of 2. If the input force were 30 lb, the output force would be

$$F_o = (\text{AMA})F_i = (2)30 \text{ lb} = 60 \text{ lb}$$

From the previous discussion, we know that in force multiplication F_i must be applied through a greater distance than the distance F_o moves. If there were no friction and $W_f = 0$, then

$$F_i d_i = F_o d_o$$

or

$$\frac{F_o}{F_i} = \frac{d_i}{d_o}$$

In this case, the mechanical advantage can be expressed in terms of the distances. Since this is an ideal situation, we define an ideal force multiplication factor, or **ideal mechanical advantage (IMA),** sometimes called theoretical mechanical advantage (TMA):

$$\boxed{\text{IMA} = \frac{F_o}{F_i} = \frac{d_i}{d_o}} \qquad \textbf{(Eq. 11.3)}$$

The IMA of a machine is the best or maximum mechanical advantage one can ever hope to achieve if friction could be eliminated completely. It is particularly convenient in designing a machine or analyzing a mechanical situation, since it can be computed from geometrical considerations (d_i and d_o) without considering the actual forces.

EXAMPLE 11.1 An engineer wants to design a machine with a 450-N force output. From geometrical considerations, it is known that the input distance is 9.0 cm and the output distance is 3.0 cm. What can be said about the input force requirements?

Solution: With $d_i = 9.0$ cm and $d_o = 3.0$ cm, the IMA of the situation is

$$\text{IMA} = \frac{d_i}{d_o} = \frac{9.0 \text{ cm}}{3.0 \text{ cm}} = 3.0$$

Notice that since the IMA is a ratio, the distances do not have to be converted to standard units. The same is true for the AMA.

Since in the absence of friction IMA = $d_i/d_o = F_o/F_i$, or

$$F_o = (\text{IMA})F_i$$

then

$$F_i = \frac{F_o}{\text{IMA}} = \frac{450 \text{ N}}{3.0} = 150 \text{ N}$$

Hence, the engineer knows that the input force must be at least 150 N. How much greater depends on the friction of the machine.

Suppose that after building the machine, it is found that the actual input force required is 180 N. Then the actual mechanical advantage of the machine is

$$\text{AMA} = \frac{F_o}{F_i} = \frac{450 \text{ N}}{180 \text{ N}} = 2.5$$

and the machine gives an actual force multiplication of 2.5. The AMA is less than the IMA, as it will be in every situation. Why?

The efficiency of a machine is closely related to its AMA and IMA. In fact, the efficiency can be written in terms of these quantities. Recall from Chapter 8 that the efficiency (Eff) is the ratio of a machine's work output (or power output) to the work input (or power input). Hence,

$$\text{Eff} = \frac{W_{\text{out}}}{W_{\text{in}}} = \frac{F_o d_o}{F_i d_i} = \frac{F_o/F_i}{d_i/d_o} = \frac{\text{AMA}}{\text{IMA}}$$

or

$$\boxed{\text{Eff} = \frac{\text{AMA}}{\text{IMA}}} \qquad \textbf{(Eq. 11.4)}$$

For example, in Example 11.1, the machine would have an efficiency of

$$\text{Eff} = \frac{\text{AMA}}{\text{IMA}} = \frac{2.5}{3.0} = 0.83 = 83\%$$

which means that 83 percent of the machine's work input would go into doing useful work and 17 percent would be lost to friction or some other cause.

11.2 Simple Machines

No matter how complex a machine is, it is made up of simple machines. **There are seven simple machines—six mechanical and one hydraulic: the lever, the pulley, the wheel and axle, the inclined plane, the wedge, the screw, and the hydraulic press.**

However, there are only three basic principles involved in simple machines—namely those of the lever, the inclined plane, and the hydraulic press. The pulley and the wheel and axle are applications of the lever principle, and the wedge and screw are basically inclined planes.

The hydraulic press involves a liquid, and it will be considered in the next section. The discussion here is concerned with the principles of the lever and the inclined plane in the six mechanical simple machines.

The Lever

The lever is a very efficient simple machine. It **consists of a rigid bar that is pivoted to rotate about a point or line called the *fulcrum* (Fig. 11.3).** The input force F_i, commonly called the *effort*, is applied to the lever to maintain or lift a load. The output force F_o must be equal to (static case) or greater than the weight of the load. The input and output lever arms L_i and L_o are the distances from the fulcrum to the effort and from the fulcrum to the load, respectively.

The ideal mechanical advantage of the lever can be calculated from work considerations (with friction neglected):

$$W_{\text{in}} = W_{\text{out}}$$

or

$$F_i s_i = F_o s_o \quad \text{and} \quad \frac{F_o}{F_i} = \frac{s_i}{s_o}$$

where s_i and s_o are the arc lengths through which F_o and F_i, respectively, move.

The two arcs subtend the same angle θ, so $s_i = L_i\theta$ and $s_o = L_o\theta$, and

$$\frac{s_i}{s_o} = \frac{L_i}{L_o}$$

Then, by Equation 11.3,

$$\text{IMA} = \frac{F_o}{F_i} = \frac{d_i}{d_o} = \frac{s_i}{s_o} = \frac{L_i}{L_o}$$

and

$$\boxed{\text{IMA} = \frac{L_i}{L_o} \quad \text{(lever)}} \qquad \textbf{(Eq. 11.5)}$$

Notice that the IMA is given by the geometry of the system. For example, in Figure 11.3, if L_i = 3 ft and L_o = 1 ft, then IMA = 3/1 = 3, and the lever ideally multiplies the force by a factor of 3, or $F_o = 3F_i$.

Also, since a lever in static equilibrium [Fig. 11.3(a)] requires that the sum of the torques be zero, $\Sigma\tau = 0 = F_oL_o - F_iL_i$, or $F_o/F_i = L_i/L_o$, the lever IMA can be derived from this static equilibrium condition as well as from work consid-

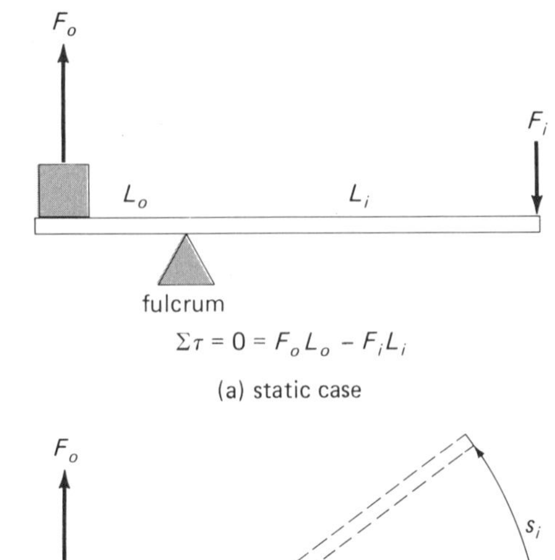

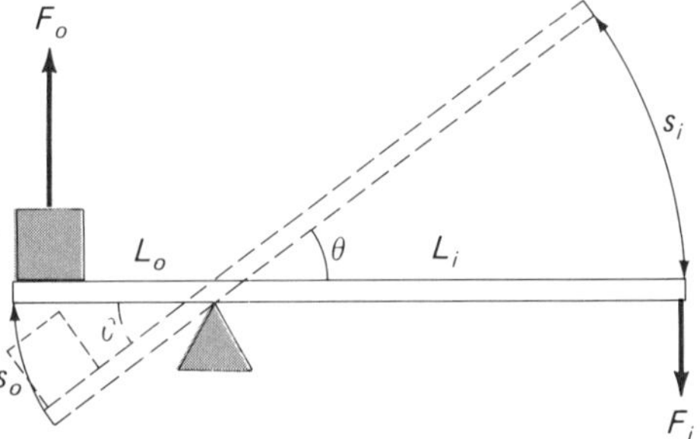

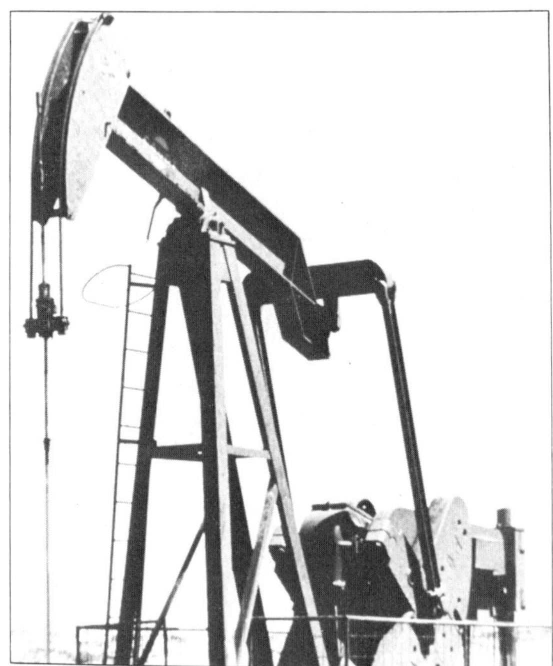

Figure 11.3
The lever. (a) In the static case of maintaining a load, the input and output torques are equal and opposite, or $\Sigma\tau = 0$ (rotational equilibrium). (b) In lifting a load, $F_is_i = F_os_o$, when friction is neglected. (c) A lever application in an oil rig. (Courtesy Chevron U.S.A., Inc.)

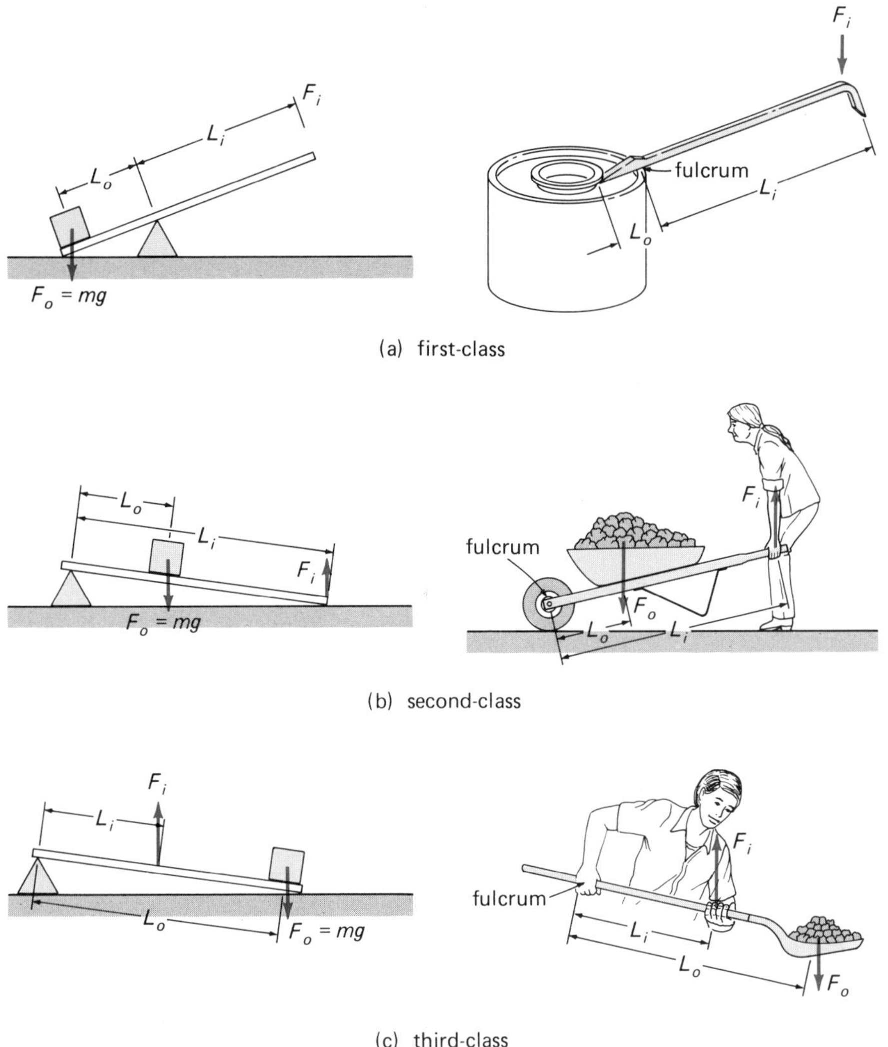

Figure 11.4
Classes of levers. (a) For a first-class lever, the fulcrum is between the load (F_o) and the effort (F_i). (b) For a second-class lever, the load is between the fulcrum and the effort. (c) For a third-class lever, the effort is between the fulcrum and the load.

erations. Frictional losses are normally quite small in the lever action, so for most practical purposes the AMA of a lever is taken to be approximately equal to the IMA.

Levers are divided into three classes, depending on the location of the fulcrum relative to the input and output forces (Fig. 11.4). In a **first-class lever,** the fulcrum is located between the effort and load. The previous derivation of the IMA of a lever was done for a first-class lever, but Equation 11.5 applies to all three lever classes. A crowbar is an example of a first-class lever.

For a **second-class lever,** the fulcrum is at one end of the lever and the effort is applied at the other end. The load is somewhere in between. A wheelbarrow is an example of a second-class lever. First-class and second-class levers are commonly used to move large loads with relatively small efforts.

For a **third-class lever,** the fulcrum is at one end of the lever and the load is at the other end.

The effort is applied somewhere in between. Lifting a load with a shovel is an example of a third-class lever. Third-class levers are commonly used to move a load through a relatively large distance at the expense of a force reduction. Notice that $L_o > L_i$.

EXAMPLE 11.2 If the handles of a wheelbarrow are 3 ft from the wheel axle, and the center of gravity of a 100-lb load (including wheelbarrow) is 1 ft from the axle, (a) what is the IMA of the wheelbarrow and (b) how much effort will have to be applied to hold the wheelbarrow legs off the ground? (See Fig. 11.4.)

Solution: (a) With $L_i = 3$ ft and $L_o = 1$ ft,

$$\text{IMA} = \frac{L_i}{L_o} = \frac{3 \text{ ft}}{1 \text{ ft}} = 3$$

(b) The IMA is the force multiplication factor, or

$$\text{IMA} = \frac{F_o}{F_i} = 3$$

so the effort (F_i) needed to balance a load $F_o =$ 100 lb is

$$F_i = \frac{F_o}{3} = \frac{100 \text{ lb}}{3} = 33.3 \text{ lb}$$

EXAMPLE 11.3 A person using a long-handled shovel lifts a shovelful of dirt. The pivot hand on the back end of the shovel handle is 4 ft from the load (includes shovel), the lifting hand is 1 ft from the pivot, and the load is 5 lb (see Fig. 11.4). (a) How much effort is required to lift the load, and (b) how much farther does the load travel than the lifting hand?

Solution: (a) This is a third-class lever with $L_o = 4$ ft and $L_i = 1$ ft:

$$\text{IMA} = \frac{L_i}{L_o} = \frac{1 \text{ ft}}{4 \text{ ft}} = \frac{1}{4}$$

Hence, with $F_o = 5$ lb,

$$\text{IMA} = \frac{F_o}{F_i} = \frac{1}{4}$$

and

$$F_o = \frac{F_i}{4}$$

or

$$F_i = 4F_o = 4(5) = 20 \text{ lb}$$

(actually slightly more, since this would be the force required for static equilibrium). Hence, there is a force reduction.

(b) Assuming the pivot hand to remain fixed, the load and lifting hand move in circular arcs, and

$$\text{IMA} = \frac{L_i}{L_o} = \frac{s_i}{s_o} = \frac{1}{4}$$

so

$$s_o = 4s_i$$

and the load moves four times the distance the effort moves (increased distance at the expense of a force reduction).

The simple lever can be used to produce large force multiplications. However, a serious limitation is the small angle of rotation about the fulcrum and the correspondingly small output distance through which the load is moved. This restriction can be overcome by allowing for the continuous rotation of the lever arms, which is the principle of the wheel and axle and of the pulley.

The Wheel and Axle **The wheel and axle consists of a wheel rigidly attached to an axle or shaft that turns with the wheel.*** As can be seen from Figure 11.5, a wheel and axle is essentially a rotating lever of unequal lever arms. If an effort force F_i is applied to the wheel, the axle has an output force F_o, which can be used to lift a load or can be applied to a particular application, as shown in the figure.

To calculate the IMA of the wheel and axle, from equilibrium considerations, $F_i R = F_o r$, and

$$\text{IMA} = \frac{F_o}{F_i} = \frac{R}{r} \quad \text{(wheel and axle)} \qquad \textbf{(Eq. 11.6)}$$

where R is the radius of the wheel and r is the radius of the axle.

* This is not to be confused with a wheel-and-axle arrangement in which the wheel (with bearings) turns on the axle.

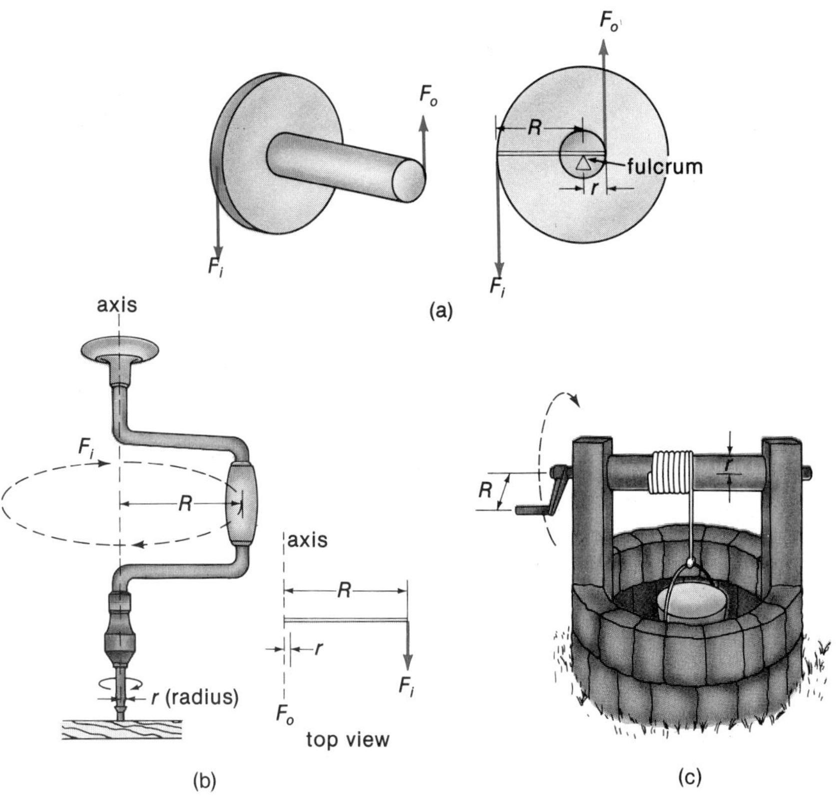

Figure 11.5
Wheel and axle. (a) The wheel and axle is essentially a rotating lever of unequal lever arms. (b) and (c) Examples of wheel and axle applications.

The Pulley Another application of the lever principle is the pulley. **A pulley is a wheel pivoted so that it can rotate freely about an axis through its center.** The wheel is usually grooved and the support rope runs in the groove (Fig. 11.6), or a chain may run on a notched wheel.

A pulley or group of pulleys in a frame with a hook used for attaching is commonly called a *block*. The ropes or supports are referred to as *tackle* or *fall*, hence the term **block and tackle.** The individual pulley wheels in the block are called **sheaves.**

As can be seen from Figure 11.6, a single fixed block (or pulley) is essentially a first-class lever with equal lever arms. In this case, $F_o = F_i$, and if an effort pulls a length of rope down a distance d_i, the load will move up an equal distance, or $d_o = d_i$. Hence, the IMA is

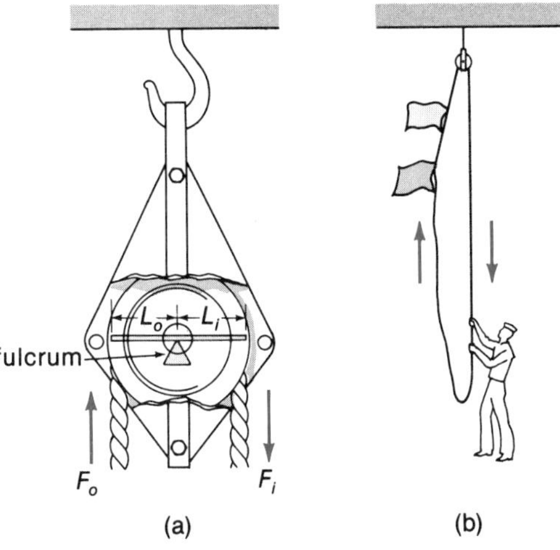

Figure 11.6
Single fixed pulley. (a) The single fixed pulley is essentially a first-class lever with equal lever arms. Hence, it has an IMA of 1. (b) Basically, a single fixed pulley is a direction changer.

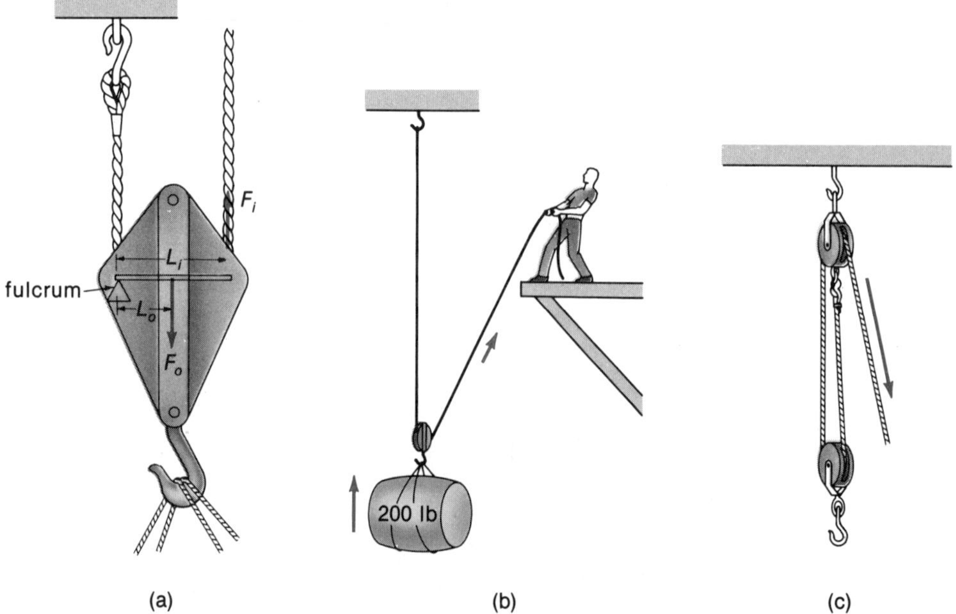

Figure 11.7
Single movable pulley. (a) The single movable pulley is essentially a second-class lever with unequal lever arms. In particular, with $L_i = 2L_o$, then IMA = 2. (b) A runner arrangement. In this case the hauling strand is a support strand. (c) A gun tackle. The top fixed pulley merely changes the direction of the pull. The hauling strand is not a support strand in this case.

$$\text{IMA} = \frac{F_o}{F_i} = \frac{d_i}{d_o} = 1 \quad \text{(single fixed pulley)} \qquad \textbf{(Eq. 11.7)}$$

and there is no force magnification. Basically, a single fixed block is a *direction changer*. A common example is a flag hoist.

How then is a pulley used to give force magnification? As in the case of a simple lever, this requires unequal lever arms; the requirement is met by a movable block.

As illustrated in Figure 11.7, a single movable block is essentially a second-class lever with the effort arm length twice that of the load arm, i.e., $L_i = 2L_o$. Also, from the figure it can be seen that if the effort pulls a length of rope d_i, the load is raised only one-half that distance, since there are two support strands to shorten, and $d_i = 2d_o$. Hence, the IMA is

$$\text{IMA} = \frac{F_o}{F_i} = \frac{d_i}{d_o} = 2 \quad \text{(single movable pulley)} \qquad \textbf{(Eq. 11.8)}$$

Notice for a movable pulley system that the weight of the movable block is part of the load.

Other block-and-tackle arrangements are shown in Figure 11.8. By similar reasoning concerning d_i and d_o and the shortening of support strands, you should be able to convince yourself that the IMA's of the arrangements are 3 and 4, respectively. Notice in the figure how the load support in each case is divided among the number of support strands, giving IMA's of 3 and 4. Further investigations of other block-and-tackle arrangements quickly lead to a general rule for finding the IMA of a block and tackle:

The number of support strands going to (and from) the movable block is equal to the IMA.

In most applications, the support strands go to the movable block, but an exception is shown in the arrangement in Figure 11.9. There are two movable blocks in this case. Block A has an IMA = 3, and block B, with three support strands going to the block and one support strand coming from it, has an IMA = 4. The total IMA of such combinations is the product of the

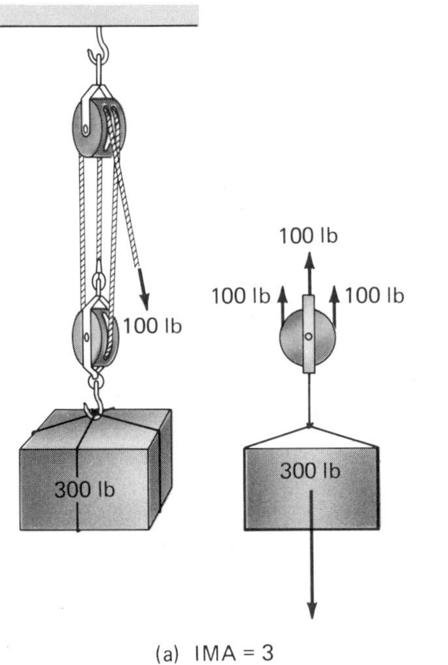

(a) IMA = 3

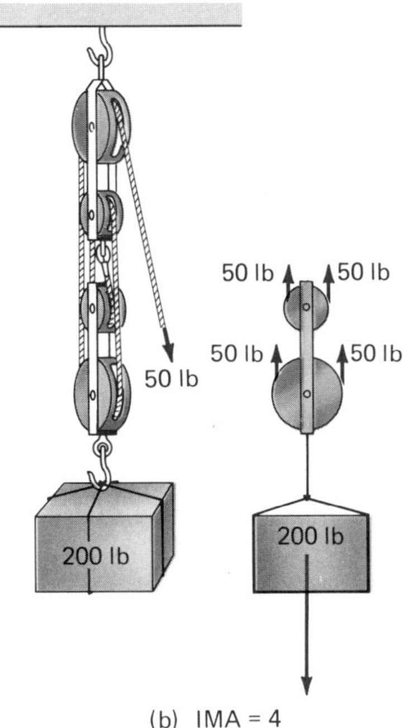

(b) IMA = 4

(c)

Figure 11.8
Block and tackle. (a) A luff tackle consists of a double fixed block and a single movable block, which gives an IMA = 3. (b) Arrangement with IMA = 4. The sheaves of a block may have different axes for ease of threading. (c) Pulleys in a crane hoist application. (Courtesy American Monorail.)

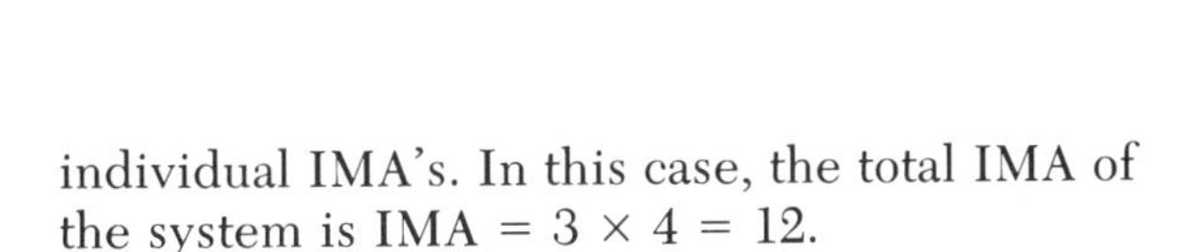

individual IMA's. In this case, the total IMA of the system is IMA = $3 \times 4 = 12$.

To find the actual mechanical advantages of block-and-tackle systems, as with any other systems, one must measure the actual input and output forces, which take friction into account.

A device based on both the wheel and axle and the block and tackle is the differential pulley and chain hoist (Fig. 11.10). The chain hoist is commonly used to lift heavy objects in factories and machine shops. The upper "block" is actually a wheel and axle. An endless chain runs over notched pulleys to prevent slipping.

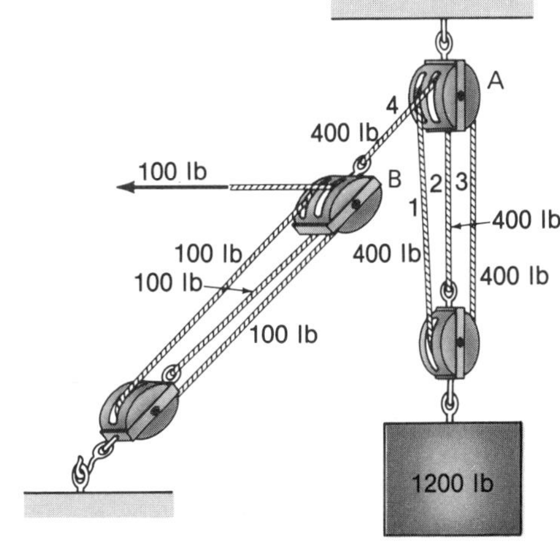

Figure 11.9
A luff upon luff. The total IMA is the product of the IMA's of the individual luffs, IMA = $3 \times 4 = 12$ for this case. (See text for description.) Ideally, a 100-lb applied force could support a 1200-lb load.

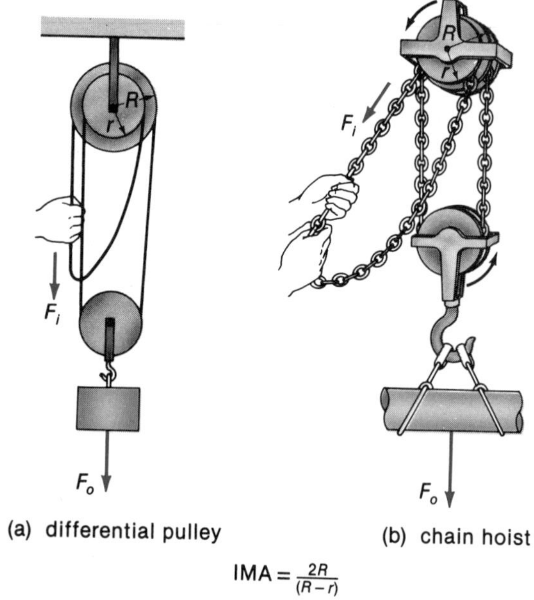

Figure 11.10
A differential pulley (a) and a chain hoist (b). A rotation of the large wheel will raise the load and at the same time the small wheel feeds the string or chain to the movable block. In one rotation of the fixed block sheaves, the load will move a vertical distance $d_o = \pi R - \pi r$. See text for description.

To find the IMA of this device, consider pulling the chain until the large wheel has made one rotation. The effort (F_i) has then moved through a distance equal to the circumference of the wheel, or $d_i = 2\pi R$. Since each of the support chains is shortened by half this amount, the load is raised a distance of πR.

However, the small wheel (axle) makes one rotation at the same time as the larger wheel does and so will feed a length of chain down to the movable block equal to its circumference $2\pi r$. Again, each of the two support chains will be increased by half this length, and the load will be lowered by πr.

The net result of these two actions is to steadily move the load upward a distance equal to the difference of the chain shortening and lengthening, i.e., $d_o = \pi R - \pi r = \pi(R - r)$. Then,

$$\text{IMA} = \frac{d_i}{d_o} = \frac{2\pi R}{\pi(R - r)}$$

$$\text{IMA} = \frac{2R}{(R - r)} \qquad \textbf{(Eq. 11.9)}$$

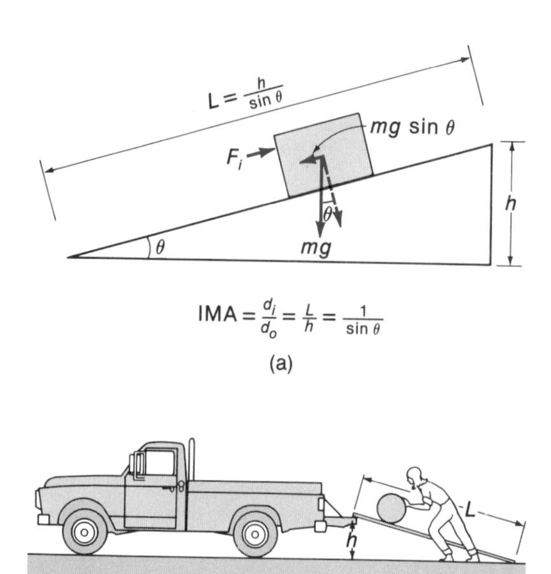

(chain hoist)

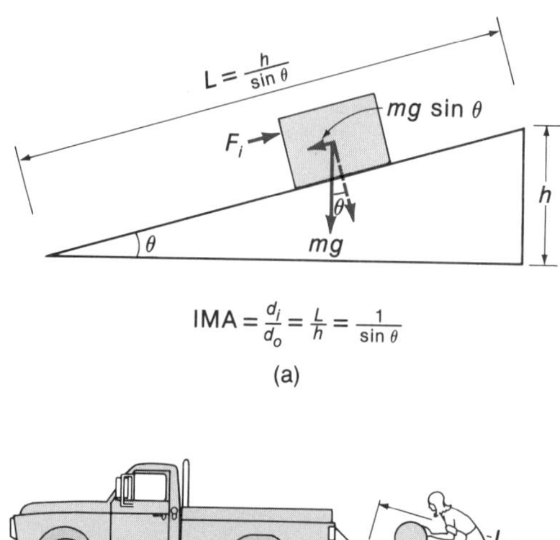

(c)

Figure 11.11
The inclined plane. Less input force is required to move a load a vertical distance (h), but load must be moved through a greater distance (L). (Photo courtesy of Sperry Vickers Tulsa Division.)

The Inclined Plane

The mechanical advantage of the inclined plane comes from applying a relatively small effort force through a greater distance than a heavy load is raised. For example, it is much easier to roll or push an object up a ramp or inclined plane than it is to lift it directly. In fact, the object may be too heavy to lift directly.

To find the IMA of an inclined plane, consider Figure 11.11. The distance through which an effort force parallel to the plane moves in moving an object up the plane is L, or $d_i = L$. The output force or the object's weight moves through a vertical distance $d_o = h$. Hence,

$$\text{IMA} = \frac{d_i}{d_o} = \frac{L}{h} = \frac{1}{\sin\theta} \quad \text{(inclined plane)} \qquad \textbf{(Eq. 11.10)}$$

Since $\sin\theta = h/L$, the IMA can also be written in terms of the angle of incline IMA $= L/h = 1/\sin\theta$.

Considering the work done in moving an object up an inclined plane, the input force must overcome the component of the object's weight parallel to the plane, $F_i = mg\sin\theta$, and acts through a distance $d_i = L = h/\sin\theta$ in moving the object up the plane. Then, $W_{\text{in}} = F_i d_i = (mg\sin\theta)(h/\sin\theta) = mgh$, which is the same as the work done in lifting the object a vertical height h with $F_o = mg$.

Steps are based on the inclined plane principle. Here, the load is moved in incremental vertical steps.

EXAMPLE 11.4 A worker must exert a force of 270 N($\approx$60 lb) parallel to the plane to move a 675-N ($\approx$150 lb) load up a 15° inclined plane. What are (a) the IMA, (b) the AMA, and (c) the efficiency of the inclined plane as a machine?

Solution: (a) With $\theta = 15°$,

$$\text{IMA} = \frac{1}{\sin\theta} = \frac{1}{\sin 15°} = \frac{1}{0.26} = 3.8$$

(b)

$$\text{AMA} = \frac{F_o}{F_i} = \frac{675\text{ N}}{270\text{ N}} = 2.5$$

(c)

$$\text{Eff} = \frac{\text{AMA}}{\text{IMA}} = \frac{2.5}{3.8} = 0.66\ (\times\ 100\%) = 66\%$$

As can be seen from Equation 11.10, the smaller the angle of incline (the smaller $\sin\theta$), the greater the mechanical advantage. Of course, for a given height, the length of incline is greater for a smaller incline angle.

Roadways are built with gradual slopes to take advantage of the inclined plane principle. The steepness of a road is described by highway engineers in terms of grade (or gradient). This is the ratio of the vertical distance, or rise or fall, to a horizontal distance of 100 ft* and is commonly expressed as a percentage.

For example, if there is a 2-ft rise for every 100 ft of horizontal distance, the grade of the road is +2 percent (2/100 × 100%). Similarly, if there is a 1.5-ft fall in a road for every 100 ft of horizontal distance, the grade is −1.5 percent. Plus and minus signs are used to indicate rise and fall, respectively.

The Wedge **The wedge** is a special application of the inclined plane principle. It usually **consists of two inclined planes set base to base** (Fig. 11.12). By driving a wedge into a material, the material can be split or cut and is forced apart a distance equal to the broad end of the wedge when the wedge is driven full length. Wedges are commonly used to split logs. Knives, hatchets, chisels, and other such cutting tools are wedges.

The IMA of such double-plane wedges is simply

$$\text{IMA} = \frac{L}{t} \quad \text{(wedge)} \qquad \textbf{(Eq. 11.11)}$$

where L and t are the length and base thickness of the wedge, respectively. Because of relatively large friction losses between a wedge and the load, the wedge has a relatively low AMA and relatively low efficiency.

A movable inclined plane can also be used as a wedge or insert to raise a large load (Fig. 11.12). The IMA in this case is the same as that of an inclined plane.

A cam used to lift valves in an internal combustion engine is a rotary wedge. If the circumference of the cam is $2L$, and R and r are the maximum and minimum distances from the rim of the cam to the pivot, then

$$\text{IMA} = \frac{L}{(R - r)} \quad \text{(cam)} \qquad \textbf{(Eq. 11.12)}$$

* As in the slope of a roof, "the rise (or fall) over the run (of 100 ft)."

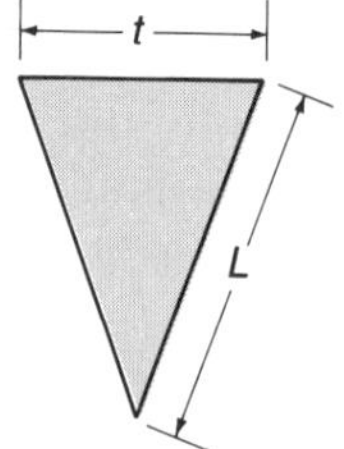

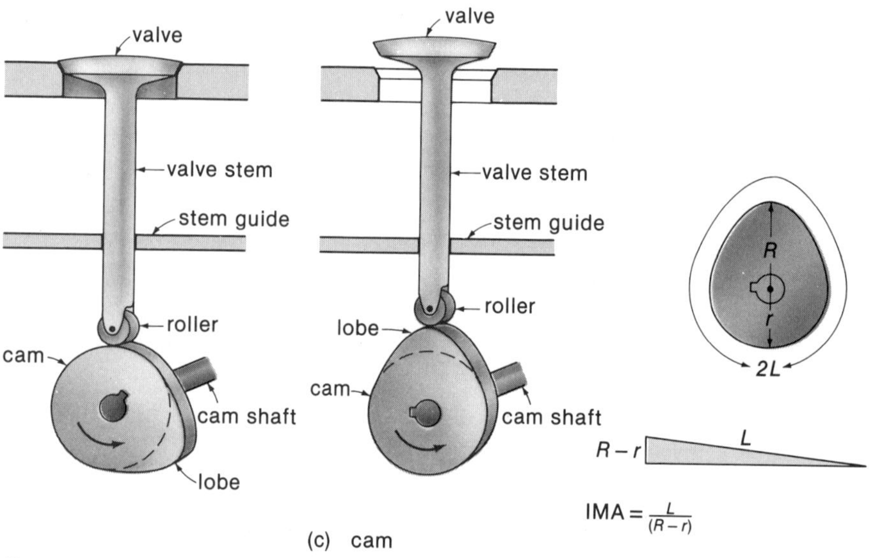

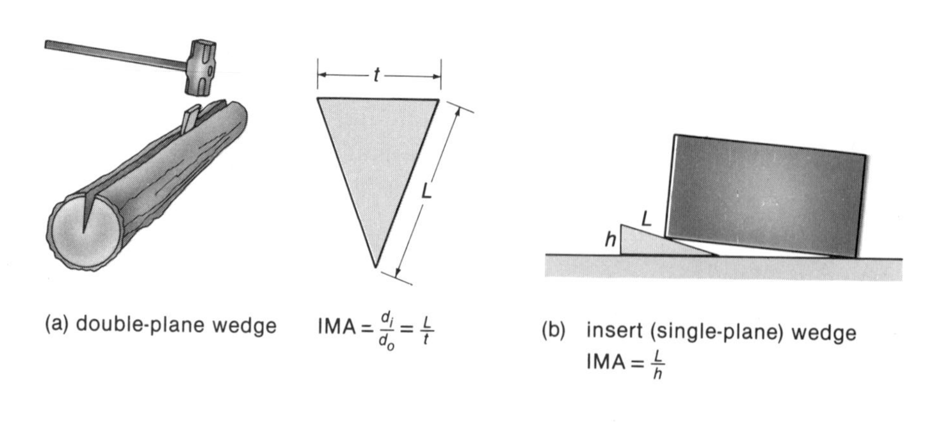

Figure 11.12
The wedge. (a) A splitting wedge consists of two inclined planes set base to base. (b) An insert wedge is a single, movable inclined plane. (c) A cam is a rotary wedge.

The Screw One of the most important and widespread applications of the inclined plane principle is the screw. **A screw is basically a spiral inclined plane wrapped around a cylinder [Fig. 11.13(a)].** The threads of a screw are usually, but not always, triangular in cross section, and the **pitch** of a screw is the distance between adjacent threads. A screw travels through its pitch distance in one complete rotation.

The spiral incline or continuous helix of threads may be cut in a clockwise or counterclockwise fashion. The terms **right-hand** and **left-hand threads,** respectively, are used to describe these screws.

A screw with right-hand threads moves away from an observer when turned clockwise. Standard screws and bolts have right-hand threads and are "screwed in" or tightened (move away) when turned clockwise.

Screws and bolts with left-hand threads are tightened when turned counterclockwise. Such screws are used in special applications. For example, the screw on one side of a turnbuckle has left-hand threads (Fig. 11.13b). As such, the turnbuckle takes up slack in a cable when turned in one direction, and slacks off when turned in the opposite direction.

A set of bicycle pedals has standard right-hand threads on the right pedal and left-hand threads on the left pedal. (You will notice an "L" and an "R" stamped on the ends of the

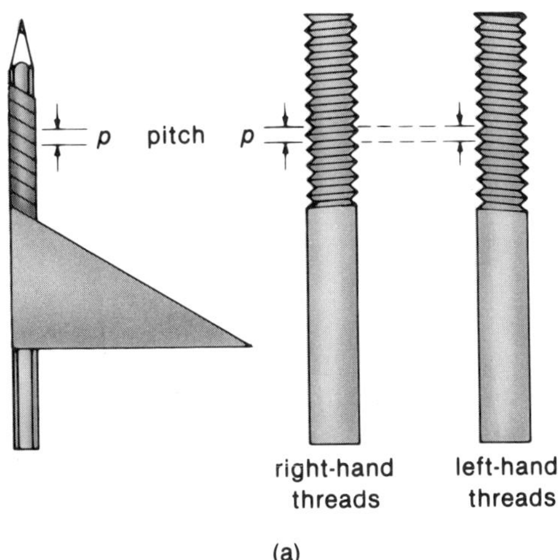

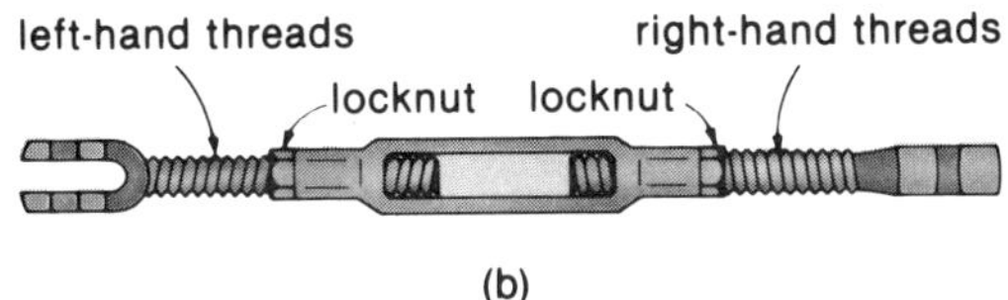

Figure 11.13
The screw. A screw is a spiral inclined plane. (a) A screw may have either right-hand or left-hand threads. The pitch p of the screw is the distance between adjacent threads. (b) A turnbuckle has right-hand threads on one screw and left-hand threads on the other.

pedal axle.) This arrangement keeps the pedals tight in the sprocket arms when the cyclist is pedaling.

The IMA of a screw depends on the lever arm used to turn it. For example, for the screw jack shown in Figure 11.14, when the jack arm of length R is turned though one complete turn, the effort moves through a distance $d_i = 2\pi R$, while the load moves through one pitch distance p, or $d_o = p$. Hence,

$$\text{IMA} = \frac{d_i}{d_o} = \frac{2\pi R}{p} \quad \text{(screw jack)} \qquad \textbf{(Eq. 11.13)}$$

Because of the smallness of p, a screw gives a large force magnification—for example, in the application of a shop vise.

Like the wedge, the screw is a relatively inefficient machine because of friction. However, this is an advantage in many cases, since the friction between the threads and the material

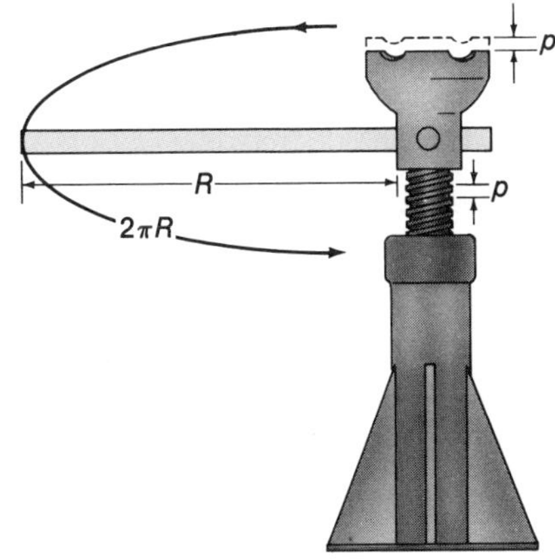

Figure 11.14
Screw jack. When the jack handle is turned through one complete turn, the effort or input force moves through a distance $d_i = 2\pi R$, while the load moves through one pitch distance p.

prevents the screw from slipping and backing out under a load.

The pitch of a screw is commonly expressed indirectly in terms of the number of threads per inch; the pitch in inches is

$$p = \frac{1}{\text{No. of threads/in.}} \qquad \textbf{(Eq. 11.14)}$$

The standard 60° thread is a triangular-cross-section thread with a 60° apex angle. The apex of the thread may be rounded or flattened to provide better stress properties. The number of threads per inch of a screw can be measured by using a thread gauge (Fig. 11.15).

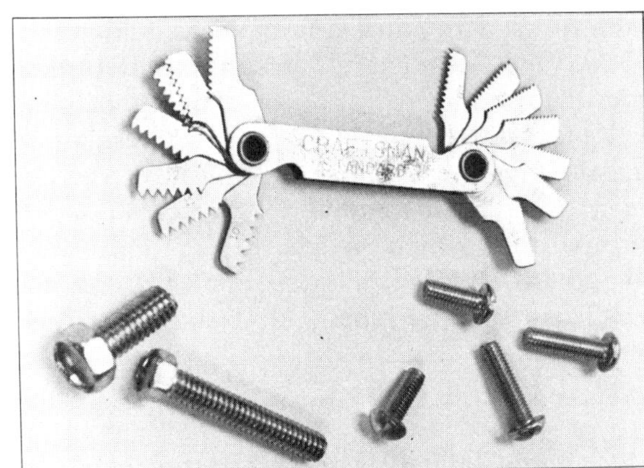

Figure 11.15
Thread gauge. The number of threads per inch of a screw can be determined by fitting the appropriate gauge in the threads.

Table 11.1
Selected Screw Sizes and Threads per Inch

Size	Basic Major Diameter (in.)	Coarse (UNC) Threads/Inch	Fine (UNF) Threads/Inch
0	0.0600	—	80
1	0.0730	64	72
2	0.0860	56	64
3	0.0990	48	56
4	0.1120	40	48
5	0.1250	40	44
6	0.1380	32	40
8	0.1640	32	36
10	0.1900	24	32
12	0.2160	24	28
$\frac{1}{4}$	0.2500	20	28
$\frac{1}{2}$	0.5000	13	20
$\frac{3}{4}$	0.7500	10	16
1	1.0000	8	12
2	2.0000	4.5	12
3	3.0000	4	12

There are a number of "standard" threads, some of which are quite specialized. The *Unified National* (UN) standard is commonly used. Screws used in most common applications are of the coarse-thread series (designated UNC) or the fine-thread series (designated UNF). Coarse-thread screws are recommended for use where vibration is not an important factor, where disassembly of parts is frequent, and where tapped holes are in metals other than steel. Fine-thread series screws (UNF) are frequently used in automotive and aircraft work, where the characteristics of coarse-thread screws are undesirable.

Some selected screw sizes and threads per inch are given in Table 11.1. The size of a screw refers to the basic diameter of the screw, or the cylinder from which the screw is cut. For example, an 8-32 screw has a diameter of 0.1640 in. and 32 threads per inch.

11.3 The Hydraulic Press

The simple hydraulic machine, the hydraulic press, is based on the relative incompressibility of liquids. Pressure is used to describe a force acting on or exerted on a liquid, which must act on a container area. The pressure p is the ratio of the force F acting perpendicularly to a surface area A, i.e.,

$$p = \frac{F}{A} \qquad \text{(Eq. 11.15)}$$

Hence, pressure has the units of lb/in^2 and N/m^2 (or pascal; 1 Pa = 1 N/m^2).

Since most liquids are practically incompressible, the pressure applied to any part of a confined liquid is transmitted to every other part of the liquid. That is, an applied pressure exerts an equal pressure on every other part of the container of the liquid. This principle, first stated by the French mathematician Blaise Pascal (1623–1662), is known as **Pascal's principle:**

> **Pressure applied to an enclosed fluid is transmitted undiminished to every portion of the fluid and to the walls of the container.**

The term *fluid* refers to both liquids and gases. Pascal's principle holds also for gases, but the pressure in a gas is not transmitted as quickly as that in a liquid.

As a demonstration of Pascal's principle, suppose some small holes were poked in a water-filled balloon—small enough so the water would not run out. Then, if pressure were applied to the balloon, water would be forced out of all the holes, as illustrated in Figure 11.16, demonstrating that the pressure is transmitted (equally) throughout the liquid.

Similarly, if a force of one pound were applied to a piston on an enclosed liquid, as in

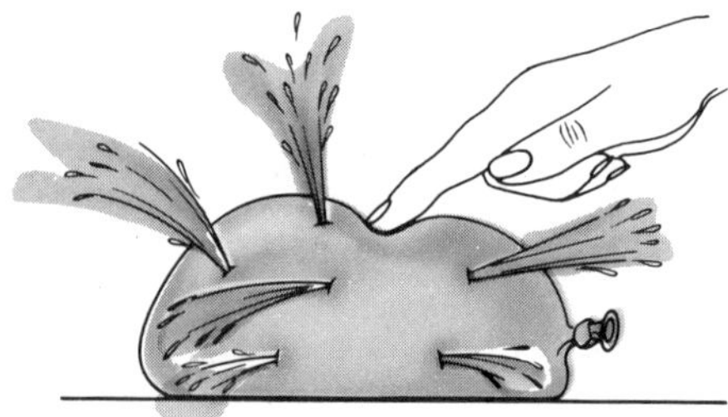

Figure 11.16
An illustration of Pascal's principle. Pressure applied to an enclosed fluid is transmitted undiminished to every portion of the fluid and to the walls of the container. This is evidenced by water being forced out of all of the holes in the balloon.

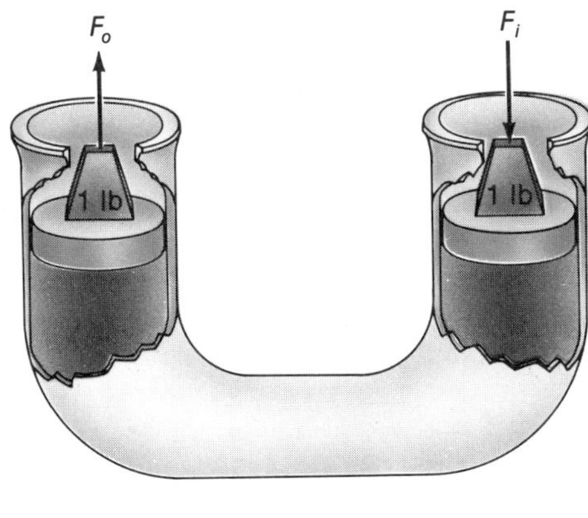

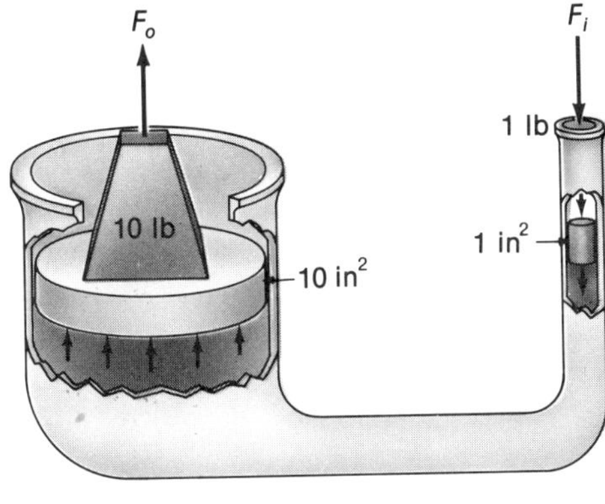

Figure 11.17

Hydraulic advantage. (a) A 1-lb weight on a piston will support another 1-lb weight on a piston of equal area, giving an IMA $= F_o/F_i = 1$ lb/1 lb $= 1$. (b) A 1-lb force applied to a piston with an area of 1 in² will support a 10-lb weight on a piston with an area of 10 in². The pressures are equal, but IMA $= F_o/F_i = A_o/A_i = 10\text{ in}^2/1\text{ in}^2 = 10$, and there is a force multiplication.

Figure 11.17, then this force would support a 1-lb weight on a piston of equal area, since the pressure is transmitted undiminished, i.e., $p_1 = p_2$. This situation is analogous to an equal arm lever. The system has an IMA = 1, since $p_1 = F_i/A = F_o/A = p_2$, and IMA $= F_o/F_i = A/A = 1$.

To obtain a force magnification, the areas of the pistons can be varied. Since the pressure is transmitted undiminished, the input pressure equals the output pressure, $p_i = p_o$, or

$$\frac{F_i}{A_i} = \frac{F_o}{A_o}$$

and

$$F_o = \left(\frac{A_o}{A_i}\right) F_i \qquad \textbf{(Eq. 11.16)}$$

Hence, if $A_o > A_i$, then $F_o > F_i$. For example, as illustrated in Figure 11.17, a 1-lb force acting on a piston with an area of 1 in² will support a 10-lb weight on a piston with an area of 10 in². The IMA in this case is

$$\text{IMA} = \frac{F_o}{F_i} = \frac{A_o}{A_i} = \frac{10\text{ in}^2}{1\text{ in}^2} = 10$$

The area is related to the distance the force would move, since the volumes of liquid displaced are equal, i.e., $V_i = V_o$ or $L_iA_i = L_oA_o$ or $L_i/L_o = A_o/A_i$, where L_i and L_o are the lengths moved by the input and output pistons, respectively.

Thus, the IMA of a hydraulic press when motion is involved in given by

$$\boxed{\text{IMA} = \frac{L_i}{L_o} \quad \text{(hydraulic press)}} \qquad \textbf{(Eq. 11.17)}$$

A diagram of a practical hydraulic press and another application of this principle are shown in Figure 11.18. Another common example of the hydraulic press principle is the garage lift used to lift automobiles, in which an air compressor supplies the input pressure. (See Special Feature 11.1.)

EXAMPLE 11.5 A hydraulic press has an output piston area of 500 in² and an input piston area of 25 in². What are (a) the IMA and (b) the minimum pressure required to support a 3600-lb load?

Solution: (a) With $A_o = 500$ in² and $A_i = 25$ in²,

$$\text{IMA} = \frac{A_o}{A_i} = \frac{500\text{ in}^2}{25\text{ in}^2} = 20$$

(b) Using IMA $= F_o/F_i$ and $F_o = 3600$ lb,

$$F_i = \frac{F_o}{\text{IMA}} = \frac{3600\text{ lb}}{20} = 180\text{ lb}$$

Then,

$$p_i = \frac{F_i}{A_i} = \frac{180\text{ lb}}{25\text{ in}^2} = 7.2\text{ lb/in}^2\ \text{(psi)}$$

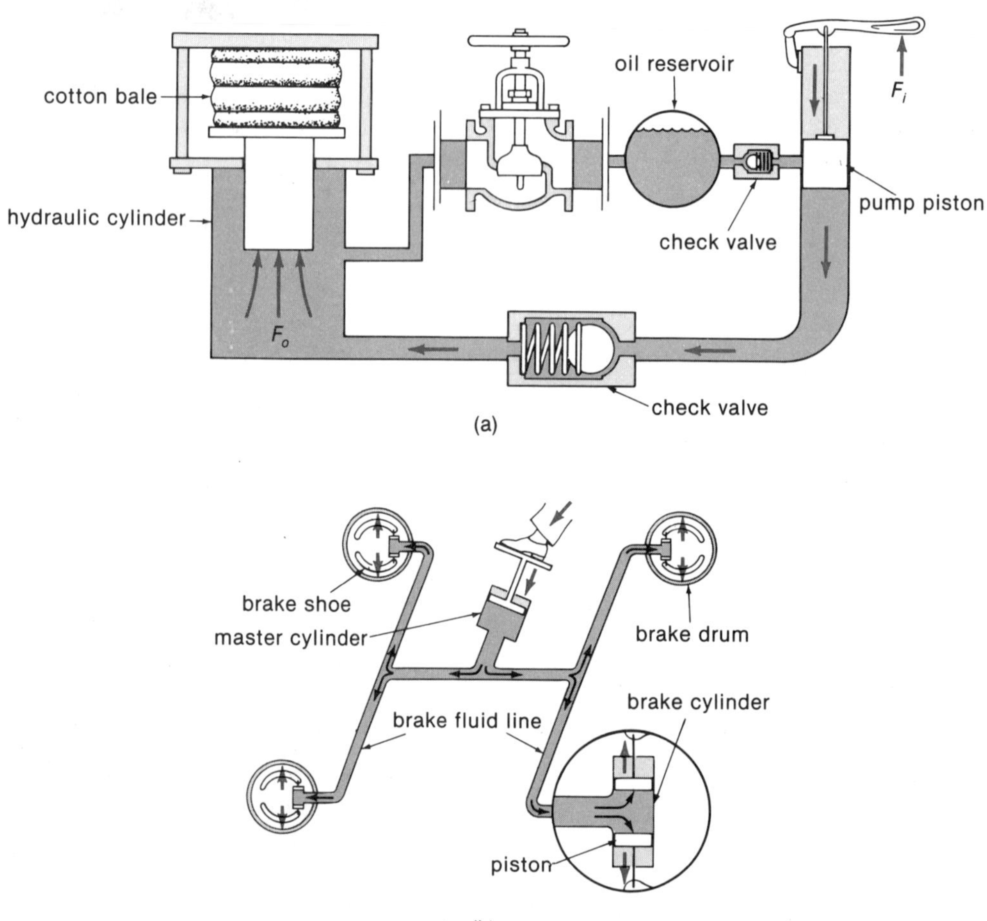

Figure 11.18
Examples of applications of Pascal's principle. (a) A hydraulic press. (b) A hydraulic brake system.

11.4 Power Transmission

Machines can be used to transfer or transmit energy. This usually involves rotary machines operating at various speeds, so it is convenient to talk about power transmission. The common forms of mechanical power transmission involve coupled simple machines, namely wheels and axles.

For example, a widely used application is the pulley and belt system [Fig. 11.20(a)]. The pulleys are actually the wheels of wheels and axles. The belt drive coupling the pulleys is commonly a V-belt that runs in the V-grooves of the pulleys so as to prevent slippage.

For rotary motion, the conservation of energy can be written in terms of the conservation of power:

$$P_{\text{in}} = P_{\text{out}}$$

or

$$\tau_i\omega_i = \tau_o\omega_o + P_f \qquad \textbf{(Eq. 11.18)}$$

where the power is expressed in terms of torque τ and angular speed ω. Here, P_f is the power (energy per time) lost to friction.

The AMA of any torque transmission system is given by the torque ratio (analogous to the force ratio in previous sections):

$$\text{AMA} = \frac{\tau_o}{\tau_i} \qquad \textbf{(Eq. 11.19)}$$

Similarly, in the absence of friction, the IMA can be written in terms of the angular speeds, since the angular speed expresses the distance (per time) through which the torque acts.

SPECIAL FEATURE 11.1

The Hydraulic Garage Lift

A common application of the hydraulic press is in the garage lift used to raise heavy automobiles and trucks (Fig. 11.19). Compressed air on the surface of an oil reservoir transmits pressure to the lift piston. The compressed air pressure doesn't have to be very great. Suppose it is 200 kPa, which is about the pressure in your automobile tires (30 lb/in^2). The pressure on the liquid in the reservoir is transmitted to the lift piston.

If the diameter of the piston is 0.30 m, then its area is 0.07 m^2, and the lift force is 14,000 N ($F = pA$), or about 3150 lb. So the next time you see a car on a lift, you'll know that Pascal's principle is in operation.

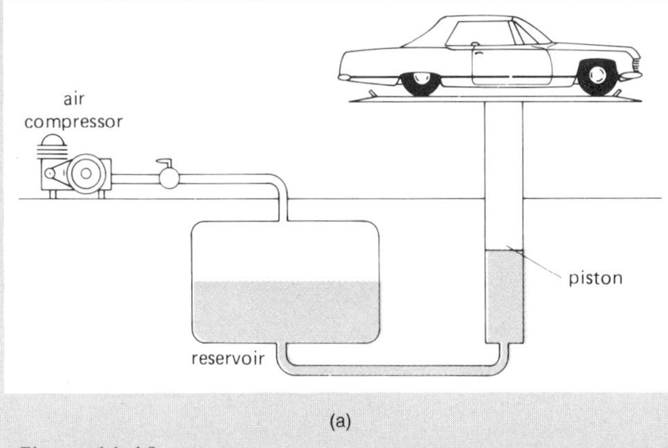

(a)

(b)

Figure 11.19
A hydraulic lift uses Pascal's principle. The air pressure on the surface of the oil in the reservoir is transmitted undiminished to the lift piston, giving a large output force.

For a belt-drive system as in Figure 11.20, the tangential speeds of the pulleys are equal if there is no belt slippage and $v_i = r_i\omega_i = r_o\omega_o = v_o$, where the r's are the radii of the pulleys. So, with ω_i and ω_o corresponding to d_i and d_o, respectively,

$$\text{IMA} = \frac{\tau_o}{\tau_i} = \frac{\omega_i}{\omega_o}$$

$$\text{IMA} = \frac{r_o}{r_i} = \frac{D_o}{D_i} \quad \text{(belt drive)} \qquad \textbf{(Eq. 11.20)}$$

where the D's are the diameters of the pulleys and are commonly expressed dimensions. The input pulley is called the drive pulley, and the output pulley is called the driven pulley.

The speed advantage of a reduction in a power transmission system is commonly expressed in terms of the **speed ratio (SR):**

$$\text{SR} = \frac{\omega_o}{\omega_i} = \frac{D_i}{D_o} = \frac{1}{\text{IMA}} \quad \text{(belt drive)} \qquad \textbf{Eq. 11.21)}$$

The speed ratio is the reciprocal of the IMA and is the speed multiplication factor. A belt-drive system may be used to decrease angular speed, which gives an increase in torque, and $D_o > D_i$ in this case.

For example, in many instances the drive pulley is coupled to a fast-rotating motor shaft, and a speed reduction is required. Conversely, a belt-drive system can be used to increase the angular speed at the expense of torque ($D_i > D_o$). Recall how a third-class lever is used to move a load through a relatively large distance at the expense of a great amount of effort.

For quick changes of speed and torque transmission, step pulleys are used [Fig. 11.20(b)]. Changes are made by shifting the belt to differ-

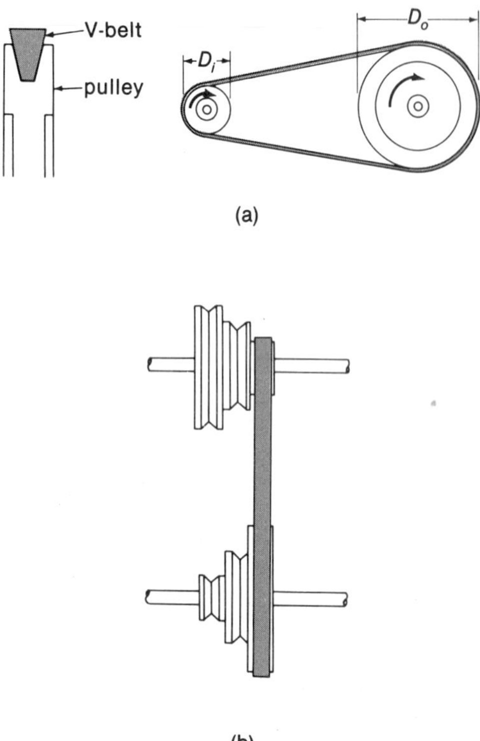

Figure 11.20
Pulley and belt systems. (a) Single pulleys and V-belt. The grooves of the pulleys in which the V-belt fits prevent slippage. With the smaller pulley as the drive pulley, there is torque multiplication but a decrease in speed. (b) A step pulley allows for quick changes in speed and torque transmission. (c) A pulley and belt system is used to transmit power from an automobile engine.

ent pulley sets on the step pulleys. Step pulleys are commonly found on drill presses and band saws, where different speeds and torques are required for working with different materials.

For example, when drilling a hard metal with a small drill, a high speed (low torque) is usually used. With a large drill, a low speed (large torque) is used, since more material is being removed and this requires a larger torque.

EXAMPLE 11.6 A motor with a speed of 1750 rpm is coupled to the drive pulley of a pulley-belt system that has a diameter of 3.0 in. If the output, or driven, pulley has a diameter of 9.0 in., (a) what is the speed ratio of the system and (b) what is the angular speed of the driven pulley?

Solution: (a) With $D_i = 3.0$ in. and $D_o = 9.0$ in., using Equation 11.20 we have

$$\text{SR} = \frac{1}{\text{IMA}} = \frac{D_i}{D_o} = \frac{3.0 \text{ in.}}{9.0 \text{ in.}} = \frac{1}{3}$$

(b) Since SR $= \omega_o/\omega_i$ (Eq. 11.21),

$$\omega_o = (\text{SR})\omega_i = (1/3)(1750 \text{ rpm}) = 583 \text{ rpm}$$

Notice that if the input pulley were larger than the output pulley, the angular speed would be increased.

Another common method of power transmission is through the use of gears. In its simplest form, a gear is a notched wheel. Gears can be used to increase or decrease the speed, magnify or reduce the applied torque, or change the direction of motion.

There are a variety of gear types. Probably the most common type is the **spur gear** (Fig. 11.21). When the mating teeth are cut so that

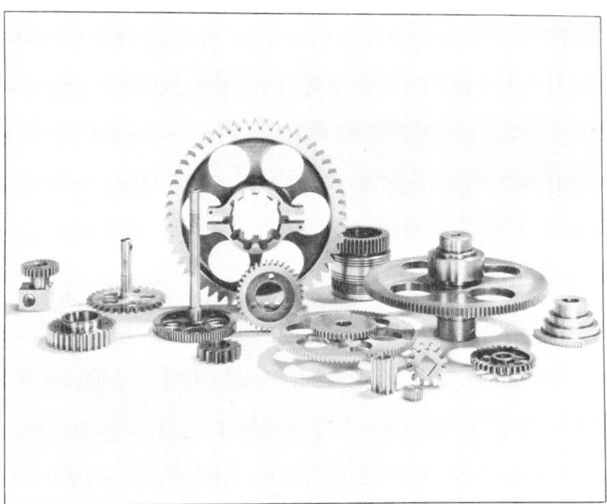

Figure 11.21
Straight spur gears. The mating teeth are cut perpendicularly to the axis of rotation. The large gears in the bottom photo are used in steel mills. (Top photo courtesy Gear Specialties, Inc. Bottom photo courtesy Illinois Gear.)

they are perpendicular to the axis of rotation, the gears are called straight spur gears. For such spur gears, the whole widths of the meshing teeth come into contact at the same time.

With helical (spiral) spur gears, the contact between two teeth starts first at the leading ends and moves progressively across the gear faces until the trailing ends are in contact. This kind of meshing action keeps the gears in constant contact, providing for less lost motion and smoother, quieter action. One disadvantage of helical spur gears is the tendency for each gear to thrust or push axially on its shaft. Special thrust bearings are necessary on the shaft to counteract this thrust.

Thrust bearings are not needed if herringbone spur gears are used. The teeth of each half of a herringbone gear are cut in different directions, and each half of the gear develops a thrust which counterbalances that of the other half. Herringbone gears are commonly used in heavy machinery.

When two gears of unequal size are meshed together, the smaller of the two is commonly called a **pinion.** Although the diameters and number of teeth on the gears are different, the teeth themselves must be of the same size for proper meshing. For example, an external gear has teeth on the inside of the wheel that point toward the axis of rotation (Fig. 11.22).

When the motion of a pinion is limited, a sector gear can be used to save space and material. In a **rack and pinion,** both gears are spur gears, but the rack is linear. The rack gear may be thought of as a piece cut from a gear of extremely large radius. The rack and pinion is useful in changing rotary motion into linear motion.

There is a wide variety of types of gears other than those discussed here.

A set of gears can be used to give a speed increase or reduction. The speed ratio of a set of gears is expressed in terms of the number of teeth on the gears. Analogous to the pulley-belt system, we have $N_i\omega_i = N_o\omega_o$, and

$$\boxed{\begin{aligned} \text{IMA} &= \frac{\omega_i}{\omega_o} = \frac{N_o}{N_i} \\ \text{SR} &= \frac{\omega_o}{\omega_i} = \frac{N_i}{N_o} \\ &\text{(gear drive)} \end{aligned}} \qquad \textbf{(Eq. 11.22)}$$

where N_i is the number of teeth on the input (drive) gear and N_o is the number of teeth on the output (driven) gear.

For example, if $N_i = 40$ and $N_o = 20$, then SR = 40/20 = 2. This gear speed ratio would be commonly expressed as 2 : 1 (read "2 to 1"), which indicates that there are twice as many teeth on one as on the other, or that the angular speed is multiplied by a factor of 2 with the

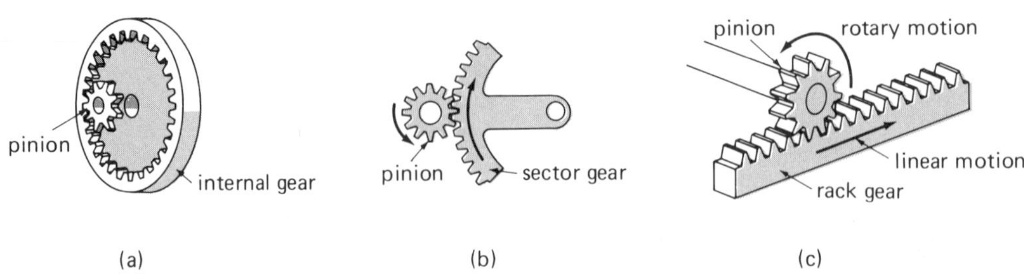

(d)

(e)

Figure 11.22
Gear arrangements. (a) An internal ring gear. (b) A sector gear used when the motion of the pinion is limited. (c) The rack and pinion changes rotary motion into linear motion. (Courtesy Illinois Gear.)

pinion gear as the output gear. If the pinion gear is the drive gear or "driver," there is a speed reduction of one half.

Several gears coupled together form a gear train. The overall speed ratio of a gear train is the product of the individual SR's of the gear pairs.

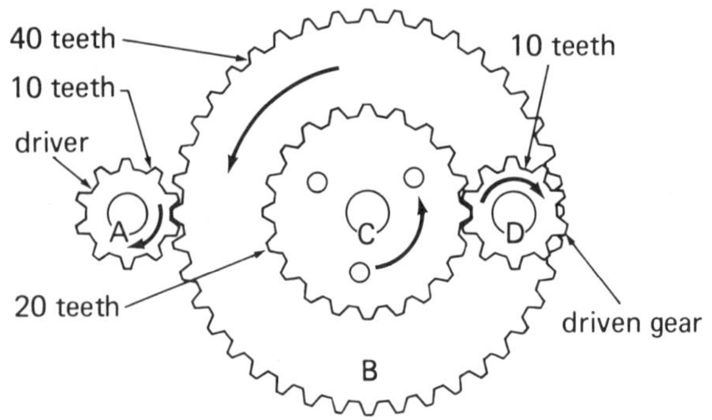

Figure 11.23
A gear train. See Example 11.7.

EXAMPLE 11.7 If the driver of the gear train in Figure 11.23 has an angular speed of 100 rpm, what is the speed of the output or driven gear?

Solution: The speed ratio of the first two gears (A and B) with $N_i = 10$ and $N_o = 40$, is

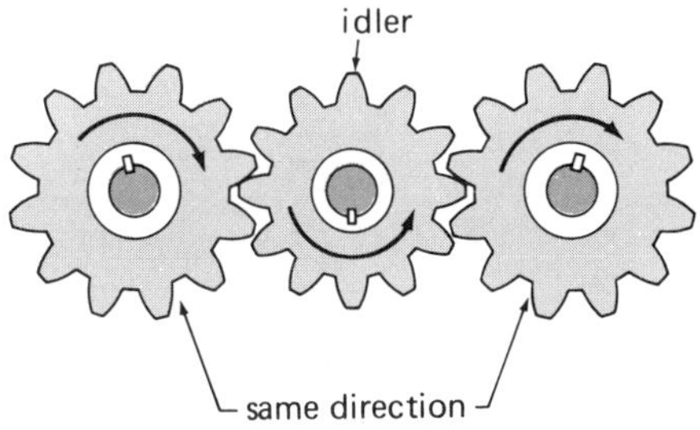

Figure 11.24
An idler gear. An idler gear is used to have drive and driven gears rotate in the same direction. An idler gear does not change the speed.

$$SR_1 = \frac{N_i}{N_o} = \frac{10}{40} = \frac{1}{4}$$

The center gears on the same axle (B and C) turn with the same speed, and the SR for the final two gears (C and D), with $N_i = 20$ and $N_o = 10$, is

$$SR_2 = \frac{N_i}{N_o} = \frac{20}{10} = 2$$

The overall speed ratio of the gear train is the product of the individual speed ratios; hence,

$$\omega_o = (SR)\omega_i = (SR_1)(SR_2)\omega_i = (\tfrac{1}{4})(2)(100 \text{ rpm}) = 50 \text{ rpm}$$

Notice that when two external spur gears mesh, they rotate in opposite directions. To maintain the same rotational direction, a third gear, called an **idler,** can be placed between the drive gear and the driven gear (Fig. 11.24). An idler does not change the speed but simply acts to make the drive gear and the driven gear turn in the same direction.

Important Terms

machine a device that multiplies force and/or changes the direction of a force

actual mechanical advantage (AMA) the ratio of the output and the input forces of a machine; the factor or magnitude of the force multiplication

ideal mechanical advantage (IMA) the mechanical advantage neglecting frictional or other losses, and hence the best or maximum advantage that can ever hope to be achieved; the IMA can be determined from geometry

efficiency the ratio of the work output and work input of a machine; also the ratio of AMA/IMA

lever a rigid bar pivoted to rotate about a point or line called the fulcrum

wheel and axle a wheel rigidly attached to an axle or shaft that turns with the wheel

pulley a wheel pivoted so that it can rotate freely about an axis through its center

inclined plane a surface elevated at an angle to the horizontal

wedge two inclined planes set base to base

screw a spiral inclined plane wrapped around a cylinder

pitch (of a screw) the distance between adjacent threads or the distance a screw travels in one rotation

Pascal's principle the principle that pressure applied to an enclosed fluid is transmitted undiminished to every portion of the fluid and the walls of the container

speed ratio (SR) the ratio of the output and input angular speeds of a power transmission system

Important Formulas

actual mechanical advantage: $AMA = \frac{F_o}{F_i}$

ideal mechanical advantage: $IMA = \frac{d_i}{d_o}$

efficiency: $Eff = \frac{W_{out}}{W_{in}} = \frac{F_o d_o}{F_i d_i} = \frac{AMA}{IMA}$

IMA's of various machines:

lever: $IMA = \frac{L_i}{L_o}$

wheel and axle: $IMA = \frac{R}{r}$

pulleys (movable): General rule: The IMA is equal to the number of the support strands going to (and from) the movable block.

chain hoist: $IMA = \frac{2R}{(R - r)}$

inclined plane: $IMA = \frac{L}{h} = \frac{1}{\sin\theta}$

wedge: $\text{IMA} = \dfrac{L}{t}$

cam: $\text{IMA} = \dfrac{L}{(R - r)}$

jackscrew: $\text{IMA} = \dfrac{2\pi R}{p}$

$$p = \frac{1}{\text{No. of threads per in.}}$$

hydraulic press: $\text{IMA} = \dfrac{A_o}{A_i} = \dfrac{L_i}{L_o}$

belt drives: $\text{IMA} = \dfrac{D_o}{D_i}$

gear drives: $\text{IMA} = \dfrac{N_o}{N_i}$

speed ratios:

belt drives: $\text{SR} = \dfrac{D_i}{D_o} = \dfrac{\omega_o}{\omega_i} = \dfrac{1}{\text{IMA}}$

gear drives: $\text{SR} = \dfrac{N_i}{N_o} = \dfrac{\omega_o}{\omega_i} = \dfrac{1}{\text{IMA}}$

Figure 11.25
A lot of lift. See Question 10.

Questions

Mechanical Advantage and Efficiency

1. Explain how machines make doing work "easier."
2. Do machines multiply work? Explain.
3. Is it possible for $F_o > F_i$? Explain.
4. Do the input and output forces have to be known to compute the IMA? Explain.
5. The efficiency, or fraction of useful work, is given by AMA/IMA. Does IMA/AMA give the fraction of lost work? Explain.

Simple Machines

6. Name and explain the principles of the six types of simple mechanical machines.
7. What is the principle of the hydraulic press?
8. Describe the different classes of levers.
9. What types of levers are (a) the "flip-top" of a soda can, (b) a pair of pliers, and (c) a nutcracker?
10. Estimate the ideal mechanical advantage of the machine shown in Figure 11.25.
11. Explain the following statement made by Archimedes, an early Greek scientist. "Give me a lever long enough and a fulcrum on which to rest it, and I will move the Earth."
12. What type of machine is a screwdriver, and how is its IMA found? (Consider uses in prying and rotating.)
13. A doorknob is a simple machine. What type of machine is it? Estimate the IMA of the common doorknob.
14. Identify and explain the operation of the machine shown in Fig. 11.26. (It was once part of a grain mill in Yugoslavia.)
15. What is the maximum IMA that can be achieved with (a) two and (b) three single pulleys?
16. What is the principle of the chain hoist, or differential pulley?

Figure 11.26
Water power. See Question 14.

17. Mathematically, the IMA of an inclined plane approaches infinity. What does this mean physically?
18. What is the IMA of a knife, and how might it be increased?
19. How practical would a turnbuckle be if it had the same type of threads on each end?
20. Explain the following screw designations: (a) 4-40, (b) 8-48, (c) ½-10.
21. What is pressure, and what is its SI unit?
22. If a different liquid is used in a hydraulic press, how is the IMA affected?

Power Transmission

23. How does the friction of a belt drive affect the AMA? Is this friction necessary? Explain.
24. What is the speed ratio and what does it indicate?
25. Explain the principle of a hand-powered egg beater.
26. What is the rack in a rack and pinion?
27. What is the speed ratio of a gear train?
28. What is the purpose of an idler gear? Does an idler contribute to the speed ratio? Explain.

Problems

Levels of difficulty are indicated by asterisks for your convenience.

11.1 Mechanical Advantage and Efficiency

1. A force of 20 lb applied to a machine produces an output force of 40 lb. What is the AMA of the machine?
2. A machine with an AMA of 3.0 has an output force of 45 N. What is the input force?

*3. A machine input force of 15 lb acts through a distance of 4.0 ft, and the 30-lb output force acts through a distance of 1.5 ft. (a) What is the AMA of the machine? (b) What is the "useful" work output? (c) How much work is done against friction?

*4. What is the IMA of the machine in Problem 3?

*5. What is the efficiency of the machine in Problem 3?

*6. A proposed machine is to have a force output of 270 N. From the blueprints it can be seen that the output distance is 4.0 cm and the input distance is 12 cm. What can be said about the magnitude of the required input force?

*7. If the machine in Problem 6 has an AMA of 2.4, what is its efficiency?

**8. A particular machine with an efficiency of 70 percent has input and output distances of 15 cm and 5.0 cm, respectively. What is the AMA of the machine?

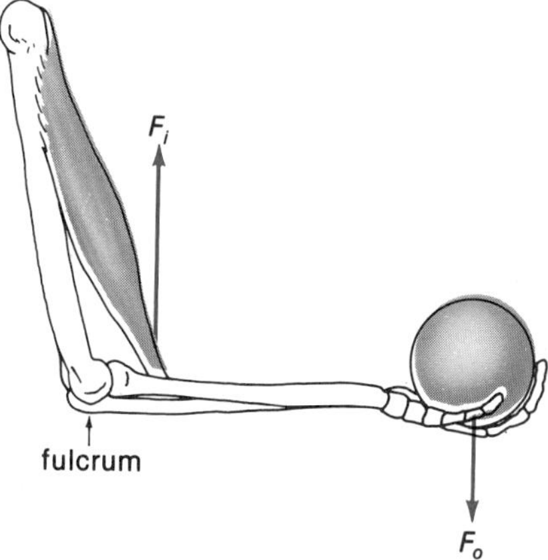

Figure 11.27
See Problem 12.

11.2 Simple Machines

9. What are the pitches of the following UNC screws? (a) 4-40, (b) 8-48, (c) ½-10 (See Table 11.1.)

*10. The oarlock on a 10-ft oar is 3.0 ft from the handle end. A person applies a force of 50 lb to the oar in rowing a boat. (a) What is the IMA of the oar? (b) What would be the ideal force output? (c) What type of lever is the oar?

*11. A 10-in. screwdriver is used to pry open a paint can lid. (a) If the blade end of the screwdriver is 1.0 in. from the lip of the can, which acts as a fulcrum, what is the IMA? (b) If an output force of 6.0 lb is required to loosen the lid, what is the required effort? (Assume negligible losses.)

*12. The human arm uses a lever principle (Fig. 11.27). (a) If the muscle is attached 6.0 cm from the joint and a 2.0-kg ball is held in the hand 36 cm from the joint, what is the vertical force supplied by the muscle? (b) What type of lever is this?

*13. The old-time catapult used to project objects operated on a lever principle (similarly to a spoon pivoted at the handle end). (a) If the input force of a particular catapult were applied 10 ft from the pivoted end and the load were 40 ft from the pivot, what would be the ideal force multiplication factor? Is the force multiplied? Explain. (b) What type of lever is this?

*14. A screwdriver has a ½-in. blade and a 2-in.-diameter handle. (a) What is the IMA? (b) If a tangential force of 3.0 lb is applied to the handle, what is the output force at the edge of the blade?

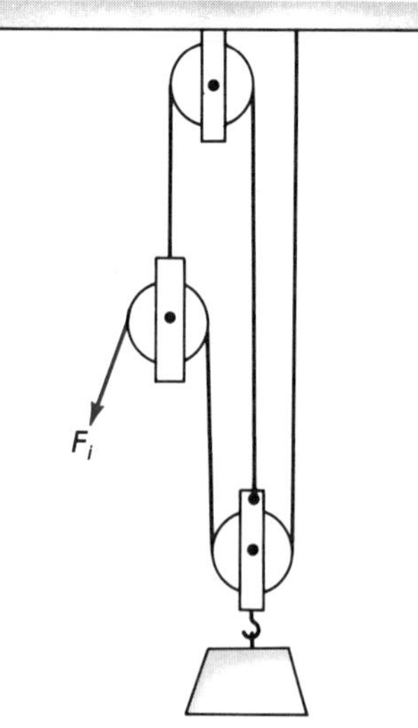

Figure 11.28
See Problem 18.

Figure 11.29
See Problem 23.

*15. A door latch mechanism has a doorknob with a radius of 2.5 cm and a shaft of 0.50-cm radius. (a) What is the IMA? (b) If a torque of 20 m-N is applied to the knob, what is the tangential output force of the shaft?

*16. A crank mechanism used to raise a bucket in a well as in Figure 11.5 has an AMA of 1.6. The handle length is 8.0 in. and the radius of the rope shaft is 2 in. (a) What is the efficiency of the mechanism? (b) What input force will hold a 24-lb bucket of water stationary?

*17. A 22-lb effort is required in raising a 20-lb load with a single fixed pulley. (a) What is the AMA of the pulley? (b) What is the efficiency?

*18. What is the IMA of the pulley arrangement in Figure 11.28?

*19. For a luff tackle arrangement as in Figure 11.8(a), an effort of 60 lb is required to raise a 150-lb crate. (a) If the movable block weighs 5.0 lb, what is the AMA? (b) What is the efficiency of the luff block?

*20. A chain hoist as in Figure 11.10 has a wheel of 6.0-cm radius and an axle of 4.0-cm radius in the upper block. (a) What is the IMA? (b) If the efficiency of the chain hoist is 55 percent, what is the effort required to lift a 300-N load (includes weight of lower block)?

*21. An inclined plane has a length of 5.0 m and a height of 1.5 m. (a) What is its IMA? (b) What is the angle of incline?

*22. A factory tractor exerts a force of 200 lb parallel to the surface of a 10° inclined plane in pushing an 800-lb crate up the plane. (a) What is the AMA? (b) What is the efficiency?

*23. (a) A uniformly sloped highway has a rise of 20 ft in a horizontal (run) of 800 ft. What are the grade and angle of the slope? (b) What is the vertical height of the hill shown in Figure 11.29? (*Hint:* $\tan \theta$ = rise/run = grade.)

*24. A wood chisel has a sloped cutting blade 2.0 cm long and 0.25 cm thick at the base. (a) What is the IMA? (b) What is the angle of incline? (A wood chisel is essentially a movable inclined plane.)

*25. A log-splitting wedge has a side length of 18 cm and a point angle of 10°. What is the IMA of the wedge?

*26. A butcher knife has a blade width of $1\frac{1}{2}$ in. and an IMA of 12. What is the thickness of the blade?

*27. A cam has a circumference length of 4.0 cm and maximum and minimum distances from the pivot to the rim of 2.5 cm and 0.50 cm, respectively. What is the IMA of the cam?

**28. A block and tackle is used to pull a 4000-lb safe up an inclined plane as shown in Figure 11.30. Neglecting friction and the weight of the block and tackle, what is the input force required to move the safe up the plane? (*Hint:* this is a compound machine.)

**29. A jackscrew has a screw thread pitch of $\frac{1}{16}$ in. and a handle length of 10 in. (a) What is the IMA of the jack? (b) If an input force of 15 lb is required to lift a 3000-lb load, what is the efficiency of the jack?

**30. Sketch a design for a block-and-tackle system with an IMA of 5.

**31. Sketch a design for a luff-upon-luff system with an IMA of 9. (See Fig. 11.9.)

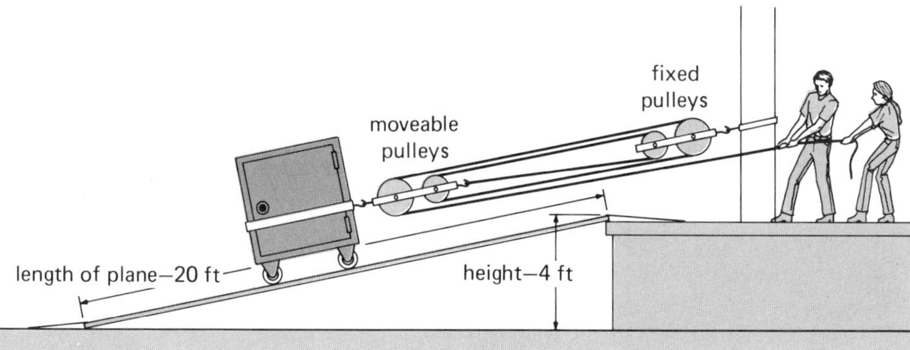

Figure 11.30
See Problem 28.

11.3 The Hydraulic Press

*32. A hydraulic lift has an output piston with an area of 800 cm^2 and an input piston area of 20 cm^2. (a) What is the IMA of the lift? (b) Assuming no frictional losses, what load could an input force of 70 lb lift?

*33. A hydraulic jack has an output piston area of 20 in^2 and an input piston area of 2.0 in^2. The handle of the jack is a lever with L_i = 8.0 in. and L_o = 2.0 in. (a) What is the IMA of the jack? (b) If an input force of 15 lb gives rise to an output force of 500 lb, what is the efficiency of the jack?

*34. A medical syringe with a plunger diameter of 2.0 cm is attached to a hypodermic needle with an inside diameter of 1.5 mm. What minimum force must be applied to the plunger to inject fluid into a vein where the blood pressure is 130 Pa?

*35. A hydraulic jack has an IMA of 6.0. (a) If a force of 100 N is applied to the driving piston, which has a diameter of 5.0 cm, what is the ideal output force of the jack? (b) What is the area of the output piston?

**36. A hydraulic press has an IMA of 10. If the pump piston supplying the driving force moves 4.0 cm each stroke, how many strokes are required to move the output piston 50 cm?

11.4 Power Transmission

*37. A V-belt pulley system has a drive pulley with a diameter of 4.0 in. and a driven, or working, pulley with a diameter of 6.0 in. (a) What is the IMA? (b) Neglecting friction, if a tangential force of 3.0 lb is applied to the drive pulley, what is the output torque?

*38. A belt drive has an input pulley with a diameter of 10 cm and an output pulley of 40 cm. If the drive pulley is connected to a motor with a shaft speed of 1000 rpm, what is the angular speed of the driven pulley?

*39. A set of step pulleys consists of pulleys with radii of 6.0 in., 4.0 in., and 2.0 in. If the input pulley is driven at a speed of 1200 rpm, what are the possible output speeds?

*40. The pulleys of a pulley V-belt system ordinarily rotate in the same direction, while intermeshed spur gears rotate in opposite directions. Sketch an arrangement for (a) two pulleys rotating in opposite directions using a single belt and (b) two spur gears rotating in the same direction using a chain "belt."

*41. A set of two spur gears has 32 teeth and 8 teeth. (a) What are the possible IMA's of the gear set? (b) If the larger gear has a rotational speed of 12 rpm, what is the rotational speed of the smaller gear?

**42. (a) What is the IMA of the bicycle drive shown in Figure 11.31? (*Hint:* this is a compound machine.) (b) If the pedal sprocket is turned with a speed of 6.0 rad/s, how far does the bicycle travel in 4.0 s, starting from rest?

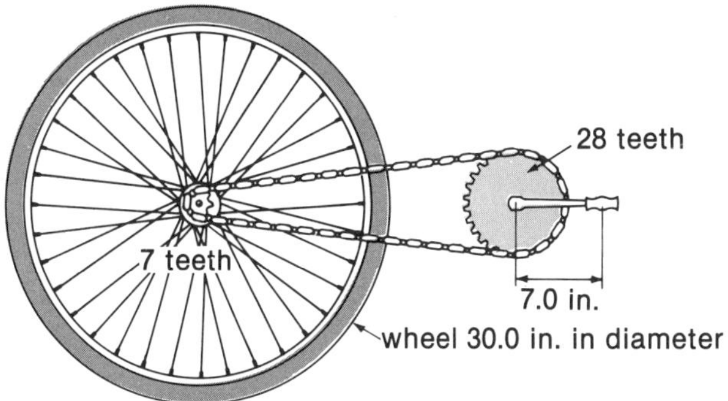

Figure 11.31
See Problem 42.

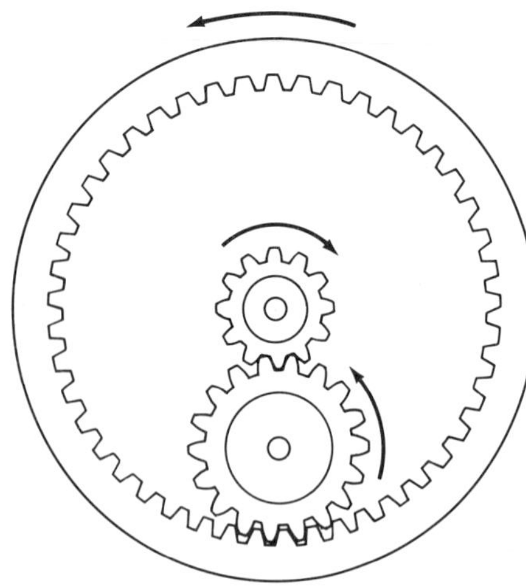

Figure 11.32
See Problem 43.

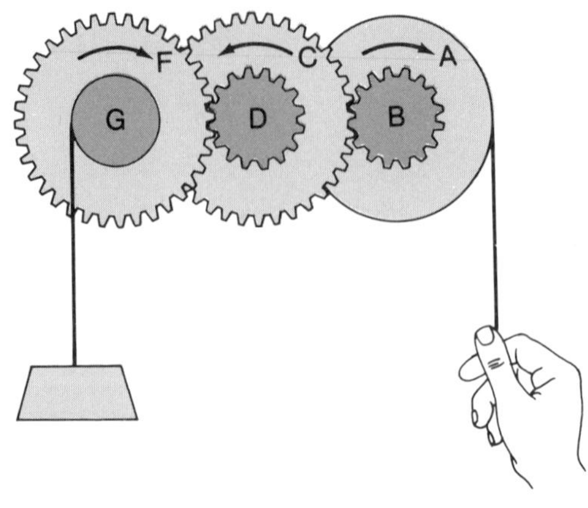

Figure 11.33
See Problem 45.

**43. The central drive gear of the so-called planetary gear arrangement shown in Figure 11.32 rotates with a speed of 100 rpm. (a) What is the IMA of the system? (b) What is the rotational speed of the external gear?

**44. If the external gear in Figure 11.32 rotates with a speed of 15.0 rpm, what is the rotational speed of the central gear?

**45. In Figure 11.33, what is the IMA of the system if the radius of wheel A is 12 in., the radius of wheel G is 6.0 in., and $N_C = N_F$? (*Hint:* consider the IMA's of the gears and the IMA of the wheels separately. The wheels are essentially a wheel and axle.)

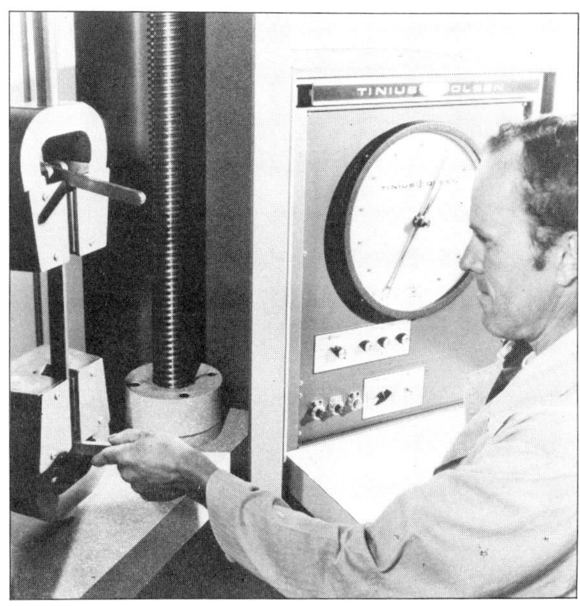

(Courtesy Tinius Olsen Testing Machine Co.)

Chapter 12

Mechanical Properties of Materials

A wide variety of materials is used in technical applications. In many instances, the applicability of a material depends on its mechanical properties. In some cases it is desirable for a material to withstand an applied force or torque, e.g., on the axle of an automobile. In other situations, a material that can be stretched or easily deformed is required, such as rubber in a rubber band or an automobile tire.

In previous chapters, forces in beams and cables and torques on *rigid* bodies were considered. However, nothing was said about the small deformations that occur in solid objects when they are subjected to forces. An important aspect of materials science is the characterization of such mechanical properties of materials.

The properties of a material depend on its composition and internal structure. The internal processes that occur in materials when they are subjected to forces are complex. However, these processes are manifested as changes in the material's bulk properties, which can be physically measured. The terms and parameters used to describe the mechanical behavior of materials will be the main focus of this chapter.

12.1 Stress and Strain

A rubber band is commonly described as being elastic, meaning that it will return to its original length after having been stretched. However, **all solid materials are elastic to some degree,** even steel. That is, a deformed solid will return to its original shape and dimensions when the deformation force is removed, *provided the force was not too great.* The elastic deformation of many solids cannot be detected visually but may be easily detected and measured by special sensitive instruments.

The cause of any deformation is an applied force, but this is expressed in terms of stress. **Stress is the ratio of the applied force F and the area A over which it acts,** i.e.,

$$\boxed{\text{Stress} = \frac{F}{A}} \qquad \textbf{(Eq. 12.1)}$$

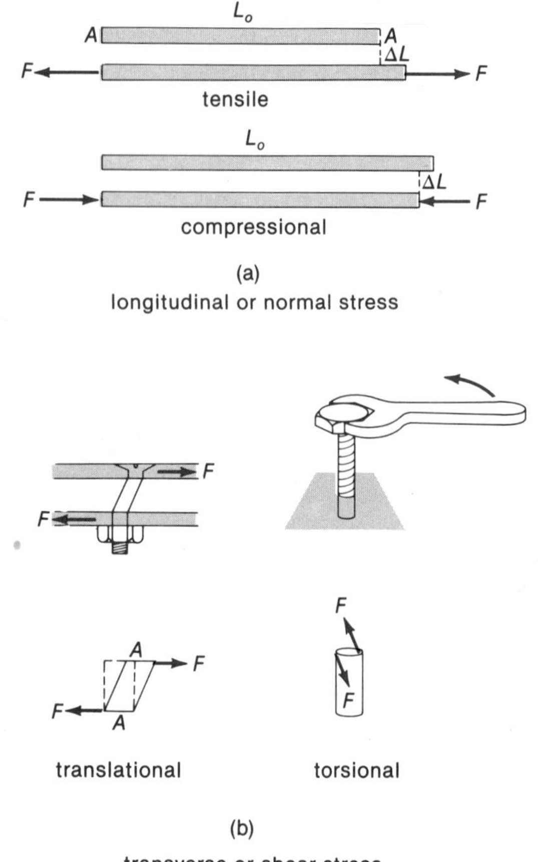

Figure 12.1
Types of stress. (a) Longitudinal or normal stress. The applied force is perpendicular or normal to the surface area. A longitudinal stress may be a tensile stress or a compressional stress. (b) Transverse or shear stress. The applied force is tangential to the surface. A shear stress may be translational stress or a torsional (twisting) stress.

For example, if a pull (force) of 100 lb is applied to a cable with a cross-sectional area of 0.50 in^2, the stress on the cable is 100 lb/0.50 in^2 = 200 lb/in^2 (or 200 psi). The unit of stress in the SI system is N/m^2.

Stress is preferred over force alone, since it takes the area of the material into account and gives the average force condition that exists within the material.

There are two types of stress: longitudinal or normal stress and transverse or shear stress (Fig. 12.1)

For normal stress, the applied force (or component thereof) is normal, or perpendicular to the surface. If a normal stress tends to elongate the material dimension, it is called a **tensile stress.** The stress in the previous cable example is a tensile stress. If a normal stress tends to compress the material dimension, it is called a **compressive stress.**

For a shear stress, the applied force is tangential, or parallel to the surface. Shear stresses are of two types. A **translational shear stress** tends to translate the surface, and a **torsional shear stress** tends to twist the material as a result of a torque (Fig. 12.1).

The result of a stress is a material deformation or a strain. **Strain is the relative change in the dimension(s) or shape of a body** and is a measure of relative distortion. For example, for a tensile stress that elongates a rod from an original length L_o to a length L, the strain is

$$\text{Tensile strain} = \frac{\text{change in length}}{\text{original length}} = \frac{\Delta L}{L_o} = \frac{L - L_o}{L_o} \qquad \textbf{(Eq. 12.2)}$$

A compressive strain is similarly defined, where the change in length is a decrease and the strain is negative.

Strain is a unitless quantity, because it is the ratio of lengths (length/length). However, the length units are sometimes retained, e.g., cm/cm or in./in. This explicitly expresses the change in length per unit length, e.g., how much longer a unit inch becomes due to the stress.

Strain may also be given as a percentage. For example, if a tensile stress increases the length of a 50-cm rod by 0.20 cm, the strain is $\Delta L/L_o$ = 0.20 cm/50 cm = 0.0040, or 0.40 percent.

The tensile and compressive properties of materials are investigated by instruments such as those shown in Figure 12.2, which supply stress-strain curves as in Figure 12.3. The initial linear relationship between the stress and strain of the material is indicative of **elastic deformation.** That is, the material returns to its original shape and dimensions when the stress is removed. **The *elastic limit* refers to the maximum stress a material can experience without becoming permanently deformed.** Some typical values

(b)

Figure 12.2
Stress testing machines. (a) Tensile testing strips of material. The chart recorder gives a stress versus strain curve. The machine can also be used for compressional testing. (Courtesy Instron Corporation, Canton, MA.) (b) Compressional stress testing. The 28-in. diameter dial allows easy reading of stress measurements. (Courtesy Tinius Olsen Testing Machine Co.)

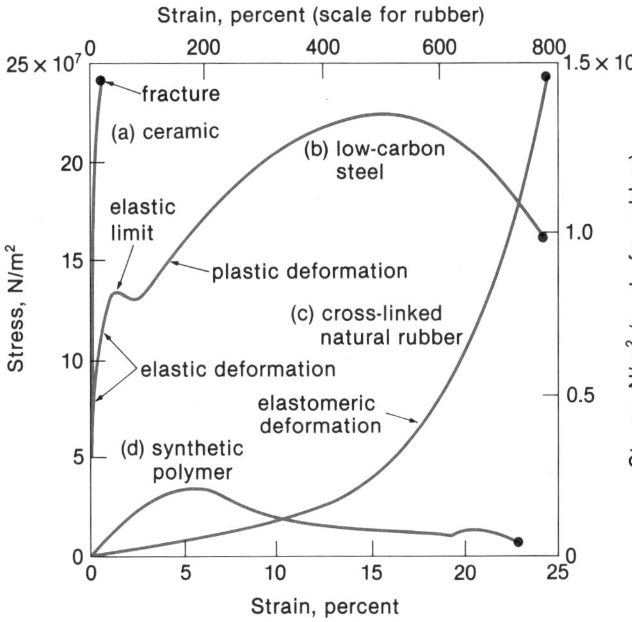

Figure 12.3
Stress-strain curves. (a) Ceramic materials fracture with little deformation. (b) Steel undergoes elastic deformation up to an elastic limit. Beyond this, permanent plastic deformation occurs until fracture. (c) Cross-linked natural rubber initially undergoes linear elastic deformation and then elastomeric deformation. Unlike plastic deformation, the material recovers when the stress is removed. (d) Synthetic polymers generally have elastic and plastic deformations.

Table 12.1
Typical Elastic Limits and Tensile Strengths for Various Materials

Material	Elastic Limit*		Tensile (Ultimate) Strength*	
	N/m^2	lb/in^2	N/m^2	lb/in^2
Aluminum	1.3×10^8	1.9×10^4	1.4×10^8	2.1×10^4
Brass	3.8×10^8	5.5×10^4	4.6×10^8	6.6×10^4
Copper	1.6×10^8	2.3×10^4	3.4×10^8	4.9×10^4
Iron	1.7×10^8	2.4×10^4	3.2×10^8	4.7×10^4
Steel				
Annealed	2.5×10^8	3.6×10^4	4.9×10^8	7.1×10^4
Spring	4.1×10^8	6.0×10^4	6.9×10^8	10×10^4

* Note: 1 lb//in² = 6.90×10^3 N/m².

of the elastic limits of materials are listed in Table 12.1.

Elastic deformation may be followed directly by fracture, as in the case of the ceramic material in Figure 12.3. However, metals and most polymers or "plastics" exhibit what is known as plastic deformation before fracturing. Unlike elastic deformation, a **plastic deformation** is permanent, and the material does not recover when the stress is removed. For example, a dent in an automobile fender is the result of a stress that produces a plastic deformation. Plastic deformation is required in forming the fender from sheet metal, but elastic deformations are usually desirable thereafter.

Some polymers exhibit various types of "beyond elastic" behavior, depending on their composition and internal structure. The cross-linked rubber in Figure 12.3 departs from the initial linear relationship between the stress and strain for sufficiently large stresses. Beyond the linear elastic range, the deformation is called elastomeric. Unlike plastic deformation, in **elastomeric deformation,** when the stress is removed the material recovers because the long molecular chains are chemically cross-linked. In the absence of cross-linking, the molecular chains of a polymer can slip relative to one another when the stress exceeds the elastic limit, and plastic deformation takes place as in the case of the synthetic polymer in Figure 12.3. The stress-strain curve for such a material varies significantly, depending on the rate of stressing and temperature.

For metals and polymer materials that show considerable plastic behavior, there are two important stress values: the **yield strength,** which is the stress at which appreciable plastic deformation begins (just beyond the elastic limit of the material), and the **tensile or ultimate strength,** which is the maximum stress a sample can support before fracturing (Fig. 12.4 and Table 12.1). The yield strength is especially signifi-

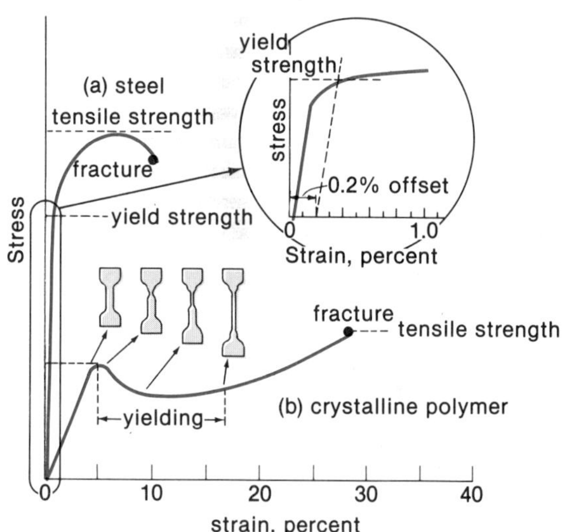

Figure 12.4
Yield and tensile strength. The yield strength is the stress at which appreciable plastic deformation begins (just beyond the elastic limit). A 0.2 percent offset yield is used as a practical designation. The tensile (or ultimate) strength is the maximum stress a sample can support before fracturing.

cant because the materials of most machines and structures are stressed in the elastic range; therefore, yield strength is a basic design factor. As a practical designation of yield strength, the 0.2 percent offset yield strength is often used. A straight line is drawn parallel to the elastic portion of the curve, which is offset by 0.2 percent along the strain axis, as illustrated in Figure 12.4.

The yield strength is then defined as the stress at which the stress-strain curve crosses the offset line. For purposes of our discussion, it should be remembered that the yield strength of a material is slightly greater than the elastic limit, which is easily determined from a stress-strain curve as the point where the curve departs from its initial linearity.

EXAMPLE 12.1 A wire made of spring steel has a diameter of $\frac{1}{8}$ in. (a) What is the maximum load that can be suspended from the wire before it is permanently deformed? (b) What is the greatest load the wire can withstand without breaking?

Solution: (a) With $d = \frac{1}{8}$ in. $= 0.125$ in., the cross-sectional area of the wire is $A = \pi r^2 = \frac{\pi d^2}{4} = \frac{\pi (0.125\text{ in.})^2}{4} = 1.23 \times 10^{-2}\text{ in}^2$. The maximum load corresponds to the elastic limit. Hence, with the stress value of the elastic limit for spring steel from Table 12.1,

$$\text{Elastic limit} = 6.0 \times 10^4\text{ lb/in}^2 = F/A$$

and

$$\begin{aligned} F &= (6.0 \times 10^4\text{ lb/in}^2)A \\ &= (6.0 \times 10^4\text{ lb/in}^2)(1.23 \times 10^{-2}\text{ in}^2) \\ &= 738\text{ lb} \qquad \text{(to three significant figures)} \end{aligned}$$

A slightly greater load would produce a stress corresponding to the yield strength, and the wire would be permanently deformed.

(b) The tensile strength is the maximum stress a material can support before fracturing. Hence, from Table 12.1,

$$\text{Tensile strength} = 10^5\text{ lb/in}^2 = F/A$$

and

$$F = (10^5\text{ lb/in}^2)(1.23 \times 10^{-2}\text{ in}^2) = 1230\text{ lb}$$

For some metals, such as low-carbon steel, and for many polymers, less stress is required after plastic deformation has begun (Fig. 12.3). This is known as *yielding*. However, some metals and polymers, especially crsytalline polymers, strengthen as a result of further plastic deformation. This is called **work-hardening** or **strain-hardening.** (It is found in all metals to some degree.)

For example, suppose you wish to sever a wire, but there are no wire cutters available. You would probably flex or bend the wire repeatedly in one place until it breaks. This is an application of strain-hardening. The repeated flexing (shear stress) makes the metal harder and more brittle. In working metals, a metal may have to be heated and cooled slowly (annealed) to remove the excessive hardness caused by work-hardening.

Some mechanical properties of materials are generally described by terms such as hardness, brittleness, ductility, and malleability.

Hardness reflects the internal resistance of a material to having its molecular particles forced farther apart or closer together. A brittle material shows little plastic behavior and is capable of withstanding only a limited amount of stress. Hardness should not be confused with brittleness. Steel and glass are both hard materials, but glass is much more brittle than steel.

A ductile material is capable of such plastic deformation that it can be drawn into a wire. Metals and polymers with extended plastic deformation ranges are good ductile materials. We have not as yet been able to produce a ductile ceramic material.

A malleable material is one that can be rolled or beaten into sheets, which requires plastic deformation. Gold, copper, lead, and aluminum are malleable materials (Fig. 12.5).

12.2 Young's Modulus

For elastic deformation, the stress is directly proportional to the strain. This relationship is referred to as Hooke's law. Recall from Chapter 7 that Hooke's law for a spring is written as $F = -kx$, where F is the spring force, x the linear displacement of the spring from its equilibrium position, and k the constant of proportionality, or spring constant.

(a)

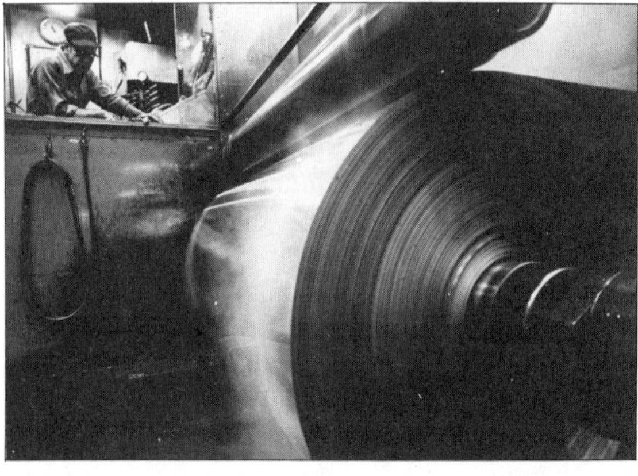

(b)

Figure 12.5
Malleable metal-working. (a) Large aluminum hot-rolling mill rolls metal sheet. (b) Successive rolling produces aluminum foil. The cooling (annealing) removes excessive hardness due to work-hardening. (Courtesy Reynolds Metals Company.)

For material bodies, Hooke's law is expressed in terms of stress and strain:

$$\text{Stress} = E(\text{strain})$$

or

$$\boxed{\text{Elastic modulus} = E = \frac{\text{stress}}{\text{strain}}} \qquad \textbf{(Eq. 12.3)}$$

where the constant of proportionality E is called the **elastic modulus.** A very rigid material would have a very high elastic modulus, since a large stress would produce a relatively small strain. Because strain is a unitless quantity, the elastic modulus has the units of stress, e.g., N/m^2, or lb/in^2.

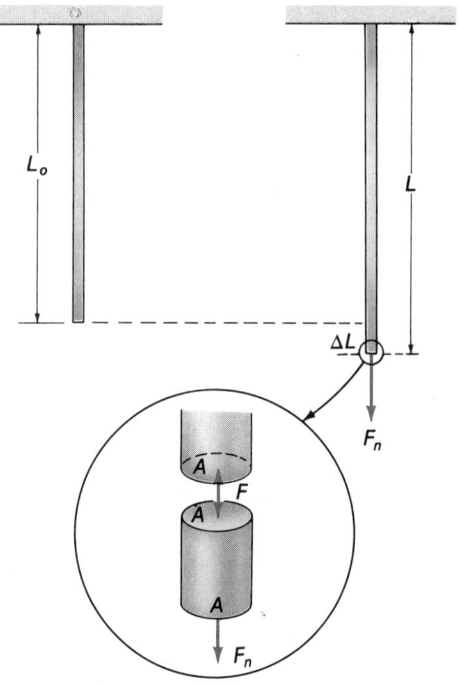

Figure 12.6
Young's modulus. The elastic constant of longitudinal stress and strain is called Young's modulus. See text for description.

In the case of longitudinal or normal stress as applied to a wire, rod, or bar, the elastic modulus is called **Young's modulus** Y,

$$Y = \frac{\text{longitudinal stress}}{\text{longitudinal strain}} \qquad \textbf{(Eq. 12.4)}$$

(Young's modulus)

where the longitudinal stress and strain may be either tensile or compressional.

For a tensile force F_n applied normally to the cross-sectional area of a wire or rod (Fig. 12.6), the longitudinal stress is

$$\text{Longitudinal stress} = \frac{F_n}{A}$$

The effect of the stress is to elongate the rod, and the longitudinal strain is

$$\text{Longitudinal strain} = \frac{\Delta L}{L_o}$$

Table 12.2
Typical Elastic Moduli for Various Materials

Material	Young's Modulus Y		Shear Modulus S		Bulk Modulus B	
	N/m^2	lb/in^2	N/m^2	lb/in^2	N/m^2	lb/in^2
Aluminum	7×10^{10}	10×10^6	2.4×10^{10}	3.4×10^6	7×10^{10}	10×10^6
Brass	9×10^{10}	13×10^6	3.5×10^{10}	5.1×10^6	6.1×10^{10}	8.5×10^6
Copper	11×10^{10}	16×10^6	4.0×10^{10}	6.0×10^6	12×10^{10}	17×10^6
Glass	5.4×10^{10}	7.8×10^6	23×10^{10}	3.3×10^6	36×10^{10}	5.2×10^6
Iron	9×10^{10}	13×10^6	7×10^{10}	10×10^6	10×10^{10}	1.45×10^6
Steel	20×10^{10}	29×10^6	8.0×10^{10}	12×10^6	16×10^{10}	23×10^6
Wood						
Longitudinal	10×10^{10}	1.45×10^6				
Radial	0.07×10^{10}	0.10×10^6				
Tangential	0.06×10^{10}	0.09×10^6				
Polyethylene	0.04×10^{10}	0.06×10^6				
Nylon	0.3×10^{10}	0.44×10^6				
Alcohol					0.11×10^{10}	0.16×10^6
Oil					0.17×10^{10}	0.25×10^6
Water					0.21×10^{10}	0.31×10^6
Mercury					2.8×10^{10}	4.0×10^6

Note: 1 $N/m^2 = 1.45 \times 10^{-4}$ lb/in^2.

Hence, Young's modulus expressed in terms of these experimental parameters is

$$Y = \frac{\text{stress}}{\text{strain}} = \frac{F_n/A}{\Delta L/L_o} = \frac{F_n L_o}{A \Delta L} \quad \textbf{(Eq. 12.5)}$$

Some typical values of Young's modulus for various materials are listed in Table 12.2.

EXAMPLE 12.2 A 60.00-cm copper rod or bar has a rectangular cross-sectional area of 2.0 cm². What load force would be required to elastically elongate the rod to a length of 60.05 cm?

Solution: The change in length of the rod is $\Delta L = L - L_o = 60.05 \text{ cm} - 60.00 \text{ cm} = 0.05$ cm. Hence, the load produces a tensile strain of

$$\text{Strain} = \frac{\Delta L}{L_o} = \frac{0.05 \text{ cm}}{60.00 \text{ cm}} = 8.3 \times 10^{-4} \text{ cm/cm}$$

That is, each original centimeter becomes 8.3×10^{-4} cm longer. Then,

$$Y = \frac{\text{stress}}{\text{strain}}$$

and

$$\text{Stress} = \frac{F_n}{A} = Y(\text{strain})$$
$$= (11 \times 10^{10} \text{ N/m}^2)(8.3 \times 10^{-4})$$
$$= 9.1 \times 10^7 \text{ N/m}^2$$

where the value of Young's modulus was obtained from Table 12.2 and 8.3×10^{-4} cm/cm $= 8.3 \times 10^{-4}$ m/m.

Since $A = 2.0 \text{ cm}^2 (10^{-4} \text{ m}^2/\text{cm}^2) = 2.0 \times 10^{-4}$ m²,

$$F_n = (9.1 \times 10^7 \text{ N/m}^2)A$$
$$= (9.1 \times 10^7 \text{ N/m}^2)(2.0 \times 10^{-4} \text{ m}^2)$$
$$= 1.8 \times 10^4 \text{ N}$$

Question: Would a stress 10 times that of the amount in this example, i.e., $F_n/A = 10 \times (9.1 \times 10^7 \text{ N/m}^2) = 9.1 \times 10^8 \text{ N/m}^2$, produce 10 times the elongation, $10\Delta L = 10(0.05 \text{ cm}) = 0.50$ cm?

Answer: If the deformation were elastic it would, since the relationship between the stress and strain in this region is linear. However, this is not the case for a stress of this magnitude for copper.

Notice in Table 12.1 that the elastic limit for copper is 1.6×10^8 N/m². Hence, the stress would exceed the elastic limit and the relationship between the stress and strain would no longer be linear. The rod would be permanently deformed.

EXAMPLE 12.3 A compressive stress of 4.0×10^4 lb/in² is applied to a 3.0-ft length of steel rail. What is the final length of the rail?

Solution: It is given that stress $= F_n/A = 4.0 \times 10^4$ lb/in², $L_o = 3.0$ ft, and $Y = 29 \times 10^6$ lb/in² (Table 12.2).* Then,

$$\frac{F_n}{A} = \frac{Y\Delta L}{L_o}$$

and

$$\Delta L = \frac{(F_n/A)L_o}{Y} = \frac{(4.0 \times 10^4 \text{ lb/in}^2)(3.0 \text{ ft})}{29 \times 10^6 \text{ lb/in}^2}$$

$$= 4.1 \times 10^{-3} \text{ ft } (= 0.0041 \text{ ft})$$

Since this is a compressive stress, the change in length is negative, $-\Delta L$, and

$$-\Delta L = L - L_o$$

or

$$L = L_o - \Delta L = 3.0 \text{ ft} - 0.0041 \text{ ft} = 2.9959 \text{ ft}$$

Note the unit cancellation. Using L_o in feet gives the answer in feet. Also note that steel is elastic and can be compressed.

12.3 Shear Modulus

When a transverse, or shear, stress is applied to a body, the elastic deformation results in a change in the shape of the body without a change in its volume. The shear stress is proportional to the shear strain, and, similarly to Young's modulus, we define a **shear modulus S** (sometimes called the modulus of rigidity):

$$\text{(Shear modulus)} \quad S = \frac{\text{shear stress}}{\text{shear strain}} \qquad \textbf{(Eq. 12.6)}$$

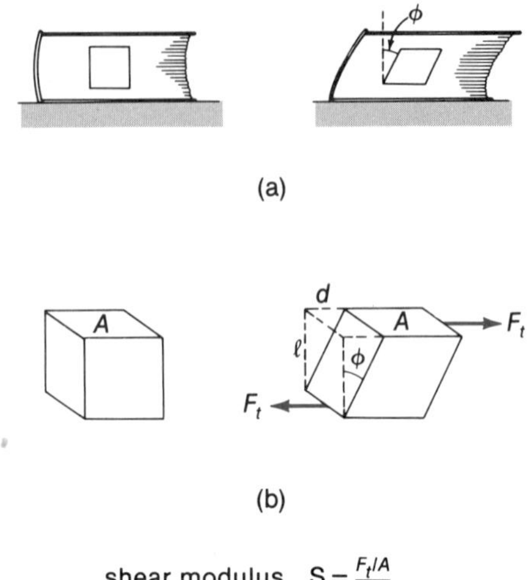

Figure 12.7
Shear modulus. (a) A shear stress changes the shape of an object but not its volume. The shear stress is translational in this case. Notice how an area is displaced by a shear angle ϕ. (b) The shear modulus S is the ratio of the shear stress F_t/A and the shear strain, which for small deformations is approximately d/ℓ.

For the case of a translational shear force applied to the surface of a block with a fixed base, as illustrated in Figure 12.7, the shear stress is F_t/A and the shear strain is defined as the angle ϕ (in radians), which is called the shear angle. The shear angle is usually quite small and hence is approximately equal to the tangent of the angle, $\phi \simeq \tan \phi = d/\ell$. In terms of these parameters the shear modulus is

$$S = \frac{F_t/A}{\phi} \simeq \frac{F_t/A}{\tan \phi} = \frac{F_t/A}{d/\ell} \qquad \textbf{(Eq. 12.7)}$$

For most materials, the value of the shear modulus is one half to one third as great as Young's modulus (Table 12.2). This means that it is easier to deform a material by shearing than by stretching or compressing.

* All quantities are assumed to be exact to any number of (significant) figures.

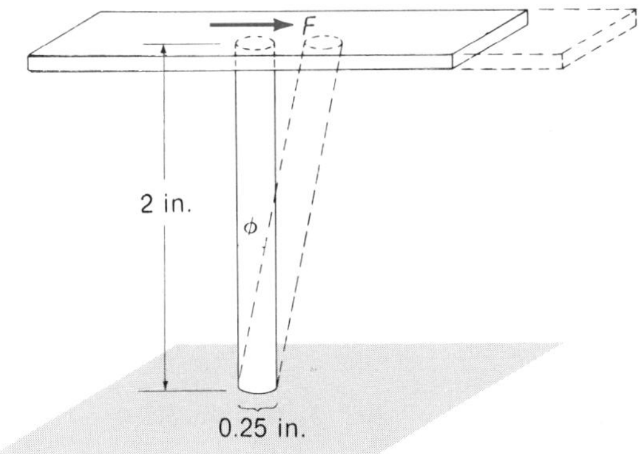

Figure 12.8
A shear stress on a metal rod. See Example 12.4.

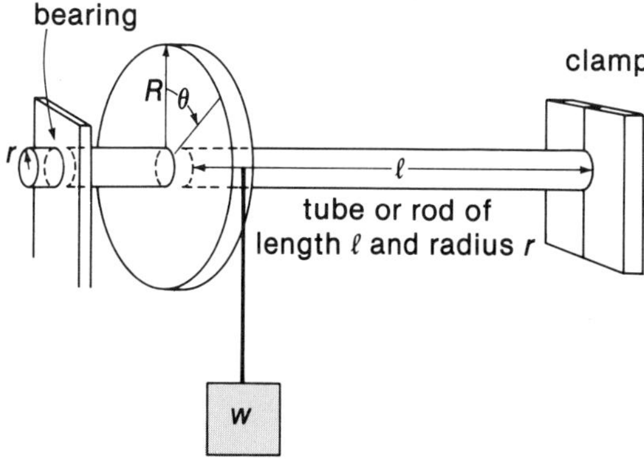

Figure 12.9
Apparatus for measurement of shear moduli. A large torque is applied to the wheel by the suspended weight. See Example 12.5.

EXAMPLE 12.4 A brass rod with a diameter of 0.25 in. supports a metal sheet 2.0 in. above a surface (Fig. 12.8). If a force of 196 lb is applied tangentially to the sheet, (a) what is the shear angle of the rod and (b) how far does the sheet move in the direction of the force?

Solution: (a) The cross-sectional area of the rod is

$$A = \pi r^2 = \frac{\pi d^2}{4} = \frac{\pi(0.25 \text{ in.})^2}{4} = 0.049 \text{ in}^2$$

The shear stress on the rod is then

$$\text{Shear stress} = \frac{F_t}{A} = \frac{196 \text{ lb}}{0.049 \text{ in}^2} = 4000 \text{ lb/in}^2$$

From Table 12.2, the shear modulus of brass is 5.1×10^6 lb/in², so the shear angle is

$$\phi = \frac{F_t/A}{S} = \frac{4.0 \times 10^3 \text{ lb/in}^2}{5.1 \times 10^6 \text{ lb/in}^2} = 7.8 \times 10^{-4} \text{ rad}$$

(b) With $\ell = 2.0$ in. and $\phi = d/\ell$,

$$d = \ell\phi = (2.0 \text{ in.})(7.8 \times 10^{-4}) = 1.6 \times 10^{-3} \text{ in.}$$

or approximately $\frac{1}{640}$ in.

A shear stress may also be of the torsional type—for example, when a torque τ is applied to a shaft. The resulting shear strain is a twisting of the shaft through twist angle θ (Fig. 12.9). It is usually easier to measure the shear modulus using a torsional shear rather than a translational shear.

For a solid cylinder fixed or clamped at one end with a length ℓ and radius r, it can be shown by calculus methods that the shear modulus for the material in terms of the twist angle θ is given by

$$S = \frac{2\tau\ell}{\theta\pi r^4} \quad \textbf{(Eq. 12.8)}$$

In using a simple apparatus as illustrated in Figure 12.9, a large torque is applied to a rod by means of a weight suspended from a wheel of radius R. The wheel is clamped near the free end of the rod to be studied (a wheel-and-axle arrangement). The torque applied to the rod is $\tau = RF = wR = mgR$.

EXAMPLE 12.5 When a weight of 75 lb is suspended from a wheel of radius 6.0 in. that is attached to a rod with a length of 30 in. and a radius of 0.50 in., a twist angle of 0.040 radians is observed (cf. Fig. 12.9). What is the shear modulus of the material?

Solution: It is given that $w = 75$ lb, $R = 6.0$ in., $\ell = 30$ in., $r = 0.50$ in., and $\theta = 0.040$ rad. The applied torque is $\tau = wR = (75 \text{ lb})(6.0 \text{ in.}) = 450$ lb-in. Then,

$$S = \frac{2\tau\ell}{\pi\theta r^4} = \frac{2(450 \text{ lb-in.})(30 \text{ in.})}{\pi(0.040)(0.50 \text{ in.})^4}$$

$$= 3.4 \times 10^6 \text{ lb/in}^2$$

Notice that the torque is expressed in lb-in. rather than the standard lb-ft. Checking with Table 12.2, we see that the rod appears to be made of aluminum.

12.4 **Bulk Modulus**

In the cases of longitudinal and shear stresses, the volume of the deformed object remains constant. However, the volume of an object does change when forces are applied uniformly over its surface. A uniform stress F/A or pressure can be applied to a solid if it is submerged in a liquid, since the pressure applied to a liquid, e.g., via a piston, is transmitted undiminished throughout the liquid and will act perpendicularly on the solid surface. (See Pascal's principle, Section 11.3.).

For example, consider a cube or a sphere acted upon by forces distributed uniformly over the object's surface (Fig. 12.10). The volume stress is then F_n/A, where F_n is the total force acting on the total area A. (In the case of a cube, this is the stress on each surface.) The volume stress causes a change in the volume of the object, $-\Delta V$, where the minus sign indicates that the change is a decrease in volume. The volume strain is then $-\Delta V/V_o$.

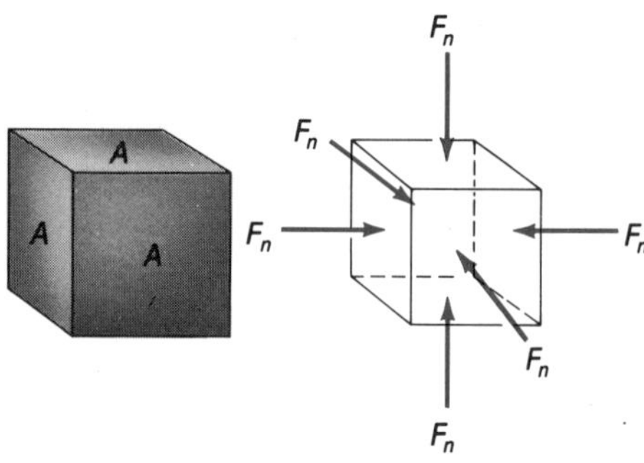

Figure 12.10.
Bulk or volume stress. Unlike longitudinal and shear stresses, a volume stress produces a change in volume. A normal stress can be applied to all surfaces of a solid by submerging it in a liquid-filled cylinder and applying pressure to a piston (Pascal's principle). The ratio of the volume stresses and the volume strain is the bulk modulus of a material.

The elastic modulus for the compressional volume change is called the modulus of volume elasticity or the **bulk modulus** $\boldsymbol{B}$**:**

(Bulk modulus) $$B = \frac{\text{volume stress}}{\text{volume strain}}$$

$$B = -\frac{F_n/A}{\Delta V/V_o} = -\frac{p}{\Delta V/V_o} \quad \textbf{(Eq. 12.9)}$$

where the volume stress F_n/A may be written as pressure p.

Liquids cannot support tensile or shear stresses, but they do support a volume compressional stress. Hence, liquids as well as solids exhibit volume elasticity. The bulk moduli for several common liquids are listed in Table 12.2. With liquids, the volume stress is usually expressed in terms of pressure, $p = F_n/A$. It should be noted that p is really a change in pressure, or Δp.

The volume elasticity is often expressed in terms of the **compressibility** $\boldsymbol{k}$**,** which is the reciprocal of the bulk modulus:

$$\text{Compressibility } k = \frac{1}{B} = -\frac{1}{p}\frac{\Delta V}{V_o} \quad \textbf{(Eq. 12.10)}$$

The greater the bulk modulus B of a material, the less compressible it is, and the smaller its compressibility k. As can be seen from Table 12.2, liquids are generally more compressible than solids since the liquids have smaller bulk moduli or greater compressibilities. Even so, large forces or pressures are required to compress liquids appreciably, and in many practical applications, liquids (and solids) are considered to be incompressible.

EXAMPLE 12.6 How many times more compressible is water than mercury?

Solution: Since $k = 1/B$, we can form a ratio, and the bulk moduli are given in Table 12.2:

$$\frac{k_{H_2O}}{k_{Hg}} = \frac{B_{Hg}}{B_{H_2O}}$$

$$= \frac{2.8 \times 10^{10}\ \text{N/m}^2}{0.21 \times 10^{10}\ \text{N/m}^2} = 13$$

Hence, water is 13 times more compressible than mercury.

EXAMPLE 12.7 How much pressure is required to compress a liter of water by 1.0 cm^3?

Solution: A liter is a volume of $V_o = 1000\ cm^3$ (or cc, cubic centimeters), and with $-\Delta V = 1.0\ cm^3$, the volume strain is

$$\frac{-\Delta V}{V_o} = \frac{1.0\ cm^3}{1000\ cm^3} = 0.001 = 10^{-3}$$

or there is a 0.1 percent reduction in volume. With $B_{H_2O} = 0.21 \times 10^{10}\ N/m^2$, by Equation 12.9,

$$p = B(-\Delta V/V_o) = (0.21 \times 10^{10}\ N/m^2)(10^{-3}) = 2.1 \times 10^6\ N/m^2$$

or $$p = 2.1 \times 10^6\ N/m^2 \left(\frac{1.45 \times 10^{-4}\ lb/in^2}{N/m^2}\right) = 3.0 \times 10^2\ lb/in^2 = 300\ lb/in^2$$

Assuming the water to be in a cubic block, each side of the cube would have an area of 0.010 m^2, since a liter has a volume of 1000 cm^3 (a cube with sides of 10 cm = 0.10 m):

$$0.010\ m^2 \left(\frac{1550\ in^2}{m^2}\right) = 15.5\ in^2$$

Hence, the force required on each side of the cube is

$$F_n = pA = (300\ lb/in^2)(15.5\ in^2) = 4650\ lb = 2.3\ \text{tons!}$$

Important Terms

stress the ratio of the applied force and the area over which it acts

longitudinal (normal) stress stress in which the applied force is normal to a surface

tensile stress a normal stress that tends to elongate a material

compressive stress a normal stress that tends to compress a material

transverse (shear) stress stress in which the applied force is tangential, or parallel, to a surface

translational stress a shear stress that tends to translate a surface

torsional stress a shear stress that tends to twist a material as a result of a torque

strain the relative change in the dimensions or shape of a body due to a stress

elastic deformation deformation in which the material returns to its original shape and dimensions when a stress is removed; characterized by a linear relationship on a stress-strain curve

elastic limit the maximum stress a material can experience without being permanently deformed

plastic deformation permanent deformation in which a material does not recover its original shape when a stress is removed

elastomeric deformation deformation exhibited by some polymers in which, due to molecular chain properties, the material recovers after a stress beyond the linear elastic range is removed

work (strain) hardening the hardening and brittling of a metal due to repeated stressing

hardness the internal resistance of a material to having its molecules forced closer together or farther apart

brittleness the material property of withstanding only a limited amount of stress before fracture

ductility the capability of a material of being drawn into a wire

malleability the capability of a material of being rolled or beaten into sheets

elastic modulus the ratio of stress and strain

Young's modulus (*Y*) the ratio of longitudinal stress and strain

shear modulus (*S*) the ratio of shear stress and strain

bulk modulus (*B*) the ratio of volume stress and strain

compressibility (*k*) the reciprocal of the bulk modulus ($k = 1/B$)

Important Formulas

stress: $$\frac{F}{A}$$

tensile strain: $$\frac{\Delta L}{L_o} = \frac{L - L_o}{L_o}$$

elastic modulus: $$\frac{\text{stress}}{\text{strain}}$$

Young's modulus: $$Y = \frac{F_n/A}{\Delta L/L_o}$$

shear modulus: (translational) $$S = \frac{F_t/A}{\phi} \simeq \frac{F_t/A}{\tan\phi} = \frac{F_t/A}{d/\ell}$$

(fixed cylinder) $$S = \frac{2\tau\ell}{\theta\pi r^4}$$

bulk modulus: $$B = -\frac{F_n A}{\Delta V/V_o} = -\frac{p}{\Delta V/V_o}$$

compressibility: $$k = \frac{1}{B} = -\frac{1}{p}\frac{\Delta V}{V_o}$$

Questions

Stress and Strain

1. What materials are not elastic?
2. How is stress defined?
3. Distinguish between longitudinal and transverse stresses.
4. Distinguish between normal and shear stresses.
5. Is there more than one type of shear stress? Explain.
6. What is strain and how is it described?
7. What are the units of (a) stress and (b) strain?
8. What is the elastic limit of a material?
9. Explain the following terms and give an example of each in a particular type of material: (a) fracture, (b) plastic deformation, (c) elastomeric deformation.
10. Distinguish between (a) yield strength and (b) tensile strength.
11. (a) Why is yield strength an important design factor, and (b) why is a 0.2 percent offset yield strength used?
12. Given two rods of equal diameter, one made of brass and the other of annealed steel, (a) which would support a greater tensile stress before becoming permanently deformed? (b) Which would fracture first?
13. How does strain-hardening affect the yield strength and tensile strength of a material?
14. Define each of the following material properties and give an example of a material for each property: (a) hardness, (b) brittleness, (c) ductility, (d) malleability.
15. Enter each of the following four solid materials into the appropriate box in Figure 12.11: lead, aluminum, diamond, and iridium (a metal used for pen points). If you are not sure of the properties, make your best guess and do a little research. What factors account for the hardness and densities of solids?

Young's Modulus

16. Express Hooke's law in terms of stress and strain.
17. A particular material has a greater Young's modulus than another material. What does this tell you?
18. What are the units of Young's modulus?
19. What is an anisotropic material, and what is an example of such a material?
20. Given two lengths of copper wire, with wire 1 twice as long as wire 2, what would be the elongation of one wire relative to the other when subjected to the same load (a) if the radius of wire 1 were twice that of wire 2, and (b) if the radius of wire 2 were twice that of wire 1?

Shear Modulus

21. Are the forces involved in a shear stress and a tensile stress the same? Explain.
22. Which would you expect to have a greater shear modulus, a piece of aluminum or a piece of wood? Explain.
23. What are the units of the shear modulus?
24. In general, would you expect the tensile strength or the shear strength of a material to be greater? Explain.
25. Given two round, solid iron support posts, post 1 having twice the diameter of post 2, what would be the displacement of one post relative to the other when subjected to the same shear force (a) if post 2 were twice as long as post 1, and (b) if post 1 were four times as long as post 2?

Bulk Modulus

26. In general, which is greater for metals: (a) Young's modulus or the shear modulus? (b) The shear modulus or the compressibility?
27. What are the units of the bulk modulus?
28. Why are there no Young's moduli and shear moduli for liquids?
29. If one material has a compressibility 1.5 times greater than another material, what can you say about the bulk moduli of the materials?

Problems

Levels of difficulty are indicated by asterisks for your convenience.

12.1 Stress and Strain

1. A 50-kg weight is suspended by means of a cable having a diameter of 1.0 cm. What is the stress in the cable?

	harder than iron	softer than iron
denser than iron		
less dense than iron		

Figure 12.11
See Question 15.

2. A tractor towing a car supplies an applied force of 0.50 ton to a cable with a diameter of 0.75 in. What is the stress in the cable?
3. A push rod in a machine receives a force of 500 N. If the rod has a square cross section 1.0 cm on a side, what is the compressional stress?
*4. A circular wire has a stress of 1270 lb/in^2 as a result of the suspension of a 10-lb picture. What is the diameter of the wire?
*5. A carpenter applies a tangential force of 80 N to the surface of a block of wood. If the surface dimensions of the block are 30 cm × 50 cm, what is the shear stress on the block?
*6. A 30-cm length of wire is stretched to a length of 30.05 cm. What is the tensile strain?
*7. A compressional strain of −0.05 percent is applied to a 20-in. rod. By how much is the length of the rod decreased?
*8. A wire made of annealed steel has a diameter of 1.0 mm. (a) What is the maximum load that can be suspended from the wire before it is permanently deformed? (b) What is the greatest load the wire can withstand without breaking?
*9. No. 24 copper wire has a diameter of 0.020 in. What is the magnitude of the normal force that will (a) permanently stretch the wire and (b) break it?
**10. An engineer wishes to suspend a 250-lb load by means of an aluminum wire of the smallest possible diameter without exceeding the elastic limit of the wire. What diameter wire should be used?

12.2 Young's Modulus

*11. When a 2.0-m length of wire of a newly developed alloy is stressed with a tension force of 400 N, its length increases by 0.080 mm. If the wire has a diameter of 0.50 cm, what is Young's modulus for the material?
*12. An aluminum wire with a diameter of 0.040 in. and a length of 20 ft is stretched by a force of 15 lb. (a) What is the increase in the length of the wire? (b) What is the maximum possible elastic elongation?
*13. A steel rod 8.0 in. long with a cross-sectional area of 0.50 in^2 is used to support a heavy load. If the rod is compressed in length by 0.00025 in., what is the magnitude of the load?
*14. A nylon strand has a length of 0.65 m and a cross-sectional area of 2.0×10^{-5} m^2. What force would be required to stretch the strand 1.0 mm?
*15. A 15-ft steel bridge girder with a cross-sectional area of 10 in^2 is subjected to a compressional load of 25 tons. What is the decrease in length of the girder?

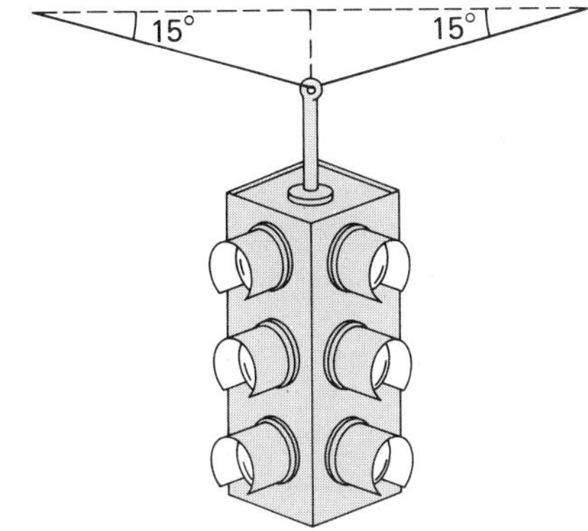

Figure 12.12
See Problem 20.

*16. A copper rod with a diameter of $\frac{1}{8}$ in. is subjected to a tensile force of 1 ton. What is the fractional increase in the length of the rod? Express the answer as a percentage.
*17. An aluminum rod and a steel rod of the same dimensions are subjected to the same tensile stress. Which rod has the greater elongation and by how much?
**18. A copper wire 50.00 cm long is stretched to a length of 50.02 cm when it supports a certain load. If an aluminum wire of the same diameter were used to support the same load, what would its initial length be so that it too would have a stretched length of 50.02 cm?
**19. How much mass would have to be suspended on a steel wire with a diameter of 0.20 cm to cause its length to increase by 0.10 percent?
**20. A 50-kg traffic light is suspended by two steel cables with radii of 1.0 cm (Fig. 12.12). If the cables form 15° angles relative to the horizontal, what is the fractional increase in the length of the cables due to the weight of the traffic light?

12.3 Shear Modulus

*21. A tangential force of 500 N is applied to the upper surface of a copper cube 20 cm on a side. What is the displacement of the upper surface of the cube?
*22. An aluminum post with a diameter of 0.50 in. supports a metal sheet 3.0 in. above a surface. If a force of 250 lb is applied tangentially to the sheet, (a) what is the shear angle of the post and (b) how far does the sheet move in the direction of the force?

*23. A rectangular form of Jell-O with a length, width, and height of 10 cm, 8.0 cm, and 4.0 cm, respectively, is subjected to a shear force of 0.40 N on its upper surface. If the surface is displaced 0.30 mm, what is the shear modulus of the Jell-O?

*24. A cube of aluminum 20 cm on a side has shear forces of 500 N applied to two opposite faces. What is the relative displacement of the cube surfaces?

*25. A torque of 100 in.-lb is applied to a cylindrical brass stud fixed at one end. If the stud has a length of 2.5 in. and a diameter of 0.50 in., what is the resulting twist angle?

*26. A 0.50-m aluminum rod fixed at one end has a diameter of 3.0 cm. What is the magnitude of the torque that would twist the rod by 1.0°?

*27. The shear modulus of a rod of a new alloy is determined using an apparatus as shown in Figure 12.9. The rod has a length of 50 cm and a diameter of 2.0 cm. If a mass of 10 kg suspended from a wheel with a radius of 30 cm produces a twist angle of 0.025 rad, what is the shear modulus of the material?

**28. For a torsion shear stress as used in the apparatus in Figure 12.9, show that the twist angle θ is related to the shear angle ϕ by $\phi = \left(\frac{r}{\ell}\right)\theta$.

**29. Two metal plates are held together by four rivets with diameters of 0.50 cm. If the maximum shear stress a single rivet can stand is 3.0×10^8 N/m^2, how much force must be applied parallel to the plates to shear off the rivets?

**30. The maximum shear stress for a type of steel is 3.5×10^8 N/m^2. How much force would have to be applied to a punch with a diameter of 2.0 cm to punch a hole in a sheet of the steel that is 1.0 cm thick? (*Hint:* draw a sketch and show that this requires a shear stress.)

12.4 Bulk Modulus

*31. A pressure of 10^8 N/m^2 is applied to the surfaces of an object with a volume of 0.450 m^2. If the volume decreases to 0.449 m^2, (a) what is the bulk modulus of the material, and (b) what is its compressibility?

*32. How much pressure would be required to decrease the volume of a glass sphere by 0.01 percent?

*33. How much pressure is required to compress a cubic foot of water by 1.0 in^3?

*34. A piston with a surface area of 20 in^2 exerts a force of 1 ton on a gallon of oil in a cylinder. What is the decrease in the volume of the oil?

*35. Which is more compressible and by how much more? (a) Glass or steel? (b) Water or oil?

*36. A brass cube 6.0 cm on a side is subjected to a uniform force of 2.5×10^7 N on each of its surfaces. What is the fractional change in the volume of the cube?

**37. Given a block of iron, what would be the relative change in volume of the block (a) in outer space and (b) on the ocean floor where the pressure is 10^8 N/m^2 (a depth of about 10 km)? Consider volume effects due to pressure changes only. (Temperature would also make a difference—thermal expansion.)

Chapter 13

Vibrations and Waves

As we learned in Chapter 12, all materials are deformed to some extent when an external force, or stress, is applied. If the stress does not exceed the elastic limit, the material recovers when the stress is removed, owing to internal molecular forces. During the deformation and the recovery, the atoms and molecules vibrate about their equilibrium positions (in addition to their normal thermal vibrations).

In some instances, the vibrational effects are not observed macroscopically. In other instances of large stresses, the material vibrates as a whole, e.g., a diving board. In any case, unless continuously driven, these mechanical vibrations eventually die out owing to the dissipation of energy through some frictional mechanism and/or the transfer of energy

Vibrations are an important technical consideration. The vibration of machinery can cause serious damage. Motor mounts and shock absorbers are examples of devices that were devised to reduce or "damp" unwanted vibrational effects. On the other hand, in some applications vibrations are promoted. For example, it is customary to put a new can of paint on a shaker or vibrator to ensure that the paint is well mixed.

To describe vibrational disturbances and motion, we speak in terms of waves and wave motion.

13.1 Waves and Wave Motion

A wave is a process involving the propagation or transfer of energy due to a physical disturbance. As a familiar example, consider dropping a stone into a calm pool of water. The stone creates a disturbance in the water, which then can be seen propagating outward in concentric circles. Or raindrops may produce many such disturbances. (See the chapter introductory photo.)

Clearly, in these examples energy is being propagated from a disturbance region. The distances the water "particles" themselves move

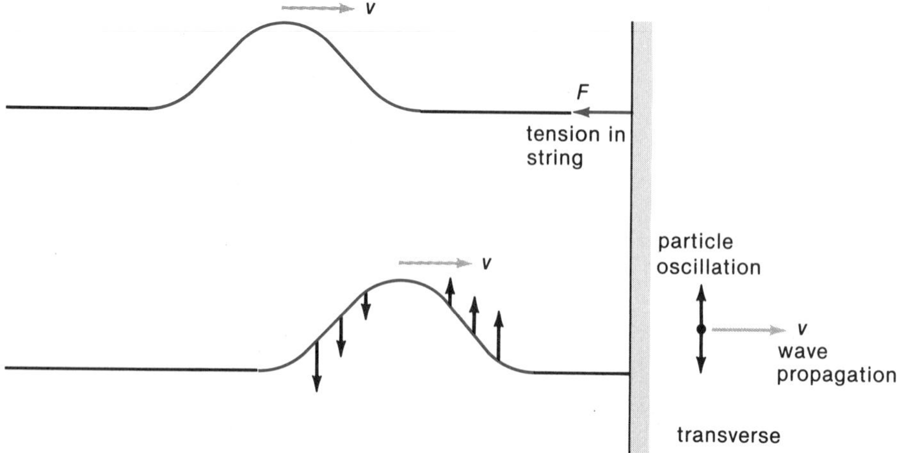

Figure 13.1

A transverse wave pulse. The pulse propagates in the stretched string with a velocity v to the right. As shown in the lower drawing, the string "particles" move up and down perpendicularly to the direction of the wave propagation.

are relatively small. Energy is transferred from one particle to another, and *the disturbance energy, rather than the medium material, is propagated.* This is called **wave motion.**

In general, we think of waves in material media. However, there is a type of wave, an electromagnetic wave, that propagates in vacuum. For example, light is an electromagnetic wave, and sunlight comes to us through the void (vacuum) of space. Electromagnetic waves are considered in Chapter 27. For the present, we are concerned with mechanical waves, which are disturbances in elastic media.

Waves are classified according to the direction of the particle motion with respect to the direction of wave propagation. Two general types of waves are transverse waves and longitudinal waves.

Transverse Waves

In transverse waves, the direction of the particle motion of the medium is perpendicular to the direction of the wave propagation.

Transverse waves are sometimes called *shear waves*. This is because of shear forces in the medium. Consider, for example, a disturbance (wave) propagating along a stretched string or rope (Fig. 13.1). Such a single disturbance is called a wave pulse. As the wave moves along the string, the individual particles move up and down in response to shear forces perpendicular to the direction of wave propagation.

The speed with which a transverse wave pulse travels along a stretched string or cord depends on the tension force F in the string and the linear mass density $\mu = m/L$ (mass per unit length) of the string. These quantities reflect measurements of elasticity and inertia. It can be shown that the speed of the wave pulse is given by the formula

$$\boxed{v = \sqrt{\frac{F}{\mu}}} \qquad \textbf{(Eq. 13.1)}$$

EXAMPLE 13.1 A 5.0-m length of rubber cord having a mass of 0.60 kg is stretched with a force of 10 N. If the cord is struck sharply near the held end, what will be the speed of the resulting wave pulse traveling down the cord?

Solution: It is given that $m = 0.60$ kg, $L = 5.0$ m, and $F = 10$ N. The linear mass density of the cord is

$$\mu = \frac{m}{L} = \frac{0.60 \text{ kg}}{5.0 \text{ m}} = 0.12 \text{ kg/m}$$

Then, using Equation 13.1 with $F = 10$ N,

$$v = \sqrt{\frac{F}{\mu}} = \sqrt{\frac{10 \text{ N}}{0.12 \text{ kg/m}}} = 9.1 \text{ m/s}$$

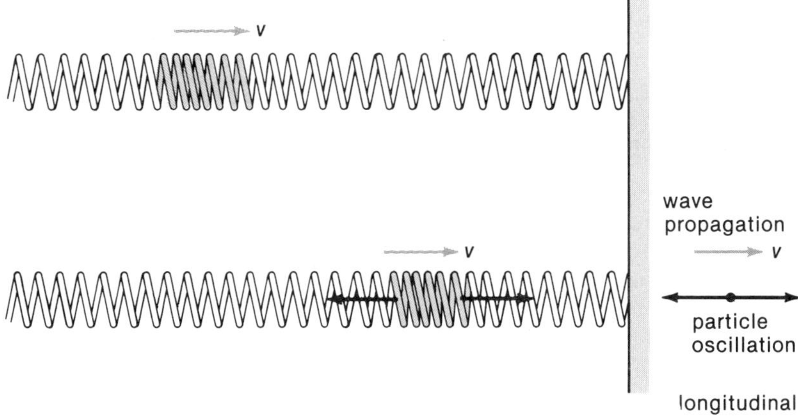

Figure 13.2

A longitudinal wave pulse. The pulse propagates in a stretched spring with a velocity v to the right. As shown in the lower drawing, the spring "particles" move back and forth parallel to the direction of the wave propagation.

Longitudinal Waves

In longitudinal waves, the direction of the particle motion of the medium is parallel to the direction of the wave propagation.

Longitudinal waves are sometimes called *compressional waves.* This is because of compressional forces in the medium. A longitudinal pulse in a coil spring is illustrated in Figure 13.2. If the coils near one end of a stretched spring are compressed together and released, a wave pulse travels down the spring. Notice that the spring "particles" move back and forth parallel to the direction of the wave propagation.

The propagation of a disturbance in a medium depends on the intermolecular forces. As we learned previously, the intermolecular forces in solids are relatively strong, and the forces become progressively weaker in liquids and gases. **Longitudinal waves propagate in all forms of matter—solids, liquids, and gases**—because of the presence of adequate intermolecular compressive forces. However, **transverse waves propagate in solids but *not* in liquids or gases.** This is because the intermolecular forces in liquids and gases will not support a shear stress.

The fact that transverse waves do not propagate in a liquid leads us to believe that a portion of the Earth's central core is a highly viscous liquid. See Special Feature 13.1.

Water waves are a combination of transverse and longitudinal motions. When a disturbance propagates through water, the water particles move in circular or elliptical orbits about their equilibrium positions, which gives rise to the commonly observed crests and troughs (Fig. 13.3). A floating object bobbing up and down in the water can be seen to describe a roughly circular path as the disturbance passes by. When water particles in a wave cannot complete the bottom parts of their path, the wave "breaks" and forms a surf.

Wave motion is one of the most important means of transmitting energy from one place to another. We communicate and receive most of the information about the world around us when our senses respond to sound waves (longitudinal waves) and light waves (transverse waves). These two important types of waves are considered in later chapters.

13.2 Periodic Motion

As we have seen, the particles of a medium vibrate or oscillate as a wave pulse passes through the medium. After the wave pulse passes by, the particles eventually return to their equilibrium positions. However, with a continuous periodic disturbance, such as the continual up-and-down shaking of a stretched string, the particles continue to oscillate.

SPECIAL FEATURE 13.1

The Earth's Liquid Outer Core

You may have heard that the Earth has a liquid center or, more correctly, a liquid outer core.* A good question to ask is, How do we know? We have never penetrated the Earth's relatively thin crust. The deepest mine shafts and drillings of a few miles have only "scratched the surface." Hot molten material comes up from the upper mantle and erupts, forming volcanoes, but this occurs only in certain areas, and overall the mantle is solid.

Waves are used to probe and investigate the interior of the Earth. We can produce minor disturbances through explosions, but nature supplies some big ones free of charge through earthquakes. Earthquakes are caused by the sudden release of built-up stress along faults, or cracks in the Earth, such as the San Andreas Fault in California. The energy from these disturbances propagates outwardly in the form of seismic waves. There are two general types of seismic waves—surface waves and body waves. The surface waves move along the Earth's surface and cause most of the earthquake damage.

Body waves travel through the Earth. There are P-waves, which are compressional (longitudinal) waves, and S-waves, which are shear (transverse) waves. The P and S stand for *primary* and *secondary,* which indicate their arrivals at a location. Primary waves travel faster than secondary waves and reach a seismic detection station first.

The longitudinal P-waves can travel through a solid or liquid, but the transverse S-waves cannot travel through a liquid. When an earthquake occurs on one side of the Earth, P-waves are detected on the opposite side and S-waves are not (Fig. 1). The absence of S-waves arriving in a "shadow zone" leads to the conclusion that the Earth must be liquid near the center. The liquid is highly viscous and metallic. When the P-waves enter and leave the liquid region, they are bent. This gives rise to a P-wave shadow zone, which tells us that only the outer part of the core is liquid.

* The Earth is divided into three regions: an outer crust about 3–30 km (5–20 mi) thick; the mantle, 2900 km (1800 mi) thick; and a core with a radius of 3450 km (2150 mi). The solid inner core has a radius of about 1200 km (750 mi).

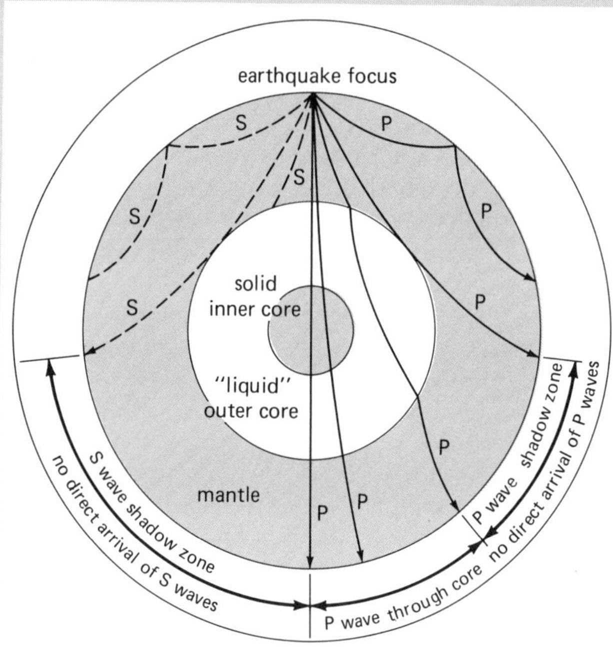

Figure 1
Evidence for the Earth's liquid outer core comes from waves generated by earthquakes. Transverse or shear waves cannot propagate through a liquid, but longitudinal or compressional waves can. Because of shadow zones where no waves are received, the Earth's outer core is believed to be a viscous liquid.

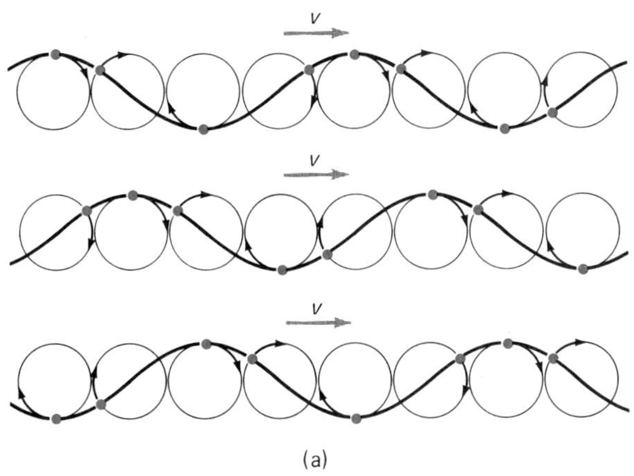

(a)

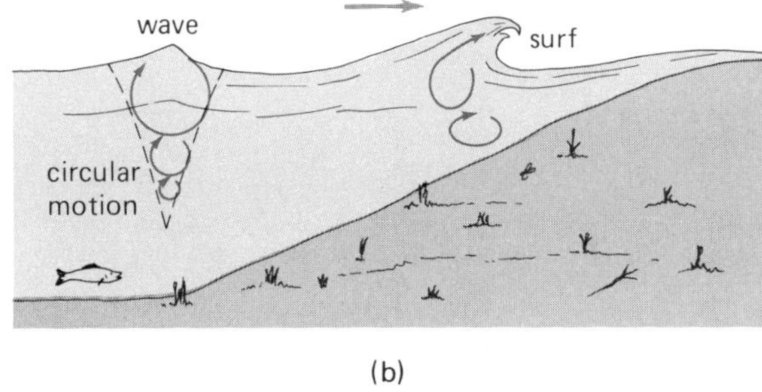

(b)

(c)

Figure 13.3
Water waves. (a) Water waves are combinations of transverse and longitudinal motions. The water "particles" move in more-or-less circular paths as the wave crest moves forward. (b) and (c) When the water particles in a wave cannot complete the bottom parts of their paths, the wave "breaks" and the crest falls to form a surf.

A periodic disturbance is one that is repetitious in time. In response to a continuous periodic disturbance, the particles are in **periodic motion.** That is, the particles continually oscillate, and the time T required for one complete oscillation is called the **period of oscillation.** Another type of periodic motion is that of a satellite in orbit about the Earth. In this case, the time required for one complete cycle of motion is the period of revolution.

The motions of the particles in a medium depend on the intermolecular forces. Of particular importance is the motion for a force that is proportional to the displacement of a particle from its equilibrium position. That is the force described by **Hooke's law,**

$$F = -kx \qquad \textbf{(Eq. 13.2)}$$

where k is the "spring" constant (cf. Section 6.3). In many instances, the intermolecular forces obey Hooke's law for small oscillations.

The minus sign in Hooke's law indicates that the force is in the direction opposite to the displacement. For example, when a spring is stretched or compressed, the spring force opposes the displacement, i.e., is a restoring force. This points up an almost obvious fact:

A wave vibration requires a restoring force.

When a particle in a medium is displaced from its equilibrium position, an intermolecular restoring force acts to return it to its equilibrium position; inertial motion causes the particle to pass through the equilibrium position, and the restoring force then acts in the opposite direction. This back-and-forth action causes the particle to vibrate about its equilibrium position.

A particle in motion under the influence of a force with the form of Hooke's law is said to be in **simple harmonic motion (SHM).** As an example of simple harmonic motion, consider a mass

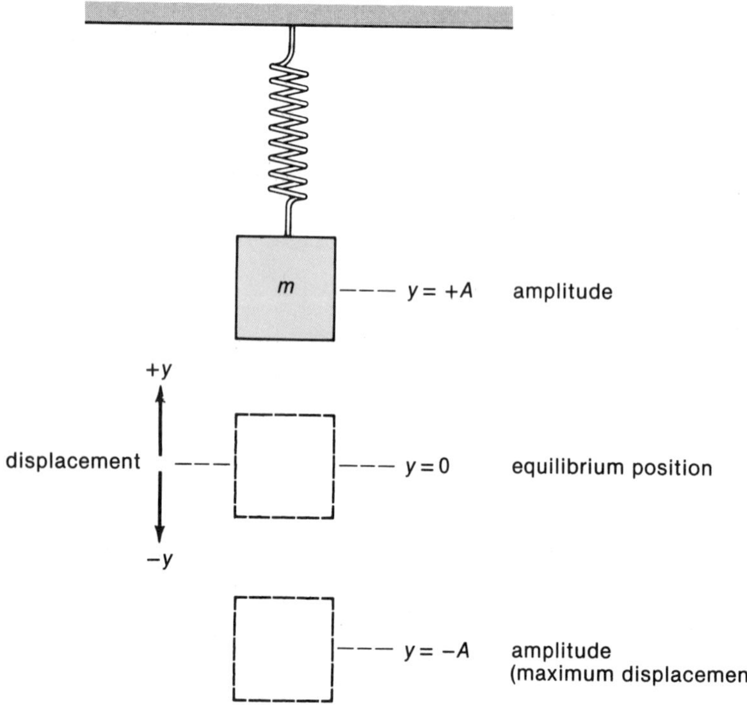

Figure 13.4
An example of simple harmonic motion. A mass on a spring undergoes SHM in oscillating up and down through its equilibrium position ($y = 0$, position of suspended mass when not in oscillation). The maximum displacement ($\pm A$) is called the amplitude of the oscillation.

suspended on a spring (Fig. 13.4). If the mass is pulled downward and released, it oscillates in simple harmonic motion.

Several parameters are used to describe the motion. The displacement y of the mass is its distance from the equilibrium position ($y = 0$) with the direction indicated by + or −. **The maximum displacement is called the *amplitude* of the oscillation.** This occurs in the motion at the positions $y = A$ and $y = -A$.

The **period** (***T***) of oscillation, or the time for one complete oscillation, is given by

$$T = 2\pi \sqrt{\frac{m}{k}} \qquad \textbf{(Eq. 13.3)}$$

(period of oscillation of a mass on a spring)

where k is the spring constant and m is the mass on the spring.

EXAMPLE 13.2 A mass of 0.10 kg oscillates on a spring with a spring constant of 16 N/m. What is the period of oscillation?

Solution: With $m = 0.10$ kg and $k = 16$ N/m,

$$T = 2\pi \sqrt{\frac{m}{k}} = 2\pi \sqrt{\frac{0.10 \text{ kg}}{16 \text{ N/m}}} = 0.50 \text{ s}$$

Simple harmonic motion occurs in other common systems—for example, in the small oscillations of a simple pendulum and a torsional (twist) pendulum (Fig. 13.5). The equation for the period of a simple pendulum undergoing small oscillations is similar to that for a mass on a spring,

$$T = 2\pi \sqrt{\frac{L}{g}} \qquad \textbf{(Eq. 13.4)}$$

(period of oscillation of a simple pendulum)

where L is the length of the pendulum (measured to the center of mass of the pendulum bob) and g is the acceleration due to gravity.

Vibrations are often described in terms of the frequency of vibration. The **frequency** (***f***) is the number of vibrations per second and is given in

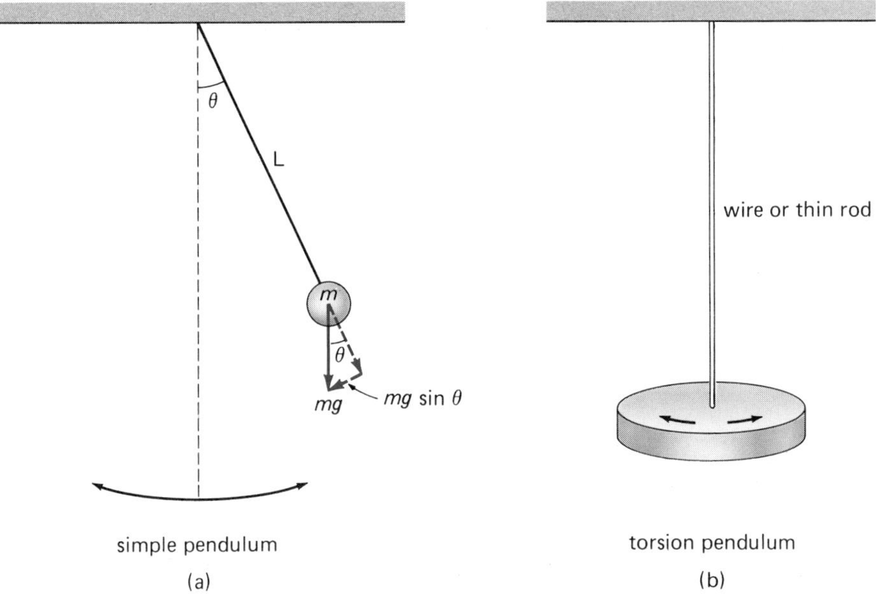

Figure 13.5
Pendula undergo simple harmonic motions. (a) A simple pendulum. For small oscillations, the restoring force, $mg \sin \theta$, is proportional to the displacement (see text). (b) A torsion pendulum. The torsional force and torque in the wire is proportional to the angular displacement θ of the disk, $\tau = -k\theta$.

units of inverse time ($1/\text{s} = \text{s}^{-1}$). As in rotational motion, the frequency is the reciprocal of the period:

$$\boxed{f = \frac{1}{T}} \qquad \textbf{(Eq. 13.5)}$$

For example, the vibrational frequency of a mass on a spring is given by

$$f = \frac{1}{T} = \frac{1}{2\pi}\sqrt{\frac{k}{m}} \qquad \textbf{(Eq. 13.6)}$$

(frequency of a mass on a spring)

The frequency is commonly expressed in cycles per second (cps), but the SI unit is the hertz (Hz), and 1 cps = 1 Hz.* In Example 13.2, the period of oscillation of the mass on a spring is $T = 0.50$ s, so the frequency of oscillation is

$$f = \frac{1}{T} = \frac{1}{0.50 \text{ s}} = 2.0 \text{ Hz}$$

* The unit is named in honor of Heinrich Hertz (1857–1894), a German physicist and an early investigator of electromagnetic waves.

That is, if the period of one oscillation is 0.50 s, then in 1.0 s the mass makes two complete oscillations and has a frequency of 2.0 cps or 2.0 Hz.

You may have been wondering what the term *harmonic* in "simple harmonic motion" refers to. **Harmonic means that the motion can be described in terms of a sinusoidal function—as a since or a cosine.** Suppose a marker is placed on an oscillating mass so that the motion is traced out on a moving paper, as illustrated in Figure 13.6. The curve traced out on the paper, and hence the path of motion of the mass, can be described by the equation of motion:

$$y = A \sin \frac{2\pi t}{T} \qquad \textbf{(Eq. 13.7)}$$

where y is the displacement, A is the amplitude, and T is the period of oscillation. Note that for $t = 0$, $y = 0$ (since $\sin 0 = 0$); for $t = T/4$, $y = A$ (since $\sin \pi/2 = 1$); and so on.

Since $f = 1/T$, Equation 13.7 can also be written

$$y = A \sin 2\pi f t \qquad \textbf{(Eq. 13.8)}$$

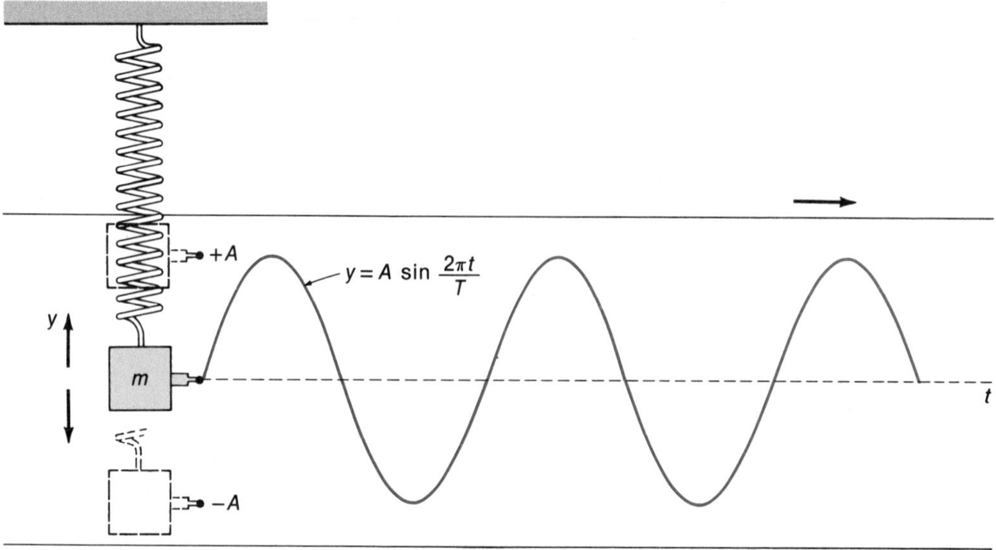

Figure 13.6
Harmonic motion. Harmonic means that the motion can be described in terms of sinusoidal functions (sines and cosines). A marker on an oscillating mass in SHM would trace out a sinusoidal curve on a moving paper.

Another common form of the equation is

$$y = A \sin \omega t \qquad \textbf{(Eq. 13.9)}$$

where ω is the angular frequency or angular speed of a particle in circular motion, where

$$\omega = \frac{2\pi}{T} = 2\pi f$$

Although linear vibration is not circular motion, the linear motion of SHM can be related to

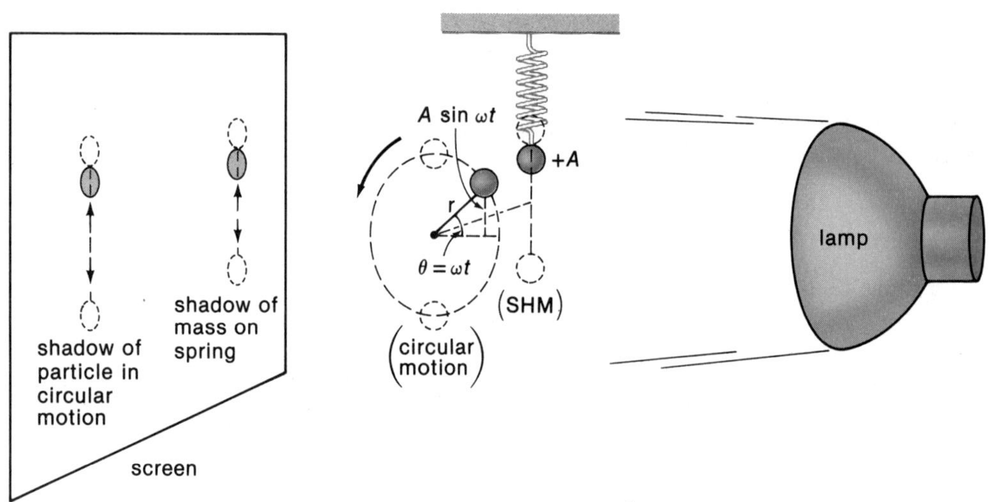

Figure 13.7
SHM and circular motion. Simple harmonic motion can be described in terms of the angular frequency of circular motion. Notice that the vertical displacement of the particle in circular motion is the same as that of a mass oscillating on a spring (with $r = A$), as indicated by the identical motions of the shadows on the screen.

circular motion, as illustrated in Figure 13.7. Notice that the vertical displacement of a particle moving on the reference circle is the same as that of a particle or mass oscillating on a spring with $r = A$. The motions of the particle shadows on the screen are identical.

Comparing Equations 13.6 and 13.9, it can be seen that for a mass on a spring $\omega = \sqrt{k/m}$.

EXAMPLE 13.3 A mass oscillates on a spring in SHM with an equation of motion of $y = (9.0 \text{ cm}) \sin 5t$. (a) What is the period of the oscillation? (b) What is the displacement of the mass at $t = 0.80$ s?

Solution: (a) Comparing the given equation of motion to the general form with the period T,

$$y = (9.0 \text{ cm}) \sin 5t$$

and

$$y = A \sin \frac{2\pi t}{T}$$

it can be seen that

$$\frac{2\pi}{T} = 5 \quad \text{or} \quad T = \frac{2\pi}{5} = 1.3 \text{ s}$$

(b) Substituting the given time of $t = 0.80$ s directly into the equation of motion,

$$y = (9.0 \text{ cm}) \sin 5t = (9.0 \text{ cm}) \sin 5(0.80)$$
$$= (9.0 \text{ cm}) \sin (4.0)$$
$$= -6.8 \text{ cm}$$

and the mass is 6.8 cm below its equilibrium position ($y = 0$). *Note well:* the angle or augment of the sine function is in radians. Remember to set your calculator to the RAD mode.

All of the preceding equations are equivalent. They describe a wave form with displacement $y = 0$ at $t = 0$. However, the general equation for simple harmonic motion is commonly written

$$\boxed{y = A \sin \left(\frac{2\pi t}{T} + \phi\right)} \quad \textbf{(Eq. 13.10)}$$

The angle of the sine function $\theta = (2\pi t/T) + \phi$ is called the *phase angle*, and ϕ is the **phase constant.**

The phase constant takes into account that the initial displacement may be other than zero. For example, suppose a mass on a spring is initially at $y = A$ (at $t = 0$), as illustrated in Figure 13.8(b). The wave form is then described by a cosine function, since $y = A \cos 2\pi t/T$ gives $y = A$ at $t = 0$. In this case, the phase constant is $\phi = 90°$. Letting $\phi = 90°$ in Equation 13.10 and using a trigonometric identity (see Appendix 3), it is easy to show that

$$y = A \sin\left(\frac{2\pi t}{T} + 90°\right) = A \cos \frac{2\pi t}{T}$$

We say that the waves described by $y = A \sin \omega t$ and $y = A \cos \omega t$ (where $\phi = 90°$) have a phase difference of 90°, or are 90° out of phase. The wave forms for $\phi = 180°$ and $\phi = 270°$ are also shown in Figure 13.8.

Essentially, the phase constant shifts the wave form in time. The phase difference for cases (b) and (d) in the figure is $\Delta\phi = 270° - 90° = 180°$, and the motions are said to be completely out of phase. This means that one mass would be going upward while the other is going downward, and vice versa. Notice that these wave forms are offset mirror images and that one has a "crest" when the other has a "trough."

If two masses oscillate together in unison, they are said to be in phase ($\Delta\phi = 0$).

Wave Energy

The total mechanical energy E of a wave disturbance is the sum of the kinetic and potential energies of the oscillating particles:

$$E = KE + PE$$

The potential energy is stored in the intermolecular "springs" as the particles move back and forth under the influence of the restoring forces.

The total energy can be expressed in terms of the amplitude of the vibration. Consider a mass oscillating on a spring. When the mass is at the maximum displacement or amplitude position ($y_{max} = A$), the total energy is all potential:

$$E = PE = \tfrac{1}{2}ky_{max}^2 = \tfrac{1}{2}kA^2 \quad \textbf{(Eq. 13.11)}$$

The total energy is proportional to the square of the amplitude. Hence,

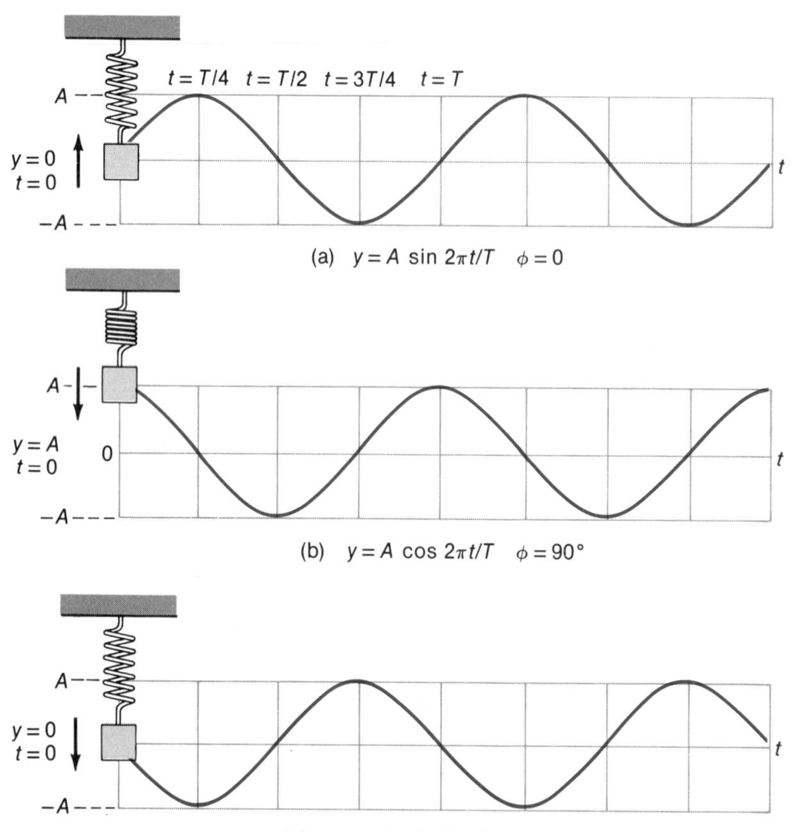

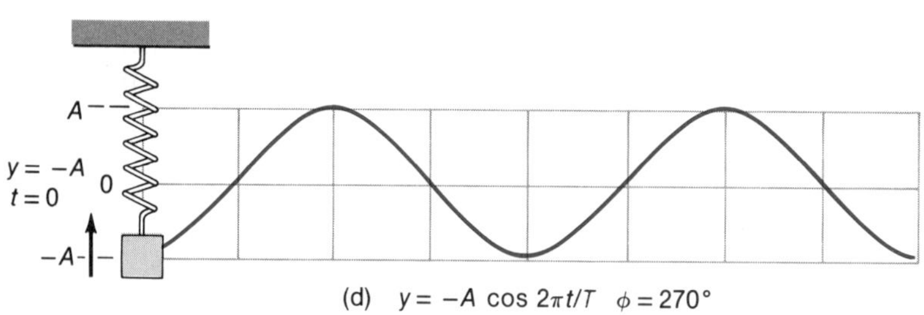

Figure 13.8
Phase constant. The phase constant essentially shifts the wave form in time and reflects different initial conditions. The waves in (a) and (c) are completely out of phase ($\Delta\phi = 180°$), as are the waves in (b) and (d).

$$E = \tfrac{1}{2}kA^2 = \tfrac{1}{2}mv^2 + \tfrac{1}{2}ky^2 = (KE) + (PE) \qquad \textbf{(Eq. 13.12)}$$

EXAMPLE 13.4 What is the magnitude of the maximum velocity of a mass of a spring oscillating in SHM in terms of the system parameters?

Solution: The maximum velocity $v_{\max}$ occurs when $y = 0$, or as the mass passes through the equilibrium position. Then, by Equation 13.12,

$$E = \tfrac{1}{2}kA^2 = \tfrac{1}{2}mv_{\max}^2$$

and

$$v_{\max} = \left(\frac{k}{m}\right)^{1/2} A = \omega A$$

[Since $f = 1/2\pi\ (k/m)^{1/2}$ and $2\pi f = \omega$, then $\omega = (k/m)^{1/2}$]

A constant periodic disturbance or driving force will cause the particles of a medium to oscillate continually. However, if the medium is no longer disturbed or driven, the vibrational motion "decays" as energy dissipates.

For example, a mass oscillating on a spring will eventually stop as energy is dissipated through some frictional mechanism. The wave motion of such processes, illustrated in Figure 13.9, is referred to as **damped harmonic motion.**

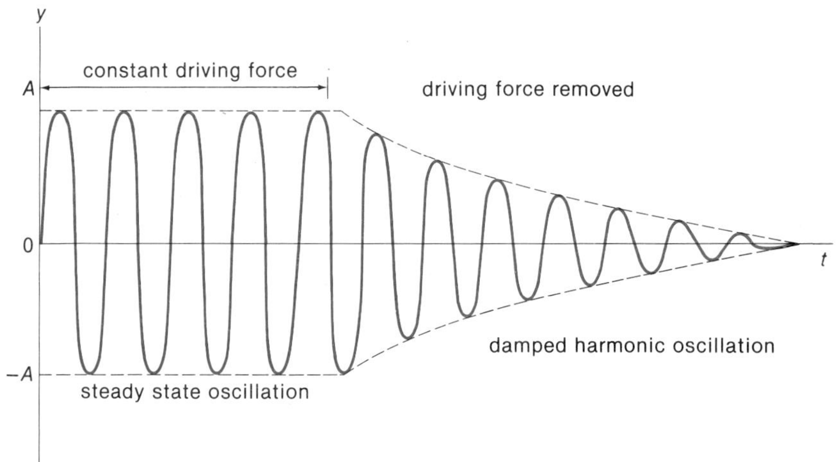

Figure 13.9
Damped harmonic oscillation. If the driving force is constant, the oscillations will be constant or in a steady state. When the driving force is removed, energy is dissipated and the oscillation is damped (decreasing amplitude).

Notice how the wave amplitude, and hence the energy, decrease with time: the more quickly the energy is lost, the greater the damping. Damping is important in many technical applications where it is desirable to prevent something from oscillating for very long. For example, on weighing scales, where quick readings are desired, the mechanism must be properly damped or the indicator will oscillate for some time.

13.3 Wave Characteristics

The speed with which a wave passes through a medium depends on the elastic properties of the material. For example, as we have seen, the speed of a wave in a stretched string depends on the tension in the string and the linear mass density of the string (Eq. 13.1). Other relationships for the speed of disturbances in materials are considered in Chapter 14 on sound. As might be expected, these involve the elastic moduli of the materials. However, the wave speed can be expressed in terms of the length and time parameters of the wave using the relationship $v = d/t$.

To characterize a length in a wave, the concept of **wavelength (λ)** is used. As illustrated in Figure 13.10, this is the distance between two adjacent crests or troughs in a wave. Notice that a sinusoidal wave form can be used to describe a longitudinal wave, in which a compression region corresponds to a crest and an expansion region to a trough. Technically,

The wavelength λ of a periodic wave is the distance between adjacent particles that are in phase.

The particles at the crests or troughs are in phase; that is, they have a zero phase difference ($\Delta\phi = 0$).

The time for a wave crest (or trough) to travel a distance λ is just the period T of the particle oscillation. Hence, we may write the **wave speed v:**

$$v = \frac{d}{t} = \frac{\lambda}{T}$$

or, with $1/T = f$,

$$\boxed{v = \lambda f} \qquad \textbf{(Eq. 13.13)}$$

EXAMPLE 13.5 A periodic wave has a wavelength of 30 cm and a frequency of 1.0 kHz. What is the wave speed?

Solution: With $\lambda = 30$ cm $= 0.30$ m and $f = 1.0$ kHz $= 10^3$ Hz ($= 10^3$/s),

$$v = \lambda f = (0.30 \text{ m})(1.0 \times 10^3/\text{s}) = 3.0 \times 10^2 \text{ m/s}$$

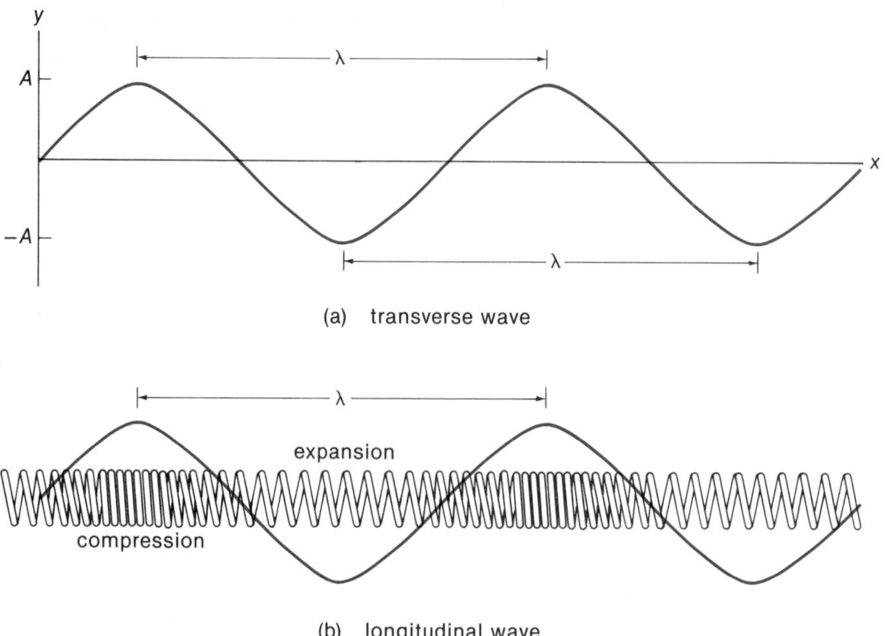

Figure 13.10
Wavelength. The wavelength is the distance between adjacent particles that are in phase. (a) For a transverse wave, this may be conveniently taken as the distance between adjacent crests or troughs. (b) For a longitudinal wave, the regions of compressions and expansions correspond to crests and troughs, respectively.

EXAMPLE 13.6 A small stick floating in a lake is observed to bob up and down 15 times in 60 seconds owing to a periodic surface disturbance. If the wave crests are 4.0 ft apart, what is the wave speed of the disturbance?

Solution: The stick undergoes 15 cycles in 60 s, so the frequency of the wave oscillation is

$$f = \frac{15 \text{ cycles}}{60 \text{ s}} = 0.25 \text{ Hz}$$

Then, with a wavelength (crest-to-crest distance) of $\lambda = 4.0$ ft,

$$v = \lambda f = (4.0 \text{ ft})(0.25 \text{ Hz}) = 1.0 \text{ ft/s}$$

Thus far, we have considered only waves propagating in media. But what happens when a wave meets a boundary or another wave in a medium? When a wave propagating in one medium strikes the boundary of another medium, the wave is partially reflected and partially transmitted. The degrees of reflection and transmission depend on the elasticity of the second medium.

Suppose the second medium is completely inelastic. The incident wave is then totally reflected. This can be illustrated by a wave in a string that has its fixed end attached to a rigid support (Fig. 13.11). The wave exerts a force on the support, and the reaction force of the support on the string causes the reflected wave to be 180° out of phase with the incident wave.

A wave reflected from a rigid boundary is 180° out of phase with the incident wave.

When the second medium is elastic, some of the wave energy is transmitted, and a disturbance propagates in the second medium. The rest of the incident disturbance energy is reflected. However, in this case, because of the elasticity of the second medium, the reflected wave is not inverted or out of phase with the incident wave. This is analogous to the reflection in a string with a movable support end (Fig. 13.11).

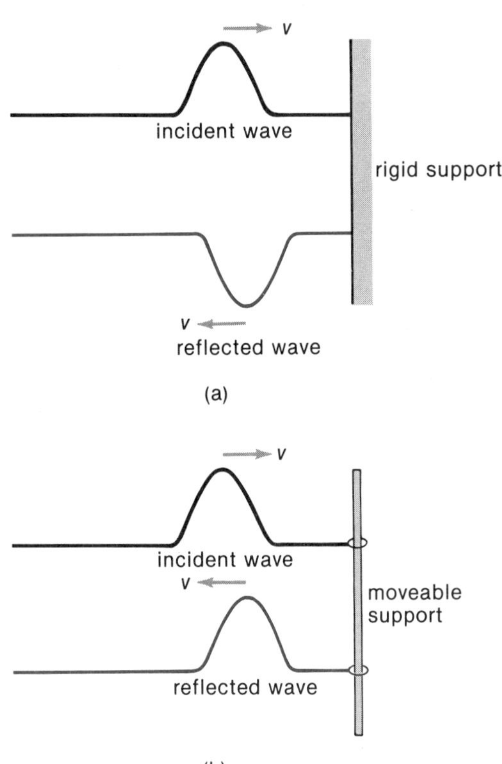

Figure 13.11
Reflection. (a) A wave reflected from a rigid support is inverted or 180° out of phase with the incident wave. (b) A wave reflected from a moveable support is not inverted and has the same phase as the incident wave.

Two or more disturbances can propagate in a medium at the same time. Often, the waves travel in directions such that they meet each other. When this occurs, we say the waves *interfere*. The combined wave form is then given by the **principle of superposition:**

> **When two or more waves interfere in a medium, the individual particle displacement is given by the vector sum of the displacements of each particle.**

Hence, the resultant wave form is the sum of the component waves.

Consider two wave pulses of equal amplitude traveling in opposite directions (Fig. 13.12). If both waves have positive displacements, the combined wave form in the superposition region is larger than either component

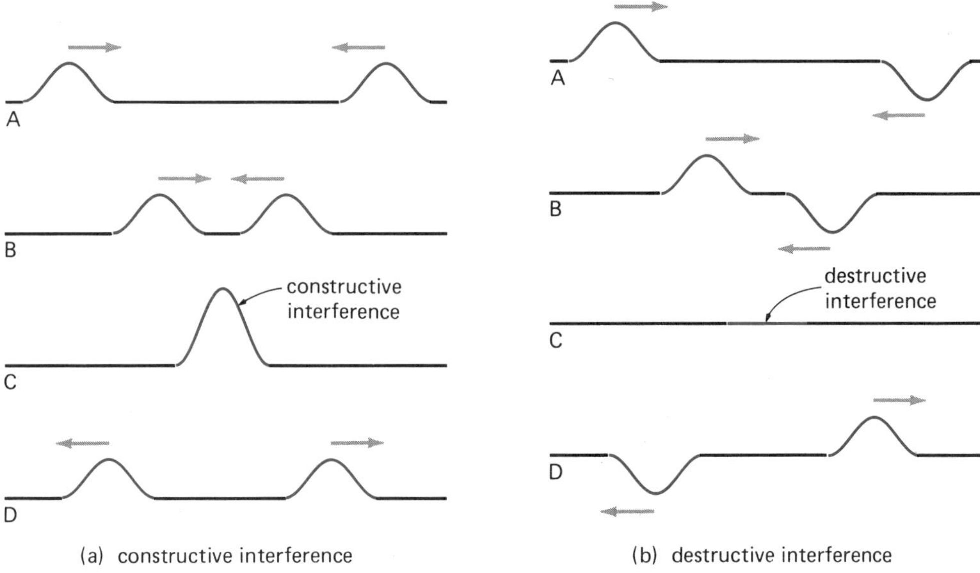

Figure 13.12
Interference and superposition. (a) When two wave pulses in phase interfere, the displacement of the combined wave is greater than either component wave, which is called constructive interference. (b) When two waves 180° out of phase interfere, the displacement of the combined wave is smaller than either component wave. This is called destructive interference.

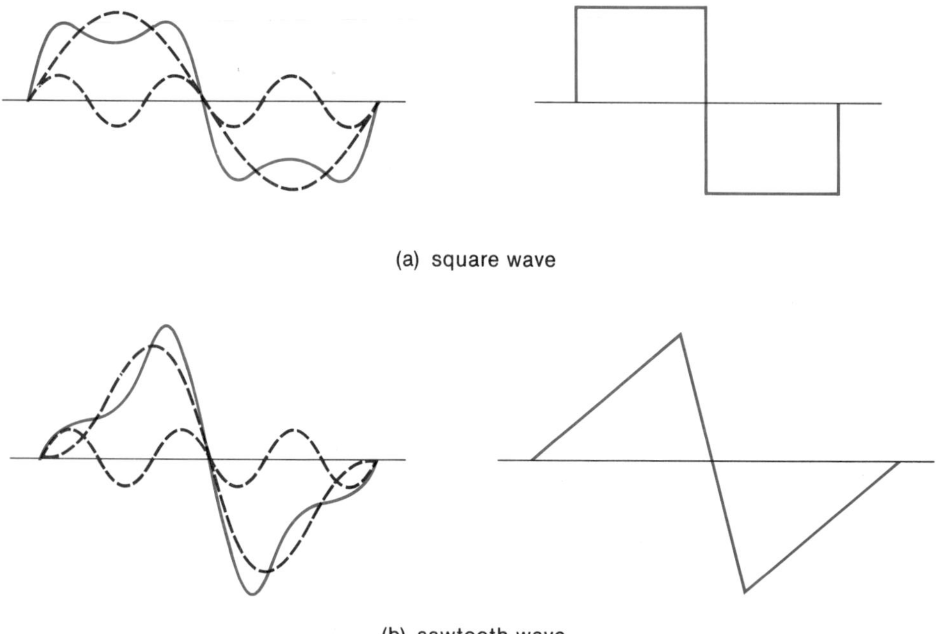

Figure 13.13
Wave forms. The diagrams illustrate how different wave forms—a square wave (a) and a double sawtooth wave (b)—can be formed by the superposition of sinusoidal waves. The superposition of two waves (dashed lines) gives the wave forms shown by the solid lines. A series of appropriately selected waves will give the desired wave form.

wave. This condition is referred to as **constructive interference.** When the wave forms are exactly superimposed, the amplitude of the combined wave is twice that of one individual wave ($A + A = 2A$). This is called *total* constructive interference.

If one of the waves has a positive displacement and the other a negative displacement, the combined wave form in the superposition region is smaller than either component. This is called **destructive interference.** When the wave forms are exactly superimposed, the wave components completely "cancel" each other in *total* destructive interference. This does not mean that energy is destroyed or lost. The medium particles still have energy. Notice that after interfering, the waves travel on as if nothing had happened.

If the frequencies and amplitudes of the interfering waves are not equal, the superposition can be quite complicated. However, by selecting many sinusoidal waves of the proper frequencies, amplitudes, and phases, the superposition principle allows the "sculpting" of various wave forms. Two such wave forms that are important in electrical applications are shown in Figure 13.13.

Question: If the total constructive interference of two identical wave pulses gives a wave form with twice the amplitude ($A + A = 2A$), does this mean that the total energy is quadrupled instead of doubled? Note: $E = E_1 + E_2 \propto A^2 + A^2 = 2A^2$, but with twice the amplitude, $E \propto (2A)^2 = 4A^2$.

Answer: No, this would violate the principle of conservation of energy. For a pulse in a string, the potential energy is associated with the stretching of the string and the kinetic energy with the vertical motion. During the process of interference or overlapping, a segment may not be stretched, but have twice the velocity. Hence, the kinetic energy is quadrupled, but the potential energy is zero. In other segments, the opposite situation occurs. For identical pulses, either the kinetic energy or the potential energy is zero in each overlap-

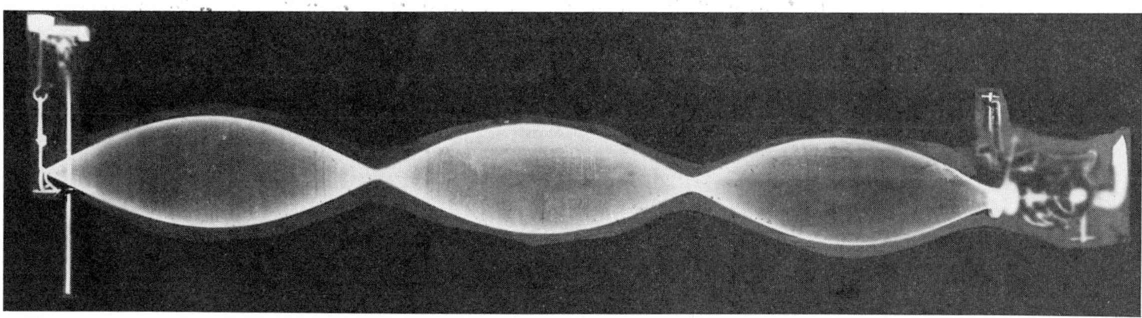

Figure 13.14
A standing wave in a string. (Courtesy of VWR Scientific.)

ping region, so the total energy can only double.

13.4 Standing Waves

When shaking the end of a stretched string or rubber cord, you may have noticed that if the string is shaken just right, the wave form appears to "stand" in the cord (Fig. 13.14). This condition arises from interference between waves traveling in opposite directions—the wave traveling away from your hand and the reflected wave (traveling in the opposite direction) from the fixed end of the string.

As we learned, the reflected wave is 180° out of phase with the incident wave. Using the principle of superposition to analyze the situation at various instants of time, we find that the displacement for certain positions along the string is always zero. The positions of zero displacement are called **nodes** (Fig. 13.15). For a string fixed at each end, the boundary conditions require a node at each end of the string. Other

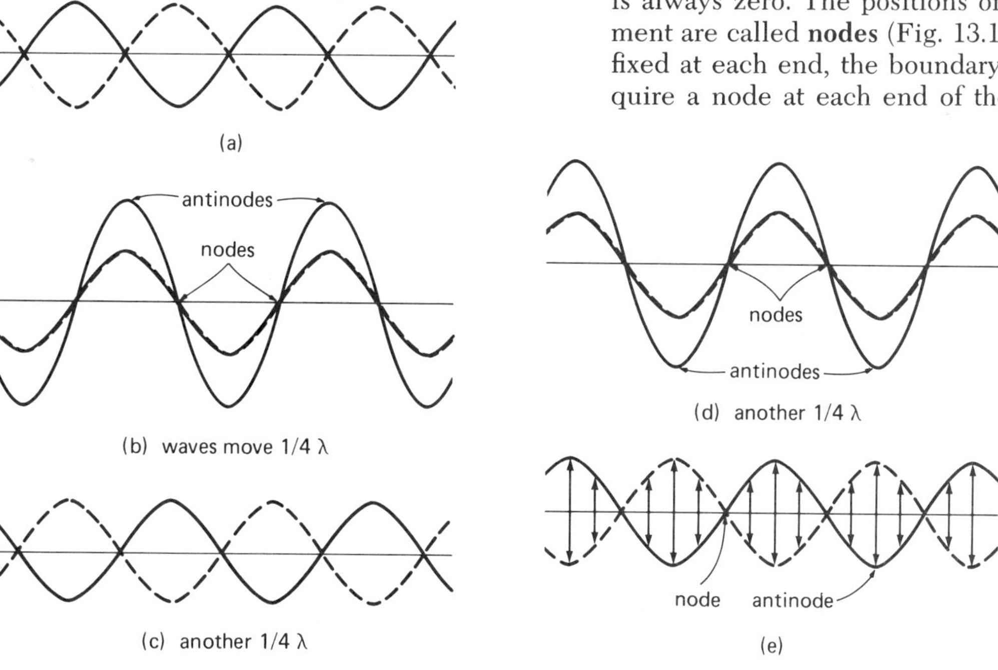

Figure 13.15
Traveling wave interference. Waves traveling in opposite directions in a string are continuously in and out of phase (a) to (d). Interfering waves of equal amplitude and wavelength give rise to standing waves (e). The positions of zero displacement are called nodes, and the positions of maximum displacement are called antinodes.

positions along the string are found to have maximum displacements. The positions of maximum displacements are called **antinodes.**

The effect of the interference of the waves with the same amplitude and frequency traveling in opposite directions is a so-called **standing wave.** Since the nodal positions are not in motion, the energy is essentially "standing" in the string as the string particles vibrate up and down in SHM between the nodal positions. Notice that the distance between adjacent antinodes or adjacent nodes is one half of the wavelength ($\lambda/2$) of the standing wave.

The standing wave shown in Figure 13.14 is just one mode of vibration for a string fixed at both ends. There are other standing wave modes that satisfy the boundary conditions of a node at each end (Fig. 13.16). As can be seen from the figure, the boundary conditions are satisfied by "fitting in" an integral number of half wavelengths $\lambda/2$ for the given length L of the string, i.e.,

$$L = 1\left(\frac{\lambda_1}{2}\right), \quad L = 2\left(\frac{\lambda_2}{2}\right),$$

$$L = 3\left(\frac{\lambda_3}{2}\right), \quad L = 4\left(\frac{\lambda_4}{2}\right), \ldots$$

Hence, the allowable wavelengths are

$$\lambda_1 = \frac{2L}{1}, \quad \lambda_2 = \frac{2L}{2},$$

$$\lambda_3 = \frac{2L}{3}, \quad \lambda_4 = \frac{2L}{4}, \ldots$$

or

$$\lambda_n = \frac{2L}{n} \qquad n = 1, 2, 3, 4, \ldots \qquad \textbf{(Eq. 13.14)}$$

Knowing the wave speed v and using the relation $v = f\lambda$, we identify the corresponding frequencies of vibration as

$$\boxed{f_n = \frac{v}{\lambda_n} = \frac{nv}{2L}} \qquad \textbf{(Eq. 13.15)}$$

$$n = 1, 2, 3, 4, \ldots$$

The speed is the same for all the frequencies, since it depends on the particular characteristic of the medium. As a result, the frequencies f_n are called the **natural or characteristic frequencies** of the normal modes of vibration.

Using Equation 13.1 for the wave speed in a stretched string, we can write the characteristic frequencies

$$f_n = \frac{v}{\lambda_n} = \frac{nv}{2L} = \frac{n}{2L}\sqrt{\frac{F}{\mu}} \qquad n = 1, 2, 3, 4, \ldots$$

$$\textbf{(Eq. 13.16)}$$

where F is the tension in the string and μ is the linear mass density ($\mu = m/L$).

The lowest frequency f_1 is called the **fundamental frequency**, and the higher frequencies, f_2, f_3, etc., are called **overtones.** The first overtone is f_2, the second overtone f_3, and so on. The complete series of characteristic frequencies is called a **harmonic series.** The first harmonic is the fundamental frequency ($n = 1$), the second harmonic is the first overtone ($n = 2$), and so on.

EXAMPLE 13.7 What are (a) the fundamental frequency and (b) the second overtone frequency for a string with a length of 2.5 m, fixed at both ends, if the wave speed in the string is 20 m/s?

Solution: It is given that $L = 2.5$ m and $v = 20$ m/s. (a) Then, for $n = 1$,

$$f_1 = \frac{v}{2L} = \frac{20 \text{ m/s}}{2(2.5 \text{ m})} = 4.0/\text{s} = 4.0 \text{ Hz}$$

(b) The second overtone is the third harmonic ($n = 3$), and

$$f_3 = \frac{3v}{2L} = \frac{3(20 \text{ m/s})}{2(2.5 \text{ m})} = 12 \text{ Hz}$$

Notice that

$$f_3 = 3f_1 = 3(4.0) = 12 \text{ Hz}$$

Standing waves are also set up in media with compressional waves. Standing waves of this type will be considered in the next chapter.

Resonance

The characteristic frequencies of a material or oscillator are the *natural* frequencies of oscillation. These frequencies depend on such factors as the restoring force and the geometry of the oscillator. In a sense, these are the frequencies at which the oscillators "want to" oscillate.

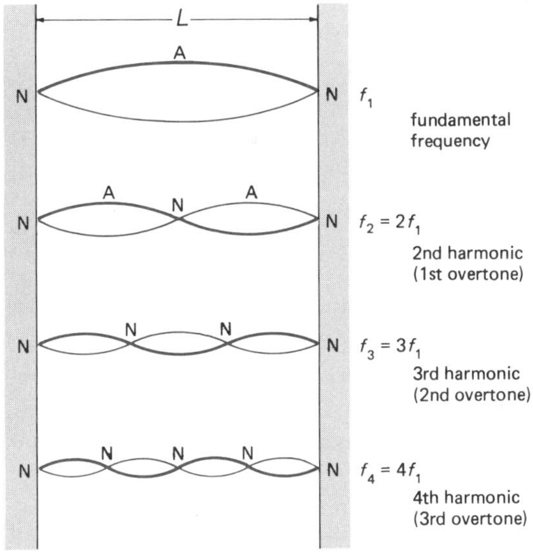

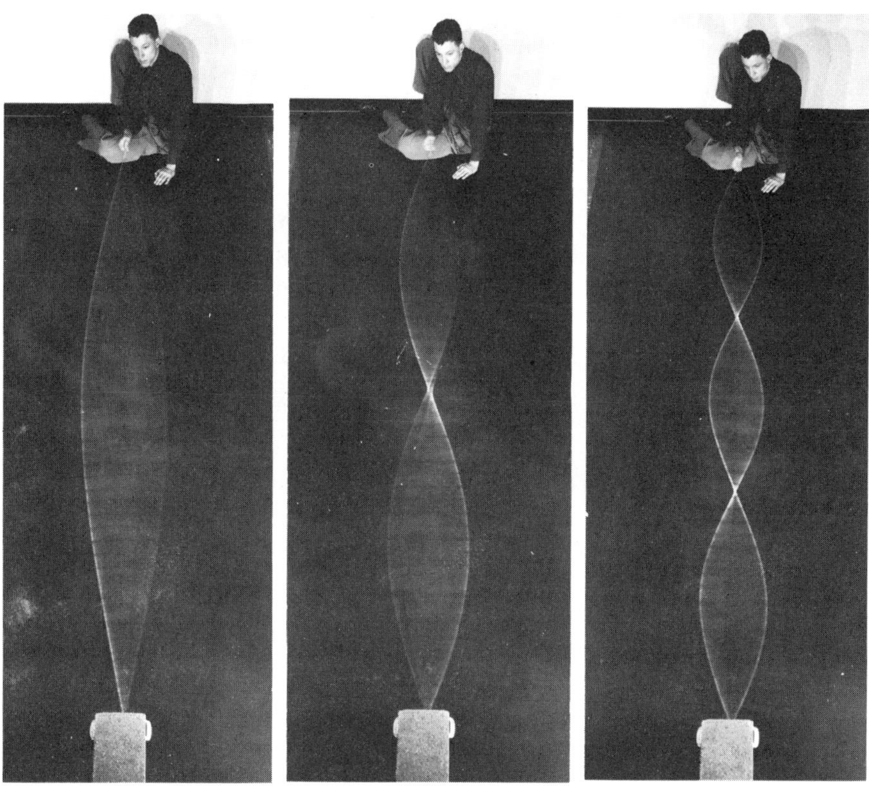

Figure 13.16

Characteristic frequencies of standing waves in a stretched string. The string must have nodes at each end. "Fitting in" different possible numbers of half-wavelengths gives rise to particular harmonic frequencies.

> **When an oscillator or medium is driven by a driving force with a frequency corresponding to a particular characteristic frequency, we say that the oscillator is driven in resonance. At resonant frequency, there is maximum energy transfer from the driving source to the oscillator.**

This is not to say that oscillations will not occur at other driving frequencies. However, the oscillations may be erratic, and maximum energy transfer will not occur. For example, if one end of a stretched string is shaken with a frequency other than one of its characteristic frequencies, it will oscillate wildly. Basically, the driving force is out of phase with the string oscillations; for example, the string might move upward when the applied force is downward. But when driven at a particular characteristic frequency, the string oscillates smoothly in standing wave form as maximum energy is transferred from the driving source to the string.

Another example of a resonance condition is provided by the proper pushing of a person on a swing. A swing is basically a simple pendulum. A simple pendulum has only one characteristic frequency—i.e., a fundamental frequency—which depends on the length of the pendulum. [Recall that for a pendulum $f = (1/2\pi)(g/L)^{1/2}$.]

When the swing is pushed with this frequency, it swings higher and higher (greater amplitude) due to maximum energy transfer. We have all experienced what happens if we change the driving frequency slightly and push the swing before it has reached the return amplitude of its swing.

An example of unwanted resonance is discussed in Special Feature 13.2. Driven in resonance by gusting winds, the Tacoma Narrows Bridge collapsed in 1940.

The maximum energy transfer of the resonance condition is also important in electrical circuits, as discussed in Chapter 25.

Important Terms

wave the propagation of energy in a medium due to a disturbance without a net transfer of matter (electromagnetic waves are an exception)

transverse (shear) wave a wave in which the particle motion of the medium is perpendicular to the direction of wave propagation

longitudinal (compressional) wave a wave in which the particle motion of the medium is parallel to the direction of wave propagation

simple harmonic motion (SHM) particle motion under the influence of a linear restoring force ($F = -kx$)

displacement the distance and direction of a particle from its equilibrium position ($y = 0$)

amplitude the maximum displacement of oscillation

period the time required for one complete oscillation

frequency the number of cycles or oscillations per second

phase constant the constant that takes into account the initial conditions of harmonic motion

damped harmonic motion wave motion with decreasing amplitude due to some frictional or loss-of-energy mechanism

wavelength the distance between adjacent particles that oscillate in phase

wave speed the time rate of change of position for a wave

principle of superposition the principle that, when two or more waves interfere in a medium, the individual particle displacement is given by the vector sum of the displacements

constructive interference the principle that the vector sum of the displacements is greater than the individual particle displacements

destructive interference the principle that the vector sum of the displacements is less than the individual particle displacements

standing wave the interference of two waves of equal amplitude and frequency traveling in opposite directions

node a point of zero displacement in a standing wave

antinode a point of maximum displacement in a standing wave

characteristic frequencies the natural or allowed frequencies of a system

fundamental frequency characteristic frequency for $n = 1$

overtones and harmonics characteristic frequencies for $n > 1$; the first overtone ($n = 2$) is the second harmonic, and so on

resonant frequency a characteristic frequency of an oscillator for which there is maximum energy transfer when driven at this frequency (condition of resonance)

SPECIAL FEATURE 13.2

Galloping Gertie: The Tacoma Narrows Bridge Collapse

The Tacoma Narrows Bridge near Tacoma, Washington, was opened to traffic on July 1, 1940. It was 2800 ft (855 m) long and 39 ft (12 m) wide. The steel girders used in its construction were 8 ft (2.4 m) tall. During its first several months in use, many transverse modes of vibration were observed.

On the morning of November 7, 1940, winds gusting up to 40–45 mi/h started the main span vibrating, with nodes, of course, at the main towers and at several places in between. The vibrational frequency was 36 vib/min with an amplitude of 1.5 ft (Fig. 13.2). At 10 A.M. the main span of the bridge began to vibrate in a torsional (twisting) mode in two segments with a frequency of 14 vib/min. This was apparently due to the loosening of a cable by which the roadway was suspended. Twisting in the wind, the bridge was nicknamed "Galloping Gertie."

The wind drove the structure near the critical velocity for the torsional mode, and this vibration built through resonance. Shortly after 11 A.M., the main span broke and collapsed.

The bridge was rebuilt using the original anchorages and tower foundations, but with a new design that stiffened the structure and increased the resonance frequency so that high winds would not set it into resonance vibrations–and the design worked.

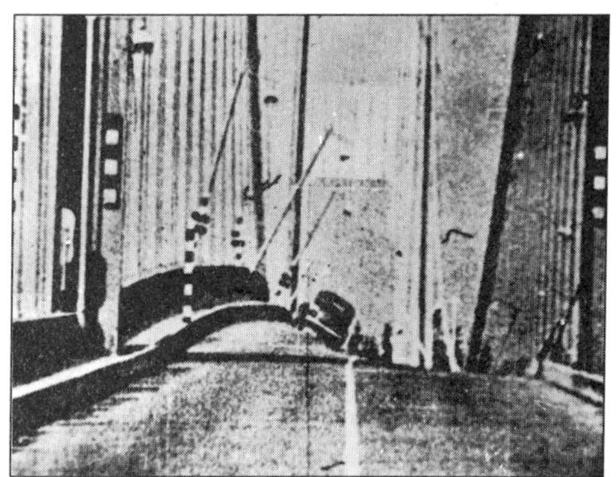

Figure 1
Tacoma Narrows bridge collapse (1940). Gale winds set the bridge into resonance (torsional) vibration. The bridge, nicknamed "Galloping Gertie," oscillated for almost an hour, then collapsed. (Courtesy of R. Stevenson and R. B. Moore.)

Important Formulas

wave speed: $v = \sqrt{\frac{F}{\mu}}$ (in stretched string)

period:

mass on spring $T = 2\pi\sqrt{\frac{m}{k}}$

simple pendulum (small angle oscillations) $T = 2\pi\sqrt{\frac{L}{g}}$

frequency: $f = \frac{1}{T}$

angular frequency: $\omega = 2\pi f = \frac{2\pi}{T}$

$\omega = \sqrt{\frac{k}{m}}$ (for spring)

SHM motion equation:

$$y = A\sin\left(\frac{2\pi t}{T} + \phi\right)$$
$$= A\sin(2\pi ft + \phi)$$
$$= A\sin(\omega t + \phi)$$

wave energy:

$$E = \tfrac{1}{2}kA^2 = \tfrac{1}{2}mv^2 + \tfrac{1}{2}ky^2 = (KE) + (PE)$$

wave speed: $v = \frac{\lambda}{T} = \lambda f$

characteristic frequencies (in stretched string):

$$f_n = \frac{v}{\lambda_n} = \frac{nv}{2L}$$

$n = 1, 2, 3, 4, \ldots$

Questions

Waves and Wave Motion

1. How are energy propagation and wave motion related in a material?
2. Why are transverse waves called shear waves, and longitudinal waves called compressional waves?
3. If two stretched strings of different mass densities have the same tension, in which string will the speed of wave propagation be greater?
4. When a musician tightens a string on a stringed instrument, what effect does this have on (a) the speed of the wave propagation and (b) the frequency?
5. Discuss whether transverse waves and/or longitudinal waves will propagate through (a) a solid, (b) a liquid, and (c) a gas.
6. What are seismic waves, and what is the difference between S and P waves?
7. Are water waves longitudinal or transverse? Explain. How does a bottle or twig bobbing up and down in a lake demonstrate that waves carry energy?

Periodic Motion

8. An object oscillates in SHM. (a) What is the requirement for SHM? (b) What does the H in "SHM" stand for and what does it mean?
9. What is the unit of frequency in terms of fundamental properties?
10. For a pendulum swinging in SHM (small arcs), what positions of the swing correspond to the amplitude and zero points of the sine curve that describes the motion? What is the total energy at these points?
11. If a mass oscillating on a spring is given more energy by a driving force, how does this affect the (a) amplitude, (b) frequency, and (c) period of oscillation? Explain.
12. Two masses oscillate up and down on springs with the same amplitude and frequency. What can you say about the spring constants of the springs? Compare the motions when they are (a) in phase, (b) one-fourth cycle out of phase, and (c) completely out of phase. (Draw some diagrams and show where the masses are relative to each other at particular times.)
13. Give some examples in which damped harmonic motion is undesirable and some examples in which such motion is desirable.

Wave Characteristics

14. When a motorboat moves across a lake, water waves lap against the shore for some time after it has passed. But few or no waves reach the shore from a rowboat. Why is this?
15. With the wave speed constant in water, how is the distance between the crests of the surface waves affected when the disturbance frequency increases?
16. Curves representing two waves are shown in Figure 13.17. Which wave has the greater (a) amplitude, (b) frequency, (c) wavelength, and (d) period? Write the general equation of motion for each wave (without phase constant).
17. What is the phase relationship between the incoming and outgoing waves reflected from (a) an elastic medium and (b) an inelastic medium?
18. Distinguish between constructive and destructive wave interference. Is energy destroyed in destructive interference? Explain. What is "destroyed"?

Standing Waves

19. How many natural frequencies are there for a mass oscillating on a spring? Explain.

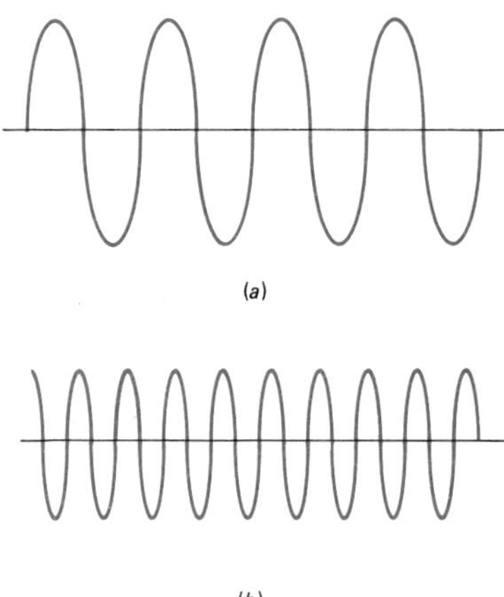

Figure 13.17
Different frequencies. See Question 16.

20. When a stretched rope is oscillated at its fourth harmonic frequency, how many standing wavelengths are formed in the rope?
21. A long, limber rod or hacksaw blade is clamped at one end, leaving the other end free to vibrate. The boundary conditions for a standing wave in the rod are a node and antinodes at the respective ends. In this case, only the odd harmonics are possible, e.g., f_1, f_3, f_5, and so on. Draw a few possible standing wave forms and demonstrate this in terms of wavelength.
22. While sitting on a swing that you have agreed to push, a fellow student informs you that the resonant frequency of the swing is 0.33 Hz. How often should you push the swing to give a smooth, high ride?
23. Explain any resonance phenomenon involved in bouncing a basketball (Fig. 13.18). What would be the case if the big fellow in the figure stooped over and bounced the ball?
24. A swing has only one characteristic frequency, f_1. However, it can be pushed smoothly at frequencies of $f_1/2$, $f_1/3$, $f_1/4$, and so on. Explain why, and describe the effects of the different driving frequencies.

Problems

Levels of difficulty are indicated by asterisks for your convenience.

13.1 Waves and Wave Motion

1. Show that $\sqrt{F/\mu}$ has the units of speed.
2. A stretched string has a linear mass density of 0.10 kg/m. If the tension in the string is 20 N, what is the speed of a transverse wave in the string?

*3. A 10-ft length of rubber cord has a weight of 1.6 lb. If the cord is stretched with a force of 8.0 lb, with what speed will a transverse disturbance travel in the cord?

*4. If it is desired for a wave pulse to travel with a speed of 30 ft/s in the cord in Problem 3, what is the required tension in the cord?

*5. A 4.0-m string stretched with a force of 20 N transmits waves at a speed of 15 m/s. What is the mass of the string?

*6. Two strings have mass densities of 0.0125 kg/m and 0.0500 kg/m, respectively. If both strings are stretched with equal tensions, in which string will the wave speed be greater, and how many times greater?

**7. Two rubber cords have weight densities of 0.75 lb/ft and 0.45 lb/ft, respectively. If the wave speed is the same in each, which cord has the greater tension, and how many times greater?

**8. By how much must the tension in a stretched string be increased in order to double the speed of waves traveling in the cord?

13.2 Periodic Motion

9. If an object oscillates with a frequency of 10 Hz, how long does it take the object to make two complete cycles?
10. Suppose it takes you 20 minutes to walk around the block twice. (a) What is the period of your motion? (b) What is your frequency in hertz? What does this mean?
11. A mass oscillates with a frequency of 5.0 Hz. How many cycles does it go through (a) in 2.0 s and (b) in 0.60 s?

*12. A mass of 0.10 kg oscillates on a spring with a spring constant of 2.3 N/m. What are (a) the period and (b) the frequency of oscillation of the mass?

*13. A 0.50-kg mass suspended from a spring stretches the spring 12 cm. If the mass is set into oscillation, what is the period of oscillation?

*14. A simple pendulum has a length of 0.50 m and a bob with a mass of 0.25 kg. What are (a) the period and (b) the frequency of the pendulum's swing for small angles of oscillation?

*15. What is the required length of a "one-second" pendulum? (Thomas Jefferson suggested that such a pendulum be used as a time standard.)

*16. If the length of a pendulum is shortened from 24 inches to 18 inches, by what factor does the period change?

*17. A mass oscillates on a spring with an amplitude of 4.0 cm. What distance does the mass travel in one period?

Figure 13.18
Dribbling in resonance. See Question 23.

*18. The equation of motion of a particle oscillation is $y = 10 \cos 2\pi t/T$ (cm). Where is the particle at (a) $t = 0$ and (b) $t = T/4$ relative to its equilibrium position?

*19. If the period of oscillation in Problem 18 is 6 s, what is the particle displacement at (a) $t = 1$ s, (b) $t = 3$ s, (c) $t = 4$ s, and (d) $t = 6$ s? *Hint:* don't forget to set your calculator on radians.

*20. The equation of motion describing an SHM is $y = 8.0 \sin 10t$ (cm). (a) What is the frequency of oscillation? (b) What is the displacement at $t = T/3$?

*21. An SHM is described by $y = -5.0 \cos \omega t$ (cm). If the frequency of oscillation is 4 Hz, what is the displacement at (a) $t = 0.50$ s, (b) $t = 1.0$ s, and (c) $t = 4.0$ s? *Hint:* don't forget to set your calculator on radians.

*22. What is the equation of motion for a mass in SHM with an amplitude of 10 cm and a frequency of 5.0 Hz when the phase constant is (a) 270°, (b) π rad, and (c) 45°?

*23. Write the equation of motion for the wave form shown in Figure 13.19. (a) What are the frequency and angular frequency of the wave? (b) What is the phase constant?

*24. Write the wave equations for the waves that are (a) +90° out of phase and (b) completely out of phase with the wave shown in Figure 13.19.

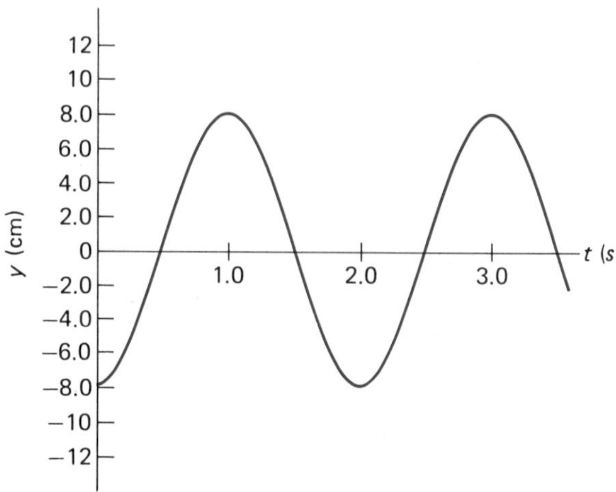

Figure 13.19
See Problem 23.

*25. The motion of a 0.15-kg mass oscillating on a spring with a spring constant of 150 N/m is described by $y = (0.12 \text{ m}) \cos \omega t$. What is the total energy of the mass?

*26. What are the position and magnitude of the maximum velocity of the mass in Problem 25?

**27. A 500-g mass on a spring with a spring constant of 240 N/m^2 is pulled down 15 cm below its equilibrium position and released. (a) What is the total energy of the oscillating mass? (b) What is its maximum velocity? (c) What is the kinetic energy of the mass at $t = 1.26$ s?

**28. The equation of motion of a 0.75-kg mass oscillating on a spring is $y = (0.25 \text{ m}) \sin 8t$. (a) What is the total energy of the mass? (b) What is its speed at $t = 1.7$ s?

**29. The pendulum of a grandfather clock is 0.35 m long, and the clock gains 5.0 min each hour (reads 65 min in a true 60-min interval). (a) Should the pendulum length be made shorter or longer? (b) By how much should the pendulum length be effectively changed so the clock reads correctly? (Consider a simple pendulum.)

**30. The period of oscillation of a torsional pendulum is given by $T = 2\pi\sqrt{I/k'}$ (where the k' is from the rotational analog of the linear case of Hooke's law, $F = -kx$, or $\tau = -k'\theta$, and I is the moment of inertia). For a particular torsional pendulum with a disk of mass 4.0 kg and a radius of 20 cm, a force of 28 N applied tangentially to the rim of the disk produces an angular displacement of 0.035 rad. When the force is removed, the pendulum oscillates in SHM. What are (a) the period and (b) the frequency of oscillation?

13.3 Wave Characteristics

31. A wave has a period of 0.40 s and a wavelength of 20 cm. What is the wave speed?

32. What is the wavelength of a wave with a frequency of 200 Hz and a wave speed of 50 m/s?

*33. Radio station frequencies (radio waves) are in the 540–1600 kHz and 88–108 MHz ranges. What are the corresponding wavelengths of these frequency ranges? (The speed of electromagnetic radio waves is 3.0×10^8 m/s.)

*34. After a motorboat passes by on a lake, an observer on the shore notices that the waves hit the shore about every 2.0 s and that the distance between the crests of the waves is about 1.5 m. (a) What is the approximate speed of the water waves? (b) Does this depend on the speed of the boat? Explain.

*35. Light waves in vacuum travel at a speed of 300,000 km/s. The frequency of visible light is on the order of 10^{14} Hz. What is the order of magnitude of the wavelength of visible light?

*36. A fisherman observes the bobber on his line to bob up and down 20 times in 30 s due to a periodic surface disturbance. If the wave crests are 60 cm apart, what is the wave speed of the disturbance?

13.4 Standing Waves

*37. The wave speed in a stretched string of length 150 cm is 60 cm/s. What are (a) the fundamental frequency and (b) the frequency of the first overtone of the string?

*38. A 10-ft rubber cord with a linear weight density of 0.080 lb/ft is stretched with a force of 20 lb. What are (a) the fundamental frequency and (b) the frequency of the third harmonic?

*39. A cord 5.0 m long has a mass of 0.25 kg and is stretched with a tension of 40 N. (a) What is the fundamental frequency of the cord? (b) What are the frequencies of the second harmonic and the second overtone?

*40. A pendulum has a length of 2.0 ft. (a) What is the fundamental frequency of the pendulum? (b) What is the second harmonic? Explain.

*41. A 2.4-m string vibrating as in Figure 13.14 has four segments. If the tension in the string is supplied by a mass of 200 g and the mass of the string is 50 g, what is the frequency of the standing wave?

**42. A thin, flexible plastic rod of length L is clamped at one end. What are the characteristic frequencies with which the rod would vibrate if the wave speed in the rod were v?

(Courtesy Memorex Corporation)

Chapter 14
Sound

What is sound? A common answer is that sound is what we hear. Physiologically, sound *is* the hearing sensation of the ear. But sound also has a broader meaning. Physically, sound is the transmission of energy or the propagation of waves in matter. In this sense, sound exists even if there is no ear to hear it. Thus, how would you answer the old question: If a tree falls in the forest and no one is there to hear it, is there sound?

In this chapter we study the nature of sound and sound phenomena. Also, some technical applications of sound are presented.

14.1 The Nature of Sound

In the physical sense, **sound is any longitudinal disturbance in an elastic medium.** For example, a vibrating tuning fork in air produces compressional or longitudinal waves that propagate outward from the source (Fig. 14.1).

The regions of compression (increased pressure and density) are called *condensations*, and the regions of expansion (decreased pressure and density) are called *rarefactions*. We hear these disturbances if the frequency of vibration is in the audible range of the human ear and if the wave has sufficient energy or intensity.

The range of the frequency response of the human ear, which is approximately from 20 Hz to 20 kHz (20,000 Hz), provides for general regions of the **sound spectrum** (Fig. 14.2). The frequency range of human hearing is called the **audible region.** Below the audible region is the **infrasonic region** (< 20 Hz), and above the audible region is the **ultrasonic region** (> 20 kHz).

Other animals have different hearing frequency responses. For example, dogs can hear sounds in the ultrasonic region. Whistles with frequencies in the ultrasonic range, which do not disturb humans, are sometimes used to call dogs. Also, in some regions cars and trucks are

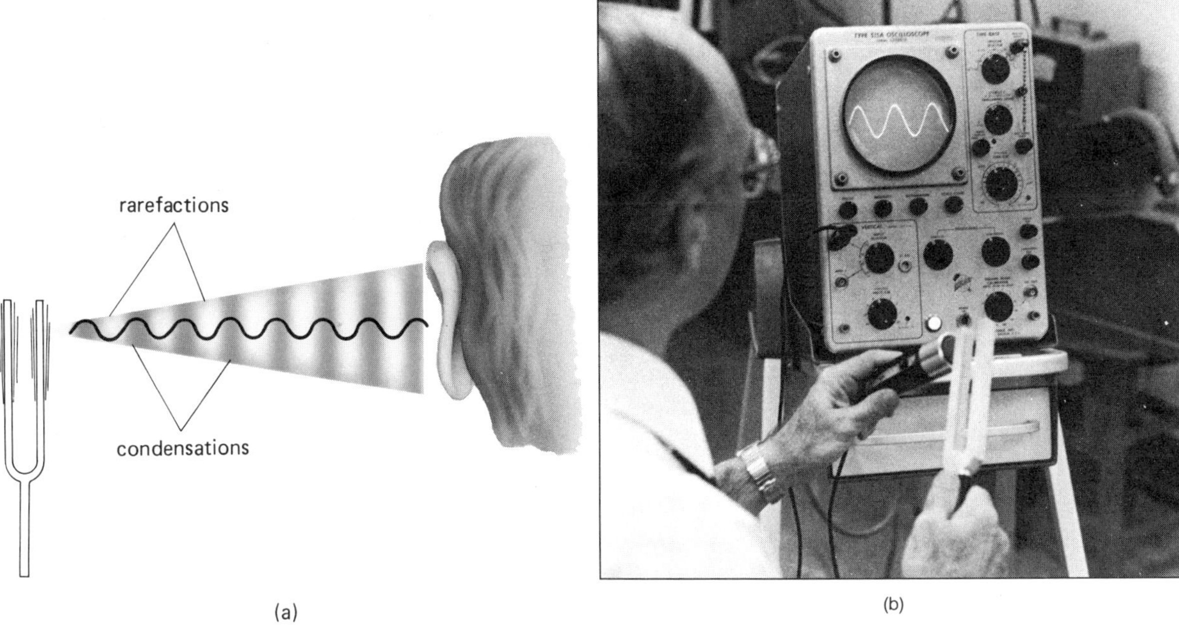

Figure 14.1

Sound waves. (a) An illustration of sound propagating from a vibrating tuning fork. The regions of increased pressure are called condensations, and the regions of decreased pressure are called rarefactions. (b) The form of the wave from a tuning fork displayed on an oscilloscope. Note the sinusoidal wave form.

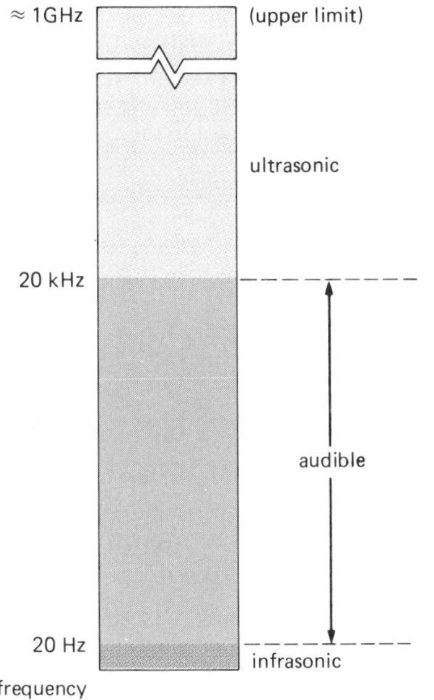

Figure 14.2

Sound frequency spectrum. We hear only sound disturbances with frequencies between about 20 Hz and 20 kHz (audible region). Below this is the infrasonic region, and above is the ultrasonic region, which has an upper limit of about a billion hertz that is set by the limit of material elasticity.

equipped with ultrasonic deer whistles. The relative air motion operates the whistle, the sound of which can be heard by deer. This has been shown to be effective in preventing deer from running across roads in front of vehicles.

Sound must also have sufficient energy or intensity to be heard. As discussed in the previous chapter, this depends on the amplitude of the wave disturbance. A large pulse disturbance, such as the explosion after shooting a gun, is heard as a loud boom. On the other hand, we often amplify soft sounds by electronic means so they can be heard.

Longitudinal (sound) waves propagate in solids and liquids as well as gases. This is evident from sounds heard through a wall and from hearing the clicking of rocks being tapped together when swimming underwater. The speed of sound in a medium depends in general on the elasticity of the medium and the inertia of its particles. Hence, we would expect the speed of sound to be different for different media. For example, it can be shown that the speed of sound waves in a thin rod is given by

$$v = \sqrt{\frac{Y}{\rho}} \quad \text{(solid rod)} \qquad \textbf{(Eq. 14.1)}$$

where Y is Young's modulus (Chapter 12) and ρ is the mass density of the material. Expressions for the speed of sound in extended solids are more complicated. In liquids, the speed of sound is given by a similar expression:

$$v = \sqrt{\frac{B}{\rho}} \quad \text{(liquid)} \qquad \textbf{(Eq. 14.2)}$$

In this case B is the bulk modulus and ρ is the mass density. The following example illustrates the differences in the speeds of sound in different forms of matter.

EXAMPLE 14.1 What is the speed of sound (a) in a steel rod and (b) in water?

Solution: Using the moduli and densities given in Tables 12.2 and 15.1, respectively:

(a) With $Y_{steel} = 20 \times 10^{10}$ N/m^2 and $\rho_{steel} = 7.8 \times 10^3$ kg/m^3,

$$v = \sqrt{\frac{Y}{\rho}} = \sqrt{\frac{20 \times 10^{10}\ \text{N/m}^2}{7.8 \times 10^3\ \text{kg/m}^3}}$$

$$= 5.1 \times 10^3\ \text{m/s} \quad (= 1.7 \times 10^4\ \text{ft/s})$$

(b) With $B_{H_2O} = 0.21 \times 10^{10}$ N/m^2 and $\rho_{H_2O} = 1.0 \times 10^3$ kg/m^3,

$$v = \sqrt{\frac{B}{\rho}} = \sqrt{\frac{0.21 \times 10^{10}\ \text{N/m}^2}{1.0 \times 10^3\ \text{kg/m}^3}}$$

$$= 1.4 \times 10^3\ \text{m/s} \quad (= 4.8 \times 10^3\ \text{ft/s})$$

The speed of sound in solids is generally greater than the speed of sound in liquids. Why?

The speed of sound in gases is given by an expression similar to Equation 14.2. However, the bulk moduli of gases vary greatly with temperature. Using the gas laws, the speed of sound in air, the most common gaseous medium, is

$$\text{at } T_C = 0°\text{C} \qquad v = 331\ \text{m/s}$$

$$(T_F = 32°\text{F}) \qquad (v = 1087\ \text{ft/s} \simeq 740\ \text{mi/h})$$

Temperature affects the speed of sound because as the temperature increases, so does the speed of the molecules of the gas. As a result, the molecules collide more often, and a disturbance is transmitted more quickly through the air.

It is found that the speed of sound increases approximately linearly with temperature over normal ranges at a rate of 0.6 m/s for each degree above 0°C (or 1.1 ft/s for each degree above 32°F). These results can be expressed in equation form by

$$v = 331 + 0.6\ T_C \quad \text{m/s}$$

or

$$v = 1087 + 1.1\ (T_F - 32°) \quad \text{ft/s}$$

$$\text{(speed of sound in air)} \qquad \textbf{(Eq. 14.3)}$$

where T_C and T_F are the Celsius and Fahrenheit temperatures, respectively.

EXAMPLE 14.2 What is the speed of sound in air at room temperature (20°C or 68°F)?

Solution: With $T_C = 20°$C or $T_F = 68°$F,

$$v = 331 + 0.6\ T_C\ \text{m/s}$$
$$= 331 + 0.6(20)\ \text{m/s}$$
$$= 343\ \text{m/s}$$

or

$$v = 1087 + 1.1(T_F - 32)\ \text{ft/s}$$
$$= 1087 + 1.1\ (68 - 32)\ \text{ft/s}$$
$$= 1127\ \text{ft/s}$$

As a general rule, the speed of sound in air at normal temperatures is on the order of $\frac{1}{5}$ mi/s.

$$\frac{1127\ \text{ft/s}}{5280\ \text{ft/mi}} = 0.21\ \text{mi/s} \simeq \tfrac{1}{5}\ \text{mi/s}$$

We are often interested in the wavelength λ of sound in a particular medium. This may be computed from the wave relationship (Eq. 13.14),

$$\lambda f = v$$

when the frequency of oscillation and the wave speed are known.

EXAMPLE 14.3 Compute the wavelengths of sound waves in the media in the previous examples if the frequency is 2000 Hz (or 2.0×10^3/s).

Solution: In a steel rod with $v = 5.1 \times 10^3$ m/s,

$$\lambda = \frac{v}{f} = \frac{5.1 \times 10^3 \text{ m/s}}{2.0 \times 10^3\text{/s}} = 2.6 \text{ m}$$

In water with $v = 1.5 \times 10^3$ m/s,

$$\lambda = \frac{v}{f} = \frac{1.5 \times 10^3 \text{ m/s}}{2.0 \times 10^3\text{/s}} = 0.75 \text{ m}$$

In air with $v = 343$ m/s,

$$\lambda = \frac{v}{f} = \frac{343 \text{ m/s}}{2.0 \times 10^3\text{/s}} = 0.17 \text{ m}$$

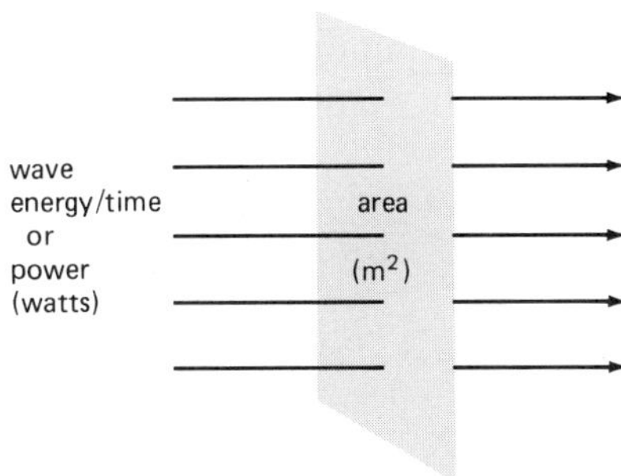

Figure 14.3

Sound intensity. The intensity is the rate of energy transfer (acoustical power, energy per time) through a cross-sectional area A.

14.2 Sound and Hearing

A great deal of communication depends on sound and hearing. Our oral communication is through voice sounds that are received by the ear. Voice sounds are produced by the disturbance of the air stream being expelled from our lungs in breathing. To demonstrate this, try to speak when you inhale as opposed to when you exhale.

The chief disturbance takes place in the larynx, or "voice box," located in the throat. Stretched across the larynx are two membrane-like bands called vocal cords, which form a slit-like opening. The flexing of the vocal cords causes the opening to widen and close. This action modulates the air stream, giving it wave form. Other voice parameters include the tongue and lips. Notice how these move as you speak.

Hearing response takes place in the ear, which is sensitive to minute pressure changes produced by the mechanical vibrations in a sound wave. The response of the human ear varies with individuals. Common physiological terms used to describe the sensations produced by the ear are **loudness, pitch,** and **quality** (or timbre). These terms are related to the physical properties of waves. The correlations of the sensory effects and physical wave properties are

Sensory Effect ↔ Physical Wave Property

loudness ↔ intensity

pitch ↔ frequency

quality ↔ wave form (overtones)

However, there is not a one-to-one correlation. The physical properties are objective and directly measurable. The sensory effects are subjective and vary with individuals.

Loudness and Intensity

The loudness of a sound is related to the energy of the disturbance. Sound, or acoustical energy, is commonly described in terms of acoustical power (in watts), which denotes the time rate of energy transfer. (Power = energy per time, $P = E/t$.)

To further characterize the flow of energy of a wave, we speak in terms of intensity. **The intensity *I* is the time rate of energy transfer through a cross-sectional area *A*,** i.e., I = energy/time/area or power/area (Fig. 14.3).

$$\text{(intensity)} \quad I = \frac{E/t}{A} = \frac{P}{A} \qquad \textbf{(Eq. 14.4)}$$

The units of intensity are then watts per square meter (W/m^2). Since the energy of a wave is proportional to the square of the wave amplitude (Chapter 13), it follows that the intensity is also proportional to the square of the wave amplitude.

The intensity of sound varies with the distance from the sound source, as can be seen by considering the intensity at various distances from a point source from which sound travels

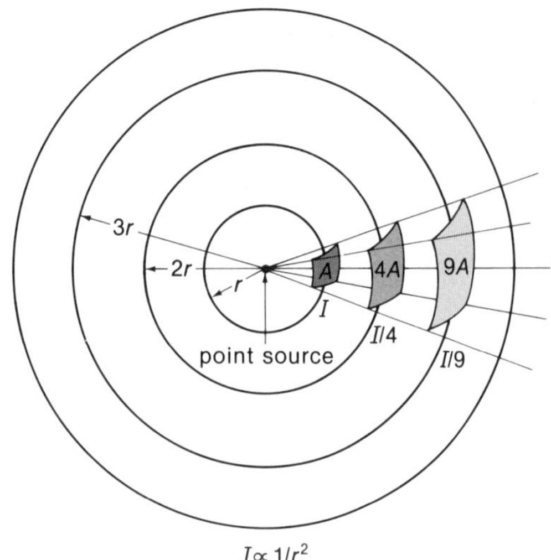

Figure 14.4
The intensity from a point source decreases as $1/r^2$. The energy from the source travels through larger and larger concentric shells (area = $4\pi r^2$). Hence, $I = P/A = P/4\pi r^2$.

outward equally in all directions (Fig. 14.4). The area through which the sound power from the source passes is the area of a sphere, $A = 4\pi r^2$; hence,

$$I = \frac{P}{A} = \frac{P}{4\pi r^2} \qquad \textbf{(Eq. 14.5)}$$

The intensity then is proportional to, or "falls off" as $1/r^2$ with, distance from the source ($I \propto 1/r^2$). This form is sometimes called an *inverse square law.* Notice in the figure that the greater the intercepting area, the greater the "spread" of energy and the smaller the intensity. Hence, the greater the distance from the source, the less the intensity.

The intensities at different distances from the source can be compared by forming the ratio of the intensities of two spherical surfaces:

$$\frac{I_2}{I_1} = \frac{P/4\pi r_2^2}{P/4\pi r_1^2}$$

or

$$\frac{I_2}{I_1} = \frac{r_1^2}{r_2^2} = \left(\frac{r_1}{r_2}\right)^2 \qquad \textbf{(Eq. 14.6)}$$

EXAMPLE 14.4 At a distance of 1 m from a point source, the sound intensity has a value of I_1. What is the intensity 3 m from the source?

Solution: With $r_1 = 1$ m and $r_2 = 3$ m,

$$\frac{I_2}{I_1} = \left(\frac{r_1}{r_2}\right)^2 = \left(\frac{1}{3}\right)^2 = \frac{1}{9}$$

or

$$I_2 = \tfrac{1}{9}I_1$$

Hence, the intensity at 3 m is $\frac{1}{9}$ of its value at a distance of 1 m.

For ordinary audible sounds, the intensity ranges from $I_o = 10^{-12}$ W/m² (the **threshold of hearing**) to about $I = 1.0$ W/m². Comparing these intensities,

$$\frac{I}{I_o} = \frac{1}{10^{-12}} \qquad \text{or} \qquad I = 10^{12}\, I_o$$

This shows that the upper normal audible intensity is a million million times the threshold intensity I_o.

To compare one sound intensity level with another, we establish an intensity level scale—much like a temperature scale. The scale is referenced to the threshold intensity of human hearing, $I_o = 10^{-12}$ W/m². To accommodate the large intensity range of audible sounds, a logarithmic scale is used, which effectively compresses the scale range.

The unit for measuring intensity levels is called the bel (in honor of Alexander Graham Bell; abbreviated B), and the sound *intensity level* (β) is defined by

$$\beta = \log \frac{I}{I_o} \quad \text{(B)} \qquad \textbf{(Eq. 14.7)}$$

(intensity level in bels)

However, the bel turns out to be a rather large unit, so the decibel (dB) is commonly used (1 bel = 10 decibels). The sound **intensity level (β) in decibels (dB)** is then given by

$$\boxed{\beta = 10 \log \frac{I}{I_o} \quad \text{(dB)}} \qquad \textbf{(Eq. 14.8)}$$

(intensity level in decibels)

The decibel scale is shown in Figure 14.5.

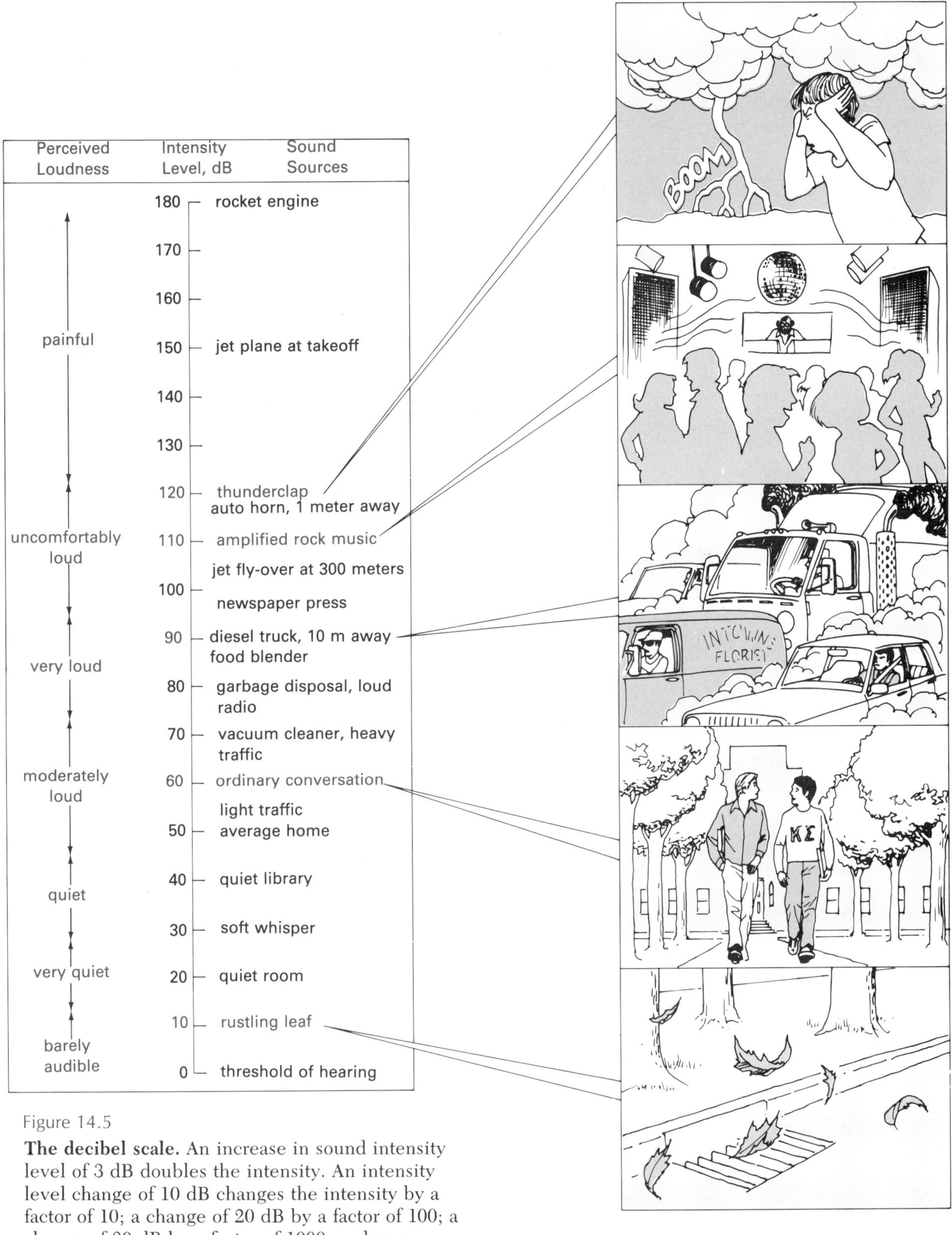

Figure 14.5

The decibel scale. An increase in sound intensity level of 3 dB doubles the intensity. An intensity level change of 10 dB changes the intensity by a factor of 10; a change of 20 dB by a factor of 100; a change of 30 dB by a factor of 1000; and so on.

EXAMPLE 14.5 What are the intensity levels in dB for the intensities (a) $I = 10^{-12}$ W/m² and (b) $I = 1.0$ W/m²?

Solution: (a) $I = I_o = 10^{-12}$ W/m² is the reference intensity, and

$$\beta = 10 \log \frac{I}{I_o} = 10 \log \frac{10^{-12}}{10^{-12}}$$

$$= 10 \log 1 = 0 \text{ dB}$$

since $\log 1 = \log 10^0 = 0$. (Recall $\log 10^x = x$.) Hence, the reference intensity is the "zero" of the dB scale.

(b) With $I = 1.0$ W/m²,

$$\beta = 10 \log \frac{I}{I_o} = 10 \log \frac{1.0}{10^{-12}}$$

$$= 10 \log 10^{12} = 10(12) = 120 \text{ dB}$$

The intensities of sounds are easily compared by their decibel differences:

> **An increase of 10 dB increases the sound intensity by a factor of 10.**
>
> **An increase of 20 dB increases the sound intensity by a factor of 100.**
>
> **An increase of 30 dB increases the sound intensity by a factor of 1000, and so on.**

Thus, a sound with an intensity level of 80 dB, for example, has an intensity 100 times greater than that of a sound with an intensity level of 60 dB.

Notice that the intensity difference in bels is the exponent of the power-of-ten factor of the difference in intensities, e.g., 80 dB − 60 dB = 20 dB = 2 B, and $10^2 = 100$. It can be shown that doubling the intensity of a sound corresponds to a 3-dB increase in intensity level.

EXAMPLE 14.6 Show that a sound with an intensity level of 70 dB has an intensity 1000 times greater than that of a sound with an intensity level of 40 dB.

Solution: The intensity level difference is 70 dB − 40 dB = 30 dB, or

$$10 \log \frac{I_{70}}{I_o} - 10 \log \frac{I_{40}}{I_o} = 30 \text{ dB}$$

where I_{70} and I_{40} are the intensities of the 70-dB and 40-dB sounds, respectively. Recalling that $\log a - \log b = \log a/b$, we have

$$10 \log \frac{\left(\frac{I_{70}}{I_o}\right)}{\left(\frac{I_{40}}{I_o}\right)} = 10 \log \frac{I_{70}}{I_{40}} = 30$$

and

$$\log \frac{I_{70}}{I_{40}} = 3$$

Since $\log (10^3) = 3$, by comparison

$$\frac{I_{70}}{I_{40}} = 10^3 \quad \text{or} \quad I_{70} = (10^3)I_{40} = (1000)\, I_{40}$$

Notice that this factor is given directly by the power of ten of the intensity difference in bels: 30 dB = 3 B, and $10^3 = 1000$.

EXAMPLE 14.7 The sound from a loudspeaker has an intensity level of 80 dB at a distance of 5.0 ft. How far from the speaker does the sound have an intensity level of 40 dB? Assume the speaker to be a point source.

Solution: The intensity level difference is

$$80 \text{ dB} - 40 \text{ dB} = 40 \text{ dB} = 4 \text{ B}$$

Hence, the intensity will be *reduced* by a factor of $10^4 = 10{,}000$, or I_2 is 1/10,000 of I_1, i.e., $I_2 = (10^{-4})I_1$. Then, using Equation 14.6 with $r_1 = 5.0$ ft,

$$\frac{I_2}{I_1} = 10^{-4} = \frac{r_1^2}{r_2^2} = \frac{(5.0 \text{ ft})^2}{r_2^2}$$

and

$$r_2^2 = 25 \times 10^4 \text{ ft}^2$$

so

$$r_2 = 5.0 \times 10^2 \text{ ft} = 500 \text{ ft}$$

Sounds with very great intensities can be dangerous. Above the intensity threshold of pain (about 120 to 125 dB), sound is painfully loud to the ear and causes hearing damage. Consequently, ear protectors must be worn in some occupations (Fig. 14.6). Longer exposure to lower sound (noise) levels can also damage hearing. See Special Feature 14.1.

Intensity is a measurable physical quantity and is related to the subjective loudness of a sound as perceived by the ear. However, the sensory loudness also depends on the frequency of the sound. Some sounds of different intensities and different frequencies are perceived as equally loud.

Figure 14.6
Ear (hearing) protectors must be worn in some occupations to protect workers from hearing loss. Notice also the safety hat. (Courtesy INCO Safety Products Co.)

SPECIAL FEATURE 14.1

Noise Exposure Limits

Sounds with very high intensities can be dangerous. Above the threshold of pain (about 120 dB), sound is painfully loud to the ear. Brief exposures to levels of 140 to 150 dB can rupture eardrums and cause permanent hearing loss. Consequently, in some occupations ear protectors or ear valves must be worn and noise intensity levels must be monitored (Fig. 1).

Figure 1
Sound level monitoring. The sound level intensity of a machine is monitored with a sound level meter. Federal regulations set limits for sound level exposure times.

Longer exposure to lower sound (noise) levels can also damage hearing. For example, there may be a hearing loss for a certain frequency range. (Have you ever noticed a temporary hearing loss after listening to the music of a loud band for a long time?)

Federal standards now set permissible noise exposure limits, as listed in Table 14.1. For example, 6 h is the maximum time that a worker can be exposed to sound-intensity levels of 92 dB. For a sound level of 95 dB, 4 h is the maximum exposure time. Notice that this is a doubling of intensity.

Table 14.1
Permissible Noise Exposure Limits*

Maximum Duration per Day (Hours)	Sound Level (dB)
8	90
6	92
4	95
3	97
2	100
$1\frac{1}{2}$	102
1	105
$\frac{1}{2}$	110
$\frac{1}{4}$ or less	115

* When the daily noise exposure is composed of two or more periods of exposure of different intensity levels, their combined effect should be considered.

Pitch and Frequency

The sensory effect of pitch is primarily related to the frequency of sound. We think of pitch in terms of the "highness" or "lowness" of sound, and in general, the greater the frequency, the greater (higher) the pitch.

For example, the faster the rotation (the greater the frequency) of a fire or police siren, the greater the pitch. Musicians refer to the pitch of a musical note. This is designated by a letter such as A, C, D, or E and by a unique position on a staff (a special graph). Each position has an associated physical frequency. Within a certain group of notes, the "A" note (440 Hz) has a lower pitch than the "E" note (660 Hz).

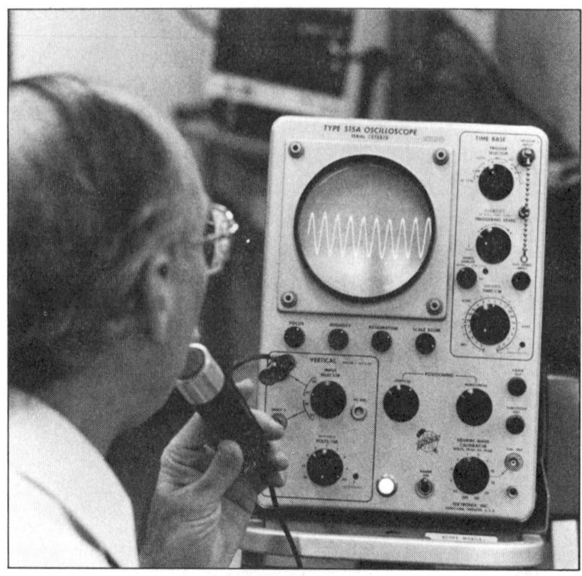

(a)

Quality and Wave Form

Musical instruments can sound notes of the same pitch, but there is a marked difference in the quality of the sounds. For example, the emissions of the same note from a piano, a guitar, and a horn sound different, as do the voices of singers.

The physiological quality of sound depends on the number and relative intensities of the overtones present. When a note is played on a musical instrument, such as a guitar, overtones are always present. The fundamental frequency is the characteristic pitch of the note, but the quality depends on overtones.

Combinations of overtones give rise to various complex wave forms (Fig. 14.7). Hence, the physical property of the wave form is associated with the quality of sound. The quality of sound is highly subjective. Some combinations of overtones or wave forms are pleasing to some people and not to others.

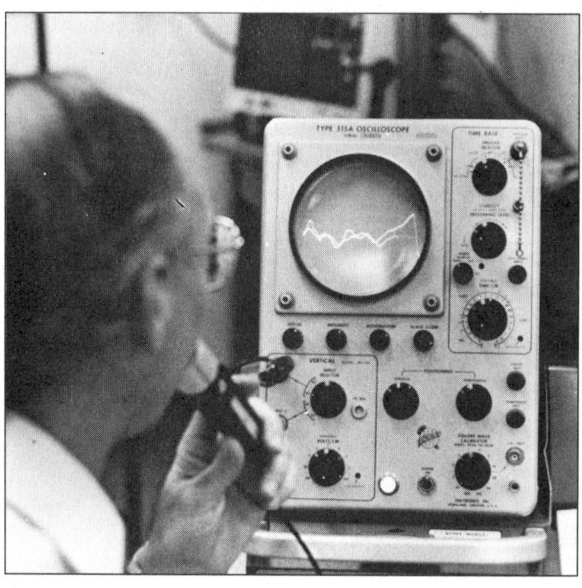

(b)

Figure 14.7
Sound quality depends on overtones and wave form. Human sounds are displayed electronically. (a) Whistling a note gives a simple sinusoidal wave form. Compare with Figure 14.1. (b) Voice sounds have complex wave forms as a result of various overtones.

14.3 Sound Phenomena

Interference

Interference effects occur for sound as they do for all other types of waves. Similar to transverse standing waves in a string, longitudinal standing waves can occur in vibrating air columns in pipes and cavities. Some possible standing wave modes for closed and open pipes are illustrated in Figure 14.8. Notice that a closed pipe has one closed end and one open end and that an open pipe has two open ends.

The closed end of a pipe is analogous to the fixed end of a string or the fixed end of a metal strip clamped in a vise. Hence, the closed end of a pipe must have a standing wave node. The

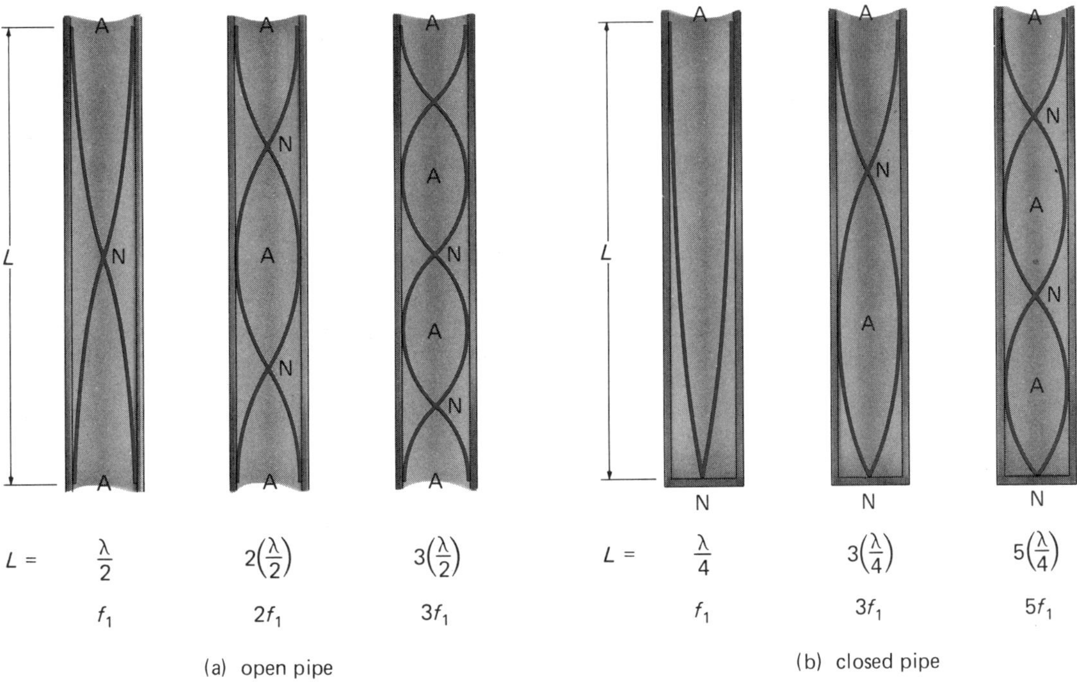

Figure 14.8
Characteristic frequencies of vibrating air columns in pipes. (a) An open pipe must have antinodes (A) at each end. As a result, all standing wave harmonics are possible in an open pipe. (b) A closed pipe must have a node (N) at its closed end and an antinode (A) at the open end. Only the odd harmonics are possible in this case.

open end of a pipe is analogous to the free end of a clamped metal strip, and it must have a standing wave antinode.

Sound traveling in pipes sets up standing waves. The characteristic frequencies of the standing waves are determined by how many wavelengths or partial wavelengths can be "fitted" into a given length of pipe consistent with the open and closed end boundary conditions. For an air column in an open pipe, antinodes are required at both ends.

This leads to standing waves in the pipe with an integral number of half wavelengths, $L = 1(\lambda_1/2)$, $L = 2(\lambda_2/2)$, $L = 3(\lambda_3/2)$, and so on. In terms of the wavelengths, these may be written collectively as

$$\text{(open pipe)} \qquad \lambda_n = \frac{2L}{n} \qquad n = 1, 2, 3, 4, \ldots$$

(Eq. 14.9)

The speed of sound is given by $v = \lambda f$, so the possible characteristic frequencies are

$$f_n = \frac{v}{\lambda_n}$$

$$\textbf{(open pipe)} \qquad \boxed{f_n = \frac{nv}{2L}} \qquad \textbf{(Eq. 14.10)}$$

$$n = 1, 2, 3, 4, \ldots$$

Thus, all of the standing wave harmonics are possible for an air column in an open pipe.

As can be seen from Figure 14.8 for a closed pipe, only an odd integral number of quarter wavelengths can be fitted into the pipe: $L = 1(\lambda_1/4)$, $L = 3(\lambda_3/4)$, $L = 5(\lambda_5/4)$, and so on; and

$$\text{(closed pipe)} \qquad \lambda_n = \frac{4L}{n} \qquad n = 1, 3, 5, \ldots$$

(Eq. 14.11)

The characteristic frequencies are then

$$f_n = \frac{v}{\lambda_n}$$

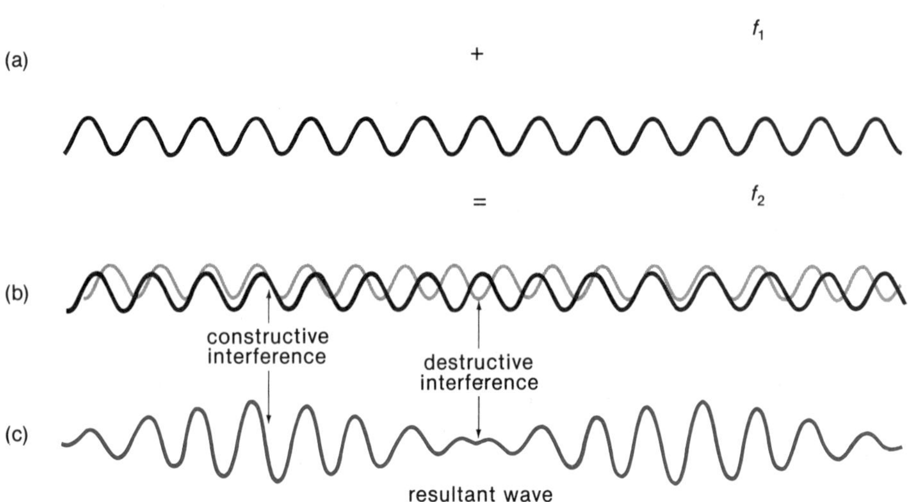

Figure 14.9
Beats. (a) When two sound waves of equal intensity (amplitude) but slightly different frequencies interfere (b), the resultant wave (c) is a pulsed disturbance with a frequency of $f_b = f_2 - f_1$.

(closed pipe) $$f_n = \frac{nv}{4L} \qquad \textbf{(Eq. 14.12)}$$

$$n = 1, 3, 5, \ldots$$

Thus, only the odd harmonics are possible for a closed pipe. The first overtone (above the fundamental frequency) is the third harmonic ($n = 3$), the second overtone is the fifth harmonic ($n = 5$), and so on.

EXAMPLE 14.8 What are the fundamental frequencies for closed and open organ pipes with lengths of exactly 1.0 m at room temperature?

Solution: From Example 14.2, the speed of sound at room temperature (20°C) is $v = 343$ m/s. Then, for a closed pipe of length $L = 1.0$ m, the fundamental frequency ($n = 1$) is

$$f_1 = \frac{(1)v}{4L} = \frac{343 \text{ m/s}}{4(1.0 \text{ m})} = 85.8 \text{ Hz}$$

Similarly, for an open pipe,

$$f_1 = \frac{(1)v}{2L} = \frac{343 \text{ m/s}}{2(1.0 \text{ m})} = 172 \text{ Hz}$$

Beats

Standing waves require a confined medium, which gives rise to boundary conditions. However, two or more waves may interfere when traveling in space. The result in this case is their combined traveling wave form as given by the superposition principle. In most instances, the wave form is quite complex. However, a special case of such interference gives rise to the phenomenon of **beats,** where the combination of sounds is heard as a fluctuating pulsation.

As illustrated in Figure 14.9, the combined or resultant wave of two sounds of equal intensity (amplitude) but slightly different frequencies is a pulsed disturbance. The amplitude of the wave form varies, and a loud tone is heard for the regions of constructive interference. The number of sound pulsations per second is called the **beat frequency.** It can be shown that the beat frequency f_b is equal to

$$f_b = f_2 - f_1 \qquad \textbf{(Eq. 14.13)}$$

where f_1 and f_2 are the slightly differing frequencies of the two interfering waves.

For example, if two tuning forks with frequencies of 512 Hz and 516 Hz vibrate simultaneously, the beat frequency of the resulting pulsation is $f_b = 516 - 512 = 4$ Hz. That is, four pulsating sounds are heard each second.

Musicians use the phenomenon of beats to tune their instruments to each other. When no beats are heard ($f_b = 0$), the instruments are in tune, or have the same pitch or frequency.

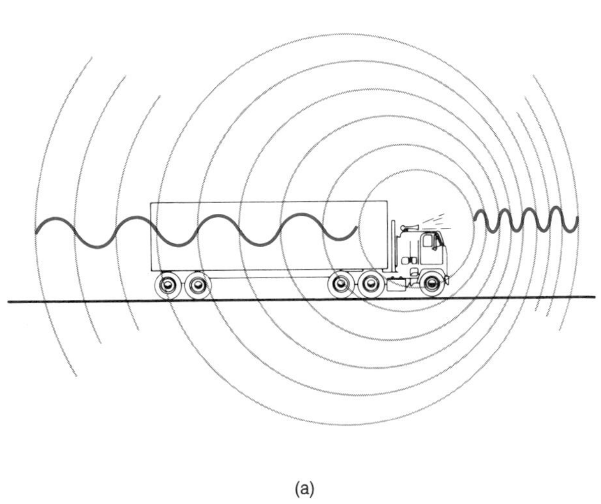
(a)

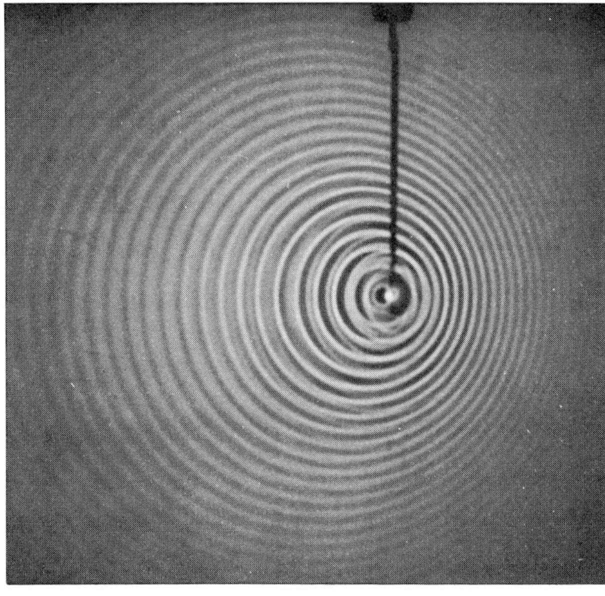
(b)

Figure 14.10
The Doppler effect. The Doppler effect is the apparent change of frequency of a sound source due to relative motion between the source and the observer. (a) Sound waves "bunch" up in front of a moving source (moving to the right), and the frequency is greater than when the source is at rest. Similarly, behind the source the waves are "spread out" and the frequency is lower. (b) The Doppler effect in water in a ripple tank. (Courtesy Educational Development Center, Newton, MA.)

The Doppler Effect

When you are standing beside a highway and a truck or car blowing its horn passes you, have you ever noticed how the pitch of the horn apparently increases as the vehicle approaches and decreases after it passes? The effect would be the same if the sound source were stationary and you were moving by it. The apparent change in frequency of a sound source due to relative motion between the source and the observer is called the **Doppler effect.***

The Doppler effect can be understood by considering the effect of motion on the wavelength of the sound (Fig. 14.10). For a moving source, the effect is to "bunch up" the waves in front of the source and hence increase the frequency. Similarly, the waves behind the source are "spread out," which decreases the frequency. In the figure, the distance between the lines represents one wavelength, e.g., the crest-to-crest distance.

* After Christian Doppler (1803–1853), an Austrian scientist, who first described the effect.

The frequency f' heard by a stationary observer from an approaching source is given by

$$f' = \left(\frac{v}{v - v_s}\right) f_s \qquad \textbf{(Eq. 14.14)}$$

(observed frequency of an approaching source)

where v is the speed of sound, v_s the speed of the source, and f_s the frequency of the source.

Similarly, the frequency heard by a stationary observer as the moving source recedes is given by

$$f' = \left(\frac{v}{v + v_s}\right) f_s \qquad \textbf{(Eq. 14.15)}$$

(observed frequency of an approaching source)

These equations are often written in combined form as

$$\boxed{f' = \left(\frac{v}{v \pm v_s}\right) f_s} \qquad \textbf{(Eq. 14.16)}$$

(moving source, stationary observer)
Approaching source, $-v_s$
Receding source, $+v_s$

SPECIAL FEATURE 14.2

Radar and the Doppler Effect

The Doppler effect is used in the determination of the velocity of a moving object by sonar (ultrasonic waves in water) and radar (radio waves in air). Let's discuss radar here, since it is more familiar. Radar stands for *ra*dio *d*etecting *a*nd *r*anging. The detecting and ranging (distance) come from reflection, the time delay, and wave speed, as discussed previously. Add the Doppler effect, and there's more.

Suppose a "Smokey Bear taking a picture" (CB radio jargon for a state trooper using radar) sitting at the side of the highway directs a beam of radio waves toward a car, and the waves are reflected back (Fig. 1). If the car being observed is not moving, the reflected waves have the same frequency as the ones directed outward. But if the car is moving toward the police car, the reflected waves have a higher frequency, or are Doppler-shifted.

In effect, the moving car acts like a moving source, and there is a double Doppler shift—coming and going. The magnitude of the shift depends on the speed of the car. A computer quickly calculates this speed and displays it for the officer to see. If the figure is above the speed limit, the next thing the driver might hear is the Doppler shift of a siren!

Figure 1
The Doppler effect in action. The Doppler shift is used in radar speed detection.

If the car is moving away from the patrol car, the reflected waves have a lower frequency. The speed is still calculated from the magnitude of the frequency shift. It's a bit more complicated, but the speed of the car can be determined even when the patrol car is moving. The computer does this easily and can "get you coming and going."

From experience, we know that the observed frequency increases, or is greater than f_s, for an approaching source, so the negative form of the equation is used. Similarly, the positive sign is used for a receding source.

EXAMPLE 14.9 The driver of a car blows the horn, which has a frequency of 500 Hz, as it passes by a stationary observer. If the car travels at 90 km/h, what is the frequency of the horn as heard by the observer (a) as the car approaches and (b) as the car recedes? (Assume the speed of sound to be 343 m/s.)

Solution: (a) With $f_s = 500$ Hz, $v_s = 90$ km/h = 25 m/s, and the speed of sound $v = 343$ m/s, for an approaching source we have

$$f' = \left(\frac{v}{v - v_s}\right) f_s$$

$$= \left(\frac{343\text{ m/s}}{343\text{ m/s} - 25\text{ m/s}}\right) 500\text{ Hz} = 539\text{ Hz}$$

(b) Similarly, when the car is receding,

$$f' = \left(\frac{v}{v + v_s}\right) f_s$$

$$= \left(\frac{343\text{ m/s}}{343\text{ m/s} + 25\text{ m/s}}\right) 500\text{ Hz} = 466\text{ Hz}$$

As has been noted, the Doppler "shift" in frequency is a relative effect and also occurs for a stationary source and a moving observer and when the source and observer are both in motion. The observed frequency of a stationary source by an observer moving with a speed v_o is given by

$$f' = \left(\frac{v \pm v_o}{v}\right) f_s \qquad \textbf{(Eq. 14.17)}$$

(moving observer, stationary source)
Approaching the source, $+v_o$
Receding from the source, $-v_o$

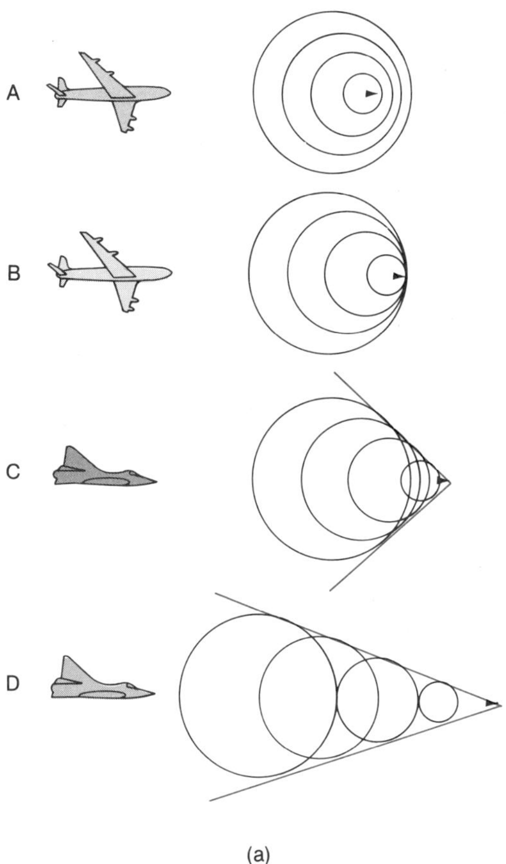

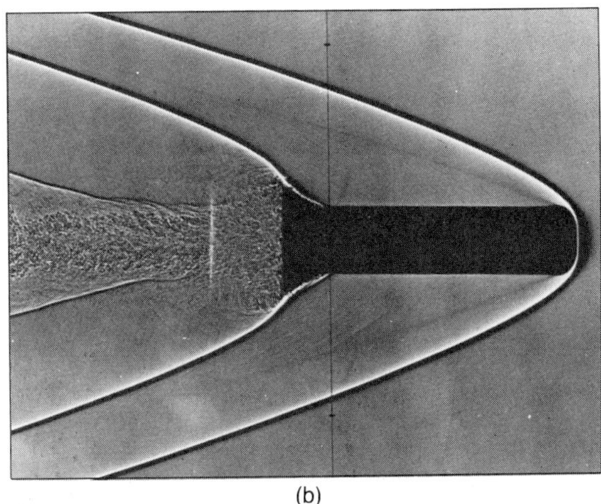

Figure 14.11
Wave patterns produced by subsonic and supersonic objects. (a) A. Subsonic. Plane flies at less than the speed of sound. B. Sonic. Plane flies at the speed of sound. C. Supersonic. Mach 1.5. Plane flies 1.5 times the speed of sound. D. Supersonic. Mach 3. Notice the buildup of the bow waves for the supersonic cases. (b) Bow wave. Strong shock waves are generated at the nose and rear surfaces of a blunt-nosed cylinder, which was gun-launched in a wind tunnel. The cylinder was moving at Mach 7 when this photograph was taken. Notice the turbulence behind the cylinder.
(Courtesy NASA.)

In the event that both the source and the observer are moving, the observed frequency is given by a combination of Equations 14.16 and 14.17,

$$f' = \left(\frac{v \pm v_o}{v \pm v_s}\right) f_s \qquad \textbf{(Eq. 14.18)}$$

where the sign conventions are as before. A practical example of the Doppler effect is given in Special Feature 14.2.

Supersonic Speeds

It is possible for sound sources to travel faster than the speed of sound; that is, a sound source can travel at a supersonic speed. When this occurs, the sound waves form two converging pressure ridges of a shock wave behind the source (Fig. 14.11).

The term *bow wave* is commonly applied to the V-shaped pressure ridge. A similar bow wave trails outward from the bow of a speedboat moving through water at a high speed, in which case the speed of the boat is greater than the wave speed in water.

Jet planes that "break the sound barrier" ($\simeq$ 770 mi/h in air) and fly at supersonic speeds produce a bow wave of air that trails outward and downward. This large air pressure disturbance is heard as a **sonic boom** as the bow wave passes over a location.

The speed of a fast-moving object is characterized by the Mach number, which is defined as

$$\text{Mach number} = \frac{v_o}{v} = \frac{\text{(speed of object)}}{\text{(speed of sound)}} \qquad \textbf{(Eq. 14.19)}$$

For example, a jet plane traveling at twice the speed of sound ($v_o = 2v$) is said to be traveling at or have a speed of "Mach 2."

For some technical applications of sound—in the use of ultrasound—see Special Feature 14.3.

SPECIAL FEATURE 14.3

Ultrasonics

Ultrasonic sound is in the frequency region above the audible region (> 20,000 Hz). This high-frequency sound has many uses and applications. For example, whereas audible-frequency sound waves have a limited propagation range in water, ultrasonic waves can travel for kilometers and are highly directional. As a result, ultrasound is used in marine applications.

The depth of the ocean is determined by depth "sounding" techniques using a fathometer. A beam of ultrasound is directed downward from a ship and is reflected from the ocean floor. The depth is computed using the ultrasound speed and elapsed time. This detection and ranging technique is called **sonar.**

Ultrasonic sonar is used to detect and determine the ranges of not only ships and submarines, but also schools of fish. In addition, combining or modulating audible sound waves with ultrasound makes underwater radio communication possible.

Ultrasound is used in many technical applications. One of the best known is ultrasonic cleaning [Fig. 1(a)]. Ultrasound is used in liquid baths to clean metal parts. Ultrasonic vibrations (small wavelengths) can loosen traces of foreign matter from otherwise inaccessible places. Jewelers make use of ultrasonic baths to clean rings and other jewelry.

Another important industrial application of ultrasound is ultrasonic drilling in the machining of very hard materials [Fig. 1(b)]. By means of an abrasive paste and an ultrasonic vibrator, the material is rapidly worn away. Since the ultrasonic drill does not rotate, the vibrator tip can be oriented to produce holes or surfaces of any shape. Ultrasonic soldering irons are also available. These are particularly useful in the soldering of aluminum. The ultrasound removes the aluminum oxide coating on the surface and eliminates the need for fluxes.

An ultrasonic beam or pulse traveling in metal can be used to detect flaws (Fig. 2). When the ultrasound strikes a flaw, which has different properties than the surrounding medium, reflection and refraction occur. The echo pattern is monitored, and an irregularity indicates the presence of a flaw in the metal. Such techniques provide a means of nondestructive testing of metal castings and other metal objects, such as airplane parts.

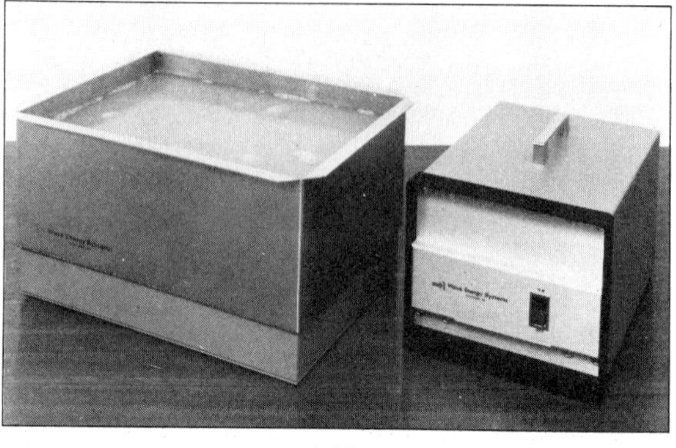

(a)

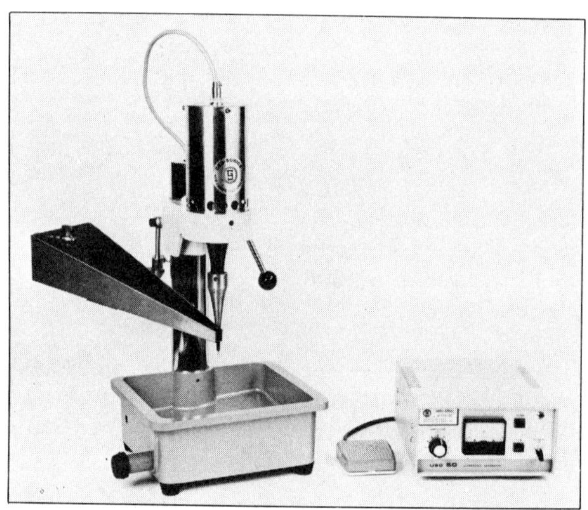

(b)

Figure 1

Ultrasonic applications. (a) An ultrasonic cleaning tank. (b) An ultrasonic drill used to make holes in glass, ceramics, and gem stones by ultrasonic vibrations rather than rotary motions, as in ordinary drilling. (Courtesy Geode Instruments, Inc.)

In the medical field, ultrasonic "sonar" can be used to view soft internal tissues and organs, such

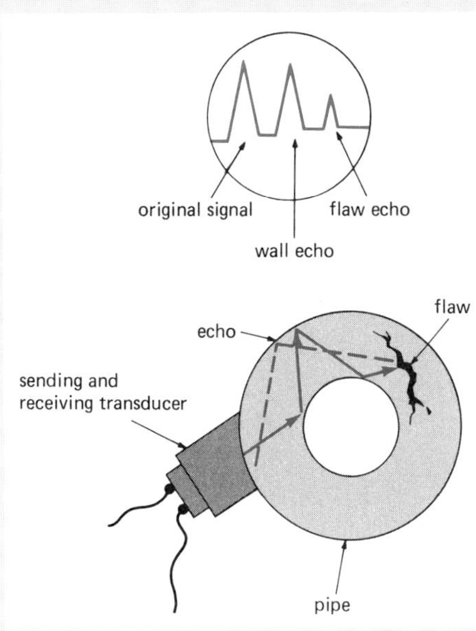

Figure 2
Ultrasonic nondestructive testing. Echoes reflected from flaws cause irregularities in the observed reflection patterns. (Courtesy Sonic Instruments, Inc.)

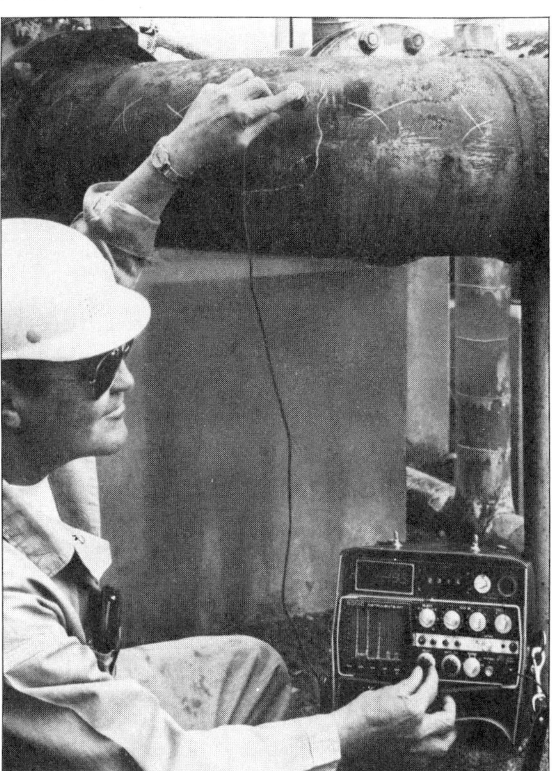

as the liver or spleen, which are nearly invisible to X-rays. Ultrasound can also be used to "view" a fetus at different stages of development without the dangerous effects presented by X-rays. The different degrees of reflection of scanned areas are monitored and stored in a computer. The computer then reconstructs an "echogram" of the region (Fig. 3).

Ultrasonic brain scans are used to detect tumors and cerebral hemorrhage. By applying the Doppler effect to reflected ultrasound, doctors can detect and monitor movements such as the actions of heart valves, the flow of blood, and the beating of fetal hearts.

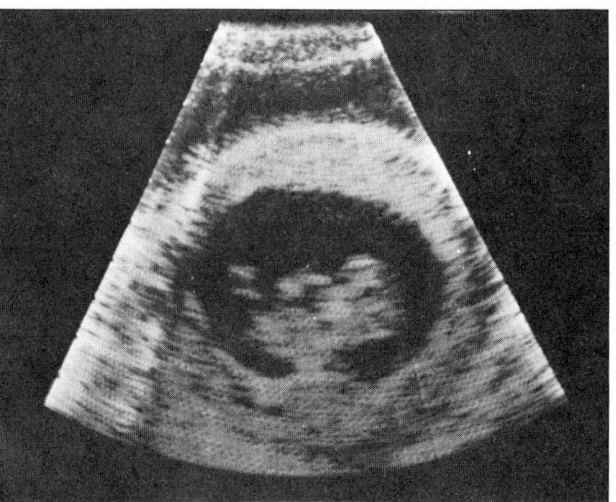

Figure 3
Medical ultrasonics. An ultrasonic scan of a human fetus. (Courtesy Picker Corporation.)

Important Terms

sound any longitudinal disturbance in an elastic medium

audible sound (human) sound in the frequency range of 20 Hz to 20 kHz

infrasonic sound sound with frequencies below 20 Hz

ultrasonic sound sound with frequencies above 20 kHz

loudness the perceived intensity of sound

pitch the perceived frequency of sound

quality (timbre) the perceived pleasantness of sound, which is related to overtones and wave form

intensity energy/time/area or power/area

intensity level the measure of sound intensity referenced to the intensity of the threshold of human hearing, $I_o = 10^{-12}$ W/m^2

decibel (dB) the unit of sound intensity level

beats fluctuating pulsations due to traveling wave interference

Doppler effect the apparent change in frequency of a sound source due to relative motion between the source and an observer

Mach number the ratio of the speed of an object and the speed of sound

Important Formulas

speed of sound:

in solid rod $$v = \sqrt{\frac{Y}{\rho}}$$

in liquid $$v = \sqrt{\frac{B}{\rho}}$$

variation with temperature $$v = 331 + 0.6\, T_C \quad \text{m/s}$$

$$v = 1087 + 1.1(T_F - 32°) \quad \text{ft/s}$$

$$v = \lambda f$$

intensity: $$I = \frac{E/t}{A} = \frac{P}{A}$$

at distance from point source $$I = \frac{P}{4\pi r^2}$$

and $$\frac{I_2}{I_1} = \left(\frac{r_1}{r_2}\right)^2$$

intensity level: $$\beta = 10 \log \frac{I}{I_o} \quad dB$$

where $$I_o = 10^{-12}\ \text{W/m}^2$$

characteristic pipe frequencies:

open pipe $$f_n = \frac{nv}{2L} \qquad n = 1, 2, 3, 4, \ldots$$

closed pipe $$f_n = \frac{nv}{4L} \qquad n = 1, 3, 5, \ldots$$

beat frequency: $$f_b = f_2 - f_1$$

Doppler frequency: $$f' = \left(\frac{v}{v \pm v_s}\right) f_s$$

(moving source, stationary observer)
Approaching source, $-v_s$
Receding source, $+v_s$

$$f' = \left(\frac{v \pm v_o}{v}\right) f_s$$

(moving observer, stationary source)
Approaching the source, $+v_o$
Receding from the source, $-v_o$

Mach number: $$M = \frac{v_o}{v} = \frac{\text{(speed of object)}}{\text{(speed of sound)}}$$

Questions

The Nature of Sound

1. Some people say there are two things that are necessary to have sound, and others say there are three things. Why the difference?
2. An alarm clock is hung in a closed glass jar. When the alarm goes off, it can be heard. Why? What would happen if the air were pumped out of the jar?
3. What is the speed of sound in vacuum? Explain.
4. In what type of medium is the speed of sound generally greatest and why?
5. The speed of sound in air is given for still air. How would wind affect the propagation of sound? What happens if you shout into the wind?
6. Sound travels faster in humid air. Why is this? (*Hint:* the major components of air are N_2 and O_2. In a volume of moist air, some of these molecules are replaced by H_2O molecules. Check their masses using a periodic table.)
7. In a popular lecture demonstration, the instructor breathes helium (He) gas. When the

Figure 14.12
Singing in the shower. See Question 14.

instructor talks, his or her voice has a high-pitched "Donald Duck" sound. What causes this? (*Hint:* the vocal cavities and wavelengths are still the same.)

8. How many times faster is the speed of light than the speed of sound in air?
9. While backpacking, you notice that there is a 4-s interval between the time you see a lightning flash and the time you hear the thunder from an approaching storm. Approximately how far away is the storm?

Sound and Hearing

10. What happens to the energy when a sound "dies out"?
11. If you moved three times as close to a point sound source, how would the sound intensity be affected?
12. After a snowfall, it "sounds" particularly quiet. Why is this?
13. Why do sounds in an empty building or room sound "hollow"?
14. Singing in the shower is a popular pastime (Fig. 14.12). Why does your voice sound fuller in the shower?
15. What are the dB levels of (a) ordinary talking, (b) a loud radio, and (c) the threshold of pain?
16. What increase in dB increases the sound intensity by a factor of (a) 2, (b) 10, and (c) 1000?
17. Is the effect the same when the sound level is increased from 10 dB to 20 dB as when it is increased from 40 dB to 80 dB? Explain.
18. Can there be a negative dB level—for example, −10 dB?
19. Estimate the dB levels that might be heard from the sound sources shown in Figure 14.13.
20. What are the federal standards for the permissible noise exposure limits for daily durations of (a) 8 h, (b) 2 h, and (c) $\frac{1}{2}$ h?

Figure 14.13
Sound intensity levels. See Question 19.

Sound Phenomena

21. Which harmonics are possible (a) in an open pipe and (b) in a closed pipe?
22. It is common to hear music, particularly the bass part, through walls or ceilings (floors if you are upstairs). What does this tell you about the natural frequency of the walls?

Figure 14.14
Vibrating strings. See Question 25.

23. When a metal hammer or wrench is dropped on a hard surface, it has a certain "ring." Why is this?
24. Do beats have anything to do with the beat of the music? Describe some practical applications of beats.
25. (a) Why are different notes obtained when the fingers are placed on different frets of a guitar (Fig. 14.14)? (b) On the low-frequency strings of an instrument, a second wire is wrapped around the main wire. What is the purpose of this?
26. If a sound source and an observer are both moving with the same velocity, is there a Doppler shift? Explain.
27. Does a sonic boom occur only when a jet plane "breaks the sound barrier"?
28. How fast would a "jet" fish have to swim to create an "aquatic boom"? Give a numerical answer.

Problems

Levels of difficulty are indicated by asterisks for your convenience.

14.1 The Nature of Sound

1. What are the wavelength limits of the audible range for human hearing? (Use 340 m/s for the speed of sound.)
2. The distance between two condensations of a sound wave is 2.0 m. Could this sound be heard by a human ear if it were intense enough? (Take the speed of sound to be 340 m/s.)
3. Show that $\sqrt{Y/\rho}$ has the units of speed.

*4. What is the speed of sound in a thin copper rod?

*5. Is the speed of sound greater in a brass rod or a steel rod? How many times greater? (*Hint:* use a ratio.)

*6. For which of the metals listed in Table 12.2 is the speed of sound in a thin rod closest to the speed of sound in a glass rod?

*7. An aluminum rod 10 m long is struck on one end. How long does it take for the disturbance to travel to the other end?

*8. A steel rail on a railroad is struck with a hammer. Assuming that the disturbance is transmitted along the track without interruption, how long will it take for the longitudinal wave to travel 1 km in one direction?

*9. What is the speed of sound in mercury?

*10. Which is greater, the speed of sound in water or the speed of sound in alcohol? How many times greater?

*11. The speed of sound in a particular liquid is 1.6×10^3 m/s. If the bulk modulus of the liquid is 0.15×10^{10} N/m^2, what is the density of the liquid?

*12. A large circular water storage tank has a diameter of 10 m. How long would it take for a disturbance to travel across the tank?

*13. What is the speed of sound in air on a warm summer day when the temperature is 86°F?

*14. What increase in temperature will cause the speed of sound in air to increase (a) by 10 m/s and (b) by 10 ft/s?

*15. A tuning fork has a frequency of 540 Hz. What is the wavelength of the sound coming from the vibrating fork when the air temperature is (a) 0°C and (b) room temperature (20°)?

*16. A hunter sees a fellow hunter 0.50 mi away fire a rifle (as evidenced by the smoke from the barrel). If the air temperature is 42°F, how long will it be until the report of the rifle is heard?

**17. A swimmer in a lake clicks rocks together every four seconds. What is the wavelength of the sound in the water?

**18. An ultrasonic cleaning bath contains alcohol. If the ultrasonic generator operates at a frequency of 50 MHz, what is the wavelength of the waves in the bath? (*Note:* the wavelengths in a cleaning bath are approximately the size of the particles to be loosened or "scrubbed" away.)

14.2 Sound and Hearing

*19. A sound has an intensity of 10^{-6} W/m^2. How much energy passes through an area of 0.15 m^2 in 2.0 s?

*20. A point source emits sound at a rate of 5.6×10^{-7} J/s. What is the sound intensity at a distance of 1.5 m from the source?

*21. If the distance from a point sound source is increased by a factor of 2.3, how is the intensity affected?

*22. The sound intensity 1.5 m from a point source is 5.0×10^{-7} W/m^2. What is the intensity 2.5 m from the source?

*23. An observer is 15 ft from a point sound source. How far toward the source must the observer move so the intensity is increased by a factor of 2.5?

*24. What are the intensity levels in dB for the following intensities? (a) 10^{-5} W/m^2, (b) 10^{-1} W/m^2, (c) 10^2 W/m^2

*25. What would be the typical reaction of a person listening to a sound with an intensity of the following? (a) 10^{-10} W/m^2, (b) 10^{-4} W/m^2, (c) 10^0 W/m^2

*26. The volume of a radio is turned up so the output is increased from 30 dB to 80 dB. By what factor is the intensity increased?

*27. If the sound intensity is decreased from 10^{-2} W/m^2 to $^{10-6}$ W/m^2, what is the corresponding decrease in dB?

*28. A factory whistle is heard by one worker at an intensity level of 60 dB and by another worker at an intensity of 80 dB. How much farther away from the whistle is one worker than the other?

*29. The average sound intensities due to continuously running machinery in different parts of a factory are 10^{-3} W/m^2 and 10^{-1} W/m^2, respectively. How long should workers be allowed to work in these areas each day?

**30. Each day a worker in a machine shop works 5 h in an area where the sound intensity is 10^{-3} W/m^2 and 2 h in another area where the sound intensity is 10^{-2} W/m^2. Considering the combined effect of the permissible noise exposure limits in each area, is this a safe industrial practice?

**31. A person 20 ft from a loudspeaker receives a certain intensity level of sound. How far toward the speaker would the person have to move for the received sound intensity to be (a) 10 times greater and (b) 100 times greater? (Assume the speaker to be a point source.)

**32. The sound from a loudspeaker has an intensity level of 100 dB at a distance of 3.0 m. Considering the speaker to be a point source, how far from the speaker will the sound have an intensity level (a) of 60 dB and (b) just barely enough to be heard?

14.3 Sound Phenomena

*33. An open organ pipe has a length of 0.80 m. What are (a) the fundamental frequency and (b) the second overtone of the pipe at room temperature (20°C)?

*34. A closed organ pipe has a length of 2.0 ft. If the air temperature is 72°F, what are the two lowest characteristic frequencies of the pipe?

*35. At room temperature (20°C), a closed organ pipe has a fundamental frequency of 256 Hz. What is the length of the pipe?

*36. What is the beat frequency of two tones with frequencies of 256 Hz and 260 Hz?

*37. A violinist with a perfectly tuned A string (f = 440 Hz) plays an A note with another violinist, and a beat frequency of 2 Hz is heard. What is the frequency of the tone from the other violin? (Is there only one possibility?)

*38. A truck approaches a stationary observer with a speed of 90 km/h. If the truck driver blows the horn, which has a frequency of 1000 Hz, what is the frequency of the horn heard by the observer (a) as the truck approaches and (b) as the truck recedes? (Assume an air temperature of 20°C.)

*39. A worker driving to work down a straight road toward the factory at a speed of 60 mi/h hears the factory whistle blow. If the frequency of the whistle is 6000 Hz and the air temperature is 72°F, what frequency is heard by the worker?

*40. A jet plane flies at a speed of 896 mi/h at an altitude where the temperature is 42°F. (a) What is the Mach number of the plane? (b) Does it cause a sonic boom?

*41. A trucker traveling along a straight road sounds the truck horn, which has a frequency of 500 Hz. If an observer beside the road in front of the truck measures the frequency to be 520 Hz, how fast is the truck moving? (Assume the speed of sound to be 340 m/s.)

**42. A car and a truck approach each other on a highway, both traveling at 60 mi/h. If the driver of the truck sounds the truck's horn, which has a frequency of 1000 Hz, what frequency is heard by the driver of the car? (Assume the speed of sound to be 1100 ft/s.)

**43. A common apparatus for measuring the speed

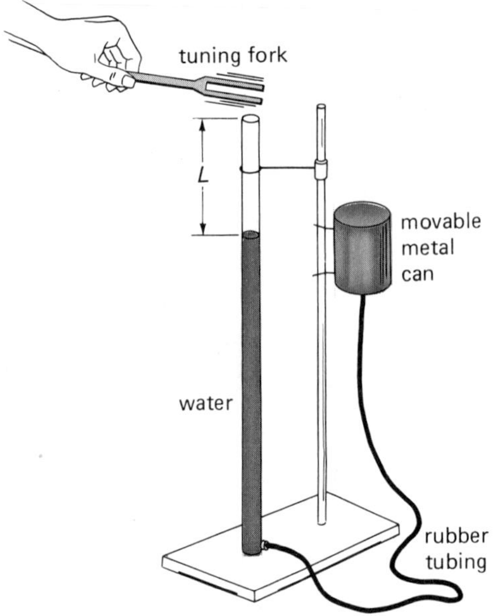

Figure 14.15
See Problem 43.

of sound in air is illustrated in Figure 14.15. Standing waves are set up in the tube, and when an antinode is near the top of the tube, a loud resonant sound is heard. By adjusting the height of the water column, or the length L of the empty resonant tube, different antinodes may be positioned at the top of the tube (longer standing wave portions in the tube). (a) Show that the length difference between two audible antinode positions is equal to $\lambda/2$. (b) If a tuning fork with a frequency of 1024 Hz is used to set up the standing wave and the tube lengths are 8.5 cm and 25.5 cm for the first and second antinodes, respectively, what is the speed of sound?

**44. The apparatus in Figure 14.15 can also be used to determine the unknown frequency of a tuning fork (cf. Problem 43). This is done by using the theoretical formula for the speed of sound in air. Suppose a tuning fork sets up a standing wave in the tube and there is a distance of 67 cm between audible antinodes. If the lab is at room temperature (20°C), what is the frequency of the tuning fork?

**45. Two identical strings on different guitars are tuned to an A note (440 Hz). If the string of one of the guitars slips so the tension is decreased by 0.90 percent, what is the beat frequency between the two strings when they are sounded?

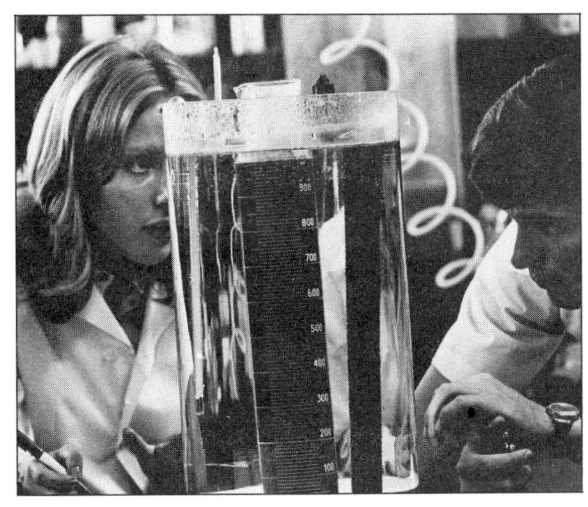

(Courtesy Reynolds Metals Company.)

Chapter 15

Fluid Mechanics

The term *fluid* is commonly thought to be synonymous with *liquid*. However, **a fluid is defined as any substance that can flow. Hence, gases as well as liquids are fluids.** Gases and liquids are different states of matter, but the common property of ability to flow gives them similar mechanical properties.

It is therefore convenient to classify matter mechanically as being either solid or fluid. The basic difference is that a solid can support a shear stress, whereas a fluid cannot. A fluid flows under the application of a shear stress instead of being elastically deformed.

There are marked differences in the compressibilities of gases and liquids. For example, air (a mixture of gases) is compressible, and water is relatively incompressible. Gases and liquids also differ in thermal properties. These properties will not be considered in the general discussion of fluids in this chapter—a simplification that makes the mechanical principles and the many diverse technical applications of fluids easier to understand.

15.1 Fluid Properties

It is convenient to consider first some of the terms used to describe the properties of fluids—namely density and specific gravity, surface tension and capillarity, and viscosity.

Density

As we learned in Chapter 1, mass (or weight) is a measure of the quantity of matter that a substance contains. We might say mercury is heavier than water. But *heavier* and *lighter* are relative terms. What is being implied is that a given volume of mercury is heavier, or contains more matter, than an *equal* volume of water. The ambiguity is removed by a density relationship, which essentially expresses the compactness of matter in a given volume.

Recall that there are two types of densities—mass density and weight density. The **mass density ρ** of a substance is defined as the mass m per unit volume V:

$$\text{Mass density} = \frac{\text{mass}}{\text{volume}}$$

$$\boxed{\rho = \frac{m}{V}} \qquad \textbf{(Eq. 15.1)}$$

The units of mass density in the metric system are kg/m^3 and g/cm^3. The unit of mass density in the British system would be slug/ft^3. However, the slug is not a common engineering unit, since the British system is a gravitational or force system. It is more common in the British system to express density in terms of weight units or to use a **weight density D,** i.e., weight w per unit volume V:

$$\text{Weight density} = \frac{\text{weight}}{\text{volume}}$$

$$\boxed{D = \frac{w}{V}} \qquad \textbf{(Eq. 15.2)}$$

The units of weight density in the British system are lb/ft^3. The mass and weight densities of some common substances are listed in Table 15.1.

Since $w = mg$, the mass and weight densities in a given system of units are related, i.e.,

$$D = \frac{w}{V} = \frac{mg}{V} = \rho g \qquad \textbf{(Eq. 15.3)}$$

Table 15.1
Densities of Some Common Substances

	Mass Density ρ		Weight Density D
Substance	***g/cm^3***	***kg/m^3***	***lb/ft^3***
Solids			
Aluminum	2.7	2700	169
Brass	8.7	8700	540
Copper	8.9	8900	555
Glass	2.6	2600	162
Gold	19.3	19,300	1204
Ice	0.92	920	58
Iron	7.9	7900	490
Lead	11.4	11,400	705
Silver	10.5	10,500	654
Steel	7.8	7800	487
Wood (oak)	0.81	810	51
Liquids			
Alcohol	0.79	790	49
Benzene	0.88	880	55
Gasoline	0.68	680	42
Glycerol	1.26	1260	79
Mercury	13.6	13,600	850
Water	1.0	1000	62.4
Sea water	1.03	1030	64.0
Gases			
Air	0.00129	1.29	0.0807
Carbon dioxide	0.00076	0.76	0.0481
Helium	0.00018	0.18	0.0111
Hydrogen	0.00009	0.09	0.0058
Nitrogen	0.00125	1.25	0.0782
Oxygen	0.00143	1.43	0.0892

Specific Gravity

It is convenient to compare the mass or weight of a substance with that of an equal volume of another substance. Water, being one of the most abundant common substances, is used as the basis of such a comparison, called *specific gravity.* The **specific gravity (sp gr) of a solid or liquid is defined as the ratio of the weight of a given volume of a substance to the weight of an equal volume of water:**

$$\text{sp gr} = \frac{\text{weight of a given volume of substance}}{\text{weight of an equal volume of water}} \qquad \textbf{(Eq. 15.4)}$$

Expressed in terms of densities,

$$\text{sp gr} = \frac{D_s}{D_w} = \frac{\rho_s}{\rho_w} \qquad \textbf{(Eq. 15.5)}$$

where the s and w subscripts refer to substance and water, respectively. (The standard density of water is taken at a temperature of 4°C [≈ 39°F], at which water has its maximum density.) The specific gravity of a liquid is measured in a special bottle called a pycnometer (Fig. 15.1).

Notice that the specific gravity is a pure number with no units. Since water in the metric cgs system has a mass density of $\rho_w = 1.0$ g/cm^3, the numerical values of the specific gravity and the mass density are equal in the cgs system, i.e.,

$$\text{sp gr} = \frac{\rho_s}{\rho_w} = \frac{\rho_s}{1.0} = \rho_s$$

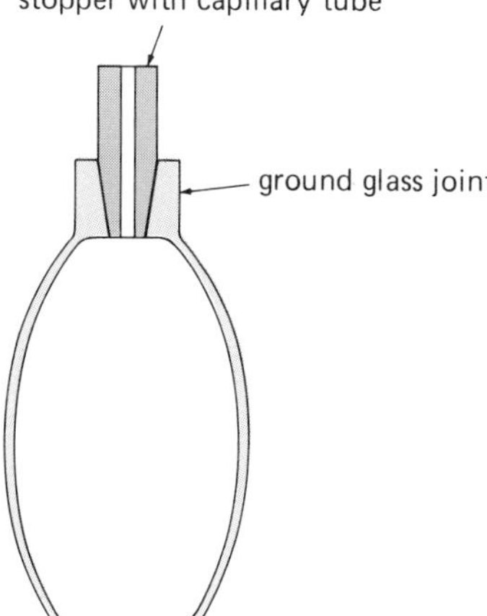

Figure 15.1

A pycnometer or specific gravity bottle. When the ground-glass stopper is inserted into the bottle filled with a liquid, the liquid level reaches the top of the capillary tube and the outside of the bottle is wiped dry. The procedure is to weigh in turn the dry, clean bottle; the bottle filled with distilled water; and the bottle filled with the liquid for which the specific gravity is to be determined.

For example, the density of mercury is 13.6 g/cm^3 and its specific gravity is 13.6, which indicates that mercury is 13.6 times denser than water.

Surface Tension and Capillarity

The molecules of a gas are separated by relatively large distances, and there is little interaction between the molecules other than collisions. In a liquid, however, attractive electrical forces exist between the relatively close molecules. Within the body of a liquid, a molecule is attracted in all directions by the surrounding neighboring molecules (Fig. 15.2).

However, a molecule near or at a free surface has a smaller force of attraction in that direction. In other words, there is a net force acting on the molecule toward the interior of the liquid.

Any system tends to move toward a condition of stable equilibrium in which its potential energy is a minimum. As a result of the net force on the molecules of a liquid at or near a free surface, the liquid adjusts its shape until its surface area and the potential energy are minimal. For example, a drop of liquid free from other

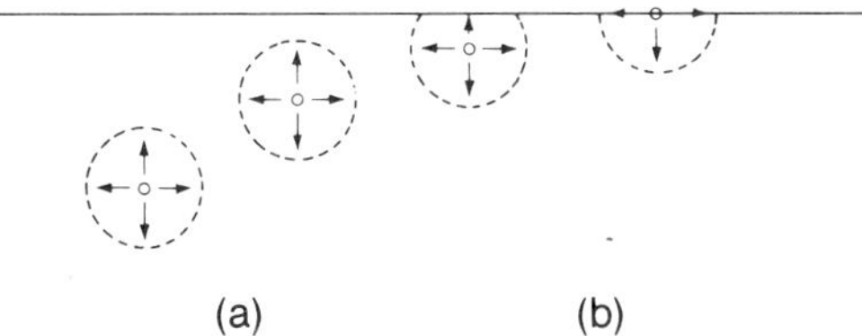

Figure 15.2

Molecular forces in a liquid. (a) Within the body of a liquid, a molecule is attracted in all directions by the surrounding molecules, and there is no net force. (b) However, at or near the surface of a liquid, there is a net force acting toward the interior of the liquid because of fewer molecules above.

forces assumes a spherical shape, since the sphere is the geometrical shape that has the least surface area for a given volume. This is observed when drops of water "bead up" on a freshly waxed automobile.

When a liquid changes shape, its surface does work and is "stretched" or in a state of tension; hence the name **surface tension.** As a result of surface tension, a razor blade or needle can be "floated" on a liquid surface (Fig. 15.3), and water bugs can scamper about on the surface of a pond.

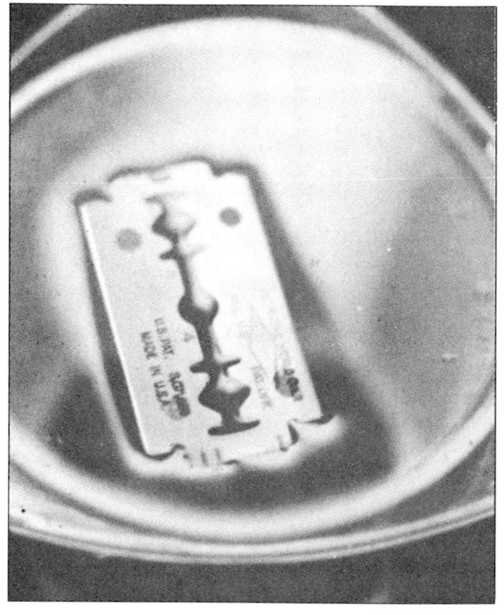

Figure 15.3

Surface tension. A razor blade floats on water because of surface tension.

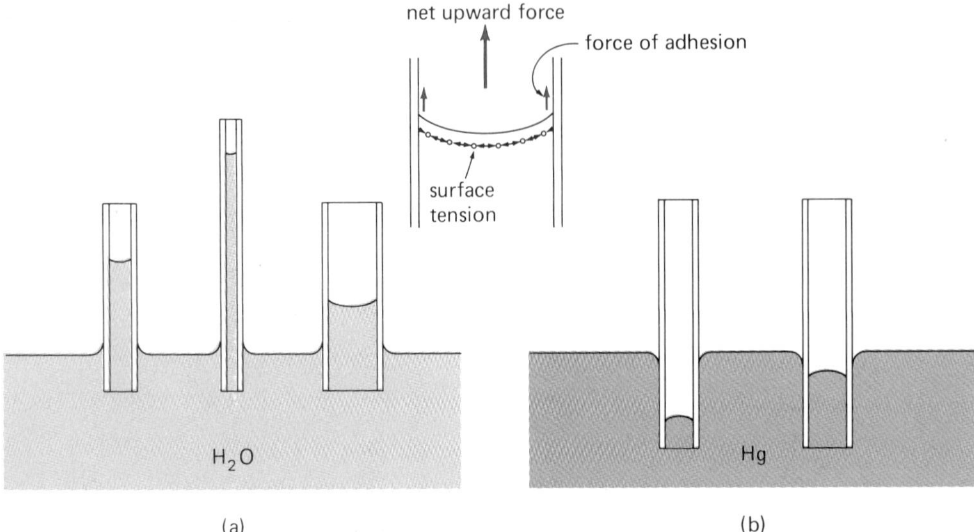

Figure 15.4
Capillary action. (a) For a liquid that wets the capillary tube, such as water on glass, the adhesion and surface tension contribute to a capillary rise in the tube with a concave meniscus or surface shape. (b) For a liquid that does not wet the capillary tube, such as mercury on glass, the surface tension causes the liquid to be depressed in the tube. In this case, the meniscus is convex.

The molecular attraction between the like molecules of a liquid is called a cohesive force. This cohesive force gives rise to *cohesion,* or the tendency of a liquid to remain as one assemblage of particles. The lack of cohesive forces between the molecules of a gas allows the gas to fill the entire space within which it is confined. **The force of attraction between the molecules of a liquid and the molecules of a container boundary (which are unlike each other) is called an adhesive force** and gives rise to *adhesion.* A liquid that *wets* a solid has greater adhesion than cohesion. This is an important property for "adhesives" such as glue or epoxy.

Surface tension and adhesive forces give rise to what is known as **capillary action,** or the rise or depression of a liquid in a small-bore capillary (Fig. 15.4). If a liquid wets the surface of a capillary (adhesion greater than cohesion), the action of surface tension in this case causes the liquid to rise vertically in the capillary tube. The adhesive wetting and surface-tension contracting of the surface essentially "draw up" or "lift" the liquid in the tube until the adhesive lifting force is balanced by the weight of the column.

The hemispherical surface of the liquid column is called its *meniscus.* In the case of capillary rise, the meniscus is concave. For liquids that do not wet the tube (cohesion greater than adhesion), the surface tension causes the liquid in the tube to be depressed, and the meniscus is convex. Mercury exhibits capillary depression.

Viscosity

Recall that a fluid is a substance that flows under the application of a shear stress. The internal resistance of a fluid during this deformation or flow is expressed in terms of **viscosity.** In a liquid, the resistance is due primarily to molecular attraction as one "layer" of the liquid slides over another. As illustrated in Figure 15.5, a shear stress causes a fluid to flow.

Viscosity relates the shear stress to the rate of angular deformation. For example, if equal shears are applied for equal amounts of time, the fluid with the smaller angle has the greater viscosity, or internal resistance to flow. In general, we tend to think of "thick" liquids, such as molasses and tar, as being highly viscous liq-

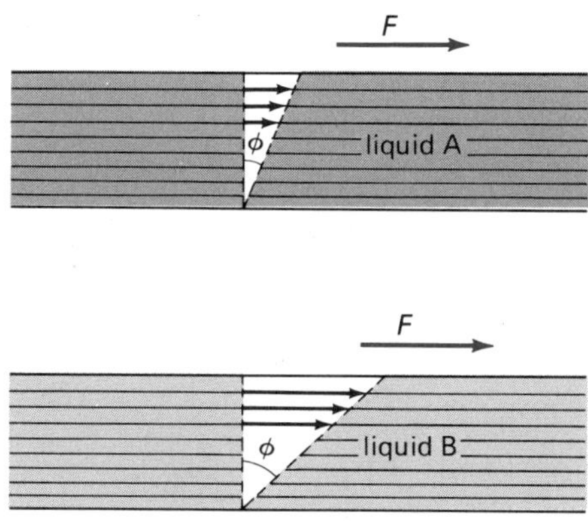

Figure 15.5
Liquid viscosity. In a liquid, the viscosity or internal resistance to flow is primarily due to molecular attraction as one "layer" of fluid slides over another. For a given shear stress, liquid A has a slower rate of change of ϕ and hence a greater viscosity than liquid B.

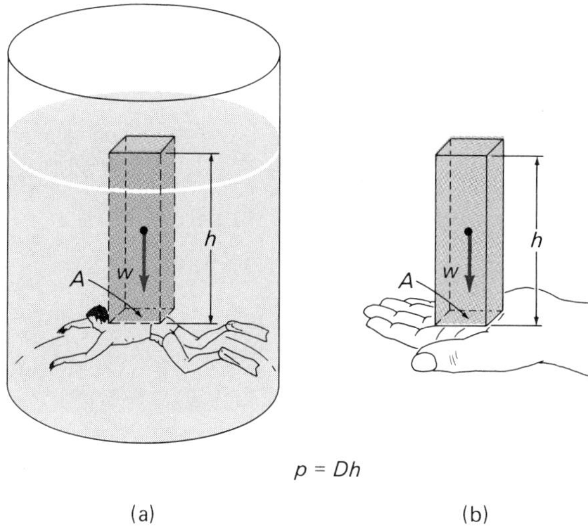

Figure 15.6
Pressure and depth. The pressure force at any depth in a fluid is equal to the weight of the column of fluid above. The pressure at depth h is given by $p = \rho gh$ (constant density). See text for description.

uids and "thin" liquids, such as alcohol and water, as being less viscous.

As might be expected, viscosity is temperature-dependent. In liquids the viscosity generally decreases with increasing temperature, and in gases the viscosity increases with increasing temperature. This is because in gases the viscosity is due to molecular collisions rather than to molecular attraction.

15.2 Pressure and Pressure Measurement

A fluid at rest exerts a force on the walls of its container that is perpendicular, or normal, to the walls, so it is convenient to express the force in terms of pressure. Recall that **pressure p** is defined as the normal force F per unit area A:

$$p = \frac{F}{A} \qquad \textbf{(Eq. 15.6)}$$

Within the volume of a fluid, the pressure depends on depth, since the force on a surface is dependent on the weight of the fluid above it (Fig. 15.6). Considering a column of height h above an area A, the mass of fluid in the column is

$$m = \rho V = \rho Ah$$

where $\rho = m/V$ is the mass density. Then the pressure due to the weight of the column is

$$p = \frac{F}{A} = \frac{w}{A} = \frac{mg}{A} = \frac{\rho gAh}{A} = \rho gh$$

or

$$\boxed{p = \rho gh = Dh} \qquad \textbf{(Eq. 15.7)}$$

where $D = \rho g$ is the weight density. Thus, the pressure within a fluid is directly proportional to the depth h. Equation 15.7 is commonly referred to as the **pressure-depth equation.**

It is assumed that the density is constant. This is a reasonable assumption for liquids and relatively small volumes of gases. In the atmosphere, however, the density of air is not constant with height, because the weight of the overlying atmosphere compresses the air near the Earth.

EXAMPLE 15.1 A cylindrical sea lab with a diameter of 10 ft and a height of 8.0 ft is lowered into the ocean so the top of the lab is 200 ft below sea level. What are (a) the pressure and force on the top of the lab and (b) the pressure on the bottom of the lab due to the sea water? (Assume significant figures.)

Solution: (a) From Table 15.1, for sea water $D = 64.0\ \text{lb/ft}^3$. Then, at a depth of $h = 200$ ft,

$$p = \rho gh = Dh = (64.0\ \text{lb/ft}^3)(200\ \text{ft})$$
$$= 12{,}800\ \text{lb/ft}^2$$

The area of the circular top of the lab is $A = \pi r^2 = \pi(5.0)^2 = 78.5\ \text{ft}^2$, and $p = F/A$, so

$$F = pA = (12{,}800\ \text{lb/ft}^2)(78.5\ \text{ft}^2)$$
$$= 1{,}004{,}800\ \text{lb} \simeq 502\ \text{tons}$$

(b) The bottom of the sea lab is $h = 208$ ft below sea level, so the pressure at this depth is

$$p = Dh = (64.0\ \text{lb/ft}^3)(208\ \text{ft}) = 13{,}312\ \text{lb/ft}^2$$

In the previous discussion the pressure was that due only to the fluid itself. In many instances, this is not the only contribution to the pressure. In an open container, a liquid is also subjected to atmospheric pressure (pressure due to the weight of the overlying atmospheric gases).

Atmospheric pressure at sea level is 14.7 lb/in^2 or $1.013 \times 10^5\ \text{N/m}^2$. By Pascal's principle, an external pressure applied to an enclosed fluid is transmitted undiminished throughout the volume of the fluid (cf. Chapter 11). Hence, the total or absolute pressure p_{abs} at a depth h in an open liquid is

$$p_{abs} = \rho gh + p_{atm}$$
$$= Dh + p_{atm} \qquad \textbf{(Eq. 15.8)}$$

where p_{atm} is the atmospheric pressure.

Pressure Measurement

The previous discussion brings up an important consideration in pressure measurement. Like any measurement, a pressure measurement is an interval or difference relative to some reference. The usual references are absolute zero pressure and atmospheric pressure (Fig. 15.7). Absolute zero pressure corresponds to a complete vacuum, or no pressure at all.

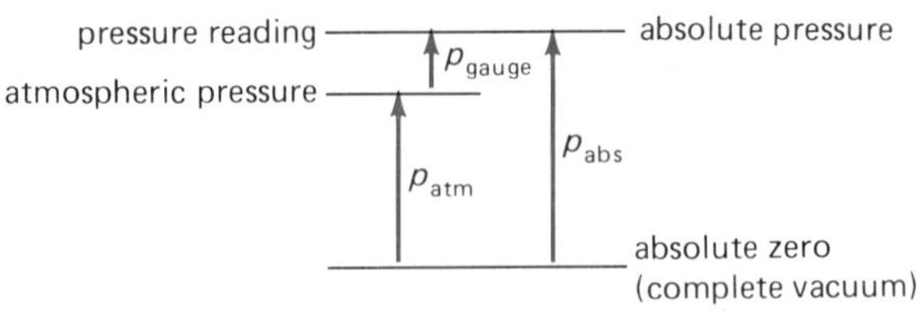

(a) positive gauge pressure
$p_{abs} = p_{gauge} + p_{atm}$

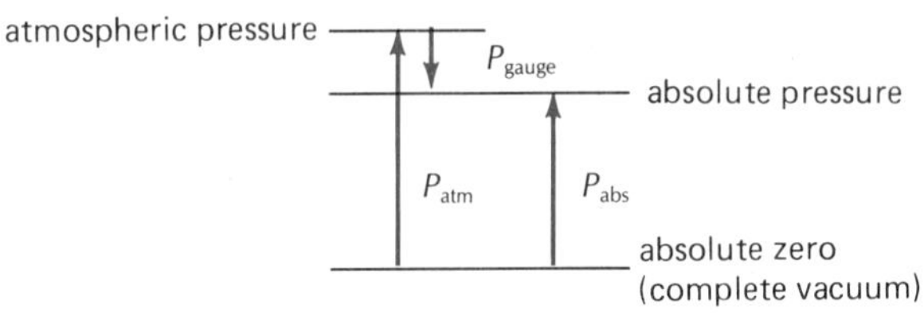

(b) negative gauge pressure (partial vacuum)
$p_{abs} = -p_{gauge} + p_{atm}$

Figure .7
Pressure measurement. Pressure measurements are usually referenced to either absolute zero pressure (complete vacuum) or atmospheric pressure. The gauge pressure is the difference between the absolute pressure and atmospheric pressure. (a) If $p_{abs} > p_{atm}$, then the gauge pressure is positive. (b) If $p_{abs} < p_{atm}$, then the gauge pressure is negative, indicating a partial vacuum.

When a pressure is expressed as a difference between a pressure value and a complete vacuum, it is called an **absolute pressure.** When a pressure is expressed as a difference between a pressure value and atmospheric pressure, it is called **gauge pressure.** As can be seen from Figure 15.7, absolute pressure = gauge pressure + atmospheric pressure, or

$$p_{abs} = p_{gauge} + p_{atm} \qquad \textbf{(Eq. 15.9)}$$

Compare this to Equation 15.8, where $p_{gauge} = \rho gh = Dh$. This would be the pressure measured on some pressure-measuring instrument—e.g., a pressure gauge—that had been zeroed at sea level.

It should be noted that gauge pressure is not always greater than atmospheric pressure. A negative gauge pressure indicates a pressure less than atmospheric pressure or the pressure of a partial vacuum.

Now let's consider some pressure measurement instruments. Atmospheric pressure is

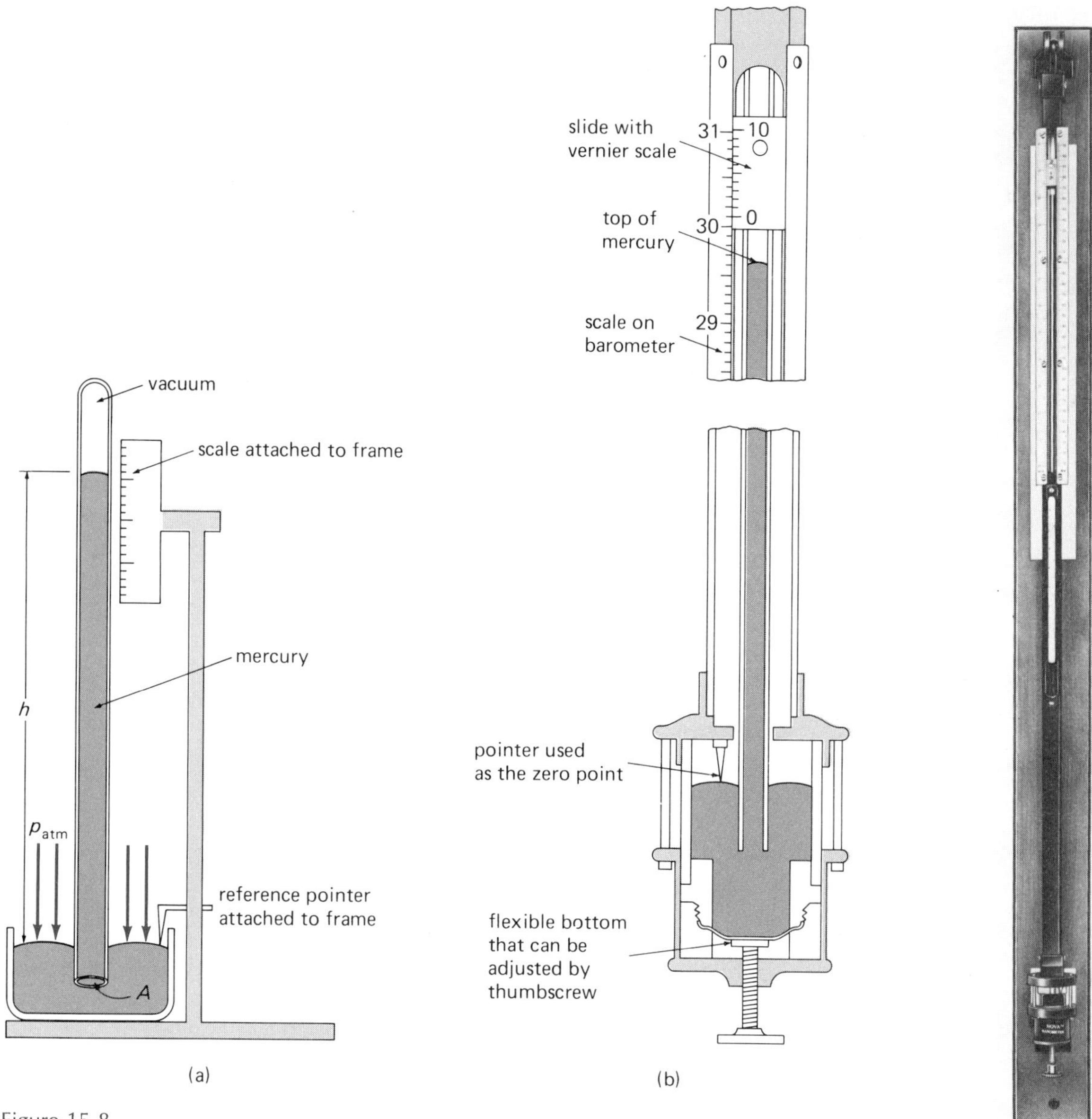

Figure 15.8
The mercury barometer. The simplified diagram (a) illustrates the principle of a mercury barometer. Atmospheric pressure supports the column of mercury in the tube, and hence the height of the column is proportional to the atmospheric pressure. Diagram (b) shows some details of an actual mercury barometer as shown in photograph (c). (Photo courtesy of Princo Instruments Co.)

measured with a **barometer.** A mercury barometer consists essentially of a tube of mercury that has been inverted in a pool of mercury (Fig. 15.8). Some of the mercury runs out of the tube, but the external air pressure on the surface of the mercury pool supports a column of mercury of height h. Since the system is in equilibrium, the external pressure is equal to the pressure due to the weight of the column of mercury, or $p = \rho gh = Dh$ (Eq. 15.7), where ρ and D are the mass and weight densities of mercury, respectively.

Standard atmospheric pressure at sea level supports a column of mercury 30 in. (actually

29.92 in.) or 76 cm = 760 mm in height. The barometric pressure is often reported in inches, centimeters, or millimeters of mercury. In honor of Evangelista Torricelli (1608–1647), who invented the mercury barometer, a pressure corresponding to one millimeter of mercury (mm Hg) is given the unit of one torr, i.e., 1 mm Hg = 1 torr. The pressure of one atmosphere (atm) in standard units is given through Equation 15.7, and

Atmospheric Pressure and Units

$$
\begin{aligned}
1 \text{ atm} &= 14.7 \text{ lb/in}^2 \\
&= 1.01325 \times 10^5 \text{ N/m}^2 \text{ (Pa)} \\
&= 1{,}013{,}250 \text{ dyn/cm}^2 \\
&= 30 \text{ in. Hg} = 76 \text{ cm Hg} \\
&= 760 \text{ mm Hg (torr)}
\end{aligned}
$$

EXAMPLE 15.2 What is the atmospheric pressure if the column height of a mercury barometer is 76 cm?

Solution: With h = 76 cm = 0.76 m and, for mercury, $\rho = 13.6 \times 10^3$ kg/m^3 (Table 15.1),

$$
\begin{aligned}
p &= \rho g h \\
&= (13.6 \times 10^3 \text{ kg/m}^3)(9.80 \text{ m/s}^2)(0.76 \text{ m}) \\
&= 1.0129 \times 10^5 \text{ N/m}^2
\end{aligned}
$$

(The value is slightly different from the accepted value because of the approximate values, or lack of significant figures, for ρ and g.)

A common pressure unit used in meteorology is the millibar (mb). One bar = 10^3 mb = 10^5 N/m^2 = 10^6 dyn/cm^2, and

$$1 \text{ atm} = 1.01325 \text{ bar} = 1013.25 \text{ mb}$$

Atmospheric pressure is also measured with an aneroid barometer (Fig. 15.9). It makes use of a sensitive metal diaphragm on an evacuated chamber to detect pressure, which is indicated mechanically.

A common device used to measure gauge pressure is the open-tube manometer (Fig. 15.10). It consists of a U-shaped tube containing a liquid, usually mercury or water depending on the magnitude of the pressure to be measured. When both ends of the tube are open to the atmosphere, the liquid levels in both arms of the U-tube are at the same height.

Figure 15.9
An aneroid barometer. An aneroid barometer is essentially a diaphragm pressure gauge which indicates atmospheric pressure mechanically. Notice that fair weather is associated with high barometric pressure and that rain is associated with low barometric pressure. (Courtesy of Taylor Instrument Division, Sybron Corp., Aden, NC 28704.)

When one end of the tube is connected to a container of gas under pressure, the liquid rises in the other arm until the pressures are equalized. The difference in the levels is a measure of gauge pressure, and

$$p = \rho g h = Dh$$

We can see that this is the gauge pressure by computing the equal pressures at the bottom of the U-tube for both columns:

$$
\begin{array}{ccc}
\text{Left column} & & \text{Right column} \\
p_{\text{abs}} + \rho g y_1 & = & \rho g y_2 + p_{\text{atm}}
\end{array}
$$

where p_{abs} is the absolute pressure of the fluid in the container. Then,

$$p_{\text{abs}} - p_{\text{atm}} = p_{\text{gauge}} = \rho g(y_2 - y_1) = \rho g h$$

A closed-tube manometer is also shown in Figure 15.10. Here the pressure is measured relative to the vacuum in the closed tube. Hence, the closed-tube manometer measures absolute pressure.

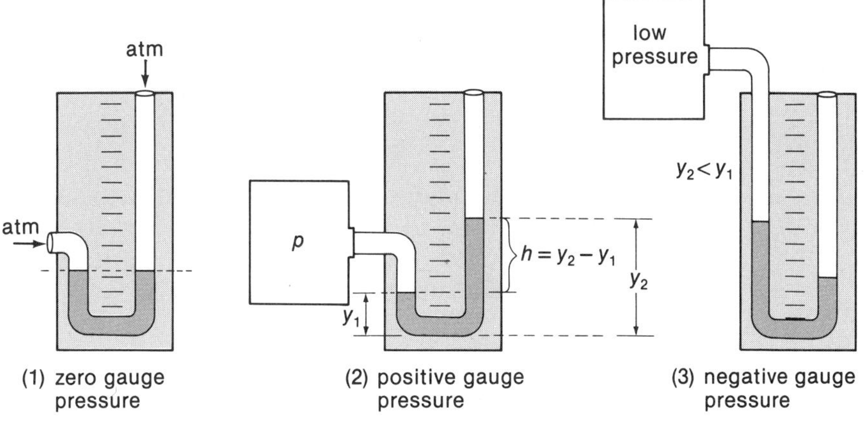

(a) open-tube manometer

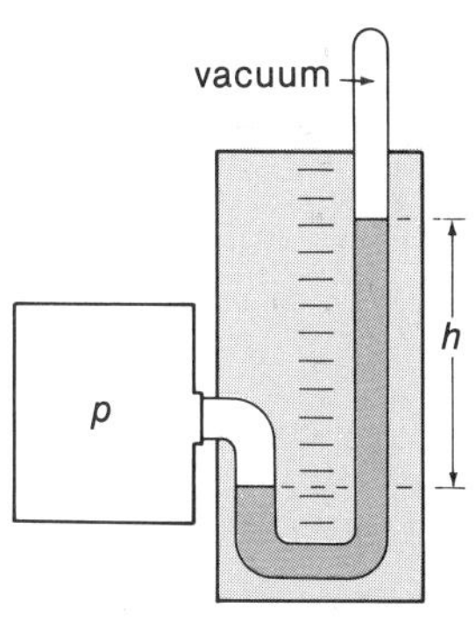

(b) closed-tube manometer

Figure 15.10

Manometers. (a) Open-tube manometer. The open-tube manometer measures pressure relative to atmospheric pressure (1). Diagrams (2) and (3) show positive and negative gauge pressures, respectively. (b) Closed-tube manometer. The manometer measures absolute pressure relative to the vacuum in the closed tube.

15.3 Buoyancy and Archimedes' Principle

When an object submerged in a liquid is lifted, it seems lighter or easier to lift (requires less force) than when it is not in a liquid. Therefore, there must be an upward force acting on the object in the liquid that balances part of the object's weight. This phenomenon is called **buoyancy,** and the upward acting force is called the **buoyant force.**

Consider an object submerged in a liquid, as illustrated in Figure 15.11. The liquid pressure acting downward on the top of the object is

$$p_1 = \rho g h_1$$

where ρ is the mass density of the liquid. Similarly, the liquid pressure acting upward on the

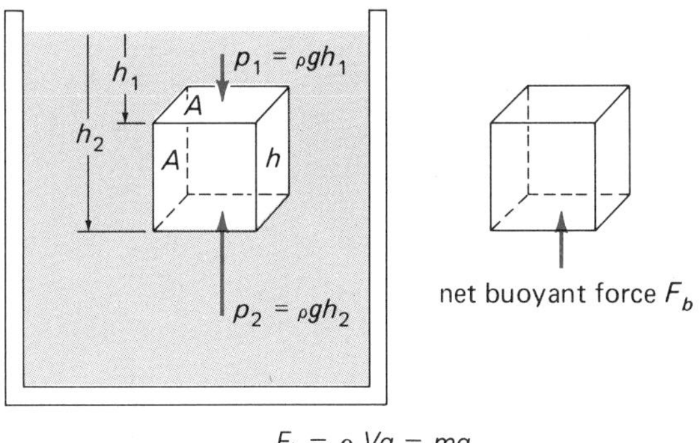

$F_b = \rho Vg = mg$
(weight of liquid displaced)

Figure 15.11

Buoyant force. The pressure on the bottom of a submerged object is greater than the pressure on the top. Hence, there is a net (buoyant) force upward. It can be shown that the buoyant force is equal to the weight of the volume of fluid displaced. See text for description.

bottom of the object is

$$p_2 = \rho g h_2$$

Since $h_2 > h_1$, the upward pressure is greater than the downward pressure. Expressing the difference of these pressures in terms of force,

$$\begin{aligned} F_b &= F_2 - F_1 = p_2 A - p_1 A \\ &= \rho g h_2 A - \rho g h_1 A \\ &= \rho g (h_2 - h_1) A \\ &= \rho g h A \end{aligned}$$

where F_b is the upward buoyant force and $h = h_2 - h_1$ is the height of the object. The volume of the object is $V = hA$; hence, the buoyant force is

$$\boxed{F_b = \rho V g = mg} \qquad \textbf{(Eq. 15.10)}$$

(buoyant force) = (weight of displaced liquid)

where $\rho = m/V$ or $m = \rho V$, which is the mass of the fluid displaced by the object, and mg is the weight of the displaced fluid. This concept was first put forth by the Greek scientist Archimedes (287–212 B.C.) and is known as **Archimedes' principle:**

> **An object in a fluid is buoyed upward with a force equal to the weight of the volume of the fluid displaced by the object.**

Archimedes' principle applies to gases as well as liquids—i.e., to fluids in general. For example, a helium balloon is buoyed up in air. Experimental proof of Archimedes' principle is illustrated in Figure 15.12.

EXAMPLE 15.3 A 10-kg block of wood 30 cm on a side is held submerged under water. What force is required to do this?

Solution: The length of a side of the block is 30 cm = 0.30 m, and the volume of the cubical block is

$$V = 0.30 \text{ m} \times 0.30 \text{ m} \times 0.30 \text{ m} = 0.027 \text{ m}^3$$

The density of water is $\rho = 1000 \text{ kg/m}^3$ from Table 15.1. Then, the weight of the volume of water displaced by the submerged block is

$$\begin{aligned} w = mg &= \rho V g \\ &= (10^3 \text{ kg/m}^3)(0.027 \text{ m}^3)(9.8 \text{ m/s}^2) \\ &= 265 \text{ N} \end{aligned}$$

By Archimedes' principle, this is the magnitude of the upward buoyant force. To keep the block submerged, this force must be balanced by an equal downward force, which is the sum of the applied force and the weight force of the 10-kg block, i.e., $F + mg = F_b$, and

$$\begin{aligned} F = F_b - mg &= 265 \text{ N} - (10 \text{ kg})(9.8 \text{ m/s}^2) \\ &= 167 \text{ N} \ (\simeq 38 \text{ lb}) \end{aligned}$$

Floating and Sinking

The magnitude of the buoyant force acting on an object determines whether the object floats or sinks. An object sinks if its weight exceeds that of the displaced fluid (buoyant force), and it rises if its weight is less than that of the displaced fluid.

Thus, a block of wood held submerged underwater and then released will rise to the surface and float. The block will adjust to a depth in the water at which its weight is equal to the weight of the water displaced ($\Sigma F = 0$, equilibrium).

This is more conveniently expressed in terms of the densities of the object ρ_o and the fluid ρ_f. The weight of the object is

$$w_o = m_o g = \rho_o g V_o$$

and the weight of the displaced fluid or buoyant force is

$$F_b = w_f = m_f g = \rho_f g V_f$$

Now suppose the object is completely immersed in the fluid and released. While it is completely submerged, $V_o = V_f$, and, dividing the two previous equations,

$$\frac{F_b}{w_o} = \frac{\rho_f}{\rho_o}$$

or

$$F_b = \left(\frac{\rho_f}{\rho_o}\right) w_o \qquad \textbf{(Eq. 15.11)}$$

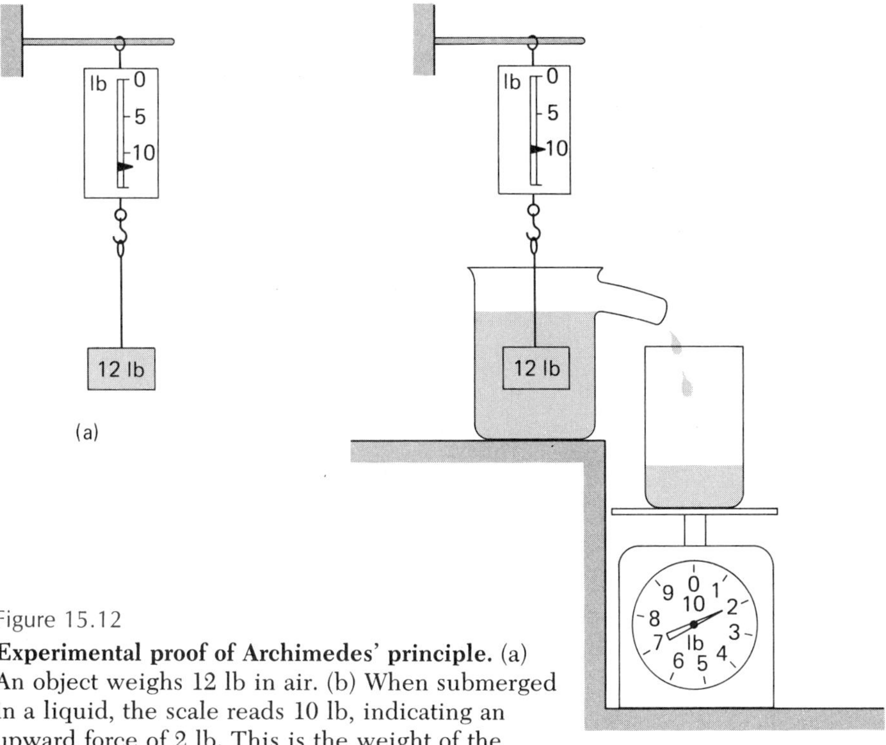

Figure 15.12
Experimental proof of Archimedes' principle. (a) An object weighs 12 lb in air. (b) When submerged in a liquid, the scale reads 10 lb, indicating an upward force of 2 lb. This is the weight of the liquid displaced.

Then,

> **(a) If $\rho_f > \rho_o$, then $F_b > w_o$, and the object will be buoyed up to the surface and will float.**
> **(b) If $\rho_f < \rho_o$, then $F_b < w_o$, and the object will sink.**
> **(c) If $\rho_f = \rho_o$, and $F_b = w_o$, the object will float in equilibrium at any submerged depth.**

(It is assumed that ρ_f is constant with depth or that the fluid is incompressible.) Hence,

> **An object will float in a fluid if the density of the object is less than or equal to the density of the fluid.**

EXAMPLE 15.4 A cube of material 20 cm on a side has a mass of 7200 g. (a) Will the cube float in water? (b) If so, what will be the height of the portion of the cube extending above the water surface?

Solution: (a) The volume of the cube is $V = L^3 = (20\text{ cm})^3 = 8000\text{ cm}^3$. Its density is then

$$\rho_c = \frac{m}{V} = \frac{7200\text{ m}}{8000\text{ cm}^3} = 0.90\text{ g/cm}^3$$

Since the density of water is $\rho_w = 1.0\text{ g/cm}^3$, then $\rho_w > \rho_c$, and the cube will float.

(b) The weight of the cube $w_c = \rho_c g V_c$ will be balanced by the buoyant force $F_b = \rho_w g V_w$, where V_w is the volume of water displaced by the *submerged portion* of the cube. That is, $F_b = w_c$, or

$$\rho_w g V_w = \rho_c g V_c$$

and

$$\frac{V_w}{V_c} = \frac{\rho_c}{\rho_w} = \frac{0.90\text{ g/cm}^3}{1.0\text{ g/cm}^3} = 0.90$$

Hence, $V_w = (0.90)V_c$, and the volume of water displaced is 0.90 of the volume of the cube, or 90 percent of the cube is submerged.

This means that 10 percent of the cube's volume is above the water, or

$$V_{\text{above}} = (0.10)V_c = (0.10)(8000\text{ cm}^3) = 800\text{ cm}^3$$

SPECIAL FEATURE 15.1

The Tip of the Iceberg

Icebergs are huge floating masses of ice that have broken off the ends of glaciers or polar ice sheets. These floating islands of ice range in size up to 30 km (19 miles) long and 60 m (200 ft) thick. See Figure 1.

The density of ice is 0.92 g/cm^3, and the density of sea water is 1.03 g/cm^3 (Table 15.1). Then, forming the density ratio, an iceberg floating in the ocean has 0.92/1.03 = 0.89, or 89 percent, of its volume below the surface. We see only the "tip of the iceberg," or 11 percent (1/9) of its volume.

Icebergs pose a serious threat to shipping. After the *Titanic* struck an iceberg and sank in 1912, an international ice patrol was established by joint action of all nations with shipping interests in the North Atlantic. The ice patrol is the responsibility of the U.S. Coast Guard, which locates icebergs, predicts their drift, and issues warnings.

Figure 1
The tip of the iceberg. Almost 90 percent of an iceberg lies below the surface. (Courtesy U.S. Coast Guard.)

The cube has a cross-sectional area of A = 20 cm × 20 cm = 400 cm^2, and $V_{\text{above}} = Ah_{\text{above}}$. Thus,

$$h_{\text{above}} = \frac{V_{\text{above}}}{A} = \frac{800 \text{ cm}^3}{400 \text{ cm}^2} = 2.0 \text{ cm}$$

and the cube extends 2.0 cm above the water surface.

See Special Feature 15.1 for a similar example.

Archimedes' principle has many applications. Large ships made of steel and weighing many tons float because of their shapes and resulting displacement volumes, whereby the average density of a ship is less than that of water. Submarines submerge and surface because of buoyancy (Fig. 15.13). When a sub's ballast tanks are flooded with sea water, the average density of the sub is greater than that of the sea water, and the sub sinks. By pumping out the tanks (removing ballast), the average density of the sub can be made less than that of the sea water, so the sub rises and floats on the surface.

A quick and common method of determining the specific gravity of a liquid is based on Archimedes' principle. This method makes use of a hydrometer, which is usually a sealed glass tube with an enclosed scale in the stem and a bulb weighted with metal shot (Fig. 15.14). The depth to which a hydrometer floats in a liquid depends on the liquid's specific gravity. The hydrometer is calibrated to read a specific gravity of 1.0 in water. In a denser liquid, a hydrometer would float higher and indicate a density greater than 1.0. Similarly, a hydrometer would float lower in a less dense liquid.

Hydrometers are used in many applications. Perhaps the most familiar uses are in battery-acid testers and automobile radiator antifreeze testers.

15.4 Fluid Flow

In the previous discussions, the conditions of fluids at rest or in equilibrium (called *hydrostatics* for liquids) were considered. The study of fluid flow (called *hydraulics* for liquids) is another important topic, and a difficult one; hence, simplifications and idealizations are in order.

We will assume that fluids in motion exhibit streamline flow. This implies that the flow is

(a)

(b)

Figure 15.13

Buoyancy. (a) A submarine submerges and surfaces by flooding and emptying ballast tanks so as to change its overall density. (Courtesy U.S. Navy.) (b) When filled with enough hot air, balloons become buoyant and rise. Why? (Courtesy E. Stampf, Lander College.)

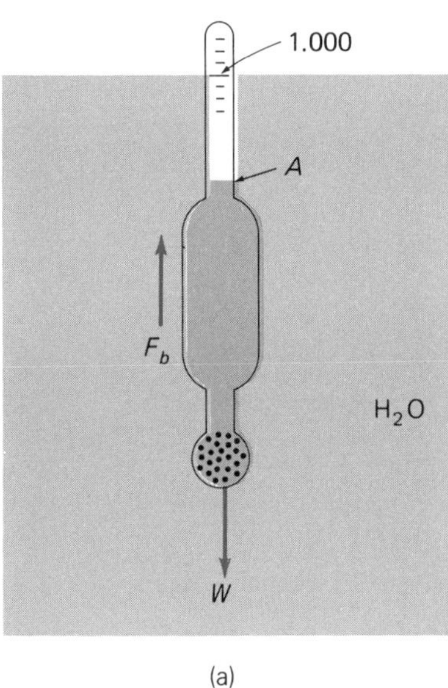

(a)

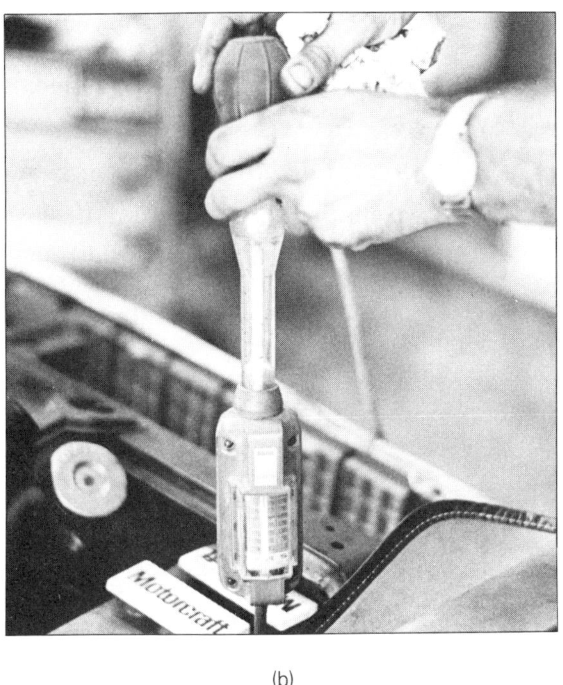

(b)

Figure 15.14

The hydrometer. (a) When a hydrometer floats in water, the stem calibration reads 1.000, the specific gravity of water. In more or less dense liquids, the hydrometer floats higher or lower, respectively, and the stem calibration indicates the specific gravities of the liquids. (b) A battery-acid tester. The hydrometer in the tester measures the specific gravity or "strength" of the battery acid solution.

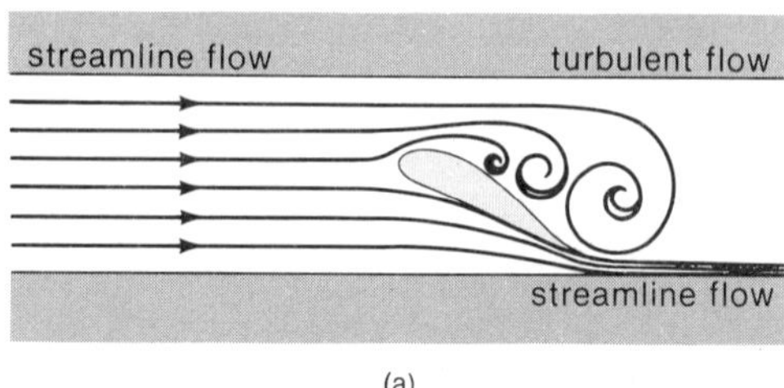

Figure 15.15
Steamline and turbulent flows. In streamline flow (a), the fluid particles follow parallel paths called streamlines. The diagram illustrates streamlines and turbulent flow around an air foil with a large angle of attack. It the photograph (b), the hot gases from a cigarette, made visible by the smoke particles, are initially in streamline flow and then become turbulent.

steady and that fluid particles follow given paths called **streamlines** (Fig. 15.15).

In streamline flow, ideally the streamlines never cross, since in such a case the particle would have the option of two paths and the flow would not be steady. (Should this occur, as it does in many practical instances, the flow is said to be turbulent and is characterized by little whirlpools and eddies that are difficult to analyze.) As a further simplification, we will also consider a fluid to be incompressible and nonviscous.

Rate of Flow

With these assumptions, certain predictions about fluid flow in smooth pipes can be made—for example, the rate of flow. The **flow rate** is defined as the volume of fluid that passes a certain cross section of its pipe per unit time.

Consider a fluid flowing in a pipe with a constant speed v. Each particle of the fluid moves a distance vt during a time t (Fig. 15.16). If the pipe has a cross-sectional area A, the volume V of fluid passing a certain point in time t is

$$V = Avt \qquad \textbf{(Eq. 15.12)}$$

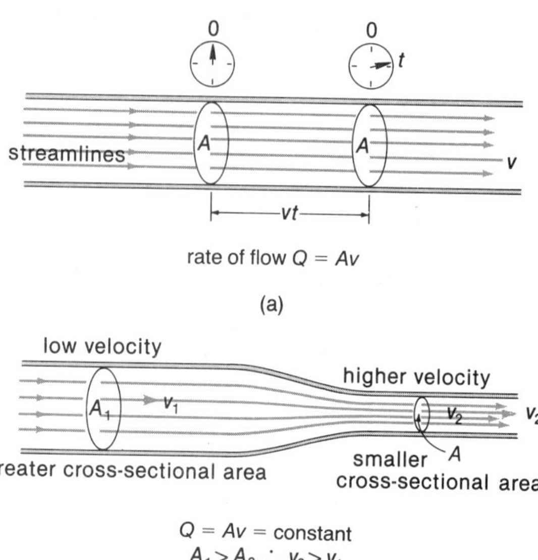

Figure 15.16
Rate of flow. (a) For a pipe with a uniform cross-sectional area A, the flow rate is $Q = Av$, where v is the average speed of the fluid. See text for description. (b) Under ideal conditions, the flow rate is constant. As a result, the speed of the fluid increases if the cross-sectional area of the pipe decreases.

Then the flow rate Q is

$$Q = \frac{V}{t} = \frac{Avt}{t} = Av$$

or

$$\boxed{Q = Av} \qquad \text{(Eq. 15.13)}$$

Notice the units of the flow rate are $(m^2)(m/s) = m^3/s$, or volume per time.

Under ideal conditions (e.g., an incompressible liquid) the flow rate will remain constant, and we have the **continuity equation:**

$$Q = A_1v_1 = A_2v_2$$
$$= Av = \text{constant} \qquad \text{(Eq. 15.14)}$$

That is, the same volume of fluid passes A_1 and A_2 in the same unit time. This is essentially a statement of conservation of mass ("What goes in must come out"), as would be expected in the practical case of a relatively incompressible liquid.

Equation 15.14 indicates that if the cross-sectional area of the pipe varies, so does the flow speed of the fluid so as to maintain the product Av constantly. Hence, if $A_2 < A_1$, then $v_2 > v_1$; that is, the speed of the fluid flow increases if the pipe is constricted. Notice in Figure 15.16 that in a region of greater flow speed, the streamlines are closer together.

EXAMPLE 15.5 A horizontal water line has a 1.0-in.-diameter pipe that reduces to a $\frac{1}{2}$-in.-diameter pipe. If the water flows through the large-diameter pipe with a speed of 6.0 ft/s, (a) what is the flow speed of the water in the $\frac{1}{2}$-in. pipe, and (b) how many gallons of water per minute will be delivered by the line?

Solution: (a) With $d_1 = 1.0$ in., $d_2 = \frac{1}{2}$ in., and $v_1 = 6.0$ ft/s, by Equation 15.14,

$$v_2 = \frac{A_1v_1}{A_2} \quad \text{or} \quad v_2 = \left(\frac{d_1}{d_2}\right)^2 v_1$$

$$= \left(\frac{1 \text{ in.}}{\frac{1}{2} \text{ in.}}\right)^2 (6.0 \text{ ft/s}) = 24 \text{ ft/s}$$

(b) The flow rate is $Q = Av$. Computing this for A_1 with $d_1 = 1.0 \text{ in.} = \frac{1}{12}$ ft,

$$Q = A_1v_1 = \frac{\pi d_1^2 v_1}{4} = \frac{\pi(\frac{1}{12}\text{ ft})^2(6.0 \text{ ft/s})}{4}$$
$$= 0.033 \text{ ft}^3/\text{s}$$

Then, using $1 \text{ ft}^3 = 7.5$ gal,

$$Q = 0.033 \text{ ft}^3/\text{s}(7.5 \text{ gal/ft}^3)(60 \text{ s/min})$$
$$= 15 \text{ gal/min}$$

Bernoulli's Equation

The flow speed of a fluid in a pipe is associated with pressure. The flow speed would also be affected if the pipe carried the fluid to a greater height; and in this case, in terms of energy, the fluid would have a greater potential energy and less kinetic energy. The density of the fluid would be important since the potential energy for a volume of fluid raised a distance h is ρgh.

A mathematical relationship involving these parameters was developed by Daniel Bernoulli (1700–1782), a Swiss mathematician, and consequently it is known as **Bernoulli's equation:**

$$p_1 + \rho gh_1 + \tfrac{1}{2}\rho v_1^2 = p_2 + \rho gh_2 + \tfrac{1}{2}\rho v_2^2$$

or

$$p + \rho gh + \tfrac{1}{2}\rho v^2 = \text{constant}$$

(Eq. 15.15)

Notice that ρgh is the potential energy *per unit volume*, i.e., $\rho gh = (m/V)gh = (mgh)/V$. Similarly, $\frac{1}{2}\rho v^2$ is the kinetic energy *per unit volume*, and pressure has the units of work *per unit volume*.

The relationship of the quantities in Bernoulli's equation is illustrated in Figure 15.17. The equation is easily derived from work-energy considerations; however, this will not be our approach. Instead, emphasis is placed on the applications of Bernoulli's equation. It should be noted that p in the equation is the absolute pressure.

First, it can be seen that for a fluid at rest ($v = 0$) in a container, the pressure difference of two depths is given by Bernoulli's equation as

$$p_2 + \rho gh_2 + p_1 + \rho gh_1$$

or

$$p_2 - p_1 = \rho g(h_1 - h_2)$$

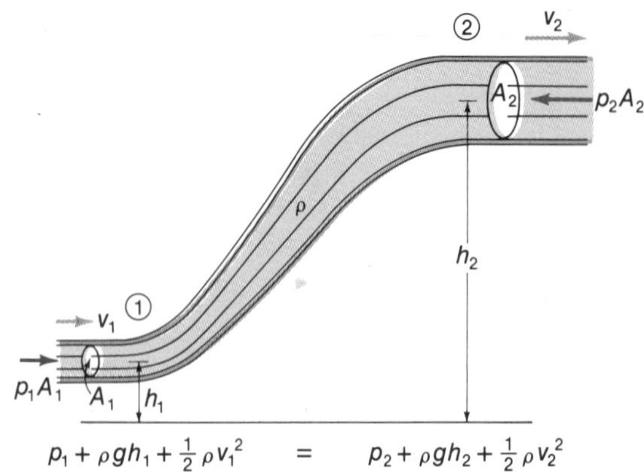

Figure 15.17
Bernoulli's equation. The diagram shows the relationships of the quantities in Bernoulli's equation for fluid flow, $p + \rho gh + \frac{1}{2} v^2$ = constant.

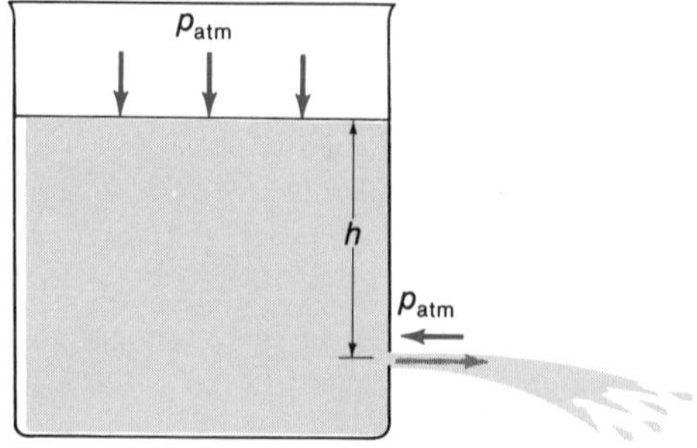

Figure 15.18
Application of Bernoulli's equation. The speed of liquid emerging from a hole in a large container can be calculated from Bernoulli's equation to be $v = \sqrt{2gh}$. See text for description.

This is the same result as that from Equation 15.7, $p = \rho gh$ or $\Delta p = \rho g \Delta h$, where h was taken as positive in the downward direction.

Bernoulli's equation can be used to calculate the speed of a liquid emerging from a hole in a container (Fig. 15.18). In this case, $p_2 = p_1 = p_{atm}$; that is, the pressure on the liquid surface is the same as that external to the hole. For a large container, the liquid level falls slowly, so, to a reasonable approximation, $v_1 = 0$. Then, applying Bernoulli's equation,

$$\rho g h_1 = \tfrac{1}{2}\rho v_2^2 + \rho g h_2$$

or

$$v_2^2 = 2g(h_2 - h_1) = 2gh$$

Thus, the magnitude of the velocity of the liquid emerging from the hole is

$$v = \sqrt{2gh} \qquad \textbf{(Eq. 15.16)}$$

Notice that this is the same as the speed of an object dropped from rest and falling through a distance h.

EXAMPLE 15.6 A small hole 0.10 cm in diameter is punched in a large tank of water 150 cm below the surface of the water. What is the flow rate of the water initially emerging from the hole?

Solution: The flow rate is Av (Eq. 15.13). The area of the hole is $A = \pi r^2 = \pi d^2/4$, and $v = \sqrt{2gh}$ (Eq. 15.16). Then, with the given data,

$$Q = Av = \frac{\pi d^2 \sqrt{2gh}}{4}$$

$$= \frac{\pi (0.10\ \text{cm})^2 \sqrt{2(980\ \text{cm/s}^2)(150\ \text{cm})}}{4}$$

$$= 4.3\ \text{cm}^3/\text{s}$$

Other common and interesting fluid applications that can be analyzed with Bernoulli's equation, e.g., airplane lift and chimney draft, are given as questions at the end of the chapter.

The Venturi Meter

The flow rates of fluids in tubes can be measured by means of Venturi meters (Fig. 15.19). A liquid-filled U-tube–type manometer is commonly used for gases, and a standpipe-type manometer is used for liquids. By Bernoulli's equation, for the flowing fluid in both types of meters,

$$p_1 + \tfrac{1}{2}\rho v_1^2 = p_2 + \tfrac{1}{2}\rho v_2^2$$

where in the horizontal tube $h_1 = h_2$. When using the equation in the British system, notice that $\rho = D/g$.

The difference in the column heights of the meter is a measure of the pressure difference, and for the standpipe-type meter,

$$p_1 - p_2 = \rho g h$$

(In the U-tube–type meter, $\rho = \rho_\ell$, where ρ_ℓ is the density of the liquid in the U-tube.) Also,

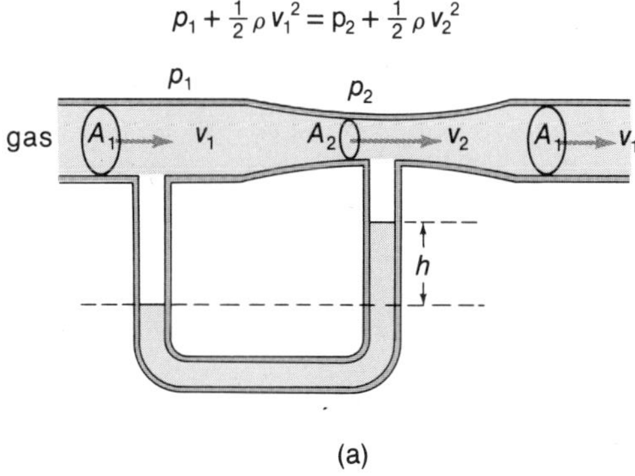

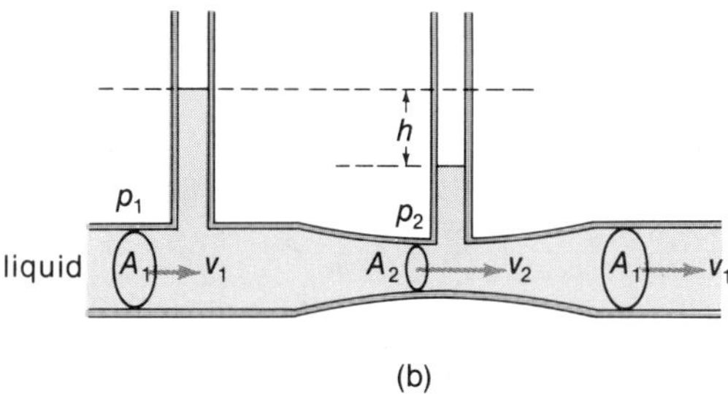

Figure 15.19
Venturi meters. The flow rates of fluids can be determined using Venturi meters, which are types of manometers. (a) U-tube–type meter for measuring the flow rates of gases. (b) Stand-pipe–type meter for measuring the flow rates of liquids. See text for descriptions.

from the continuity equation,

$$A_1 v_1 = A_2 v_2$$

From these three equations, the flow rate in the main pipe, after some algebra, can be shown to be

$$Q = A_1 v_1$$

$$= A_1 \left[\frac{2gh}{(A_1^2/A_2^2) - 1} \right]^{1/2} \qquad \textbf{(Eq. 15.17)}$$

(The flow rate expression for a U-tube meter gas measurement includes the mass densities of the gas and of the tube liquid, since they do not cancel out. In the standpipe meter the standpipe liquid is the same as that in the main tube.)

See Special Feature 15.2 for an application of a Venturi tube.

Viscous Flow and Poiseuille's Law

Real fluids have internal resistance to flow, which is described by viscosity. Such flow is

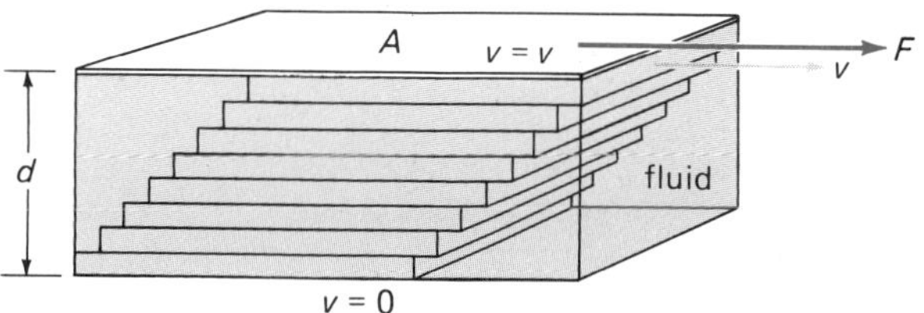

Figure 15.20
Laminar flow. Internal friction (viscosity) gives rise to a layered structure motion in response to a shear stress. Laminar flow is characteristic of streamline flow for viscous liquids at low speeds.

difficult to analyze, but a general insight into viscous flow can be obtained by considering laminar flow (Fig. 15.20). Internal friction causes layers to move over each other in response to a shear stress. This layered-structure motion, or laminar flow, is characteristic of streamline flow of viscous fluids at low velocities. At large velocities the flow becomes turbulent.

In terms of laminar flow, the **viscosity** η is given by

$$\eta = \left(\frac{F}{A}\right)\left(\frac{d}{v}\right) \qquad \textbf{(Eq. 15.18)}$$

where the quantities are identified in Figure 15.20.

The SI unit of viscosity is the poise (P), in honor of the French physicist Jean Poiseuille (1799–1869), who studied fluid flow, particularly blood flow. From Equation 15.18, the poise can be seen to be equal to

$$(\text{N/m}^2)(\text{m/m/s}) = \text{N-s/m}^2 \text{ (or Pa-s)}$$

The smaller unit of centipoise (cP) is often used (1 P = 1000 cP). The values of the viscosities of some liquids are listed in Table 15.2.

Table 15.2
Viscosities of Various Liquids

Liquid	Viscosity (P) at 20°C
Blood	2.5×10^{-3}
Glycerin	1.5
Mercury	1.6×10^{-3}
Oil (light machine)	0.11
Water	1.0×10^{-3}

SPECIAL FEATURE 15.2

The Carburetor

Notice in Figure 15.19 that at the construction of the Venturi tube the speed of the fluid increases and the pressure is reduced. This principle is used in the gasoline carburetor, which mixes air and fuel for combustion in an engine (Fig. 1).

Gasoline is maintained in the float chamber by a float-actuated valve at a level slightly below the opening of the outlet jet. When the throttle valve is open, air flows downward through a Venturi throat into the intake manifold. Because of the constriction in the throat, the air speed is increased and the pressure reduced in this region. Because of the pressure reduction, fuel flows from the outlet jet and mixes with the air.

The air flow is controlled by the throttle valve. An increase in the air flow by further opening the throttle valve will increase the pressure difference, and more fuel will flow.

When the engine is idling the throttle valve is closed. The vacuum on the manifold side of the valve then causes the fuel to flow past the idle adjusting screw to the outlet just below the throttle valve. When the throttle valve is opened, the vacuum at the idle fuel outlet is reduced, and fuel is supplied through the main jet.

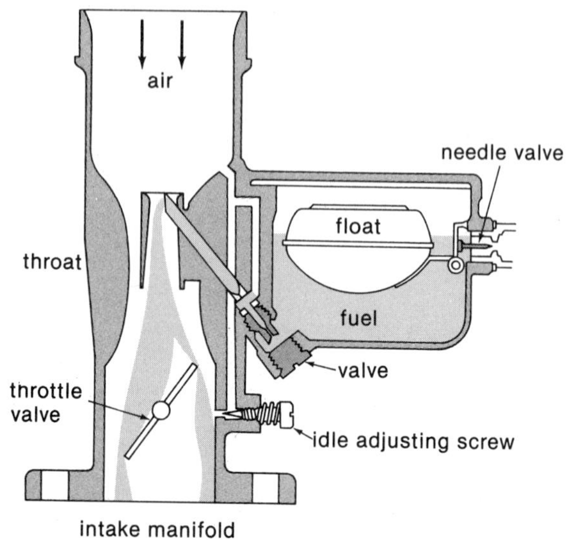

Figure 1

The gasoline carburetor. The simplified diagram illustrates the principle of the carburetor, which mixes fuel and air for combustion. A pressure reduction occurs in the venturi throat. See text for description.

Poiseuille's work led to the relationship describing the average flow rate, $\overline{Q} = A\overline{v}$, in terms of fluid and pipe or conduit parameters. We know that when the area of a pipe changes, the pressure changes (cf. Fig. 15.19), so the flow rate would be expected to be directly proportional to the pressure difference, $\Delta p = p_1 - p_2$ (where $p_1 > p_2$), and the radius r of the pipe (Fig. 15.21). Also, the flow rate would be inversely proportional to the length L of the pipe (longer path of resistance) and the viscosity (the greater the viscosity, the smaller the flow rate).

The final relationship is given by what is known as **Poiseuille's law:**

$$\overline{Q} = \frac{\pi r^4 \Delta p}{8\eta L} \qquad \textbf{(Eq. 15.19)}$$

Note that the flow rate depends on the fourth power of the radius (r^4) of the pipe, which makes this a very sensitive parameter.

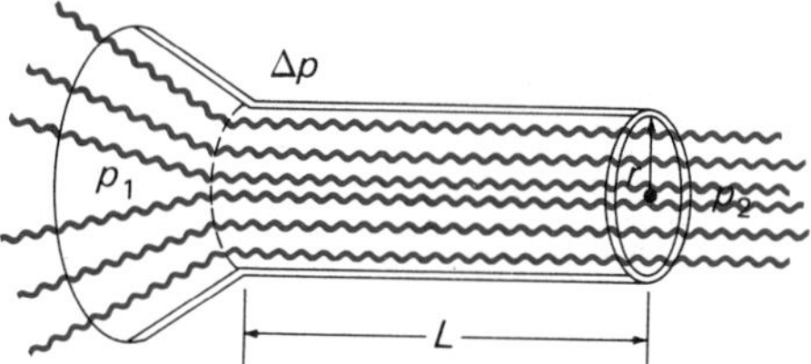

Figure 15.21

Poiseuille's law. The flow rate of a viscous fluid through a pipe or conduit is proportional to the pressure difference $\Delta p = p_1 - p_2$, and also the fourth power of the radius of the pipe, r^4. See text for description.

EXAMPLE 15.7 The radius of a particular length of water pipe is reduced by $\frac{1}{4}$ due to mineral deposits. By what factor is the flow rate reduced if the pressure difference remains the same?

Solution: It is given that $r_2 = \frac{3}{4}r_1$ (one-fourth reduction). Then, forming a ratio with Poiseuille's law,

$$\frac{\overline{Q}_2}{\overline{Q}_1} = \left(\frac{r_2}{r_1}\right)^4 = \left(\frac{3}{4}\right)^4 = 0.32$$

and $$\overline{Q}_2 = 0.32\,\overline{Q}_1$$

That is, the flow rate is decreased by more than $\frac{2}{3}$, or 67 percent. This example illustrates the effect of arterial deposits on blood circulation.

Important Terms

density the mass or weight per unit volume

specific gravity the ratio of the weight of a given volume of a substance to the weight of an equal volume of water

surface tension the tendency of a free surface of a liquid to contract and behave like a stretched membrane due to a net force on molecules at the free surface

cohesive force molecular attraction between like molecules

adhesive force molecular attraction between unlike molecules of different substances

capillary action the rise or depression of a liquid in a capillary due to surface tension and adhesive forces

meniscus the hemispherical shape of the surface of a liquid column

viscosity the internal resistance of a fluid to flow under the application of a shear stress

pressure the normal force per unit area

absolute pressure pressure referenced to complete vacuum

atmospheric pressure the pressure due to the gases of the atmosphere

gauge pressure the difference between the absolute and atmospheric pressures

barometer instrument used to measure atmospheric pressure

Archimedes' principle an object in a fluid is buoyed up with a force equal to the weight of the volume of the fluid displaced by the object

floating criterion the principle that an object will float in a fluid if the density of the object is less than or equal to the density of the fluid

streamline flow ideal fluid flow in which the fluid particles follow nonintersecting paths

Important Formulas

density:

mass	$\rho = m/V$
weight	$D = w/V = \rho g$
specific gravity:	$\text{sp gr} = \dfrac{D_s}{D_w} = \dfrac{\rho_s}{\rho_w}$
	$\text{sp gr} = \rho_s$ (in cgs system where $\rho_w = 1.00\ \text{g/cm}^3$)
pressure:	$p = \dfrac{F}{A}$
pressure-depth equation:	$p = \rho g h = Dh$
absolute pressure:	$p_{abs} = \rho g h + p_{atm} = Dh + p_{atm} = p_{gauge} + p_{atm}$
atmospheric pressure:	$p_{atm} = 1.01325 \times 10^5\ \text{N/m}^2$ (Pa) $= 14.7\ \text{lb/in}^2$ $= 30$ in. Hg $= 760$ mm Hg (torr) $= 1013.25$ mb
buoyant force:	$F_b = \rho V g = mg$ (where m is mass of fluid displaced)
flow rate:	$Q = Av$
continuity equation:	$Q = Av = \text{constant}$
Bernoulli's equation:	$p + \rho g h + \frac{1}{2}\rho v^2 = \text{constant}$
viscosity:	$\eta = \left(\dfrac{F}{A}\right)\left(\dfrac{d}{v}\right)$
Poiseuille's law:	$\overline{Q} = \dfrac{\pi r^4\,\Delta p}{8\eta L}$

Questions

Fluid Properties

1. Fluids are commonly thought of as liquids. Why are gases also classified as fluids?
2. (a) What is one of the densest metals? (b) What is one of the densest liquids?
3. Which has a greater specific gravity: (a) copper or steel? (b) Alcohol or gasoline?
4. The specific gravity is equal to the magnitude of the mass density in the cgs system. Is it equal to the magnitudes of the mass densities in the SI and British systems? Explain.
5. (a) Why can water bugs walk on water? (b) What causes raindrops to "bead up" on a freshly waxed car?

6. Why does a liquid rise higher in a smaller capillary tube? (See Fig. 15.4.) What limits the height of a column of liquid in a capillary tube?
7. Why is glue called an "adhesive"? Is its adhesive or cohesive force stronger? Why won't glue stick to some surfaces—for example, a greasy surface?
8. How does viscosity generally vary with temperature (a) in liquids and (b) in gases?

Pressure and Pressure Measurements

9. Distinguish between absolute pressure and gauge pressure. What is meant by a negative pressure, and how would such a pressure be indicated on an open-tube manometer?
10. Explain the principle of a barometer. Does it measure absolute or gauge pressure?
11. A common method of transferring liquids is through siphoning action, where the outlet of the siphon tube has to be below the level of the liquid in the initial container. Give an explanation of this action. Can siphoning be done in a vacuum?
12. Explain the principle of the vacuum cleaner. How does it "pick up" dirt?
13. Refer to Figure 15.22. (a) Water leaks through a hole in the bottom of a cup. What would happen if you placed the palm of your hand tightly over the mouth of the cup, and why? (b) A glass is filled with water and an index card is placed on the top. If you hold the card and turn the glass over, you can remove your hand without causing the card to fall off and the water to drain out. Explain why. Could you turn the glass sideways or horizontally with the same result?
14. How do suction cups work? Would they work on the outside of a spaceship on the moon?
15. A "plumber's helper," or plunger, is used to unplug drains. When a plunger is used on a sink, is the obstruction pushed or pulled in the drain?
16. Gasoline cans with spouts commonly have a plastic-capped vent hole on the top. What is the purpose of the vent, and what happens if you forget to remove the cap when pouring?
17. Automobile tires are inflated to pressures on the order of 200 kPa (30 lb/in^2). Yet bicycle tires are inflated to pressures more than twice this. Why such a greater pressure for a lighter vehicle? (How many more times greater is the typical automobile tire pressure than atmospheric pressure?)

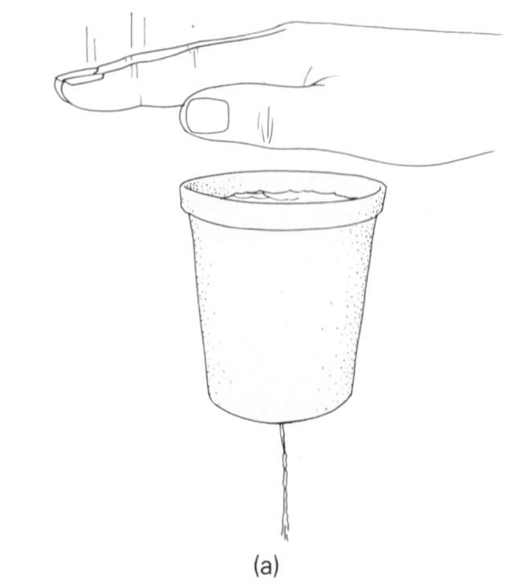

(a)

(b)

Figure 15.22
Atmospheric pressure in action. See Question 13.

18. What would be the height of a column of a barometer that contained (a) a liquid one-half as dense as mercury, and (b) water?
19. In a mercury or liquid barometer, does the height of the column depend on the diameter or the cross-sectional area of the tube? Explain.

Buoyancy and Archimedes' Principle

20. Why do helium and hot air balloons rise? How is ballast (weights) used in hot air balloon travel?
21. (a) If water and glycerol are poured together, which floats on top? (b) Will a solid steel ball float in any liquid? Explain.

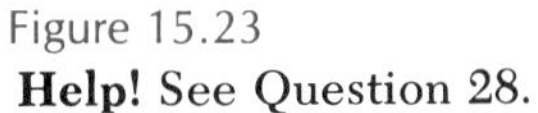

Figure 15.23
Help! See Question 28.

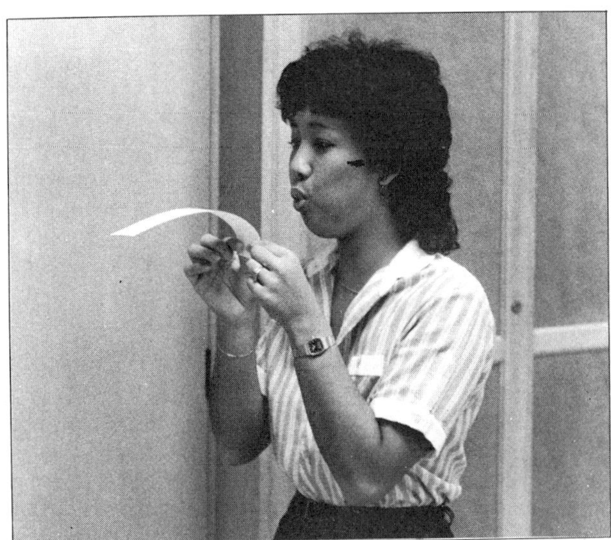

Figure 15.24
Blow and the paper rises. See Question 32.

22. Cruise ships weigh on the order of 70,000 tons. Explain how such heavy ships can float. How much water does a 70,000-ton ship displace? Does the ship float higher or lower in the water when it is loaded? Explain.
23. (a) Why can humans float in water? Can you affect your floating depth in any way? Explain. (b) Why can people float easily when swimming in the Great Salt Lake in Utah?
24. If a Ping Pong ball is held well below the surface of a body of water and then released, it rises and "pops" out of the water. What causes the ball to rise above the surface?
25. How could you use a plastic milk jug in a swimming pool to illustrate the density floating and sinking rules given in the chapter?
26. The setup shown in Figure 15.12 is used to illustrate Archimedes' principle. Could you also use it to determine the density of an irregularly shaped heavy (sinking) object like a rock? How about to find the density of an irregularly shaped floating object? Explain.
27. A container of water sitting on a scale has a certain weight. If a block of wood is placed in the water and it floats due to the upward buoyant force, does the scale reading change? Explain.
28. Referring to Figure 15.23, (a) what is the principle of a life preserver? (b) Explain, in terms of water displacement, why the boat sinks with a hole in the bottom. How does bailing help?

Fluid Flow

29. Why is the flow rate equation a continuity equation for incompressible fluids?
30. Explain why water coming from a hose goes a greater distance when the hose is fitted with a nozzle than when it is not.
31. Show that $\frac{1}{2}\rho v^2$ and p have units of energy per volume and work per volume, respectively.
32. Explain why a piece of paper rises when a person blows across the top of it, as shown in Figure 15.24.
33. (a) The curvature of an airplane wing causes air to flow faster over the top surface of the wing than beneath the bottom surface. Explain how this causes the airplane's "lift." (b) Explain why tall chimneys and smokestacks provide better draft than short ones.
34. (a) Why does the top on a convertible (automobile) "pop up" or bulge when the car goes at a fast speed? (b) When you drive your car on an interstate and a large truck with "the hammer down" goes by in the passing lane, the car sometimes seems to sway or veer toward the truck. What causes this?
35. Fluid flow requires some driving force, commonly supplied by pumps. Explain the operation of the pumps shown in Figure 15.25, (a) a lift pump, and (b) a cylinder-piston pump.
36. Discuss the effects of blood flow from arteries into capillaries in terms of Poiseuille's law.

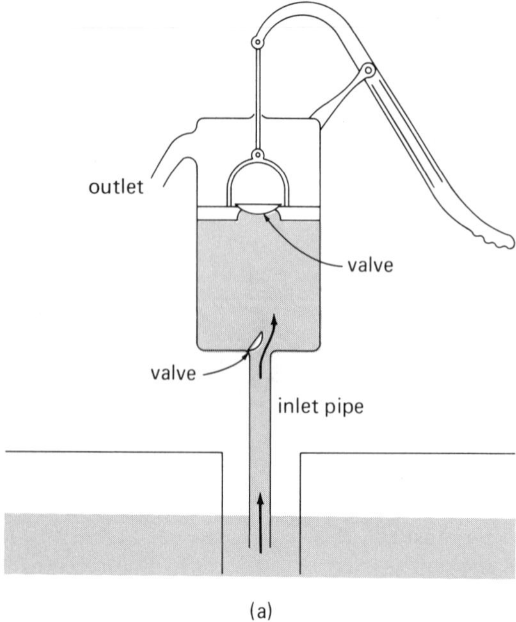

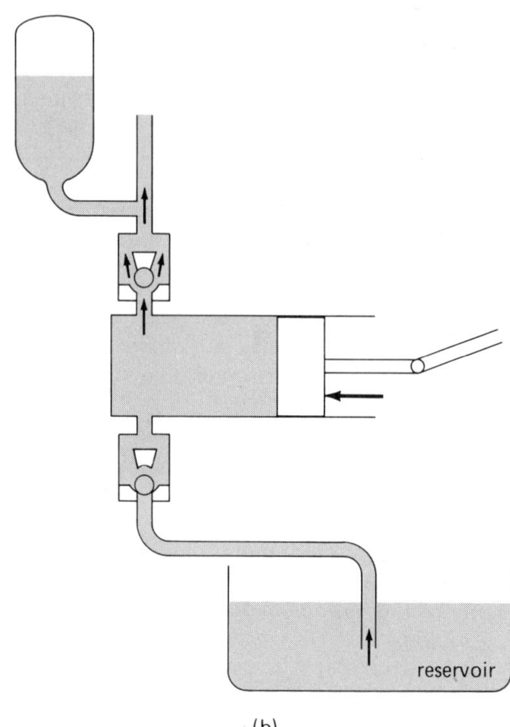

Figure 15.25
Pumps. See Question 35.

Problems

Levels of difficulty are indicated by asterisks for your convenience.

15.1 Fluid Properties

1. A 5.0-kg object has a volume of 2.0×10^{-3} m^3. What are its (a) mass density and (b) weight density in lb/ft^3?
2. What would be the volume of 2.0 lb of copper?
*3. Given 10 lb of aluminum and 2.4 lb of lead, which has the greater volume, and how many times greater?
*4. A pure gold nugget with a volume of 1.5 cm^3 is placed on a double-pan balance. What would be the volume of the brass weights needed to balance the nugget?
*5. A 100-mL graduated cylinder is half filled with alcohol and half with gasoline. What is the total mass of the combined liquids?
*6. A particular substance has a weight of 6.0 lb and a volume of 0.020 ft^3. What is its specific gravity?
*7. A piece of metal alloy is found to have a mass of 500 g and a volume of 75 cm^3. What is its specific gravity?
*8. The specific gravity of a particular liquid is 6.5. What would be the weight of 0.50 ft^3 of the liquid?
**9. Suppose alcohol were used as the standard for specific gravity. In that case, what would be the alcoholic specific gravities of (a) iron and (b) water?
**10. A capillary tube has a diameter of 1.0 mm. (a) If water rises to a height of 8.0 mm in the tube, what is the net upward adhesive force? (b) Assuming the same force, how far would blood rise in the tube? (Density of blood: 1.1 g/cm^3.)

15.2 Pressure and Pressure Measurement

11. A swimmer dives to a depth of 20 ft. What is the pressure, in lb/in^2, experienced by the swimmer due to the water above him?
12. What is the total force, in (a) newtons and (b) pounds, due to the atmosphere on the body of a person with a total body area of 0.20 m^2?
13. A table top measures 1 m $\times$ 1 m. What is the force on the table top due to the atmosphere in (a) newtons and (b) pounds? (c) Why can such a table be easily picked up with such a force on its top?
*14. A coin lies on the bottom of a pool under 2.50 m of water. What is the total (absolute) pressure on the coin?
*15. What is the gauge pressure, in atmospheres, on a deep-sea diver at a depth of 50 fathoms? (1 fathom = 6 ft)
*16. The absolute pressure on an object submerged in glycerol is 1.063×10^5 N/m^2. What is the depth of the object?

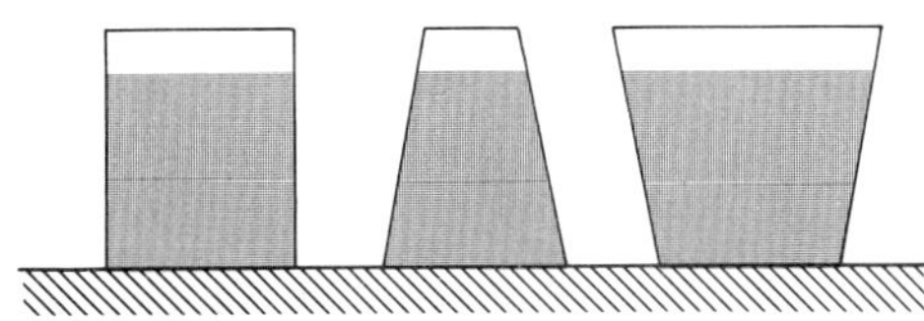

Figure 15.26
Same heights, different pressures? See Problem 24.

*17. On a particular day, the weather report gives the barometer reading as 30.15 in. of Hg. What is the atmospheric pressure in lb/in^2?

*18. What is the height of the column of a mercury barometer if the atmospheric pressure is 998 mb? (*Hint:* use a ratio.)

*19. What would be the height of the column of a water barometer at normal atmospheric pressure?

*20. An open-tube mercury manometer is used to measure the pressure of a gas in a container. If the height difference of the columns is 15 cm, what is the gauge pressure of the gas?

*21. The gauge pressure of a gas is 0.60 lb/in^2. If a closed-tube mercury manometer were connected to the gas container, what would be the height difference of the columns in the manometer arms? (*Hint:* be careful of units.)

*22. An open-tube mercury manometer has a height difference in the column arms of -10.0 cm. What are (a) the gauge pressure and (b) the absolute pressure?

**23. The pressure of a gas in a container is measured with both an open-tube mercury manometer and a closed-tube mercury manometer. If the gauge pressure of the gas is 100 N/m^2, what is the reading for the height difference of the unconnected arms for each manometer? (Is an open-tube manometer practical for measuring pressures of this order?)

**24. Three open containers as shown in Figure 15.26, with the same circular base areas, are filled with water to the same height. (a) Justify that the total force on the base of each container is the same. (b) The total force on the base of each container is the same, yet it is obvious that each container holds a different volume of water and that the *weight* of the water in each container is different. Explain this "hydrostatic paradox."

15.3 Buoyancy and Archimedes' Principle

*25. What is the force required to hold a cubic block of ice with sides of 0.20 m submerged under water?

*26. If the cube of ice in Problem 25 is released, what will be the height of the portion of the cube extending above the water surface when the ice floats in equilibrium?

*27. A cylindrical piece of polymer material with a mass of 650 grams has a radius of 5.0 cm and a length of 15 cm. If the cylinder is placed in a container of water, will it float?

*28. A steel machine part with a volume of 0.20 ft^3 weighs 20 lb. When it is suspended from a scale and submerged in benzene, what does the scale read?

*29. A crane lifts a rectangular iron bar, 10 ft long and 0.50 ft $\times$ 0.50 ft on an end, from the bottom of a lake. What is the minimum upward force the crane must exert when the bar is (a) in the water and (b) above the water?

*30. A submarine weighs 800 tons. How much water must be displaced for the sub to be in equilibrium at some submerged depth in the ocean? (Assume a constant sea water density.)

**31. An ocean-going barge is 50 m long and 20 m wide, and has a mass of 50 metric tons. Will the barge clear a sandbar 0.50 m below the surface of the water?

**32. A block of iron readily sinks, but ships constructed of iron float. A solid cubic block of iron 1.0 m on a side, with a density of 8.0×10^3 kg/m^3, is made into iron sheets, and from them an open cube is made that does not sink. (a) What should be the minimum volume of such a cube? (b) What is the length of the sides of the cube?

15.4 Fluid Flow

*33. Show that the pressure-depth equation (Eq. 15.7) can be derived from Bernoulli's equation.

*34. Water flows from a 1.0-in.-diameter pipe with a speed of 1.2 ft/s. How long will it take to fill a 100-gallon container?

*35. A gasoline pump delivers gasoline into a flexible rubber hose with a diameter of 1.5 in. at a rate of 4.0 gal/min. The nozzle of the hose has a diameter of 0.75 in. (a) What is the flow rate of the gasoline from the nozzle in ft^3/s? (b) What is the flow speed of the gasoline from the nozzle?

*36. In a hydraulic system, oil flows through a 2.0-cm-diameter pipe with a speed of 25 cm/s. If the pipe couples into another pipe where the flow speed is 10 cm/s, what is the diameter of the second pipe?

*37. An ideal liquid flows with a velocity of 20 cm/s through a horizontal pipe with a cross-sectional area of 10 cm^2. A constriction in the pipe narrows the cross section to 2.0 cm^2. (a) What is the flow velocity in the constriction? (b) What is the difference between the pressure at the constriction point and the pressure in the nonconstricted portion of the pipe? (Assume the liquid to be water.)

*38. A water tower springs a leak 10 m below the water level. (a) At what speed does the water initially emerge from the hole? (b) If the hole has an area of 0.20 cm^2, what is the initial flow rate?

*39. Water flows through a horizontal pipe with a speed of 2.0 ft/s. The pipe constricts to a smaller cross section, and a Venturi meter across the constriction shows a difference of 0.25 ft in the heights of the columns of the standpipes. What is the speed of the water in the smaller pipe?

*40. Gasoline flows through a horizontal pipeline with a flow rate of 30 ft^3/min. A pressure gauge measures the pressure to be 300 lb/ft^2 in a section of the pipeline with a diameter of 4.0 in. If the line expands to a pipe with a diameter of 6.0 in., what are (a) the flow speed in the larger pipe and (b) the pressure in this pipe?

*41. A Venturi meter is used to measure the flow rate of a liquid in a pipe with a cross-sectional area of 0.050 m^2. If the cross-sectional area of the meter constriction is 0.0080 m^2 and the height difference of the standpipe columns is 0.15 m, what is the flow rate?

*42. Fluid A flows three times as fast as fluid B when the tangential force per unit area on fluid A is twice that on fluid B. All other things being equal, which fluid has the greater viscosity, and how many times greater?

*43. The radius of a length of pipe carrying a liquid is decreased by 2.0 percent due to deposits on its inner surface. (a) By how much must the pressure difference between the ends of the pipe be increased to maintain the same flow through the pipe? (b) If the "pipe" were a blood vessel, what effect would this have on the heart?

**44. Show that Equation 15.17 can be derived from the equation of continuity, Bernoulli's equation, and the pressure-depth equation.

**45. How much blood flows each second through an artery that is 10 cm long and has a radius of 0.50 cm if the pressure difference across the artery is 30 mm Hg (torr)?

**46. A hypodermic needle has an inner diameter of 0.10 cm and a length of 4.0 cm. It is attached to a syringe with a plunger end area of 5.0 cm^2. The syringe is filled with water, and a force of 100 N is applied to the plunger. What is the flow rate from the needle when the water is squirted into the air? (*Hint:* find the pressure difference using Pascal's principle. See Chapter 11.)

(Courtesy of Corning Glass Works)

Chapter 16

Temperature and Heat

The temperature of a substance and how much heat it gains or loses are important considerations in numerous technical applications. The terms *temperature* and *heat* are used frequently, and we all understand their general meanings. We turn up the "heat" in the house to make it warmer—raise the temperature. We put things in the refrigerator to cool them off—lower their temperatures. However, most people find it difficult to give precise definitions of temperature and heat. (Can you define them?)

To better understand and describe temperature and heat, one must look closely at the physical situations of these quantities.

16.1 The Difference Between Temperature and Heat

It is evident from experience that temperature is associated with how hot or cold something is. We associate temperature with the sensations of "hot" and "cold" based on our sense of touch. For example, if you had two bowls of water at sufficiently different temperatures, you could tell which was hotter, or had a higher temperature, by comparing the bowls with your hands. Notice that this is comparison, or *relative* measure. You are comparing the hotness and coldness of the water in the bowls relative to the (body) temperature of your hands. Hence, we can say

Temperature is a relative measure or indication of hotness or coldness.

With regard to heat, we know that it is associated with energy transfer. When you put your hand in a bowl of water, your hand feels warmer or cooler when heat is transferred to or from it. Heat energy always "flows" from a substance with a higher temperature to one with a lower temperature.* Thus, we can say

> **Heat is energy transferred from one body to another because of a temperature difference.**

Heat, then, is energy in transit, so to speak.

It is common observation that when a hot object (or fluid) is brought into contact with a colder object, the objects eventually come to the same temperature, or to **thermal equilibrium.** For example, thinking of the two bowls of water again, if you dumped the water from one bowl into the other, all the water would eventually come to the same temperature (with proper mixing). This would be somewhere between the temperatures of the individual bowls of water.

We'll return to these ideas in a later section, after defining some temperature scales and heat units.

16.2 Temperature Measurement and Heat Units

Although our sense of touch gives a relative indication of hot and cold, it does not provide a good means of measuring temperature since our temperature sense is subjective and can be fooled. Also, the measurement range is limited. Instead, we use some physical property that depends on temperature and construct a thermometer, a device that measures relative hotness or coldness by a change in some physical property.

Fortunately, there are many physical properties that can be used. Probably the most obvious, and by far the most commonly used, property is thermal expansion (and contraction, i.e., negative expansion).†

Almost all substances expand with increasing temperature, but by different amounts (see Chapter 17). Conversely, most substances contract with decreasing temperature. Metals are a good example. The thermal expansion is small, but it can be made evident by using two metals in the form of a bimetallic strip (Fig. 16.1). Two metal strips, such as brass and iron, are bonded together. Because the metals expand differ-

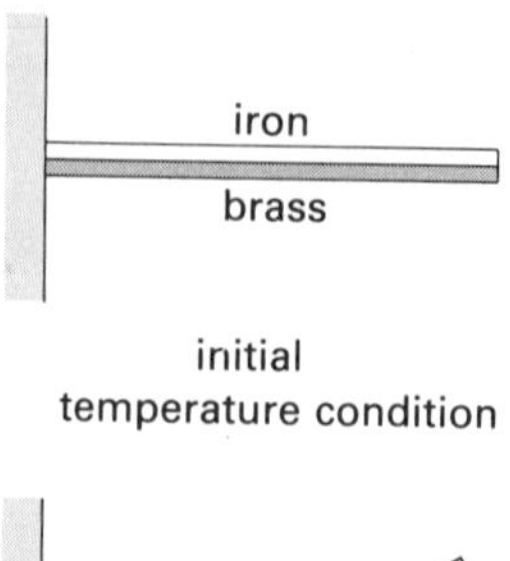

initial temperature condition

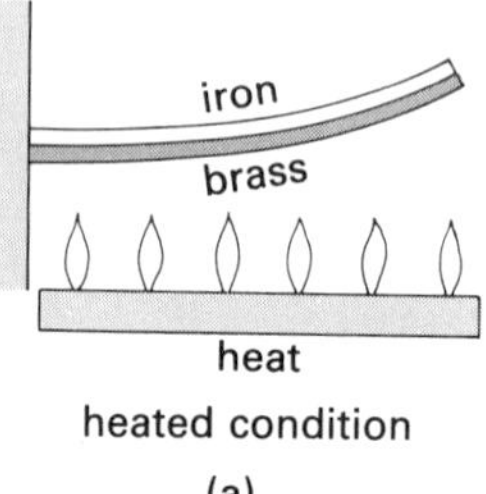

heated condition

(a)

(b)

Figure 16.1

Bimetallic strip. Because metals expand differently, the bimetallic strip bends toward the metal with the smaller linear expansion.

* An early theory of heat considered it to be a fluid-like substance called caloric (Latin *calor*, meaning heat) that could be made to flow in and out of a body. Even though this theory has long since been abandoned, we still commonly say that heat "flows" from one body to another.

† See the Chapter Supplement on thermometry for other properties used to measure temperature.

Figure 16.2
Bimetallic thermometer. A bimetallic strip wound in a coil or helix is used in a dial thermometer.

ently, the bimetallic strip bends toward the strip with the smaller linear expansion when heated.

A bimetallic strip could be used as a thermometer by calibrating its deflection, since each position of the deflecting end corresponds to a different temperature. A more convenient form of bimetallic thermometer is a strip wound in a coil or helix. The deflection is indicated by means of a dial (Fig. 16.2). Such dial thermometers are commonly used as cooking thermometers in ovens. A common application of a bimetallic coil for temperature control is given in Special Feature 16.1.

By far the most common temperature-measuring device is the **liquid-in-glass thermometer,** with which we are all familiar. This thermometer makes use of the thermal expansion of a liquid. It consists of a glass bulb, usually containing mercury or alcohol (colored with a dye to make it more visible), which is connected to a glass tube with a small capillary bore. An increase in the temperature of a liquid in the bulb causes it to expand up the bore, and the change in the height of the liquid column in the tube provides a means of measuring a change in temperature.

As with any measurement, we need a scale with numerical or unit values. One way to do this is to use two reference temperatures defined by some physical phenomena to establish two reference points, or marks on the thermometer. (Recall that the meter and the yard are defined practically by two reference marks on metal bars.) Numerical temperature values may then be assigned to these reference points.

This assignment is arbitrary, but one of the most widely used schemes is to assign the number zero to the temperature at which ice melts at standard atmospheric pressure (the ice or melting point) and the number 100 to the temperature at which water boils at standard atmospheric pressure (the steam or boiling point). This assignment forms the basis of the **Celsius temperature scale.**

To calibrate the thermometer, the interval between the ice and steam points is divided into 100 equal subintervals, or degrees (much like dividing a meter into centimeters). The temperature is then read in degrees Celsius (°C),* as indicated by the liquid level in the capillary bore. Thus, we refer to the temperature at which water freezes (or ice melts) as 0°C and the temperature at which water boils (or steam

* This scale is sometimes called the centigrade (Latin *centi-*, 100, and German *grade*, degree) scale because it has 100 units between the ice and steam points. Officially it is the Celsius scale, in honor of its inventor, A. C. Celsius (1701–1744), a Swedish astronomer.

SPECIAL FEATURE 16.1

The Thermostat

When we want to adjust the temperature in our homes, we usually turn the thermostat up or down. This common wall device controls the temperature by turning heating and cooling systems on and off.

If you pull the cover off a thermostat, you will ordinarily see a mechanism such as that shown in Figure 1. The temperature sensor is a bimetallic coil to which a glass vial containing mercury is attached. As the coil expands or contracts with temperature changes, the vial is tilted, and the mercury moves from one end to the other. Within the vial are electrical contacts. In moving to different ends of the vial, the mercury makes or breaks electrical contact to turn heating and cooling systems on or off, depending on the temperature for which the thermostat is set.

Setting a thermostat for a desired temperature tilts the vial to the appropriate position. Remove the cover of a thermostat set in the "off" position and observe the tilting of the vial and movement of the mercury when you adjust the temperature control.

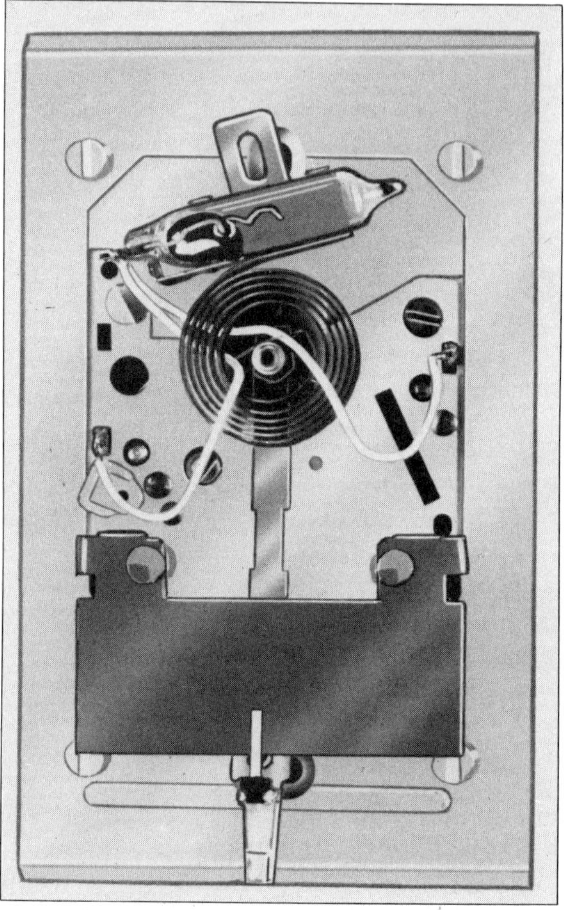

Figure 1
Thermostat. The expansion of a bimetallic coil causes the vial of mercury to tip and make or break electrical contact, which turns heating and cooling systems on and off. (Courtesy Honeywell, Inc.)

condenses) as 100°C (Fig. 16.3). Room temperature on the Celsius scale is 20°C.

Another scale in wide use in the United States for everyday temperature measurements is the **Fahrenheit temperature scale.*** On this scale, as you well know, the ice point is 32°F and the steam point is 212°F (Fig. 16.3). Hence, there are 180 degrees between the two reference points. Since there are 100 degrees Celsius (100°C) as compared to 180 degrees Fahrenheit (180°F) over this interval, each Celsius degree is nearly twice as large as a Fahrenheit degree or, more exactly, $\frac{180}{100} = \frac{9}{5} = 1.8$ times as large. (That is, $1°\text{C} = \frac{9}{5}°\text{F}$, or $5°\text{C} = 9°\text{F}$, and $1°\text{F} = \frac{5}{9}°\text{C} = 0.56°\text{C}$—equivalent degree *intervals,* not temperature readings.) Notice that the temperature readings on the Celsius and Fahrenheit scales are equal at −40°, i.e., −40°C = −40°F.

To convert from one temperature scale reading to the other, both the difference in the magnitudes of degree intervals and the difference in values of the lower reference temperatures (0°C and 32°F) must be considered. The formulas for conversion are

$$T_F = \tfrac{9}{5}T_C + 32° \quad (°\text{F})$$
$$T_C = \tfrac{5}{9}(T_F - 32°) \quad (°\text{C}) \qquad \textbf{(Eq. 16.1)}$$

* Named for G. D. Fahrenheit (1686–1736), a German physicist who invented the liquid-in-glass thermometer. Fahrenheit originally used the temperature of an ice-salt mixture for a low temperature reference and normal body temperature as an upper reference point. On his scale, the ice and steam points of water turned out to be 32°F and 212°F, respectively.

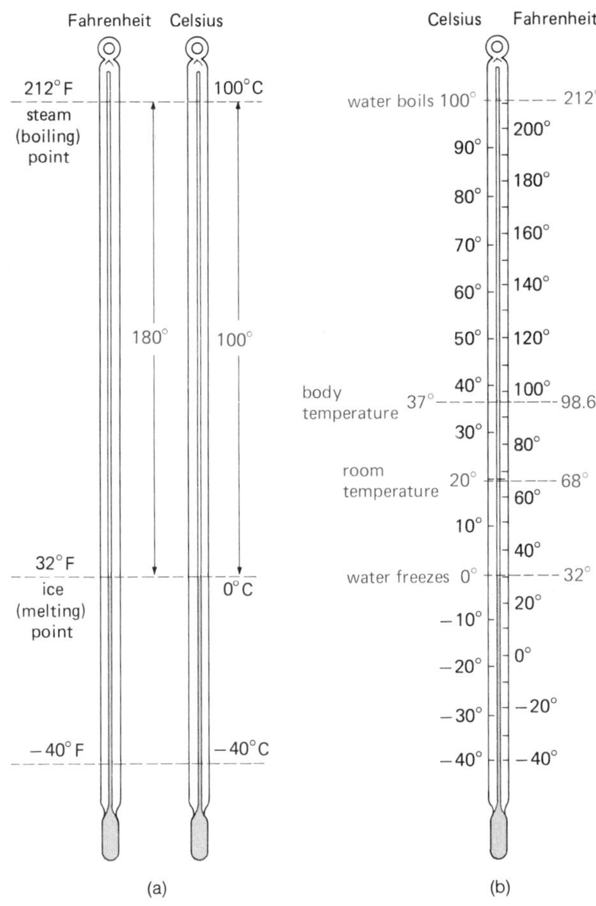

Figure 16.3
The Celsius and Fahrenheit temperature scales.

where T_C and T_F are the Celsius and Fahrenheit temperatures, respectively.

EXAMPLE 16.1 What are the equivalent Fahrenheit and Celsius temperatures, respectively, of (a) 20°C, room temperature, and (b) 98.6°F, normal body temperature?

Solution:

(a) $T_F = \frac{9}{5}T_C + 32° = \frac{9}{5}(20°) + 32° = 68°F$

(b) $T_C = \frac{5}{9}(T_F - 32°) = \frac{5}{9}(98.6° - 32°) = 37°C$

Absolute Temperature Scales

The Celsius and Fahrenheit scales are used for everyday temperature measurements, and negative temperatures (below zero) are common. Upon reflection, one might ask if there is a lower limit, or absolute zero temperature. Because of the thermal properties of gases (dis-

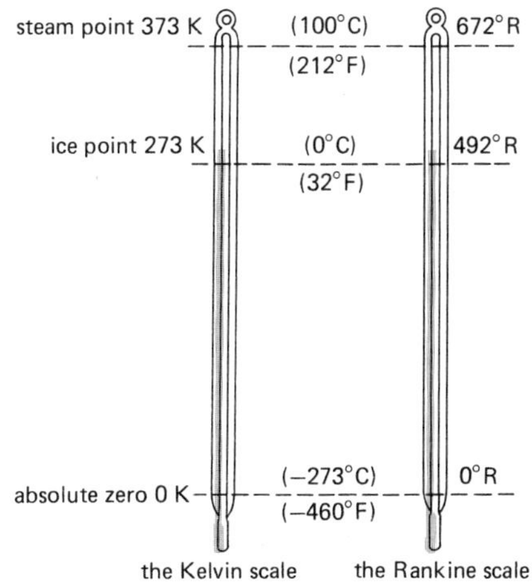

Figure 16.4
The Kelvin and Rankine absolute temperature scales. The lowest reading on the absolute temperature scales is absolute zero. (Liquid-in-glass thermometers are used for illustrations only; they cannot be used to measure extremely low temperatures.)

cussed in the next section), an absolute zero is indicated at a temperature of −273°C or −460°F.

This absolute zero of temperature is analogous to the zero end of a measuring stick, with the ice and steam point references defining a length interval along the measuring stick. For scientific reasons, discussed in the next section, it is convenient to define absolute temperature scales based on absolute zero. Two such absolute scales are the **Kelvin temperature scale** and the **Rankine temperature scale** (Fig. 16.4), named after Lord Kelvin (William Thomson, 1824–1907), a British scientist, and William Rankine (1820–1872), a Scottish engineer, who proposed the respective scales.

The Kelvin scale unit is the kelvin (K; *not* degree Kelvin) and is associated with the Celsius scale. For example, the freezing point of water on the Kelvin scale is 273 K. The Rankine scale uses degrees Rankine (°R) and is associated with the Fahrenheit scale. The respective absolute temperatures are related to the Celsius and Fahrenheit temperatures by the relationships

$$T_K = T_C + 273 \quad \text{(K)}$$
$$T_R = T_F + 460° \quad \text{(°R)}$$

Eq. 16.2)

SPECIAL FEATURE 16.2

Temperature Conversion Mnemonic Procedure

A mnemonic (memory-aid) procedure is sometimes used in the conversion of Celsius and Fahrenheit temperature readings, as follows. The acronyms C-FAMS and F-CADS are remembered. The first two letters of each acronym indicate the direction of the conversion, e.g., C-F (Celsius-Fahrenheit). The last three letters stand for Add-Multiply-Subtract (AMS) and Add-Divide-Subtract (ADS), respectively. The number 40 is added and subtracted, and $\frac{9}{5}$ (or 1.8) is what is multiplied and divided in each case.

Let's take an example. Suppose we want to convert 100°C to the corresponding Fahrenheit reading. Applying C-FAMS stepwise,

$$\underset{\text{(Add)}}{100°\text{C} + 40 =} \quad \underset{\text{(Multiply)}}{140° \times \tfrac{9}{5}} \quad \underset{\text{(Subtract)}}{= 252° - 40} = 212°\text{F}$$

which is the correct answer, as we know. Going the other way with F-CADS,

$$\underset{\text{(Add)}}{212°\text{F} + 40 =} \quad \underset{\text{(Divide)}}{252°/(\tfrac{9}{5})} \quad \underset{\text{(Subtract)}}{= 140° - 40} = 100°\text{C}$$

It is easy to show that these methods are equivalent to the expressions in Equation 16.1. Let's look at C-FAMS. Starting with T_C, we first Add,

$$(T_C + 40)$$

then Multiply,

$$\tfrac{9}{5}(T_C + 40)$$

then Subtract:

$$\tfrac{9}{5}(T_C + 40) - 40$$

or

$$\tfrac{9}{5}T_C + 72 - 40 = \tfrac{9}{5}T_C + 32 = T_F$$

Notice in F-CADS that dividing by $\frac{9}{5}$ is equivalent to multiplying by $\frac{5}{9}$.

---

where T_K and T_R are the absolute temperatures in kelvins and degrees Rankine, respectively. The Kelvin scale is used mainly in scientific work and the Rankine scale in engineering applications.

Notice that the Celsius and Kelvin scales have equivalent temperature intervals. That is, a temperature change of one degree Celsius corresponds to a change of one kelvin. Similarly, the Fahrenheit and Rankine scales have equivalent temperature intervals—a change of one degree Fahrenheit corresponds to a change of one degree Rankine.

EXAMPLE 16.2 What are the equivalent absolute temperatures of (a) 100°C and (b) −15°C on the Kelvin and Rankine scales?

Solution: (a) By Equation 16.2, for 100°C,

$$T_K = T_C + 273 = 100° + 273 = 373 \text{ K}$$

and, since by definition 100°C = 212°F,

$$\begin{aligned} T_R &= T_F + 460° \\ &= 212° + 460° \\ &= 672°\text{R} \end{aligned} \qquad \text{(cf. Fig. 16.4)}$$

(b) For −15°C,

$$T_K = T_C + 273 = -15° + 273 = 258 \text{ K}$$

To find the temperature in degrees Rankine, −15°C is first converted to degrees Fahrenheit:

$$\begin{aligned} T_F &= \tfrac{9}{5}T_C + 32° = \tfrac{9}{5}(-15°) + 32° \\ &= -27° + 32° = 5°\text{F} \end{aligned}$$

and

$$T_R = T_F + 460° = 5° + 460° = 465°\text{R}$$

Absolute zero is believed to be the lower limit of temperature, although we have never physically been able to obtain this temperature. (In fact, there is a physical law stating that absolute zero cannot be reached, as will be presented in Chapter 18.)

No upper limit of the temperature scale is known. In terms of our measuring-stick analogy, we have a temperature "measuring stick" starting at zero and extending indefinitely, or at least with no upper end (limit) in sight.

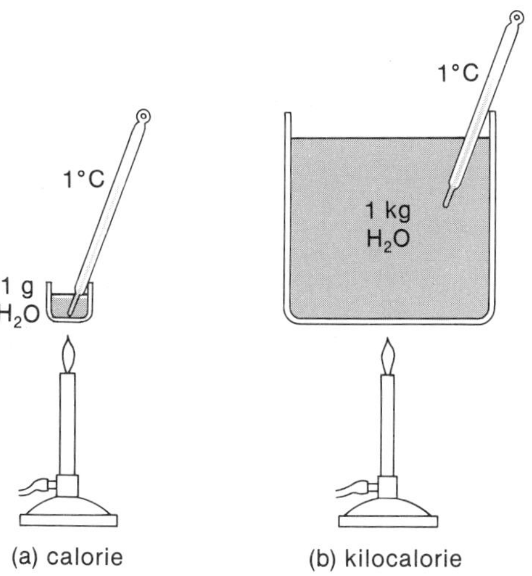

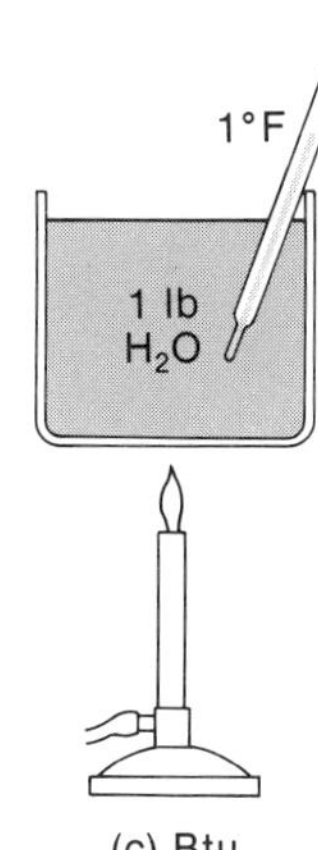

Figure 16.5
Units of heat energy. (a) A calorie raises the temperature of one gram of water 1°C. (b) A kilocalorie raises the temperature of 1 kg of water 1°C. (c) A Btu raises the temperature of one pound of water 1°F.

Heat Units

Since heat is energy, a quantity of heat may be expressed in a standard energy unit: joule or ft-lb. However, the commonly used units of heat energy are defined in terms of temperature changes. These units, the calorie, kilocalorie, and British thermal unit, are defined as follows (Fig. 16.5):

One calorie (cal) is the quantity of heat required to raise the temperature of one gram of water one degree Celsius (14.5°C to 15.5°C).

One kilocalorie (kcal) is the quantity of heat required to raise the temperature of one kilogram of water one degree Celsius (14.5°C to 15.5°C), and 1 kcal = 1000 cal.

One British thermal unit (Btu) is the quantity of heat required to raise the temperature of one pound of water one degree Fahrenheit (63°F to 64°F).

The degree intervals are specified in the definitions—e.g., a "15° calorie"—because the amount of heat required to raise the temperature of a given quantity of water one degree varies slightly with temperature. For most practical purposes, this variation may be neglected, as will be done in our discussion. The relationships between the heat units are

$$\mathbf{1\ Btu = 252\ cal = 0.252\ kcal}$$

The kilocalorie is used as the unit of food energy. Unfortunately, the food kilocalorie has come to be commonly called a "calorie," so confusion sometimes results. To help avoid such confusion, the kilocalorie is often referred to as a "big calorie" and abbreviated Cal, in contrast to the "little calorie" (cal).

Although we commonly use the Calorie unit, other countries use the standard energy unit. See Fig. 16.6.

The Btu is commonly used in engineering applications. For example, furnaces and air conditioners are rated in "Btu's." The rating indicates the amounts of heat in Btu (per hour) that the heating and cooling units can supply or remove.

As a form of energy, heat energy can be transformed into mechanical energy, and vice versa. For example, heat engines, such as the internal combustion engine, convert heat energy into mechanical energy, and machines lose mechanical energy to the "heat of friction."

A quantitative relationship between thermal energy and mechanical energy was established by James Joule (after whom the joule unit is named) in 1843, using the conservation of energy. His measurements were not very accurate, and more modern techniques using electrical energy have established the so-called **mechanical equivalent of heat** to be

$$\mathbf{1\ cal = 4.186\ J}\ \ (\mathbf{= 4.2\ J})$$

$$\mathbf{1\ kcal = 4186\ J}$$

$$\mathbf{1\ Btu = 778\ ft\text{-}lb = 1054\ J}$$

Figure 16.6
It's a joule. (a) In Australia diet drinks are labeled as being low "joule." (b) In Germany the labeling is a bit more specific. The label reads: "less than 4 kilojoules (1 kcal) in 0.3 Liter." How does this compare to our diet drinks?

EXAMPLE 16.3 The efficiency of a machine is 75 percent, where the lost energy is lost to the heat of friction. If the machine is supplied with 12 kW of power, how much frictional heat develops each second?

Solution: Since $12 \text{ kW} = 12 \times 10^3 \text{ W} = 12 \times 10^3 \text{ J/s}$, there are 12×10^3 J of energy delivered to the machine each second. With an efficiency of 75 percent, one fourth (25 percent) of this amount, or 3.0×10^3 J, is lost to friction. Hence,

$$3.0 \times 10^3 \text{ J} \left(\frac{1 \text{ cal}}{4.2 \text{ J}}\right) = 7.1 \times 10^2 \text{ cal} = 0.71 \text{ kcal}$$

or

$$3.0 \times 10^3 \text{ J} \left(\frac{1 \text{ Btu}}{1054 \text{ J}}\right) = 2.8 \text{ Btu}$$

which is the heat energy developed each second due to friction.

16.3 Kinetic Theory and the Perfect Gas Law

The statements on temperature and heat given at the beginning of the chapter are really operational definitions. To obtain a better insight, let's look at matter on a molecular level. Recall that matter is made up of molecules—either single atoms or combinations of atoms or ions. (Ions are electrically charged atoms or molecules.) The molecules are continuously jiggling around in motion.

In a solid, the molecules or atomic ions are attracted to each other and are held together by relatively strong forces. As a result, **a solid has a definite shape and volume.** The molecules vibrate over short distances about their equilibrium positions (Fig. 16.7). It is sometimes instructive to visualize the molecules as being connected by tiny springs, which represent the intermolecular forces. The greater the temperature of a solid, the faster the molecules vibrate.

In a liquid, the intermolecular forces are not strong enough to hold the molecules in fixed positions. So **a liquid has a definite volume, but it is without a definite shape and assumes the shape of its container.**

In a gas, at ordinary pressures, the molecules are relatively far apart and are in rapid, random motion, interacting primarily through collisions since the intermolecular forces are so weak that the gas molecules do not stay together. As a result, **a gas has neither a definite shape nor volume, and it completely fills the volume of its container.**

The energy associated with random translational motion (motion of a molecule as a whole)

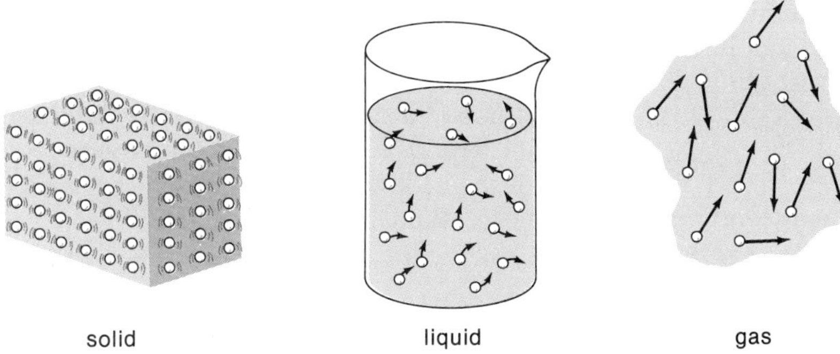

Figure 16.7
States of matter as described by molecular theory. A solid has definite shape and volume. A liquid has a definite volume without a definite shape. A gas has neither a definite shape nor volume.

is often called **thermal energy.*** However, in diatomic and other complicated gases, liquids, and solids there may also be rotational and/or vibrational motions of the atoms within the molecules (Fig. 16.8). Then, too, there is potential energy associated with the molecular and atomic "springs." The *total* (kinetic plus potential) energy contained within a body is called its **internal energy.**

In terms of these considerations, temperature is associated with the thermal energy, or random motions of molecules, in a substance. Thermal energy is the "temperature" energy that gives a temperature reading. (For example, a small amount of heat is transferred to the thermometer bulb, which comes to thermal equilibrium with the substance.)

When heat is added to a substance, it may go into thermal energy of the molecules, which increases the temperature. It may also go into the internal rotational and vibrational energies, which do not raise the temperature. Generally, there is a combination of both. Hence,

> **Heat is internal energy that is added to or removed from a body.**

* *Thermal energy* is a general term rather than a scientific one. It is used in different contexts by different people. We will use it here in association with the *random* translational motions of molecules. An object moving as a whole, such as a thrown baseball, has *ordered,* rather than random, translational motion.

In the kinetic theory of gases, this idea is simplified by considering a gas to be made up of solid particles (no intramolecular energy) that interact only through collisions (no intermolecular forces). Hence, the thermal energy is the

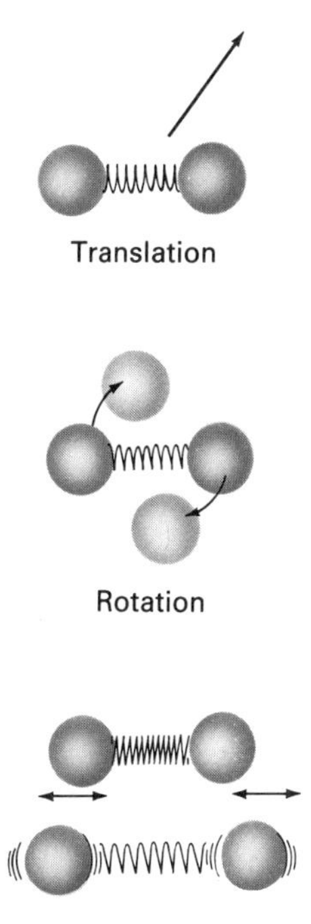

Figure 16.8
Molecular motions. In translational motion, the molecule moves as a whole. There may also be intramolecular rotational and vibrational motions of the atoms.

internal or total energy for such a *perfect* or *ideal* gas. Detailed analysis of the kinetic theory shows that the average kinetic energy per molecule of a perfect gas is related to the *absolute* temperature by the equation

$$\overline{KE} = \tfrac{1}{2}m\bar{v}^2 = \tfrac{3}{2}kT \qquad \textbf{(Eq. 16.3)}$$

where m is the mass of a gas molecule, and because of the numerous molecules of a gas with a distribution of speeds, a special average molecular speed $\bar{v}$ is used. The constant k is called Boltzmann's constant and is equal to 1.38×10^{-23} J/K.

EXAMPLE 16.4 Give an estimate of the average speed of nitrogen molecules (the major constituent of air) at room temperature (20°C). The mass of a nitrogen molecule is 4.7×10^{-26} kg.

Solution: Assuming that the nitrogen approximates a perfect gas, and considering a single molecule,

$$\overline{KE} = \tfrac{1}{2}m\bar{v}^2 = \tfrac{3}{2}kT$$

and

$$\bar{v} = \sqrt{\frac{3kT}{m}}$$

Then, with $k = 1.38 \times 10^{-23}$ J/K and $T = T_K = T_C + 273 = 20 + 273 = 293$ K,

$$\bar{v} = \left[\frac{3(1.38 \times 10^{-23}\ \text{J/K})(293\ \text{K})}{(4.7 \times 10^{-26}\ \text{kg})}\right]^{1/2}$$

$$= 5.1 \times 10^2\ \text{m/s} \qquad (= 1100\ \text{mi/h})$$

As the example shows, the gas molecules move with tremendous average speeds. The pressure of a gas results from collisions between the gas molecules and the walls of the container. An increase in pressure is due to an increased number of molecular collisions with the container walls. This requires an increase in the average speed or kinetic energy of the gas molecules, which results from an increase in temperature (Eq. 16.3). Thus, the pressure of a gas is directly proportional to its temperature. This is why we are warned not to throw aerosol cans into open flames, because the pressure increase might cause the can to explode.

The volume of a gas can also vary with temperature. In a nonrigid container, such as a balloon, the volume (and pressure) can increase with increasing temperature. Thus, the condition or state of a gas is determined by three factors: pressure p, volume V, and temperature T.*

In terms of kinetic theory, it can be shown that the relationship between these factors is given by

$$\boxed{pV = NkT} \qquad \textbf{(Eq. 16.4)}$$

where N is the number of molecules (reflecting the mass or quantity of gas) and T is the *absolute* temperature. This expression is called the **perfect** or **ideal gas law.**

No real gas is ideal, but the behavior of most gases at low densities generally follows the perfect gas law relationship within reasonable limits of temperature and pressure that are not close to the liquefaction point.

For a given quantity or mass of gas, Nk is a constant, and the perfect gas law may be written in the form $pV/T = Nk$ (a constant), or

$$\boxed{\frac{p_1V_1}{T_1} = \frac{p_2V_2}{T_2}} \qquad \textbf{(Eq. 16.5)}$$

For a gas at constant temperature and mass, we have

$$pV = C \quad \text{(a constant)}$$

or

$$p_1V_1 = p_2V_2 \qquad \textbf{(Eq. 16.6)}$$

This relationship was discovered empirically by Robert Boyle (1629–1691), an English chemist, and is commonly known as **Boyle's law.**

Also, for a gas at constant pressure and mass,

$$\frac{V}{T} = C' \quad \text{(a constant)}$$

or

$$\frac{V_1}{T_1} = \frac{V_2}{T_2} \qquad \textbf{(Eq. 16.7)}$$

* Because there are three variables, measurements are often taken at 0°C and 1 atm of pressure, i.e., standard temperature and pressure (STP).

This relationship was discovered empirically by Jacques Charles (1747–1832), a French physicist, and is known as **Charles' law.**

EXAMPLE 16.5 A cylinder of acetylene gas has a gauge pressure of 200 lb/in^2 at a temperature of 17°C. If the temperature of the gas is raised to 37°C, what is the gauge pressure?

Solution: It is given that $p_1 = 200$ lb/in^2, $T_1 = 17°C + 273 = 290$ K, and $T_2 = 37°C + 273 = 310$ K. (Remember that T is *absolute* temperature.) Since the cylinder is rigid, the volume is constant, i.e., $V_1 = V_2$, and

$$\frac{p_1V_1}{T_1} = \frac{p_2V_2}{T_2}$$

or

$$\frac{p_1}{T_1} = \frac{p_2}{T_2}$$

and

$$\begin{aligned} p_2 &= p_1\left(\frac{T_2}{T_1}\right) \\ &= (200\ \text{lb/in}^2)\left(\frac{310\ \text{K}}{290\ \text{K}}\right) \\ &= 214\ \text{lb/in}^2 \end{aligned}$$

The linear relationship of p and T for a gas maintained at constant volume [$p = (Nk/V)T$] allows the determination of the value of absolute zero (Fig. 16.9). By plotting p versus T for different gases, straight lines are obtained over normal temperature ranges. Real gases liquefy at low temperatures, so there is a deviation from the straight-line relationship as the temperature is lowered. However, by extrapolating the lines to zero pressure, the value of absolute zero (−273°C) is determined.

The linear relationship also forms the basis of the constant-volume gas thermometer. Once the p versus T curve is determined for a particular volume of gas, an arbitrary temperature may be measured from the pressure reading of the gas in the thermometer.

16.4 Specific Heat

As is well known, when a quantity of heat ΔQ is added to a substance, its temperature normally increases. Similarly, if a substance loses heat, its temperature decreases. Hence, we may write

$$\Delta Q \propto \Delta T$$

which is read "The quantity of heat ΔQ added (or removed) is proportional to the change in temperature ΔT." This relationship can be expressed in equation form as

$$\Delta Q = C\Delta T \qquad \textbf{(Eq. 16.8)}$$

where C is the **heat capacity** of the substance.

The heat capacity is not a very useful quantity, since it is different for objects of the same material—for example, different amounts of water. It should be evident that the amount or mass of a substance is an important consideration in describing the temperature change resulting from the addition or removal of heat. Being more specific and taking mass into account,

$$\Delta Q \propto m\Delta T$$

and, in equation form,

$$\boxed{\begin{aligned} \Delta Q &= mc\Delta T \\ \text{or} \quad c &= \frac{\Delta Q}{m\Delta T} \end{aligned}} \qquad \textbf{(Eq. 16.9)}$$

where c is the *specific heat capacity,* which is commonly referred to simply as **specific heat,** and $c = C/m$ (heat capacity per unit mass).

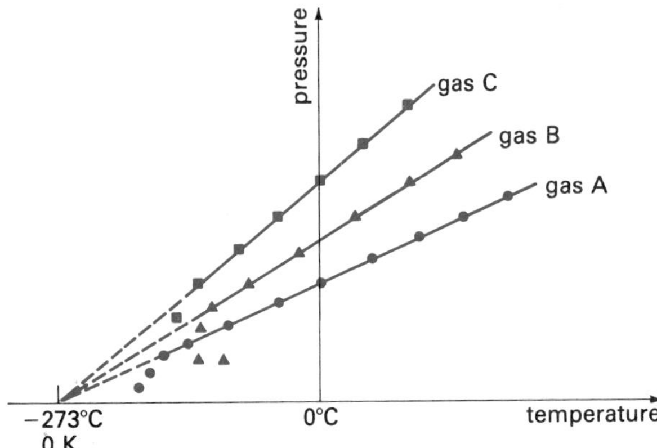

Figure 16.9
Absolute zero. The pressures of gases follow linear relationships with temperature (at constant volume). Extrapolating to zero pressure gives the value for "absolute" zero temperature, 0 K (or −273°C). Actual gases liquefy at low temperatures.

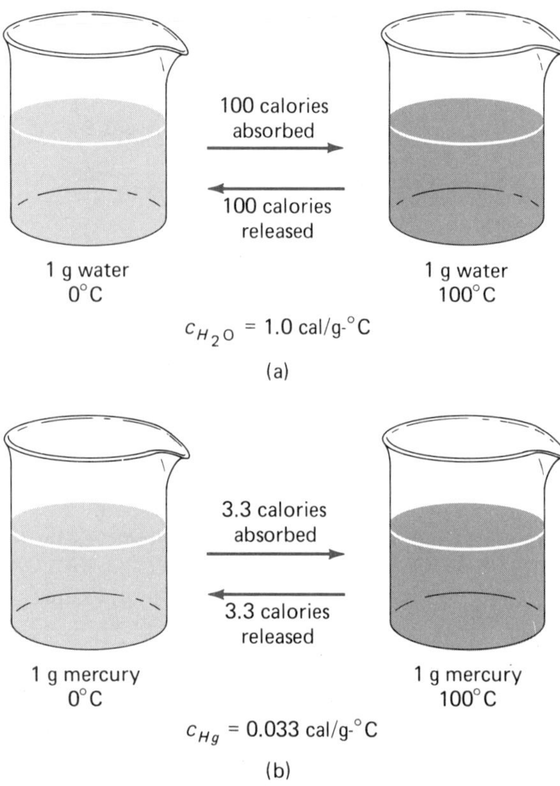

Figure 16.10

Specific heat. The specific heat of a substance is specific for that substance and is the amount of heat required to raise the temperature of 1 gram (or 1 kg) of a substance 1°C. For example, water has a specific heat of c_{H_20} = 1 cal/g-°C or 1 kcal/kg-°C. To raise 1 gram (or 1 kg) of water from 0°C to 100°C requires 100 cal (or 100 kcal). But, for 1 gram (or 1 kg) of mercury with a specific heat of c_{Hg} = 0.033 cal/g-°C or 0.033 kcal/kg-°C, only 3.3 cal (or 3.3 kg) must be supplied for the same temperature change.

Table 16.1
Specific Heats of Various Substances

Substance	c (cal/g-°C, kcal/kg-°C, or Btu/lb-°F)
Solids	
Aluminum	0.21
Asbestos	0.20
Brass	0.094
Copper	0.093
Glass (typical value)	0.16
Ice	0.50
Iron	0.113
Lead	0.031
Silver	0.056
Soil (typical value)	0.25
Wood (typical value)	0.40
Liquids	
Ethyl alcohol	0.60
Benzene	0.41
Gasoline	0.50
Mercury	0.033
Turpentine	0.42
Water	1.00
Gases	
Air	0.17
Steam	0.48

The specific heat is the amount of heat required to raise the temperature of one gram of a substance one degree Celsius.

From the definition of the calorie (raises 1 g of water 1°C), the **specific heat of water** is c = 1 cal/g-°C, i.e.,

$$c = \frac{\Delta Q}{m\Delta T} = \frac{1\text{ cal}}{(1\text{ g})(1°\text{C})} = 1\text{ cal/g-°C}$$

Thus, unlike the heat capacity, the specific heat of c = 1 cal/g-°C is *specific* for water, as are the specific heats for other substances (see Fig. 16.10).

The specific heat of a substance is a material property, and in general, from Equation 16.9,

The specific heats of various substances are given in Table 16.1. As a result of the analogous definition of the Btu to that of the calorie, the numerical values of the specific heat are the same in both systems of units for a given material.

It should be noted that the pound unit is used for m in Equation 16.9 since we have been working in the British system. More correctly, we should write $c = \Delta Q/w\Delta T$, where w is the weight in pounds. However, we will continue to use the form of Equation 16.9, keeping in mind that m is "in pounds."

Other units of specific heat are kcal/kg-°C (which is also 1.0 for water) and, in the SI system, J/kg-K.

EXAMPLE 16.6 Given 2.0 kg of aluminum and 2.0 kg of water at 20°C, how much heat is required to raise the temperature of each to 30°C?

Solution: From Table 16.1, the specific heats of aluminum and water are $c = 0.21$ kcal/kg-°C and $c = 1.0$ kcal/kg-°C, respectively. With $\Delta T = 30°\text{C} - 20°\text{C} = 10°\text{C}$, for aluminum,

$$\Delta Q = mc\Delta T = (2.0 \text{ kg})(0.21 \text{ kcal/kg-°C})(10°\text{C}) = 4.2 \text{ kcal}$$

For water,

$$\Delta Q = mc\Delta T = (2.0 \text{ kg})(1.0 \text{ kcal/kg-°C})(10°\text{C}) = 20 \text{ kcal}$$

Notice from the example that it takes much more heat to raise an amount of water through 10°C than to raise an equivalent amount of aluminum. Water has one of the highest specific heats of common substances. As a result, it is frequently used to store thermal energy in solar heating systems. (See Chapter 32.)

The specific heat capacity is descriptive of how much heat a substance will "hold," or its "capacity" per unit mass, for a given temperature change. Given equal masses of two different substances, the one with the greater specific heat will require more heat to raise its temperature by a given amount.

We often say that certain materials "hold their heat longer." This occurs because such materials have relatively greater specific heat capacities and, hence, have more heat energy per unit mass to start with. Since it takes more heat per unit mass to raise the temperature, such substances have more "stored" heat, and take a longer time to cool off, or lose their greater amounts of heat.

This effect of substances with high specific heats is sometimes painfully evident, as when eating a baked potato or the cheese on a pizza. Have you ever burned your mouth when eating these foods? That is because they have higher specific heats and more "stored" heat per unit mass than other foods that cool off quickly (and lull us into thinking that the potato and cheese have also cooled).

Calorimetry

As the name implies, calorimetry involves the measurement of heat. Through the use of the known values of specific heats of materials and temperature measurements, we can readily calculate the amount of heat absorbed or given up by specific substances, as shown in the preceding section. First, however, accurate heat measurements must be made experimentally to determine the value of the specific heat of a material.

The basic principle of calorimetry is the conservation of energy. If a hot body and a cold body are brought into thermal contact, they will eventually come to thermal equilibrium at the same temperature because of the transfer or "flow" of heat. If no heat is lost to the surroundings, then, by the conservation of energy,

$$\underset{\text{(by the hot body)}}{\text{Heat lost}} = \underset{\text{(by the colder body)}}{\text{Heat gained}}$$

Measurements of the quantities of heat transferred from one substance to another are carried out in specially designed vessels called *calorimeters*. A common type of calorimeter used in the physics laboratory is shown in Figure 16.11. It is basically an insulated container in which heat transfer takes place and temperature can be measured.

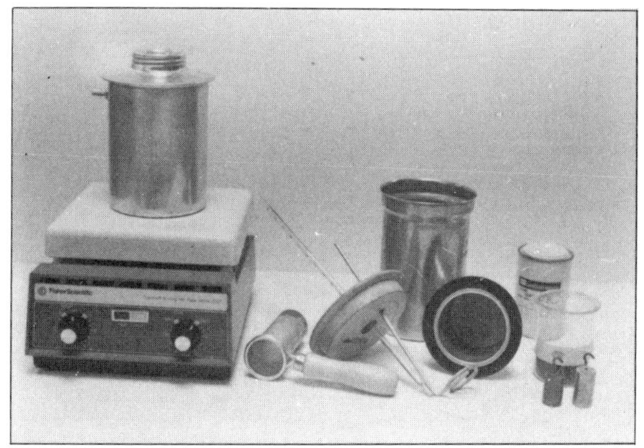

Figure 16.11
Calorimetry. The common laboratory calorimeter (center) consists of a cup (with black insulating ring) that sits in another can. A thermometer and stirrer extend into the cup through the can lid. Metal shot or slugs are heated in the wood-handled can that fits into the steam generator on the hot plate.

The specific heat of a substance is ordinarily measured by a calorimetry procedure called the *method of mixtures*. A hot substance of known temperature and mass is placed in or mixed with a colder substance of known temperature and mass, e.g., a metal in a liquid or one liquid in another liquid.

Let's assume we wish to determine the specific heat of aluminum. Neglecting the calorimeter cup for the moment, suppose 300 g of aluminum at 85.0°C is placed in 500 g of water at 15.0°C and that the final temperature of the mixture in thermal equilibrium is 22.7°C. Then, assuming no heat loss to the surroundings,

$$\text{Heat lost by Al} = \text{heat gained by } H_2O$$

or

$$\Delta Q_{\text{lost}} = \Delta Q_{\text{gained}}$$

and

$$(mc\Delta T)_{\text{Al}} = (mc\Delta T)_{H_2O}$$

Solving for the aluminum specific heat, c_{Al}, and using the given data with $c_{H_2O} = 1.0$ cal/g-°C,

$$c_{\text{Al}} = \frac{(mc\Delta T)_{H_2O}}{(m\Delta T)_{\text{Al}}}$$

$$= \frac{(500 \text{ g})(1.0 \text{ cal/g-°C})(22.7\text{°C} - 15.0\text{°C})}{(300 \text{ g})(85.0\text{°C} - 22.7\text{°C})}$$

$$= 0.21 \text{ cal/g-°C}$$

Of course, in practice the calorimeter cup (and stirrer) must be taken into account, since it too is involved in the heat transfer process. Calorimeter cups are commonly made of aluminum, the specific heat of which is well known.

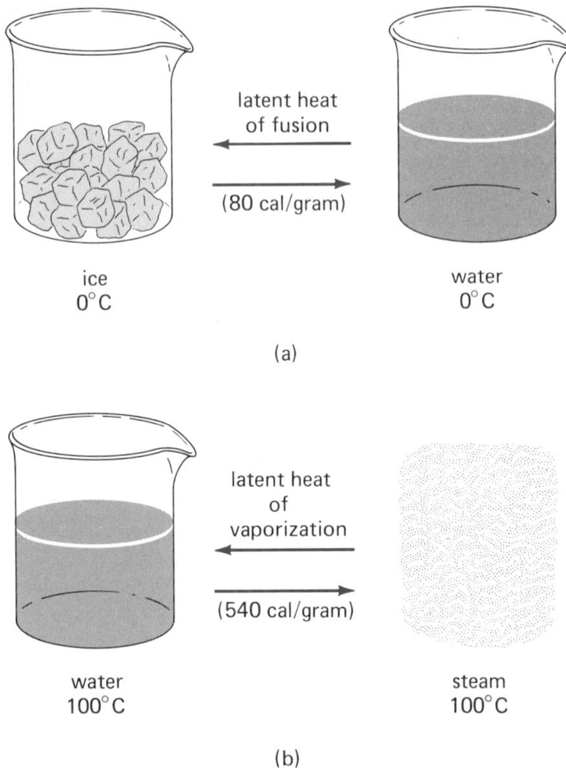

Figure 16.12
Phase changes. (a) Ice absorbs 80 cal/g at 0°C to melt. When water freezes at 0°C, 80 cal/g are removed. (b) Water absorbs 540 cal/g at 100°C to vaporize. When steam condenses at 100°C, 540 cal/g are released.

16.5 **Phase Changes and Latent Heat**

As described in Section 16.3, according to the kinetic theory, the form of matter depends on the degree of interaction among the molecular particles. The three common forms of matter—solid, liquid, and gas—are known as phases, and **the transition from one form to another is called a phase transition or phase change.***

Phase changes may be effected by supplying heat energy, which goes into the work of separating the molecules to be in the next phase, e.g., changing a liquid to a gas. In the reverse process, internal energy is given up (Fig. 16.12). The quantity of heat involved in a phase change per unit mass is called the **latent heat**—latent implying "hidden," since the heat is applied or removed without a temperature change.

In the solid-liquid phase change there is melting or fusion, and the temperature at which this occurs is commonly called the melting point, e.g., 0°C for water (at 1 atm). The heat added or removed in this process is referred to as the **latent heat of fusion,** and

The latent heat of fusion L_f is the heat energy per unit mass (or weight) involved in a solid-liquid phase change.

In equation form,

$$\boxed{L_f = \Delta Q/m \qquad \text{or} \qquad \Delta Q = mL_f} \quad \textbf{(Eq. 16.10)}$$

* The term *states of matter* is sometimes used. However, in physics *phases of matter* is preferred, as *state* has a different meaning (Chapter 18).

Table 16.2
Heats of Fusion and Vaporization (at Atmospheric Pressure)

	Latent Heat of Fusion, L_f		Latent Heat of Vaporization, L_v	
Substance	***cal/g (kcal/kg)***	***Btu/lb***	***cal/g (kcal/kg)***	***Btu/lb***
Alcohol				
Ethyl	24.9	45	204	367
Methyl	22.0	40	263	473
Ammonia	1.08	195	327	465
Helium	1.25	2.25	5.0	9.0
Lead	6.3	10.6	222	315
Mercury	2.7	5.0	71	128
Nitrogen	6.2	11	47.8	85
Oxygen	3.3	5.9	51	92
Silver	2.1	38	558	1004
Water	80	144	540	970
Zinc	24	43	475	855

The units of latent heat are cal/g (or kcal/kg) and Btu/lb. For example, as listed in Table 16.2, the latent heat of fusion for water is 80 cal/g or 144 Btu/lb. That is, at 0°C, 80 cal of heat are required to melt 1 g of ice, or 144 Btu are required to melt 1 lb of ice. Unlike specific heats, the latent heats are not the same in the different systems of units.

The variation of temperature with heat energy for the phase change of ice into water is illustrated in Figure 16.13. As heat is being absorbed by ice at 0°C, the ice and melted water coexist in equilibrium at the same temperature (horizontal line). When the ice is completely melted, part of the additional heat energy goes into raising the temperature of the water. Inversely, when water at 0°C freezes to ice, 80 cal/g must be extracted from the water.

After a solid substance has melted, some of the added heat energy goes into the kinetic energy of the liquid molecules, which raises the temperature of the liquid. When the liquid particles acquire sufficient energy to overcome the intermolecular forces, another phase change occurs.

In the liquid-gas phase change there is a vaporization or condensation, and the temperature at which this occurs is commonly called the boiling point, e.g., 100°C for water (at 1 atm). The heat added or removed in the process is referred to as the **latent heat of vaporization,** and

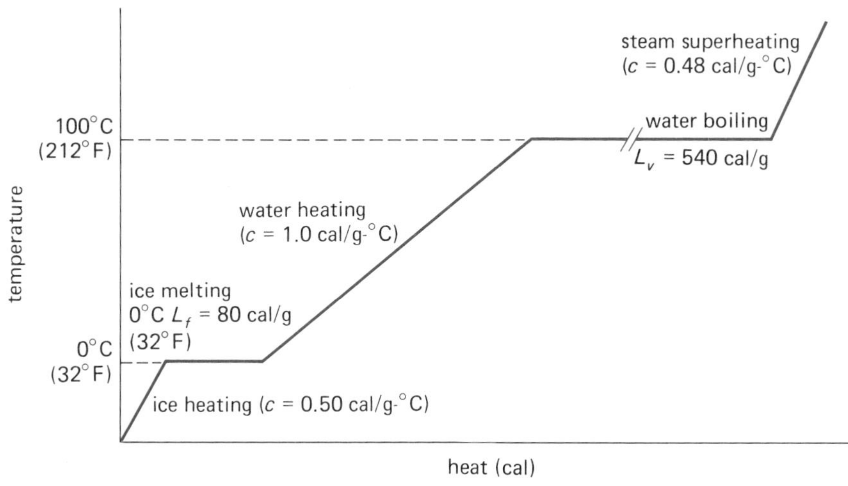

Figure 16.13
A graph of temperature versus heat for water. Follow the graph from left to right. As heat is added to ice below 0°C, the temperature of the ice increases. At 0°C, the ice melts and the added heat goes into the latent heat of fusion. The temperature remains constant (horizontal line), assuming all of the heat goes into melting the ice. The temperature of the water then increases to 100°C where the added heat goes into the latent heat of vaporization.

The latent heat of vaporization L_v is the heat energy per unit mass (or weight) involved in a liquid-gas phase change.

In equation form,

$$L_v = \Delta Q/m \qquad \text{or} \qquad \Delta Q = mL_v \qquad \textbf{(Eq. 16.11)}$$

The latent heat of vaporization of water is 540 cal/g or 970 Btu/lb. This phase change is illustrated graphically in Figure 16.13.

Under proper conditions of temperature and pressure, usually low pressures, it is possible for a substance to change from the solid phase directly to the gaseous phase without passing through the liquid phase. This process is referred to as *sublimation*. Some substances, such as dry ice (solid CO_2) and mothballs (*p*-dichlorobenzene), sublime at normal temperatures and pressures.

EXAMPLE 16.7 How much heat energy is required to change 100 g of water at 80°C to steam at 110°C?

Solution: To raise the temperature of water from 80°C to 100°C requires

$$\begin{aligned}\Delta Q_1 &= mc\Delta T \\ &= (100\text{ g})(1.0\text{ cal/g-°C})(100\text{°C} - 80\text{°C}) \\ &= 2000\text{ cal}\end{aligned}$$

At this point, latent heat is required ($L_v = 540$ cal/g):

$$\Delta Q_2 = mL_v = (100\text{ g})(540\text{ cal/g}) = 54{,}000\text{ cal}$$

Then, to raise the steam to 110°C ($c_{\text{steam}} = 0.48$ cal/g-°C),

$$\begin{aligned}\Delta Q_3 &= mc\Delta T \\ &= (100\text{ g})(0.48\text{ cal/g-°C})(110\text{°C} - 100\text{°C}) \\ &= 480\text{ cal}\end{aligned}$$

The total heat energy required is then

$$\begin{aligned}\Delta Q_t &= \Delta Q_1 + \Delta Q_2 + \Delta Q_3 \\ &= 2000\text{ cal} + 54{,}000\text{ cal} + 480\text{ cal} \\ &= 56{,}480\text{ cal}\end{aligned}$$

The vaporization of liquids also occurs at temperatures below their boiling points, by evaporation. The average kinetic energy of the liquid molecules is determined by a distribution of molecular velocities—some fast, some slow. When a fast-moving particle approaches the liquid surface with sufficient energy to overcome the retarding intermolecular forces that give rise to surface tension (Chapter 15), it escapes from the liquid, becoming a gas particle.

Since the evaporating particles carry away energy, **evaporation is a cooling process.** The cooling mechanism of our bodies depends on the evaporation of perspiration. The heat of vaporization or evaporation is supplied by our body heat.

Boiling and Freezing

The boiling and freezing points of substances vary widely; see Table 16.3. This is due to differences in the molecular interactions or bonding. Notice that the phase-change temperatures in the table are given for a pressure of 1 atm (standard pressure). Pressure makes a difference to boiling and freezing points.

To understand this, let's first see why a liquid boils. When heat is added to a liquid and its temperature increases, the average kinetic energy of its molecules increases. At the boiling point temperature, the average molecular en-

Table 16.3
Melting Points and Boiling Points (at Atmospheric Pressure)

	Melting Point		Boiling Point	
Substance	**°C**	**°F**	**°C**	**°F**
Alcohol				
Ethyl	−117	−179	78	172
Methyl	−98	−144	65	149
Ammonia	−75	−103	−34	−29
Helium	−272	−458	−269	−452
Lead	327	620	1744	3171
Mercury	−39	−38	357	675
Nitrogen	−210	−346	−196	−321
Oxygen	−219	−362	−183	−297
Silver	960	1760	2212	4014
Water	0	32	100	212
Zinc	419	786	907	1665

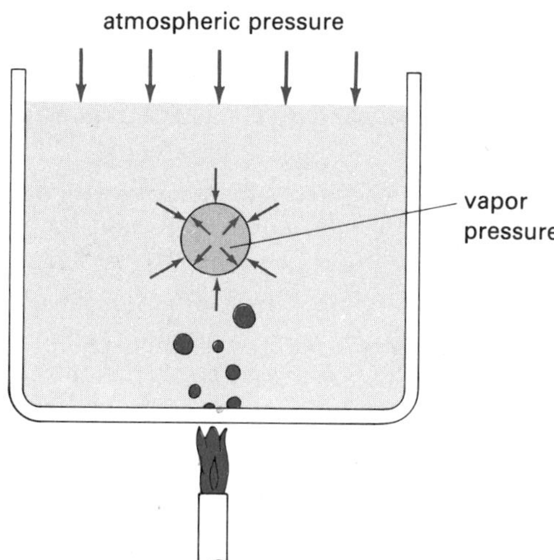

Figure 16.14
Boiling. When the vapor pressure in bubbles is greater than the external pressure, boiling starts. The bubbles rise, break the surface, and the gas escapes.

ergy is equal to the work necessary to change the liquid into a gas, and bubbles begin to form. More heat produces vapor bubbles in the liquid, usually near the bottom of the container where it is being heated and is hottest (Fig. 16.14).

If the vapor (gas) pressure in the bubbles is greater than the pressure above due to the atmosphere (and the weight of the liquid*), boiling starts. The bubbles rise and break through the surface, and the gas escapes. Thus, we can keep a liquid from boiling, or raise its boiling point temperature, by increasing the pressure on the liquid. A practical application of this is the pressure cooker, in which the gas pressure in the enclosed container raises the boiling point and allows higher temperatures for faster cooking.

At reduced pressures, the boiling points of liquids are lowered. For example, at high altitudes, where the atmospheric pressure is less (than at sea level), the boiling point of water is less than 100°C. At Pike's Peak the atmospheric pressure is about 600 torr (mm Hg), and water boils at about 94°C.

* In most cases this is negligible compared with atmospheric pressure.

By sufficiently reducing the pressure on a liquid, you can have boiling without external heating. This can be demonstrated by placing water in a vacuum jar. As air is evacuated from the jar, the pressure and hence the boiling point are lowered. Eventually, the boiling point is lowered to the temperature of the water and it boils "without heating." If this pressure is maintained, the boiling quickly stops because the latent heat of vaporization is taken from the internal energy of the water. This cools the water and lowers its temperature below the boiling point.

If the pressure is further reduced, the boiling and cooling continue until the temperature is reduced to the freezing point and ice forms on the surface of the boiling water! Here water coexists in all three phases—solid, liquid, and gas. This is known as the triple point; it occurs at a pressure of 4.6 torr and a temperature of 0.01°C. This fixed point is used in the definition of the kelvin unit and temperature scale (see Table 1.3).

How about the freezing point and pressure? Most liquids contract on freezing or solidifying. Increased pressure helps this process along and thereby raises the freezing point. However, there are some exceptions, most notably water. The open molecular structure of freezing water causes an expansion, or increase in volume (Chapter 17), and ice is less dense than liquid water. In this case, increased pressure tends to lower the freezing (melting) point temperature and make the ice melt, or have a smaller volume. The melting point isn't lowered much, only about 0.0075°C per atmosphere of pressure increase.

As a final topic, let's consider the boiling and freezing points of aqueous (water) solutions. For example, when you dissolve table salt in water, how does this affect the boiling and freezing points? If you check the boiling point temperature of salt water, you will find that it boils slightly above 100°C, so the boiling point is increased by salt. This is because salt ions are attracted to the water molecules and more energy is required to vaporize them.

On the freezing side, the dissolved salt lowers the freezing point. Here, the salt ions get in the way of the water molecules forming ice-crystal structures. Only when the water molecules are slowed down sufficiently (lower temperature) are the attractive forces large enough to cause freezing. In general, adding anything to water has this effect.

We use salt in making homemade ice cream and on icy sidewalks and roads. When salt is added to ice, it mixes with the outer layers of the ice. The freezing or melting point of the ice-salt mixture is below 0°C and it melts, taking heat from the ice. This effect causes ice to melt on sidewalks. In the case of ice cream making, the melting lowers the temperature of the salt-ice-water mixture around the container to below 0°C, and more heat can be conducted from the ice cream mix, causing it to freeze—the result we all wait for impatiently.

Important Terms

temperature a relative measure or indication of hotness or coldness

heat energy transferred from one body to another because of a temperature difference; the internal energy that is added to or removed from a body

thermal equilibrium a condition of having the same temperature throughout a system, and hence no net heat transfer

thermometer a device that measures relative hotness or coldness by a change in some physical property

Celsius temperature scale a scale with 100 degrees between the ice point (0°C) and the steam point (100°C)

Fahrenheit temperature scale a scale with 180 degrees between the ice point (32°F) and the steam point (212°F)

absolute zero the lower limit of temperature

Kelvin temperature scale an absolute temperature scale with the lowest temperature being absolute zero (0 K = −273°C)

Rankine temperature scale an absolute temperature scale with the lowest temperature being absolute zero (0°R = −460°F)

calorie (cal) the quantity of heat required to raise the temperature of one gram of water one degree Celsius

kilocalorie (kcal) the quantity of heat required to raise the temperature of one kilogram of water one degree Celsius

British thermal unit (Btu) the quantity of heat required to raise the temperature of one pound of water one degree Fahrenheit

mechanical equivalent of heat the standard energy unit equivalent of heat units (1 cal = 4.186 J and 1 Btu = 778 ft-lb)

thermal energy a general term used to describe the energy of the random translational motions of molecules associated with temperature

internal energy the total (kinetic plus potential) energy contained within a body

perfect (ideal) gas a theoretical gas made up of solid particles (no intramolecular energy) that interact only through collisions (no intermolecular forces)

perfect (ideal) gas law the relationship between the pressure, volume, and temperature of a perfect gas, $pV = NkT$

Boyle's law the pressure and volume of a gas are inversely proportional (at constant temperature)

Charles' law the volume and absolute temperature of a gas are directly proportional (at constant pressure)

heat capacity the ratio of heat transfer and the resulting temperature change of a body

specific heat (capacity) the amount of heat required to raise the temperature of a unit mass (e.g., one gram) of a substance one unit (e.g., one degree Celsius)

phase change the transition from one state of matter to another, e.g., solid to liquid

latent heat the energy per unit mass involved in a phase change

triple point the point at which a substance can coexist in all three phases, e.g., for water, 4.6 torr and 0.01°C

Important Formulas

Celsius-Fahrenheit conversion:	$T_F = \frac{9}{5}T_C + 32°$	(°F)
	$T_C = \frac{5}{9}(T_F - 32°)$	(°C)
Kelvin and Rankine absolute temperatures:	$T_K = T_C + 273$	(K)
	$T_R = T_F + 460°$	(°R)
average kinetic energy per molecule of a perfect gas:	$\overline{KE} = \frac{1}{2}m\bar{v}^2 = \frac{3}{2}kT$	
perfect (ideal) gas law:	$pV = NkT$ (T absolute)	
Boltzmann's constant	$k = 1.38 \times 10^{-23}$ J/K	
heat capacity:	$C = \dfrac{\Delta Q}{\Delta T}$	
or	$\Delta Q = C\Delta T$	
specific heat (capacity):	$c = \dfrac{\Delta Q}{m\Delta T}$	
or	$\Delta Q = mc\Delta T$	
latent heat:	$L = \dfrac{\Delta Q}{m}$	
or	$\Delta Q = mL$	
conversion factors and constants:	1 cal = 4.186 J (= 4.2 J)	
	1 kcal = 4186 J	
	1 Btu = 778 ft-lb = 1054 J	
latent heats of water	L_f = 80 cal/g (kcal/kg)	
	L_v = 540 cal/g (kcal/kg)	

Questions

Temperature and Heat

1. Is the temperature sense of our bodies accurate? Explain.
2. Why does the Fahrenheit scale have such strange numbers for the freezing and boiling points of water?
3. How does a degree Celsius compare with a degree Fahrenheit?
4. What are (a) room temperature and (b) normal body temperature on the Celsius scale?
5. Which is larger, a degree Fahrenheit or a kelvin, and by how much?
6. On what temperature scale would a temperature change of 10 units be the largest?

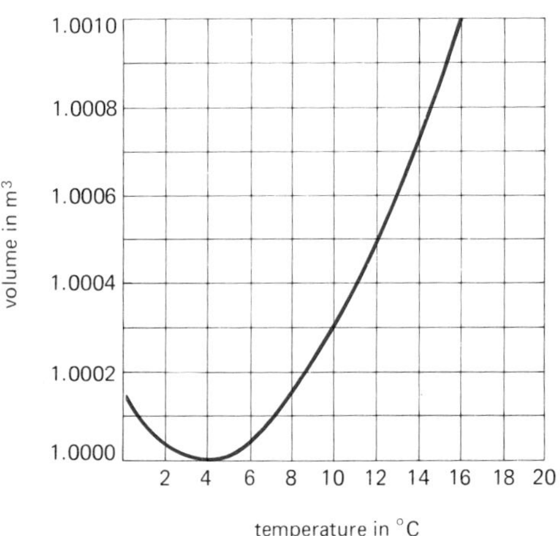

Figure 16.15
Volume versus temperature. See Question 10.

7. What is the interval between the ice and steam points of water on the Kelvin scale?
8. What are the minimum and maximum temperatures on the absolute temperature scale?
9. Is the magnitude of one kelvin the same as that of one degree Rankine? Explain.
10. Water has unique thermal expansion properties (Fig. 16.15). As the temperature increases between 0°C and 4°C, the volume *decreases*. Would water be a good liquid for a liquid-in-glass thermometer? How would such a thermometer behave near the ice point?
11. (a) Heat might be thought of as one of the "middle men" of energy. Why is this? (b) When you leave an outside door open on a cold day, does cold come in or does heat go out?
12. Define (a) calorie, (b) kilocalorie, and (c) Btu, and tell which is the largest heat unit.
13. A person buys a "25,000-Btu" window air conditioner. Explain this rating.
14. Distinguish between a gram calorie and a food calorie.
15. What is meant by the mechanical equivalent of heat?

Kinetic Theory and the Perfect Gas Law

16. How does the kinetic theory view (a) a solid, (b) a liquid, and (c) a gas?
17. What is the difference between the thermal energy and the internal energy of a perfect gas?
18. How is temperature related to a gas according to kinetic theory?
19. If equal masses of helium and oxygen gases are at the same temperature, do they have equal internal energies? Explain.
20. Heat always flows from a substance with a higher temperature to one with a lower temperature.

Does it always flow from a substance with more thermal energy to one with less thermal energy? How about internal energy? (*Hint:* consider dropping a hot BB into a tube of water at room temperature.)

21. A dilute (perfect) gas at a constant pressure decreases in volume as the temperature is lowered. What does the perfect gas law predict the volume to be at absolute zero? Does this actually happen? Explain.

Specific Heat

22. (a) What is heat capacity, and why is it not a very useful quantity? (b) What is specific heat capacity and why is it "specific"?
23. In using a liquid to store thermal energy in a solar heating system, would you choose one with a large or small specific heat? What liquid would be a likely candidate?
24. Given equal amounts (masses) of ethyl alcohol and gasoline at room temperature, which would require more heat energy to raise its temperature by five degrees?
25. If equal amounts of heat are added to two containers of water and the temperature change of the water in one of the containers is twice that of the other, what can you say about the quantities of water in the containers?
26. A piece of aluminum and a piece of copper with the same mass are placed in a laboratory oven and heated to a given temperature. Assuming the same, constant cooling rate for both, would both pieces of metal cool to room temperature in the same amount of time after being removed from the oven? Explain.
27. Blocks of metal with equal masses are heated to the same temperature and placed on a block of paraffin (Fig. 16.16). Explain why they melt to different depths when they were initially at the same temperature. Give an estimate of the specific heat of zinc.

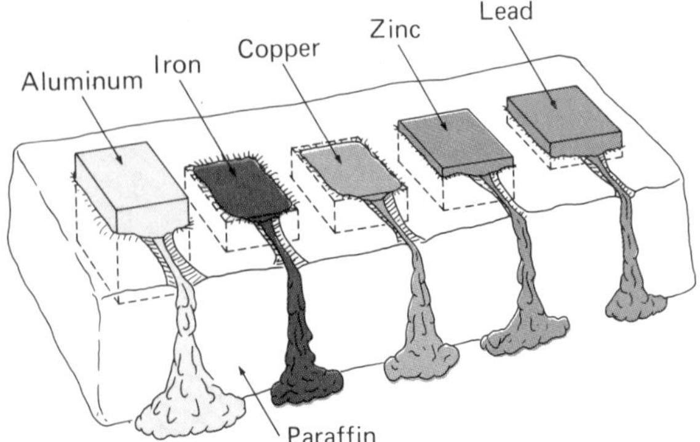

Figure 16.16
Melting to different depths. See Question 27.

Phase Changes and Latent Heat

28. Explain what is meant by "latent" heat. Why is there no temperature change associated with a phase change?
29. Explain why the latent heat of vaporization of water is almost seven times the latent heat of fusion.
30. The latent heat of alcohol is less than the latent heat of water. What does this imply?
31. (a) Why does the mirror in a bathroom fog up when you take a shower? (b) Why can you "see" your breath on a cold day?
32. A pan of water on a stove boils "faster" when the burner is on high heat, and boils "more slowly" when the burner is on low heat. Is the temperature of the water greater for a fast boil? Explain.
33. Why does covering a pot of water with a lid help the water to boil more quickly?
34. Automobile cooling systems operate under pressure. (a) What is the purpose of this? (b) What would happen if you removed the radiator pressure cap immediately after turning off a hot engine, and why? (Don't try this—it's very dangerous.)
35. We often say that a fan blowing on us feels "cool," yet it blows air at room temperature. Explain.
36. What phase change is associated with frost-free refrigerators? Explain.

Problems

Levels of difficulty are indicated by asterisks for your convenience.

16.2 Temperature Measurement and Heat Units

*1. Convert the following temperatures to degrees Celsius.
(a) 41°F, (b) −13°F, (c) 0°F, (d) 180°F

*2. Convert the following temperatures to degrees Fahrenheit.
(a) 60°C, (b) −25°C, (c) 4°C, (d) 150°C

*3. Convert the following temperatures to kelvins.
(a) 43°C, (b) −15°C, (c) 50°F, (d) 450°R

*4. Convert the following temperatures to degrees Rankine.
(a) 100°F, (b) 58°F, (c) 25°C, (d) 318 K

*5. Prove that −40°F is equal to −40°C.

*6. A 1.5-kg hammer falls from a height of 5.0 m. If all of the potential energy lost could be converted into heat energy, how many calories would be generated?

*7. Assuming that 5.0 Btu of heat could be converted completely into mechanical energy (which is never the case) and that the energy is used in doing work in lifting a 100-lb object, to what height could the object be lifted?

*8. The efficiency of a machine is 60 percent, where the energy lost goes into the heat of friction. If the work input of the machine is 2000 J, how many kilocalories of energy are lost?

*9. Plot temperature in degrees Celsius (T_C) versus temperature in degrees Fahrenheit (T_F) on a Cartesian (rectangular) graph. Describe the relationship in terms of the graph parameters.

16.3 Kinetic Theory and the Perfect Gas Law

10. What is the average kinetic energy of the molecules of a quantity of perfect gas at 23°C?

11. What is the average speed (in mi/h) of the molecules in a dilute oxygen gas at room temperature? (The mass of an oxygen molecule, O_2, is 5.31×10^{-26} kg.)

*12. An ideal gas at 27°C is heated to a temperature of 627°C. By how many times is the internal energy of the gas increased?

*13. A quantity of ideal gas is at room temperature (20°C). If enough heat were added to double its internal energy, what would be the Celsius temperature of the gas?

*14. If the pressure of a quantity of a perfect gas is doubled and its volume is decreased by one half, what is the effect on its temperature?

*15. If the temperature of a quantity of perfect gas is increased from 20°C to 80°C, how is its internal energy affected?

*16. Two equal quantities of perfect gas are at temperatures of 20°C and 20°F. (a) If the temperatures of the gases are raised to 80°C and 80°F, respectively, which gas has the greater increase in internal energy? (b) How many times greater is this increase than the other?

*17. A quantity of perfect gas has a temperature of 0°C. An equal quantity of perfect gas is twice as hot or energetic. What is its temperature in degrees Celsius?

*18. A constant-volume gas thermometer has a pressure of 1000 Pa at 20°C. If the pressure increases to 2000 Pa, what is the temperature in degrees Celsius?

*19. A quantity of perfect gas occupies 4.0 L at 20°C and has a pressure of 1.6×10^5 Pa. (a) What is the volume at the same temperature when the pressure is decreased to 1.2×10^5 Pa? (b) What is the pressure at the same temperature if the volume is compressed to 1.5 L? (c) What is the temperature (in °C) at a pressure of 1.2×10^5 Pa and a volume of 1.5 L?

*20. The temperature of a quantity of ideal gas is 80°F and occupies a volume of 0.25 m^3 at a pressure of 20 lb/in^2. If the temperature is decreased to 20°F and the pressure is decreased to 15 lb/in^2, what is the volume of the gas if it is in a flexible container?

16.4 Specific Heat

*21. Show that one cal/g-°C is equivalent to one Btu/lb-°F.

*22. How much heat is required to raise the temperature of 50 g of water by 30°C?

*23. In an annealing process, a 2.5-lb glass plate is slowly cooled from 220°F to 70°F. How much heat is given up by the glass plate?

*24. The temperature of 0.50 lb of aluminum and 0.50 lb of iron, both initially at room temperature, is increased to 100°F. Which metal absorbs more heat, and how much more?

*25. Twenty calories of heat are added to 50 g of silver at room temperature. What is the final temperature of the silver?

*26. In a solar heating system with 500 gal of water initially at 70°F, 20,000 Btu of solar energy is absorbed. What is the final temperature of the water? (One gallon of water weighs 8.3 lb.)

*27. A 50-g aluminum rod at room temperature is immersed in an 86°F water bath. (a) How much heat does the rod absorb from the water bath in coming to the bath temperature? (b) How much heat is lost by the bath?

*28. A 1.0-lb copper fitting at 212°F is placed in 1.0 gal of water at room temperature. Neglecting the container and any losses, what is the final equilibrium temperature? (*Hint:* one gallon of water weighs 8.3 lb. See Problem 31.)

*29. In a calorimetry experiment, 250 g of aluminum shot at 100°C is carefully poured into 0.20 L of water at room temperature. Neglecting the container and any losses, what is the final equilibrium temperature? (*Hint:* see Problem 31.)

**30. Equal amounts of heat are added to different quantities of brass and lead. The temperature of the brass increases by 10°C, and the temperature of the lead increases by 5°C. Which piece of metal has the greater mass, and how many times more?

**31. In calorimetry, a solid substance, usually at a higher temperature, is mixed with the calorimeter liquid. The mixture then comes to some final equilibrium temperature. Neglecting the calorimeter cup, show that the final temperature is given by

$$T_f = \frac{(mcT_i)_s + (mcT_i)_\ell}{(mc)_s + (mc)_\ell}$$

where the s and ℓ subscripts refer to the solid substance and liquid, respectively.

**32. A liter of ethyl alcohol at 60°C is mixed with 200 cc of water at 10°C. Neglecting any losses, what is the final equilibrium temperature of the mixture?

**33. In a calorimetry experiment, 500 g of metal at 100°C is mixed with 0.50 kg of water at room temperature in an aluminum calorimeter cup. The cup has a mass of 250 g. If the final temperature of the mixture is 25°C, what is the specific heat of the metal?

16.5 Phase Changes and Latent Heat

34. How much heat is required to melt 50 g of ice at 0°C?

35. How many Btu's are required (a) to melt 1.0 kg of ice at 32°F and (b) to vaporize 1 gal of water at 212°F?

36. A quantity of steam (500 g) at 212°F is condensed. How much heat is liberated?

*37. How much heat is required to raise the temperature of 10 g of ice at −10°C to water at 20°C?

*38. A gallon of water at 200°F is converted to superheated steam at 250°F. How much heat is required?

*39. How much heat is required to convert 100 g of ice at −5°C to steam at 110°C?

*40. In an industrial process, 2.0 lb of steam at 220°F is refrigerated and converted to ice at 30°F. How much heat is extracted?

**41. How much heat must be added to 100 mL of ethyl alcohol at room temperature to produce an alcohol vapor of 78°C?

**42. Four ice cubes, each with a mass of 10 g at 0°C, are placed in 200 mL of water at room temperature. Assuming no heat loss, what is the final temperature after the ice melts and the water comes to equilibrium?

**43. A 200-g piece of ice at −10°C is placed in 100 cm^3 of water at 50°C. Assuming no heat loss, how much liquid is there when the ice and water are in thermal equilibrium? (Neglect any density variations.)

Chapter Supplement

Thermometry—Temperature Measurement

Temperature is measured by reproducible changes in the physical properties of materials. As we have seen in this chapter, thermal expansion is a commonly used property in bimetallic and liquid-in-glass thermometers, but it is not the only one. For example, in addition to thermal expansion, electrical and radiation properties are also used.

Various physical properties have different ranges over which they are applicable for temperature measurement or thermometry. Some of the common practical temperature-measuring instruments and their ranges are listed in Table 16S.1.

In this supplement, some of the not-so-common temperature measuring devices will be discussed. Although some of the electrical and radiation principles involved will be discussed in later chapters, you should be able to grasp the general ideas by drawing on your practical experience.

Table 16S.1
Temperature-Measuring Instruments

Instrument	Approximate Range
Liquid-in-glass thermometer	
Alcohol	−80°C–100°C
Mercury	−38°C–350°C
Bimetallic thermometer	−40°C–500°C
Electrical resistance thermometer	−272°C–1600°C
Thermocouple	−260°C–1600°C
Optical pyrometer	600°C upward
Infrared pyrometer	−20°C–1700°C

Electrical Resistance Thermometers

The electrical resistance of most metals increases with temperature. (Electrical resistance is the opposition of a material to the flow of electric current. See Chapter 21.) The increase is rather large. For example, in platinum a 39

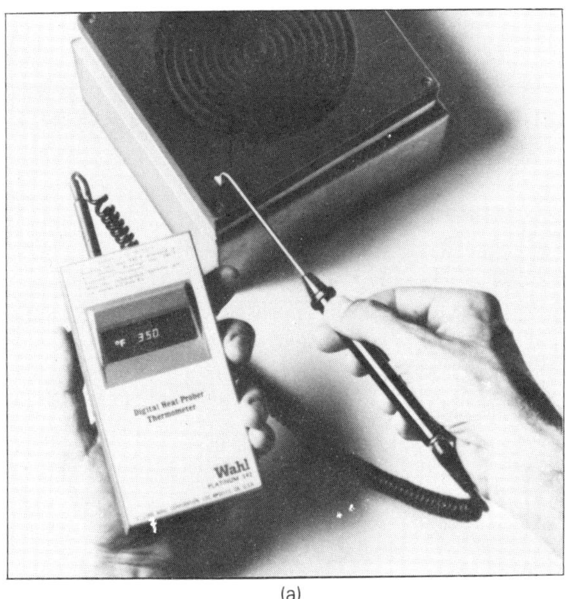

(a)

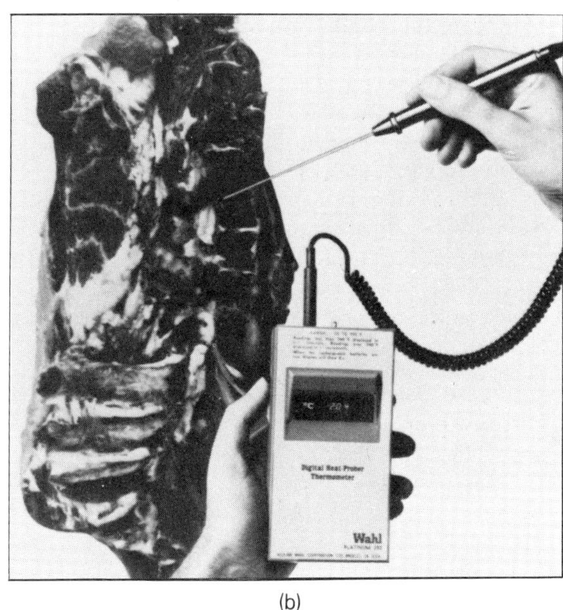

(b)

Figure 16S.1

Resistance thermometers. Based on the change of electrical resistance with temperature, resistance thermometers (here with digital readouts) are versatile instruments and may be used to measure (a) surface temperatures, and (b) interior temperatures of solids and liquids.

percent change occurs between 0°C and 100°C. Hence, changes in the electrical resistance of a metal can be calibrated to temperature and used as a resistance thermometer. Platinum, which can be used to measure in a temperature range of −260°C to 1100°C, is especially suitable.

If the metal in a circuit is connected to the appropriate meters, the temperature can be read. The metal temperature probe is convenient in many applications (Fig. 16S.1).

Thermocouple Thermometers

The thermocouple thermometer is based on an electrical effect. If two different metal wires, such as copper and iron, are joined together to form a closed loop, an electrical current flows in the loop if one junction is kept at a different temperature from the other (Fig. 16S.2).

The electrical current is commonly measured in terms of voltage with a millivoltmeter. The voltage is proportional to the difference in the temperatures of the junctions. If the temperature of one of the junctions (reference junction) is known, the temperature of the other junction

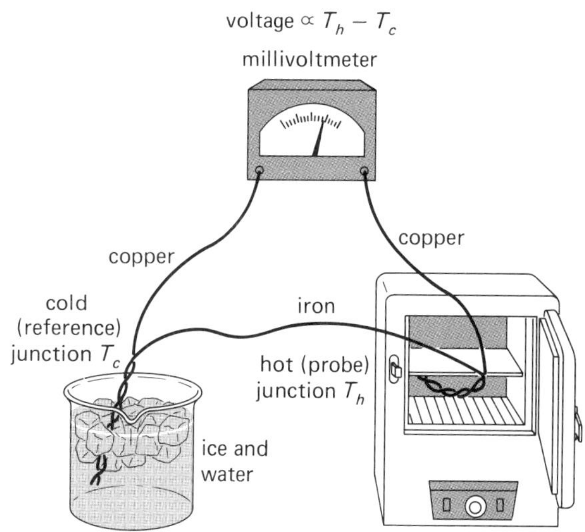

Figure 16S.2

The thermocouple thermometer. When the junctions of a thermocouple are at different temperatures, a voltage develops that is proportional to the temperature difference. One of the junctions is used for a reference at a known temperature.

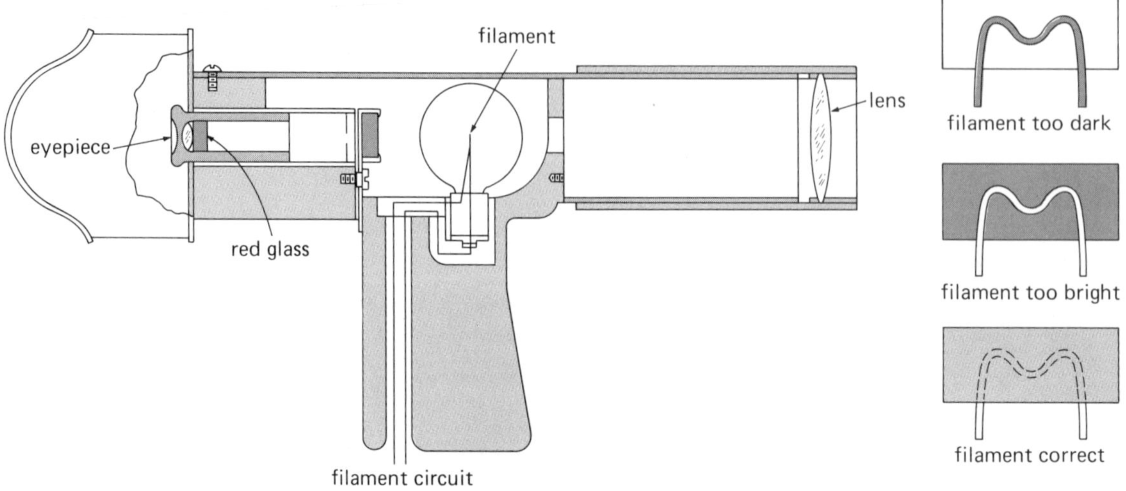

Figure 16S.3
Optical pyrometer. (a) The filament provides a visual brightness comparison that is matched to a temperature "background." The filament current is proportional to the temperature. (b) An optical pyrometer being used to measure the temperature of molten steel.

can be measured. For example, a cold (reference) junction in ice and water is at 0°C, since ice and water coexist at this temperature.

The thermocouple is particularly convenient for many applications because of the small size of the metal wire junction. It can be inserted wherever a temperature measurement is desired. The reference junction and electrical meter can be some distance away.

Pyrometers

Pyrometers make use of the radiation of light emitted by an object. The wavelength or color of the light is proportional to the temperature of the body. For example, when metals are heated to high temperatures they glow with different colors (emit different radiations) with increasing temperature. The approximate relationships are

Minimum visible red	475°C
Dull red	600°C
Cherry red	700°C
Light red	850°C
Orange	900°C
Yellow	1000°C
Blue-white	1150°C and higher

Thus, the temperatures of very hot bodies can be judged visually by their color and brightness (intensity). Direct visual estimation of temperature is subjective and may be quite inaccurate. An optical pyrometer is an instrument designed to improve visual estimates of temperature by providing the eye with a source of brightness for comparison (Fig. 16S.3). The comparison source is a lamp filament. Varying the current in the filament varies its brightness

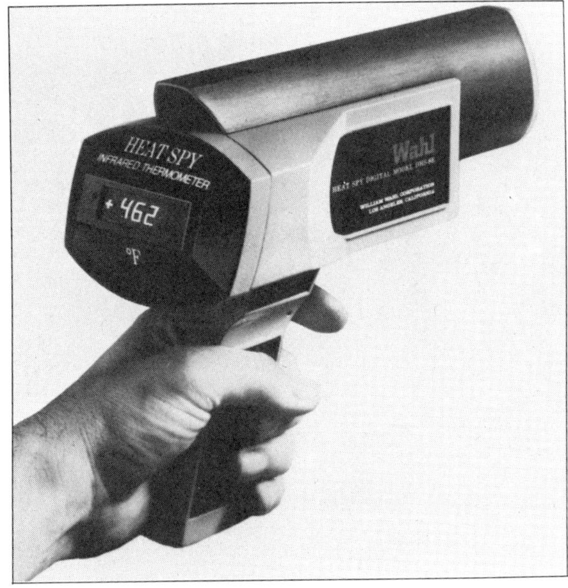

(a)

(c)

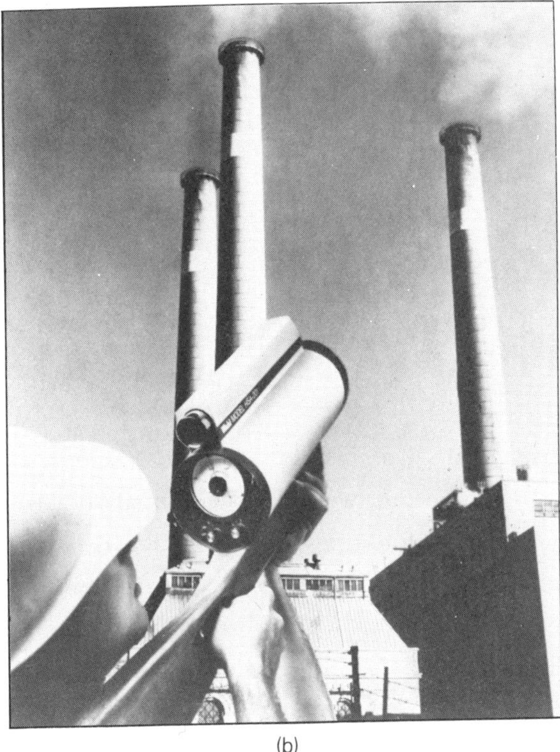

(b)

Figure 16S.4
An infrared thermometer (a) and some applications (b) and (c). The temperature of a surface can be determined from the emitted infrared radiation merely by pointing the instrument toward and focusing it on the surface.

so as to match the brightness of the image of a hot body as seen through the pyrometer. The filament current is calibrated in terms of temperature.

All warm bodies emit infrared radiation. The characteristics of this radiation depend on a body's temperature. Infrared radiation cannot be seen with the human eye, so photocells and other special detectors must be used. An infrared pyrometer is used to detect the infrared radiation from a body, and the radiation is then related to its temperature. These instruments have more precision and accuracy and a greater range than optical pyrometers. Infrared "thermometers" are now used quite extensively (Fig. 16S.4). You don't even have to be close to take a body's temperature; you can just point the pyrometer at it.

(Courtesy Bethlehem Steel Corp.)

Chapter 17

Thermal Properties of Materials

Having gained an insight into the nature of heat and temperature, we will now look at some of the basic thermal properties of matter. Heat energy may be thought of as the common denominator of energy on which our society depends. Either directly or indirectly, the energy used domestically and in industry generally involves heat. Home heating, electrical power generation, steel-making, and many other industrial processes use thermal energy.

As a result, the thermal properties of materials are important technical considerations. Heat must be transferred; its effect on materials must be known; and temperature, the chief measurable parameter of heat, must be monitored and accurately measured. These processes and effects form the basis of this chapter.

17.1 Heat Transfer

In the previous chapter, general reference was made to heat being transferred or "flowing" from one body to another. However, little was said about the methods of heat transfer or the *rate* at which heat is transferred. These considerations are of some major importance.

In some instances we wish to retard heat flow—for example, in preventing the loss of heat from hot water or steam pipes and in preventing the thawing of frozen foods. In other instances, we wish to promote the transfer of heat—for example, in the cooling of automobile engines.

There are three methods of heat transfer: conduction, convection, and radiation.

Conduction

The method of heat transfer easiest to describe quantitatively is conduction, which is the primary method of heat transfer in solids. Whenever there is a temperature difference within a body or between two bodies in thermal contact, a spontaneous flow of heat takes place from the region of higher temperature to that of lower temperature—down a "temperature hill," so to speak.

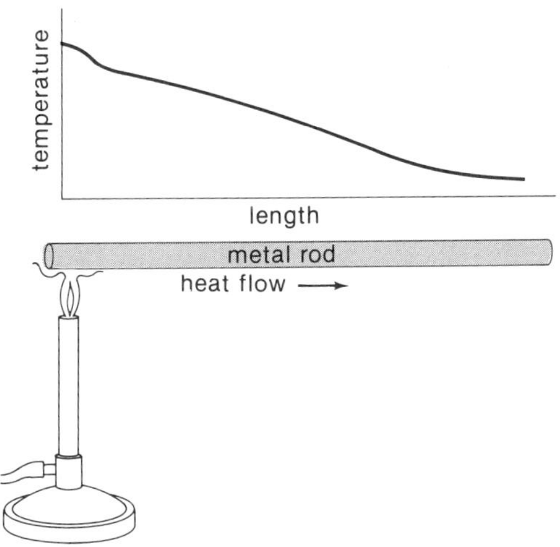

Figure 17.1
Heat transfer by conduction. A spontaneous flow of heat in a metal rod takes place from a region of higher temperature to a region of lower temperature by molecular interaction. The thermal gradient or "hill" between the two regions is illustrated in the graph.

Such a temperature variation with position or length is referred to as a **thermal or temperature gradient,** and the greater the thermal gradient, the steeper the "temperature hill" (Fig. 17.1). The average temperature gradient is the change in temperature per unit length, $\Delta T/L$, where L is the length or thickness of the material and ΔT is the temperature difference between the ends or surfaces of the material.

A qualitative understanding of the microscopic basis of the conduction of heat can be gained from the kinetic theory, in which heat is described in terms of the mechanical energy of the atoms and molecules of a substance. In a solid, the molecules are held together in a fixed array and vibrate about their equilibrium positions (Section 16.3). The molecules of a higher temperature region have, on the average, more kinetic energy than those of a neighboring region of lower temperature, and energy is transferred by molecular interaction to the region of lower temperature (less molecular motion).

Also, in some materials, particularly metals, electrons are not tightly bound to the atoms and are more or less free to move throughout the material. The electrons carry energy from one region to another. This is usually the most important mechanism for heat conduction in metals. As will be learned in a later chapter, the mobility of electrons in metals makes them good conductors of electricity. Hence, good conductors of electricity are generally good conductors of heat.

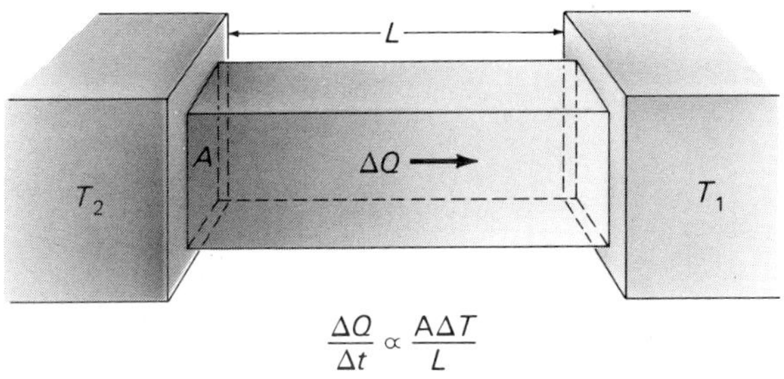

Figure 17.2
Rate of heat conduction. The rate $\Delta Q/\Delta t$ at which heat is conducted through a bar is proportional to the cross-sectional area A of the bar and the temperature difference between the ends of the bar, $\Delta T = T_2 - T_1$. The conduction rate is inversely proportional to the length L of the bar.

To describe the rate of heat conduction quantitatively, it is convenient to consider a bar or slab of material, as illustrated in Figure 17.2. The bar has a uniform cross section of area A and a length L, and its ends are kept at constant temperatures T_2 and T_1, where $T_2 > T_1$. It is assumed that no heat is lost from the sides of the bar; i.e., the sides are thermally insulated. Then, the quantity of heat ΔQ that is conducted through the bar in a time Δt defines the average rate of heat flow $\Delta Q/\Delta t$.

In analyzing the situation, it should be apparent from experience that

1. The rate of heat flow is directly proportional to the cross-sectional area A. (A bar with a greater cross-sectional area conducts more heat per unit time.)
2. The rate of heat flow is directly proportional to the temperature difference $\Delta T = T_2 - T_1$ between the ends of the bar. (The greater the temperature difference, the greater the amount of heat transferred per unit time.)
3. The rate of heat flow is inversely proportional to the length L of the bar. (The longer the bar, the less heat transferred per unit time, since the heat energy must travel a greater distance.)

Thus,

$$\frac{\Delta Q}{\Delta t} \propto \frac{A \Delta T}{L}$$

or, in equation form,

$$\boxed{\frac{\Delta Q}{\Delta t} = \frac{kA\Delta T}{L}} \quad \text{(Eq. 17.1)}$$

where k is a constant of proportionality characteristic of the particular material and is called the **thermal conductivity.**

The thermal conductivity of a material is characteristic of its ability to conduct heat.

It should be noted in Equation 17.1 that $\Delta T/L$ is the change in temperature per unit length, or the thermal gradient. If heat is conducted such that the temperature difference is equalized, i.e., $\Delta T = 0$, then there is no thermal gradient and the bar is in thermal equilibrium at a constant temperature.

From Equation 17.1 in the form

$$k = \Delta QL/A\Delta t\Delta T$$

the thermal conductivity can be seen to have metric units of cal-cm/cm²-s-°C = cal/cm-s-°C (or kcal/m-s-°C) and, in the British system, Btu-in./ft²-h-°F. The mixed units in the British system are geared for thermal conduction across materials with thicknesses in inches and areas in square feet. Also, the unit time is practically expressed in hours.

A list of the thermal conductivities of various materials is given in Table 17.1. A material with a relatively large thermal conductivity is a good thermal conductor, and a material with a low thermal conductivity is a poor thermal conductor, or a *thermal insulator.*

Some applications of thermal conductivities are shown in Figure 17.3. We use Styrofoam coolers with poor thermal conductivity to keep things cold. Also, we often say that a tile floor is cold to the bare feet, and we use a rug to avoid stepping on the tiles. The rug and the floor are at the same temperature, but their thermal conductivities are different (see Table 17.1 and assume the rug is cotton). The title floor only *feels* colder because it conducts heat from your feet faster.

Table 17.1
Thermal Conductivities

Material	k	
	*cal/cm-s-°C**	*Btu-in./ft²-h-°F*
Good conductors		
Aluminum	0.50	1.5×10^3
Brass	0.25	0.75×10^3
Copper	0.92	2.8×10^3
Iron	0.11	0.33×10^3
Lead	0.08	0.24×10^3
Silver	1.0	3.0×10^3
Steel	0.11	0.33×10^3
Average conductors		
Asbestos	1.4×10^{-3}	4.2
Brick	1.7×10^{-3}	5.1
Concrete	4.0×10^{-3}	12
Floor tile	1.6×10^{-3}	4.8
Glass (typical value)	2.0×10^{-3}	6.0
Ice	5.3×10^{-3}	16
Water	1.4×10^{-3}	4.2
Good insulators (poor conductors)		
Air	0.55×10^{-4}	0.17
Corkboard	1.0×10^{-4}	0.30
Cotton	1.8×10^{-4}	0.54
Fiberboard	1.4×10^{-4}	0.42
Foam plastic	1.0×10^{-4}	0.30
Glass wool	0.90×10^{-4}	0.27
Vacuum	0	0
Wood (typical value)	2.5×10^{-4}	0.75

* 1 kcal/m-s-°C = 10 cal/cm-s-°C

See Special Feature 17.1 for an old question that involves thermal conductivity.

EXAMPLE 17.1 The bottom of an iron tank used to heat water for an industrial process is 0.50 in. thick and has an area of 12 ft². If the tank contains boiling water at 212°F and is heated at the bottom by a row of gas flames that maintain an average temperature of 400°F, (a) what is the rate of thermal conduction through the bottom of the tank? (b) How much heat passes through the bottom of the tank in 30 minutes?

Solution: It is given that $L = 0.50$ in., $A = 12$ ft², and $\Delta T = T_2 - T_1 = 400°\text{F} - 212°\text{F} = 188°\text{F}$. From Table 17.1, for iron $k = 0.33 \times 10^3$ Btu-in./ft²-h-°F. Then, by Equation 17.1,

(a)

(b)

Figure 17.3
Thermal conductivity. (a) Materials such as Styrofoam, which have a lot of air spaces, are poor thermal conductors. This property helps keep objects inside the cooler cold (or warm) for long periods of time. (b) Is a tile floor colder than a rug? See text for description.

(a) $$\frac{\Delta Q}{\Delta t} = \frac{kA\Delta T}{L}$$

$$= \frac{(0.33 \times 10^3 \text{ Btu-in./ft}^2\text{-h-°F})(12 \text{ ft}^2)(188\text{°F})}{0.50 \text{ in.}}$$

$$= 1.5 \times 10^6 \text{ Btu/h}$$

(b) In $\Delta t = 30 \text{ min} = 0.50 \text{ h}$,

$$\Delta Q = (1.5 \times 10^6 \text{ Btu/h})\Delta t$$

$$= (1.5 \times 10^6 \text{ Btu/h})(0.50 \text{ h})$$

$$= 7.5 \times 10^5 \text{ Btu}$$

SPECIAL FEATURE 17.1

Does Hot Water Freeze Before Cold Water?

This is a common question, and the answer depends on the conditions. Suppose that two identical *covered* pans with equal amounts of water at temperatures of 20°C and 40°C, respectively, are placed in a freezer at 0°C. All other things being equal (except the cooling rates), the pan with the lower initial temperature, 20°C, would freeze first. The pan at the higher temperature would lose heat at a faster rate (greater temperature difference), but it has more heat to lose. It would cool down to 20°C, then cool at the same rate as the other pan did during this first period of time.

Think of it this way: suppose you were going to run 20 m and another person were going to run 40 m on the same track. The other runner might run at a faster rate to your 20-m mark, but then if he slowed down and ran the second 20 m at the same rate as you, he would never be able to catch up, and you'd get to the finish line first.

It might be possible for the hot water to freeze first under some special conditions. Suppose the pans were *uncovered* and the hot water were very hot. In cooling down, more of the hot water would evaporate, which would leave less water (mass) in the pan to cool to freezing. Also, a hot pan might melt into the frost layer on the freezer shelf, and with better contact would have greater thermal conductivity and a faster cooling rate.

EXAMPLE 17.2 Ice is stored in a refrigerator-freezer at 0°C, and the outside ambient temperature is 20°C. The refrigerator door is 81 cm wide, 62 cm high, and 10 cm thick, with plastic foam insulation. If the refrigerator is unplugged, how much ice melts per hour as a result of the heat conducted through the door?

Solution: It is given that $\Delta T = T_2 - T_1 = 20°\text{C} - 0°\text{C} = 20°\text{C}$, $L = 10$ cm, $A = w \times h = 81 \text{ cm} \times 62 \text{ cm} = 5.0 \times 10^3 \text{ cm}^2$, and for plastic foam $k = 1.0 \times 10^{-4}$ cal/cm-s-°C (Table 17.1). Then,

$$\frac{\Delta Q}{\Delta t} = \frac{kA\Delta T}{L}$$

$$= \frac{(1.0 \times 10^{-4} \text{ cal/cm-s-°C})(5.0 \times 10^3 \text{ cm}^2)(20°\text{C})}{10 \text{ cm}}$$

$$= 1.0 \text{ cal/s}$$

or

$$\frac{\Delta Q}{\Delta t} = 1.0 \text{ cal/s } (3600 \text{ s/h})$$

$$= 3.6 \times 10^3 \text{ cal/h}$$

Hence, $\Delta Q = 3.6 \times 10^3$ cal are conducted through the door each hour.

With the latent heat of fusion for ice $L_f = 80$ cal/g and $\Delta Q = mL_f$, the mass of ice melted per hour is

$$m = \frac{\Delta Q}{L_f} = \frac{3.6 \times 10^3 \text{ cal}}{80 \text{ cal/g}} = 45 \text{ g}$$

(Question: After all the ice is melted, will the rate of heat transfer remain the same?)

R-Value

In these energy-conscious times, the conduction or lack of conduction of heat is quite important. For building materials, the lack of thermal conduction (i.e., thermal resistance, which describes the insulating properties of a material) is a main consideration. The **thermal resistivity** ρ is the reciprocal of the thermal conductivity:

$$\rho = \frac{1}{k} \qquad \textbf{(Eq. 17.2)}$$

Using k in British units gives the insulation property of a material per inch of thickness.

The effective thermal insulation of a material is often rated in terms of its R-value. The **R-value** is a practical measure of thermal resistance and is given by

$$\boxed{\text{R-value} = \rho L = \frac{L}{k} \quad (L \text{ in inches})} \qquad \textbf{(Eq. 17.3)}$$

where the conductivity k is in *British units* and L is the thickness of the material *in inches*. For example, 3.0 in. of foam plastic would have an R-value of (with $k = 0.30$ Btu-in./ft²-h-°F; Table 17.1).

$$\text{R-value} = \frac{L}{k} = \frac{3.0}{0.30} = 10$$

which is expressed as R-10.

Notice that the heat flow equation (Eq. 17.1) can be written in terms of an R-value:

$$\frac{\Delta Q}{\Delta t} = \frac{kA\Delta T}{L} = \frac{A\Delta t}{(L/k)} = \frac{A\Delta T}{R} \qquad \textbf{(Eq. 17.4)}$$

Hence, the greater the R-value of a material, the smaller the rate of heat flow, or the greater the insulating value.

A given R-value of insulation requires different thicknesses of different materials. For example, each of the following material thicknesses has an R-value of 19 (R-19):

Fiberglas*	(6.0 in.)
Wood	(14 in.)
Brick	(7.9 ft)

For example, for wood ($k = 0.75$; Table 17.1), using Equation 17.3,

$$L = k(\text{R-value}) = (0.75)(19) = 14 \text{ in.}$$

A common use of R-values is in specifying the thermal insulation requirements in building and construction. The recommended insulation standards in R-values for ceiling, walls, and floors in different parts of the country are shown in Figure 17.4.

* Registered trademark of Owens-Corning Fiberglas Corp.

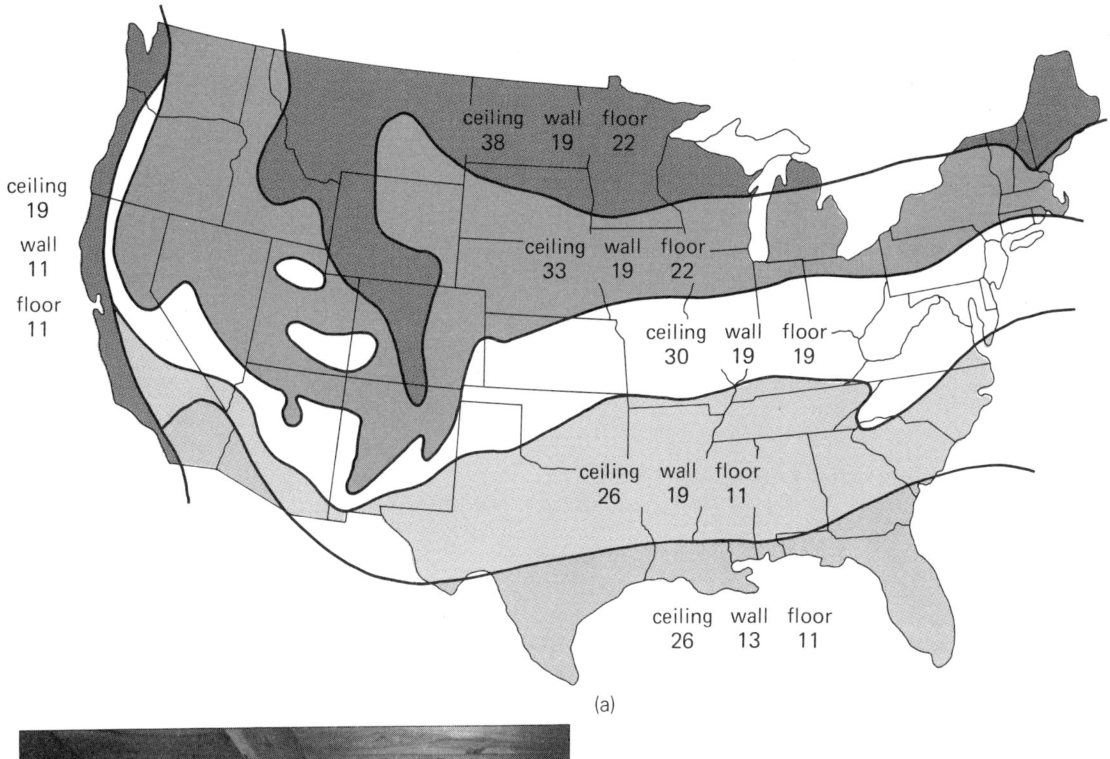

(a)

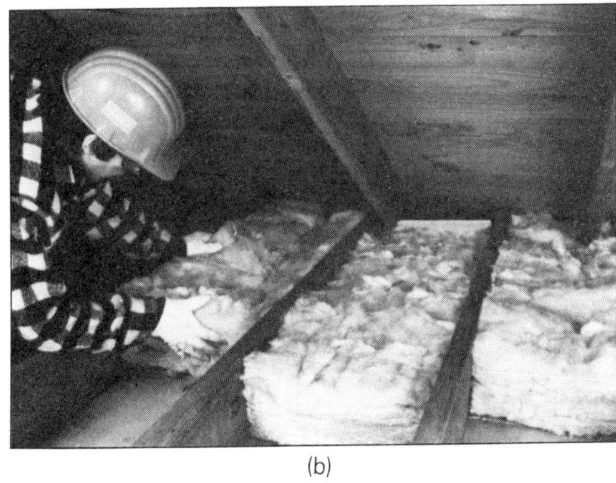

(b)

Figure 17.4
Insulation. (a) Recommended insulation R-values for ceiling, walls, and floors in different parts of the country. (b) Fiberglas insulation being installed in an attic. (Courtesy Owens-Corning Fiberglas Corp.)

Properly insulated heating and cooling ducts, ceilings, walls, and floors provide significant savings in fuel costs. For a given R-value, less insulating thickness is required for a material having a smaller thermal conductivity (higher thermal resistivity). As noted above, 6.0 in. of Fiberglas provides the same insulation as 14 in. of wood. Fiberglas and similar commercial materials have superior resistance to thermal conduction because of millions of tiny air cells in the material. (Why?) The R-values of such commercial materials are commonly found printed on the package or material.

For a composite of several layers of different materials, the R-value of the composite is the sum of the R-values of the individual layers.

Convection

In a convection process, heat is transferred by mass transfer. That is, heat energy is carried along by a moving substance, such as in a water- or an air-circulating heating system. Since matter must be free to move in the process, convection is an important effect in fluids—gases and liquids. Obviously, convection does not generally occur in a solid.

Figure 17.5
Convection. Heat transfer by convection involves mass transfer. Rising hot air sets up a convection cycle that distributes heat around the room.

When a portion of a gas or liquid is heated, a density and pressure difference develops within the fluid due to thermal expansion. The heated portion is less dense than the material in the surrounding regions. As a result, the material in the warmer portion of the fluid rises or is "buoyed up" (Archimedes' principle, Chapter 15), and heat is transferred. The rising motion of the fluid is referred to as a *convection current.* This is the basis of the common saying "Hot air rises" (Fig. 17.5).

Convection currents are readily observed in water being heated in a glass container or in air rising from a hot road in the summer. As a mass of fluid rises and cools, fluid to the side moves in to take the place of the rising fluid, and the cool fluid descends so as to set up a convection cycle. The quantitative description of convection is extremely complex, as might be expected.

The expansion and rising of a heated portion of a fluid is called *natural convection.* Years ago, home heating relied primarily on natural convection cycles, as is still the case today with space heaters. However, in modern heating and cooling systems, the fluid is moved by the action of a fan or pump, which is heat transfer through *forced convection.* Common examples are the cooling systems of automobile engines and the forced-air heating (and cooling) of homes and buildings (Fig. 17.6).

Convection currents are also important to the burning of a fire in a fireplace. The fire gets fresh air needed for combustion from convection currents, which are set up by the rising of hot air in the chimney and aided by the "draft" of the Bernoulli effect (Chapter 15). Air blowing across the top of the chimney reduces the pressure in this area and creates a pressure difference between the top and bottom of the chimney, and we commonly say the chimney has a good draft or "draws" well.

The air flow for convection may take heated air up the chimney. That air can come from other portions of the house. It may be drawn from the heated air in a closed room, and that air is replaced by colder air coming through cracks around windows or under doors. (A fireplace would not operate in a completely leakproof house.)

In either case, fireplaces are not very efficient. Their efficiency may be increased with "heatolators." In one type, air pipes installed in the chimney are heated and the warmed air is circulated back into the room. In modular, set-in fireplaces, air from the room is blown by a fan through space behind the firewall and back into the room.

Radiation

The third method of heat transfer, radiation, is quite different from conduction and convection, inasmuch as radiant energy can move through the empty space of a vacuum. Radiant energy, or radiation, is a form of electromagnetic radia-

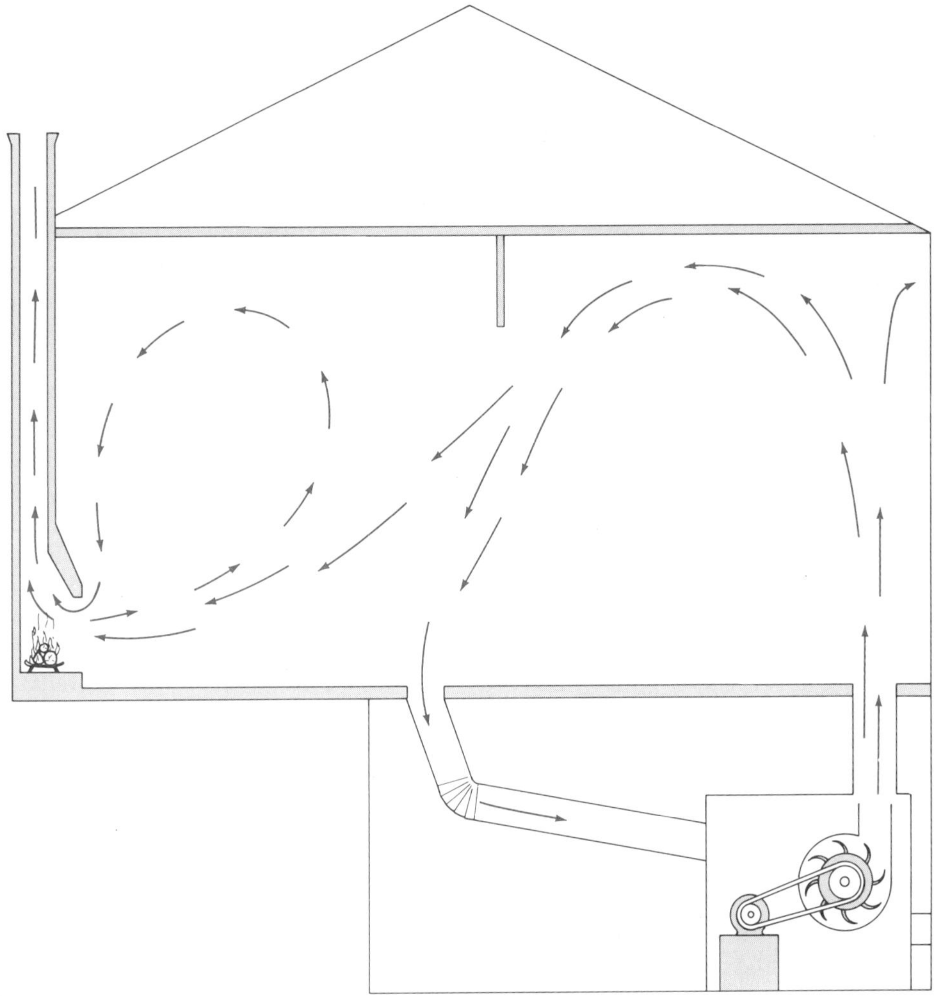

Figure 17.6

Forced-air convection. Heat transfer by convection in most homes is assisted by mechanical means. Convection provides fresh air for combustion in fireplaces, but a lot of energy is lost up the chimney.

tion. More will be said about the electromagnetic nature of radiation in a later chapter. For the present discussion, we are concerned with the general properties of commonly observed radiation heat transfer.

Electromagnetic radiation has many representative forms. Light is a form of radiation, and light and *heat* are transferred from the Sun to Earth through the void (vacuum) of space. Other forms of radiation include radio waves, microwaves, and X-rays. The production of these types of radiation involves special processes that will be discussed later. For the present we will consider thermal radiation, which is emitted by a substance by virtue of its temperature.

All objects emit radiant energy, and heat transfer is effected by the emission and absorption of radiation by matter. It is found that the rate at which thermal energy is radiated from the surface of an object is directly proportional to the fourth power of the *absolute* temperature of the body, i.e.,

$$\frac{\Delta Q}{\Delta t} \propto T^4$$

The total radiation emitted depends on the surface area A of the object, so we write

$$\frac{\Delta Q}{\Delta t} \propto eAT^4$$

where e is the **emissivity,** which is a measure of the ability of a body to emit thermal radiation. The emissivity is a unitless quantity varying from 0 to 1 and reflects the fact that some bodies are better emitters than others.

Radiation emission is commonly expressed in terms of the intensity I, which is the power emitted per unit area: $I = P/A = (\Delta Q/\Delta t)/A$, where $\Delta Q/\Delta t$ is power (heat energy per time). The units of intensity are watt/square meter (W/m^2) in the SI system. Thus, in equation form,

$$\boxed{I = \frac{P}{A} = e\sigma T^4} \qquad \textbf{(Eq. 17.5)}$$

where σ is the constant of proportionality. This relationship is known as the **Stefan-Boltzmann law,** and σ is called the Stefan-Boltzmann constant; $\sigma = 5.7 \times 10^{-8}\ W/m^2\text{-}K^4$.

EXAMPLE 17.3 What is the intensity of the radiation being emitted by an object with an emissivity of 0.50 at room temperature (20°C)?

Solution: With $e = 0.50$ and $T_K = T_C + 273 = 20 + 273 = 293$ K,

$$I = e\sigma T^4 = (0.50)(5.7 \times 10^{-8}\ W/m^2\text{-}K^4)(293\ K)^4$$
$$= 210\ W/m^2$$

Suppose the object has an area of $0.20\ m^2$; then the radiant power being emitted is ($I = P/A$)

$$P = IA = (210\ W/m^2)(0.20\ m^2) = 42\ W$$

or 42 joules of energy are being emitted each second (watt = joules per second).

Radiation emitted by objects at room temperature is usually a negligible mode of heat transfer. However, at high temperatures, such as in the filament of an incandescent light bulb, radiation is usually far more important than either conduction or convection, owing to the strong dependence of the emission of radiation on temperature (T^4). The same is true for the heating effect of a fire in a fireplace (without a heatolator). The fire "feels" warm because of radiation transfer, since convection takes heat up the chimney.

As can be seen from Equation 17.5, the other factor on which the radiation from a body depends is its emissivity e. As a general rule, a good emitter is also a good absorber. That is, the emissivity of a body is generally equal to its absorptivity, $e = a$. The absorptivity is a measure of the amount of radiation a body absorbs, and it varies from 0 to 1. The idealized case of a perfect emitter and absorber is called a **black body** ($e = a = 1$).

The emissivity depends in large part on the surface characteristics. A body with a black, nonshiny surface usually has a high emissivity and is also a good absorber of radiation. A shiny surface generally has a low emissivity and is a poor absorber, since the surface reflects radiation. These absorption properties can be demonstrated by painting the bulb of one thermometer with aluminum paint and another with black paint. When placed in sunlight (radiation), the black-bulb thermometer will read a higher temperature because of greater absorption.

There are many applications and processes that involve heat transfer or the lack of it. An example of the latter, that takes into account all three methods of heat transfer, is the vacuum bottle, which is used to keep liquids and foods hot or cold (Fig. 17.7). The inner double-walled bottle is made of glass, a poor conductor, and the space between the walls is evacuated to prevent heat transfer by convection. The external surface of the inner wall is silvered to reduce radiation losses through the partial vacuum between the walls.

On the other hand, efficient heat transfer is essential in automotive cooling systems (Fig. 17.8). In the common liquid-cooled system, heat from the internal combustion in the cylinders is conducted through the engine block and is removed by a forced convection cycle of water circulating through the engine cooling system. A water pump driven by the engine via the fan belt provides the forced convection that carries the water through the radiator. Here, the water circulates through tubes with fins that provide more surface area. Heat is removed by forced airflow through the radiator from the engine fan.

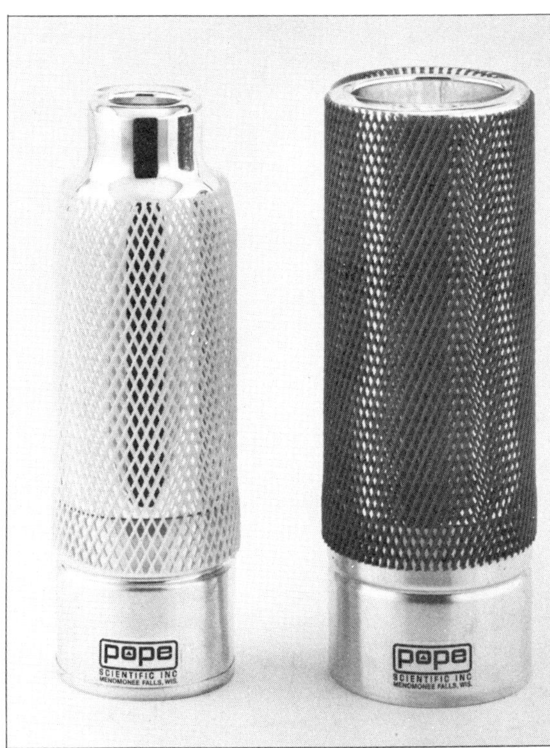

Figure 17.7
Vacuum bottles. Commonly called Dewar flasks, these double-walled, evacuated, silvered glass containers are highly effective in preventing heat transfer. Notice the protective matting on the bottles. This allows handling, reduces breakage, and prevents flying glass should the bottle be broken. If a bottle is broken, it "implodes" because of the partial vacuum, but glass could come flying outward. Why? (Courtesy Pope Scientific, Inc.)

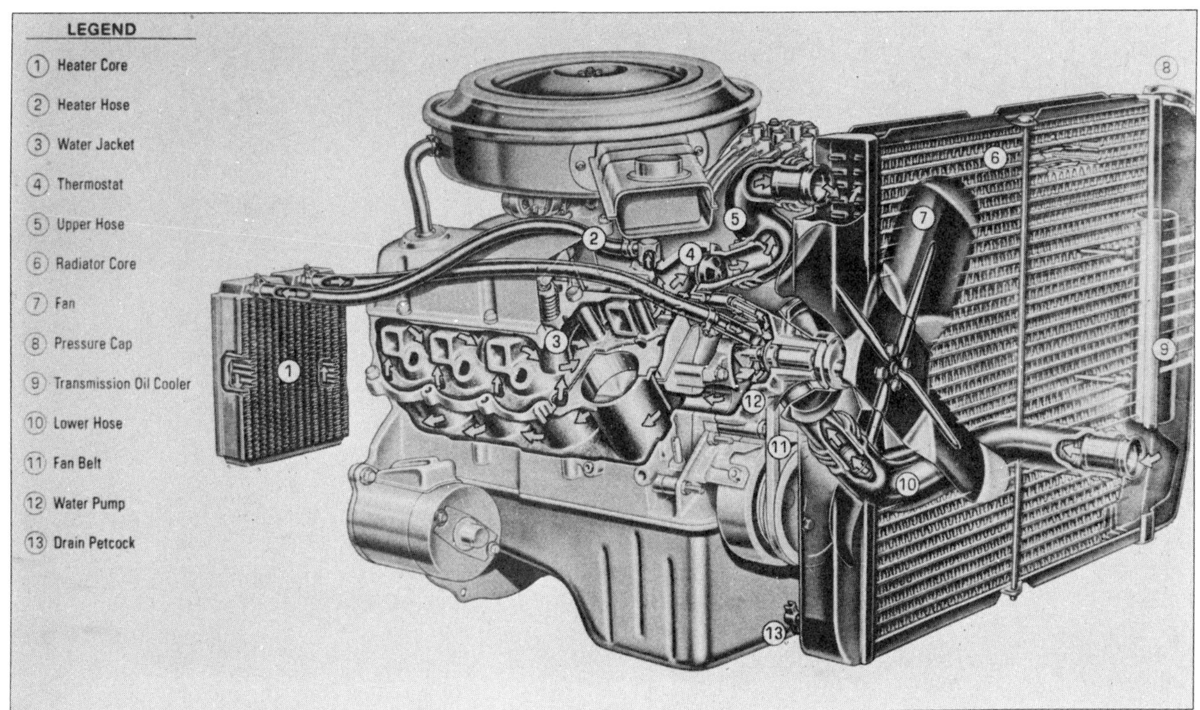

Figure 17.8
The automotive cooling system. Most of the heat is removed by convection. The radiator is better called a heat exchanger. (Courtesy Ford Motor Co.)

SPECIAL FEATURE 17.2

The Microwave Oven

Heat transfer by radiation, in the form of the microwave oven, is a relatively new boon to cooking. Microwaves are a type of electromagnetic radiation that is absorbed by water molecules as a result of molecular resonance in the microwave frequency range.

In a microwave oven (Fig. 1), the microwaves are generated by a device called a *magnetron*. On leaving the magnetron, the microwaves are routed to the oven via a metal channel, or wave guide. A metal stirrer disperses the waves in the oven chamber.

Microwave radiation does not affect all materials in the same way. Metals reflect microwaves, similarly to the way a mirror reflects visible light. That is why the metal walls of the oven do not get hot. The see-through door panel is shielded with a fine metal grid that allows light to pass through but reflects the longer microwaves. Some materials, however, are relatively transparent to microwaves; these include glass, ceramics, paper, and plastic. Microwaves pass through plastic wrap and glass or ceramic cooking dishes, but are absorbed by the food, causing it to be heated.

Water, fat, and sugar are common components of food that readily absorb microwaves. The molecules of these substances have resonance vibrations in the microwave frequency region (cf. Chapter 13). The radiation penetrates the food only to a depth of about 2–3 cm (about 1 in.), depending on the density of the food. However, because the air in the oven is relatively cool, the food loses heat, which gives rise to the false impression that it is being cooked from the inside out. Foods are cooked by the conduction of heat from the surface to the interior, much as in a conventional oven. It is advisable to let large food items sit for a while after the oven has shut off so they will be heated or cooked throughout.

Microwaves are absorbed rather uniformly, and a food warms quickly without burning. In conventional ovens, in contrast, the high surface temperature causes foods to brown and burn.

We are often told that metal utensils and foil should not be used in microwave ovens. Yet some units are equipped with a metal grill or shelf, and there is a technique of using a piece of aluminum foil to keep certain parts of a food cooler. However, reflections from a deep metal pan would keep the microwaves from reaching and cooking the food. Also, if the metal is not positioned properly, the radiation reflected from it could cause an electrical arc or spark, which could start a fire or, over time, damage the oven. The best policy is to follow the manufacturer's recommendations for proper use.

An important safety feature of a microwave oven is the automatic shut-off when the door is opened. If you were able to reach into the oven while it was running, you could be injured, since the molecules in your hand would absorb the radiation.

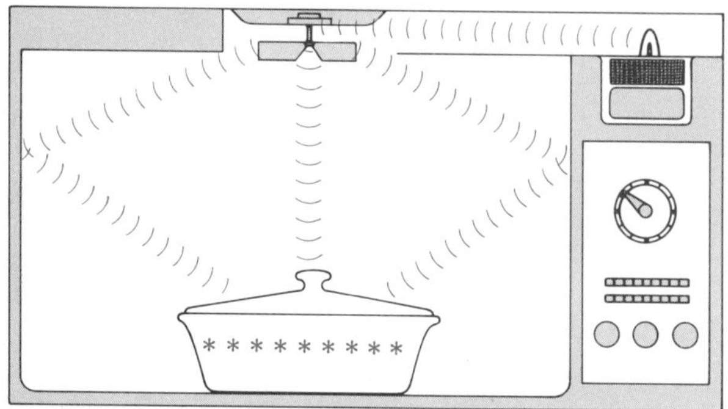

Figure 1

Heat transfer by radiation. In a microwave oven, microwaves (a type of electromagnetic radiation) are absorbed by water and fat molecules in food, causing it to become hot.

Radiator is somewhat of a misnomer, since most of the heat is removed by forced air convection (after being conducted to the air). A more appropriate name for the auto radiator would be a **heat exchanger.** Of course, some heat energy is removed from the hot radiator by radiation, but it is a relatively small amount. If the cooling system depended solely on radiation heat transfer, the engine would quickly overheat.

An example of heat transfer by radiation is given in Special Feature 17.2.

17.2 Thermal Expansion of Materials

It is a common observation that substances expand or change in size with changes in temperature. For example, houses are heard creaking as they relieve stresses due to the thermal expansion associated with seasonal temperature changes. With few exceptions, all substances increase in size with increasing temperature, and correspondingly decrease in size with decreasing temperature.

In terms of the kinetic theory of matter, an increase in temperature causes a greater vibration of the molecular particles, which increases the average distance between the particles in a solid or liquid. This is manifested as an overall change in the dimensions of the substance.

Linear Expansion

For a solid object, which has a definite shape, thermal expansion can be analyzed in terms of the specific dimensions of the object. The change in any one dimension of a solid is called **linear thermal expansion.**

For relatively small temperature changes, it has been found experimentally that the change ΔL in any linear dimension of a solid is proportional to the temperature change ΔT (Fig. 17.9). It is also proportional to the initial length L_o. For example, if two bars are made of the same material, but one is twice as long as the other, the longer bar expands twice as much for the same temperature change.

These relationships can be expressed by the equation

$$\boxed{\Delta L = \alpha L_o \Delta T} \qquad \textbf{(Eq. 17.6)}$$

where the constant of proportionality α is called the **coefficient of linear thermal expansion,** and $\Delta L = L - L_o$, or the difference between the final and initial lengths, respectively.*

The coefficient of linear thermal expansion $\alpha = \Delta L/L_o\Delta T$ can be seen to be the change in unit length per length per degree change in temperature. Since the ratio $\Delta L/L_o$ is dimensionless, the units of α are inverse degrees, i.e., 1/°C or 1/°F (°C^{-1} or °F^{-1}). As a material property, the coefficient of linear thermal expansion is different for different materials. The approximate values of α for several common materials are given in Table 17.2.

* For a temperature decrease, ΔL is negative owing to a thermal contraction (negative expansion).

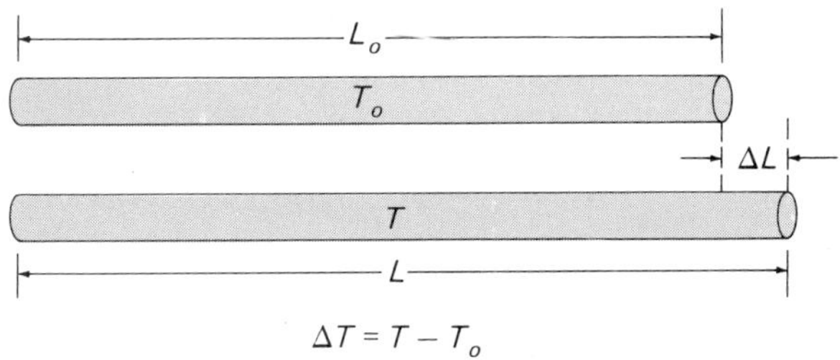

Figure 17.9
Linear thermal expansion. The change in length L of a rod is proportional to its initial length L_o and the change in temperature ΔT.

Table 17.2
Typical Coefficients of Thermal Expansion

	Linear Thermal Coefficient, α	
Substance	**°C^{-1}**	**°F^{-1}**
Aluminum	2.4×10^{-5}	1.3×10^{-5}
Brass	1.9×10^{-5}	1.0×10^{-5}
Concrete (varies)	0.9×10^{-5}	0.5×10^{-5}
Copper	1.7×10^{-5}	0.94×10^{-5}
Glass		
Ordinary (varies)	0.8×10^{-5}	0.4×10^{-5}
Pyrex	0.3×10^{-5}	0.16×10^{-5}
Ice	5.1×10^{-5}	2.8×10^{-5}
Iron	1.2×10^{-5}	0.66×10^{-5}
Lead	2.9×10^{-5}	1.6×10^{-5}
Steel	2.0×10^{-5}	1.1×10^{-5}
Wood, oak		
Along grain	0.5×10^{-5}	0.27×10^{-5}
Across grain	5.4×10^{-5}	2.9×10^{-5}
	Volume Thermal Coefficient, β	
	°C^{-1}	**°F^{-1}**
Alcohol		
Ethyl	1.12×10^{-3}	0.60×10^{-3}
Methyl	1.22×10^{-3}	0.66×10^{-3}
Benzene	1.24×10^{-3}	0.69×10^{-3}
Gasoline	1.08×10^{-3}	0.58×10^{-3}
Mercury	0.18×10^{-3}	0.10×10^{-3}
Water (20°C)	0.21×10^{-3}	0.12×10^{-3}

If it is desired to find the expanded length L of an object directly, we have, from Equation 17.6,

$$\Delta L = L - L_o = \alpha L_o \Delta T$$

or

$$\boxed{L = L_o(1 + \alpha\Delta T)} \quad \textbf{(Eq. 17.7)}$$

EXAMPLE 17.4 A 90-ft steel rail is laid in the spring of the year when the temperature is 65°F. On a hot summer day when the temperature is 100°F, what are (a) the change in length and (b) the length of the rail?

Solution: With L_o = 90 ft, ΔT = 100°F − 65°F = 35°F, and $\alpha = 1.1 \times 10^{-5}$/°F (Table 17.2),

(a) $\Delta L = \alpha L_o \Delta T = (1.1 \times 10^{-5}/°\text{F})(90 \text{ ft})(35°\text{F})$
$= 0.035 \text{ ft } (= 0.42 \text{ in.})$

(b) The new length is then

$$L = L_o + \Delta L = 90 \text{ ft} + 0.035 \text{ ft} = 90.035 \text{ ft}$$

which could have been found directly from Equation 17.7.

Although the linear thermal expansion may seem quite small, engineers must consider this effect in allowing for thermal stresses that could cause damage (Fig. 17.10).

EXAMPLE 17.5 Suppose the steel rails in Example 17.4 had a cross-sectional area of 10 in² and were butted together when they were laid. How much longitudinal stress would develop in a rail when the temperature increased to 100°F?

Solution: Recall, from Section 12.2,

Longitudinal stress =
Young's modulus × longitudinal strain

or

$$\frac{F_n}{A} = Y\left(\frac{\Delta L}{L_o}\right)$$

Then, $Y = 29 \times 10^6$ lb/in² for steel (Table 12.2) and, using the values of ΔL and L_o from Example 17.4,

$$\frac{F_n}{A} = Y\left(\frac{\Delta L}{L_o}\right)$$

$$= (29 \times 10^6 \text{ lb/in}^2)\left(\frac{0.035 \text{ ft}}{90 \text{ ft}}\right)$$

$$= 1.1 \times 10^4 \text{ lb/in}^2$$

With a cross-sectional area of A = 10 in², this corresponds to a longitudinal force of

$$F_n = (1.1 \times 10^4 \text{ lb/in}^2)A$$

$$= (1.1 \times 10^4 \text{ lb/in}^2)(10 \text{ in}^2)$$

$$= 1.1 \times 10^5 \text{ lb} \qquad (= 55 \text{ tons!})$$

which illustrates the importance of considering stresses due to thermal expansion.

Area Expansion

Area expansion is simply linear expansion in two dimensions. Consider a rectangular surface of L_o and width w_o, which has an initial area of $A_o = L_o w_o$. If the temperature of the object is increased by an amount ΔT, the expansions of length and width dimensions are given by Equation 17.7.

$$L = L_o(1 + \alpha\Delta T)$$

$$w = w_o(1 + \alpha\Delta T)$$

The expanded area is then

$$A = Lw = L_o w_o(1 + \alpha\Delta T)^2$$

$$= L_o w_o(1 + 2\,\alpha\Delta T + \alpha^2\Delta T^2)$$

Since α is on the order of 10^{-5} (Table 17.2), the last term with α^2 can be neglected to a good approximation, and

$$\boxed{A = A_o(1 + 2\,\alpha\Delta T)} \quad \textbf{(Eq. 17.8)}$$

where $A_o = L_o w_o$. In terms of the incremental change in area $\Delta A = A - A_o$, Equation 17.8 can be written

$$\boxed{\Delta A = 2\alpha A_o \Delta T} \quad \textbf{(Eq. 17.9)}$$

The **coefficient of area thermal expansion** is thus 2α, or twice the coefficient of linear thermal expansion.

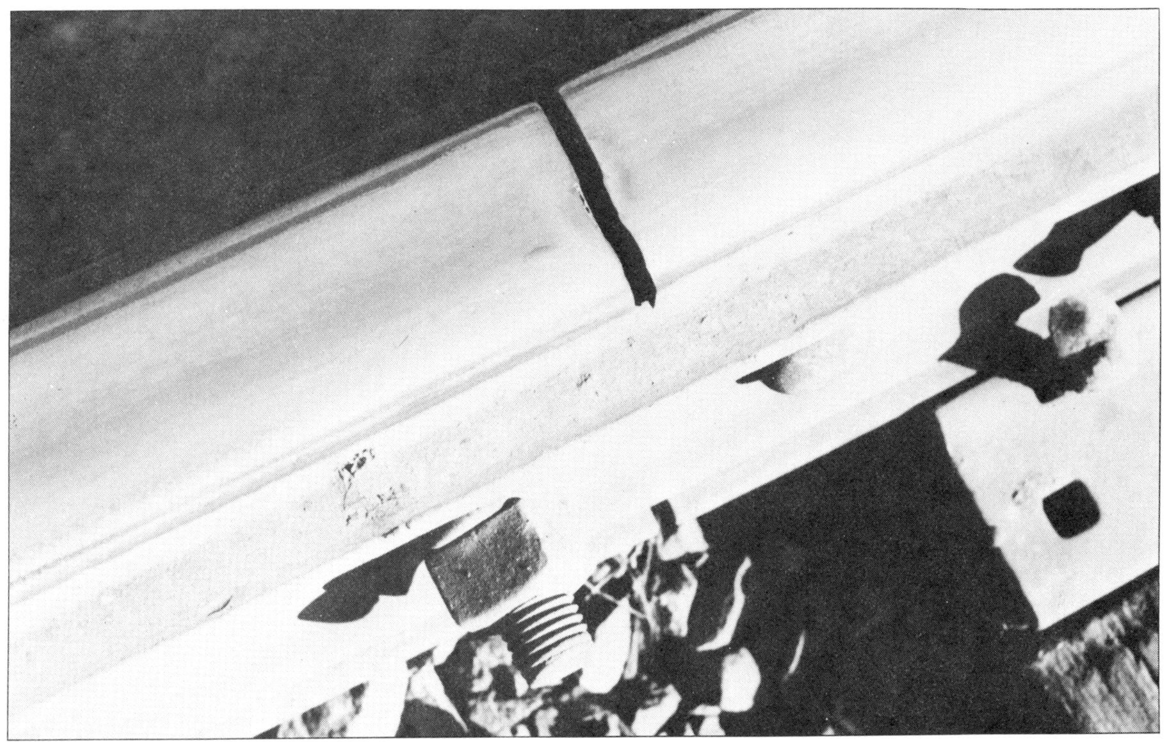

(a)

(b)

(c)

Figure 17.10
Allowances for thermal expansion. (a) Expansion gaps are left between railroad rails to allow for thermal expansion. (b) Not enough gap. These railroad rails in Asbury Park, NJ, buckled on a hot July day. (c) Expansion joints allow bridges to expand safely.

EXAMPLE 17.6 A flat circular disk cut from a piece of aluminum sheet at room temperature has a diameter of 20.0 cm. If the disk is heated to 150°C, what is the change in the area of the disk?

Solution: With a radius of 10.0 cm, the initial area of the circular disk (at room temperature, 20°C) is

$$A_o = \pi r^2 = \pi(10.0)^2 = 314 \text{ cm}^2$$

From Table 17.2, $\alpha_{\text{Al}} = 2.4 \times 10^{-5}\ °\text{C}^{-1}$ and $\Delta T = 150°\text{C} - 20°\text{C} = 130°\text{C}$; hence,

$$\begin{aligned}\Delta A &= 2\alpha A_o \Delta T \\ &= 2(2.4 \times 10^{-5}\ °\text{C}^{-1})(314 \text{ cm}^2)(130°\text{C}) \\ &= 2.0 \text{ cm}^2\end{aligned}$$

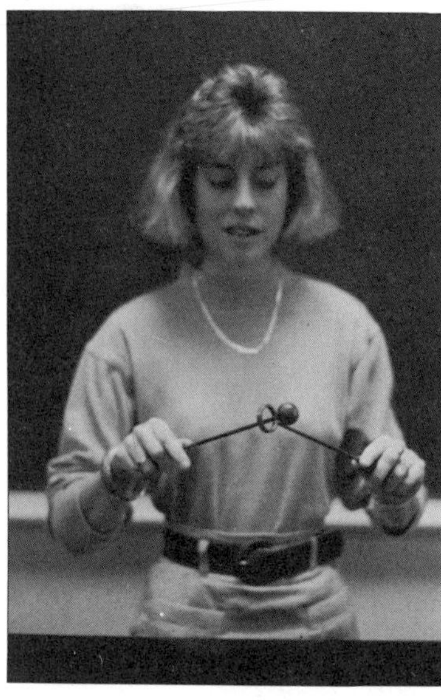
(a)

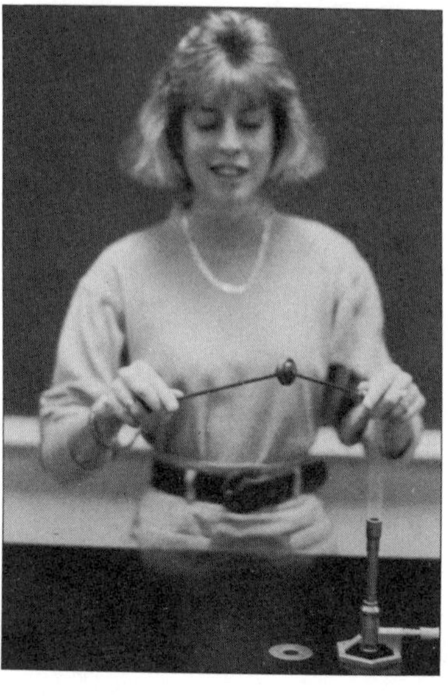
(b)

Figure 17.11
Area thermal expansion. (a) At room temperature, the ball fits easily through the ring. (b) However, if the ball is heated, it expands and will not fit through the ring. If both the ball and ring are heated, the ball will fit through the ring because the area of the ring hole also gets larger as a result of thermal expansion.

Suppose the aluminum sheet from which the disk in the preceding example was cut is placed in an oven. Does the hole in the sheet get larger or smaller? A common incorrect answer is that the hole gets smaller, but actually the hole gets larger. The hole behaves the same as if the disk of material were still there, and hence it gets larger since the disk expands (see Fig. 17.11).

Volume Expansion

With the consideration of the linear expansion of a solid in three dimensions, it can be shown that the change in volume due to thermal expansion is given, to a good approximation, by

$$\Delta V = \beta V_o \Delta T$$

or

$$V = V_o(1 + \beta \Delta T) \qquad \textbf{(Eq. 17.10)}$$

where $\Delta V = V - V_o$, V_o is the initial volume, and for a solid the **coefficient of volume thermal expansion** $\beta = 3\alpha$.

Liquids follow the same relationship for the volume thermal expansion of a solid. However, β is generally larger for liquids (Table 17.2) and is not related to the linear coefficient of thermal expansion α for solids.

EXAMPLE 17.7 A 100-gal steel gasoline tank on a truck is filled to the top on a cold night when the temperature is 35°F. During the day, the temperature increases to 95°F. Assuming that the truck has not run and no gasoline has been used, how much gasoline overflows from the tank?

Solution: With $V_o = 100$ gal, $\Delta T = 95°\text{F} - 35°\text{F} = 60°\text{F}$, and $\beta = 0.58 \times 10^{-3}/°\text{F}$ (Table 17.2), the change in the volume of the gasoline would be

$$\begin{aligned}\Delta V_g &= \beta V_o \Delta T \\ &= (0.58 \times 10^{-3}/°\text{F})(100 \text{ gal})(60°\text{F}) \\ &= 3.48 \text{ gal}\end{aligned}$$

However, the steel gasoline tank would also increase in volume. With $\alpha_{\text{steel}} = 1.1 \times 10^{-5}/°\text{F}$ (Table 17.2),

$$\begin{aligned}\Delta V_t &= 3\alpha V_o \Delta T \\ &= 3(1.1 \times 10^{-5}/°\text{F})(100 \text{ gal})(60°\text{F}) \\ &= 0.20 \text{ gal}\end{aligned}$$

Hence, the amount of gasoline overflow would be

$$\begin{aligned}\Delta V &= \Delta V_g - \Delta V_t \\ &= 3.48 \text{ gal} - 0.20 \text{ gal} \\ &= 3.28 \text{ gal}\end{aligned}$$

SPECIAL FEATURE 17.3

Water Density and Freezing at the Top

The densities of most liquids increase as the liquids are cooled and their temperatures reduced. This is the case for water as it is cooled to 4°C (Fig. 1). However, when water is cooled below 4°C to its freezing point, its density decreases. This implies that the formation of the open lattice structure that is characteristic of ice occurs over the temperature range of 4°C to 0°C, rather than solely at the freezing point.

This unusual property accounts for the fact that open containers of water freeze at the top first. Most of the cooling takes place at the open surface. As the temperature of the top layer of water is lowered toward 4°C, the cooler, denser water sinks to the bottom. However, below 4°C, the water at the top is less dense than the water below and remains at the top, where it freezes when the freezing point is reached.

Think of the environmental effects if this were not the case. Lakes, ponds, and rivers would freeze from the bottom up, and much of the aquatic animal and plant life would be destroyed—not to mention the effect on ice skating.

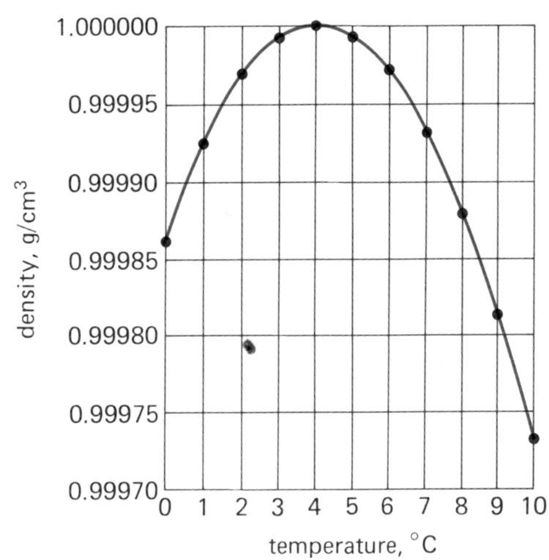

Figure 1
Density of water versus temperature. The maximum density of water occurs near 4°C. Below 4°C, water is less dense due to the formation of an open hexagonal structure of the molecular units that is characteristic of ice (e.g., six-sided snowflakes).

Most liquids, like solids, expand when the temperature is increased and contract (negative expansion) when the temperature is decreased. One notable exception is water in a certain temperature range. As water is cooled from normal temperatures to 4°C, it contracts, or decreases in volume, as would be generally expected. However, as water is cooled below 4°C to the freezing point (0°C), it expands! Inversely, when water is heated from 0°C to 4°C it contracts, and above 4°C it expands.

Expressed in terms of density (mass/unit volume), this means that water has its maximum density at 4°C (actually 3.98°C). This somewhat peculiar behavior accounts for the fact that water freezes at the top first (Special Feature 17.3).

17.3 Heat of Combustion

We burn fossil fuels to obtain energy for heating. In a similar sense, foods are body fuels. Both fossil-fuel burning and digestion involve the conversion of chemical energy to heat energy.

The intrinsic heat values of fuels and foods are expressed in terms of the **heat of combustion,** which is the heat produced per unit mass (or weight) of the substance when it is burned in oxygen. The units of the heat of combustion are thus cal/g, kcal/kg, and Btu/lb.

For example, the heat of combustion of gasoline is approximately 20,500 Btu/lb. This means that when one pound of gasoline reacts with an amount of oxygen necessary to ensure its complete combustion, 20,500 Btu of heat energy are liberated. Compared to the heat of combustion of 11,500 Btu/lb for alcohol, it is readily seen that gasoline has a greater intrinsic heat value than alcohol.

Similarly, a weight-watcher is interested in knowing that the heat of combustion of a boiled egg is 1600 kcal/kg as compared with 2100 kcal/kg for a scrambled egg. (Recall that the kilocalorie is the food "Calorie." The Calories

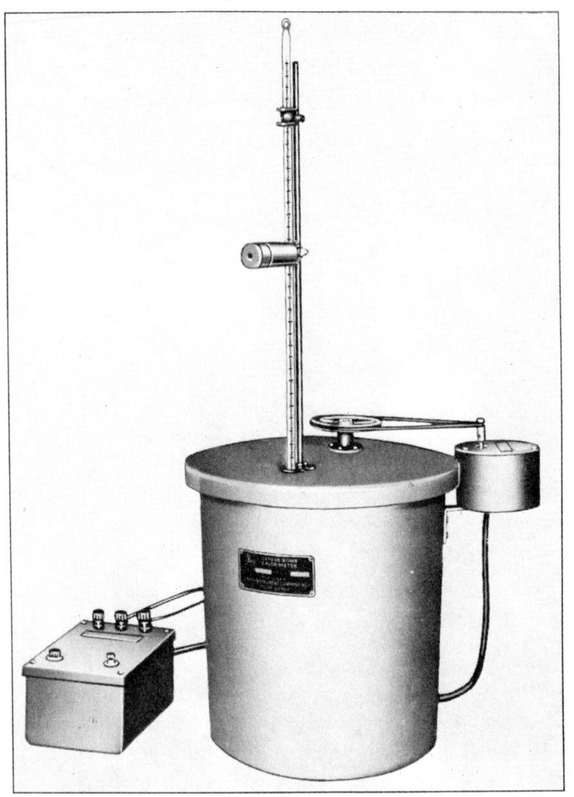

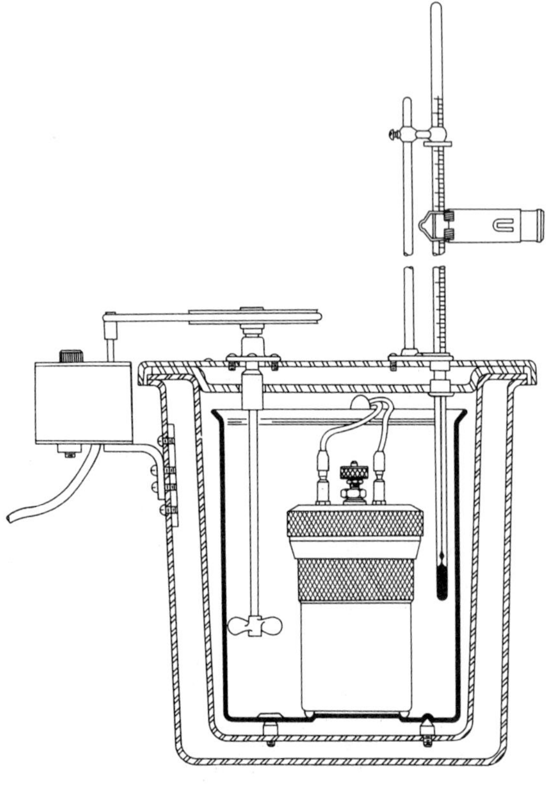

Figure 17.12

A bomb calorimeter. Combustion takes place in the inner bomb container, and the heat of combustion is determined by the method of mixtures.
(Courtesy Parr Instrument Company.)

listed in weight charts are given for average amounts, e.g., for an average-sized egg.)

Heats of combustion are ordinarily measured in a bomb calorimeter (Fig. 17.12). The "bomb" is a heavy steel cylinder fitted with a gas-tight screw cover. A known mass of a substance is placed in the bomb calorimeter cup in an atmosphere of pure oxygen. Upon ignition by means of an electric current, combustion takes place in the form of an explosion. The heat produced by the combustion of the substance is determined by the method of mixtures, as in a regular calorimeter (Section 16.4).

Typical heats of combustion of various substances are given in Table 17.3. (The numerical value is the same for cal/g and kcal/kg. Why?) The heats of combustion of gaseous fuels, such as natural gas, are sometimes expressed as heat content per unit volume of gas, e.g., Btu/ft^3, at standard temperature and pressure (STP; i.e., 0°C and 1 atm).

The heat liberated by a quantity of substance on complete combustion is given by

$$\Delta Q = mH \qquad \textbf{(Eq. 17.11)}$$

where H is the heat of combustion and m is the mass. When the heat of combustion is expressed in the British system, weight (w) is substituted for mass (m).

EXAMPLE 17.8 How much thermal energy is released with the complete combustion of 1 ton of soft (bituminous) coal?

Solution: From Table 17.3, the heat of combustion of bituminous coal is $H = 13{,}500$ Btu/lb, and with $w = 1$ ton $= 2000$ lb,

$$\begin{aligned}\Delta Q &= wH = (13{,}500 \text{ Btu/lb})(2000 \text{ lb}) \\ &= 27 \times 10^6 \text{ Btu} \\ &= 27 \text{ million Btu}\end{aligned}$$

Table 17.3
Typical Values of Heats of Combustion

Substance	cal/g or kcal/kg	Btu/lb
Fuel		
Alcohol	6400	11,500
Coal		
Anthracite (hard coal)	8000	14,400
Bituminous (soft coal)	7500	13,500
Coke	6000	11,000
Diesel oil	10,500	18,900
Fuel oil	10,300	18,500
Gasoline	11,400	20,500
Natural gas	10,000	18,000
	(10,500 kcal/m³)*	(12,000 Btu/ft³)*
Wood (pine)	4500	8000
Foods		
Bread (white)	2000	3600
Butter	8000	14,400
Eggs		
Boiled	1600	2800
Scrambled	2100	3880
Ice cream	2100	3800
Meat (lean)	1200	2150
Milk	700	1250
Potatoes (white, boiled)	970	1750
Sugar (white)	4000	7200

* Volume at STP.

Important Terms

conduction a method of heat transfer due to molecular interactions

thermal conductivity (k) a measure of a material's ability to conduct heat

thermal resistivity (ρ) the reciprocal of thermal conductivity; a measure of a material's resistance to conducting heat

R-value a rating of the insulating value of materials given by the product of the thermal resistivity and material thickness

convection a method of heat transfer involving mass transfer

radiation a method of heat transfer by electromagnetic radiation, which requires no material medium

emissivity a measure of the ability of a body to emit radiation

black body an ideal emitter and absorber

heat of combustion (H) the intrinsic heat value per unit mass of a substance, usually a food or fuel

Important Formulas

thermal conduction rate: $\frac{\Delta Q}{\Delta t} = \frac{kA\,\Delta t}{L}$

thermal resistivity: $\rho = \frac{1}{k}$

R-value: $\text{R-value} = \rho L = \frac{L}{k}$ (L in inches)

Stefan-Boltzmann's law: $I = \frac{P}{A} = e\sigma T^4$

Stefan-Boltzmann's constant: $\sigma = 5.7 \times 10^{-8}\ \text{W/m}^2\text{-K}^4$

linear thermal expansion: $\Delta L = \alpha L_o \Delta T$
or $L = L_o(1 + \alpha\Delta T)$

area thermal expansion: $\Delta A = 2\alpha A_o \Delta T$
or $A = A_o(1 + 2\alpha\Delta T)$

volume thermal expansion: $\Delta V = \beta V_o \Delta T$

or $V = V_o(1 + \beta \Delta T)$

($\beta \simeq 3\alpha$ for a solid)

heat of combustion: $H = \dfrac{\Delta Q}{m}$

or $\Delta Q = mH$

Questions

Heat Transfer

1. Give examples of situations in which heat transfer is (a) desired and (b) undesired.
2. On what parameters does the rate of conductive heat transfer depend?
3. Why do underground water pipes sometimes freeze only after it has been very cold for several days?
4. Is baked food more likely to burn on the bottom in an aluminum baking pan or in a glass baking dish? Which has a greater thermal resistance: (a) copper or iron? (b) Concrete or brick?
5. How does the rate of thermal conduction of the bottom of a metal cooking pan vary with (a) the area and (b) the thickness?
6. Thermal underwear has a knitted structure with lots of holes (Fig. 17.13). Wouldn't a material without holes be a better insulator? Explain.
7. Thermopane windows have double panes of glass separated by a small air space. Why are these windows better for insulation than single-pane windows? How do storm doors and windows help reduce heating bills?
8. Foam insulation is sometimes blown between the outer walls and inner walls of a house. If air is a poor conductor, why bother with the insulation?
9. The outside coils of window air conditioners have a fin network on them. What is the purpose of the fins?
10. Why are the recommended insulation R-values for ceilings greater than those for walls and floors? (See Fig. 17.4.)
11. A student defines thermal resistance as the R-value per unit thickness. Is this correct?
12. In some states, highway signs warn: "Bridge freezes before road" (Fig. 17.14). Why is this?
13. (a) Why do we generally wear dark clothes in the winter and light-colored clothes in the summer? (b) In the summer, why are blacktop roads hotter than the ground by the side of the road? (c) What is the purpose of painting mobile home roofs with aluminum paint?
14. A big roaring fire in a fireplace is only about 10 percent efficient in heating a room. Why the low efficiency?
15. Discuss the energy balance and the average temperature of the Earth if its only heat-loss mechanisms were conduction and convection.
16. (a) When your skin is hot, the blood vessels in the skin dilate, or get larger in diameter. When the skin is cold (below 37°C), the blood vessels constrict. What is the purpose of this action? (b) Alcohol (taken internally) causes the blood ves-

Figure 17.13
Holey long johns. See Question 6.

Figure 17.14
Why not together? See Question 12.

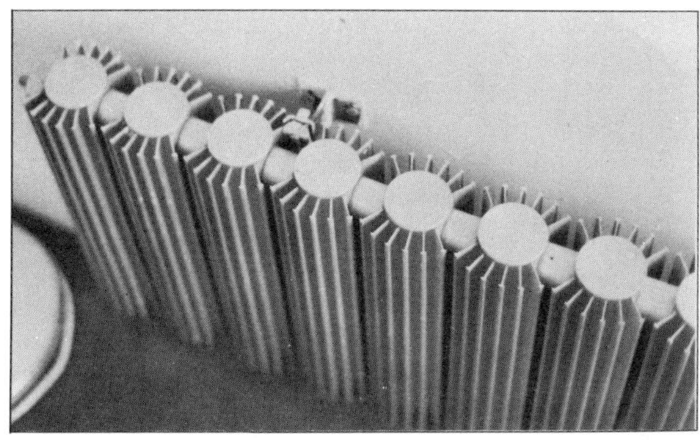

Figure 17.15
Hot-water heater. See Question 20.

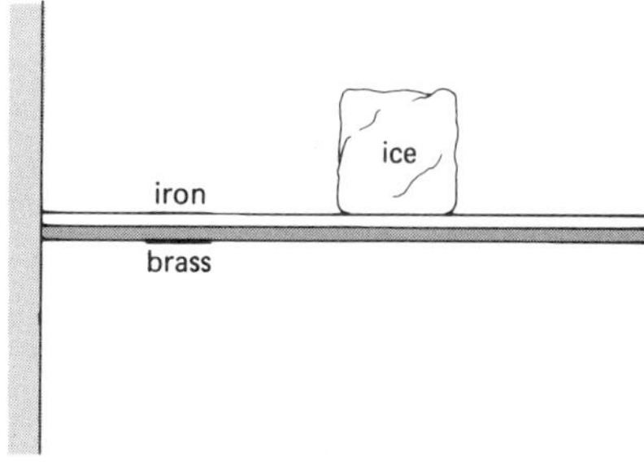

Figure 17.16
A bending situation. See Question 25.

sels in the skin to dilate, and drinkers feel a "warm glow." Is this really a warming process for the body? Explain.

17. Suppose you take cream in your coffee. If you have a hot cup of coffee you want to drink later, which method would keep the coffee hot longer: to put the cream in it right away or wait until you are ready to drink it? (*Hint:* think about temperature differences and cooling or heat transfer.)
18. Some questions about microwave ovens: (a) In heating frozen foods in sealed pouches, why do you first poke holes in the pouch? (b) Why are microwave ovens built so they will not operate with the doors open? (c) What is the purpose of the metal grating on the inside of the glass in the door?
19. What is the emissivity of a body? Can it have a value of 2.0?
20. Explain the purpose of the fins on the hot-water heater shown in Figure 17.15.

Thermal Expansion of Materials

21. Why do ordinary glass containers crack more readily than Pyrex containers when filled with hot liquids?
22. Bonded materials often break at the bond because of "mismatch expansion." What does this mean?
23. On a hot afternoon or cold night you may hear creaking noises in a house's attic. What causes this?
24. Why are concrete highways and driveways poured in sections rather than in one long strip? Sidewalks are sometimes poured in strips, but grooved joints are made. What does this do?
25. A piece of ice is placed on a bimetallic strip as shown in Figure 17.16. What happens in this case?
26. When one drinking glass is stuck inside another, an old trick to unstick them is to put water in one of them and run water of a different temperature over the outside of the other. Which water should be hot and which should be cold?
27. If you turn on a hot-water faucet or the hot water in a shower to a moderate flow, you may observe that the flow decreases after awhile, but this is not observed with cold water. What causes the flow of hot water to change?
28. (a) Railroad rails are laid with some space between them. What would happen if the rails were laid on a cold day and butted up against each other? (b) On a hot afternoon you may observe gasoline dripping from a recently filled automobile gas tank. Is the owner trying to waste fuel and money? Explain.
29. When a thermometer is inserted into hot water, the mercury or alcohol is sometimes observed to fall slightly before it rises. Why?
30. Do all liquids expand with a temperature increase? Explain.

Heat of Combustion

31. In Table 17.3, the heats of combustion are given in cal/g or kcal/kg. Are these ratios the same? Explain.
32. Which has more energy content, hard coal or soft coal?
33. Gasohol is a mixture of gasoline and alcohol. Does gasohol have a greater intrinsic energy value than gasoline?
34. Is the listed heat of combustion of a substance released when the substance is burned ordinarily? Explain.
35. Why would scrambled eggs have a greater heat of combustion than boiled eggs? (See Table 17.3.)

Problems

Levels of difficulty are indicated by asterisks for your convenience.

17.1 Heat Transfer

1. The temperature varies uniformly from 40°C to 30°C between the ends of a 20-cm metal bar. What is the thermal gradient in the bar?
2. If the bar in Problem 1 is made of copper and has a cross-sectional area of 5.0 cm^2, what is the rate of heat flow through the bar?
3. The emissivity of an object is 0.67. How many times greater would the radiation intensity be from a black body if the object and the black body were both at the same temperature?
4. If the absolute temperature of a radiator is doubled, how is the emitted intensity affected?

*5. A wall is made up of equal areas of wood and glass. The glass is 0.50 cm thick and the wood is 4.0 cm thick. For the same temperature difference, which conducts more heat, and how many times more?

*6. What is the rate of heat conduction through a 4.0 ft × 8.0 ft piece of fiberboard 0.50 in. thick, if the temperature difference of the opposite surfaces is 30°F?

*7. A 4.0 ft × 2.5 ft window in a building has a single pane of glass $\frac{1}{8}$ in. thick. If it is 70°F on the inside and 20°F on the outside, how much heat is conducted through the window in 24 hours?

*8. A plastic foam ice chest has dimensions of 0.50 m × 0.25 m × 0.25 m ($\ell \times w \times h$) and has a foam thickness of 2.5 cm. If the chest is filled with ice and the outside temperature on a hot day is 25°C, (a) what is the rate of heat conduction into the chest? (Assume equal conduction for all parts.) (b) How much ice will melt in 4.0 min?

*9. The bottom of a round copper teakettle has a radius of 10 cm and a thickness of 3.0 mm. The kettle sits on a stove burner at a temperature of 150°C and is full of boiling water. (a) What is the rate of heat conduction? (b) Ideally, how much water is "boiled away" in 5.0 min? Is the answer reasonable on a practical basis? Explain.

*10. (a) Which of the metals in Table 17.1 have the greatest and least thermal resistivities, and what are they? (b) What are the R-values for these metals for sheets 1.0 in. thick?

*11. Compare the R-values for equal thicknesses of concrete and floor tile. How many times greater is one than the other?

*12. What thicknesses of (a) fiberboard and (b) glass wool will give an R-value of R-30?

*13. What thickness of (a) brick, (b) concrete, and (c) air would have the same insulating properties as 4.0 in. of foam plastic? Explain why the best of these insulators is not used exclusively in walls.

*14. What is the difference in the thicknesses of foam plastic used when following recommended insulation standards for ceiling insulation in Minneapolis and in San Francisco? (See Fig. 17.4.)

*15. Contractors in Chicago and Miami both follow the recommended insulation standards for wall insulation for their regions. What is the difference in the thicknesses of the insulations if Fiberglas is used?

*16. Compare the R-values of a 4.0-in. space for (a) foam plastic, (b) air, and (c) vacuum. Which would be best for home insulation? Is it practical?

*17. The space between an inner wall and outer wall that measures 10 ft × 8.0 ft is insulated with foam having an R-value of R-13. (a) What is the rate of heat transfer through the insulated space when the temperature difference is 20°F? (b) How thick is the insulation?

*18. The temperature of an object at room temperature (20°C) is doubled to 40°C. What is the effect on the radiation intensity?

*19. What is the intensity of the radiation being emitted by an object with an emissivity of 0.40 at a temperature of 100°C?

*20. What is the temperature in degrees Celsius of a black body that radiates with an intensity of 462 W/m^2?

**21. Compare the R-values of a thermopane window with $\frac{1}{8}$-in. glass panes and a $\frac{1}{4}$-in. air space with that of a single-pane window with a thickness of $\frac{3}{16}$ in.

**22. Thermal *resistance* R takes into account the area of a material and is written $R = \rho L/A$. (a) Why is the thermal resistance inversely proportional to the area? (b) Show that the rate of conductive heat flow is $\Delta Q/\Delta t = \Delta T/R$ and (c) that R-value $= RA$. (In Chapter 21 you will find that the electrical resistance $R = \rho L/A$, where ρ is the electrical resistivity, and $\Delta Q/\Delta t = I = \Delta V/R$, where $\Delta Q/\Delta t$ is the flow of electrical charge or current and ΔV is a voltage difference.)

17.2 Thermal Expansion of Materials

(Assume any number of significant figures.)

23. The temperature of an aluminum rod 40 cm long is increased from room temperature to 80°C. What is the increase in the length of the rod due to thermal expansion?

*24. What temperature increase is required to increase the length of a 1.0-ft iron bar by 0.021 in.?

*25. A 15-cm Pyrex glass stirring rod at room temperature is dropped into boiling water. What is the length of the rod when in thermal equilibrium with the water?

*26. Two glass rods of equal length, one of Pyrex and the other of ordinary glass, are heated so they expand linearly by the same amount. Which has the greater temperature change, and how many times greater?

*27. A block of oak wood has a rectangular surface of 10 in. × 6.0 in., where the longer dimension is along the grain. If the block undergoes a temperature change of 75°F, what is the increase in the surface area?

*28. In a region of the northern United States, the difference between the average yearly high and low temperatures is 150°F. How wide a gap between 90-ft steel railroad rails would just safely allow for this total temperature variation? Would this be practical? Explain.

*29. A copper plate at 10°C is heated to 70°C. What is the percent increase in the area of the plate?

*30. A cube of steel undergoes a temperature change of 200°C. What is the percent increase in the volume of the cube?

*31. A brass cylinder with a radius of 3.0 cm and a length of 15 cm at room temperature is heated to 300°C. What is the volume of the cylinder at this temperature?

*32. A liter of water is measured out at room temperature. (a) What is the volume of the water at 100°C? (b) What is the minimum possible volume reachable by lowering the temperature?

**33. A brass rod with a radius of 5.5 mm fits through a circular hole with a 1.0 percent area tolerance at room temperature. To what temperature would the rod have to be heated so it would not fit in the hole?

**34. An aluminum sheet 1.0 ft × 1.0 ft at room temperature has a circular hole with a diameter of 2.5 in. near its center. The sheet is placed in an oven and heated to 200°F. What are (a) the area of the hole and (b) the lengths of the sides of the sheet at this temperature?

**35. An automobile's 16-gal steel gasoline tank is filled to capacity on a cool morning when the temperature is 4°C. If one pint of gasoline is used during the drive to work and the temperature rises to 18°C that afternoon, will gasoline overflow from the tank, and if so, how much? (Neglect expansion of tank. Is this a reasonable assumption? Explain.)

17.3 Heat of Combustion

*36. How much energy is released with the complete combustion of 1 gal of gasoline?

*37. How much anthracite (hard) coal would be required to give an energy output equivalent to that of the complete combustion of 100 lb of pine wood?

*38. What is the Calorie content of a slice of bread (0.50 oz) spread with butter (0.01 oz)?

*39. What is the energy content of a gallon of gasohol (90 percent gasoline and 10 percent alcohol by volume)? How does this compare to a gallon of gasoline?

*40. (a) A "Count Your Calories" book lists the calories of a large raw or boiled egg as 80 Cal. What is the mass of the egg? (b) If the egg is scrambled, the calorie content is listed as 115 Cal. Explain the difference between a boiled egg and a scrambled egg, and compute the mass of the additional ingredient.

**41. A fuel oil salesperson tries to convince the management of a company that has an industrial process requiring a heat input of 1 million Btu/h to convert from (soft) coal to fuel oil. Give some figures about the quantities of fuel needed to support this argument and other practical or economical considerations.

Chapter 18

Thermodynamics, Heat Engines, and Heat Pumps

As the name implies, **thermodynamics** deals with the transfer and actions (dynamics) of heat (Greek *therme*, meaning heat). In general, it is a broad and comprehensive branch of science that is concerned with all types of energy aspects, but chiefly the relationship between heat and mechanical energy. The formal development of thermodynamics began less than 200 years ago, primarily growing out of efforts to produce heat engines—devices for converting heat energy into mechanical work. These include steam engines, gasoline engines, diesel engines, jet engines, and any device that converts heat into work.

Some of the general aspects of thermodynamics have been presented in previous chapters. However, there are other important principles that govern the utilization of heat and work. These laws of thermodynamics are basic in the operation and design of heat engines and heat pumps, which are topics of this chapter.

A heat pump is a device that uses mechanical energy or work to transfer heat from a lower-temperature source to a higher-temperature region. Can you think of a common heat pump? How about a refrigerator or an air conditioner? An air conditioner takes heat from a cool room and transfers it to the hot outdoors. It takes work or energy to do this. Check your electric bills in the summer.

18.1 The First Law of Thermodynamics and Thermodynamic Processes

The first law of thermodynamics is simply a statement of the conservation of energy for a thermodynamic system. It is expressed mathematically as

$$\boxed{\Delta Q = \Delta U + W} \qquad \textbf{(Eq. 18.1)}$$

This equation expresses the general fact that when an amount of heat ΔQ is added to or removed from a system, there is a change in the internal energy ΔU of the system, and/or work W is done by or on the system.

To see how the conservation of energy — the first law — applies to thermodynamic systems, we must first be able to describe changes in a system.

The working substance in thermodynamic applications involving the conversion of thermal and mechanical energies is commonly a fluid — a gas or liquid. We say that a particular quantity of matter comprises a **thermodynamic system.** In general, a system interacts with its surroundings. One such interaction involves the transfer of heat into and out of the system.

For example, a quantity of gas confined in a cylinder with a movable piston constitutes a thermodynamic system. Heat may be added to the system and the gas may expand to do mechanical work on its surroundings (moving the piston). In another process, the piston may compress (do work on) the gas, causing its internal (heat) energy and temperature to increase.

Clearly, the state (condition) of the system changes with such interactions. In order to analyze these changes, or thermodynamic processes, we must be able to describe the various states of the system. This is done practically in terms of measurable macroscopic quantities. For example, a quantity of gas is described by its pressure, volume, and temperature.

Such quantities that characterize the state of a system are called ***thermodynamic coordinates.*** For an ideal gas, the relationship between the thermodynamic coordinates is given by the perfect gas law (Eq. 16.4):

$$pV = NkT$$

Any such equation relating the thermodynamic coordinates of a system is called an **equation of state.** Different systems have different equations of state, some of which are quite complicated.

The notion of thermodynamic coordinates results from graphs of these variables. For a given quantity of gas, the thermodynamic coordinates (p, V, T) correspond to the (x, y, z) coordinates on a Cartesian graph. Each point or equilibrium state is characterized by three thermodynamic coordinates, which satisfy the ideal gas law relationship.

A *process* is a change in the state of the system or a change in thermodynamic coordinates.

In three dimensions, the thermodynamic coordinates define a p-V-T thermodynamic surface. Such a thermodynamic surface for a substance like carbon dioxide (CO_2) is shown in Figure 18.1. Recall that real substances undergo

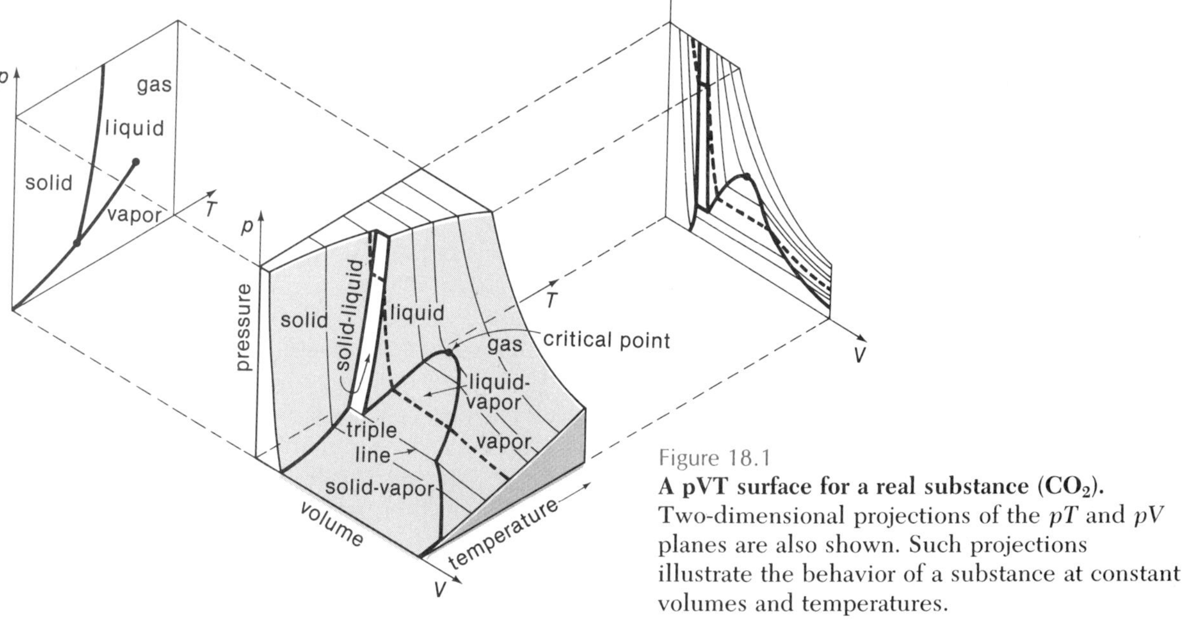

Figure 18.1
A pVT surface for a real substance (CO_2). Two-dimensional projections of the pT and pV planes are also shown. Such projections illustrate the behavior of a substance at constant volumes and temperatures.

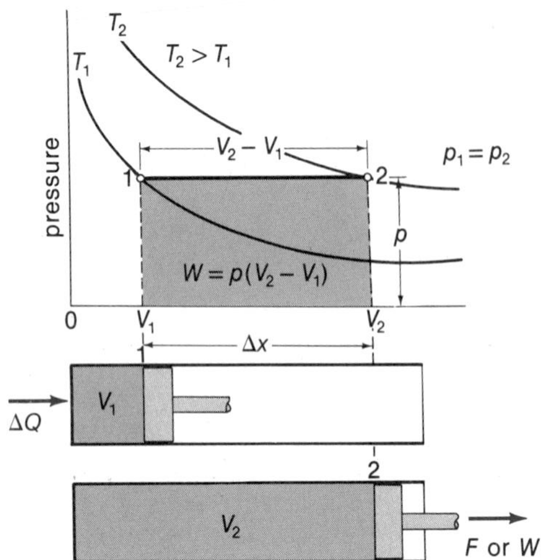

Figure 18.2
An isobaric (constant-pressure) process. The gas takes in an amount of heat Q while doing mechanical work on the piston during the expansion in going from point 1 (p_1, V_1, T_1) to point 2 (p_2, V_2, T_2). Notice that the temperature and the internal energy increase during the process. The curved lines are lines of constant temperature (isotherms).

phase changes and that there are three distinct phase regions.

Because there are three thermodynamic variables for a given mass of gas, it is convenient to consider "iso-" (Greek *isos*, equal) processes, in which one of the coordinates is kept constant.

(a) Isobaric Process A constant-pressure, or isobaric, process for an ideal gas is illustrated in Figure 18.2.

An isobaric process is one in which the pressure of the system remains constant.

The p-V diagram is analogous to a two-dimensional x-y plot on Cartesian axes. The isobaric process path is called an **isobar.** In the illustrated process, an amount of heat ΔQ is added to a given mass of gas in a cylinder. If the pressure is kept constant, the piston must move so as to increase the volume. In this case, we say that the work is done *by* the system (the gas) *on* the surroundings (the piston).

Notice that the temperature of the gas increases during the process, since from the equation of state, $pV = NkT$, where p is constant for the isobaric process. Recall that the internal energy of an ideal gas is directly proportional to its temperature. Hence, an increase in temperature means that the internal (kinetic) energy also increases ($+\Delta U$).

The work done by the gas in this expansion process is equal to the area under the curve. Recall that work W is equal to the force F times the distance Δx through which the force acts:

$$W = F\Delta x$$

In terms of pressure ($p = F/A$), $F = pA$, where A is the area of the piston, and

$$W = pA\Delta x$$

But $A\Delta x$ is just the change in the volume of the gas, $A\Delta x = \Delta V$; hence,

$$\boxed{W = p\Delta V = p(V_2 - V_1)} \qquad \textbf{(Eq. 18.2)}$$

and $p\Delta V$ can be seen to be the area under the process path curve.

Since $V_2 > V_1$, the work done by the system on its surroundings is positive ($+W$). In terms of the first law,

$$\Delta Q = \Delta U + W = \Delta U + p\Delta V$$
(isobaric process)

That is, the heat energy added to the system ($+\Delta Q$) is equal to the increase in the internal energy of the system plus the work done by the system.

For the reverse process of Figure 18.2, an external force applied to the system in state 2 could take the system to state 1 in a compression process along the reverse process path. In this case, work would be done *on* the system (the gas) *by* the surroundings (the piston). With the final volume being less than the initial volume ($V_1 > V_2$), the work done on the system is negative: $p(V_2 - V_1) = -W$. In the process, heat ($-\Delta Q$) would be ejected from the system since its temperature and internal energy would be decreased ($T_1 < T_2$, and $-\Delta U$).

To move a system along a given process path, the process must take place slowly so the system is in thermal equilibrium at each point along the process curve. The process can then be reversed along a *known* process path as just suggested. The ability to do this exactly is called *reversibility*.

HEAT AND LIGHT

Heat and phase change. Casting molten high-purity iron.

A one-minute exposure captures multiple lightning flashes illuminating Kitt Peak National Observatory in Arizona. [© Gary Ladd, 1972]

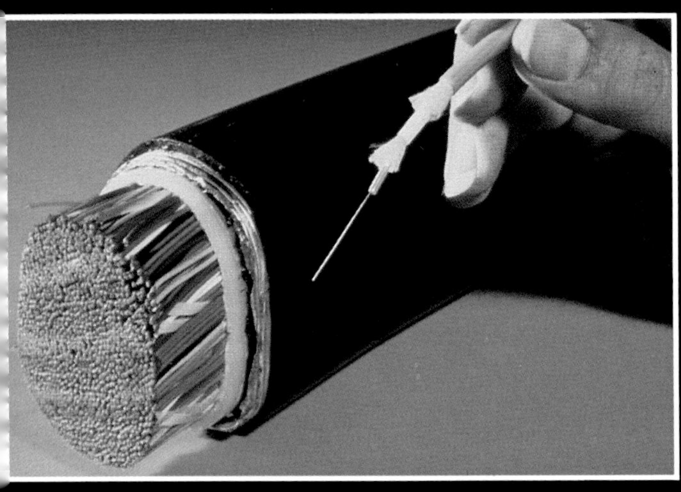

Ultra-pure glass optical fibers carry voice, video, and data signals in high-capacity telecommunications networks. (Courtesy Corning Incorporated)

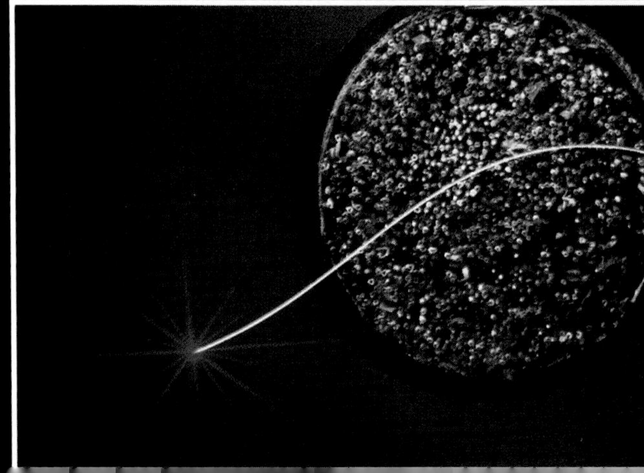

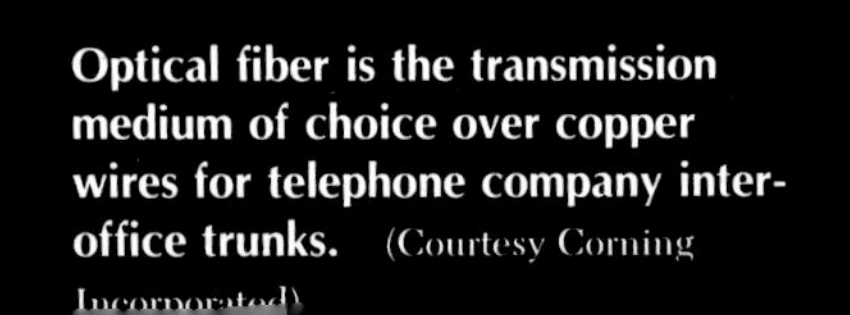

Optical fiber is the transmission medium of choice over copper wires for telephone company inter-office trunks. (Courtesy Corning Incorporated)

Light produced during laser machining is analyzed in order to study the interaction between laser materials and to help develop improvements in laser welding and cutting. (Courtesy Department of Energy)

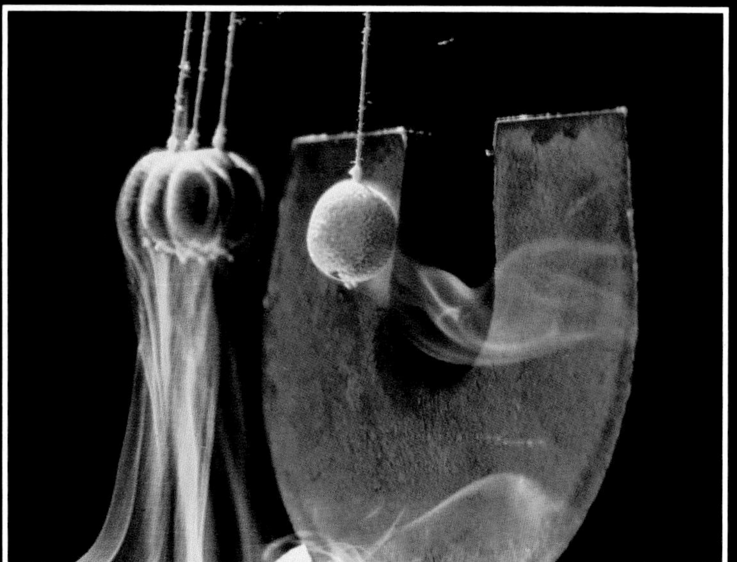

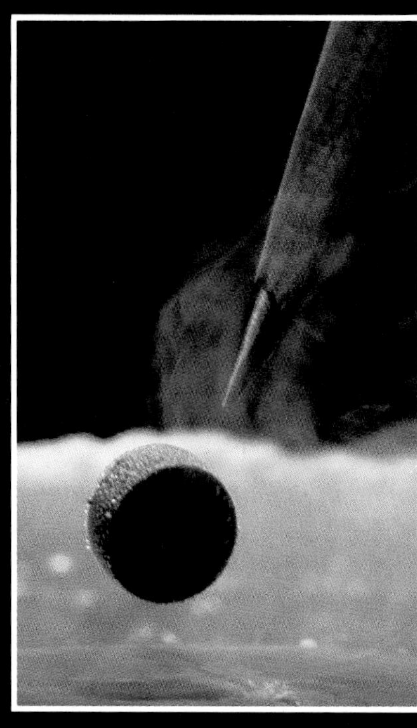

Superconductivity and levitation. When ceramic materials are superconducting, they repel magnetic fields and are repelled by magnets. This stops the ceramic pendulum bob (left) or causes a piece of material to levitate (right). [*Left:* Courtesy Bill Pierce, *Time Magazine*. *Right:* Courtesy IBM Research Division, Thomas J. Watson Research Center]

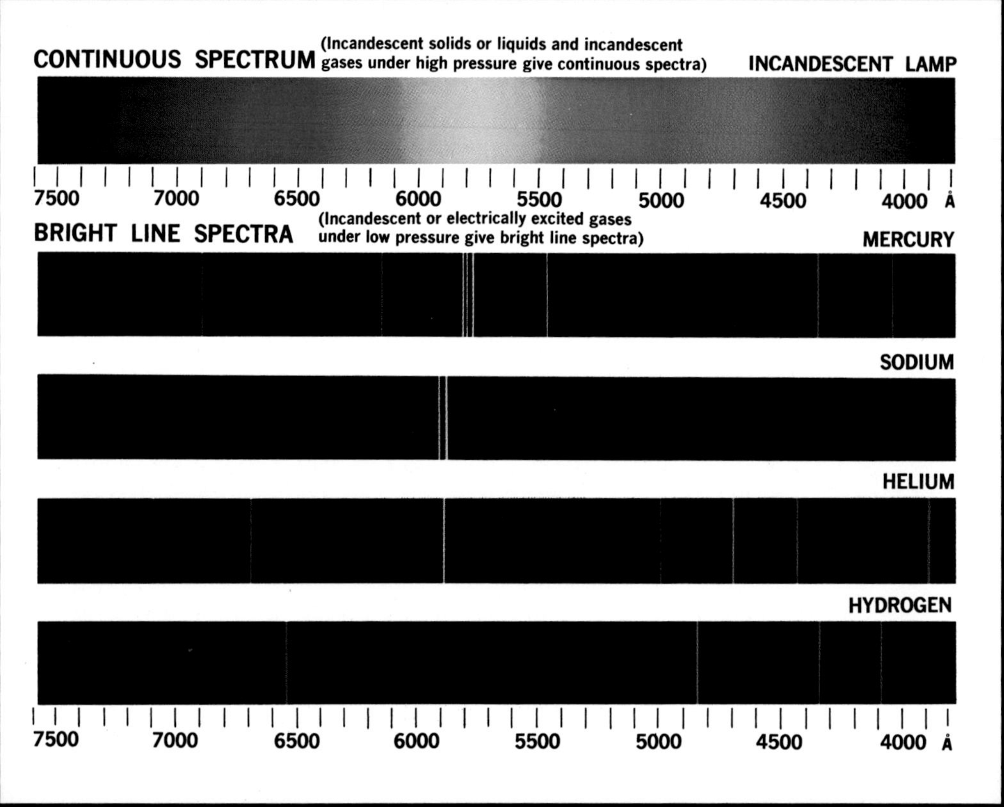

CONTINUOUS SPECTRUM (Incandescent solids or liquids and incandescent gases under high pressure give continuous spectra) **INCANDESCENT LAMP**

7500 7000 6500 6000 5500 5000 4500 4000 Å

BRIGHT LINE SPECTRA (Incandescent or electrically excited gases under low pressure give bright line spectra)

MERCURY

SODIUM

HELIUM

HYDROGEN

7500 7000 6500 6000 5500 5000 4500 4000 Å

Continuous and line spectra.

Spectra and solar research. To improve efficiency, a "tandem" solar cell using two photoelectric materials is tested. A spectrum-splitting filter is used to divide sunlight into two beams—one containing wavelengths best suited to produce a current in silicon, and the other containing wavelengths more suited to gallium arsenide. (Courtesy Department of Energy)

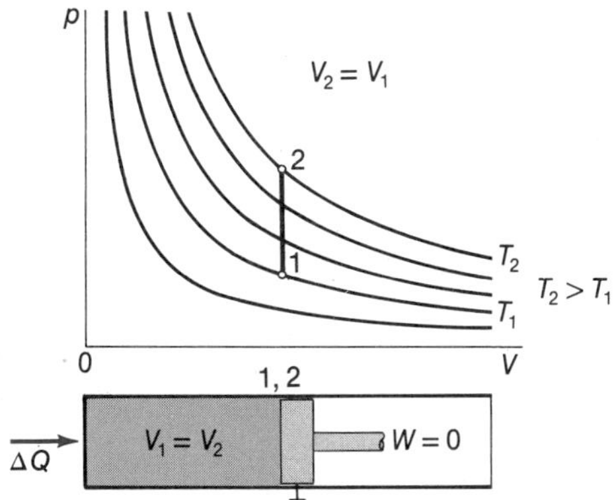

Figure 18.3
An isometric (constant-volume) process. When heat is added to the system and the volume is kept constant, the temperature of the gas must increase. In an isometric process, only heat is accepted (or rejected), since no mechanical work is done ($\Delta V = 0$).

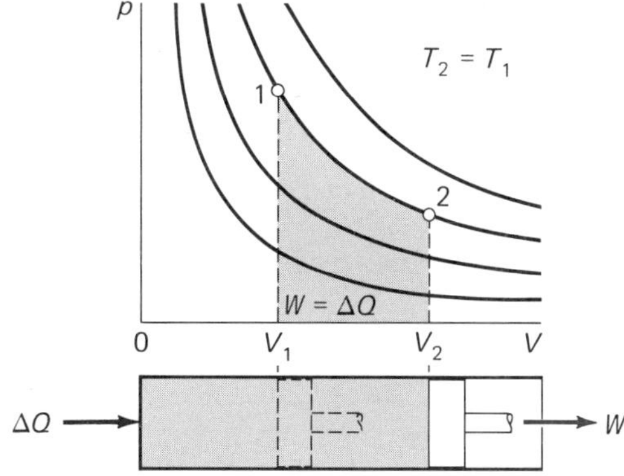

Figure 18.4
An isothermal (constant-temperature) process. When heat is added to the system, both the pressure and volume must change to keep the temperature constant. Since the temperature of the ideal is constant, so is its internal energy, and the work done by the gas is equal to the heat input.

This is somewhat of an ideal condition, since in most real processes only the initial and final states are known. Between these states, the gas is in a chaotic condition and the thermodynamic coordinates are unmeasurable and unknown; hence, the process is irreversible. This does not mean that the system cannot be taken from the final state back to the initial state, but only that the process cannot be reversed along the previous process path since it is unknown. Even so, it is instructive to consider ideal, reversible processes.

(b) Isometric Process Another "iso-" process—an isovolumetric (isometric for short), or constant-volume, process—is illustrated in Figure 18.3.

An isometric process is one in which the volume of the system remains constant.

The process path is called an **isomet.** In this case, heat energy is added to the system. And if the volume is maintained constantly (the piston is not allowed to move), the pressure must increase as the system goes from state 1 to state 2. The temperature of the system must also increase. An increase in temperature means an increase in the internal energy ($+\Delta U$) of the ideal gas system. Why?

For an isometric process, no work is done, since the area under the curve is zero ($\Delta V = 0$, and $W = p\Delta V = 0$). Hence, by the first law,

$$\Delta Q = \Delta U + W = \Delta U + p\Delta V = \Delta U + 0$$

or

$$\Delta Q = \Delta U$$
(isometric process)

and all the heat goes into the internal energy of the gas.

(c) Isothermal Process A third "iso-" process is an isothermal, or constant-temperature, process, as illustrated in Figure 18.4.

An isothermal process is one in which the temperature of the system remains constant.

In this type of process, the system is "moved along" an **isotherm.** As the system goes from state 1 to state 2, heat energy is added to it, and both the pressure and volume change to keep the temperature constant (the pressure decreases and the volume increases for the process in Fig. 18.4). Again, the work done is equal to the area under the process curve, but it is more difficult to calculate than in the case of a constant-pressure process.

In an isothermal process, the internal energy of the ideal gas system is unchanged ($\Delta U = 0$), since the temperature is constant. Then, by the

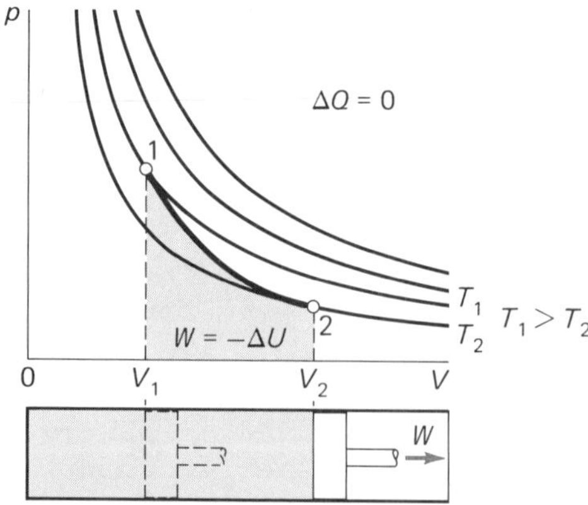

Figure 18.5
An adiabatic (no heat transfer) process. All of the thermodynamic coordinates change in an adiabatic process. Since no heat enters or leaves the system ($\Delta Q = 0$), the work must be done at the expense of the internal energy of the system.

first law, the added heat energy is equal to the work done by the system:

$$\Delta Q = W$$
(isothermal process)

Thus, an isothermal process is one of energy transformation—heat energy to mechanical energy and vice versa.

(d) Adiabatic Process Another important process in which all of the thermodynamic coordinates vary is an adiabatic process.

> **An adiabatic process is one in which no heat energy enters or leaves the system ($\Delta Q = 0$).**

The graph in Figure 18.5 illustrates an adiabatic expansion process. The process curve for an adiabatic process is called an **adiabat.** Notice that the temperature of the ideal gas system decreases in the process and, hence, there is a decrease in internal energy ($-\Delta U$). This is also reflected in the first law, with work being done by the system ($+W$) and $\Delta Q = 0$:

$$\Delta Q = 0 = -\Delta U + W$$
(adiabatic expansion)

where ΔU is required to be negative. Thus, for an adiabatic expansion, work is done by the system at the expense of its internal energy.

For an adiabatic compression (the reverse process), work done on the system goes into increasing the internal energy of the system, as evidenced by an increase in the temperature of the system or as predicted by the first law.

18.2 The Second and Third Laws of Thermodynamics

From experience we know that heat is transferred spontaneously, or "flows," from a hotter body to a colder body when the bodies are in thermal contact. However, if heat were to flow from a colder to a hotter body, the first law would not be violated, since the heat gained would equal the heat lost in either case. Therefore, another thermodynamic principle is needed that specifies the *direction* in which a process can take place. This is embodied in the **second law of thermodynamics.** A simple form of this law is

> **Heat will not flow spontaneously from a colder body to a hotter body.**

For example, a cold drink always warms up to room temperature. It has never been observed that the drink gets colder while the room or surroundings become warmer. The empirical second law reflects our experience; it is a "law" because no case to the contrary has ever been observed.

A more practical form of the second law, which applies to the heat engines (and heat pumps) discussed in the following section, is

> **Heat energy cannot be completely transformed into mechanical work (and vice versa) in a cyclic process.**

The key words here are "cannot be completely" and "cyclic process." The first phrase implies that there will always be some heat lost or wasted when that energy is transformed to mechanical work in a cyclic process, e.g., the conversion of combustion heat energy to mechanical work by the internal combustion engine of an automobile.

A "cyclic" process means that an engine goes through a cycle of processes that brings it back to its original state so a continuous work output can be maintained by the repeating cycle. As we have seen, heat may be converted completely to work in a *single* isothermal process, but this would not make for a practical heat engine with continuous work output.

As in the case of simple mechanical machines (Chapter 11), because energy is lost or wasted the efficiency of a "thermal" machine can never be 100 percent. If it were, we would be able to construct a perpetual-motion machine. Thus, the second law tells us that it is impossible to construct a thermal perpetual-motion machine. This point will be considered further in the next section.

By telling us what *does not happen,* **the second law indicates the *direction* in which a thermodynamic process can take place.** Theoretical microscopic analyses of many possible processes show that a system naturally moves from a more orderly state toward a more disorderly state. The same is true in a mechanical sense. For example, the natural tendency of one's room is to become disorderly. A thermodynamic measure of disorderliness is a quantity called **entropy.** In the context, the second law can be stated

The total entropy of the universe increases in every natural process.

Mathematically, the change in entropy (ΔS) of a system is given by

$$\boxed{\Delta S = \frac{\Delta Q}{T}} \qquad \textbf{(Eq. 18.3)}$$

where T is the absolute temperature. As can be seen, the SI unit of entropy is J/K. Without the use of calculus, this expression can be used only for isothermal processes.

EXAMPLE 18.1 What is the change in entropy when 1.0 g of water freezes to ice at 0°C?

Solution: It is given that $m = 1.0$ g and $T_C = 0$°C. The amount of heat energy involved in the phase change is the latent heat of fusion (as discussed in Chapter 17), and

$$\Delta Q = -mL_f = -(1.0 \text{ g})(80 \text{ cal/g}) = -80 \text{ cal}$$

or

$$\Delta Q = -80 \text{ cal } (4.2 \text{ J/cal}) = -336 \text{ J}$$

ΔQ is negative because (latent) heat is given up by or removed from the system. Since $T_K = T_C + 273 = 0 + 273 = 273$ K, we have

$$\Delta S = \frac{\Delta Q}{T} = \frac{-336 \text{ J}}{273 \text{ K}} = -1.23 \text{ J/K}$$

But this is a *decrease* in entropy, and the molecules are more orderly in ice than in water. Is there something wrong?

No, the entropy of a system can decrease, but the *total* entropy of the universe must increase. (Water does not naturally or *spontaneously* freeze.) That is, the energy for doing work on the system to remove the heat from the water came from somewhere else (an entropy increase), and overall the *total* entropy of the universe increased.

Suppose, for example, the water froze as a result of being in contact with a "heat sink" at a temperature of −10°C. With $\Delta Q = +336$ J (heat added to sink system),

$$\Delta S \text{ (heat sink)} = \frac{\Delta Q}{T} = \frac{+336 \text{ J}}{263 \text{ K}} = +1.28 \text{ J/K}$$

Hence,

$$\begin{aligned}\Delta S \text{ (universe)} &= \Delta S \text{ (heat sink)} + \Delta S \text{ (water)} \\ &= 1.28 \text{ J/K} - 1.23 \text{ J/K} \\ &= +0.05 \text{ J/K}\end{aligned}$$

In effect, entropy is a measure of a system's capability to do work or transfer heat. A system at a high temperature *naturally* tends to do work on and/or transfer heat to its surroundings. In the process, the entropy of the system is increased. Hence, the greater the entropy of a system, the less *available* energy the system has with which to do work.

In a sense, the heat energy is more orderly when it is concentrated in the system. When the heat energy is transferred from the system (in a natural process), it is more "spread out" or disorderly, and there is an increase in entropy (Fig. 18.6).

Since natural processes continually occur with heat transfer from hotter to colder bodies, the entropy of the universe is continually increasing, and the universe should thermodynamically "run down." The entropy should

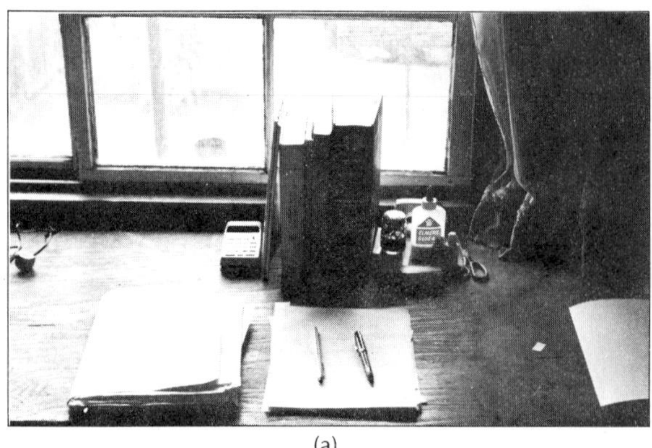

(a)

(b)

Figure 18.6

Entropy of the increase. Entropy in effect is a measure of disorder, which increases in every natural process. Straightening up or making a messy desk orderly (a decrease in entropy) requires work or an expenditure of energy, which creates entropy such that there is always a net entropy increase in the universe. (Courtesy J. and A. Turk.)

reach a maximum when everything is at the same temperature. This limit is called the **heat death of the universe.** The final temperature is estimated to be a few degrees above absolute zero. Not any too warm, but this would occur billions of years from now, so don't worry.

Thermodynamics is an empirical science, and its "laws" are not based on absolute impossibilities of things happening but on the experience that those things have never been observed. Because of this, we can say with great certainty, "Entropy can be created but not destroyed" and "Energy can be neither created nor destroyed," which are the essences of the second and first laws of thermodynamics.

The **third law of thermodynamics** states:

> **It is impossible to attain a temperature of absolute zero.**

The third law has never been violated experimentally, although in cryogenic (low-temperature) experiments, absolute zero has been approached to within 0.00000001 (10^{-8}) degree. Theoretically, a temperature of absolute zero would allow a heat engine to have an ideal thermal efficiency of 100 percent, as we discuss in the following section.

18.3 Heat Engines

A heat engine is a device that converts heat energy into mechanical work. There are many practical examples of heat engines: the gasoline engine, the steam engine, the jet engine, and even the human body, in the sense that we use heat energy obtained from food to do work.

Thermodynamically, it is convenient to consider a heat engine in terms of a representative diagram, as in Figure 18.7. This approach considers only the involved processes and not the

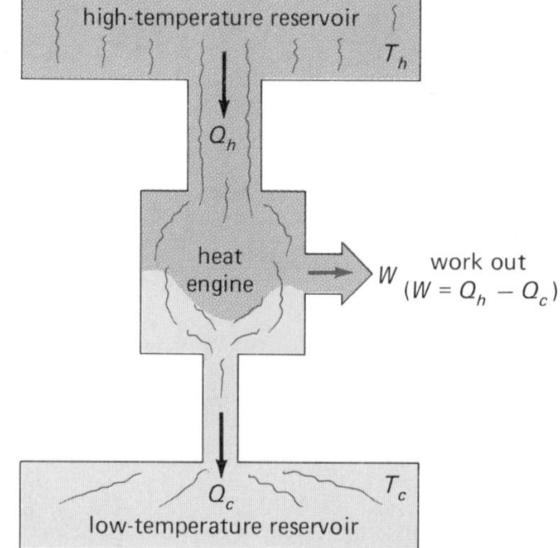

Figure 18.7

Representative diagram of a heat engine. An amount of heat Q_h is supplied to the engine from the high-temperature reservoir. Part of this energy goes into mechanical work and the rest is expelled to the low-temperature reservoir, i.e., $Q_h = W + Q_c$.

SPECIAL FEATURE 18.1

The Drinking Bird Heat Engine

A novel example of a cyclic heat engine is the toy drinking bird (Fig. 1). It doesn't look much like a heat engine, but it falls under the definition of one. To start the "engine," you wet the absorbent flock material on the head and beak of the bird with water. The water evaporates, lowering the temperature (T_c) and removing heat (Q_c).

The liquid inside the body is ether, which has a low boiling point and readily vaporizes at room temperature T_h. The evaporation of ether in the lower part of the body creates pressure above the liquid. The ether in the tube does not evaporate as readily because the head is cooled by the evaporation of water from the flock covering, and there is less vapor pressure in the head. The pressure difference forces the ether up the tube into the head.

The rising liquid raises the center of gravity of the bird above the pivot point, and the bird pitches forward for a "drink" (to rewet the flock). In this position, the pressures in the head and body are equalized, and the ether drains back into the body. The bird pivots back and the cycle begins again, with the liquid in the body being warmed by heat from the atmosphere (Q_h).

The bird could be hooked up so its motion could do some useful work. This heat engine is cheap to run, but it is doubtful if "birdmobiles" will ever replace piston-engine automobiles.

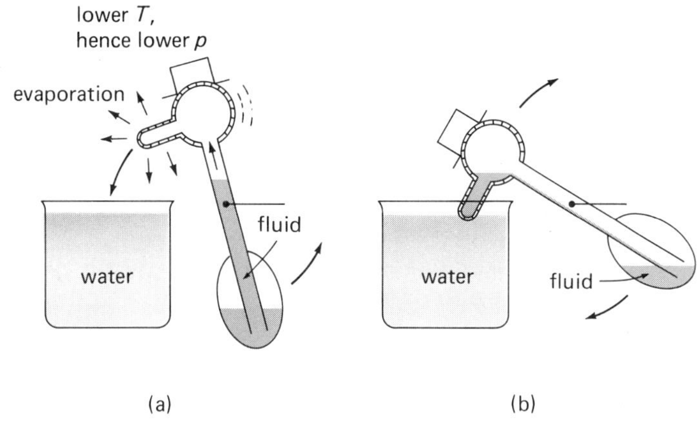

Figure 1
The "drinking bird" heat engine. A strange-looking heat engine perhaps, but it does the job with work output.

actual components of the engine. The basic processes of a heat engine are

1. An amount of heat Q_h is supplied to the engine from a high-temperature heat reservoir at temperature T_h.
2. Mechanical work W is performed by the engine by using a portion of the heat input.
3. The remaining heat Q_c is rejected to a low-temperature reservoir (e.g., the surroundings) at temperature T_c.

A rather simple heat engine is considered in Special Feature 18.1.

From a practical standpoint, an important consideration of a heat engine is its efficiency. Recall from Chapter 11 that the mechanical efficiency (Eff) of a simple machine is defined as the ratio of the work output W_{out} to the work input W_{in}:

$$\text{Eff} = \frac{W_{out}}{W_{in}}$$

The thermal efficiency ε_{th} of a heat engine is similarly defined:

$$\varepsilon_{th} = \frac{W_{out}}{Q_{in}} \qquad \textbf{(Eq. 18.4)}$$

where Q_{in} is the input heat. A practical heat engine operates in a cycle so that a continuous work output is maintained. Hence, Equation 18.4 is the efficiency *per cycle.*

Since in one cycle the system comes back to its initial state, the internal energy is the same, and $\Delta U = 0$ for the cycle. By the first law, $\Delta Q = \Delta U + W$, and for the cycle,

$$\Delta Q = 0 + W_{out}$$

or

$$Q_h - Q_c = W_{out}$$

Then, the **thermal efficiency** of the cycle with $Q_{in} = Q_h$ is

$$\varepsilon_{th} = \frac{W_{out}}{Q_{in}} = \frac{Q_h - Q_c}{Q_h} = 1 - \frac{Q_c}{Q_h} \quad \textbf{(Eq. 18.5)}$$

EXAMPLE 18.2 A heat engine absorbs 250 kcal of heat energy and rejects 150 kcal while doing work in a cycle. (a) How much work is done? (b) What is the thermal efficiency of the engine?

Solution: (a) With $Q_h = 250$ kcal and $Q_c = 150$ kcal, for a cyclic process ($\Delta U = 0$),

$$W = \Delta Q = Q_h - Q_c$$
$$= 250 \text{ kcal} - 150 \text{ kcal} = 100 \text{ kcal}$$

or, in mechanical work units,

$$100 \text{ kcal } (4186 \text{ J/kcal}) = 41.86 \text{ J}$$

(b) The thermal efficiency is

$$\varepsilon_{th} = 1 - \frac{Q_c}{Q_h}$$
$$= 1 - \frac{150 \text{ kcal}}{250 \text{ kcal}} = 0.40 \quad (= 40\%)$$

There are many possible cyclic thermodynamic processes through which a heat engine could be operated. But is there one cycle that is more efficient than all the others? Sadi Carnot (1796–1832), a French engineer, suggested such a cycle, which can be shown to be the theoretical limit for the maximum possible efficiency of an engine cycle. It consists of two adiabats and two isotherms (Fig. 18.8).

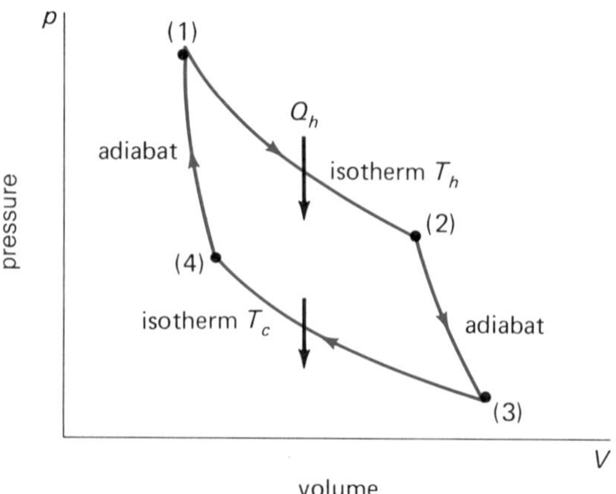

Figure 18.8
The Carnot cycle. The Carnot cycle, consisting of two adiabats and two isotherms, is the cycle for the theoretical limit of maximum heat engine efficiency.

Since there is no heat transfer in adiabatic processes, such a cycle is unachievable for practical engines. But the cycle does set a theoretical limit for the best efficiency one can hope to achieve with a heat engine. **A theoretical engine operating through a Carnot cycle is called an *ideal* or *Carnot engine.***

An important result from the detailed analysis of a Carnot cycle is the relationship between the heat Q_h transferred from the high-temperature reservoir at T_h and the heat Q_c rejected to the low-temperature reservoir at T_c, namely,

$$\frac{Q_c}{Q_h} = \frac{T_c}{T_h} \quad \textbf{(Eq. 18.6)}$$

Using this relationship for a Carnot cycle and Equation 18.5, we have a **Carnot efficiency ϵ_c** for an ideal heat engine:

$$\varepsilon_c = 1 - \frac{Q_c}{Q_h} = 1 - \frac{T_c}{T_h}$$

or

$$\varepsilon_c = \frac{T_h - T_c}{T_h} = 1 - \frac{T_c}{T_h} \quad \textbf{(Eq. 18.7)}$$

Note that the temperatures of the reservoirs in this case are *absolute* temperatures (either Kelvin or Rankine).

EXAMPLE 18.3 An engineer wishes to design a heat engine operating between reservoirs with temperatures of 70°F and 300°F with an efficiency of 30 percent. Is this possible?

Solution: Using the Rankine temperature scale, $T_c = 70°\text{F} + 460° = 530°\text{R}$ and $T_h = 300°\text{F} + 460° = 760°\text{R}$. Then, the Carnot efficiency of an engine operating between these temperatures is

$$\varepsilon_c = 1 - \frac{T_c}{T_h} = 1 - \frac{530°\text{R}}{760°\text{R}}$$

$$= 1 - 0.7 = 0.3 \quad (= 30\%)$$

Since the Carnot efficiency is the theoretical limit, which can never be achieved practically, a practical heat engine operating between these temperatures is not possible.

The Carnot efficiency (Eq. 18.7) has several important implications. It points out that **the greater the temperature difference between the operating reservoir temperatures of a heat engine, the greater the efficiency of the engine.** For example, if T_h is twice T_c, or $T_c/T_h = \frac{1}{2} = 0.5$, then the Carnot efficiency is 50 percent. But, if T_h is three times T_c or $T_c/T_h = \frac{1}{3}$, then ε_c is $\frac{2}{3}$, or 67 percent.

Also, the third law of thermodynamics is contained in Equation 18.7. To have 100 percent efficiency would require $T_c = 0$ K. But, by the second law, a heat engine with 100 percent efficiency is forbidden. Hence, it is impossible to have a temperature of absolute zero, which is the third law of thermodynamics.

Two-Stroke Cycle Engines

To illustrate an engine cycle, let's consider the two-stroke cycle gasoline engine. Such internal combustion engines are used in some lawn mowers, motorcycles, outboard motors, and chain saws. A single cylinder of a two-stroke cycle engine is illustrated in Figure 18.9. Such piston-driven engines are referred to as *reciprocating engines.*

The cyclic series of thermodynamic processes through which the engine operates is called the **Otto cycle,** after the German engineer Nikolas Otto (1832–1891), who designed and built one of the first successful two-stroke gasoline engines. In the figure, the series of reversible thermodynamic processes in the *p-V* diagrams are theoretical representations that help analyze the cycle.

Starting with the compression stroke, the piston advances and compresses the air-fuel mixture in an adiabatic compression (1–2). The spark plug ignites the mixture, and the pressure and temperature of the system increase (2–3), giving the "high-temperature reservoir" condition of the system. The combustion gases then expand, driving the piston downward for work output (3–4).

In the last process, the valves open, the combustion products are ejected, and the fresh charge of air-fuel enters the cylinder (4–1). In the process the system is cooled to the "low-temperature reservoir" condition. This completes the two-stroke (up and down) cycle of the engine piston.

A second type of internal combustion engine that may also operate in a two-stroke cycle is the diesel engine, developed by the German engineer Rudolf Diesel (1858–1913). It operates on a cycle similar to the Otto cycle, but with some important differences (Fig. 18.10). Physically, the marked difference between diesel and gasoline engines is the absence of an ignition system in the diesel engine. After the compression process (1–2 in Fig. 18.9), fuel oil is injected into the cylinder, and combustion occurs because of the high temperature of the compressed air-fuel mixture (2–3). Most two-stroke diesel engines use a supercharger to provide to the cylinder a new charge of air that is free of burned gases from the previous cycle.

The major differences between the Diesel cycle and the Otto cycle are the compression ratios and the constant-pressure processes. It can be shown that diesel engines have higher efficiencies than Otto engines as a result of the much higher compression ratios they can attain. The thermal efficiencies of Diesel cycle engines are about 40 percent, as compared with about 30 percent for Otto cycle engines.

Four-Stroke Cycle Engines

Two-stroke cycle engines have one power stroke per cycle; i.e., power is produced each

spark plug
fuel and air being compressed
exhaust port
valve
fuel and air from carburetor
crankcase

(1) compression stroke

(2) ignition

expanding gases

(3) power stroke

exhaust gases
new charge

(4) exhaust and recharge

p
pressure
(3)
expansion
heating
(4)
Q_h
cooling
(2)
Q_c
compression
(1)
volume
V

(a)

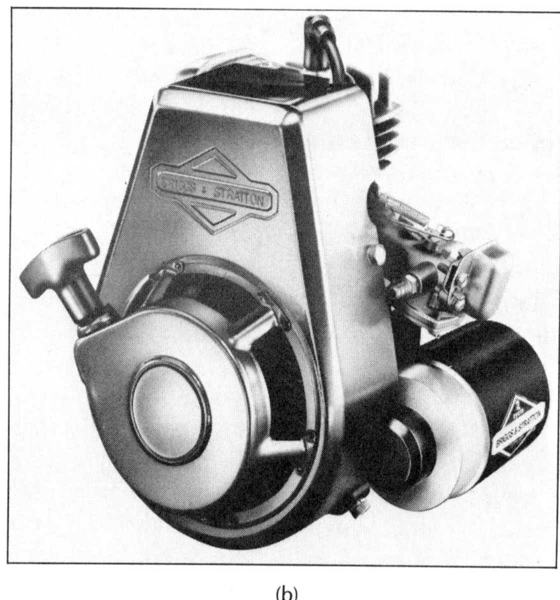

(b)

Figure 18.9

The two-stroke-cycle engine. (a) The engine operates through a cyclic series of thermodynamic processes called an Otto cycle, as shown theoretically on the graph: (1–2) is an adiabatic compression process; (2–3) is an isometric heating process; (3–4) is an adiabatic expansion process; (4–1) is an isometric cooling process. (b) A practical two-stroke-cycle engine. (Courtesy Briggs & Stratton Corp.)

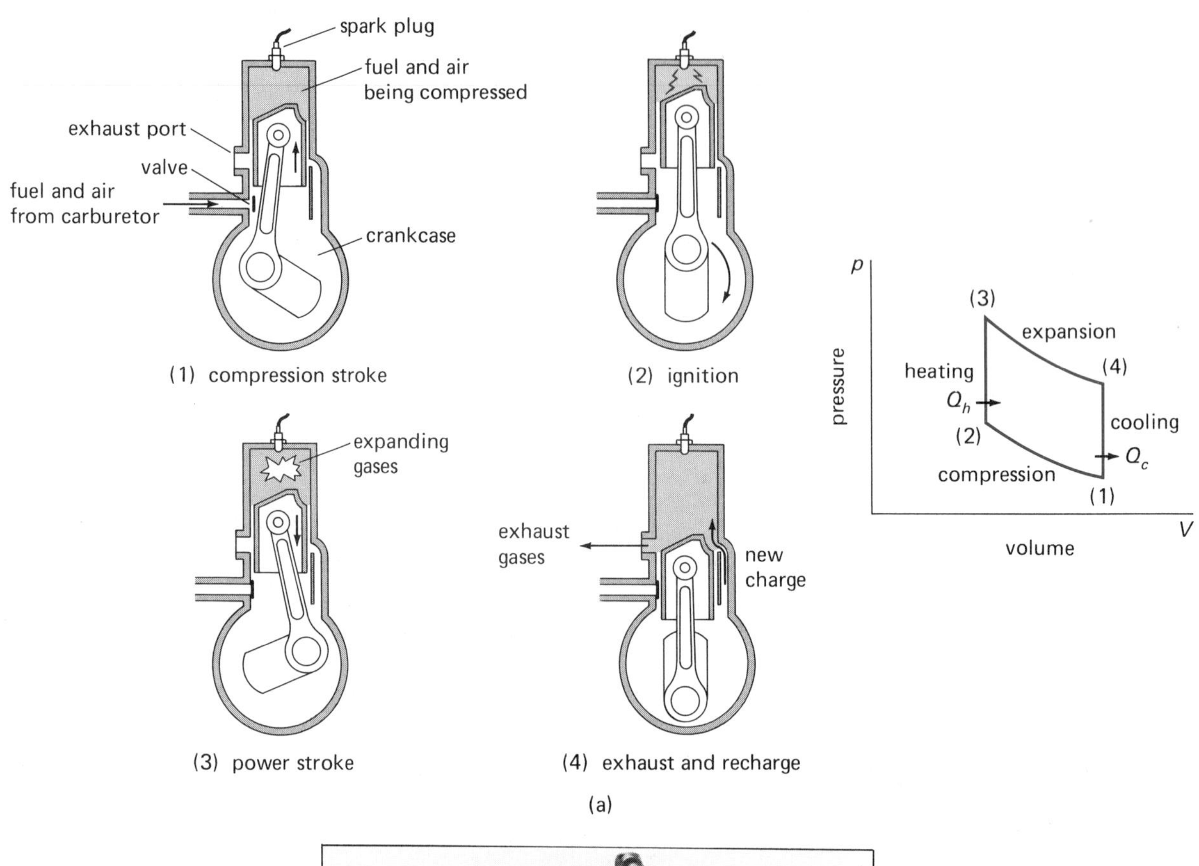

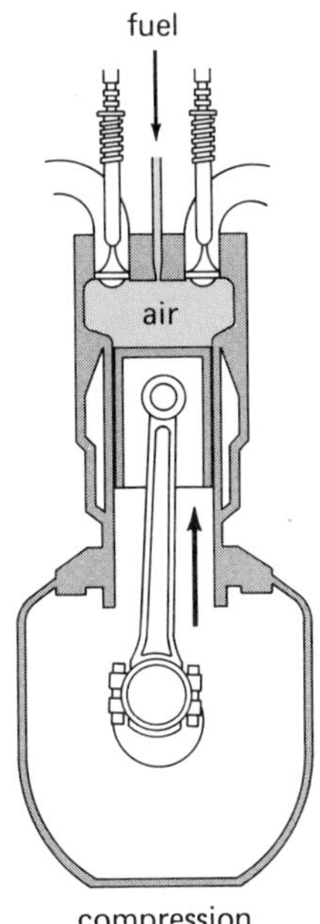

Figure 18.10
The diesel engine. The diesel engine is a two-stroke-cycle engine that does not use a spark plug. Near the end of the compression stroke (as shown here), fuel is injected into the chamber and is ignited by the heat of compression.

time the piston goes up and down. This is an advantage for lightweight engine applications in which a large horsepower-per-pound ratio is desired. However, one disadvantage of the two-stroke cycle engine is the waste or loss of fuel with the exhaust gases (process 4–1, Fig. 18.9).

Most of our automobiles use four-stroke cycle gasoline internal combustion engines modeled after an engine built by Otto around 1876. The four-stroke cycle engine has one more piston cycle (up and down strokes) than the two-stroke cycle engine. The additional cycle prevents the loss of fuel with the exhaust gases. Diesel engines can also be operated in a four-stroke cycle.

As illustrated in Figure 18.11 for a gasoline engine, the four-stroke cycle adds an isobaric leg (intake and exhaust strokes). Compare Figure 18.11 with Figure 18.9. The work for the extra cycle is supplied by the rotational energy of a flywheel. During the intake stroke, the entering fuel mixture does a small amount of work on the moving piston (the area under the intake isobar). During the exhaust stroke, the energy stored in the flywheel moves the piston to push out the hot gases through the exhaust valve. This work equals the area under the exhaust isobar.

To obtain greater work output, multicylinder engines are used. For example, automobile engines have four, six, eight, or even twelve cylinders. In each of these, all of the cylinders go through their four-stroke cycle in two rotations of the crankshaft (to which the piston rods are attached). The cylinder firing, or ignition, occurs in a regular sequence for smoother power output. For example, in a four-cylinder engine (Fig. 18.12), the four cylinders are displaced in timing by a quarter cycle so the crankshaft gets a power stroke every half revolution.

Rotary Engines

Despite the good record of reciprocating engines, efforts have continually been made to develop engines of other designs, in particular ones that would reduce the number of moving parts so as to minimize the need for maintenance and adjustment. One of the serious contenders is the rotary engine. James Watt, the developer of the steam engine, tried to build a rotary steam engine in the late 1700's but could not overcome the problem of sealing between the rotary member and its casing.

A practical internal combustion rotary engine was invented by Felix Wankel, a German engineer, in the early 1950's. The principle of the **Wankel rotary engine** is illustrated in Figure 18.13.

An important feature of the engine is that it has only two moving parts: (1) a triangular rotor with convex sides and an internal gear and (2) a geared output shaft. Inside a casing with a water-cooled jacket, the three apexes of the rotor rub over the casing. The space between the rotor and the casing is divided into three compartments. The air-fuel mixture enters through an inlet port and is compressed and spark-ignited. The combustion gases expand to deliver power and are finally exhausted through the outlet port.

In the process, the charge in the compartment system goes through a cycle of four thermodynamic processes similar to those in a reciprocating engine. However, in the rotary engine, three *separate* series of processes are

spark plug
spark
air
fuel
piston
valves
intake
compression
power
exhaust

p
expansion
compression
intake
atmospheric pressure
exhaust
V
(a)

(b)

Figure 18.11

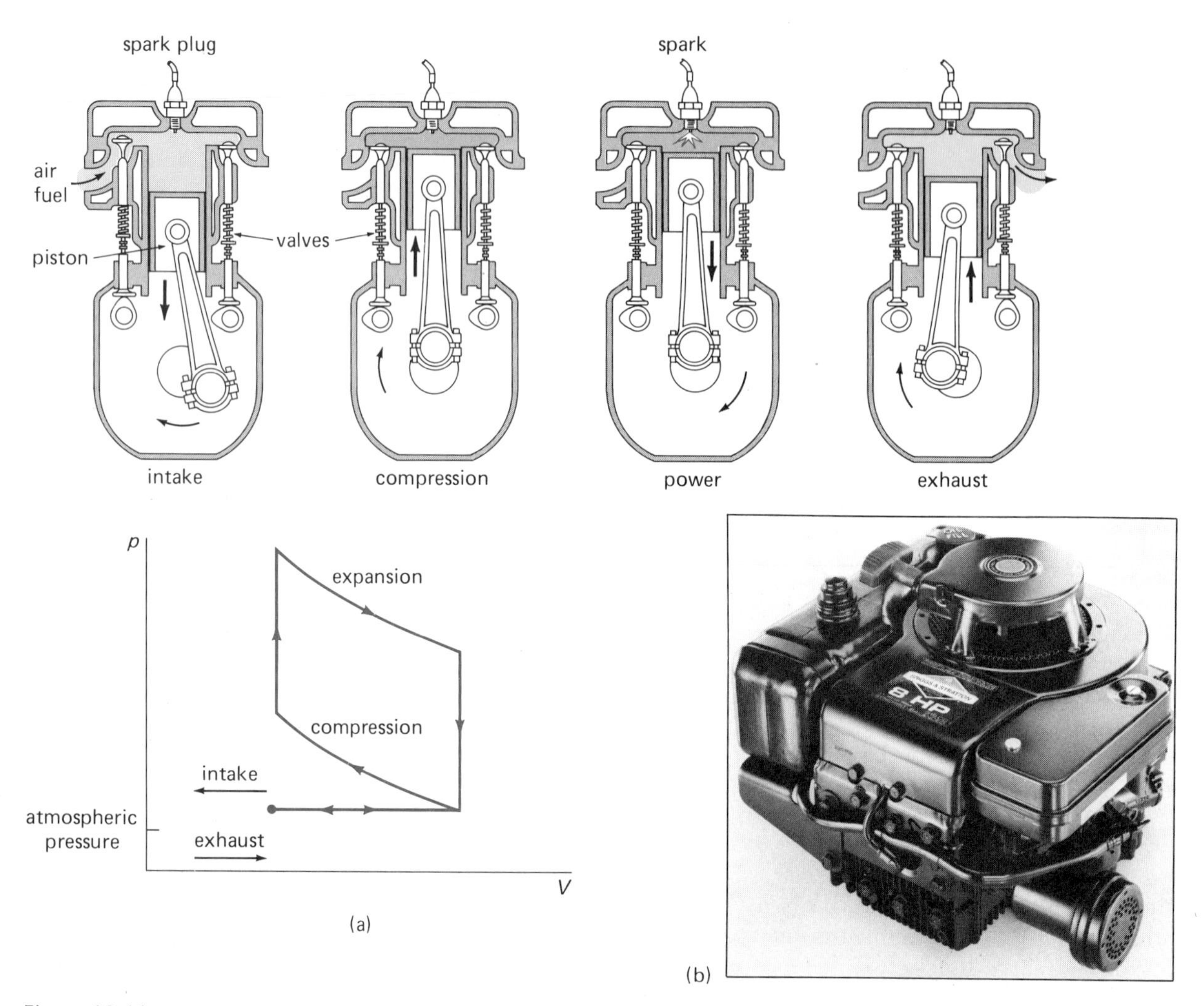

The four-stroke-cycle engine. This engine has one more piston cycle (up and down — two strokes) than the two-stroke-cycle engine. The additional cycle avoids loss of fuel with the exhaust gases and adds an isobaric leg to the theoretical thermodynamic Otto cycle, as shown in the graph. (b) A practical four-stroke-cycle engine. (Courtesy Briggs & Stratton Corp.)

Figure 18.12

Four-cylinder engine. In a four-cylinder engine the cylinders are displaced in timing by a quarter cycle, so the crankshaft gets a power stroke every half revolution. (Courtesy American Motors Corp.)

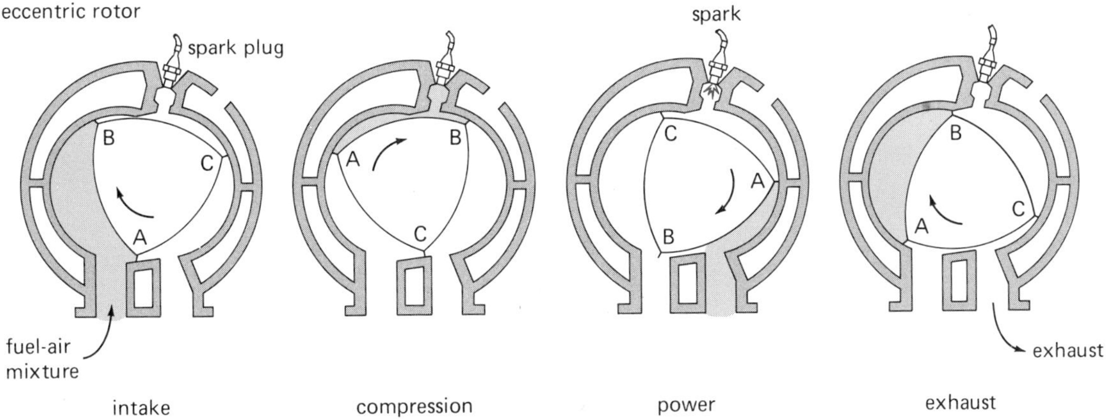

Figure 18.13
The Wankel engine. This rotary engine undergoes a four-stroke Otto cycle in each of the three variable chambers in one revolution of the rotor. That is, three separate processes are taking place simultaneously. The four-stroke process is shown here for only one chamber.

taking place simultaneously—one at each rotor face. The processes are out of phase—i.e., in different stages of progress. But, since each four-process cycle takes place during one revolution of the rotor, there are three power impulse processes per revolution. In terms of the number of power impulses per revolution of the output shaft, the rotary engine is equivalent to a six-cylinder, four-cycle reciprocating engine.

Elimination of a crankshaft and piston rods makes the rotary engine relatively compact. It has been used in automobiles but does not have good fuel economy.

18.4 Heat Pumps and Refrigerators

Technically, **a heat pump is a device that transfers heat from a low-temperature reservoir to a high-temperature reservoir.** By the second law of thermodynamics, this requires work input. Theoretically, a heat pump may be thought of as a heat engine operated in reverse (Fig. 18.14), and a common example is a refrigerator.

Heat is taken from a low-temperature reservoir (the refrigerator), which requires mechanical work on the system. The heat is then rejected to a high-temperature reservoir (the surroundings). In a descriptive sense, a better name for a refrigerator is a heat pump, since it is essentially "pumping" heat from a low-temperature reservoir to a high-temperature reservoir.

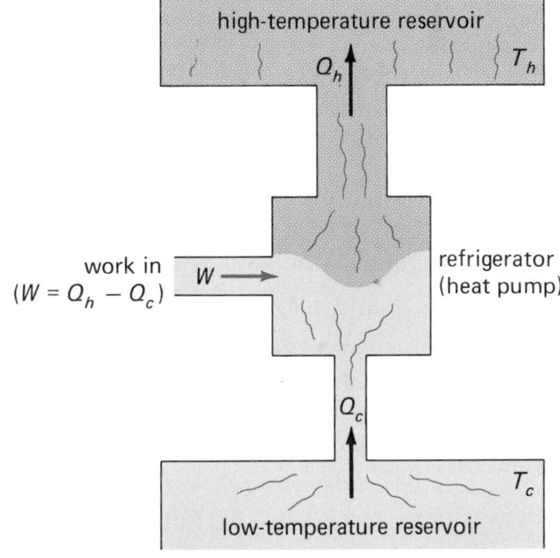

Figure 18.14
Representative diagram of a heat pump or refrigerator. Heat Q_c is taken from the low-temperature reservoir and rejected to a high-temperature reservoir. This requires work input, and $W = Q_h - Q_c$.

A schematic diagram of a typical refrigeration system is shown in Figure 18.15. The working substance, called the *refrigerant*, is used to absorb and transport heat. It is a fluid that readily undergoes a phase change (liquid-gas) at operating temperatures. Common refrigerants are ammonia (boiling point [b.p.] 28°F at 1 atm of pressure); Freon-12, a fluorocarbon compound

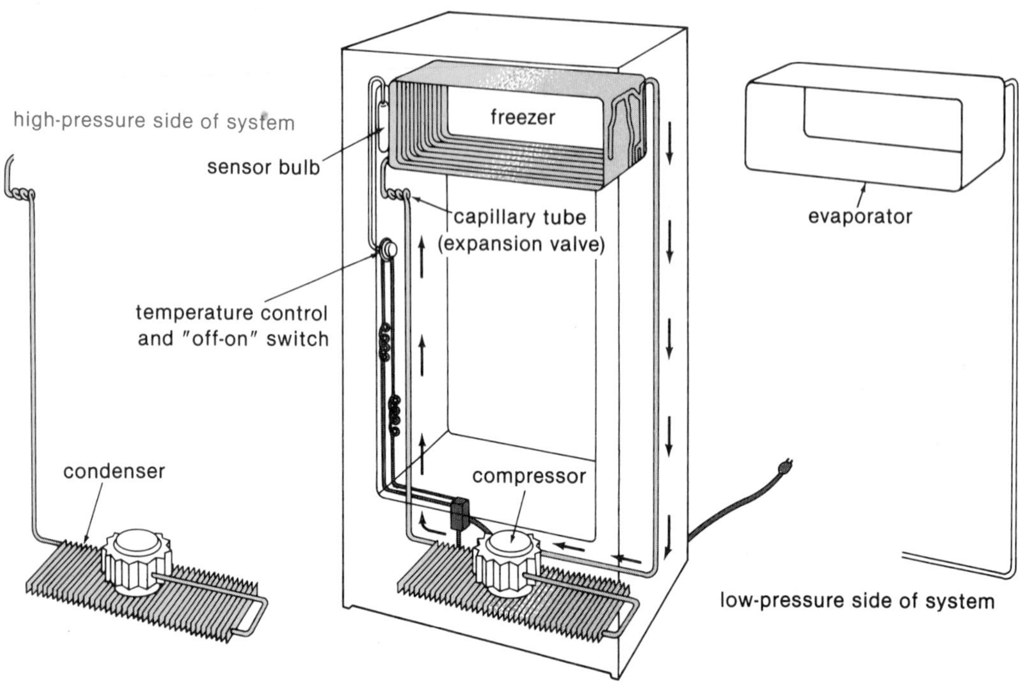

Figure 18.15

Simplified refrigeration system. Liquid refrigerant evaporates in the evaporator and removes heat. The gas is compressed by the compressor and liquefies in the condenser, where heat is given up to the surroundings. See text for further description. (Courtesy Frigidaire Company, Division of White Consolidated Industries.)

(b.p. 22°F); and sulfur dioxide (b.p. 13.8°F). Ammonia is the most common industrial refrigerant (e.g., in ice plants) and Freon is the most common household refrigerant (e.g., in refrigerators and air conditioners).

The basic components of a refrigeration system are a compressor, a condenser, an expansion valve, and an evaporator. The compressor supplies the necessary work to move the refrigerant through the system. The gaseous refrigerant is compressed by the compressor and emerges at high temperature and pressure. It then passes into the condenser, where it is cooled until it liquefies. The condenser is cooled by forced air or water, which carries away the rejected heat. During the process, the quantity of heat Q_h is rejected from the system (to the high-temperature reservoir, the surroundings).

The liquid refrigerant then passes through an expansion valve metering device, which admits the proper quantity of refrigerant to the low-pressure side of the system. In passing through

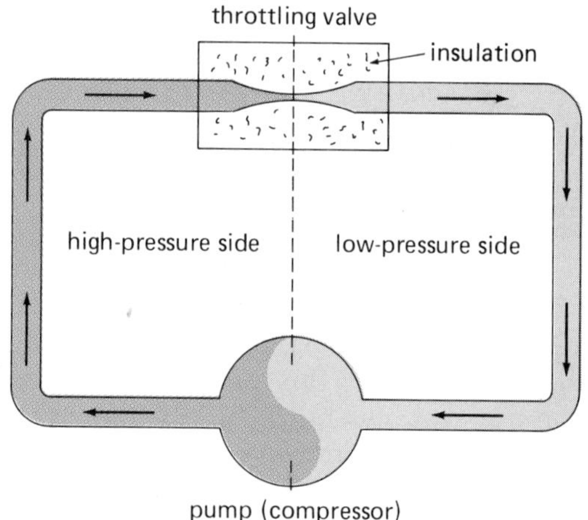

Figure 18.16

Throttling process. As a liquid under high pressure expands through the throttling expansion valve, particle vaporization occurs and the temperature is lowered at the expense of the internal energy. See text for description.

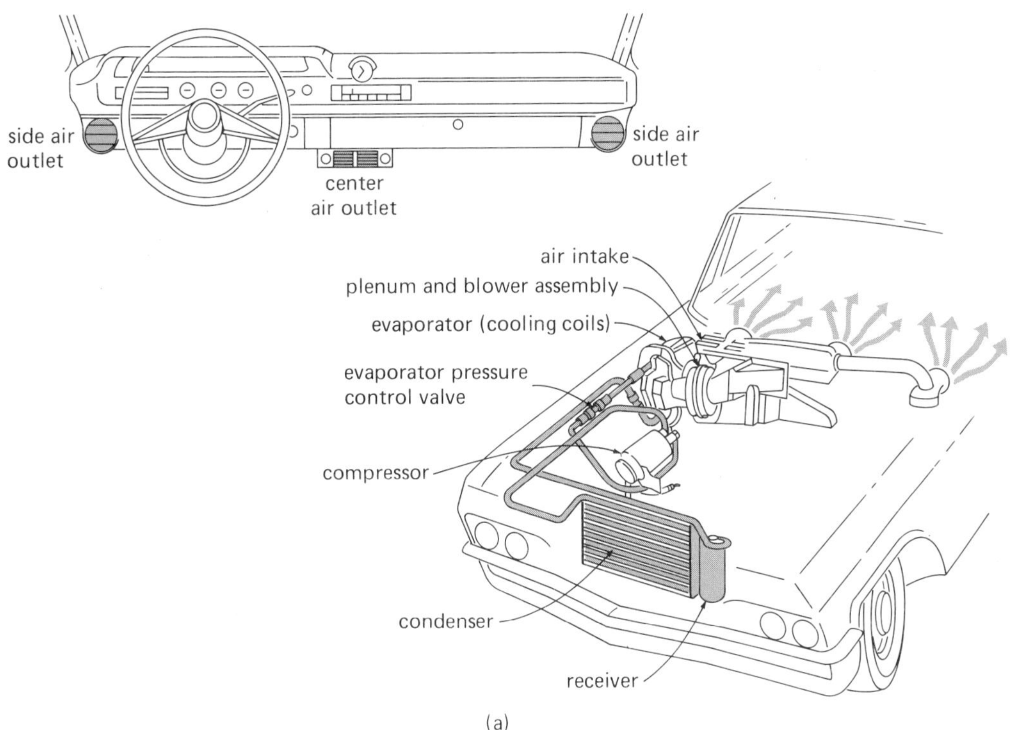

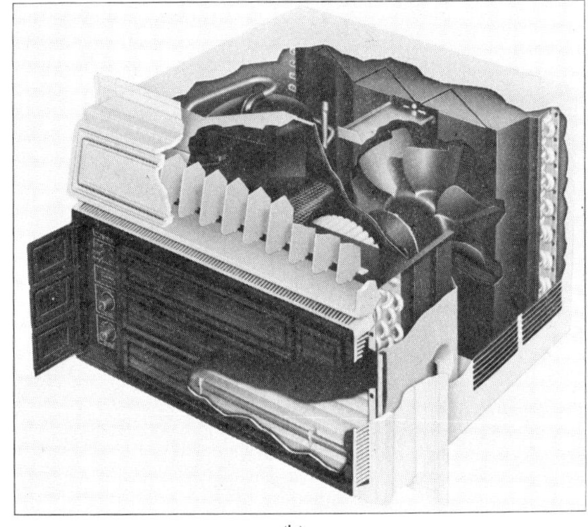

Figure 18.17

Air conditioners. An air conditioner is a form of refrigerator or heat pump. (a) Elements of an automotive air conditioning system. (Courtesy Frigidaire Company, Division of White Consolidated Industries.) (b) Exposed view of a window air conditioner. (Courtesy Carrier Corporation.) In these applications, the inside of the car and the inside of the house, respectively, correspond to the inside of a refrigerator.

the expansion valve, the refrigerant undergoes a so-called **throttling process** (Fig. 18.16). The liquid refrigerant experiences a drop in temperature, and partial vaporization occurs as it expands in the throttling process.

The valve is insulated to prevent the loss of heat, meaning the refrigerant undergoes a near-adiabatic process, $\Delta Q = 0$. Then, by the first law, $\Delta Q = 0 = \Delta U + W$, or $W = -\Delta U$. This means that the work done by the refrigerant in passing through the valve is done at the expense of its internal energy ($-\Delta U$), which effectively lowers the temperature of the fluid. As shown in Figure 18.16, it is convenient to divide the system into a high-pressure side and a low-pressure side. The same is true for the refrigeration system.

The cooled liquid refrigerant then flows through the evaporation coils, where it absorbs a quantity of heat Q_c from the low-temperature reservoir (interior of the refrigerator). The absorbed heat causes the liquid refrigerant to boil, and the heat is carried away as the latent heat of vaporization as the refrigerant is drawn into the compressor, where the cycle begins again.

An air conditioner, which is a form of refrigerator, goes through a similar cycle (Fig. 18.17). In this case, the low-temperature reservoir is usually the inside of a house or building, and heat is pumped to the outside environment (the

SPECIAL FEATURE 18.2

Heat Pump Heating and Cooling

As we have seen, a heat pump is synonymous with a refrigeration system. However, the term is now more generally applied to a year-round heating and cooling system. Such heat pumps are now commonly used in domestic heating and cooling.

When operating as a cooling system, the heat pump extracts heat from inside the home and expels it, together with the heat equivalent of the compression work, to an outside *heat sink* (the outside air).

Conversely, when the heat pump is operating as a heating system, the heat is taken from a *heat source* such as the outside air and delivered to the inside heated space, along with the heat equivalent of the work of compression (Fig. 1).

The heat pump cycle is effectively the same as the vapor-compression refrigeration cycle studied previously. The main difference between the systems is the objective of the application. A refrigeration system is chiefly concerned with the low-temperature effect produced at the evaporator coils, while a heat pump is concerned with both the cooling effect produced at the evaporator coils and the heating effect produced at the condenser coils.

The common design of heat pumps for residential and smaller commercial applications uses air as the heat source–sink and as the heating and cooling medium. Other designs use water as the heat source. Here, the heat source is usually an underground reservoir of water, e.g., a well or storage tank. With a higher specific heat capacity and higher temperature than the outside air, the heat pump cop is much greater. Why?

In the basic air design (Fig. 1), heating and cooling are accomplished by changing the direction of the refrigerant flow. Two independent air circuits are employed—an outdoor coil circuit and a conditioner coil circuit. The circuit interchange is controlled by means of a four-way valve.

During the cooling cycle, when the heat pump is functioning as an air conditioner, the compressor delivers hot compressed refrigerant gas to the outdoor coil where it is condensed, giving up the latent heat of condensation to the outside air. From the outdoor coil, the now-liquid refrigerant flows to a liquid receiver and then through a throttling valve to the conditioner coils.

Here the refrigerant is vaporized by absorbing the heat of vaporization from the air passing through the conditioner coils to the conditioned space. The refrigerant gas then returns to the compressor, completing the cycle. During a heating cycle, when the heat pump is functioning as a heating system, the process cycle is reversed.

The performance and application of any year-round heating-cooling system depends on several factors. In particular, the selection and design of an air-source heat pump are affected by the outdoor temperature variation in a given locality and by frost formation on the heat source coils. The indoor heating requirements are greatest when the outdoor heat source temperatures are the lowest. In cold and humid climates, frost may form on the coils, which, if allowed to accumulate, will interfere with the heat transfer. You may have experienced the "frosting up" of air conditioner coils on a humid summer day. Heat pumps are usually equipped with automatic defrost mechanisms.

Similarly, the outside coils of a heat pump operating in the heating cycle can become frosted when the outdoor temperature is low. These problems can be overcome by proper design and the use of supplemental electrical heating, particularly during very cold periods in winter.

---

high-temperature reservoir). The process can be reversed in a commercial heat pump for both heating and cooling. See Special Feature 18.2.

The cooling efficiency of a refrigerator or heat pump depends on the amount of heat Q_{out} extracted from the low-temperature reservoir and the expended work input W_{in}. This is expressed in the **coefficient of performance (cop):**

$$\text{cop} = \frac{Q_{out}}{W_{in}} \qquad \textbf{(Eq. 18.8)}$$

Since a practical heat pump system operates in a cycle to provide a continuous removal of heat, $\Delta U = 0$ for the cycle, as in the case of a heat engine. Then, by the conservation of energy (first law),

$$W_{in} = \Delta Q = Q_h - Q_c$$

The **cop per cycle** is then

$$\boxed{\text{cop} = \frac{Q_{out}}{W_{in}} = \frac{Q_c}{Q_h - Q_c}} \qquad \textbf{(Eq. 18.9)}$$

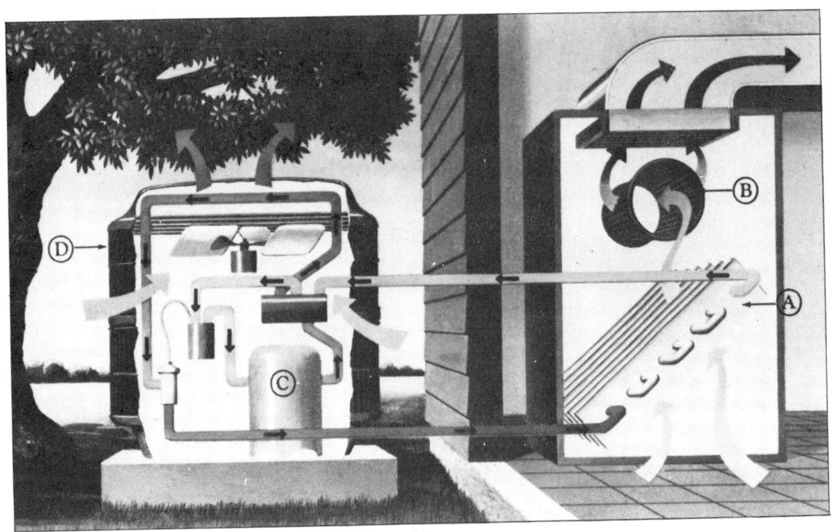

(a)

(b)

Figure 1

Heat pump operation. (a) Cooling cycle. In the summer, the heat pump works like a conventional air conditioner. Refrigerant in the indoor coils (A) absorbs heat from the air inside the house, and fan (B) circulates the cooled air. Compressor (C) pressurizes the refrigerant and sends it to the outdoor coils (D), where heat is expelled. (b) Heating cycle. In the winter, the heat pump works like an air conditioner in reverse, extracting heat that is present even in cold winter outside air. The heat is absorbed by the refrigerant in the outdoor coils (A). The compressor (B) pressurizes the refrigerant and sends it to the indoor coils (C), where it gives up heat to the air being circulated through the house by the fan (D). (Courtesy Carrier Corporation.)

Comparing Equation 18.9 with Equation 18.8, we see that the cop in words is

$$\text{cop} = \frac{\text{net refrigeration effect}}{\text{heat equivalent of the mechanical work input}}$$

In terms of an ideal or Carnot heat pump, which uses the Carnot cycle of a heat engine in reverse with the relationship $Q_c/Q_h = T_c/T_h$ (Eq. 18.6), the Carnot cop_c is given by

$$\text{cop}_c = \frac{T_c}{T_h - T_c} \qquad \textbf{(Eq. 18.10)}$$

where the T's are absolute temperatures. Hence, pumping heat when there is a large temperature difference requires a large work input, as expressed by a relatively small cop, or less cooling per unit work input.

Notice that the cop is greater than 1 (or 100 percent) for normal refrigerator temperature operation, since $T_h - T_c$ is less than T_c. For ideal cycles, cop's up to about 7.0 are possible. However, actual refrigerators do not have such high cop's because there are many losses that the ideal cycle does not take into account. The cop's of actual refrigerator cycles range from 1.5 to 4.0, depending on the operating conditions.

EXAMPLE 18.4 The working substance of an ideal refrigerator operating in a Carnot cycle removes heat from the freezing compartment at −28°C and rejects it at room temperature (20°C). What is the coefficient of performance?

Solution: With $T_c = -28°C + 273 = 245$ K and $T_h = 20°C + 273 = 293$ K,

$$\text{cop}_c = \frac{T_c}{T_h - T_c} = \frac{245\text{ K}}{293\text{ K} - 245\text{ K}} = 5.1$$

Thus, the net refrigerating effect is ideally 5.1 times as great as the mechanical work input; i.e., by Equation 18.8,

$$\text{cop} = \frac{Q_{\text{out}}}{W_{\text{in}}} = 5.1$$

or

$$Q_{\text{out}} = (5.1)\ W_{\text{in}}$$

Refrigeration Ratings

Refrigeration systems are rated in terms of the amounts of heat they will remove in a given time, just as heating units are rated in terms of the amounts of heat they will deliver per time.

Small refrigeration systems are commonly rated in Btu per hour (or "Btu's" for short). For example, window air conditioners range from 5000 to 25,000 Btu's.

Large refrigerating systems are rated in tons of refrigeration. A "ton" of heat energy is defined as the quantity of heat that must be removed from a ton of water at 32°F to cause it to freeze. Recall that the latent heat of fusion for water is 144 Btu/lb (Chapter 16). Hence, a "ton" of heat energy (1 ton = 2000 lb) is

$$144\text{ Btu/lb} \times 2000\text{ lb} = 288{,}000\text{ Btu}$$

In terms of a ton of refrigeration, which is heat removed *per time* ($\Delta Q/\Delta t$), the rating is specified over a 24-hour period. This time defines the standard commercial ton of refrigeration. That is, a one-"ton" refrigeration system will remove 288,000 Btu in 24 hours. Hence,

$$\textbf{1 ton of refrigeration} = \textbf{288,000 Btu/24 h}$$
$$= \textbf{12,000 Btu/h}$$

Typical commercial cold-storage refrigeration units range from 20 to 50 tons.

EXAMPLE 18.5 A commercial refrigeration system removes 480,000 Btu of heat energy in 2.0 h. What is its rating in standard tons of refrigeration?

Solution: The system removes

$$\Delta Q/\Delta t = 480{,}000\text{ Btu/2.0 h} = 240{,}000\text{ Btu/h}$$

Then

$$240{,}000\text{ Btu/h (1 ton/12,000 Btu/h)} = 20\text{ tons}$$

Important Terms

thermodynamics the branch of physics that deals primarily with the transfer and actions (dynamics) of heat (Greek *therme,* heat)

first law of thermodynamics a statement of the conservation of energy for a thermodynamic system; i.e., the change in heat is equal to the change in internal energy plus work

thermodynamic coordinates the variables that describe a thermodynamic system, e.g., p, V, and T for a gas

thermodynamic process a change in the state of a system or a change in the thermodynamic coordinates

isobaric process one in which the pressure of the system remains constant

isometric (isovolumetric) process one in which the volume of the system remains constant

isothermal process one in which the temperature of the system remains constant

second law of thermodynamics heat will not flow spontaneously from a colder body to a hotter body, or heat cannot be completely transformed into mechanical work (and vice versa) in a cyclic process, or the total entropy of the universe increases in every natural process

entropy a thermodynamic property that indicates the direction of a process; a measure of disorderliness or of a system's capability to do work or transfer heat

heat death of the universe the condition of maximum entropy, when everything would be at the same temperature

third law of thermodynamics it is impossible to attain a temperature of absolute zero

heat engine a device that converts heat energy into mechanical work

thermal efficiency the ratio of the work output and heat input of a heat engine

Otto cycle the thermodynamic cycle of common internal combustion piston engines

Carnot cycle the theoretical cycle for maximum heat engine efficiency, consisting of two adiabats and two isotherms

Carnot efficiency the theoretical maximum (ideal) efficiency of a cyclic heat engine operating in a Carnot cycle

diesel engine an internal combustion piston engine that uses the temperature of compression for fuel ignition (no spark plugs)

Wankel rotary engine an internal combustion engine that uses a rotary member instead of pistons

heat pump a device that transfers heat from a low-temperature reservoir to a high-temperature reservoir with work input

coefficient of performance (cop) the ratio of the heat output and work input of a heat pump; a measure of cooling efficiency

ton of refrigeration a refrigeration rating; 1 ton of refrigeration equals 12,000 Btu/h

Important Formulas

first law of thermodynamics:
$$\Delta Q = \Delta U + W = \Delta U + p\Delta V$$

entropy:
$$\Delta S = \frac{\Delta Q}{T}$$

thermal efficiency:
$$\varepsilon_{th} = \frac{W_{out}}{Q_{in}} = \frac{Q_h - Q_c}{Q_h} = 1 - \frac{Q_c}{Q_h}$$

Carnot efficiency:
$$\varepsilon_c = \frac{T_h - T_c}{T_h} = 1 - \frac{T_c}{T_h}$$

coefficient of performance:
$$\text{cop} = \frac{Q_{out}}{W_{in}} = \frac{Q_c}{Q_h - Q_c}$$

Carnot coefficient of performance:
$$\text{cop}_c = \frac{T_c}{T_h - T_c}$$

(1 ton of refrigeration = 288,000 Btu/24 h = 12,000 Btu/h)

Questions

The First Law of Thermodynamics and Thermodynamic Processes

1. What is meant by the internal energy of a system? Identify the internal energy of a real gas.
2. Describe several "iso-" processes and the advantages of such processes.
3. For a gas system, how much work is done in an isometric process? Explain.
4. How is the work involved in a process represented on a p-V diagram?
5. Distinguish between reversible and irreversible processes.
6. How is the internal energy of an ideal gas affected in an isothermal process?
7. Give the names of the process paths for the following processes: (a) $\Delta P = 0$, (b) $\Delta V = 0$, (c) $\Delta T = 0$, (d) $\Delta Q = 0$.
8. What gas process has the net effect of transforming heat energy into mechanical energy?
9. The p-T phase diagram for water is shown in Figure 18.18. (a) What is the condition of the water (i) when a series of processes takes the system along the curves and (ii) when a curve is crossed? What type of "iso-" processes would these be? (b) Comparing to Figure 18.1, what is the difference between substances that have oppositely slanting fusion curves (slanting to the left and right of the vertical)? (c) Is it possible, by varying the pressure, to have water (i) boil at room temperature or (ii) freeze at room temperature with this constant volume?
10. Referring to Figure 18.18, what are (a) the triple point and (b) the critical point? (The latter may require a little research.)

The Second and Third Laws of Thermodynamics

11. Does heat flowing spontaneously from a colder body to a hotter body violate the first law? Explain.
12. Is it possible to have heat energy completely transformed into mechanical energy in a single process? Explain.
13. State a form of the second law in terms of the efficiency of a cyclic heat engine.
14. Does a quantity of water have more entropy when it is a liquid or when it is frozen? Explain.
15. How does the entropy of gas in the cylinder of a heat engine change during the compression stroke of the piston?

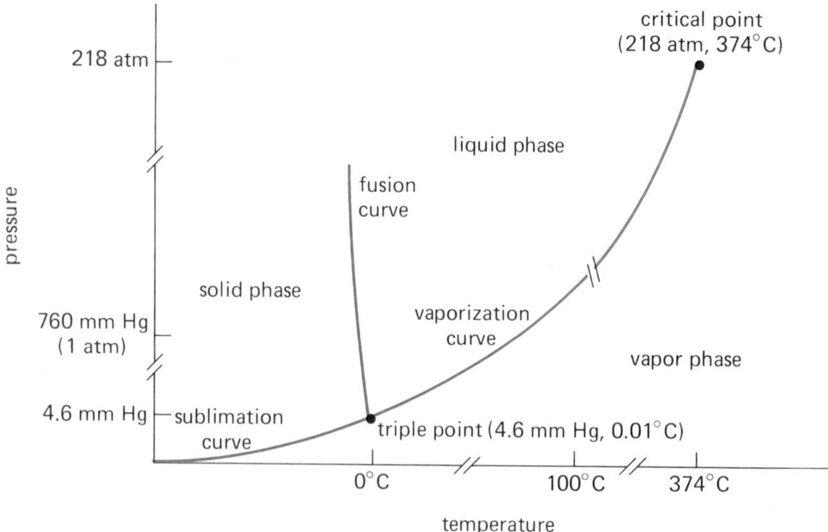

Figure 18.18
Phase changes. See Question 9.

16. Is heat energy generally becoming less available? Explain.
17. Suppose a temperature below absolute zero could be achieved. What would be the ideal efficiency of a heat engine in this better-than-ideal (impossible) situation?
18. Some common sayings are "You can't get something for nothing," "You can't even break even," and "I'll never sink that low." Are these expressions in any way descriptive of the laws of thermodynamics? Explain.

Heat Engines

19. What is the difference between the thermal efficiency and the Carnot efficiency of a heat engine?
20. According to the Carnot cycle, how can the efficiency of a heat engine be increased?
21. Draw a representative diagram of a heat engine with 100 percent efficiency.
22. In Figure 18.11, the power output of the engine can be seen to be 8 hp. What would be the engine rating in the SI system?
23. An automobile crankshaft is shown in Figure 18.19. Explain the purpose of the somewhat eccentric shape. What type of multicylinder is this crankshaft from? The camshaft is also shown. As it rotates, the cam lobes lift the valves. Why is there one lobe for each valve? Compare with Figure 18.11.
24. Why is the number of power outputs per revolution of the output shaft of a single-rotor Wankel

Figure 18.19
Pistons and cam. See Question 23.

engine equivalent to that of a six-cylinder piston engine?

Heat Pumps and Refrigerators

25. The heat out of a heat pump is greater than the heat in. Does this violate the first law?
26. (a) Would leaving the refrigerator door open be a practical way to air-condition the kitchen? Explain. (b) Is it energy-efficient to put hot leftovers in a refrigerator? Explain.
27. Prior to refrigerators, iceboxes with blocks of ice in a top compartment were used to cool foods and keep them from perishing. Is the old icebox a heat pump? Why was the ice compartment at the top and not at the bottom?
28. On hot, humid days the coils on home and auto air conditioners sometimes frost and "freeze up." The conditioned air is then less cool. Why? Why does the air conditioner compressor get hot when this happens (sometimes to the point that it burns out)?
29. Sketch a design for a simple compressor—i.e., its components—and describe the principle of operation.
30. Design a heat pump that uses water as a heat transfer agent.

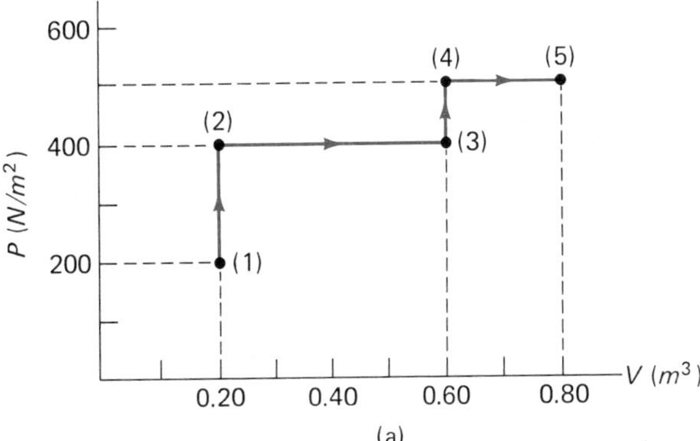

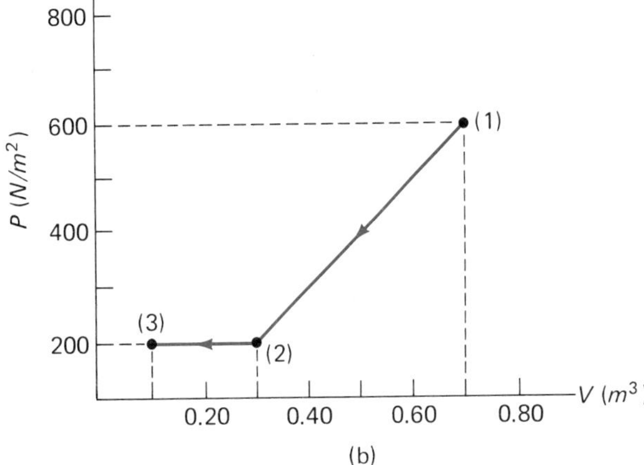

Figure 18.20
pV graphs. See Problem 4.

Problems

Levels of difficulty are indicated by asterisks for your convenience.

18.1 The First Law of Thermodynamics and Thermodynamic Processes

1. A system receives 100 cal of heat energy. If 25.0 cal go into internal energy, how many joules of mechanical work are done by the system?
2. On a p-T diagram, draw (a) an isobar, (b) an isotherm, and (c) an isomet.
3. On a V-T diagram, draw (a) an isotherm, (b) an isomet, and (c) an isobar.

*4. How much work is done in the ideal gas processes shown in the graphs in Figure 18.20? Identify the type of process for each path.

*5. A quantity of gas in a piston-cylinder arrangement is compressed isobarically at a pressure of 1 atm with a volume decrease of 0.0150 m^3. If the internal energy of the gas increases by 500 J, how much heat (in calories) is given up to the surroundings?

*6. A system of gas absorbs 10 Btu of heat in an isometric process as the pressure varies from 1.0 atm to 1.1 atm. How is the internal energy of the system affected by the process?

*7. In an isothermal process, a perfect gas does 250 J of work on its surroundings. (a) What is the change in the internal energy of the gas? (b) Does the system absorb or give up heat? How much?

*8. An ideal gas in a vertical piston-cylinder absorbs 0.50 Btu of heat during an isothermal expansion. If the piston and load combined weigh 1800 lb, how far will the piston be moved upward?

*9. In an adiabatic compression, 450 J of work is done on a gas system. (a) How much heat is given up by the system? (b) How is the internal energy of the gas affected? (c) What is the change in entropy of the system?

**10. A system of ideal gas with an internal energy of 147 J is at a temperature of 200 K. If 90.0 cal of heat is added to the system and 20.0 cal of this goes into work done by the system, what is the final temperature of the gas?

**11. In an isobaric process, 25 kcal of heat is absorbed by a system of perfect gas. If the pressure of the gas is 2.0×10^5 N/m^2 and the volume increases by 0.20 m^3, what is the change in the internal energy of the gas?

18.2 The Second and Third Laws of Thermodynamics

*12. What is the change in entropy when 0.50 kg of water freezes to ice at 0°C? What is the change in entropy when the ice melts at this temperature?

*13. One hundred grams of water is converted to steam at 100°C. What is the change in entropy?

*14. What is the change in entropy when 50 g of ethyl alcohol vaporizes at its boiling point of 78°C? (L_v = 200 cal/g)

*15. Which process has the greater change in entropy—200 g of ice changing to water at 0°C, or 40.0 g of steam condensing to water at 100°C? Explain any negative changes of entropy.

*16. A system of perfect gas at STP undergoes an isothermal expansion and does 2500 J of work on its surroundings. What is the change in the entropy of the system?

*17. A thermodynamic system undergoes an isothermal process at 27°C with an entropy change of 50 J/K while 3.0×10^3 J of work is done on the system. What is the change in the internal energy?

**18. (a) The variables of temperature and entropy may be plotted on a *T-S* diagram. An *isentropic* process is one in which the entropy of the system remains constant. Plot an isentropic process on a *T-S* diagram. What type of process is this in terms of the processes studied in the text? (b) Plot an isothermal process on a *T-S* diagram and show that the area under the curve is equal to the heat added to or removed from the system.

18.3 Heat Engines

*19. A heat engine absorbs 500 cal of heat and rejects 325 cal while doing work in a cycle. (a) How many joules of work are done? (b) What is the thermal efficiency of the engine?

*20. The thermal efficiency of a particular heat engine is 35 percent. If the engine absorbs 600 Btu of heat, (a) how much heat is rejected per cycle? (b) How much work is done?

*21. For a particular auto with a four-cylinder engine, the tachometer indicates that the engine is turning at a rate of 3000 rpm at a normal speed. (a) How many power strokes or outputs does the engine have each minute? (b) Would the tach rate be the same for a six- or eight-cylinder engine for the same power output? Explain.

*22. A heat engine has a thermal efficiency of 20 percent. If the heat input is supplied to the engine by the condensation of 2.0 lb of steam at 212°F each cycle, (a) what is the work output? (b) How much heat is rejected to the surroundings?

*23. The working gas of a heat engine absorbs 8.0 Btu of heat each cycle. If the gas expands 0.30 ft^3 isobarically at 1 atm pressure, what is the efficiency of the engine? (*Hint:* be careful of units.)

*24. A steam engine does 1.6×10^4 J of useful work each cycle, but loses 2.3×10^4 J to friction and exhausts 5.0 kcal of heat. What are the engine's (a) heat input and (b) efficiency?

*25. A heat engine operates with 25 percent efficiency. If the power output of the engine is rated at 2.0 hp, how much energy per hour must be supplied to the engine?

*26. Assuming the engine in Problem 19 goes through a Carnot cycle, what is the Carnot efficiency?

*27. An engineer designs a heat engine with an efficiency of 40 percent that rejects heat to a low-temperature reservoir at room temperature. Above what temperature, in degrees Celsius, must the high-temperature reservoir be?

*28. An ideal heat engine has an efficiency of 33 percent. How many times hotter is the high-temperature reservoir than the low-temperature reservoir?

*29. A Carnot engine with an efficiency of 40 percent takes in the working substance at 200°C. At what temperature, in degrees Celsius, is it exhausted?

*30. Which has the greater theoretical efficiency, a heat engine operating between heat reservoirs at 300°C and 100°C, or one operating between 300 K and 100 K? What is the efficiency in each case?

*31. A steam engine takes in superheated steam at 300°F and exhausts it at 212°F. What is the Carnot efficiency?

*32. It is possible to use the heat of the ocean to supply the mechanical energy to turn a generator and produce electricity. A diagram of such a system is shown in Figure 18.21. In certain tropical areas, the surface layer in the ocean is about 25°C and the lower layers are about 5°C. Answer the following questions about the system: (a) Is the system a heat engine? (b) Why is ammonia used as a working fluid, and how is it used to produce mechanical energy? (c) What would be the ideal efficiency of the system? (d) Do you think ocean thermal energy conversion will ever be economically feasible?

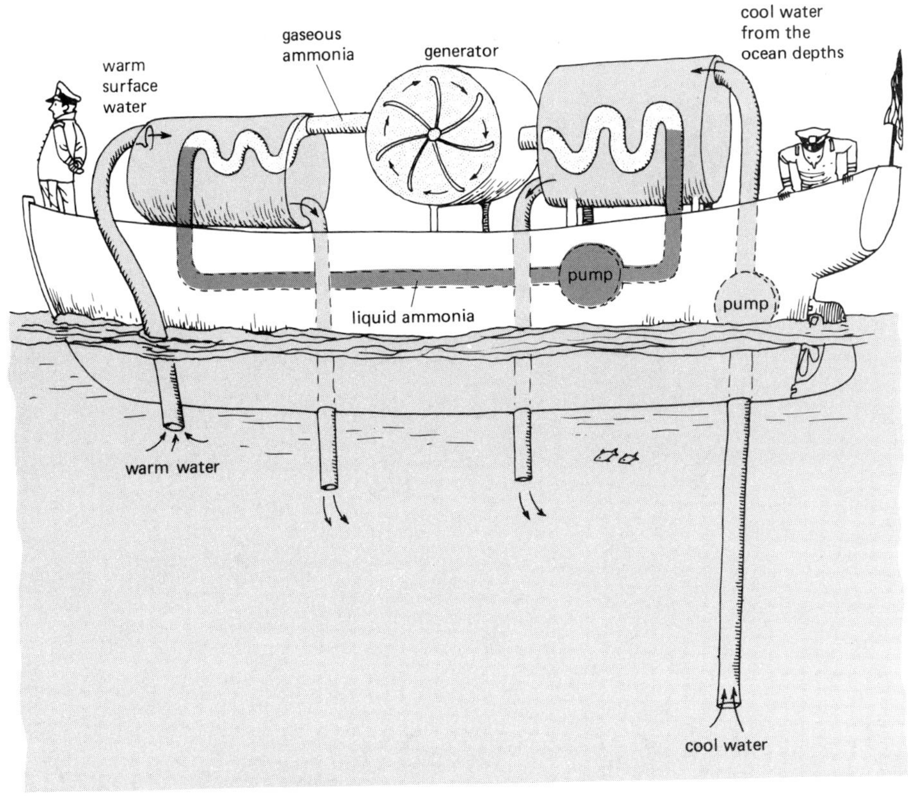

Figure 18.21
See Problem 32.

*33. An engineer wishes to design a heat engine that operates between reservoirs with temperatures of 200°C and 100°C with an efficiency of 30 percent. Is this possible?

**34. During a Carnot cycle, 300 cal of heat is absorbed by a gas while it undergoes an isothermal expansion at 177°C. If 200 cal are exhausted, (a) what is the temperature, in degrees Celsius, of the isothermal compression of the cycle? (b) What is the efficiency? (c) How much work, in joules, is done by the gas during the cycle?

18.4 Heat Pumps and Refrigerators

35. What is the coefficient of performance of a refrigerator that removes 150 cal of heat for a work input of 150 J?

*36. A refrigerator takes 1.8 Btu of heat from a low-temperature reservoir and exhausts 3.0 Btu to a high-temperature reservoir each cycle. (a) What is the cop? (b) What is the work input each cycle?

*37. A refrigerator with a cop of 4.0 removes 100 cal from its storage area each cycle. (a) How much heat is exhausted each cycle? (b) What is the work input in joules?

*38. The working substance of an ideal refrigerator removes heat from a compartment at 0°F and rejects it at room temperature. What is the coefficient of performance of the refrigerator?

*39. A commercial refrigeration system removes 360,000 Btu of heat in 4.0 h. What is the rating of the system in tons of refrigeration?

*40. An ice plant refrigeration unit is rated at 25 tons of refrigeration. How much water at 0°F could the system freeze in one hour? (Express the amount in weight tons.)

*41. A refrigeration system that removes 8000 Btu/h of heat has a work input of 2.0 hp. (a) What is the horsepower input per ton of refrigeration? (b) Express the cop in terms of hp/N (horsepower per ton). What is the cop?

**42. A refrigerator has a cop of 1.5. What is the horsepower input required for the refrigerator to remove 12,000 Btu in 30 minutes?

**43. (a) Show that the ideal cop for normal refrigerator operating temperatures is greater than 1. (b) The actual or thermal cop of a refrigerator is about 2.5. What does this tell you about the amount of heat pumped per work input? (c) On moderately cold days, a heat pump used to heat a house has a thermal cop of about 3.0. If the work input is 4000 J, how much heat would be delivered to the home? In terms of energy, how much of this would be cost-free? (d) How does the cop of a heat pump used for home heating vary with outside temperature?

(© Gary Ladd)

Chapter 19

Electrostatics

The study of electricity begins with electrostatics. As the name implies, electrostatics deals with static situations—the statics of electric charges. As in mechanical statics in Chapter 3, there is no charge motion, but there is a force involved. It is the electrical force, which is the basis of electrical phenomena, or electricity.

In this chapter you will find that many of the principles studied previously apply to electricity: for example, force, field, work, and energy. Electricity is an integral part of our society, with many technical applications. Therefore, a basic understanding of electricity is highly important.

19.1 Electric Charge and Force

Electric charge is the basis of electrical phenomena. But what is electric charge? Basically, we don't know what electric charge is, only what it does in terms of fundamental force interactions. Along with mass, length, and time, **electric charge is a fundamental property.**

Electric charge is involved in the fundamental interaction called electric force, just as mass is involved in the fundamental gravitational interaction (Chapter 7). But on a comparative scale the electric force is billions upon billions, even trillions upon trillions, of times stronger than gravity.

It has been found that electric charge is associated with certain atomic particles. There are two types of charges, arbitrarily characterized as positive (+) and negative (−). The smallest unit of charge is associated with electrons and protons,* the particles that make up atoms.

A simplified model of the atom pictures the protons to be at the center of the atom as a "nu-

* There is experimental evidence for the existence of subatomic particles with a fractional electronic charge, called *quarks*. A quark can have a charge that is 1/3 or 2/3 of the electron's charge, but quarks are not believed to exist outside the nucleus in a free state.

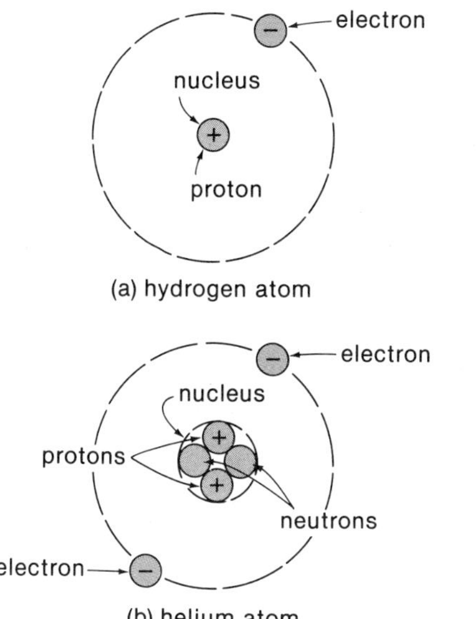

Figure 19.1
Atomic models of (a) hydrogen and (b) helium. The negatively charged electrons of an atom orbit about a nucleus containing positively charged proton(s) and neutral neutrons (for other than the hydrogen atom shown).

cleus" and the electrons to be orbiting about the nucleus (Fig. 19.1). In this so-called "solar system" model, the electrical force of attraction between an electron and the nuclear protons supplies the centripetal force necessary to keep the electron in orbit, as gravity does between the Sun and a planet. (The nuclei of atoms other than ordinary hydrogen atoms also contain neutral particles, called neutrons. They are discussed in Chapter 31.)

Table 19.1
Basic Electric Charges

Electron	$e^- = -1.6 \times 10^{-19}$ C
	$(m_e = 9.11 \times 10^{-31}$ kg)
Proton	$p^+ = +1.6 \times 10^{-19}$ C
	$(m_p = 1.67 \times 10^{-27}$ kg)

The standard unit of charge in the SI system is the coulomb (C), named in honor of Charles A. de Coulomb (1736–1806), a French physicist, for his work in electrostatics. The magnitudes of the charges of the electron and proton are given in Table 19.1. As can be seen, the charges are equal with opposite polarity (+ and −).

The electronic charge is quite small. In actual electrostatic situations, we work with larger charges (given the symbol q), which are an accumulation of electronic charges. This is really a "net" charge. Equal numbers of protons and electrons cancel each other electrically, so an object with equal numbers of electrons and protons is electrically neutral.

If an electron is removed from a neutral atom, the atom has a net charge of $q = +1.6 \times 10^{-19}$ C because of the "uncancelled" proton. Such an electrically charged atom or molecule is called an *ion* (Fig. 19.2).

An object has a net charge q because it has an excess n of electrons or protons. This may be represented as

$$q = ne \qquad \textbf{(Eq. 19.1)}$$

where e is the electronic charge (a positive charge in the absence of electrons).

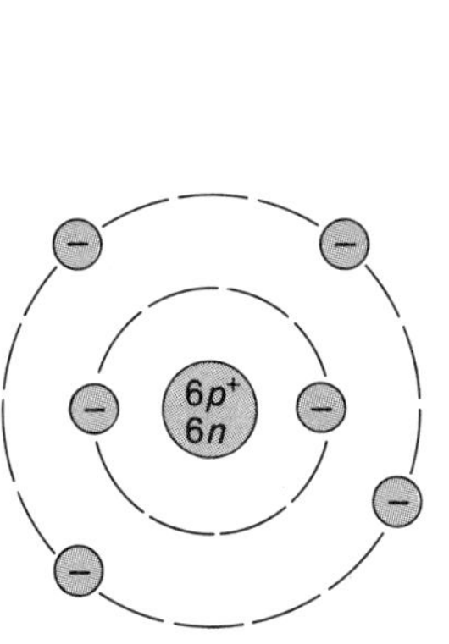

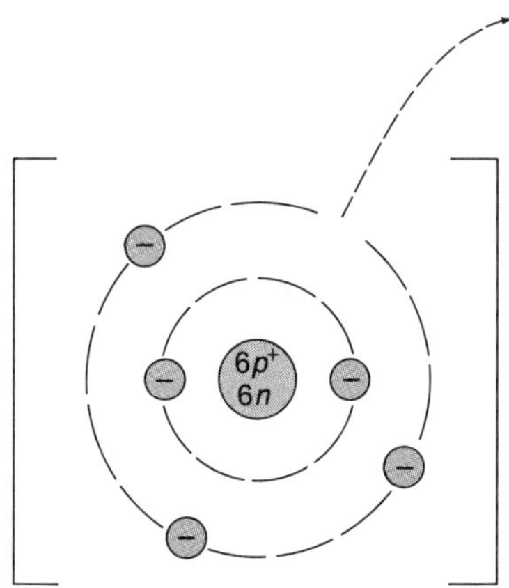

Figure 19.2
Ionization. (a) A neutral atom contains an equal number of electrons and protons. (An atom with six protons is a carbon atom.) (b) If an electron is removed, the atom becomes a positively charged ion, since there is one more positive charge than there are negative charges.

EXAMPLE 19.1 A piece of cloth has a net charge of −0.50 C on a dry day because of "static" electricity. How many excess charges does the cloth have?

Solution: It is given that $q = -0.50$ C; hence, the excess charges are electrons. Then

$$q = ne$$

or

$$n = \frac{q}{e} = \frac{-0.50\text{ C}}{-1.6 \times 10^{-19}\text{ C}}$$
$$= 3.1 \times 10^{18} \text{ electrons}$$

Charges are divided into two types because of two types of mutual interactions—attractive and repulsive. This property is summed up in the **law of charges:**

> **Like charges repel, and unlike charges attract.**

As illustrated in Figure 19.3, two positive charges or two negative charges experience mutually repulsive forces (Newton's third law), and a positive charge and a negative charge experience mutually attractive forces.

Since charges experience forces, it might be expected that they would move in fixed objects. This depends on the particular material. Just as there are thermal conductors and insulators, there are also electrical conductors and insulators. Recall from Chapter 17 that metals are good thermal conductors because some of the electrons in a metal are "free" to move around (not permanently bound to a particular molecule or atom). Such electron mobility in materials also makes them good electrical **conductors.**

> **A conductor is a material in which electrons are relatively free to move.**

In other materials, such as wood or plastic, there is little electron mobility. These materials are called **insulators** (poor conductors).

> **An insulator is a material in which there is relatively little electron mobility.**

A material with electron mobility between that of a good conductor and that of a poor conductor (an insulator) is called a **semiconductor.**

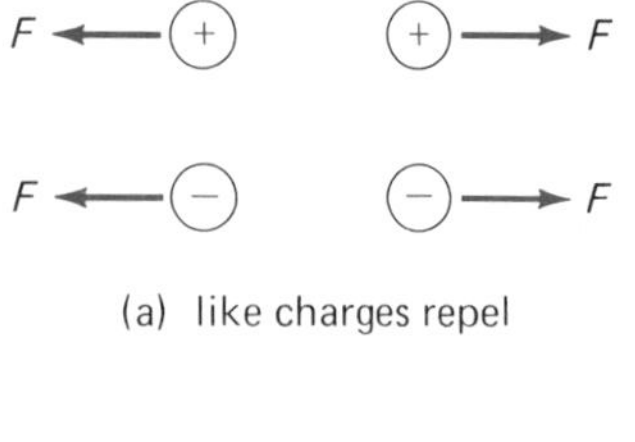

Figure 19.3
The law of charges. Like charges repel (a) and unlike charges attract (b).

Coulomb's Law

The magnitude of the electrostatic force F between two charges q_1 and q_2 is given by an expression known as **Coulomb's law:**

$$\boxed{F = \frac{kq_1q_2}{r^2}} \qquad \textbf{(Eq. 19.2)}$$

where r is the distance between the charges and k is a constant. In the SI system,

$$k = 9.0 \times 10^9 \text{ N-m}^2/\text{C}^2$$

The constant k is sometimes written $k = \frac{1}{4}\pi\varepsilon_o$, where ε_o is called the permittivity of free space (vacuum) and $\varepsilon_o = 8.85 \times 10^{-12}\text{ C}^2/\text{N-m}^2$.

The direction of the force is given by the law of charges. As illustrated in Figure 19.4, the mutual forces on two charges are equal and opposite (Newton's third law). If there are more than two charges, the net force on a particular charge

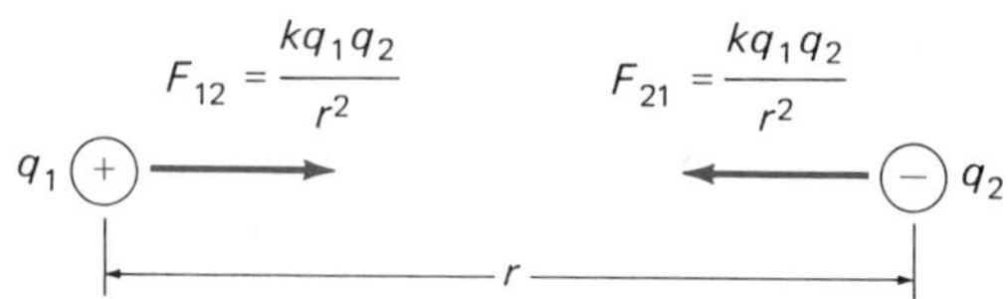

Figure 19.4
Coulomb's law. The magnitude of the electrostatic force on one charge due to another is given by Coulomb's law. The mutual forces on the charges are equal and opposite, with the direction of the forces being given by the law of charges.

is the vector sum of the forces acting on the charge.

EXAMPLE 19.2 Two point charges of 2.0×10^{-6} C and 8.0×10^{-6} C, respectively (often written 2.0 μC and 8.0 μC [μC, micro-coulomb; 1 μC = 10^{-6} C]), are 0.30 m apart. What is the force on the 2.0-μC charge?

Solution: It is given that $q_1 = 2.0 \times 10^{-6}$ C, $q_2 = 8.0 \times 10^{-6}$ C, and $r = 0.30$ m. Then

$$F = \frac{kq_1q_2}{r^2}$$

$$= \frac{(9.0 \times 10^9 \text{ N-m}^2/\text{C}^2)(2.0 \times 10^{-6} \text{ C})(8.0 \times 10^{-6} \text{ C})}{(0.30 \text{ m})^2}$$

$$= 1.6 \text{ N, away from the other charge}$$

What are the magnitude and direction of the force on the other charge?

EXAMPLE 19.3 On a relative scale, how much stronger is the electrical force than the gravitational force?

Solution: To get a comparison, consider the situation of the simple (hydrogen) atom, as illustrated in Figure 19.1. Then the electrical and gravitational forces are

$$F_e = \frac{kp^+e^-}{r^2} \quad \text{and} \quad F_g = \frac{Gm_pm_e}{r^2}$$

where the m's are the masses of the particles and r is their separation distance. When we form a ratio, the r's cancel, and

$$\frac{F_e}{F_g} = \frac{kp^+e^-}{Gm_pm_e}$$

Putting in the values of k and G and the magnitudes of the charges and masses from Table 19.1, we find that

$$\frac{F_e}{F_g} \approx 10^{40}$$

or

$$F_e \approx 10^{40} F_g$$

This means that the electrical force (F_e) is 10,000 trillion, trillion, trillion times the gravitational force (F_g)!

19.2 Electrostatic Charging

Electrostatic effects are readily observable. Have you ever shuffled across a carpeted floor and been zapped by a spark when you reached for the doorknob? This happens because you are electrically charged and there is enough electric force to cause the air molecules to be ionized. There is then a charge flow, giving rise to a spark.

The outer electrons of the atoms in some materials are loosely bound and can be freed and transferred to another object. For example, when a hard rubber rod is rubbed with a piece of fur, electrons are transferred to the rod from the fur. With an excess of electrons, the rod is negatively charged. (How about the fur?)

Similarly, when a glass rod is rubbed with silk, the glass gives up electrons to the silk and becomes positively charged (deficiency of electrons). This is called *charging by friction.* In the same manner, you become charged when you shuffle across a carpet.

Since rubber and glass (and you) are poor conductors, the charge stays on the rods for some time before leaking off, if the air is dry. In moist air an invisible film of water condenses on objects, which makes the surface conductive. Charges quickly leak off and there is no charge build-up.

The electric force and the polarity of charges can be demonstrated with charged rods using an electroscope (Fig. 19.5). This consists of a metal insulated bulb connected to metal foil leaves. For example, if you bring a negatively charged rod near the metal bulb of the electroscope, electrons in the bulb are repelled by the electric force and law of charges. Since the metal is a conductor, the electrons move as far away as possible—to the metal foil leaves. The leaves, each of which now has an excess of negative charge, are repelled apart or diverge. When the rod is removed, the electrons redistribute over the conductor, and the leaves come together.

If a positively charged rod is brought near the bulb, a similar effect occurs. In this case the leaves are positively charged. Why? The electroscope tells us that the rods are charged, but not how they are charged (positively or negatively). However, this can be determined by charging the electroscope with a known type of

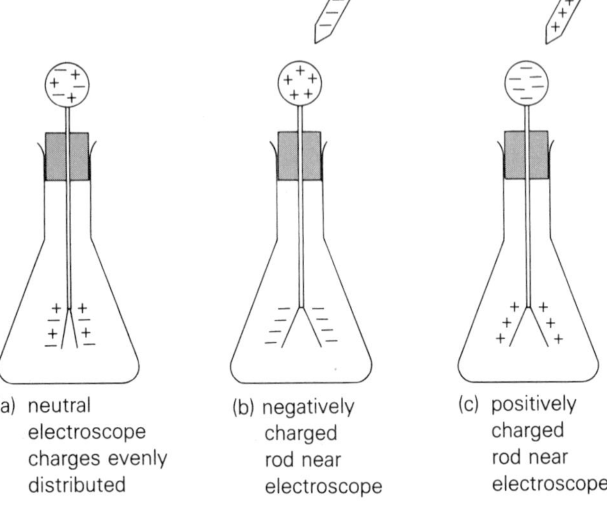

Figure 19.5

The electroscope. A common type of electroscope consists of an insulated metal ball connected to metal foil "leaves." The electroscope can be used to demonstrate a net charge on an object and the electrical force. See text for description.

charge. For example, if the bulb were touched with a negatively charged rod, some electrons would be transferred to the bulb, the electroscope would be negatively charged, and the leaves would diverge. Why? This is called *charging by contact.*

A positively charged rod brought near the bulb would then cause the leaves to collapse, and a negatively charged rod would cause the leaves to further diverge. Why? (Think of electron flow and the effect of the electric force.)

An object can be charged without friction or contact of a charged rod, by bringing a charged rod close by. For example, suppose a charged rod is brought near an electroscope bulb, as shown in Figure 19.6. Touching the bulb with the finger "grounds" the electroscope.* This provides a path for electrons to get farther away from the charged rod, and some of them are conducted to ground.

When the finger is removed, the electroscope is left with a deficiency of electrons and it is positively charged. How could you test to see if it were positively charged?

Charging by induction need not involve a removal of charge. An object can have regions of charge. For example, if a charged rod is brought near an insulated metal ball, as shown in Figure 19.6(b), there are a separation of charge and a net electrical force, even though the net charge of the ball as a whole is zero. Induction was also used in "charging" the electroscope in Figure 19.5. The induction results in a **polarization** or separation of charge. All materials, including insulators, can be polarized by induction to some degree. Here there is a polarization of electric charges in molecules. A molecule with a separation of charge is called an electric dipole. Some materials have permanent dipoles, and molecular dipoles can be induced in others.

Some common examples of combinations of charging by friction and induction are shown in Figure 19.7. When you rub an inflated balloon on your hair (or sweater), it will stick to a wall or

* *Electrical ground* refers to earth (ground) or some other large conductor that can receive or supply electrons without significantly affecting its own electrical condition. An electrical ground is analogous to a thermal reservoir in thermodynamics. By another analogy, this is like adding to or taking a cup of water from the ocean.

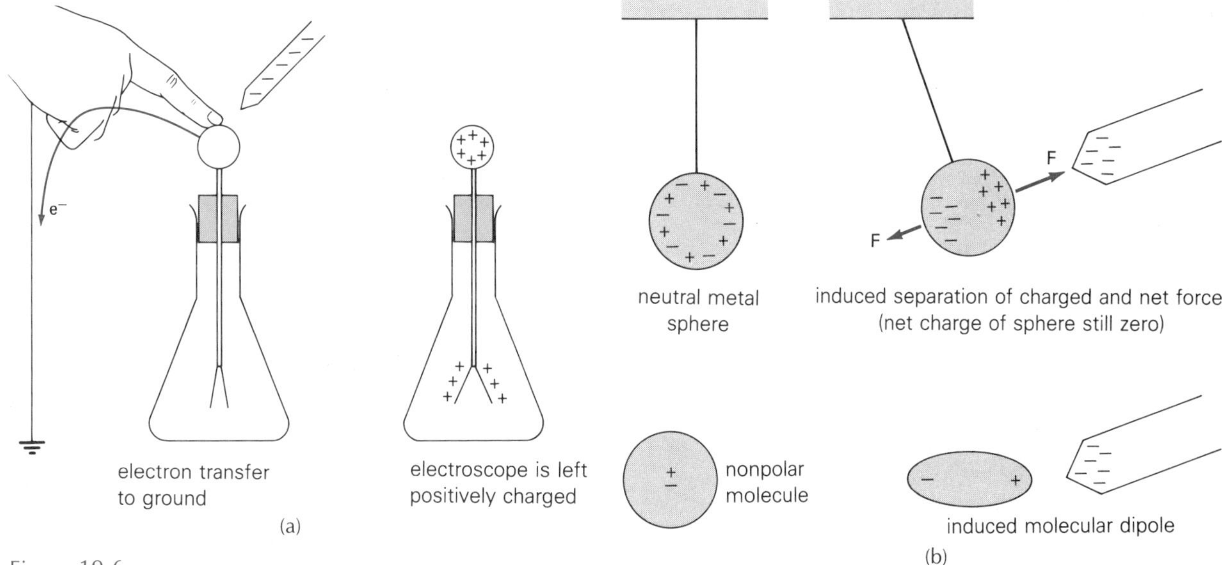

Figure 19.6
Charging by induction. (a) The electroscope is left with a net charge. (b) Induction causes regions of charge. See text for description.

ceiling. The balloon is charged by friction, and the charged balloon induces an opposite charge on the wall, which provides the attractive electrical force. Similarly, a hard rubber comb attracts bits of paper after being run through the hair.

"Static electricity" can be undesirable as well as practically useful. You may have noticed a chain or conductive strip dangling to the ground from a gasoline truck. This is intended to prevent charge build-up and sparking.* Similarly, medical personnel in operating rooms

* Tires in contact with the road develop a negative charge. By induction, parts of the metal body of the vehicle near the tires are positively charged, and sparking could occur between the truck and a nearby grounded object or oppositely charged body. A chain to ground would conduct some electrons from the truck's body, but this would leave it positively charged and still prone to sparking.

(a)

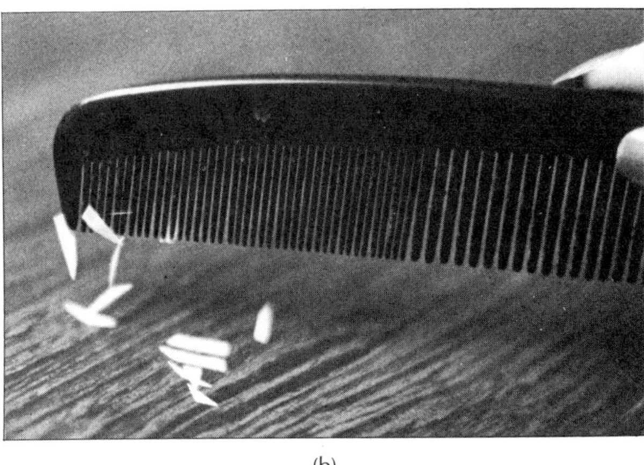

(b)

Figure 19.7
Examples of electrostatic charging and induction. (a) Balloons stick to walls after being charged by friction. (b) Bits of paper are picked up by a charged rubber comb.

(a)

(b)

Figure 19.8
Static cling. Products to prevent undesirable electrostatic effects.

wear shoe coverings with conducting strips to prevent electrostatic build-up and sparking in the presence of flammable gases. You can now buy products to prevent "static cling" in clothes (Fig. 19.8). The electrostatic precipitator discussed later in Special Feature 19.1 is an example of a practical use of electrostatics. You'll find another example, the electrostatic copier, in Chapter 26.

19.3 Electric Field

A convenient quantity in electrostatics is the electric field.

> **The electric field *E* is defined as the force per unit charge.**

The electric field at a particular point in space due to a charge q is determined by placing (or imagining) a positive test charge q_o at the point. A positive charge is used by convention. Then, the magnitude of the electric field is given by

$$E = \frac{F}{q_o} = \frac{kq_oq}{q_or^2} = \frac{kq}{r^2}$$

or

$$\boxed{E = \frac{F}{q_o} = \frac{kq}{r^2}} \qquad \textbf{(Eq. 19.3)}$$

The units of E can be seen to be newtons/coulombs (N/C).

The direction of the electric field is the direction in which a positive charge would experience a force (positive test charge). For example, the E field vectors around positive and negative charges are illustrated in Figure 19.9. It is common to connect the vector arrows to form field lines, or **lines of force.**

The electric field is a force "mapping" of space due to a charge configuration. It is analogous to knowing the price per pound of an article. Given the price per pound and how many pounds are desired, the price is easily determined. Similarly, given the electric field (N/C) and the magnitude of the charge (C), the force acting on the charge is easily found. A positive charge would experience a force (and move if released) in the direction of and along a field line. The force on a negative charge would be in the opposite direction.

EXAMPLE 19.4 (a) What is the electric field at the origin for the charge configuration shown in Figure 19.10? (b) What force would be experienced by a charge of $-6.0\ \mu\text{C}$ placed at the origin?

Solution: (a) It is given that $q_1 = +4.0\ \mu\text{C} = +4.0 \times 10^{-6}$ C, $q_2 = -4.0\ \mu\text{C} = -4.0 \times 10^{-6}$ C,

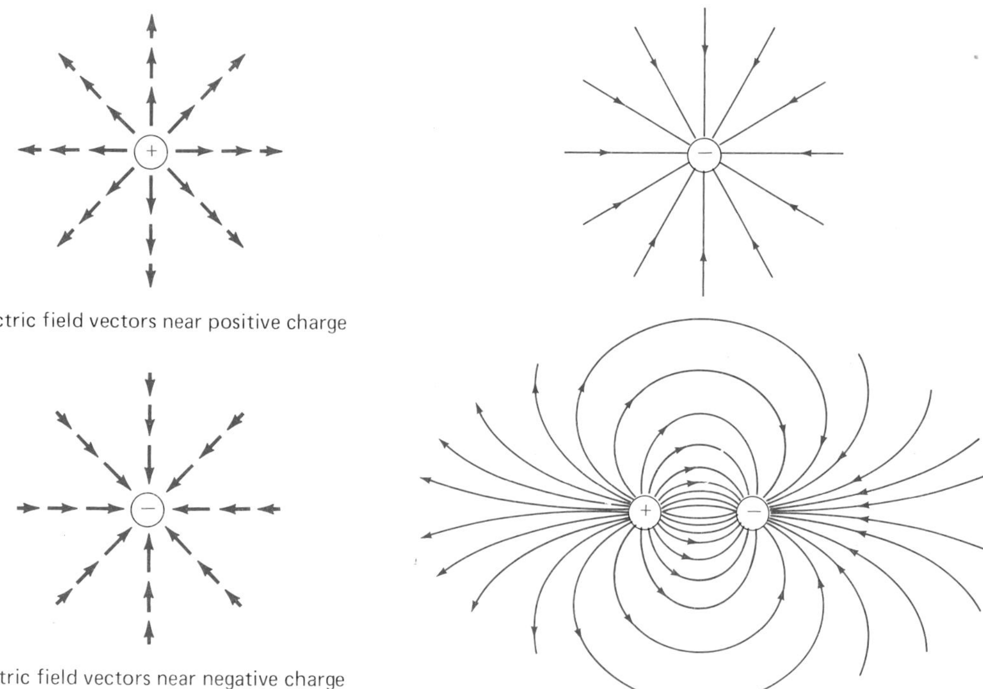

Figure 19.9
Electric field and lines of force. (a) The electric field is determined at points in space using a positive test charge, giving the force per unit charge (N/C) at different points as indicated by the vector arrows. (b) The vectors are connected to form lines of force which show the path and direction of a positive charge released in the field.

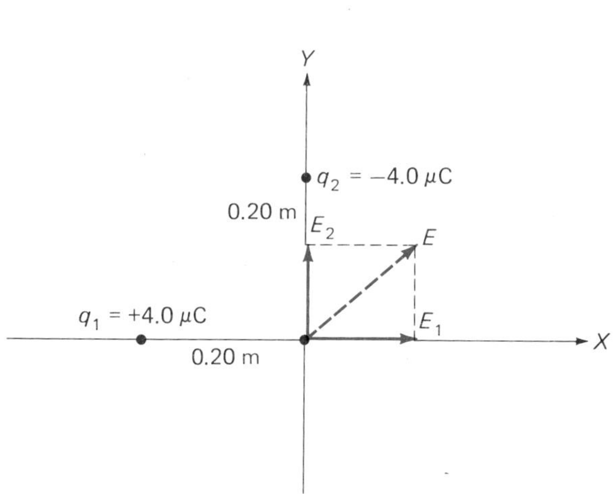

Figure 19.10
See Example 19.4.

and $r_1 = r_2 = 0.20$ m. Since the quantities are identical in magnitude,

$$E_1 = E_2 = \frac{kq}{r^2}$$

$$= \frac{(9.0 \times 10^9 \text{ N-m}^2/\text{C}^2)(4.0 \times 10^{-6} \text{ C})}{(0.20 \text{ m})^2}$$

$$= 9.0 \times 10^5 \text{ N/C}$$

and E_1 is in the x direction and E_2 in the y direction (as determined by a positive test charge). Then, the vector sum or resultant has a magnitude of

$$E = (E_1^2 + E_2^2)^{1/2} = (2E_1^2)^{1/2}$$

$$= [2(9.0 \times 10^5 \text{ N/C})^2]^{1/2} = 1.3 \times 10^6 \text{ N/C}$$

at an angle of

$$\theta = \tan^{-1}\left(\frac{E_2}{E_1}\right) = \tan^{-1}(1) = 45°$$

(b) If a charge $q = -6.0\ \mu\text{C} = -6.0 \times 10^{-6}$ C is placed at the origin, it experiences a force of

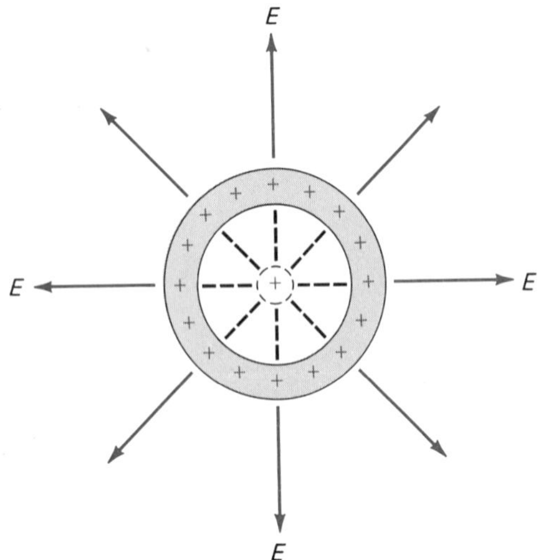

Figure 19.11
Charged spherical conductor. All the charges on the conductor reside on the outer surface. The electric field outside the sphere (hollow or solid) is the same as if all the charges were concentrated at its center.

$$F = qE = (6.0 \times 10^{-6}\text{ C})(1.3 \times 10^{6}\text{ N/C})$$
$$= 7.8\text{ N}$$

in the direction opposite the E field.

The electric field of a charged, isolated conductor is a bit more difficult to determine. One simple case is a charged spherical conductor (Fig. 19.11). The electric field outside the conductor is the same as if all the charge on the surface of the sphere were concentrated at its center (similar to the center of mass).

Since the excess charges of the conductor are mutually repelled, they tend to get as far away from each other as possible. Hence,

> **The excess charge on a conductor resides on the outer surface of the conductor.**

This is true for any isolated conductor. The electric field inside a hollow or solid charged conducting sphere, or any charged conductor, is zero, since there is no excess charge within the conductor.

The electric field of a charged, irregularly shaped conductor varies, depending on the shape of a particular portion of the conductor. In talking about the electric field of a charged conductor, it is convenient to express the E field in terms of the surface charge density ($\sigma = q/A$) and to use permittivity (ε_o) notation. The electric field at the outside surface of a charged sphere with $A = 4\pi r^2$ is

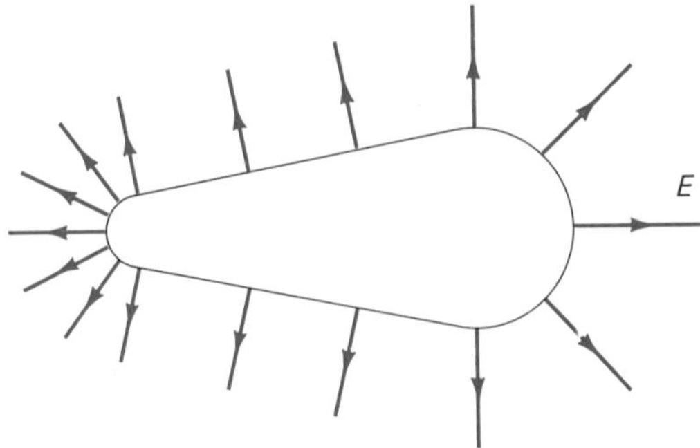

Figure 19.12
A charged irregularly shaped conductor. The electric field is greater near sharp surfaces.

$$E = \frac{kq}{r^2} = \frac{q}{4\pi\varepsilon_o r^2} = \frac{q}{\varepsilon_o(4\pi r^2)}$$
$$= \frac{q}{\varepsilon_o A} = \frac{\sigma}{\varepsilon_o}$$

and

$$\boxed{E = \frac{\sigma}{\varepsilon_o}} \qquad \textbf{(Eq. 19.4)}$$

where E is perpendicular to the surface. The electric field near *any* charged conducting surface is proportional to the surface charge density.

The electric field of an irregularly shaped conductor is greater near sharp surfaces (Fig. 19.12). If the electric field becomes large enough, the molecules of the air are ionized, and a spark discharge occurs. This is called a *corona discharge* and is most likely to happen between two conductors with sharp edges in close proximity. (See Special Feature 19.1.)

19.4 Electric Potential Energy and Electric Potential

Electric Potential Energy

With an electric force between two charges, there is also a mutual potential energy of the charges. This **electric potential energy U** is given by the relationship

$$\boxed{U = \frac{kq_1q_2}{r}} \qquad \textbf{(Eq. 19.5)}$$

The Electrostatic Precipitator

A practical application of electric field and electric force is the Cottrell or electrostatic precipitator, which is used to remove particulate matter from flue gases. The developmental work on it was done by F. G. Cottrell, an American physical chemist, in the early 1900's. The precipitator was used in the smelting industry in 1912.

The basic principle of an electrostatic precipitator is illustrated in Figure 1. The positively charged plates are called discharge plates, and the negatively charged plates are called collection plates. As a result of a large electric field between the plates, a corona discharge occurs around the needle projections on the positive plates. (Some precipitator designs use wires as positive discharge electrodes.)

The particulate matter, such as soot, in flue gases passing through the precipitator chamber is ionized. The positively charged particles then move toward and collect on the negative collection plates. The particulate matter accumulates and falls to the bottom of the chamber, where it is removed.

Electrostatic precipitators are capable of removing more than 90 percent of the particulate matter from flue gases (Fig. 2). They are commonly used in fossil-fuel electrical generating plants and industrial operations in which there is a large amount of particulate matter in the flue gases resulting from incomplete combustion and impurities.

Figure 2
Electrostatic precipitation in action. The two smokestacks on the left are equipped with precipitators; the two on the right are not. (Courtesy P. Highsmith.)

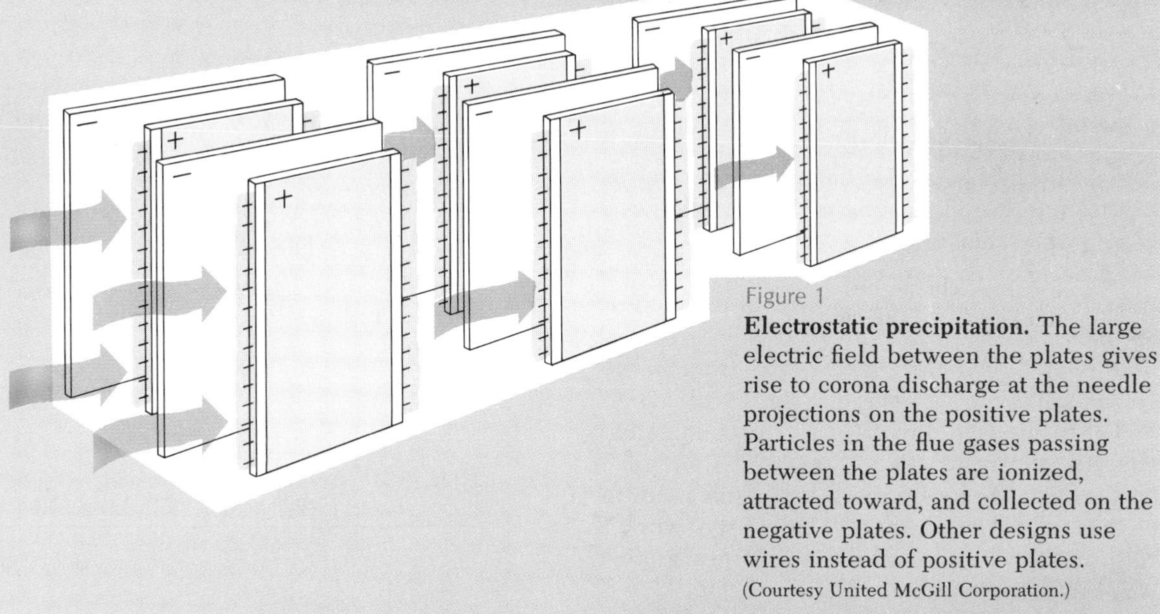

Figure 1
Electrostatic precipitation. The large electric field between the plates gives rise to corona discharge at the needle projections on the positive plates. Particles in the flue gases passing between the plates are ionized, attracted toward, and collected on the negative plates. Other designs use wires instead of positive plates. (Courtesy United McGill Corporation.)

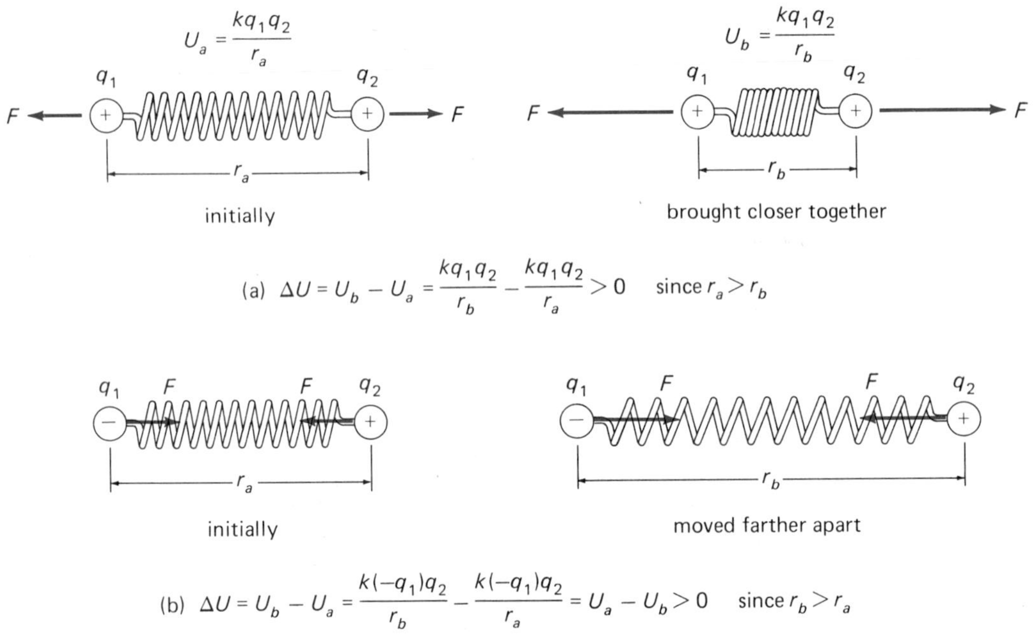

$U_a = \frac{kq_1q_2}{r_a}$ initially

$U_b = \frac{kq_1q_2}{r_b}$ brought closer together

(a) $\Delta U = U_b - U_a = \frac{kq_1q_2}{r_b} - \frac{kq_1q_2}{r_a} > 0$ since $r_a > r_b$

initially

moved farther apart

(b) $\Delta U = U_b - U_a = \frac{k(-q_1)q_2}{r_b} - \frac{k(-q_1)q_2}{r_a} = U_a - U_b > 0$ since $r_b > r_a$

Figure 19.13
Electric potential energy. It is convenient to imagine the electrical force as arising from a spring between two charges. (a) When two like charges are brought closer together (spring compressed), work is done and there is an increase in the potential energy $\Delta U > 0$. (b) Similarly, if two unlike charges are moved farther apart, work is done and the potential energy increases.

where r is the separation distance between charges q_1 and q_2. Essentially, this is the energy gained or lost in bringing the charges together.

The potential energy may be positive or negative, depending on whether the force between the charges is repulsive or attractive. Using the signs of the charges mathematically indicates whether the potential energy is positive or negative, e.g., $k(+q_1)(+q_2)/r = +U$ and

$$k(-q_1)(+q_2)/r = -U$$

Since energy is a scalar quantity, **the total potential energy of a configuration of charges is simply the scalar sum, $U_1 + U_2 + U_3 + \cdots$.**

Of particular interest is the change in potential energy ΔU when one or both of the charges are moved. It is sometimes convenient to think of the electrical force as arising from a tiny spring between the charges (Fig. 19.13).

To bring two positive (or two negative) charges closer together (compression of a spring) would require work and result in an increase in potential energy. Similarly, work is required to separate two unlike charges, and the potential energy is increased.

How would the potential energy of two like charges be affected if the charges were moved farther apart? It should be apparent that this would result in a decrease in potential energy, i.e., $\Delta U < 0$, or negative.

EXAMPLE 19.5 Two charges, $q_1 = -2.0\ \mu C$ and $q_2 = +3.0\ \mu C$, are 0.20 m apart. What are (a) the mutual potential energy of the charges and (b) the change in potential energy if the separation distance is decreased to 0.15 m?

Solution: It is given that $q_1 = -2.0\ \mu C = -2.0 \times 10^{-6}$ C, $q_2 = +3.0\ \mu C = +3.0 \times 10^{-6}$ C, $r_a = 0.20$ m, and $r_b = 0.15$ m.

(a) Then

$$U_a = \frac{kq_1q_2}{r_a}$$

$$= \frac{(9.0 \times 10^9\ \text{N-m}^2/\text{C}^2)(-2.0 \times 10^{-6}\ \text{C})(3.0 \times 10^{-6}\ \text{C})}{0.20\ \text{m}}$$

$$= -0.27\ \text{J}$$

(b) $\Delta U = U_b - U_a = \dfrac{kq_1q_2}{r_b} - \dfrac{kq_1q_2}{r_a}$

$$= \frac{(9.0 \times 10^9 \text{ N-m}^2/\text{C}^2)(-2.0 \times 10^{-6} \text{ C})(3.0 \times 10^{-6} \text{ C})}{0.15 \text{ m}} - (-0.27 \text{ J})$$

$$= -0.36 \text{ J} + 0.27 \text{ J} = -0.09 \text{ J}$$

and the potential energy decreases.

Electric Potential

In electrical applications, a very important quantity is the electric potential, or energy per unit charge. The electric potential energy at any point due to a charge q is determined by using a positive test charge q_o (as in the case of the E field), i.e.,

$$U = \frac{kq_oq}{r}$$

Then the **electric potential (or voltage, V)** is defined

$$\boxed{V = \frac{U}{q_o} = \frac{kq}{r}} \qquad \textbf{(Eq. 19.6)}$$

$$\text{Electric potential} = \frac{\text{potential energy}}{\text{charge}}$$

and, as energy/charge, has units of J/C. However, the unit of joule/coulomb is given the name **volt (V)** in both the SI and British systems in honor of Alessandro Volta (1745–1827), an Italian scientist who invented one of the first practical batteries. As with electric potential energy, the sign of the electric potential is determined by the signs of the charges.

Since the electric potential energy and the work done in bringing charges together are equal in magnitude, an alternative form of the electric potential equation is

$$V = \frac{W}{q_o} \qquad \textbf{(Eq. 19.7)}$$

Of particular interest is the **potential** or **voltage difference between two points.** This is equal to the work done in moving a test charge between two points, i.e.,

$$\Delta V = V_b - V_a = \frac{W_{ab}}{q_o} \qquad \textbf{(Eq. 19.8)}$$

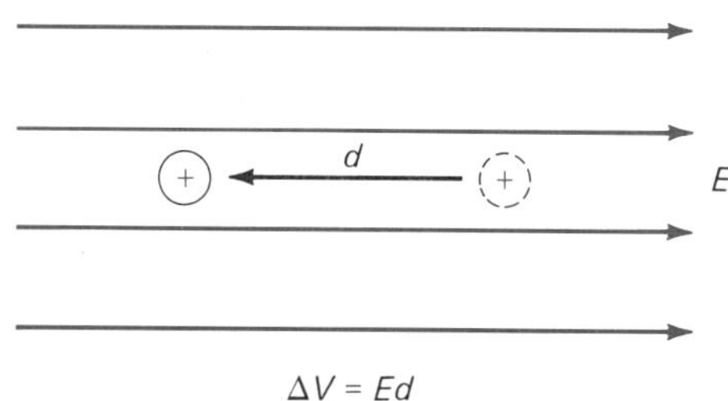

Figure 19.14
Electric potential. When a positive test charge is moved against a uniform electric field, work is done and the electric potential (work per charge) is increased.

The potential difference can be related to the electric field. Consider the case of a uniform electric field (Fig. 19.14). In moving a positive test charge from point a to point b against the field, the work done is

$$W_{ab} = Fd = q_oEd$$

since $E = F/q_o$. Then

$$\Delta V = \frac{W_{ab}}{q_o} = Ed$$

which is commonly written

$$V = Ed \qquad \textbf{(Eq. 19.9)}$$

with ΔV being understood, as in the case of x for Δx. Notice that the electric field $E = V/d$ has units of V/m, which are equivalent to N/C ($E = F/q_o$).

Since the work is positive (done against the field) in this case, the potential difference is positive. This means that the charge was moved to a higher potential, which is analogous to moving to a higher potential in a uniform gravitational field. For the general case,

$$\Delta V = -Ed \cos \theta \qquad \textbf{(Eq. 19.10)}$$

where θ is the angle between the electric field vector and the displacement ($\theta = 180°$ in Fig. 19.14).

If $\theta = 90°$ ($\cos 90° = 0$), then $\Delta V = W = 0$, and no work is done in moving a charge at right angles to the electric field lines. Such paths are called *equipotentials,* since the electric potential must be constant along the paths ($\Delta V = 0$; see Fig. 19.15).

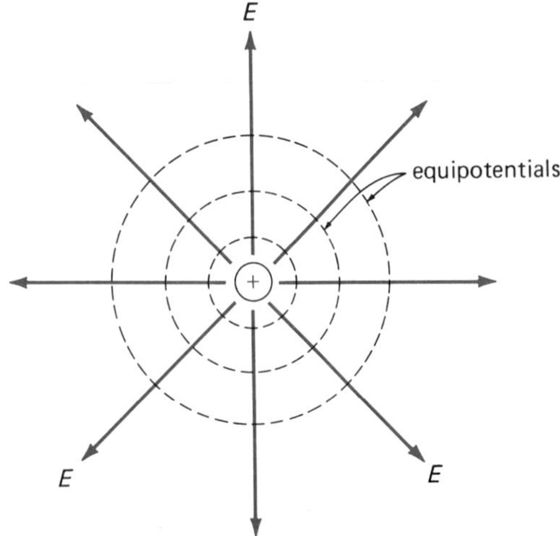

Figure 19.15
Equipotentials. When a charge is moved at right angles to the electric field, no work is done, since the potential is the same everywhere along the path, i.e., $\Delta V = W/q = 0$.

The Electron Volt

A common unit of energy used in modern physics is the electron volt (eV).

> **An electron volt is the energy acquired by an electron accelerating through a potential difference of one volt.**

In terms of the joule unit,

$$\Delta U = e\Delta V = (1.6 \times 10^{-19}\ \text{C})(1\ \text{volt})$$
$$= 1.6 \times 10^{-19}\ \text{J} = 1\ \text{eV}$$

and

$$\mathbf{1\ eV = 1.6 \times 10^{-19}\ J}$$

Larger eV units are obtained by using metric prefixes, e.g., keV (kilo-eV; 1 keV = 10^3 eV) and MeV (mega-eV; 1 MeV = 10^6 eV).

EXAMPLE 19.6 A doubly ionized particle is released in a uniform electric field of 1000 V/m. (a) What is the potential difference between the point of release and the location of the particle after it has traveled 0.25 m? (b) How is the energy of the particle affected? (c) If the particle has a mass of 3.0×10^{-26} kg, what is its acceleration?

Solution: It is given that $q = +2e = +2(1.6 \times 10^{-19}\ \text{C})$, since two electrons were removed from the particle, $E = 1000$ V/m $= 10^3$ V/m, and $m = 3.0 \times 10^{-26}$ kg.

(a) $\Delta V = Ed = (10^3\ \text{V/m})(0.25\ \text{m}) = 250\ \text{V}$

This is actually a negative voltage difference, i.e., -250 V, since the particle had a decrease in potential energy ($V = -Ed\cos\theta = -Ed\cos 0 = -Ed$).

(b) $$\Delta U = q\Delta V = 2(1.6 \times 10^{-19}\ \text{C})(-250\ \text{V})$$
$$= -8.0 \times 10^{-17}\ \text{J}$$

Remember, electric potential energy is charge times voltage. The decrease in potential indicates an equivalent increase in the kinetic energy of the particle. Since $\Delta U = q\Delta V = (1e)(1\ \text{V}) = 1$ eV for one electronic charge through a potential difference of one volt, we can write in this case

$$\Delta U = q\Delta V = (2e)(-250\ \text{V}) = -500\ \text{eV}$$
$$(= -8.0 \times 10^{-17}\ \text{J})$$

(c) $F = qE = ma$; hence,

$$a = \frac{qE}{m} = \frac{2(1.6 \times 10^{-19}\ \text{C})(10^3\ \text{V/m})}{3.0 \times 10^{-26}\ \text{kg}}$$
$$= 1.1 \times 10^{10}\ \text{m/s}^2$$

A way to electrostatically generate millions of volts is discussed in Special Feature 19.2 on page 408.

Important Terms

electric charge the fundamental property associated with electric force

coulomb (C) the unit of electric charge

law of charges like charges repel, and unlike charges attract

Coulomb's law the relationship that gives the electrostatic force between two charges

conductor a material in which electrons are relatively free to move

insulator a material in which there is relatively little electron mobility

semiconductor a material with intermediate electron mobility

electrostatic induction the redistribution of charge in an object due to the influence of a charged body nearby but not in contact

electrical ground the earth (ground) or some other large conductor that can receive or supply electrons without significantly affecting its own electrical condition

polarization a separation of charge so as to give regions of net charge

electric field the force field surrounding a charged body; electric field = force/charge, in the direction a positive charge would experience a force

electric potential energy the mutual potential energy of charges due to the electrical force

electrical potential (voltage) the potential energy per charge, commonly called voltage and measured in volts

electron volt (eV) a unit of energy; the energy acquired by an electron accelerating through a potential difference of one volt

Important Formulas

net charge: $q = ne$

$e = 1.6 \times 10^{-19}$ C

Coulomb's law: $F = \dfrac{kq_1q_2}{r^2}$

$k = 9.0 \times 10^9$ N-m²/C²

electric field: $E = \dfrac{F}{q_o} = \dfrac{kq}{r^2}$

(for positive test charge)

at the outside surface of a charge sphere $E = \dfrac{\sigma}{\varepsilon_o}$

$\sigma = \dfrac{q}{A}$

surface charge density

$\varepsilon_o = 8.85 \times 10^{-12}$ C²/N-m²

permittivity of free space

electric potential energy: $U = \dfrac{kq_1q_2}{r}$

electric potential (voltage): $V = \dfrac{U}{q_o} = \dfrac{kq}{r}$

$V = \dfrac{W}{q_o}$

potential or voltage difference $\Delta V = V_b - V_a = \dfrac{W_{ab}}{q_o}$

relation to electric field $\Delta V = -Ed \cos \theta$

constants and conversion factors:

$k = 9.0 \times 10^9$ N-m²/C² $= \dfrac{1}{4\pi\varepsilon_o}$

$\varepsilon_o = 8.85 \times 10^{-12}$ C²/N-m²

1 eV $= 1.6 \times 10^{-19}$ J

Questions

Electric Charge and Force

1. About how many electrons would it take to make up a charge of one coulomb? How many protons would it take?
2. Two negative charges of −1 C each are placed at opposite ends of a meterstick. (a) Could a free electron be placed somewhere on the meterstick so it would be in static equilibrium? (b) What would happen if an electron and/or a proton were placed on the meterstick other than at the 50-cm position?
3. A charge of +1 C is placed at one end of a meterstick and a charge of −1 C at the other end. Could a free electron be placed somewhere on the meterstick so it would be in static equilibrium? How about a proton?
4. The gravitational force is weaker than the electric force. It is easy to feel or experience the gravitational force—for example, when you pick up a heavy object. Electrons and protons are all around. Why don't we generally feel the electric force?
5. A large charge of $+Q$ and a small charge of $-q$ are a short distance apart. How do the electric forces on the two charges compare?
6. Is the Coulomb constant k a universal constant like G? Explain.
7. An electron is a certain distance from a positive charge. If the electron is moved twice as far away from the other charge, how is the electric force affected?
8. When thin plastic food wrap is pulled from its roll in a box and cut off, it often sticks to itself. Why?
9. Distinguish between electrical conductors and insulators. What is a semiconductor?

Electrostatic Charging

10. What occurs (a) when a positively charged insulator is brought near an electroscope bulb? (b) When the charged insulator touches the bulb? (c) Whan a negatively charged insulator is then brought near the bulb?
11. What is (electrical) ground?
12. Given an object and an electroscope, how can you tell if the object is charged? If it is, how can you tell what type of net charge it has?

SPECIAL FEATURE 19.2

The Van de Graaff Generator

An application that uses the principles discussed in this chapter is the Van de Graaff electrostatic generator, which was invented by the American physicist Robert Van de Graaff around 1930. It generates electrostatic potentials of *millions* of volts. The high voltages are used to accelerate charged particles that can be used to probe the nucleus and to produce X-rays for medical and industrial applications.

The basic design of the Van de Graaff generator is shown in Figure 1. A motor-driven pulley (P) causes a belt to move past the "teeth" of a positively charged metal comb (A). Negative charges are transferred to the comb, leaving the left side of the belt positively charged. At the top comb (B), electrons are transferred to the belt, leaving the metal sphere positively charged. Continuous operation results in the build-up of a huge charge on the sphere and the "stored" electric potential of millions of volts.

When the sphere is charged to a high enough potential, corona discharge can occur through the ionization of air (Fig. 2). At STP (standard temperature and pressure, 0°C and 1 atm), the electric field required for the ionization of air is about 3 million volts per meter (V/m). Although the corona discharge is rather spectacular, it is usually undesirable, since it limits the voltage that can be accumulated on the sphere. The discharge can be suppressed and higher voltages achieved by surrounding the charged sphere with a gas of high ionization potential under high pressure. Electric fields greater than 10 million V/m can then be achieved.

Some other effects of the high voltage of a Van de Graaff generator are shown in Figure 3. Notice the human electroscope, with the hair taking the place of the leaves.

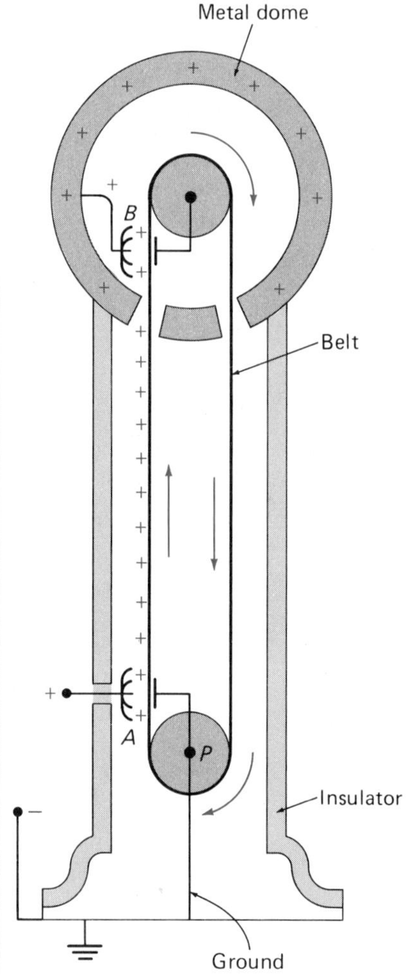

Figure 1
A diagram of a Van de Graaff generator. The accumulated charge on the sphere results in an electric potential of millions of volts. (Courtesy R. Serway and J. Faughn.)

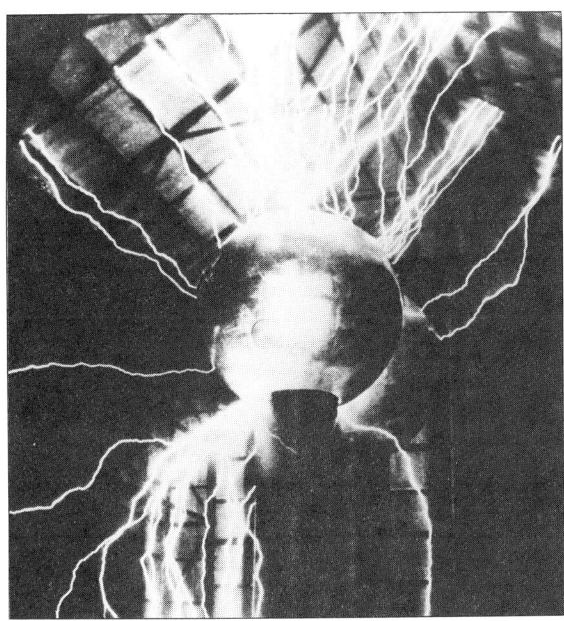

Figure 2
Big volts. Ionization discharge in the air from a Van de Graaff generator. The ionization of air requires an electric field of about 3 million volts per meter.

(a)

(b)

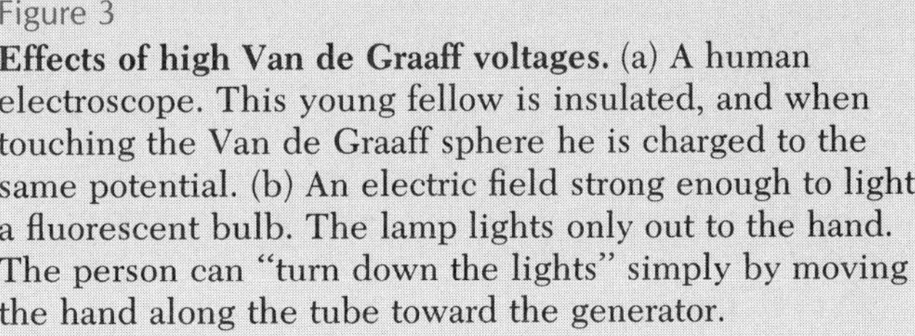

Figure 3
Effects of high Van de Graaff voltages. (a) A human electroscope. This young fellow is insulated, and when touching the Van de Graaff sphere he is charged to the same potential. (b) An electric field strong enough to light a fluorescent bulb. The lamp lights only out to the hand. The person can "turn down the lights" simply by moving the hand along the tube toward the generator.

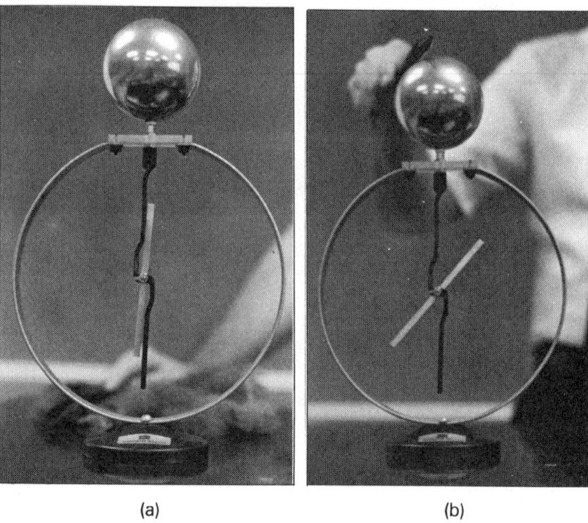

(a) (b)

Figure 19.16
A Braun electroscope. (a) Electroscope uncharged. (b) Electroscope charged. See Question 13.

13. Another type of electroscope, a Braun electroscope, is shown in Figure 19.16. Explain its principle of operation. Does this type of electroscope have any practical advantages over the foil-leaf electroscope?
14. An electroscope is negatively charged, and its leaves diverge. What would happen to the leaves (and why) if you touched the bulb with (a) your finger and (b) a glass rod that had been rubbed with silk?
15. How could you charge an electroscope negatively by using induction? How could you prove it is negatively charged?
16. (a) What causes static cling in clothing? Why is this more of a problem on a dry day? (b) Why is dust so difficult to get off a phonograph record?
17. (a) We commonly rub balloons on our hair to charge them electrostatically. How could a bald-headed person charge a balloon so it would stick to a wall? (b) What determines whether a charged balloon will stick to a wall or ceiling?
18. What would happen in Figure 19.6(b) if the rod touched the metal sphere? What would happen if the sphere were an insulator?
19. Two metal spheres on insulating rods are in contact, as shown in Figure 19.17. How could you charge both spheres without directly touching them? Would the spheres be charged positively or negatively?

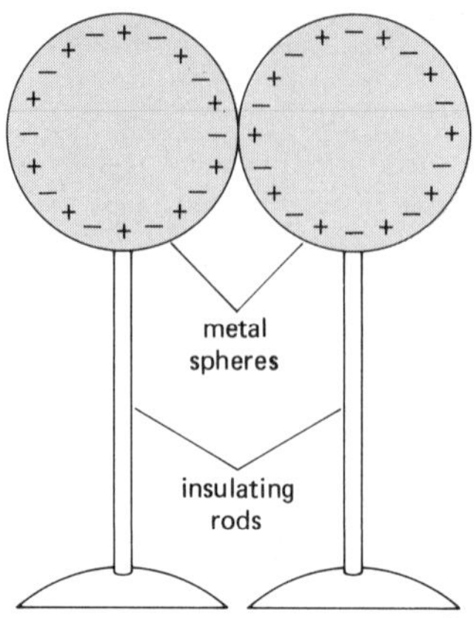

Figure 19.17
Contact. See Question 19.

Electric Field

20. What is an electric field, and how is it determined? Is the electric field a vector or scalar quantity?
21. Why do electric field lines always begin on positive charges and end on negative charges?
22. Sketch the electric field in the vicinity of two isolated (a) positive charges and (b) negative charges. (Don't forget the directional arrowheads.)
23. What happens to charge placed on an isolated conductor?
24. When a closed, hollow metal object, such as a metal box, is charged by induction by outside charges or is given a net charge by contact, the electric force or electric field inside the object is zero. Why?
25. The arrangement in Question 24 is an example of electrostatic shielding. A closed metal container acts as an electrostatic shield. If there are charges inside the shield, they will set up their own electric field. But the field inside will be unaffected by charges outside. Would it be possible to construct a gravitational shield? Explain. (Electrostatic shielding is common in electronic equipment—for example, a radio set, in which it is desirable to keep the electric fields originating in one part of the set from interfering with components in another part.)
26. Would it be safe to get inside a Van de Graaff generator sphere and then have it charged? (See Question 24 and Special Feature 19.2.)
27. Explain why the boy's hair in Figure 3 of Special Feature 19.2 stands on end. Is he insulated from ground?

Electric Potential Energy and Electric Potential

28. What gives rise to electric potential energy?
29. What are the units of (a) electric field, (b) electric potential energy, and (c) electric potential?
30. Is the electric potential energy increased or decreased (a) when two like charges are separated

and (b) when two unlike charges are brought closer together?

31. One charged object has twice the electric potential energy of another. Does it necessarily have twice the electric potential? Can it have?
32. What is the sign of the change in potential difference when a negative charge is moved (a) in the same direction as an electric field, (b) in the opposite direction from an electric field, and (c) perpendicular to an electric field?
33. What is an equipotential, and how might one be determined in an electric field?
34. What is an electron volt, and how does it compare in magnitude to the other common units of the same property?

Problems

Levels of difficulty are indicated by asterisks for your convenience.

19.1 Electric Charge and Force

1. On a dry day, a person picks up a static charge of -8.0×10^{-12} C from a rug. How many excess electrons does the person have?
2. It is desired to give an insulated conductor a net charge of +9.6 μC. How many electrons would have to be removed?

*3. In nuclear physics an alpha particle has a net charge of +2 (two protons). How many alpha particles would give a net charge of +6.4 μC?

*4. Two particles with electric charges of +2.0 μC and -4.0 μC are 0.30 m apart. What is the force acting on each particle?

*5. If an electron is placed 20 cm from a charge of +0.80 C, (a) what is the force experienced by the electron? (b) If the electron were free to move, what would be its initial acceleration?

*6. How far apart would an electron and a proton have to be to have a mutual electrical attraction of 1.0 N?

*7. A charge of -3.0 μC experiences a repulsive force of 2.7 N due to another charge located 40 cm away. What are the sign and magnitude of the other charge?

*8. Two charges of +1.0 μC and +4.0 μC are located on the x-axis at positions (40 cm, 0) and (-20 cm, 0), respectively. (a) Could a third charge placed at $x < -20$ cm or $x > 40$ cm be in static equilibrium? If so, where? (b) Could a third charge be placed between the two charges and be in static equilibrium? If so, where? (*Hint:* draw a sketch of the situation.)

*9. Charges of +4.0 μC and -9.0 μC are located on the y-axis at positions (0, 10 cm) and (0, -30 cm), respectively. (a) Could a third charge be placed between the charges and be in static equilibrium? If so, where? (b) Could a third charge be placed at an off-axis position and be in static equilibrium? If so, where? (c) Could a third charge be in static equilibrium anywhere? If so, where? (*Hint:* draw a sketch of the situation.)

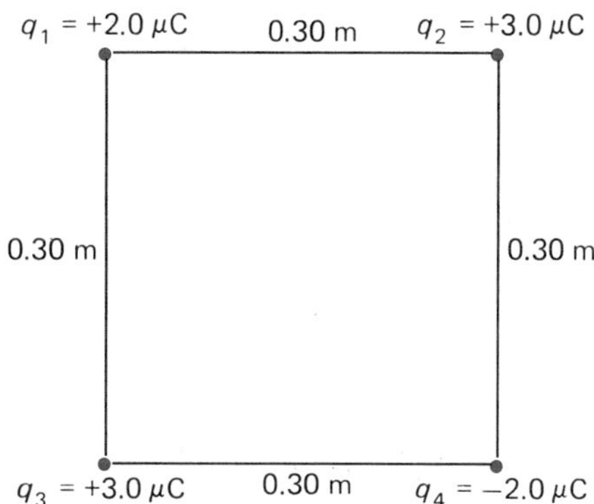

Figure 19.18
See Problem 10.

**10. Four charges are located at the corners of a square, as illustrated in Figure 19.18. (a) What is the net force on q_1? (b) What is the net force on q_4?

**11. Three identical charges are located at the corners of an equilateral triangle, as shown in Figure 19.19. (a) What is the net force on q_1? (b) What are the net forces on q_2 and q_3?

19.3 Electric Field

*12. What is the electric field 60 cm away from a charge of +3.0 μC? (Remember to give magnitude and direction.)

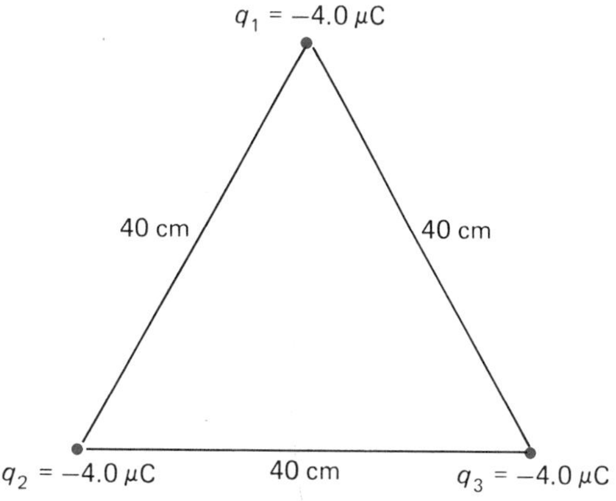

Figure 19.19
See Problem 11.

*13. A charge configuration produces an electric field of 6.0×10^5 N/C in the $+x$ direction at the point $(x, y) = (4.0$ cm, 5.0 cm$)$. What force would be experienced by a charge of (a) +2.0 μC and (b) −3.0 μC located at this position?

*14. Two like charges of −3.5 μC are located at positions (x, y) of (20 cm, 0) and (−20 cm, 0), respectively. Where is the electric field due to these charges zero?

*15. A charged, hollow spherical conductor with an area of 0.80 m^2 has a net charge of +2.5 μC. (a) What is the electric field near the outside surface of the sphere? (b) What is the electric field 5.0 cm from the center of the sphere?

*16. Consider the charge configuration described in Problem 9. (a) What is the electric field at the origin? (b) Where is the electric field zero?

*17. Referring to Figure 19.10, what is the electric field at the point (0.20 m, 0)?

*18. What are the magnitude and direction of a vertical electric field that would support the weight of an electron?

**19. What is the electric field at the center of the square in Figure 19.18?

**20. What is the electric field at the center of the triangle in Figure 19.19?

19.4 Electric Potential Energy and Electric Potential

*21. (a) What is the electric potential energy of two electrons 1.0 mm apart? (b) What would be the change in the potential energy if the separation distance were increased to 2.0 mm?

*22. Two charges of +5.0 μC and −4.0 μC are located at positions (x, y) of (−20 cm, 0) and 10 cm, 0), respectively. (a) What is the electric potential energy of the charges? (b) How much work is required to move the −4.0 μC charge to the position (20 cm, 0)?

*23. The electric potential energy between two identical charges of +10 μC is 1.50 J. If work is done so the potential energy is increased by 0.50 J, what is the final separation distance of the charges?

*24. What is the electric potential energy of the charge configuration in Figure 19.18?

*25. What is the electric potential energy of the charge configuration in Figure 19.19?

*26. The electric potential at a particular location is 9.0×10^4 J/C as determined by using a test charge of +3.0 μC. What is the electric potential energy of the charge at this point?

*27. What is the electric potential at a point 15 mm from a charge of −6.0 μC?

*28. A charge of +8.0 μC is located at the origin. (a) What is the electric potential at the position (30 cm, 0)? (b) What is the potential difference between the points (30 cm, 0) and (45 cm, 0)?

*29. A uniform electric field of 2.5×10^2 V/m exists between two charged parallel plates. How much work is required to move a charge of −4.0 μC (a) a distance of 10 cm in the direction of the field and (b) 20 cm in the direction opposite the field?

*30. How much energy, in joules, is acquired by an electron when accelerated through a potential difference of 1 kV?

*31. An alpha particle has a charge of +2 and a mass of 2.4×10^{-27} kg. (A charge of +2 means the particle has an equivalent charge of two protons.) How much energy, in (a) electron volts and (b) joules, is acquired by the particle when it accelerates through a potential of 50 kV?

**32. An alpha particle (see Problem 31) is released in an electric field of 50 V/m. (a) What is the acceleration of the particle? (b) What is the kinetic energy of the particle, in eV, when it has traveled 0.30 m?

**33. A charge is moved in a 0.30–0.40–0.50 m triangle in a uniform field of 3.0×10^3 N/C in the $+x$ direction (see Fig. 19.20). (a) If the 0.30-m leg of the triangle is in the $+x$ direction, what is the change in potential for each side of the triangle? (b) What is the total change in potential when the charge is back to its starting point?

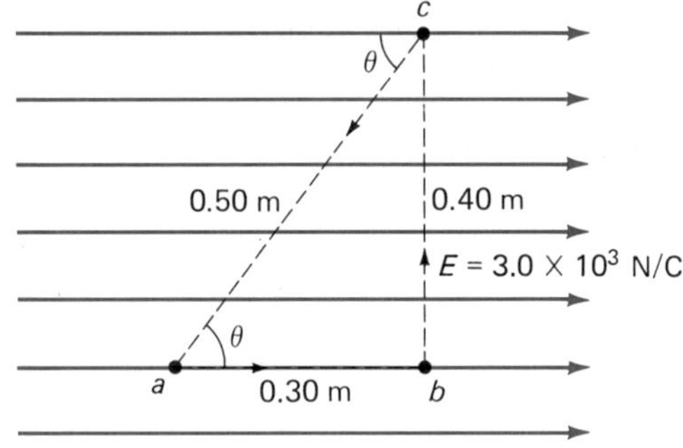

Figure 19.20
See Problem 33.

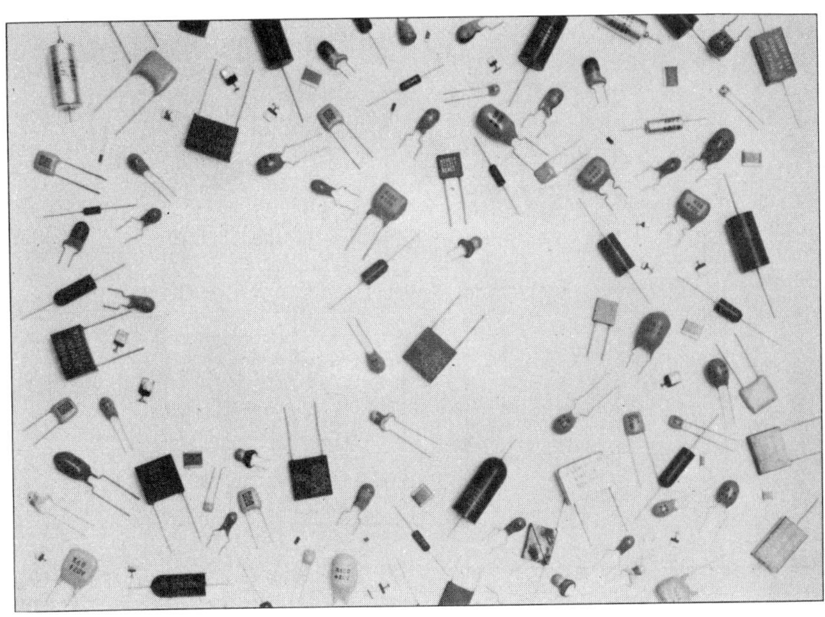

(Courtesy Union Carbide)

Chapter 20

Capacitance and Dielectrics

A charged conductor has electric potential by virtue of its charge, since work is done in charging a conductor (electric potential = work/charge). Hence, a charged conductor has energy, or the capacity to do work.

A measure of the charge on a conductor at a given potential is given by a quantity called *capacitance,* and electrical energy can be stored in a device called a *capacitor*. The energy storage capacity of a capacitor is enhanced by dielectric materials—the topic of discussion in this chapter.

20.1 Capacitance and Capacitors

When a conductor becomes charged (due to a transfer of charge), another conductor somewhere becomes equally and oppositely charged. Two closely separated conductors, each with equal and opposite charge Q, are called a capacitor. Of particular practical interest is the **parallel-plate capacitor** (Fig. 20.1). It consists of two parallel-plate conductors, each of area A, separated by a distance d.

When a capacitor is connected to a battery, the battery supplies the energy to transfer electronic charge from one plate to the other through chemical action (Chapter 21). Suppose a negative charge is transferred from one plate to the other. The additional transfer of charge then requires progressively more work, since the charge already on the plate repels additional charges. This is analogous to inflating a tire with a hand pump. As more air is put into the tire and the pressure builds, it becomes increasingly difficult (requires more work) to put additional air in the tire.

The total charge Q that is transferred depends on the electric potential, or voltage V, of

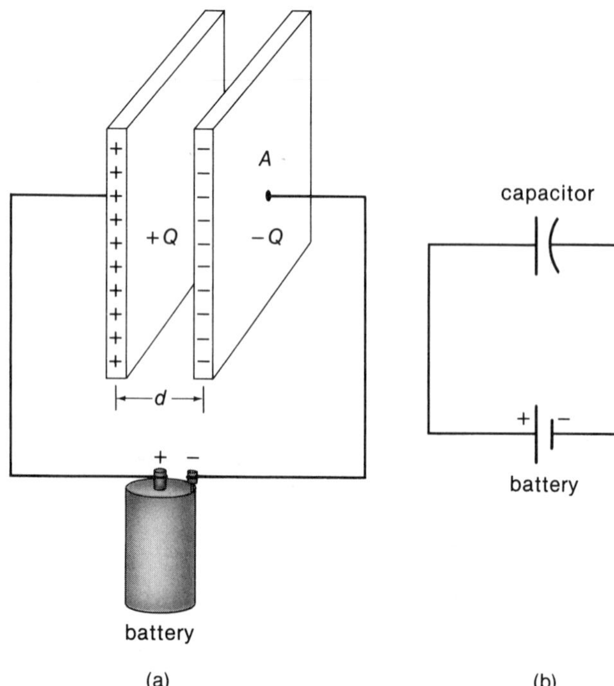

Figure 20.1
Parallel-plate capacitor. (a) Two parallel plates of equal area A separated by a distance d form a capacitor. When connected to a battery, the plates obtain a charge Q. The battery supplies the work to charge the capacitor. (b) Equivalent circuit symbols.

the battery. The greater the voltage, the greater the charge on the plates, i.e.,

$$Q \propto V$$

and

$$\boxed{Q = CV \quad \text{or} \quad C = \frac{Q}{V}} \qquad \textbf{(Eq. 20.1)}$$

where C is called the **capacitance.** The capacitance is essentially the charge "capacity" of a capacitor for a given voltage.

From the equation, the unit of capacitance can be seen to be coulomb/volt. However, this unit is given the name **farad** (F) in honor of the English scientist Michael Faraday (1791–1867), an early investigator of electrical phenomena. The farad is too large a unit for practical applications, so the smaller units microfarad and picofarad are commonly used:

$$1 \text{ microfarad } (\mu\text{F}) = 10^{-6} \text{ farad (F)}$$

$$1 \text{ picofarad (pF)} = 10^{-12} \text{ farad (F)}$$

The microfarad is often called a "mike" and the picofarad a "mickey-mike," "micro-mike," or "puff" by persons working in electrical applications.

When the plates of a capacitor are charged with a charge Q ($+Q$ and $-Q$ on opposite plates), it can be shown that there is a uniform electric field between the plates (Fig. 20.2), which has a magnitude

$$E = \frac{\sigma}{\varepsilon_o} = \frac{Q}{\varepsilon_o A} \qquad \textbf{(Eq. 20.2)}$$

where $\sigma = Q/A$ is the surface charge density of the plates, and the constant ε_o is the permittivity

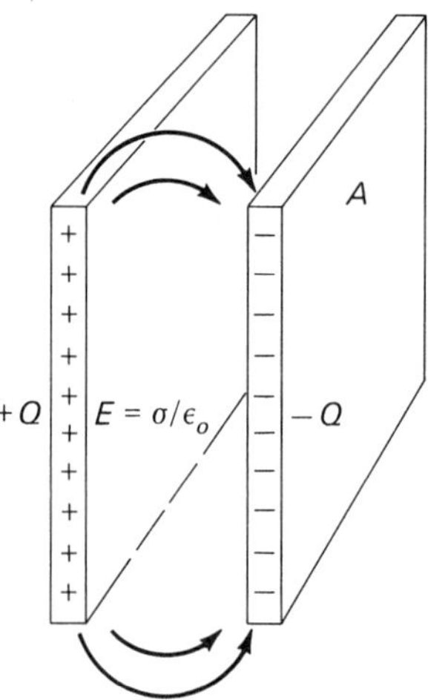

Figure 20.2
Electric field of a parallel-plate capacitor. The magnitude of the uniform electric field between two charged plates is given by $E = \sigma/\varepsilon_o = Q/\varepsilon_o A$, where $\sigma = Q/A$. For an ideal case, the field fringing at the edges of the plates is neglected.

of free space (vacuum), $\varepsilon_o = 8.85 \times 10^{-12}$ $C^2/N\text{-}m^2$ (cf. Chapter 19.1). For the ideal case, the E field fringing at the edges of the capacitor is considered negligible.

The potential difference between the plates is then

$$V = Ed = \frac{Qd}{\varepsilon_o A}$$

where d is the separation distance of the plates. The capacitance in terms of the capacitor geometry is then given by

$$\boxed{C = \frac{Q}{V} = \frac{\varepsilon_o A}{d}} \qquad \textbf{(Eq. 20.3)}$$

(parallel plate capacitor)

Hence, the greater the area A of the plates and the closer they are, the greater the capacitance (more charge on the plates for a given voltage).

EXAMPLE 20.1 A parallel-plate capacitor has a plate area of 0.70 m^2 and a separation distance of 0.50 mm. (a) What is the capacitance of the capacitor? (b) If a voltage of 12 V is placed across the plates, what is the electric field between the plates, and how much excess charge is on each plate?

Solution: It is given that $A = 0.70\ m^2$ and $d = 0.50\ mm = 5.0 \times 10^{-4}$ m. Then,

$$\text{(a)}\ C = \frac{\varepsilon_o A}{d} = \frac{(8.85 \times 10^{-12}\ C^2/N\text{-}m^2)(0.70\ m^2)}{5.0 \times 10^{-4}\ m} = 1.2 \times 10^{-8}\ F = 120\ \mu F$$

(b) With $V = 12$ V,

$$E = \frac{V}{d} = \frac{12\ V}{5.0 \times 10^{-4}\ m} = 2.4 \times 10^4\ V/m$$

and

$$Q = CV = (1.2 \times 10^{-8}\ F)(12\ V) = 1.4 \times 10^{-7}\ C = 0.14\ \mu C$$

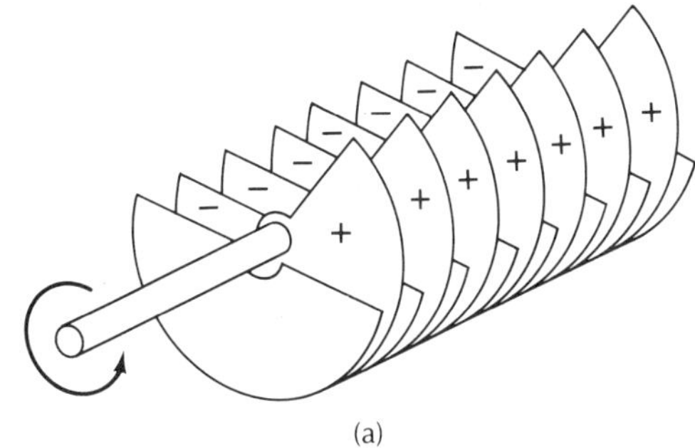

(a)

(b)

Figure 20.3
Variable air capacitor. The capacitance is varied by rotating the shaft, which varies the common effect area of two sets of semicircular plates—one fixed and the other movable. (b courtesy of P. Highsmith.)

Capacitors with a fixed geometry and a constant capacitance are called fixed capacitors. Variable capacitors are also available. The most common type of variable capacitor, the variable air capacitor, is shown in Figure 20.3. It is used in radio tuning circuits.

Capacitance is varied by rotating a shaft, which varies the common area of a set of fixed and movable plates. The greater the common area of the plates, the greater the capacitance. The maximum common area is the area of one plate times the number of intervening spaces between the plates. That is, if there are N plates of area A, the maximum common area is $(N - 1)A$.

EXAMPLE 20.2 A variable air capacitor is designed with half-circular plates that have radii of 1.7 cm and separation distances of 0.20 cm. How many plates would be required to make a variable capacitor with a maximum capacitance of 50 pF?

Solution: It is given that $r = 1.7\text{ cm} = 1.7 \times 10^{-2}\text{ m}$, $d = 0.20\text{ cm} = 2.0 \times 10^{-3}\text{ m}$, and $C = 50\text{ pF} = 50 \times 10^{-12}\text{ F} = 5.0 \times 10^{-11}\text{ F}$. The area A of a half-circular plate is

$$A = \tfrac{1}{2}(\pi r^2) = \tfrac{1}{2}\pi(1.7 \times 10^{-2}\text{ m})^2$$
$$= 4.5 \times 10^{-4}\text{ m}^2$$

The capacitance with the maximum common area of N plates is $C = \varepsilon_o(N-1)A/d$, and

$$N - 1 = \frac{Cd}{\varepsilon_o A}$$
$$= \frac{(5.0 \times 10^{-11}\text{ F})(2.0 \times 10^{-3}\text{ m})}{(8.85 \times 10^{-12}\text{ C}^2/\text{N-m}^2)(4.5 \times 10^{-4}\text{ m}^2)}$$
$$= 25$$

and

$$N = 25 + 1 = 26\text{ plates}$$

20.2 Energy of a Charged Capacitor

As we have learned, work is required to charge a capacitor. Hence, a charged capacitor has electric potential energy. As the charge on the plates goes from zero to a total charge Q in the charging process, the potential difference, or voltage, across the plates goes from zero to V. Work is done in moving charge (from the negative plate to the positive plate) through the increasing potential difference.

To compute this work, we use an average potential difference $\overline{V}$,

$$\overline{V} = \frac{V_{\text{final}} + V_{\text{initial}}}{2} = \frac{V + 0}{2} = \tfrac{1}{2}V$$

where $V_{\text{initial}} = 0$. This equation is similar to the average speed in Chapter 4.2, $\bar{v} = (v_f + v_o)/2$. Then the work or, equivalently, the potential energy of the charged capacitor is

$$W = U = Q\overline{V} = Q(\tfrac{1}{2}V)$$
$$U = \tfrac{1}{2}QV$$

We refer to this as the energy "stored" in a charged capacitor. Using the defining equation for capacitance, $Q = CV$, the potential energy can be written in three forms:

$$\boxed{U = \tfrac{1}{2}QV = \frac{Q^2}{2C} = \tfrac{1}{2}CV^2} \qquad \textbf{(Eq. 20.4)}$$

(energy stored in a capacitor)

The last form is usually the most convenient in practical applications.

EXAMPLE 20.3 A 2.0-μF capacitor is connected to a 6.0-V battery. How much energy is stored in the fully charged capacitor?

Solution: It is given that $C = 2.0\ \mu\text{F} = 2.0 \times 10^{-6}\text{ F}$ and $V = 6.0\text{ V}$. Then,

$$U = \tfrac{1}{2}CV^2 = \tfrac{1}{2}(2.0 \times 10^{-6}\text{ F})(6.0\text{ V})^2$$
$$= 3.6 \times 10^{-5}\text{ J}$$

Notice that the energy is proportional to the square of the applied voltage. If a 12-V battery were used, the capacitor would have four times the energy as with a 6.0-V battery.

Energy Density

A charged capacitor has potential energy by virtue of the work done in assembling charges on the plates. Another way of looking at this is that a charged capacitor has energy because an electric field has been established in the region between the plates. In this case, the energy is associated with the field itself, and we say that the energy is stored in the electric field.

It is therefore convenient to define an electric **energy density u_e**, or the electric energy per unit volume, $u_e = U/V'$, where V' is the volume of the region between the plates (not voltage).

Considering the energy of a parallel-plate capacitor and related equations,

$$U = \tfrac{1}{2}CV^2 = \tfrac{1}{2}\left(\frac{\varepsilon_o A}{d}\right)(Ed)^2$$

The volume of the region between the capacitor plates is $V' = Ad$, hence

$$U = \tfrac{1}{2}\varepsilon_o E^2(Ad) = \tfrac{1}{2}\varepsilon_o E^2 V'$$

and

$$\boxed{u_e = \frac{U}{V'} = \tfrac{1}{2}\varepsilon_o E^2} \qquad \textbf{(Eq. 20.5)}$$

(energy density)

Notice that the energy density is proportional to the square of the electric field. This result is general and is valid for any electric field, even electromagnetic waves in space.

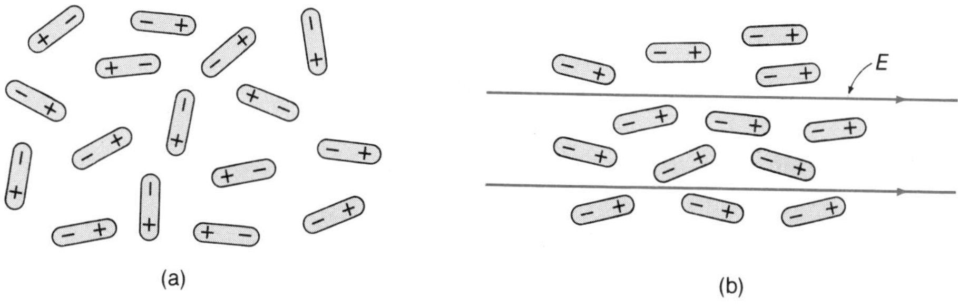

Figure 20.4
Molecular dipoles. (a) Polar molecules have regions or centers of positive and negative charges. Ordinarily, the molecules are randomly oriented. (b) When placed in an electric field, the dipoles tend to line up with the field. Molecular dipoles may also be induced in ordinary nonpolar molecules (cf. Section 19.2).

20.3 Dielectrics and Dielectric Constants

So far in our discussion, the region between the plates of a capacitor has contained air (or vacuum). It is important that the plates do not come into contact. This may be prevented by placing an insulating material between the plates.

In an insulator, the atomic electrons generally are not free to move. However, the molecules of an insulating material, although electrically neutral, may have "centers" of positive and negative charges and hence are termed *polar* molecules (Fig. 20.4). Such a separation of charge is called an **electric dipole.** When placed in an electric field, the dipoles tend to align with the field that is under the influence of the electric forces on their centers of charge. The same effect can occur for materials with nonpolar molecules. The E field polarizes the nonpolar molecules by shifting or displacing the electrons of the molecule. These molecules have induced dipoles (cf. Chapter 19.2).

What then happens to the properties of a capacitor when an insulating material, called a **dielectric,** is placed between the plates? As illustrated in Figure 20.5, when a dielectric is placed in a charged capacitor, its dipole polar-

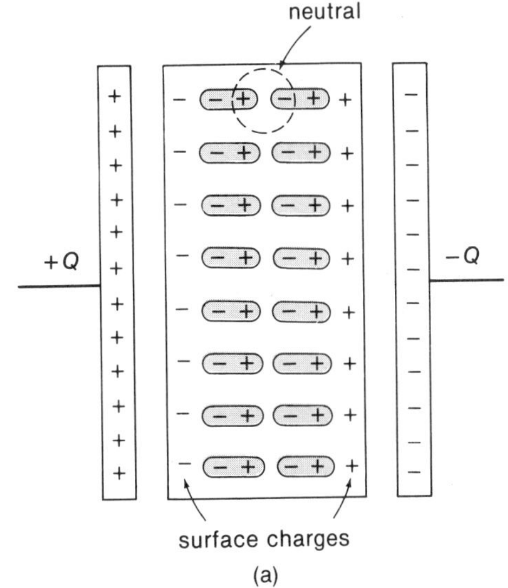

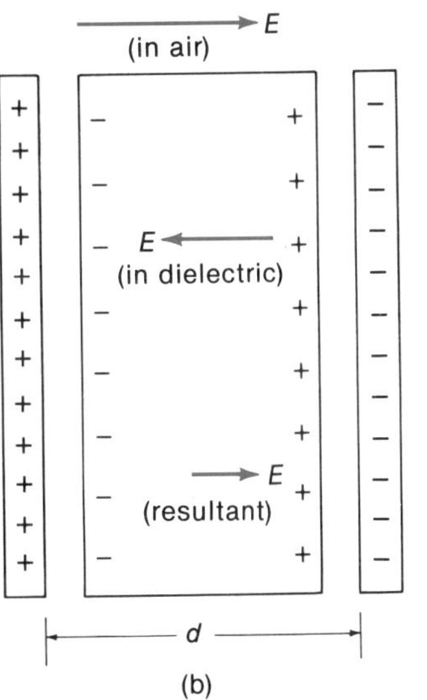

Figure 20.5
Dielectric in a charged capacitor. (a) When a capacitor with a dielectric is charged, the alignment of the dipoles gives rise to surface charges on the dielectric. (b) The electric field due to the surface charges on the dielectric is opposite to that of the electric field of the capacitor with a dielectric (in air). As a result, the electric field (vector resultant) is reduced, as is the voltage, $V = Ed$. But with the dielectric, the capacitance is increased, since $C = Q/V$.

Table 20.1
Typical (Average) Dielectric Constants and Dielectric Strengths of Some Materials

Material	Dielectric Constant K	Dielectric Strength *volts/mil**	Dielectric Strength *volts/meter*
Vacuum	1.0000	0	0
Air	1.0006	75	3.0×10^6
Liquids			
Ethyl alcohol	25	—	—
Silicone oil	2.5	500	20×10^6
Transformer oil	4.0	400	16×10^6
Water	80	—	—
Solids			
Glass	3.8–6.8	—	—
Mica	7.0	5000	200×10^6
Paper, general	3.5	1100	45×10^6
Polyethylene	2.3	450	18×10^6
Polystyrene	2.6	550	22×10^6
Porcelain	6.0–10	50–400	$2.0–16 \times 10^6$
Teflon	2.1		
Titanates (Ba, Ca, Mg, etc.)	15–12,000	50–300	$2.0–12 \times 10^6$
Titanium dioxide (TiO_2)	14–110	100–200	$4.0–8.0 \times 10^6$

* 1 mil = 0.001 inch

ization has the net effect of inducing charges on its surfaces. This results in an electric field within the insulator that is opposite to that of the plates. The electric field of the capacitor is effectively decreased, as is the voltage across the capacitor, since $V = Ed$. But since $C = Q/V$, the capacitance increases.

The increase in capacitance due to a dielectric is characterized by the **dielectric constant** $\boldsymbol{K}$ of the material:

$$K = \frac{C}{C_o} \qquad \textbf{(Eq. 20.6)}$$

where C and C_o are the capacitances of a capacitor with and without the dielectric, respectively. The dielectric constants of several materials are listed in Table 20.1. Notice that the dielectric constant of air is almost one. This is why air can be used instead of vacuum, which has a K of exactly one.

The dielectric property of a material can also be expressed in terms of the permittivity ε. The capacitance of a capacitor in air is $C_o = \varepsilon_o A/d$. (We now distinguish the capacitance in air or vacuum as C_o. Notice how the C_o and ε_o go together.) Then, with $C = KC_o$ (Eq. 20.6), the capacitance with a dielectric is

$$C = KC_o = \frac{K\varepsilon_o A}{d} = \frac{\varepsilon A}{d}$$

and

$$\boxed{\varepsilon = K\varepsilon_o} \qquad \textbf{(Eq. 20.7)}$$

(permittivity)

The effect of a dielectric on a particular property can be found by replacing ε_o (the permittivity of free space) in an equation with ε (the dielectric permittivity). Equivalently, ε_o is multiplied by K. For example, with a dielectric the energy density of a capacitor is

$$u_e = \tfrac{1}{2}\varepsilon E^2 = \tfrac{1}{2}K\varepsilon_o E^2$$

EXAMPLE 20.4 A parallel-plate capacitor with a plate area of 0.50 m^2 has a polystyrene plastic film, 0.030 mm thick, sandwiched between the plates. (a) What is the capacitance? (b) If the capacitor is connected to a 12-V battery, how much energy is stored in the "cap"? (A capacitor is sometimes called "cap" for short.)

Solution: It is given that $A = 0.50\ \text{m}^2$, $d = 0.030\ \text{mm} = 3.0 \times 10^{-5}\ \text{m}$, and, from Table 20.1, $K = 2.6$ for polystyrene.

$$\text{(a) } C = KC_o = \frac{K\varepsilon_o A}{d}$$

$$= \frac{(2.6)(8.85 \times 10^{-12}\ \text{C}^2/\text{N-m}^2)(0.50\ \text{m}^2)}{3.0 \times 10^{-5}\ \text{m}}$$

$$= 3.8 \times 10^{-7}\ \text{F} = 0.38\ \mu\text{F}$$

$$\text{(b) } U = \tfrac{1}{2}CV^2 = \tfrac{1}{2}(3.8 \times 10^{-7}\ \text{F})(12\ \text{V})^2$$

$$= 2.7 \times 10^{-5}\ \text{J}$$

As we have seen, dielectric materials serve two purposes in capacitors: (1) they insulate the plates from one another, and (2) they increase the capacitance by a factor of K. Plastic films and paper allow the construction of tubular capacitors, in which the dielectric is placed between strips of aluminum foil and the "parallel-plate" capacitor is rolled into a cylindrical tube (Fig. 20.6).

To maximize the energy storage capabilities of a capacitor of a given plate area, a material of high dielectric constant should be used, the dielectric should be as thin as possible, and a large voltage should be applied to the capacitor ($U = \frac{1}{2}KC_oV^2$ and $C_o \propto 1/d$). The latter two items tend to work against each other, since the thinner the dielectric, the more likely it is to break down at a high voltage.

The voltage at which electrical breakdown (electrical discharge through the material) occurs is called the **dielectric strength** of the material. It is commonly expressed in volts/mil (1 mil = 0.001 in.) or volts/meter. The dielectric strengths of some materials are listed in Table 20.1. Dielectric strength depends on the material properties and on material defects and impurities.

Throughout the foregoing discussion, it was assumed that there is no flow of electrons until the breakdown voltage is reached. This is an ideal case. In actual practice, every insulator conducts charge to some degree. Although this flow of charge, called **leakage current,** is very small, a charged, isolated capacitor will eventually lose its charge as a result of electron conduction through the material from one plate to the other. In some capacitors, such as the electrolytic type (see the following discussion), the leakage current is quite large and a charged electrolytic capacitor rapidly loses its charge.

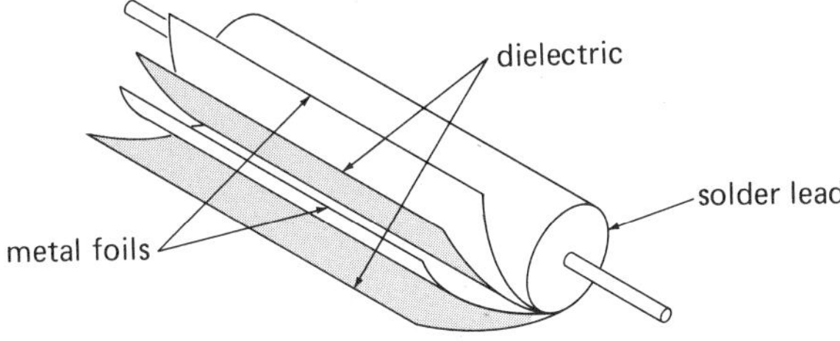

Figure 20.6

Tubular capacitor. A dielectric sheet or film is placed between sheets of aluminum foil and rolled into a cylindrical tube. The total capacitance is the sum of the rolled "parallel-plate" capacitors. (The film may be metalized by vapor deposition so separate metal foils are not needed.)

Types of Capacitors

There are many types of capacitors, and they come in a variety of sizes and shapes (see the chapter introductory photo). The variable air capacitor has already been discussed. Capacitors with constant capacitances are referred to as *fixed capacitors,* and they are further divided according to the dielectric. For example, common types of fixed capacitors are paper and (plastic) film tubular, ceramic, and electrolytic capacitors.

The basic construction of a paper or film tubular capacitor is illustrated in Figure 20.6. In a paper tubular capacitor, the paper may be impregnated with wax or oil to enhance the capacitance (increase K). Some actual tubular capacitors are shown in Figure 20.7.

Ceramic capacitors use a ceramic dielectric (porcelain, TiO_2, or titanates) coated on each side with a metal that acts as the plate conductor. Ceramic capacitors come in a variety of shapes and sizes (Fig. 20.8). The leads protrude through a protective insulating coating (usually plastic). Ceramic capacitors have very low leakage currents and can be designed with very large capacitances because of the relatively high dielectric constants of ceramic materials (see Table 20.1).

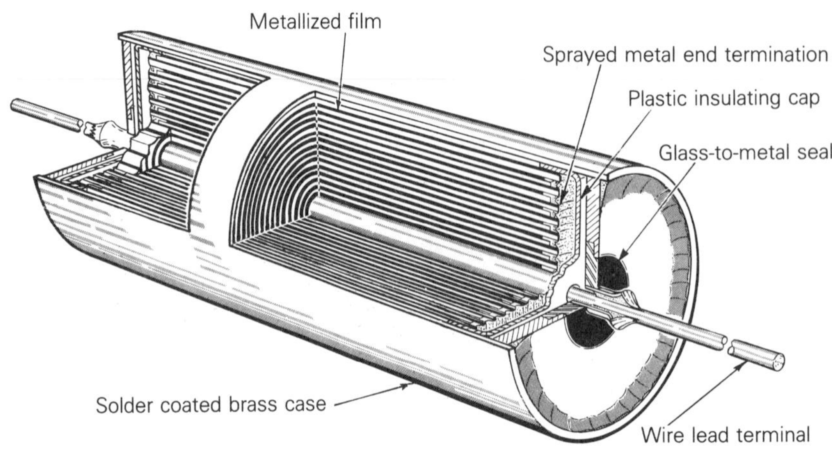

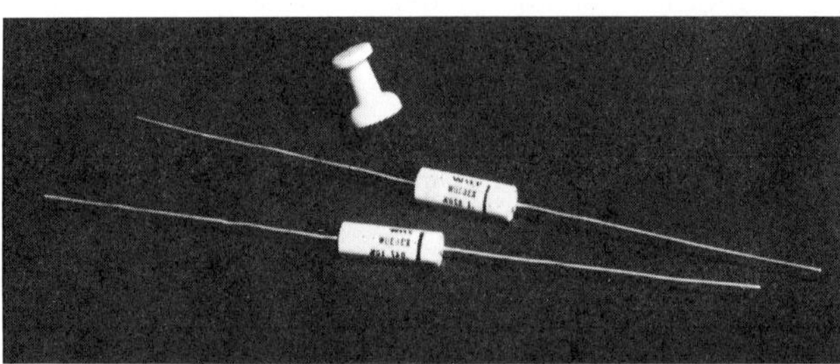

Figure 20.7
Tubular film capacitors and exposed view of construction. (Diagram courtesy Electronics Division, Union Carbide Corporation; photo courtesy TRW, Inc.)

Electrolytic capacitors can be designed with capacitances of several thousand microfarads (Fig. 20.9). The dielectric is a thin oxide film (only a few millionths of an inch thick) that is formed electrolytically on the surface of one plate; the surface is usually a foil of aluminum or tantalum. An electrolytic coating or a layer of paper saturated with an electrolyte is placed over the oxide layer on the foil. Another foil is placed over the electrolyte layer for the negative plate.

Figure 20.8
Ceramic capacitors. (Courtesy Electronics Division, Union Carbide Corporation.)

The purpose of the electrolyte is to electrolytically heal flaws that invariably occur in the thin oxide film. The polarity (+ or −) of the electrolytic capacitor is important because the oxide foil will not oxidize unless it is positive. Electrolytic capacitors are characterized by low breakdown voltages and high leakage currents.

20.4 Capacitors in Series and in Parallel

In electrical circuits, two or more capacitors may be connected. One such arrangement of **capacitors in series** is shown in Figure 20.10. In a series connection, the negative plate of one capacitor is connected to the positive plate of another. Thus, the charge on each capacitor is the same as that transferred by the battery, i.e.,

$$Q = Q_1 = Q_2 = Q_3$$

(capacitors in series)

In series, the sum of the electric potentials, or voltage "drops," across the capacitors is equal to the voltage "rise" of the battery:

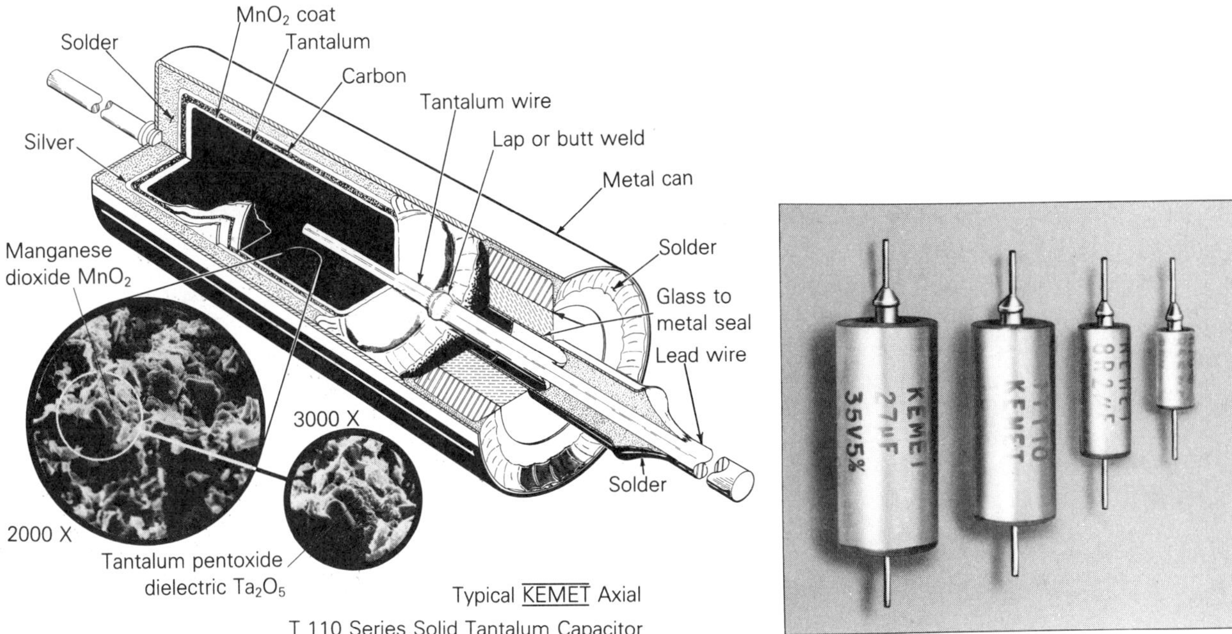

Figure 20.9
Electrolytic capacitors and exposed view of their construction. (Courtesy Electronics Division, Union Carbide Corporation.)

$$V = V_1 + V_2 + V_3$$
(capacitors in series)

and with $V = Q/C$,

$$\frac{Q}{C_s} = \frac{Q_1}{C_1} + \frac{Q_2}{C_2} + \frac{Q_3}{C_3}$$

$$= Q\left(\frac{1}{C_1} + \frac{1}{C_2} + \frac{1}{C_3}\right)$$

Hence,

$$\boxed{\frac{1}{C_s} = \frac{1}{C_1} + \frac{1}{C_2} + \frac{1}{C_3}} \qquad \textbf{(Eq. 20.8)}$$

(equivalent series capacitance)

where C_s is the equivalent capacitance of the three capacitors in series. That is, the three capacitors in series could be replaced by one capacitor of value C_s.

Equation 20.8 may be extended to any number of capacitors in series. For two capacitors in series, the equation can be written

$$C_s = \frac{C_1C_2}{C_1 + C_2} \qquad \textbf{(Eq. 20.8a)}$$

It is interesting to note, as can easily be shown with a few examples, that **the equivalent capacitance of capacitors in series is always less than that of the smallest capacitance.**

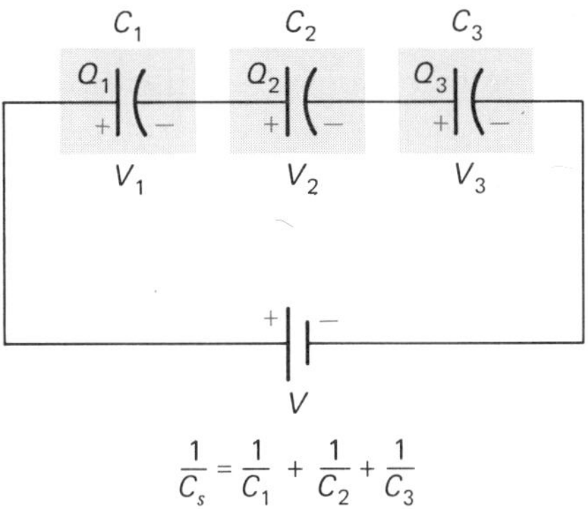

Figure 20.10
Capacitors in series. The charge Q on each capacitor is the same, but the voltage "rise" of the battery is equal to the sum of the voltage "drops" across the capacitors, i.e., $V = V_1 + V_2 + V_3$. The reciprocal of the total equivalent capacitances C_s is equal to the sum of the reciprocals of the individual capacitors.

EXAMPLE 20.5 Two capacitors, 4.0 μF and 8.0 μF, are connected in series with a 12-V battery. What are (a) the equivalent capacitance of the series combination and (b) the charge on the capacitors?

Solution: It is given that $C_1 = 4.0\ \mu\text{F}$, $C_2 = 8.0\ \mu\text{F}$, and $V = 12$ V.

(a) In series,

$$C_s = \frac{C_1 C_2}{C_1 + C_2}$$

$$= \frac{(4.0\ \mu\text{F})(8.0\ \mu\text{F})}{4.0\ \mu\text{F} + 8.0\ \mu\text{F}} = 2.7\ \mu\text{F}$$

(In this case the μF unit was used instead of converting to farads. Why?)

(b) With $C_s = 2.7\ \mu\text{F} = 2.7 \times 10^{-6}$ F,

$$Q = C_s V = (2.7 \times 10^{-6}\ \text{F})(12\ \text{V})$$

$$= 3.2 \times 10^{-5}\ \text{C} = 32\ \mu\text{C}$$

and, in series, this is the charge on each capacitor.

Another circuit arrangement in which the charge is shared between two or more capacitors is when the capacitors are connected in parallel. For **capacitors in parallel,** the positive plates of all the capacitors are connected, as are all the negative plates (Fig. 20.11). In this case, each of the voltage drops across the capacitors is equal to the voltage rise of the battery, i.e.,

$$V = V_1 = V_2 = V_3$$

(capacitors in parallel)

The total charge transferred from the battery is equal to the sum of the charges on the individual capacitors:

$$Q = Q_1 + Q_2 + Q_3$$

(capacitors in parallel)

Then it follows ($Q = CV$) that

$$C_p V = C_1 V_1 + C_2 V_2 + C_2 V_3$$

$$= V(C_1 + C_2 + C_3)$$

where the last step uses the fact that all the voltage drops are equal. Hence,

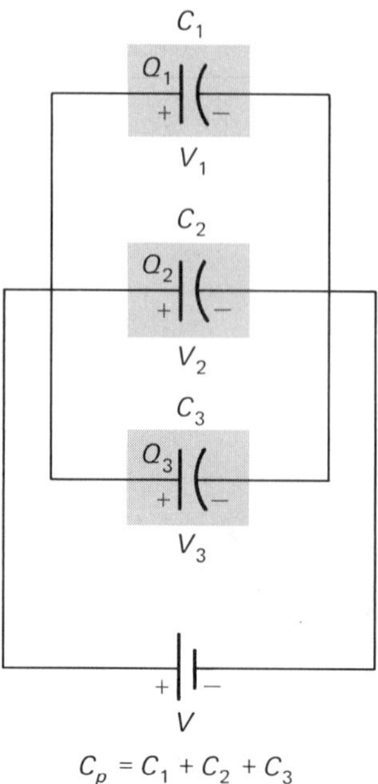

Figure 20.11
Capacitors in parallel. The voltage across each capacitor is the same as that of the battery, but the total charge transferred by the battery is equal to the sum of the charges on the individual capacitors, $Q = Q_1 + Q_2 + Q_3$. The total equivalent capacitance C_p is equal to the sum of the individual capacitances.

$$\boxed{C_p = C_1 + C_2 + C_3} \qquad \textbf{(Eq. 20.9)}$$

(equivalent parallel capacitance)

where C_p is the equivalent capacitance of three capacitors in parallel. This sum may be extended to any number of capacitors in parallel.

Capacitors may be connected in series-parallel combinations such as that shown in Figure 20.12. The combination may be reduced to a single, equivalent capacitance.

EXAMPLE 20.6 Three capacitors are connected in a circuit, as shown in Figure 20.12. What are (a) the equivalent capacitance of the circuit and (b) the charge transferred by the battery?

Solution: (a) The equivalent capacitance of C_1 and C_2 in parallel is

$$C_p = C_1 + C_2 = 2.0\ \mu\text{F} + 5.0\ \mu\text{F} = 7.0\ \mu\text{F}$$

Then the total equivalent capacitance is the combination of C_p in series with C_3, and

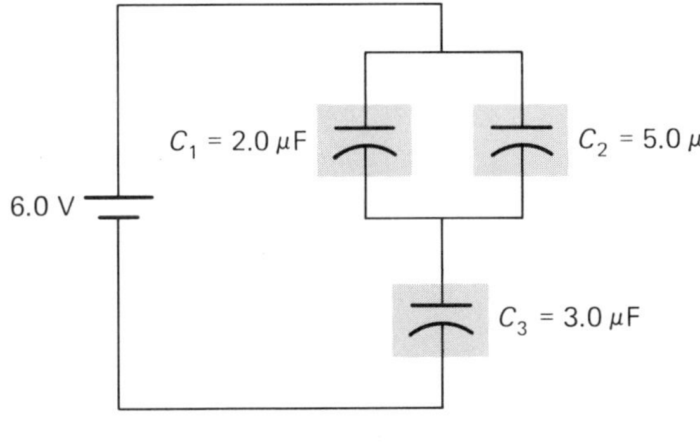

(a) actual circuit

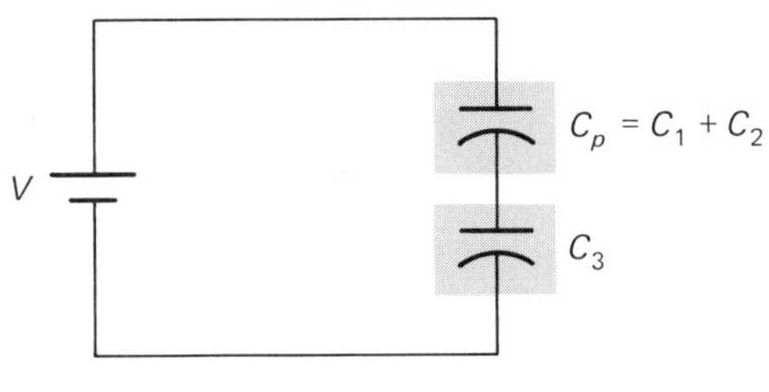

(b) equivalent circuit

Figure 20.12
See Example 20.6.

$$C_T = C_s = \frac{C_p C_3}{C_p + C_3}$$

$$= \frac{(7.0\ \mu\text{F})(3.0\ \mu\text{F})}{7.0\ \mu\text{F} + 3.0\ \mu\text{F}} = 2.1\ \mu\text{F}$$

(b) With $V = 3.0$ V, the charge transferred by the battery is

$$Q = C_T V = (2.1 \times 10^{-6}\text{ F})(3.0\text{ V})$$

$$= 6.3 \times 10^{-6}\text{ C}$$

$$= 6.3\ \mu\text{C}$$

20.5 Capacitor Charging and Discharging

When a capacitor is charged in a dc circuit, a transient current flows until the capacitor is fully charged. A sustained current does not flow since the capacitor gap is an "open" in the circuit.

A capacitance C is charged through a resistance R (in series) at an exponential rate, and the charge Q on the voltage V across the capacitor at time t is given by*

$$Q = Q_{max}(1 - e^{-t/RC})$$

and

$$V = V_{max}(1 - e^{-t/RC}) \qquad \textbf{(Eq. 20.10)}$$

(capacitor charging)

where Q_{max} and V_{max} are the maximum charge and voltage, respectively.

The constant RC is called the **time constant,** $\tau = RC$. In one time constant, $t = \tau$, a capacitor is charged to 0.63 (63%) of its maximum value of charge or voltage; in two time constants, $t = 2\tau$, to 0.87 of its maximum value; and so on (Fig. 20.13).

Similarly a capacitor discharges according to the relationship

$$Q = Q_{max}\, e^{-t/RC}$$

and

$$V = V_{max}\, e^{-t/RC} \qquad \textbf{(Eq. 20.11)}$$

(capacitor discharging)

In this case the capacitor loses its charge exponentially, and after one time constant, $t = \tau$, a capacitor loses 0.63 of its charge, or its voltage is reduced to 0.37 (37%) of its maximum value (Fig. 20.13).

EXAMPLE 20.7 A 1.0-MΩ resistor is in series with a 1.5-μF capacitor. (a) What is the time constant of the arrangement? (b) If the series RC arrangement is connected to a 90-V battery, what is the voltage across the capacitor in one time constant?

Solution: It is given that $R = 1\text{ M}\Omega = 10^6\ \Omega$ (a one-"meg" resistor) and $C = 1.5\ \mu\text{F} = 1.5 \times 10^{-6}$ F (a 1.5-"mike" capacitor).

(a) $\tau = RC = (10^6\ \Omega)(1.5 \times 10^{-6}\text{ F}) = 1.5$ s

(b) With $V_{max} = 90$ V, and $t = \tau = RC$,

$$V = V_{max}(1 - e^{-t/RC}) = (90\text{ V})(1 - e^{-RC/RC})$$

$$= (90\text{ V})(1 - e^{-1}) = (90\text{ V})(1 - 0.37)$$

$$= (90\text{ V})(0.63) = 57\text{ V}$$

* Electrical resistance, which is measured in ohms, will be discussed in detail in following chapters.

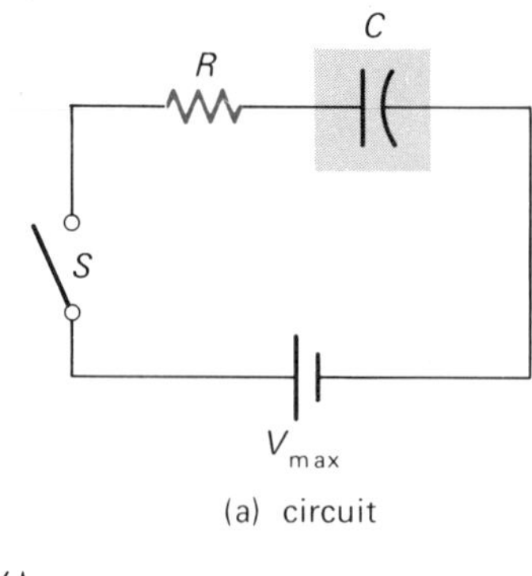

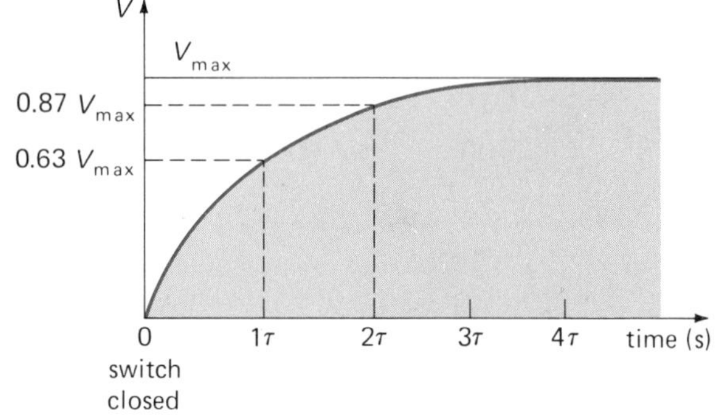

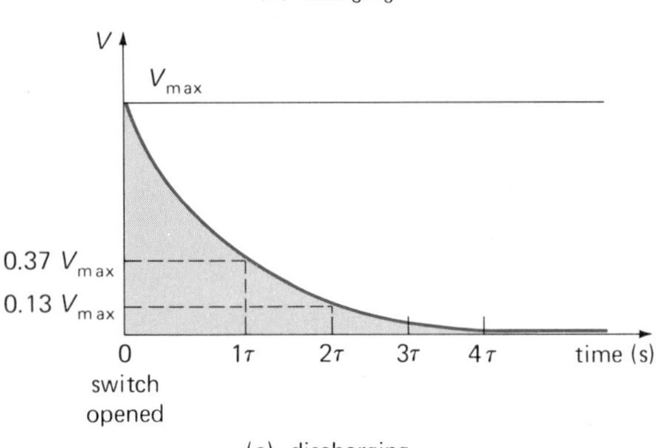

Figure 20.13
Capacitor charging and discharging. When a capacitor is charged through a resistance (a), the charge and the voltage across the capacitor build up exponentially with time (b), as shown here for voltage. Similarly, a capacitor discharges exponentially with time (c). The charge and discharge rates depend on the time constant RC. After one time constant, the voltage across and charge on the capacitor rises or falls 0.63 of its maximum value.

(To find e^{-1} using a calculator, enter -1 and push the e^x button; or enter 1, push the e^x button, and take the reciprocal, $1/e^x = e^{-x}$.)

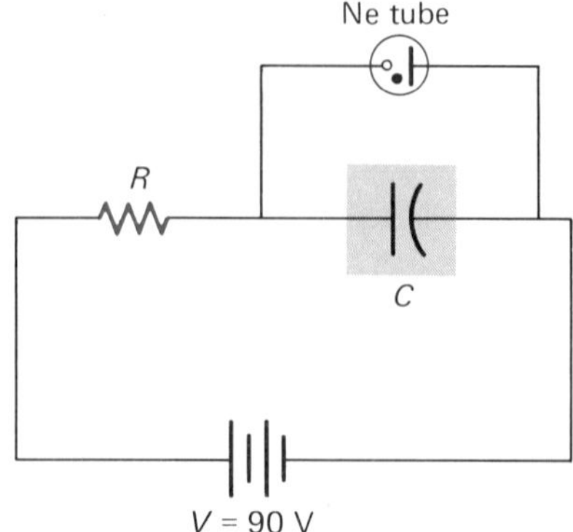

Figure 20.14
RC "flasher" circuit. When the voltage across the capacitor builds up to the firing potential of the neon tube, it discharges through the tube, causing it to glow momentarily, or flash. The flash rate depends on the charging rate or the RC time constant of the circuit.

The *RC* "Flasher" Circuit

The effect of the *RC* time constant can be seen for the so-called *RC* flasher circuit (Fig. 20.14). In this circuit, a small neon tube is placed across (in parallel with) the capacitor. The neon tube has a "firing potential" of about 70 V. That is, the neon tube conducts very little, or has a very large resistance, when the voltage across it is less than 70 V. However, above 70 V an ionization discharge occurs. The tube's resistance suddenly drops, and the tube conducts (as evidenced by a visible glow).

In the circuit with a battery of about 90 V, the voltage across the capacitor and neon tube is controlled by the capacitor. The capacitor charges according to the *RC* time constant. When the voltage across the capacitor reaches about 70 V [see Fig. 20.13(b)], the tube begins to conduct, and the capacitor discharges through the tube. The sudden surge of energy causes the tube to "flash."

When the current from the capacitor can no longer sustain the discharge, the tube stops conducting, with the voltage having fallen to an intermediate value. Then the voltage again rises to the breakdown voltage, and the cycle repeats. This charging-discharging process takes place again and again, and the neon tube flashes intermittently at a rate determined by the *RC* time constant of the circuit. By decreasing *R* and/or

C, the time constant is decreased, and the neon tube will flash at a faster rate.

Important Terms

capacitance (C) the ratio of the charge on a conductor and the applied voltage, $C = Q/V$

farad (F) the unit of capacitance (coulomb/volt)

energy density the electric energy per unit volume of capacitor space stored in the electric field

electric dipole a separation of charge, such as in a polar molecule, which may be permanent or induced

dielectric an insulating material used between capacitor plates

dielectric constant (K) the ratio of dielectric capacitance and vacuum capacitance, $K = C/C_o$

dielectric strength the voltage at which electrical breakdown (electric discharge through an insulating material) occurs

time constant (τ) the product of resistance and capacitance, $\tau = RC$

Important Formulas

capacitance: $C = \dfrac{Q}{V}$

electric field between parallel plates: $E = \dfrac{\sigma}{\varepsilon_o} = \dfrac{Q}{\varepsilon_o A}$

$\varepsilon = 8.85 \times 10^{-12}\ \text{C}^2/\text{N-m}^2$

parallel plate capacitance (vacuum): $C_o = \dfrac{\varepsilon_o A}{d}$

energy stored in a capacitor: $U = \frac{1}{2}QV = \dfrac{Q^2}{2C} = \frac{1}{2}CV^2$

energy density: $u_e = U/V' = \frac{1}{2}\varepsilon_o E^2$

dielectric constant: $K = \dfrac{C}{C_o} = \dfrac{\varepsilon}{\varepsilon_o}$

series capacitance: $\dfrac{1}{C_s} = \dfrac{1}{C_1} + \dfrac{1}{C_2} + \dfrac{1}{C_3} + \cdots$

for two capacitors $C_s = \dfrac{C_1 C_2}{C_1 + C_2}$

parallel capacitance: $C_p = C_1 + C_2 + C_3 + \cdots$

capacitor charging: $Q = Q_{\max}(1 - e^{-t/RC})$

or $V = V_{\max}(1 - e^{-t/RC})$

capacitor discharging: $Q = Q_{\max}\, e^{-t/RC}$

or $V = V_{\max}\, e^{-t/RC}$

Questions

Capacitance and Capacitors

1. How does the capacitance of a parallel-plate capacitor change with voltage?
2. Why is charging a capacitor analogous to inflating a tire with a hand pump?
3. Explain why a charged capacitor has energy even though the plates are conductors.
4. For a parallel plate capacitor, $C = \varepsilon_o A/d$. Show that the units of C are farads. What are a "mike" and a "puff"?
5. How many picofarads are in 1.0 microfarad?
6. How does the electric field between capacitor plates vary with the separation distance?
7. Two parallel-plate capacitors have the same charge, but one of the capacitors has plates of smaller area. In which capacitor is the electric field greater?
8. For a parallel-plate capacitor, how does the capacitance vary with (a) the area of the plates, (b) the distance between the plates, and (c) the applied voltage?
9. (a) Distinguish between fixed and variable capacitors. (b) Describe the principle of a variable air capacitor—in particular, the effect of the number of plates.

Energy of a Charged Capacitor

10. Two capacitors with capacitances of 1.0 μF and 0.15 μF are charged to a given voltage. (a) Which has more charge? (b) Which has more energy? How much more in each case?
11. Two capacitors with capacitances of 5 pF and 10 pF are charged so they both have the same amount of charge on their plates. (a) Which is at a lower voltage? (b) Which has more energy? What are the factors in each case?
12. A variable air capacitor is charged to a certain voltage. If the capacitance is doubled, what is the effect on the charge of the capacitor?
13. How does the energy density of a capacitor vary with the electric field?
14. What effects can be used to maximize the energy storage capacity of a capacitor of fixed area? Explain how these combined effects set a practical limit.

Dielectrics and Dielectric Constants

15. (a) Distinguish between polar and nonpolar molecules. (b) What is an induced dipole?
16. What is the effect of an electric field on an electric dipole?
17. Is a dielectric material a conductor or an insulator? Explain.
18. What are the units of the dielectric constant?
19. Why does the dielectric constant vary for different materials?
20. A dielectric is placed in an isolated charged capacitor (fixed Q). What is the effect on (a) the electric field, (b) the voltage across the capacitor, and (c) the capacitance?

21. A dielectric is placed in a capacitor connected to a battery (fixed V). What is the effect on (a) the voltage across the capacitor, (b) the charge on the capacitor, and (c) the capacitance?
22. Give two major reasons for using dielectric materials in capacitors.

Capacitors in Series and in Parallel

23. Distinguish between series and parallel connections.
24. Two sets of identical capacitors are connected, with one set in series and the other in parallel. (a) For a given voltage, which set has more charge on the capacitor plates? (b) Is the voltage the same across the capacitors in each set?
25. How can two capacitors of different capacitances be connected so that the equivalent capacitance is (a) greater than either capacitance and (b) less than either capacitance?
26. What is meant by the dielectric strength of a material, and on what does it depend?
27. What is leakage current?
28. Compare the electrical characteristics of (a) polymer-film, (b) ceramic, and (c) electrolytic capacitors.

Capacitor Charging and Discharging

29. Explain why the charging and discharging of a capacitor varies exponentially with time rather than varies linearly.
30. What is the effect of an electrical resistance in a circuit when charging and discharging a capacitor?
31. (a) From the charging equations (Eq. 20.10), what is the charge on a capacitor at $t = 0$? (b) According to the discharge equations (Eq. 20.11), how long would it take for a capacitor to completely discharge? Is this a practical result?
32. Capacitance is defined as $C = Q/V$. Using the charging equations (Eq. 20.10), how does the capacitance vary when a capacitor is charged?
33. What would happen in a flasher circuit if the batteries started to "run down" and the terminal voltage fell below 70 V?

Problems

Levels of difficulty are indicated by asterisks for your convenience.

20.1 Capacitance and Capacitors

1. A parallel-plate air capacitor with a capacitance of 3.0 μF is connected to a 12-V battery. How much charge is transferred to the capacitor?
2. When a capacitor is connected to a 120-V source, it is found that there is a charge of 6.0×10^{-5} C on the plates. What is the capacitance?
3. A capacitor of 1.5 μF is connected to a voltage supply of 500 V. (a) How much charge is transferred to the plates? (b) Considering the charge on both plates, what is the total net charge of the capacitor?

*4. An air capacitor has a charge of 5.0×10^{-8} C on its plates, each of which has an area of 0.75 m^2. (a) What is the magnitude of the electric field between the plates? (b) If the separation distance of the plates is 2.0 mm, what are the capacitance and the potential difference of the plates?

*5. A parallel-plate air capacitor has plates 100 cm by 75 cm that are separated by a distance of 1.0 mm. What is its capacitance?

*6. Each of the plates of a capacitor has an area of 15 m^2. What separation distance would give a capacitance of 0.25 μF in air?

*7. To show why the farad is an impractically large unit, compute the area of the plates of a 1.0-F capacitor with a plate separation distance of 1.0 mm in air. Express the result in square miles.

*8. A variable air capacitor has plates with areas of 3.0×10^{-4} m^2 and a separation distance of 0.60 mm. How many plates would be required if the capacitor had a range of 0 to 124 pF?

**9. If a 20-plate variable air capacitor has half-circular plates with radii of 2.5 cm and separation distances of 0.75 mm, what is its maximum capacitance?

20.2 Energy of a Charged Capacitor

10. A 20-pF air capacitor is charged to 12 V. How much energy is stored in the capacitor?
11. If the voltage applied to a capacitor is increased from 10 V to 30 V, by what factor would the energy of the capacitor increase?

*12. A parallel-plate air capacitor is charged to 30 V. (a) How many times more energy would the capacitor have if it were charged to 120 V? (b) By what factor would the energy of the capacitor change if the separation distance were decreased by one half and the voltage decreased to 80 V from 120 V?

*13. A parallel-plate air capacitor with a plate area of 0.50 m^2 and a separation distance of 1.4 mm is charged to 12 V. (a) How much energy is stored in the capacitor? (b) What is the energy density?

*14. A 4.0-μF capacitor is connected to a variable voltage source. To what voltage must the capacitor be charged so that it has 0.0020 J of stored electrical energy?

*15. A parallel-plate air capacitor with a separation distance of 2.0 mm is charged to 120 V. What is the energy density of the capacitor?

**16. A parallel-plate air capacitor is charged to 20 V. If the plate area is 0.80 m^2 and the separation distance is 1.0 mm, (a) how much energy is stored in the capacitor? (b) What is the energy density? (c) What is the force on the plates?

20.3 Dielectrics and Dielectric Constants

17. A capacitor has a capacitance of 75 pF in air and a capacitance of 200 pF with a dielectric material between its plates. What is the dielectric constant of the material?

18. A fixed parallel-plate capacitor has a capacitance of 0.15 μF in air. A polymer film sandwiched between the plates increases the capacitance to 0.39 μF. What is the material? (See Table 20.1.)

19. What is the permittivity of mica?

*20. If the capacitor described in Problem 13 had a material with a dielectric constant of 4.0 between its plates, (a) what would be its capacitance? (b) How much energy would be stored in the capacitor?

*21. Two identical parallel-plate capacitors, one with polystyrene dielectric and the other with Teflon dielectric, are charged to the same voltage. (a) Which capacitor has greater capacitance? (b) Which has greater energy? How many times more in each case?

*22. Two parallel plates, 10 cm^2 in area, are separated by 1.5 mm. (a) What is the capacitance in vacuum? (b) What is the capacitance if a sheet of polyethylene 1.5 mm thick is inserted between the plates? (c) What is the capacitance if the capacitor is immersed in transformer oil?

*23. A tubular capacitor contains a film of polystyrene 1.0 mm thick. Electrical breakdown occurs when the capacitor is charged to 250 V. What is the dielectric strength of the film in (a) volts/meter and (b) volts/mil?

20.4 Capacitors in Series and in Parallel

24. What is the total capacitance of two capacitors of 2.0 μF and 4.0 μF connected (a) in series and (b) in parallel?

25. What are the maximum and minimum capacitances that can be obtained by combinations of any of three capacitors with capacitances of 0.50 μF, 1.0 μF, and 1.5 μF?

*26. An electronics technician has three individual capacitors available with capacitances of 5.0 pF, 10 pF, and 20 pF. What different values of capacitance can he get using only series or parallel combinations of two or more capacitors?

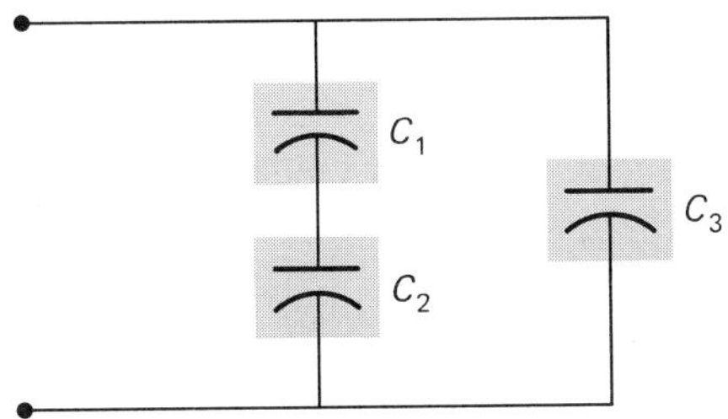

Figure 20.15
See Problem 28.

*27. Two capacitors have capacitances of 2.0 μF and 6.0 μF. (a) If they are connected in parallel and charged to a potential difference of 10 V, what is the charge on each capacitor, and what is the total charge? (b) If they are connected in series (after being discharged) and each receives a charge of 4.0 μC when connected to a battery, what is the potential difference across each capacitor, and what is the voltage of the battery?

*28. Three capacitors, with $C_1 = 0.20$ μF, $C_2 = 0.60$ μF, and $C_3 = 0.40$ μF, are connected as shown in Figure 20.15. (a) What is the total equivalent capacitance of the arrangement? (b) If a 12-V battery is connected to the terminals, how much charge is drawn from the battery?

*29. For the arrangement in Figure 20.15 with $C_1 =$ 20 pF and $C_2 = 30$ pF, what value of C_3 would give a total equivalent capacitance of 50 pF?

**30. For the arrangement in Figure 20.15 with $C_1 =$ 2.0 μF and $C_3 = 3.0$ μF, what value of C_2 would give a total equivalent capacitance of 4.0 μF?

**31. Three capacitors are connected as shown in Figure 20.15, with $C_2 = C_3 = 1.0$ μF. If the total equivalent capacitance is $\frac{4}{3}$ μF, what is the capacitance of C_1?

**32. Are there other possible values of capacitances using two or more of the capacitors in Problem 26 in different combinations? Explain, and compute any possible values.

20.5 Capacitor Charging and Discharging

33. What is the time constant of a 2.0-μF capacitor being charged through a resistance of 200 Ω?

34. What value of resistance should be used with a 0.25-μF capacitor to give a time constant of 1.0 s?

*35. What percentage of the maximum voltage would (a) a charging capacitor and (b) a discharging capacitor have in three time constants?

*36. What is the range of a variable capacitor that could be used with a 500-kΩ resistor to give a time constant between 1.0 s and 2.0 s?

*37. A 100-kΩ resistor is in series with a 0.50-μF capacitor. (a) What is the time constant? (b) If the *RC* series arrangement is connected to a 120-V source, what is the voltage across the capacitor in two time constants? (c) In a time of 0.18 s?

*38. A 1.5-μF capacitor with a voltage of 50 V is discharged through a 2.0-MΩ resistor. What is the charge on the capacitor in (a) one time constant and (b) 2.5 time constants? (c) In a time of 5.0 s?

(Courtesy Jean-Claude Lejeune, Stock, Boston)

Chapter 21

Current, Resistance, and Power

In our discussion of electrostatics in the last chapter, the electric charges were generally fixed or motionless. Static electricity is important, but the vast majority of electrical applications involve electric current, or charges in motion. When you turn on a flashlight or a car's headlights, there is a flow of electrons in a wire circuit. The lamp filament (a wire too) is part of the circuit, and here electrical energy is turned into heat and light (radiant energy). Similarly, when you switch on a light in the home, there is a flow of electrons in a circuit, but in this case the electrons move alternately back and forth. More will be said about this alternating current in Chapter 24. First a little history.

Early investigators thought electricity was the result of different fluids—positive and negative. That is probably why we speak about a "flow" of charges, similarly to a "flow" of heat, which was also originally thought to be a fluid.

Ben Franklin advanced a single-fluid theory of electricity. He postulated the existence of a tenuous, imponderable fluid. In his analysis, all bodies normally contained a certain quantity of the fluid; a surplus or deficit gave rise to electrical properties. With an excess, a body was positively excited, and with a deficit it was negatively excited. This theory was a basis for the later idea that it was the positive electric charges that flowed or moved. Today, we know it is the electrons that move in a conductor and give rise to a current.

21.1 Electric Current

In solid conductors, particularly metals, it is the outer atomic electrons that are relatively free to move. The atomic nuclei with the positively charged protons are fixed in the lattice structure of the material. (In ionic liquid conductors, both positive and negative *ions* can move.) But what

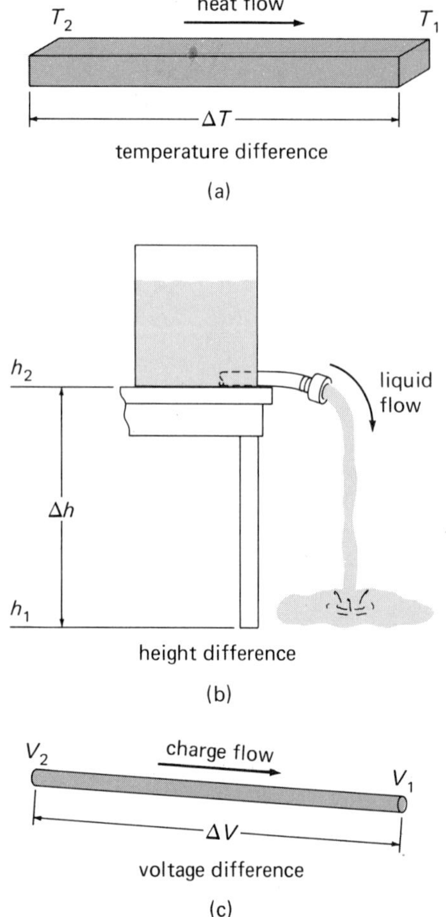

Figure 21.1
"Flows" generally require differences in conditions.

causes the electrons to move or flow, or causes an electric current in a conductor?

In the case of heat flowing in a thermal conductor or water flowing downhill in a conduit (pipe), we know the cause is a temperature *difference* or a height (potential energy) *difference* (Fig. 21.1). In the electrical case, there is also a difference—an electric potential, or voltage (energy/charge), *difference*. In a sense, the electron flow or current is down an electrical "hill," from a higher to a lower potential.

Of course, the "flows" in Figure 21.1 would only be temporary. When thermal equilibrium (equal temperatures), equal heights, and equal potentials were reached, or when there were no differences, the flows would cease. A sustained heat flow would require a heat pump and a system of thermal conductors, or a *circuit*, to maintain the temperature difference. Similar situations are needed for water flow and electric current (Fig. 21.2).

In the water "circuit," the path for water flow is complete and the pump supplies the work to maintain the potential difference for continuous flow. The water flow can do work, such as turning a paddle wheel. Similarly,

A sustained electric current requires a complete circuit and a voltage source.

Conducting wires usually complete an electric circuit, and the voltage source may be a battery. (See the Chapter Supplement on voltage sources.) The electric current can do work and light a bulb in the circuit. A circuit can be "opened" and "closed" by means of a switch. When you turn a light on, you are closing a switch and completing a circuit. When you turn a light off, you are opening the switch and interrupting the circuit.

Electric current describes the flow rate of charge:

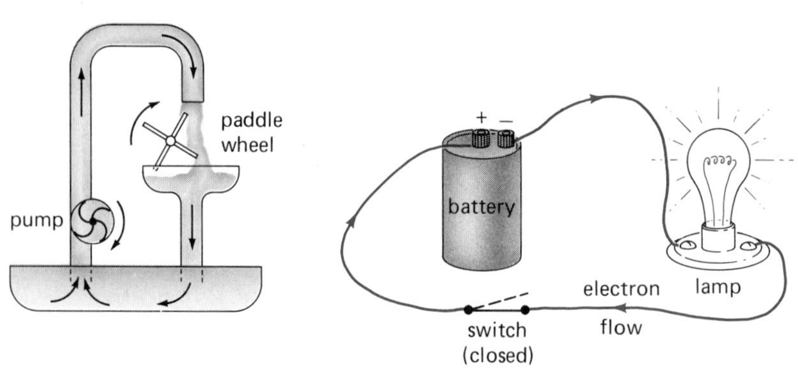

Figure 21.2
Circuit analogies. For sustained flow, a completed circuit and a method of maintaining a potential difference are required for both liquid and electrical circuits.

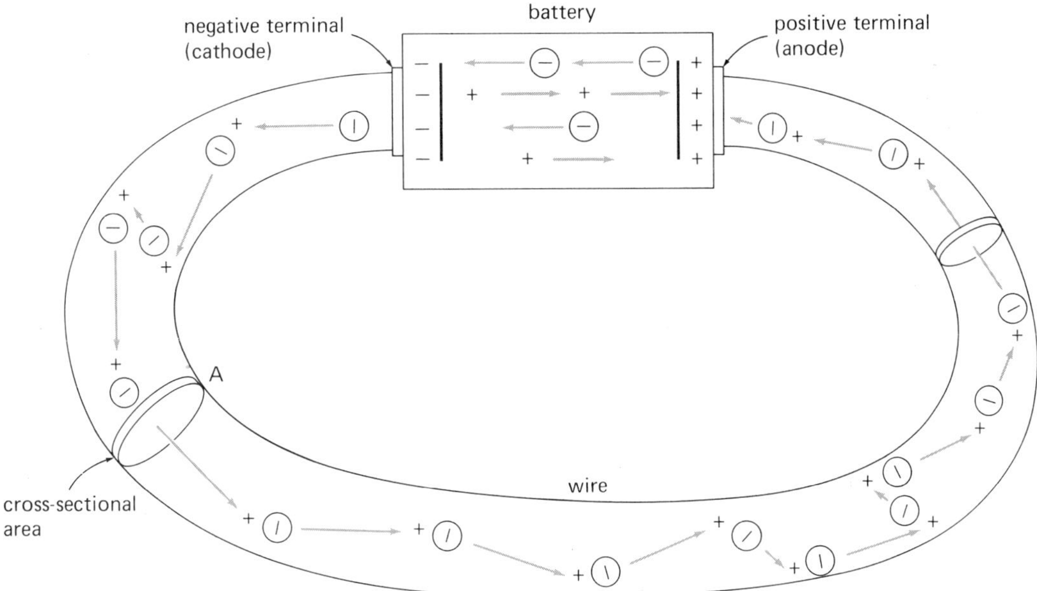

Figure 21.3
Electric current. An illustration of the charge flow in a battery circuit. The electric field set up in a wire by the battery causes electrons to move toward the positive terminal, giving rise to an electron flow or current. The electron motion is random and chaotic as a result of collisions, but there is a net "drift" of electrons in the wire toward the positive terminal.

An electric current I is the net charge q passing through a cross-sectional area of a conductor per unit time.

$$\boxed{I = \frac{q}{t}} \qquad \textbf{(Eq. 21.1)}$$

The unit of current can be seen to be coulomb per second. However, this unit is given the name **ampere** (A, or "amp" for short) in honor of the French physicist Andre Ampere (1775–1836).

$$I\ (\text{amp}) = \frac{q\ (\text{coulomb})}{t\ (\text{second})}$$

EXAMPLE 21.1 A current of 1.0 A flows in a conductor. How many electrons pass through a cross-sectional area of the conductor in 1.0 s?

Solution: It is given that $I = 1.0$ A and $t = 1.0$ s. Then $I = q/t$, or

$$q = It = (1.0\ \text{A})(1.0\ \text{s}) = 1.0\ \text{C}$$

of charge passes through the cross-sectional area. Then $q = ne$, where e is the electronic charge, and

$$n = \frac{q}{e} = \frac{1.0\ \text{C}}{1.6 \times 10^{-19}\ \text{C}} = 6.3 \times 10^{18}\ \text{electrons}$$

In a battery circuit the *net* motion of the electronic charge carriers is in one direction around the circuit (Fig. 21.3). Such a current is called a **direct current** (dc). The electron flow is away from the negative terminal of the battery toward the positive terminal.

The idea of electron flow is convenient in describing electrical current, but don't get the idea that the electrons flow around the circuit like water through pipes. Water circuits are helpful analogies, but they can be misleading, as in the case of "current flow."

In the absence of a voltage difference, the electrons in a metal wire move around randomly and chaotically at very high speeds—on the order of 1000 km/s (about 2 million miles per hour). With the electrons moving about in all directions, there is no net flow one way or the other.

With a potential difference from a battery across the wire, the electrons still collide and move chaotically, but there is a general charge movement along the wire toward the lower potential. This electron movement has an average drift velocity on the order of 0.1 cm/s—not much of a flow. In the case of alternating current (discussed in Chapter 24), the electrons move periodically back and forth and there is *no* net electron flow.

Yet, when you switch on a light it comes on instantaneously. Telephone conversations in the form of electric signals travel hundreds of miles in telephone lines. If this travel depended on actual electron flow, telephone conversations would be pretty slow. What, then, travels through the circuit? It is the electric field. The circuit wires act as guides for the electric force field that drives the conducting electrons. The electric field travels through the circuit with a speed near the speed of light (186,000 mi/s), so the results are almost instantaneous. The energy of the electric field does the work in the circuit.

So it's really energy that flows in a circuit, producing only a slow net motion of the electrons. However, we still talk in terms of current and charge flow because they are convenient terms for describing the overall electrical effects.

21.2 Ohm's Law

Motion is generally accompanied by resistance—for example, frictional resistance between objects and internal resistance, or viscosity, in fluid flow. Similarly, when there is charge flow, there is electrical resistance. On the atomic level, this opposition to charge flow or current arises from collisions between electrons and the lattice atoms or ions of the material. Hence, **electrical resistance (*R*)** is a material property.

The current I in a conductor is directly proportional to the potential difference, or voltage

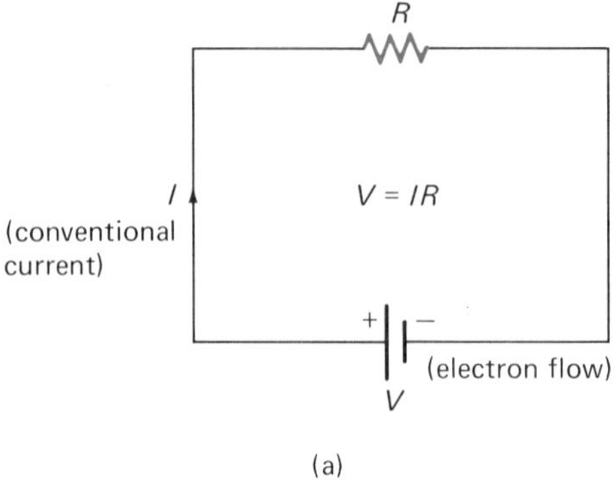

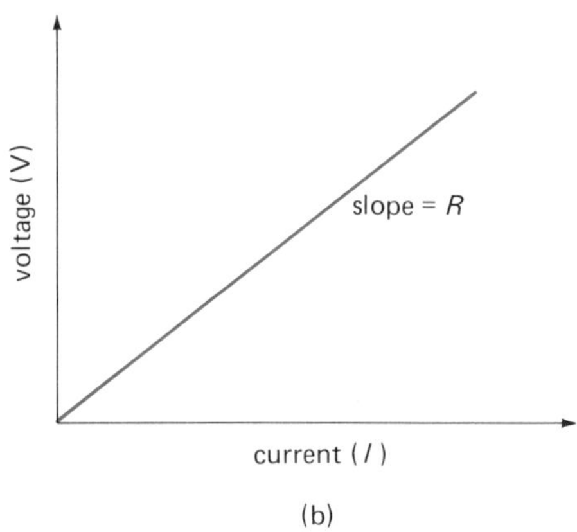

Figure 21.4
Ohm's law. (a) The relationship between voltage, current, and a resistance in many circuits is given by Ohm's law, $V = IR$. The resistance R might be a light bulb, as in Figure 21.2. (b) On a graph of V versus I for an ohmic conductor, the resistance is the slope of the straight line.

V, across the conductor, i.e.,

$$V \propto I$$

For many conductors, particularly metals, there is a linear relationship between current and voltage (Fig. 21.4), and we may write

$$\boxed{V = IR} \qquad \textbf{(Eq. 21.2)}$$

where R is the resistance and has the unit of ohm (Ω, Greek omega). This expression, known as **Ohm's law** (for Georg Ohm, 1789–1854, the German physicist who discovered it), is one of the most important relationships in electrical work.

A conductor that obeys Ohm's law is called an **ohmic conductor.** (It should be kept in mind that some materials are non-ohmic and do not follow Ohm's law.) As can be seen from Equation 21.2, for a given (constant) voltage V, the smaller the resistance, the greater the current flow, and vice versa ($I = V/R$). The resistance in a circuit element is represented by the symbol ⩘⩘.

EXAMPLE 21.2 When a particular lamp is connected to a 12-V battery, a 0.20-A current flows in the circuit. (a) What is the resistance of the lamp? (b) If the lamp were replaced with one having twice the resistance, what would be the current in the circuit?

Solution: It is given that $V = 12$ V and $I = 0.20$ A.

(a) Then, by Ohm's law,

$$R = \frac{V}{I} = \frac{12 \text{ V}}{0.20 \text{ A}} = 60 \ \Omega$$

(b) Since the resistance of the replacement lamp is twice as great, or $R = 2(60 \ \Omega) = 120 \ \Omega$, the current is half as much, as you might expect; or, directly from Ohm's law,

$$I = \frac{V}{R} = \frac{12 \text{ V}}{120 \ \Omega} = 0.10 \text{ A}$$

The actual flow of charge in a circuit is in the direction of the net electron motion. However,

By convention, the current direction in a dc circuit is taken to be the direction in which positive charge carriers would move.

Thus, the direction of the **conventional current** is opposite to that of the actual electron flow, or away from the positive terminal of the battery (Fig. 21.4).

The conventional current direction is based on historical reasons. As pointed out in the chapter introduction, it was once thought that current was associated with positive charge flow. Today we know it is the electrons that move in a circuit, but the current flow is still represented in the direction of conventional current in circuit diagrams.

21.3 Resistance and Resistivity

Let's take a closer look at the property of electrical resistance. Resistance is an intrinsic material property, and a variety of materials are used to make resistors (Fig. 21.5). However, resistance also depends on the geometry of the material. In general, the resistance of any material with a uniform cross-sectional area depends on four factors (Fig. 21.6):

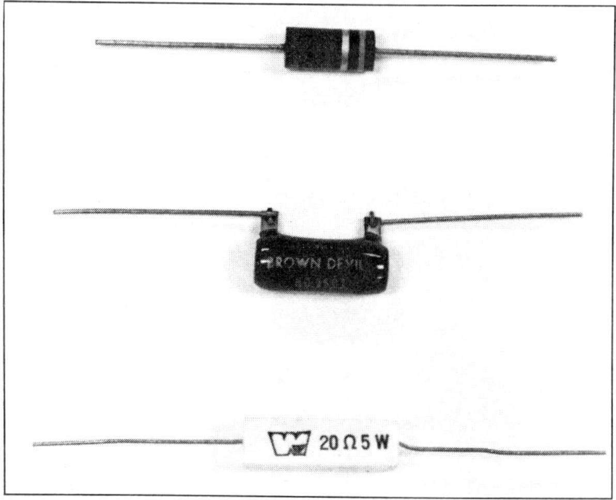

Figure 21.5
Commercially available resistors. The value of the tip resistor is given by color-coded bands.

1. Length
2. Cross-sectional area
3. Kind of material
4. Temperature

The resistance of a conductor, say in the form of a wire, depends on the wire's length L and its cross-sectional area A:

$$R \propto \frac{L}{A}$$

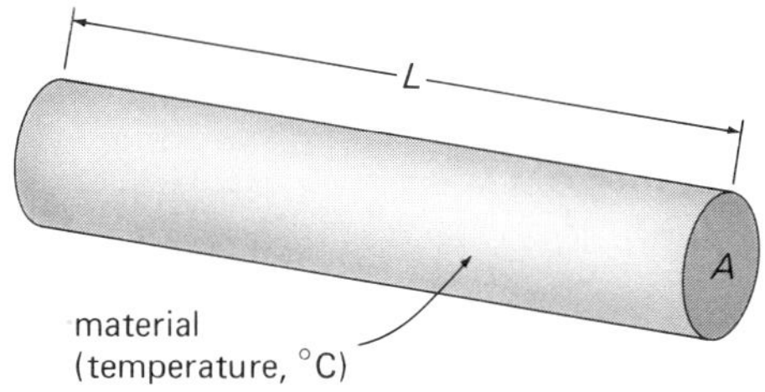

Figure 21.6
Resistance factors. The electrical resistance of a conductor depends on (1) its length, (2) its cross-sectional area, (3) the kind of material, and (4) temperature.

That is, the longer the wire, the greater the resistance, and the greater the cross-sectional area, the smaller the resistance (greater area for current flow). A wire conductor and resistance are analogous to a pipe with fluid flow and viscosity.

Different materials offer different resistances to electrical currents as a result of their compositions. The dependence of resistance on the kind of material and on temperature is characterized by the **resistivity** $\boldsymbol{\rho}$, and

$$\boxed{R = \frac{\rho L}{A}} \qquad \textbf{(Eq. 21.3)}$$

The units of resistivity from the equation are

$$\rho = \frac{R\ (\text{ohm})\ A(\text{m}^2)}{L(\text{m})} = \text{ohm-m } (\Omega\text{-m})$$

The resistivities of some common materials at room temperature (20°C or 68°F) are given in Table 21.1.

Table 21.1

Resistivities and Temperature Coefficients of Resistance for Various Materials

	ρ (20°C)		α
Material	**Ω-*m***	**Ω-*CM/ft***	**°C^{-1}**
Conductors			
Aluminum	2.8×10^{-8}	17.0	3.9×10^{-3}
Carbon	3.6×10^{-8}	21.8	-0.5×10^{-3}
Copper	1.7×10^{-8}	10.3	3.9×10^{-3}
Gold	2.4×10^{-8}	14.5	3.4×10^{-3}
Iron	10×10^{-8}	60.6	5.0×10^{-3}
Manganin (alloy of Cu, Mn, and Ni)	44×10^{-8}	267	0.01×10^{-3}
Mercury	96×10^{-8}	582	0.89×10^{-3}
Nichrome (alloy of Ni and Cr)	100×10^{-8}	606	0.44×10^{-3}
Nickel	7.8×10^{-8}	42.3	6.0×10^{-3}
Platinum	10×10^{-8}	60.6	3.0×10^{-3}
Silver	1.6×10^{-8}	9.70	3.8×10^{-3}
Tungsten	5.6×10^{-8}	34.0	4.5×10^{-3}
Insulators (typical orders of magnitude)			
Glass	10^{12}	10^{19}	—
Rubber	10^{15}	10^{22}	—
Wood	10^{10}	10^{17}	—

EXAMPLE 21.3 A copper wire has a diameter of 0.25 cm. What is the resistance of a 10-m length of this wire at room temperature?

Solution: Given: $d = 0.25$ cm $= 2.5 \times 10^{-3}$ m, $L = 10$ m, and, from Table 21.1, $\rho_{Cu} = 1.7 \times 10^{-8}$ Ω-m. The cross-sectional area of the wire is

$$A = \pi r^2 = \frac{\pi d^2}{4} = \frac{\pi(2.5 \times 10^{-3}\ \text{m})^2}{4}$$

$$= 4.9 \times 10^{-6}\ \text{m}^2$$

Then,

$$R = \frac{\rho L}{A} = \frac{(1.7 \times 10^{-8}\ \Omega\text{-m})(10\ \text{m})}{4.9 \times 10^{-6}\ \text{m}^2} = 0.035\ \Omega$$

This example shows why the resistance of the connecting wires in an ordinary electrical circuit is generally considered negligible.

Notice in Table 21.1 that the resistivities are also given in units of Ω-CM/ft, where CM is for circular mil. These are practical units that are particularly applicable to circular wires. Wires used in electrical applications generally have diameters that are in fractions of inches. (See Section 21.5 on wire gauges.) As a result, they are measured in mils (1 mil = 0.001 in.).

A wire with a diameter of 1 mil (0.001 in.) has an area of 1 CM (circular mil).

The area in CM is given simply by

$$A_{CM} = (d_{mils})^2 \qquad \textbf{(Eq. 21.4)}$$

where d is the diameter of the wire in mils (Fig. 21.7).

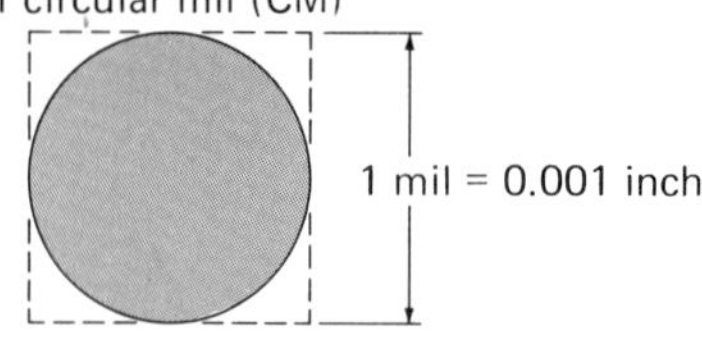

Figure 21.7

Circular mil. A wire with a diameter of 1 mil has an area of 1 CM.

EXAMPLE 21.4 A copper wire has a diameter of 0.064 in. What is the resistance of 1000 ft of this wire at room temperature?

Solution: It is given that $L = 1000$ ft and $d = 0.064$ in. $= 64$ mil. (Inches are converted to mils by moving the decimal point three places to the right.) The area of the wire in CM is then

$$A_{CM} = (d_{mils})^2 = (64)^2 = 4.1 \times 10^3 \text{ CM}$$

From Table 21.1, $\rho_{Cu} = 10.3$ Ω-CM/ft, and

$$R = \frac{\rho L}{A} = \frac{(10.3\ \Omega\text{-CM/ft})(10^3\ \text{ft})}{4.1 \times 10^3\ \text{CM}} = 2.5\ \Omega$$

Temperature Dependence of Resistance

The resistivity, and hence the resistance ($R \propto \rho$), are functions of temperature. **For most metallic conductors, the resistance increases with increasing temperature.** This is due to the increased molecular motion in the conductor, which hinders the flow of charge. For liquid conductors and some semiconductors, the resistance decreases with increasing temperature.

The temperature dependence of resistance is generally linear for normal temperature ranges. An equation similar to that of linear thermal expansion (Section 17.2) can be written for this relationship:

$$\Delta R = \alpha R_o \Delta T \quad (\text{or } \Delta\rho = \alpha\rho_o\Delta T) \qquad \textbf{(Eq. 21.5)}$$

where α is the **temperature coefficient of resistance** and $\Delta R = R - R_o$ is the difference between the final and initial resistances, respectively, for a temperature change ΔT. (Resistance is usually used instead of resistivity, since it can be measured directly in applications.)

Since $\Delta R/R_o$ is dimensionless, α has the units of 1/°C, or °C^{-1}. The α's for some conductors are listed in Table 21.1. Notice that carbon (a semiconductor) has a negative temperature coefficient of resistance. This means that the resistance decreases with increasing temperature.

Equation 21.5 can also be written

$$\Delta R = R - R_o = \alpha R_o \Delta T$$

and

$$R = R_o(1 + \alpha\Delta T) \quad [\text{or } \rho = \rho_o(1 + \alpha\Delta T)] \qquad \textbf{(Eq. 21.6)}$$

EXAMPLE 21.5 A piece of aluminum metal has a resistance of 65 Ω at 20°C. If the metal is heated to 100°C, what is its resistance at this temperature?

Solution: It is given that $R_o = 65\ \Omega$ and $\Delta T = T - T_o = 100°\text{C} - 20°\text{C} = 80°\text{C}$. From Table 21.1, $\alpha_{Al} = 3.9 \times 10^{-3}/°\text{C}$. Then the increase in resistance is

$$\Delta R = \alpha R_o \Delta T = (3.9 \times 10^{-3}/°\text{C})(65\ \Omega)(80°\text{C}) = 20\ \Omega$$

and

$$R = R_o + \Delta R = 65\ \Omega + 20\ \Omega = 85\ \Omega$$

(R could have been found directly from Eq. 21.6.)

Because of the linear relationship of resistance and temperature for metallic conductors, metals are used in electrical resistance thermometers (see the Chapter 16 Supplement on thermometry). See Special Feature 21.1 for a special condition of superconductivity, or zero resistance.

21.4 Electric Power

The resistance of materials to electric current results in the conversion of electrical energy to heat. We have all experienced an electrical component becoming hot—for example, a light bulb. Inversely, heat energy is converted to electrical energy in electrical generation processes.

These energy conversion processes with *time* are expressed in terms of **electric power.** Recall that power P is the work W per time t,

$$(\text{power}) \quad P = \frac{W}{t} \qquad \textbf{(Eq. 21.7)}$$

But electrical work is $W = qV$, so

$$P = \frac{W}{t} = \frac{qV}{t} = IV$$

SPECIAL FEATURE 21.1

Superconductivity

The electrical resistance of a metal conductor can be decreased by cooling. You may wonder how far one can go with this decrease. Strangely enough, for certain materials you can go all the way. At relatively low temperatures some materials exhibit **superconductivity,** and their electrical resistance vanishes, or goes to zero!

Above a certain critical temperature T_c, such a material is an ordinary conductor with some electrical resistance. At or below the critical temperature, however, its resistance falls abruptly to zero, as shown in the graph in Figure 1. The critical temperature is a property of a particular substance, just as the freezing point of water is 0°C.

In a superconducting state, an electric current established in a loop would persist indefinitely, with no resistive losses. Currents of several hundred amps introduced in a superconducting ring have been observed to remain constant for several years. In a sense, the conducting electrons of a superconductor never collide with the material lattice, while a conductor at normal temperatures always has some resistance and energy loss. The trick for superconduction is to maintain materials at very low temperatures.

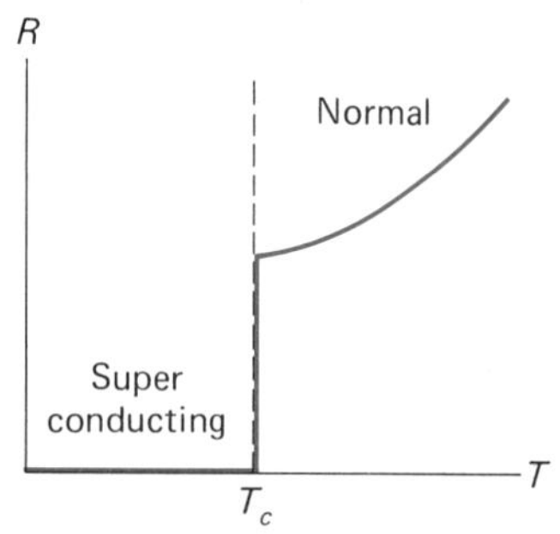

Figure 1
Resistance versus temperature for a pure superconductor. For metals such as mercury, lead, or aluminum, the resistance decreases with decreasing temperature down to the critical temperature T_c, below which the resistance drops to zero.

Superconductivity was discovered by a Dutch physicist, Heike Kammerlingh-Onnes, in 1911 and was first observed in solid mercury at a temperature of about 4 K (4 kelvins above *absolute* zero, −269°C or −452°F — pretty cold!). The mercury was cooled to this temperature using liquid helium. The boiling point, or the temperature at which helium condenses to a liquid, is about −267°C at 1 atm of pressure. (Liquefaction is accomplished at higher temperatures by increased pressure; cf. Chapter 16. Reduced pressure lowers the boiling point temperature.) Lead also exhibits superconductivity when cooled to liquid helium temperatures.

Liquid helium is relatively expensive, \$3 to \$6 a liter, depending on the amount you buy. So there has been a search for other materials that become superconducting at higher temperatures. Over the years, other superconducting metals and alloys were found, and the critical temperature crept upward to about 18 K (−255°C or −427°F). In 1973, it went up to 23 K (−250°C or −418°F).

In 1986 there was a major breakthrough and a new class of superconductors was discovered with higher critical temperatures. These were ceramic "alloys," or mixtures of rare earth elements such as lanthanum and yttrium. The new superconductors

---

or

$$\boxed{P = IV} \qquad \textbf{(Eq. 21.8)}$$

where $I = q/t$. Hence, the electric power is given by the product of the current and the voltage. Recall that the unit of power in the SI system is the watt (W).

Using Ohm's law, $V = IR$, Equation 21.8 can be written in other forms by substituting for V and I:

$$\boxed{P = IV = \frac{V^2}{R} = I^2R} \qquad \textbf{(Eq. 21.9)}$$

The power ratings of electrical components may be given directly in watts or in terms of voltage and current (Fig. 21.8). Notice that the can opener operates on 120 V (VAC — volts alternating current) and draws 1.4 A of current. Its watt power rating is $P = IV = (1.4\text{ A})(120\text{ V}) = 168\text{ W}$; that is, the can opener uses 168 J of electrical energy each second.* Some of this energy goes into doing useful work (opening a

* Although our initial discussion is concerned with direct current, the general forms of the equations apply to alternating current conditions (Chapter 24) and are used here for familiar illustration.

were prepared by grinding a blend of metallic elements and heating them at a high temperature, which produces a ceramic material. For example, one of the new superconductors was a mixture of yttrium, barium, and copper oxide. In 1986 the critical temperature rose to 57 K (−216°C or −357°F), and in 1987 a critical temperature of 98 K (−175°C or −283°F) was attained. Since then, a new thallium compound has been reported that achieves superconductivity at 125 K (−148°C or −234°F). There have also been reports of bits of materials becoming superconducting at room temperature (293 K), but these are questioned.

The 98-K material was a major breakthrough. Such "high-temperature" superconductors can be cooled using liquid nitrogen (Fig. 2). Liquid nitrogen has a boiling point of 77 K (−196°C or −321°F) and costs only about 25 to 40 cents per liter. (Nitrogen is readily plentiful as the major constituent of air.) This is an important, probably revolutionary scientific discovery. However, the revolution is going to take a while, which many reports fail to point out.

Figure 2

Magnetic levitation. A small magnet levitates above a superconducting ceramic disk, which is at a temperature of 77 K.

One of the main applications of superconductors is in superconducting magnets. Electromagnets are used in motors (Chapter 24), and the greater the strength of the magnet, the more powerful the motor. The magnet's strength depends on the current in the windings (wires). Without resistance, no loss occurs and there is greater current. (Superconducting magnets have been used on ship motors for some years, but with liquid helium.)

Superconducting magnets make certain processes more efficient and enable larger currents to be conducted. One potential application is in magnetically levitated trains, which are repelled off the track and ride on cushions of air. Superconducting magnets could also be used for propulsion. (Experimental "MagLev" trains have already been built using low-temperature superconducting magnets; see Chapter 24.) Other possible applications are underground electrical transmission cables and electric cars.

However, it is generally thought that such applications are some distance in the future. An application that is probably more imminent is computer interconnects. Interconnects are the metallic connections between computer chips by which they "talk" to each other. Superconducting interconnects would decrease power dissipation and possibly speed up the signal transfer, which makes for faster computers. The absence of electrical resistance opens up many possibilities.

can) and some energy is lost due to resistive heating, which is commonly called **joule heat** or **I^2R losses.**

The latter is particularly true for a light bulb. Before discussing this heating effect, let's see how the current requirement and resistance of a light bulb can be determined.

EXAMPLE 21.6 What is the current requirement and resistance of a 60-W household light bulb?

Solution: It is given that $P = 60$ W. Taking an operating voltage of $V = 120$ V (see Fig. 21.8),

$$I = \frac{P}{V} = \frac{60 \text{ W}}{120 \text{ V}} = 0.50 \text{ A}$$

and, using Ohm's law,

$$R = \frac{V}{I} = \frac{120 \text{ V}}{0.50 \text{ A}} = 240 \ \Omega$$

A 60-W light bulb uses or dissipates 60 J of electrical energy each second. Unfortunately, more than 95 percent of this is not in the form of visible light and is lost in the forms of nonvisible radiation and thermal energy. The filament

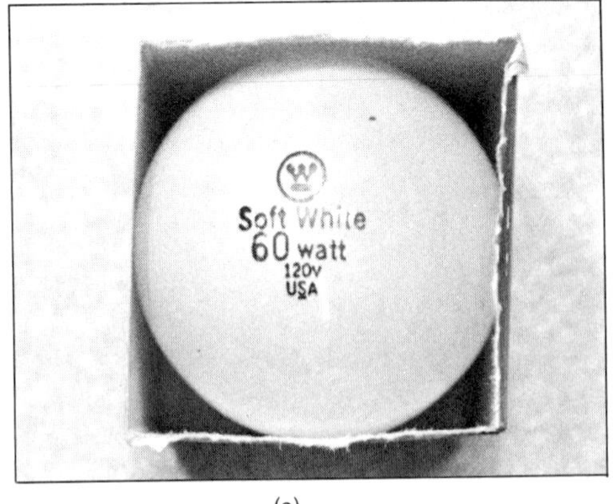

(a)

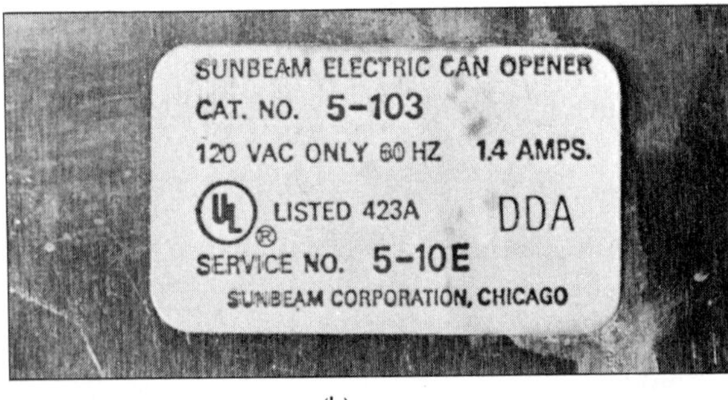

(b)

Figure 21.8
Power ratings. (a) The light bulb uses 60 watts (60 J/s) at 120 V. (b) The power rating for this can opener is expressed in terms of current and voltage ($P = IV$).

temperature of a common light bulb is on the order of 2500 to 3000°C. This is an example of a necessary but unwanted heating effect.

In applications in which heating effects are wanted, such as hair dryers and coffee makers, joule heating is promoted. This is done by using a heating element of relatively low resistance (but much higher than that of the connecting wires) so there will be a large current to take advantage of the square of the current in the I^2R losses. Another way of seeing this is through the equation for power in the form of $P = V^2/R$. For a constant voltage, when the resistance is decreased, the power increases.

EXAMPLE 21.7 An automatic coffee maker is rated at 1625 W at 120 V. (a) How much current does the coffee maker draw? (b) What is its resistance?

Solution: It is given that $P = 1625$ W and $V = 120$ V.

(a) $P = IV$, or

$$I = \frac{P}{V} = \frac{1625 \text{ W}}{120 \text{ V}} = 13.5 \text{ A}$$

(b) $P = \frac{V^2}{R}$, or

$$R = \frac{V^2}{P} = \frac{(120 \text{ V})^2}{1625 \text{ W}} = 8.9 \ \Omega$$

or, alternatively by Ohm's law, $V = IR$, or

$$R = \frac{V}{I} = \frac{120 \text{ V}}{13.5 \text{ A}} = 8.9 \ \Omega$$

Compare the resistance of the coffee maker with that of the light bulb in the previous example.

The power requirements of some common appliances and power tools are listed in Table 21.2.

The electrical energy is given by Equation 21.7 in the form

$$\text{(energy or work)} \quad \boxed{W = P \cdot t} \quad \textbf{(Eq. 21.10)}$$

where the unit is the watt-second (W-s). It is too small a quantity for most practical purposes, so the larger unit of kilowatt-hour (kWh) is used. The electric company bills us for the use of a certain number of kilowatt-hours, or electrical energy. The usage is monitored and recorded by a kilowatt-hour meter, found near the electrical service entrance of a home or business.

21.5 Wire Gauges

Miles of conductor wire are used in electrical applications. In an effort to standardize wire sizes, manufacturers have listed the sizes of wires in wire tables according to gauge numbers or gauge sizes.

Unfortunately, different wire tables assign different diameters to different gauge numbers. For example, the diameter of a No. 00 gauge wire (commonly written 2/0 and pronounced "two-oh" or "double-oh") in different wire tables varies from 0.3648 in. (Brown & Sharpe) to

Table 21.2
Typical Power Requirements of Some Common Household Appliances and Power Tools

Appliance	Power Requirement	Tool	Power Requirement
Air conditioner		Arc welder,	
Room	1500 W	portable	68,000 W
Central	4500 W	Belt sander	600 W
Blanket, electric	180 W	Bench grinder	650 W
Blender	800 W	Drill press	1250 W
Coffee maker,		Motor, ½ hp	500 W
automatic	1625 W	Saw	
Cooker, slow	300 W	Circular,	
Dishwasher	1200 W	portable	550 W
Food processor	330 W	Radial arm	1325 W
Heater		Table	1500 W
Portable	1400 W		
Water	4500 W		
Iron	1100 W		
Microwave oven	625 W		
Refrigerator			
Regular	400 W		
Frost-free	500 W		
Stove			
Range-top	6000 W		
Oven	4500 W		
Television			
Black-and-white	50 W		
Color	100 W		
Toaster	950 W		

0.380 in. (Stub's), to 0.3310 in. (Washburn & Moen), to 0.348 in. (British Standard).

In the United States, gauge sizes have come into dominant use and are now commonly referred to as **American Wire Gauges (AWG).** These vary from the hair-thin No. 60 to No. 4/0 (0000), which is almost half an inch in diameter.

The diameters and cross-sectional areas of some of the wires with lower AWG numbers are listed in Table 21.3. Notice that

The lower the gauge number, the larger the cross section of the wire and the smaller the resistance per length.

In wire tables, various data are given for a particular type of metal wire. For example, commonly listed items are the ohms per 1000 feet, the feet per ohm, and the pounds per 1000 feet (see Table 21.4).

Wires of different gauge sizes are used in different applications (Fig. 21.9). Some consideration is given to mechanical properties, but cost and current requirements are of prime importance. The latter is determined by the length and cross-sectional area of the wire. Gauge sizes are based on the cross-sectional area, which is inversely proportional to the resistance ($R = \rho L/A$).

Notice in Table 21.3 that

A decrease of ten gauge numbers is approximately equivalent to increasing the cross-sectional area by a factor of 10, which decreases the resistance by $\frac{1}{10}$ (other factors kept constant).

For example, compare AWG No. 11 (8234.0 CM) with AWG No. 1 (83,690 CM). Analysis of Table 21.3 reveals that

The cross-sectional area is approximately doubled for a decrease of three gauge numbers, and doubling the area decreases the resistance by one half (other factors kept constant).

For example, compare AWG No. 11 (8234.0 CM) with AWG No. 8 (16,510 CM).

EXAMPLE 21.8 In a household electrical circuit, 150 ft of AWG No. 14 solid copper wire is used. (a) What is the resistance of the wire at room temperature? (b) If AWG No. 12 wire were used instead, how would this affect the resistance?

Solution: (a) From Table 21.4, AWG No. 14 copper wire has a resistance of 2.525 Ω/1000 ft (R/L) at 20°C. Then, for L = 150 ft,

$$R_{14} = (R/L)L = (2.525\ \Omega/1000\ \text{ft})(150\ \text{ft})$$
$$= 0.379\ \Omega$$

(b) For No. 12 wire, the resistance is 1.588 Ω/1000 ft, and

$$R_{12} = (1.588\ \Omega/1000\ \text{ft})(150\ \text{ft}) = 0.238\ \Omega$$

So,

$$\frac{R_{12}}{R_{14}} = \frac{0.238\ \Omega}{0.379\ \Omega} = 0.63$$

and the resistance would be 0.63 as great.

Table 21.3
American Wire Gauge (AWG) Sizes

	British			Metric	
		Cross Section at 20°C			
Gauge No.	*Diameter in mils at 20°C*	*Circular Mils*	*Sq. Inches*	*Diameter in mm at 20°C*	*Cross Section in mm² at 20°C*
0000	460.0	211600	0.1662	11.68	107.15
000	409.6	167800	0.1318	10.40	85.03
00	364.8	133100	0.1045	9.266	67.43
0	324.9	105500	0.08289	8.252	53.48
1	289.3	83690	0.06573	7.348	42.41
2	257.6	66370	0.05213	6.544	33.63
3	229.4	52640	0.04134	5.827	26.67
4	204.3	41740	0.03278	5.189	21.15
5	181.9	33100	0.02600	4.621	16.77
6	162.0	26256	0.02062	4.115	13.30
7	144.3	20820	0.01635	3.665	10.55
8	128.5	16510	0.01297	3.264	8.366
9	114.4	13090	0.01028	2.906	6.634
10	101.9	10380	0.008155	2.588	5.261
11	90.74	8234	0.006467	2.305	4.172
12	80.81	6530	0.005129	2.053	3.309
13	71.96	5178	0.004067	1.828	2.624
14	64.08	4107	0.003225	1.628	2.081
16	50.82	2583	0.002028	1.291	1.309
18	40.30	1624	0.001276	1.024	0.8231
20	31.96	1022	0.0008023	0.8118	0.5176
22	25.35	642.4	0.0005046	0.6438	0.3255
24	20.10	404.0	0.0003173	0.5106	0.2047
26	15.94	254.1	0.0001996	0.4049	0.1288
28	12.64	159.8	0.0001255	0.3211	0.08098
30	10.03	100.5	0.00007894	0.2546	0.05093

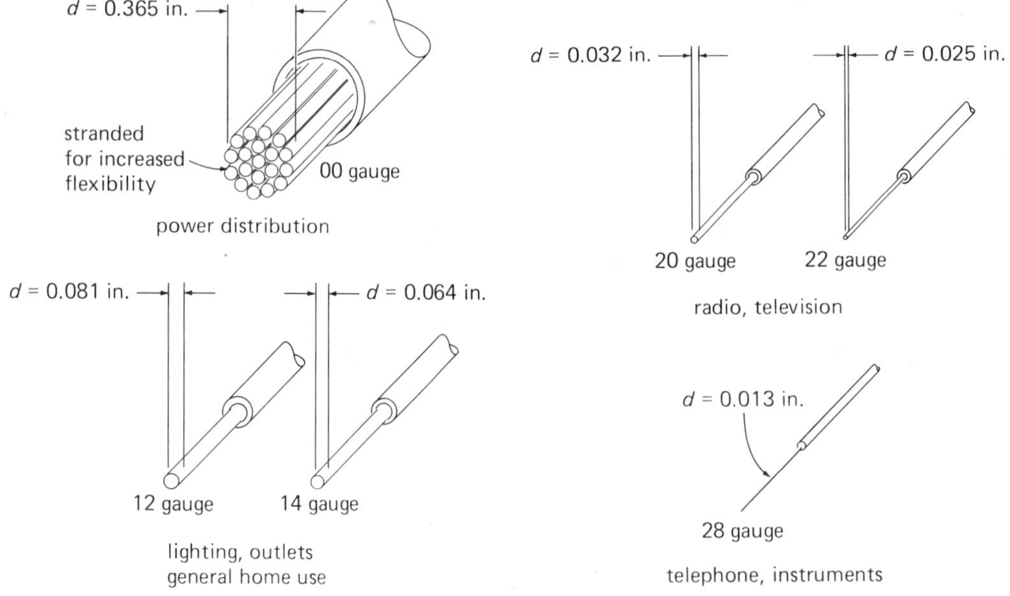

Figure 21.9
Applications of different wire gauges (AWG).

Table 21.4
Properties of Annealed Copper Wire at 20°C

	British			Metric		
Gauge No.	*ohms/1000 ft*	*feet/ohm*	*pounds/1000 ft*	*ohm/km*	*meters/ohm*	*kg/km*
0000	0.04901	20400	640.5	0.1608	6219	953.2
000	0.06180	16180	507.9	0.2028	4932	755.9
00	0.07793	12830	402.8	0.2557	3911	599.5
0	0.09827	10180	319.5	0.3224	3102	475.4
1	0.1239	8070	253.3	0.4066	2460	377.0
2	0.1563	6400	200.9	0.5127	1951	299.0
3	0.1970	5075	159.3	0.6465	1547	237.1
4	0.2485	4025	126.4	0.8152	1227	188.0
5	0.3133	3192	100.2	1.028	972.9	149.1
6	0.3951	2531	79.46	1.296	771.5	118.2
7	0.4982	2007	63.02	1.634	611.8	93.78
8	0.6282	1592	49.98	2.061	485.2	74.37
9	0.7921	1262	39.63	2.599	384.8	58.98
10	0.9989	1001	31.43	3.277	305.1	46.77
11	1.260	794.0	24.92	4.132	242.0	37.09
12	1.588	629.6	19.77	5.211	191.9	29.42
13	2.003	499.3	15.68	6.571	152.2	23.33
14	2.525	396.0	12.43	8.282	120.7	18.50
15	3.184	314.0	9.858	10.45	95.71	14.67
16	4.016	249.0	7.818	13.17	75.90	11.63
18	6.358	156.6	4.917	20.95	47.74	7.37
20	10.15	98.50	3.092	33.31	30.02	4.602
22	16.14	61.95	1.945	52.96	18.00	2.894
24	25.67	38.96	1.223	84.21	11.87	1.820
26	40.81	24.50	0.7692	133.9	7.468	1.145
28	64.90	15.41	0.4837	212.9	4.697	0.7199
30	103.2	9.691	0.3042	338.6	2.954	0.4527

Important Terms

electric current (*I*) the net charge q passing through a cross-sectional area of a conductor per unit time, $I = q/t$

ampere (A) the unit of electric current

Ohm's law for an ohmic conductor, the ratio of the voltage and current is equal to a constant, $R = V/I$, where R is the resistance

conventional current the current direction in which positive charge carriers would flow

resistivity (*ρ*) a measure of the resistive properties of a material, taking into account the geometrical considerations of cross-sectional area and length, $\rho = RA/L$

circular mil (CM) a relative measure of area for wires of circular cross section; a wire with a diameter of 1 mil (0.001 in.) has an area of 1 CM

temperature coefficient of resistance (*α*) the fractional change in resistance per unit temperature change, $\alpha = (\Delta R/R_o)/\Delta T$

electric power (*P*) the electric energy per unit time, $P = W/t = IV$

joule heat (I^2R losses) the energy lost due to resistive heating

Important Formulas

current: $I = \dfrac{q}{t}$

Ohm's law: $V = IR$

resistivity: $\rho = \dfrac{RA}{L}$ or $R = \dfrac{\rho L}{A}$

area in circular mils: $A_{CM} = (d_{mils})^2$

change in resistance [resistivity] with temperature:

$\Delta R = \alpha R_o \Delta T$
$[\Delta\rho = \alpha\rho_o\Delta T]$

or

$R = R_o(1 + \alpha\Delta T)$
$[\rho = \rho_o(1 + \alpha\Delta T)]$

electric power: $P = IV = \dfrac{V^2}{R} = I^2R$

Questions

Electric Current

1. If a wire is connected to two objects and current flows in the wire, what do you know about the electric potentials of the objects? Which way does the current flow?
2. Voltage sources "pump" energy into a circuit. Where does this energy go?
3. Distinguish between a transient current and a steady current. Give an example of each.
4. When a capacitor is connected to a battery, a current flows in charging the capacitor. How can this be when a capacitor is an open circuit?
5. When room lights are turned on by means of a wall switch, what occurs in the circuit?
6. (a) Why do we refer to a "flow" of charge? (b) If the drift velocity of electrons in a conductor is so small, why does an auto battery influence the starter as soon as you turn on the ignition switch?

Ohm's Law

7. What is a non-ohmic conductor? Give an example of such a conductor.
8. In a conductor, what are the actual charge carriers? What is the direction of the conventional current in a dc circuit?
9. If the resistance in a simple battery circuit is doubled, how is the current affected?
10. If the voltage in a circuit is doubled, how can the current in the circuit be maintained constantly?

Resistance and Resistivity

11. On what factors does the resistance of a material depend?
12. How does the resistance of a wire depend on the dimensions of the wire?
13. What does the resistivity of a material represent, and what are the units of resistivity?
14. Explain the advantage of using the unit of circular mil (CM) for wire areas.
15. In large current transfer applications, bus bars are used. These are metal bars of large cross-sectional areas. What is the advantage of such bars?
16. Why are copper and aluminum wires commonly used in electrical circuits rather than, say, iron wires?
17. If resistance were plotted versus temperature, what would be the slope of a straight-line graph? What type of resistance would this be?
18. An aluminum wire and a copper wire have the same dimensions and length. How do their resistances compare?
19. Compare the resistances of two copper wires of equal length, but one with a diameter of 1 mil and the other with a diameter of 3 mils.
20. Given two copper wires. If the radius of one wire is twice the other and the larger wire is three times as long, how do the resistances of the wires compare?

Electric Power

21. What is the difference between joule heating and I^2R losses?
22. If the current in a circuit is doubled, how is the power affected?
23. Which has a greater resistance, a 75-W or a 100-W light bulb? If the tungsten filament in each bulb were the same length, how would they differ otherwise?
24. What does the electric company bill us for, and what unit is used?
25. What six items are the biggest users of electrical energy in a typical home?
26. A hot dog can be cooked by "plugging it in," so to speak (Fig. 21.10). (*Note:* this is a lecture demonstration to be undertaken by the instructor and should *not* be attempted at home. It can be dangerous.) It takes only a minute or so to heat the hot dog. There are now commercially available hot dog cookers that operate on the same principle and cook several dogs at one time. What causes the hot dog to cook?

Wire Gauges

27. How does the resistance per length vary with AWG numbers?
28. Which is bigger around, AWG No. 000 or AWG No. 0 wire?
29. Given equal lengths of No. 10 copper wire and No. 22 copper wire, which has greater resistance?

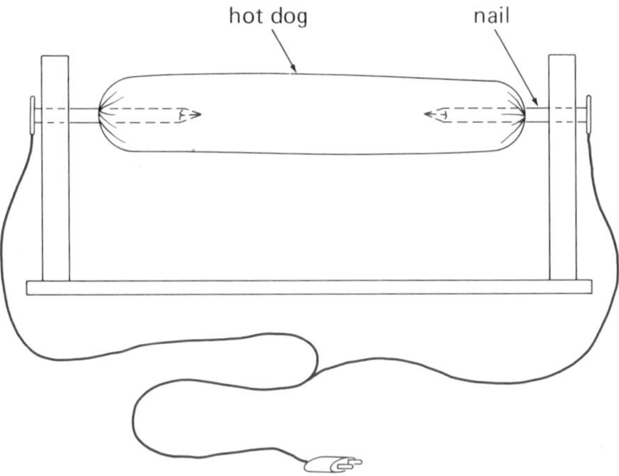

Figure 21.10
Hot dog! See Question 26.

30. A circuit is wired with No. 14 wire. What gauge wire of the same material should be used to decrease the wire resistance by (a) one half and (b) three fourths?
31. A circuit originally wired with No. 20 copper wire is rewired with No. 10 copper wire. What is the effect on the resistance and power loss in the circuit?

Problems

Levels of difficulty are indicated by asterisks for your convenience.

21.1 Electric Current

1. A net charge of 3.0 C flows through a cross-sectional area of a conductor in 0.75 s. What is the current?
*2. A current of 100 mA flows in a conductor. What is the net number of electrons passing through a cross-sectional area of the conductor in 0.50 s?
*3. There is a net movement of 5.0×10^{16} electrons in one direction in a wire in 2.0 s. What is the current in the wire?
*4. If a current of 1.0 A flows in a wire, what is the net number of electrons passing through a cross-sectional area of the wire in 3.0 s?

21.2 Ohm's Law

5. An appliance has a resistance of 200 Ω and operates on 110 V. How much current does the appliance draw?
6. When a trouble light is connected to a 12-V battery, a current of 500 mA flows in the circuit. What is the resistance of the circuit? What accounts for most of this resistance?
7. A circuit component with a resistance of 150 Ω requires a current of 0.80 A for proper operation. What is the voltage requirement for the component?
*8. An ohmic resistance in a circuit draws 150 mA from a 12-V source. How much current would be drawn from a 60-V source by the same resistance?
*9. When a resistor is connected to a 12-V source, it draws 0.250 A of current. When it is connected to a 90-V source, it draws 1.88 A. Is this an ohmic resistance?
*10. If a 12-V battery causes the current flow in Problem 1, what is the resistance of the circuit?

21.3 Resistance and Resistivity

11. High-purity copper has a resistivity of 1.6×10^{-8} Ω-m at room temperature. What is the resistance of a copper wire 1.0 m in length and 1.0 mm in diameter?
12. In a particular application, it is desired that a 50-m length of aluminum wire have a resistance of 2.0 Ω. What diameter wire should be used?
13. Given wires of the same cross-sectional area, how much greater is the resistance per unit length (a) for iron than for copper, and (b) for nichrome than for aluminum?
14. A large wire has a diameter of $\frac{1}{2}$ in. What is the cross section of the wire in circular mils?
15. A wire has a cross-sectional area of 4.1×10^5 CM. What is the diameter of the wire in millimeters?
*16. Show that 1 CM = $\pi/4$ square mils, where a square mil is the actual area unit.
*17. What would be the order of the resistance of a glass cube with a volume of 8.0 cm^3?
*18. A conductor has a square cross section with a length of 0.150 in. on a side. What is the cross-sectional area in CM? (Cf. Problem 16.)
*19. A copper wire has a diameter of 0.460 in. What is the resistance of 500 ft of this wire at room temperature?
*20. What length of nichrome wire with a diameter of 0.201 in. will provide a resistance of 10 Ω at room temperature?
*21. What length of manganin wire with a diameter of 10 mm has a resistance of 5.0 Ω at room temperature?
*22. A square iron bar, 1.5 in. on a side and 2.0 ft in length, is used as a conductor. What is the resistance of the bar at room temperature?
*23. A copper bar has a resistance of 0.50 Ω at room temperature. What would be the resistance of the bar at 80°C?
*24. A carbon resistor has a resistance of 240 Ω at room temperature. What would be the resistance at 100°C?
*25. An aluminum conductor and a similar tube of mercury have the same resistance at room temperature. Which has the greater change in resistance, and how many times greater, if the temperature of each is increased by the same amount?
*26. What is the temperature coefficient of resistance, in $°F^{-1}$, for (a) copper and (b) tungsten?
**27. In a particular circuit design, a metal conductor is to have a resistance of 0.150 Ω at room temperature. When the circuit is in operation, its ambient temperature is 120°C. What metal should be used to have a minimum resistance change, and what would be the resistance change?

**28. A tungsten wire is 0.010 in. in diameter and 15 ft long. How much more current would be conducted at room temperature than at a temperature of 1000°C in a circuit with a constant 50-V source?

21.4 Electric Power

29. A 60-W light bulb operates on 120 V. (a) How much current flows through the bulb? (b) What is the resistance of the bulb?
30. A portable electric heater is rated at 1400 W at 120 V. (a) How much energy is dissipated by the heater in 5.0 min? (b) What is the resistance of the heater?
31. A circuit component with a resistance of 250 Ω operates on 110 V. What is the power requirement of the component?
*32. A carbon resistor has a resistance of 15.0 Ω at room temperature. It is used in an application with an ambient temperature of 300°C and a voltage source of 12.0 V. Compare the power expenditures at these temperatures, with the same voltage source.
*33. A 100-Ω resistor is rated at 2.0 W. (a) What is the maximum safe voltage that could be applied to the resistor? (b) What would be the current at this voltage?
*34. How many kilowatt-hours of electrical energy does a 100-W light bulb use during 5.0 h of operation?
*35. A hair dryer rated at 1200 W operates on 120 V. (a) How much current does the dryer draw? (b) If the dryer cord has a resistance of 0.020 Ω, what fraction of the total wattage is dissipated in the cord as joule heat?
*36. Which would draw more current, a color TV or a black-and-white TV? How many times more? (*Hint:* see Table 21.2.)
**37. A water heater automatically operates for 2.0 h during the course of the day. If the cost of electricity is 12 cents per kWh, what is the cost of operating the heater for a month (30 days)? (See Table 21.2.)
**38. In the preparation of a meal, a food processor is used for 5.0 min, a coffee maker for 10 min, the range top of the stove for 30 min, and the stove oven for 45 min. During the cleanup, a dishwasher is used for 30 min. Assuming the power requirements in Table 21.2 for these appliances, what is the cost of the electricity for this use if the local rate is 9 cents per kWh?

21.5 Wire Gauges

*39. A wiring application calls for 500 m of AWG No. 10 copper wire. (a) What is the resistance of the wire at room temperature? (b) If AWG No. 12 were mistakenly used, how would this affect the resistance? (Give the factor of the effect.)
*40. A circuit uses AWG No. 14 copper wire. The resistance in the wire is 1.5 Ω. What is the length of the wire in meters?
*41. In a household electrical circuit, 400 ft of AWG No. 16 copper wire is used. (a) What is the resistance of the wire at room temperature? (b) What is the weight of the wire? (c) If AWG No. 13 were used, what would be the resistance and weight?
**42. An electrical power line contains a 5-mi length of AWG No. 0000 aluminum wire. (a) What is the resistance of the wire at room temperature? (b) What is the weight of the wire? (For AWG No. 0000 aluminum wire, cross-sectional area = 2.116×10^5 CM, and $\rho = 17.0$ Ω-CM/ft.)
**43. A 100-ft length of AWG No. 5 copper wire is used in a circuit application where the temperature of the wire is 70°C. (a) What is the resistance of the wire? (Neglect thermal expansion of the wire.) (b) If the potential difference between the ends of the wire is 2.0 V, how much current flows in the wire?

Chapter Supplement

dc Voltage Sources

To have a direct current, we must have a source of dc voltage, i.e., an electric potential difference of fixed polarity (+ and −). The positive terminal of the source is at a higher potential than the negative terminal.

Batteries

One of the most common dc voltage sources is the battery. A battery converts chemical energy into electrical energy in a chemical cell. The term *battery* refers to a "battery" of cells.

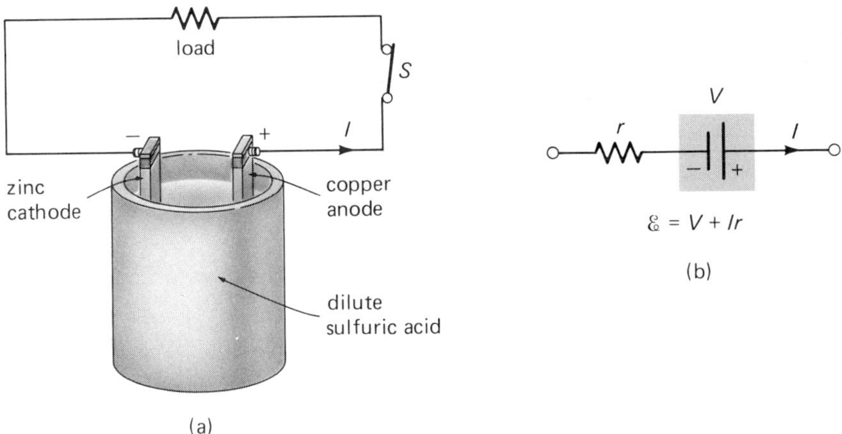

Figure 21S.1
A simple battery. (a) Charge flows in a circuit as a result of the chemical action of the cell. (b) Circuit symbol of a cell showing explicitly the internal resistance r. The emf of a battery is equal to the terminal voltage V plus the voltage drop across the internal resistance.

All battery cells have the same basic components: two electrodes—an anode (+) and a cathode (−)—and an electrolyte in which ionic charges are transferred between the electrodes. A simple voltaic cell is illustrated in Figure 21S.1.

A battery cell is said to be the seat of an **emf (electromotive force).*** The emf ($\mathscr{E}$) is the potential difference developed by the chemical action of the battery. But, because of the internal resistance r of the cell, there is an internal voltage drop, Ir, and the measured ***terminal* voltage *V*** of the battery is

$$V = \mathscr{E} - Ir \qquad \textbf{(Eq. 21S.1)}$$

Note that the emf is the open circuit voltage ($I = 0$ and $\mathscr{E} = V$).

A battery cell has a particular terminal voltage. To increase the voltage, several cells or batteries can be connected in series (Fig. 21S.2). In this case the voltages of the batteries add, and the current through each of the batteries is the same.

When batteries are connected in parallel (Fig. 21S.2), the voltage across all the batteries is the same, and each battery contributes a fraction of the total current.

* This is a misleading name. The electromotive force is *not* a force, but a voltage, or potential difference.

Batteries are divided into primary and secondary types. **A secondary battery can be recharged, while a primary battery is not designed to be.** The chemical reaction in a primary cell may be reversible to some degree, but recharging may cause such a battery to explode (due to gas formation) or to leak corrosive liquid.

One of the most common primary batteries is the carbon dry-cell battery (Fig. 21S.3). The negative electrode is a zinc cylinder, which also serves as the battery casing. The positive electrode is a carbon rod. The electrolyte is a paste of zinc chloride ($ZnCl_2$), ammonium chloride (NH_4Cl), and manganese dioxide (MnO_2).

The newer "alkaline" dry-cell batteries use potassium hydroxide (KOH, an alkali) instead of NH_4Cl. The reaction in the NH_4Cl cell produces ammonia gas (NH_3), which impairs the chemical reaction, causing the cell voltage to drop. This is not the case in alkaline cells, which have more constant outputs and longer lives.

A common secondary or rechargeable battery is the automobile battery (Fig. 21S.4). The 12-V auto battery is actually a series connection of six 2.0-V cells. Such batteries are called storage batteries. In the common lead storage battery, the positive and negative electrodes are porous lead dioxide (PbO_2) and spongy lead (Pb), re-

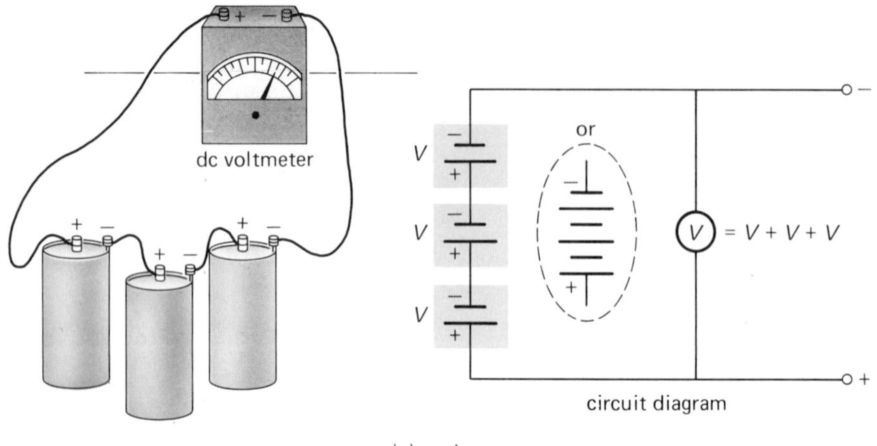

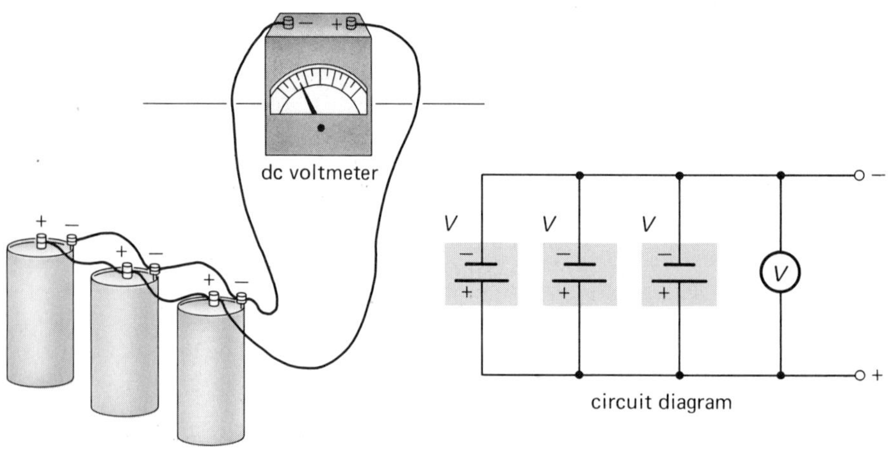

Figure 21S.2
Batteries in series and parallel. (a) In series, the voltages of the batteries add, and the current through each of the batteries is the same. (b) In parallel, the voltage across all of the (identical) batteries is the same, and each battery contributes a fraction of the total current.

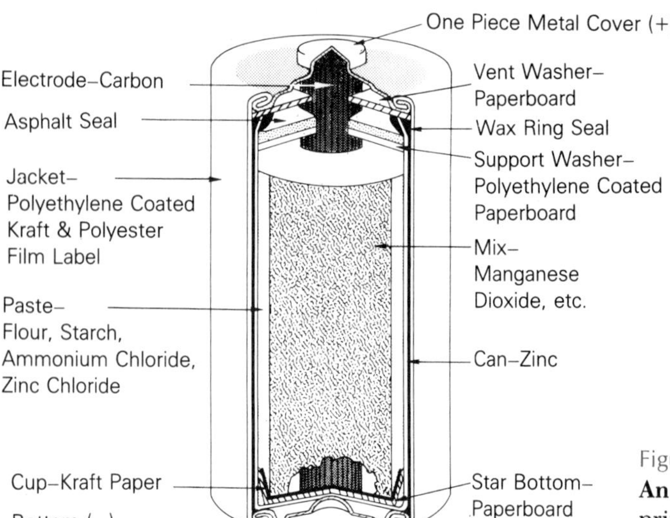

Figure 21S.3
An exposed view of the common dry-cell battery (a primary D-cell battery). The electrolyte is a paste. (Courtesy Union Carbide Co., Battery Products Division.)

(a)

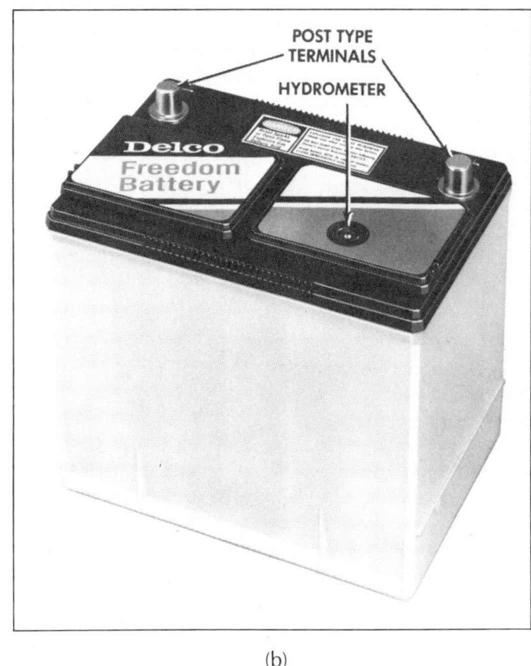

(b)

Figure 21S.4
Rechargeable automotive batteries. (a) Exposed view of a typical automotive battery. There are six 2.0-V cells in series to produce a 12-V output. (b) A "Freedom" battery, which requires no addition of water. The battery has a built-in hydrometer for a "state of charge" check. (Courtesy Delco Remy Division, General Motors Corporation.)

spectively. The electrolyte is sulfuric acid (H_2SO_4), which decomposes into H^+ ions and sulfate ions $(SO_4)^{2-}$.

The overall chemical reaction of the battery is given by

$$\underset{\text{(anode +)}}{PbO_2} + \underset{\text{(cathode −)}}{Pb} + \underset{\text{(electrolyte)}}{H_2SO_4}$$

$$\text{charge} \rightleftarrows \text{discharge}$$

$$\underset{\text{(lead sulfate on electrodes)}}{2PbSO_4} + \underset{\text{(water)}}{2H_2O}$$

In the discharge process, the concentration and density of the sulfuric acid electrolyte drop because of the formation of water. The "charge" of a battery can thus be checked by testing the density of the electrolyte with a hydrometer. This is done externally in many batteries, but some of the newer sealed batteries have built-in hydrometers (Fig. 21S.4).

The capacity rating of a battery is given in ampere-hours (A-h or amp-h) for a particular discharge current. For example, a battery might have a capacity rating of 60 amp-h at 3.0 amp. This means that with 3.0 amps being drawn from the battery, it will have a life of 60 amp-h/3.0 amp = 20 h. In general,

$$\text{Battery life (hours)} = \frac{\text{amp-hour rating}}{\text{rated amps}} \qquad \textbf{(Eq. 21S.2)}$$

The amp-hour rating is decreased if more than the rated amp current is drawn from the battery. The amp-hour rating also decreases with temperature. The rate of the chemical reaction of a battery falls with falling temperature, and consequently the current is also decreased. That is why automobile engines turn over slowly on cold days.

If the battery is connected to a voltage source that reverses the discharge current flow, the chemical action is reversed and the battery is recharged. This reverse action is the basis of electrolysis used in electroplating. For example, if a battery is connected between two electrodes in a copper sulfate ($CuSO_4$) solution, cop-

(a)

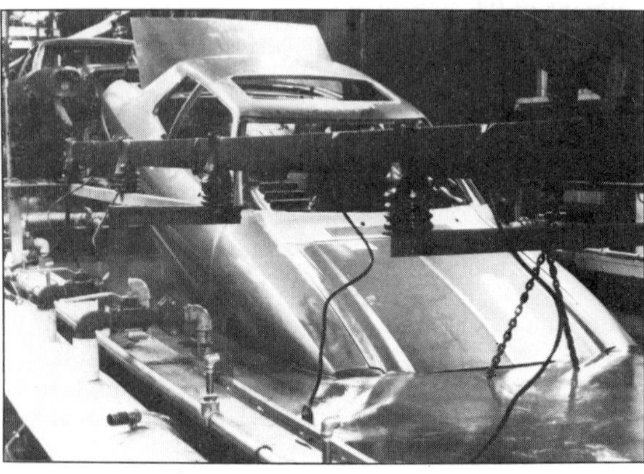

(b)

Figure 21S.5

Automotive electrocoat primer process. (a) In the process, a phosphate cleaning spray first strips any oil or dust from the car bodies, and a zinc-phosphate film is deposited on the sheet metal for good adhesion of the primer paint. (b) The car bodies are then dipped into an electrically charged vat of special primer paint. The bodies receive an opposite polarity from attached electrodes, causing paint to adhere instantly to the sheet metal. An average of almost 6 pounds of paint adheres to each vehicle. (Courtesy Ford Motor Co.)

per ions (Cu^{2+}) are attracted to the cathode and become copper atoms, and a layer of copper is "plated" on the electrode. Electroplating is used in industrial processes to apply a metal coating to objects—for example, chrome plating and other protective coatings (Fig. 21S.5).

Generators

Another source of dc voltage and current is the dc generator. In this case, mechanical energy is converted into electrical energy. The principle of the dc generator is discussed in Chapter 24.

Rectifiers

The normal electrical outlets in the home and laboratory are sources of alternating current (ac) and voltage. The polarity of ac voltage and current changes back and forth from plus to minus with time. With dc power supplies (Fig. 21S.6) it is possible to rectify ac current and convert it to dc current. Alternating current and rectification are discussed in Chapters 25 and 26.

Summary

The emf (electromotive force, $\mathscr{E}$) is the potential difference developed by the chemical action of a battery. The terminal voltage V is given by

$$V = \mathscr{E} - Ir$$

where r is the internal resistance of the battery.

The capacity rating of a battery is given in amp-hours at a rated discharge current. The bat-

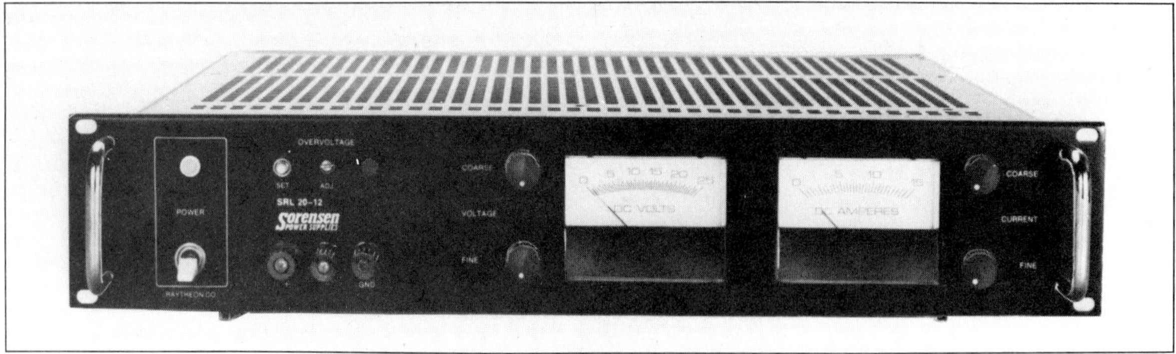

Figure 21S.6

A dc power supply. The supply converts or rectifies ac voltage and current to dc voltage and current. (Courtesy Sorensen Co., a unit of Raytheon Co.)

tery life is then given by

$$\text{Battery life (h)} = \frac{\text{amp-h rating}}{\text{rated amps}}$$

Secondary batteries are rechargeable and primary batteries are not.

Questions

1. Distinguish between the emf and the terminal voltage of a battery.
2. What are the different effects of batteries connected in series and those connected in parallel?
3. What is the difference between primary and secondary batteries?
4. Explain the principle of the common dry-cell battery used in flashlights.
5. What are the plate materials and the electrolyte of automotive batteries?
6. Why does a battery turn an automobile engine slowly on a cold day?
7. How are storage batteries rated?
8. Explain the principle of electroplating.

Problems

1. A battery has an internal resistance of 0.25 Ω. When it is connected to a resistance of 20 Ω, a current of 0.60 A flows in the circuit. What is the emf of the battery?
2. A battery is rated at 120 amp-h at 4.0 A. What is the rated battery life?

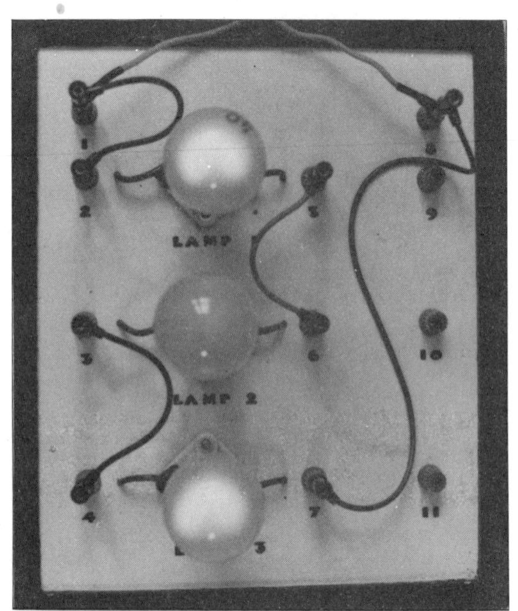

Chapter 22

Basic dc Circuits

Thousands of different circuits are used in various applications. Although the circuits are physically different and some circuits may be quite complex, the same basic principles apply to all. The essential relationship used in analyzing simple circuits is that of Ohm's law—the relationship between voltage and current.

As we learned in Chapter 21, the relationship between voltage and current depends on resistance. In this chapter the effects of the resistances of components in dc circuits will be studied. Major questions in circuit analysis concern how much current is "drawn" from the voltage source and how much power is expended in the circuit by various components.

22.1 **Resistances in Series and in Parallel**

Any dc circuit component other than a voltage source can be represented as a resistance. Even the connecting wires of a circuit have resistances, but these are usually considered negligible. In general there are two basic arrangements of resistances in circuits: series and parallel.

Resistances in Series

Figure 22.1 shows a circuit diagram for three resistances connected in series (resistances connected end to end). We generally say that the battery in a circuit is the source of a "voltage rise" and that "voltage (IR) drops" occur across the resistances. One may compare the series electric circuit in Figure 22.1 with the liquid circuit in Figure 22.2, in which the water height h is equal to the sum of the water "drops," h_1, h_2, and h_3, or $h = h_1 + h_2 + h_3$.* Similarly, in the electrical circuit

* This is a limited analogy—a resemblance or an explanation based on certain similarities between two things. Keep in mind that there are many differences between electrical and liquid circuits.

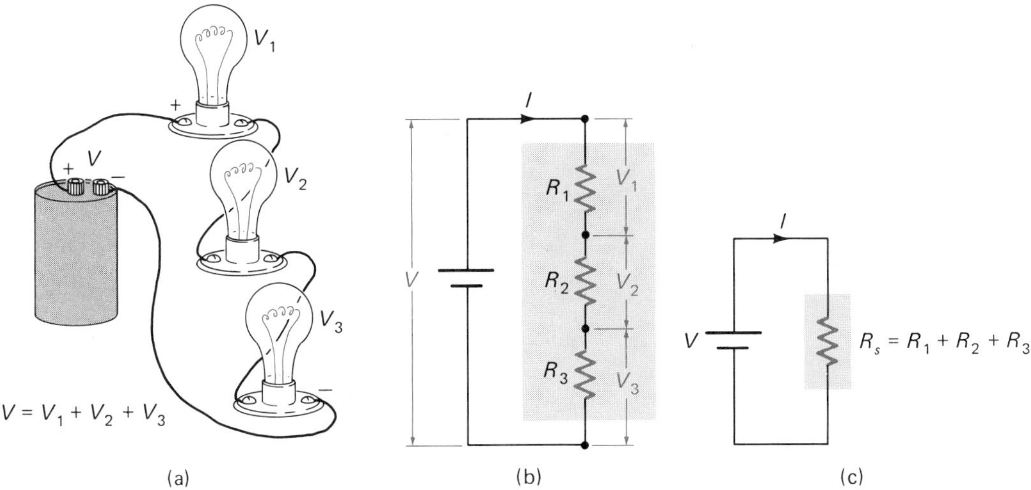

Figure 22.1
Resistances in series. (a) A battery circuit with three light bulbs (resistances) connected in series. (b) A circuit diagram for a battery and three resistances in series (connected end to end). Compare diagram (a) with diagram (b). The voltage rise of the battery is equal to the sum of the voltage drops across the individual resistances, $V = V_1 + V_2 + V_3$, and the current through each resistance is the same. (c) A circuit with the three resistances replaced by a single equivalent resistance R_s.

$$V = V_1 + V_2 + V_3$$

or the voltage "rise" of the battery equals the sum of the voltage "drops" of the resistances.

In a circuit with resistances in series, the voltage rise V of the voltage source (e.g., a battery) is equal to the sum of the voltage drops across the individual resistances.

Also,

In a series circuit, the current through each resistance is the same.

$$I = I_1 = I_2 = I_3$$

This is analogous to liquid flow through a continuous (series) pipe. Thus we may write, with $V = IR$,

$$V_s = V_1 + V_2 + V_3$$

$$IR_s = IR_1 + IR_2 + IR_3 = I(R_1 + R_2 + R_3)$$

and

$$\boxed{R_s = R_1 + R_2 + R_3} \qquad \textbf{(Eq. 22.1)}$$

(series resistances)

where R_s is the equivalent resistance of the series arrangement (Fig. 22.1). The formula can be extended to any number of resistances.

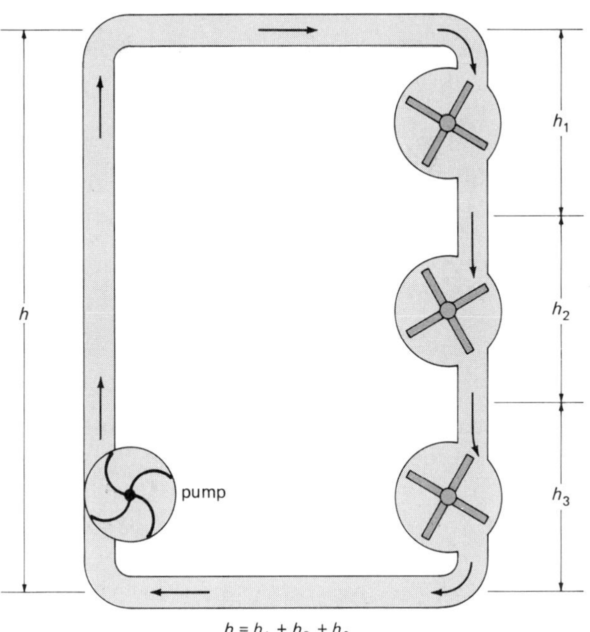

Figure 22.2
A liquid circuit with paddle wheels in series. Compare with the electrical circuit in Figure 22.1 for analogies.

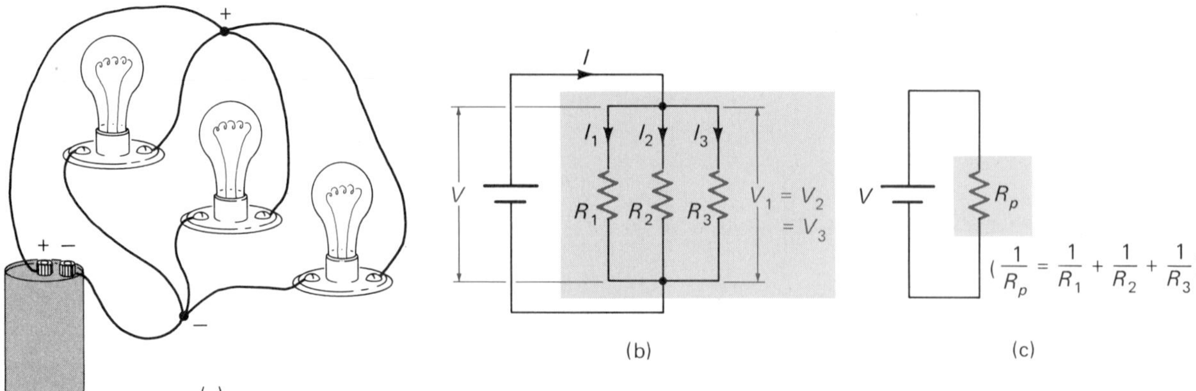

Figure 22.3
Resistances in parallel. (a) A battery circuit with three light bulbs (resistances) connected in parallel. (b) A circuit diagram for battery and three resistances in parallel (one side of each resistance connected together, and the other sides connected together). Compare diagram (a) with diagram (b). The voltage drop across each resistance is the same as the voltage rise of the battery, but the current from the battery divides among the resistances, and $I = I_1 + I_2 + I_3$, where $I_i = V/R_i$. (c) A circuit with the three resistances replaced by a single equivalent resistance R_p.

EXAMPLE 22.1 Suppose each of the resistances in Figure 22.1 is 10 ohms and the battery has a terminal voltage of 12 V. (a) What is the equivalent resistance? (b) How much current is drawn from the battery?

Solution: It is given that $R_1 = R_2 = R_3 = 10\ \Omega$ and $V = 12$ V. Then,

(a)
$$R_s = R_1 + R_2 + R_3 = 10\ \Omega + 10\ \Omega + 10\ \Omega = 30\ \Omega$$

That is, the three 10-Ω resistances could be equivalently replaced with a 30-Ω resistance.

(b) With $V = IR_s$,

$$I = \frac{V}{R_s} = \frac{12\text{ V}}{30\ \Omega} = 0.40\text{ A}$$

Hence, a current of 0.40 A flows through each of the resistances.

Note that the voltage drop across each resistance is $V_1 = IR_1 = (0.40\text{ A})(10\ \Omega) = 4.0$ V and, similarly, $V_2 = V_3 = 4.0$ V, such that

$$V = V_1 + V_2 + V_3 = 4.0\text{ V} + 4.0\text{ V} + 4.0\text{ V} = 12\text{ V}$$

(sum of voltage drops equals voltage rise)

Resistances in Parallel

Figure 22.3 shows a circuit diagram for three resistances connected in parallel (one side of each resistance connected together and the other sides connected together).

In a circuit with resistances in parallel, the voltage drop across each resistance is equal to the voltage rise of the battery.

That is,

$$V = V_1 = V_2 = V_3$$

This is analogous to a liquid parallel circuit where $h = h_1 = h_2 = h_3$ (Fig. 22.4).

But notice that the current from the battery divides at the junction, and

$$I = I_1 + I_2 + I_3$$

For resistances in parallel, the current entering a junction is equal to the sum of the currents leaving the junction.

Naturally, the current divides according to the resistances in the parallel branches, with the greatest amount of current taking the "path of least resistance."

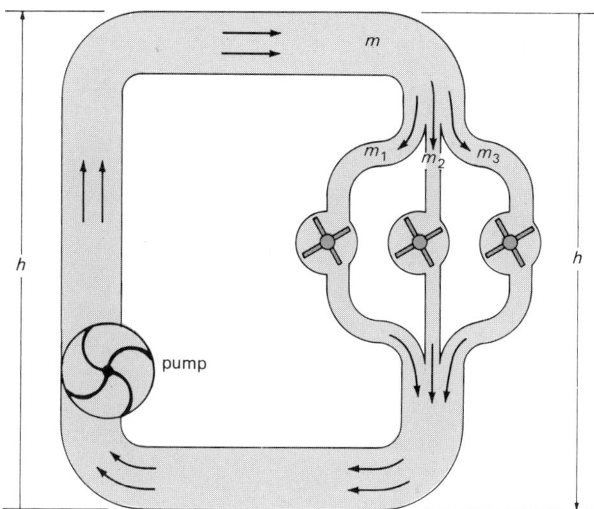

Figure 22.4
A liquid circuit with paddle wheels in parallel. Compare with the electrical circuit in Figure 22.3 for analogies.

Thus, we may write, with $I = V/R$,

$$I = I_1 + I_2 + I_3$$

$$\frac{V}{R_p} = \frac{V}{R_1} + \frac{V}{R_2} + \frac{V}{R_3} = V\left(\frac{1}{R_1} + \frac{1}{R_2} + \frac{1}{R_3}\right)$$

and

$$\boxed{\frac{1}{R_p} = \frac{1}{R_1} + \frac{1}{R_2} + \frac{1}{R_3}} \qquad \textbf{(Eq. 22.2)}$$

(parallel resistances)

where R_p is the equivalent resistance of the parallel arrangement (Fig. 22.3). The formula can be extended to any number of resistances.

In the case of two resistances in parallel, we have the more convenient form

$$\frac{1}{R_p} = \frac{1}{R_1} + \frac{1}{R_2} = \frac{R_1 + R_2}{R_1 R_2}$$

and

$$\boxed{R_p = \frac{R_1 R_2}{R_1 + R_2}} \qquad \textbf{(Eq. 22.3)}$$

(two resistances in parallel)

EXAMPLE 22.2 Suppose the resistances in Figure 22.3 are two 10-Ω resistances and a 20-Ω resistance and that the battery has a terminal voltage of 12 V. (a) What is the equivalent resistance? (b) What are the current drawn from the battery and the current through each resistance?

Solution: It is given that $R_1 = R_2 = 10\ \Omega$, $R_3 = 20\ \Omega$ and $V = 12$ V. Then,
(a)

$$\frac{1}{R_p} = \frac{1}{R_1} + \frac{1}{R_2} + \frac{1}{R_3} = \frac{1}{10\ \Omega} + \frac{1}{10\ \Omega} + \frac{1}{20\ \Omega}$$

$$= \frac{2}{20\ \Omega} + \frac{2}{20\ \Omega} + \frac{1}{20\ \Omega}$$

$$= \frac{5}{20\ \Omega} = \frac{1}{4\ \Omega}$$

and

$$R_p = 4.0\ \Omega$$

Alternatively, combining the two 10-Ω resistances,

$$R_{p_1} = \frac{R_1 R_2}{R_1 + R_2} = \frac{(10\ \Omega)(10\ \Omega)}{10\ \Omega + 10\ \Omega} = 5.0\ \Omega$$

Then, combining this equivalent resistance with the 20-Ω resistance,

$$R_p = \frac{R_{p_1} R_3}{R_{p_1} + R_3} = \frac{(5\ \Omega)(20\ \Omega)}{5\ \Omega + 20\ \Omega} = 4.0\ \Omega$$

(b) The current drawn from the battery is

$$I = \frac{V}{R_p} = \frac{12\ \text{V}}{4.0\ \Omega} = 3.0\ \text{A}$$

Then, since the voltage drop across each resistance is the same,

$$I_1 = \frac{V}{R_1} = \frac{12\ \text{V}}{10\ \Omega} = 1.2\ \text{A}$$

$$I_2 = \frac{V}{R_2} = \frac{12\ \text{V}}{10\ \Omega} = 1.2\ \text{A}$$

$$I_3 = \frac{V}{R_3} = \frac{12\ \text{V}}{20\ \Omega} = 0.6\ \text{A}$$

Notice that

$$I = I_1 + I_2 + I_3 = 1.2\ \text{A} + 1.2\ \text{A} + 0.6\ \text{A} = 3.0\ \text{A}$$

Also, the current I from the battery divides at the parallel junction inversely proportionately to the resistances in the parallel branches. For example, one half as much current flows through the 20-Ω resistance as through a 10-Ω resistance.

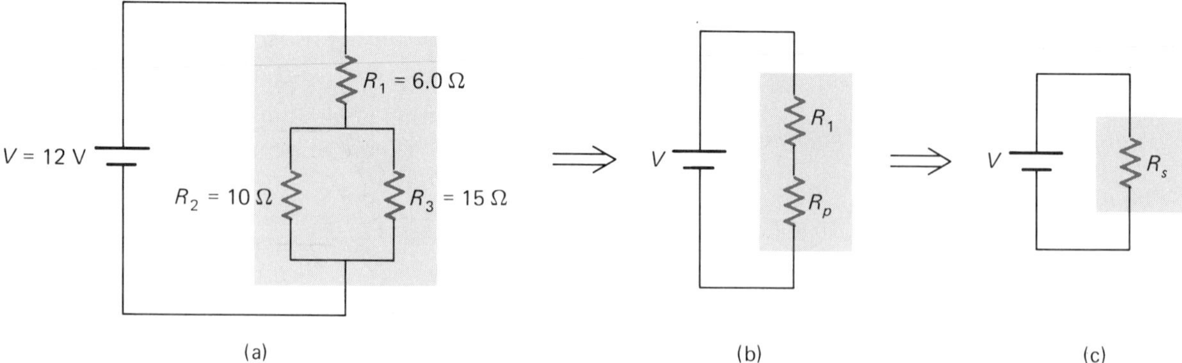

Figure 22.5
Series-parallel resistances and circuit reduction. (a) The resistance R_1 in series with resistances R_2 and R_3, which are in parallel. (b) R_2 and R_3 can be equivalently replaced by R_p, and R_1 and R_p are in series. (c) R_1 and R_p can be equivalently replaced with a single resistance R_s. The current from the battery is then given by $I = V/R_s$.

Another interesting fact for resistances in parallel is that

> **The equivalent resistance R_p is always less than the smallest resistance in the parallel arrangement.**

For instance, see the alternative method of resistance combinations in the previous example.

Series-Parallel Resistances

Circuits may have resistances in a variety of series-parallel arrangements. For example, in Figure 22.5, R_1 is in a series with the parallel arrangement of R_2 and R_3.

To find out how much current is drawn from the battery by such a series-parallel arrangement, the resistances are "reduced" to a single equivalent resistance. Combining the two parallel resistances, we see that they could be equivalently replaced by a resistance of

$$R_p = \frac{R_2 R_3}{R_2 + R_3} = \frac{(10\ \Omega)(15\ \Omega)}{10\ \Omega + 15\ \Omega} = 6.0\ \Omega$$

Then R_1 and R_p in series are combined:

$$R_s = R_1 + R_p = 6.0\ \Omega + 6.0\ \Omega = 12\ \Omega$$

Hence, the three resistances in the series-parallel arrangement could be equivalently replaced by a 12-Ω resistance. The current supplied by the battery is then

$$I = \frac{V}{R_s} = \frac{12\ \text{V}}{12\ \Omega} = 1.0\ \text{A}$$

EXAMPLE 22.3 (a) What is the current through each of the resistances in Figure 22.5? (b) How much power is expended by each?

Solution: (a) It was shown that 1.0 A is drawn from the battery. Since R_1 is in a series with the battery, a current of 1.0 A flows through R_1. The voltage drop across R_1 is then

$$V_1 = I_1 R_1 = (1.0\ \text{A})(6.0\ \Omega) = 6.0\ \text{V}$$

The voltage drop across R_p is

$$V_p = I_1 R_p = (1.0\ \text{A})(6.0\ \Omega) = 6.0\ \text{V}$$

(Notice that $V = V_1 + V_p = 6.0\ \text{V} + 6.0\ \text{V} = 12\ \text{V}$, or voltage drops equal voltage rise.)

V_p is the voltage drop across both R_2 and R_3, and

$$I_2 = \frac{V_p}{R_2} = \frac{6.0\ \text{V}}{10\ \Omega} = 0.60\ \text{A}$$

$$I_3 = \frac{V_p}{R_3} = \frac{6.0\ \text{V}}{15\ \Omega} = 0.40\ \text{A}$$

(Notice that $I_1 + I_2 + I_3 = 0.6\ \text{A} + 0.4\ \text{A} = 1.0\ \text{A}$. That is, the current divides between the two resistances at the junction.)

(b) With the preceding values,

$$P_1 = I_1 V_1 = (1.0\ \text{A})(6.0\ \text{V}) = 6.0\ \text{W}$$

$$P_2 = I_2 V_p = (0.60\ \text{A})(6.0\ \text{V}) = 3.6\ \text{W}$$

$$P_3 = I_3 V_p = (0.40\ \text{A})(6.0\ \text{V}) = 2.4\ \text{W}$$

Notice that the power expended by the total equivalent resistance $R_s = 12\ \Omega$ is $P = IR_s = (1.0\ \text{A})(12\ \Omega) = 12\ \text{W}$ and that $P = P_1 + P_2 + P_3 = 6.0\ \text{W} + 3.6\ \text{W} + 2.4\ \text{W} = 12\ \text{W}$.

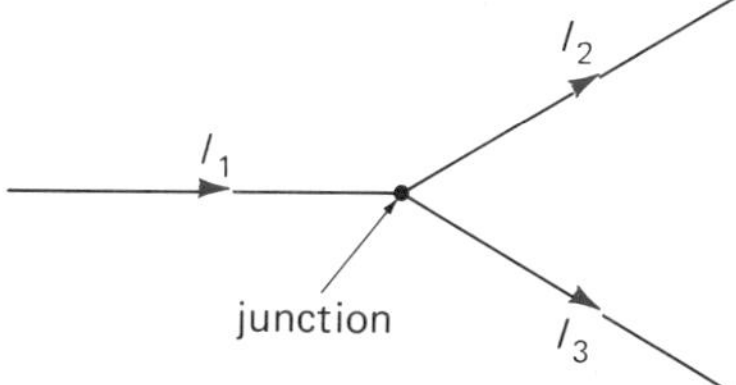

Figure 22.6
Kirchhoff's current or junction rule. The sum of the currents entering and leaving a junction is zero, $\Sigma I = 0$, i.e., $I_1 - I_2 - I_3 = 0$, where the currents leaving a junction are taken to be negative. Equivalently, the currents "into" a junction are equal to the currents "out," $I_1 = I_2 + I_3$.

22.2 Kirchhoff's Rules and Multiloop Circuits

The elements of circuit analysis are summed up in two general statements known as **Kirchhoff's rules:** the current rule and the voltage rule.

Kirchhoff's current rule states:

> **The sum of the currents entering and leaving a junction is zero, i.e.,**

$$\boxed{\Sigma I = 0} \qquad \textbf{(Eq. 22.4)}$$

or

$$\Sigma I_{in} = \Sigma I_{out}$$

A **junction** is a point where three or more connecting wires of a circuit meet. A **branch** is a circuit segment between two junctions that contains one or more circuit elements.

The current rule is essentially a statement of the conservation of charge. For example, for the junction in Figure 22.6, taking currents entering a junction as positive and those leaving as negative,

$$I_1 - I_2 - I_3 = 0$$

$$(\text{or} \quad \underset{\text{in}}{I_1} = \underset{\text{out}}{I_2 + I_3})$$

In applying the current rule to a junction, the current directions are assumed. Should the assumed direction be incorrect, the solved value of the current will be negative, indicating that it is actually in the opposite direction.

Kirchhoff's voltage rule states:

> **The sum of the voltage rises and the voltage drops around a closed circuit loop is zero, i.e.,**

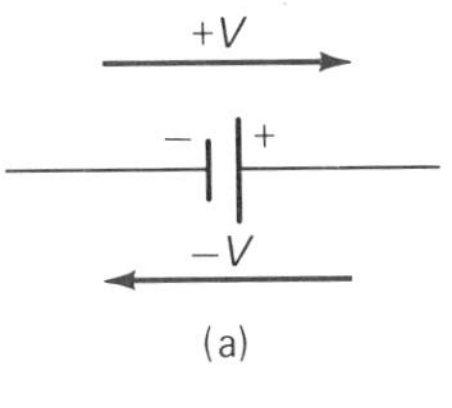

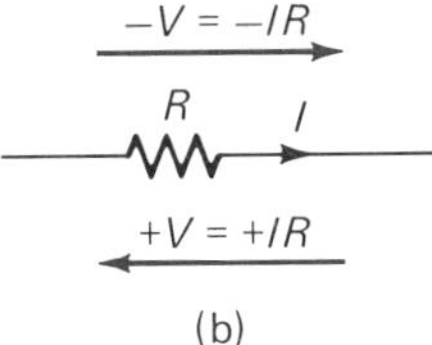

Figure 22.7
Sign convention for Kirchhoff's rules. See text for description.

$$\boxed{\Sigma V = 0} \qquad \textbf{(Eq. 22.5)}$$

A closed loop is any continuous path around a circuit that leaves a point in one direction and returns to the same point. An equivalent statement of the voltage rule is that the sum of the voltage rises equals the sum of the voltage drops around a closed circuit loop. This is essentially a statement of the conservation of energy.

A sign convention is used for voltage changes as a loop is traversed, since the direction around the loop can be either clockwise or counterclockwise:

1. The voltage change of a battery is taken to be positive when the battery is traversed in the direction of its positive (+) terminal, and negative when it is traversed in the direction of its negative (−) terminal. See Fig. 22.7(a).
2. When passing through a resistance, the voltage change across the resistance is taken to be negative if it is traversed in the direction of the assigned current flow, and positive if the resistance is traversed in the opposite direction. See Fig. 22.7(b).

For example, for the circuit in Figure 22.8, Kirchhoff's voltage rule gives

$$V - IR_1 - IR_2 - IR_3 = 0$$

(Compare this with the development of Eq. 22.1.) The direction in which a circuit loop is traversed does not make any difference. If the

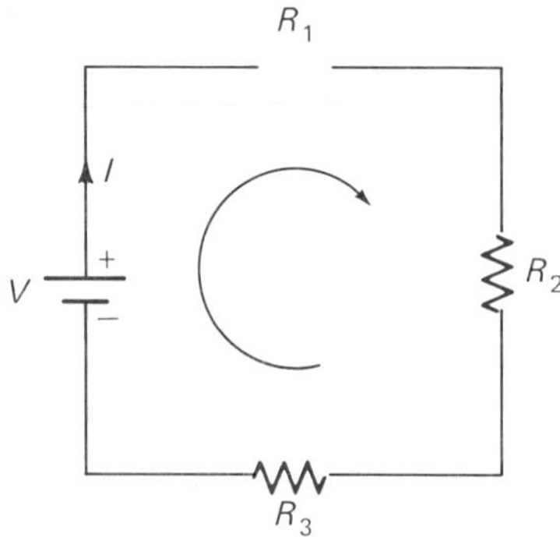

Figure 22.8
Kirchhoff's voltage or loop rule. The sum of the voltage rises and voltage drops around a closed circuit loop is zero, $\Sigma V = 0$, i.e., $V - IR_1 - IR_2 - IR_3 = 0$. If the loop is traversed in the opposite direction, the signs in the equation would be reversed. See Figure 22.7.

loop is traversed in the opposite direction, we have, by the sign convention,

$$-V + IR_1 + IR_2 + IR_3 = 0$$

which is the same as the previous equation.

The important thing in applying Kirchhoff's rules to a circuit is to make certain that all the circuit elements are traversed at least once in making the closed circuit loops. In some cases, traversing a particular loop may be unnecessary in solving for the currents, as the following example shows.

EXAMPLE 22.4 Find the current in each of the resistances in Figure 22.5, using Kirchhoff's rules (see Fig. 22.9).

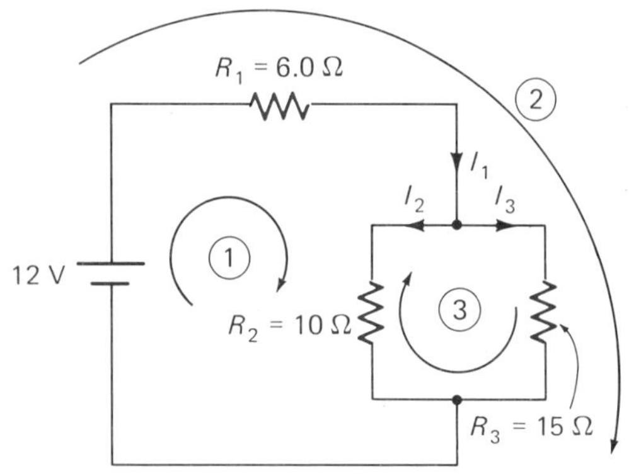

Figure 22.9
Application of Kirchhoff's rules. See Example 22.4. This circuit is the same as in Figure 22.5 (slightly redrawn) with circuit loops indicated.

Solution: Applying the current rule at the junction indicated in Figure 22.9, $\Sigma I = 0$,

$$I_1 - I_2 - I_3 = 0$$

Then, applying the voltage rule around loop 1 and loop 2,

$$\text{Loop 1:}\quad V - I_1R_1 - I_2R_2 = 12 - 6I_1 - 10I_2 = 0$$

$$\text{Loop 2:}\quad V - I_1R_1 - I_3R_3 = 12 - 6I_1 - 15I_3 = 0$$

This gives three equations with three unknowns (I_1, I_2, and I_3) from which the values of the currents may be found.

Subtracting the loop 2 equation from the loop 1 equation, we have

$$-10I_2 + 15I_3 = 0$$

(Notice that this is the same expression that would be obtained if the parallel branch loop were traversed—loop 3.) Hence,

$$I_2 = 1.5I_3$$

Substituting into the initial current equation,

$$I_1 = I_2 + I_3 = 1.5I_3 + I_3 = 2.5I_3$$

Then, substituting these current values into the loop 1 equation,

$$12 - 6(2.5I_3) - 10(1.5I_3) = 0$$

$$I_3 = 12/30 = 0.40 \text{ A}$$

and $\quad I_2 = 1.5I_3 = 1.5(0.40 \text{ A}) = 0.60 \text{ A}$

Also,

$$I_1 = I_2 + I_3 = 0.60 \text{ A} + 0.40 \text{ A} = 1.0 \text{ A}$$

Compare these results with those of Example 22.3.

The use of Kirchhoff's rules in the previous example may be likened to cracking a nut with a sledgehammer. It is more convenient to analyze such simple circuits directly, as in Example 22.3, by finding the equivalent resistance. The purpose of Example 22.4 was to show how Kirchhoff's rules and the sign convention are applied.

The real value of Kirchhoff's rules comes in multiloop circuits with more than one voltage source, as in Figure 22.10. Here, the equivalent resistance method cannot be used.

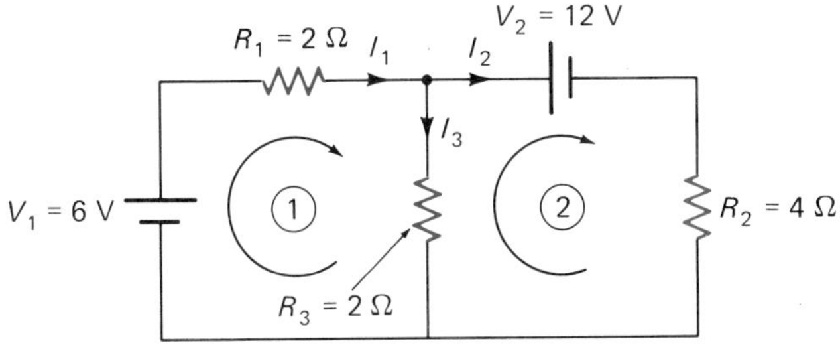

Figure 22.10
Multiloop circuit with two voltage sources. Circuits with more than one voltage source in different circuit loops are conveniently analyzed by Kirchhoff's rules. See text for analysis (Example 22.5).

EXAMPLE 22.5 Find the currents in the branches in the circuit in Figure 22.10.

Solution: Applying Kirchhoff's rules,

Current rule: $\quad I_1 - I_2 - I_3 = 0$

Voltage rule:

Loop 1:

$$V_1 - I_1R_1 - I_3R_3 = \quad 6 - 2I_1 - 2I_3 = 0$$

Loop 2:

$$-V_2 - I_2R_2 + I_3R_3 = \underline{-12 - 4I_2 + 2I_3 = 0}$$

Adding the equations: $\quad -6 - 2I_1 - 4I_2 = 0$

and $\quad I_1 = -2I_2 - 3$

Substituting this into the current rule equation,

$$I_1 - I_2 - I_3 = (-2I_2 - 3) - I_2 - I_3 = 0$$

and

$$I_2 = \frac{-I_3}{3} - 1$$

Putting this into the loop 2 equation,

$$-12 - 4\left(\frac{-I_3}{3} - 1\right) + 2I_3 = 0$$

and

$$\frac{10}{3} I_3 = 8$$

or

$$I_3 = 2.4 \text{ A}$$

Then,

$$I_2 = \frac{-I_3}{3} - 1 = \frac{-2.4}{3} - 1 = -0.8 - 1 = -1.8 \text{ A}$$

The minus sign indicates that I_2 is in the direction opposite to that assumed in the figure. Finally,

$$I_1 = I_2 + I_3 = -1.8 + 2.4 = 0.6 \text{ A}$$

Hence, I_1 and I_2 flow into the junction and $I_1 + I_2 = I_3$.

22.3 Voltage Sources in Series

When a circuit has several voltage sources, it is often helpful in analyzing the circuit to theoretically or physically replace two or more dc voltage sources *in series* in one branch with a single, equivalent voltage source.

> **The equivalent voltage source of two or more dc sources in series in one circuit branch is found by summing the sources according to directional polarity.**

For example, in the first circuit in Figure 22.11, the polarities are in the same direction, so the two batteries may be replaced by an 18-V battery ($V_s = 6\text{ V} + 12\text{ V} = 18\text{ V}$). The 6-V and 12-V batteries "work together," so to speak.

However, in the second circuit in the figure, the batteries have polarities in opposite directions, so the equivalent would be a 6-V battery with the polarity as shown ($V_s = 12\text{ V} - 6\text{ V} = 6\text{ V}$). In this case, the 6-V and 12-V batteries "work against" each other.

Care must be taken to make certain that the batteries are in the same branch (segment between two junctions). The third circuit in Figure 22.11 illustrates this situation. The 4-V and 2-V batteries are in the same branch and can be replaced by an equivalent 8-V battery. However, this procedure cannot be carried further, since the 6-V battery and the equivalent 8-V battery are not in series but in parallel.

The methods of replacing batteries in series and resistances in series and parallel by equiva-

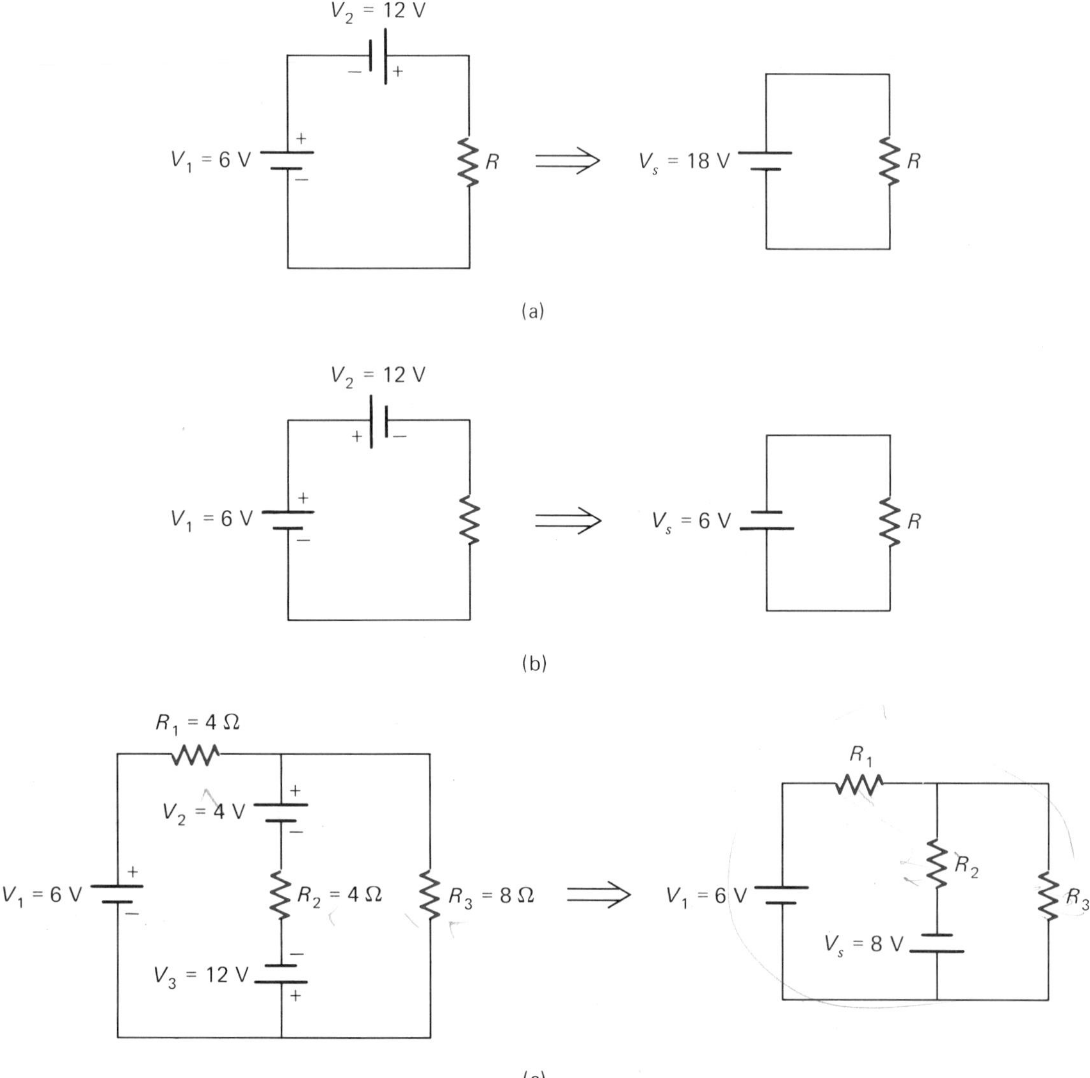

Figure 22.11
Voltage sources in series. (a) With polarities in the same direction, the two (or more) batteries may be equivalently replaced with a single battery having a voltage equal to the sum of the individual sources. (b) With polarities in opposite directions, the two (or more) batteries may be equivalently replaced with a single battery having a voltage equal to the absolute difference of their individual voltages. (c) V_2 and V_3 may be replaced by an equivalent battery since they are in series, but V_1 and V_s may not since they are in parallel.

lent batteries and resistances can be used to greatly simplify the analyses of circuits in appropriate cases. For example, consider the circuit and the equivalent circuit in Figure 22.12.

The three resistances in the right side of the circuit are in parallel and can be replaced by an equivalent resistance R_p. The top parts of the resistances have a common junction even though they are separated by wires (of negligible resistance). The two batteries in series in the left loop of the circuit can be replaced by an equivalent voltage source V_s. Then Kirchhoff's rules can easily be applied to the reduced equivalent circuit.

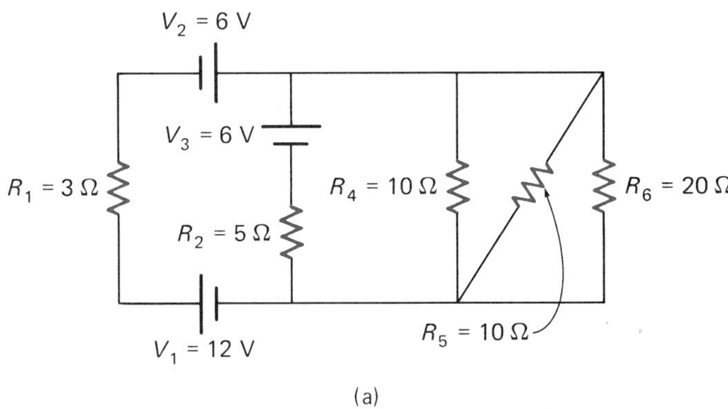

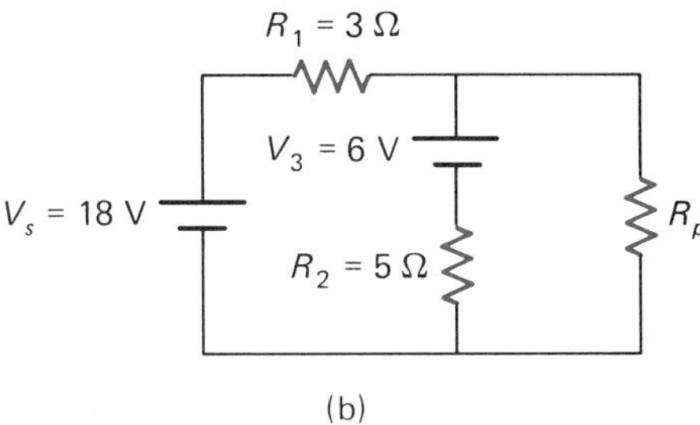

Figure 22.12
Circuit reduction. By combining batteries and resistances into their equivalents, circuits can sometimes be simply reduced. (b) An equivalent circuit of (a). The reduced circuit may be conveniently analyzed by Kirchhoff's rules.

22.4 Circuit Applications

The Wheatstone Bridge Circuit

The Wheatstone bridge circuit, as illustrated in Figure 22.13, is used to determine the values of unknown resistances. (The circuit is named in honor of Sir Charles Wheatstone, an English physicist.)

In the circuit R_1, R_2, and R_3 are known resistances and R_x is the unknown resistance. When the switch in the center branch is closed, the values of one or more of the known resistances are adjusted (by using a resistance box, for example) until there is no current through the center branch, as indicated by an ammeter (or galvanometer).

It is then said that the meter is in a "null" condition (zero reading) and that the bridge is "balanced." In this condition, the voltage drops across the adjacent resistances in the opposite arms of the bridge are equal, i.e., $V_{AB} = V_{AC}$ and $V_{BD} = V_{CD}$. This can readily be shown by applying Kirchhoff's rules to the two resistance loops:

$$\text{Loop 1:} \quad I_2R_x - I_1R_1 = 0$$

$$\text{Loop 2:} \quad I_2R_3 - I_1R_2 = 0$$

Dividing one equation by the other to cancel the I's and solving for R_x,

$$R_x = \frac{R_3R_1}{R_2} \qquad \textbf{(Eq. 22.6)}$$

In the case of a slide-wire Wheatstone bridge, where R_1 and R_2 are sections of a uniform wire attached to a meterstick, the resistances are proportional to the lengths L_1 and L_2 of the wire, and Equation 22.6 can be written

$$R_x = R_3\left(\frac{L_1}{L_2}\right) \qquad \textbf{(Eq. 22.7)}$$

since $R_1/R_2 = L_1/L_2$.

EXAMPLE 22.6 In a slide-wire Wheatstone bridge circuit, a fixed-value resistor, $R_3 = 150\ \Omega$, is used. When the bridge circuit is balanced, the tap key is at the 40-cm position on the meterstick. What is the value of the unknown resistance?

Solution: It is given that $R_3 = 150\ \Omega$, $L_1 = 40$ cm, and $L_2 = 60$ cm. Then,

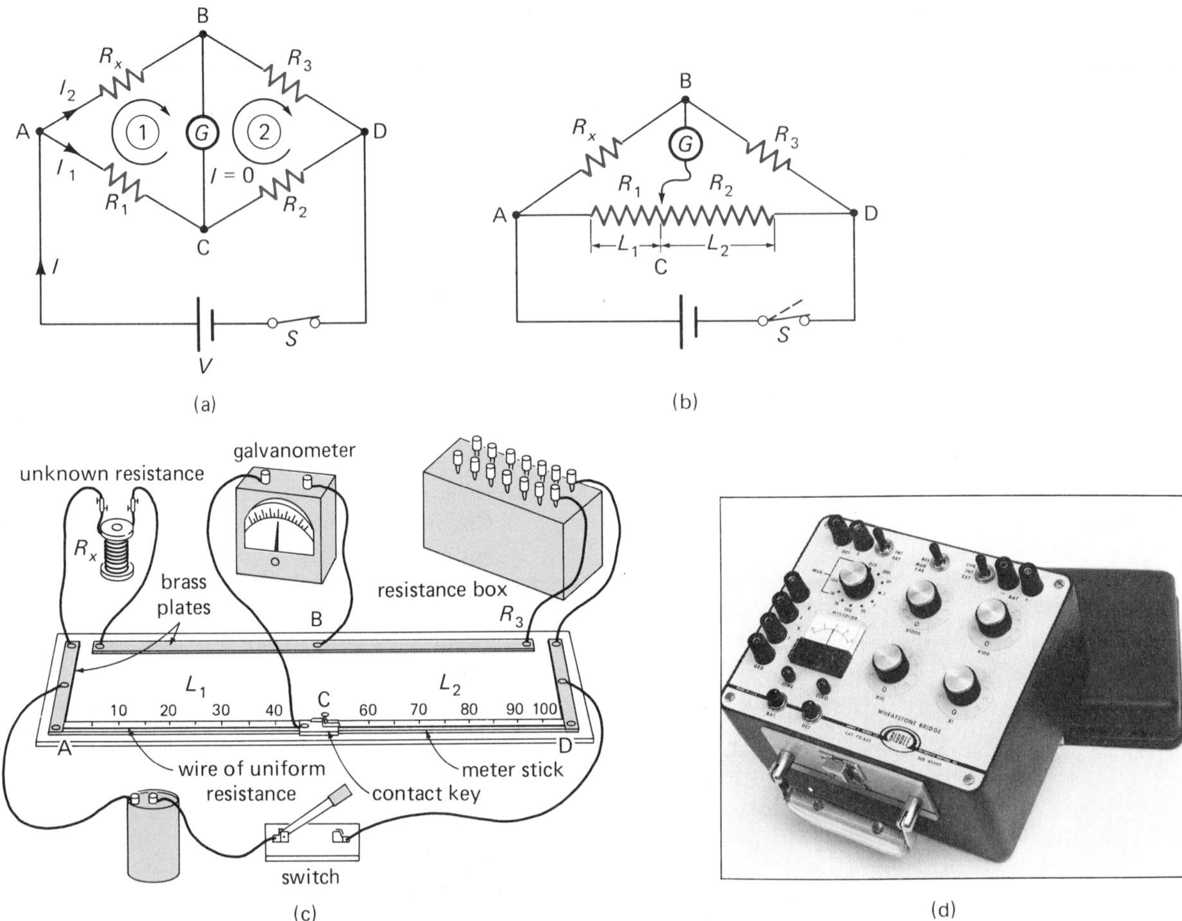

Figure 22.13
Wheatstone bridge circuit for measuring resistance. (a) When the values of one or more of the known resistances are adjusted so there is no current through the galvanometer G, the value of the unknown resistance is given by $R_x = R_3R_1/R_2$. (b and c) A slide-wire Wheatstone bridge. Here, resistances may be measured in terms of the lengths of the wire segments, and $R_1/R_2 = \ell_1/\ell_2$. The resistance box allows adjustment of the range of the bridge. (d) A portable Wheatstone bridge. (Courtesy James G. Biddle Co.)

$$R_x = R_3\left(\frac{R_1}{R_2}\right) = R_3\left(\frac{L_1}{L_2}\right)$$

$$= (150\ \Omega)\left(\frac{40\text{ cm}}{60\text{ cm}}\right) = 100\ \Omega$$

The Potentiometer Circuit

A three-contact variable resistor is termed a *potentiometer* or a *voltage divider,* depending on how it is used (Fig. 22.14). If one stationary contact and the movable contact are used, the variable resistor is a **rheostat.** The single resistance (and voltage drop) can be varied from zero to a maximum value. If all three contacts are used, the variable resistor is a **voltage divider.** The movable contact controls how the supply voltage V is divided between V_{ab} and V_{bc}.

An important use of a voltage divider is in a potentiometer circuit used to measure the emf of a battery. A voltmeter connected across the terminals of a battery will measure the terminal voltage, which is smaller than the emf because of the internal resistance of the battery (see the Chapter 21 Supplement). However, for a potentiometer circuit as in Figure 22.15, which is based on zero current being drawn from the bat-

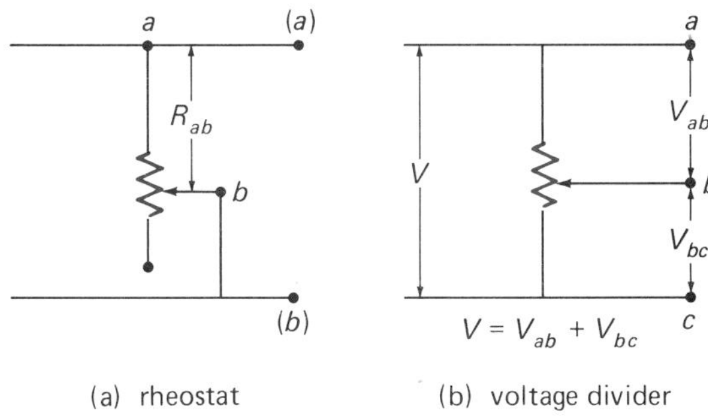

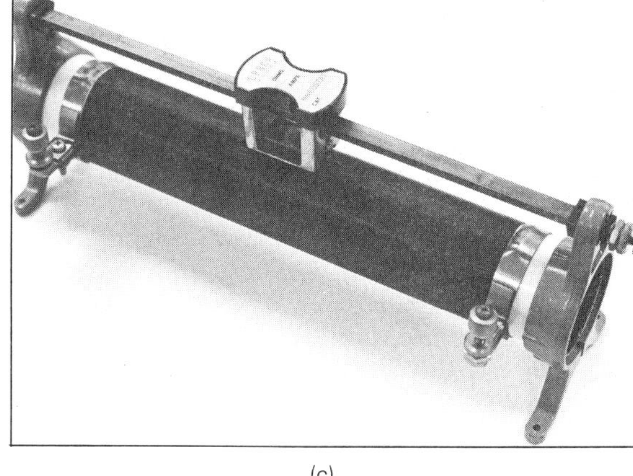

(c)

Figure 22.14

Rheostat and voltage divider. A three-contact variable resistor is a rheostat and a voltage divider, depending on how it is used. (a) As a rheostat using two contacts, the resistance can be varied from zero to the maximum value. (b) Using three contacts, the variable resistor acts as a voltage divider. The movable contact controls the division of the voltage between V_{ab} and V_{bc}. (c) A common type of wire-wound rheostat/voltage divider used in the physics laboratory.

tery of unknown emf, the internal resistance is not important, since there is no voltage drop due to this resistance.

In Figure 22.15, $\mathscr{E}_s$ is a standard cell of known emf and $\mathscr{E}_x$ is the cell or battery of unknown emf. A double throw switch S allows connection to either $\mathscr{E}_s$ or $\mathscr{E}_x$. If the switch is thrown to the $\mathscr{E}_s$ terminal, current will flow from this cell, as indicated by the ammeter A. However, by adjusting the variable resistor contact C, it is possible to locate a position where no current flows through the ammeter. (The lower "working" battery works against and cancels out $\mathscr{E}_s$, so to speak.) Applying Kirchhoff's voltage rule to the loop with $\mathscr{E}_s$, we have

$$\mathscr{E}_s - IR_s = 0 \qquad \text{or} \qquad \mathscr{E}_s = IR_s$$

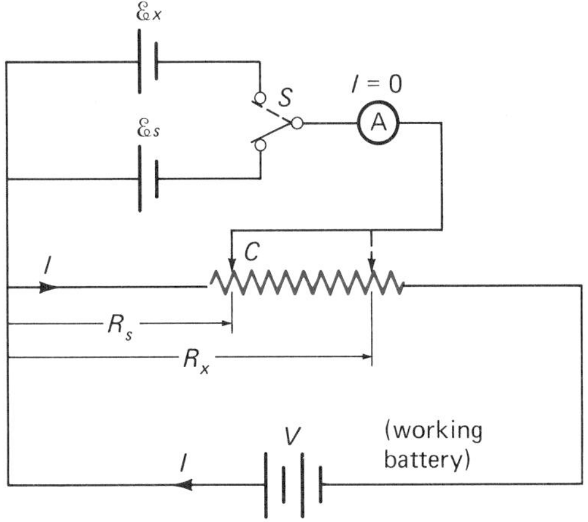

Figure 22.15

Potentiometer battery circuit. The circuit allows the measurement of unknown battery emf's ($\mathscr{E}_x$) by comparing against a standard cell ($\mathscr{E}_s$). See text for description.

Then, throwing the switch to the $\mathscr{E}_x$ terminal and repeating the zero-current procedure, around the loop with $\mathscr{E}_x$ we have

$$\mathscr{E}_x - IR_x = 0 \qquad \text{or} \qquad \mathscr{E}_x = IR_x$$

(The current is the same in each case, since the working battery loop has the same resistance in each case.) Combining the equations for the two cases, we have

$$\mathscr{E}_x = \left(\frac{R_x}{R_s}\right)\mathscr{E}_s \qquad \textbf{(Eq. 22.8)}$$

As in the case of the Wheatstone bridge, if a uniform slide-wire potentiometer is used, this equation can be written in terms of the wire lengths, which are easily measured, i.e.,

$$\mathscr{E}_x = \left(\frac{L_x}{L_s}\right)\mathscr{E}_s \qquad \textbf{(Eq. 22.9)}$$

Important Terms

junction a point where three or more connecting wires of a circuit meet

branch a circuit segment between two junctions that contains one or more circuit elements

Kirchhoff's rules:

current rule the sum of the currents entering and leaving a junction is zero

voltage rule the sum of the voltage rises and the voltage drops around a closed circuit loop is zero

Wheatstone bridge a circuit for measuring unknown resistances

rheostat a two-contact variable resistor, commonly wire-wound, or a slide-wire

potentiometer (voltage divider) a three-contact variable resistor in which the voltage across two end contacts is divided by the center contact

potentiometer circuit a circuit used to measure the electromotive forces (emf's) of batteries

Important Formulas

series resistance: $R_s = R_1 + R_2 + R_3 + \cdots$

parallel resistance: $\dfrac{1}{R_p} = \dfrac{1}{R_1} + \dfrac{1}{R_2} + \dfrac{1}{R_3} + \cdots$

(for two resistances) $R_p = \dfrac{R_1 R_2}{R_1 + R_2}$

Kirchhoff's rules:

current rule (at junction): $\Sigma I = 0$

or $\Sigma I_{in} = \Sigma I_{out}$

voltage rule (around closed circuit loop): $\Sigma V = 0$

sign convention:
(a) The voltage change of a battery is taken to be positive when the battery's positive (+) terminal is in the direction of motion when a loop is traversed, and negative when the motion is in the direction of the battery's negative (−) terminal.
(b) When passing through a resistance, the voltage change across the resistance is taken to be negative if it is traversed in the direction of the assigned current flow, and positive if the loop is traversed in the opposite direction.

Wheatstone bridge equation: $R_x = \dfrac{R_3 R_1}{R_2} = R_3\left(\dfrac{L_1}{L_2}\right)$ (slide wire)

potentiometer equation: $\mathscr{E}_x = \left(\dfrac{R_x}{R_s}\right)\mathscr{E}_s = \left(\dfrac{L_x}{L_s}\right)\mathscr{E}_x$ (slide wire)

Questions

Resistances in Series and in Parallel

1. For resistances in a series circuit, how are the voltage drops across the resistances related to the voltage rise of the battery?
2. In a series circuit, what is the current through each resistance?
3. Is the equivalent series resistance always greater than the smallest resistance in the series arrangement? Explain.
4. For resistances in parallel, how are the voltage drops across the resistances related to the voltage rise of the battery?
5. How are the individual currents through resistances in parallel related to the current drawn from the battery?
6. Can resistors be connected in parallel to achieve a greater equivalent resistance? Explain.
7. How does the current divide at a parallel junction in a circuit with respect to the resistances in the parallel branches?

Kirchhoff's Rules and Multiloop Circuits

8. What are the physical principles involved in Kirchhoff's (a) voltage rule and (b) current rule?
9. Distinguish between a circuit junction and a branch.
10. As you solve for the currents in the branches of a multiloop circuit, suppose a current comes out to be negative. What does this mean? What if all the currents come out to be negative?
11. What is the effect on an equation obtained from Kirchhoff's voltage rule if the current loop is traversed in the opposite direction?
12. What is an important consideration in applying Kirchhoff's rules to a circuit with regard to circuit elements? What would happen if this were not observed?

Voltage Sources and Circuit Applications

13. What is the effect of two batteries in series with opposite directional polarity?
14. What is the effect of two batteries in series with the same directional polarity?
15. Why can't the polarity rule for summing voltage sources in series be applied to voltage sources in parallel in a multiloop circuit (with a load)? Would the rule apply if only voltage sources were connected in parallel? Explain.
16. What are the conditions when a Wheatstone bridge is "balanced"?
17. Does it make any difference in which arm of a Wheatstone bridge an unknown resistance is placed?
18. Is a variable resistor used in a slide-wire Wheatstone bridge? Explain. Suppose only fixed resistances were used in a Wheatstone bridge circuit. Would it be functional for measurement?
19. Distinguish between a rheostat and a voltage divider.
20. Explain why a potentiometer is said to act as a voltage divider.
21. Explain how a potentiometer circuit is used to measure the emf of a battery.

Problems

Levels of difficulty are indicated by asterisks for your convenience.

22.1 Resistances in Series and in Parallel

*1. What resistance should be connected in series with a 10-Ω resistor in a circuit with a 120-V source so that a current of 2.4 A flows in the circuit?

*2. A circuit with a 6.0-V battery has a 3-Ω resistor and a 15-Ω resistor in series. (a) What is the current through each resistor? (b) What is the voltage drop across each resistor?

*3. Three resistors of 10 Ω, 20 Ω, and 30 Ω are connected in series with a 12-V battery. (a) What is the voltage drop across each resistor? (b) How much power is dissipated in each resistor?

*4. A circuit is made up of a 12-V battery with a 4-Ω resistor and a 6-Ω resistor in parallel. (a) What is the voltage across each resistor? (b) How much current flows through each resistor?

*5. What resistance should be connected in parallel with a 5-Ω resistor in a circuit with a 12-V source so that a current of 3.0 A is drawn from the source?

*6. (a) What is the equivalent resistance of the arrangement shown in Figure 22.16? (b) If the circuit is connected to a 12-V battery, how much current flows through each resistance?

*7. If you were given three 10-Ω resistors, what different values of resistance could you put into a circuit, using any number of the three resistors?

*8. Consider the circuit in Figure 22.17. (a) What is the current through each resistor? (b) What is the voltage drop across each resistor? (c) What is the total power expended in the circuit?

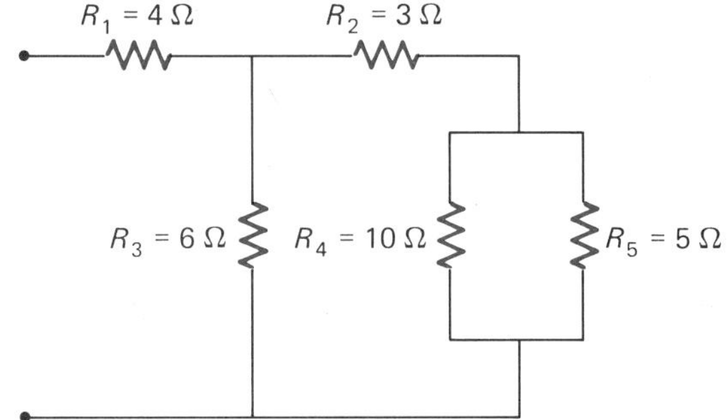

Figure 22.16
See Problem 6.

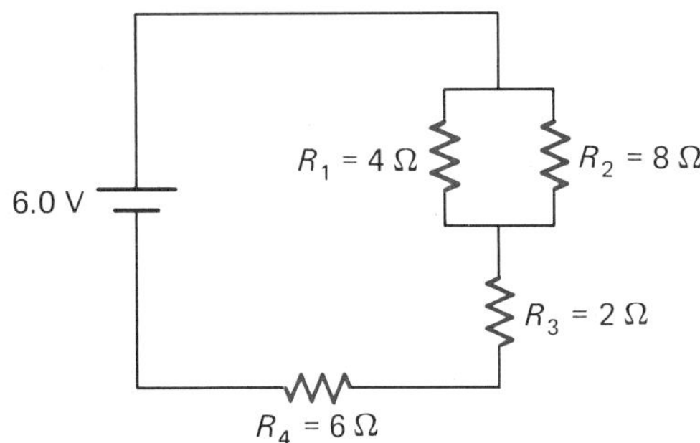

Figure 22.17
See Problem 8.

**9. Prove that the equivalent resistance of resistors in parallel is always less than that of the smallest resistor in the arrangement.

**10. What is the equivalent resistance of the arrangement shown in Figure 22.18 if all of the resistors are 10 Ω?

**11. In Figure 22.18, if all the odd-numbered resistors are 5 Ω and all the even-numbered resistors are 10 Ω, and the circuit is connected to a 12-V battery, (a) how much current is drawn from the battery? (b) What is the current through each resistor?

22.2 Kirchhoff's Rules and Multiloop Circuits

*12. Compute the currents in a circuit as in Figure 22.10 with $V_1 = 20$ V, $V_2 = 6.0$ V, $R_1 = 4.0\ \Omega$, $R_2 = 1.0\ \Omega$, and $R_3 = 5.0\ \Omega$.

*13. What is the current through each of the resistors in the circuit in Figure 22.11(c)?

*14. For the circuit shown in Figure 22.19, (a) compute the current through each resistor and (b) compute the power dissipated by each resistor and the power drawn from each battery.

*15. (a) Find the current through each of the resistors in the circuit in Figure 22.12(b). (b) Then find the current through each of the resistors in the R_p parallel arrangement.

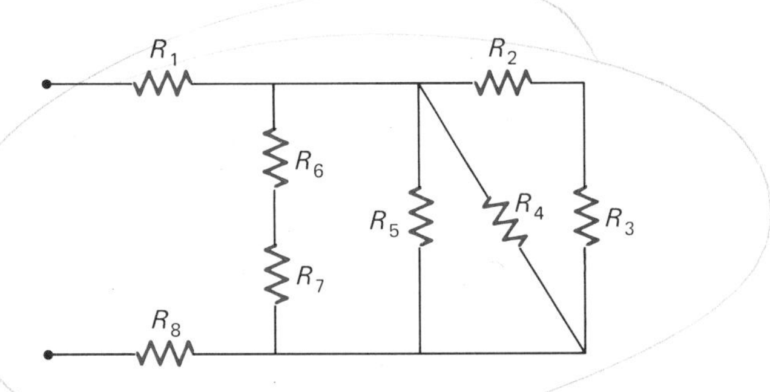

Figure 22.18
See Problems 10 and 11.

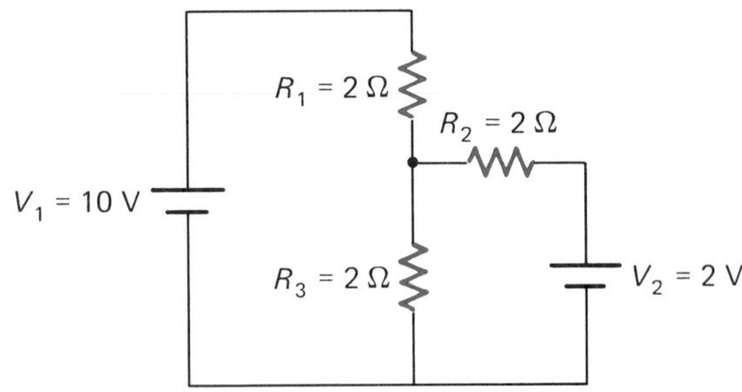

Figure 22.19
See Problem 14.

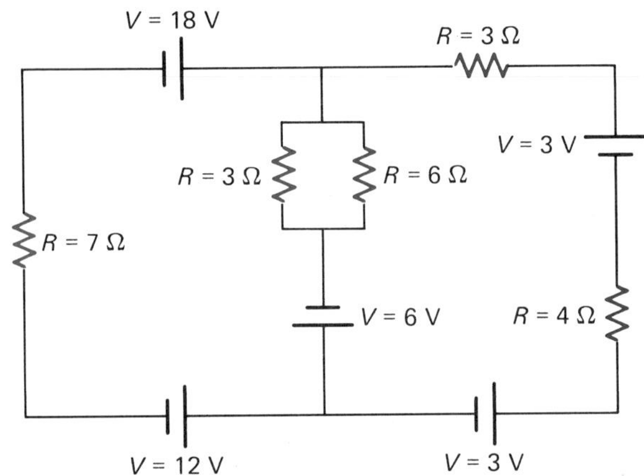

Figure 22.20
See Problem 16.

*16. In the circuit in Figure 22.20, what is the current through the 6-Ω resistance in the center parallel arrangement?

22.4 Circuit Applications

17. A Wheatstone bridge in a balanced condition has three known resistances $R_1 = 10\ \Omega$, $R_2 = 3.0\ \Omega$, and $R_3 = 4.5\ \Omega$. What is the value of the unknown resistance in the bridge circuit?

18. In a slide-wire Wheatstone bridge circuit as in Figure 22.13, a fixed-value resistor, $R_3 = 200\ \Omega$, is used. When the bridge circuit is balanced, the key is at the 80-cm position on the meterstick. What is the value of the unknown resistance?

*19. A slide-wire Wheatstone bridge with a 1-m wire is used to measure the value of an unknown resistance. When a fixed resistor $R_3 = 120\ \Omega$ is used, the value of the unknown resistance is computed to be 40 Ω. What is the position of the tap key when the bridge is balanced?

*20. A variable, wire-wound resistor is used as a voltage divider. If a 12-V source is connected across the resistor and the variable contact is one third of the length from one end, how is the voltage divided?

*21. A potentiometer uses a standard cell with 6.0-V emf. With the switch in the standard circuit, a resistance of 8.0 Ω gives a zero ammeter reading. With the switch in the unknown cell circuit, a resistance of 2.0 Ω gives a zero ammeter reading. What is the emf of the unknown cell?

*22. A slide-wire potentiometer has a wire length for the standard cell that is four times the wire length for an unknown cell for the balanced conditions. (a) If a standard cell with an emf of 12.0 V is used, what is the emf of the unknown cell? (b) What would be the case if the unknown cell had an emf of 12.0 V?

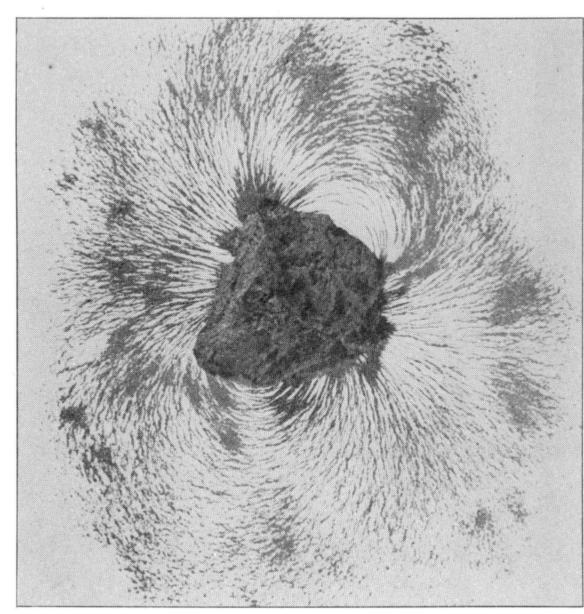

Magnetic lodestone and iron filings.

Chapter 23

Magnetism

Electricity and magnetism are closely related. In fact, we will find they are inseparable, although we talk about electric forces and magnetic forces. There is a distinction between the static electrical couloumb force and the magnetic force, which involves moving electric charges. Since both forces ultimately arise from electric charges, the general term *electromagnetic force* is sometimes applied to both.

Electromagnetic forces are basic in the operation of electric meters, motors, and generators and in many other important technical applications. The basic principles and applications of electromagnetism are the subject of this and the next chapter.

23.1 Magnets and Magnetic Fields

The first known magnets were *natural* magnets (see the chapter introductory photo). These were lodestones found in the ancient province of Magnesia in the mid-East, from which magnetism derives its name. These natural magnets were observed to attract bits of iron and to exhibit attractive and repulsive forces between each other. The Chinese were perhaps the first to use natural magnets to make induced magnetic needles for compasses.

The magnets with which we are most familiar are the common bar magnet, the horseshoe magnet (a bar magnet bent in the form of a horseshoe), and those used to stick things on the refrigerator. If a bar magnet is dipped into iron filings, the filings cling to the magnet in concentrations near the ends of the bar (Fig. 23.1).

These regions of apparently concentrated magnetic strengths are called **magnetic poles.** The poles are distinguished as the north (seeking) and south (seeking) poles. This terminology comes from the compass properties of a suspended magnet. The north pole "seeks," or points toward, the north direction on Earth, and the south pole points toward the south.

Figure 23.1
Magnetic poles. Iron filings collect about the poles of a magnet. (Courtesy of K.F. Kuhn and J.S. Faughn.)

Unlike electrical charges, we have never been able to isolate a single magnetic pole. That is, we have never found a magnetic monopole to occur by itself—opposite poles always occur together.* If we break a magnet in two, we have two smaller magnets. The reason for this will be discussed later.

When two magnets are brought close together, it is quickly observed that the magnets attract each other in some cases and repel each other in other cases. This action is described by a **law of poles**, similar to the law of changes:

> **Like magnetic poles repel, and unlike magnetic poles attract.**

That is, N-N (north-north) and S-S repel, and N-S (or S-N) attract, as illustrated in Figure 23.2.

The magnitudes of the attractive and repulsive forces depend on the so-called pole strengths, and there is a formula, similar to Coulomb's law for electric forces, by which the forces may be computed. Although the concept of poles is useful in describing magnetic interactions, poles do not exist in modern electromagnetic theory. Instead, we will be concerned with **magnetic field strength *B*,** which we will see is the magnetic force "per moving electric charge."

A magnetic field surrounds a magnet and is represented by magnetic field lines (of force), as illustrated in Figure 23.3. This is analogous to the electric field and lines of force around electric charges. The direction of the magnetic field indicates the direction of the force that would be experienced by a north magnetic pole at that point in space. As a result, a compass needle will line up with a field line, with the north pole pointing in the direction of the field. The closer together, or denser, the field lines in a particular region, the greater the magnitude of the magnetic force in that region.

The fact that a compass needle lines up with a magnetic field indicates that there is a magnetic field associated with the Earth (Fig. 23.4). It is as though there were a large bar magnet in the Earth, although this is not physically possible. By the law of poles, the north (seeking)

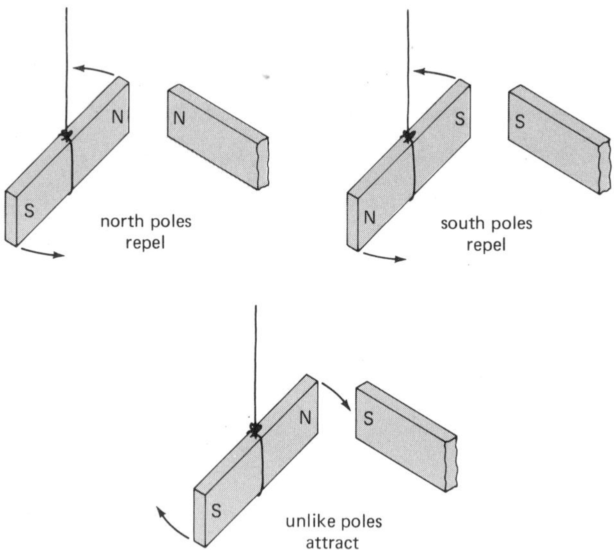

Figure 23.2
The law of poles. Like magnetic poles repel, and unlike poles attract.

* The existence of magnetic monopoles has been predicted, and at the time of this writing, two reports of possible monopole "events" in research experiments have been made. However, the existence of magnetic monopoles is far from being confirmed.

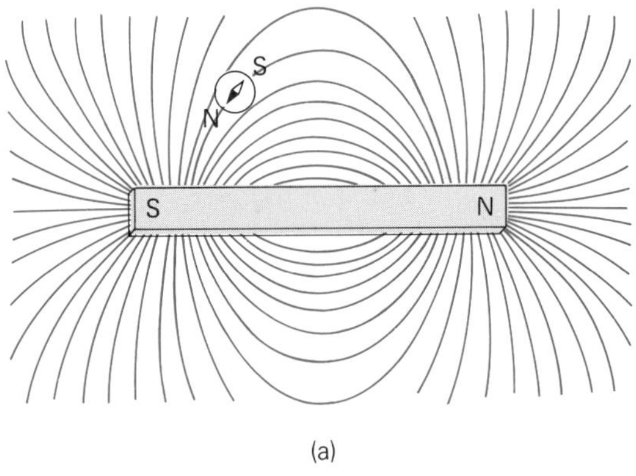

(a)

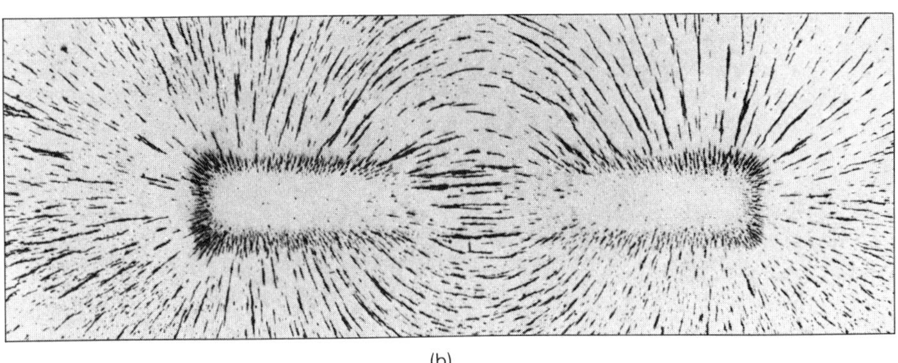

(b)

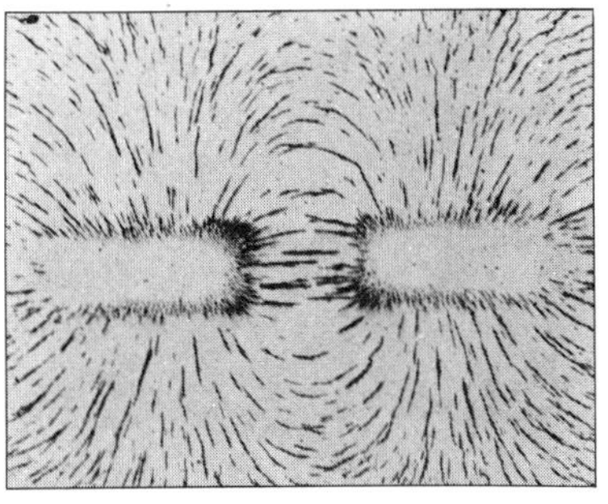

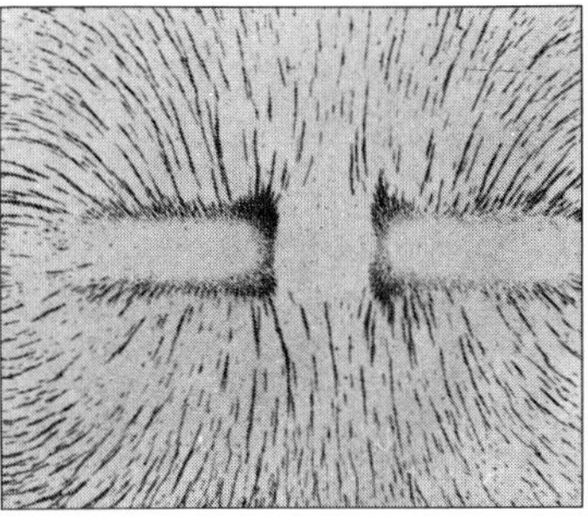

(c)

Figure 23.3

Magnetic field and lines of force. (a) Schematic illustration. In the vicinity of a magnet, a compass needle lines up with the magnetic field with the north pole of the compass needle pointing in the direction of the field. (b) Iron filings show the geometry of the magnetic field. (c) Patterns of attractive and repulsive poles.

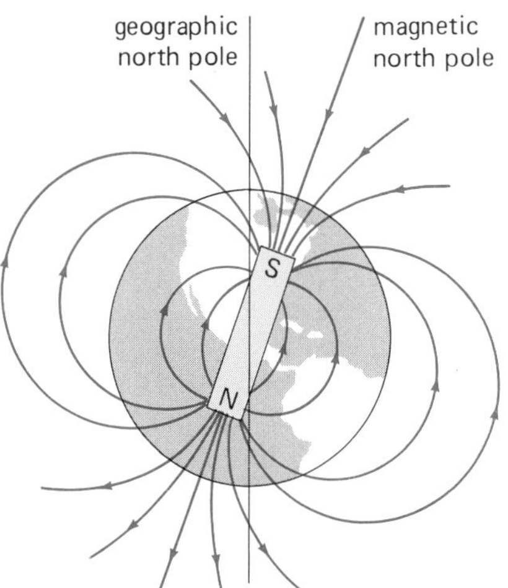

Figure 23.4

The Earth's magnetic field. The field is like that of a bar magnet inside the Earth (although this is not the case). The Earth's magnetic south pole is near the north geographic pole—about 1600 km (1000 mi) from the geographic pole (true north).

SPECIAL FEATURE 23.1

The Earth's Magnetic Field

We don't really know what causes the Earth's magnetic field. But the fact that a current produces a magnetic field leads scientists to speculate that there may be a flow of charge in the Earth's liquid outer core, which is chiefly iron, that gives rise to the magnetic field. The flow of charge is thought to be associated with the Earth's rotation.

Some of the other planets have magnetic fields, whereas others have no or very weak magnetic fields. For example, the magnetic fields of Jupiter and Saturn are much larger than that of the Earth. These planets are largely gaseous and have fast rotations of 10 to 11 hours. Venus and Mercury, on the other hand, have very weak magnetic fields. These planets are about as dense as Earth but rotate relatively slowly (243 and 58 days, respectively).

Another theory is that the Earth's magnetic field arises from currents in thermal convection cycles in the liquid outer core due to heat coming up from the central core. If the heat flow subsided and the directions of the convection cycles were reversed, with subsequent heat flow, this could account for the Earth's magnetic pole reversals, which have occurred at various intervals over geologic time.

The movements in such models could also account for fluctuations in the Earth's magnetic field. The magnetic north pole and the geographic North Pole do not coincide. (We generally refer to the magnetic north pole when talking about "magnetic north" even though it is actually a south magnetic pole. See Fig. 23.4.) The magnetic north pole is about 1500 km (930 mi) south of the geographic North Pole and is displaced some 13° of latitude. The south magnetic pole is displaced even farther, so the magnetic axis does not lie along the Earth's rotational axis as defined by the geographic poles, and is not even along a straight line through the Earth's center.

Also, the Earth's magnetic field changes such that the magnetic poles "wander." The north pole, on which we have more data, moves about 1° of latitude (about 110 km or 70 mi) in a decade. For some unknown reason, it has progressively moved northward from its 1904 position at latitude 69° N and also westward, crossing the 100° W longitudinal meridian. There are also sometimes daily shifts by as much as 80 km (50 mi), with the pole returning to its starting position. This is thought to be due to bursts of charged particles from the Sun that disturb the Earth's magnetic field.

pole of the compass needle lines up with the field and points toward the geomagnetic south pole, which is in the vicinity of the geographic North Pole. (The Earth's magnetic field is believed to be due to internal electrical currents arising from its liquid outer core and rotation. See Special Feature 23.1 and the principles of electromagnetism in the following sections.)

23.2 Electromagnetism

In electrical experiments carried out in 1820, the Danish physicist Hans Christian Oersted discovered a relationship between electricity and magnetism. It was found that

> An electric current gives rise to a magnetic field.

This can be shown by the arrangement illustrated in Figure 23.5. When current is flowing in the circuit, the compass needle is deflected, indicating the presence of a magnetic field. When there is no current in the circuit, the compass needle points in its normal north-south direction.

When a compass is used to map out the magnetic field around a straight wire segment, it is found that the field is in concentric circles around the wire (Fig. 23.6). The circular sense of the magnetic field around a current-carrying wire is given by a **right-hand (thumb) rule:**

> If a current-carrying wire is grasped with the right hand with the thumb in the direction of the conventional current, the curled fingers indicate the circular sense (direction) of the magnetic field.

The actual direction of the B field at a particular point is tangential to the circular field line. The magnitude of the magnetic field at a distance d from a long straight wire is given by

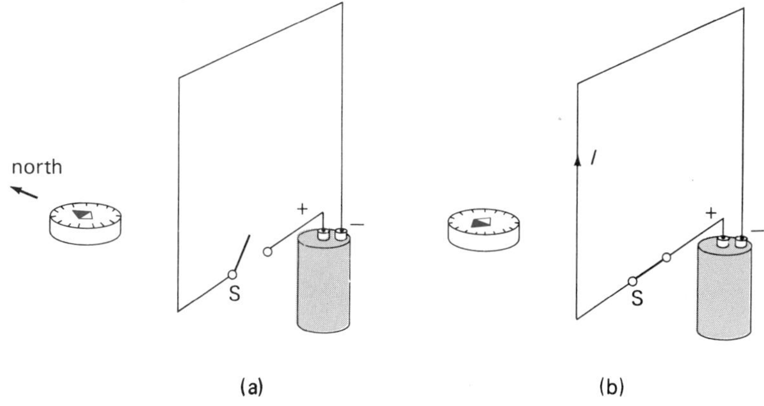

Figure 23.5

An electric current gives rise to a magnetic field. (a) With the switch open and no current in the circuit, the compass needle points north. (b) When the switch is closed and there is current in the circuit, the compass needle is deflected, indicating the presence of a magnetic field (other than that of the Earth).

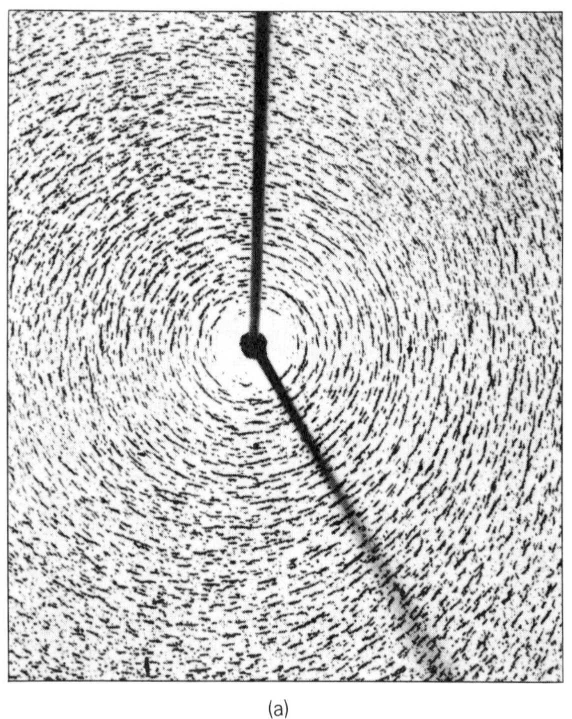

Figure 23.6

The magnetic field around a current-carrying wire. (a) The magnetic field is found to be a series of concentric circles around the wire. (b) The direction of the field is given by a right-hand rule. With the thumb of the right hand in the direction of the conventional current, the curled fingers indicate the circular sense of the magnetic field. The direction of the field is actually tangential to the circle at a given point. [(a) Courtesy Educational Development Center, Newton, MA.]

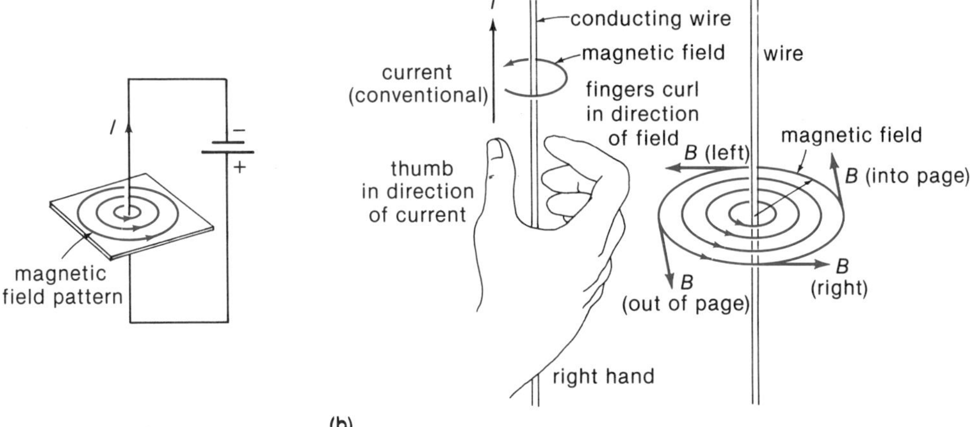

$$B = \frac{\mu_o I}{2\pi d} \qquad \text{(Eq. 23.1)}$$

(magnetic field around a long straight wire)

where μ_o is a constant called the permeability of free space (vacuum) and

$\mu_o = 4\pi \times 10^{-7}$ weber/ampere-meter (Wb/A-m)

This is analogous to the permittivity of free space ε_o in the electrical case.

As can be seen from Equation 23.1 and the units of μ_o, the units of the magnetic field are weber/(meter)^2, or Wb/m^2. The Wb/m^2 has been given the special name **tesla (T)** in the SI system, and 1 Wb/m^2 = 1 T.* The horizontal component of the Earth's magnetic field at the equator is approximately 10^{-5} T.

EXAMPLE 23.1 A wire carrying a current of 2.0 A runs along the right outer edge of the page of this text (I toward the top of the page). What is the magnetic field (magnitude and direction) at a point 15 cm from the edge of the page, toward the binding, in the plane of a flat page?

Solution: It is given that I = 2.0 A and d = 15 cm = 0.15 m. The magnitude of the magnetic field at a point 0.15 m perpendicular from the wire (Eq. 23.1) is

$$B = \frac{\mu_o I}{2\pi d} = \frac{(4\pi \times 10^{-7}\ \text{Wb/A-m})(2.0\ \text{A})}{2\pi\ (0.15\ \text{m})}$$
$$= 2.7 \times 10^{-6}\ \text{Wb/m}^2 \text{ or T}$$

The direction of the field, as given by the right-hand rule, is upward or out of the plane of the page.

Other current-carrying wire configurations for which the magnetic fields are easily calculated are given in the following paragraphs.

* The weber is in honor of Wilhelm Weber, a German physicist, and the tesla is in honor of Nikola Tesla, a Yugoslavian inventor who worked in the United States. The unit of magnetic field strength in the cgs system is the gauss (G), in honor of the German mathematician Carl Gauss, who worked with Weber on magnetic research (1 T = 10^4 G).

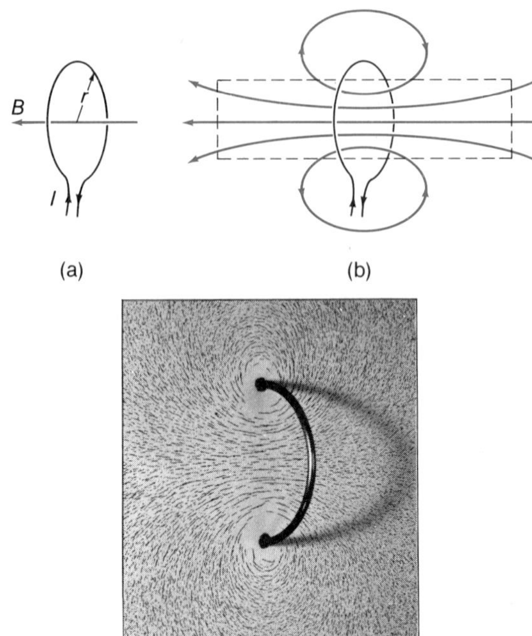

Figure 23.7
Magnetic field at the center of a current-carrying circular loop. (a) The magnetic field at the center of the loop is uniform or constant. (b) The complete field of the loop is similar to that of a bar magnet.

The magnitude of the magnetic field at the center of a circular current loop is given by

$$B = \frac{\mu_o I}{2r} \qquad \text{(Eq. 23.2)}$$

(magnetic field at the center of a circular loop)

where r is the radius of the loop (Fig. 23.7). Notice that the complete magnetic field for a circular loop is similar to that for a bar magnet. Hence, a circular loop can produce the magnetic field of a bar magnet without the concept of poles.

If there are N adjacent similar loops, then the magnitude of the field at the center of the coil of loops is found by replacing I with NI in Equation 23.2:

$$B = \frac{\mu_o NI}{2r} \qquad \text{(Eq. 23.3)}$$

(magnetic field at the center of a coil of N circular loops)

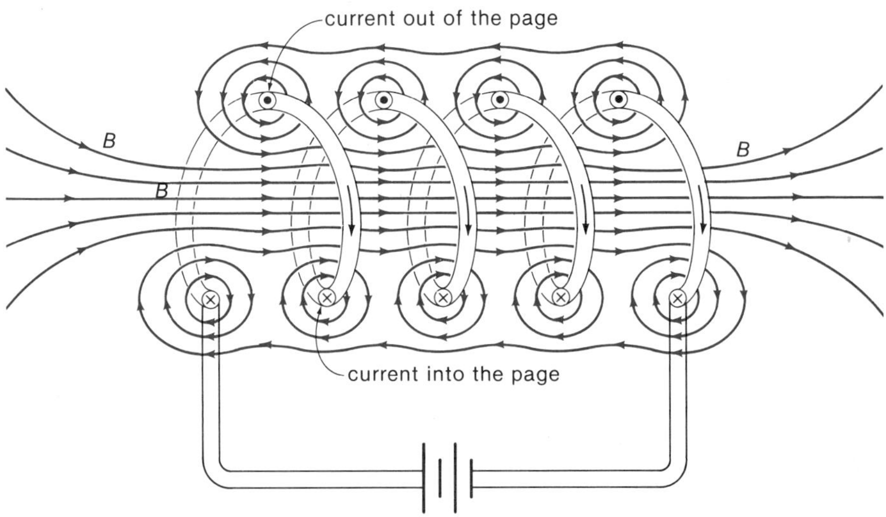

Figure 23.8
The magnetic field of a solenoid. A solenoid is a coil or helix of many loops of wire. Each loop contributes to the field such that the magnetic field is relatively constant across most of the interior of the coil. The current directions are indicated by dots and crosses. A dot indicates that the current is out of the page, and a cross indicates that the current is into the page. Think of an arrow approaching and receding.

A solenoid is a long coil of many circular turns of wire closely wound in the shape of a cylinder (Fig. 23.8). A solenoid is considered long if its length is several times greater than its diameter. The many loops of wire produce a rather uniform magnetic field in the interior of the solenoid. The magnitude of the magnetic field in the interior of a solenoid is given by

$$\boxed{B = \frac{\mu_o NI}{L} = \mu_o nI} \qquad \textbf{(Eq. 23.4)}$$

(magnetic field inside a solenoid)

where L is the length of the solenoid and $n = N/L$ is the turn density, or number of turns per unit length. A solenoid with a material core is used in many technical applications (see the Chapter Supplement). The purpose of the core is considered in the following section.

23.3 Magnetic Materials

Some materials are "magnetic" and others are nonmagnetic. For example, iron is magnetic; a piece of iron experiences an interaction with a permanent magnet. Wood and aluminum do not, and are called nonmagnetic.

It might be thought that the magnetic properties of some materials arise from orbiting electrons in the atoms. Such an electron, a moving electronic charge, would constitute a current and give rise to a magnetic field. This does occur; however, analysis shows that the magnetic field produced by one atom is often canceled by an oppositely moving electron in the same atom. As a result, the magnetic field due to orbiting electrons is either zero or very small for most materials.

The magnetic properties of many materials is explained by electron "spin." Classically, we think of an electron spinning on its axis, which would also give rise to a current. However, electron spin is purely a quantum mechanical effect (which will not be discussed), and the classical model of a spinning electron is not the case. In atoms containing many electrons, the electrons usually pair up with their spins opposite each other. The fields cancel each other, and hence most substances are nonmagnetic.

However, the magnetic fields produced by the electron spins do not cancel completely in some materials, such as iron, and these materials can be magnetized. Strong coupling exists between neighboring atoms, and large groups of atoms with aligned spins are formed. These

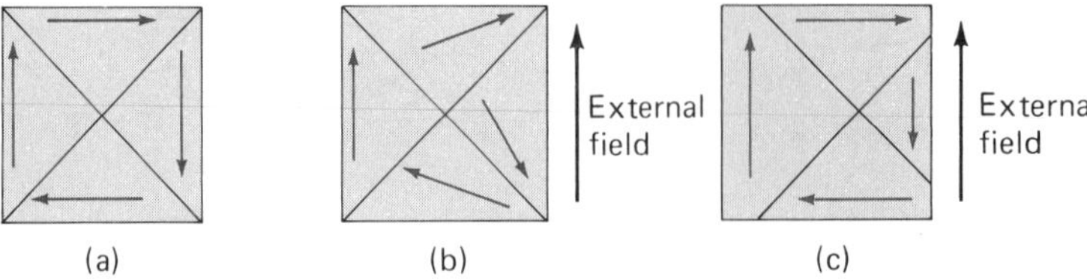

Figure 23.9

Magnetic domains. (a) A simple illustration of an unmagnetized sample with four domains. (b) Magnetization due to rotation of the domains. (c) Magnetization due to a shift in domain boundaries.

groups are called *domains*. In an unmagnetized substance, the domains are randomly oriented [Fig. 23.9(a)]. When an external magnetic field is applied, the fields of the domains tend to align with the external field, and the substance is magnetized.

The magnetization of material results from two effects. The domains may rotate slightly to become more aligned with the external field [Fig. 23.9(b)]. Also, in some instances, domains that are already aligned with the external field grow at the expense of other domains [Fig. 23.9(c)].

When the external magnetic field is removed, some domains remain aligned in the field direction, but many return to their random orientation. The remanent magnetism is usually small. Hence, the material can be made a magnet at will by turning the current in the surrounding coil on and off. This is the principle of the **electromagnet.** Electromagnets are used in many applications, such as industrial lifting (Fig. 23.10).

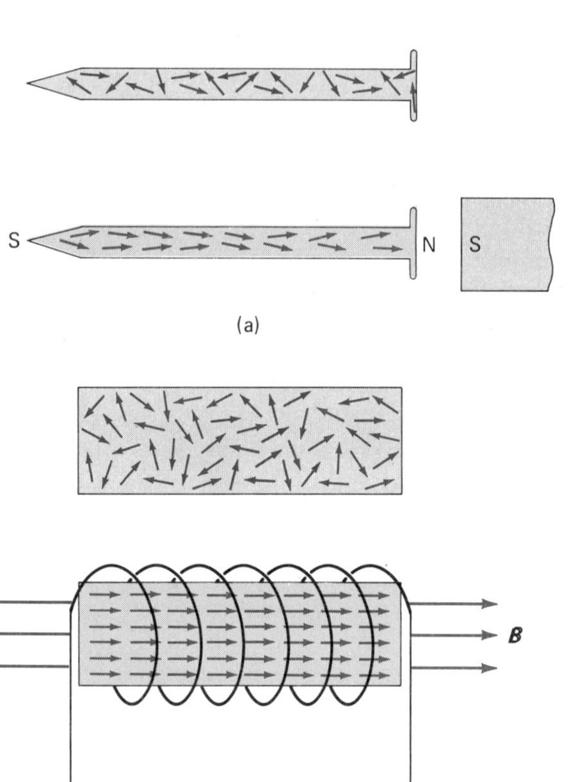

Figure 23.10

Magnetic domains. Ordinarily, the magnetic domains in a magnetic material are randomly oriented, and their effects cancel. A magnetic field from a magnet (a) or from a current-carrying wire (b) causes the domains to align and the iron becomes magnetized. When the switch in (b) is opened and the magnetic field is removed, the magnetic domains again become randomly oriented. This is the principle of an electromagnet (c). Pellets of iron oxide used in a direct reduction steel-making process are lifted by an electromagnet. (Courtesy Koff Industries, Inc.)

By heating and cooling the material in the external magnetic field, the domain alignment can be "locked in," and a permanent magnet is produced.

A quantity used to indicate the "magnetizability" of a material is the **relative permeability K_m**, which is defined as

$$\boxed{K_m = \frac{B_m}{B}} \qquad \textbf{(Eq. 23.5)}$$

Here, B is the magnetic field in a circular coil with an air (vacuum) core, and B_m is the magnetic field in the coil with a material core. This is analogous to the dielectric constant of a material ($K = C/C_o$; cf. Chapter 20).

For example, if a current-carrying solenoid has a material core, the magnetic field in the core (Eq. 23.5) is

$$B_m = K_m B = K_m \mu_o n I = \mu n I$$

where μ is the permeability of the material, and

$$\mu = K_m \mu_o \qquad \text{or} \qquad K_m = \frac{\mu}{\mu_o} \qquad \textbf{(Eq. 23.6)}$$

(This is analogous to the dielectric permittivity $\varepsilon = K\varepsilon_o$.)

Magnetic materials are classified as being ferromagnetic, paramagnetic, or diamagnetic. **Ferromagnetic materials have high relative permeabilities.** The basic ferromagnetic materials are iron, nickel, cobalt, and some alloys. In general, $K_m > 100$ for ferromagnetic materials, but the relative permeability of iron can be as high as 8000, and that of permalloy, an alloy of about 80 percent nickel and 20 percent iron, can be up to 80,000. Above a certain temperature known as the Curie temperature, the materials lose their ferromagnetic properties. The Curie temperature for iron is 770°C.

Paramagnetic materials have a K_m slightly greater than one. Aluminum is an example. The relative permeability of a paramagnetic material indicates that the magnetic moment alignment in a magnetic field of such a material is small. A strictly nonmagnetic material would have a K_m of one.

Diamagnetic materials have a K_m slightly less than one. Such materials have the property of being feebly repelled by a strong magnet. Bismuth is an example of a diamagnetic material.

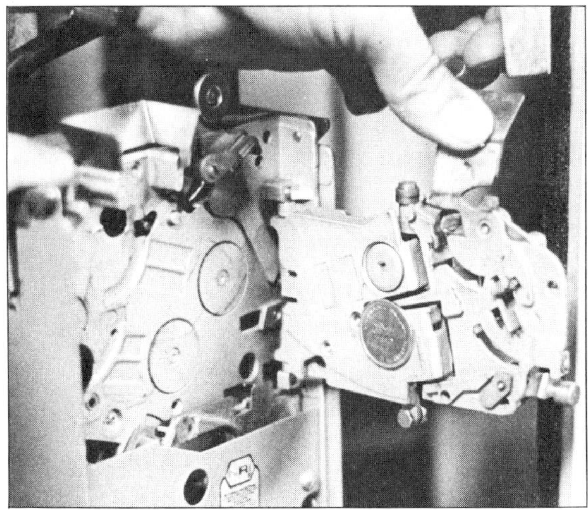

Figure 23.11

A coin machine mechanism. Magnets are used to stop slugs and washers that may damage the mechanism. Some Canadian coins are also attracted by magnets, as shown here by the suspended Canadian quarter.

EXAMPLE 23.2 To prevent damage and economic losses, coin-operated machines commonly are equipped with magnetic devices to prevent items such as metal slugs or washers from entering the coin mechanism. Some Canadian coins are also stopped by the device (Fig. 23.11). Why is this?

Solution: The magnetic devices have small permanent magnets that attract and prevent ferromagnetic objects, e.g., ferrous washers, from entering the coin mechanism. Some Canadian coins, unlike U.S. coins, still contain nickel and hence are stopped by the magnets. (Nickel is ferromagnetic.) The coin-reject mechanism, which is designed to mechanically eject bent coins or an improper number of coins, will generally eject the ferromagnetic coins, since they are attracted to the permanent magnets.

23.4 Magnetic Force and Torque on Current-Carrying Wires

An electrically charged particle moving in a magnetic field experiences a force.

Consider the case of a positively charged particle moving in a uniform magnetic field, as illus-

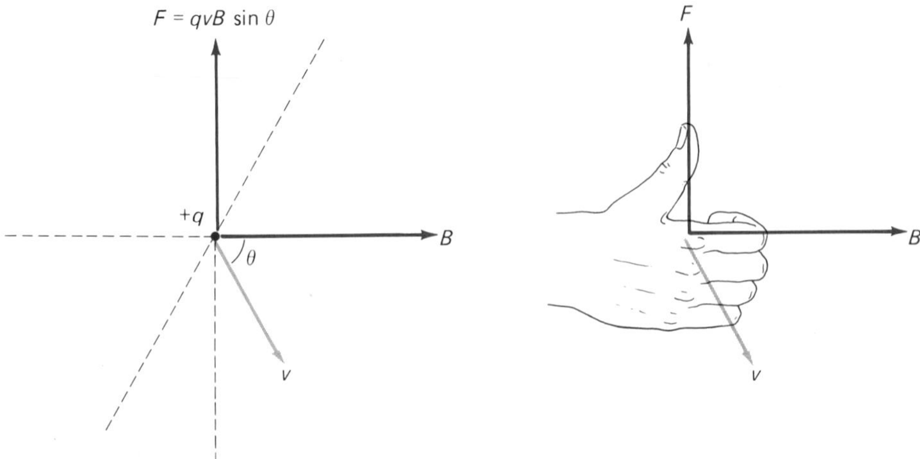

Figure 23.12
Force on a charged particle moving in a magnetic field. The magnitude of the force is $F = qvB \sin \theta$. The direction of F is perpendicular to the plane of v and B. If the right hand is used to turn or cross the v vector into the B vector, the thumb points in the direction of F.

trated in Figure 23.12. The magnitude of the force experienced by the charge is given by

$$\boxed{F = qvB \sin \theta} \qquad \textbf{(Eq. 23.7)}$$

(magnetic force on a moving charged particle)

where θ is the angle between the velocity vector v and the magnetic field vector B (Fig. 23.12). Note that for $\theta = 0$,

$$B = F/qv$$

That is, the magnetic field is the force per "moving electric charge."

The direction of the force on the charged particle is given by a right-hand rule:

> **If the fingers of the right hand turn or "cross" the v vector into the B vector, then the thumb points in the direction of the force F.**

If the particle is negatively charged, the force is in the opposite direction. As can be seen in Figure 23.12, the direction of F is perpendicular to the plane of v and B in either case.

The magnetic force on a *beam* of charged particles is important in nuclear applications, which are discussed in Chapter 31 on modern physics. For the present discussion, the more important aspect is the force on a current-carrying wire in a magnetic field. This is more applicable to practical technical applications.

Consider a wire carrying a (conventional) current I in a uniform magnetic field (Fig. 23.13). It is convenient to think of the current as

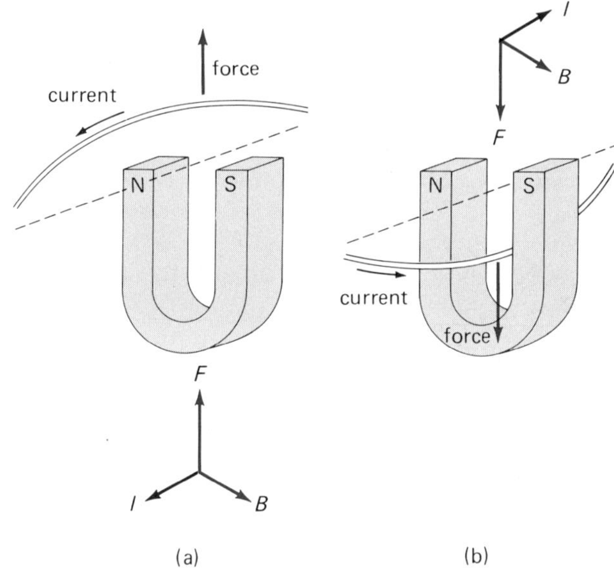

Figure 23.13
Force on a section of current-carrying wire. The magnitude of the force is $F = I\ell B \sin \theta$. The direction of F is perpendicular to the plane of I and B, as if by a right-hand rule I were turned or crossed into B.

a beam or flow of positive (conventional) charges in the wire. Then the magnetic force on the charges or the wire is

$$F = qvB \sin\theta = q\left(\frac{\ell}{t}\right) B \sin\theta = \left(\frac{q}{t}\right) \ell B \sin\theta$$

or

$$\boxed{F = I\ell B \sin\theta} \qquad \textbf{(Eq. 23.8)}$$

(magnetic force on current-carrying wire)

The direction of the force on the wire is similar to that on a moving charge. In this case, the direction of the conventional current I is crossed or turned into B with the fingers of the right hand, and the thumb points in the direction of F.

EXAMPLE 23.3 A long, straight wire running along the right outer edge of the page of this book carries a current of 2.5 A in a conventional direction from the bottom of the page to the top. If there is a uniform magnetic field of 2.0×10^{-4} Wb/m^2 downward (perpendicular to the plane of the page), what is the force per unit length on the wire?

Solution: It is given that $I = 2.5$ A, $B = 2.0 \times 10^{-4}$ Wb/m^2, and $\theta = 90°$. Then, by Equation 23.8,

$$\frac{F}{\ell} = IB \sin 90°$$

$$= (2.5\ \text{A})(2.0 \times 10^{-4}\ \text{Wb/m}^2)(1.0)$$

$$= 5.0 \times 10^{-4}\ \text{N/m}$$

since $\sin 90° = 1.0$. The direction of F is given by the right-hand rule of crossing the direction of I into B, and this is in the plane of the page toward the binding of the book.

Definition of the Ampere

The magnetic force on a current-carrying wire provides a mechanical means for the definition of the ampere. Consider two parallel wires separated by a distance d, carrying currents I_1 and I_2, respectively. The magnitude of the magnetic field due to I_1 at the second wire is given by

$$B_1 = \frac{\mu_o I_1}{2\pi d}$$

The direction of B_1 is at right angles ($\theta = 90°$) to the second wire, so the magnitude of the force per unit length on the second wire is

$$\frac{F_2}{\ell} = I_2 B_1 = \frac{\mu_o I_1 I_2}{2\pi d}$$

It can be shown that a similar expression gives the force per length on the first wire. This follows directly from Newton's third law.

The magnetic force between such a parallel wire arrangement is then used to define the ampere:

> **The ampere is defined as the current which, if maintained in each of two long parallel wires separated by a distance of 1.0 m in free space, would produce a force between the wires (due to their magnetic fields) of 2×10^{-7} N for each meter of length.***

Torque on a Current-Carrying Wire Loop

An important application of the magnetic force is in the production of a torque on a current-carrying wire loop (Fig. 23.14). The loop is pivoted so it can rotate in a uniform magnetic field. Consider the forces on each section of the rectangular loop. The forces on the sides are in the direction of the axis of rotation, so they produce no torque. Moreover, they cancel, so there is no net force in this direction. Since the currents in the end segments and the magnetic field are always perpendicular, the magnitude of the forces is $F = IBw$, where w is the width of the loop.

Referring to the side-view figure, we see that the moment or lever arm is $r_\perp = (\ell/2) \cos\alpha$, where α is the angle between the plane of the loop and the magnetic field and ℓ is the side length of the loop. Or, in terms of the complementary angle ϕ, this becomes $r_\perp = (\ell/2) \sin\phi$. The torque τ_1 on each end segment is then

$$\tau_1 = Fr_\perp = (IBw)(\ell/2) \sin\phi$$

* National Institute of Standards and Technology.

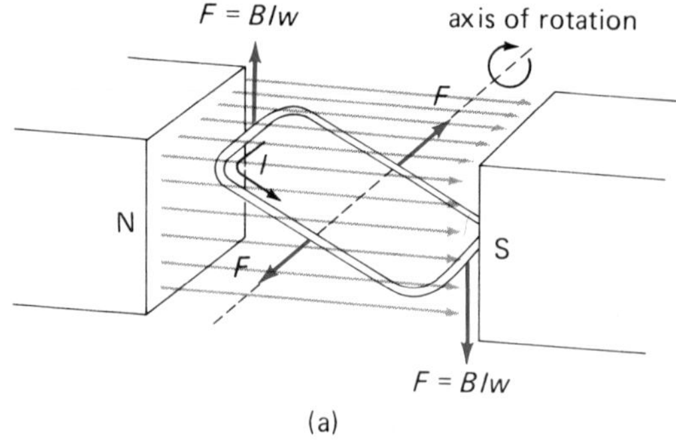

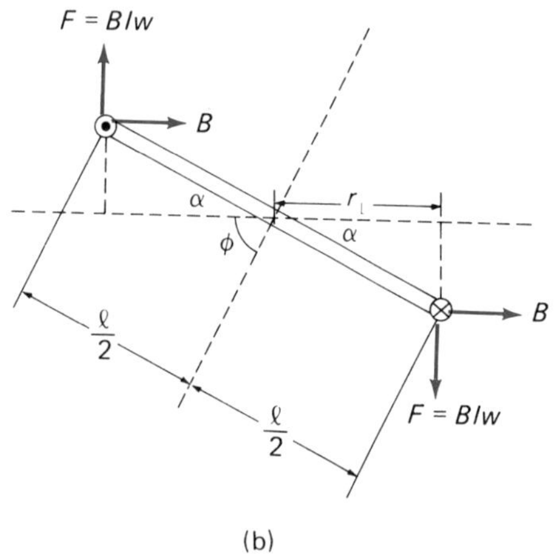

Figure 23.14
Torque on a current-carrying loop. (a) When a current-carrying loop is in a magnetic field, it experiences a torque due to the magnetic force on the loop. (b) Geometry to analyze torque. See text for description.

The total torque is twice this, since there is an equal force on each end:

$$\tau = 2\tau_1 = BIw\ell \sin \phi$$

or

$$\boxed{\tau = BIA \sin \phi} \qquad \textbf{(Eq. 23.9)}$$

(torque on current-carrying wire loop)

where $A = w\ell$ is the area of the loop and ϕ is the angle between a normal to the plane of the loop and the magnetic field. The direction of the torque is such that it would produce a clockwise rotation in the figure.

Notice that the torque is maximum where the plane of the loop is parallel to the field ($\phi = 90°$). As the loop rotates, the torque decreases, and it is zero when the plane of the loop is perpendicular to the magnetic field (no net force). If the loop rotates past this equilibrium position, the torques act to restore it to this position. Hence, the loop rotates until it is in stable equilibrium.

23.5 The dc Motor

The torque due to the magnetic force on a current-carrying loop is the basis of the direct-current (dc) motor. A simple, single-loop dc motor is illustrated in Figure 23.15. A current loop ro-

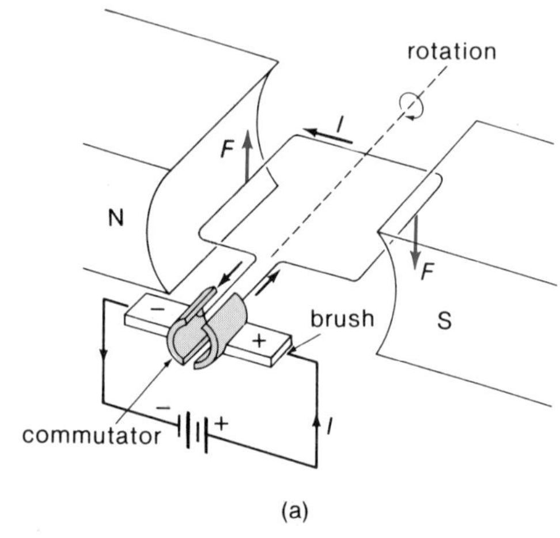

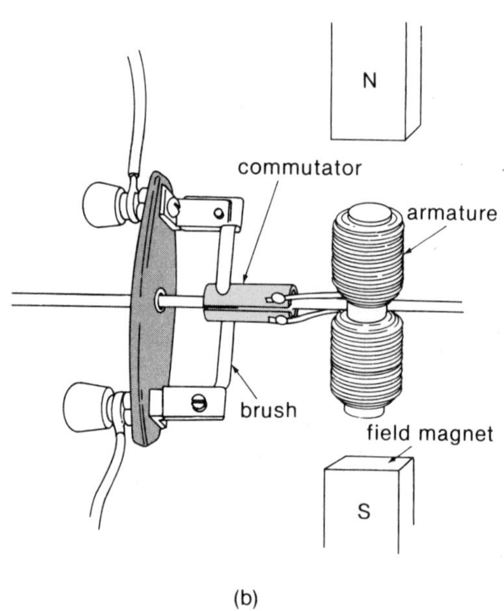

Figure 23.15
Simple dc motor. (a) The single loop armature rotates as a result of the magnetic torque. (b) The basic parts of a dc motor. The commutator is a split cylinder, each half of which makes contact with one brush, then the other. This reverses the current in the armature coils and the forces on them.

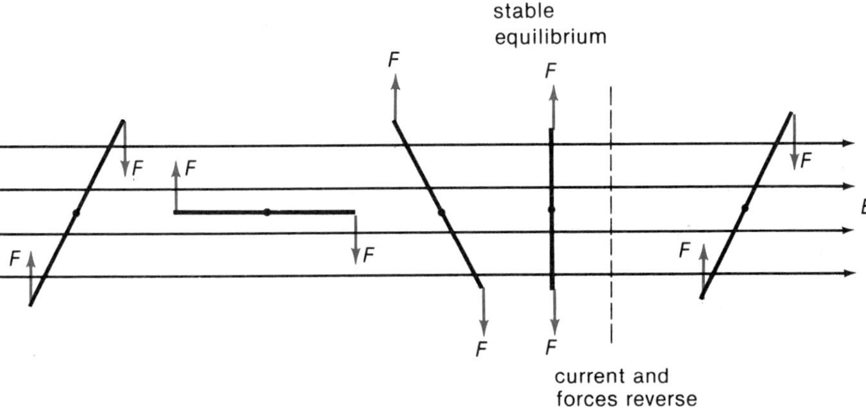

Figure 23.16
Armature rotation. End view of a single loop armature. The forces cause the loop to rotate until it is in stable equilibrium. The current in the loop then reverses due to the split-ring commutator, and the loop rotates through the next half-cycle. This action gives continuous rotation.

tates in a magnetic field. The output shaft of the loop (motor) then has rotational mechanical energy that can be used to do useful work. Hence, as a basic function,

> A motor converts electrical energy into mechanical energy.

As we learned in the last section, a current loop rotates to a position of stable equilibrium, e.g., through one half turn or cycle (180°). To provide a continuous rotation of the loop or motor armature, either the current or the magnetic field must be reversed. It is easier to reverse the current in the loop, and this is accomplished with a split-ring commutator (Fig. 23.15). The commutator consists of two metal half-rings insulated from each other. The ends of the loop are fixed to respective halves of the ring. Current is supplied to the commutator and loop by means of contact brushes.

To understand the current reversal in the loop, notice that during a half-cycle one of the half-rings is positive and the other is negative. During the next half-cycle, the polarity of the half-rings is reversed. Hence, the current in the loop is reversed each half-cycle. The effect of this is illustrated in Figure 23.16.

During one half-cycle, the loop rotates toward the position of stable equilibrium. At this position, the current reverses because of the change in polarity of the split ring halves. As a result, the forces on the loop are reversed, and a condition of unstable equilibrium exists. The rotational inertia of the armature carries it through this position, and the loop rotates through another half-cycle. This process repeats, and the motor armature rotates continuously.

The simple dc motor discussed here shows the basic principle of the motor. There are many motor designs. Smoother operation and greater torques can be obtained by adding more coils to the armature and by increasing the magnetic field, e.g., by replacing the permanent magnets with electromagnets.

23.6 The Ammeter and Voltmeter

The Galvanometer

The galvanometer forms the basis of measuring current and voltage in the form of an ammeter and voltmeter, respectively. A **galvanometer is an instrument that detects small currents.** A common type of galvanometer is shown in Figure 23.17.* Here, any small current in a pivoted coil (called a d'Arsonval movement) between permanent pole faces produces a torque. This causes the coil to rotate, and the pointer needle is deflected. A restoring spring returns the needle to its zero position when there is no current in the coil.

* Only the moving-coil-type galvanometer, which is suitable for measuring dc currents, is considered here. Modifications may be made for measuring average alternating currents (ac). See Chapter 25.

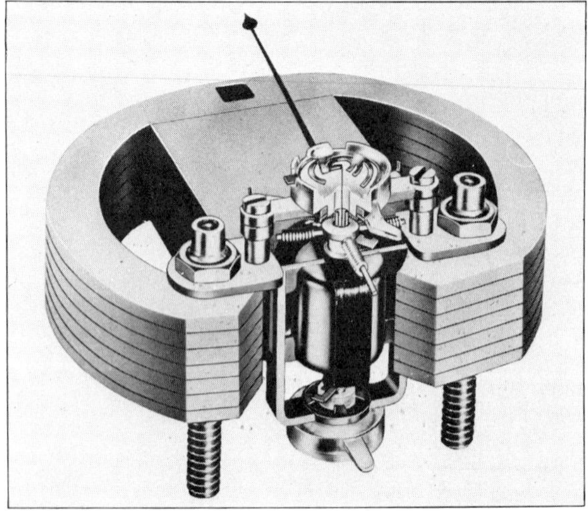

Figure 23.17
A galvanometer movement. When current passes through the coil between the pole faces of a magnet, a torque causes the coil to move. The deflection, as indicated by a pointer, is proportional to the current. (Courtesy Simpson Electric Co.)

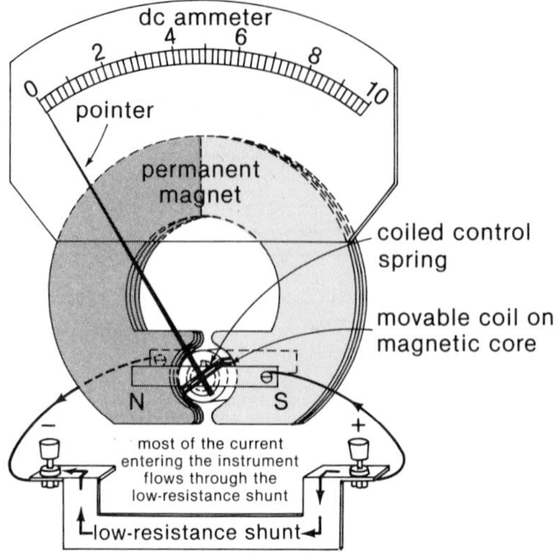

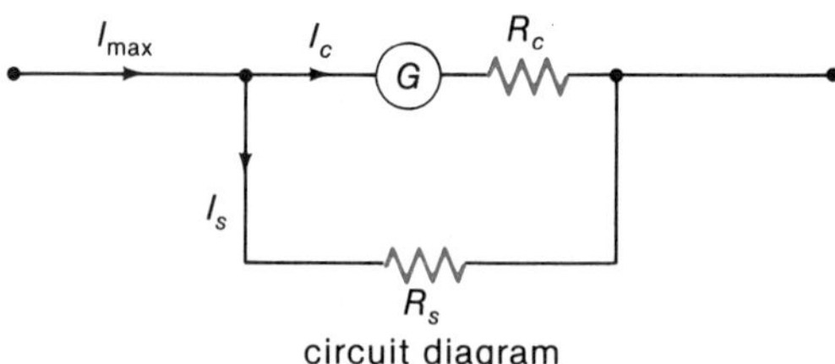

Figure 23.18
The dc ammeter. By connecting a shunt resistance in parallel with the galvanometer, appreciable dc currents may be measured. Most of the current flows through the shunt, while only a small fraction flows through the galvanometer coil.

The galvanometer is essentially an ammeter, and the needle scale can be calibrated in current units. However, the resistance of the coil R_c is very small, and the I^2R losses limit the current that can be carried by the coil. For example, a coil might have an $R_c = 25$ ohms and maximum allowable current of 1.0 mA. By mechanically designing the galvanometer so that there is a full-scale deflection (maximum pointer deflection) of 1.0 mA, the galvanometer becomes a milliammeter. At half the maximum current, the pointed needle will indicate 0.5 mA, and so on.

The Ammeter

The galvanometer can be used to measure larger currents by connecting an appropriate resistance, called a shunt resistance R_s, in parallel with the coil (Fig. 23.18). Suppose we wish to have an ammeter capable of reading up to 1.0 A (full-scale deflection). Then, a shunt resistance is chosen so that only 1.0 mA (0.001 A) flows through the galvanometer coil. The other 0.999 A passes through the shunt resistance.

Since R_c and R_s are in parallel, the voltage across these resistances is the same (Kirchhoff's voltage rule), and

$$I_cR_c = I_sR_s$$

Also, by Kirchhoff's current rule,

$$I_{max} = I_c + I_s$$

where I_{max} is the maximum current of the desired range of the instrument (full-scale deflection). Then,

$$I_cR_c = (I_{max} - I_c)R_s$$

and

$$\boxed{R_s = \frac{I_cR_c}{I_{max} - I_c}} \qquad \textbf{(Eq. 23.10)}$$

(ammeter shunt resistance)

EXAMPLE 23.4 A galvanometer has a coil resistance of 25 ohms, and the coil current for full-scale deflection is 1.0 mA. What is the required shunt resistance to convert the galvanometer into a 1.0-A meter?

Solution: It is given that $R_c = 25\ \Omega$, $I_c = 1.0$ mA = 0.001 A = 10^{-3} A, and $I_{max} = 1.0$ A. Then, by Equation 23.10,

$$R_s = \frac{I_c R_c}{I_{max} - I_c}$$

$$= \frac{(10^{-3}\ \text{A})(25\ \Omega)}{1.0\ \text{A} - 0.001\ \text{A}} = 0.025\ \Omega$$

A multirange ammeter can be designed with different shunt resistances for more versatility (Fig. 23.19). An important point to remember in using an ammeter is that

> **An ammeter is connected in series with the circuit element through which the current is to be measured.**

That is, the ammeter must be connected directly "in line" with the circuit element. This is analogous to measuring fluid flow through a pipe. The measuring device must be connected "in line." If the ammeter were mistakenly connected in parallel, the current would be shorted through the ammeter because of its extremely low resistance. This could burn out the shunt resistor or blow a fuse in the instrument.

The Voltmeter

A voltmeter is a galvanometer with a scale calibrated in volts. For example, for a galvanometer with R_c = 25 ohms and I_c = 1.0 mA = 0.001 A (full-scale deflection), by Ohm's law the full-scale voltage deflection V_c is

$$V_c = I_c R_c = (0.001\ \text{A})(25\ \Omega)$$

$$= 0.025\ \text{V} = 25\ \text{mV}$$

Hence, the galvanometer alone can be used as a millivoltmeter.

For an extended-range voltmeter, a large "multiplier" resistance R_m is placed in series with the galvanometer coil (Fig. 23.20). The multiplier resistance must be appropriately chosen. Suppose it is desired to have a full-scale range of V_{max}. Then, with the maximum allowable current I_c through the coil,

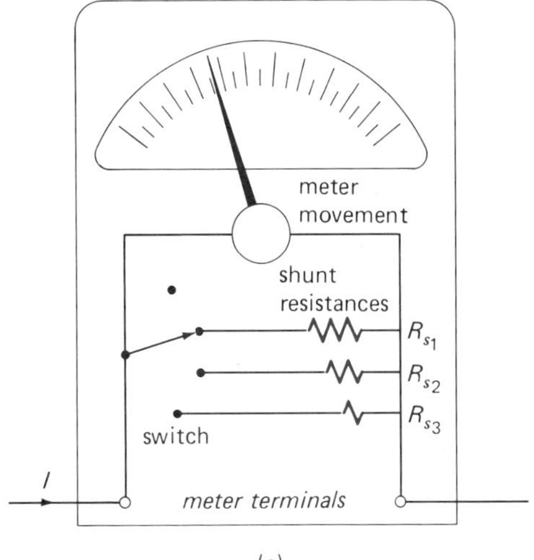

(a)

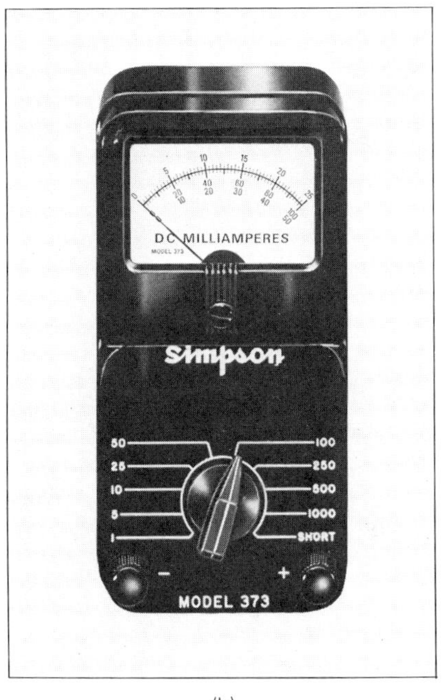

(b)

Figure 23.19
Multirange dc ammeter. (a) A meter can be adapted to read various ranges of current by selecting different shunt resistance. (b) A multirange milliammeter. (Courtesy Simpson Electric Co.)

$$V_{max} = I_c R_c + I_c R_m$$

and

$$R_m = \frac{V_{max} - I_c R_c}{I_c} \qquad \textbf{(Eq. 23.11)}$$

(voltmeter multiplier resistance)

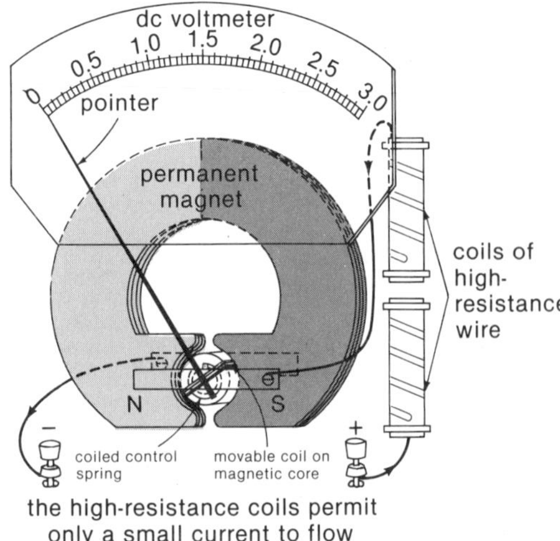

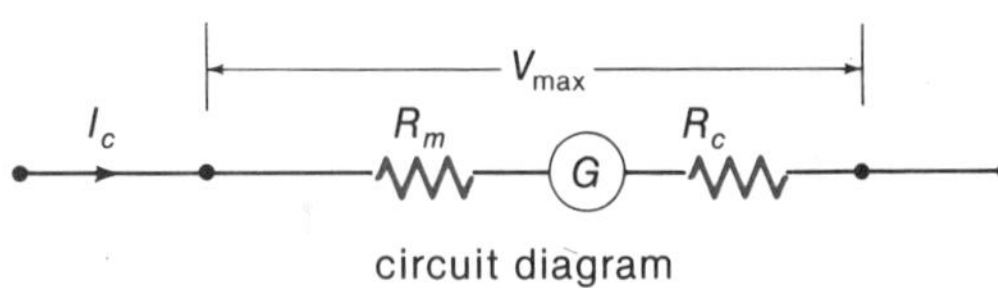

Figure 23.20
The dc voltmeter. By connecting a "multiplier" resistance in series with a galvanometer, dc voltages may be measured. The high-resistance multiplier permits only a small current to go through the galvanometer.

EXAMPLE 23.5 A galvanometer has a coil resistance of 25 ohms, and the coil current for full-scale deflection is 1.0 mA. What is the required multiplier resistance to convert the galvanometer into a 1.0-V meter?

Solution: It is given that $R_c = 25\ \Omega$, $I_c = 1.0$ mA $= 10^{-3}$ A, and $V_{max} = 1.0$ V. Then, by Equation 23.11,

$$R_m = \frac{V_{max} - I_c R_c}{I_c}$$

$$= \frac{1.0\ \text{V} - (10^{-3}\ \text{A})(25\ \Omega)}{10^{-3}\ \text{A}} = 975\ \Omega$$

Like an ammeter, a voltmeter can be made multirange by having several multiplier resistances (Fig. 23.21). However, unlike an ammeter,

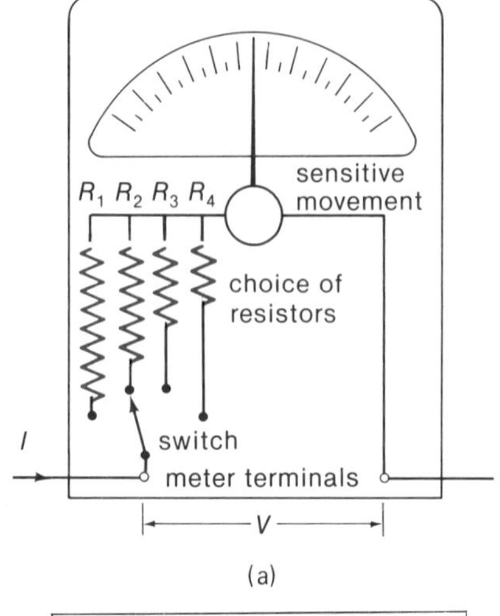

(a)

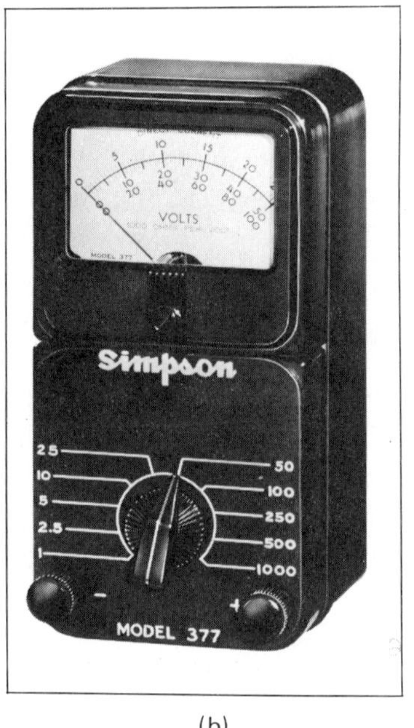

(b)

Figure 23.21
Multirange dc voltmeter. A meter can be adapted to read various ranges of voltages by selecting different multiplier resistances. [(b) courtesy Simpson Electric Co.]

A voltmeter is connected in parallel with the circuit element across which the voltage is to be measured.

If a voltmeter were mistakenly connected in series, the large resistance of the meter would affect the current and the voltage drops in the circuit. In parallel with a circuit element, only the small current I_c is drawn from the circuit.

Notice that a galvanometer is used in direct current and voltage measurements. If the current were alternating, the meter needle would oscillate back and forth. To measure alternating current and voltage, the current must first be rectified. More is said about rectification in Chapter 26.

Multimeters

For versatility, the elements of the multirange ammeter and voltmeter are often combined in a single unit to form a multimeter (Fig. 23.22). A multimeter also contains an ohmmeter for the direct measurement of resistances. Digital multimeters with electronic-display readouts are coming into increasing use.

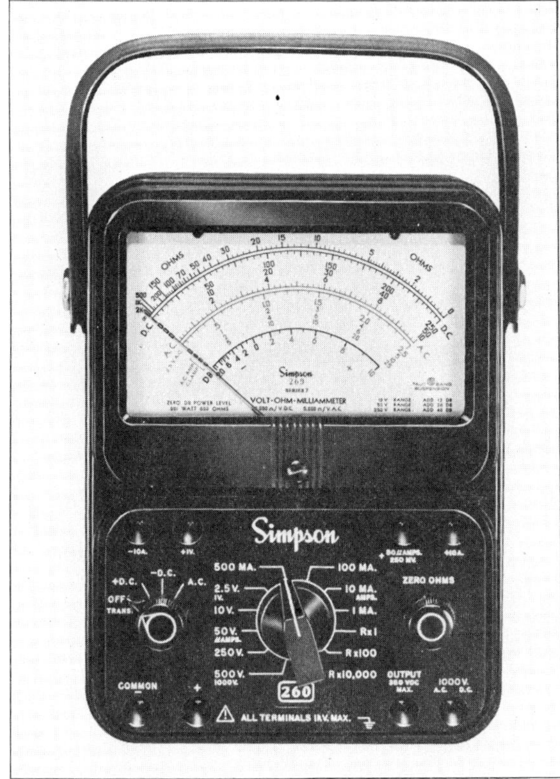

(a)

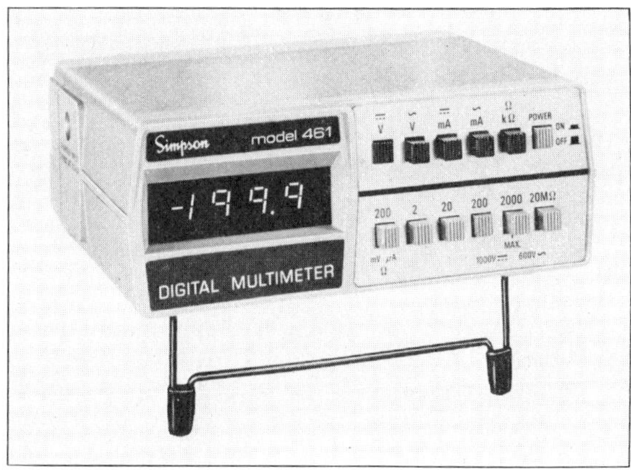

(b)

Figure 23.22
Multimeters. Multimeters contain an ammeter and a voltmeter for both dc and ac measurements, as well as an ohmmeter for resistance measurements. (Courtesy Simpson Electric Co.)

Important Terms

magnetic poles regions of apparently concentrated field strengths in a magnet, designated as north and south poles

law of poles like magnetic poles repel, and unlike magnetic poles attract

magnetic field (*B*) the magnetic force per moving electric charge

tesla (T) the SI unit of magnetic field (1 T = 1 Wb/m^2)

magnetic moment the vector representation of an atomic magnet

magnetic domain a group of aligned magnetic moments

relative permeability (K_m) a relative measure of the "magnetizability" of a material ($K_m = B_m/B = \mu/\mu_o$)

ferromagnetic material a material with a high relative permeability

paramagnetic material a material with a relative permeability slightly greater than one

diamagnetic material a material with a relative permeability slightly less than one

ampere (definition) the current which, if maintained in each of two long parallel wires separated by a distance of 1.0 m in free space, would produce a force between the two wires (due to their magnetic fields) of 2×10^{-7} N for each meter of length

motor a device that converts electrical energy into mechanical energy

galvanometer an instrument that detects small currents

Important Formulas

magnetic field:

around a long, straight, current-carrying wire $\quad B = \dfrac{\mu_o I}{2\pi d}$

at the center of a circular, current-carrying wire $B = \frac{\mu_o I}{2r}$

inside a solenoid $B = \frac{\mu_o NI}{L} = \mu_o nI$

$(n = N/L)$

$\mu_o = 4\pi \times 10^{-7}$ T-m/A (Wb/A-m)

right-hand thumb rule: If a current-carrying wire is grasped with the right hand with the thumb in the direction of the conventional current, the curled fingers indicate the circular sense (direction) of the magnetic field.

relative permeability: $K_m = \frac{B_m}{B} = \frac{\mu}{\mu_o}$

force on a moving charged particle: $F = qvB \sin \theta$

right-hand rule: If the fingers of the right hand turn or "cross" the v vector into the B vector, then the thumb points in the direction of F.

force on a current-carrying wire: $F = I\ell B \sin \theta$

right-hand rule: If the fingers of the right hand turn or "cross" a section of wire in the direction of the conventional current I into the B vector, the thumb points in the direction of F.

torque on a current-carrying wire: $\tau = BIA \sin \phi$

ammeter shunt resistance: $R_s = \frac{I_c R_c}{I_{max} - I_c}$

voltmeter multiplier resistance: $R_m = \frac{V_{max} - I_c R_c}{I_c}$

Questions

Magnets, Magnetic Fields, and Electromagnetism

1. What are natural magnets?
2. What are magnetic poles, and why are they called north and south poles?
3. Could magnetic poles be given designations of + and − equally as well as north and south? Explain.
4. If the poles of a magnet were labeled in reverse, would this change the law of poles? Explain.
5. Describe the principle of a compass.
6. Do magnetic poles have definite points of location? Discuss the concept of poles in terms of electromagnetism.
7. Sketch the magnetic field in the vicinity of the north pole ends of two bar magnets in close proximity, with the lengths of the magnets on a common line.
8. Sketch the magnetic field in the vicinity of the north pole of one bar magnet and the south pole of another in close proximity, with the lengths of the magnets along a common line.
9. A long, straight horizontal wire carries a direct current to the north. In what general directions would a compass needle point if the compass were placed (a) above the wire, (b) below the wire, and (c) on either side of the wire? What would happen in each case if the wire carried alternating current?
10. You are given four items—a piece of string, a needle, and two identical iron bars, one of which is a permanent magnet. Using the string and the bars, the needle and the bars, and only one of the bars alone, give three ways you could identify which bar is the magnet.
11. If a current-carrying solenoid were suspended on a string so it could rotate freely, what would happen, and which way would the end of the solenoid left by the current point? (Draw a sketch to illustrate this.)

Magnetic Materials

12. What are, and what gives rise to, atomic magnetic moments?
13. Distinguish between permeability and relative permeability.
14. (a) Are nails attracted to either pole of a bar magnet? Why or why not? (b) Several nails are suspended head-to-tail from a bar magnet. Identify the poles of each nail. Consider suspensions from both ends of the magnet. (Draw a sketch to illustrate this.)
15. Distinguish between and give examples of ferromagnetic, paramagnetic, and diamagnetic materials.
16. (a) How are permanent magnets made? Can any material be used? (b) What is the principle of industrial electromagnets?
17. A common doorbell continues to ring as long as the button is depressed. Referring to Figure 23.23, explain the operation of such a doorbell.
18. (a) Why does striking a magnet on a hard surface weaken its magnetism? How about heating the magnet? (b) Why doesn't breaking a magnet in two give single magnetic poles? What if you kept breaking it into smaller pieces?
19. Two materials have relative permeabilities of 200 and 1.5, respectively. Which would make a better magnet? Explain.

Magnetic Force and Torque

20. Describe the interaction between a moving electric charge and a magnetic field.

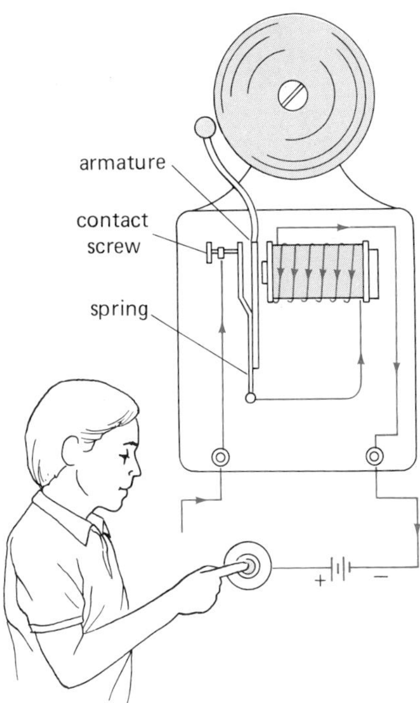

Figure 23.23
A doorbell. See Question 17.

21. Does a stationary electric charge in a magnetic field experience a force? Explain.
22. An electric field can be used to distinguish between positively and negatively charged particles. Could you do this with a magnetic field? Explain.
23. Describe the right-hand rule for finding (a) the direction of the magnetic field around a current-carrying wire and (b) the force on a moving charge and a current-carrying wire in a magnetic field.
24. How is the ampere defined?
25. Two long parallel wires carry currents. What is the direction of the force on each wire (a) if the currents are in the same direction and (b) if the currents are in the opposite directions?
26. Is there a torque on a current-carrying wire loop (a) when the plane of the loop is parallel and (b) when it is perpendicular to a magnetic field? Explain.

The dc Motor

27. What is the basic function of a motor?
28. A dc motor has a solid ring commutator. Would you buy the motor? Explain.
29. When is the torque on an armature loop in a dc motor (a) the greatest and (b) the least?

The Ammeter and Voltmeter

30. Can a galvanometer be used directly as an ammeter or a voltmeter? Explain.
31. What is the purpose of (a) a shunt resistance in an ammeter and (b) a multiplier resistance in a voltmeter?
32. How are ammeters and voltmeters connected in circuits? Explain why.
33. The scale of a multirange ammeter is changed so as to read a larger current. How is the shunt resistance affected?
34. The scale of a multirange voltmeter is changed so as to read a smaller voltage more accurately. How is the multiplier resistance affected?

Problems

Levels of difficulty are indicated by asterisks for your convenience.

23.1–2 Magnetic Fields and Electromagnetism

1. A long, straight, vertical wire carries a current of 20 A in the upward direction. What is the magnitude of the magnetic field 50 cm east of the wire?
2. A circular loop of wire with a diameter of 20 cm carries a current of 10 A. What is the magnitude of the magnetic field at the center of the loop?

*3. A long, horizontal, straight wire runs north and south. It is desired to produce a magnetic field of 2.0×10^{-6} T vertically upward at a point 40 cm directly east of the wire. What should be the magnitude and direction of the current in the wire?

*4. Consider a straight wire parallel to the top of the page of this book carrying a current of 10 A to the left. (a) What is the magnetic field 20 cm from the wire toward the bottom of the page? (b) What is the magnetic field 20 cm in the opposite direction from the wire?

*5. Two long, straight, parallel wires 30 cm apart both carry a current of 5.0 A as illustrated in Figure 23.24. What are the magnetic fields at points A, B, and C? (Note that the symbols for currents into and out of a page are a cross [×] and a dot [·], respectively. Think of the notched end and the tip of an arrow.)

*6. Suppose in Figure 23.24 that the current I_1 is into the page. What would be the magnetic fields at points A, B, and C in this case?

*7. The magnetic field 10 cm from a straight, current-carrying wire is 5.0×10^{-6} T. If the current in the wire is doubled and reversed in direction, how does this affect the magnetic

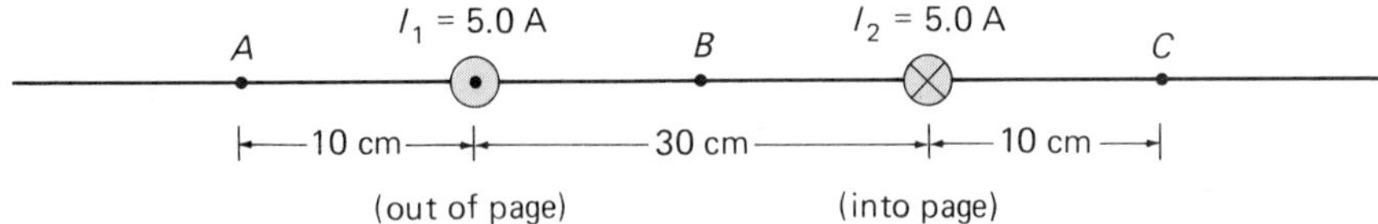

Figure 23.24
See Problems 5, 6, and 24.

field at distances of (a) 10 cm and (b) 5.0 cm from the wire?

*8. The planes of two circular loops of wire with radii of 10 cm are parallel, and the loops are close together. What is the approximate magnitude of the magnetic field on a center axis and between the loops if both loops carry a current of 4.0 A (a) in the same direction and (b) in opposite directions?

*9. Five adjacent circular loops with radii of 15 cm carry a current of 2.0 A. What is the approximate magnitude of the magnetic field at the center of the coil?

*10. An air solenoid, 12 cm long and with a diameter of 3.0 cm, has 60 loops of wire. What is the magnitude of the magnetic field inside the solenoid if it carries a current of 5.0 A?

*11. The turn density of an air solenoid is 10 turns/cm. What current would produce a magnetic field of 1.0×10^{-5} T in the solenoid?

*12. It is desired to make an air solenoid 15 cm in length so that a current input of 20 A produces a magnetic field of 5.0×10^{-2} T inside the solenoid. (a) What is the required turn density? (b) How many turns are there on the solenoid?

*13. The magnetic field in a circular coil in air is 3.0×10^{-5} T. When a particular material is placed in the coil, it is found that the magnetic field is 7.5×10^{-3} T. What are (a) the relative permeability and (b) the permeability of the material?

*14. A solenoid with a turn density of 50 turns/cm has an iron core with a relative permeability of 8000. (a) If the solenoid carries a current of 2.0 A, what is the magnetic field near the end of the core? (b) If the core is removed, what is the magnetic field?

**15. Two straight wires carry currents as illustrated in Figure 23.25. (a) What is the magnetic field at a point midway between the wires? (b) Would it be possible to make the magnetic field zero at this point by varying the currents? Explain.

23.4 Magnetic Force and Torque

*16. What is the magnitude of the force on an electron moving westward with a velocity of 2.5×10^4 m/s in a horizontal magnetic field of 6.0×10^{-5} T, if the electron moves (a) parallel to the field and (b) perpendicular to the field?

*17. A particle with a charge of $+3.0\ \mu C$ moves due east with a speed of 4.0×10^3 m/s. A horizontal magnetic field of 1.5×10^{-4} T is directed northeast. What is the magnetic force on the particle when it enters the field?

*18. In a particle accelerator, a horizontal beam of protons travels with a speed of 1.0×10^6 m/s. Looking in the direction of motion of the beam, a turning magnet deflects the beam to the right with a force of 8.0×10^{-8} N. What are the magnitude and direction of the magnetic field of the turning magnet?

*19. A vertical wire carries a current of 10 A upward in a horizontal field of 0.025 T directed eastward. What is the force per length on the wire?

*20. A long, horizontal wire in a north-south direction carries a current of 2.0 A north. If a horizontal magnetic field of 5.0×10^{-3} T is directed 30° east of north, what is the force per length on the wire?

*21. An air solenoid with a turn density of 20 turns/cm carries a current of 6.0 A. A wire running along the central axis of the solenoid carries a current of 1.0 A. What is the force per length on the wire?

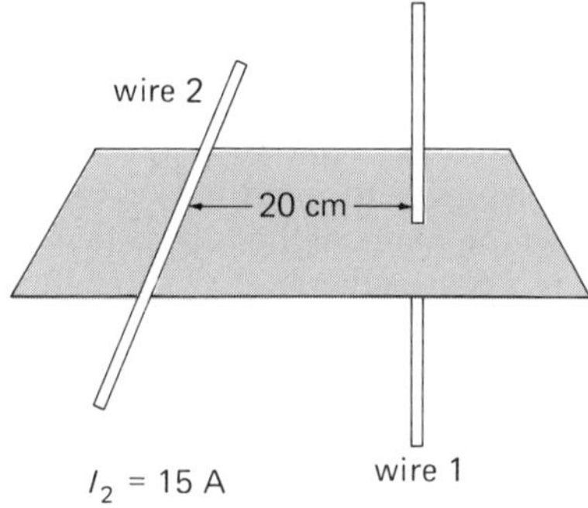

Figure 23.25
See Problem 15.

*22. A square, pivoted loop 10 cm on a side carries a current of 0.50 A in a uniform magnetic field with a magnitude of 2.0×10^{-4} T. What is the magnitude of the torque on the loop (a) when the plane of the loop is parallel to the field and (b) when the plane of the loop is perpendicular to the field?

*23. A pivoted rectangular loop with dimensions of 5.0 cm × 10 cm carries a current of 2.0 A in a uniform magnetic field with a magnitude of 5.0×10^{-2} T. (a) What is the magnitude of the torque on the loop when the angle between the normal to the plane of the loop and the magnetic field is 30°? (b) What is the torque when the angle between the plane of the loop and the magnetic field is 60°? (c) When it is 90°?

**24. Consider the parallel wires illustrated in Figure 23.24. (a) What is the force per length on wire 2? (b) What is the force per length on wire 1? (c) If the current in wire 1 were into the page, what would be the force per length on each wire?

23.6 The Ammeter and Voltmeter

*25. A galvanometer has a coil resistance of 10 Ω, and the coil current for a full-scale deflection is 0.50 mA. What is the required shunt resistance to convert the galvanometer into a 3.0-A ammeter?

*26. An ammeter has a shunt resistance of 0.50 Ω for a 10-A scale. If the coil resistance of the galvanometer is 25 Ω, what is the coil current for full-scale deflection?

*27. A multirange ammeter is designed with 1.0-A, 3.0-A, and 30-A ranges. If a galvanometer with a coil resistance of 5.0 Ω has a coil current of 2.5 mA for full-scale deflection, what are the required shunt resistances?

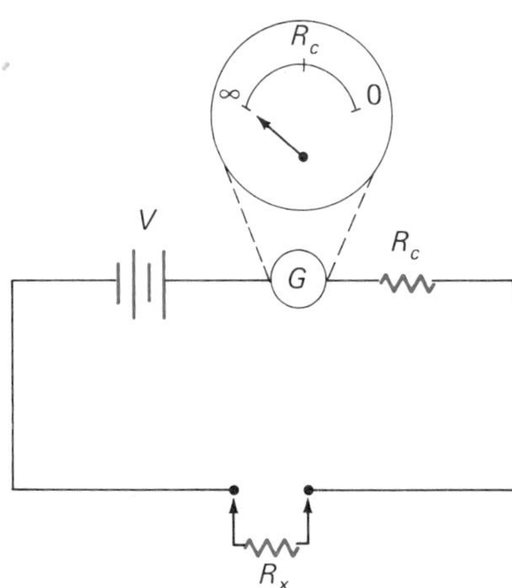

Figure 23.26
See Problem 30.

*28. If the galvanometer in Problem 25 is used in a voltmeter, what is the required multiplier resistance to convert the galvanometer to a 10-V meter?

*29. A multirange voltmeter is designed with 1.0-V, 3.0-V, and 30-V ranges. If the galvanometer described in Problem 27 is used in the voltmeter, what are the required multiplier resistances?

**30. A basic circuit for an ohmmeter using a galvanometer is shown in Figure 23.26. (a) As indicated on the scale, zero resistance is at the extreme right and infinite resistance is at the extreme left. When an external resistance R_x is exactly equal to the galvanometer coil resistance R_c, the reading is in the center of the scale. Explain why these scale positions are as they are. (b) Is the ohmmeter scale linear? Explain.

Chapter Supplement

Relays and Solenoids

Two electromagnetic devices that have numerous technical applications are relays and solenoids. Basically, they convert electromagnetic energy into mechanical energy or motion.

Relays

A relay is essentially an electromagnetic switch (Fig. 23S.1). It uses a coil with a fixed core, i.e., an electromagnet. When the coil is energized, the magnetic force attracts a contact arm, called an armature, toward the coil. The motion of the armature then either opens or closes ("breaks or makes") an auxiliary circuit, depending on the arrangement. The relay in Figure 23S.1 opens the circuit when activated.

One of the main advantages of a relay-switching circuit is that it allows low-voltage control of

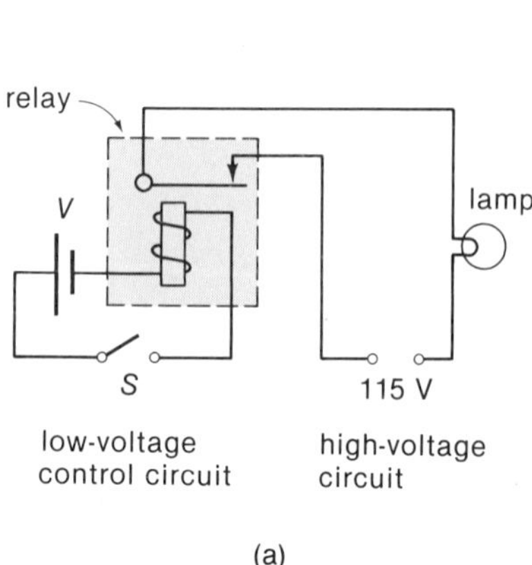

Figure 23S.1
Relay. (a) A relay switching circuit. When switch S is closed in the low-voltage control circuit, the electromagnet attracts the armature contact and opens the high-voltage circuit. (b) A general-purpose relay. (Courtesy Magnetcraft Electric Co.)

larger, high-voltage circuits. The low-voltage control of lights in the home is coming into increasing use. Large-current motors and machines can be controlled from a distance without the need to run heavy wires to the controlling switch. This provides convenience and safety.

Through design considerations of the coils and cores, relays provide a sensitive means of control. Relays have a wide variety of applications in electronic circuits such as those in radios and television sets. Also, electromechanical relays have been used for many years in current-voltage regulators in automobiles. Such regulators provide control of generator output and circuit voltage so as to meet various battery and operating requirements.

A cut-out relay is used to prevent current flow from the battery to the generator, which would discharge the battery. Electronic regulators that employ transistors for the switching functions are being used on most new cars because of their size and economy.

Another common use of an electromagnetic switching device is in the circuit breaker, which is used to prevent circuit overloads (Fig. 23S.2). Circuit breakers are now commonly used in household electrical circuits in place of fuses. If the current in the circuit exceeds a predetermined value, the breaker is activated and the relay opens the circuit. The circuit breaker may be "reset," or closed manually, by a cam switch. If a circuit breaker repeatedly opens, steps should be taken to find out why and to remedy the situation. Remember that a circuit breaker is a safety device, installed for protection from electrical overload.

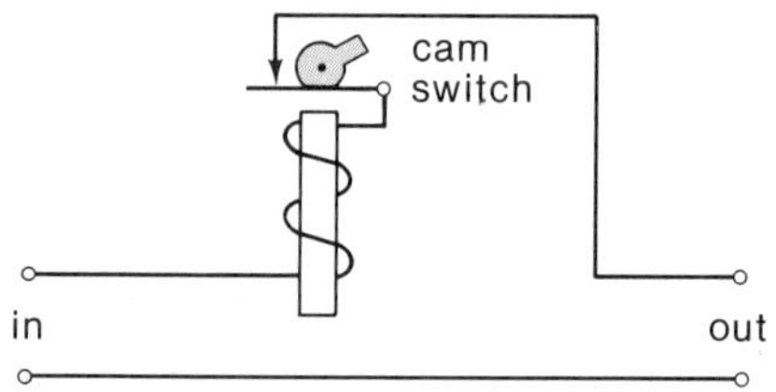

Figure 23S.2
Simplified circuit breaker diagram. When the current in the circuit exceeds a predetermined value, the relay (circuit breaker) opens the circuit. The circuit breaker may be reset (closed) manually by the cam switch. (See also Fig. 25.20.)

Solenoids

Solenoids are similar to relays, except that in the solenoid the electromagnet contains a movable core. Basically, a solenoid is an electromechanical device that converts electrical energy into linear motion (Fig. 23S.3). When the electromagnet is energized, the magnetic force draws the core into the coil, providing the mechanical action. A spring may be used to return the core when the coil circuit is opened.

Solenoids are used for the electrical control of mechanical actions. A common example is the starting system of an automobile (Fig. 23S.3). When the starter switch is closed, a low-current starter solenoid on the cranking motor moves a plunger, which mechanically engages the starter gear and closes a switch that connects the motor to the high-current battery. When the starter switch is released, the solenoid is de-energized, and a spring within the solenoid assembly pulls the gear out of mesh and interrupts the current flow to the cranking motor.

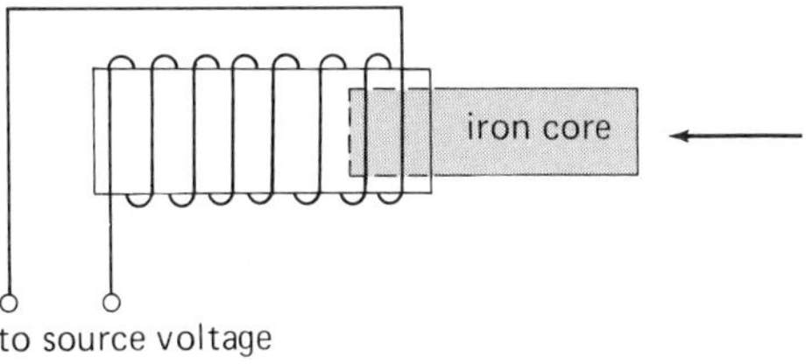

(a) solenoid principle

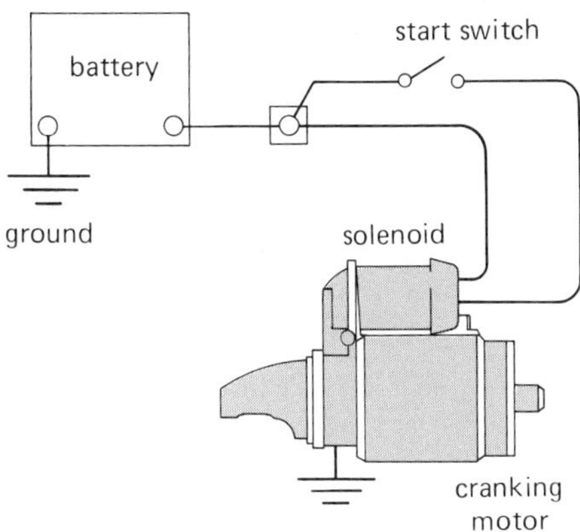

(b) auto starting system

Figure 23S.3
The solenoid. (a) A magnetic solenoid is similar to a relay, except that the electromagnet contains a movable core. (b) Auto starting system. When the switch is closed, the solenoid is activated and the plunger core engages the starter gear. When the main motor starts, the solenoid is deactivated and the starter gear is disengaged.

Turbine generators. (Courtesy Bureau of Reclamation, United States Department of Interior.)

Chapter 24

Electromagnetic Induction

The discovery that an electrical current produces a magnetic field prompted a question on the reverse situation: Can a magnetic field produce a current? Several scientists worked on this problem in the early 19th century, and Michael Faraday (in honor of whom the capacitance unit is named) is given credit for the discovery of electromagnetic induction.

This effect forms the basis for the production and transmission of electrical power through generators, transformers, and many other electromagnetic applications. In previous chapters, we considered batteries primarily as voltage sources of direct current. This involved the conversion of chemical energy to electrical energy. In this chapter we study how electric currents, both alternating and direct, are generated through electromagnetic induction. In this case, mechanical energy is converted to electrical energy.

24.1 Faraday's Law of Induction

Experimenting in the early 1830's, Faraday observed that

> **An emf (voltage) is induced in a conducting loop when there is relative motion between the conductor and a magnetic field, such that there is a change in the number of field lines through the loop.**

For example, as illustrated in Figure 24.1, if a magnet is moved toward a loop of wire, an induced current is detected in the circuit by a galvanometer or an ammeter. The same effect is observed by moving a coil of wire near a stationary magnet. In either case, the number of field lines through the loop changes. Other observations made by Faraday were that

1. The magnitude of the induced emf (or current) is proportional to the number of loops N of wire.
2. If the direction of the relative motion is reversed, the induced current is reversed.

As an example of the latter condition, if the magnet were moved away from the loop in Figure 24.1, the galvanometer needle would be deflected in the other direction.

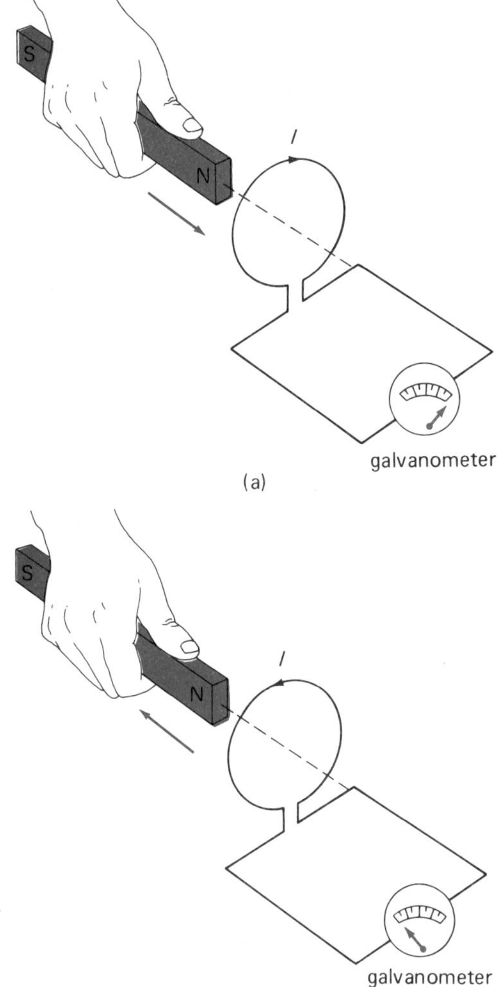

Figure 24.1
Magnetic induction. (a) When the magnet is moved toward the loop of wire and there is a change (increase) in the number of field lines through the loop, a current is induced, as indicated by the deflection of the galvanometer. (b) When the magnet is moved away from the loop and number of field lines through the loop decreases, the induced current is reversed.

Another observation that is critical to the explanation of electromagnetic induction is that

3. The magnitude of the induced emf is proportional to the strength of the magnetic field and the rate of change of the number of field lines through the loop.

For example, the greater the pole strength of the magnet and the faster it is moved, the greater the induced emf in the stationary loop.

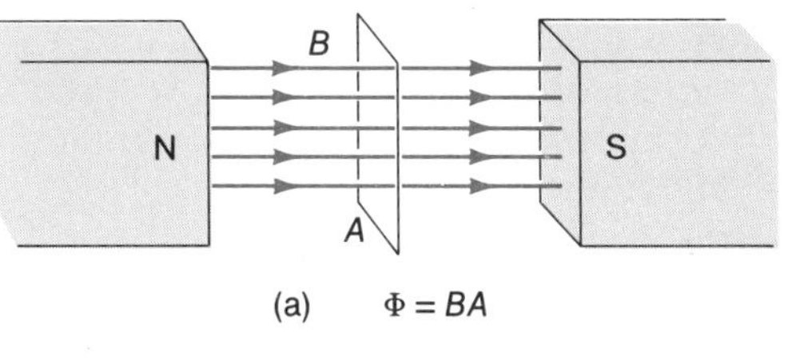

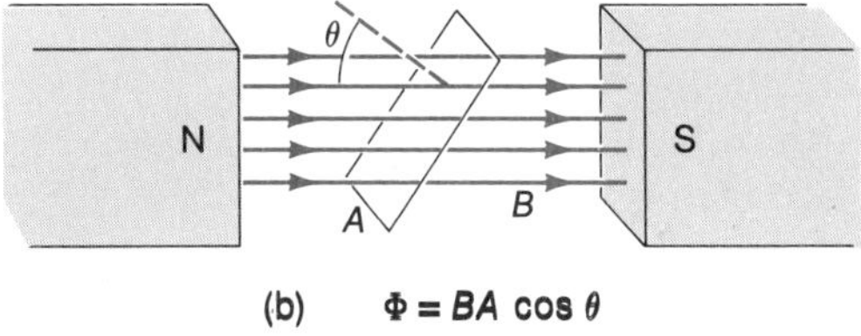

Figure 24.2
Magnetic flux. The magnetic flux is a measure of the number of field lines through an area. (a) With the magnetic field perpendicular to the plane of the loop, $\phi = BA$. (b) In general, $\phi = BA \cos \theta$.

To describe this, it is convenient to consider the **magnetic flux Φ:**

$$\boxed{\Phi = BA} \qquad \textbf{(Eq. 24.1)}$$

(magnetic flux)

where the magnetic field B is perpendicular to the plane of the area A of the loop (Fig. 24.2). The unit of the magnetic flux can be seen to be the weber, $B\ (\text{Wb/m}^2)\ A\ (\text{m}^2) = \Phi\ (\text{Wb})$.*

The magnetic flux is a measure of the number of magnetic field lines through an area.

In general,

$$\Phi = BA \cos \theta \qquad \textbf{(Eq. 24.2)}$$

where θ is the angle between the magnetic field lines and a normal to the plane of the loop (Fig. 24.2). $A \cos \theta$ is the area of the loop that is "open," or exposed to the magnetic field lines.

The above observations can be expressed in a relationship known as **Faraday's law of induction:**

$$\boxed{\mathscr{E} = -N\left(\frac{\Delta\Phi}{\Delta t}\right)} \qquad \textbf{(Eq. 24.3)}$$

* The magnetic field B is sometimes called the *flux density* because of its units, Wb/m^2.

where $\mathscr{E}$ is the average induced emf and N is the number of loops of wire. The $\Delta\Phi/\Delta t$ is the time rate of change of flux through a loop and reflects item 3 given previously. The minus sign indicates the polarity of the induced voltage, or the direction of the induced current with regard to the conversion of energy, as will be explained in the next section.

Hence, to produce an emf, one must have a change in the flux, $\Delta\Phi$. Generally, a change in flux (the number of field lines through a loop) can occur in two ways:

1. $\Delta\Phi = (\Delta B)A$ changing magnetic field, constant area

and/or

2. $\Delta\Phi = B(\Delta A)$ constant magnetic field, changing area

Actually, there are three ways to produce a change in flux. The angle θ in Equation 24.2 may also be varied with time. That is, the loop may be rotated. This in effect changes the open or exposed area of the loop to the magnetic field lines, so we include this in number 2 as a ΔA. Then Equation 24.3 may be generalized to

$$\mathscr{E} = -N\left[\left(\frac{\Delta B}{\Delta t}\right)A + B\left(\frac{\Delta A}{\Delta t}\right)\right] \quad \textbf{(Eq. 24.4)}$$

The cases in Figure 24.1 are examples of the $(\Delta B/\Delta t)A$ component. The magnitude of the magnetic field (number of field lines) through the loop varies with time, owing to the relative motion.

Another example of this case is illustrated in Figure 24.3. Here, when the switch is closed in the battery circuit, the current goes from zero to a constant value. This changing current produces a magnetic field that increases with time through the adjacent loop. The result of this $\Delta B/\Delta t$ is detected by the galvanometer as an induced current. This is an example of **mutual induction.** After a constant value of the current is reached in the battery circuit, the induced current drops to zero. Why? What will happen when the switch is opened?

An example of the $B(\Delta A/\Delta t)$ component of Equation 24.4 would be a loop that is adjustable so as to change its area. Think of such a loop on a balloon that is being blown up. A more practical example of $(\Delta A/\Delta t)$ is given in the next section.

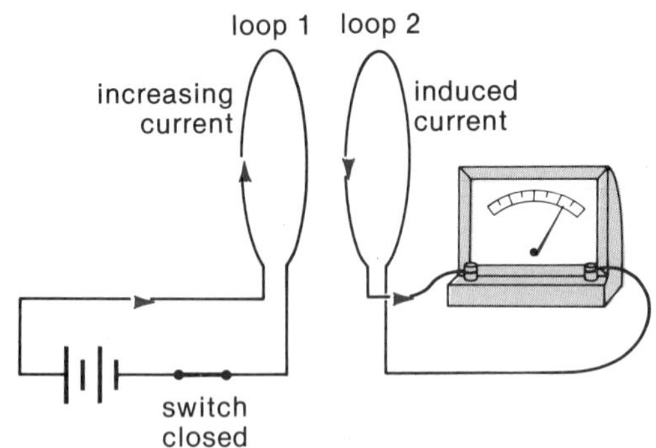

Figure 24.3
Magnetic induction due to changing current. When the switch is closed, the current builds up from zero to a constant value in loop 1. The resulting time-varying magnetic field $\Delta B/\Delta t$ through the nearby loop 2 induces a current in that loop. When a steady current flows in loop 1, the magnetic field is constant ($\Delta B/\Delta t = 0$) and the induced current in loop 2 goes to zero. When the switch is opened, the process is reversed.

EXAMPLE 24.1 When the switch is closed in the battery circuit in Figure 24.3, the magnetic field through loop 2 goes from zero to a constant value of 1.6 Wb/m^2 in 0.02 s. If the area of the loop is 0.05 m^2, (a) what is the average emf produced during $0 < t < 0.02$ s, and (b) what is the induced current in loop 2 after 0.02 s?

Solution: It is given that $B_1 = 0$, $B_2 = 1.6$ Wb/m^2, $A = 0.05$ m^2, and $\Delta t = 0.02$ s.

(a)
$$\begin{aligned}\mathscr{E} &= -N\left(\frac{\Delta\Phi}{\Delta t}\right)\\ &= -\left(\frac{\Delta B}{\Delta t}\right)A\\ &= -\frac{(B_2 - B_1)A}{\Delta t}\\ &= -\frac{(1.6\ \text{Wb/m}^2)(0.05\ \text{m}^2)}{0.02\ \text{s}}\\ &= -4\ \text{V}\end{aligned}$$

where $N = 1$.

(b) Since after 0.02 s the current in loop 1 and, hence, the magnetic field are constant, $\Delta B/\Delta t = 0$ and $\mathscr{E} = 0$.

Notice that if the switch were opened, the average induced voltage due to the decreasing magnetic field would be $\mathscr{E} = +4$ V. The signs of the voltages indicate that the polarity of the voltages, or direction of the current, is reversed for the two cases.

24.2 Lenz's Law and Back emf

An example of the $B(\Delta A/\Delta t)$ component of Faraday's law is illustrated in Figure 24.4. As the metal bar is moved along the metal frame with a speed v in a uniform magnetic field, the change in area is given by

$$\Delta A = \ell \Delta d = \ell v \Delta t$$

Hence, $\Delta A/\Delta t = \ell v$ and

$$\mathscr{E} = -\frac{\Delta \Phi}{\Delta t} = -B\left(\frac{\Delta A}{\Delta t}\right) = -B\ell v \qquad \textbf{(Eq. 24.5)}$$

This is referred to as a **motional emf.**

However, it is important to determine the direction of the induced current. Let's assume that it is to the right in the figure. The point to keep in mind is that a current, induced or otherwise, gives rise to a magnetic field. Then, using the right-hand thumb rule (Section 23.2), the magnetic field of the induced current would add (be in the same direction as the applied uniform B field). This would increase the magnetic field through the loop and give rise to an additional change in flux by the $(\Delta B/\Delta t)A$ component of Faraday's law.

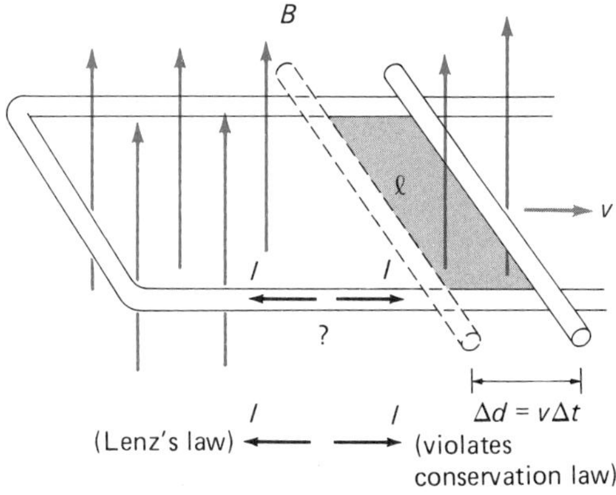

Figure 24.4
Motional emf and Lenz's law. When the bar is moved on a metal frame in a constant magnetic field, the $B(\Delta A/\Delta t)$ component of Faraday's law gives rise to an induced emf and a current in the circuit. By Lenz's law, the induced current is in the direction indicated to produce a magnetic field that opposes the increase in the flux. Otherwise, the conservation of energy would be violated.

Thus, the induced current would increase, which would increase the induced magnetic field, and so on, such that we would have a runaway condition. That is, we would violate the conservation of energy rule and have an electromagnetic perpetual-motion machine, so to speak.

A statement for the conservation of energy in electromagnetic induction was formulated by the Russian physicist H. Lenz shortly after Faraday put forth his law of induction. The statement is known as **Lenz's law:**

> An induced current is in the direction such that its effect opposes the change that produces it.

By means of this law, the conservation of energy is applied to induced voltages and currents.

For the case in Figure 24.4, the induced current is then to the left, as indicated. The *change* producing the current is an *increase* in flux (due to an increase in area). The induced current produces a magnetic field opposite to the existing magnetic field in the loop, which decreases the net magnetic field through the loop. This gives rise to a *decrease* in flux, which opposes the change as required by Lenz's law.

Back emf in dc Motors

In Section 23.5, it was learned that the principle of the dc motor is a current-carrying loop rotating in a uniform magnetic field. The rotation is due to the magnetic torque on the loop. However, as the loop rotates in the magnetic field, there is a change of flux through the loop (cf. Fig. 24.2).

Since the rotation of the loop and the change in flux are due to the current delivered to the motor, by Lenz's law an induced current or emf must oppose the delivered current (Fig. 24.5). As a result, the induced emf is called a **back emf (or counter emf).** The net voltage is then the

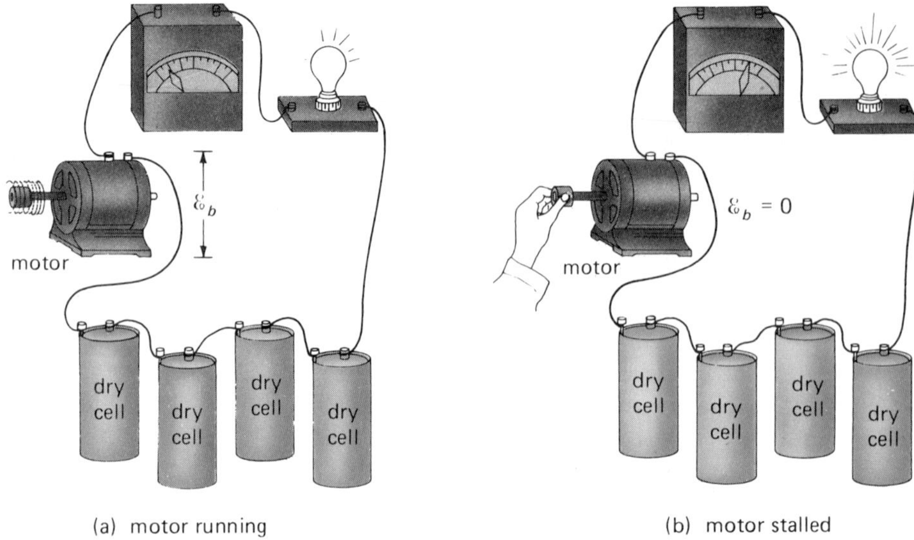

Figure 24.5
Back emf. (a) The induced (back) emf in a dc motor opposes the current delivered by the voltage source, as indicated by the decreased brightness of the bulb. (b) If the motor is stalled, the back emf is zero and the current to the circuit is greater, as indicated by the increased brightness of the bulb.

applied voltage V minus the back emf $\mathscr{E}_b$, and, by Ohm's law,

$$V - \mathscr{E}_b = IR \qquad \textbf{(Eq. 24.6)}$$

where R is the resistance of the motor windings.

The back emf depends on the rotational speed of the motor armature, since the greater the speed, the greater the flux change. But the back emf tends to reduce the armature current and, hence, the torque on the armature. The back emf then controls the speed of the armature to some extent.

24.3 **Electrical Generators**

A generator is a device that converts mechanical energy into electrical energy. Essentially, a generator performs the reverse function of a motor. The principle of the generator involves electromagnetic induction.

Consider the simple electrical generator illustrated in Figure 24.6. A loop is mechanically rotated in a magnetic field. The ends of the loop are connected to slip rings, which are in contact with conducting brushes. As the loop is turned through one half-cycle, the flux increases during the first quarter-cycle and decreases during the next quarter-cycle. The induced emf has the same polarity, and the induced current flows in one direction during this half-cycle. The emf and current go through a maximum, as shown in the graph.

During the next half-cycle, a similar effect occurs, but the polarity of the induced emf and the direction of the current are reversed. The polarity of the emf and the directions of the current alternate each half-cycle. The induced current is therefore an **alternating current (ac),** and the voltage is referred to as an *ac voltage.* Such an ac generator is sometimes called an *alternator.* In a regular generator, many rotating loops are used to increase the emf output.

It can be shown that the **instantaneous emf produced by a simple ac generator** is given by

$$\mathscr{E} = NBA\omega \sin \theta \qquad \textbf{(Eq. 24.7)}$$

where N is the number of loops, A is their common area, ω is the rotational or angular frequency in rad/s, and θ is the angle between the magnetic field and the normal to the plane of the loop (Fig. 24.7).

Hence, when $\theta = 0$ and $\theta = 180°$, $\mathscr{E} = 0$. Also, when $\theta = 90°$ and $\theta = 270°$, $\mathscr{E} = \mathscr{E}_{max}$ (see Fig. 24.6). The equation may be conveniently written

$$\mathscr{E} = \mathscr{E}_{max} \sin \theta \qquad \textbf{(Eq. 24.8)}$$

where $\mathscr{E}_{max} = NBA\omega$. Or, since $\theta = \omega t = 2\pi ft$,

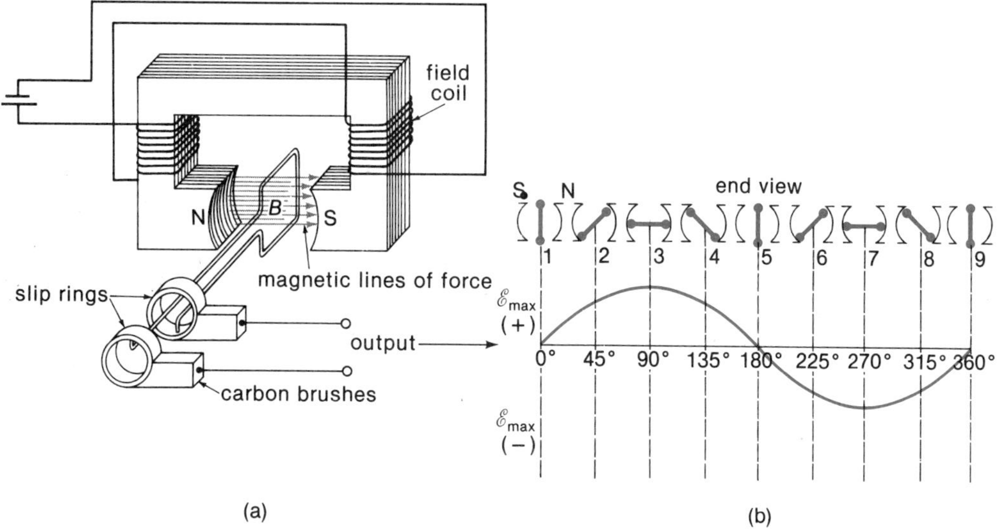

Figure 24.6

ac generator. (a) The elements of a simple ac generator. (b) A graph of voltage (or current, $I = \mathscr{E}/R$) output with respect to armature rotation. The polarity of the voltage and direction of the current reverses with each half-cycle.

$$\mathscr{E} = \mathscr{E}_{max} \sin \omega t$$
$$\mathscr{E} = \mathscr{E}_{max} \sin 2\pi f t \qquad \textbf{(Eq. 24.9)}$$

The electricity used in the home and industry is primarily ac. The generating frequency f in the United States is 60 Hz (60 cycle), and the voltage is generally 110–120 VAC (or 220–240 VAC in some applications).*

In specifying the amount of ac voltage and current, a special time-averaged value is used. It would be misleading to say the ac voltage is equal to $\mathscr{E}_{max}$, since the emf or voltage reaches this value only momentarily each half-cycle. Alternating quantities are expressed in *effective* values in terms of the average power delivered to a resistor. Recall that $P = V^2/R$ and

$$P = \frac{V^2}{R} = \frac{\mathscr{E}^2}{R} = \frac{(\mathscr{E}_{max} \sin \omega t)^2}{R}$$
$$= \frac{\mathscr{E}^2_{max} (\cos 2\omega t + 1)}{2R}$$

where the trigonometric identity $\sin^2 \theta = \frac{1}{2}(\cos 2\theta + 1)$ is used.

The time average of the cosine term is zero, since it is equally positive and negative during the cycle. Hence,

$$P_{eff} = \frac{V^2_{eff}}{R} = \frac{\mathscr{E}^2_{max}}{2R}$$

Thus, we define an average or effective value of voltage, V_{eff}:

$$V^2_{eff} = \frac{\mathscr{E}^2_{max}}{2}$$

and

$$V_{eff} = \frac{1}{\sqrt{2}} \mathscr{E}_{max} = 0.707\ V_{max} \qquad \textbf{(Eq. 24.10)}$$

(effective ac voltage)

where $\mathscr{E}_{max} = V_{max}$ and $1/\sqrt{2} = 0.707$. This effective value is also called the *root-mean-square (rms) value.*

* The accepted abbreviation for alternating current is ac. Alternating voltage is commonly written as VAC and is read "volts ac."

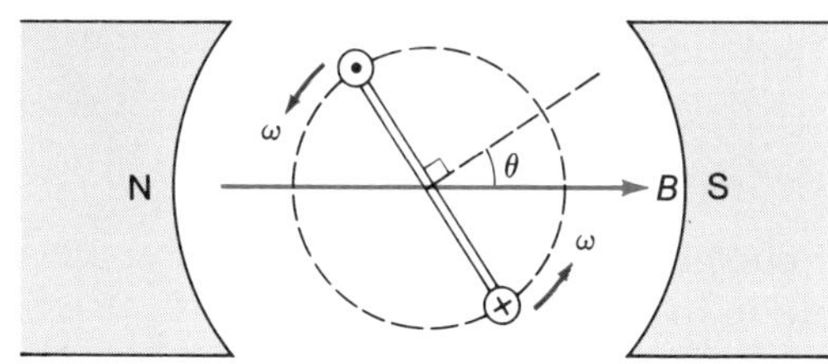

Figure 24.7

Generator geometry. The generator output depends on the rotational frequency ω of the loop or armature.

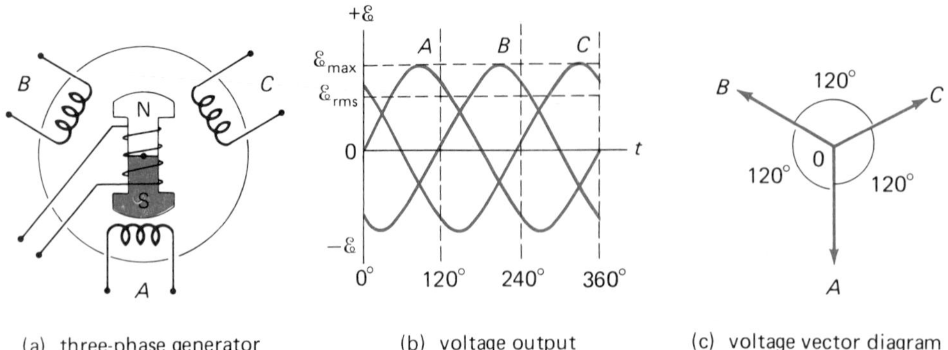

Figure 24.8

Three-phase ac generator. (a) The magnetic field can be made to rotate and the output coils (A, B, and C) remain stationary. (b) The voltage (and current) outputs are 120° out of phase (c).

There is an expression similar to Equation 24.10 for current:

$$I_{\text{eff}} = 0.707\ I_{\text{max}}$$

Using the effective voltage and effective current, expressions for Ohm's law ($V = IR$) and power ($P = IV$) can be used in ac applications. The ranges in the effective 110–120 VAC and 220–240 VAC are due to fluctuations in V_{max} in generation. One commonly sees 110 VAC, 115 VAC, and 120 VAC listed as normal household voltages; 220–240 VAC is used in applications with greater power requirements, e.g., electric stoves and central air conditioners.

We define the ac amp in terms of power, which is independent of the current direction.

> One ampere of ac current is said to flow in a circuit if the current produces the same average joule heating effect as one amp of dc current under the same conditions.

EXAMPLE 24.2 An ac generator (alternator) has 90 turns on its armature with an area of 0.01 m². (a) If the armature is turned with a frequency of 60 Hz in a magnetic field of 0.50 Wb/m², what is the effective voltage output? (b) If the output load of the generator is 480 ohms, what is the effective current in the circuit?

Solution: It is given that $N = 90$, $A = 0.01\ \text{m}^2$, $f = 60$ Hz, and $B = 0.50\ \text{Wb/m}^2$. Then,

(a) $$\mathscr{E}_{\text{max}} = NBA\omega = NBA\ 2\pi f$$

$$= (90)(0.50\ \text{Wb/m}^2)(0.01\ \text{m}^2)2\pi(60\ \text{Hz})$$

$$= 170\ \text{V}$$

and

$$V_{\text{eff}} = 0.707\mathscr{E}_{\text{max}} = (0.707)(170\ \text{V}) = 120\ \text{V}$$

(b) With $R = 480\ \Omega$ and $V = IR$, where V and I are the effective voltage and current, respectively, then

$$I = \frac{V}{R} = \frac{120\ \text{V}}{480\ \Omega} = 0.25\ \text{A}$$

An ac generator may be wound with more than one set of loops or output coils. The magnetic field can be made to rotate and the coils remain stationary (Fig. 24.8). The induction effect is the same as that of a rotating loop and stationary magnetic field. A generator with three sets of coils 120° apart is a three-phase alternator with three ac voltages and currents that are 120° out of phase.

In commercial electrical generation, heat from the combustion of fossil fuels or from nuclear reactions is used to generate steam, which powers a turbine (Fig. 24.9). The turbine supplies the mechanical energy for the generator.

The dc Generator

A dc current can also be generated by electromagnetic induction. The dc generator is essentially a dc motor operated in reverse (Fig. 24.10). With the use of a split-ring commutator, the output voltage has a constant polarity, and the current is in a constant direction. However, both pulsate with time. The fluctuations can be

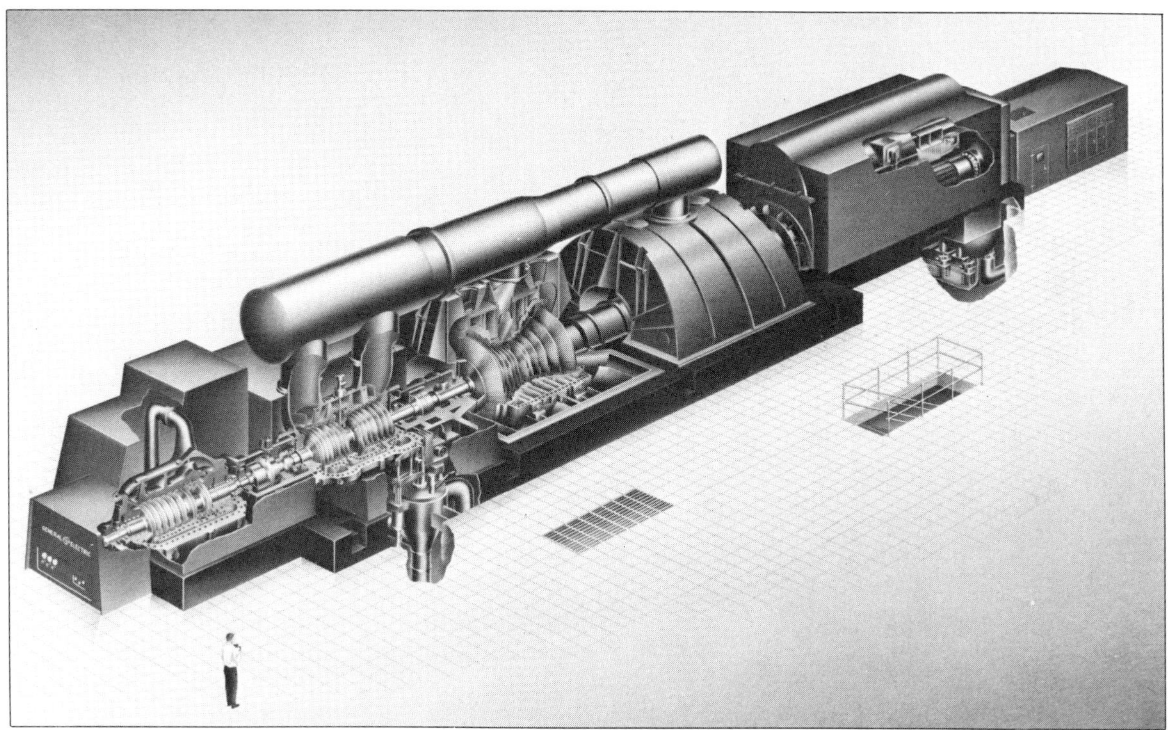

Figure 24.9
Steam-turbine generator unit. A cutaway drawing showing a typical fossil-fueled steam-turbine generator unit as it would appear installed in a major power-generating station. The unit is rated at more than 500,000 kW and is capable of producing enough electricity to supply the demands of a city of more than a half-million people. (Courtesy General Electric Co., Schenectady, N.Y.)

decreased by using a two- or three-phase generator. In these cases, the voltage and the current never go to zero.

At one time, dc generators were widely used in automobiles as a source of current for electrical systems and battery recharge (Fig. 24.11). To get appreciable output, the generator armature had to be rotated at relatively high speeds—in other words, the generator was most effective when the motor was running at a

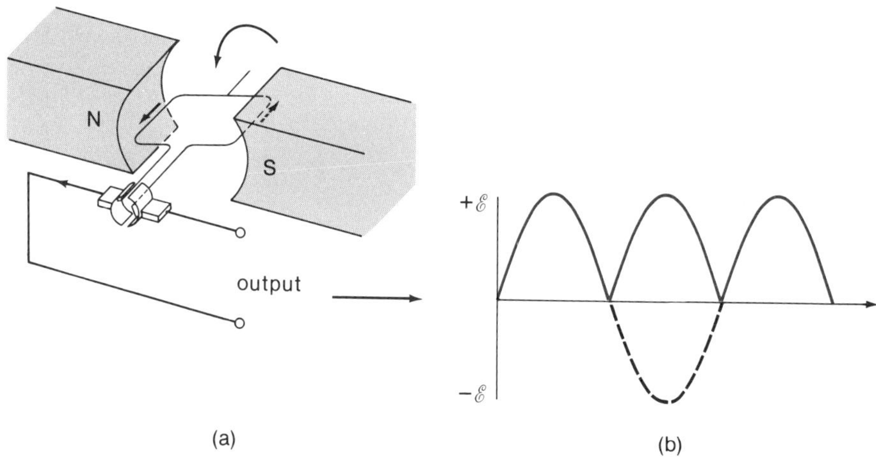

Figure 24.10
dc generator. (a) The elements of a simple dc generator. Because of the split-ring commutator, the output voltage (b) has a constant polarity and the current is in one direction.

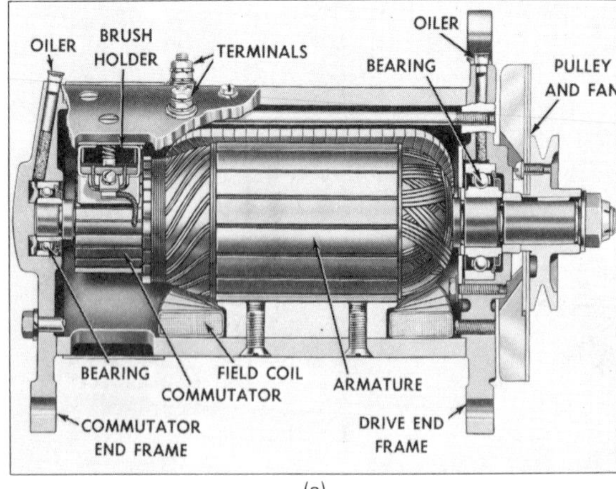

(a)

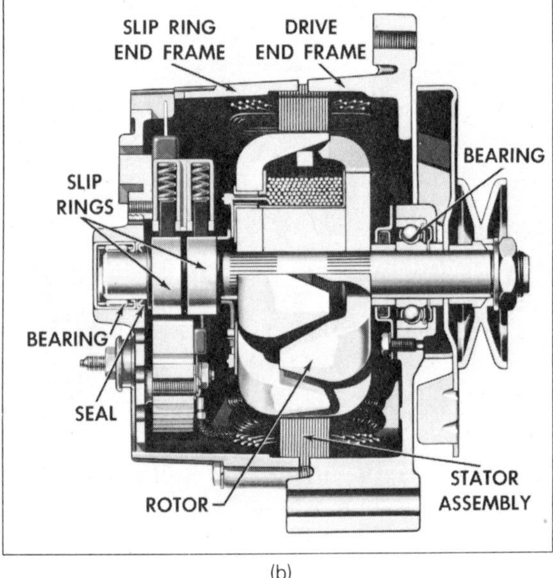

(b)

Figure 24.11
Automobile generator. (a) An exposed view of a generator with commutator for dc output. (b) An exposed view of an ac alternator. A rectifier bridge changes the ac voltage to dc voltage. (Courtesy Delco Remy Division, General Motors Corp.)

high speed. (Power transmission was by means of the fan belt.) These dc generators have been replaced, for the most part, by ac alternators, which have greater outputs at low speeds. The ac output is changed (rectified) to dc before use. (Rectification is discussed in Chapter 26.)

24.4 ac Motors

Motors as well as generators can be ac or dc. In general, dc motors are used in applications where speed control is required, whereas ac motors are used mainly in constant-speed applications. Also, ac motors are usually more trouble-free than dc motors. A few common types of ac motors are discussed here.

Universal Motors

The universal motor operates on either ac or dc. Recall that in dc operation the current direction changes because of the split-ring commutator and because the magnetic field is in one direction. If the input is ac, the current also alternates directions; but in this case the polarities of the brush contacts alternate as well. Hence, in a unidirectional magnetic field, a condition of stable equilibrium would arise for the armature.

However, since the same current flows through the armature coils and the field coils, the magnetic field also alternates directions. As a result, the motor continues to run in a single direction.

Universal motors are not generally used for heavy-duty applications because of excess sparking at the brushes. They are widely used in small appliances and power tools such as fans, vacuum cleaners, food mixers, and portable drills and saws (Fig. 24.12).

Induction Motors

A common type of ac motor is the induction motor. Like all motors, it contains a rotating part, the rotor (armature), and a stationary part, the stator. However, an induction motor does not use a commutator or brushes. In an induction motor, the stator rather than the rotor is connected to the ac current supply. That is, current is not fed to the rotor as in previously discussed cases.

The ac current in the stator windings (field coil) produces a time-varying magnetic field that induces an ac current in the rotor windings. The interaction of the induced rotor current with the stator field then causes the rotor to turn.

The principle of a three-phase induction motor is illustrated in Figure 24.13. A three-phase alternating current represents three separate currents that are 120° out of phase (see Fig. 24.8). The three-phase current is used to estab-

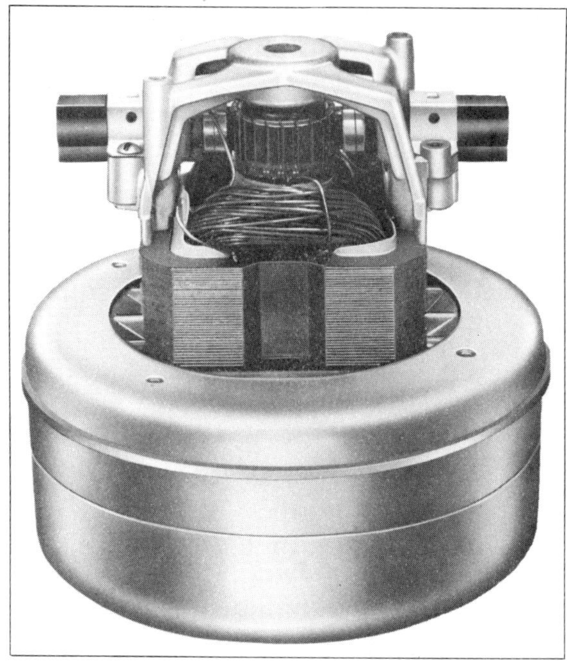

Figure 24.12

Universal motor. This motor is used to drive a centrifugal fan designed for use in vacuum cleaners. The motor windings are cooled by "working air," which is directed over the windings after leaving the fan system. (Courtesy Ametek, Lamb Electric Division.)

lish a "rotating" magnetic field in the stator. As the field rotates, the rotor follows the field around and the motor shaft rotates.

The rotor of such a motor consists of a circular, laminated iron cylinder with copper bars or heavy copper wire in slots around the core. Because the rotor appears somewhat like a circular cage, it is called a squirrel-cage rotor.

Three-phase electricity is commonly available in industry. However, in homes the current is mainly single-phase. Squirrel-cage motors can be run on a single-phase source, but some means of starting the motor is required. In this case, the rotor currents induced by the stator field can only keep a rotating motor going. They cannot start it.

Single-phase induction motors may be started by what is known as phase splitting. One method of phase splitting involves the use of a capacitor. As is discussed in Chapter 25, capacitance in an ac circuit causes the current in the circuit to lead the applied voltage in phase. With the connection of a starting coil and a capacitor across the running (stator) windings, a phase shift in the stator and rotor fields occurs (Fig. 24.14).

This type of motor is known as a *capacitor-start induction motor*. The capacitor and starting coil are taken out of the circuit by a centrifugal switch when the motor approaches full speed. Induction motors are used to power all types of medium and heavy machinery.

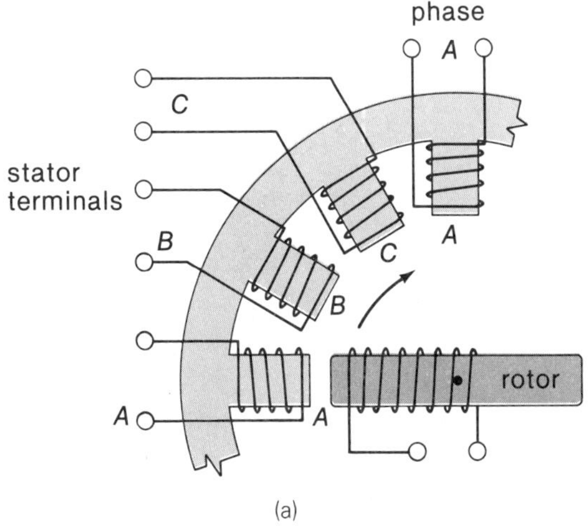

(a)

(b)

Figure 24.13

Three-phase induction motor. (a) The three phases of the current that are 120° out of phase establish a "rotating" magnetic field in the stator. The interaction of the induced current in the rotor windings causes the rotor to follow the field and rotate. (b) Exposed view of a 5-hp, 1800-rpm, three-phase motor. (Courtesy Delco Products Division, General Motors Corp.)

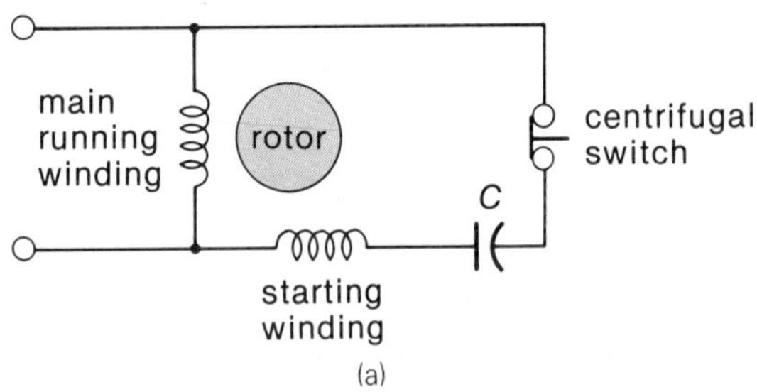

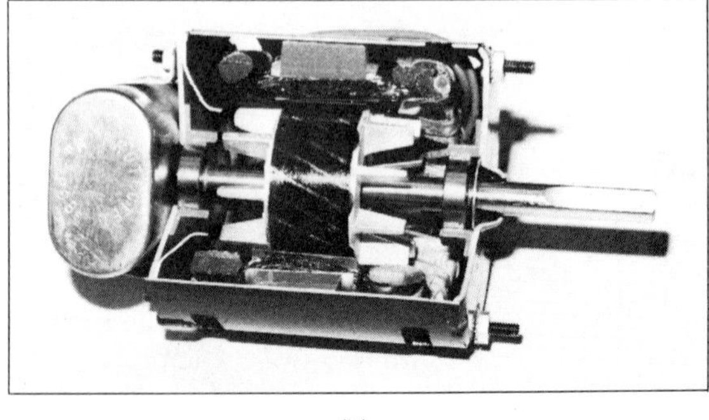

Figure 24.14
Capacitor-start, split-phase induction motor. See text for description. (Courtesy Fasco Industries, Motor Division.)

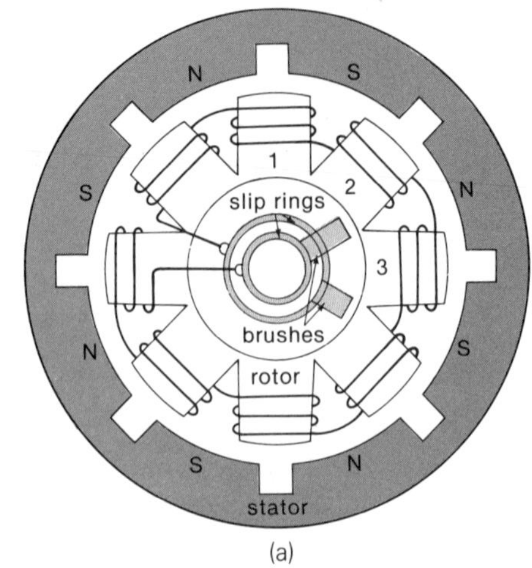

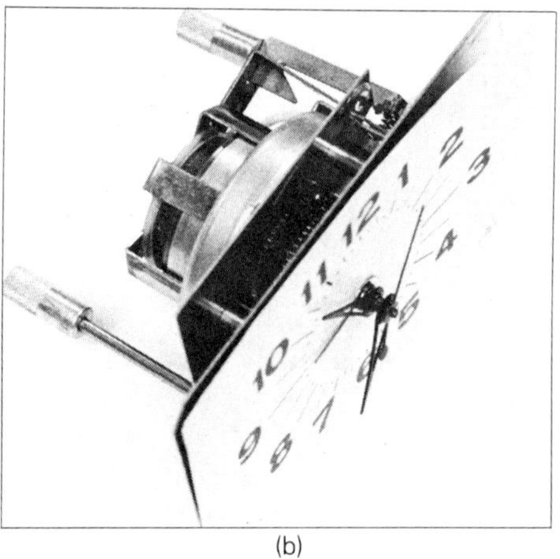

Figure 24.15
Synchronous motor. (a) The stator poles are ac-excited and continually change polarity. As a result, the constant dc-excited rotor poles are attracted and repelled, causing the rotor to rotate. As long as the frequency of the applied ac voltage is constant, the speed of the motor is constant. As a result, synchronous motors are commonly used in electric clocks (b).

Synchronous Motors

A synchronous motor operates in step, or in "sync," with the frequency of the applied ac voltage (usually 60 Hz). The synchronous motor is essentially an ac generator or alternator used as a motor.

The rotor of a synchronous motor consists of a series of alternate poles that are dc-excited (Fig. 24.15). As a result, the rotor poles do not change polarity. The stator poles arise from coils in slots in the motor frame. Alternating current is applied to these coils, and the alternate stator poles continually change polarity (N and S). Hence, the constant rotor poles are attracted toward and repelled from alternate stator poles according to the law of poles. During each cycle of the applied ac voltage, the rotor moves the distance of two stator poles.

The speed of a synchronous motor depends on the frequency of the applied ac voltage and the number of poles in the motor. As long as the frequency of the applied ac voltage is constant, the speed of the motor does not vary. Because of this, synchronous motors find many applications in timing mechanisms, e.g., electric clocks (Fig. 24.15).

A special application of electromagnetic induction and a *linear* synchronous motor is given in Special Feature 24.1.

24.5 Transformers

The transformer is a device used to increase or decrease ac voltage. This is done by mutual electromagnetic induction, a distinct advantage of alternating current over direct current.

Recall the mutual induction of two loops (Fig. 24.3). This is the principle of a transformer, but in the transformer case there are two

SPECIAL FEATURE 24.1

MagLev: The Train of the Future

A train that levitates a half-inch off the track and travels at 300 mi/h ($\simeq$ 480 km/h)? This may well be the train of the future, using the principles of electromagnetic induction—the MagLev (*mag*netic *lev*itation) train (Fig. 1). Conventional trains have wheels that ride directly on rails with wheel flanges against the rails for guidance; this arrangement gives rise to a great deal of frictional loss. The maximum practical speed is about 185 mi/h (300 km/h).

The electromagnetic principles of the experimental MagLev train are involved in (a) levitation, (b) guidance, and (c) propulsion. The on-board magnets are powerful superconducting magnets. At very low temperatures, certain metals show the property of superconductivity—a condition of zero electrical resistance (see Special Feature 21.1). In the absence of resistance, very large currents and magnetic fields are possible. (Recall $B \propto I$, Chapter 23.) Simplified diagrams of the operating principles of the MagLev train are shown in Figure 2.

1. **Levitation.** When a magnet on board the train approaches and passes over coils on the guideway, a current is induced in a coil and it becomes an electromagnet with the same polarity as the on-board magnet (Lenz's law). The repulsion of the magnets causes the train to "levitate" off the guideway [Fig. 2(a)].

Figure 1
MagLev train. The MagLev train levitates one-half inch off the guideway and is guided and propelled by electromagnetic principles. (Courtesy Budd Company.)

2. **Guidance.** When the train deviates to either the right or left, a current is induced in coils along the sides of the guideway (which are also used for propulsion). The magnetic forces bring the train back to the center of the guideway [Fig. 2(b)]. This is a null (zero) flux guidance system.
3. **Propulsion.** The train is propelled by a linear synchronous motor (cf. Fig. 24.15), which uses the interaction of on-board magnets and the

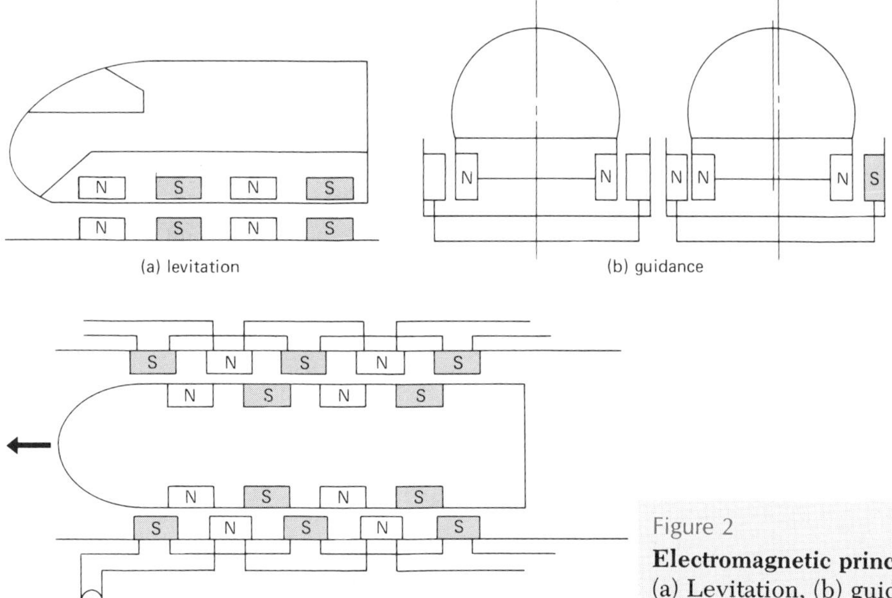

Figure 2
Electromagnetic principles for a MagLev train. (a) Levitation, (b) guidance, and (c) propulsion. See text for description.

changing polarities of magnetic coils along the sides of the guideway [Fig. 2(c)]. The on-board magnets are attracted by the coil magnets of different polarity immediately ahead of them and repulsed by those immediately behind. The polarity of the guideway magnets are reversed at regular, synchronous intervals, and this propels the train forward. The time interval between the guideway coil reversals establishes the speed of the train's movement.

Causing the polarities of the guideway coils to be "out of sync" would slow the train, giving a braking action, which may be supplemented by frictional skids.

American and Japanese companies have built prototype MagLev trains. Speeds in excess of 300 mi/h have been achieved on experimental guideways. The MagLev train's greatest potential is currently in heavy, short-distance commuter transit.

sets of coils, a primary coil and a secondary coil, wound near or over each other on a core with a high permeability, such as iron (Fig. 24.16).

With an ac source connected to the primary, a changing magnetic flux is established within the core of the transformer. The iron core serves to concentrate and contain the flux.

When an input ac voltage V_p is applied to the primary coil, the applied voltage is related to the changing flux, $\Delta\Phi/\Delta t$, as

$$V_p = -N_p \left(\frac{\Delta\Phi}{\Delta t}\right)$$

where N_p is the number of turns of the primary.

Similarly, assuming the same flux passes through the secondary, the induced emf in this coil is

$$V_s = -N_s \left(\frac{\Delta\Phi}{\Delta t}\right)$$

where N_s is the number of turns of the secondary. Then, dividing one equation by the other, we have

$$\boxed{\frac{V_s}{V_p} = \frac{N_s}{N_p}} \qquad \textbf{(Eq. 24.11)}$$

(transformer voltages and turns)

Hence, by varying the number of turns of the secondary N_s relative to the number of turns on the primary N_p, the voltage can be "stepped up" or "stepped down."

For example, if $N_s = 100$ turns, $N_p = 20$ turns, and the primary voltage is 120 V,

$$V_s = \left(\frac{N_s}{N_p}\right) V_p = \left(\frac{100}{20}\right) 120 \text{ V} = 600 \text{ V}$$

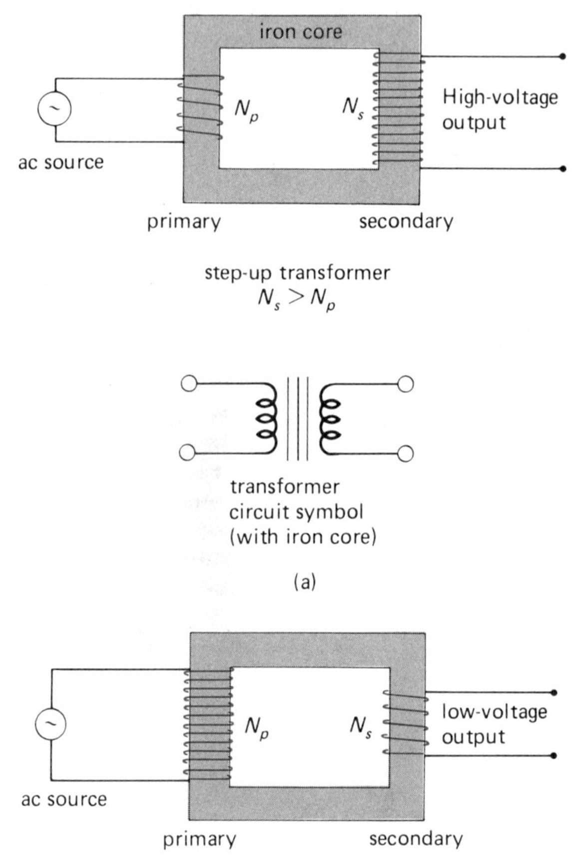

Figure 24.16
The transformer. (a) Step-up transformer. The number of turns on the secondary is greater than the number of turns on the primary. (b) Step-down transformer. The number of turns on the primary is greater than the number of turns on the secondary.

MODERN PHYSICS

Irradiated reactor assemblies exhibit a characteristic glow when immersed in water, known as Chrenkov radiation. Light is emitted because the speed of charged radioactive decay particles is greater than the speed of light in water.
(Courtesy Department of Energy)

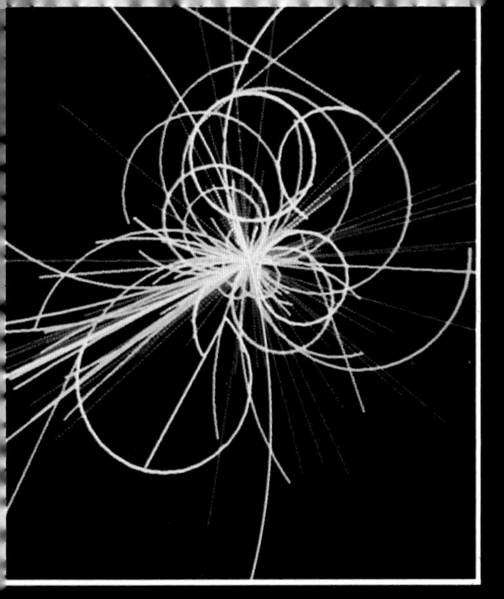

Particle accelerators aim at the heart of matter. The new Superconducting Super Collider (SSC) will be 84 km (52 mi) in circumference, with a tunnel ten feet in cross section. Approximately 10,000 superconducting magnets will focus and guide two beams of protons in opposite directions around the tunnel. The beam speeds will approach the speed of light, and they will collide with an energy of 40 trillion electron volts. The collisions will create new subatomic particles that will be detected and analyzed. Note the size of the SSC in comparison with that of Manhattan Island and other accelerators. (Courtesy Department of Energy)

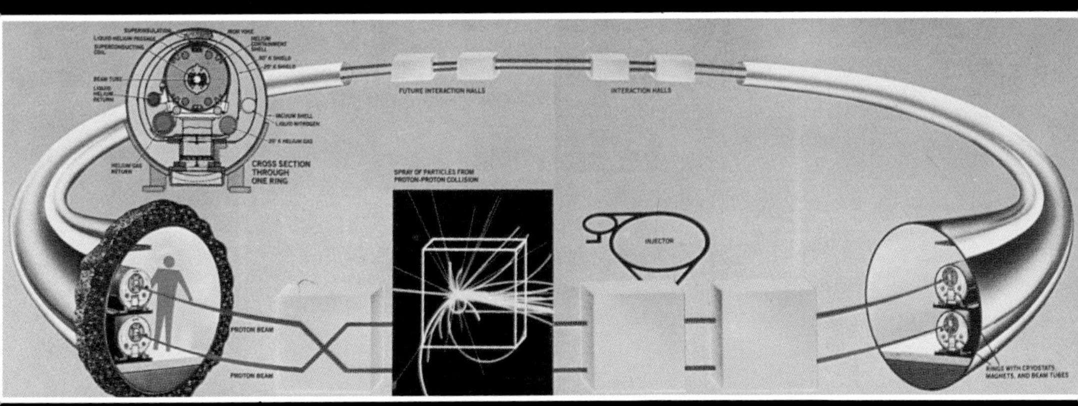

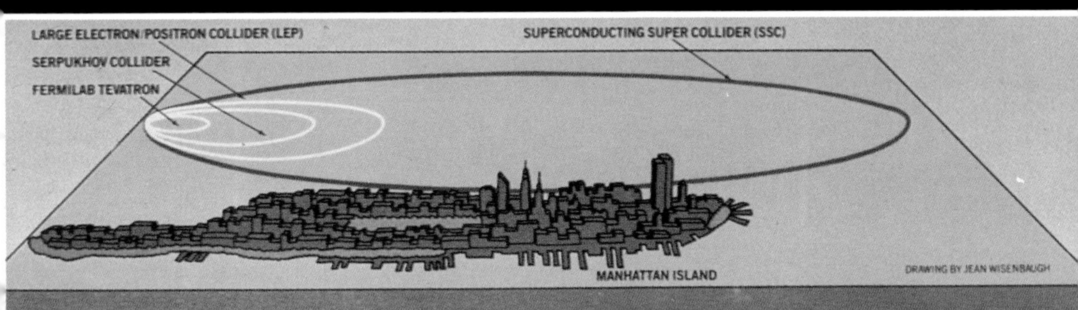

A spectacular effect is achieved when the phases of a lunar eclipse are superimposed over the Space Shuttle Columbia. (Courtesy NASA)

An artist's conception of Space Station Freedom, a proposed permanently-manned configuration that is scheduled for late 1999. Three sets of solar arrays will provide more than

Portable photovoltaic units were developed as one source of power for remote areas or underdeveloped countries. A 1 kW irrigation pump illustrates the versatility of such systems. (Courtesy Department of Energy)

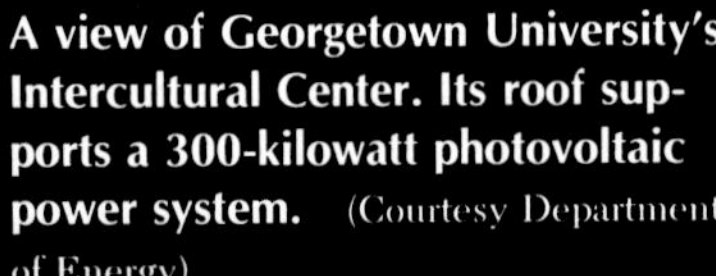

A view of Georgetown University's Intercultural Center. Its roof supports a 300-kilowatt photovoltaic power system. (Courtesy Department of Energy)

Views of heliostats and the central receiver pilot plant in California. A 72-acre field of more than 1800 heliostats (mirrors), each about 20 feet square, surrounds the central receiver, which is installed in a high tower. Mirrors reflect the Sun's rays up to the receiver where water boils and becomes superheated steam. The steam is piped to ground level and turns turbines that drive generators. An automatic control system keeps each mirror tracking the Sun's arc through the sky for maximum efficiency. A thermal storage system saves excess heat for use after sundown or during cloudy periods. (Courtesy Department of Energy)

In the case $N_s > N_p$, the transformer produces a larger voltage output and is called a **step-up transformer.**

Inversely, the primary may have more turns than the secondary (Fig. 24.16). This can be accomplished simply by making the secondary the primary and the primary the secondary. Then—with $V_p = 600$ V, $N_s = 20$ turns, and $N_p = 100$ turns—by Equation 24.11,

$$V_s = \left(\frac{N_s}{N_p}\right) V_p = \left(\frac{20}{100}\right) 600 \text{ V} = 120 \text{ V}$$

In the case $N_s < N_p$, the transformer produces a lower output voltage and is called a **step-down transformer.**

The efficiency of a transformer is defined as the ratio of the power output to the power input ($P = IV$).

$$\text{Eff} = \frac{\text{power output}}{\text{power input}} = \frac{I_s V_s}{I_p V_p} \quad \textbf{(Eq. 24.12)}$$

(transformer efficiency)

where I_p and I_s are the primary and secondary currents, respectively. Well-designed transformers can have efficiencies in excess of 90 percent. Types of commercial transformers are shown in Figure 24.17.

Two sources of power loss in transformers are stray flux and eddy currents. In the case of stray flux, if all of the magnetic flux generated by the primary does not pass through the secondary, then the efficiency is reduced. Also, since iron is a conductor, the emf due to the changing flux in the volume of the transformer core sets up current loops within the core known as **eddy currents.** These eddy currents give rise to I^2R losses and lower transformer efficiency. They can be minimized by laminated iron cores, which tend to break up the eddy current loops.

It should be realized that even though the transformer can be used to greatly increase the voltage, this is not a case of "getting something for nothing." To understand why, let's assume the case of an ideal or perfect transformer (100 percent efficiency). For this ideal case,

$$\text{Power input} = \text{power output}$$

or

$$I_p V_p = I_s V_s$$

and, with Equation 24.12,

(a)

(b)

Figure 24.17

Transformers. (a) Typical dry-type transformers. (b) A high-voltage transformer with a fan cooling system. Using fans, the power rating of the transformer is increased by over 33 percent. [(b) Courtesy Square D Company.]

$$\boxed{\frac{I_p}{I_s} = \frac{V_s}{V_p} = \frac{N_s}{N_p}} \quad \textbf{(Eq. 24.13)}$$

(transformer currents)

Hence, when the voltage is stepped up, the current is reduced. Inversely, when the voltage is stepped down, the current is increased.

EXAMPLE 24.3 In the previous example of a step-up transformer with $V_p = 120$ V, $N_p = 20$ turns, and $N_s = 100$ turns, if the current in the primary is 5.0 A, what is the current in the secondary? (Assume an ideal transformer.)

Solution: It is given that $I_p = 5.0$ A. Then,

$$I_s = \left(\frac{N_p}{N_s}\right) I_p = \left(\frac{20}{100}\right) 5.0 \text{ A} = 1.0 \text{ A}$$

In this case, the voltage is stepped up (by a factor of 5 from 120 V to 600 V), but the current is stepped down (by a factor of $\frac{1}{5}$). It can easily be shown that if the voltage is stepped down, the current is stepped up proportionally.

The Induction Coil

As a final consideration of electromagnetic induction in this chapter, let's examine the induction coil (Fig. 24.18). The induction coil is essentially a transformer that operates on dc current.

When the switch is closed in the primary circuit, the iron core acts as a magnet and attracts the armature, which breaks or opens the circuit. The field in the core then collapses, and the armature is released and again makes contact. (The capacitor diminishes the armature's arcing.) In this manner, a current *alternately* builds up and diminishes in the coil, giving rise to a changing magnetic flux in the core.

As the flux changes through the many turns of the secondary coil, a high voltage is induced across the gap and a spark "jumps" the gap. Such induction coils are used to produce the high voltages (12,000 V or more) required for the production of sparks in automobile ignition systems. See Special Feature 24.2.

Another application of electromagnetic induction is described in the Chapter Supplement.

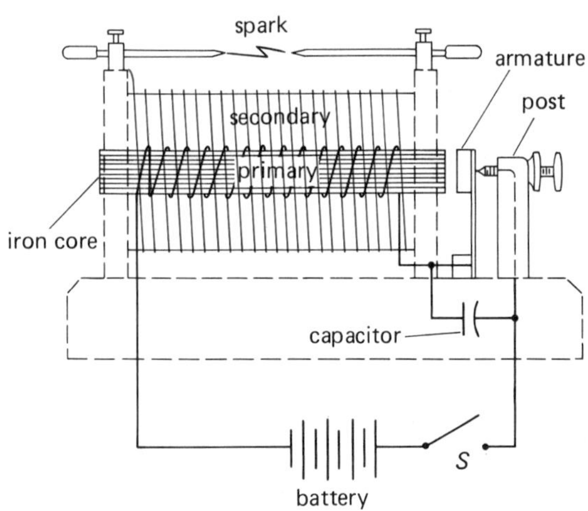

Figure 24.18
An induction coil will produce a high voltage from a low voltage dc source. When switch S in the primary circuit is closed, the iron core acts as a magnet, attracts the armature, and opens the circuit. This repeated action builds up a flux in the secondary coil, which produces a high voltage and causes a spark to jump the gap in the circuit.

Important Terms

magnetic flux a measure of the number of the magnetic field lines passing through an area $\Phi = BA \cos \theta$

magnetic induction the induced emf in a conducting loop when there is relative motion between the conductor and a magnetic field, such that there is a change in the number of field lines through the loop

mutual induction the induced emf in one conductor as a result of a change in the magnetic field from another conductor

Faraday's law of induction the law that the induced emf is proportional to the time rate of change of flux

Lenz's law the law that an induced current is in the direction such that its effects oppose the change that produces it

back (counter) emf the induced emf in a motor winding that opposes the applied voltage

generator a device that converts mechanical energy into electrical energy

alternating current (ac) current that alternates back and forth due to changes in voltage polarity

transformer a device used to increase (step up) or decrease (step down) ac voltage by mutual induction

eddy currents currents induced in a transformer core due to a changing magnetic flux

Important Formulas

magnetic flux: $\Phi = BA \cos \theta = BA$ $(\cos \theta = 90°)$

Faraday's law of induction:

$$\mathscr{E} = -N \left(\frac{\Delta \Phi}{\Delta t}\right)$$

$$= -N \left[\left(\frac{\Delta B}{\Delta t}\right) A + B \left(\frac{\Delta A}{\Delta t}\right)\right]$$

back (counter) emf: $\mathscr{E}_b = V - IR$

SPECIAL FEATURE 24.2

The Auto Spark Ignition System

The common ignition system that "fires" spark plugs in an automobile engine has an ignition "coil," which is actually a transformer (Fig. 1). The 12 V from the battery would not cause the spark discharge in the spark-plug gap to ignite the air-fuel mixture. When a spark plug is supposed to fire, a switch (the breaker points, or "points") breaks the circuit between the battery and the primary of the coil. (The points are located in the distributor, which distributes the secondary voltage to the spark plugs in turn.) Voltages of 12,000 V or more are generated in this manner.

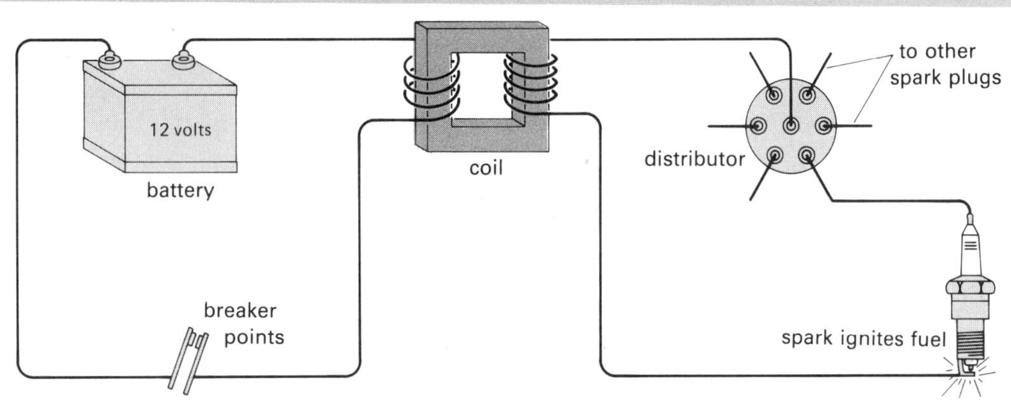

Figure 1
Spark ignition system. A simplified diagram of an auto spark ignition system shows the ignition coil, or step-up transformer.

instantaneous ac generator emf:
$$\mathscr{E} = NBA\omega \sin\theta = \mathscr{E}_{max}\sin\omega t = \mathscr{E}_{max}\sin 2\pi ft \quad (\mathscr{E}_{max} = NBA\omega)$$

effective ac power:
$$P_{eff} = \frac{V_{eff}^2}{R} = \frac{\mathscr{E}_{max}^2}{2R}$$

effective ac voltage:
$$V_{eff} = \frac{1}{\sqrt{2}}\mathscr{E}_{max} = 0.707\ V_{max}$$

transformer equations:
$$\frac{V_s}{V_p} = \frac{N_s}{N_p} = \frac{I_p}{I_s}$$

Questions

Faraday's Law of Induction

1. A current produces a magnetic field. Can a magnetic field produce a current? Explain.
2. Does the orientation of a loop with respect to a magnet make any difference in inducing a voltage in the loop through relative motion? Explain.
3. A voltage is induced in a loop when it is stationary and a magnet is brought toward it or when the loop is brought toward a stationary magnet. What would be the effect if (a) both the magnet and the loop were brought toward each other? (b) If both the magnet and loop had the same velocity?
4. Where does the induced voltage in a loop come from? Is this a case of getting something for nothing? Explain.
5. In inducing a current in a loop by bringing a magnet toward the loop, would it make any difference in either the magnitude or the direction of the induced current if the north or south pole were used? Explain.
6. Can an induced emf be produced only by a time-varying magnetic field? Explain.

Lenz's Law and Back emf

7. What is Lenz's law, and on what principle is it based?
8. Suppose the induced current in a loop were in a direction that enhanced the change that produced it. What would be the result?
9. What is meant by back emf, and how does it affect the speed of a motor?
10. When a magnet is brought toward a loop or coil, the magnetic field from the induced current is

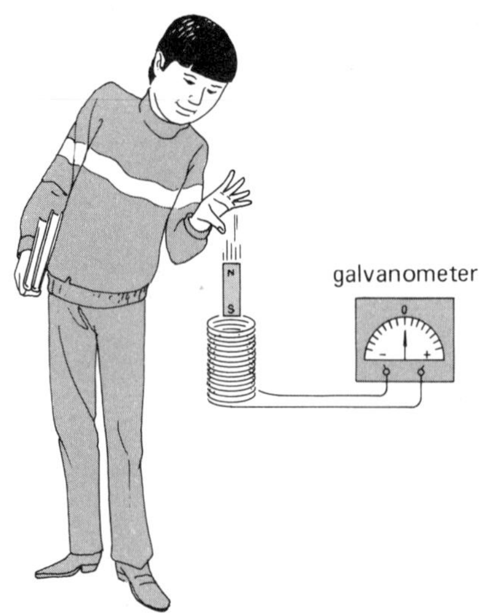

Figure 24.19
Galvanometer deflection. See Question 11.

Figure 24.20
A new invention? See Question 20.

like that of a repelling magnet. As a result, work must be done in moving the magnet toward the coil. Why is this?

11. A bar magnet is dropped through a coil (Fig. 24.19). (a) What will be observed on the galvanometer? (b) As the magnet enters the coil, will the acceleration be greater or less than the acceleration due to gravity? How about as it leaves? Explain.

Electrical Generators

12. What is the overall function of a generator?
13. Explain the principle and main parts of an ac generator. Give three ways in which the voltage output of a generator could be increased.
14. What is meant by 60-cycle (Hz) 110 VAC?
15. Suppose the rotational frequency of a generator is increased from 60 Hz to 120 Hz. What effect would this have on the voltage output?
16. How are the amounts of ac voltage and current specified when the polarity and direction of each are continually changing? How is the ac ampere defined?
17. Can Ohm's law and the power relation be used in ac applications? Explain.
18. What is meant by rms values of voltage and current?
19. Why is the range of ac voltage sometimes specified as 110–120 VAC or 220–240 VAC?
20. An inventor proposes to revolutionize electrical generation with a new type of generator, as shown in Fig. 24.20. (a) What type of voltage does the generator produce? (b) The inventor states that the internal friction losses of the spring are practically negligible, so the magnet will oscillate indefinitely and therefore provide a cheap source of electrical power. Would you financially back this invention? Explain.

ac Motors

21. Explain how a universal or series motor can operate on both ac and dc.
22. What is the difference between a rotor and a stator in a motor? Which receives the current input? Explain.
23. Explain the principle of a three-phase induction motor. Can single-phase induction motors be started by rotor currents alone? Explain.
24. (a) What is the principle of a synchronous motor? (b) Explain the operation of the linear synchronous motor system of the MagLev train.

The Transformer

25. Would a transformer operate (a) on a battery? (b) On a dc generator?
26. What would be the effect(s) if the primary and secondary of a transformer had the same number of windings? What kind of "step" transformer would this be?
27. Some transformers have one or more terminals, or "taps," on the secondary so that connecting to different taps puts different fractions of the total number of secondary windings into a circuit. What is the advantage of this?
28. A fluorescent lamp needs an initial voltage of 12,000 V to ionize the lamp gases. This is supplied by a transformer with a 120-V input. How many times more windings are on the secondary than on the primary?
29. An ac source has a 10-amp output. A particular circuit requires a 2.0-amp input. How could you accomplish this? Explain.

30. A person has a single transformer with 50 windings on one part of the core and 500 windings on the other. Is this a step-up or step-down transformer? Explain.
31. Does a transformer operate on self-induction or mutual induction? Does it have both? If so, what are the effects?
32. How efficient are transformers, and what are some sources of power loss?
33. Explain how an induction coil can be operated on dc current. Is an induction coil a transformer? Explain.

Problems

Levels of difficulty are indicated by asterisks for your convenience.

24.1 Faraday's Law of Induction

1. A circular wire loop with a radius of 10 cm is in a uniform magnetic field of 6.0×10^{-4} T. What is the flux through the loop if the plane of the loop is (a) perpendicular to the field and (b) parallel to the field?
2. A square loop 20 cm on a side is in a uniform magnetic field of 4.5×10^{-2} Wb/m^2. What is the flux through the loop if the normal to the plane of the loop makes an angle of (a) 30° and (b) 60° with the magnetic field?

*3. Two loops with areas of 0.15 m^2 and 2.0 m^2 are in a uniform magnetic field of 0.30 T. If the normal to the plane of the smaller loop is at an angle of 45° to the magnetic field and the normal of the larger loop is at an angle of 55°, through which loop is the flux greater, and how many times greater?

*4. The flux through a coil with 50 turns decreases from 0.08 Wb to 0.03 Wb in 0.10 s. (a) What is the average induced emf in the coil? (b) If the coil has a resistance of 0.25 Ω, what is the average induced current in the coil during this time?

*5. A loop of wire with an area of 0.050 m^2 has a magnetic field equal to 0.75 T perpendicular to its plane. In a time interval of 1.0 ms, the magnetic field increases to 1.50 T. (a) What is the average induced emf in the loop during this time? (b) If the loop were replaced with a coil of wire with 50 turns of the same area, what would be the emf?

*6. The magnetic field perpendicular to the plane of a horizontal wire loop with an area of 0.50 m^2 increases from 1.5 T to 2.5 T in $\frac{1}{8}$ s. (a) What is the average emf induced in the loop during this time? (b) If the loop has a resistance of 0.16 Ω, what is the average induced current in the loop? (c) If the field is vertically upward, what is the direction of the induced current, looking down on the loop from above?

*7. The normal to the plane of a circular wire loop with a radius of 5.0 cm makes an angle of 30° with the field lines of a uniform magnetic field of 1.8 T. If the magnetic field decreases to zero in 0.10 s, what is the average voltage induced in the loop during this time?

*8. A coil of wire has a cross-sectional area of 80 cm^2. A uniform magnetic field parallel to the coil axis increases from 2.0 T to 7.0 T in 0.25 s, and an average emf of 16 V is induced in the coil. How many loops are in the coil?

*9. A horizontal coil of wire with 30 turns has a cross-sectional area of 0.040 m^2. A magnetic field directed downward at an angle of 30° relative to the normal of a coil increases from 1.0 T to 2.5 T in 0.20 s. (a) What is the average induced voltage in the coil? (b) If the coil has a resistance of 2.0 Ω, what is the average induced current? (c) Looking down on the coil from above, what is the direction of the induced current?

*10. A square coil of wire with 20 turns and an area of 0.010 m^2 is placed perpendicular to a magnetic field of 0.75 Wb/m^2. The coil is flipped so that its plane is parallel with the field in 0.30 s. What is the average induced emf in the coil?

*11. A wire 10 cm in length is moved with a constant velocity of 4.0 m/s perpendicular to a magnetic field of 0.80 T. What is the induced voltage across the length of the wire?

24.2 Lenz's Law and Back emf

*12. A dc motor has a resistance of 2.0 Ω. When connected to a 12-V source, with the armature moving at its operational speed, a back emf of 4.0 V is generated. (a) What is the current in the armature coil at the operational speed? (b) What is the starting current?

*13. A 110-V dc motor has an armature resistance of 5.0 Ω and draws 6.0 A of current when it is operating. (a) What is the starting current of the motor? (b) What is the back emf when it is operating?

*14. A series wound 220-V dc motor has a field coil resistance of 50 Ω and an armature resistance of 10 Ω. When in operation, the motor draws 2.0 A of current. (a) What is the total resistance of the motor? (b) What is the back emf generated? (c) How much power is lost to the back emf?

**15. A shunt (parallel) wound 120-V dc motor has a field coil resistance of 160 Ω and an armature resistance of 10 Ω. In operation, a back emf of

90 V is generated. (a) What is the initial starting current of the motor? (b) What is the operating current? (c) What is the power loss due to the back emf?

24.3 Electrical Generators

16. Normal household voltage is (a) 110–120 V and (b) 220–240 V. What are the peak voltages of these ranges?

*17. A simple ac generator has a coil of 50 turns and an area of 0.020 m^2, which rotates with a speed of 100 rpm between the pole faces of a permanent magnet. (a) If the magnetic field between the pole faces is 2.0 T, what is the maximum instantaneous emf generated? (b) What is the effective voltage?

*18. Suppose the generator described in Problem 17 were turned with a frequency of 60 Hz. (a) What would be the instantaneous emf at t = 2.0 s? (b) If the generator delivered power to a 500-Ω load, what would be the effective current in the load?

**19. An alternator has 4000 turns, with each turn having an area of 250 cm^2. The coil rotates at 50 rps in a uniform magnetic field of 1.5 T. (a) What is the maximum value of the generated emf? (b) What is the instantaneous emf $\frac{3}{8}$ of a revolution after the maximum value? (c) If the alternator has an output load of 100 Ω, what is the effective power dissipated by the load?

24.5 Transformers

20. A step-up transformer with 100 turns on the primary and 300 turns on the secondary is operated on a 120-V line that supplies 6.0 A of current. Assuming 100 percent efficiency, (a) what are the voltage and current in the secondary? (b) What is the power output? (c) What is the power input?

21. A step-down transformer has 80 turns on its input winding. If a 240-V input is stepped down to 60 V, how many turns are on the output winding? (Assume 100 percent efficiency.)

*22. An ac generator with an output of 440 V and a resistance of 10 Ω supplies power to a 24-kV line. If the primary of the step-up transformer has 120 turns, (a) how many turns are there on the secondary? (b) What is the line current? (Assume 100 percent efficiency.)

*23. An industrial furnace requires 200 A at 240 V. If power is supplied to the plant location by a 6-kV line, (a) what is the required turn ratio for a step-down transformer for the furnace? (b) What is the current in the line? (Assume 100 percent efficiency.)

*24. A utility pole transformer steps down 12 kV to 240 V. (a) What is the ratio of the secondary turns to the primary turns? (b) If the current input is 1.0 A, what is the current output? (Assume 100 percent efficiency.)

*25. It is desired to step up a 1.0-A ac current to 6.0 A. If an ideal transformer with 90 turns on the input winding is used, (a) how many turns are on the output winding? (b) How is the voltage affected?

*26. A doorbell transformer steps down 120 V to 12 V, and a current of 0.50 A is delivered to the bell. (a) What is the ratio of the transformer windings? (b) What is the current input? (Assume 100 percent efficiency.)

**27. A step-down transformer has an input of 120 V at 0.50 A and an output of 11 V at 4.8 A. What is the transformer efficiency?

**28. A step-up transformer with an efficiency of 90 percent has 75 primary turns and 600 secondary turns. If the primary draws 10 A of current at 115 V, what are the current and voltage of the secondary? Assume only I^2R losses.

Chapter Supplement

Induction and Dielectric Heating

Electromagnetic induction provides a means of heating conductive materials. The material to be heated is essentially made the secondary of a transformer, which is then a shorted secondary (Fig. 24S.1). As with a regular transformer, an ac source input to the primary induces a current in the secondary. The secondary material is then heated due to I^2R (joule heat) losses. More specifically, the induced currents are eddy currents within the conductive material.

When the frequency of the ac source is increased, the eddy currents have a tendency to

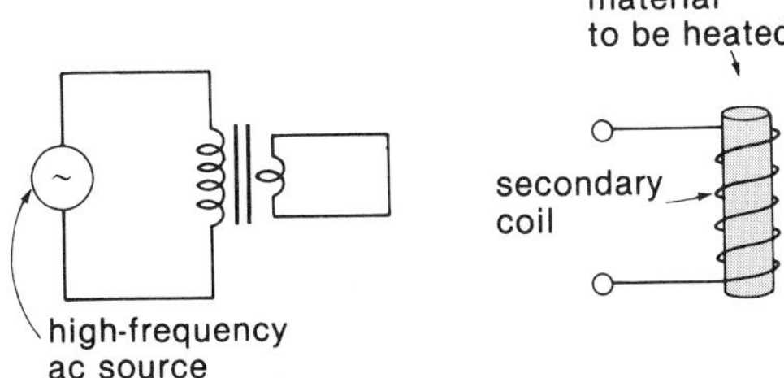

Figure 24S.1

Principle of induction heating. The material to be heated serves as the secondary of a transformer. As a shorted secondary, eddy currents set up within the material produce joule heat.

flow on the surface of the material. This phenomenon is known as the *skin effect.* Hence, the depth of the heating may be controlled by varying the input frequency. This is useful in industrial processes where it is necessary to heat the surface of a material to harden it. Induction heating is used in many industrial heat-treating processes (Fig. 24S.2).

Induction heating is also now being used for cooking. Glass induction cooktops on kitchen stoves have a high-frequency coil just below the surface. This induces a current in any magnetic utensil of ferrous material such as cast iron or steel. The induced current in a metal pan causes joule heating and boils water or cooks food in the pan. The only heat on the cooktop is that absorbed from the pan.

It is even possible to place a paper towel between the cooktop's surface and a pan of water and bring the water to a boil without burning the paper.

It should be noted that induction heating is limited to conductive materials. Nonconductive materials (insulators) may be heated by dielectric heating. In this case, the material serves as the dielectric of a capacitor.

When a high-frequency ac voltage is applied to the capacitor plates, the material is heated as a result of internal friction. In the material the molecules undergo vibrational motions as the molecular dipoles attempt to follow the rapidly changing field (high frequency). Energy is dissipated as heat as a result of these molecular motions. An application of dielectric heating is the gluing together of plywood sheets to form a board.

(a)

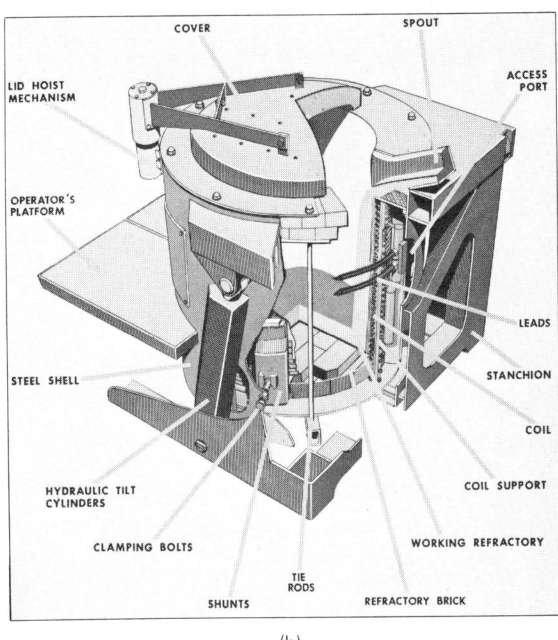

(b)

Figure 24S.2

Induction furnace. (a) An induction furnace and (b) parts of a "coreless" induction furnace, which essentially uses a transformer action between the induction coil as the primary and the metal to be heated as the secondary. (Courtesy Inductotherm Corp.)

Watt-hour meter. (Courtesy Westinghouse Electric Corporation.)

Chapter 25

Basic ac Circuits

Alternating-current applications constitute the lion's share of the electrical industry. The majority of the electricity generated in the United States is of the alternating variety, and this is transmitted for use in the home and industry.

In dc circuits, the resistance of a circuit determines the current for a fixed voltage source. However, in ac circuits there are other considerations because of the alternating nature of the current and voltage. In this chapter we investigate these considerations in order to gain a basic understanding of ac circuits. In addition, we consider high-voltage power transmission and basic household circuits, along with the all-important subject of electrical safety.

25.1 Reactance

In ac circuits there are three basic elements that determine the amount of current flow in a circuit for a given alternating voltage source. These are resistance, capacitance, and inductance.

In a dc circuit or an ac circuit containing only a pure resistance, the relationship between effective current and voltage is given by the common form of Ohm's law, $V = IR$. In such an ac circuit, the current and voltage have maximum and zero values at the same time and are said to be *in phase* (Fig. 25.1).

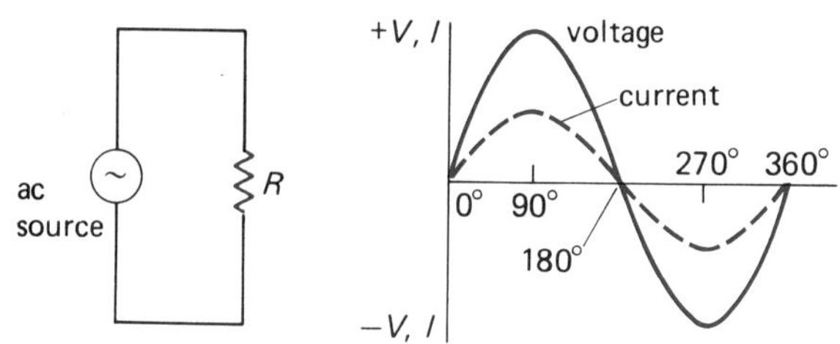

Figure 25.1

Pure resistance circuit. In a circuit with only resistance, the current and the voltage are in phase.

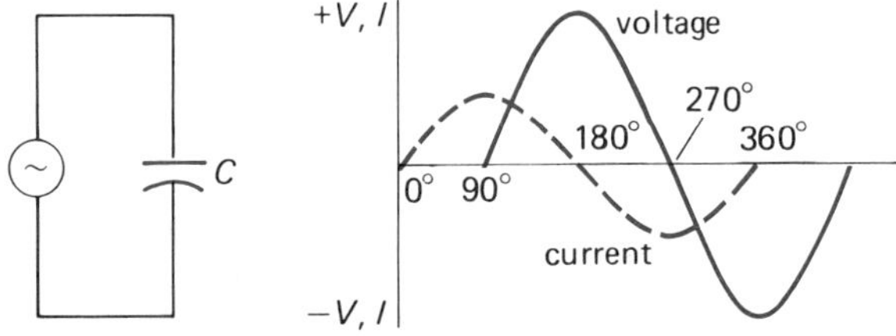

Figure 25.2
Pure capacitive circuit. In a circuit with only capacitance, the current leads the voltage by 90°.

In a pure resistive ac circuit, the current and voltage are in phase.

Many household appliances, such as heaters and lights, approximate this condition.

Capacitive Reactance

The situation is different, however, in an ac circuit with only a capacitor (Fig. 25.2). In this case, as the voltage increases, charge builds up on the capacitor, giving rise to an opposition to the charge flow. The current then goes to zero when the capacitor voltage is a maximum (similar to the dc charging case). In the discharging part of the half-cycle, the current is a maximum when the voltage is zero. These changes repeat during the next half-cycle, when the voltage is in the opposite (alternate) direction.

In terms of the phase-angle relationship of the wave forms (Chapter 13), the current and the voltage are out of phase by 90°, and

In a pure capacitive ac circuit, the current leads the voltage by 90°.

The effect of a capacitor is to prevent charge flow by causing an opposing voltage to build up on the plates. The opposition to alternating current in a capacitor circuit is expressed in terms of the **capacitive reactance X_C**, which is given by

$$X_C = \frac{1}{2\pi fC} = \frac{1}{\omega C} \qquad \textbf{(Eq. 25.1)}$$

where f is the frequency of the alternating voltage source (and circuit current), C is the capacitance in farads (F), and in terms of angular frequency, $\omega = 2\pi f$. The reactance is in ohms.

In a pure capacitive circuit, the Ohm's law relationship has the form

$$V = IX_C \qquad \textbf{(Eq. 25.2)}$$

where V and I are the effective voltage and current, respectively.

EXAMPLE 25.1 If in Figure 25.2 the voltage source is 120 V–60 Hz and the capacitor is 30 μF, what are (a) the capacitive reactance and (b) the current in the circuit?

Solution: It is given that $V = 120$ V, $f = 60$ Hz, and $C = 30\ \mu\text{F} = 30 \times 10^{-6}$ F.

(a) $$X_C = \frac{1}{2\pi fC} = \frac{1}{2\pi(60\text{ Hz})(30 \times 10^{-6}\text{ F})} = 88\ \Omega$$

(b) $V = IX_C$, and

$$I = \frac{V}{X_C} = \frac{120\text{ V}}{88\ \Omega} = 1.4\text{ A}$$

Inductive Reactance

Another important consideration in ac circuits is inductance, which arises from an inductor. An **inductor** is simply a coil of wire (Fig. 25.3). When current flows through an inductor in an ac circuit, the associated alternating magnetic field gives rise to a changing flux in the coil and **self-induction.** As a result, a self-induced back emf will exist in accordance with Lenz's law. This back emf opposes the change and, hence, the current flow in the circuit.

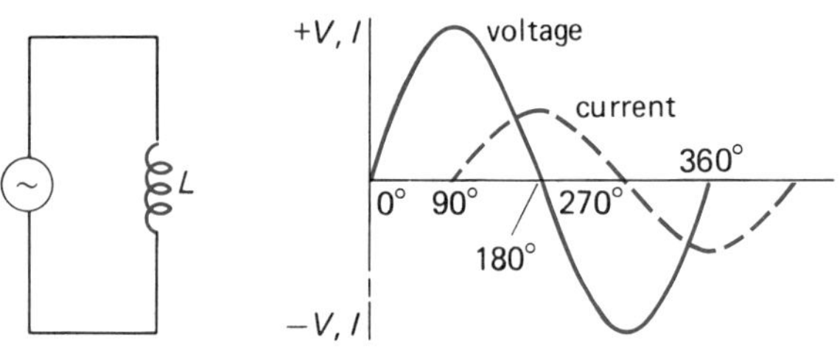

Figure 25.3
Pure inductive circuit. In a circuit with only inductance, the voltage leads the current by 90°.

The self-induced back emf is given by Faraday's law of induction, $\mathscr{E}_b = -N\Delta\Phi/\Delta t$. But, since the geometry of the coil is fixed, the time rate of change of flux, $\Delta\Phi/\Delta t$, is proportional to the average time rate of change of current in the coil circuit, $\Delta I/\Delta t$. Hence, we may write

$$\mathscr{E}_b = -L\left(\frac{\Delta I}{\Delta t}\right)$$

where L is the inductance of the coil with the unit of the **henry (H)**.* The inductance characterizes the opposition to the current flow in the circuit. The smaller unit millihenry (mH) is often used (1 mH = 10^{-3} H).

In a circuit with only an inductor (Fig. 25.3), the back emf opposes and "delays" the circuit current so that it lags behind the driving voltage. The voltage reaches a maximum when the current is zero, and the current is a maximum when the voltage is zero. Hence, as in the capacitive case, the current and voltage are out of phase by 90°, but

In a pure inductive ac circuit, the voltage leads the current by 90°.

The phase relationships of pure inductive and pure capacitive circuits are opposite. A helpful device with which to remember these phase relationships is

E̲L̲I̲ the I̲C̲E̲ man

From *ELI* (with E representing the voltage V), for an inductance L, the E (voltage) leads or comes before I (current). Similarly from *ICE*, for a capacitance C, the I (current) leads or comes before E (voltage).

The opposition to alternating current in an inductor circuit is expressed in terms of the **inductive reactance X_L**, which is given by

$$\boxed{\begin{aligned} X_L &= 2\pi fL \\ &= \omega L \end{aligned}} \qquad \textbf{(Eq. 25.3)}$$

* In honor of Joseph Henry (1797–1879), an American physicist, for his work in electromagnetic induction. Henry discovered electromagnetic induction independently during the same period as Michael Faraday, but Faraday published his findings first.

where f is the frequency of the alternating voltage source (and circuit current), L is the inductance in henrys, and the reactance is in ohms.

Similarly to the capacitive case, in a pure inductive circuit the Ohm's law relationship has the form

$$V = IX_L \qquad \textbf{(Eq. 25.4)}$$

where V and I are the effective voltage and current, respectively.

EXAMPLE 25.2 If the voltage source in Figure 25.3 is 120 V–60 Hz and the inductance is 125 mH, what is the current in the circuit?

Solution: It is given that $V = 120$ V, $f = 60$ Hz, and $L = 125$ mH $= 0.125$ H. Then,

$$X_L = 2\pi fL = 2\pi(60 \text{ Hz})(0.125 \text{ H}) = 47\ \Omega$$

and $V = IX_L$, so

$$I = \frac{V}{X_L} = \frac{120 \text{ V}}{47\ \Omega} = 2.6 \text{ A}$$

25.2 Impedance, Resonance, and Power Factor

In the previous section, it was instructive to consider *pure* capacitive and *pure* inductive circuits. In such circuits there is a reactance opposition to current flow, and

The reactance of an ac circuit is the *nonresistive* opposition to current flow arising from capacitance and/or inductance.

However, it is impossible to have such pure circuits, since in any circuit there must be some resistance R. Therefore, in practical applications we must consider the effects of current opposition from both resistance and reactance (capacitive and/or inductive).

The Series *RC* Circuit

Suppose an ac circuit has only a resistance and a capacitance, as illustrated in Figure 25.4. The phase difference between the current and the voltage varies with the circuit element. As a result, the effective opposition to the current flow in the circuit must be found by a special means: a **phase diagram** (Fig. 25.4).

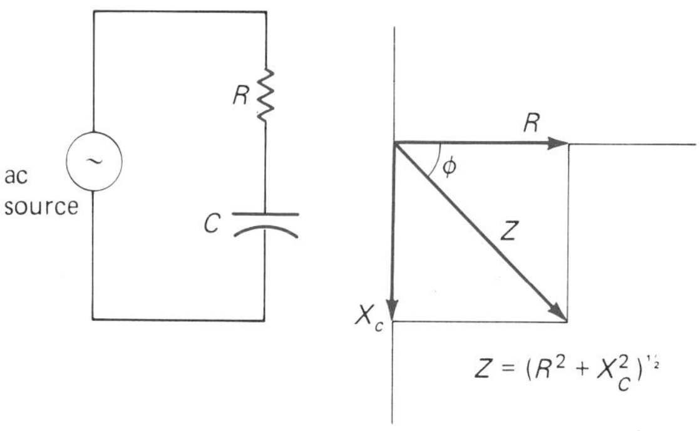

Figure 25.4
Series RC circuit. In a series *RC* circuit (a) the impedance Z is the phasor (vector) sum of the resistance and capacitive reactance (b).

In a phase diagram, the current oppositions of the circuit elements are given vector-like properties and are called **phasors.** The resistance is plotted along the horizontal axis, since a resistive element has a phase difference of zero ($\phi = 0$). The capacitive reactance is plotted along the negative vertical axis. This indicates a phase difference of $\phi = -90°$. A negative phase angle implies that the voltage lags behind the current, as is the case for capacitance.

Then the phasor sum is the effective opposition or impedance Z to the current flow, and

$$Z = [R^2 + X_C^2]^{1/2} \qquad \textbf{(Eq. 25.5)}$$

The Ohm's law relationship for the circuit is

$$V = IZ \qquad \textbf{(Eq. 25.6)}$$

EXAMPLE 25.3 A series *RC* circuit has a resistance of 100 Ω and a capacitance of 30 μF. If the voltage source operates at 120 V–60 Hz, what is the current in the circuit?

Solution: It is given that $R = 100\ \Omega$, $C = 30\ \mu\text{F}$, $V = 120$ V, and $f = 60$ Hz. From Example 25.1, the capacitive reactance is $X_C = 88\ \Omega$. Hence,

$$Z = [R^2 + X_C^2]^{1/2}$$
$$= [(100\ \Omega)^2 + (88\ \Omega)^2]^{1/2} = 133\ \Omega$$

and $V = IZ$, so

$$I = \frac{V}{Z} = \frac{120\ \text{V}}{133\ \Omega} = 0.90\ \text{A}$$

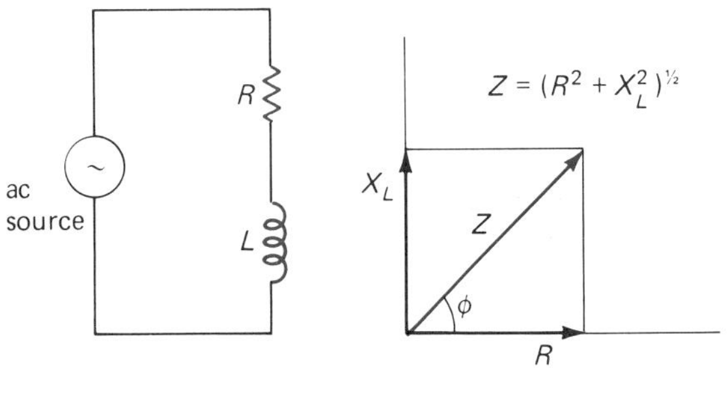

Figure 25.5
Series RL circuit. In a series *RL* circuit (a), the impedance is the phasor (vector) sum of the resistance and the inductive reactance.

The Series *RL* Circuit

The analysis of the series *RL* circuit is similar to that of the *RC* circuit (Fig. 25.5). However, in this case the inductive reactance is plotted along the positive vertical axis. Here the positive phase angle ($\phi = +90°$) implies that the voltage leads the current. In this case, the impedance is

$$Z = [R^2 + X_L^2]^{1/2} \qquad \textbf{(Eq. 25.7)}$$

and, as before, $V = IZ$.

The Series *RLC* Circuit

In the general case, an ac circuit contains all three circuit elements (Fig. 25.6). Here again,

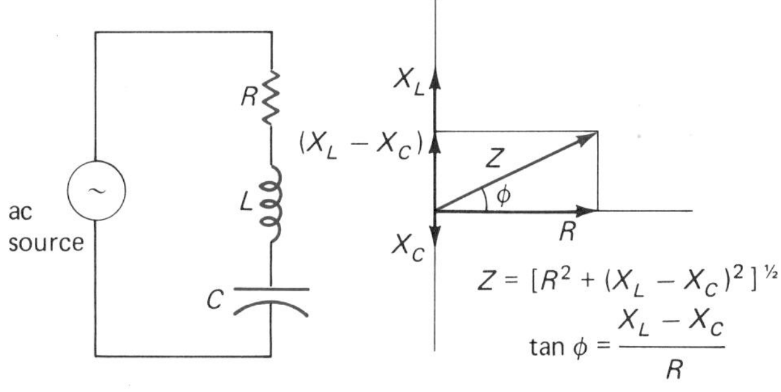

Figure 25.6
Series RLC circuit. In a series *RLC* circuit (a), the impedance is the phasor sum of the resistance and the total reactance $X = X_L - X_C$ (b).

phasor addition is used to determine the impedance. The phasor addition of the reactances, $X_L - X_C$, gives the total reactance. **The impedance is the phasor sum of the total reactance and the resistance.** From the diagram, we see that

$$\boxed{Z = [R^2 + (X_L - X_C)^2]^{1/2}} \qquad \textbf{(Eq. 25.8)}$$

and the **phase angle ϕ** is

$$\boxed{\tan \phi = \frac{X_L - X_C}{R}} \qquad \textbf{(Eq. 25.9)}$$

If $X_L > X_C$, then the phase angle is positive $(+\phi)$ and the circuit is said to be *inductive* (as shown in the figure). If $X_C > X_L$, then the phase angle is negative $(-\phi)$ and the circuit is said to be *capacitive*.

EXAMPLE 25.4 A series *RLC* circuit with a voltage source of 120 V–60 Hz has a resistance of 100 Ω and capacitive and inductive reactances of 88 Ω and 47 Ω, respectively (as in Examples 25.1 and 25.2). What is the current in the circuit?

Solution: It is given that $V = 120$ V, $R = 100\ \Omega$, $X_C = 88\ \Omega$, and $X_L = 47\ \Omega$. Then,

$$Z = [R^2 + (X_L - X_C)^2]^{1/2}$$
$$= [(100\ \Omega)^2 + (47\ \Omega - 88\ \Omega)^2]^{1/2} = 108\ \Omega$$

and

$$I = \frac{V}{Z} = \frac{120\ \text{V}}{108\ \Omega} = 1.1\ \text{A}$$

Notice that the circuit is capacitive $(X_C > X_L)$, and the phase diagram would have a negative phase angle:

$$\phi = \tan^{-1}\left(\frac{X_L - X_C}{R}\right) = \tan^{-1}\left(\frac{47\ \Omega - 88\ \Omega}{100\ \Omega}\right)$$
$$= \tan^{-1}(-0.41) = -22°$$

Circuit Resonance

As can be seen from the phase diagram in Figure 25.6, the combined effect of capacitive reactance and inductive reactance is a tendency to cancel each other. The reactances cancel completely when $X_L - X_c = 0$, or $X_L = X_C$.

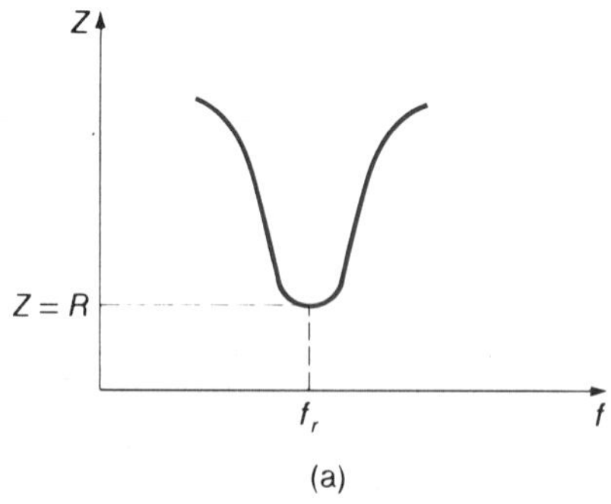

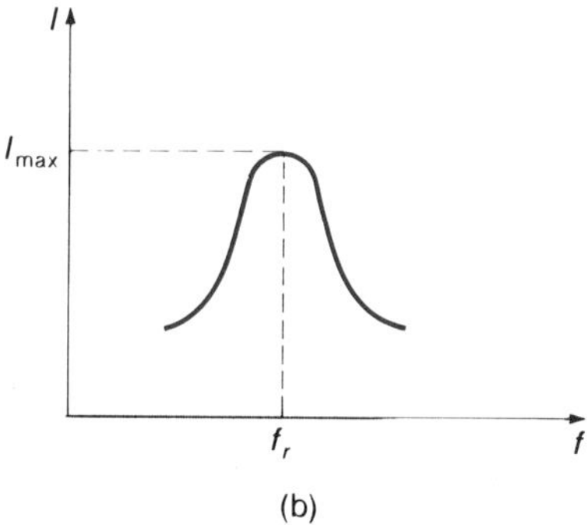

Figure 25.7
Resonance. In a series *RLC* circuit, at the resonance frequency f_r, the reactance is zero ($X_L = X_c$) and the impedance Z is completely resistive (a). The impedance is then a minimum and the current in the circuit is a maximum (b).

For fixed L and C, this condition occurs for a particular **resonant frequency f_r:**

$$X_L = X_C$$

or

$$2\pi f_r L = \frac{1}{2\pi f_r C}$$

Solving for f_r,

$$\boxed{f_r = \frac{1}{2\pi\sqrt{LC}} = \frac{0.159}{\sqrt{LC}}} \qquad \textbf{(Eq. 25.10)}$$

When the voltage source drives an *RLC* circuit at the resonant frequency, the circuit is said to be *in resonance*. **In this case, the impedance is completely resistive and is a minimum** (Fig. 25.7). **The current in the circuit will be a maximum when it is in resonance.**

EXAMPLE 25.5 What is the resonant frequency of a series *RLC* circuit that has a resistance of 100 Ω, an inductance of 50 mH, and a capacitance of 0.10 μF?

Solution: It is given that $L = 50$ mH $= 50 \times 10^{-3}$ H and $C = 0.10\ \mu$F $= 0.10 \times 10^{-6}$ F. Then,

$$f_r = \frac{0.159}{\sqrt{LC}} = \frac{0.159}{[(50 \times 10^{-3}\text{ H})(0.10 \times 10^{-6}\text{ F})]^{1/2}}$$

$$= 2.2 \times 10^3 \text{ Hz} = 2.2 \text{ kHz}$$

Notice that the resonant frequency does not depend on the resistance of the circuit.

Analogous to mechanical wave resonance (Section 13.4), **there is a maximum power transfer to the circuit when it is in resonance.**

An important application of circuit resonance is in the radio receiver. With the use of a variable capacitor, the circuit becomes "tunable" and the resonant frequency of the circuit can be varied. When you turn the tuning knob of a radio, you are turning the shaft of a variable air capacitor. When the resonant frequency of the circuit is the same as that of a radio station signal, the power transfer of radio waves to the circuit is a maximum and the receiver has the greatest response to the incoming signal.

The Power Factor

When an *RLC* circuit is in resonance, the impedance is completely resistive and the current is a maximum, as we have learned. Thus, the power consumption of the circuit is a maximum and equal to the I^2R losses, $P = IV = I^2R$. However, when there is a nonzero total reactance, the current and the power consumption are less (Fig. 25.7).

In the nonzero reactance case, no power is consumed by the capacitance or the inductance. Energy is merely stored and released in these circuit elements, since the current and voltage are out of phase in the circuit. The effective power loss in a circuit with resistance and reactance is given by

$$P = IV \cos \phi \qquad \textbf{(Eq. 25.11)}$$

where I is the current in the circuit as measured by an ammeter, V is the voltage of the driving source, and ϕ is the phase angle.

The quantity $\cos \phi$ is called the **power factor.** Note that the power loss is a maximum when $\cos \phi = 1$ ($\phi = 0°$, pure resistance, resonance condition) and a minimum when $\cos \phi = 0$ ($\phi = 90°$, pure reactance, ideal condition). From the phase diagram in Figure 25.6, it can be seen that the power factor is given by

$$\cos \phi = \frac{R}{Z} = \frac{R}{[R^2 + (X_L - X_C)^2]^{1/2}} \qquad \textbf{(Eq. 25.12)}$$

EXAMPLE 25.6 A series *RLC* circuit with a 125-Ω resistance, a 50-μF capacitance, and a 0.30-H inductance is driven by a 220-V–60-Hz source. What is the power loss of the current?

Solution: It is given that $R = 125\ \Omega$, $C = 50\ \mu\text{F} = 50 \times 10^{-6}$ F, $L = 0.30$ H, $V = 220$ V, and $f = 60$ Hz. Then the reactances are

$$X_C = \frac{1}{2\pi fC} = \frac{1}{2\pi(60\text{ Hz})(50 \times 10^{-6}\text{ F})} = 53\ \Omega$$

$$X_L = 2\pi fL = 2\pi(60\text{ Hz})(0.30\text{ H}) = 113\ \Omega$$

Hence, the circuit is inductive ($X_L > X_C$). The impedance of the circuit is

$$X = [R^2 + (X_L - X_C)^2]^{1/2}$$

$$= [(125\ \Omega)^2 + (113\ \Omega - 53\ \Omega)^2]^{1/2} = 139\ \Omega$$

The current in the circuit (as measured by an ammeter) is

$$I = \frac{V}{Z} = \frac{220\text{ V}}{139\ \Omega} = 1.6\text{ A}$$

The power factor of the circuit is

$$\cos \phi = \frac{R}{Z} = \frac{125\ \Omega}{139\ \Omega} = 0.90$$

The power loss is then

$$P = IV \cos \phi = (1.6\text{ A})(220\text{ V})(0.90) = 317\text{ W}$$

25.3 Electrical Power Transmission

One of the most important ac circuits is that used in the transmission of electrical power from the generating station to the consumer. The distribution of electrical power is quite complicated, so only the general features are considered here.

With few exceptions, nearly all of the electrical power generated in the United States today is in the form of 60-Hz, three-phase alternating current (cf. Section 24.3). Three-phase power is more economical in material requirements, e.g., wire. However, one of the chief considerations in power transmission and distribution of any phase power is power losses. The two main sources of power losses are line losses and leakage losses.

Line Losses

Line losses are due to the I^2R ohmic resistive losses of the power lines themselves. The resistance of a line is relatively constant and depends on the wire gauge and the length of the line. For example, an AWG 3/0 aluminum wire has a resistance on the order of 0.50 Ω/mi. Aluminum is used almost exclusively in long-distance transmission lines because of its lightness, strength, and economy.

The important point to notice in the $P = I^2R$ relationship is that the power loss varies as the square of the current. Hence, it is desirable to keep the current as small as possible.

The greatest advantage in using ac power is the ability to step up the voltage for transmission and to step it down to whatever voltage is needed by the consumer through the use of transformers. Since the transformer is an electromagnetic induction device, it cannot be used for this purpose with dc power. Recall from Section 24.5 that when the voltage is stepped up, the current is stepped down proportionately. Hence, the I^2R losses are reduced for high-voltage–low-current transmission. This is illustrated in the following example.

EXAMPLE 25.7 Suppose there is a power input of 1.1 kW to a 60-mi line with a resistance of 0.50 Ω/mi. Assuming a power factor of 1, compare the power loss of the transmission at (a) 220 V and (b) 2200 V.

Solution: (a) It is given that $P_{in} = IV$,

$$I = \frac{P_{in}}{V} = \frac{1100 \text{ W}}{220 \text{ V}} = 5.0 \text{ A}$$

and there is a current of 5.0 A in the line. That is, the power source supplies 5.0 A at 220 V. Then, with $R = (60 \text{ mi})(0.50 \ \Omega)/\text{mi} = 30 \ \Omega$,

$$P_{loss} = I^2R = (5.0 \text{ A})^2(30 \ \Omega) = 750 \text{ W}$$

and the percent loss is

$$\% \text{ loss} = \frac{P_{loss}}{P_{in}} (\times 100\%)$$

$$= \frac{750 \text{ W}}{1100 \text{ W}} (\times 100\%) = 68\%$$

(b) With $V = 2200$ V (stepped up by a factor of 10),

$$I = \frac{P_{in}}{V} = \frac{1100 \text{ W}}{2200 \text{ V}} = 0.50 \text{ A}$$

where it is assumed that the transformer losses are negligible. Hence, the line current is stepped down by a factor of 10 to 0.50 A, and

$$P_{loss} = I^2R = (0.50 \text{ A})^2(30 \ \Omega) = 7.5 \text{ W}$$

In this case,

$$\% \text{ loss} = \frac{P_{loss}}{P_{in}} (\times 100\%)$$

$$= \frac{7.5 \text{ W}}{1100 \text{ W}} (\times 100\%) = 0.68\%$$

Thus, the power loss is reduced for high-voltage–low-current transmission.

Another point to consider is the voltage drop, ΔV, in the transmission line.

(a) At 220 V input with 5.0 A and 30 Ω,

$$\Delta V = IR = (5.0 \text{ A})(30 \ \Omega) = 150 \text{ V}$$

Thus, there will be only a voltage of

$$220 \text{ V} - \Delta V = 220 \text{ V} - 150 \text{ V}$$

$$= 70 \text{ V}$$

at the end of the line.

(b) For 2200 V input with 0.50 A and 30 Ω,

$$\Delta V = IR = (0.50 \text{ A})(30 \ \Omega) = 15 \text{ V}$$

and

$$2200 \text{ V} - \Delta V = 2200 \text{ V} - 15 \text{ V}$$

$$= 2185 \text{ V}$$

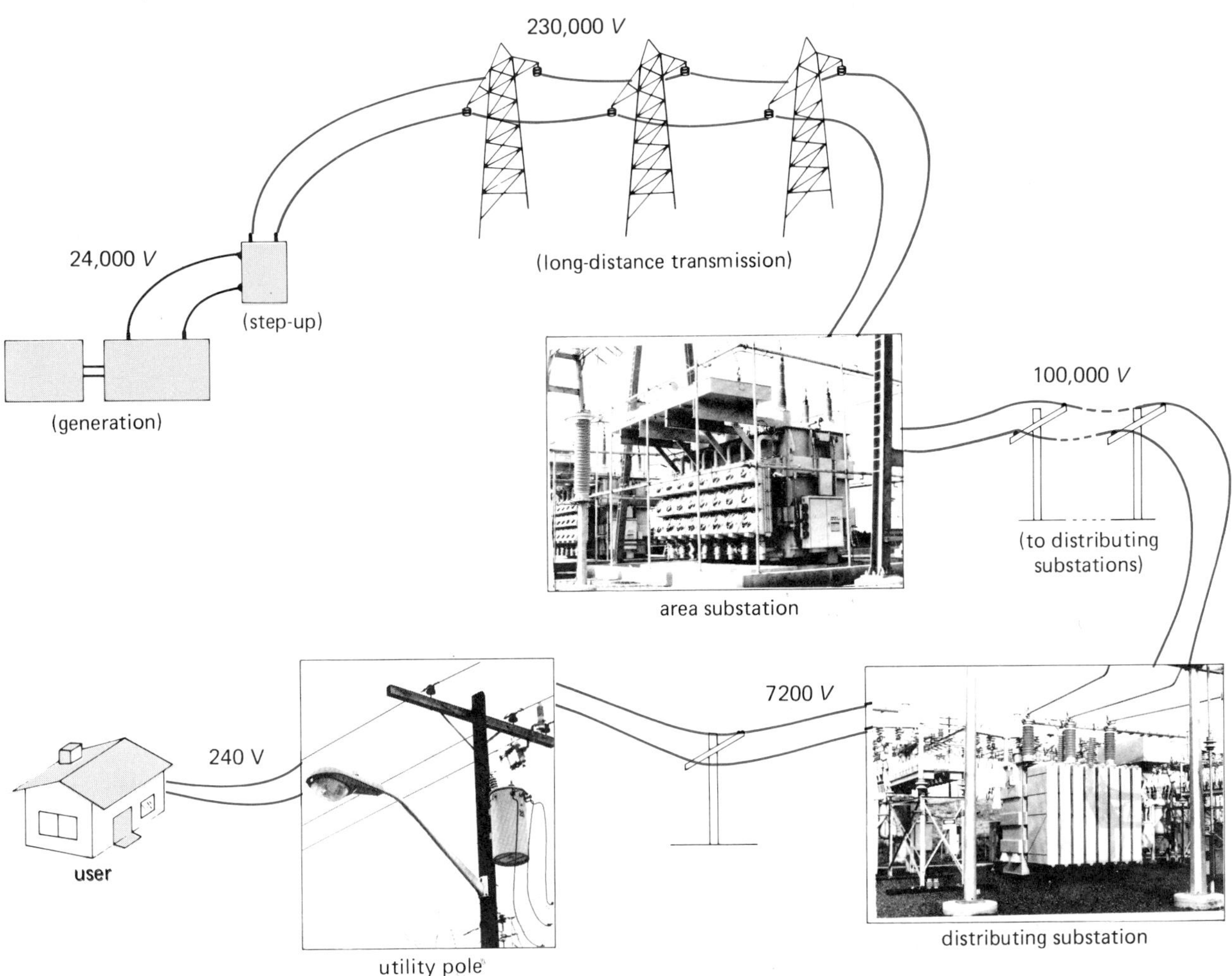

Figure 25.8
A diagram of an ac power transmission system. The voltage is stepped-up at the generating station for transmission to an area substation, where it is stepped-down (notice the cooling fans in the photo). The voltage is further stepped-down at the distributing substation and at the home utility pole.

At the very large transmission voltages used in commercial power transmission, the voltage drop along the line is practically negligible.

Typical voltage values of an electrical distribution system are illustrated in Figure 25.8. Notice that the main transmission line voltage is on the order of hundreds of kilovolts. The voltage is stepped down at area substations, and feeder lines supply power to smaller distributing substations. From these, main lines, or "mains," carry the power to utility pole transformers where the voltage is stepped down to 240 V for consumer use.

Leakage Losses

At high transmission voltages, the air surrounding the line may be ionized. This provides a conduction path to nearby trees, buildings, or the ground and gives rise to leakage losses. Long insulators are used at power line towers to prevent leakage losses (Fig. 25.9).

Leakage losses are generally greater during wet weather because moist air is more easily ionized than dry air. Under certain atmospheric conditions, the extreme ionizing of air near power lines with very large voltages gives rise to an arcing or corona loss. The corona discharge can sometimes be seen and the accompanying crackling noise heard.

Figure 25.9
Power line insulators. Long insulators are used at power line towers to prevent leakage losses.

Figure 25.10
Power capacitors. Power capacitors are used to balance the inductive reactance of the line. The round object at the left is a transformer used to operate the switching device that switches the capacitance in and out of the line circuit.

Load Balancing

In a power distribution system, a great deal of attention must be given to "balancing" the load. This refers to balancing the capacitive or inductive reactances of the load. It is desired to make the power factor as large, or as resistive, as possible. The greatest power will then be delivered to the load.

Residential loads are generally resistive and may have a power factor of 0.90 to 0.95. However, the increasing use of motors for heating and cooling and for appliances provides inductance and tends to lower the power factor. This is particularly true for industrial loads that include large motors.

In order to balance the inductive reactance of such loads, banks of capacitors are installed in the line. You have no doubt seen these power capacitors on utility poles (Fig. 25.10). A capacitor bank is usually switched into and out of the line by means of a time switch. In the evenings, the inductive load generally decreases. If the capacitor bank were left in the line, the line load would then become unbalanced by capacitive reactance.

EXAMPLE 25.8 A small shop has a resistive load of 250 Ω and an inductive load of 300 Ω. (a) What is the power factor of the total load? (b) What capacitance would have to be placed in the line to balance the load completely?

Solution: It is given that $R = 250\ \Omega$ and $X_L = 300\ \Omega$.

(a) The impedance of the load is

$$Z = [R^2 + X_L^2]^{1/2}$$
$$= [(250\ \Omega)^2 + (300\ \Omega)^2]^{1/2} = 391\ \Omega$$

and

$$\cos\phi = \frac{R}{Z} = \frac{250\ \Omega}{391\ \Omega} = 0.64$$

(b) To balance the load completely, or have $\cos\phi = 1$, requires that $X_C = X_L$ (then $Z = R$ and $\cos\phi = R/Z = R/R = 1$). Hence,

$$X_C = \frac{1}{2\pi fC} = 300\ \Omega$$

and

$$C = \frac{1}{2\pi f(300\ \Omega)}$$

$$= \frac{1}{2\pi(60\ \text{Hz})(300\ \Omega)}$$

$$= 8.8 \times 10^{-6}\ \text{F} = 8.8\ \mu\text{F}$$

Devices are now sold to do this in the home.

*25.4 Household Circuits

Most ac circuits are no doubt found in homes. Even though these circuits are used by almost everyone, their principles are unfamiliar to most people because the circuits are hidden behind walls.

The standards and safety of household wiring are of prime concern and are the chief functions of two organizations. The National Fire Protection Association issues a book, *The National Electrical Code*, every three years. "The Code," as it is commonly called, is a set of rules, which have been established over many years, for the proper use of materials and for the correct, safe installation of electrical wiring.

The National Fire Protection Association has no authority to enforce the rules of the Code. That is left to local authorities, who generally follow the Code, with occasional local exceptions.

Materials standards are the main concern of the Underwriters' Laboratories (UL). This is a nonprofit organization that tests and establishes standards for electrical equipment and materials. You may have seen the UL seal or stamp on electrical appliances.

Manufacturers submit their products to the Underwriters' Laboratories for testing. If a product is "listed by UL," this means that it has passed UL tests and meets minimum safety standards. A UL listing does not mean that the product can be used for any purpose, but that it can be safely used for the purpose for which it is intended.

Power is supplied to the home from the utility pole transformer by a three-wire system (see Fig. 25.8). The transformer is tapped so that there is a 240-V potential difference between the two "hot" wires (Fig. 25.11). A hot wire is a current-carrying wire at a potential other than ground.

The third wire of the system is grounded at the transformer and at the service entrance to the home. ("Grounded" means that the wire is connected to a driven ground rod or a metal cold-water pipe that goes into the ground and hence is at the same potential as the ground, or earth.) This third wire between the transformer and service entrance is called a neutral wire and carries no current. In the three-wire color system, the grounded wire is always white or gray. The hot wires are usually black and red, but black may be used for both hot wires.

A potential of 120 V, needed for most household appliances, can be obtained by making a connection between either of the hot wires and the grounded wire. However, note that in a

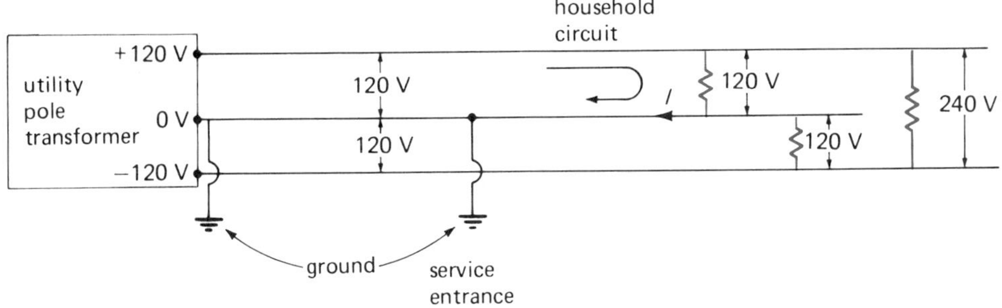

Figure 25.11
Household circuit and entrance. The utility pole transformer is tapped so there is a 240-V potential difference between two hot wires. A third neutral wire is grounded at the transformer and service entrance. The 120 V for household appliances is obtained by connecting across either of the hot wires and the ground wire. The grounded wire is then no longer "neutral" since it is part of the circuit and carries current.

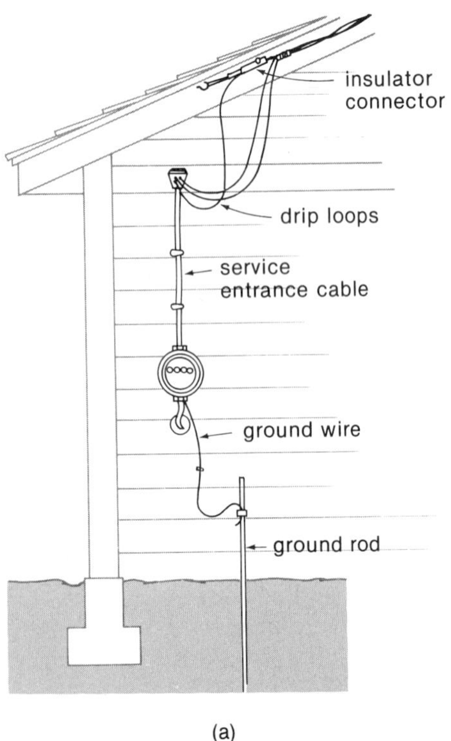

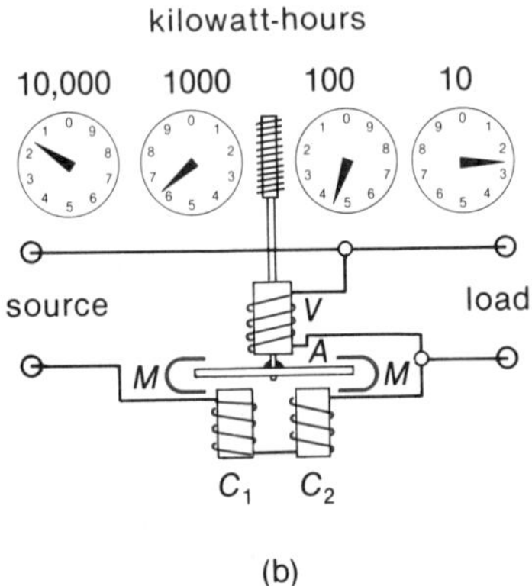

Figure 25.12
Home service entry. (a) Power is brought to the home through a watt-hour meter (see introductory photo of this chapter). (b) The watt-hour meter, used to record electrical energy consumption, is essentially a small single-phase induction motor. A coil V acts as a voltage winding. The coils C_1 and C_2 act as current windings. An aluminum disk A turns like an armature to follow the field set up by the coils. Permanent magnets M induce eddy currents in the rotating disk that oppose the motion of the disk and produce the slippage required for the induction motor operation.

household two-wire circuit, the grounded wire is no longer a neutral wire. It is a grounded, *current-carrying* wire.

A diagram of a typical home service entrance is shown in Figure 25.12. At the building entry, the wires have drip loops to prevent moisture from entering the service entrance. The wires pass through a watt-hour meter and into a service panel box inside the house. For modern wiring, No. 1/0 or No. 3/0 three-wire is preferred for 150-A or 200-A service panels, respectively. Many homes have 150-A service panels, but 200-A service panels are required for homes with electrical heating. With increasing electrical use and larger homes, 300-A service panels are sometimes used.

A simplified diagram of circuits for household electrical distribution is shown in Figure 25.13. In the service panel, the circuits are fused or have circuit breakers. No. 14 wire is a common wire size in household circuits, but the selection of wire size depends on the ampere rating of the circuit. As can be seen in Figure 25.13, common circuits are 120 V with 15 A and 20 A for lighting and appliances, and 240 V with 30 A and 50 A for water heaters and electric ranges. Notice that elements in a circuit are connected in parallel. Why?

Light Circuits

To illustrate the wiring of a specific element, let's consider some light circuits. Electrical cable used for wiring may have two or three wires. For plastic-sheathed cable, the wire insulation colors are black and white for two-wire cable,

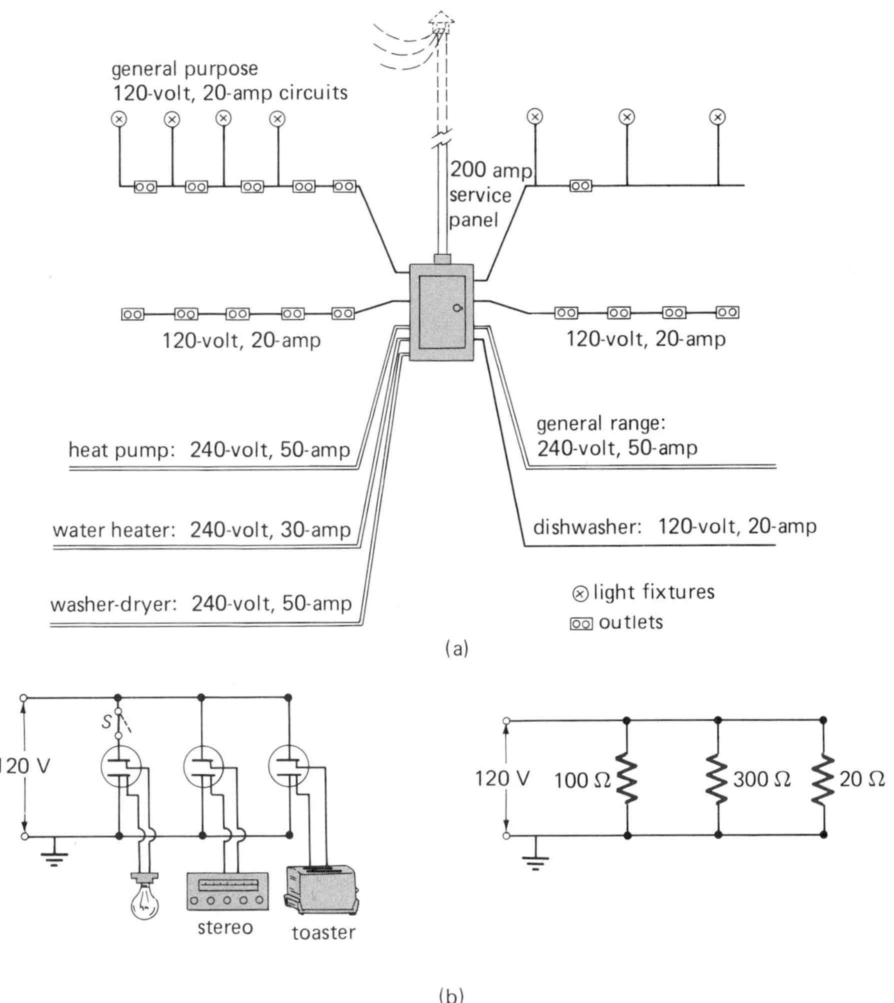

Figure 25.13
Simplified home service panel and circuits. (a) The voltage and ampere ratings of a circuit depend on its use. (b) Expanded view of a household circuit. Notice that the circuit elements are connected in parallel.

and for three-wire cable the colors may be black, white, and red; black, white, and green; or black, white, and a bare wire.

Most codes require that all junction and receptacle boxes be grounded. Figure 25.14 shows a properly grounded junction box using a three-wire cable with a bare *grounding* wire (otherwise usually green). A grounding wire is a "dedicated" ground and carries no current.

The wires go from the junction box to a pull-chain light. Solderless wire connectors, or "wire nuts," are used on the splices. At the light fixture, the black wire is connected to the darker brass screw and the white wire to the silver-colored screw.

Figure 25.15 shows a diagram for a ceiling light controlled by a wall switch through a thin-wall conduit. (Grounding wires for junction box and receptacle are not shown, for convenience.) From this and the previous figure, the applications of two general rules of wiring can be seen:

1. The white (ground) wire should always run without interruption from the source to the consumption element, e.g., a light fixture. This means that a switch must interrupt, or be placed in, the hot (black wire) side of the circuit. More is said about this with regard to safety in the next section.
2. Never connect a black (hot) wire to a white

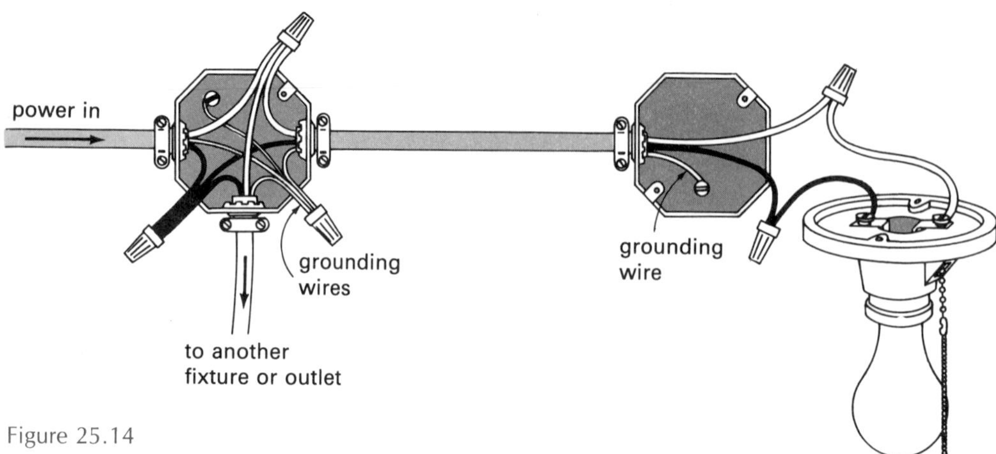

Figure 25.14
Grounded junction boxes. Most codes require all junction and receptacle boxes to be grounded, which is achieved by means of a grounding wire. If the hot black wire should come into contact with the box, it would short the circuit and open a circuit breaker or blow a fuse.

(grounded) wire. That is, black wires are connected to black, and white wires are connected to white. If a hot wire makes contact with a grounded wire, either because of an improper connection or accidentally because of faulty insulation, a short circuit results. The circuit then has only the negligible resistance of the wires, and a large current flow results.

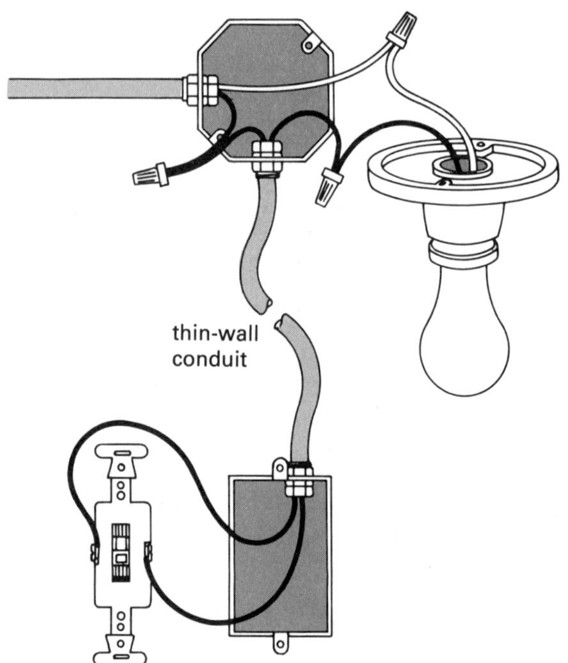

Figure 25.15
Ceiling light controlled by a wall switch. The switch is placed in the black hot line that runs to the switch through a thin-wall conduit. The switch is in series with the circuit element. (Grounding wires not shown here for convenience.)

As a final example, consider the common case of turning a light on and off at two different locations, e.g., a garage light from the house and from the garage, or a hall light from each end of the hall or from upstairs and downstairs. The switch used to control a light from a single point, as in Figure 25.15, is called a single-pole switch.

Control of a light from two locations requires the use of two three-way switches. The name "three-way switch" tends to be confusing. Such a switch will not control a light three ways or from three places, but three-way switches do have three terminals (Fig. 25.16). The common terminal screw is usually identified by being a darker color than the others.

Figure 25.17 shows diagrams for a light that can be controlled by two switches at different locations. Trace the circuit in the simplified diagrams and prove to yourself that each switch controls the light independently.

25.5 Electrical Safety

The importance of electrical safety cannot be overstressed. Electrical accidents can result in property damage, personal injury, and death.

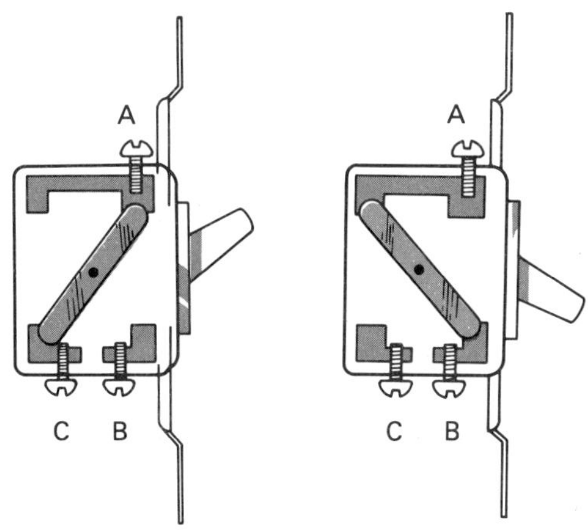

Figure 25.16
A three-way switch. The switch has three terminals and two switching actions. The common terminal *A* is either in contact with terminal *B* or terminal *C*.

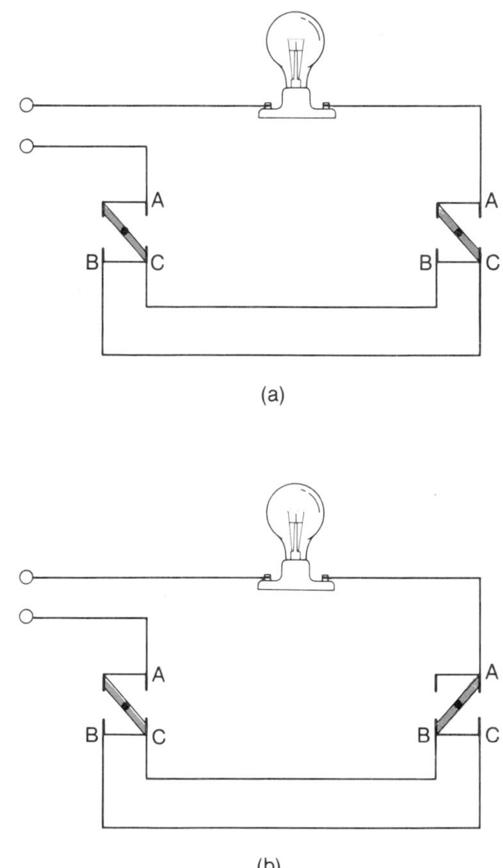

Figure 25.17
Circuit for controlling a light from two locations. (a) Positions of three-way switches when light is off. (Both switches could be in *AB* contact too.) (b) The light can be turned on or off by using either switch.

Fuses and Circuit Breakers

The common methods of protecting against property damage through circuit overloads and overheating are fuses and circuit breakers. Every hot wire must be protected by a fuse or a circuit breaker. A **fuse** is essentially a short piece of metal with a low melting point and a resistance that is relatively large compared with an equal length of circuit wire. When the current in a fused circuit exceeds the fuse rating, e.g., 15 or 20 A, the joule heat vaporizes or melts the fuse strip. The fuse "blows," and the circuit is open. Either a circuit overload or a short circuit will cause a fuse to blow.

There are various types of fuses (Fig. 25.18). Most common in older homes is the Edison-base type. Its base is the same as that of an ordinary light bulb. There are also "slow-blow" types (trade name, Fusetron), which handle temporary overloads for a few seconds before blowing. This type of fuse accommodates sudden surges of current, e.g., when a motor starts up.

Edison-base fuses are physically interchangeable—for instance, a 30-A fuse can be placed in a 15-A circuit. Because of the danger of this, Edison-base fuses are no longer permitted in new installations. They can be used only as replacements. To prevent interchanging when replacing these fuses, a non-tamperable fuse, called a Type-S (trade name, Fustat), can be used. It will not fit an ordinary Edison-base socket. An unremovable adapter is installed in the socket. Each rated Type-S fuse has an adapter with different-size threads. For example, a 30-A Type-S fuse will not fit a 15-A adapter (Fig. 25.18).

Circuit breakers are now used in place of fuses as required by the Code (Fig. 25.19). In a circuit breaker, a predetermined value of current will cause to breaker to "trip," or open, the circuit. Common circuit breakers are either thermally or magnetically activated (Fig. 25.20). In the former, the circuit current passes through a bimetallic strip (cf. Chapter 16). If too much current flows, the I^2R heating causes the strip to bend and open the circuit. Similarly, an excess of current in a conductor causes a magnetically activated circuit breaker to open (recall $B \propto I$). A circuit breaker may be reset by means of a reset switch.

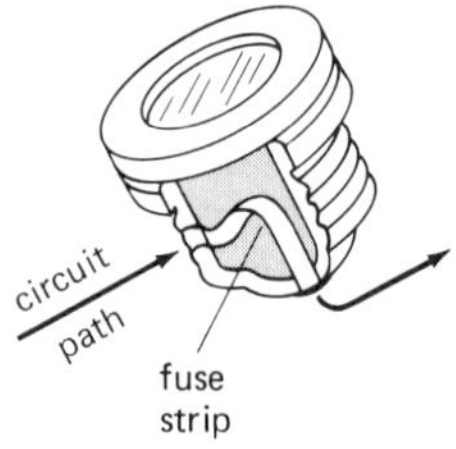

Figure 25.18
Fuses. A fuse "blows," or opens, when joule heating from a current larger than the rated value melts the fuse strip. Edison-base fuses (lower left in photo) are interchangeable, and any value of fuse can be put into a socket. Type-S fuses (top in photo) have an adapter installed in the socket specific for a rated fuse. Notice the different threads on the fuses.

Solid-state trip circuit breakers are coming into more and more use. They offer much greater reliability and accuracy in sensing overcurrents. Through the use of transformers and solid-state components, current levels are measured and timed, then compared to predetermined values. When the predetermined current and time levels have been reached, the solid-state trip unit sends a signal to an internal tripping solenoid, which trips the circuit breaker.

In either case—whether a circuit is opened when a fuse blows or when a circuit breaker trips—steps should be taken to find out why and to remedy the problem. Remember, these are safety devices, and when they open a circuit they are telling you something—namely, that the circuit is overloaded or shorted.

Figure 25.19
A circuit breaker panel. The circuit breakers shown here are magnetically activated. A current in excess of its rated value will cause the circuit breaker to open the circuit. The circuit breaker may be reset by means of a switch.

Three-Prong and Polarized Plugs

Switches, fuses, and circuit breakers are always placed in the hot (high-voltage) side of the line, so as to interrupt power flow to the circuit element. However, fuses and circuit breakers may not always protect you from electrical shock. An

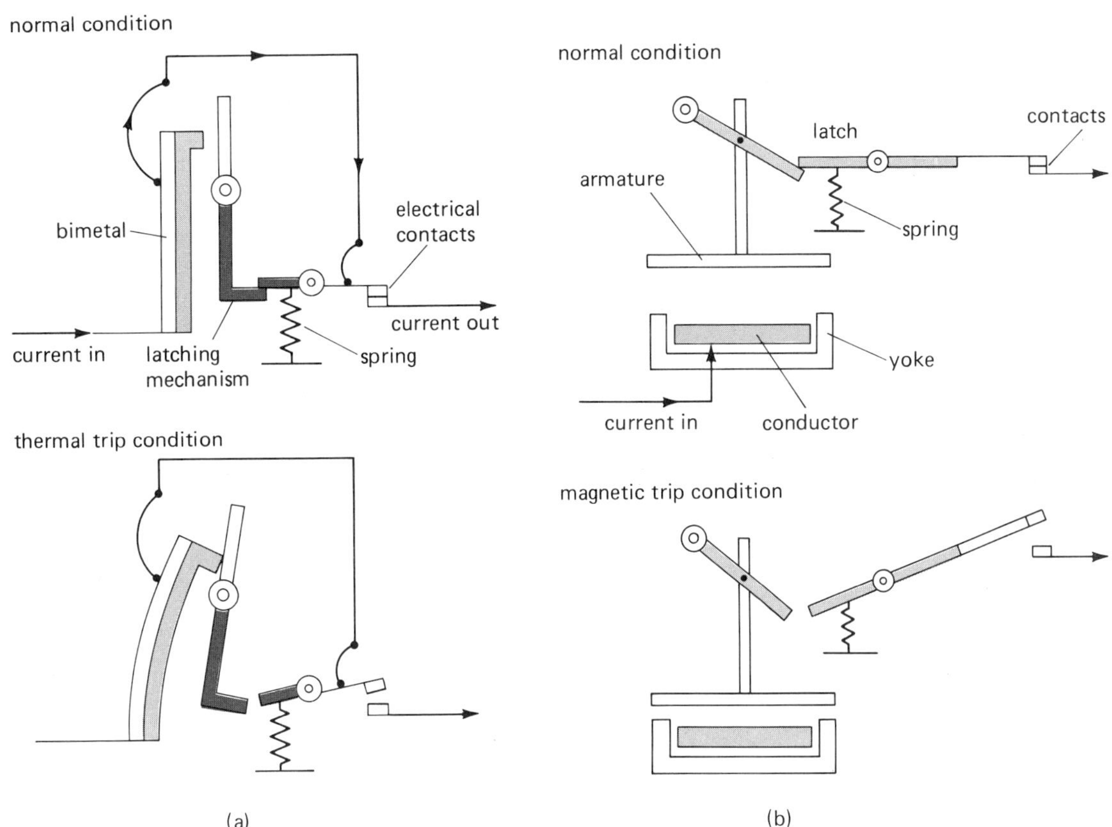

Figure 25.20
Circuit breaker diagrams. (a) Thermal trip. The joule heat generated by the current passing through the bimetallic strip causes it to bend. With enough current and heating, the strip trips the circuit. (b) Magnetic trip. When the current is large enough, the magnetic force draws the armature down to the yoke and causes the latch to release. (Courtesy Square D Company.)

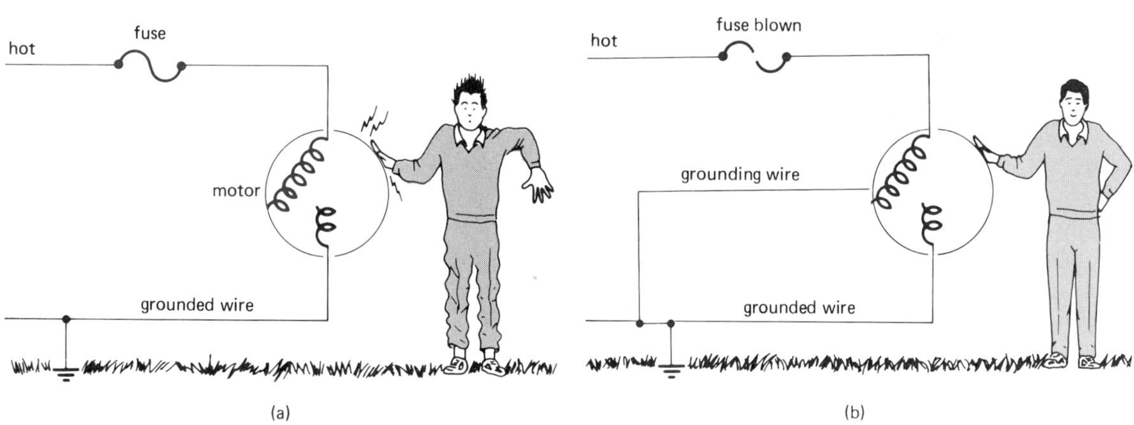

Figure 25.21
Grounding safety. (a) A dangerous situation. If the winding of a motor, with a casing not properly grounded, comes into contact with the casing, a shock is received when it is touched. (b) If the casing is grounded with a grounding wire, the short circuit causes the fuse to blow and prevents a shock.

SPECIAL FEATURE 25.1

Personal Safety and Electrical Effects

Personal safety in working with electricity involves common sense and a fundamental knowledge of electricity. The main thing to remember is not to become part of a circuit yourself. As with any other circuit component, the current drawn by the human body depends on its resistance and the voltage source, $I = V/R_{body}$.

If the skin is dry, there is usually a high resistance, and an electrical shock may only be an uncomfortable and surprising tingle. However, if the skin is wet—for example, with perspiration—the resistance may be low enough to allow injurious and fatal current flows. Table 25.1 shows that only milliamps are needed.

If you come into contact with a "hot" wire, say at 120 V, current could flow through parts of your body. If the circuit is completed through your hand (finger-to-finger), a shock and perhaps a burn can result. However, if the circuit is completed through the body (hand-to-hand or hand-to-foot), there can be serious results, depending on the amount of current flow. You may lose muscle control and not be able to let go. (Muscles are controlled by nerves, which are activated by electrical impulses.)

Larger amounts of current can cause breathing difficulties, and the heart may have uncontrolled contractions (ventricular fibrillation). If the current is greater than 100 mA (0.1 A), death can result. This isn't much current on a relative basis. A 60-watt light bulb operating on 120 V draws $I = P/V = 60/120 = 0.5$ A of current. Keep in mind that it is the current that kills, not the high voltage itself.

When working with electrical equipment or on a circuit, the best policy is to unplug or open the circuit at the source (service panel).

Table 25.1
The Effects of Electric Currents on Humans*

Current (mA)	Effect
1	Barely perceptible
5–10	Mild shock
10–15	Difficult to let go
15–25	Muscular freeze; cannot let go
25–50	Breathing difficulty
50–100	Breathing may stop; ventricular fibrillation
>100	Death

* Varies with individuals.

example is illustrated in Figure 25.21. A hot wire inside a tool (e.g., an electric drill) or appliance—for example, from the winding of a motor—may come into contact with the metal frame or housing. If you touch the frame, watch out. See Special Feature 25.1 for some possible effects.

To prevent this, a grounding wire is used. The circuit is then completed (shorted) to ground, and the fuse in the circuit is blown. This is why many electrical tools and appliances have three-prong plugs (Fig. 25.22). The grounding prong is usually U-shaped. In the wall receptacle, this connection runs to ground. The use of a three-prong plug in an older-type, "two-hole" socket requires an adapter. The grounding lug, or in some cases a wire, of the adapter should be connected to the grounded receptacle by means of the plate-fastening screw. Many times people do not ground the grounding lug, defeating the whole purpose of the safety grounding system.*

Have you ever tried to plug something in and the plug wouldn't go, but when you turned it over, it fit? If you take a look at a wall receptacle, you'll find that one of the slits is bigger than the other. And if you look at the plug, you will find that one prong is bigger than the other (Fig. 25.23). This is called a **polarized plug.**

Polarizing in the electrical sense refers to a method of identification by which proper connections can be made. The original intent in making this type of plug was safety. The small slit in the receptacle is the hot side and the

* When connecting the grounding lug to the plate-fastening screw, it is assumed that the receptacle box is grounded by means of a third dedicated grounding wire (not the zero potential *ground* wire, which is a current-carrying wire in the circuit). If this is not done, the three-prong plug does not serve its intended purpose.

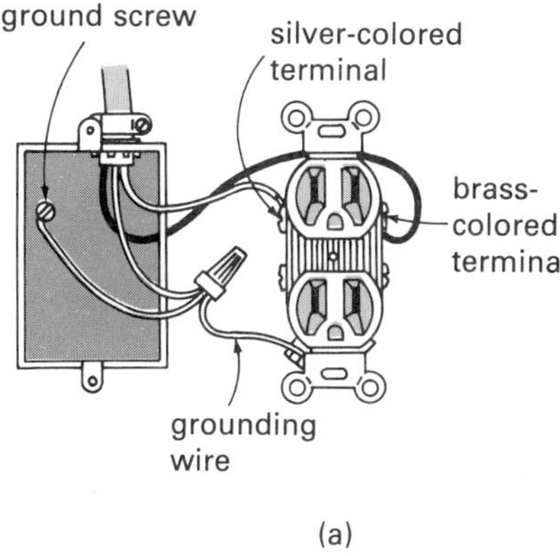

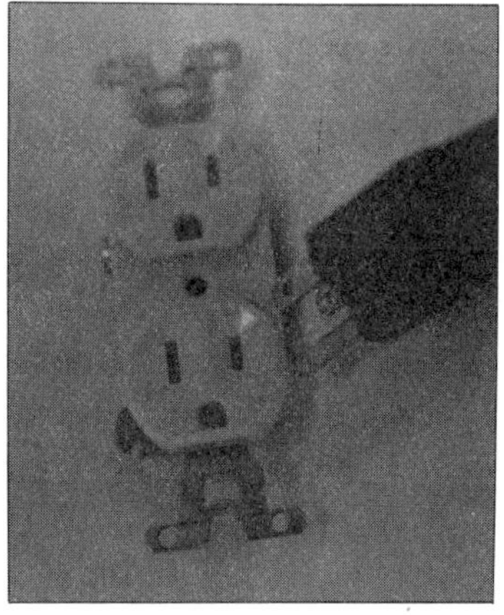

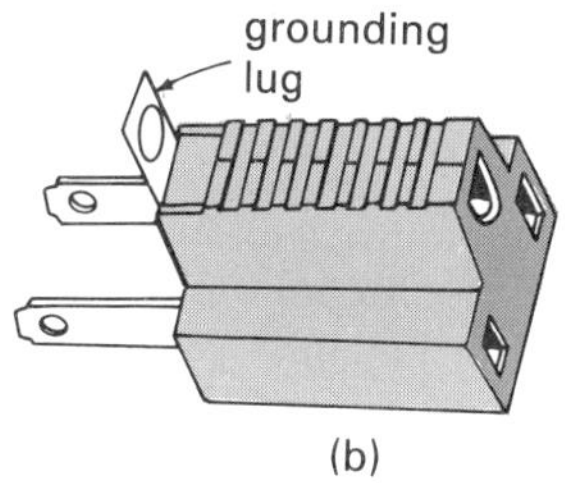

Figure 25.22
A three-prong plug and adapter. The adapter has a grounding lug or wire that should be connected to a grounded receptacle by means of the plate-fastening screw. Otherwise, the whole purpose of the grounding safety system is defeated.

large slit the neutral or ground side, if the wires are properly connected. Theoretically, the housing of an appliance can always be connected to the ground side by means of the polarized plug. The effect is similar to that of a three-prong plug. If a condition as in Figure 25.21 occurs, the hot wire is shorted to ground, which blows a fuse or trips a circuit breaker.

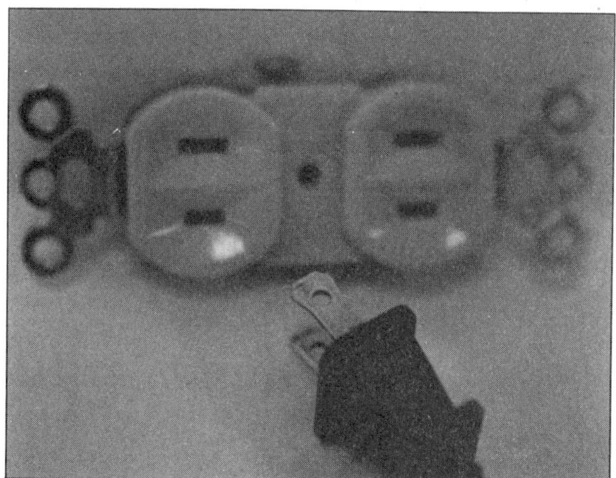

Figure 25.23
A "polarized" plug. This is an early attempt at electrical safety that is still used. It's a good backup system if the receptacle and appliance are properly wired.

However, this system leaves too much to human error. If the receptacle or the appliance is not wired (polarized) properly, a dangerous situation can exist. Polarization is ensured with a "dedicated" third grounding wire as in a three-prong plug arrangement, which is the accepted safety system. The two-prong polarized plug system remains as a general backup safety system, *provided* it is wired properly. However, the use of polarized plugs is decreasing.

Before leaving plugs, look at Figure 25.24. The photo shows a converter set for foreign electrical systems. Plugs with flat parallel blades as we use in the United States (along with Japan, and Central and parts of South America) are not used in many other countries, just as the common 120-V voltage is not. In Europe and many other places, ordinary appliances and lamps operate on 240 V. Therefore, if you travel with your own appliances (electric razor, hair dryer, etc.), it is important to take a converter set. You can imagine what might happen if you plugged your electrical equipment into a 240-V line, but you couldn't do this because the plug would not fit.

The rectangular-shaped object in Fig. 25.24 is a transformer for stepping down voltage from 240 V to 120 V. One of the plugs is plugged into

Figure 25.24
Strange-looking plugs? Not so strange in foreign countries in which other types of plugs are used, as well as 240 V for common appliances. This is a converter set with a transformer for stepping down voltage from 240 V to 120 V and adapter plugs. See text for description.

the socket of the appropriate country. There are two holes on the back of a plug (not seen in the photo) into which the round pins on the back of the transformer are inserted. The U.S. two-blade plug then goes into the front of the transformer.

The plug with the flat angled blades on the left is the type used in Australia, New Zealand, and China. The round pin plug on the right is the type used throughout Europe, the Middle East, and most of Asia. The three-prong blade plug in the center is the type used in Great Britain, Ireland, and parts of Africa. Notice how it resembles the type plug we use for 240-V appliances such as an electric stove.

Ground-Fault Interrupter

When a hot wire touches a grounded component other than the grounded (white) circuit wire, a **ground-fault** occurs. Its effect is the same as that of a short circuit. A fuse will blow or a circuit breaker will trip; the grounding wire system already discussed ensures this to prevent dangerous situations. However, millions of two-wire electrical tools and appliances are in use in which ground-faults can occur. Also, some people foolishly cut off the grounding prong of a three-prong plug so the plug will fit a two-wire receptacle.

In some instances, e.g., due to poor contact, a ground-fault current may be too small to blow a fuse or trip a circuit breaker. But even such a ground-fault current is potentially dangerous. To protect against ground-faults, a safety device called a **ground-fault interrupter (GFI)** may be used. A GFI is particularly desirable for use of electrical equipment outdoors, when a person is easily grounded to the earth, especially wet earth.

The operation of the GFI depends on the fact that, under normal operation, the currents in the hot (black) wire and the grounded (white) wire are essentially the same. When a ground-faurlt occurs, some of the current leaks to ground. The GFI senses any difference in the currents of the circuit wires. If this exceeds a preset trip level, the GFI trips and opens the circuit. A typical GFI can open a circuit within 0.025 ($\frac{1}{40}$) s after detecting a ground-fault leakage current of 5 mA.

A GFI will not prevent a person in contact with the ground-fault circuit from receiving a shock. But it reacts quickly enough so that the shock current is not fatal.

It should be noted that GFI's are not intended to replace circuit breakers or fuses. (GFI's are relatively expensive.) Circuit breakers and fuses are used to protect a circuit from overloading and overheating. GFI's protect individuals from ground-fault shocks.

Important Terms

reactance current opposition in an ac circuit due to a capacitance (X_C) or an inductance (X_L).

ELI the ICE man for an inductance (L), the voltage (E) leads the current (I); and for a capacitance (C), the current (I) leads the voltage (E)

impedance Z the total current opposition in an ac circuit due to resistance and capacitive and/or inductive reactances

phase diagram vector-like phasor diagram of R, X_L, and X_C that gives the impedance and phase angle

phase angle (ϕ) the angle of a phase diagram that gives the phase difference between the voltage and current

circuit resonance the condition when a circuit is completely resistive ($X_L = X_C$) and there is maximum power transfer

resonant frequency the circuit frequency for circuit resonance ($X_L = X_C$) for which there is maximum power transfer

power factor ($\cos \phi$) factor of effective power loss ($P = IV \cos \phi$)

line losses losses in power transmission lines due to resistance

leakage losses losses in power transmission lines due to the ionization of air

grounded wire a wire at zero or ground potential

grounding wire a grounded wire that is not ordinarily current-carrying

fuse a protective device that opens a circuit when an excess of current melts a metal strip due to joule heating

circuit breaker a protective device that mechanically opens a circuit by using magnetic or thermal effects when an excessive current flows

ground-fault interrupter (GFI) a protective device that senses differences in currents and opens a circuit as a result

Important Formulas

capacitive reactance: $X_C = \dfrac{1}{2\pi f C} = \dfrac{1}{\omega C}$

and $V = IX_C$

inductive reactance: $X_L = 2\pi f L = \omega L$

and $V = IX_L$

impedance: $Z = [R^2 + (X_L - X_C)^2]^{1/2}$

and $V = IZ$

phase angle: $\tan \phi = \dfrac{X_L - X_C}{R}$

If $X_L > X_C$, then $+\phi$ and circuit is inductive; if $X_C > X_L$, then $-\phi$ and circuit is capacitive.

resonant frequency: $f_r = \dfrac{1}{2\pi\sqrt{LC}} = \dfrac{0.159}{\sqrt{LC}}$

power factor: $P = IV \cos \phi$

$$\cos \phi = \frac{R}{Z} = \frac{R}{[R^2 + (X_L - X_C)^2]^{1/2}}$$

Questions

Reactance

1. For an ac circuit, when are the current and voltage in phase?
2. When are the phase relations between voltage and current the same for a circuit in ac and dc applications?
3. What is capacitive reactance, and what causes the opposition to current flow?
4. Does more current flow in a capacitor circuit when the frequency of the voltage source is increased? Explain.
5. What is the phase relation between the voltage and the current in a pure capacitive circuit?
6. In a pure capacitive circuit, the capacitance is doubled. How does this affect the current for a given voltage?
7. What is inductive reactance, and what causes the opposition to current flow?
8. What is inductance, and what are the units?
9. What is the phase relation between voltage and current in a pure inductive circuit? How does this compare with a pure capacitive circuit?
10. Does the current flow in an inductor circuit increase when the frequency of the voltage source is increased? Explain.
11. In a pure inductive circuit, the inductance is doubled. How does this affect the current for a given applied voltage?

Impedance, Resonance, and Power Factor

12. What is impedance, and how is it represented on a phasor diagram?
13. What are the units of (a) impedance, (b) capacitive reactance, and (c) inductive reactance?
14. How does the phase angle in a phasor diagram tell whether the circuit is inductive or capacitive?
15. Is a series *RLC* circuit always either inductive or capacitive? Explain.
16. When is an *RLC* circuit in resonance?
17. How is the resonance frequency of an *RLC* circuit affected when (a) the capacitance is increased, (b) the inductance is decreased, and (c) the inductance is increased and the capacitance is decreased?
18. Why is there maximum power transfer to an *RLC* circuit when it is driven in resonance?
19. Can series *RL* and *RC* circuits be in resonance? Explain.
20. What is the power factor of an *RLC* circuit, and what is its significance? Are there power factors for *RL* and *RC* circuits?

Electrical Power Transmission

21. What are the major sources of losses in electrical power transmission?
22. What kind of wire is predominantly used for electrical power lines? Could a better type of conductor be used, and if so, why isn't it?
23. Why is three-phase power transmission more economical than single-phase power transmission?
24. The most economical ac power transmission occurs for high-voltage–low-current transmission. Why is this, and how are these conditions accomplished?
25. Explain what is meant by "load balancing."

Figure 25.25
Voltage or current? See Question 34.

Household Circuits

26. What are (a) the Code and (b) UL certification?
27. (a) Explain the three-wire system and common color code used in household service and circuits. (b) Does a truly "neutral" wire carry current? Explain.
28. Distinguish between "grounded" and "grounding" wires in household circuits.
29. Which wire should run without interruption from the source to the consumption element in household circuits?
30. Can a three-way switch control a circuit element in three ways or from three places? Explain.

Electrical Safety

31. (a) What is the principle of operation of a fuse? (b) What are slow-blow fuses? (c) Distinguish between Edison-base fuses and Type-S fuses. What is the safety advantage of the latter?
32. A very *dangerous* practice of some ignorant or foolish people is to place a penny in a fuse socket with the blown fuse when a replacement is not available. Why should this never be done?
33. How do magnetic and thermal circuit breakers work?
34. Warning signs commonly state, "Danger—High Voltage" (Fig. 25.25). Wouldn't it be more appropriate to say, "Danger—High Current"? Explain.
35. What determines the electrical shock a person receives, and what are the effects of (a) 10–25 mA, (b) 25–50 mA, and (c) 50–100 mA currents?
36. Why should switches, fuses, and circuit breakers always be placed on the hot side of a line?
37. What is electrical polarization? Explain its application in three-prong plugs and polarized plugs.
38. What is wrong with the wiring arrangement shown in Figure 25.26, and what is the proper arrangement?

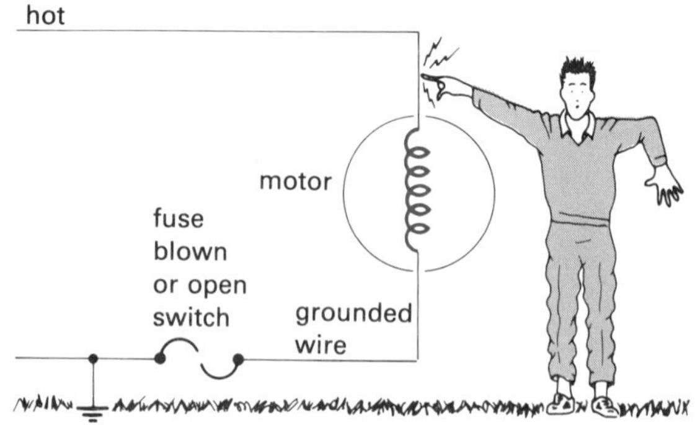

Figure 25.26
Shocking situation. See Question 38.

Problems

Levels of difficulty are indicated by asterisks for your convenience. (All currents and voltages in the problems are effective or rms values.)

25.1 Reactance

*1. A 10-μF capacitor is connected to a 120-V–60-Hz source. (a) What is the capacitive reactance? (b) What is the current in the circuit? (Neglect wire resistance.)

*2. A 3.0-μF capacitor is connected to a 90-V source. If the current in the circuit is 0.15 A, what is the frequency of the source? (Neglect wire resistance.)

*3. It is desired to have 0.10 A of current flow in a pure capacitive circuit with a 120-V–60-Hz source. What is the required capacitance of the circuit?

*4. A variable capacitor in a circuit with a 120-V–60-Hz source is initially at 0.25 μF. The capacitance is increased to 0.50 μF. (a) Is the current in the circuit increased or decreased? (b) By how much (by what factor) does the current change?

*5. A 100-V–50-Hz voltage source is in a circuit with a 1.5-μF capacitor. (a) How much current is drawn from the source? (b) If the capacitor is connected to a 120-V–60-Hz source, by what factor would the current change (Neglect wire resistance.)

*6. A coil of wire has a back emf of 2.0 V when the average time rate of change of current in the coil is 25 A/s. What is the inductance of the coil?

*7. An inductor has a reactance of 90 Ω when driven at a frequency of 60 Hz. What is the inductance?

*8. A 250-mH coil of negligible resistance is connected to a 120-V–60-Hz source. (a) What is the reactance of the coil? (b) What is the current in the circuit?

*9. In a pure inductive circuit, 1.5 A is drawn from a 120-V–60-Hz source. What is the inductance of the circuit?

*10. It is desired to have 0.50 A flow in a coil of negligible resistance when it is connected to a 24-V–50-Hz source. What should be the inductance of the coil?

**11. A coil of negligible resistance with an inductance of 300 mH is connected to a 100-V–50-Hz source and then to a 120-V–60-Hz source. What is the difference in the currents drawn from the two sources?

25.2 Impedance, Resonance, and Power Factor

*12. A 200-Ω resistor and a 25-μF capacitor are connected in series to a 120-V–60-Hz source. (a) What is the impedance of the circuit? (b) What is the current in the circuit?

*13. A series circuit with a 50-Ω resistor and a 350-mH inductor is driven by a 120-V–60-Hz power supply. (a) What is the impedance of the circuit? (b) What is the current in the circuit?

*14. A series *RLC* circuit with a 150-Ω resistor, a 7.5-μF capacitor, and a 500-mH inductor is connected to a 120-V–60-Hz source. (a) Is the circuit inductive or capacitive? (b) What are the impedance and phase angle between the voltage and the current? (c) What is the current in the circuit?

*15. If the circuit described in Problem 14 were connected to a 100-V–50-Hz source, would the circuit have more current than for a 120-V–60-Hz source? Justify your answer.

*16. An *RLC* circuit has a capacitance of 10 μF and an inductance of 100 mH. What is the resonant frequency of the circuit?

*17. What is the resonant frequency of the circuit in Problem 14?

*18. It is desired for a series *RLC* circuit with a capacitance of 50 μF to have a resonant frequency of 1000 Hz. (a) What value of resistance should be used? (b) What value of inductance should be used?

*19. An *RLC* circuit with a resistance of 400 Ω has capacitive and inductive reactances of 300 Ω and 150 Ω, respectively. (a) Is the circuit capacitive or inductive? (b) What is the power factor of the circuit?

*20. What is the power loss of the circuit in Problem 12?

*21. What is the power loss of the circuit in Problem 13?

*22. What is the power loss of the circuit in Problem 14?

**23. A series *RLC* circuit with a resistance of 150 Ω, a 75-μF capacitance, and a 0.40-H inductance is driven by a 220-V–60-Hz source. (a) What is the power loss of the circuit? (b) At what frequency would the power loss be a maximum?

**24. A 100-W light bulb is connected in series with a 3.0-μF capacitor and a 1.0-H inductor. If the circuit is plugged into a 120-V–60-Hz outlet, what is the current in the circuit?

**25. What value of inductance for the circuit in Problem 24 would allow the light bulb to glow the brightest (maximum current)?

**26. A 12-V–40-Hz source is used in a series circuit with a 250-Ω resistance, a 4.8-μF capacitance, and a 1.5-H inductance. (a) What is the power loss of the circuit? (b) At what frequency would the power loss be a maximum? (c) What capacitance in the circuit would give maximum power loss for the 40-Hz source?

25.3–5 Electrical Power Transmission and Safety

*27. A 70-mi power line with a resistance of 0.50 Ω/mi and a line voltage of 12 kV has a power input of 6.5 kW. What is the percent power loss of the line? (Assume a power factor of unity and neglect leakage losses.)

*28. A generating station supplies 6.0 A at 440 V to a step-up transformer with 100 primary turns and 8000 secondary turns. The transformer output is fed into a 100-mi transmission line with a resistance of 0.50 Ω/mi. Assuming an ideal transformer and a line power factor of unity, (a) what would be the percent power loss of the line? (b) What would be the percent power loss if the voltage were not stepped up? (Neglect leakage losses.)

*29. An industrial plant with a 240-V–60-Hz service has a resistive load of 450 Ω and an inductive load of 250 Ω. (a) What is the power factor of the load? (b) What capacitance would balance the load?

*30. A 12-kV–60-Hz power line has a resistive load of 150 Ω and an inductive load of 200 Ω. (a) What is the power factor of the line? (b) What capacitance would balance the load?

*31. A person comes in contact with a 120-V line and receives a mild shock. What is the order of resistance of the person's body?

*32. A careless house painter carrying an aluminum ladder hits a 220-V service line with the ladder. If the painter's body resistance is 10 kΩ, what is the result?

Light-sensitive photoresist material used to make chip circuits. Taken with an electron microscope, the photo's magnification is 9,800X. The pattern shown has dimensions as narrow as one micrometer—a hundred times smaller than the width of a human hair.

Chapter 26

Electronics and Solid State Physics

Probably the fastest-growing area of modern technology is electronics. Since the discovery of the electron vacuum tube in the late 1800's, electronics has developed so much that electronic devices affect everyone's life. They are found in the home and in industry and are coming into increasing use in medicine. These devices include radio, television, computers, monitors, and a multitude of controlling and switching devices.

In the 1940's another discovery was made that revolutionized electronics. Work in solid state physics showed that the functions of vacuum tubes could be performed by solid state devices. At the same time, a great reduction in power consumption and size was realized. For example, we are all familiar with small "transistorized" radios that operate on small batteries.

In this chapter we look at some of the principles of electronics and solid state physics, along with some common devices. Although solid state components have for the most part replaced vacuum tubes, we begin our study by considering the operation of electron tubes. Their principles are more readily understood and contribute to the understanding of analogous solid state components.

26.1 The Electron Tube

Thomas Edison, the inventor of the incandescent lamp, made a discovery that was the basis of the electron tube. In experimenting with his lamp, Edison placed a metal plate in the bulb and connected the plate to a battery (Fig. 26.1). He observed that current flowed in the plate circuit if the positive terminal of the battery was connected to the plate, but if the negative terminal of the battery was connected to the plate, no current flowed.

Edison recorded and even patented this so-called **Edison effect,** but he did not apply it. Without the benefit of modern electron theory, the Edison effect was unexplained. However, with our present knowledge, this effect is read-

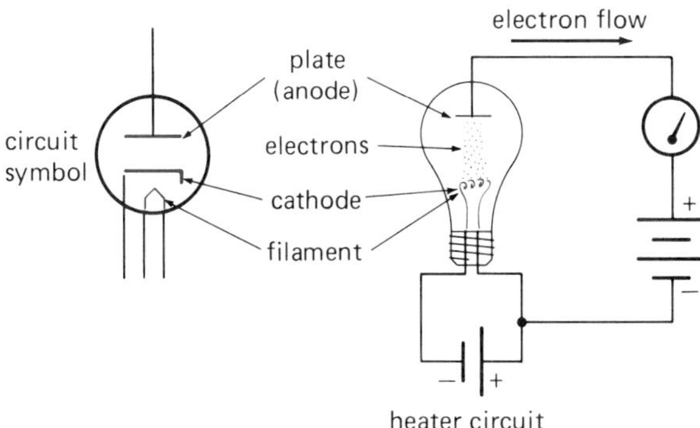

Figure 26.1

The Edison effect. When the plate of a tube is connected to the positive terminal of a battery, a current is observed in the circuit. The circuit is completed by thermionic emission of electrons from the heated filament. The circuit symbol for a tube diode is shown at the left. (The heater symbol is often omitted for convenience.)

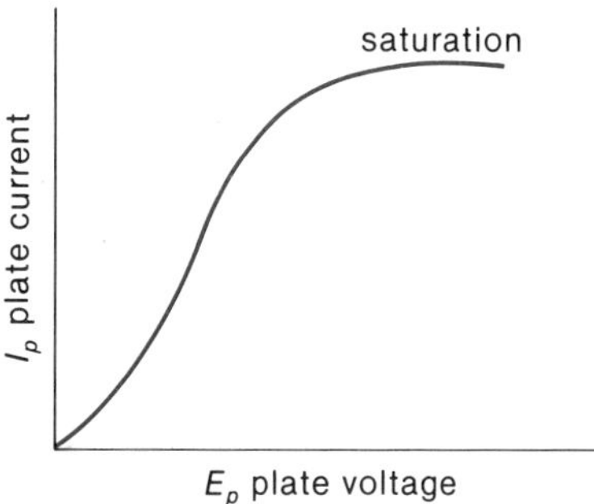

Figure 26.2

Plate current versus plate voltage for a vacuum diode. As the plate voltage is increased, the plate current increases up to a certain (saturation) value. Beyond this, any further increase in plate voltage will not increase the current.

ily understood. Since there is a current in the circuit, there must be a flow of electrons. In the tube the electrons are attracted toward the positive plate (anode) and, hence, must be emitted from the hot filament (the negative cathode of the tube).

The current flowing in the heater circuit heats the filament. This energy causes electrons to be released from, or "boiled off," the filament. This process is called **thermionic emission.** The free electrons are then attracted toward the positive plate, and a current is set up in the plate circuit. If the plate were negative, the electrons would be repelled by the plate, and no current would flow in the circuit.

With two electrodes, an anode and a cathode, such tubes are called **diodes** (*di* meaning "two"). There is usually a good partial vacuum in an electron tube, since air molecules would hinder the electron flow and cause the filament to oxidize. As a result, electron tubes are commonly referred to as *vacuum tubes.*

Saturation

In an actual tube, the plate is usually a cylindrical piece of metal that surrounds the filament. At a given operating temperature, the heater emits a certain average number of electrons. With no plate voltage, a "space charge," or cloud of electrons, exists around the heater.

When a positive voltage is applied to the plate (E_p),* many of the electrons are attracted to the plate, giving rise to a plate current (I_p). If the plate voltage is increased, more electrons are drawn to the plate and the current increases (Fig. 26.2). Finally, a voltage is reached at which all of the emitted electrons are attracted to the plate. This is called the **saturation point.** Any further increase in plate voltage will not increase the current.

Rectification

One of the great advantages of the diode is that it can be used for **rectification**—that is, changing ac to dc. As we have seen, when the plate is positive, current flows. If an ac source is used instead of a battery, the polarity of the plate changes with the frequency of the ac source. As a result, the current flow in the circuit is turned "on" and "off." In a sense, the diode tube acts as a valve (in Great Britain, electron tubes are called valves).

The effect of this valve action is illustrated in Figure 26.3. The diode conducts only during the positive half-cycle. The output voltage is then always positive and hence is dc. Since every half-cycle is missing, this process is called **half-wave rectification.**

* In electronics, voltages are commonly represented by the symbol *E*.

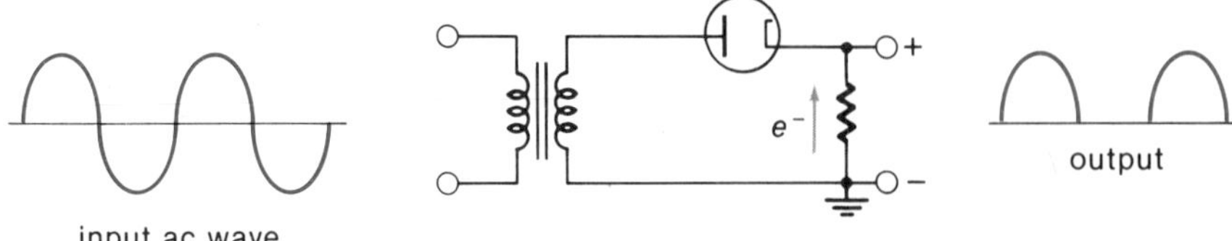

Figure 26.3
Half-wave rectification. The diode is like a "valve" and conducts only during the positive half-cycle. The output voltage then always has a positive polarity and hence is dc. Negative half-wave rectification may be obtained by reversing the diode in the circuit.

By using two diodes, **full-wave rectification** can be achieved (Fig. 26.4). Notice in the figure that the voltages are supplied to the tubes from a center-tapped transformer. The current through the load resistance is in the same direction during each half-cycle and hence is dc.

Full-wave rectification can also be achieved using a single tube containing two plates—a duodiode (see Problem 3). A bridge rectifier circuit may also be used; it is considered in the next section.

The Triode and Amplification

In 1907, Lee DeForest, an American scientist, made an important improvement to the vacuum tube by adding a third element called a **grid.** With three elements, the tube is called a **triode.** The grid consists of a fine wire coil or mesh located between the plate and cathode (Fig. 26.5). Electrons from the cathode can easily pass through.

The purpose of the grid is to control the plate current (hence the common name *control grid*). If a negative "bias" voltage is applied to the grid, the electron flow from the cathode to the plate is impeded, and only a limited number of electrons pass on to the plate. The more negative the grid voltage (E_g) is made, the smaller the plate current becomes (Fig. 26.6). Finally, a point is reached at which the plate current goes to zero. This is called the *cutoff voltage* (or bias) of the tube.

The control grid permits the amplification of a signal. That is, a small input signal can be amplified into a relatively large output signal (Fig. 26.7). This is very important in many applications. For example, radio waves received by a radio antenna produce only minute voltage signals in the receiving circuit. These must be amplified for the radio to function properly.

There are many other types of tubes that are used for various applications, such as tetrodes (four elements), pentodes (five elements), and gas-filled tubes, which are not considered here.

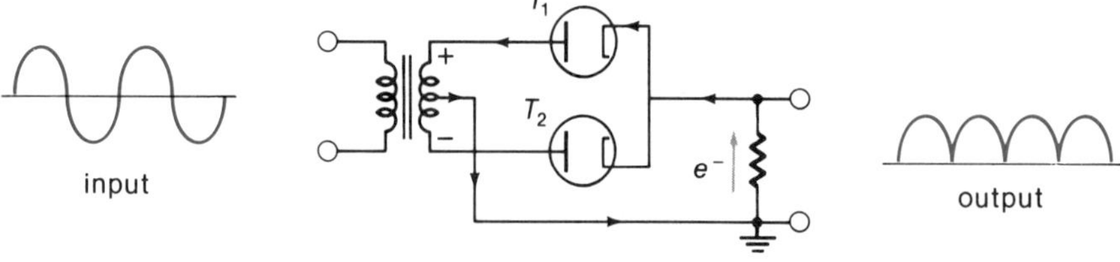

Figure 26.4
Full-wave rectification. During one half-cycle, the secondary of the center-tapped transformer has the polarity as shown. The plate of T_1 is then positive and conducts, while the plate of T_2 is negative and does not conduct. During the next half-cycle, the polarity is reversed; T_2 conducts and T_1 does not. In either case, the electron current through the load is in the same direction, and the polarity of the voltage output is the same (pulsating dc).

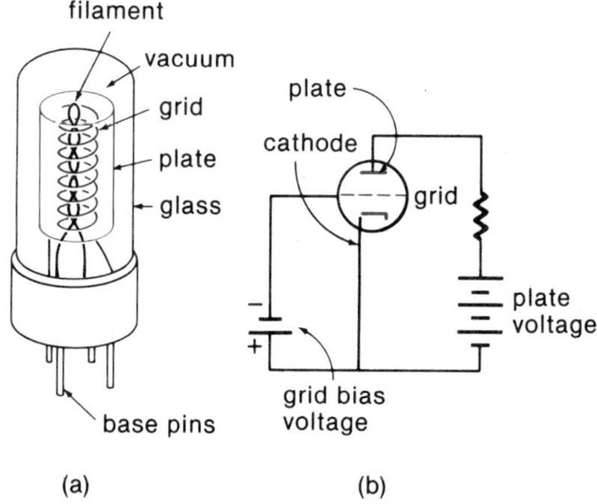

Figure 26.5
The triode vacuum tube. (a) The three elements of a triode are the plate, cathode (filament), and grid. The latter is located between the plate and the filament. (b) Circuit diagram for the triode. The negative-bias voltage on the grid can be used to control the tube (plate) current.

26.2 Solid State Diodes and Transistors

As the name implies, solid state physics is the study of solid materials. Of particular interest in electronic applications are semiconductors. Recall that semiconductors are materials with mediocre electrical conductivity.

Two of the most important semiconductor materials are silicon and germanium. The atoms of these elements have four outermost electrons, which are shared with the other atoms of the element. This is illustrated for silicon in Figure 26.8(a). The silicon atoms are said to be in a crystalline lattice structure, which means they are in an orderly array.

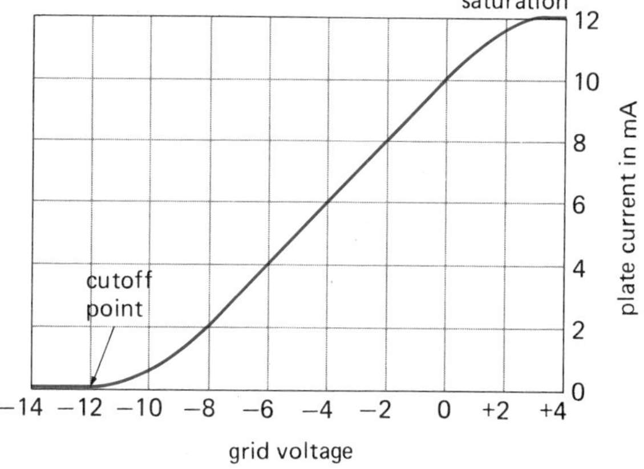

Figure 26.6
Grid voltage versus plate current. With a positive grid voltage or "bias," the tube is quickly saturated. With a negative bias, the cutoff voltage is reached, at which the plate current is zero.

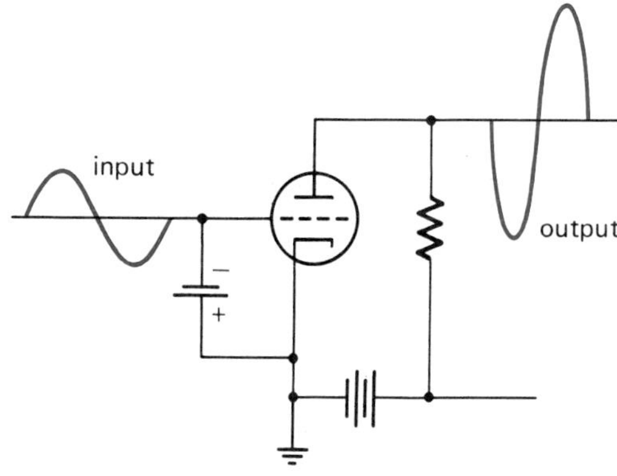

Figure 26.7
Basic amplifier circuit. A small variation of the grid voltage produces a proportionately large change in the plate current and voltage. This gives the triode the ability to amplify a signal.

In pure crystals of silicon, all the electrons are bound, which leaves none free to conduct electricity. Hence, pure silicon (or germanium) is a good insulator. However, there are usually impurity atoms in the crystalline structure, which make it a semiconductor. This is done purposefully by **doping,** or adding impurity atoms to, the pure material. For example, phosphorus impurity atoms might be added to the silicon [Fig. 26.8(b)].

Phosphorus atoms have five outermost electrons, so there is one extra electron per impurity atom in the doped lattice structure. If a voltage is applied to the material, these extra electrons can move in the crystal, and it is semiconducting.

> **An impurity that contributes extra electrons is called a do*n*or impurity**

and

> **A semiconductor in which the charge carriers are *n*egative electrons is called an n-type semiconductor.**

A pure silicon crystal may also be doped with boron impurity atoms [Fig. 26.8(c)]. Boron atoms have three outermost electrons, so one

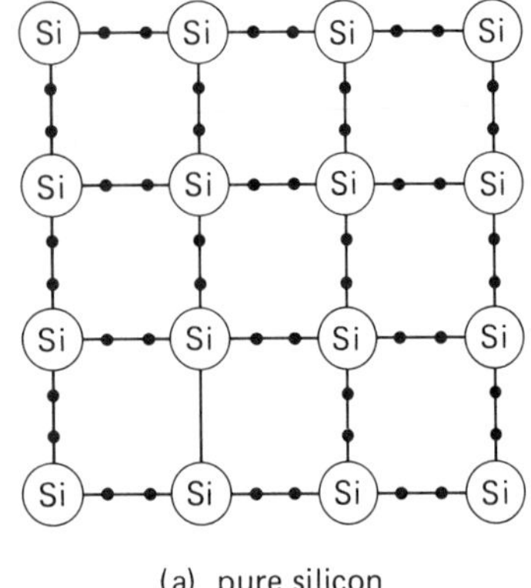

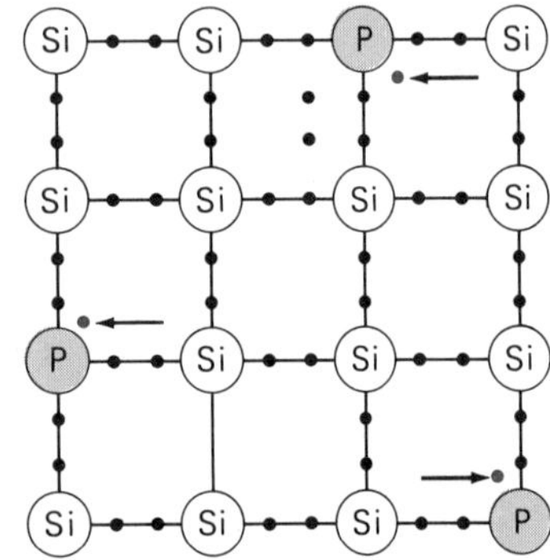

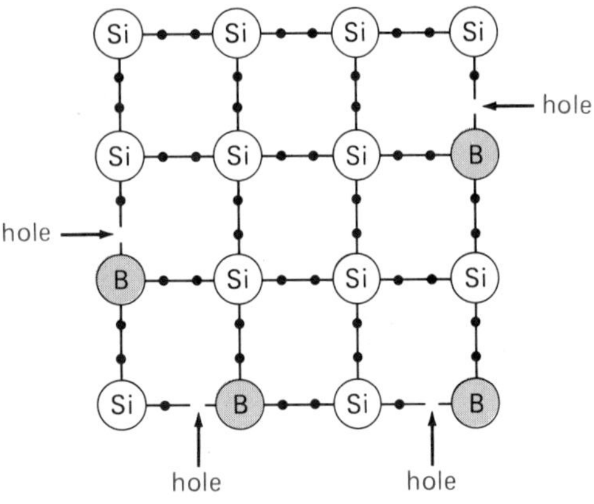

Figure 26.8
Doped semiconductors. (a) Pure silicon is a good insulator. (b) Impurity atoms that donate extra electrons add to the conduction and give an n-type semiconductor. (c) Impurity atoms with less valence electrons create "holes" that add to the conduction and give a p-type semiconductor.

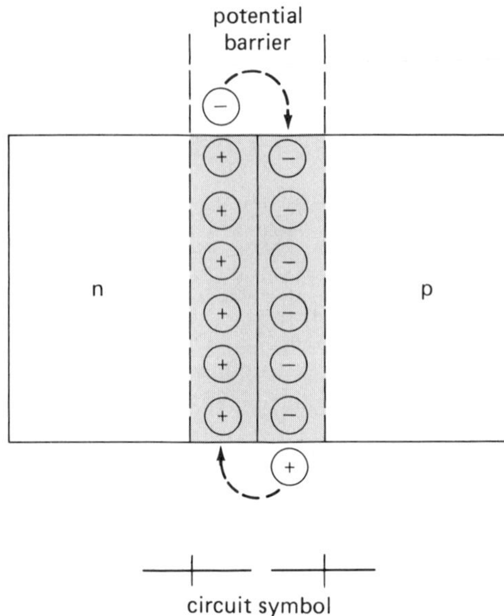

Figure 26.9
A junction diode. The diffusion of electrons and holes in the material sets up a potential barrier across the junction.

electron is lacking, and there is a "hole," or vacancy, near each impurity atom in the lattice. When a voltage is applied to the crystal, electrons can transfer to fill holes, thereby leaving other holes behind. Hence, a flow of electrons in one direction is equivalent to a flow of holes in the opposite direction. In effect, the holes behave like positive charges, and we refer to **positive holes.**

> **An impurity that creates a "hole," or vacancy, is called an acce*p*tor impurity**

and

> **A semiconductor in which the charge carriers are considered to be *p*ositive "holes" is called a p-type semiconductor.**

The Junction Diode

When a thin piece of p-type semiconductor is placed in contact with a piece of n-type semiconductor, a **junction diode** is formed (Fig. 26.9), its name derived from the earlier vacuum diode it replaced. At the junction between the two crystals, the charge carriers of each tend to diffuse into the other crystal. Electrons move

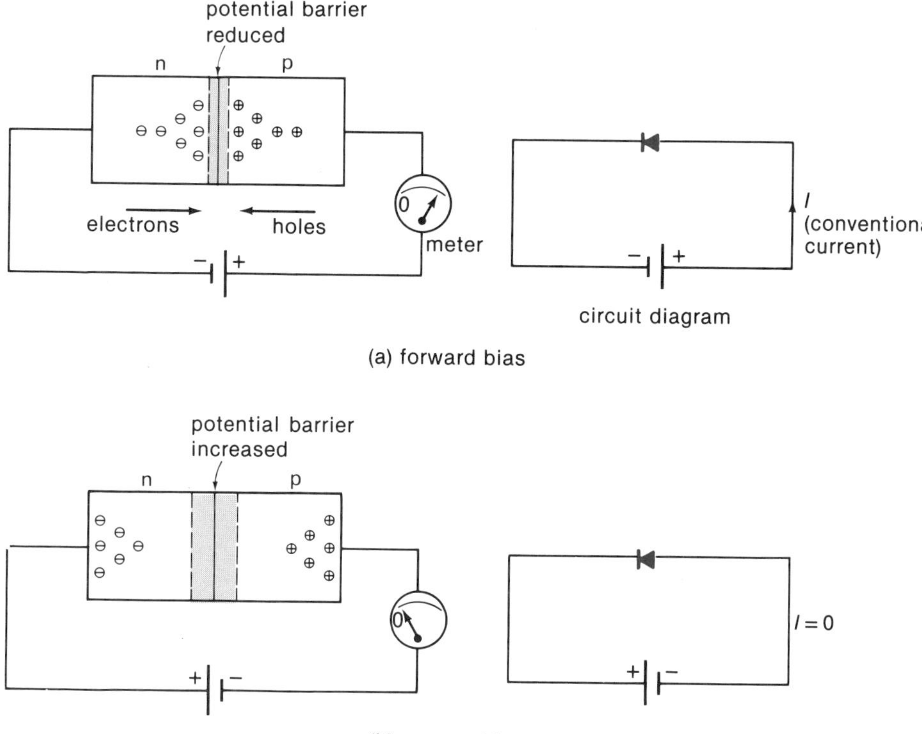

Figure 26.10
Diode bias. (a) When biased in the forward direction, the potential barrier is reduced and there is current in the circuit. (b) When biased in the reverse direction, the potential barrier is increased and there is little or no current in the circuit.

across the junction into the vacancies of the p-type crystal and, in effect, holes move into the n-type crystal. This action sets up a potential barrier across the junction. That is, a small voltage, or potential, exists between the regions close to the junction, which prevents further diffusion.

The potential barrier gives the junctioned n-p crystals the unidirectional current flow effect of a diode. If an external voltage is applied to the crystals (p-type positive polarity and n-type negative polarity), the potential barrier is reduced and current flows in the circuit (Fig. 26.10). With this polarity, the crystal diode is said to be biased in the forward direction, or to have **forward bias.** The arrowhead circuit symbol points in the direction of the conventional current flow (positive charge carriers).

However, if the polarity is reversed (n-type positive polarity and p-type negative polarity), the potential barrier is increased, and little or no current flows across the junction. In this case, the diode is said to have **reverse bias.** Hence, the crystal diode has the same unidirectional flow characteristics as the vacuum tube diode.

Crystal diodes have all but replaced vacuum tubes in many applications. For example, diode circuits are used for rectification. The basic circuits are the same as those in Figures 26.3 and 26.4, with the vacuum tubes replaced by diodes.

A bridge-type circuit for full-wave rectification is shown in Figure 26.11. Notice that the bridge circuit for full-wave rectification does not require a center-tapped transformer. Crystal diodes offer great advantages in size, power consumption, and material economy. No heating element is required for a crystal diode as it was for thermionic emission in the vacuum

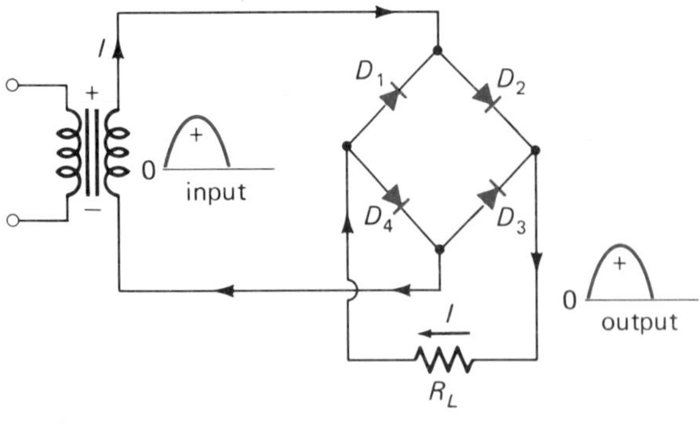

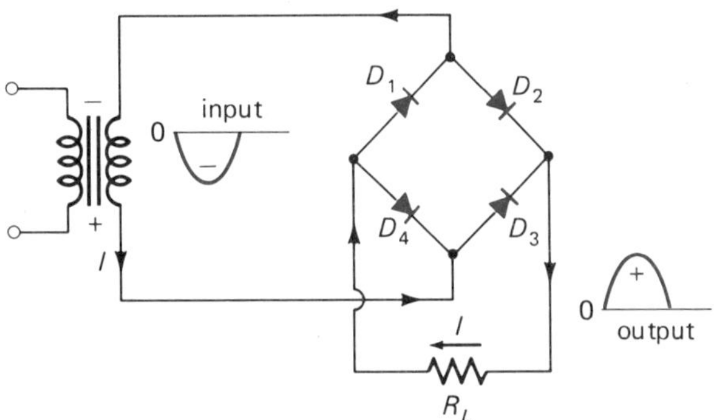

Figure 26.11
Diode bridge circuit for rectification. (a) During one half-cycle, the conventional current is as shown. (Trace the circuit, remembering that the diode arrowhead points in the direction of conventional current flow.) (b) During the next half-cycle, the current through the load resistance R_L is in the same direction; hence the output is direct current.

tube. This gives a difference of milliwatts versus watts in power consumption.

Another application of a solid state diode is given in Special Feature 26.1.

The Transistor

The solid state analog of the triode vacuum tube is the transistor. In a transistor, a thin piece of one type of crystal is sandwiched between two pieces of the other type of crystal (Fig. 26.12). This gives rise to npn and pnp transistors. In either case, one of the outer crystals is the **emitter,** the center crystal the **base,** and the other outer crystal the **collector.** These correspond to the cathode, grid, and plate, respectively, in a vacuum tube.

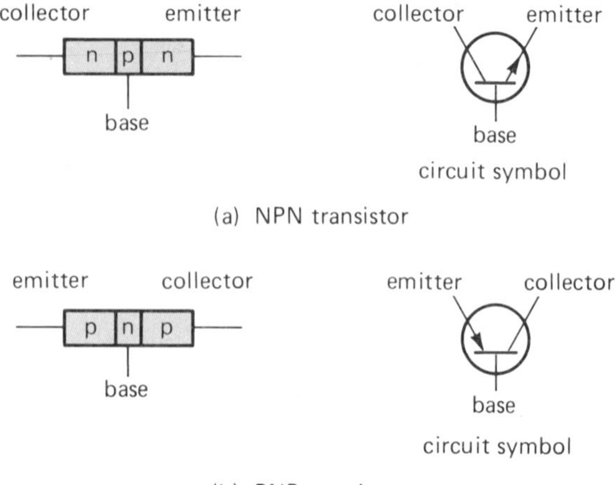

Figure 26.12
Transistors. (a) An npn transistor. (b) A pnp transistor.

Analogously to the vacuum tube, the base (grid) current can be used to control the current flow, and the transistor can be used to amplify signals. There are various circuit configurations for transistors. They depend on which element is common to both the input and output circuits. Each configuration has different circuit characteristics, which you will study if you take an electronics course.

Another solid state application involving electrostatics—the electrostatic copier—is discussed in Special Feature 26.2.

26.3 Printed Circuits and Integrated Circuits

Printed Circuits

With the development of solid state components such as diodes and transistors, there have been rapid strides toward circuit miniaturization. This has been particularly true, and a necessity, in aerospace applications where size and weight are of prime importance. It is also noticeable in everyday commercial applications, e.g., pocket calculators and radios.

One method of assembling the miniature components in a convenient package is on a printed circuit (Fig. 26.13). The "printing" of the circuit connections (taking the place of wires) is done on a copper-clad fused glass cloth board. The circuit is painted on the board by

SPECIAL FEATURE 26.1

LED's—Light-Emitting Diodes (and Solar Cells)

The LED displays used for hand calculators, cash registers, digital clocks, gas pumps, and so on are applications of a special type of semiconductor diode. When a semiconductor diode is biased in the forward direction, electrons and holes flow in opposite directions, as we have seen. Some of the electrons and holes combine in the process, and when they do, energy is released in the form of light (Fig. 1).

If one of the semiconducting materials is transparent, such as gallium phosphide, light is emitted from the diode (LED). LED's can be arranged in the shapes of numbers and letters for use in readout displays. (LED's have been replaced in large part by LCD's, or liquid-crystal displays, in watches, calculators, and the like because they take less power. This type of display will be discussed in Chapter 27.)

Solar cells are made by using the reverse process. If a diode has a thin, transparent p-type layer, incoming light will pass through the crystal to the junction and strike the n-type (electron carrier) material. This can give electrons enough energy to be kicked across the junction into the p-type material. As a result, a voltage develops across the diode, similarly to what happens in a battery, and a "solar current" flows in a circuit connected to the diode. (See Chapter 32 for more uses of solar cells.)

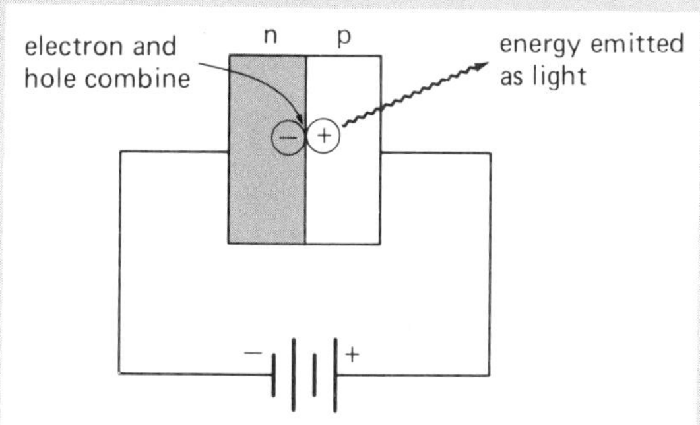

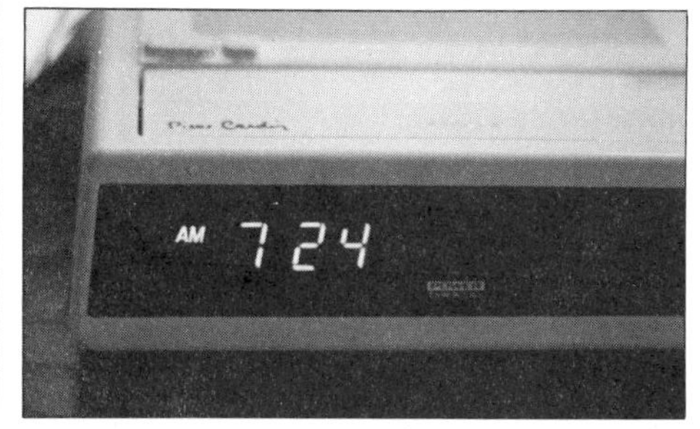

Figure 1
LED (light-emitting diode). In an LED, the combination of electrons and holes results in the release of energy in the form of light. Such diodes are used in LED displays, for example, in digital clocks.

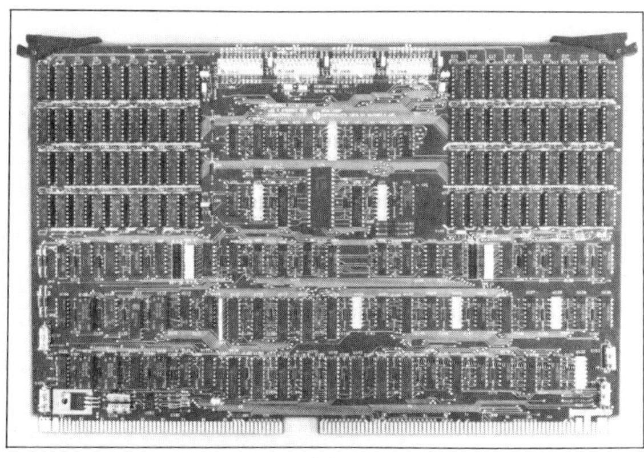

(a)

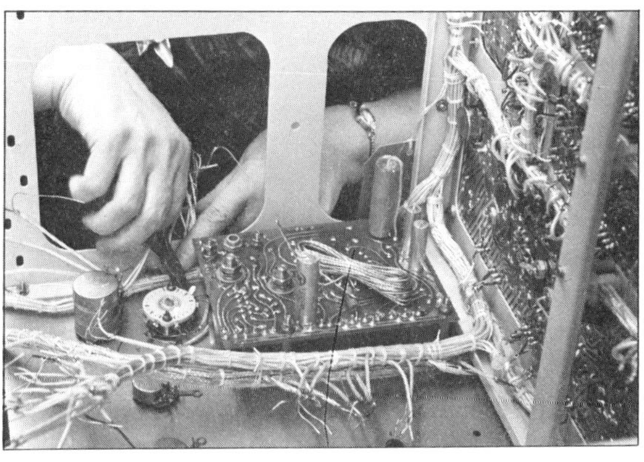

(b)

Figure 26.13
Printed circuits. (a) This printed circuit board is a memory module used in a microprocessor. (Courtesy Motorola Semiconductor Products.) (b) Printed circuit boards being used in macro circuits. (Courtesy General Electric Co.)

SPECIAL FEATURE 26.2

Xerography and Electrostatic Copiers

Xerography is a dry process by which anything written, typed, printed, or photographed can be copied. The word was coined from the Greek *xeros*, meaning "dry," and *graphein*, meaning "to write."

The heart of xerography is the photoconductor. Some semiconductors are photoconductors. In the dark, a photoconductor such as selenium is a good insulator and hence can be electrostatically charged. However, when light strikes a photoconductor, it becomes conductive and permits the electrical charge to leak away from the part hit by the light.

How Xerography Works

1 Surface of a plate coated with a photo-conductive metal is electrically charged as it passes under wires.

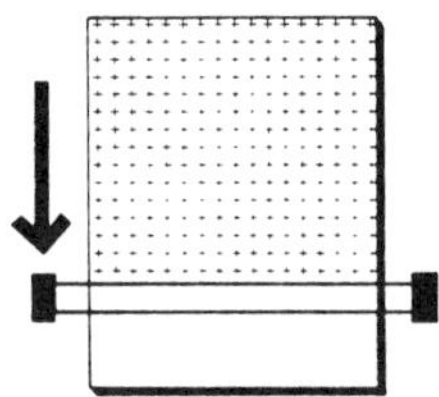

2 Plus marks represent positively charged plate.

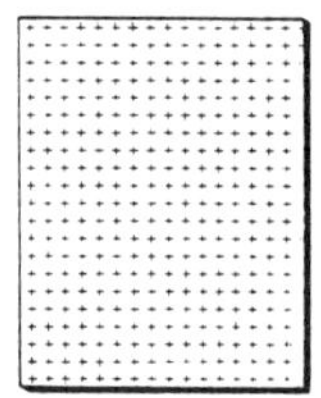

3 Original document is projected through a lens. Plus marks represent latent image retaining positive charge. Charge is drained in areas exposed to light.

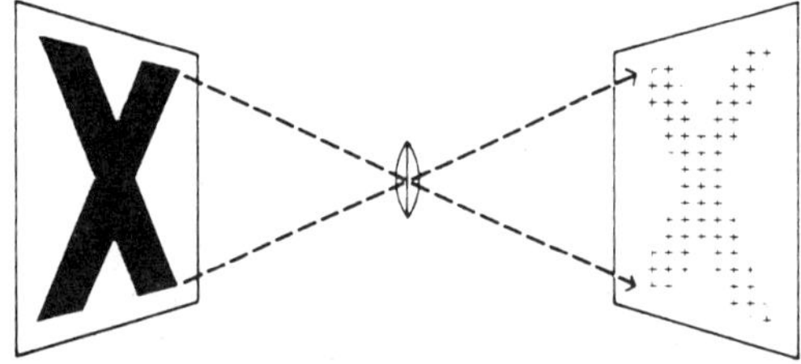

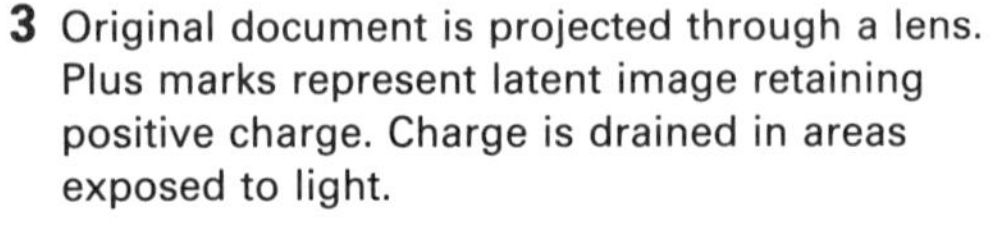

4 Negatively charged powder (toner or "dry ink") is applied to the latent image, which now becomes visible.

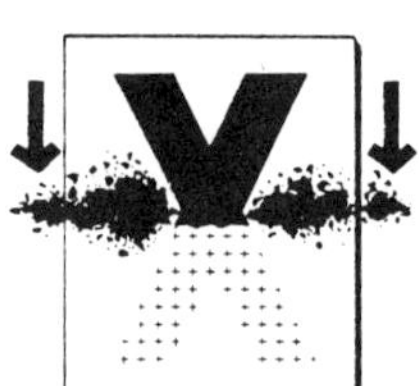

Figure 1
Electrostatic copying. The steps of the xerography process are shown in the diagram, and an electrostatic copier is shown in the photo. This Xerox copier/duplicator uses a photoconductor-coated belt, rather than a drum, as used commonly in older models. (Courtesy Xerox Corporation.)

hand or applied by a screen-printing process using an acid-resistant paint. When the board is placed in an acid bath, the unpainted portions of the copper are etched away, leaving the desired circuit connections.

A photographic method may also be used. The copper of the board is coated with a photosensitive material. A mask of the desired circuit is placed on the board, which is then exposed to light. In the developing process, the light-exposed surfaces are etched away.

After the circuit has been etched, small holes are drilled at the proper locations in the circuit for mounting the leads of the circuit elements, e.g., resistors, capacitors, and sockets, which are soldered in place. Printed circuit plug-in modules make the assembly of the electronic systems used in technology quite easy.

Integrated Circuits

Even further miniaturization has come about through the use of integrated circuits (IC's). An integrated circuit consists of thousands of diodes, transistors, resistors, and capacitors, which are fabricated on a silicon chip, usually only a few millimeters square (Fig. 26.14 and chapter introductory photo).

In transfer electrostatic copying, a photoconductor-coated plate, drum, or belt is electrostatically charged and then illuminated with a projected image of the writing or object to be copied (Fig. 1). The illuminated portions of the plate become conducting and discharge.

The plate then comes into contact with a negatively charged black powder called toner or "dry ink." The charge-retaining portions of the plate attract the toner, and it adheres to them. Paper is then placed over the plate and a positive charge is delivered to the back of the paper. This attracts the toner from the plate surface to the paper. The image is then fused permanently into the paper by heat, and a copy of the original image is produced.

5 Paper (or other material) is placed over plate and given positive charge.

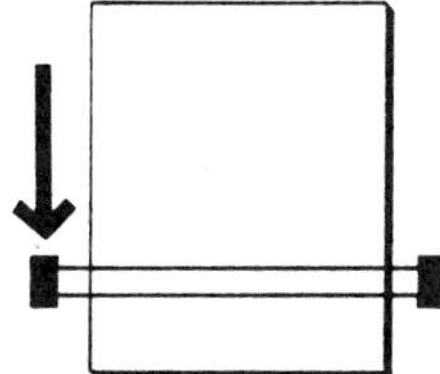

6 Positively charged paper attracts dry ink from plate, forming direct positive image.

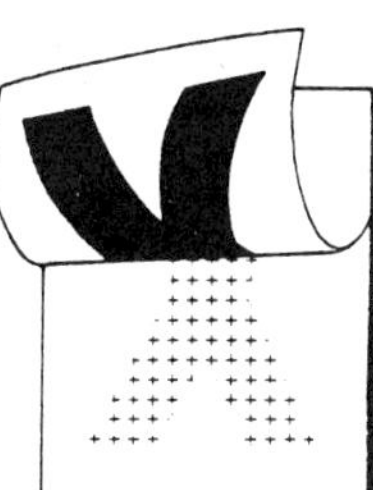

7 Image is fused into the surface of the paper or other material by heat for permanency.

Single crystals of lightly doped silicon are drawn from molten material (about 1100°C). These are cleaved into thin wafers. Then hundreds of identical IC chips are fabricated on the surface of the wafer in a series of steps. The chips are separated and packaged for use.

Components are fabricated on the wafer by several techniques. For example, diodes and transistors can be made by vapor deposition. Successive deposits of n- and p-type impurities diffuse into the wafer to create diodes—the pnp and npn structures needed for transistors. A focused beam of ion impurities can also be used in a process called *ion implant*.

The basic means of creating the interconnecting circuit is by a photosensitive technique. The wafer is covered with a thin layer of photosensitive plastic called *photoresist*. A photographic negative with the desired circuit pattern is placed on the resist layer and then is exposed to ultraviolet light. Where the light strikes the resist, it hardens and remains in place when the unexposed portion is washed away. The hardened plastic can be dissolved later.

A thin layer of metal is then deposited over the resist pattern. When the resist is dissolved, the metal is removed with it. This leaves a

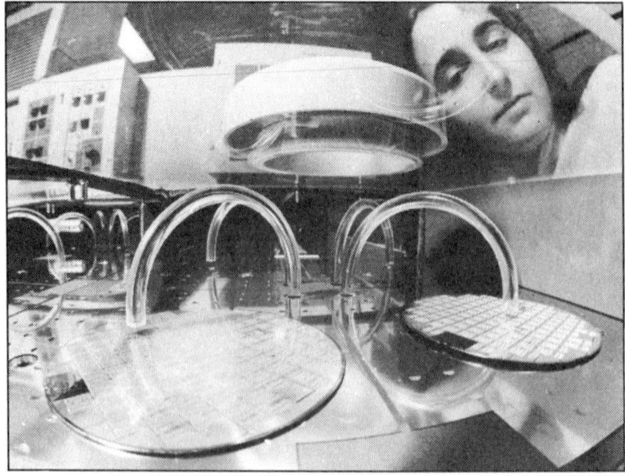

(a)

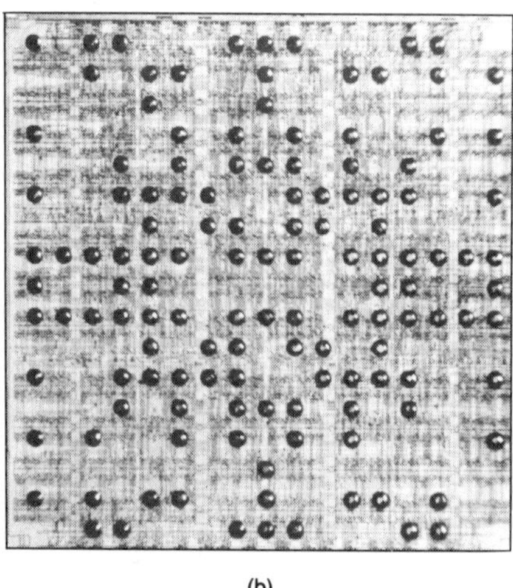

(b)

Figure 26.14

Integrated circuits. (a) The processing of razor-thin slices of silicon. The thin waves float on a cushion of air and the tube-like optical sensor loops control their movement. (b) The finished product—a new high-performance logic chip for a computer. The chip contains over 700 logic circuits. (See also chapter introductory photo.)

metal pattern for interconnecting the components on the surfaces where the resist was originally removed.

Resistors can be made from n- or p-type regions, but they have a wide range of tolerances. For precision applications, a small amount of material such as nichrome is deposited on the wafer, and a laser is used to burn away the nichrome until the desired resistance is obtained.

26.4 Radio and Television

Electronics has many applications. Because radio and television (TV) are so common, we will look at their principles of operation. The circuits and techniques used in radio and TV are quite complicated, but a general understanding of their operations can be gained from the basic theories presented here.

Radio

As we all know, radio is used to transmit sounds over long distances. Sound waves themselves do not travel very far. For example, someone shouting may be heard up to distances of several hundred feet. Recall from Chapter 14 that sound waves are mechanical vibrations. In air, these waves travel approximately $\frac{1}{5}$ mi/s and are readily attenuated, or damped.

Radio waves, on the other hand, are electromagnetic waves that need no medium of transport and travel at the speed of light. Early radio was called "wireless" because no wires were needed for transmission, in contrast to telegraphy. Radio waves are generated by electrons oscillating in a wire or antenna. The orientation of the E field of the wave with respect to the

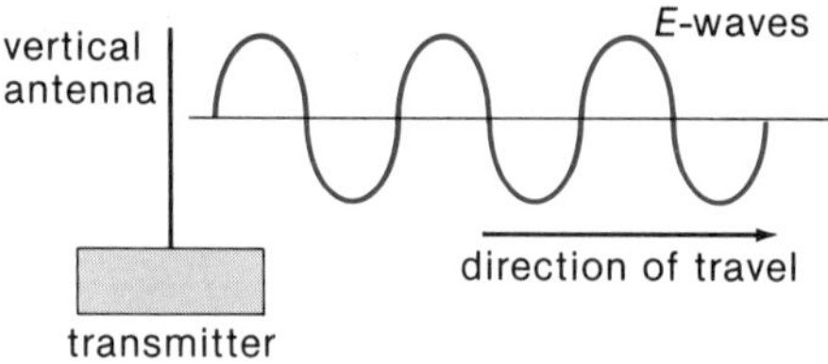

(a) vertical polarization

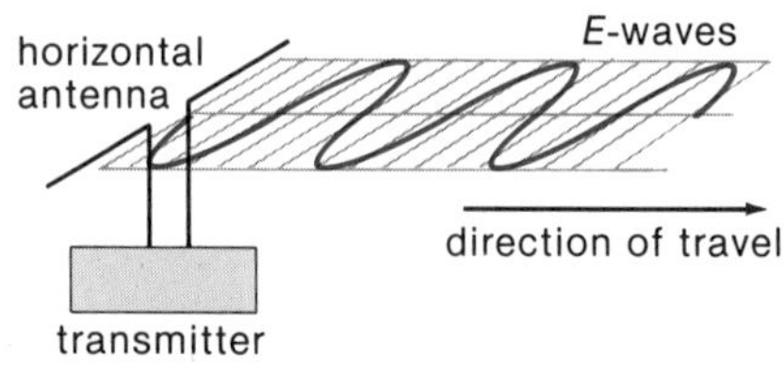

(b) horizontal polarization

Figure 26.15

Radio wave propagation. The E field of electromagnetic radio waves is in the same plane as the antenna. The E wave may be (a) vertically polarized or (b) horizontally polarized, depending on the orientation of the antenna.

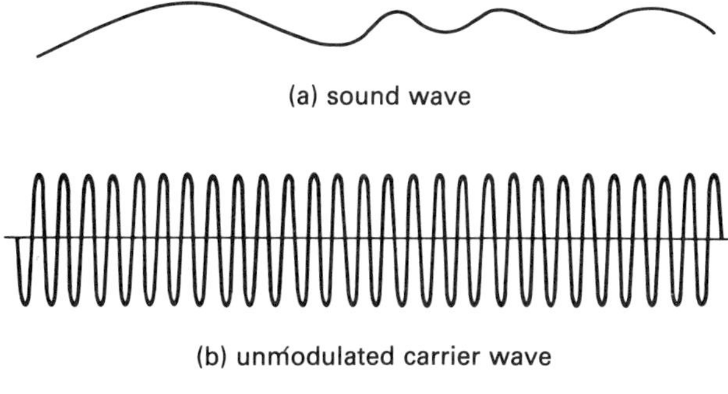

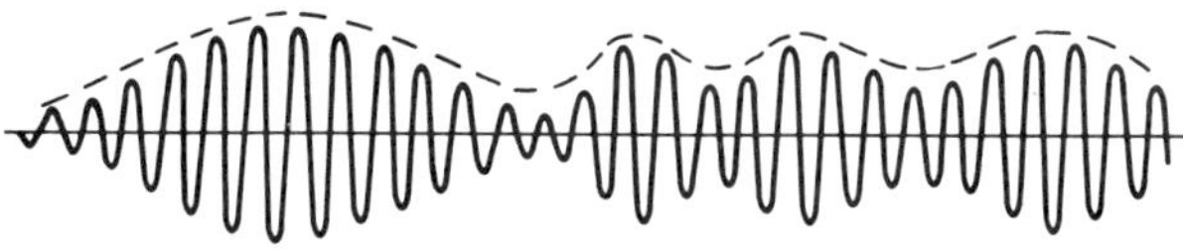

(c) amplitude modulated (AM) wave

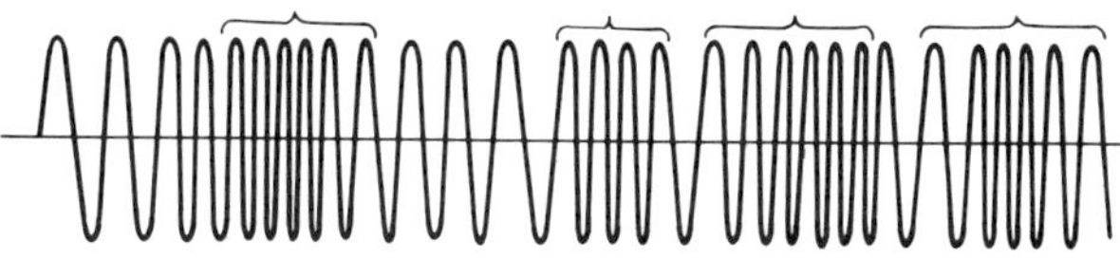

(d) frequency modulated (FM) wave

Figure 26.16
Modulation. A sound wave (a) can be impressed on an unmodulated carrier wave (b) so that it is (c) amplitude modulated (AM) or (d) frequency modulated (FM).

Earth's surface is called *polarization.* The E waves are in the same plane as the antenna. Hence, depending on the orientation of the antenna, radio waves may be vertically or horizontally polarized (Fig. 26.15, B field component neglected for convenience).

You may be wondering how radio waves carry sound, since in the kHz and MHz ranges they are above the audible frequencies. This is accomplished by a process called ***modulation.*** The radio wave acts as a "carrier" wave for the sound waves. The sound waves are electrically impressed on the radio waves, so they modulate the carrier (Fig. 26.16). The carrier wave may be amplitude modulated (AM) or frequency modulated (FM). In either case, the variations of the audible sound wave are impressed on the carrier wave. At the receiver, the modulated radio waves are demodulated. **Demodulation** (or **detection**) is the separation of the audio portion of the signal from the carrier wave.

The AM frequency spectrum for radio is from 550 to 1750 kHz, and the FM frequency spectrum is from 88 to 108 MHz (check your radio dial). Each station is assigned a particular individual band, or channel. For AM stations, the band width is 10 kHz, which is 5 kHz on each side of the station's assigned frequency. For FM stations, the channel band is 200 kHz wide.

With greater band widths, a greater range of frequencies can be transmitted by FM. This adds to the fidelity of the sound, particularly for music. Also, static is less of a problem. In AM applications, static also modulates the wave and is carried along. This unwanted amplitude modulation is eliminated by FM. However, because of the high frequency, FM has a relatively short range and depends primarily on line-of-sight transmission.

There are several different types of radio receivers. One of the simplest is a tuned-circuit receiver, which uses a single diode detector. A diagram of this type of receiver circuit is shown in Figure 26.17.

The receiving antenna picks up the radio signals and is inductively coupled to a tuning circuit. The parallel LC tuning circuit (sometimes called a tank circuit) has a resonant frequency given by the same equation as for a series RLC circuit (see Section 25.2). By adjusting the variable capacitor, a particular radio frequency can be selected.

The weak signal is amplified and fed into the diode detector. This input signal contains an rf (radio frequency) component, the carrier, and an af (audio frequency) component, the sound signal. The diode detector is known as a peak detector because it produces a dc voltage equal to the peak values of the input signal. During each carrier cycle, the diode turns on briefly and charges the capacitor C_1 to the peak voltage of the particular carrier cycle.

Between cycles, the capacitor would ordinarily discharge through the resistor R. However, by making the RC time constant much greater than the period of the carrier wave, there is little discharge between cycles. As a result, most of the carrier signal is removed and the output has the form of the upper part of the modulated wave, which is the form of the original modulat-

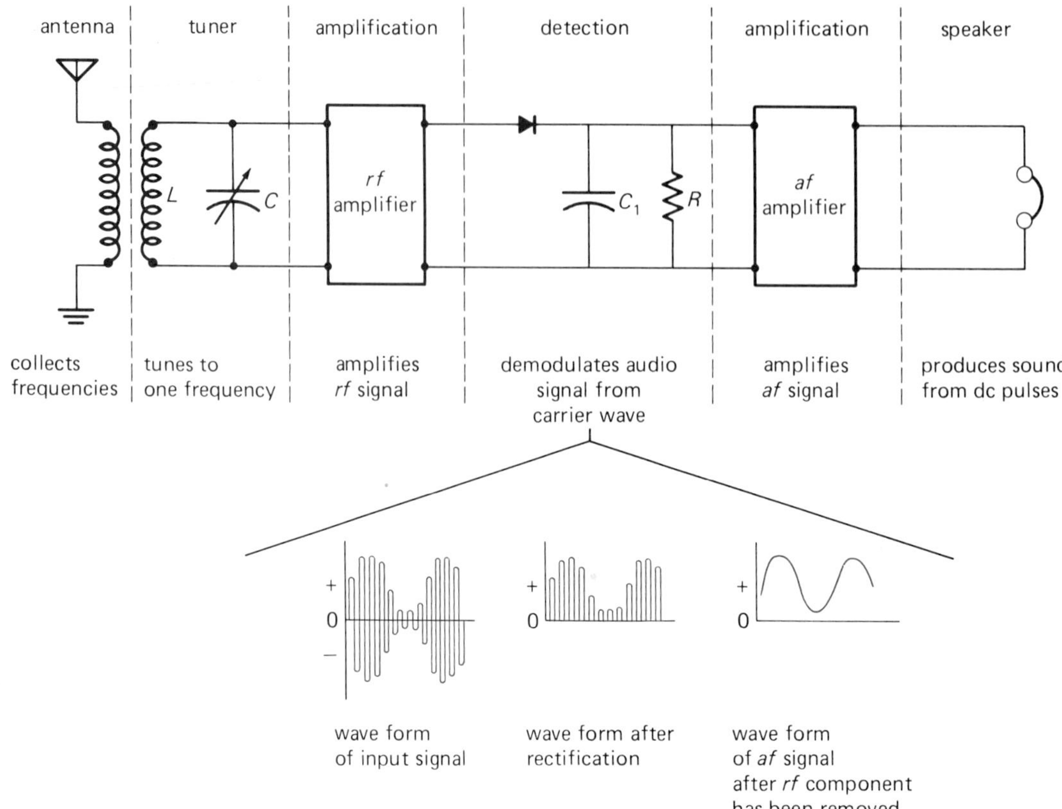

Figure 26.17
Basic radio-receiving circuit with diode detector. The different stages serve the functions indicated. (See text for description.)

ing sound wave. The *RC* time constant must be carefully selected. If it is too long, the circuit cannot detect the next carrier peak.

The demodulated af signal is then amplified and converted back into sound by a speaker.

Woofers and Tweeters

For you stereo buffs, we might say a word about speakers. To provide optimum response for different audio frequency bands, three general types of speakers are used. Low-frequency speakers are called **woofers,** and high-frequency speakers are called **tweeters.** There are also intermediate speakers for intermediate frequencies.

Special crossover networks are used in speaker combinations so that signals in a specific frequency range are channeled to the proper speaker. A simple crossover network is illustrated in Figure 26.18.

Notice the woofer (low f) is connected through an inductance L, and the tweeter (high f) is connected through a capacitance C. For high-frequency signals, the reactance of the coil, $X_L = 2\pi f L$, is large, and the signals go to the tweeter through the capacitance, which has

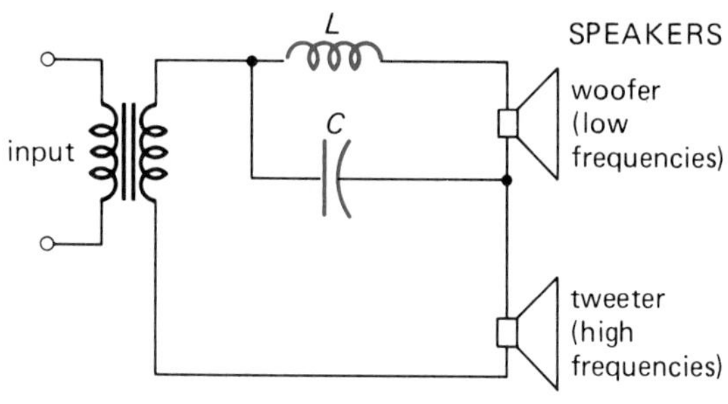

Figure 26.18
Simplified crossover speaker network. The high and low frequencies "cross over" between woofer and tweeter speakers. See text for description.

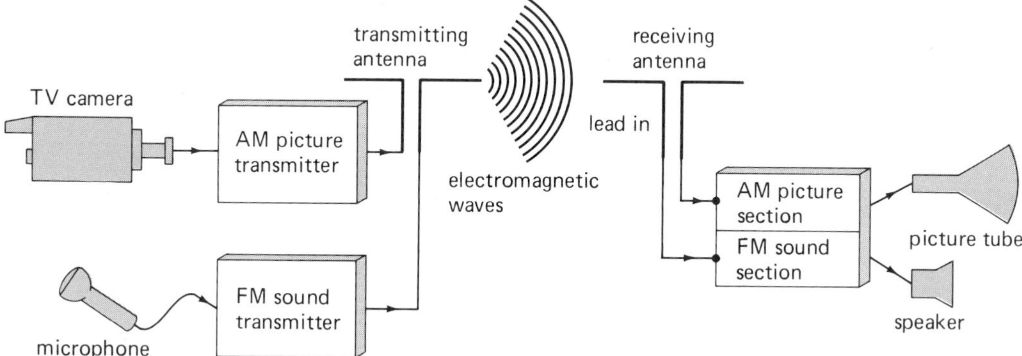

Figure 26.19
Simplified diagram of TV transmission and reception. TV combines both a picture AM signal and an audio FM signal.

a low reactance for high frequencies, $X_C = 1/2\pi fC$.

By similar reasoning, low-frequency signals are channeled to the woofer. (The principles of microphones and [loud]speakers are considered in the Problem Section.)

Television

Television (TV) combines both sight and sound, using an AM picture signal and an FM audio signal. Detailed analysis of the various TV circuits is quite complicated. We consider here only some of the basic components. A basic diagram of TV transmission and reception is shown in Figure 26.19.

The picture signal in the TV camera is formed by means of an **image-orthicon tube** (Fig. 26.20). Light from objects activates an array of photosensitive cells, which release electrons. These are attracted to a target screen. As a result, there is an "electron image" of the objects on the screen. This is scanned by an electron beam, which converts the image information to electrical signals.

These signals are transmitted, along with the accompanying FM sound, to the receiver. The picture tube of the receiving set is a cathode ray tube (CRT). The electron beam of the tube scans the fluorescent screen in a 525-line pattern in a fraction of a second (Fig. 26.21). The signals from the television camera reproduce the camera electron image on the screen in a mosaic picture of light and dark spots, and a black-and-white picture is formed.

Color television is somewhat more involved. With filters the color camera separates the light into three beams, and correspondingly most color picture tubes have three electron beams, one for each of the primary additive colors (Fig.

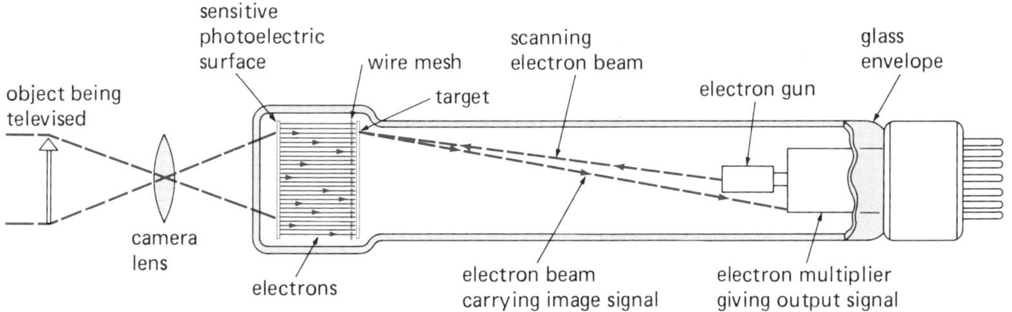

Figure 26.20
An image orthicon tube. This tube is the basic element of a television camera. (See text for description.)

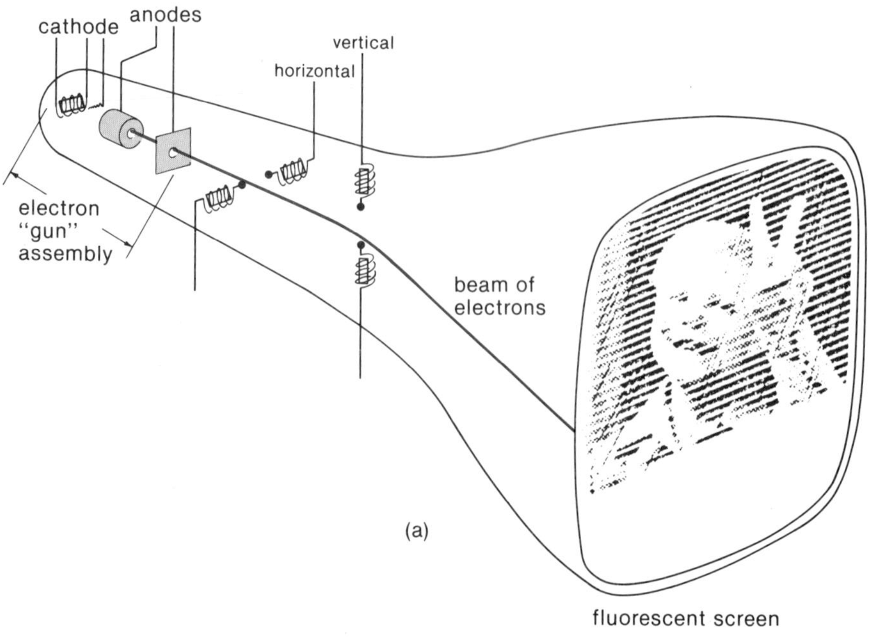

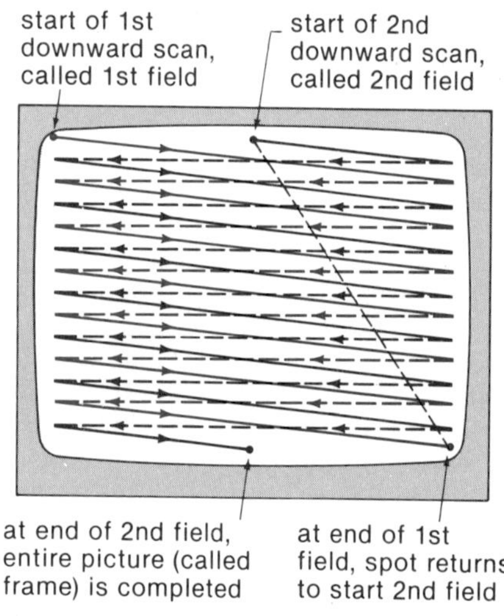

Figure 26.21
The television receiver. (a) Diagram of a CRT (cathode ray tube). Signals to the different deflecting coils control the electron beam on the screen. (b) The beam scans the tube screen in hundreds of horizontal lines in a fraction of a second.

26.22). In the United States, we have a compatible color television system. Compatibility means that a color TV receiver can reproduce both color and black-and-white programs. Of course, black-and-white programs are in black and white on a color set. Black-and-white sets can reproduce color programs without adjustment, but in black and white.

In the three-beam picture tube, the beams are directed through a shadow mask onto the screen. On the screen are thousands of phosphor dots arranged in triads. Each triad has one dot for each of the three primary colors (Fig. 26.23). Each dot glows with its specific color when hit by an electron beam. The emissions from the three dots of the triad cannot be distin-

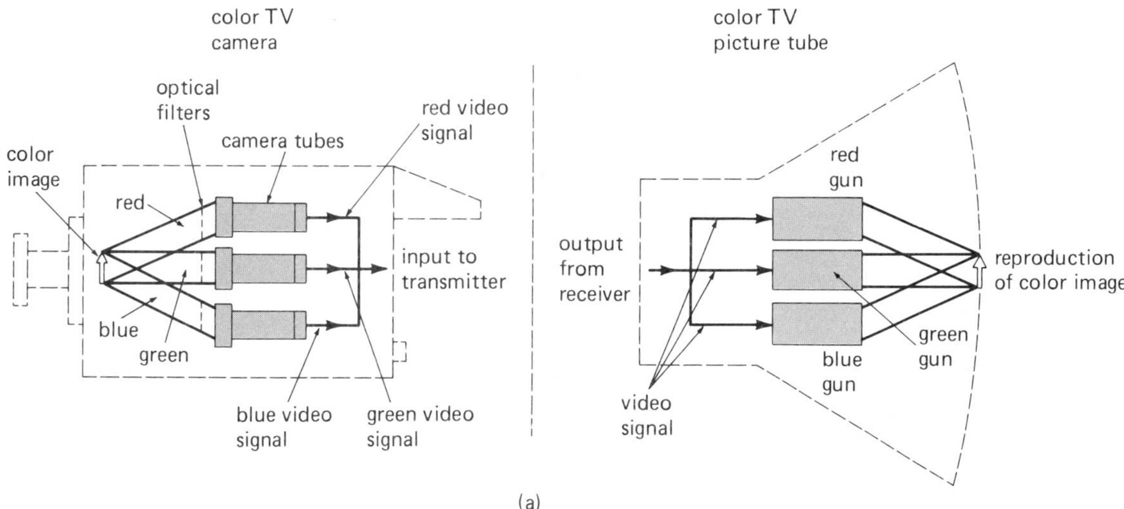

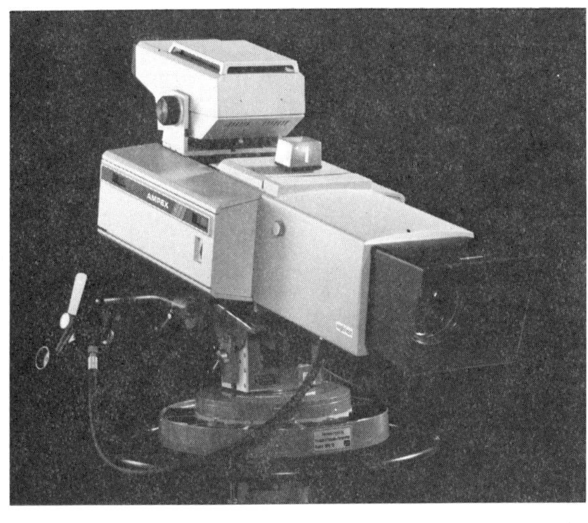
(b)

Figure 26.22
Elements of color television. (a) In the color camera, the incoming light beam is separated into three light beams of primary colors. In the reception set, the beams are recombined to produce a colored image. (b) An actual television camera. (Courtesy Ampex Corporation.)

guished separately at normal viewing distances. The eye sees the light from these primary color sources as a single color, which depends on the color-additive properties of the light components (see Chapter 29).

For either black-and-white or color picture tubes, high voltages on the order of thousands of volts (up to 20,000 volts) are used on the picture screen anode of the tube. The deceleration of electrons that have accelerated through such high potentials can produce X-rays, particularly in color picture tubes, which must be properly shielded. Even when the set is shut off, a dangerous potential can remain because of capacitive effects. Extreme caution should be used when working on a TV set. In most cases, repairs should be left to an experienced technician.

Some color sets use a tube with a single electron beam. This is known as a chromatron tube. The chromatron tube uses the principle of sequential dot display. Sets of primary-color phosphor dots are excited sequentially in rapid succession by the single beam. The beam is modulated so that each color dot contributes to producing the right color.

Television channels are divided into two groups: VHF (very high frequency, Channels 2 through 13) and UHF (ultra high frequency, Channels 14 through 83). The TV channels have a relatively wide band width of 6 MHz since they must carry more information than radio.

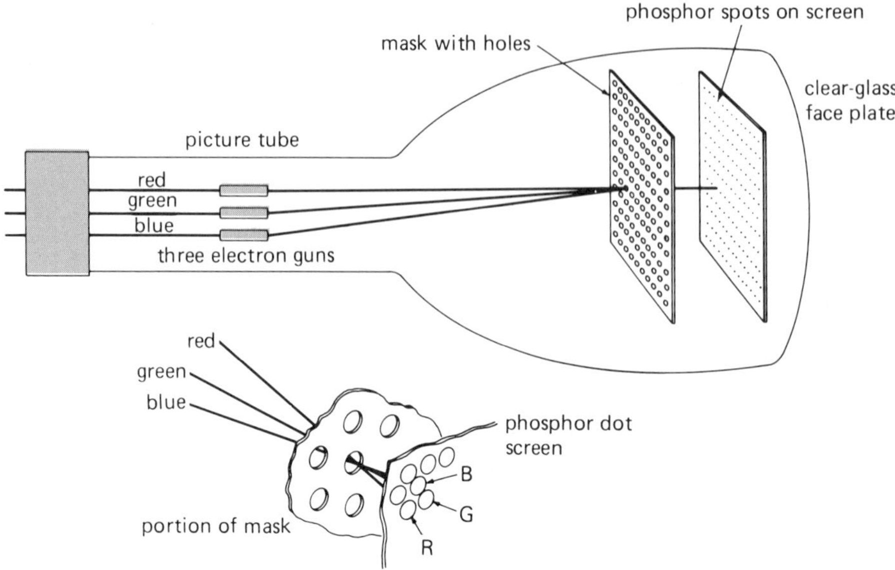

Figure 26.23
Color reproduction. In many color TV receivers, the color beams are directed though a screen mask and excite phosphor-dot triads on the screen.

Important Terms

Edison effect the flow of electrons in a lamp bulb from the filament to a positive plate installed in the bulb

thermionic emission the release, or "boiling off," of electrons from a hot tube filament

diode an electronic element with two electrodes, a positive anode and a negative cathode

saturation the condition of maximum current flow in an electronic element

rectification the changing of ac to dc

triode an electronic element with three electrodes, one of which (the grid or base) is used to control the current flow

grid the control element in a tube triode

cutoff voltage the negative bias grid voltage which gives zero plate current

doping impurity atoms added to an insulator to make it a semiconductor

donor impurity atoms that contribute conducting electrons to a material

n-type semiconductor a semiconducting material in which the charge carriers are negative electrons

acceptor impurity atoms that create holes or vacancies in a material

p-type semiconductor a semiconducting material in which the charge carriers are considered to be positive "holes"

junction diode a diode made of n- and p-type semiconductors

bias voltage applied to an electronic element

forward bias applied voltage that enhances current flow

reverse bias applied voltage that retards current flow

transistor a solid state triode with an emitter, a collector, and a base (corresponding to a vacuum tube filament, plate, and grid)

printed circuits circuits "printed" or etched on copper-clad glass cloth board

integrated circuits (IC's) miniature circuits on semiconductor chips

modulation the superposition of a voice or sound wave on a radio carrier wave, by either amplitude modulation (AM) or frequency modulation (FM)

image-orthicon tube the image-forming tube of a TV camera

Questions

The Electron Tube

1. What is the Edison effect, and how is it related to thermionic emission?
2. What are some good thermionic emitters? Comment on carbon and tungsten.
3. Why does an electron tube have a partial vacuum?

4. What are the "-odes" of a diode?
5. What is the saturation point of a vacuum tube?
6. Explain how a diode acts as a "valve" and how it rectifies.
7. Distinguish between half-wave and full-wave rectification.
8. How does a triode amplify? Can it also rectify?
9. What is meant by the cutoff voltage of a triode?

Solid State Diodes and Transistors

10. Explain what is meant by "doping."
11. What is a "hole," and how can it be considered to be a charge carrier?
12. When a solid state diode is biased in the forward direction, is the p-type material positive or negative? Why?
13. When a solid state diode is conducting, which way do the holes flow across the junction?
14. If a piece of an n-type or a p-type semiconductor were placed in a battery circuit, would there be conduction in each case? Explain. What if the polarity were reversed?
15. What are some advantages of solid state diodes and transistors over vacuum tubes?
16. Explain how a transistor is analogous to a triode vacuum tube in components and operation.
17. Why is a transistor like two diodes "back to back"? Would "head to head" be a better description?
18. How are printed circuits "printed" or made?
19. What are IC's, and how are they made?

Radio and Television

20. How are radio waves polarized?
21. Explain how radio waves in the kHz and MHz ranges can carry sound.
22. What are the band widths of (a) an AM radio station and (b) an FM radio station? Why is there a difference?
23. When you tune in a station on the radio, what are you actually doing?
24. What is the energy conversion associated with (a) microphones and (b) speakers?
25. A dog is a woofer and a bird is a tweeter. Is there any correlation between the sounds of these animals and woofer and tweeter speakers? What is the function of a crossover network?
26. There can be several AM radio stations in one area. Theoretically, how many different 10-kHz-wide channels, or radio stations, can the AM frequency band accommodate in one area?
27. FM radio channels are 200 kHz wide. How many FM stations could the FM frequency band theoretically accommodate in one area?
28. Explain how a picture is produced on a TV screen.
29. What is a compatible color TV system?
30. What would be the result if all of the dots in a color triad in a TV set were equally excited?
31. Why is there danger of X-rays from TV sets?
32. Distinguish between VHF and UHF television.

Problems

1. Draw a diode half-wave rectifier circuit using a solid state diode.
2. It is sometimes desired to have a negative output (with respect to ground) from a diode half-wave rectifier circuit. Such a negative rectifier is called an inverted diode. Draw an inverted diode circuit.
3. Instead of using two diode tubes for full-wave rectification as in Figure 26.4, a single duodiode tube (one with two anodes and a single cathode filament) may be used. Draw such a duodiode symbol in a circuit (such as that in Fig. 26.4) that would give full-wave rectification.
4. Draw a circuit with two solid state diodes that gives full-wave rectification using a center-tapped transformer.
5. A technician in an electronics shop must build a bridge rectifier but finds that he is out of solid state diodes. However, some old vacuum tube diodes are available. Draw a bridge rectifier circuit with an input transformer showing how the technician should connect the tubes.

A microphone converts sound energy (waves) into electrical energy (signals or impulses). Problems 6 and 7 are concerned with two types of microphones.

6. A diagram of a carbon button microphone is shown in Figure 26.24. A button contains loosely packed carbon granules. Explain how the microphone converts sound waves into electrical signals.

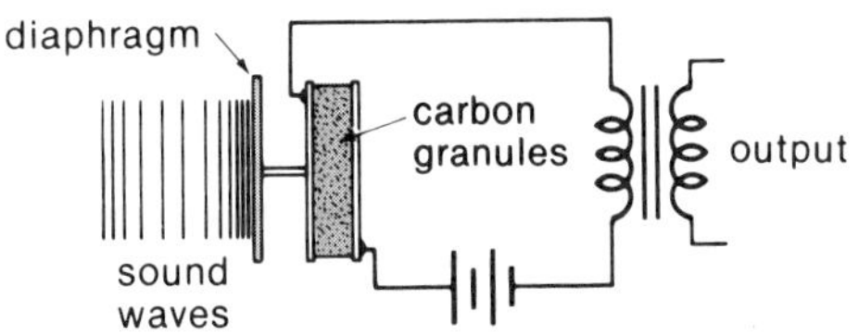

Figure 26.24
Carbon button microphone. See Problem 6.

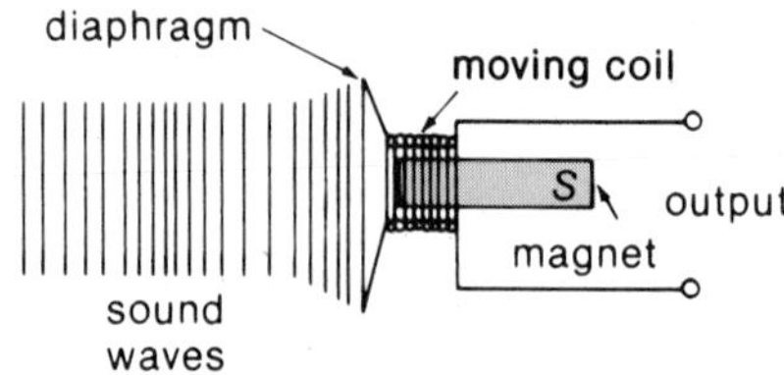

Figure 26.25
Dynamic microphone. See Problem 7.

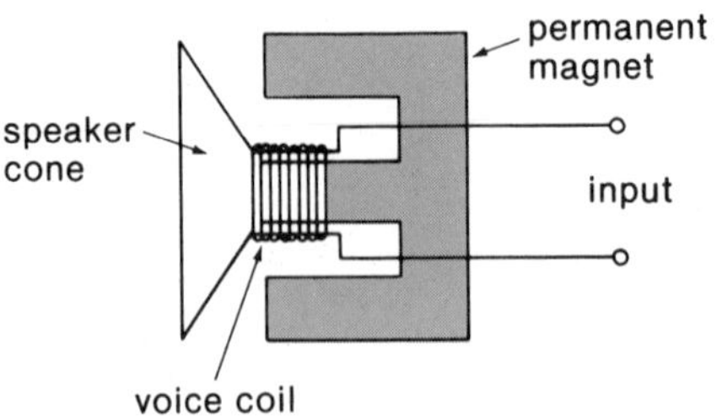

Figure 26.26
Speaker for sound reproduction. See Problem 9.

7. A dynamic microphone uses a moving coil in a fixed magnetic field (Fig. 26.25). Explain how the microphone converts sound waves into electrical signals. Why is no voltage source (e.g., a battery) required for the operation of the microphone?
8. An electrical signal in the form of a sine wave with constant amplitude is fed into a diode detector circuit. Sketch the form of the output.

A speaker converts electrical energy (signals or impulses) into sound energy (waves). Problems 9 and 10 are concerned with two types of speakers.

9. A diagram of a permanent magnet speaker is shown in Figure 26.26. The voice coil is attached to the speaker cone (diaphragm). Explain how the speaker converts electrical signals into sound waves. (An electrodynamic speaker uses an electromagnet instead of a permanent magnet.)
10. Headphones and earphones are used for private listening to stereos and radios. A diagram of a typical headphone speaker is shown in Figure 26.27. It contains a soft iron diaphragm and a permanent magnet. Explain the principle of operation.

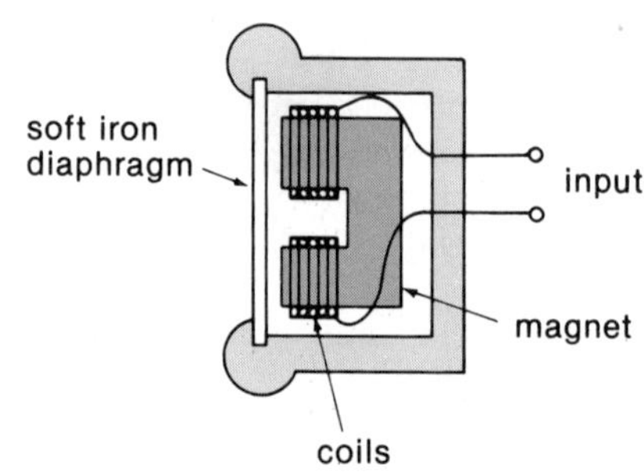

Figure 26.27
Another type speaker. See Problem 10.

(Courtesy General Electric Co.)

Chapter 27

Light and Illumination

Light is to the eye as sound is to the ear. We need proper illumination to see and obtain visual information. Automobiles are equipped with headlights and various other lights to provide illumination and to convey information—for example, turn signals and tail lights. Proper lighting is important in the classroom, in the home, and in industry.

Also like sound, light is a disturbance; hence, it may be characterized as a wave. But light is distinctly different from the waves we have studied so far. No medium is required for the propagation of light. It will travel through a vacuum. For example, we receive light from the Sun through the void (vacuum) of space.

As *acoustics* refers to the general study of sound and hearing, *optics* refers to the general study of light and vision. The physical properties of light provide for many technical applications. In this chapter some of the important properties of light are discussed.

27.1 The Nature of Light

In the classical sense, light is an electromagnetic wave.* Light waves are characterized by electric and magnetic fields (Fig. 27.1). The electric field vector E and magnetic field vector B are perpendicular to each other and are perpendicular to the direction of propagation. Hence, light is a transverse wave.

Electromagnetic radiation spans a wide range of frequencies (Fig. 27.2), but light is generally taken to mean electromagnetic waves in or near the visible region. For example, radio waves are electromagnetic waves but are not generally thought of as light. To produce radio waves, electrons are oscillated back and forth in metal antennas hundreds of thousands of times a second. The oscillation frequencies are controlled by crystals—e.g., quartz crystals—which vibrate naturally at radio frequencies. The resonant frequencies depend on the elasticity of the crystal and on its size and shape.

In order for light to radiate from an antenna, the electrons would have to oscillate many billions or trillions of times each second. This is

* In later chapters you will learn that in some instances light must be characterized as a particle to explain certain phenomena, giving rise to the "dual nature" of light.

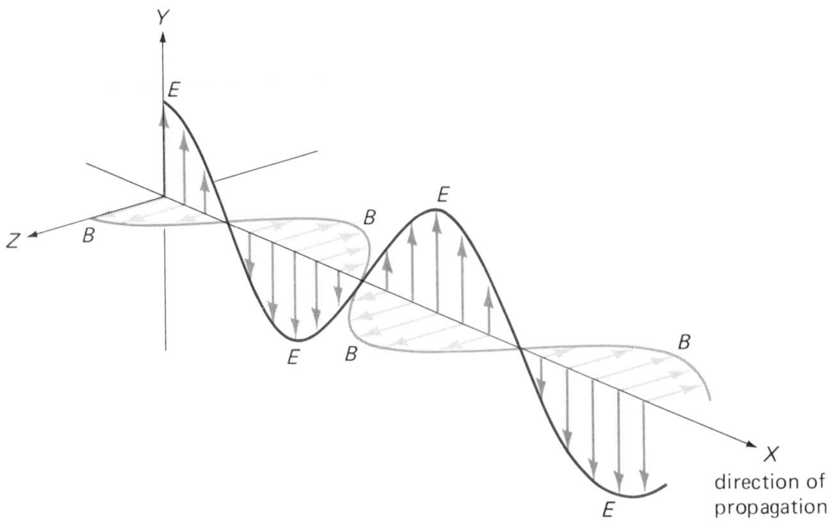

Figure 27.1
Light is an electromagnetic wave. The electric field vector E and the magnetic field vector B are perpendicular to each other and are perpendicular to the direction of propagation.

beyond the range of crystal vibrations in terms of both size and elasticity. The "antennas" of visible light emissions are atoms themselves. In our simplified model of the atom (Chapter 19), we speak of transitions rather than vibrations. When atoms become excited, electrons are raised to higher orbits, or energy levels, from the nucleus. The electrons generally remain in the higher energy levels for only fractions of seconds and then spontaneously return to their normal orbits. When they do so, energy is emitted as light or electromagnetic radiation, with

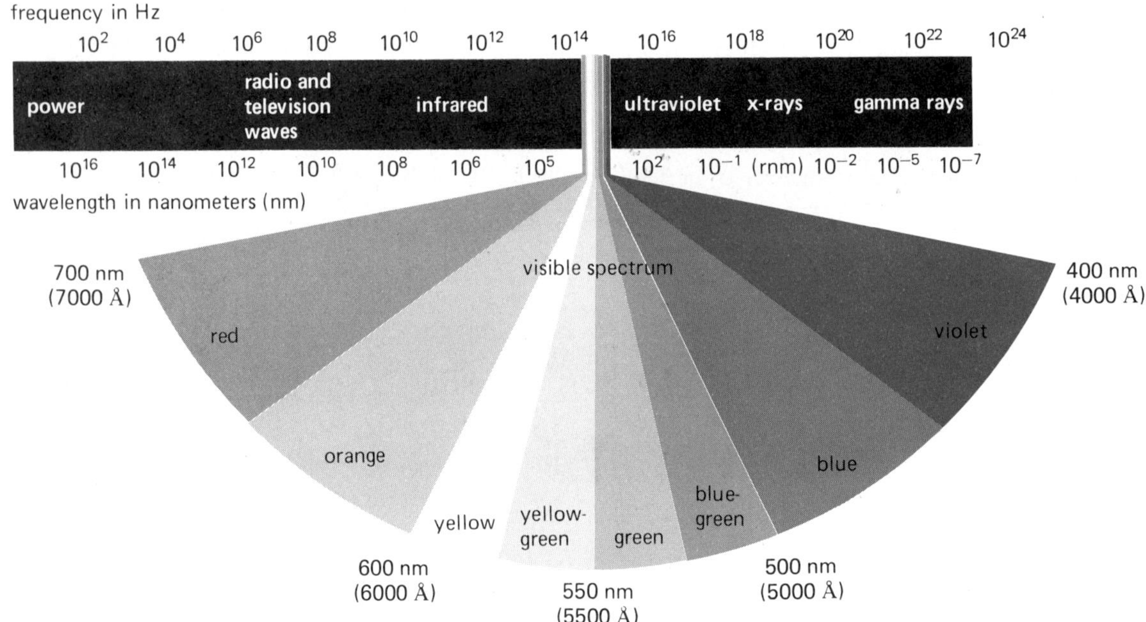

Figure 27.2
The electromagnetic spectrum. The visible region comprises wavelengths between approximately 400 and 700 nm (or 4000 and 7000 Å).

different frequencies of light corresponding to transitions between different energy levels or orbits. This concept will be discussed in more detail in Chapter 30.

Similar to the acoustical frequency spectrum, different regions of the electromagnetic (EM) spectrum are given different names (Fig. 27.2). Notice that the visible region is only a small part of the EM spectrum. Above and below the visible region are the ultraviolet and infrared regions, respectively. The radiations in these and other regions cannot be perceived by the human eye.

In the visible region, light of different frequencies (or wavelengths) is perceived by the eye as different colors. Color is the physiological frequency response of the eye, analogous to the physiological frequency response of the ear to the pitch of sound waves. More will be said about color in Chapter 29.

The frequency and wavelength of light are related by the equation

$$\boxed{\lambda f = c} \qquad \textbf{(Eq. 27.1)}$$

where c in this case is the speed of light. In vacuum, the speed of light is

$$c = 3 \times 10^8 \text{ m/s} = 3 \times 10^{10} \text{ cm/s}$$

(speed of light in vacuum)

All electromagnetic radiations travel at this speed in vacuum or in "free space," which is approximately 186,000 mi/s. The speed of light is different in different media, as we shall learn.

EXAMPLE 27.1 Light traveling in vacuum has a frequency of 6×10^{14} Hz. What is the wavelength of the light?

Solution: With $f = 6 \times 10^{14}$ Hz (or s^{-1}) and $c = 3 \times 10^{10}$ cm/s,

$$\lambda = \frac{c}{f} = \frac{3 \times 10^{10} \text{ cm/s}}{6 \times 10^{14}\text{/s}} = 5 \times 10^{-5} \text{ cm}$$

Notice from Figure 27.2 that this light is in the blue-green portion of the visible spectrum.

The example shows that light waves have relatively short wavelengths. A small length unit sometimes used to express the wavelengths of electromagnetic radiation is the **angstrom***:

$$1 \text{ angstrom (Å)} = 10^{-10} \text{ m} = 10^{-8} \text{ cm}$$

Hence, the wavelength of the light in the previous example, in angstroms, is

$$\lambda = 5 \times 10^{-5} \text{ cm} \left(\frac{1 \text{ Å}}{10^{-8} \text{ cm}}\right) = 5000 \text{ Å}$$

Another appropriate, commonly used small unit is the nanometer (nm; 1 nm = 10^{-9} m = 10^{-7} cm). In this case,

$$\lambda = 5 \times 10^{-5} \text{ cm} \left(\frac{1 \text{ nm}}{10^{-7} \text{ cm}}\right) = 500 \text{ nm}$$

Light of a single frequency or wavelength is called **monochromatic light** (monochromatic—one color). Most of the light we experience is polychromatic, or composed of light of several frequencies. Light having all the frequencies of the visible spectrum is called "white" light. Sunlight is an example of white light.

Since light is an electromagnetic wave, its wave properties can be used to explain various phenomena. The study of phenomena using the wave nature of light is called **physical optics** or **wave optics.**

When a beam of white light is incident on a glass prism (Fig. 27.3), the emergent beam is spread out in a continuous band, or spectrum, of colors. Isaac Newton got this result from an experiment using sunlight. The phenomenon is explained by the wave nature of light. It arises because the speed of light for different wavelengths is slightly different in transparent media (see Chapter 28). The separation of light into its component wavelengths (colors) is called **dispersion.**

In vacuum, all the wavelength components of white light have the same speed. However, in media such as the glass of a prism, light with a longer wavelength travels faster than light with a shorter wavelength (e.g., red light travels faster than blue light). As a result, a white beam of light is spread out into a band of colors.

The wave nature of light can be demonstrated by several experiments, which are also used to measure the wavelength of light.

* Named after a Swedish physicist, Anders Angström.

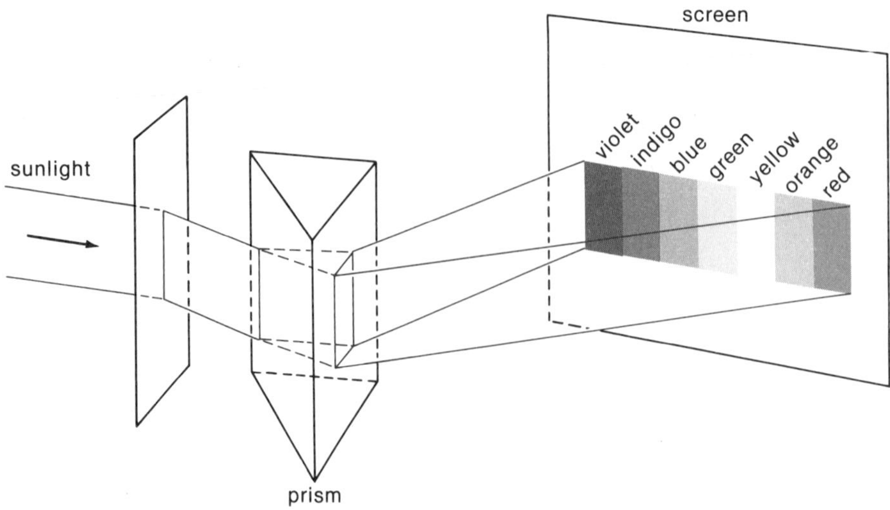

Figure 27.3
Prism dispersion. Isaac Newton showed that a prism separates a beam of sunlight into its component colors and produces a continuous spectrum.

27.2 Interference

As learned in Chapter 13, waves interfere with each other, and we refer to constructive and destructive interference, as determined by the principle of superposition. Recall that to have complete constructive interference, the waves must have the same frequency and be exactly in phase, i.e., have a phase difference of zero. To have complete destructive interference, the waves must be exactly out of phase, or have a phase difference of 180°.

Consider waves of the same frequency from two point sources, as illustrated in Figure 27.4. If the waves arrive at point P in phase, they interfere constructively. Similarly, if they arrive exactly out of phase, the waves interfere destructively.

Such sinusoidal waves in space can be described by the equation

$$y = A \sin\left(\frac{2\pi x}{\lambda} + \phi\right) \qquad \textbf{(Eq. 27.2)}$$

Plotting y versus x gives a sinusoidal curve, similar to plotting y versus t for $y = A \sin(2\pi t/T + \phi)$ in Chapter 13. The graph of Equation 27.2 essentially gives the spatial form of the wave at a given instant—like taking a photograph of the wave. The ϕ in the equation is again a phase constant that determines whether the wave has a sine or cosine form.

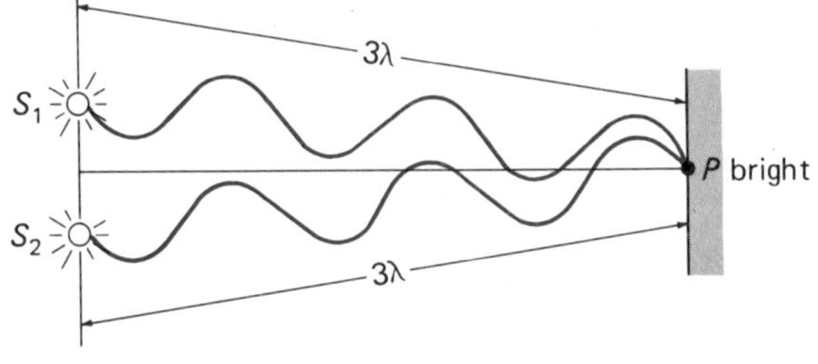

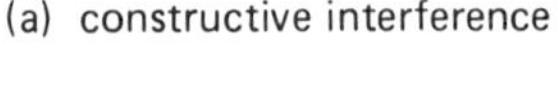

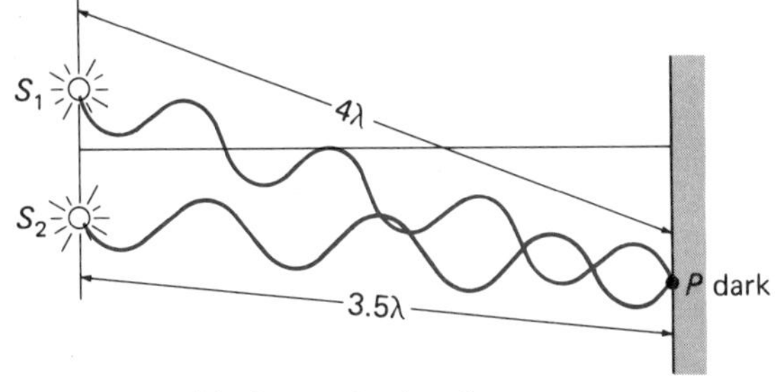

Figure 27.4
Interference. (a) Total constructive interference occurs when the waves arrive at a point exactly in phase. In the figure, the path difference of the waves from the two point sources is zero, $PD = 3\lambda - 3\lambda$, and the waves arrive exactly in phase. (b) Total destructive interference occurs when the waves arrive at a point completely out of phase. In the figure, the path difference is $PD = 4\lambda - 3.5\lambda = \lambda/2$, and the phase difference is $\Delta\theta = \pi$ rad or completely out of phase.

The waves in Figure 27.4 have the form*

$$y = A \sin \frac{2\pi x}{\lambda} \qquad \textbf{(Eq. 27.3)}$$

where $\phi = 0$.

The phase angle θ of the wave is $\theta = 2\pi x/\lambda$ (where $\sin \theta = \sin 2\pi x/\lambda$). Hence, the phase difference $\Delta\theta$ of two waves of the same frequency (or wavelength) is given by

$$\boxed{\Delta\theta = \frac{2\pi \Delta x}{\lambda}} \qquad \textbf{(Eq. 27.4)}$$

(phase difference)

This shows that the phase difference can be expressed in terms of the path difference $\Delta x = x_2 - x_1$. The path difference is simply the difference in the path lengths traveled by the waves.

It is convenient to express the path lengths in wavelength units. For example, in Figure 27.4(a), the path difference is zero, $\Delta x = 3\lambda - 3\lambda = 0$, and hence the phase difference is zero. If the path lengths from the sources to point P were 6λ and 5λ, respectively, then the path difference would be $\Delta x = 6\lambda - 5\lambda = \lambda$. The phase difference would then be $\Delta\theta = 2\pi\Delta x/\lambda = 2\pi\lambda/\lambda = 2\pi$.

With a little thought, you should be able to convince yourself that the waves arrive at P exactly in phase and that

Total constructive interference occurs if the path difference is zero or an integral number of wavelengths.

Conditions for total constructive interference are

Phase difference (rad)	Path difference
$\Delta\theta = 0, 2\pi, 4\pi, 6\pi, \ldots$	when $\Delta x = 0, \lambda, 2\lambda, 3\lambda, \ldots$,

respectively. Here, $\Delta\theta = 4\pi$ $(= 2\pi + 2\pi)$ rad is equivalent to $\Delta\theta = 0$, since 2π rad is a complete cycle, and so on.

For the case in Figure 27.4(b), the path difference is $\Delta x = 4\lambda - 3.5\lambda = 0.5\lambda = \lambda/2$. The phase difference is then $\Delta\theta = 2\pi\Delta x/\lambda = 2\pi(\lambda/2)/\lambda = \pi$ rad (or 180°), and the waves are completely out of phase. By reasoning as before, we can see that the waves arrive at P exactly out of phase, and

Total destructive interference occurs if the path difference is an odd integral number of half wavelengths.

Conditions for total destructive interference are

Phase difference (rad)	Path difference
$\Delta\theta = \pi, 3\pi, 5\pi, \ldots$	when $\Delta x = \lambda/2, 3\lambda/2, 5\lambda/2, \ldots$,

respectively. Here, $\Delta\theta = 3\pi$ $(= \pi + 2\pi)$ rad is equivalent to $\Delta\theta = \pi$ rad (or 180°), and so on.

EXAMPLE 27.2 What is the phase difference for two waves of the same frequency arriving at a point P with the respective path lengths of 6λ and 4.5λ?

Solution: The path difference of the waves is

$$\Delta x = 6\lambda - 4.5\lambda = 1.5\lambda = \frac{3\lambda}{2}$$

Hence, the waves are out of phase by a phase difference of

$$\Delta\theta = \frac{2\pi\Delta x}{\lambda} = \frac{2\pi(3\lambda/2)}{\lambda} = 3\pi$$

and total destructive interference occurs.

When light waves interfere constructively at a point, a bright stationary interference pattern will be formed on a screen placed at the point, as indicated in Figure 27.4(a). Similarly, when light waves interfere destructively, a dark stationary interference pattern is formed on the screen [Fig. 27.4(b)]. *Stationary* refers to the constant intensity of the interference pattern; that is, it does not vary with time.

Stationary interference patterns require that the phase differences of the interfering waves be constant in time. If waves have the same frequency and their phase differences are constant in time, they are said to be *coherent*. Light waves from ordinary sources may have the same

* This is sometimes written $y = A \sin kx$, where $k = 2\pi/\lambda$ is called the *wave number*. Note the analogy to

$$y = A \sin 2\pi t/T = A \sin \omega t.$$

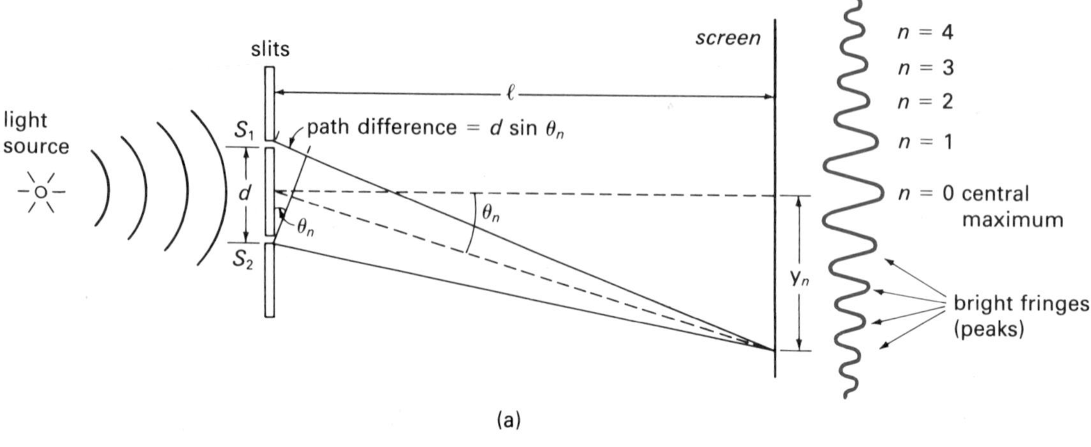

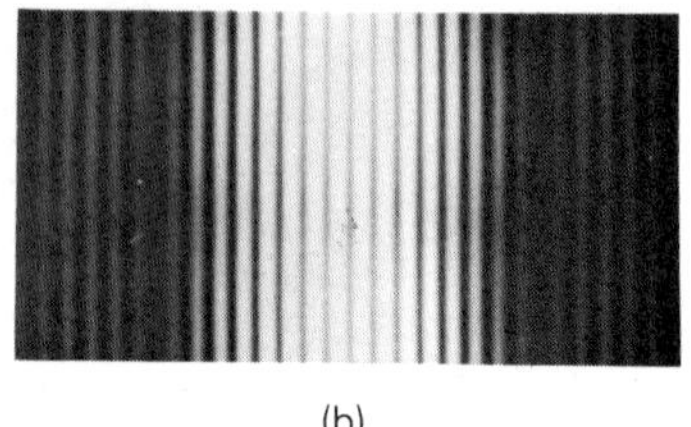

Figure 27.5

Young's interference experiment. Light from two slits interferes constructively and destructively on a screen (a), producing alternate bright and dark fringes, as shown in the photo (b). By analyzing the geometry, the wavelength of the light can be computed. See text for description.

frequency, but their phase "constants" (ϕ), and hence their phase differences, vary with time. Light waves brought together from two different light sources are almost never coherent and will not produce stationary interference patterns. (An exception is laser light, which is discussed in Chapter 30.)

Young's Interference Experiment

One of the earliest demonstrations of the interference of light was performed in 1801 by Thomas Young, an English scientist. It is often referred to as **Young's double-slit experiment** because Young used light from a single source passing through two small slits to obtain coherent light waves. The two slits act as "sources," and since they "emit" the same light, the waves from the slits are coherent (Fig. 27.5). The waves interfere, and an interference pattern of alternate bright and dark fringes is observed on a screen.

The path difference between the two light rays from the sources is $d \sin \theta$, where d is the distance between the slits. Hence, constructive interference will occur at locations on the screen when

$$d \sin \theta = n\lambda \qquad n = 0, 1, 2, 3, \ldots \qquad \textbf{(Eq. 27.5)}$$

Applied to the interference pattern, $n = 0$ corresponds to the central maximum, $n = 1$ to the first bright fringe (called the first-order fringe) on either side of the central maximum, and so on.

From the geometry of the experiment, a particular bright fringe at point P located a distance y_n from the center of the central maximum is

$$y_n = L \tan \theta = L \frac{\sin \theta}{\cos \theta}$$

The angle θ is normally quite small, so $\cos \theta \approx 1$, and to a good approximation

$$y_n = L \sin \theta \qquad \textbf{(Eq. 27.6)}$$

Combining Equations 27.5 and 27.6, we have

$$\boxed{\lambda = \frac{y_n d}{nL} \qquad n = 1, 2, 3, \ldots} \qquad \textbf{(Eq. 27.7)}$$

Hence, Young's experiment provides a method of measuring the wavelength of light from the geometry of the experiment.

EXAMPLE 27.3 A monochromatic light source is used in Young's double-slit experiment where the distance between the slits is 0.05 mm and the screen is at a distance of 1.5 m from the slits. The center of the second-order bright fringe is located 3.0 mm from the center of the central maximum on the screen. What is the wavelength of the light?

Solution: With $d = 0.05$ mm, $L = 1.5$ m $= 1.5 \times 10^3$ mm, and $n = 2$ with $y = 3.0$ mm,

$$\lambda = \frac{y_n d}{nL} = \frac{(3.0 \text{ mm})(0.05 \text{ mm})}{2(1.5 \times 10^3 \text{ mm})}$$

$$= 0.05 \times 10^{-3} \text{ mm}$$

$$= 5.0 \times 10^{-5} \text{ cm } (= 500 \text{ nm})$$

The light appears blue-green since the wavelength is in that region of the visible spectrum.

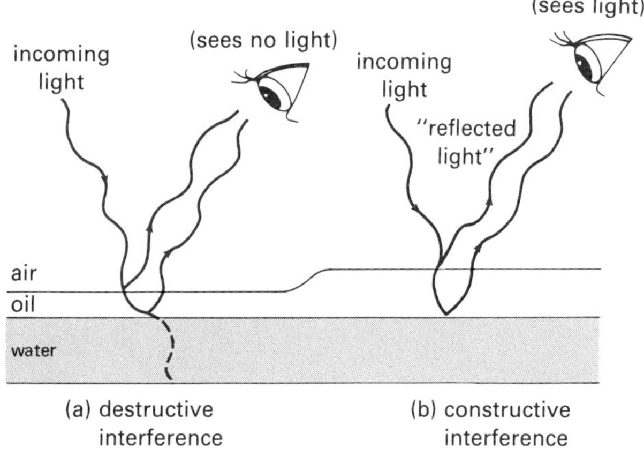

Figure 27.6

Thin film interference. Reflections from the top and bottom surfaces of the oil film give rise to interference. (a) If the film thickness is such that the reflective waves are completely out of phase, total destructive interference occurs, and no light is seen. (The light is transmitted, not destroyed.) (b) With film thicknesses for which in-phase constructive interference occurs, light is seen (reflected).

Thin Film Interference

Have you ever noticed the multicolored patterns on an oil or gasoline slick on a wet pavement, or in soap bubbles, and wondered what caused them? These are common examples of interference—thin film interference. Even though these films are transparent, there is some reflection at the surfaces, and interference occurs for light reflected from the top and bottom surfaces. When the film thickness is such that the reflected waves are in phase, constructive interference occurs and light is reflected from the film (Fig. 27.6). When the film thickness is such that the reflected waves are out of phase, destructive interference occurs, and the light is transmitted (not destroyed).

The path difference of the reflected waves depends on the angle at which the light strikes the film; however, there is another consideration in this case. Light waves reflected from the surface of an optically denser medium, e.g., an air-film interface, undergo a 180° phase shift. This is analogous to a wave in a stretched string being reflected from a rigid support (Chapter 13). Similarly, waves reflected from a less optically dense medium, e.g., a film-air interface, undergo no phase shift.

For example, suppose light with a wavelength λ is normally incident on a thin film in air (air on both sides as in the case of a soap bubble) and the film thickness is $\frac{1}{4}\lambda$. Then the path difference for the waves reflected from the top (air-film) and the bottom (film-air) surfaces is $\frac{1}{4}\lambda$ down and $\frac{1}{4}\lambda$ back, or $\lambda/2$. It might be expected that destructive interference would occur. However, the wave reflected from the top air-film surface is phase shifted by 180°, and constructive interference occurs in this case (no phase shift at the film-air surface).

But consider the case in Figure 27.6 with the oil film being $\frac{1}{4}\lambda$ thick. Here too the path difference is $\lambda/2$ for normally incident reflected waves. However, the reflected waves from both surfaces (air-oil and oil-water) undergo 180° phase shifts. The reflected waves are then out of phase for this path difference, and destructive interference or transmission occurs. An application of this type of destructive thin film interference is described in Special Feature 27.1.

In any case, when white light falls on a thin film, some wavelengths are reflected and others are transmitted. For a film of relatively uniform thickness, the reflection may be selective for a particular wavelength region of the visible spectrum, and the film is seen to be one particular color. However, oil and soap films generally vary in thickness, and a vivid display of different colors is seen.

SPECIAL FEATURE 27.1

Nonreflecting Lenses

Thin films are used to make nonreflective glass lenses such as those used on cameras and binoculars. Interference properties allow for the reduction of reflected light from the lenses or the greater transmission of light, which is needed for exposing camera film and for binocular viewing. A nonreflective lens has a thin film coating with a thickness of $\frac{1}{4}\lambda$ for a particular light component. This is usually chosen from the yellow-green portion of the visible spectrum (about 550 nm wavelength), for which the eye is most sensitive. Other wavelengths are still reflected and give the glass a somewhat bluish appearance. You have probably noticed this bluish hue in coated lenses for cameras and other optical instruments.

The film coating serves a double purpose. Not only does it promote nonreflection from the front of the lens, but it also cuts down on back reflection (Fig. 1). Some of the light transmitted through the lens is reflected from the back surface. It could be reflected again from the front surface of an uncoated lens and give rise to a poor image on the photographic film in a camera. However, the reflections from the two film surfaces of a coated lens interfere destructively, and there is no reflection.

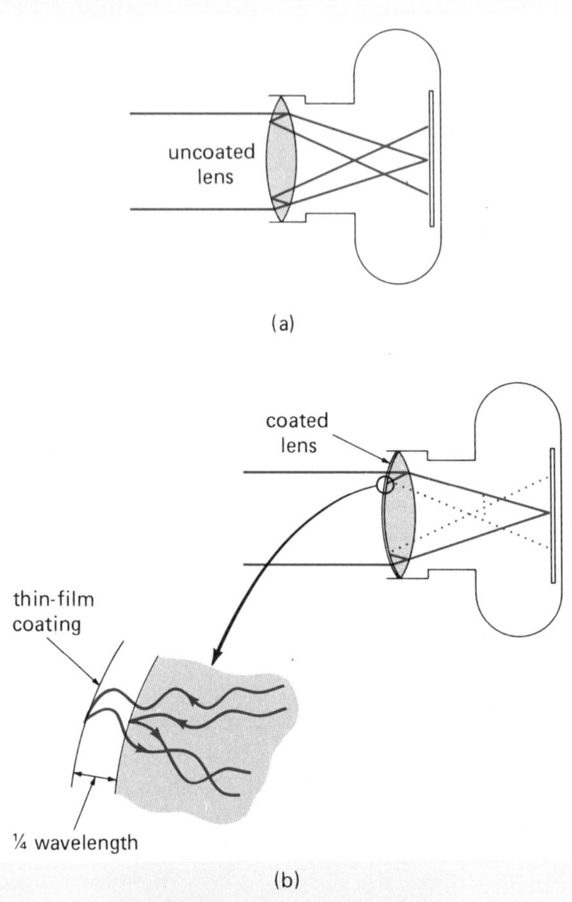

Figure 1
Nonreflective coatings for lenses. (a) Reflection from the interior surface of a lens can give a poor focus or image on the camera film. (b) With a thin film coating, the reflections from the film surfaces interfere destructively, and reduce the reflection.

The principle of thin film interference is used to compare and measure the thicknesses of objects. Glass plates, called **optical flats,** are ground and polished so they are as flat as possible. Test plates can be polished optically flat with a tolerance of approximately 5×10^{-7} cm. If there is a uniform air film, or air "wedge," between the flats, regular interference patterns are observed when the flats are illuminated (Fig. 27.7).

An optical flat can also be used to determine the flatness of a reflecting surface, such as a mirror or metal plate. Irregularities on the surface would give rise to air films of different thicknesses and an irregular interference pattern. Similar techniques can be used to determine irregularities in the shapes of lenses (see the Chapter Supplement).

Figure 27.7
Optical flats. When a uniform air film or wedge exists between the flat glass plates, regular interference fringes occur. They can be used to measure the thickness of objects or to determine whether the reflecting surfaces are optically flat. (Courtesy Edmund Scientific Co.)

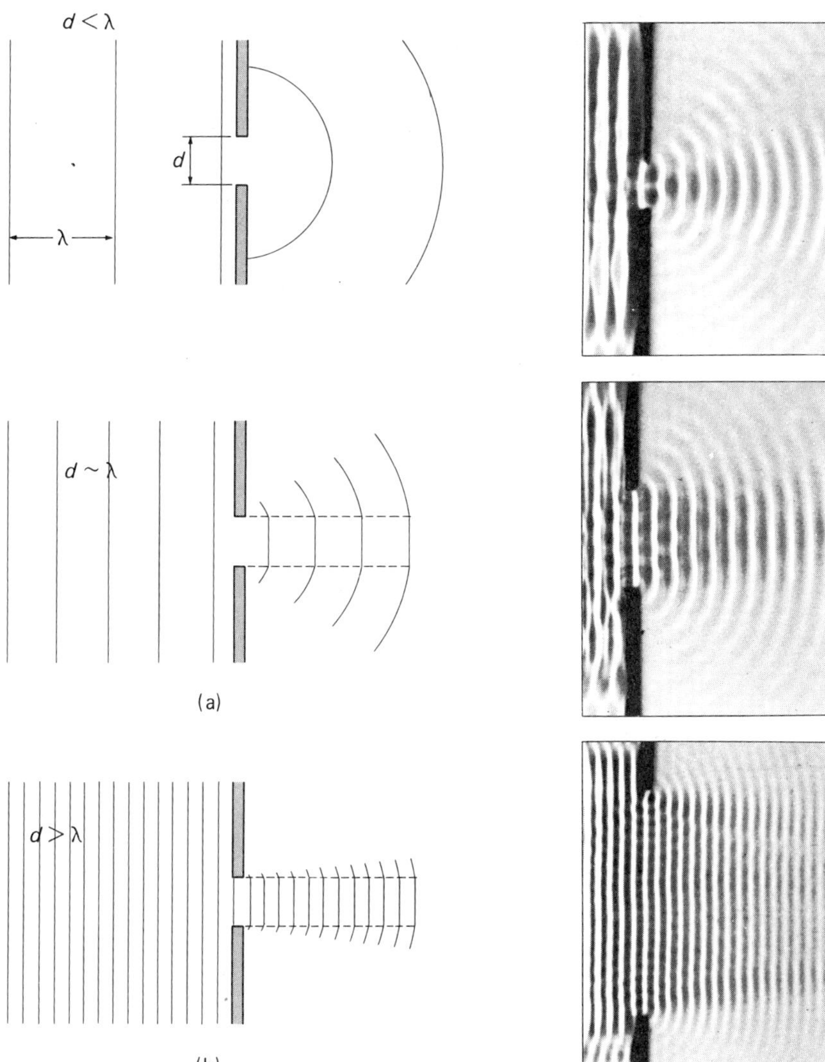

Figure 27.8
Diffraction. Diffraction is the bending of waves around the corner or edge of an opaque object, e.g., the corners of a slit. To have appreciable diffraction, the slit width *d* must be smaller than or on the same order as the wavelength of the light (a). If the slit width is greater than the wavelength, little diffraction occurs (b). The photos show these effects for water waves. (Courtesy Educational Development Center, Newton, MA.)

27.3 Diffraction

Diffraction refers to the bending of waves around the corner or edge of an opaque object. We normally think of the boundary of a shadow as being quite distinct. But on close observation, the boundary is somewhat fuzzy owing to diffraction effects. The diffraction of sound waves is more evident. We can hear sound coming from around a corner; hence, the sound waves must be bent, or diffracted.

It is diffraction that gives the slits in Young's experiment the character of point sources. The amount of diffraction depends on the slit width and wavelength of the light. To have appreciable diffraction, the slit width must be on the same order or less than the wavelength (Fig. 27.8). When the slit width is much less than the

Figure 27.9
Diffraction grating. A master diffraction grating being ruled by a diamond-point scribe. (Courtesy J. Williams, F. Trinklein, and H. Metcalfe.)

wavelength, spherical waves spread out from the slit as though from a point source. Notice in the figure the diffraction around the edge of the slit. If the slit width is greater than the wavelength, little diffraction occurs.

A **diffraction grating** consists of a large number of equally spaced parallel lines—commonly up to 10,000 per centimeter—and comes in two types. A transmission grating has the lines cut on a piece of plastic or glass (Fig. 27.9). The transparent spaces between the grooved lines act as a series of parallel slits. More common is the reflection grating, which may be made by depositing a thin film of aluminum on an optically flat surface, then removing some of the metal in a regularly spaced, parallel line pattern. Precision diffraction gratings are also made by using laser beams. The beams expose a photosensitive material, which is then etched.

It is more convenient to determine the wavelength of light using a diffraction grating because the interference fringes are narrower and more intense than those obtained with a double slit. A diffraction grating extends the interference from a two-slit situation to interference from a large number of slits, or "sources."

The analysis of the diffraction grating is more complicated than that of double-slit diffraction. However, a bright fringe occurs at a point when the light from *all* of the slits interferes constructively. As before, this requires a path difference from adjacent slits, $d \sin \theta$, to be an integral number of wavelengths. Thus, **bright fringes occur at angles θ_n when**

$$d \sin \theta_n = n\lambda \qquad n = 0, 1, 2, 3, \ldots \qquad \textbf{(Eq. 27.8)}$$

Hence, d is the distance between the lines, or the slit width. This is given by the reciprocal of the number of lines N (per inch or centimeter), $d = 1/N$. Equation 27.8 is called the **grating equation.**

EXAMPLE 27.4 A diffraction grating has 7000 lines/cm. When monochromatic light is diffracted, it is found that the first-order fringe is at an angle of 20°. What is the wavelength of the light?

Solution: With $N = 7000$ lines/cm, $d = 1/N = 1/7000 = 1.4 \times 10^{-4}$ cm. Then, for $n = 1$ and $\theta = 20°$,

$$\lambda = \frac{d \sin 20°}{n} = \frac{(1.4 \times 10^{-4}\ \text{cm})(0.34)}{1}$$

$$= 4.8 \times 10^{-5}\ \text{cm}\ (= 480\ \text{nm})$$

Hence, the light appears blue.

Notice that the angle θ depends on the wavelength. If the incoming light consisted of many wavelengths, e.g., white light, each component wavelength would be deviated differently. From Equation 27.8 we can see that the red end of the visible spectrum (long wavelengths) would be diffracted more than the violet end (shorter wavelengths). As a result, the interference fringes would consist of a spectrum of colors.

Of course, the zeroth-order fringe would consist of white light, since when $n = 0$, then $\theta = 0$ for all wavelengths. However, the first-order fringe ($n = 1$) and subsequent fringes consist of spectra that are increasingly dispersed, or spread apart.

Because of this property, diffraction gratings are used in some instruments to select specific wavelengths (Fig. 27.10). When the grating is rotated slightly, the various wavelengths "parade" by a given point. The rotation control can be calibrated in wavelength units, and a narrow wavelength region can be selected. Prisms can also be used, but they disperse light into relatively wide spectra. Instruments used to ob-

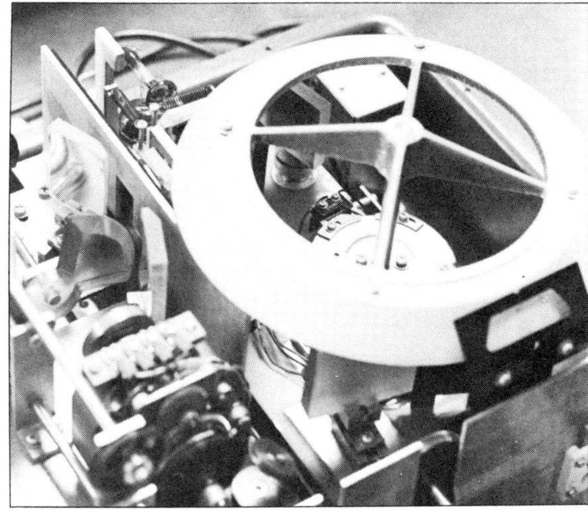

Figure 27.10
Diffraction spectrometer. In this instrument, infrared light enters from the left and strikes the reflection diffraction grating (at top center, see at the left of the second quadrant of the divided circle). Wavelengths of the infrared spectrum that may be absorbed by a sample can be selected by rotating the grating.

serve and analyze spectra are called **spectrometers.**

27.4 Polarization

The transverse nature of light can be demonstrated by polarization. Recall that the electric and magnetic components of electromagnetic waves are mutually perpendicular and that these vectors are perpendicular to the direction of propagation.

In most cases, the electric and magnetic vectors of the waves in a light beam are completely random in their plane of oscillation. The light is then said to be unpolarized. This condition is represented by a random orientation of vectors, as shown in Figure 27.11. By convention, only the electric field vectors are used to illustrate the orientation of the field vectors.

If for some reason there is a partial preferential orientation of the field vectors, the light is said to be **partially polarized.** Similarly, when the field vectors all lie in a plane, the light is said to be **plane polarized** or **linearly polarized.** For example, radio (electromagnetic) waves are generated by accelerating electric charges in an antenna. The electric field of the wave is preferentially oriented parallel to the antenna, and the waves are best received when the receiver antenna is parallel to, or aligned with, the polarized waves. On the other hand, longitudinal waves, such as sound waves, cannot be polarized.

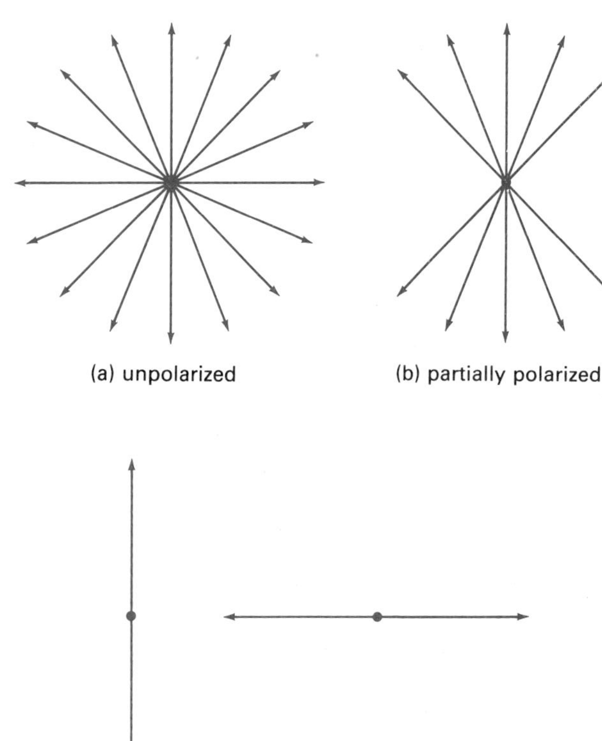

Figure 27.11
Polarization. The polarization of light is represented by the electric field vectors of a wave. (a) Unpolarized (vectors randomly oriented). (b) Partially polarized (preferential orientation). (c) Plane or linearly polarized (vectors in a plane).

The polarization of light can be effected by several means. Three of these are (1) selective absorption by crystals, (2) reflection, and (3) scattering.

1. **Selective absorption.** Certain crystals have the ability to selectively absorb field vector components with certain orientations, and light passing through such crystals is polarized. Around 1930 Edwin Land, an American scientist, found a way to align tiny needle-shaped crystals in sheets of transpar-

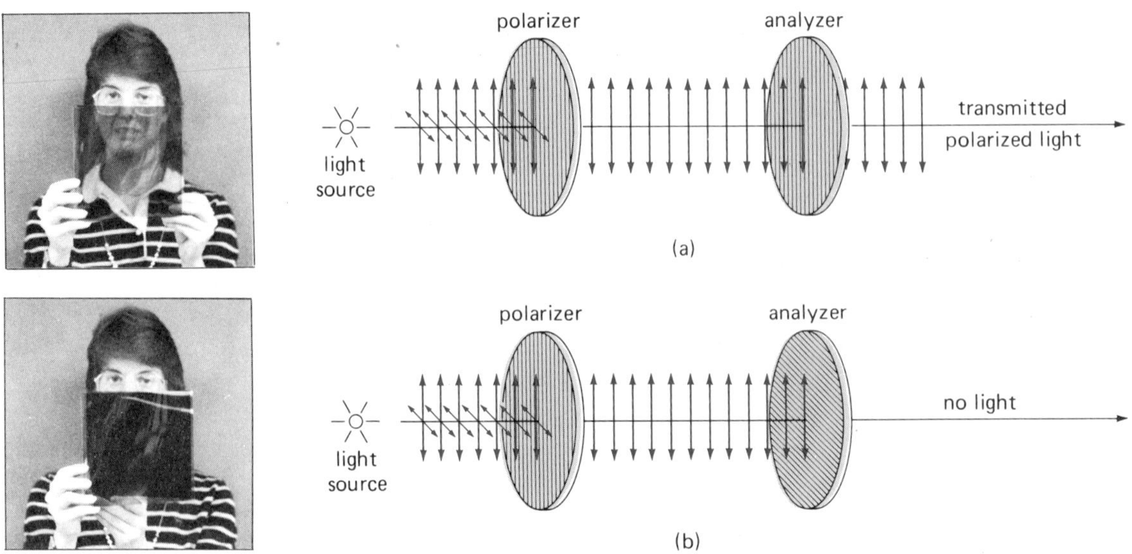

Figure 27.12
Polarized light. Light is polarized by Polaroid film, but since the human eye cannot detect polarized light, an analyzer is needed. (a) Light is transmitted when the planes of polarization of the polarizer and the analyzer are in the same direction. (b) When the planes of polarization are at right angles ("crossed Polaroids"), little, or ideally, no light is transmitted.

ent celluloid, thereby producing a thin, polarizing material now known commercially as Polaroid.

Improved Polaroid films have been developed using polymer materials. During the manufacturing process, the film is stretched so as to align the long chain molecules of the polymer. With proper treatment, the outer (valence) molecular electrons can move along the oriented chains. As a result, the molecules readily absorb light with E vectors parallel to the oriented chains and transmit light perpendicular to the chains.

The direction perpendicular to the oriented chain molecules is commonly called the *transmission axis* or *polarization direction*. Hence, when unpolarized light falls on a polarizing sheet, the sheet acts as a polarizer and polarized light is transmitted (Fig. 27.12).*

The human eye cannot distinguish between polarized and unpolarized light. Hence, we need an analyzer to detect polarized light. The analyzer can be another sheet of polarizing film. When the analyzer has the same orientation as the polarizer, light is transmitted. But when it is rotated 90°, or when the polarizing sheets are "crossed," the analyzer absorbs the polarized light and, ideally, no light is transmitted (Fig. 27.12).

You can demonstrate this yourself with two pairs of polarizing sunglasses. Try it the next time you are at a store that sells polarizing sunglasses (if you don't personally have two pairs of polarizing sunglasses).

2. **Reflection.** Light is also partially polarized by reflection. The direction of polarization is parallel to the reflecting surface, and the degree of polarization depends on the material and the angle at which the light strikes the surface. That is why we use polarizing sunglasses (Fig. 27.13). Light reflected from roads, water, and other surfaces has great intensity and glare. By properly aligning the polarizing lenses of the glasses (with the transmission axis vertical), some of the partially polarized reflected light is absorbed, which reduces the glare.

* A common analogy is to view a polarizer as a picket fence through which a random wave in a rope will pass only parallel to the pickets and be polarized in this direction. However, keep in mind that in a polarizing sheet, the blocking "pickets" (absorbing molecular chains) are perpendicular to the transmission axis.

(a)

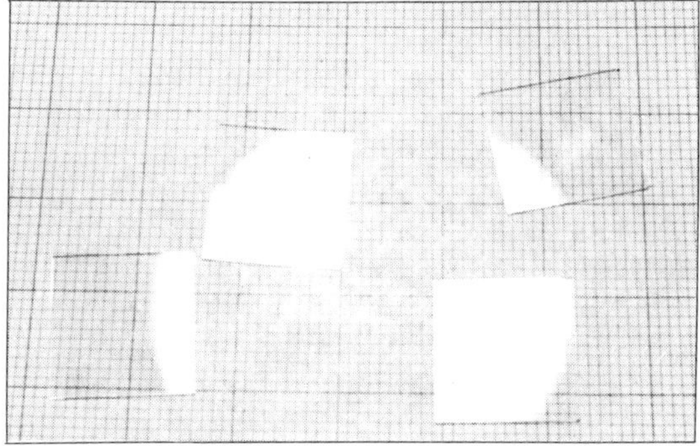

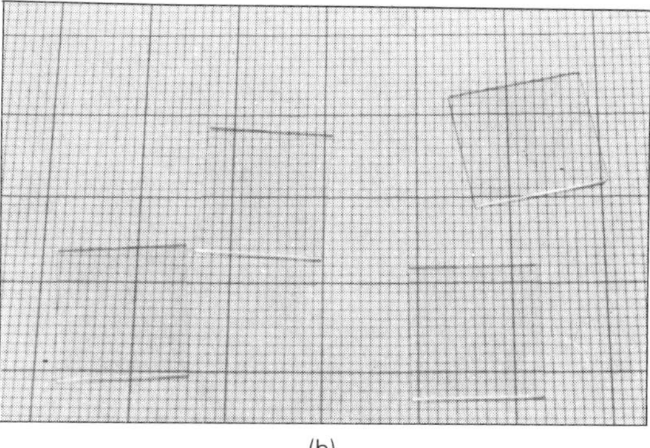

(b)

Figure 27.13
Glare reduction. Light reflected from surfaces is partially polarized parallel to the surface (a). Polarizing sunglasses with vertical planes of polarization reduce the intensity or glare by blocking out the polarized light, as shown in (b) for glass surfaces. [(b) Courtesy L. Greenberg.]

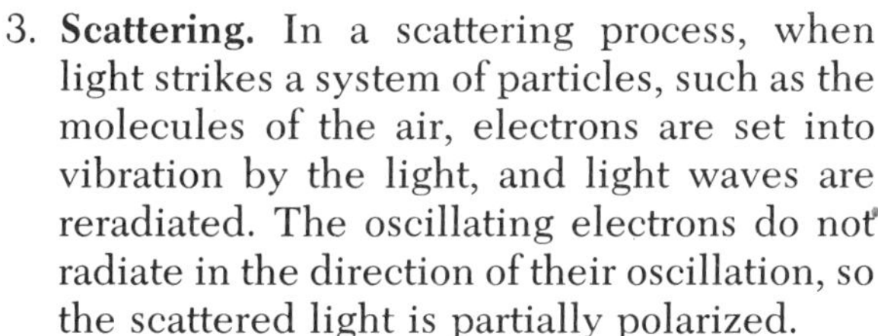

3. **Scattering.** In a scattering process, when light strikes a system of particles, such as the molecules of the air, electrons are set into vibration by the light, and light waves are reradiated. The oscillating electrons do not radiate in the direction of their oscillation, so the scattered light is partially polarized.

   As a result, sunlight scattered by the atmosphere and received at the Earth's surface is partially polarized. You can check this yourself by looking at a blue sky through polarizing sunglasses and then rotating the glasses. Light from different areas of the sky will show different degrees of transmission depending on the degree of polarization, which depends on the relative direction of the Sun.

   It is believed that some insects, such as bees, use the polarized sky light for navigation. Scientists have put bees in a box with a Polaroid film over the top. When the film is rotated, the flying bees change direction.

There are many applications of polarizing films and filters. For example, most transparent materials have a polarizing effect when under stress. When a sample under stress is placed between crossed polarizers (so light is not ordinarily transmitted), patterns such as those

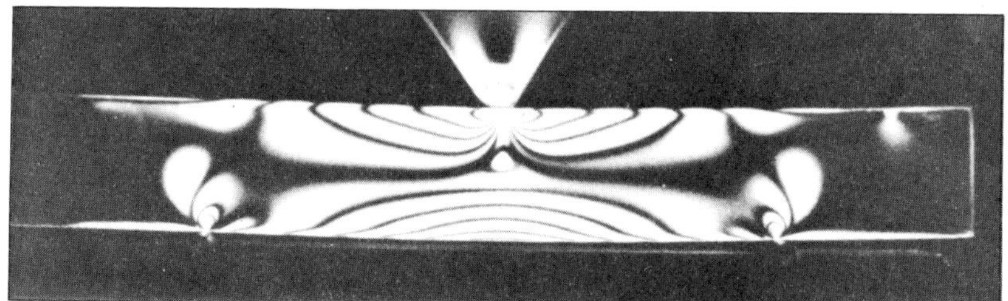

Figure 27.14
Optical stress analysis. Strain distribution in a plastic beam subjected to stress. The sample is viewed between crossed polarizers. The strain pattern disappears when the stress is removed. (Courtesy R. Serway.)

SPECIAL FEATURE 27.2

LCD's—Liquid Crystal Displays

A now very common application of polarization is in LCD's, or liquid crystal displays. These displays are found on calculators, wrist watches, and even gas pumps. One type of LCD makes use of the light-polarizing properties of some liquid crystals. (Liquid crystals are liquids that show some degree of molecular order.)

Such crystals have the ability to rotate, or "twist," the polarization direction of polarized light by 90° (Fig. 1). In a so-called twisted nematic display, a liquid crystal is sandwiched between crossed polarizer sheets and backed by a mirror. Light falling on the surface of the LCD is polarized, twisted, reflected, and twisted again, and then it leaves the LCD. Hence, the display appears light when illuminated.

But when a voltage is applied to the crystal, it loses its twisting property, and the crystal appears dark when illuminated because the light is blocked by the polarizing sheet and not reflected. Such voltage-induced dark regions of the crystal are used to form the numbers and letters of the display. You can show that the light coming from such an LCD is polarized by a Polaroid sheet (or polarizing sunglasses), as shown in Figure 1.

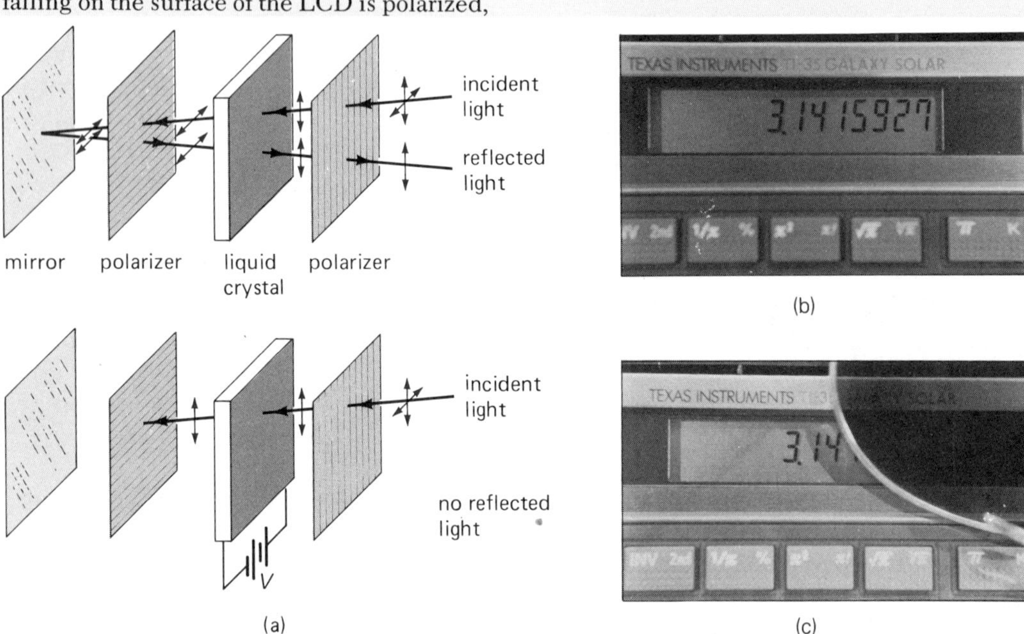

Figure 1
LCD (liquid crystal display). (a) When no voltage is applied to the liquid crystal, it "twists" the light polarization through 90°. When voltage is applied, there is no twisting, and light does not pass through the second polarizer and is not reflected. The crystal then appears dark. (b) The bright areas of an LCD are due to reflected polarized light, as can be shown with a polarizing sheet (analyzer—polarizing sunglasses in this case).

shown in Figure 27.14 are observed. This provides a means for analyzing strain distributions in transparent samples. The method is called optical stress analysis. See also Special Feature 27.2.

27.5 **Illumination**

Proper lighting, or illumination, is important in the home, classroom, factory, and any working area where visual observation is used. Photometry deals with the measurement of light in the visible region. This is the wavelength region approximately between 400 nm and 700 nm (or 4000 Å and 7000 Å).

Light Sources

Our natural light source is the Sun. However, many kinds of artificial light sources have been

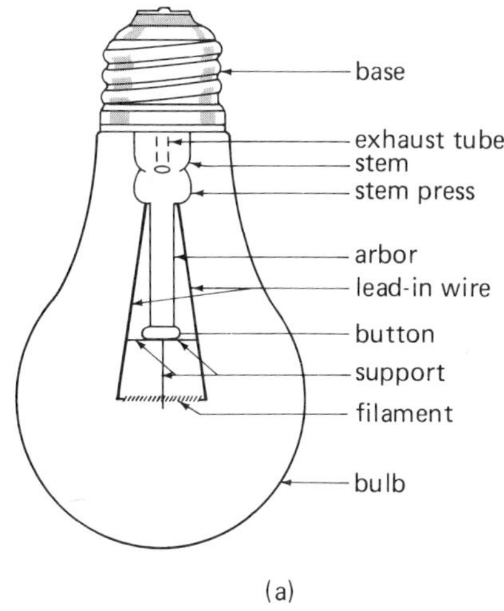

(a)

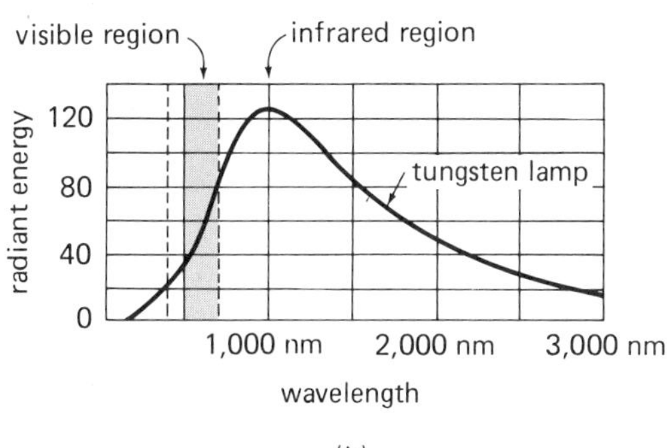

(b)

Figure 27.15
The incandescent lamp and light spectrum. (a) Parts of a lamp, and (b) the energy spectrum. Notice that most of the radiation is in the infrared region.

developed. The most commonly used light sources are described in the following paragraphs.

Incandescent Lamps The common light bulbs around the house are incandescent lamps. An incandescent lamp consists of a coiled wire filament usually made of tungsten, sealed in a glass bulb (Fig. 27.15). The bulb is filled with a mixture of argon and nitrogen gases at low pressure. This prevents oxidation of the filament and retards its evaporation, which would decrease the life of the lamp. (The gray spot that develops on the bulb after long use is due to the evaporation of metal from the filament.)

The tungsten filament becomes white-hot when carrying an electrical current. The filament temperatures of bulbs of common wattage ratings range from 2500 to 3000°C in normal operation. Light from incandescent lamps has a continuous spectrum, but most of the radiation lies in the infrared part of the spectrum (Fig. 27.15), and more than 95 percent of the energy is radiated as heat. This makes the incandescent lamp a relatively inefficient light source.

Even so, several billion incandescent lamps are used in the United States each year. The lamp bulbs commonly are frosted to diffuse the light and soften shadows, providing pleasanter illumination.

Halogen lamps have greater efficiency than common incandescent lamps. (*Halogen* refers to the chemical family of elements that includes fluorine, chlorine, and iodine.) A tungsten filament can burn hotter in a halogen (typically iodide) atmosphere. This produces more light in the visible region, or whiter light, and requires less power input for a given light output. Because of the chemical effect of the iodide on glass, the filament must be placed in quartz rather than glass, which increases the cost. The chief applications of halogen lamps are automobile headlights and projector bulbs.

Fluorescent Lamps A fluorescent lamp is basically a gas discharge tube. It contains mercury vapor and an inert gas, such as argon, under low pressure. During operation, some of the mercury atoms are ionized, and the ionized gas gives rise to a current discharge in the tube (Fig. 27.16). In the process, the mercury atoms are excited and emit light, primarily in the ultraviolet region.

The glass tube is coated with a phosphor material such as calcium tungstate. When the ultraviolet radiation strikes the phosphor, it fluoresces and emits visible light. **Fluorescence** is the process of absorption of light at one wavelength and the emission of light at another wavelength. The type of light emitted depends on the type of phosphor coating used. With the proper selection of phosphor coatings, the majority of the light from fluorescent lamps is emitted in the visible region. Because of the fluorescent conversion and lower operating temperatures, fluorescent lamps are considerably more efficient than incandescent lamps.

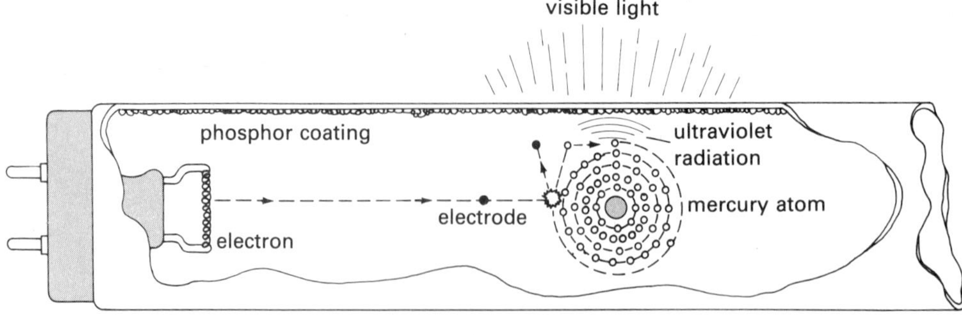

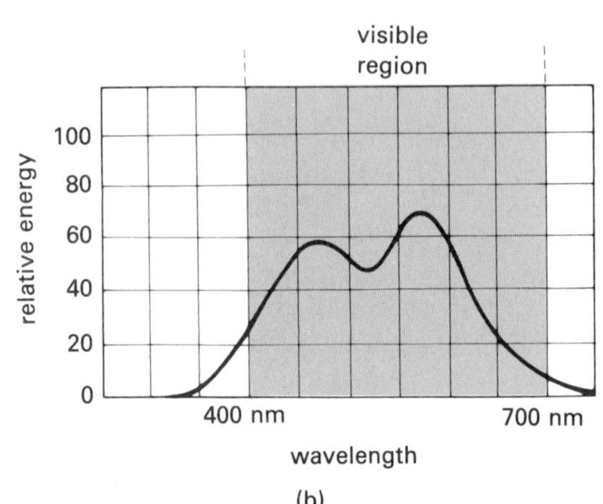

Figure 27.16
The fluorescent lamp. (a) Mercury atoms are ionized and excited. They emit light in the ultraviolet region. The ultraviolet light is absorbed by a phosphor coating on the inside of the tube, and visible light is emitted. (b) Energy spectrum. Notice that most of the radiation is in the visible region. Compare to Figure 27.15 spectrum.

Certain types of fluorescent lamps have phosphor coatings that emit in the near-ultraviolet region. These lamps produce very little visible radiation and are called "black lights." They are intended for use in activating fluorescent and phosphorescent materials in advertising and other displays.*

Other gas discharge tubes are used for special types of lighting. For example, neon discharge tubes emit red light, and these tubes are commonly used in neon signs. The color depends on the type of gas used.

Luminous Flux

Illumination involves the irradiation of a surface with visible light. This, in turn, depends on the amount of light energy traveling from a source or sources to the surface. To express this, we refer to **luminous flux,** which is the time rate of flow of visible light energy. Being a measure of energy per time, luminous flux has the units of joule/second, or watt. Luminous flux is therefore a measure of light power.

The human eye is not equally sensitive to all colors. Some colors evoke a greater brightness response than others (see Section 29.2). The response of the human eye to various wavelength colors is illustrated by a spectral efficiency curve, which depicts the wavelength of maximum visual sensitivity, about 555 nm or yellow-

* Phosphorescence is the process in which electrons remain in excited states for relatively long times before de-excitation and the emission of light. As a result, a phosphorescent material can "glow in the dark" for some time after being exposed to light and excited.

Table 27.1
Approximate Wavelength-Color Responses

Color	Wavelength (nm)
Red	610–700
Orange	590–610
Yellow	550–590
Green	500–550
Blue	450–500
Violet	400–450

green color (Table 27.1), which is assigned a value of 1.0, or 100 percent.

The relative efficiencies of other visible wavelengths are expressed as decimal fractions or percentages of this value. Thus, according to the efficiency curve, given a yellow-green light source and a blue light source (470 nm) emitting equal light fluxes, the yellow-green source would appear approximately five times as bright as the blue source. (For $\lambda = 470$ nm, $e_r = 0.20$, where e_r is the relative efficiency [Fig. 27.17].)

In order to associate and standardize the visual sensitivity of various wavelengths to the light power in watts, the special luminous flux unit of the **lumen (lm)** is defined.

One lumen is equivalent to 1/680 W of light with a wavelength of 555 nm.

EXAMPLE 27.5 Light from a yellow-green source and light from a blue source both have luminous fluxes of 1/680 W. What are the fluxes in lumens?

Solution: By definition, the yellow-green light (555 nm) has a flux of $F = 1.0$ lm. From the spectral sensitivity curve, blue light has a relative sensitivity of about $e_r = 0.20$ (20 percent), so the equivalent flux of the blue light is

$$F_e = e_r F = (0.20)(1.0 \text{ lm}) = 0.20 \text{ lm}$$

Hence, the lumen unit relates the luminous flux to the visual brightness perceived by the eye. If the lights in the previous example were combined, the eye would perceive a brightness of 1.2 lm, and the specific light power would be $2/680 = 1/340$ W.

Since illumination is concerned with visual response, the lumen is the preferred practical photometric unit of luminous flux. Moreover, most common light sources, such as electric light bulbs, emit visible light composed of many wavelengths. A 100-watt light bulb emits a total flux of about 1700 lumen.

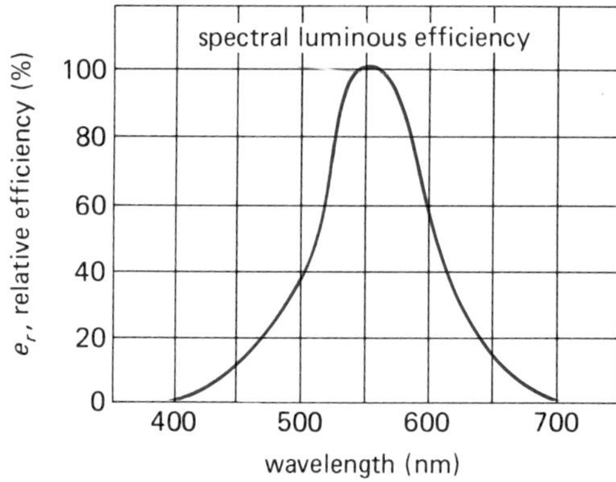

Figure 27.17
Spectral efficiency curve. The curve gives the response of the average eye to equal amounts of energy at various wavelengths relative to yellow-green light at 500 nm (5000 Å) taken as 100 percent.

Illuminance

A surface is said to be illuminated when it is irradiated with visible light. For example, the page of this book must be illuminated if you are to read it. The **illuminance** E (also called illumination) of a given surface is defined as the flux falling on it divided by the area of the surface:

$$\text{(illuminance)} \quad E = \frac{F}{A} \qquad \textbf{(Eq. 27.9)}$$

For several sources, the total illuminance is given by the sum of the individual source illuminations.

The unit of illuminance in the British system is the lumen/ft², or **foot-candle (fc)**; 1 fc = 1 lm/ft². The SI unit of illuminance is the **lux (lx)**, and 1 lx = 1 lm/m². A hectolux (hlx) is 100 lx.

Proper illumination is an important consideration (Fig. 27.18). Good lighting reduces eyestrain, accidents, and poor workmanship. Work efficiency and output go hand in hand with proper illumination. However, too much lighting, which produces glare, is as undesirable as not enough. The recommended illuminations for various seeing tasks are given in Table 27.2.

Illumination can be conveniently expressed in terms of the intensity (I) of a source and the distance from the source.

$$E = \frac{F}{A} = \frac{I}{r^2}$$

or

$$E = \frac{I}{r^2} \qquad \textbf{(Eq. 27.10)}$$

where r is the distance from the source or the radius of a sphere with area A. The luminous intensity in a particular direction is commonly called the *candle power* of the source. The accepted SI unit of this intensity is the candela

(a)

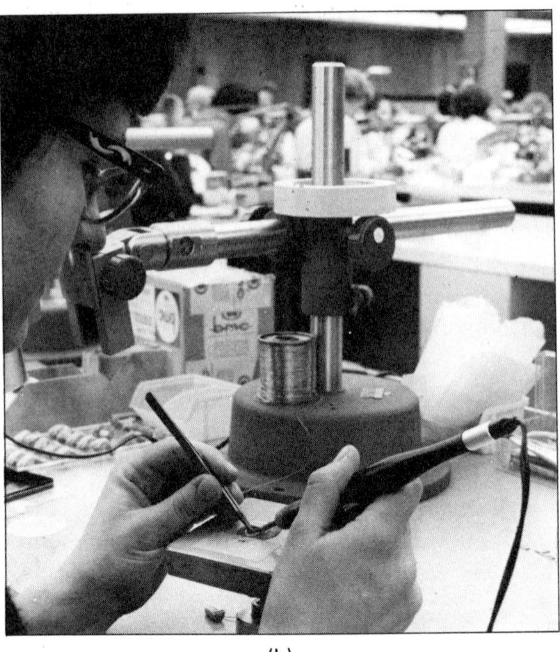

(b)

Figure 27.18

Proper illumination is important in working conditions. (a) Medium tasks are done more accurately with a minimum of 100 fc (11 hlx). (b) Extra-fine tasks are easier to see when low contrast is offset with 1000 fc (110 hlx). (Courtesy General Electric Co.)

(cd). Notice from the equation that 1 fc is equal to 1 cd/ft^2, and 1 lx is equal to 1 cd/m^2.*

Hence, the illumination falls off as $1/r^2$ from the source. Recall the similar "inverse square" relationship for the intensity of sound in Chapter 14. For a constant source intensity I, we may write

$$\frac{E_2}{E_1} = \frac{r_1^2}{r_2^2} = \left(\frac{r_1}{r_2}\right)^2 \qquad \textbf{(Eq. 27.11)}$$

* The luminous intensity is defined in terms of the solid angle Ω (see Table 1.2) where $I = F/\Omega$. For a spherical surface, $\Omega = A/r^2$.

Equation 27.10 assumes the incident light to be normal to the surface. However, if the light strikes the surface at an angle of incidence θ (Fig. 27.19), which is often the case, the illumination is given by

$$\boxed{E = \frac{I \cos \theta}{r^2}} \qquad \textbf{(Eq. 27.12)}$$

EXAMPLE 27.6 A machine casting with a flat surface area of 0.50 ft^2 is illuminated by two

Table 27.2

Seeing Tasks and Minimum Recommended Lighting Levels

Seeing Task	Description	Recommended *fc*	Recommended *hlx*
Casual	Noncritical tasks	30	3.2
Rough	Large, easily identifiable objects	50	5.4
Medium	Most manufacturing activities	100	11
Fine	Intricate tasks, detailed assembly	500	54
Extra fine	Very small parts, inspection	1000	110

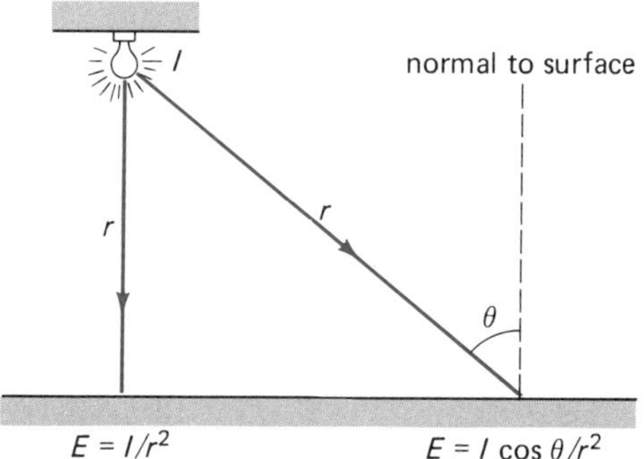

Figure 27.19

Illumination. The illuminance is given by $E = I/r^2$ when the incident light is normal to the surface. When the incident light is at an angle θ relative to a normal to the surface, then $E = I \cos \theta / r^2$.

light sources, as illustrated in Figure 27.20, with $I_1 = 3400$ cd and $I_2 = 1000$ cd. (a) What is the total illumination of the surface? (b) If source S_2 is turned off and the casting is elevated 3.0 ft, what is the illumination?

Solution: (a) The illumination from S_1 is

$$E_1 = \frac{I_1}{r_1^2} = \frac{3400 \text{ cd}}{(6.0 \text{ ft})^2} = 94 \text{ cd/ft}^2 = 94 \text{ fc}$$

The illumination from S_2 is

$$E_2 = \frac{I_2 \cos 60°}{r_2^2} = \frac{(1000 \text{ cd})(0.50)}{(3.0 \text{ ft})^2} = 56 \text{ fc}$$

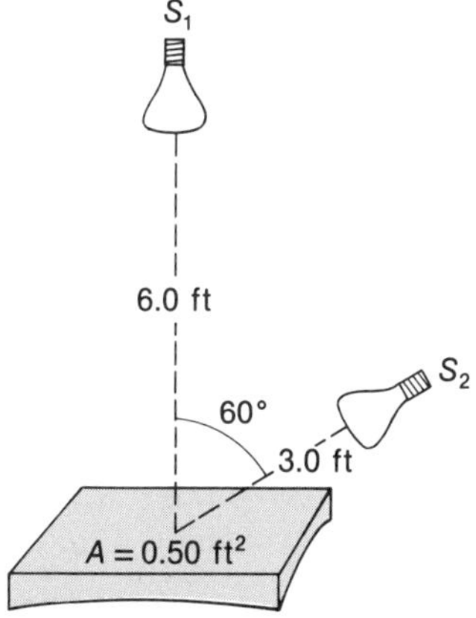

Figure 27.20

See Example 27.6.

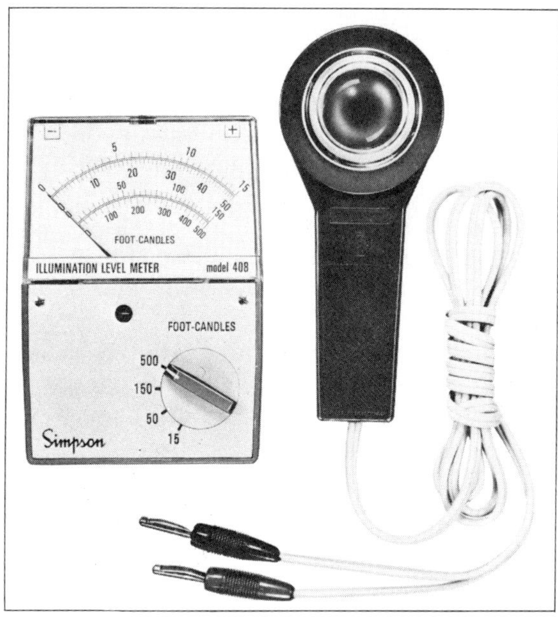

Figure 27.21

A light or illumination level meter. The current produced by light falling on the photoelectric "eye" is measured in proportionate foot-candles on the meter. (Courtesy Simpson Electric Co.)

Hence,

$$E_t = E_1 + E_2 = 94 \text{ fc} + 56 \text{ fc} = 150 \text{ fc}$$

(b) Since $E_1 = 94$ fc at $r_1 = 6.0$ ft, then at a distance of $r_2 = 3.0$ ft, by Equation 27.11,

$$\frac{E_2}{E_1} = \left(\frac{r_1}{r_2}\right)^2 = \left(\frac{6.0 \text{ ft}}{3.0 \text{ ft}}\right)^2 = 4$$

or

$$E_2 = 4E_1 = 4(94 \text{ fc}) = 376 \text{ fc}$$

Illumination can be measured by several means. One common device is a light meter (Fig. 27.21). This type of instrument makes use of the so-called photoelectric effect (see Chapter 30). When light falls on certain metals, such as selenium, electrons are emitted and an electric current flows in an appropriate circuit. The current is proportional to the light flux. The current meter reading of the instrument is calibrated in illumination units. Photographic exposure meters are specialized light meters of this type.

Important Terms

light electromagnetic waves, generally taken to be in or near the visible region of the electromagnetic spectrum

physical (wave) optics the study of phenomena using the wave nature of light

dispersion the separation of light into component wavelengths (colors) due to the speed of light being slightly different for different wavelengths in transparent media

optical flats flat glass plates used to test the flatness of surfaces by thin film (air wedge) interference

diffraction the bending of waves around the corner or edge of an opaque object

polarization preferential orientation of the field vectors of electromagnetic waves

luminous flux the time rate of flow of visible light energy or light power

illuminance (illumination) the luminous flux falling on a surface divided by the area of the surface

Important Formulas

electromagnetic wave characteristics:

$$\lambda f = c$$

$$(c = 3 \times 10^8 \text{ m/s} = 3 \times 10^{10} \text{ cm/s})$$

wave equation:

$$y = A \sin\left(\frac{2\pi x}{\lambda} + \phi\right)$$

phase difference (spatial):

$$\Delta\theta = \frac{2\pi\,\Delta x}{\lambda}$$

conditions for interference:

total constructive:

Phase difference (rad) $\Delta\theta = 0, 2\pi, 4\pi, \ldots$ Path difference when $\Delta x = 0, \lambda, 2\lambda, \ldots$

total destructive:

Phase difference (rad) $\Delta\theta = \pi, 3\pi, 5\pi, \ldots$ Path difference when $\Delta x = \frac{\lambda}{2}, \frac{3\lambda}{2}, \frac{5\lambda}{2}, \ldots$

Young's interference experiment:

$$d \sin\theta = n\lambda \qquad n = 0, 1, 2, 3, \ldots \text{ (bright fringe)}$$

$$\lambda = \frac{y_n d}{nL} \qquad n = 1, 2, 3, \ldots$$

grating equation:

$$d \sin\theta_n = n\lambda \qquad n = 0, 1, 2, 3, \ldots \text{ (bright fringe)}$$

where $d = \frac{1}{N}$

illuminance (illumination):

$$E = \frac{F}{A} \quad \frac{\text{(luminous flux)}}{\text{(area)}}$$

$$= \frac{I}{r^2}$$

$$E = \frac{I \cos\theta}{r^2} \quad \text{(non-normal incidence)}$$

Questions

The Nature of Light

1. What is the range of the wavelength of visible light? Which end of the spectrum has the longer wavelength?
2. Give the order of magnitude of the frequency of visible light.
3. What are the electromagnetic radiations that lie on both sides of the visible spectrum, and which has the greater wavelength?
4. An acronym to remember the colors of the visible spectrum (or rainbow) is ROY G. BIV. What do the letters of the name stand for?
5. How long does it take light from the Sun to reach Earth?
6. Is the speed of light the same in all media? Explain.

Interference

7. Distinguish between the phase and phase constant of a wave.
8. Compare the wave equations for a sinusoidal wave for temporal (time) and spatial (space) representations.
9. How is the phase difference of two waves related to their path difference?
10. What are the conditions for (a) total constructive interference and (b) total destructive interference?
11. What is meant by a stationary interference pattern? Does light from ordinary light sources form stationary interference patterns? Explain.
12. When are light waves coherent? How are coherent light "sources" obtained for Young's experiment?
13. Is light "destroyed" in destructive interference? Explain. How about the light for the dark fringes or destructive-interference regions in the interference pattern of Young's experiment?
14. When washed dishes aren't rinsed well, they often show a display of colors after they dry. What causes this?
15. What would be the effect if a thin film were illuminated with monochromatic light instead of white light? Consider films of uniform and nonuniform thicknesses.

16. Will a thin film always be nonreflecting (i.e., will destructive interference occur) for normal incidence when the film thickness is one fourth of a wavelength of the light? Explain.
17. Explain why the film thickness on nonreflecting glass is one fourth of the wavelength of yellow-green light. What would happen if the film thickness were half a wavelength?
18. The film on most nonreflecting lenses is just the thickness to produce destructive interference for yellow-green light. If you wanted to have destructive interference primarily for blue light, would you make the film thinner or thicker? Explain.
19. Suppose you want to use thin-film interference to make "reflecting" glass or a mirror. How could you do this? (*Hint:* discuss in terms of film thickness *and* type of material; e.g., what if the film material were optically denser than glass?)
20. Explain why a uniform air wedge between two optical flats produces an interference pattern of alternate bright and dark lines when illuminated with monochromatic light. What would be seen if white light were used?

Diffraction

21. How is the slit width of a diffraction grating determined?
22. How are diffraction gratings used in spectrometers?
23. Some of the vivid colors of the fanned tail of a peacock are due to diffraction. Why is this?
24. Radio and television waves can be diffracted. Which would be expected to show the greater diffraction for common-sized objects that might be encountered: (a) AM or FM radio waves, and (b) VHF or UHF television waves? (See Chapter 26.)
25. What size objects or openings would be good to diffract (a) AM radio waves and (b) FM radio waves? (See Chapter 26.)

Polarization

26. Describe some methods by which light may be polarized. Can light be unpolarized? Explain.
27. How do polarizing sunglasses reduce glare?
28. Does it make any difference how the plane, or direction of polarization, of the lenses of sunglasses is oriented? Explain.
29. Given two pairs of sunglasses, how could you tell if they were nonpolarizing or polarizing, or could you?
30. Three-dimensional pictures are projected on the screen by two projectors with slightly different images, polarized in opposite directions (Fig. 27.22). One image is for one eye and the other for the other eye, so the brain will interpret this as depth, or a third dimension. Explain how this is accomplished when wearing the required 3-D glasses. (Without the glasses, the image on the screen appears fuzzy or blurred.)

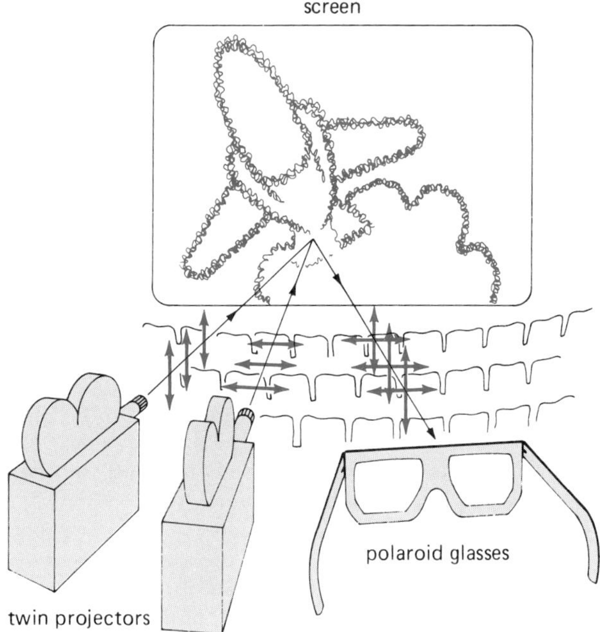

Figure 27.22
3-D movie. See Question 30.

Illumination

31. Mercury-vapor street lights have a bluish hue, while the newer sodium arc lamps have a yellow appearance. Explain this difference.
32. At a disco, the black light over the band blows out and there are no more bulbs available. The proprietor decides to use an infrared lamp instead to get the fluorescent effects. Is this a good idea? Explain.
33. Why can we still see things when black lights are used?
34. Some soap manufacturers add a blue dye to their detergents that fluoresces with a slight blue color. What is the purpose of this? (*Hint:* "whiter whites" have a cleaner appearance when they are slightly bluish in color.)
35. During the summer, "brownouts," or voltage reductions, are experienced in some large cities due to electrical demand exceeding the supply. How would this affect the light emitted from an incandescent lamp? (*Hint:* see the Chapter 16 Supplement.)
36. Distinguish between luminous flux and illuminance, and give the units of each.

37. How are illumination, working conditions, and efficiency related?
38. What is the principle of a light meter, and what does it measure?

Problems

Levels of difficulty are indicated by asterisks for your convenience.

27.1 The Nature of Light

1. Light traveling in a vacuum has a frequency of 5.0×10^{14} Hz. What is the wavelength of the light?
2. What are the frequency limits of visible light?
3. Ultraviolet light has a wavelength of 300 nm and infrared light has a wavelength of 800 nm. What are the light frequencies?

*4. The general speed of light in water is 2.3×10^8 m/s. (a) What is the wavelength of green light ($f = 6.0 \times 10^{14}$ Hz) in water in nanometers? (b) How does this compare with the wavelength of the light in vacuum?

*5. The wavelength of monochromatic light in a particular type of glass is 4000 Å. (a) What is the speed of this light in the glass if the frequency is 5.0×10^{14} Hz? (The frequency is the same in both a medium and a vacuum.) (b) What is the ratio of c/c_m, where c_m is the speed of light in the glass medium?

27.2 Interference

*6. What is the equation of a spatial sinusoidal wave with a phase constant of (a) $\pi/2$ rad, (b) π rad, (c) $3\pi/2$ rad, and (d) 2π rad?

*7. A wave in space is described by the equation $y = 10 \sin \pi x$ (cm). (a) What is the wavelength of the wave? (b) What is the displacement of the wave after it has traveled a distance of 5λ?

*8. What is the wave equation of a transverse light wave in vacuum with an amplitude of 0.25 cm (at $x = 0$) and a frequency of 6.0×10^{14} Hz?

*9. What is the phase difference for two waves of the same frequency arriving at a point after having traveled the respective path lengths of 8.0λ and 7.5λ? Do they interfere constructively or destructively?

*10. Two waves with the same frequency interfere after traveling 20λ and 22λ, respectively. (a) What is the phase difference of the waves? (b) Is the interference constructive or destructive?

*11. Two waves have identical wave forms given by $y = 12 \sin 2\pi x$ (cm). What is the amplitude of the combined waves if they interfere after traveling (a) 9.0λ and 8.0λ, respectively, and (b) 7.0λ and 5.5λ, respectively?

*12. Young's double-slit experiment is performed with monochromatic light, with a distance between the slits of 1.0 mm and with the screen a distance of 1.2 m from the slits. If the center of the second-order bright fringe is located 1.5 mm from the center of the central maximum on the screen, what is the wavelength of the light in nanometers, and what is its color?

*13. When orange light ($\lambda = 600$ nm) is used for Young's experiment, the first-order bright fringe is observed 2.5 mm from the center of the central maximum on a screen 1.0 m from the slits. What is the slit width used in this experiment?

*14. If light having two component wavelengths of 400 nm and 700 nm is used in the experimental setup in Problem 12, how far apart are the first-order fringes of each component?

*15. If the slit width in Young's experiment were doubled and all else remained the same, where would the second-order bright fringe of monochromatic light be located with respect to the original interference pattern?

*16. Show that the distance between adjacent bright fringes in Young's experiment using monochromatic light is given by $L\lambda/d$.

*17. A thin soap film has a uniform thickness. What would be the thicknesses (in terms of wavelength units in the film) for reflected monochromatic light to interfere (a) constructively and (b) destructively? Assume normal incidence and justify your answers.

*18. Two camera lenses are coated with thin films to make them "nonreflecting" for (a) blue light (500 nm) and (b) red light (700 nm). What are the film thicknesses (in terms of wavelength units in the films)? Assume normal incidence.

27.3 Diffraction

19. A diffraction grating has 25,000 lines/in. What is the slit width of the grating?

*20. Monochromatic light is diffracted by a grating with 10,000 lines/cm. It is found that the first-order fringe is diffracted by an angle of 15°. What is the wavelength of the light?

*21. A diffraction grating has 2500 lines/cm. What is the angle of the second-order fringe when light with a wavelength of 5000 Å is diffracted?

*22. A diffraction grating has 10,000 lines/cm and is illuminated with light having a frequency of 5.0×10^{14} Hz. What are the angles of (a) the first-order bright fringe and (b) the second-order bright fringe?

*23. What is the angular difference between the ends of the first-order spectrum when white light is diffracted by a grating with 5000 lines/cm?

27.4 Polarization

*24. Suppose you had two sheets of Polaroid film and you held one in front of the other while looking through the sheets (see Fig. 27.12). (a) If you rotated one of the sheets through one complete rotation, how many times would the Polaroids be crossed and darken? Start with uncrossed (light-transmitting) Polaroids. (b) If both sheets were rotated through one complete rotation at the same time but in opposite directions, how many times would the Polaroids be crossed and darken? (c) What would be the case if the Polaroids were rotated in the same direction with the same speed?

27.5 Illumination

25. Light from a yellow-green light source (λ = 555 nm) has a luminous flux of 1/272 W. What is the flux in lumens?

*26. Light from a yellow-green source (λ = 555 nm) and light from an orange source (λ = 600 nm) both have luminous fluxes of 1/680 W. What are the perceived fluxes in lumens? (*Hint:* see Fig. 27.17.)

*27. The light from a green source (λ = 510 nm) has a luminous flux of 1/340 W, and light from a yellow-green source (λ = 555 nm) has a luminous flux of 1/680 W. (a) What are the fluxes in lumens? (b) What are the perceived brightness and specific light power of the combined lights?

*28. Green light with a wavelength of 520 nm has a luminous flux of 1/340 W. What would be the flux power of red light (630 nm) that could be perceived with the same brightness?

*29. The proper lighting level required for working on an assembly line is 100 fc. What is the luminous flux on a 0.50 ft × 0.80 ft metal sheet on the line?

*30. A light source with an intensity of 1200 cd is 0.50 m from a workbench where intricate welding is done. Does the source supply the recommended lighting level?

*31. A light source provides an illumination of 800 lx at a distance of 2.0 m from the source. What is the illumination 4.0 m from the source?

**32. A lamp with an intensity of 2000 cd is 4.0 ft directly above a radio on a repair bench. A television set is 3.0 ft to the side of the radio. What is the illumination of each?

**33. Suppose the light source S_1 in Figure 27.20 has an intensity of 2500 cd. (a) What would be the required intensity of S_2 so that it would produce an equal illumination on the surface? (b) What would be the total flux on the surface?

Chapter Supplement

Newton's Rings

High-quality optical instruments (Chapter 29) require lenses that have been ground and polished smoothly enough so they do not produce distortions. The quality and perfection of a lens can be inspected by interference techniques. A common method involves Newton's rings. When a curved lens surface is placed on an optical flat, a uniform, curved air wedge is formed between the lens and the flat (Fig. 27S.1). As a result, an interference pattern is produced.

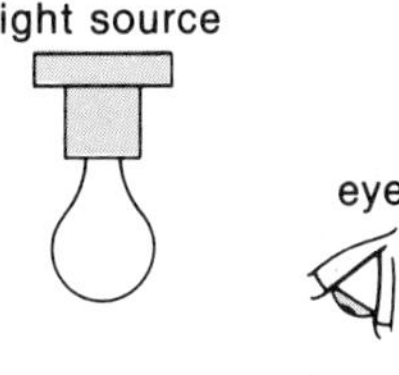

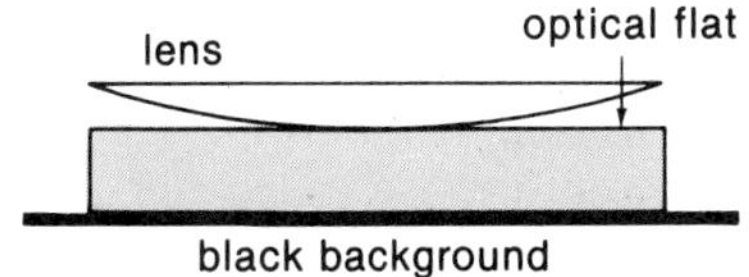

Figure 27S.1

Lens inspection. The arrangement for optical testing for irregularities in a lens surface. The air wedge between the lens and the optical flat gives rise to interference fringes.

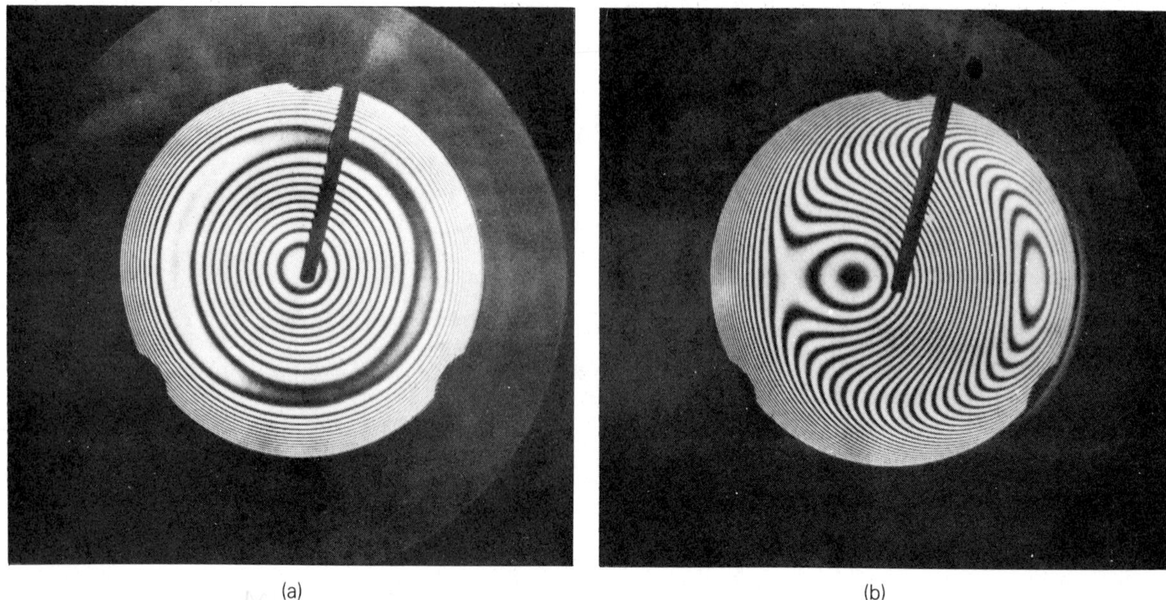

Figure 27S.2
Newton's rings. (a) Circular interference fringes (rings) indicate the lens is of high quality and smoothly ground. (b) A distorted interference pattern indicates irregularities in the lens' surface. [From Physical Science Study Committee (PSSC).]

The interference pattern in this case is a set of bright and dark circular fringes, or rings (Fig. 27S.2). This effect was described by Isaac Newton in 1708; hence the name *Newton's rings.* If the lens is perfectly ground, a set of concentric rings is observed, indicating the high quality of the lens. However, if there are imperfections in the lens's surface, a distorted interference pattern is observed.

Chapter 28

Mirrors and Lenses

Reflection

Two of the most common optical components are mirrors and lenses. Hardly a day goes by that each of us does not look into a mirror, and many of us wear lenses (glasses). There are also numerous technical applications of mirrors and lenses.

The prominent physical properties of mirrors and lenses are reflection and refraction, respectively. As we analyze these properties, it is convenient to ignore the wave properties of light and consider rays of light, or ray optics. After defining this concept and considering the physics of reflection and refraction, we shall look at the properties and characteristics of certain common types of mirrors and lenses and the images they form.

28.1 Light Rays

Waves can be characterized by wave fronts. A **wave front** is defined by the adjacent portions of a wave that are in phase. For example, in water, waves spread out from a single point disturbance in a succession of circular wave fronts (Fig. 28.1). All of the particles on a crest (or in a trough) are in phase and define circular wave fronts.

At a very great distance from the point source, the wave fronts are nearly parallel to each other, and they approximate plane wave fronts. A long disturbance source, such as a meterstick dashed up and down in water, would produce plane wave fronts.

Light waves can be characterized by wave fronts; however, it is often convenient to talk about rays of light. A **ray** is defined by a line drawn perpendicular to the wave fronts in the direction of the wave propagation (Fig. 28.1). This is consistent with the common observation that light travels in straight lines. A beam of light can be thought of as a parallel group of rays, although we usually draw a single ray to represent a light beam.

The analysis of light phenomena using rays is called **geometrical optics** or **ray optics.** This is

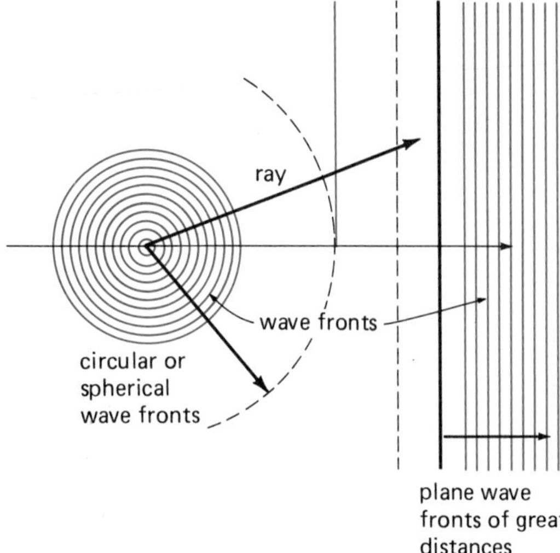

Figure 28.1
Wave fronts and rays. The adjacent portions of a wave that are in phase form a wave front. A ray is a line perpendicular to the wave fronts and in the direction of wave propagation. At large distances from a point source, the wave fronts approximate plane waves.

because we neglect the wave nature of the light and consider only the geometry of the rays. Two particular light phenomena, reflection and refraction, are easily represented by ray optics.

28.2 **Reflection and Refraction**

Reflection

When a wave strikes the boundary of another medium, the wave is reflected and/or transmitted. In the case of sound, the reflected wave is heard as an echo. The direction of the reflected wave depends on the direction of the incident wave. As stated in the **law of reflection,**

> **The angle of incidence is equal to the angle of reflection; i.e., $\theta_i = \theta_r$.**

This is illustrated in Figure 28.2 for a light ray reflected from a flat surface, such as that of a plane mirror. The angle of incidence θ_i and the angle of reflection θ_r are measured from the normal, which is a line perpendicular to the reflecting surface. The incident and reflected rays and the normal to the surface lie in the same plane.

The reflection from a smooth surface is called **regular** or **specular reflection.** A regularly reflecting surface is called a *mirror surface.* Light is also reflected from rough surfaces, such as the page of this book. In this case the law of reflection still applies, but the light is reflected in many directions (Fig. 28.2). This type of reflection is called **irregular** or **diffuse reflection.**

Refraction

When light is transmitted from one medium to another at an angle, the light is "bent," or refracted. That is, the light rays are diverted from their original direction (Fig. 28.3). This happens because the speed of light is different in different media. The frequency of the light is the same in both media, so the wavelengths are different in the different media ($\lambda f = v$).

For the case illustrated in Figure 28.3, the speed of light is less in the second medium than in the first, and the light rays are bent toward the normal. An example is light going from air into water or glass.

> **Light passing into an optically denser medium is refracted *toward* the normal.**

In general, the greater the density of a material, the slower the speed of light in the material.*

Similarly, for the reverse situation,

> **Light passing into a less optically dense medium is refracted *away* from the normal.**

Notice that this condition is illustrated by reverse ray tracing in the diagram in Figure 28.3. In ray optics, situations may be reversed by reverse ray tracing.

The refraction of light gives rise to many interesting phenomena. An example of the apparent change of position of an object due to refraction is shown in Figure 28.4. Also, some atmospheric refraction effects are discussed in Special Feature 28.1.

The speed of light in a medium is characterized by the **index of refraction *n*** of the medium. This is defined as

$$n = \frac{c}{c_m} \qquad \textbf{(Eq. 28.1)}$$

* Optical density and mass density are not the same, although there is a general correlation. An optically denser material, or one that bends light more, may have a smaller mass density than another material.

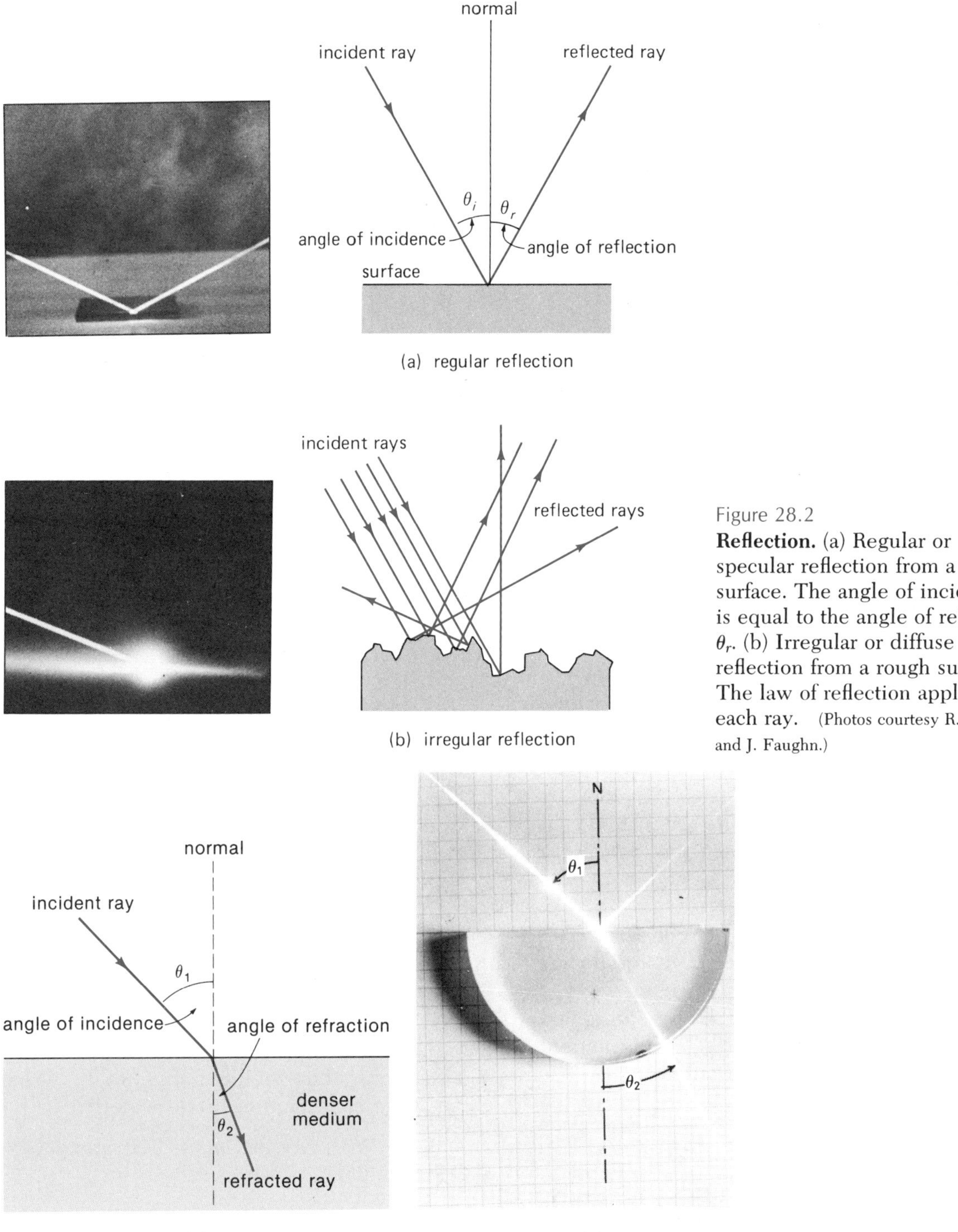

Figure 28.2
Reflection. (a) Regular or specular reflection from a mirror surface. The angle of incidence θ_i is equal to the angle of reflection θ_r. (b) Irregular or diffuse reflection from a rough surface. The law of reflection applies to each ray. (Photos courtesy R. Serway and J. Faughn.)

Figure 28.3
Refraction. When light passes from one medium into another at non-normal incidence, it is deviated from its original path, or refracted. A light ray is bent toward the normal when passing into an optically denser medium as shown here in (a) and (b). A light ray passing into a less optically dense medium is bent away from the normal, as may be seen by reverse ray tracing.

Figure 28.4
Refraction in action. The pencil appears to be almost severed because of refraction. (Courtesy J. and A. Turk.)

where c is the speed of light in vacuum and c_m is the speed of light in the medium. Since the speed of light is greatest in vacuum, the index of refraction of a material is always greater than one ($c > c_m$, and $n = c/c_m > 1$). Typical indices of refraction of some common materials are listed in Table 28.1.

For practical purposes, the index of refraction of air is taken to be unity (1.00), which means the speed of light in air is essentially the same as that in vacuum. This is a good approximation since $n_{air} = 1.00029 = c/c_{air}$, and $c \simeq c_{air}$.

EXAMPLE 28.1 What is the speed of light in water?

Solution: Table 28.1 indicates that water has an index of refraction $n = 1.33$. Then,

$$n = \frac{c}{c_{H_2O}}$$

and

Table 28.1
Typical Indices of Refraction for Different Materials

Material	n (Index of Refraction)
Gases	
Air	1.00029
Carbon dioxide	1.00045
Liquids	
Benzene	1.50
Carbon tetrachloride	1.46
Ethyl alcohol	1.36
Water	1.33
Solids	
Diamond	2.42
Glass	
Ordinary crown	1.50
Dense flint	1.70
Ice	1.31
Polystyrene	1.58

$$c_{H_2O} = \frac{c}{n} = \frac{3 \times 10^8 \text{ m/s}}{1.33}$$

$$= 2.26 \times 10^8 \text{ m/s}$$

The angle of incidence θ_1 and the angle of refraction θ_2 (Fig. 28.3) are related by an expression known as **Snell's law:**

$$\boxed{n_1 \sin \theta_1 = n_2 \sin \theta_2} \qquad \textbf{(Eq. 28.2)}$$

where n_1 and n_2 are the indices of refraction in the respective media. Snell's law can be used to compute the angle of refraction in a medium.

EXAMPLE 28.2 A light beam in air enters water in a container at an incident angle of 30°. What is the angle of refraction in the water?

Solution: With $\theta_1 = 30°$, $n_{air} = 1.00$, and $n_{H_2O} = 1.33$ (Table 28.1),

$$\sin \theta_2 = \frac{n_1 \sin \theta_1}{n_2}$$

$$= \frac{(1.00) \sin 30°}{1.33} = \frac{(1.00)(0.500)}{1.33} = 0.376$$

Then

$$\theta_2 = \sin^{-1}(0.376) = 22°$$

SPECIAL FEATURE 28.1

Atmospheric Refractions: Mirages, Hot Air, and Twinkling Stars

Atmospheric refraction effects are common. One of these is the mirage. The term *mirage* generally brings to mind the image of a thirsty person in a desert "seeing" a pool of water that really isn't there. This is not the mind playing tricks, but is an optical illusion. A more common mirage is the "wet spot" we often see on a hot highway in the summer that we can never seem to reach. What we are actually seeing is a reflection of the sky [Fig. 1(a)]. The warm air layer near the road is less dense than the cooler air above it. This density difference causes sky light to be refracted, or bent, to the eye. In the desert, a water image may be due to the refraction of light from the leaves of a tall, distant palm tree.

It is also refraction that allows us to "see hot air" rising from hot surfaces such as a road or a hot plate. If you stop and think for a moment, you'll know that you can't see air. What you are seeing is the refraction of light waves passing through the different-density regions of the rising convection currents. Similarly, atmospheric motions and density changes refract incoming starlight, causing stars to twinkle.

When watching a sunset (or moon-set), you may have noticed that the Sun near the horizon appears to be flattened [Fig. 1(b)]. This is also due to atmospheric refraction. The density of the atmosphere decreases with altitude, so the rays from the top and bottom portions of the Sun on the horizon are refracted by different degrees. This produces the observed flattening. Rays from the sides of the Sun on a horizontal plane are generally refracted by the same amount, so the Sun still appears circular along its sides. Also, because of refraction the Sun can be seen for several minutes after it actually sinks below the horizon.

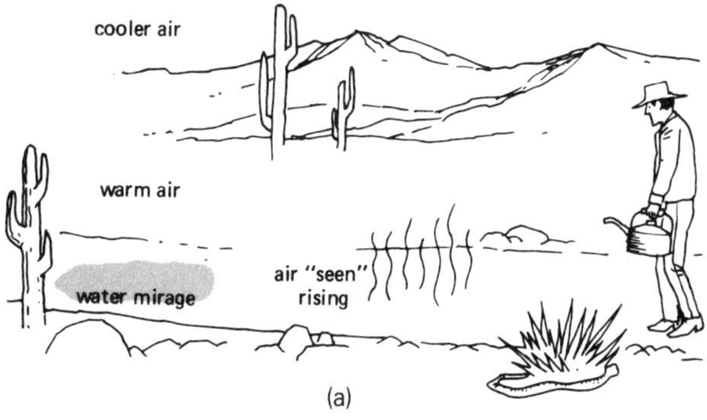

Figure 1
Atmospheric refraction. (a) Layers of air at different temperatures and densities act like different media and refract or bend skylight so we see a water mirage on hot road surfaces. Refraction also accounts for being able to "see" hot air rising. (b) The Sun appears flattened because of different refraction of light rays from the top and bottom portions. [(b) Courtesy G. Holton, F. Rutherford, and F. Watson.]

The index of refraction varies slightly with the wavelength of light and the temperature of the material. The variation of the index of refraction with wavelength gives rise to the dispersion property of prisms, discussed in the preceding chapter. Note from Snell's law (as applied in Example 28.2) that different indices of refraction give quite different refraction angles or a spreading of wavelength components.

Snell's law also provides a means of measuring the index of refraction of a transparent material. The test material is placed in contact with a material of known index of refraction. A light beam passing through the materials is at a given

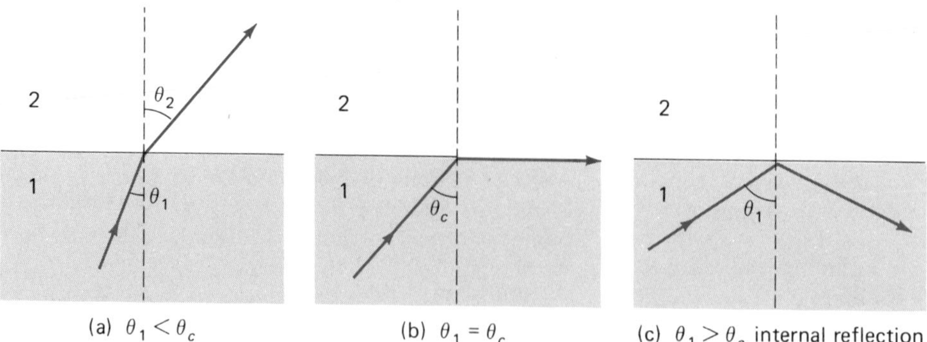

Figure 28.5
Internal reflection. (a) A light ray passing into a less optically dense medium is bent away from the normal. (b) If the angle of incidence θ_1 is increased to a critical angle θ_c, the refracted ray is parallel to the surface. (c) If $\theta_1 > \theta_c$, total internal reflection occurs.

angle of incidence to the surface of the test material. By measuring the angle of refraction, we can obtain the index of refraction of the test material. An instrument used to measure indices of refraction is called a *refractometer.*

Total Internal Reflection

As already learned, when light passes from a denser medium into a less dense medium, e.g., from water to air, the refracted beam is bent away from the normal (Fig. 28.5). If the angle of incidence increases, the angle of refraction also increases according to Snell's law.

At some critical angle of incidence, $\theta_1 = \theta_c$, the angle of refraction is 90° and the refracted beam just grazes the surface of the water. Then, for incident angles greater than the critical angle, the light does not pass into the air at all but is totally reflected—just as though the surface were a mirror. This phenomenon is called **total internal reflection.**

The critical angle depends on the index of refraction of the material. When $\theta_1 = \theta_c$, then $\theta_2 = 90°$, and by Snell's law,

$$\frac{\sin \theta_1}{\sin \theta_2} = \frac{\sin \theta_c}{\sin 90°} = \frac{n_2}{n_1} \qquad \text{where } n_2 < n_1$$

and

$$\sin \theta_c = \frac{n_2}{n_1} \qquad \textbf{(Eq. 28.3)}$$

since $\sin 90° = 1$. When the second medium is air ($n_2 = 1.00$), the equation becomes

$$\boxed{\sin \theta_c = \frac{1}{n}} \qquad \textbf{(Eq. 28.4)}$$

(critical angle of total internal reflection, air incidence)

where n is the index of refraction of the material in which the light is internally reflected.

EXAMPLE 28.3 What is the critical angle for total internal reflection in crown glass?

Solution: From Table 28.1, for crown glass $n = 1.50$. Then,

$$\sin \theta_c = \frac{1}{n} = \frac{1}{1.50} = 0.67$$

and

$$\theta_c = \sin^{-1}(0.67) = 42°$$

The property of total internal reflection allows specially cut prisms to reflect light in optical instruments in place of mirrors. For example, a right-angle prism can be used to reflect light through 90° (Fig. 28.6). Since the critical angle for glass is less than 45°, the light rays are totally reflected. Two such prisms can be used to construct a periscope.

Using total internal reflection, light can be "piped" from one location to another in glass or plastic rods (Fig. 28.7). On entering the light pipe, the light undergoes repeated internal re-

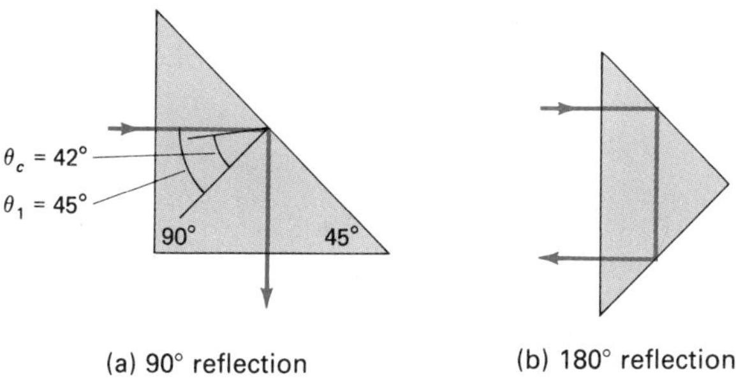

Figure 28.6
Internal reflections and right angle prisms. (a) Since $\theta_1 > \theta_c$, the light rays are internally reflected. When the incident beam is normal to a surface adjacent to the right angle of the prism, it is reflected through 90°. (b) If the incident beam is normally incident on the face opposite the right angle, it is reflected through 180° by two internal reflections.

flections and follows the contour of the pipe. Flexible light pipes can be constructed by fusing bundles of transparent fibers together. This has given rise to the relatively new field of **fiber optics.**

Flexible light guides can be used to transmit light to otherwise inaccessible places. Also, by using two sets of fibers in a well-randomized fiber bundle, light can be transmitted through one set and returned by the other. This arrange-

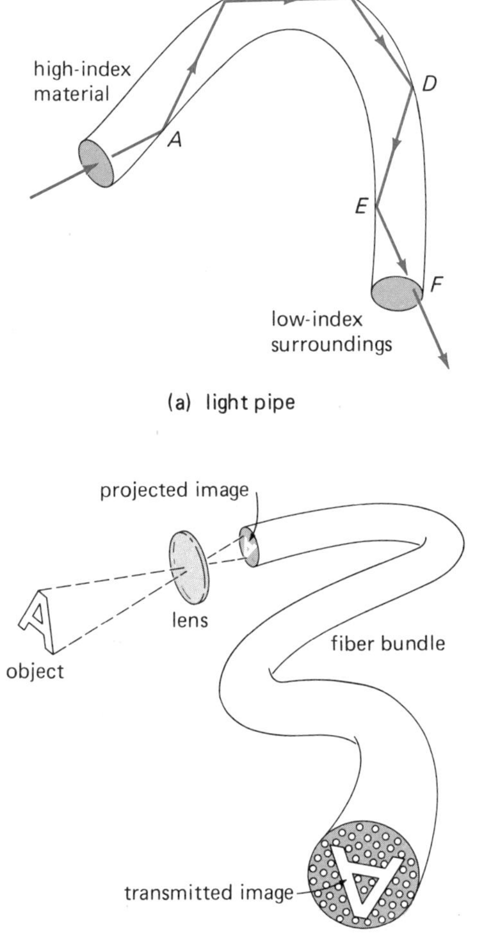

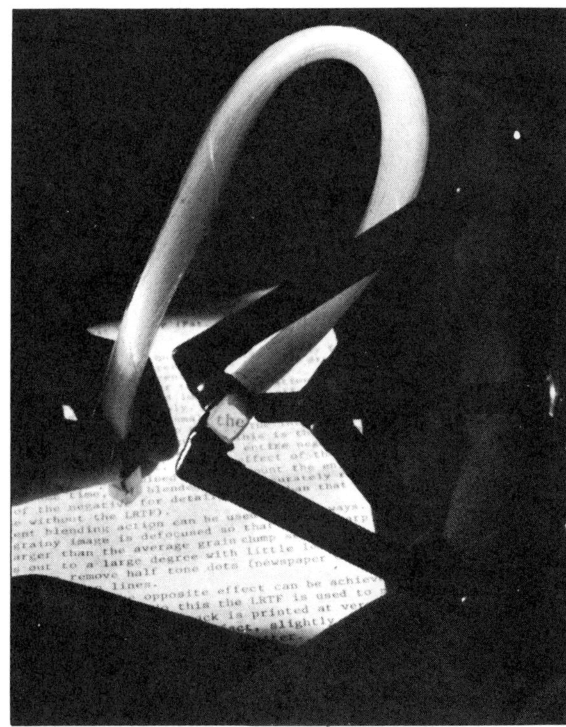

Figure 28.7
Fiber optics. (a) A light pipe. By internal reflections, light is "piped" around the contour of the rod. (b and c) By using fused bundles of transparent fibers, flexible light pipes can be constructed and images transmitted. (Courtesy L. Greenberg.)

(a)

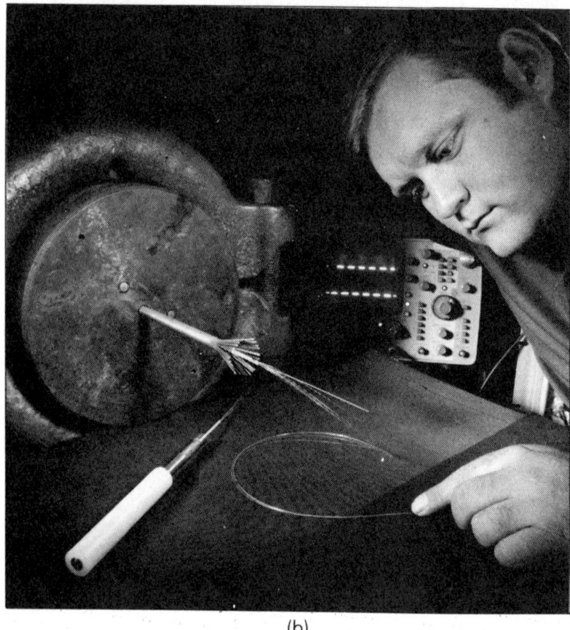

(b)

Figure 28.8
Fiber optics applications. (a) A portable, battery-powered fiberscope allows visual inspections of places that cannot be viewed directly. Here, a technician inspects a valve inside a pipe. (b) Inspection of optical fibers that will be used in fiber optic telephone communications. [(A) courtesy American Optical. (b) Courtesy AT&T Bell Laboratories.]

ment is used in a fiberscope, which allows visual inspection of places that cannot be viewed directly (Fig. 28.8). Such techniques with fiber optics have made it possible for physicians to view internal portions of the human body, such as the stomach and heart valves. Optical fibers are the basis of a newly developed form of telephone communication that uses transmitted light signals instead of electronic signals in wires.

A natural effect involving internal reflection and dispersion is discussed in Special Feature 28.2.

28.3 Mirrors

A mirror is any smooth surface that regularly, or specularly, reflects light. As a result, a mirror produces images of objects. A still pool of water can act as a mirror. Reflections can be seen in a glass windowpane in the evening or at night when the light transmitted from the outside is reduced.

Commercially, mirrors are available as polished metal surfaces and, most commonly, glass surfaces that have been "silvered" with compounds of tin or mercury. (Silver is still applicable as a mirror coating but is more expensive than tin or mercury.) A glass mirror may be front-coated or back-coated, depending on the application. Common glass mirrors are usually back-coated, but mirrors for telescopes are front-coated. Two common types of mirrors are plane mirrors and spherical mirrors.

Plane Mirrors

Reflecting surfaces that are flat, or planar, form plane mirrors. The common wall mirror in a bathroom is a plane mirror. We use plane mirrors to view images— often our own. Figure 28.9 illustrates how the image of an object is found for a plane mirror using ray optics. Each ray from an object obeys the law of reflection, $\theta_i = \theta_r$, and to an observer the light appears to come from an image "inside" or behind the mirror. Notice that the image is the same distance behind the mirror as the object is in front of the mirror, i.e., $d_i = d_o$.

There is also a right-left reversal of the object and the image. As we all have observed when

SPECIAL FEATURE 28.2

The Rainbow

What causes a rainbow? Rain, of course — or, more correctly, water droplets in the air after a rain — and the Sun must be there too. But we do not see a rainbow after every rain, so there must be some special conditions. Basically, whether or not we see a rainbow depends on the relative positions of the Sun and the observer. You may have noticed that the Sun is behind you when you see a rainbow.

The rainbow is produced by refraction, dispersion, and internal reflection in water droplets. Occasionally, more than one rainbow can be seen. The primary rainbow is sometimes accompanied by a fainter and higher secondary rainbow arc. Have you ever seen a secondary rainbow? For the primary rainbow, light is reflected once inside a water droplet (Fig. 1). Being refracted and dispersed, the sunlight is spread out into a spectrum of colors.

Because of the conditions on the refraction and internal reflection, the angles between the incoming and outgoing visible-light colors lie in the narrow range of 40° for violet to 42° for red. This means that you see a rainbow only when the Sun is positioned so that the dispersed light is reflected to you through these angles. With millions of water droplets in the air, a colorful arc is seen, running vertically in color from violet to red. (Below the rainbow arc, the light from the droplets combines to form a brightly illuminated region.)

The less frequently seen secondary rainbow is caused by a double reflection in the water droplets. This gives rise to a sequence of colors in the secondary rainbow that is an inversion of that in the primary rainbow.

As the Sun rises, less of the rainbow arc is seen. In fact, an observer on the ground cannot see a primary rainbow when the Sun's altitude (angle above the horizon) is greater than 42°. The primary rainbow is then below the horizon. As an observer is elevated, more of a rainbow arc is seen. It is common to see a completely circular rainbow from an airplane (no rainbow's end, no pot of gold). You have probably seen a circular "rainbow" in the fine spray of a garden hose.

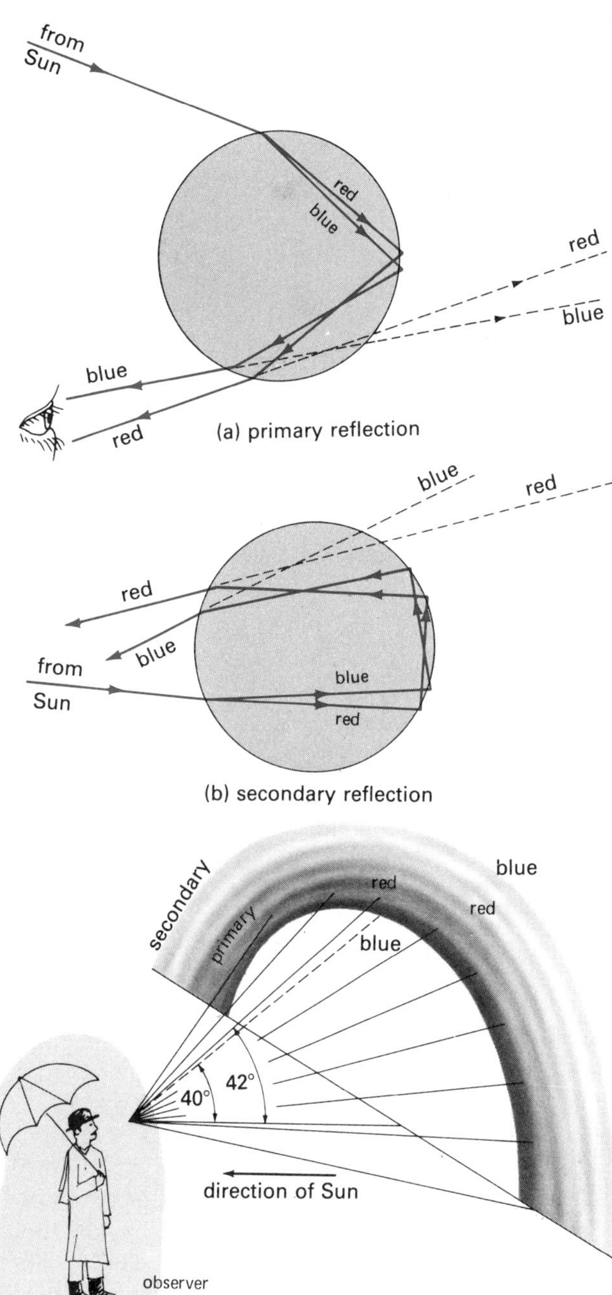

Figure 1
Dispersion, refraction, internal reflection, and rainbows. (a) Single and (b) double internal reflections can occur in water droplets, giving rise to primary and secondary rainbows. (c) Dispersion separates the component colors of sunlight, and an observer sees different colors coming from different altitudes or a rainbow of colors.

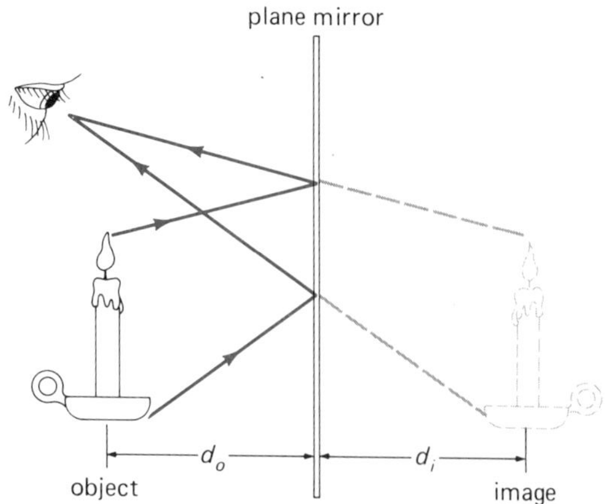

Figure 28.9
Plane mirror image. The image is found by ray tracing and the law of reflection. The image is the same size as the object and "behind" the mirror the same distance as the object is in front of the mirror, $d_i = d_o$.

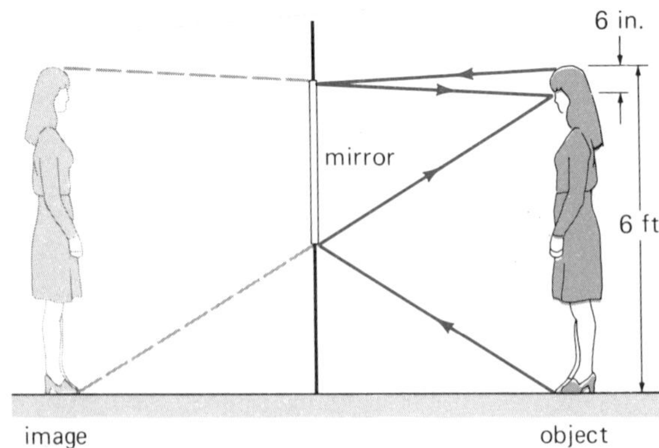

Figure 28.10
See Example 28.4.

looking into a plane mirror, if you raise your right hand, your image will raise its left hand. One other characteristic of the image formed by a plane mirror is that it is the same size as the object. This is not necessarily the case for spherical mirrors, as we shall discuss.

EXAMPLE 28.4 What is the minimum length of a plane mirror that can be used by a person 6 ft tall to view her complete image (see Fig. 28.10)?

Solution: With the eyes 6 in. below the top of the person's head, the top edge of the mirror is at a height of 5 ft 9 in., since the normal to the point of reflection bisects the distance (6 in.) from the top of the head to the eyes ($\theta_i = \theta_r$).

Similarly, the normal of the reflected ray from the feet to the eyes bisects the distance from the feet to the eyes (5 ft 6 in. = 66 in.), and 66 in./2 = 33 in., which is the height of the bottom of the mirror. Hence, the mirror is 69 in. − 33 in. = 36 in. = 3 ft in length, or one-half the person's height.

Spherical Mirrors

Spherical mirrors are regularly reflecting spherical surfaces (Fig. 28.11). This allows for two types of spherical mirrors: concave and convex. For a **concave mirror,** the reflecting surface is the inside of the spherical surface. (Like a *cave*, a con*cave* mirror is recessed.) For a **convex mirror,** the reflecting surface is the outer portion of the spherical surface.

Both concave and convex mirrors are characterized by a **radius of curvature *R*,** which is the radius of the sphere of which the spherical surface is a part. A line through the **center of curvature *C*** and the center of vertex *V* of the mirror surface is called the **mirror axis.** The distance from the vertex *V* to the center of curvature *C* is equal to the radius of curvature *R*.

Another important point on a spherical mirror is the **focal point.** It can readily be shown that the focal point lies midway between *V* and *C*; hence,

$$f = \frac{R}{2} \qquad \textbf{(Eq. 28.5)}$$

(focal length for spherical mirrors)

That is, the focal length is one-half the radius of curvature.

The significance of the focal point is illustrated in Figure 28.11. For a concave mirror, rays parallel to the mirror axis are reflected through, and hence are "focused" (converge) at, the focal point. (The law of reflection applies to each ray.) As a result, **a concave mirror is called a *converging mirror*.**

Similarly, rays parallel to the axis of a convex mirror diverge as though they come from a point source at the focal point. Hence, **a convex mirror is called a *diverging mirror*.** Practical exam-

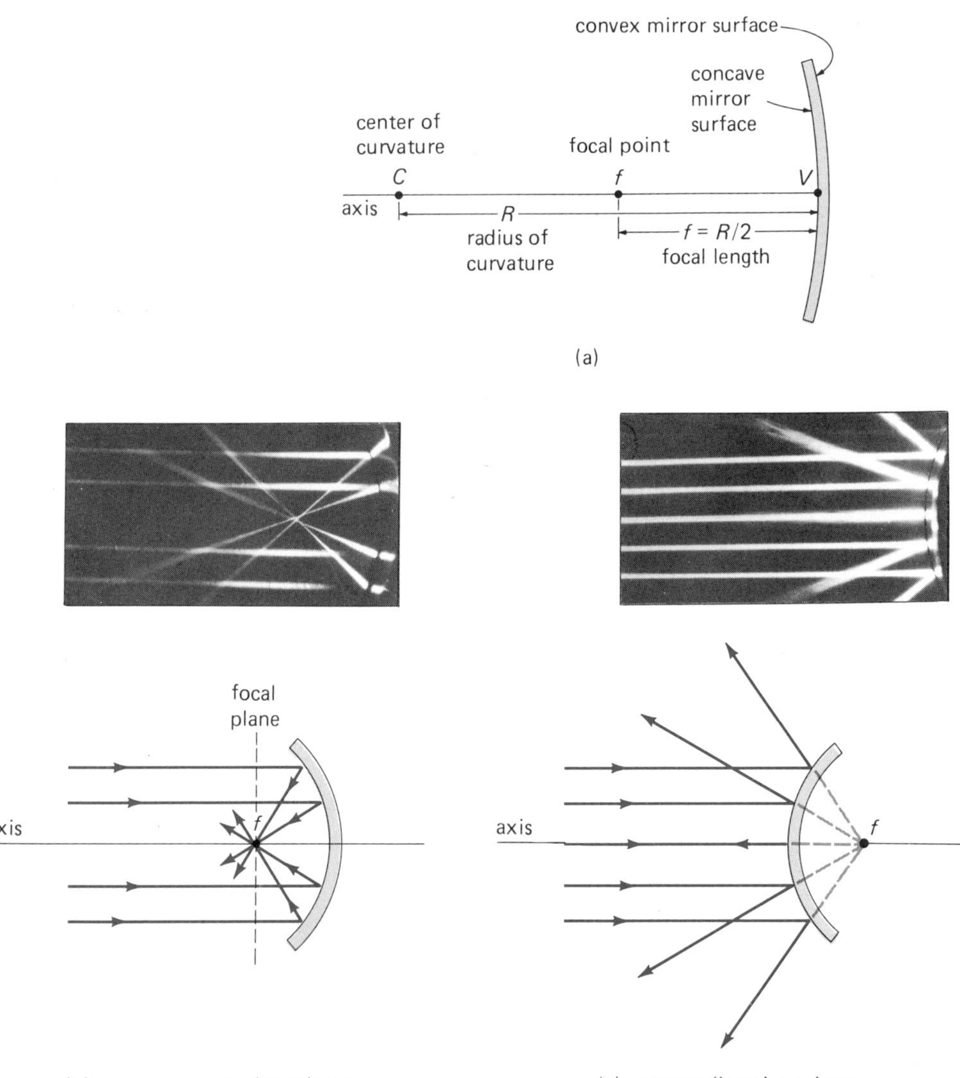

Figure 28.11
Spherical mirrors. (a) Mirror geometry and nomenclature. (b) Rays parallel to the mirror axis of a concave mirror are reflected through or converge at the focal point. (c) Rays parallel to the mirror axis of a convex mirror are reflected as though they diverge from the focal point. (Courtesy P. Highsmith.)

ples of converging and diverging mirrors are shown in Figure 28.12.

The images formed by spherical mirrors can be determined by ray diagrams in which two (or three) particular rays are used. As illustrated in Figure 28.13, these are

1. **Parallel ray.** A ray parallel to the mirror axis is reflected through the focal point (concave mirror) or appears to come from the focal point (convex mirror).
2. **Chief ray.** A ray through the center of curvature is reflected back along the original ray path.
3. **Focal ray.** A ray through the focal point is reflected parallel to the mirror axis.

Actually, only two rays are needed to define the image, e.g., rays 1 and 2 in Figure 28.13. It is customary to use an arrow for an object. The point at which the rays from the tip of the object arrow intersect defines the location of the tip of the image arrow.

The image is described by its characteristics, which are

Real or virtual
Upright or inverted (in orientation)
Enlarged or reduced (in size)

Figure 28.12
Spherical mirror applications. (a) A bulb at the focal point of a spherical flashlight reflector gives a parallel beam. (b) A spherical "mirror" for converging radio and television waves from satellites. The detector is located at the focal point of the TV "dish" antenna. (c) Convex spherical mirrors provide a wide-angle view for store monitoring. (d) A convex mirror is used to give wide-angle views in mountainous regions in order to see approaching traffic around curves, here in Germany. Note the distances in kilometers.

With regard to real and virtual images,

A real image is formed by the convergence of light rays and can be formed on a screen (Fig. 28.14).

A virtual image is one from which light rays appear to diverge and which cannot be formed on a screen.

Another way of thinking of a virtual image is one that is formed "inside" or behind the mir-

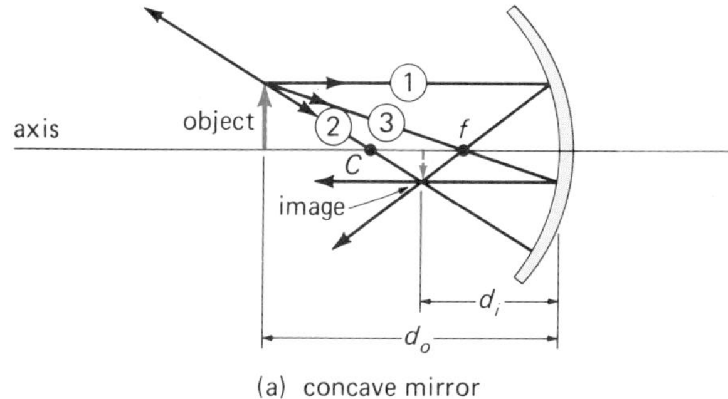

(a) concave mirror

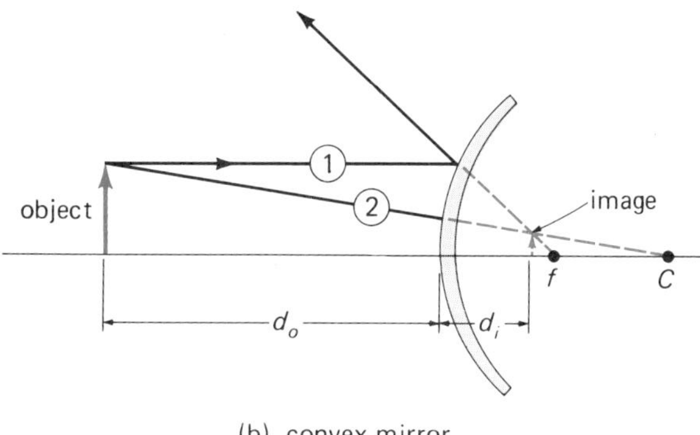

(b) convex mirror

Figure 28.13
Ray diagrams. (a) For a concave or converging mirror. (b) For a convex or diverging mirror. See text for descriptions.

ror. From experience and after drawing a few ray diagrams, it becomes evident that

> Plane mirrors and convex (diverging) spherical mirrors always form virtual images.

In particular, all the images for a convex mirror are virtual, upright, and reduced in size. *Note:* we are able to see a virtual image in a mirror because the real image of the object is formed on the retina "screen" of our eye (see Chapter 29).

A converging mirror can form both real and virtual images, depending on the position of the object. There are three object regions and two boundary locations along the mirror axis, as illustrated in Figure 28.15. The image characteristics for these regions and point locations are also given in the figure.

The Spherical Mirror Equation The image distance d_i and the magnification (enlargement or reduction) factor can be found from ray diagrams if they are drawn to scale. However, the same information can be obtained analytically from the spherical mirror equation and the magnification equation. These are:

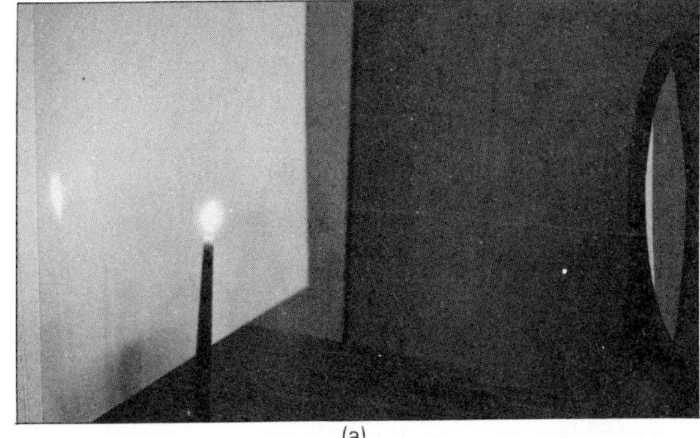

(a)

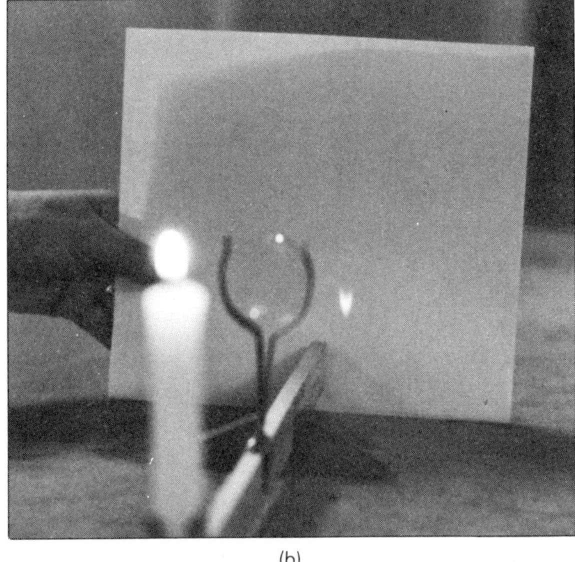

(b)

Figure 28.14
Real images can be formed on a screen. The image of the candle flame is formed on a paper screen by a small concave spherical mirror (a) and a biconvex spherical lens (b). Notice that the images are inverted.

$$\frac{1}{d_o} + \frac{1}{d_i} = \frac{1}{f} \qquad \textbf{(Eq. 28.6)}$$

(spherical mirror equation)

$$M = -\frac{d_i}{d_o} \qquad \textbf{(Eq. 28.7)}$$

(magnification equation)

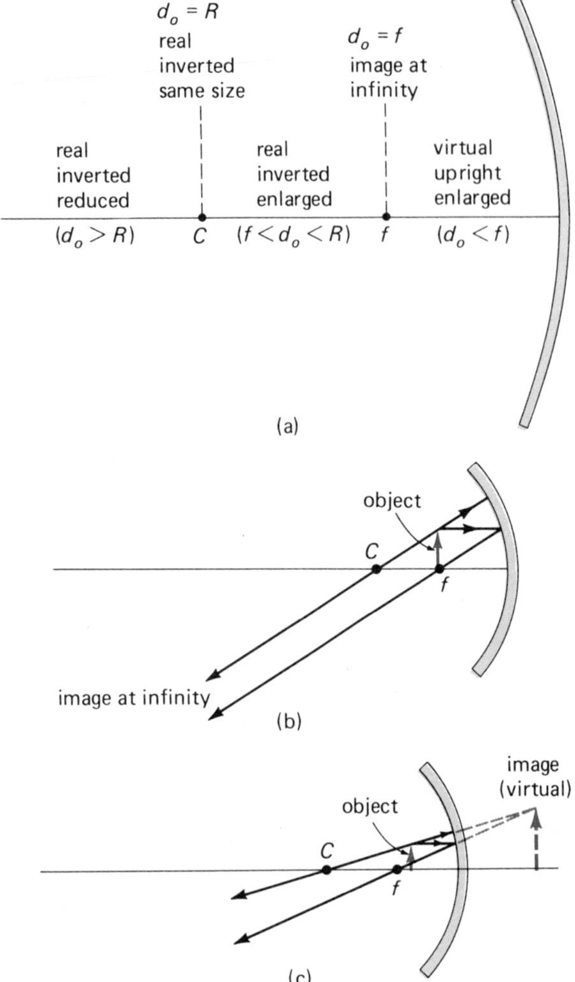

Figure 28.15
Image characteristics for spherical converging mirrors. (a) Regions and points for given object distances with image characteristics. (b) Ray diagrams for object at focal point. Reflected rays are parallel, and we say they converge or form an image at infinity. By reverse ray tracing, the image of an object at infinity is formed in the focal plane of the mirror (cf. Fig. 28.11). (c) Ray diagram for object inside the focal point. The rays appear to come from a virtual image inside or behind the mirror.

The magnification equation gives the linear or height **magnification factor M** of the image. For example, if $M = 2 = y_i/y_o$, then $y_i = 2y_o$, or the image is twice as tall as the object.*

* A helpful way to remember that M is given by the ratio of d_i/d_o is that it is i over o, or in alphabetical order.

Table 28.2
Sign Convention for Spherical Mirror Equation

d_o is always positive		Concave (converging) mirror: f is positive Convex (diverging) mirror: f is negative	
If d_i is	**Image is**	**If M is**	**Image is**
+	Real	+	Upright
−	Virtual	−	Inverted

The minus sign arises from a sign convention used for spherical mirrors, which gives the characteristics of an image. This convention is given in Table 28.2.

Other useful forms of the mirror equation (Eq. 28.6), depending on what is being solved for, are

$$d_i = \frac{d_o f}{d_o - f} \qquad \textbf{(Eq. 28.6a)}$$

or

$$d_o = \frac{d_i f}{d_i - f} \qquad \textbf{(Eq. 28.6b)}$$

or

$$f = \frac{d_i d_o}{d_i + d_o} \qquad \textbf{(Eq. 28.6c)}$$

EXAMPLE 28.5 An object is placed 15 cm in front of a concave mirror that has a radius of curvature of 20 cm. Where is the image formed, and what are its characteristics?

Solution: It is given that $d_o = 15$ cm and $R = 20$ cm; hence, $f = R/2 = 20$ cm$/2 = 10$ cm. Also, since the mirror is concave, f is positive. Using Equation 28.6 in one of its alternative forms,

$$d_i = \frac{d_o f}{d_o - f} = \frac{(15 \text{ cm})(10 \text{ cm})}{15 \text{ cm} - 10 \text{ cm}} = 30 \text{ cm}$$

Hence, the image is real ($+d_i$) and is formed 30 cm in front of the mirror. Then,

$$M = -\frac{d_i}{d_o} = -\frac{30 \text{ cm}}{15 \text{ cm}} = -2$$

and the image is inverted ($-M$) and is twice as tall as the object, $|M| = 2$ (enlarged by a factor of two).

EXAMPLE 28.6 An object is placed 30 cm in front of a diverging mirror that has a focal length of 10 cm. Where is the image, and what are its characteristics?

Solution: It is given that $d_o = 30$ cm and $f = -10$ cm (minus because the mirror is convex; cf. Table 28.2). Then,

$$d_i = \frac{d_o f}{d_o - f} = \frac{(30 \text{ cm})(-10 \text{ cm})}{30 \text{ cm} - (-10 \text{ cm})} = -7.5 \text{ cm}$$

Hence, the image is virtual ($-d_i$, as would be expected) and appears to be 7.5 cm behind the mirror. Then,

$$M = -\frac{d_i}{d_o} = -\frac{(-7.5 \text{ cm})}{30 \text{ cm}} = +0.25$$

and the image is upright ($+M$) and is reduced by a factor of 0.25, or one fourth.

EXAMPLE 28.7 Using the spherical mirror equation, show that the image distance is equal to the object distance for a plane mirror.

Solution: A plane mirror is essentially a spherical mirror with a radius of curvature of infinity, $R = \infty$. (Infinitely far from the center of curvature, the surface of a sphere approximates a plane.) Then $f = \infty$ also, since $R = f/2 = \infty/2 = \infty$, and

$$\frac{1}{d_o} + \frac{1}{d_i} = \frac{1}{f} = \frac{1}{\infty} = 0$$

and

$$\frac{1}{d_o} = -\frac{1}{d_i}$$

or

$$d_i = -d_o$$

The minus sign indicates that the image is virtual. Also notice that

$$M = -\frac{d_i}{d_o} = -\left(\frac{-d_o}{d_o}\right) = 1$$

EXAMPLE 28.8 Where is the image formed for a concave mirror for an object at infinity?

Solution: With $d_o = \infty$,

$$\frac{1}{d_o} + \frac{1}{d_i} = \frac{1}{\infty} + \frac{1}{d_i} = \frac{1}{f}$$

and

$$d_i = f$$

Hence, the image is real ($+d_i$) and is formed at the focal point (or in the focal plane for an extended object).

This case provides a method for determining the focal length (and radius of curvature) of a concave mirror, since the distance from the vertex of the mirror to the screen (focal plane) is the focal length. Objects far from the mirror may be considered at infinity.

28.4 Lenses

Lenses are usually made of glass (sometimes of plastic) and make use of the refractive properties of the material. Recall that refraction is the bending of light as it passes from one medium into another. For example, light rays are bent, or deviated from their original paths, by a prism.

It two prisms are placed base to base, they tend to converge an incident light beam parallel to their common base line (Fig. 28.16). As such, the prisms approximate a spherical biconvex lens. The spherical biconvex lens is a type of converging lens. Similarly, if two prisms are placed point to point, they approximate a spherical biconcave lens, which is a type of diverging lens.

Meniscus lenses are the type commonly used in corrective glasses. They include several types of converging and diverging lenses (Fig. 28.17). In general, a converging lens is thicker at its center than at its periphery, and a diverging lens is thinner at its center than at its periphery.

Analyses of lenses can become quite complicated, so we will limit our discussion to thin biconvex and biconcave lenses. The geometries of these lenses are shown in Figure 28.18. Because of symmetry, there is a center of curvature for each surface an equal distance on each side of a lens on an axis through the lens.

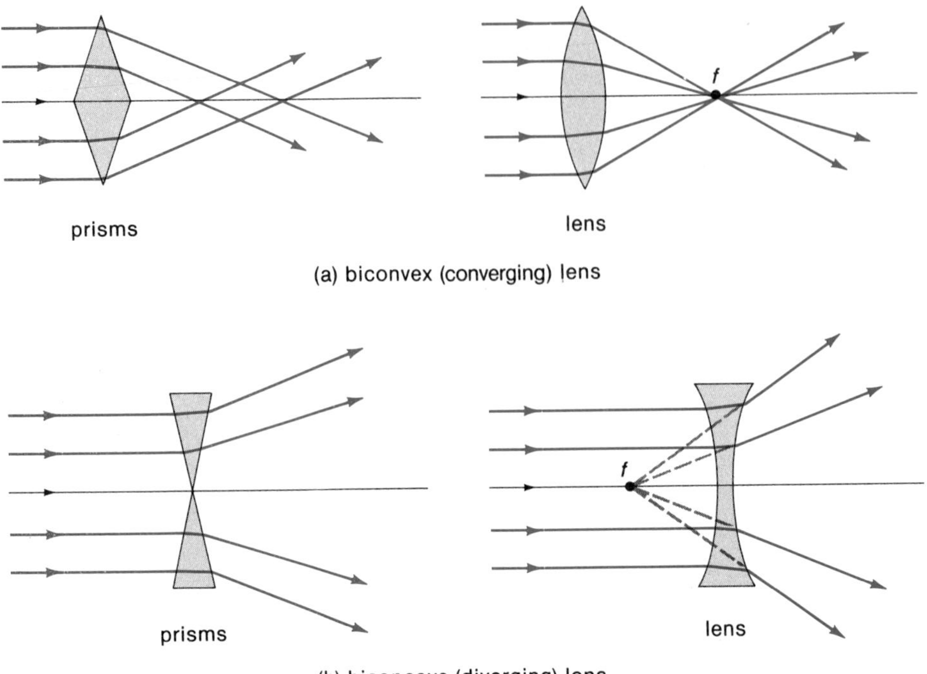

Figure 28.16
Spherical lenses. (a) A biconvex lens converges light rays at a common focal point. This is approximated by two prisms base-to-base. (b) A biconcave lens is a diverging lens and is approximated by two prisms tip-to-tip.

However, the focal length is *not* equal to one-half the radius of curvature, as for a spherical mirror. For a lens, the focal length of a surface depends on the index of refraction of the lens material. (See the Chapter Supplement.)

For a convex lens, incident rays parallel to the lens axis converge, or are "focused," at the focal point on the opposite side of the lens (see Fig. 28.16). Rays parallel to the axis of a concave lens appear to diverge from the focal point on the incident side of the lens.

Similar to spherical mirrors, the images formed by spherical lenses can be determined by ray diagrams in which two (or three) particular rays are used. As illustrated in Figure 28.19, these are

1. **Parallel ray.** A ray parallel to the lens axis is refracted through the focal point (converging convex lens) or appears to come from the focal point (diverging concave lens).
2. **Chief ray.** A ray through the center of the lens is undeviated.

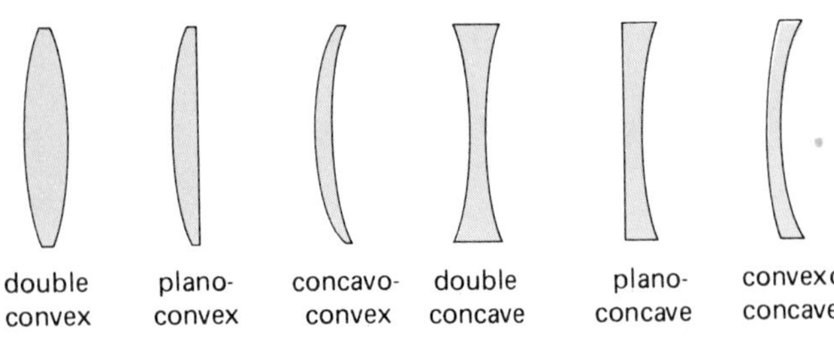

Figure 28.17
Types of lenses. In general, a converging lens is thicker at its center than at its periphery, and a diverging lens is thinner at its center than at its periphery.

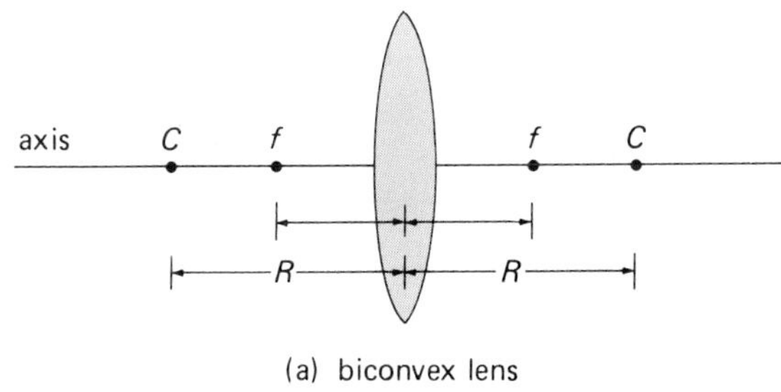

(a) biconvex lens

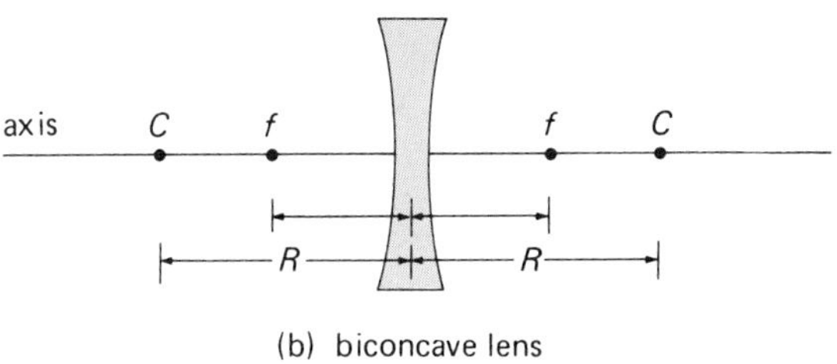

(b) biconcave lens

Figure 28.18

Lens geometry. (a) For a biconvex lens and (b) for a biconcave lens. The nomenclature is the same as that for a spherical mirror. However, because of symmetry, lenses have centers of curvature and focal points on each side. The condition of $R = f/2$ for spherical mirrors *does not* hold for spherical lenses, but varies with the lens material. (See Chapter Supplement.) As a result, C and R are commonly omitted in ray diagrams.

3. **Focal ray.** A ray through the focal point is refracted parallel to the lens axis.

As can be seen from the figure,

> A convex lens always forms a virtual image when the object is inside the focal point.

Also,

> A concave lens always forms a virtual image.

It is convenient to refer to the opposite sides of a lens as the object side and the image side. When the image is formed on the image side (side opposite the object), it is real and can be formed on a screen as shown in Figure 28.14. When the image is formed on the object side, it is virtual.

The Spherical Thin Lens Equation

For spherical lenses there are a thin lens equation and a magnification equation, which are

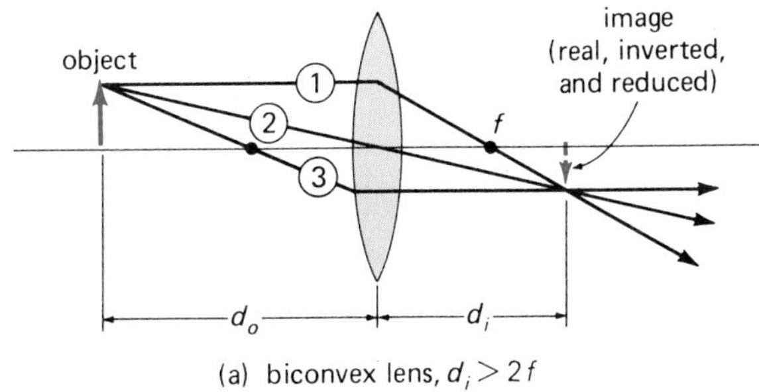

(a) biconvex lens, $d_i > 2f$

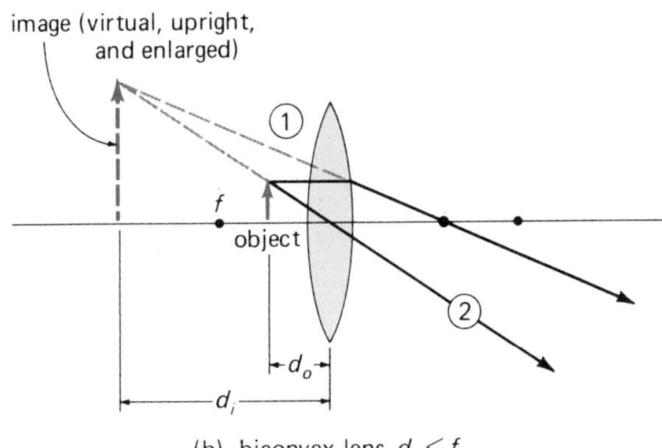

(b) biconvex lens, $d_i < f$

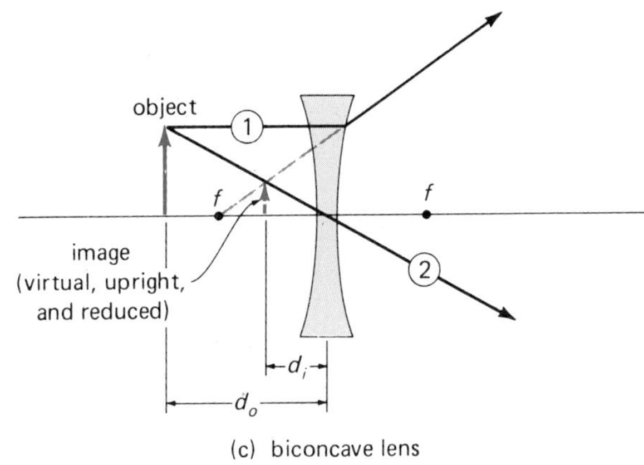

(c) biconcave lens

Figure 28.19

Ray diagrams. (a and b) For a biconvex lens with $d_o > 2f$ and $d_o < f$, respectively. (c) For a biconcave lens. See text for descriptions.

identical to the equations for spherical mirrors:

$$\frac{1}{d_o} + \frac{1}{d_i} = \frac{1}{f} \qquad \textbf{(Eq. 28.8)}$$

(thin lens equation)

The alternative forms given as Equations 28.6(a–c) still apply.

Table 28.3
Sign Convention for Thin Lenses

d_o is always positive		Convex (converging) lens: f is positive Concave (diverging) lens: f is negative	
If d_i is	**Image is**	**If M is**	**Image is**
+	Real	+	Upright
−	Virtual	−	Inverted

Also,

$$M = -\frac{d_i}{d_o} \qquad \textbf{(Eq. 28.9)}$$

(magnification equation)

The similar sign convention is given in Table 28.3.

EXAMPLE 28.9 An object is placed 25 cm in front of a biconvex converging lens that has a focal length of 10 cm. Where is the image formed, and what are its characteristics?

Solution: It is given that $d_o = 25$ cm and $f = 10$ cm. Also, since the lens is biconvex, f is positive. Then,

$$d_i = \frac{d_o f}{d_o - f} = \frac{(25 \text{ cm})(10 \text{ cm})}{25 \text{ cm} - 10 \text{ cm}} = 16.7 \text{ cm}$$

Hence, the image is real ($+d_i$) and is formed 16.7 cm from the lens on the image side (the side opposite the object). Also,

$$M = -\frac{d_i}{d_o} = -\frac{16.7 \text{ cm}}{25 \text{ cm}} = -0.67$$

and the image is inverted ($-M$) and reduced to two-thirds the size of the object, $y_i = |M|y_o = (0.67)y_o$.

Lens Combinations

In many optical instruments, such as microscopes and telescopes (discussed in the next chapter), a combination of lenses called a compound lens system is used. When two or more lenses are used in combination, the image of the system can be found by considering each lens individually. The image of one lens is used as the object of the next, and so on. This is illustrated in Figure 28.20 for a two-lens system.

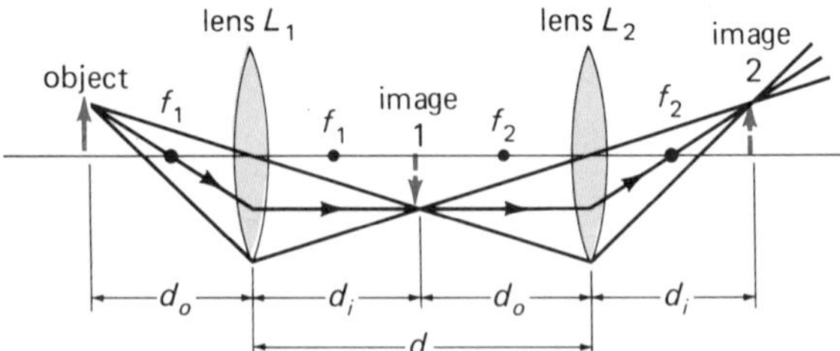

Figure 28.20
Lens combination. The image is found by treating the lenses individually. The image of the first lens is used as the object for the second lens. See text for description.

The lens equation can be used to determine the location of the final image by applying it to each lens. This is straightforward for the case in Figure 28.20, where the real image of the first lens becomes a real object for the second lens.

However, if the lenses are close enough together so that the image of the first lens is not formed before the rays are intercepted by the second lens, a modification must be made in the sign convention. In this case, the image of the first lens is treated as a *virtual* object for the second lens, and the object distance of the virtual object is taken to be *negative* in the lens equation.

The total magnification M_t of a compound lens system is given by the product of the magnification factor for each lens. For example, for a two-lens combination as in Figure 28.20,

$$M_t = |M_1| \times |M_2| \qquad \textbf{(Eq. 28.10)}$$

Aberrations

Spherical mirrors and lenses may not form perfect images owing to material defects *and* to various effects called **aberrations.** The latter are natural, inherent effects. Two will be considered here—spherical aberration, which applies to both mirrors and lenses, and chromatic aberration, which applies to lenses.

Spherical Aberration The previous discussion on the image formations of mirrors and lenses actually applies only for objects on or near the mirror or lens axis. If the object is not near the axis, a blurred, out-of-focus image is formed because the rays near the periphery of the con-

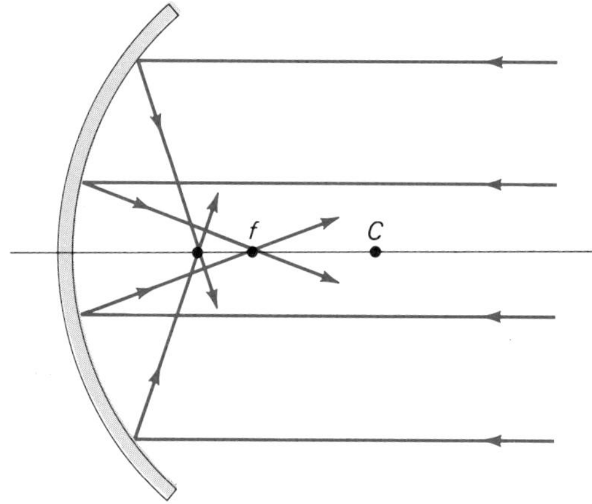

Figure 28.21

Spherical aberration. Rays near the periphery of a spherical converging mirror (or lens) are brought to focus nearer the mirror than those rays near the axis, giving a blurred or distorted image.

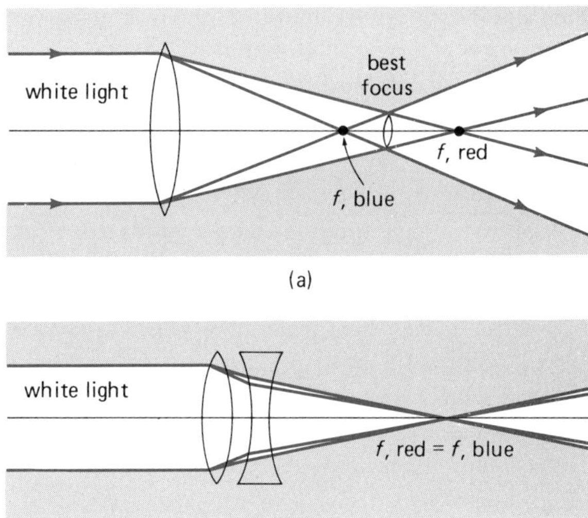

Figure 28.22

Chromatic aberration. (a) Because of dispersion, a lens has different focal points for different wavelengths or colors. (b) To eliminate chromatic aberration, an achromatic doublet is used, which focuses the colors at the same point.

verging spherical mirror or spherical lens are brought to focus nearer the mirror or lens than are those rays near the axis (Fig. 28.21).

Spherical aberration may be eliminated through the use of parabolic (parabola-shaped) mirrors. All the incident rays parallel to the axis of a parabolic mirror have a common focus. The spherical aberration of a lens can be reduced by placing a diaphragm in front of the lens. A diaphragm has an adjustable aperture, or hole, and the outer rays can be blocked off by reducing the aperture. Of course, this also reduces the intensity.

Chromatic Aberration As we have learned, the index of refraction is slightly different for different wavelengths of light. Hence, there is a dispersion effect as white light passes through a lens. As a result, the rays of the different wavelength components do not come together in a common focus (Fig. 28.22).

This aberration effect can be eliminated by using a lens combination called an *achromatic doublet* (achromatic—without color). The lenses of the achromatic doublet are made of two different materials with different indices of refraction, e.g., crown glass and flint glass. Essentially, one lens refractively counteracts the dispersion of the other lens, so a sharp image is formed.

Important Terms

wave front adjacent portions of a wave that are in phase

ray a line perpendicular to wave fronts and in the direction of wave propagations

geometrical (ray) optics the analysis of light phenomena using rays

law of reflection the angle of incidence is equal to the angle of reflection, i.e., $\theta_i = \theta_r$

regular (specular) reflection reflection from a smooth mirror surface

irregular (diffuse) reflection reflection from a rough or irregular surface

refraction the deviation of light from its original direction when transmitted from one medium to another at an angle due to different wave speeds in the media, $n = c/c_m$

index of refraction (*n*) the ratio of the speed of light in vacuum and the speed of light in a medium

Snell's law a relationship between indices of refraction and angles of incidence and refraction:

$$n_1 \sin \theta_1 = n_2 \sin \theta_2$$

total internal reflection reflection at a surface when the incident angle is greater than the critical angle of the medium ($n_1 > n_2$)

fiber optics transmission of light in transparent pipes or fibers due to internal reflection

mirror any smooth surface that regularly or specularly reflects light

plane: a mirror with a flat or plane surface

spherical:

concave (converging)—a mirror with a reflecting surface that is the inside of a spherical surface

convex (diverging)—a mirror with a reflecting surface that is the outside of a spherical surface

real image one formed by the convergence of light rays and which can be formed on a screen

virtual image one from which light rays appear to diverge and which cannot be formed on a screen

focal point the point at which parallel rays are focused or appear to be focused for spherical mirrors and lenses

lens a transparent object that converges or diverges light due to refractive properties and shape

converging lens in general, one that is thicker at its center than at its periphery

diverging lens in general, one that is thinner at its center than at its periphery

aberrations inherent effects in mirrors and lenses that give rise to blurred or out-of-focus images, e.g., spherical aberration and chromatic aberration of reflection and refraction, respectively, due to off-axis effects

Important Formulas

law of reflection: $\theta_i = \theta_r$

index of refraction: $n = \dfrac{c}{c_m}$

Snell's law: $n_1 \sin\theta_1 = n_2 \sin\theta_2$

critical angle for total internal reflection: $\sin\theta_c = \dfrac{n_2}{n_1}$ $(n_2 < n_1)$

$= \dfrac{1}{n}$ (where n_2 is for vacuum of air)

focal length of spherical mirror: $f = \dfrac{R}{2}$

spherical mirror and thin lens equation and magnification factor:

$$\frac{1}{d_o} + \frac{1}{d_i} = \frac{1}{f}$$

$$M = -\frac{d_i}{d_o}$$

(other useful forms) $d_i = \dfrac{d_o f}{d_o - f}$

$$d_o = \frac{d_i f}{d_i - f}$$

$$f = \frac{d_i d_o}{d_i + d_o}$$

sign convention: d_o is always positive

Converging (concave mirror or biconvex lens): f is positive

Diverging (convex mirror or biconcave lens): f is negative

If d_i is	Image is	If M is	Image is
+	Real	+	Upright
−	Virtual	−	Inverted

Questions

Light Rays

1. How are wave fronts defined?
2. What is a ray, and how are rays used to represent light?
3. What is the difference between physical optics and geometrical optics?

Reflection and Refraction

4. A statement supplementary to the law of reflection that is often omitted is that the incident and reflected rays lie in a plane. Discuss what this means.
5. If the moon had a very smooth surface, how would a full moon appear?
6. Why are only parts of the laser beam visible in Figure 28.23? (Blackboard erasers are being clapped together in the background.)

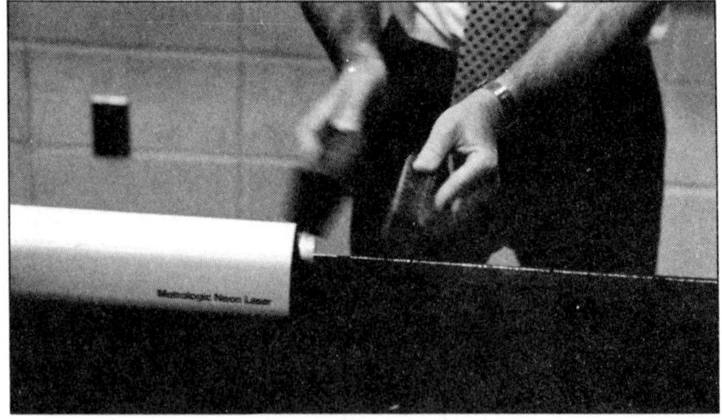

Figure 28.23
Visible laser beam. See Question 6.

Figure 28.24
Hot spot? See Question 8.

7. Why does the glass of a windowpane act as a mirror at night and not during the day? At night, two similar images, one behind the other, are often seen. Why is this?
8. What's going on in Figure 28.24? Is the finger being held over a candle flame? Explain. (*Hint:* there's a glass plate between the candles.)
9. How is light refracted when (a) entering a less dense medium at an angle, (b) entering a denser medium at an angle, and (c) in both cases for normal incidence?
10. Why should you expect the speed of light to be slower in a denser transparent medium?
11. In a lab experiment, the index of refraction of a transparent material is measured to be 1.01. Is this possible? Explain.
12. Is the speed of light greater in water or in ice?
13. How can the angle of refraction be computed?
14. A design or picture on the bottom of a swimming pool has varying distortions when viewed through the water in a filled pool. What causes these distortions?
15. While sitting by a swimming pool on a fixed bench in the spring, before the pool is filled, a person notices that he can see only a limited portion of the design on the bottom of the pool (Fig. 28.25). However, when the pool is filled, he notices that he can see another portion of the design that was previously hidden by the pool wall. How is this possible?
16. Does atmospheric refraction have any effect on the length of the day? That is, would the daylight hours be longer or shorter if we had no atmosphere?
17. Is it possible to have total internal reflection with light in air incident on water? Explain.
18. Consider light in air incident on a glass surface. What is the maximum angle of refraction that can be achieved?
19. Design a periscope to be used to see over and around objects, using (a) plane mirrors and (b) prisms.
20. Explain how refraction causes the pencil in Figure 28.4 to appear almost severed.

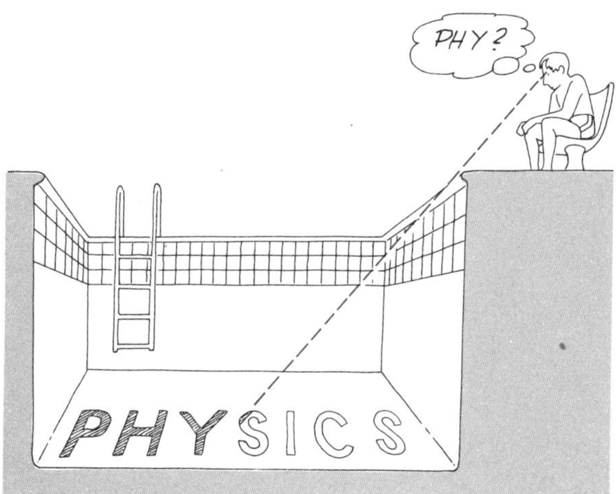

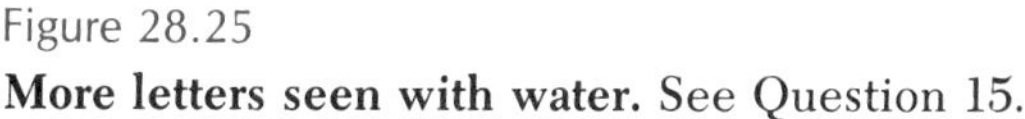

Figure 28.25
More letters seen with water. See Question 15.

21. (a) How much of the underwater world could you see, looking at various angles of incidence? (b) How would a fish see the above-water world when looking up at various angles?
22. Refraction properties are sometimes described with a marching band analogy. Consider the rows of a band marching at an angle into a muddy field with the band members in the dry and muddy regions maintaining the same cadence. Describe the analogies to the refraction of light in terms of frequency, wavelength, speed, and change of direction.
23. (a) Is there an absence of rainbows after some rains, or is there always a rainbow and an observer may not be in the proper location to see it? Explain. (Rains after dark are excluded.) (b) Do two observers at different locations see exactly the same rainbow? Explain your answer.

Mirrors

24. What are the characteristics of the image of an object for a plane mirror?
25. If you walk toward a plane, full-length mirror, what does your image do? How fast does the image move? Is the image in step with you (like marching in a band)?
26. In detective movies and also in the study of children's activities, one-way mirrors are used for secret observations. (A one-way mirror is seen as a mirror from one side but can be seen *through* from the other side.) Reflecting sunglasses are another example. How do one-way mirrors work? (*Hint:* at night, a glass windowpane is a one-way mirror.)
27. Operating rooms and dentists' lamps have large spherical reflectors. Also, flashlights have spherical reflectors. What is their purpose?
28. A dentist sometimes uses a small converging spherical mirror to examine cavities when filling them. What is the advantage of a converging mirror over a plane mirror?
29. Referring to Figure 28.26, (a) why is the wording on the ambulance painted backward? (Did the painter make an error in this case? Check it out with a plane mirror.) (b) What is the purpose of the dual mirror on the truck? (c) Explain why in some auto side-view mirrors "objects in mirror are closer than they appear."

Lenses

30. How are converging and diverging lenses approximated by prisms? Are prisms as effective as lenses? Explain.
31. Given two converging lenses of different focal lengths, which would give greater magnification for a fixed object distance?

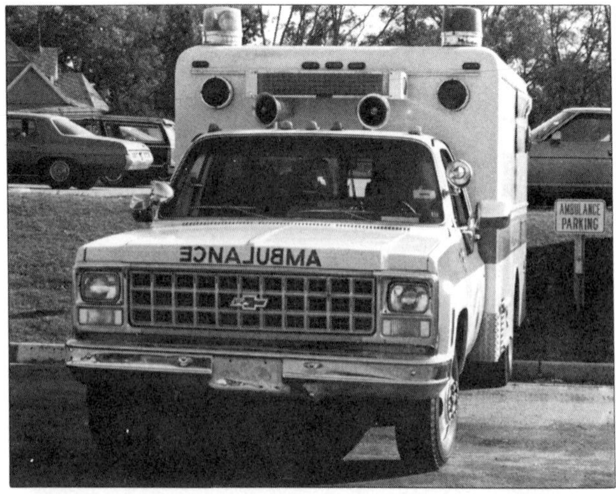

(a)

(b)

(c)

Figure 28.26
Mirror applications. See Question 29.

32. What would be the effect of a small light source placed at the focal point of a converging lens? How about at the focal point of a diverging lens?
33. A magnifying glass is a converging lens. Sunlight can be focused to a small spot using such a lens. What is the small spot an image of? Why does the small bright spot burn holes in paper or leaves when it is focused on them? (If you've ever focussed the spot on your skin, you know it gets hot.)
34. Why are distortions caused by (a) spherical aberration and (b) chromatic aberration? How may these aberrations be reduced or eliminated?
35. Discuss the types of aberrations that may affect the images formed by (a) front-coated spherical converging mirrors and (b) back-coated spherical converging mirrors.
36. What would the index of refraction of glass have to be in order to make a spherical, biconvex lens so that $f = R/2$? (See the Chapter Supplement.) Would this be possible with ordinary glass? Explain.

Problems

Levels of difficulty are indicated by asterisks for your convenience.

28.2 Reflection and Refraction

1. A light ray strikes a plane mirror surface at an incident angle of 35°. At what angle relative to the surface is the ray reflected?
2. Two light rays strike a common point on a mirror surface at angles of 25° and 40°, respectively. What is the angle between the reflected rays?
3. What is the speed of light in ice?
4. The speed of light in a transparent substance is 1.5×10^{10} cm/s. What is the index of refraction of the substance?

*5. What percentage of the speed of light in vacuum is the speed of light in air?
*6. Which is greater, the speed of light in flint glass or the speed of light in diamond, and by how much?
*7. A light beam in air is incident on a water surface at an angle of 45° relative to the normal of the water surface. What is the angle of refraction of the light in the water?
*8. When light strikes the surface of one of the liquids in Table 28.1 at an angle of incidence of 30° in air, it is found that the angle of refraction in the liquid is 20°. What is the liquid?
*9. An open container made of crown glass is full of water. If light in air strikes the water surface at an angle of incidence of 50° and passes through the water into the glass, (a) what is the angle of refraction in the glass? (b) What would be the effect without water?
*10. What are the critical angles of (a) diamond and (b) ethyl alcohol?
*11. (a) At what angle does a submerged diver have to shine a light toward the surface of a lake so that the beam is just refracted parallel to the surface? (b) What is the angle of reflection if the angle of incidence on the water-air interface is 55°?
*12. (a) What is the critical angle in crown glass at a glass-water interface? (b) How does this differ from the critical angle for a glass-air interface?
*13. A light pipe is made of polystyrene. What is the condition for light to be "conducted" in the pipe?
**14. A flint glass plate in air is 10 cm thick. Light strikes one surface of the plate at an angle of 45°, passes through the plate, and goes out the other side. (a) At what angle does the beam emerge from the plate? (b) How far is the emergent beam laterally displaced from the direction of the original incident beam?

28.3 Mirrors

15. A person stands 1.0 m in front of a plane mirror. Where is the image formed, and what are its characteristics?
16. What is the minimum length of a plane mirror that can be used by a person 5 ft 6 in. tall to view his or her complete image? Assume the person's eyes to be 4.0 in. below the top of the head.
17. Draw ray diagrams for a concave spherical mirror for the general object distances of (a) $d_o > R$, (b) $f < d_o < R$, and (c) $d_o < f$.

*18. A concave mirror has a radius of curvature of 40 cm. If an object 6.0 cm tall is placed 30 cm in front of the mirror, where is the image formed, and what are its characteristics?
*19. An object is placed 10 cm from the vertex of a concave mirror with a radius of curvature of 30 cm. (a) Is the image larger or smaller than the object, and by how much? (b) Can the image be formed on a screen?
*20. A convex spherical mirror has a radius of curvature of 20 cm. If an object is placed 20 cm from the mirror, where is the image formed, and what are its characteristics?
*21. A large convex mirror in a department store has a focal length of 0.50 m. The manger of the store stands 3.0 m in front of the mirror. Where is the manager's image, and what are its characteristics?
*22. An image is formed on a screen located 50 cm from a concave mirror with a radius of curva-

ture of 60 cm. (a) Where is the object located? (b) What is the magnification factor?

**23. (a) An inverted image is formed on a screen located 50 cm in front of a concave mirror. If the image is twice the size of the object, what is the radius of curvature of the mirror? (b) Where should the object be located to have a magnification factor of 1?

**24. A large diverging mirror with a focal length of 40 cm is used to monitor the aisles in a supermarket. A person 6.0 ft tall stands 4.0 cm from the mirror. (a) What is the percentage reduction of the image formed by the mirror? (b) What is the height of the image?

**25. A good reflecting surface such as a plane mirror has a reflectivity of about 95 percent, or 5 percent of the incident light is absorbed on reflection. If four mirrors are set up so that an incident beam is reflected from one to the other, what percentage of the incident light energy will be reflected from the fourth mirror? (*Hint:* the answer is *not* 80 percent.)

**26. Two plane mirrors are set side to side with an angle of 50° between their reflecting surfaces. If a light ray is incident on one of the mirrors at an angle of 40° and is reflected to the other mirror, what is the angle of reflection of the ray from the second mirror?

**27. Using Figure 28.27, derive (a) the mirror equation and (b) the magnification factor. (*Hint:* use similar triangles.)

28.4 Lenses

28. Draw ray diagrams for a biconvex lens for the general object distances (a) $d_o = 2f$, (b) $f < d_o < 2f$, and (c) $d_o < f$.

29. Draw ray diagrams for a biconvex lens for a long object arrow that extends on both sides of the lens axis for (a) $d_o > 2f$ and (b) d_o slightly greater than f.

*30. An object is placed 30 cm in front of a biconvex lens that has a focal length of 20 cm. (a) Where is the image formed, and what are its characteristics? (b) If the object is 10 cm tall, how tall is the image?

*31. What is the magnification factor of the image formed by a converging lens that has a focal length of 30 cm if an object is placed 20 cm from the lens?

*32. A biconcave lens has a focal length of 10 cm. If an object is placed 25 cm from the lens, where is the image and what are its characteristics?

*33. In a particular application, it is desired to have a biconcave lens form an image one-half the size of an object located 20 cm from the lens. What is the focal length of the required lens?

*34. An image of an object is magnified by a factor of 2.5 and formed on a screen 20 cm from the lens. What is the focal length of the lens?

*35. A biconvex lens with a focal length of 10 cm forms an image on a screen 25 cm from the lens. (a) Is the image upright or inverted? (b) Where is the object located?

**36. Using Figure 28.28, derive (a) the thin lens equation and (b) the magnification factor. (*Hint:* use similar triangles.)

**37. Two biconvex lenses with the same focal length (10 cm) are positioned 40 cm apart. If an object is placed 30 cm in front of one of the lenses, (a) where is the final image formed, and what are its characteristics? (b) If the object is 9.0 cm tall, how tall is the final image? (Draw a sketch or ray diagram to help visualize the situation.)

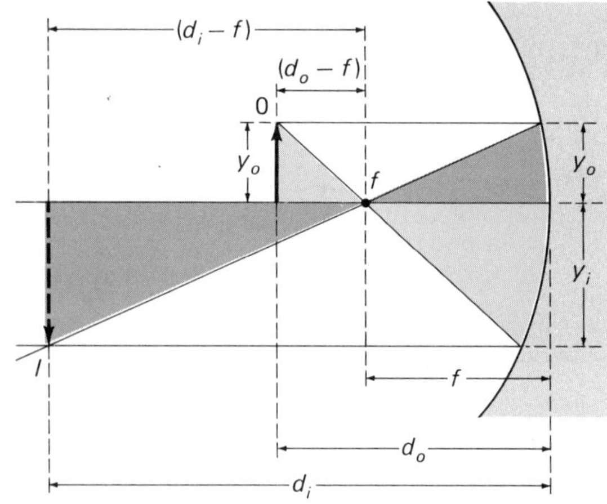

Figure 28.27
See Problem 27.

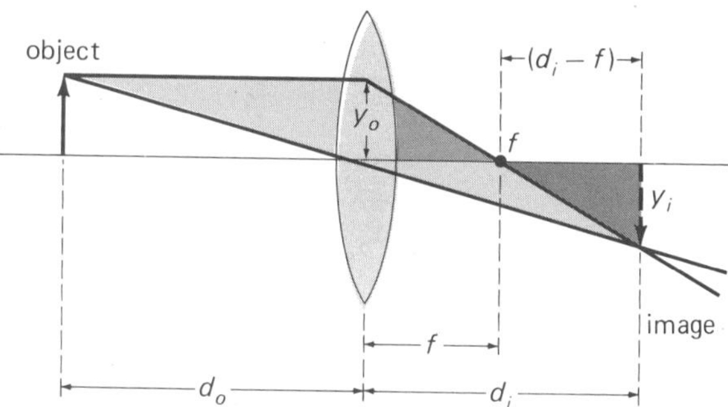

Figure 28.28
See Problem 36.

**38. A combination of two biconvex lenses with focal lengths of $f_1 = 20$ cm and $f_2 = 30$ cm, respectively, are separated by a distance of 25 cm. An object is placed 40 cm in front of the first lens. (a) Where is the final image formed? (b) What are the image characteristics?

Chapter Supplement

The Lens Maker's Equation and Lens Power

The biconvex and biconcave spherical lenses considered in this chapter had common focal lengths for the sides of the lens because of equal radii of curvature. However, other types of spherical converging and diverging lenses (Fig. 28.17 and Fig. 28S.1) are commonly used, and the focal lengths of these lenses are important considerations. The refractive index of the lens materials was not considered in the previous discussion.

In general, the focal length of a thin lens in air is given by the so-called **lens maker's equation:**

$$\frac{1}{f} = (n - 1)\left(\frac{1}{R_1} - \frac{1}{R_2}\right) \qquad \textbf{(Eq. 28S.1)}$$

where n is the index of refraction of the lens material and R_1 and R_2 are the radii of curvature of the first and second lens surfaces, respectively. The equation locates the focal point at which light passing through a lens would converge (converging lens) or the focal point from which transmitted light appears to diverge (diverging lens).

The sign convention for the radii of curvature in the lens maker's equation is as follows:

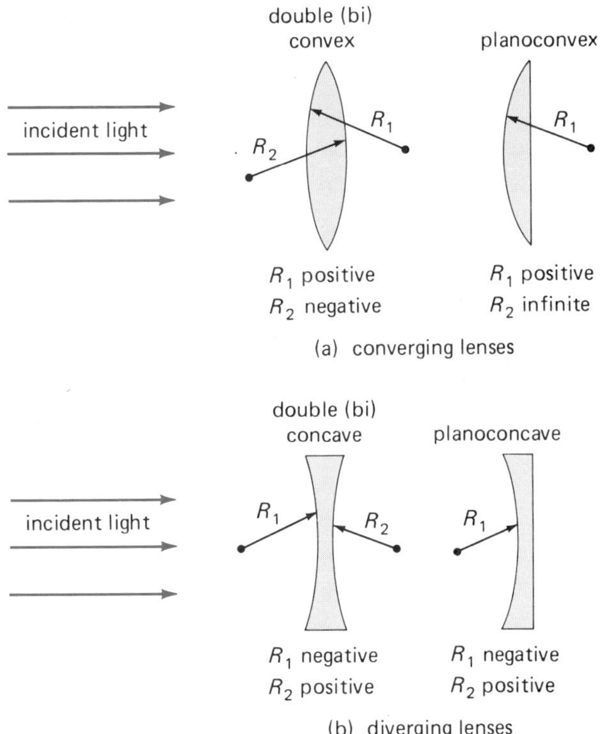

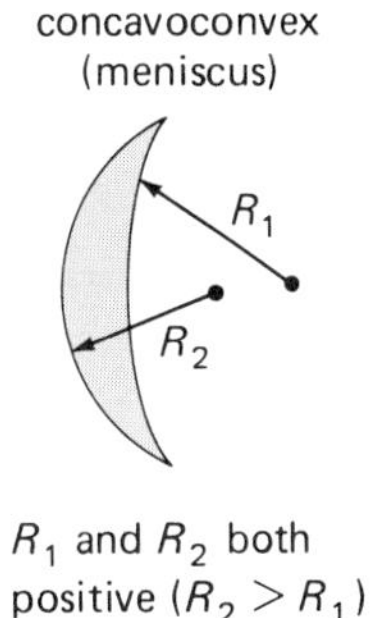

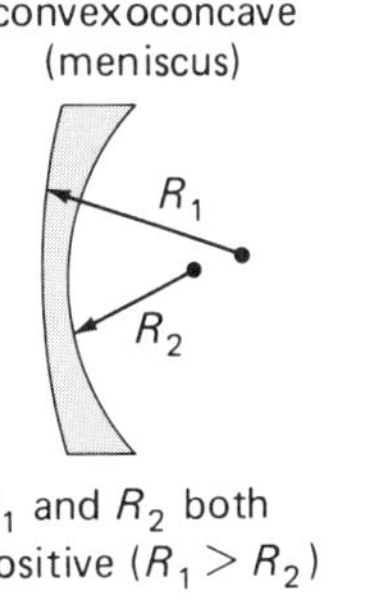

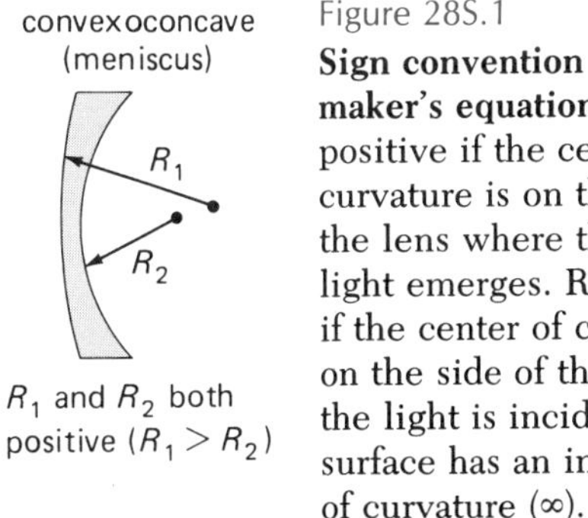

Figure 28S.1
Sign convention for the lens maker's equation. R is positive if the center of curvature is on the side of the lens where the refracted light emerges. R is negative if the center of curvature is on the side of the lens where the light is incident. A plane surface has an infinite radius of curvature (∞).

1. R is positive if the center of curvature is on the side of the lens where the refracted light emerges.
2. R is negative if the center of curvature is on the side of the lens where the light is incident.
3. A plane (flat) surface has an infinite radius of curvature (∞).

This sign convention is illustrated for converging and diverging lenses in Figure 28S.1.

EXAMPLE 28S.1 A plano-convex lens is made of crown glass, and the radius of curvature of the convex surface is 30 cm. What is the focal length of the lens for light incident on the convex surface?

Solution: $n = 1.5$ (Table 28.1), $R_1 = +30$ cm, and $R_2 = \infty$. Then,

$$\frac{1}{f} = (n-1)\left(\frac{1}{R_1} - \frac{1}{R_2}\right)$$

$$= (1.5-1)\left(\frac{1}{30 \text{ cm}} - \frac{1}{\infty}\right)$$

$$= \frac{0.5}{30}$$

and

$$f = \frac{30 \text{ cm}}{0.5} = 60 \text{ cm}$$

Hence, the lens is converging ($+f$) and a beam of light normally incident on the convex surface would converge at a point 60-cm on the opposite side of the lens.

The reciprocal of the focal length ($1/f$) is called the **lens power** (***P***) and is expressed in units of **diopters (D)** when the focal length is in meters.

$$\underset{\text{(diopters)}}{\underset{\text{lens power}}{P}} = \frac{1}{f \text{ (in meters)}} \qquad \textbf{(Eq. 28S.2)}$$

Optometrists usually prescribe corrective lenses for glasses in diopters.

EXAMPLE 28S.2 An optometrist prescribes for a patient a corrective convex-concave lens with a power of -2.0 D. The lens maker uses a crown glass blank with a convex front surface that has a radius of curvature of 40 cm to make the lens. What radius of curvature should the rear surface of the lens have?

Solution: $P = 1/f = -2.0$ D (m^{-1}), $R_1 = 40$ cm $= 0.40$ m, and $n = 1.5$ (Table 28.1). Then,

$$\frac{1}{f} = (n-1)\left(\frac{1}{R_1} - \frac{1}{R_2}\right)$$

and

$$-2.0 = (1.5-1)\left(\frac{1}{0.40} - \frac{1}{R_2}\right)$$

$$\frac{1}{R_2} = \frac{1}{0.40} + \frac{2.0}{0.5} = 2.5 + 4.0 = 6.5$$

$$R_2 = \frac{1}{6.5} = 0.15 \text{ m} = 15 \text{ cm}$$

A positive R_2 indicates that the center of curvature of the rear surface is on that side of the lens (the same as for R_1), and with $R_1 > R_2$, the lens is a negative meniscus (see Fig. 28S.1).

Problems

1. A biconvex lens made of crown glass has radiis of curvature of 20 cm for one surface and 30 cm for the other. What is the focal length of the lens if the light is incident from (a) the $R = 20$ cm side and (b) the $R = 30$ cm side?
2. A plano-concave crown glass lens has a concave surface with a radius of curvature of 40 cm. (a) What is the focal length of the lens? (b) What is the lens power?
3. A positive-meniscus flint glass lens has a front surface with a radius of curvature of 15 cm and a rear surface with a radius of curvature of 20 cm. (a) What is the focal length of the lens? (b) What is the lens power?
4. Prove that the different-shaped lenses in Figure 28S.1 are converging or diverging regardless of the direction of the incident light. (*Hint:* turn the lens in Fig. 28S.1 around.)
5. An optometrist prescribes for a patient a corrective meniscus lens with a power of +1.5. The lens maker uses a crown glass blank with a convex front surface having a radius of curvature of 20 cm to make the lens. What is the radius of curvature of the rear surface for the prescribed lens power?

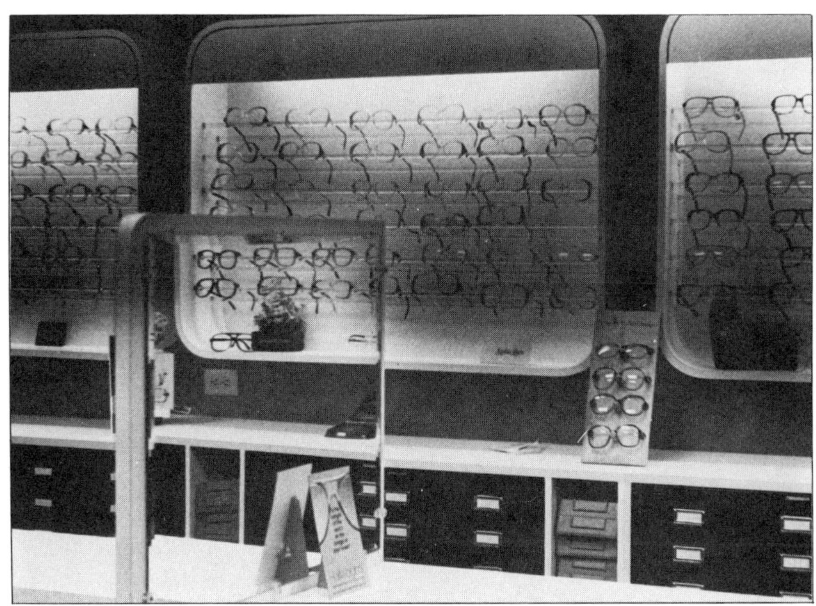

Chapter 29

Vision and Optical Instruments

Vision, or sight, is perhaps our most important sense. We tend to take vision for granted—until things start to appear blurred. Then we can often correct the problem with lenses (glasses). Even when we don't have vision defects, we use optical instruments containing mirrors and lenses to improve or extend our visual observations. For example, with microscopes and telescopes we are able to see very small and very distant objects that are not visible to the unaided eye.

In this chapter we'll take a look at these things so you can understand how you "see" the world.

29.1 The Human Eye

The eye is a unique optical instrument and one of our most valuable possessions. Without it, we would not see the beauty of the sunset or be able to read this and other books to gain knowledge.

The anatomy of the eye is shown in Figure 29.1. The eyeball is nearly spherical, with a white outer covering called the sclera (the "white of the eye"). Light enters the eye through a curved, transparent tissue called the cornea. Behind the cornea is a circular diaphragm, the iris, which has a central hole called the pupil. Our knowledge of everything we see depends on the information conveyed through the tiny pupils of our eyes. The size of the pupil aperture, or opening, is adjusted by muscle action and controls the amount of light entering the eye. For example, in very bright light the iris diaphragm closes, and the pupil becomes very small. How about in dim light?

A converging crystalline lens composed of glassy fibers is situated behind the iris. The curvature of the crystalline lens is controlled by the ciliary muscles. By adjustments and changes in the curvature of the lens, which is called **accommodation,** the images of objects at different dis-

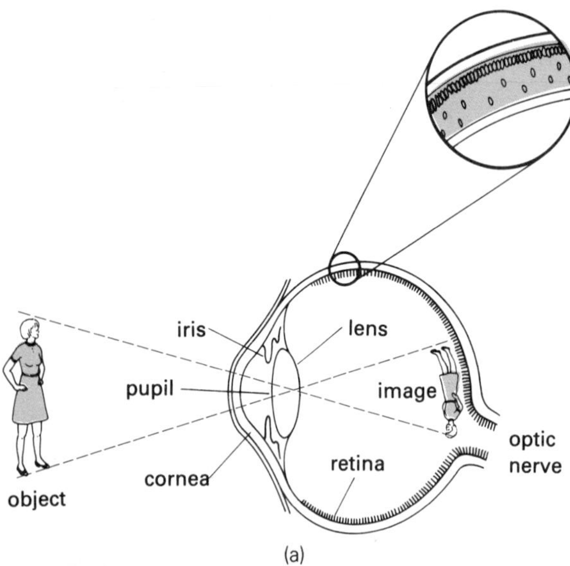

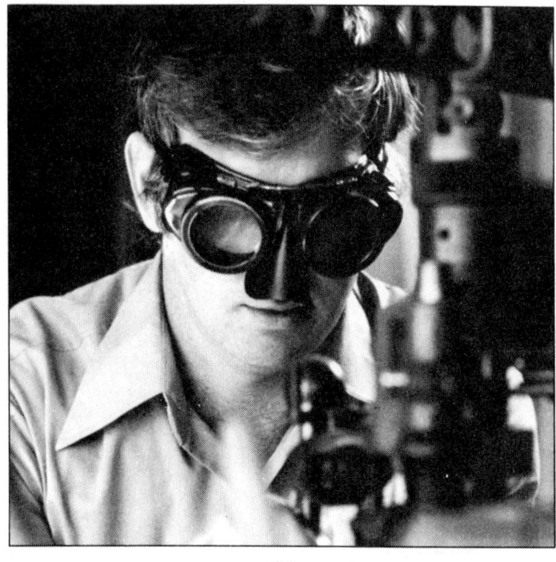

Figure 29.1

The human eye. (a) Structure of the eye. An inverted, real image is formed on the retina. See text for description. (b) Proper eye protection should always be worn. (Courtesy INCO Safety Products Co.)

tances from the eye can be focused on the **retina** on the back wall of the eyeball. The eyeball contains a fluid in front of the lens and a gelatinous material in the space behind the lens.

The operation of the eye is similar in several ways to that of a simple camera (Fig. 29.2). Each has a lens, but the curvature of the camera lens cannot be changed. Instead, the camera lens is moved back and forth so as to focus images on the film. Each has a variable diaphragm and a shutter. The shutter of the eye is the eyelid, which, unlike the shutter of a camera, is open for continuous exposure. Because of the refractive properties of the converging lenses, both the camera and the eye form inverted, or upside-down, images on light-sensitive surfaces (the film of the camera and the retina of the eye).

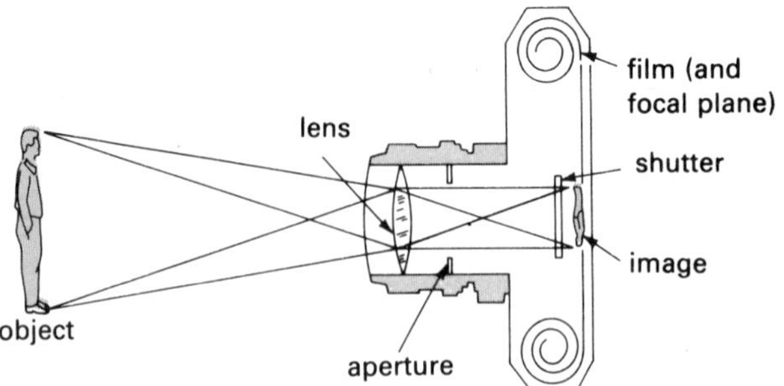

Figure 29.2

A simple camera. The operation of a camera is similar in several ways to that of the eye. Compare with Fig. 29.1.

Question: If the images in our eyes are upside down, why don't we see the world that way?

Answer: By some means, the brain learns early in life to interpret the inverted image of the world right side up. Experiments have been done with persons wearing special glasses that give them an inverted view of the world. After some initial run-ins, they become accustomed to and function quite well in their upside-down world.

The retina "film" of the eye is composed of two types of photosensitive cells called **rods** and **cones** (their names are descriptive of their shapes). The more numerous rods have a greater sensitivity to light and can distinguish between low light intensities for twilight (black-and-white) vision. The cones respond selectively to certain colors of light, some to one color and others to other colors. Cones are considerably less sensitive to light than are rods. That is why we cannot see color in very dim light. The rods and cones of the retina are connected to optic nerve fibers, which relay the light-stimulated signals to the brain.

Figure 29.3

Demonstration of blind spot. Hold the book at arm's length and, with your right eye closed, look intently at the black cross. Slowly bring the book toward your face. What happens to the square and the dot?

In the region where the optic nerve enters the eyeball (Fig. 29.1), there are no rods or cones. As a result, your eye has a "blind spot" for which there is no optical response. Figure 29.3 can be used to demonstrate this blind spot. Hold the book at arm's length and, with your right eye closed, look intently at the black cross. Then slowly bring the book toward your face. At certain points, you will see the square and dot alternately disappear and reappear. Where did they go? Nowhere. Their images were just crossing your eye's blind spot. The blind spot is not noticed in ordinary vision because of eye and object movements and binocular vision.

Binocular (two-eyed) vision also accounts for some of our depth perception, which depends on slight differences between the shapes and positions of the images on the retinae of the two eyes. The differences occur because the eyes are set several centimeters apart and get slightly different views of objects. Some other visual effects are discussed in Special Feature 29.1.

Visual Defects

Optical illusions as discussed in Special Feature 29.1 aren't visual defects. Defects generally occur because the eye is abnormal for some reason. This can be due to a variety of reasons, including disease and injury. Let's take a look at several common visual defects that can be corrected with glasses.

The points between which the eye can see distinctly are called the far point and the near point. The far point is normally without limit (infinity), and the near point depends on the accommodation of the crystalline lens. Bring your finger slowly toward your nose. At some point (your near point), you will see the finger blur or see two images. The normal eye produces sharp images on the retina for objects between the near and far points. However, a person may be nearsighted or farsighted and see blurred images (Fig. 29.4). This results from images not being focused on the retina because

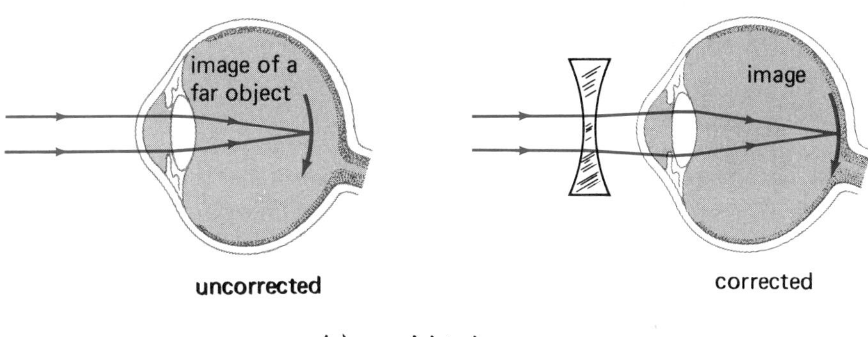

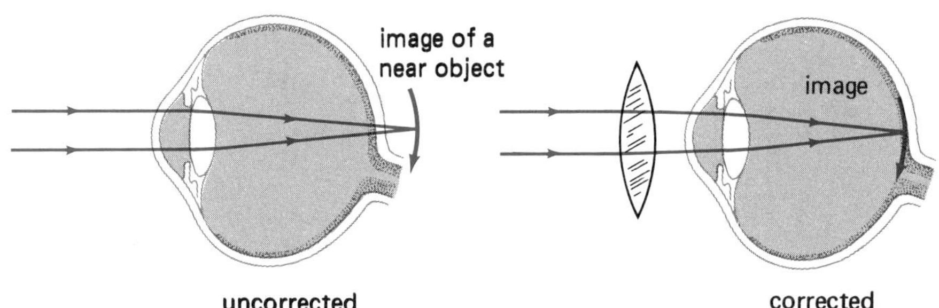

Figure 29.4

Common visual defects. When the image is not formed on the retina, corrective lenses bring things back into focus for (a) nearsightedness and (b) farsightedness.

SPECIAL FEATURE 29.1

Optical Illusions

"Seeing is believing," goes the old saying. But this is not always the case; at least you shouldn't believe everything you see. We can be visually fooled by several means. For example, fooling our depth perception causes objects in 3-D movies to seem to "jump" out of the screen, and atmospheric refraction produces mirages (Chapter 28).

Some optical illusions have to do with the eye itself and interpretations by the brain. For example, in Figure 1 you may see fleeting patches of gray on the white areas between the black diamonds (spots before your eyes!). This is because the stimulation of one area of the retina can affect the sensations in an adjacent region, producing an illusion.

Also in Figure 1 is a design that can be used to illustrate an *afterimage*. When a particular area of the retina is stimulated, it may "remember" the stimulation after you have shifted your gaze to another object. Look intently at the white dot on the black Y for 15 to 20 seconds (quite a long time), then transfer your gaze to the white dot between the E's. You'll see a ghostly white afterimage Y.

Other optical illusions are shown in Figure 2.

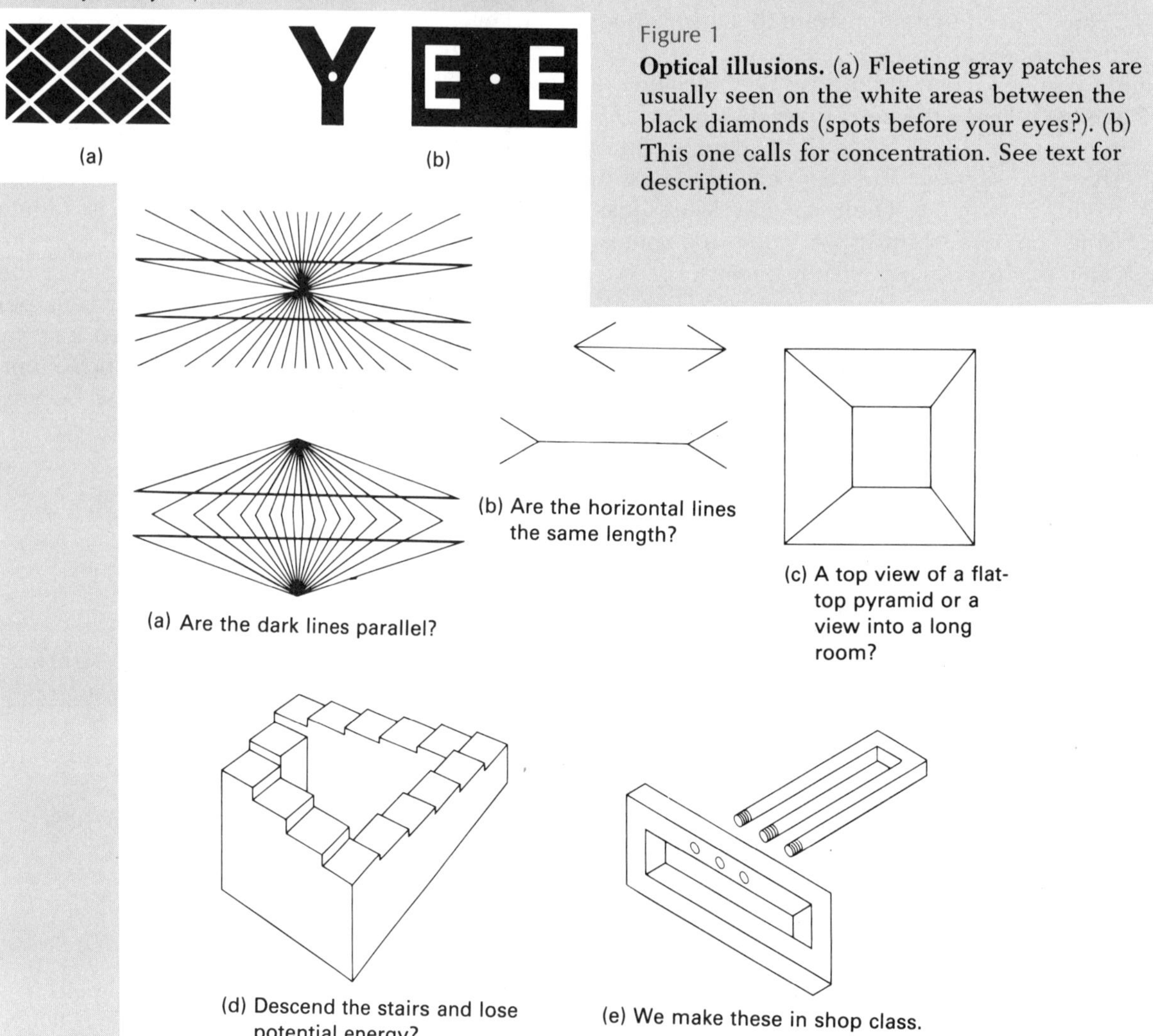

Figure 1
Optical illusions. (a) Fleeting gray patches are usually seen on the white areas between the black diamonds (spots before your eyes?). (b) This one calls for concentration. See text for description.

Figure 2
Some geometrical illusions.

Table 29.1
Approximate Near Points of the Normal Eye

Age (yr)	Near Point (cm)
10	10
20	12
30	15
40	25
50	40
60	100

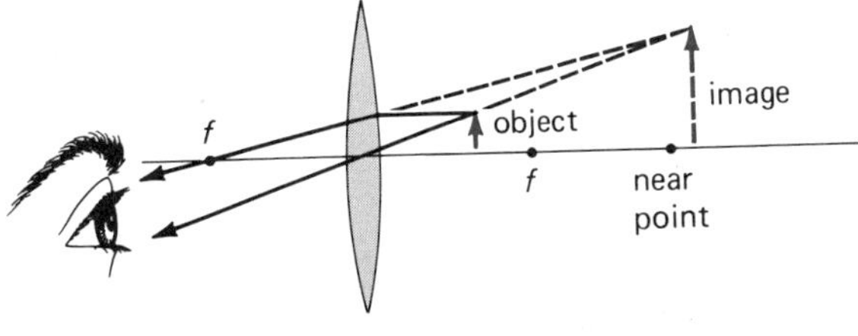

(a)

(b)

Figure 29.5
Near-point correction. The near point of the eye recedes with age, and closer objects appear blurred. (a) A converging lens "projects" the image of a close object beyond the near point, and it can be seen clearly. (b) People with this normal farsightedness commonly wear glasses for reading.

the size and shape of the eyeball and the crystalline lens are not properly matched.

Nearsightedness, or myopia, arises when the image is formed in front of the retina. A nearsighted person can see near objects clearly, but not distant objects. This defect is corrected by wearing glasses with a diverging lens for the nearsighted eye. The lens diverges the incoming rays so that the image formed by the crystalline lens is moved backward to the retina [Fig. 29.4(a)].

Farsighted persons can see far objects clearly, but near objects are blurred or out of focus. **Farsightedness,** or hyperopia, is due to the image being formed behind the retina. A converging lens will correct this by converging the incoming rays so that the image is moved forward to the retina [Fig. 29.4(b)].

Farsightedness (not being able to see close objects clearly) occurs naturally with age. You may have noticed an older person without glasses holding reading material away from him- or herself, even at arm's length. Children can see objects clearly as close as 10 cm from the eye (Table 29.1). However, as a person grows older the ciliary muscles weaken, and the crystalline lens loses its elasticity, or hardens, limiting the eye's accommodation. As a result, the near point recedes with age. When a person's "arms get too short," he or she has to get "reading glasses." The converging lens forms an image of a close object outside the near point, which the eye sees clearly (Fig. 29.5). Notice that the image is larger than the object. Reading glasses magnify things slightly.

This type of farsightedness may be called a *normal* defect, since it occurs naturally with age. So the next time you see someone holding something at arm's length to read, don't laugh. You'll be there someday.

Another common defect, **astigmatism,** occurs when the cornea and/or the crystalline lens of the eye are not perfectly spherical. As a result, the light rays have different focuses in different planes or directions. That is, a viewed object may be distinct in one direction and blurred in another. Astigmatism may be corrected with a lens having a greater curvature in the plane in which the cornea or crystalline lens has deficient curvature.

Bifocal glasses are sometimes used to correct a combination of defects. The bifocal lens was invented by Ben Franklin and consists of two lenses on the same piece of glass (Franklin glued two lenses together). For example, a small lower lens can correct for farsightedness (for reading), and the upper lens can be used to correct for nearsightedness or astigmatism. In some cases, trifocals are used.

29.2 Color

Color is the physiological sensation of the brain in response to the light excitation of cone receptors in the retina. Many animals have no cone cells and are color-blind, living in a black-and-white world. For example, a color TV would be no better than a black-and-white set in a dog's world.

In the human eye, the cones are sensitive to light with frequencies between about 7.5×10^{14} and 4.3×10^{14} Hz (400 nm to 700 nm wavelength). Different frequencies of light are perceived by the brain as having different colors (see Table 27.1). The association of colors with particular light frequencies is subjective. Monochromatic light has a particular physical frequency, but as pitch is to hearing, color is to vision—it may vary from person to person.

The concept of color vision is not well understood. One of the most popular theories is that three different types of cones are contained in the retina, each of which responds to light from a different part of the visible spectrum (Fig. 29.6). The "blue" cones have maximum response for light with a wavelength around 430 nm, the "green" cones for wavelengths around 550 nm, and the "red" cones for wavelengths around 580 nm.*

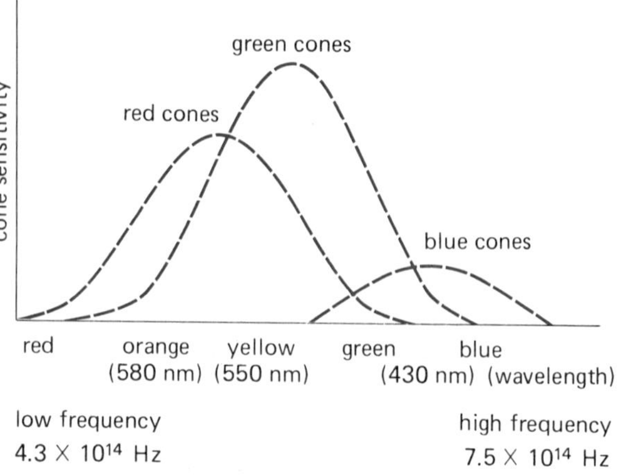

Figure 29.6
Cone sensitivity. Different cones of the eye are believed to respond to different frequencies of light to give three basic color responses.

Combinations and different degrees of cone stimulation give rise to intermediate colors. For example, when red and green cones are equally stimulated by light of a particular frequency, the brain interprets this as yellow in color. But when the red cones are stimulated more strongly than the green cones, the brain "sees" orange.

Color-blindness results when one type of primary cone is lacking. Occasionally this occurs because of failure to inherit the appropriate gene for the cone formation. The color genes are found in the female sex chromosome, so almost all color-blind people are male (about 4 percent of the male population). For example, if a person completely lacks red cones, he can see green through orange-red colors by use of his green cones. However, he is not able to distinguish among these colors satisfactorily because he has no red cones to contrast with the green ones. A similar condition exists if the green cones are missing. In either case, it is difficult or impossible to distinguish colors of the larger wavelengths, and the condition is called red-green color blindness.

Color Mixing

White light is a mixture of all visible frequencies or colors, although perhaps not of equal intensities. For example, sunlight has a predominant yellow-green component. Even so, the composite colors of white light can be easily demonstrated by dispersion with a prism.

When colors of light are mixed together, it is found that one doesn't need all colors to make white light—just red, green, and blue. When light beams of these colors are projected on a screen so the beams overlap, additive mixtures of colors and white are produced (Fig. 29.7). Evidently the excitation of all types of cones in the eye causes the combination of signals to be interpreted as white and other combinations as different colors. This is referred to as the **additive method of color production.**

By adding varying amounts of red, green, and blue light, we find that any color of the visible

* The cone "color" indicates only a general response region of the spectrum. For example, light with a wavelength of 580 nm is yellow-orange in color, but the major cone response in this end of the spectrum is from a "red" cone.

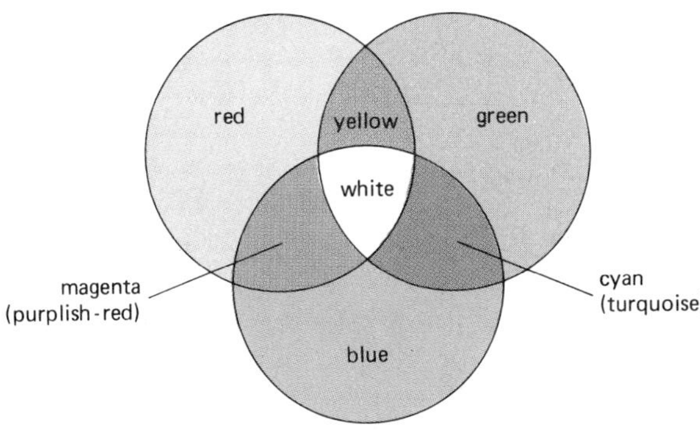

Figure 29.7
Additive method of color production. Light mixing of the primary colors (additive primaries) produces the color as illustrated in the regions of overlapping light beams.

spectrum can be generated. Recall that the triad dots on a color TV screen have red, green, and blue phosphors (Chapter 26). As a result, **red, green,** and **blue** are called the **additive primaries** or **primary colors.**

Not only does a mixture of the additive primaries appear white to the eye, but many pairs of color combinations do also. The colors of such a pair are said to be **complementary colors.** For example, the complement of blue is yellow, of red is cyan (turquoise), and of green is magenta (purplish red). This is not surprising, since from Figure 29.7 we can see that a combination of red and green is interpreted by the brain as yellow. Presumably, yellow light stimulates the red and green cones, which along with blue cone stimulation are a "white" combination.

Thus we see that an object has color because of the light coming from it. Other than light sources, objects have color when illuminated with white light because they reflect the wavelengths of the color they appear to be. The light coming from an object can be looked at in terms of selective reflection or selective absorption. When white light strikes a colored surface, certain light frequencies cause resonance electronic oscillations in the surface atoms, somewhat as sound causes a tuning fork to resonantly vibrate. The electrons of a particular atom have a narrow range of vibrational frequencies. The light components in this range are absorbed and re-emitted (reflected). The other light frequencies are absorbed by the material but go into heat (internal energy) rather than being re-emitted.

The paper of this page has atoms with a great range of resonance frequencies. When white light strikes the page, enough frequencies are reflected to your eye so that it appears white. The black print, on the other hand, has very few atoms with vibrations in the visible range of frequencies, so the light is almost totally absorbed, and the print appears black. (The light goes into the internal energy of the print ink. Recall how a thermometer with its bulb painted black registers a higher temperature than one with an unpainted bulb.) When white light strikes transparent red glass or a red rose, only red light is transmitted through the glass or reflected from the rose. All of the other colors are absorbed.

Selective absorption is important in the mixing of pigments for color production, such as in making paints and dyes for clothing. The additive method discussed previously for light does not apply here. You probably know or can guess, or any artist will tell you, that if you mix red, green, and blue pigments or paints you won't get white as in the case of light. You'll get some sort of dark brown color.

But how about mixing blue and yellow paints? You might know that this produces green. The same effect is obtained with light by passing white light through blue and yellow filters (Fig. 29.8). This is because the blue pigment in the paint or filter absorbs the wavelengths or colors except those in the blue region of the spectrum. Likewise, the yellow pigment in the paint or filter selectively absorbs the wavelengths of light except those in the yellow region. The wavelengths in the intermediate green region are not strongly absorbed by either pigment, and hence green light is transmitted through the filters or is reflected from the paint mixture.

This is an example of the **subtractive method of color production or mixing.** A mixture of absorbing pigments results in the subtraction of colors, and the eye sees the color that is not subtracted or absorbed. Three particular pigments—**cyan, magenta,** and **yellow**—are called the **subtractive primaries** or **primary pigments** (colors). Various combinations of two of

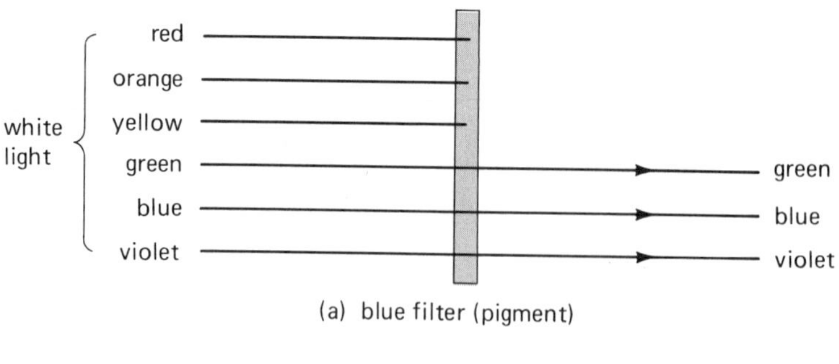

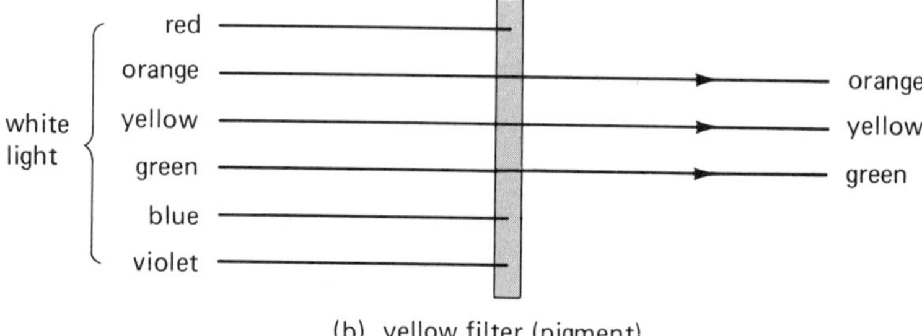

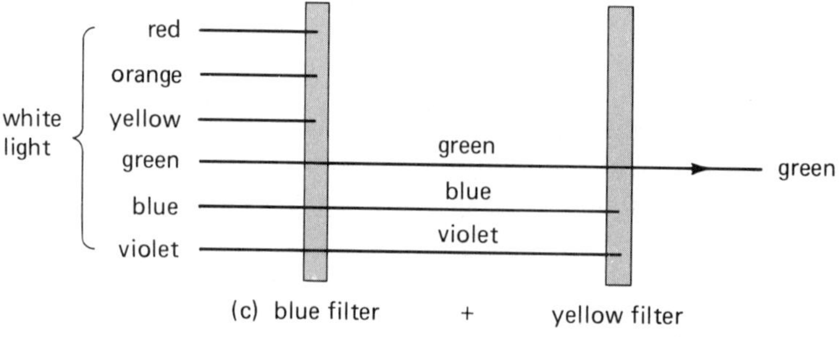

Figure 29.8
Selective absorption. A particular filter (or paint pigment) absorbs wavelengths or colors in a particular region. By subtracting (absorbing) component colors of white light, a particular color is produced.

the subtractive primaries produce the three additive primary colors (red, blue, and green), as illustrated in Figure 29.9. Notice that the magenta pigment "subtracts" the green color from the cyan-yellow overlap. As a result, magenta is sometimes referred to as "minus green." (Think about adding a magenta filter to Figure 29.8.) Similarly, cyan is "minus red" and yellow is "minus blue."

When all of the primary pigments are mixed in the proper proportions, almost all of the wavelengths of the visible spectrum are absorbed or subtracted, and the mixture appears black. People in the paint and dye businesses commonly refer to the subtractive primaries as being red, yellow, and blue, because magenta is a shade of red and cyan is a shade of blue. In paint stores, paint of any color can be produced by mixing proper combinations of subtractive primaries.

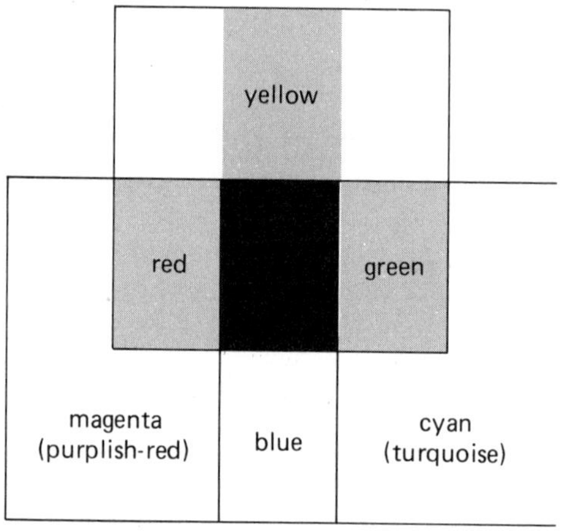

Figure 29.9
Subtractive primaries. When the subtractive primaries or pigments—cyan, magenta, and yellow—are mixed, the colors (or lack of color) are obtained as illustrated in the overlapping regions.

EXAMPLE 29.1 Suppose a yellow banana and a red apple were illuminated with green light. How would they look?

Solution: The banana would appear green. Since green is close to yellow, some green light would be reflected by the "yellow" oscillators in the banana peel. The apple would appear dark because most of the green light would be absorbed. If there were some leaves on the apple stem, you'd see green leaves and a dark apple. See Special Feature 29.2 for another color phenomenon.

Question: If the scattering by atmospheric gases is greater for visible light of greater frequency, why isn't the sky violet? Violet light should be scattered more than blue light.

Answer: Violet light *is* scattered more than blue light, but the sky is blue for a couple of reasons. First, and more important, the eye is more sensitive to blue light than to violet light. See Figure 29.6. Second, sunlight contains more blue light than violet light. The greatest color component is yellow-green, and the distribution generally decreases toward the ends of the spectrum.

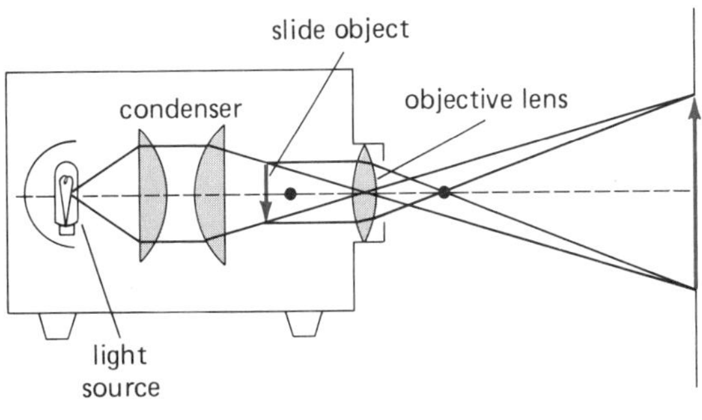

(a)

(b)

Figure 29.10
Projectors. (a) Diagram of a simple projector, and (b) various kinds of projectors. (Courtesy Eastman Kodak Company.)

29.3 **Projectors**

The projection of the images of slides and films on a screen is commonplace. We all like to show slides of our vacation or last week's party, or show a home movie. We do this by means of a projector. The basic components of a slide projector are shown in Figure 29.10, along with some common types of projectors.

A pair of "condensing" lenses concentrates light from a source onto a slide. The slide is placed just beyond the focal point of the converging lens, and a magnified image of the slide is formed on the screen. Because of the refractive properties of the lens, which are indicated by the light rays in the diagram, the image of an upright object is inverted. Therefore, slides are placed in the projector upside down so the image on the screen is seen right side up.

Motion picture projectors have basically the same components, along with the machinery to advance the film at a given speed. By quick advancement of one slightly different picture frame after another, the projected image is given the illusion of motion.

A sound motion picture also involves optical methods. When a motion picture is filmed, a recording amplifier drives a special lamp that exposes a narrow "sound track" along the side of the film. This track is a variation in optical density (light and dark regions) that follows the variations in the lamp voltage induced by sounds.

SPECIAL FEATURE 29.2

Why the Sky Is Blue (and Sunsets Are Red)

If we lived on the moon, which has no atmosphere, the sky would appear black except in the vicinity of the Sun. That is what astronauts on the moon see. On Earth, however, our sky is blue as a result of the scattering of sunlight in the atmosphere. As sunlight passes through the atmosphere, the nitrogen and oxygen molecules of the air absorb some of the light and re-emit it. The light is scattered in all directions from the free gas molecules.

This scattering is selective, with the resonant frequencies of the small molecular oscillators in the ultraviolet region. The frequencies of light in the visible region are below the resonant frequencies but are close enough to be absorbed and scattered somewhat, particularly in the blue end of the spectrum. Thus, the light in the visible spectrum is preferentially scattered, with the light at the blue end of the spectrum being scattered about ten times more than light at the red end. This is called **Rayleigh scattering** after Lord Rayleigh, the British scientist who explained it. Some of the scattered light reaches the Earth, and we see it as blue sky light (Fig. 1).

Keep in mind that *all* colors are present in sky light, but the dominant wavelength or color lies in the blue. You may have noticed that the sky light is more blue directly overhead or high in the sky and less blue toward the horizon, becoming white just above the horizon. This is because there are fewer scatters along a path through the atmosphere directly overhead (your zenith position) than toward the horizon, and multiple scattering along the horizon path gives rise to the white appearance. By analogy, if you add a drop of milk to a glass of water and illuminate the suspension with intense white light, the scattered light has a bluish hue. And yet, a glass of milk is white (due to multiple scatterings). Atmospheric pollution may enhance the milky white appearance of the sky.

The scattering of sunlight by the atmospheric gases *and* small particles gives rise to red sunsets. Generally it might be thought that since the distance sunlight travels through the atmosphere is greater to an observer at sunset, then most of the higher-frequency colors of the visible spectrum would be scattered from the sunlight and only light in the red end of the spectrum would reach the observer. However, it has been shown that the dominant color of this light, due solely to molecular scattering, is orange.*

Hence, there must be scattering by small particles in the atmosphere that shifts the light from the setting (or rising) sun toward the red. Foreign particles in the atmosphere are not necessary to give a blue sky; they even detract from it. Yet, such parti-

* Bohren, C. F., and Fraser, A. B., Colors of the Sky, *The Physics Teacher*, Vol. 23, No. 5, p. 267.

To reproduce the sound when the film runs through the projector, an optical system sends a beam of light through the sound track to a photocell or phototube. The variations in the light and dark portions of the track cause fluctuations in the light beam that are monitored by the phototube, and the electrical impulses are fed into an amplifier-speaker system that reproduces the original sounds.

29.4 Microscopes

The Simple Microscope (Magnifying Glass)

When we want to see small objects better, we might use a magnifying glass or a simple microscope, which is a biconvex lens. The lens makes an object appear larger or magnified because the refractive bending of light widens the angle of view (Fig. 29.11). The eye sees light as if it had traveled in straight lines.

Ordinarily, an object cannot be brought closer than the near point of the eye in order to form a sharp image. However, when using a magnifying glass (or eyeglasses with converging lenses; see Fig. 29.5), an object can be brought closer than the near point, with the object inside the focal point of the lens (Fig. 29.12). As a result, a magnified, virtual image is viewed by the eye.

The magnification of a magnifying glass is given by the magnification equation

$$M = \frac{d_i}{d_o} \qquad \textbf{(Eq. 29.1)}$$

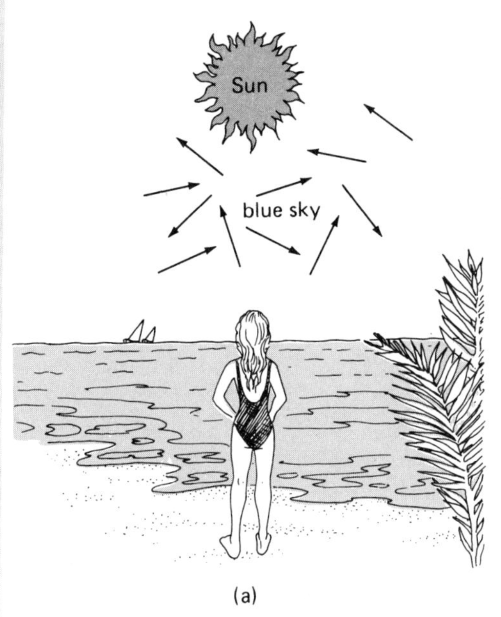

(a)

Figure 1
Atmospheric scattering accounts for (a) the blueness of the sky and (b) along with particulate matter, red sunset.

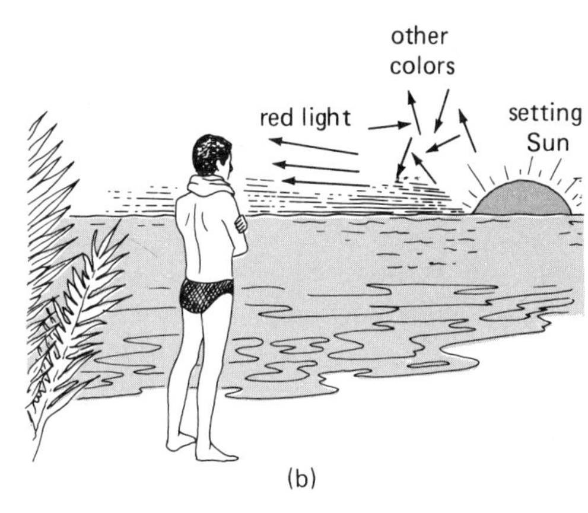

(b)

cles are necessary for deep red sunsets and sunrises, which occur most often where there is a high-pressure air mass to the west (for sunsets) or to the east (for sunrises), since the particle concentration is generally higher in a high-pressure air mass than in a low-pressure air mass.

The beauty of red sunrises and sunsets is often made more spectacular by layers of pink-colored clouds. The cloud color is due to the reflection of red light. Since the water droplets of clouds scatter visible light of all wavelengths about equally, the clouds do not affect the color of the light, but merely diffusely reflect the incident red light.

(minus sign omitted). For maximum magnification, the object is brought just inside the focal point of the lens and $d_o \simeq f$. The image is at the near point of the eye (d_i = near point). Taking an average near point to be 25 cm (d_i = 25 cm), we have for this approximation

$$\boxed{M \simeq \frac{25 \text{ cm}}{f}} \qquad \textbf{(Eq. 29.2)}$$

(magnification of magnifying glass)

EXAMPLE 29.2 A draftsman views the fine details of a blueprint with a magnifying glass that has a focal length of 10 cm. What is the maximum magnification?

Solution: With f = 10 cm, by Equation 29.2,

$$M \simeq \frac{25 \text{ cm}}{f} = \frac{25 \text{ cm}}{10 \text{ cm}} = 2.5$$

It would seem that by using a single lens of a sufficiently short focal length any magnifying power could be obtained. However, practical magnifying glasses are limited to sharp image magnifications of about 3× (3 times; read "three X") because of spherical aberration. Hand lenses with double and triple lenses are used to give greater magnifications.

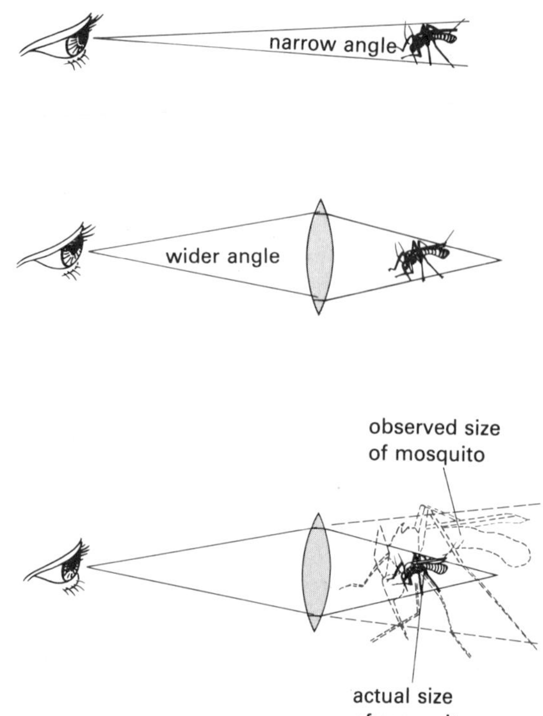

Figure 29.11
Magnification. A single lens or simple microscope expands the angle of view so that object appears larger.

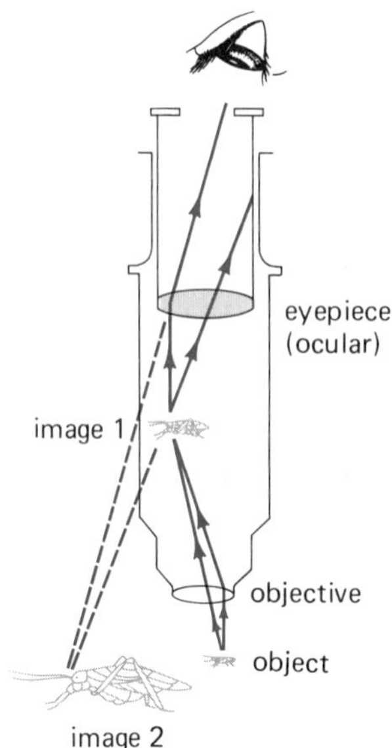

Figure 29.13
The compound microscope. The objective lens forms a real, magnified image just inside the focal point of the eyepiece, which produces a magnified, virtual image that is viewed by the observer.

The Compound Microscope

To obtain greater magnification than that given by a single lens or magnifying glass, a compound microscope may be used. Here more than one lens is used, as shown for a basic compound microscope in Figure 29.13. The objective lens with a relatively short focal length forms a magnified image of the object. The object is just beyond the focal point of the objective lens, which forms a real, inverted, and magnified image.

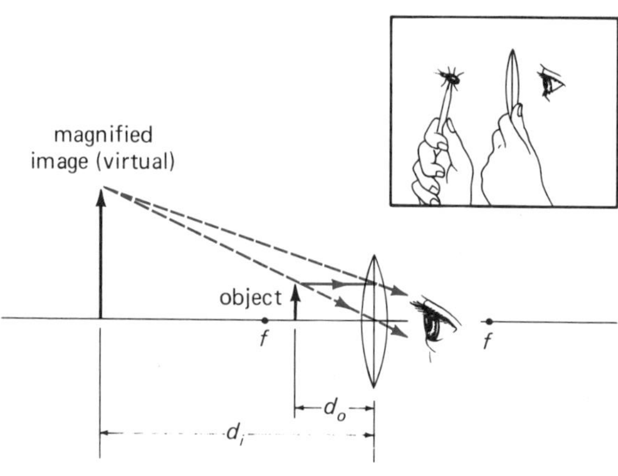

Figure 29.12
The magnifying glass or simple microscope. When an object is brought inside the focal point of the converging lens, the eye sees a large, virtual image.

This image is formed just inside the focal point of the eyepiece (or ocular) lens, which has a greater focal length than the objective lens. The eyepiece acts as a magnifying glass and produces a magnified virtual image that is viewed by the observer.

As discussed in the previous chapter, the total magnification (M_t) of a compound lens system is given by the product of the component lenses' magnifications. In this case,

$$M_t = M_O M_E = \left(\frac{d_i}{d_o}\right)_O \left(\frac{d_i}{d_o}\right)_E \quad \textbf{(Eq. 29.3)}$$

where M_O and M_E are the magnifications of the objective and eyepiece, respectively.

For a typical microscope, the object is placed just outside the focal point of the objective ($d_o \simeq f_o$) and the image distance might be about 20 cm ($d_i = 20$ cm); then $(d_i/d_o)_O \approx (20 \text{ cm}/f_o)$. For the intermediate (object) image of the eyepiece, the object is just inside the focal point of the lens ($d_o \simeq f_e$), and the image distance is at the average near point of the eye ($d_i \simeq 25$ cm), so $(d_i/d_o)_E \simeq (25 \text{ cm}/f_e)$. Hence, for these approximations

$$\boxed{M_t \simeq \frac{(20)(25)\text{ cm}^2}{f_o f_e}} \qquad \textbf{(Eq. 29.4)}$$

(magnification of compound microscope)

EXAMPLE 29.3 A technician uses a low-power compound microscope to view the etched circuits on a printed circuit board. The objective and eyepiece of the "scope" have focal lengths of 20 mm and 50 mm, respectively. What is the magnification?

Solution: It is given that $f_o = 20$ mm $= 2.0$ cm and $f_e = 50$ mm $= 5.0$ cm. Then, using Equation 29.4,

$$M_t \simeq \frac{(20)(25)\text{ cm}^2}{f_o f_e} = \frac{(20)(25)\text{ cm}^2}{(2.0\text{ cm})(5.0\text{ cm})} = 50\times$$

An exposed view of a modern microscope is shown in Figure 29.14. Nowadays microscopes are usually binocular (two oculars, i.e., an eyepiece for each eye) as opposed to the older monocular type. The compound-system eyepieces are interchangeable, and different eyepieces with different magnifications can be used, e.g., 5× or 10×. There are commonly three objectives of different magnifications on a rotating turret.* The limit of magnification of a compound microscope is about 2000×. The microscope is also equipped with a light-condensing system and an iris diaphragm below the microscope platform, or stage; the diaphragm focuses and controls the illumination from an external or internal light source.

Greater magnification can be achieved by electron microscope techniques (up to 200,000×). The electron microscope is discussed in the Chapter 30 Supplement.

* Some newer microscopes have zoom lenses for variable magnification, similar to the zoom lens on a camera.

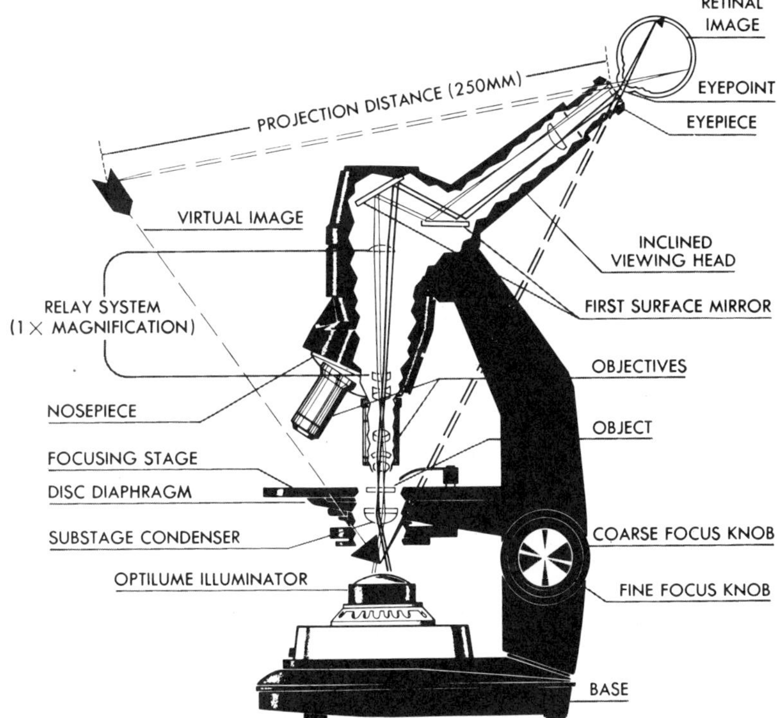

Figure 29.14
A modern microscope. The exposed view shows the lens system and light path. (Courtesy Bausch & Lomb.)

29.5 Telescopes

Telescopes are used to make distant objects appear closer. Essentially, telescopes collect and concentrate light energy to form images. This is particularly the case for astronomical telescopes used to view distant stars and galaxies. Smaller telescopes are used to view terrestrial objects—for example, the transit telescope used by a surveyor. There are two general types of telescopes: refracting telescopes and reflecting telescopes. They are based on lens refraction and mirror reflection, respectively.

The Refracting Telescope

The basic components of a refracting telescope are shown in Figure 29.15. The refracting telescope is similar to the compound microscope inasmuch as it has objective and eyepiece lenses. However, a telescope uses a large objective lens with a long focal length. The parallel rays from a distant object form an inverted, real image in the focal plane of the objective lens (at f_o). This is just inside the focal point of the eyepiece (f_e), which acts as a magnifier. As a result, a final magnified, virtual image is viewed by the observer.

The magnification of a telescope is defined in terms of angular distances (instead of linear distances) and can be shown to be

$$M = \frac{f_o}{f_e} \qquad \textbf{(Eq. 29.5)}$$

(magnification of refracting telescope)

where f_o and f_e are the focal lengths of the objective and eyepiece, respectively.

EXAMPLE 29.4 A refracting telescope has an objective lens with a focal length of 100 in. and an eyepiece with a focal length of 2.0 in. What is the magnification of the telescope?

Solution: It is given that f_o = 100 in. and f_e = 2.0 in. Then,

$$M = \frac{f_o}{f_e} = \frac{100\text{ in.}}{2.0\text{ in.}} = 50\times$$

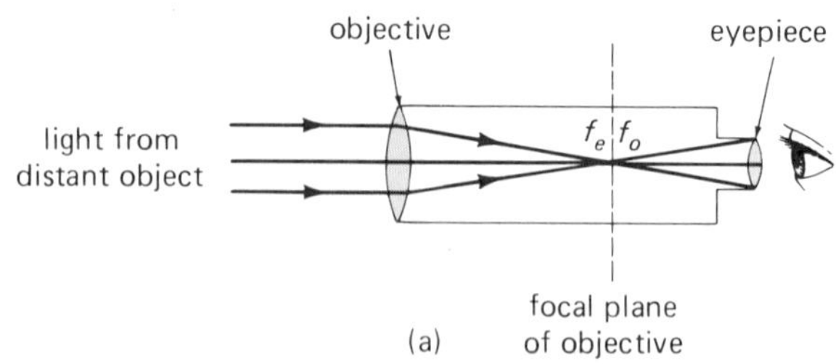

Figure 29.15
The refracting telescope. (a) The basic elements of a refracting telescope. A large objective lens forms a real image of a distant object in its focal plane, which is just inside the focal point of the eyepiece. The observer sees a magnified, virtual image. (b) The 40-inch (1.0 m, objective diameter) refracting telescope at the Yerkes Observatory. (Courtesy Yerkes Observatory.)

The refracting telescope just described is commonly referred to as an **astronomical telescope.** The final image is inverted relative to the original object. This poses no problem for astronomical work, but for viewing objects on Earth it is desirable to have the final image right side up. This may be accomplished by using a diverging lens for an eyepiece or by using an intermediate "erecting" lens between the objective and the eyepiece. Such telescopes with right-side-up images are called **terrestrial telescopes** (Fig. 29.16).

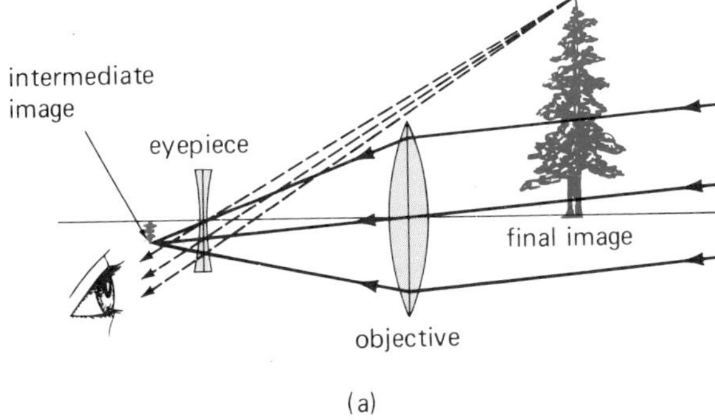

(a)

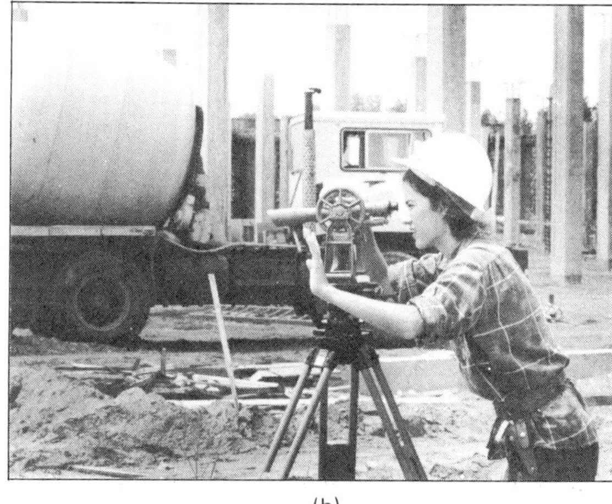

(b)

(c)

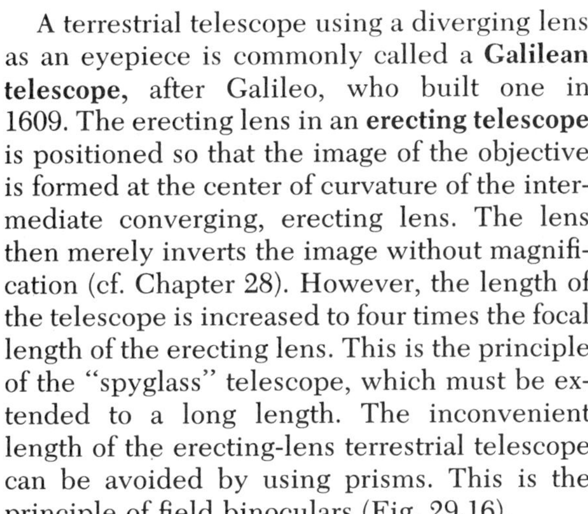

Figure 29.16

Terrestrial telescopes. (a) The principle of a Galilean telescope using a diverging lens for an eyepiece. (b) A surveyor's transit is a type of terrestrial telescope. (c) Prisms are used in binoculars to produce an erect image without the increased length required for an erecting-lens telescope.

A terrestrial telescope using a diverging lens as an eyepiece is commonly called a **Galilean telescope,** after Galileo, who built one in 1609. The erecting lens in an **erecting telescope** is positioned so that the image of the objective is formed at the center of curvature of the intermediate converging, erecting lens. The lens then merely inverts the image without magnification (cf. Chapter 28). However, the length of the telescope is increased to four times the focal length of the erecting lens. This is the principle of the "spyglass" telescope, which must be extended to a long length. The inconvenient length of the erecting-lens terrestrial telescope can be avoided by using prisms. This is the principle of field binoculars (Fig. 29.16).

The Reflecting Telescope

To form images of distant stars, sufficient light energy must be collected or received by the telescope. The amount of light entering a refracting telescope can be increased by increasing the size of the objective lens. However, there are physical limitations to this approach in the grinding of the lens, as well as material defects and aberration effects.

Another approach is to use a reflecting telescope. The basic components of this type of telescope are shown in Figure 29.17. In this case, the objective is a concave mirror; it collects and focuses the light. To make the image conveniently accessible to the eye or a camera, the rays may be deflected by a plane mirror to the side of the tube and observed, with the image magnified by an eyepiece.

However, if the telescope is large enough, the observation may be made inside the telescope tube. This is the case for the 200-in. or 5.1-m (mirror diameter) Hale Observatories telescope on Palomar Mountain in California (Fig. 29.17). The largest reflecting telescope, which

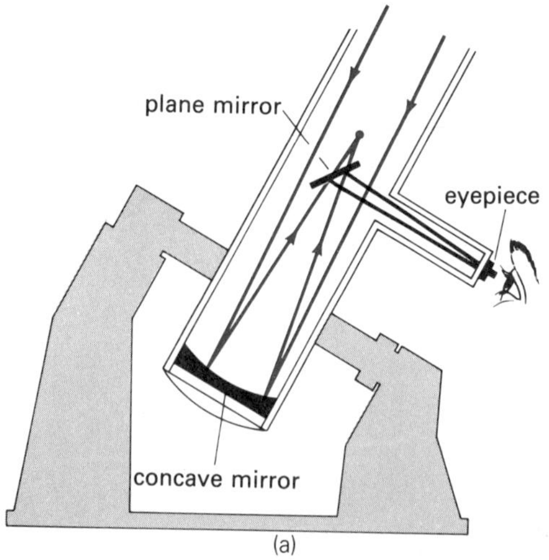

Figure 29.17

The reflecting telescope. (a) A concave mirror collects and focuses the light. A mirror may be used to bring the rays to a side eyepiece. (b) The Hale 200-inch (5.1 m, mirror diameter) telescope. The glass mirror weighs 15 tons and is 24 inches thick at the edge and 20 inches thick at the center. A thin coating of aluminum provides the reflecting surface. Observations are made at the prime focus position at the center of the telescope inside a cylindrical cage at the top. (Courtesy Hale Observatories.)

is in the Soviet Union, has a 236-in. (6-m) reflector.

Large mirrors can be constructed more easily than large lenses, since only one surface need be ground and silvered. Reflecting telescopes are also free from chromatic aberrations. Reflecting telescopes usually use parabolic mirrors to make them free of spherical aberrations as well.

The trick in building large mirrors is to make them light enough so they don't sag under their own weight. New telescope designs have mirror segments that are computer-positioned or thin mirrors that are supported in segments. Telescopes with 8-m and 10-m mirrors, using such designs, are now under construction.

Another type of telescope is discussed in Special Feature 29.3.

Important Terms

accommodation adjustment of the curvature of the crystalline lens of the eye for focusing

retina the light-sensitive part of the eye on which images are formed

rods photosensitive cells of the retina that can distinguish between low light intensities for twilight vision

cones the photosensitive cells of the retina that are responsible for color vision

color a physiological sensation of the brain in response to light excitation of the cone receptors of the retina

primary colors (additive primaries) red, green, and blue

complementary colors combinations of colored lights that produce white

subtractive primaries (primary pigments) cyan, magenta, and yellow

Rayleigh scattering the preferential scattering of light by molecules that is responsible for the blueness of the sky

simple microscope (magnifying glass) a biconvex converging lens

compound microscope a lens system of two or more converging lenses

refracting telescope a telescope that uses the refractive properties of lenses

astronomical telescope a telescope that gives an inverted final image

terrestrial telescope a telescope that gives a right-side-up final image

Galilean telescope a terrestrial telescope that uses a diverging lens as an eyepiece

erecting telescope a terrestrial telescope that uses an intermediate converging lens

reflecting telescope a telescope that uses a mirror for the collection and converging of light

Important Formulas

magnification of a magnifying glass (simple microscope): $M \simeq \dfrac{25 \text{ cm}}{f}$

magnificaton of a compound microscope: $M_t \simeq \dfrac{(20)(25)\text{ cm}^2}{f_o f_e}$

magnification of a refracting telescope: $M = \dfrac{f_o}{f_e}$

Questions

The Human Eye

1. Many people wear "contacts." On what part of the eye are contact lenses worn?
2. How are the operations of a camera shutter and of the eyelid generally different? Can you think of an instance when a camera shutter is operated like an eyelid?
3. If upside-down images are formed on the retina, how do we know that the world isn't really upside down when we see it?
4. Compare the focusing processes of the eye and a simple camera. Does the eye "camera" have black-and-white or color film?
5. What is meant by "twilight vision"?
6. Moonlight is reflected sunlight. Why don't we see color when viewing objects in moonlight?
7. Describe the following parts of the eye: (a) sclera, (b) iris, and (c) pupil.
8. Answer the questions in Figure 29.18.
9. Why do older people sometimes hold reading material at arm's length?
10. A person wears bifocals to correct for nearsightedness and astigmatism. How might the bifocal lens be ground for these vision defects?

Color

11. In very rare instances of color-blindness, a person lacks blue cones (called "blue weakness"). What colors would this person have difficulty in distinguishing?
12. Is white a color? Is black? Explain.
13. Suppose that the beams of red and blue spotlights overlap each other on a wall. (a) What color will the wall appear to be if it is white? (b) How about a blue wall? (c) A green wall?
14. In a department store window display, white light passing through a red filter falls on cyan- and yellow-colored objects. What colors are seen by someone looking at the display?
15. How do different-colored spotlights influence the appearance of a performer's clothes—for example, a red spotlight on a performer in a dark blue suit? What color spotlight would you use if you wanted to make the performer really stand out?
16. What color would a green chalkboard appear to be when viewed through yellow sunglasses? How about when viewed through rose-colored (red) glasses? If the chalkboard had writing on it

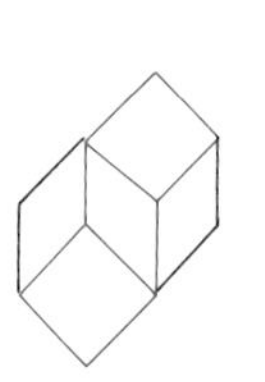

(a) Which cube does not have all of its sides?

(b) Are the diagonal lines parallel?

(c) Are all the men the same size?

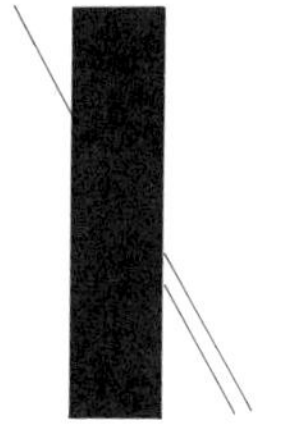

(d) Is the lower line on the right an extension of the line on the left?

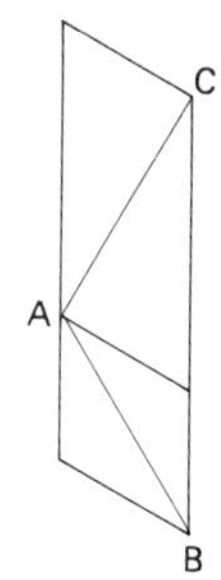

(e) Is line AB equal in length to line AC?

Figure 29.18
Seeing is believing. See Question 8.

SPECIAL FEATURE 29.3

Radio Telescopes

Although not an optical instrument, the radio telescope is an important astronomical tool. Stars and galaxies emit radio waves as well as light waves. This fact was discovered accidentally by an electrical engineer named Carl Jansky in 1931 while he was working on a static problem in intercontinental radio communications. He found that an annoying static hiss was coming from a fixed direction in space, which later was found to be the center of our galaxy (the Milky Way).

Thus, Jansky discovered radio waves coming from space. This was quickly recognized as another source of astronomical information, and radio telescopes were built (Fig. 1). A radio telescope operates similarly to reflecting light telescopes, inasmuch as radiation is collected by a large-area reflector and focused to form an "image." However, the parabolic collector of a radio telescope does not look like a mirror surface, since it is covered with wire mesh. To prevent confusion, the collector of a radio telescope is referred to as a "dish" instead of a mirror. Satellite TV antenna dishes are common.

The radio telescope dish is not a mirror for light waves, but it is for radio waves. This is because electromagnetic waves cannot detect any hole or surface irregularity that is smaller than the wavelength of the radiation. Since radio waves range from about 1 cm to several meters in wavelength, the wire-mesh surface on the metal dish acts as a good reflecting surface for such waves.

Another noticeable difference between optical and radio telescopes is that in a radio telescope there is no film or eyepiece. Instead, the radio waves are detected by an antenna positioned at the focal point of the dish. The signals are amplified, and the information received by the telescope is displayed on a recorder so it can be "seen."

Radio telescopes have added a new dimension to astronomy. They supplement optical telescopes and offer some definite advantages. Radio waves pass freely through the huge clouds of dust that ex-

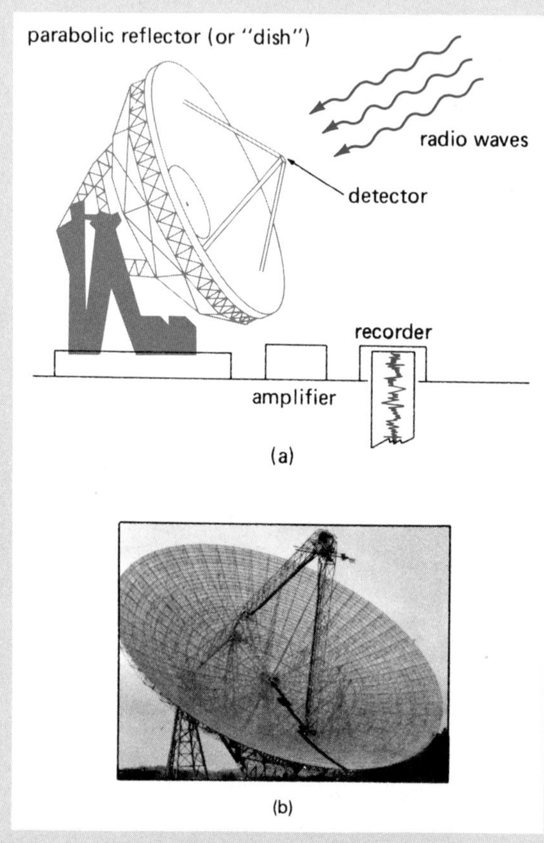

Figure 1
The radio telescope. (a) The telescope "dish" reflects radio waves to a focus at the antenna or detector, which feeds the signal into an amplifier and recorder. (b) The radio telescope at Green Banks, West Virginia, which collapsed due to structural failure, but will be rebuilt.

in white chalk, what color would the writing appear to be in these cases?

17. How would the American flag appear when illuminated with (a) red light, (b) blue light, and (c) green light?
18. Two complementary colors produce white, but if a colored light from a spotlight falls on an object with its complementary color, it appears black. Why?
19. (a) Why does a white piece of paper appear white, red, blue, or the color of whatever type of light illuminates it? (b) Can white be obtained by the subtractive method of color production? Explain.
20. By what process of color production could red and cyan colors be combined to get white?
21. What colors are obtained from the following mixtures: (a) blue light and green light, (b) yellow

Figure 2
The 305-m (1000 ft) diameter telescope at Arecibo, Puerto Rico. The dish was made by placing wire mesh in a natural bowl in the mountains. The antenna is mounted on a trolley suspended by cables 500 ft above the valley floor. Although the dish is fixed, signals are received from many directions as a result of the Earth's rotation and revolution.

ist in our galaxy and hide a large part of it from optical view. Radio waves are easily able to penetrate the Earth's atmosphere, whereas a large part of the incoming light is reflected and scattered.

Radio astronomy has extended the dimensions of the known universe almost twofold. With radio telescopes astronomers can detect galaxies that are two times farther away than those detected by optical telescopes. Although the emission of radio waves may be less intense, the penetration of the radio waves and construction of large dishes permit their detection (Fig. 2). Visual light from distant galaxies is blanketed by background illumination of the night sky. Radio telescopes can be operated around the clock, and observations can be made day or night.

pigment and cyan pigment, (c) red light and blue light, (d) magenta pigment and cyan pigment?

22. Is color TV based on the additive or subtractive method of color production? Explain.

Projectors

23. Why are slides placed in projectors upside down?
24. Could slides be put in a projector right side up and give a right-side-up image on a screen if other lenses were added to the system? Explain.
25. How are animated cartoons made?

Microscopes

26. What is a magnifying glass, and how is it positioned for maximum magnification?

27. Is there a practical limit to the magnification of a magnifying glass? Explain.
28. What are the basic components of a compound microscope? Explain how the image of an object is magnified.

Telescopes

29. Why are upside-down images no problem in astronomical telescopes?
30. A detective uses a small astronomical telescope to view a crime committed in a distant apartment building. How would he describe what he saw? If you were a defense attorney, how would you try to discredit the detective's testimony at a trial?
31. Distinguish between (a) astronomical and terrestrial telescopes and (b) reflecting and refracting telescopes.
32. Why do astronomers want to put telescopes in high-altitude orbits above the Earth?
33. What determines how far telescopes can be used to see objects? Do reflecting telescopes or refracting telescopes offer any advantage in this respect?

Problems

Levels of difficulty are indicated by asterisks for your convenience.

29.3 Projectors

*1. An old "lantern" projector has a lens with a focal length of 30 cm. If a slide is placed 35 cm from the lens and an image is formed on a screen, where is the screen and what is the magnification?

*2. A slide is positioned in a carousel projector 41.0 mm from the lens, and an image is formed on a screen at a distance of 1.64 m. What is the focal length of the lens?

**3. It is desired to have the complete picture on a 35-mm slide fill a screen 1.4 m tall when the image is projected. (a) If the screen is located 4.0 m from the projector, what is the focal length of the required lens? (b) What is the power of the lens? (See the Chapter 28 Supplement.)

29.4 Microscopes

*4. A machinist examines a metal piece with a magnifying glass having a focal length of 5.0 cm. (a) What is the approximate maximum magnification? (b) Is the observed image sharp?

*5. The gears of a watch are examined by a jeweler with a jeweler's (magnifying) glass. If a maximum magnification of 2.5× is observed by the jeweler, what is the power of the lens? (See the Chapter 28 Supplement.)

*6. A compound microscope has an objective that magnifies the object on the microscope stage by 10× and an eyepiece with 20× magnification. What is the total magnification?

*7. A specimen is observed under a compound microscope that has an objective with a focal length of 15 mm and an eyepiece with a focal length of 45 mm. What is the approximate total magnification?

*8. A compound microscope has an objective lens with a focal length of 20 mm and an eyepiece with a magnification of 15×. What is the approximate total magnification?

29.5 Telescopes

*9. The Palomar Mountain Hale Observatories reflecting telescope's mirror is 200 inches in diameter. (a) What is the diameter of the mirror in meters? (b) How does the circular area of the mirror surface compare to the floor area of your room?

*10. The objective and eyepiece of a refracting telescope have focal lengths of 100 cm and 5.0 cm, respectively. What is the magnification of the telescope?

*11. Given four converging lenses with focal lengths of 75 cm, 50 cm, 10 cm, and 5.0 cm, respectively, what are the possible magnifications of refracting telescopes you could construct with the lenses?

*12. In designing a refracting astronomical telescope having an objective lens with a power of +1.25 D, it is desired to have a magnification of 50×. What should the focal length of the required eyepiece be?

**13. Prove that using a third erecting lens to make a terrestrial telescope increases the length of the telescope to four times the focal length of the erecting lens.

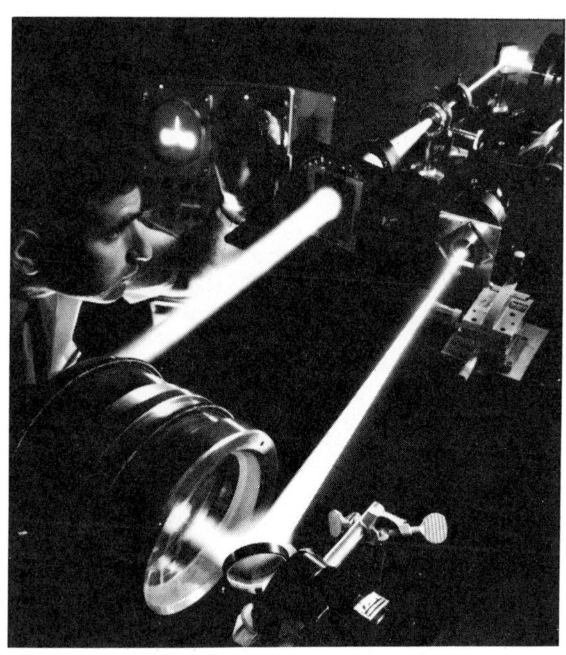

Chapter 30

Quantum Physics and the Dual Nature of Light

The study of physics that developed around 1900 and afterward is known as modern physics, as distinguished from the earlier "classical" physics. Basically, classical physics is concerned with macroscopic phenomena, and modern physics with submicroscopic phenomena.

Modern physics involves some new considerations. One of these is the quantum theory of light. Certain phenomena could not be explained by the classical wave nature of light. They required the introduction of a new concept—the quantum—and their investigation led to the recognition of the dual nature of light.

30.1 The Ultraviolet Catastrophe and Quantization

The wave nature of light (electromagnetic waves) satisfactorily explains such phenomena as interference and polarization (Chapter 27). Classical physics provided empirical relationships such as the Stefan-Boltzmann law, which adequately describes the intensity of black-body radiation (Chapter 17). However, when standing wave principles were applied to the black-body cavity radiators, it was predicted that the intensity I should be proportional to the inverse fourth power of the wavelength ($1/\lambda^4$) of the radiation.

This classical prediction gave rise to what is known as the **ultraviolet catastrophe** (Fig. 30.1). There was good agreement between the theory and experimental data for long wavelengths. However, at short wavelengths there was a major problem. Classical theory predicted that, at shorter (ultraviolet) wavelengths, the energy emitted by a black-body radiator would be infinitely large (the area under the curve), rather than energy going to zero as λ approached zero, as experimentally observed. This was a catastro-

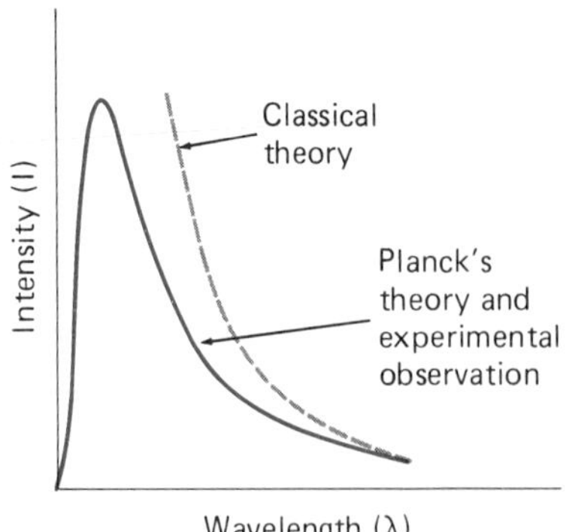

Figure 30.1
Blackbody radiation spectrum. An intensity versus wavelength graph showing the agreement with Planck's theory and experimental observation, as compared to classical theory, which predicted an ultraviolet catastrophe.

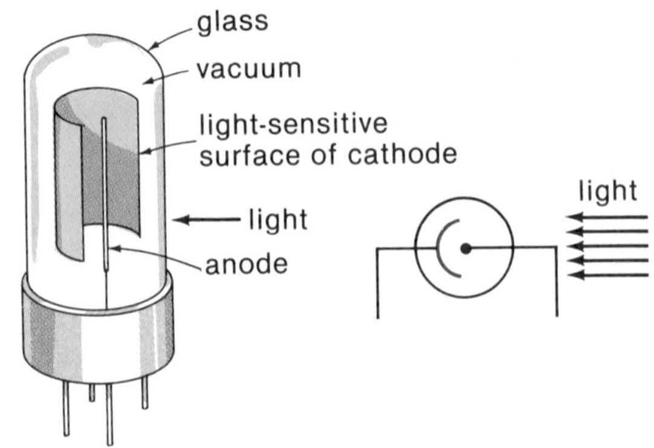

Figure 30.2
The phototube. When light falls on the photoelectric material used as the cathode, electrons are emitted and are attracted to the anode. The photocurrent is the result of light energy being converted into electrical energy.

phe because it violated energy principles. Another way of looking at this is that the theory predicted that the short-wavelength radiation would be most intense, so all objects would glow blue.

The dilemma was resolved by a German physicist, Max Planck (pronounced "Plonk"), who postulated that the radiators, or oscillators, within the walls of the body have discrete energies. **Planck's hypothesis** states that the energy of an oscillator depends on its frequency according to the relation

$$\boxed{E = hf} \qquad \textbf{(Eq. 30.1)}$$

(quantum energy)

where h is **Planck's constant,** which has a value of

$$h = 6.63 \times 10^{-34} \text{ J-s}$$

When Planck's hypothesis was applied, the theory conformed to experimental observations, thus satisfying the scientific method. The discrete amounts of energy of the oscillators were referred to as quanta (singular: quantum). Planck's quantum hypothesis set the stage for the development of modern quantum physics. Some of the important applications of quantum physics are described in the following sections, particularly the quantum nature of light.

The quantization of light gave rise to what is known as **the dual nature of light.** In order to explain classical phenomena, light is treated as a wave. Yet, in other, modern phenomena such as the photoelectric effect, light must be viewed as a "particle" or quantum. Further examples of the quantum nature of light are given in the following sections.

30.2 The Photoelectric Effect

Certain materials—for example, metallic oxides such as selenium oxide and cesium oxide—are photosensitive, or light-sensitive. When these materials are exposed to light, electrons are emitted from their surfaces. The emission of electrons from a material when it is exposed to light is called the **photoelectric effect.** By using a photoelectric material as the cathode of a vacuum tube, we form a phototube (Fig. 30.2). When the phototube is exposed to light, a current flows in its circuit. The photoelectric effect also occurs in semiconducting diodes called photocells. They are discussed in Chapter 32 as solar cells.

In a phototube, the "photocurrent" is proportional to the intensity of light (Fig. 30.3). That is, the greater the light intensity, the greater the photocurrent. (This effect is used in light meters to measure the intensity of light.) Classically, one would expect this to be the case since the greater the intensity (energy/area-time), the more energy is available to free electrons from the material.

However, there are other effects that cannot be explained by classical wave theory. It is

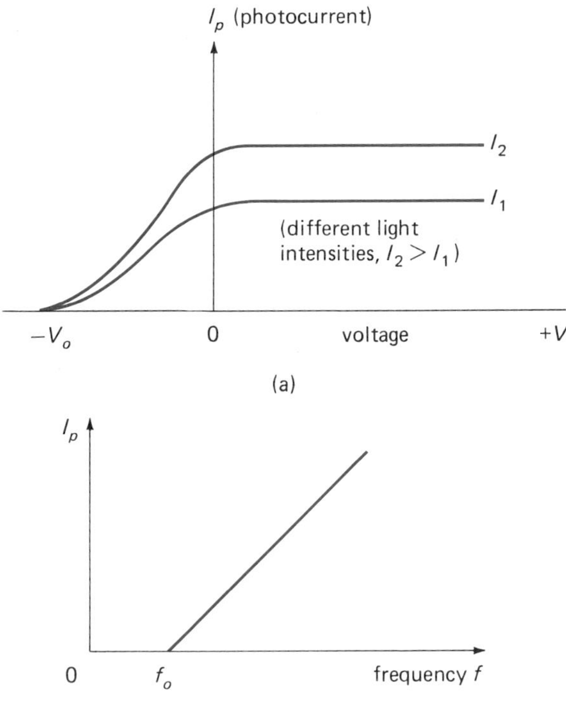

Figure 30.3
Phototube characteristics. (a) The photocurrent is proportional to the intensity of the light—the greater the light intensity, the greater the photocurrent. If the phototube is reverse-biased, the photocurrent goes to zero at a "stopping potential," $-V_o$. (b) No photoemission is observed for light having a frequency below a certain cutoff frequency f_o.

found that a photocurrent is observed immediately on illumination of the photomaterial. Also, no photoemission occurs for light having a frequency below a certain cutoff frequency f_o.

Using classical wave theory, calculations show that an appreciable time would be required for an electron to gain enough energy from continuous light waves to be freed from the material. The energy of a wave depends on the wave amplitude and not on the frequency. A light wave with sufficient intensity, even with a frequency below f_o, should be able to free electrons from the material. Hence, classical theory fails to explain the observed effects.

An explanation of the photoelectric effect was put forth by Albert Einstein in 1905. He applied Planck's quantum hypothesis to light. In his theory, the energy of a light beam is in discrete quanta called **photons,** with a single photon having an energy of $E = hf$.

EXAMPLE 30.1 A beam of yellow light with a wavelength of 6000 Å is emitted from a source. What is the energy of a photon, or quantum, of this light?

Solution: It is given that

$$\lambda = 6000 \text{ Å} \times (10^{-8} \text{ cm/Å}) = 6.0 \times 10^{-5} \text{ cm}$$

Recall that 1 Å = 10^{-10} m = 10^{-8} cm. Then, $\lambda f = c$, and

$$f = \frac{c}{\lambda} = \frac{3.0 \times 10^{10} \text{ cm/s}}{6.0 \times 10^{-5} \text{ cm}} = 5.0 \times 10^{14} \text{ Hz (s}^{-1})$$

And

$$E = hf = (6.63 \times 10^{-34} \text{ J-s})(5.0 \times 10^{14} \text{ s}^{-1}) = 3.3 \times 10^{-19} \text{ J}$$

Hence, a photon carries a small amount of energy, but a beam of light contains many photons.

Einstein applied the conservation of energy to the situation. If an electron in the photomaterial did absorb a photon and was emitted, then

energy of photon = energy needed to free electron + kinetic energy of emitted electron

or

$$E = hf = \phi + K \qquad \textbf{(Eq. 30.2)}$$

where K is the kinetic energy of the freed electron and ϕ is the amount of work needed to free the electron. ϕ is called the **work function** of the material.

Quantizing the work function, we may write

$$\phi = hf_o \qquad \textbf{(Eq. 30.3)}$$

where f_o is called the **threshold frequency** and is the same as the cutoff frequency. Hence, light with a frequency less than the threshold frequency will not have enough energy in its photons to free the electrons. Combining these equations, we have

$$hf = hf_o + K \qquad \textbf{(Eq. 30.4)}$$

Notice that there is no time delay in this theory. In absorbing a photon with a frequency greater than f_o, the photoelectron receives a definite amount of energy all at once.

EXAMPLE 30.2 A beam of monochromatic light with a frequency of 5.0×10^{14} Hz is incident on a photoelectric material that has a work function of 2.0 eV. (a) What is the kinetic energy of the photoelectrons? (b) What is the threshold frequency of the material?

Solution: It is given that $f = 5.0 \times 10^{14}$ Hz and $\phi = 2.0$ eV $(1.6 \times 10^{-19}$ J/eV$) = 3.2 \times 10^{-19}$ J.

(a) With $hf = \phi + K$,

$$K = hf - \phi = (6.63 \times 10^{-34} \text{ J-s})(5.0 \times 10^{14} \text{ Hz}) - 3.2 \times 10^{-19} \text{ J}$$

$$= 0.1 \times 10^{-19} \text{ J}$$

(b) $\phi = hf_o$, and

$$f_o = \frac{\phi}{h} = \frac{3.2 \times 10^{-19} \text{ J}}{6.63 \times 10^{-34} \text{ J-s}} = 4.8 \times 10^{14} \text{ Hz}$$

There are many applications of phototubes and photocells. One common application is the so-called "electric eye" circuit (Fig. 30.4). When light falls on the phototube, current flows in the battery-relay circuit. Interrupting the light opens the circuit and the relay. The opening and closing of the relay acts as a switch that controls some other device.

Common applications of such circuits include automatic door openers in supermarkets and other stores and counters that count the number of people passing through the door, e.g., in a library. In burglar alarm detection systems, a beam of nonvisible ultraviolet light is used. When someone breaks the beam, an alarm sounds. Photocells are used to turn on streetlights when the Sun goes down.

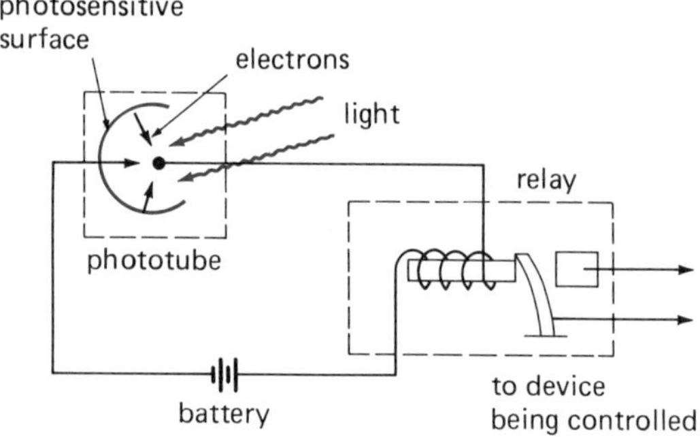

Figure 30.4
Electric eye circuit. When light falls on the phototube, there is a current in the battery-relay circuit. Interrupting the light opens the circuit, and the relay controls some device.

Photoelectric relays are used for monitoring and control in various types of commercial processes (Fig. 30.5). Photocells are also used in one type of smoke detector, since smoke particles scatter light. (An ionization-type smoke detector is described in Chapter 31.)

30.3 The Bohr Theory of the Hydrogen Atom

The quantum theory played an important role in the development of our present model of the atom. Recall that we now visualize a planetary model of the atom. Electrons are held in orbit around the nucleus of the atom by the attractive electrical (Coulomb) forces between them and the nuclear protons.

As we learned in Chapter 27, when sunlight or light from an incandescent source is analyzed by a prism or grating, a continuous spectrum is observed. However, when light from a gas discharge tube is analyzed, a discrete, or line, spectrum is observed. That is, only spectral lines of certain frequencies or wavelengths are found (Fig. 30.6).

This puzzled scientists in the early 1900's. An explanation was put forth in 1913 by the Danish physicist Niels Bohr in his theory of the hydrogen atom. Bohr obviously chose the hydrogen atom as a start because it is the simplest—one electron and one proton. He assumed the electron to be in a circular orbit around the nuclear proton, much the same as a planet is in orbit about the Sun.

In analyzing this situation, Bohr applied classical principles, but he also postulated that the angular momentum of the electron was *quantized.* That is, the electron could have only discrete values of angular momentum in units of $nh/2\pi$, where $n = 1, 2, 3, \ldots$. These n's are called principal quantum numbers, and h is Planck's constant.

This analysis restricted electrons to orbits with discrete radii (see Fig. 30.7). (Recall that the angular momentum, $L = mvr$, contains the radius of a particle in angular motion.) In a par-

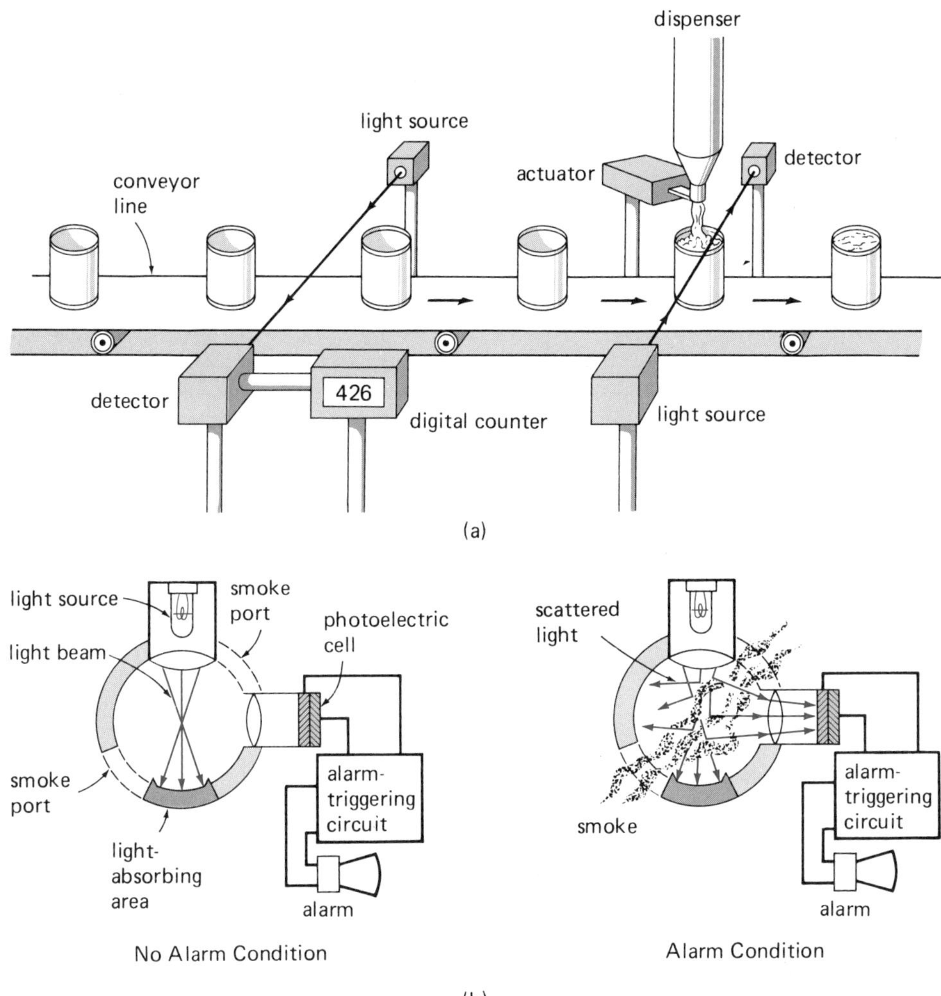

Figure 30.5

Applications of photocells. (a) Counting and control. Interrupting the light beam counts the containers and controls the fill level of the dispensed material. (b) Photoelectric smoke detection. When smoke enters the detector, the scattered light is detected by the photocell.

ticular orbit, the hydrogen electron has a particular energy. Bohr's theory showed this to be

$$E_n = \frac{-13.6}{n^2}\text{ eV} \quad n = 1, 2, 3, \ldots \qquad \textbf{(Eq. 30.5)}$$

where the unit of energy is the electron volt (eV).

The "energy levels" corresponding to the particular orbits are illustrated in Figure 30.8. The lowest energy level ($n = 1$) is called the **ground state.** All other energy levels are called **excited states** ($n = 2, 3, 4, \ldots$).

Because the energy is negative, we say that the electron is in a negative potential energy well, similar to the gravitational case (Chapter 8). A particle normally at the bottom of the well

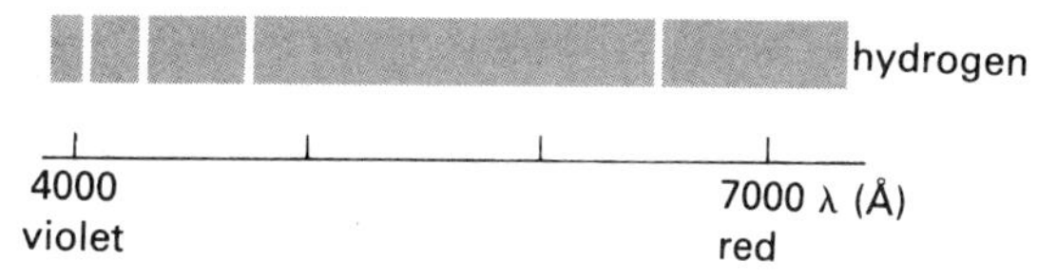

Figure 30.6

Line spectrum. The visible line spectrum of hydrogen. Only spectral lines of certain wavelengths or frequencies are observed.

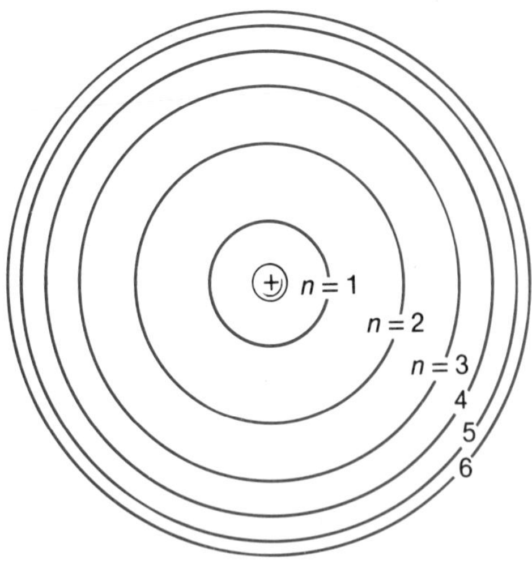

Figure 30.7
Bohr model of the hydrogen atom. In the Bohr theory, the hydrogen electron can be only in discrete orbits characterized by n quantum numbers. The electron has a different energy in each orbit.

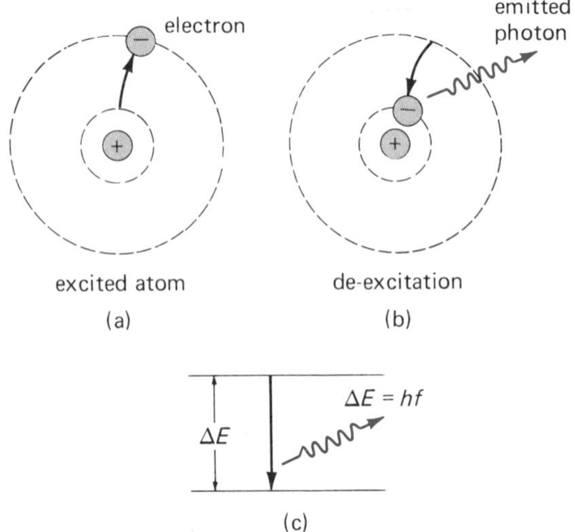

Figure 30.9
Photoemission. (a) When an electron is excited to a higher orbit or energy level, it eventually "decays" (b), with emission of a photon of discrete amount of energy at a certain frequency determined by the difference in the energy levels (c).

(ground state) must be given energy, or be "excited," to raise it up in the well. However, the hydrogen electron can be excited only by discrete amounts.

It is sometimes convenient to consider the energy levels to be like steps on an energy "ladder." As in the case of a person on a real ladder, a particle going up or down must do so in discrete steps on the ladder "rungs." Notice that the "rungs" in the hydrogen atom are not evenly spaced. If the hydrogen electron is given 13.6 eV of energy or more, it is raised to the top of the well, and the electron is no longer bound to the nucleus; i.e., the atom is ionized.

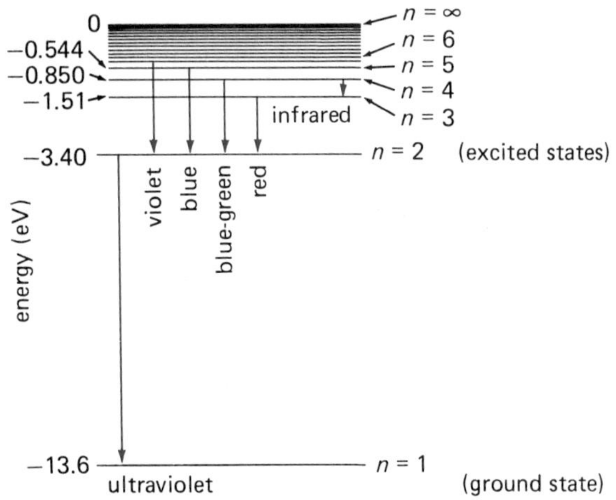

Figure 30.8
Energy level diagram for hydrogen. The $n = 1$ level is called the ground state, and all other levels are excited states. In making transitions to lower states, energy is emitted as photons of light. For $n_f = 2$, the wavelengths of the photons correspond to the wavelengths of the lines of the observed spectrum in the visible region for some transitions.

Consider now an electron making transitions downward on the energy ladder (not necessarily one rung at a time). The energy differences between higher and lower energy levels are given by

$$\begin{aligned}\Delta E &= E_i - E_f \\ &= 13.6\left(\frac{1}{n_f^2} - \frac{1}{n_i^2}\right)\text{ eV}\end{aligned} \qquad \textbf{(Eq. 30.6)}$$

where the i subscript represents an arbitrary initial state and the f represents an arbitrary lower final state.

Just as energy is given to the electron to raise it in the potential well, energy is released when the electron returns to a lower state. This energy is emitted in the form of electromagnetic radiation as a quantum, or photon, of light, since there is a discrete energy difference between the energy levels (Fig. 30.9).

According to Bohr, the frequency of the photon is $f = \Delta E/h$ or, in terms of the wavelength ($\lambda f = c$),

$$\lambda = \frac{c}{f} = \frac{hc}{\Delta E}$$

Then, expressing the constants in convenient units,

$$\lambda = \frac{12{,}400}{\Delta E}\ \text{Å} = \frac{1240}{\Delta E}\ \text{nm} \qquad \textbf{(Eq. 30.7)}$$

where λ is in angstroms or nanometers with ΔE in electron volts.

EXAMPLE 30.3 What is the wavelength of the emitted light when a hydrogen electron makes a transition from $n_i = 4$ to $n_f = 2$?

Solution: Using Equation 30.6, we see the energy difference between these levels is

$$\Delta E = 13.6\left(\frac{1}{n_f^2} - \frac{1}{n_i^2}\right) = 13.6\left(\frac{1}{(2^2)} - \frac{1}{(4^2)}\right)$$

$$= 13.6\left(\frac{3}{16}\right) = 2.55 \text{ eV}$$

Then,

$$\lambda = \frac{12{,}400}{\Delta E} = \frac{12{,}400}{2.55}\ \text{Å}$$

$$= 4863\ \text{Å} = 4.863 \times 10^{-5} \text{ cm}$$

which is the wavelength of blue light. (Significant figures assumed)

Hence, according to the Bohr theory, radiation is emitted only for transitions between discrete energy levels. This gives rise to a discrete (line) spectrum. Some transitions for the lines of $n_f = 2$ for hydrogen are illustrated in Figure 30.8. The theory agrees with experiment and predicts the wavelengths of the observed visible spectral lines (Fig. 30.6)—the so-called Balmer series. Transitions to other final states produce lines in the nonvisible regions. For example, transitions to the ground state ($n_f = 1$) from nearby states produce lines in the ultraviolet region.

Thus, when a current passes through a gas discharge tube, gas atoms are excited. In the deexcitation process, photons are emitted with the characteristic spectral lines of the particular gas. In a solid-filament incandescent lamp, the atoms interact strongly and the electrons in the conductor are not associated with a single atom. In that case, the electron energies have a continuous range and hence give rise to a wide, continuous spectrum.

30.4 The Laser

We hear more and more about lasers. Laser beams have been sent to the moon and have been used in eye surgery, in surveying, and to drill holes. We now hear talk of laser weapons capable of destroying satellites. Because of such publicity, the laser is sometimes thought of as an incredible source of energy. That is not the case. The laser is a light source, and, like other light sources, it converts one form of energy to another—always at a loss. The uniqueness of the laser as a light source is in the special properties of the light it produces.

The development of the laser was one of the crowning successes of the modern approach to science. In many instances, scientific discoveries have been made accidentally, and even though these discoveries were applied, no one fully understood how or why they worked. Often a trial-and-error, or Edisonian, approach has been used for inventing. (In developing the incandescent lamp, Edison tried many filament materials until he found one that worked—a carbonized thread from his wife's sewing basket.)

However, the laser was first developed "on paper" and then built as predicted. Atomic theory allowed scientists to "look" into the atom for previously unobserved and unapplied phenomena. As a result of quantum theory, we now have the laser.

The theory behind the laser was developed during the middle and late 1950's by the American scientists Arthur Schalow and Charles Townes, and the first working device was constructed in 1960. The term **laser** is an acronym for *l*ight *a*mplification by *s*timulated *e*mission of *r*adiation. The laser was the first device capable of amplifying light waves themselves. You may think that the amplification of electromagnetic radiation is not new. For example, radio waves are amplified in a radio. But here the current signals are amplified *after* the electromagnetic energy has been converted to the vibrational energy of electrons in the antenna (Chapter 26).

Electrical circuits are incapable of handling the frequencies of light (a million times those of radio waves), because the circuit electrons cannot oscillate at these frequencies. The laser, in which light is amplified using energy stored in atoms, overcomes this limitation.

The principle of the laser was first developed for *m*icrowave frequencies, and the first device was called a *m*aser. The laser is an optical maser. The same principles are involved, but the laser involves higher frequencies. The first maser used ammonia as an amplifying medium. Other substances, such as ruby crystals, and various gases, have been found to be suitable for higher light frequency "lasing" action, or amplification. To understand this action, let's take a look at the "aser" part of a laser—*a*mplification by *s*timulated *e*mission of *r*adiation.

When an electron is excited to a higher energy level in an atom, it generally "decays" after a short time (on the order of 10^{-8} s) and returns to a lower energy level with the emission of a photon. This is called **spontaneous emission** (Fig. 30.10). However, an excited atom can be *stimulated* to emit a photon. In a **stimulated emission** process, an excited atom is struck by a photon of the same energy of the allowed transition, and the atom emits an identical photon. After emission, the two photons have the same frequency, are emitted in the same direction, and are in phase. Since one photon goes in and two come out, this is an amplification process (which was predicted by Einstein in 1917).

The monochromatic (single-frequency), phase, and directional properties of the light of the lasing action give laser beams some unique properties. Light from common sources, such as an incandescent lamp, is incoherent. That is, the waves have no particular relationship to each other. Similarly, if you threw a handful of gravel into a pond, the resulting waves would be incoherent. In light sources, excitation occurs randomly, and atoms emit randomly and at different frequencies (different transitions). As a result, an incoherent light beam is "chaotic" (Fig. 30.11). Such a beam spreads out and becomes less intense.

To get monochromatic light, all but one frequency must be filtered out of the beam, which is then weak and still incoherent. When the

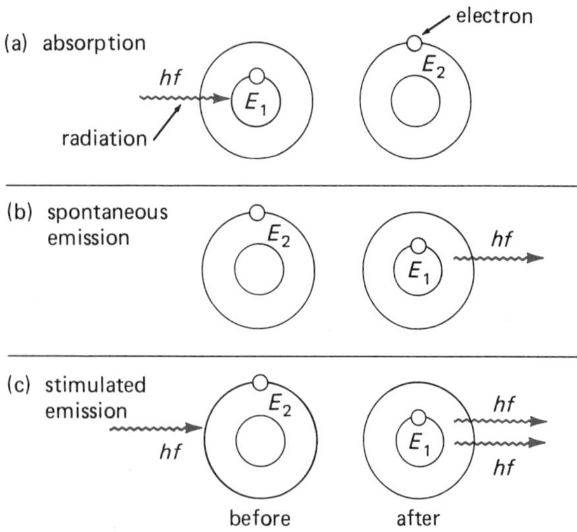

Figure 30.10
Radiation emission. (a) An atom absorbs energy and becomes excited. (b) It generally decays spontaneously after a short time, with the emission of a photon. But, if another photon strikes the excited atom (c), it is stimulated into emission.

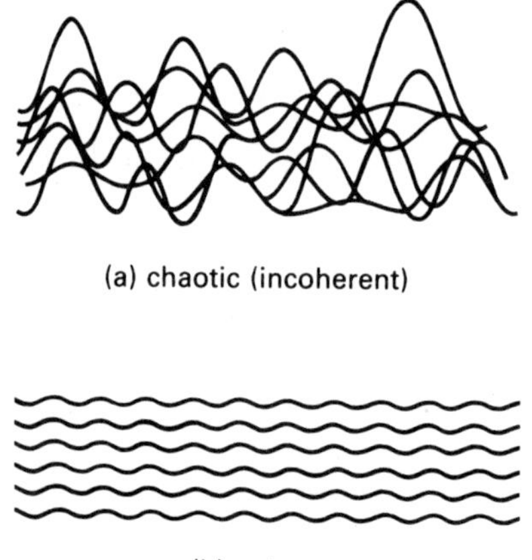

Figure 30.11
Laser light is coherent. (a) Light from common sources is chaotic or incoherent, with the waves having no particular relationship to each other. (b) Coherent laser light has the same frequency, phase, and direction.

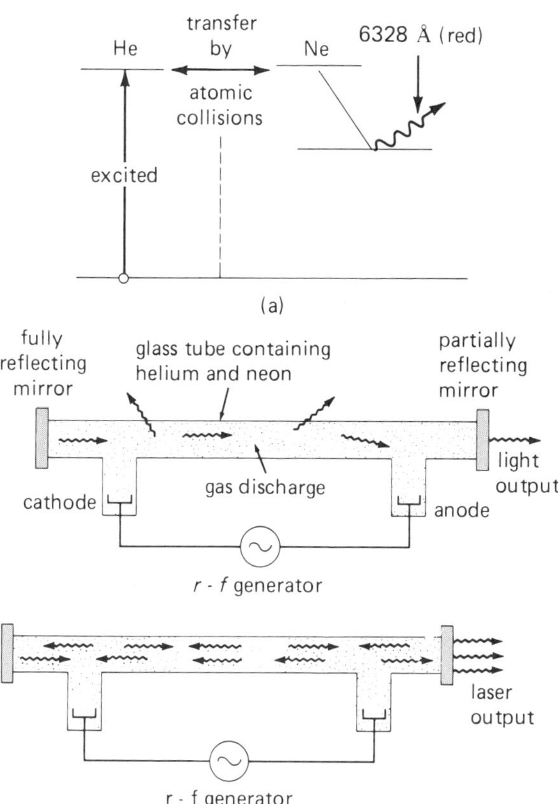

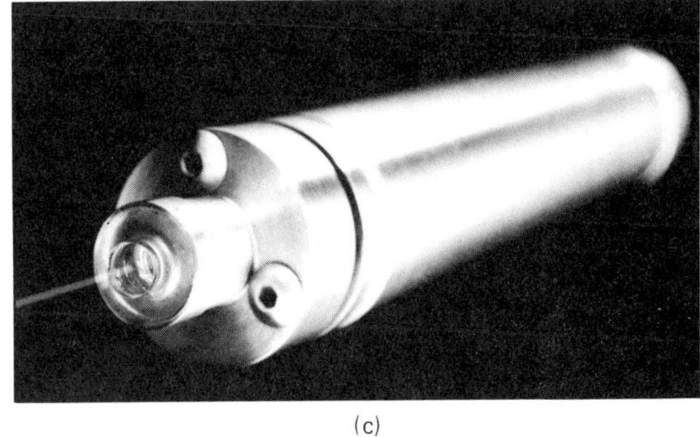
(c)

Figure 30.12
He-Ne laser. (a) The laser light comes from a neon atom transition. (b) A schematic diagram of a gas laser. (c) A laser. Steel-ceramic tubes, which are much more reliable than glass tubes, are now commonly used. (Courtesy Metrologic Instruments, Inc.)

waves (or photons) of a beam of light have the same frequency, phase, and direction, the beam is **coherent.** It is possible to make a coherent beam that does not spread out appreciably and that, if amplified, can be very intense.

There are pulsed lasers, which produce pulses of coherent light, and continuous lasers. Let's focus on the continuous-beam helium-neon (He-Ne) gas laser, which you are most likely to see in the classroom and in some common applications. The gas mixture is subjected to a high voltage in a laser tube, and an electrical discharge is induced (Fig. 30.12). Electron collisions efficiently bring He atoms to an excited state that has a relatively long lifetime ($\sim 10^{-4}$ s). The Ne atom also has an excited state with the same energy. Hence, there is a good chance that before an excited He atom emits a photon spontaneously, it will collide with an unexcited Ne atom and transfer energy to it. (There are about ten He atoms to one Ne atom in the gas mixture.) The excited Ne atoms decay after a short time, and the photon of the transition has a frequency that is in the visible red region.

The amplification of the light emitted by the Ne atoms is accomplished by reflections from mirrors placed at the ends of the laser tube [Fig. 30.12(b)]. Reflecting back and forth in the tube, the photons cause stimulated emissions, and an intense beam of photons builds up along the direction of the tube axis. Part of the beam emerges through one of the end mirrors, which is only partially silvered. The outgoing beam is highly intense, coherent, and directional.

A common He-Ne laser application is given in Special Feature 30.1. The laser light seen in this case is scattered, or indirect. An intense laser beam can be dangerous to the eye, and you should never look directly into a laser because eye damage can occur. In the medical field, a *slight* amount of such damage is used to "weld" detached retinas in eyes (Fig. 30.13).

The list of laser uses is quite long and continues to grow. The distance of the moon from the Earth has been accurately measured using a

SPECIAL FEATURE 30.1

Lasers in the Supermarket

You may have noticed the pink glow of He-Ne laser light in a supermarket checkout that uses a scanner for the optical reading of the Universal Product Code (UPC), or "bar code," on prepackaged items (Fig. 1). The checkout person moves an item across an opening through which a laser beam is directed toward the bar code label. The dark areas of the code absorb light, and the light areas reflect. A light-sensitive detector reads the pattern of bars from the reflection pattern. This information goes into a central computer, which identifies the product and sends the programmed price for that item to the electronic cash register.

(a)

(b)

Figure 1
Bar code scanning system. A He-Ne laser is used in the optical scanning system (a) that reads product codes (b) on groceries and other items. The zero to the left of the bar code indicates a grocery item. The next five digits identify the manufacturer (13000, H.J. Heinz Co.), and the next five digits identify the product (45090, 16-oz can of pork and beans). (Courtesy H.J. Heinz Co.)

laser pulse reflected from a mirror placed on the moon by astronauts. Surveyors use laser beams to "line up" measurements, as in drilling tunnels. Long-distance communication via laser beams in optical fibers is another application (Chapter 28).

High-intensity lasers can cut metals (Fig. 30.14) as well as cloth (laser knife or scissors) in commercial garment-making. Metal welding is also done with lasers. Computer printers now print with lasers. Solid state lasers are used in compact disc (CD) players instead of needles as in conventional record players. Also, laser videodiscs are available. The videodisc is about the same size as an LP album. More than 50,000 separate images can be stored on each side of a disc, along with two-channel audio. Information is coded in tracks of microscopic indentations. As the disc spins, a laser "needle" reads the tracks for playback on a conventional TV receiver.

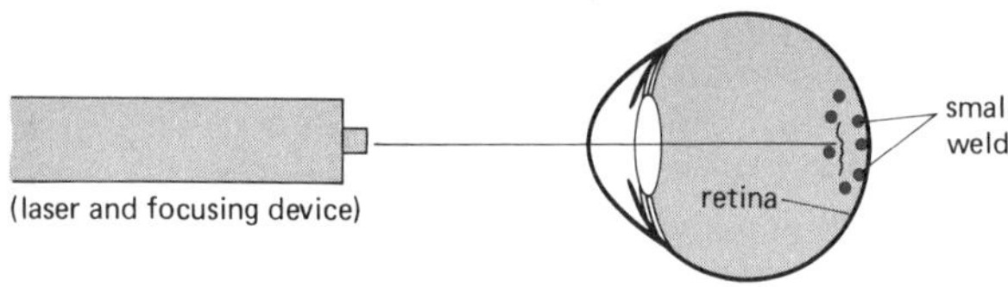

Figure 30.13
Medical laser application. Lasers are used to "weld" detached retinas in eyes.

The coherent property of laser light has made the production of three-dimensional images possible through **holography** (Greek *holos,* meaning "whole"). A hologram is made by using reference and interference beams (Fig. 30.14). Their interference patterns on a photographic film record the information from "in-depth" parts of the object. When the film is developed, the interference pattern appears as a meaningless array of light and dark areas. However, when the film is illuminated with a laser beam (or another light source, for some holograms), the recorded information is reproduced and perceived as a three-dimensional image. Someday holographic, 3-D TV sets may be common.

Various new types of lasers have been developed—glass, chemical, and semiconductor. The produced light ranges from infrared through ultraviolet, and some lasers are "tunable" and can be adjusted to different frequencies. X-ray lasers are also being developed. They hold the promise of in-depth, possibly three-dimensional (holographic), views of microscopic structures. Also, as will be learned in the next chapter, experimental work is being done in the use of lasers to induce nuclear fusion.

30.5 X-Rays

X-rays were discovered by the German scientist Wilhelm Roentgen (Fig. 30.15) in 1895 in the course of his work with high-voltage discharge tubes. The discovery was made accidentally when Roentgen noticed that a photographic plate near a discharge tube had been exposed, evidently by some unknown "X" radiation, or X-rays.

Most of us have had an X-ray to check for a broken bone or some other medical condition. The X-ray film shows internal structures that cannot normally be seen (Fig. 30.16). X-rays are highly penetrating electromagnetic radiation

(a)

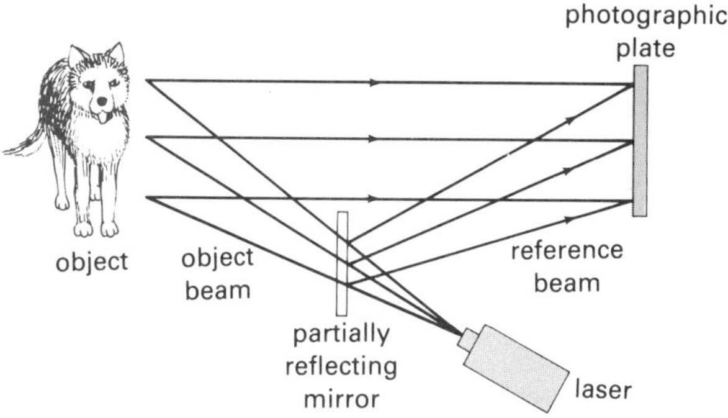

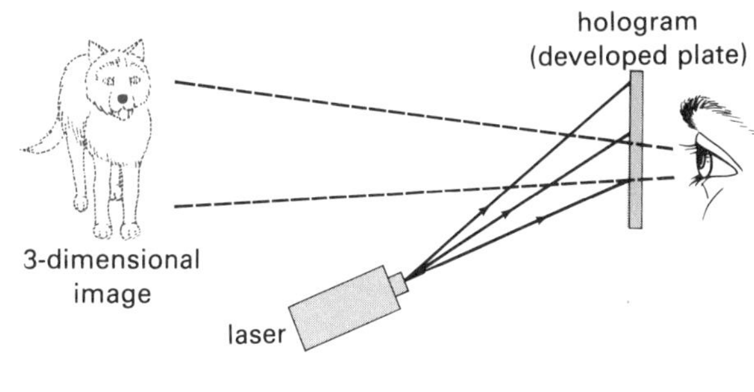

Figure 30.14

Laser uses. (a) A laser being used to cut metal. (b) An arrangement for making and viewing a hologram. See text for description.

Figure 30.15
Wilhelm Roentgen. A bust of Roentgen in the Deutsches Museum in Munich. (Photo by author.)

that can pass through many solid bodies that are opaque to other types of radiation. Denser, more solid objects absorb more of the X-rays than less dense materials. In the human body, the bones absorb more X-rays than the surrounding tissues. As a result, a contrasted picture of the internal body structures is obtained.

The basic principle of an X-ray tube is illustrated in Figure 30.16(b). Thermionic electrons emitted from the cathode are accelerated toward the anode "target" by a high applied voltage (typically 50 to 150 kV). The target material is usually copper or tungsten. As the cathode rays "collide" with the target, they interact electrically with the electrons of the atoms of the target material.

The repulsive electrical force causes the accelerated cathode electrons to slow down. In doing so, they lose energy in the form of electromagnetic radiation, and X-rays are emitted. In Germany the name for X-rays is *Bremsstrahlung*, or "braking rays."

Because of the high voltages applied to the tube, the emitted radiation is highly energetic,

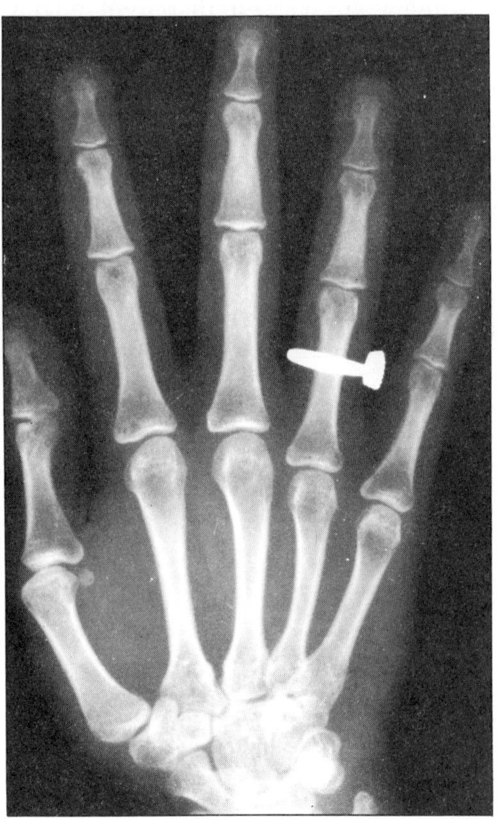

(a)

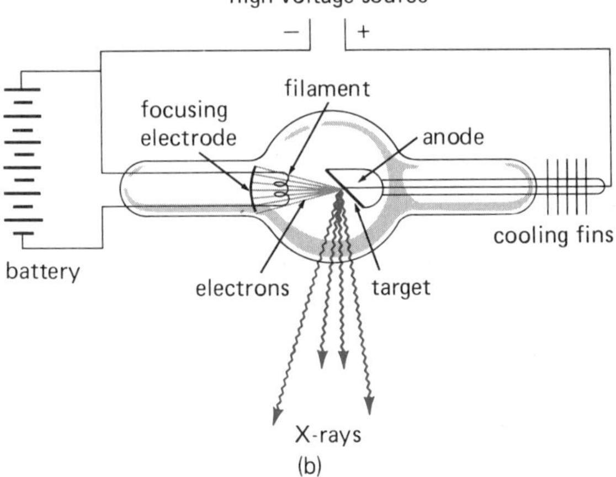

(b)

Figure 30.16
X-rays. (a) Dense, solid objects absorb more X-rays than do less dense materials. The bones absorb more X-rays than the surrounding tissue, and a contrasted picture of a hand is obtained on a photographic film. Notice the ring on the finger. (Courtesy K.F. Kahn and J.S. Faughn.) (b) Diagram of an X-ray tube. Electrons emitted by the heated cathode filament are accelerated by a high voltage. On striking the target, they are decelerated by repulsive electrical interactions with the electrons in the atoms of the target material, and X-rays are emitted.

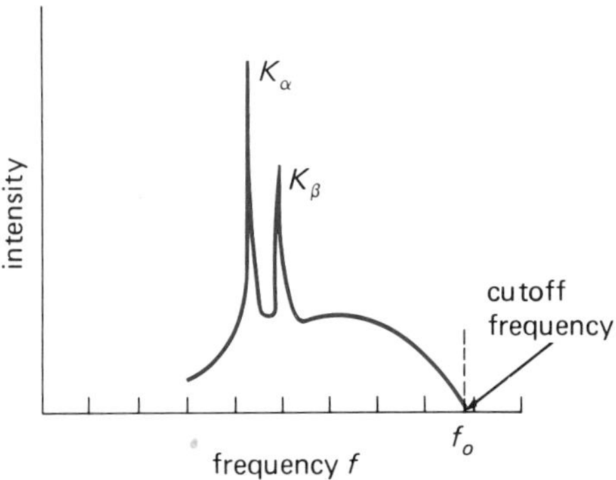

Figure 30.17

A typical X-ray spectrum. The spiked lines are caused by the filling of inner shell vacancies in target atoms from which electrons have been dislodged. The continuous background spectrum results from general electron deceleration. The cut-off frequency f_o corresponds to an electron being completely stopped and giving up all of its energy.

i.e., has a high frequency ($E = hf$). When the X-ray spectrum is analyzed, two types of spectra are observed—a continuous spectrum and a discrete line spectrum (Fig. 30.17). As you might expect, the line spectrum is produced by electron transitions between energy levels. However, in this case the spectral lines are due to transitions in "inner" energy levels.

In multi-electron atoms, as those in the target material, we refer to energy "shells" rather than levels. There are the K, L, M, N, etc., shells, corresponding to the quantum numbers $n = 1$, 2, 3, 4, etc., respectively. The inner shells are filled with electrons, so normally no states are available for transitions in these shells. (The outer electrons of multi-electron atoms are responsible for the optical spectra of these elements.)

On colliding with the target, some of the accelerated electrons may have sufficient energy to dislodge some of the inner electrons of the target atoms, say in the K shell. The vacancies are immediately filled by electrons from higher shells—either the L, M, or N shells—and high-energy quanta are emitted. The emission lines associated with filling K shell vacancies are called K_α, K_β, and K_γ, corresponding to electron transitions from the L, M, and N shells, respectively. The spectral lines are referred to as **characteristic X-rays** because they are characteristic of the energy shells of a particular target material.

The continuous-background spectrum arises from the deceleration of the cathode rays that do not dislodge target electrons. This can occur over a wide range with the release of varying amounts of energy, so the spectrum is continuous. Notice that there is a **cutoff frequency** f_o of the continuous spectrum. This short wavelength (high frequency) corresponds to the case in which a decelerated electron is completely stopped and gives up all of its energy.

Quantum theory allows the calculation of this cutoff wavelength for a given tube voltage. Recall that the energy acquired by an electron on being accelerated through a potential difference V is $E = eV$. If this is all given up in a quantum of energy $E = hf_o$, then

$$hf_o = eV$$

and the **cutoff wavelength** is given by

$$\lambda_o = \frac{c}{f_o} = \frac{hc}{eV} = \frac{12{,}400}{V}\ \text{Å} = \frac{1240}{V}\ \text{nm} \qquad \textbf{(Eq. 30.8)}$$

Since X-rays are highly energetic radiation, overexposure to them can be dangerous in that burns and other medical damage may result. You may have noticed when getting an X-ray that the technician always goes behind a shield. Lead is commonly used as a shielding material because it absorbs X-rays. Color television tubes, which operate at high voltages, can produce X-rays when electrons in the cathode ray tube are decelerated at the screen. As a result, the tube is shielded, and people are advised not to sit too close to the television set.

As in the medical field, X-rays are used in industry to "see inside" materials, e.g., to check castings and welds for flaws (Fig. 30.18). A new application of X-rays and the idea of the wave-like nature of particles are considered in the Chapter Supplement.

Important Terms

quantum a discrete packet of energy, $E = hf$

dual nature of light to explain different phenomena, light is sometimes treated as a wave and sometimes as a particle or quantum

Figure 30.18
X-ray application. X-ray inspection of aluminum diecast automotive wheels. (Courtesy Seifert X-ray Corp.)

photoelectric effect the emission of electrons from certain materials when exposed to light

photon a quantum of light

ground state the lowest energy level ($n = 1$) for an atomic electron

excited state an energy level ($n = 2, 3, \ldots$) above the ground state ($n = 1$)

spontaneous emission the emission from the natural decay or transition of an electron in an excited state

stimulated emission emission resulting from an excited atom being struck by a photon of the same energy as an allowed transition, with the emission of two identical photons

laser (*l*ight *a*mplification by the *s*timulated *e*mission of *r*adiation) a device that amplifies light by stimulated emission

coherent light a light beam having waves with the same frequency, phase, and direction

holography the production of three-dimensional pictures through the use of coherent light and interference properties

X-rays high-frequency electromagnetic radiation that is produced by the deceleration of electrons and that can penetrate opaque materials

characteristic X-rays discrete frequency X-rays resulting from inner shell vacancies being filled

cutoff frequency the maximum frequency of an X-ray spectrum, determined by the total energy transfer of the decelerating electrons

Important Formulas

quantum energy:
$E = hf$
$h = 6.63 \times 10^{-34}$ J-s (Planck's constant)

photoelectric equations: $E = hf = \phi + K$
where $\phi = hf_o$
and $hf = hf_o + K$

Bohr theory equations:

energies of electron states:
$$E_n = \frac{-13.6}{n^2}\ \text{eV} \qquad n = 1, 2, 3, \ldots$$

energy difference between energy levels:
$$\Delta E = E_i - E_f$$
or
$$\Delta E = 13.6\left(\frac{1}{n_f^2} - \frac{1}{n_i^2}\right)$$

wavelength of emitted photon:
$$\lambda = \frac{hc}{\Delta E} = \frac{12{,}400}{\Delta E}\ \text{Å} = \frac{1240}{\Delta E}\ \text{nm}$$

cutoff wavelength of X-ray spectrum:
$$\lambda_o = \frac{12{,}400}{V}\ \text{Å} = \frac{1240}{V}\ \text{nm}$$

Questions

The Ultraviolet Catastrophe and Quantization

1. Distinguish between modern physics and classical physics.
2. Describe the ultraviolet catastrophe and why it was a "catastrophe."

3. How did Planck solve the ultraviolet catastrophe?
4. What is a quantum?

The Photoelectric Effect

5. According to classical theory, what should be observed in the photoelectric effect for (a) light intensity and photocurrent, (b) the time for the initiation of photocurrent, and (c) a cutoff frequency?
6. How did Einstein explain the photoelectric effect? (Einstein's Nobel prize was awarded for his theory of the photoelectric effect, not for his theory of relativity.)
7. If a photoelectric material has a cutoff frequency, why does white light always give a photocurrent in a light meter?
8. A photographer uses a light meter to check out a modeling set that is illuminated with red light, and gets a zero reading. Yet the meter operates fine with white light. What is wrong?
9. Which has more energy, (a) a quantum of red light or a quantum of blue light? (b) Infrared-radiation quanta or visible-light quanta?
10. Can you have a photon of white light? Explain.
11. If electromagnetic radiation is made up of quanta, why don't we hear the radio intermittently with the arrival of discrete packets of energy?

The Bohr Theory of the Hydrogen Atom

12. What was Bohr's nonclassical quantum postulate in his theory of the hydrogen atom? What did it lead to?
13. Why is the electron in a Bohr orbit accelerated if it is traveling with a constant speed?
14. Classically, why should an electron in a circular orbit emit radiation? Why was it necessary for Bohr to assume that a bound electron in an orbit did not emit radiation?
15. Is a satellite in circular orbit about the Earth as the hydrogen electron is in circular orbit about the nuclear proton? Give differences and/or similarities.
16. Does the light from neon signs have a continuous spectrum? Explain.
17. An absorption spectrum is obtained by passing light through a cool gas. The gas atoms absorb light, but they also re-emit it, so why are there dark absorption lines?
18. In which transition is the photon of greater energy emitted, $n = 3$ to $n = 1$, or $n = 4$ to $n = 2$?
19. A hydrogen electron is in the excited state $n = 4$. How many photons of different frequencies could possibly be emitted in the electron's return to the ground state?

The Laser

20. What does the term *laser* stand for? How is light amplified in a laser?
21. Why is laser light monochromatic?
22. Describe some industrial applications of lasers.
23. Incoherent light is sometimes likened to a crowd of people. How would the people be moving in this case? Could a different group of people be analogous to coherent light? How about a marching band? Explain.
24. In addition to stimulated emission, another condition for laser operation is "population inversion." That is, more atoms must be in an excited state than in the ground state. Why is this necessary?
25. Would a He-Ne laser operate if one end of the tube were covered with a transparent piece of glass? Explain.
26. Explain the general meaning of the series of numbers in a product code. How are they read?
27. What is holography and what is its principle? Could holograms be made with an incandescent lamp? Explain.

X-Rays

28. Why is an X-ray tube called a cathode ray tube?
29. Electromagnetic waves are generally radiated by accelerating electric charges. Yet X-rays are called "braking rays" instead of "accelerating rays." Explain why.
30. X-ray emission is sometimes described as an inverse photoelectric effect. Explain why.
31. What gives rise to (a) a continuous X-ray spectrum and (b) characteristic X-rays?
32. What does the labeling of emission lines as K_α, K_β, and K_γ signify? What do L_α and L_β mean?
33. Why does the X-ray spectrum have a cutoff frequency?

Problems

Levels of difficulty are indicated by asterisks for your convenience.

30.1 The Ultraviolet Catastrophe and Quantization

1. A black-body oscillator has a frequency of 10 kHz. What is the energy of the oscillator according to Planck's hypothesis?
2. If a black-body oscillator has an energy of 3.3×10^{-30} J, what is the frequency of the oscillator?

*3. What is the energy difference between the oscillators of two black bodies, one with a frequency of 5.0 kHz and the other with a frequency of 7.5 kHz?

30.2 The Photoelectric Effect

4. What is the threshold frequency of a photoelectric material with a work function of 4.0×10^{-19} J?
5. A photoelectric material has a work function of 1.5 eV. (a) What is the threshold frequency of the material? (b) What is the cutoff frequency of its photoemission?

*6. When a photomaterial is illuminated with monochromatic light with a frequency of 6.5×10^{14} Hz, electrons are emitted from it with a kinetic energy of 1.3×10^{-19} J. (a) What is the work function of the material? (b) What is the threshold frequency?

*7. When a phototube is reverse-biased, it is found that the photocurrent goes to zero for a voltage of -2.5 V when the tube is illuminated with monochromatic light of a particular frequency. What is the kinetic energy of the photoelectrons (a) in electron volts and (b) in joules?

*8. If the monochromatic light used in Problem 7 has a frequency of 7.3×10^{14} Hz, what is the work function of the material in the phototube in electron volts?

*9. A beam of monochromatic light with a wavelength of 650 nm is incident on a photoelectric material with a work function of 0.50 eV. (a) What is the kinetic energy of the photoelectrons? (b) Below what frequency of light would there be no photoemission?

*10. A phototube with a material having a work function of 2.0 eV is illuminated with a beam of monochromatic light with a frequency of 3.5×10^{14} Hz. What is the kinetic energy of the emitted electrons?

*11. A beam of monochromatic light with a wavelength of 6500 Å is emitted from a laser. What is the energy of the photons?

*12. How many photons with a frequency of 5.0×10^{14} Hz will supply 1.0 J of energy?

**13. A beam of light with an intensity of 3.0×10^{-3} J/m^2-s is normally incident on a piece of photomaterial. (a) If the frequency of the light is 5.5×10^{14} Hz, how many photons strike a square centimeter of the material each second? (b) If the beam totally illuminates a 2.0 cm $\times$ 1.5 cm section of the material, which has a work function of 0.25 eV, what is the photocurrent?

30.3 The Bohr Theory of the Hydrogen Atom

14. Show that $nh/2\pi$ has units of angular momentum.
15. According to the Bohr theory, what are the three lowest possible values of angular momentum of the electron of a hydrogen atom?
16. (a) What is the value of the energy of a hydrogen electron in the $n = 3$ state? (b) How much energy is needed to ionize the atom when it is in this state?
17. (a) How much energy does a hydrogen electron have when it is in the $n = 4$ state? (b) How much energy was required to excite the electron from the ground state?

*18. A hydrogen electron is in the fifth excited state. (a) How much energy was required to excite the electron from the ground state? (b) How much more energy would be necessary to ionize the atom?

*19. A hydrogen electron makes a transition from $n = 3$ to $n = 1$. (a) What is the energy difference between the levels? (b) What is the wavelength of the emitted photon?

*20. An electron in a hydrogen atom makes a transition from the $n = 6$ state to the $n = 4$ state. (a) What is the frequency of the emitted photon? (b) Does a photon emitted by a transition from the $n = 4$ state to the $n = 2$ state have the same frequency? Justify your answer.

*21. What are the wavelengths of the hydrogen spectrum lines in the visible region, i.e., in the Balmer series?

*22. In making a transition to the ground state, a hydrogen atom emits a photon with a wavelength of 1.215×10^{-5} cm. What was the initial state of the electron?

*23. A photon with a wavelength of 102.564 nm is absorbed by a hydrogen atom in the ground state. To what state is the atom excited?

**24. A photon with a wavelength of 6561 Å is incident on a hydrogen atom. What will happen? Justify your answer.

30.5 X-Rays

*25. What is the cutoff wavelength of an X-ray spectrum from a tube with a voltage of 75 kV?

*26. An X-ray tube has an applied voltage of 150 kV. What is the cutoff frequency of the tube spectrum?

*27. The cutoff wavelength of a spectrum from an X-ray tube is 2.48×10^{-9} cm. What is the voltage of the tube?

*28. Above what voltage should an X-ray tube be operated so that the continuous spectrum has wavelengths greater than 6.20×10^{-9} cm?

Chapter Supplement

The Electron Microscope

In some instances, electromagnetic waves seem to have a particle-like nature—the dual nature of light. Conversely, it has been shown that particles exhibit a wave-like nature. For example, a beam of electrons can be diffracted, and this characteristic is described by a wave relationship. A comparison of X-ray and particle diffraction is shown in Figure 30S.1.

The spacing between the atoms in a lattice of a crystalline material is of the same order as the wavelength of the X-rays, so a crystal diffracts X-rays ($d \simeq \lambda$, Chapter 27). As a result, X-rays are used to study the internal structures of crystals. Similarly, a beam of electrons with an appropriate energy shows a diffraction pattern, so the electrons too must have a wave-like nature. These "matter waves" are believed to guide, or pilot, the particles and are a topic in the advanced field of quantum mechanics.

This "dual nature of particles" is applied in the electron microscope. A magnetic coil can focus an electron beam just as a glass lens can focus a light beam. Such magnetic "lenses" are used to focus the electron beam in an electron microscope in an arrangement similar to that of glass lenses in a light microscope (Fig. 30S.2).

The electron beam is directed on a very thin specimen. Different numbers of electrons pass through different parts of the specimen, depending on its structure. The transmitted beam is then brought to focus, forming an image on a film or a fluorescent screen.

Specimens that cannot be thin-sectioned can be viewed by other means. A specimen is coated with a thin layer of metal. The electron beam is then made to scan across the specimen. The surface irregularities cause variations in the intensity of the reflected electrons, and the

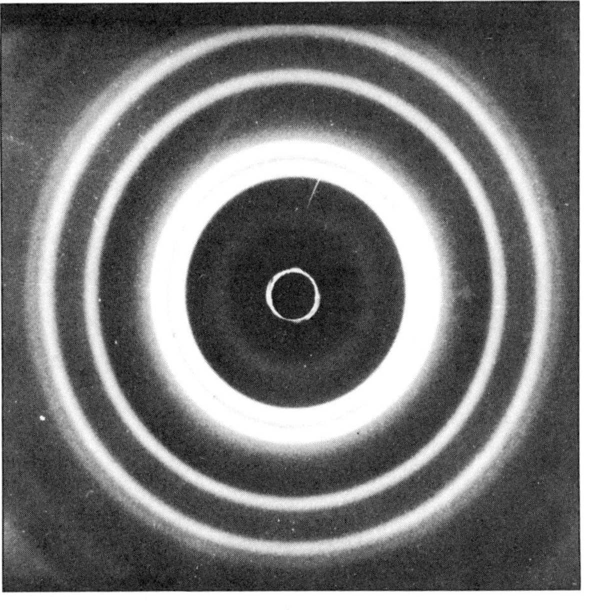

(a)

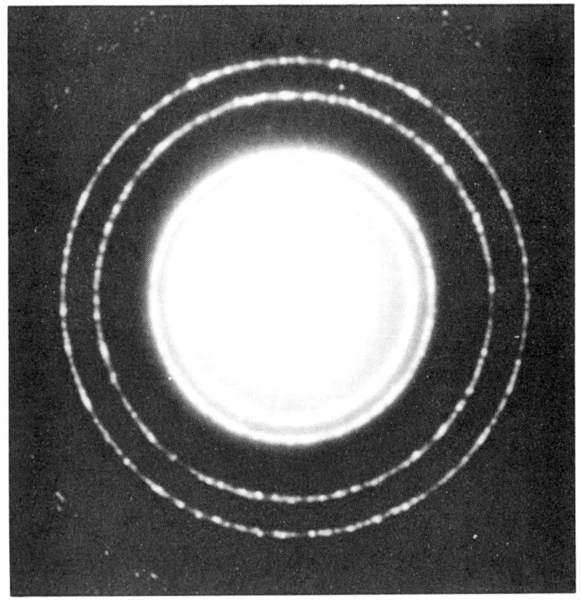

(b)

Figure 30S.1

Diffraction patterns. Patterns made by (a) X-rays passing through aluminum foil and (b) a beam of electrons passing through the same foil. (Center pattern in X-ray diffraction is black because the center of the film is blocked off to prevent overexposing the film.)

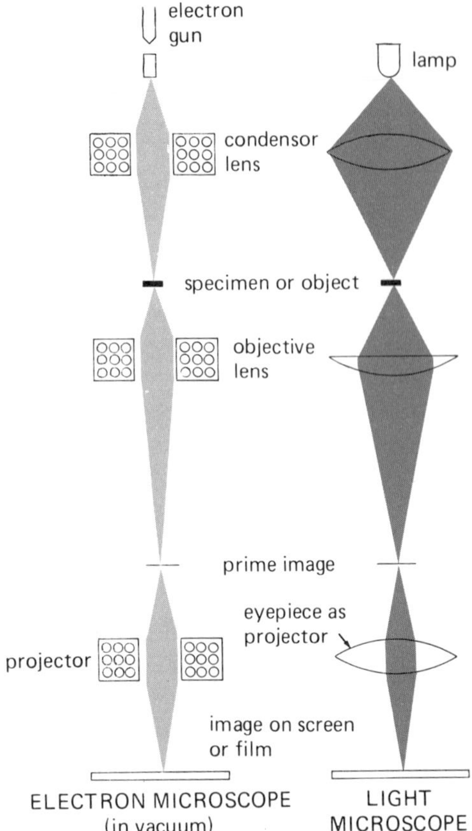

Figure 30S.2
Electron and light microscopes. A comparison of the general components of an electron microscope and a light microscope. The electron microscope uses magnetic coil "lenses," while the light microscope uses glass lenses.

reflected beam is focused to form an image. The metal coating makes the specimen electrically conducting. Otherwise, a nonuniform charge would build up on the specimen and cause the image to be distorted.

Both the transmission electron microscope and the scanning electron microscope are enclosed in a high-vacuum chamber so that the electrons are not deflected by air molecules. As a result, an electron microscope looks nothing like a light microscope (Fig. 30S.3). However, magnifications up to 100,000 times can be achieved with an electron microscope, whereas a light microscope is limited to a magnification of about 2000 times.

A new microscope technique is being used to enhance the study of such things as living cells

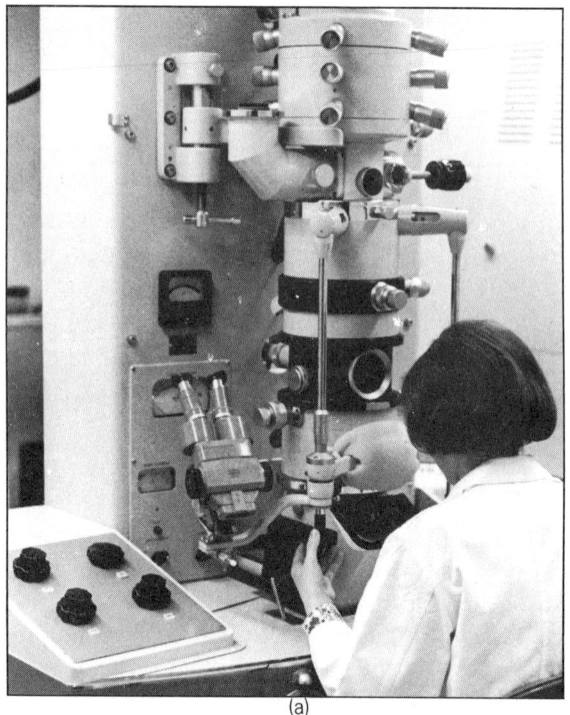

(a)

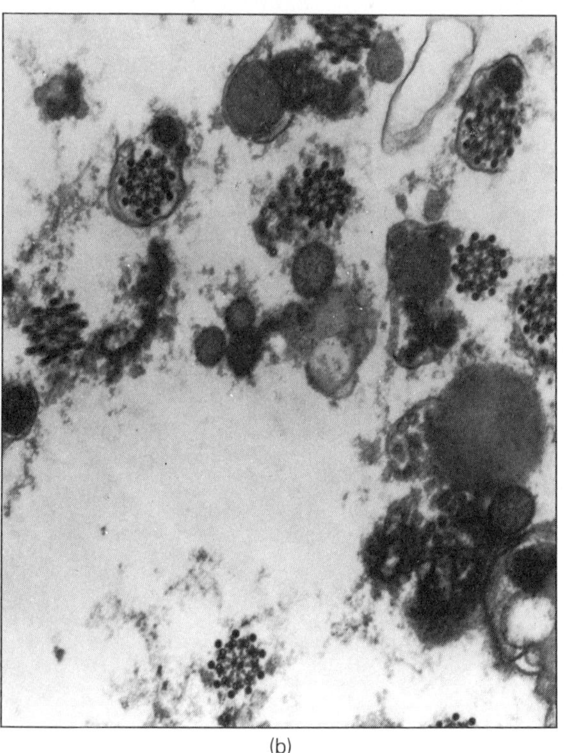

(b)

Figure 30S.3
Electron microscopy. (a) An electron microscope. (b) An electron micrograph of biological tissue at 33,000× magnification. (Courtesy L. Greenberg.)

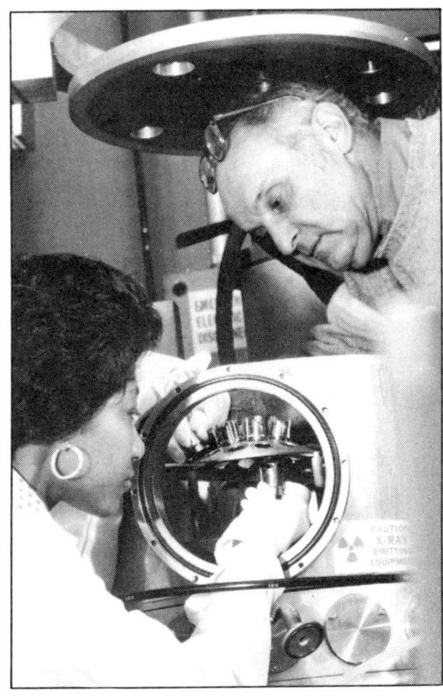

(a)

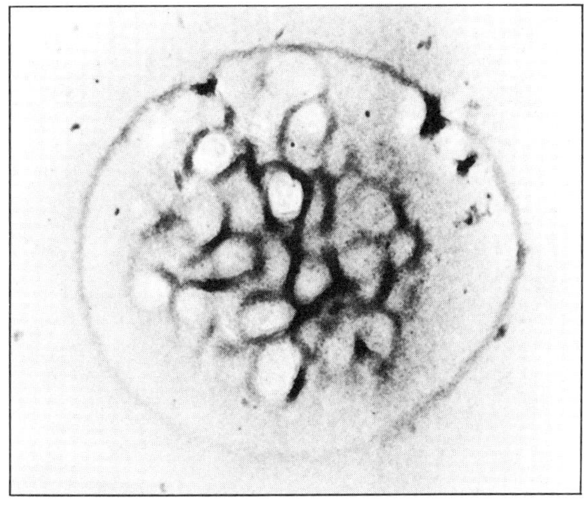

(b)

Figure 30S.4
A new technique. (a) X-rays are used to form a replica of an object in photosensitive material, and the replica is then observed with an electron microscope. (b) Electron micrograph of an X-ray replica of a blood platelet, which is only about 3 μm (3×10^{-6} m) in diameter. (Courtesy IBM.)

(Fig. 30S.4). This method uses X-rays to form a replica of the cell in a photosensitive material called resist (Chapter 26). The replica is then observed in an electron microscope. Resist, which is used in fabricating electronic circuits, can record much smaller features than can the photographic films used for conventional X-ray pictures.

This method makes it possible to see cell features only 50 Å (50×10^{-10} m) wide, which is about 1/100 the wavelength of visible light. The blood platelet shown in Figure 30S.4 is about 3 μm (~1/10,000 in.) in diameter. The dark filaments are believed to be a kind of skeleton in the cell, probably formed of muscle filaments. This structure is of great interest to biologists and cannot be seen in conventional electron micrographs.

Three Mile Island Nuclear Plant.

Chapter 31

The Nucleus and Nuclear Energy

One of the main considerations of modern physics is the nucleus. In the early 1900's, Lord Rutherford performed some experiments which indicated that the protons of an atom are concentrated at its center in a "nucleus." Since then, a great deal of nuclear research has been done, and in the 1940's we learned how to "split the atom [nucleus]." Announcement of this feat was made to the world with the dropping of atomic bombs on Hiroshima and Nagasaki near the end of World War II. This brought terms like "fission," "radioactive fallout," and "the bomb" into common usage.

Since 1959, vast amounts of energy released from atomic nuclei have been harnessed for the production of electricity, bringing into use such terms as "reactor" and "meltdown." As you know, a great deal of controversy surrounds the use of nuclear energy. However, nuclear technology has many important noncontroversial applications. Nuclear principles and applications are discussed in this chapter.

31.1 The Nucleus and the Nuclear Force

As we learned in the preceding chapter, our simplistic "planetary" model of the atom views the electrons of an atom to be in orbit about the nuclear protons. The simplest atom, hydrogen, has one proton and one electron. However, to account for the masses of all other atoms, there must be another nuclear particle—the neutron.

Unlike the electron and proton, the **neutron** has no electrical charge, and it is slightly more massive than the proton (Table 31.1). The existence of the neutron was not verified until 1932. The nuclear protons and neutrons are referred to collectively as **nucleons**, and they make up more than 99 percent of the mass of an atom (Fig. 31.1).

In a neutral atom, the number of nuclear protons is equal to the number of orbital electrons. However, **the number of protons in the nucleus determines the type of atom,** i.e., the atom of a particular element. Electrons may be removed

Table 31.1
Atomic Particles

Particle	Symbol	Charge (coulomb)	Mass	
			kg	u
Electron	${}_{-1}^{0}e$	-1.6×10^{-19}	9.11×10^{-31}	5.49×10^{-4}
Proton	${}_{1}^{1}p$ or ${}_{1}^{1}H$	$+1.6 \times 10^{-19}$	1.67265×10^{-27}	1.00728
Neutron	${}_{0}^{1}n$	0	1.67495×10^{-27}	1.00867

(or added) in ionization processes, but the number of protons in an atom remains the same. For example, all atoms with six protons in their nuclei are carbon atoms, and they make up that particular element.

However, the atoms of a particular element can contain different numbers of neutrons in their nuclei, forming "species" of the element called **isotopes.** Isotopes of hydrogen and carbon are illustrated in Figure 31.2. Usually, one isotope of an element is naturally more abundant than the others. A particular nuclear species, or isotope, is referred to as a **nuclide.**

The nuclei of atoms are designated symbolically by a nuclear notation using the chemical symbol of an element (Fig. 31.3). The left superscript, called the **mass number *A*,** indicates the number of nucleons (protons + neutrons) in the nucleus. The left subscript, called the **proton** or **atomic number *Z*,** indicates the number of protons in the nucleus.* The right subscript, the **neutron number *N*,** indicates the number of neutrons. Hence, $A = Z + N$.

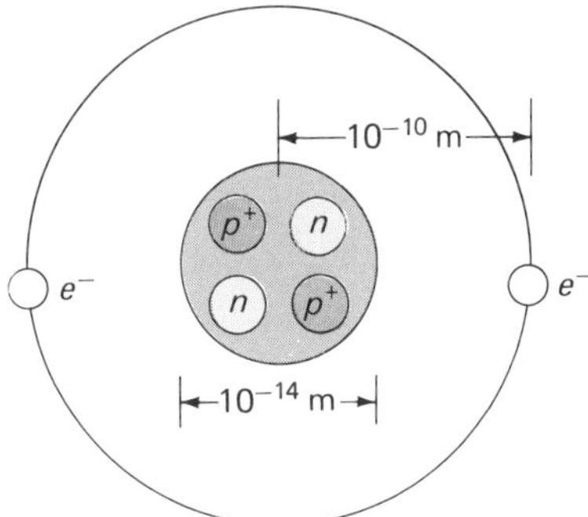

Figure 31.1
Atomic model. A simplified schematic diagram of a helium atom shows atomic divisions. The nucleons (protons and neutrons) make up over 99 percent of the mass of the atom and are very compact. (Drawn here separated for convenience.)

In the case of the carbon isotopes shown in Figure 31.2, we then write ${}_{6}^{12}C_6$, ${}_{6}^{13}C_7$, and ${}_{6}^{14}C_8$. The neutron number is somewhat redundant, since it can be obtained from $N = A - Z$. As a result, it is often omitted in the nuclear notation, i.e., ${}_{6}^{12}C$, ${}_{6}^{13}C$, and ${}_{6}^{14}C$. In words, these isotopes are referred to as carbon-12, carbon-13, and carbon-14. An alphabetical listing of the elements with their proton numbers is given in Appendix 4.

In nuclear notation, the proton, neutron, and electron are written ${}_{1}^{1}p$ (or sometimes ${}_{1}^{1}H$), ${}_{0}^{1}n$, and ${}_{-1}^{0}e$, respectively. Notice in Table 31.1 that two mass units are listed. Since the kilogram unit is inconveniently large for nuclear particles, another unit called the **unified atomic mass unit (u)** is defined. This is based on the carbon-12 atom. One u is defined as $\frac{1}{12}$ of the mass of a neutral carbon-12 atom, which is defined to have a mass of 12.0000 u, and

$$1 \text{ u} = 1.66 \times 10^{-27} \text{ kg}$$

The atomic weights (masses) of the elements given in tables or on periodic charts are based on the percentages of naturally occurring isotopes, e.g., the atomic weight of carbon is 12.01115 u and reflects the presence of carbon-13 and carbon-14 in a sample.

* The atomic number originally referred to the number of electrons an atom contains. In neutral atoms, the number of electrons and protons are equal, so the atomic number was also applied to the number of protons. However, proton number is a more descriptive term because of the ionization of electrons, as mentioned earlier.

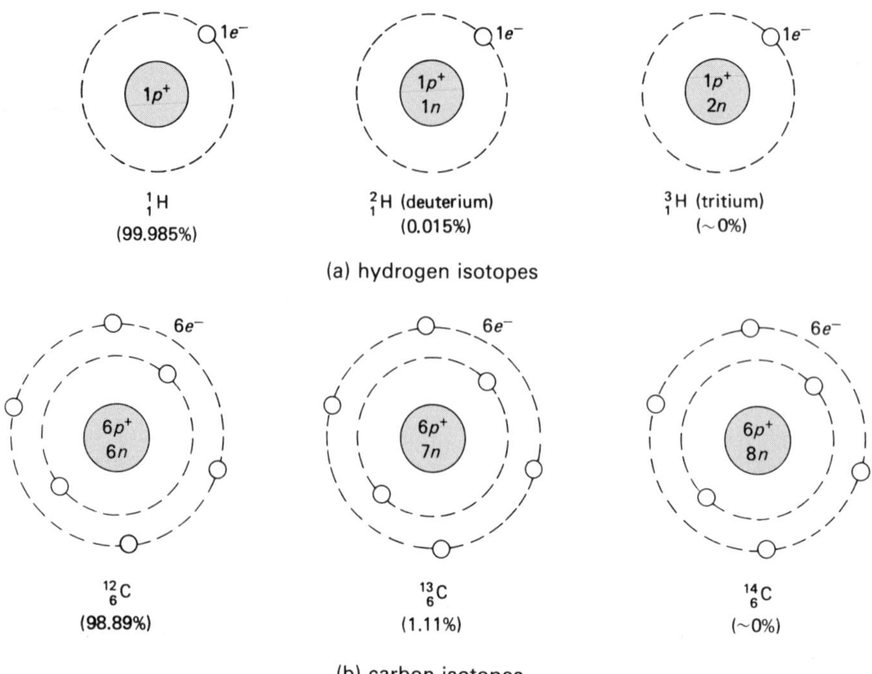

Figure 31.2
Isotopes. (a) Hydrogen isotopes. Hydrogen has three isotopes. The $^1_1$H isotope is referred to simply as hydrogen. The $^2_1$H and $^3_1$H isotopes are called deuterium and tritium, respectively. The percentage of natural abundance is listed below each isotope. (b) Carbon isotopes. Three of several carbon isotopes are referred to as carbon-12, carbon-13, and carbon-14.

The Nuclear Force

Within the small confines of the nucleus, the nucleons are clustered together, with each nucleon taking up about the same amount of space (Fig. 31.4). The nucleons are on the order of 10^{-15} m apart. This means that there are large repulsive electrical forces between the positively charged protons ($F = kq_1q_2/r^2$). What then holds the nucleus together? Since the nucleons are mass particles, the ever-present gravitational force is there ($F = Gm_1m_2/r^2$). However, as was shown in Chapter 19, the gravitational force is a factor of 10^{-40} smaller than the electrical force, so this is negligible.

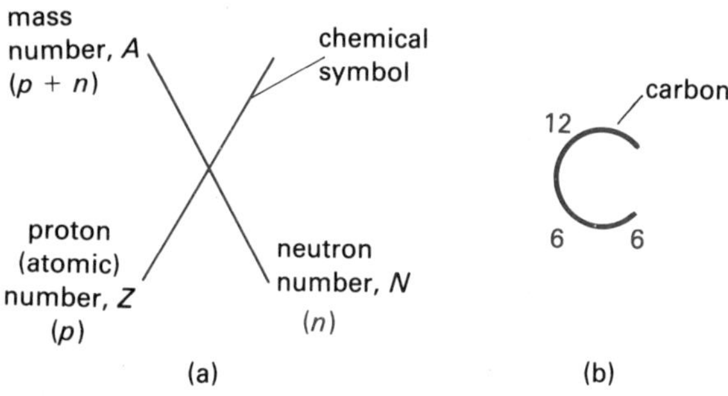

Figure 31.3
Nuclear notation. (a) A general case and (b) for a carbon nucleus. The notation conveys that this carbon nuclide has a mass number of 12 or 12 nucleons ($A = p + n$), made up of 6 protons ($Z = 6$) and 6 neutrons ($N = 6$).

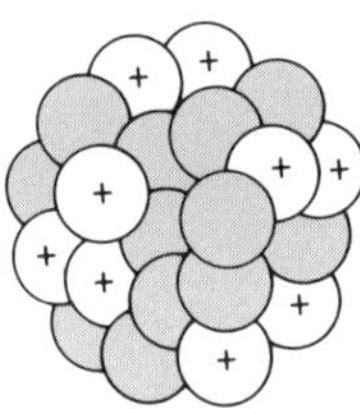

Figure 31.4
Nucleons are clustered together in the nucleus. Each nucleon takes up about the same amount of space.

Evidently, there is some strong attractive force in the nucleus that overcomes the repulsive electrical forces and holds the nucleus together. We call this attractive force acting between nucleons the nuclear force or the strong interaction.* The **nuclear force** is a fundamental force like the electrical and gravitational forces, but it is more complicated and not completely understood.

Experiments show that the nuclear force acts between any two pairs of nucleons. That is, there are nuclear forces between proton-proton, proton-neutron, and neutron-neutron pairs. Also, the nuclear force is very short-ranged. It does not act outside the nucleus and so does not act between nucleons with separation distances greater than approximately 10^{-12} cm ($r > 10^{-12}$ cm).

Summarizing these properties, **the nuclear force**

1. **Is strongly attractive**
2. **Acts between any nucleon pair**
3. **Is short-ranged (zero for separation distances of $r > 10^{-12}$ cm)**

Since the nuclear force acts between any pair of nucleons, it appears that one of the functions of the neutrons in the nucleus is to provide added nuclear force to overcome the electrical repulsion of the protons. This is evidenced by the increasing number of neutrons in the nuclei of heavier elements, which have more protons (see Fig. 31.5). As the graph in Figure 31.5 shows, the Z and N numbers of stable nuclei are about the same up to about Z = 20 (calcium). Heavier nuclei (greater Z) have an increasingly greater number of neutrons than protons, which provide more attractive nuclear force.

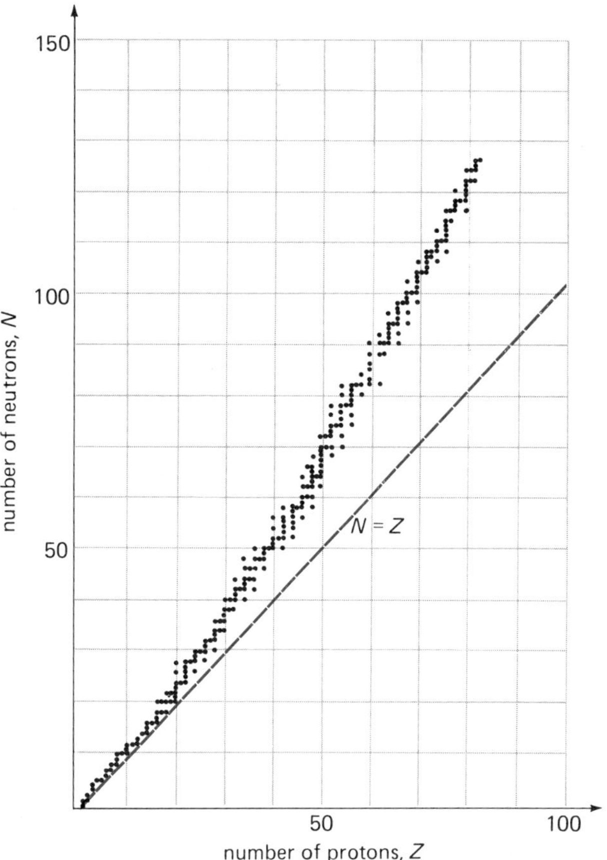

Figure 31.5

Neutron number (N) versus proton number (Z) for stable nuclides. Nuclides with proton numbers greater than 20 (or mass number $A > 40$) have more neutrons than protons. The excess neutrons act as a nuclear "glue" that holds a nucleus together.

31.2 Radioactivity

There are more than 250 stable nuclei making up the graph in Figure 31.5. However, there are approximately 1400 known nuclides. The big difference reflects the known unstable nuclides. Unstable nuclei spontaneously "decay" with the emission of energetic particles and are said to be radioactive. The "radio-" part refers to the emitted radiation, which in the modern context may be a particle or a wave. So a radioactive nuclide or isotope is "active" in emitting radiation. Notice that the graph in Figure 31.5 terminates at proton number 83. There are no stable nuclei with proton numbers greater than 83.

Of the nearly 1200 known unstable nuclides, only a small number occur naturally. The other radioactive nuclei are made artificially. The unstable nuclides found in nature decay with the emission of alpha particles or beta particles, which are sometimes accompanied by gamma rays. Using a magnetic field, which deflects electrically charged particles, it was found that alpha "rays" were positively charged particles and beta "rays" were negatively charged parti-

* There is also a weak nuclear force, or weak interaction, which will be considered later in this chapter.

SPECIAL FEATURE 31.1

Marie and Pierre Curie

Marie Sklodowska (1867–1934) was born in Warsaw, Poland, and received her early scientific training from her father. She became involved in a students' revolutionary organization and found it advisable to leave Warsaw. In Paris she earned a science degree, and in 1895 she married Pierre Curie, who was a physicist well known for his work on crystals and magnetism (Fig. 1). One of his most important discoveries was that the magnetic properties of substances change at a certain temperature, the "Curie temperature."

After their marriage, Marie did doctoral research on radioactivity, and Pierre joined his wife in this work. In 1898 they announced their discovery of two new radioactive elements that they had isolated, radium and polonium (named after Marie Curie's native country). From 8 tons of uranium pitchblende ore, they painstakingly isolated 10 mg of radium and a smaller amount of polonium. In 1903 they were awarded the Nobel prize in physics for their discovery in radioactivity. The prize was shared with Henri Becquerel, who had discovered the radioactive properties of uranium in 1896. Pierre Curie was killed in a horse-drawn vehicle accident in 1906. Marie Curie (commonly known as Madame Curie) was appointed to her husband's professorship at the Sorbonne, the first woman to have this post.

In 1911 Mme. Curie was awarded the Nobel prize in chemistry for her work on radium and the study of its properties. She was the first person to win two Nobel prizes in science. The rest of her career was spent in establishing and supervising laboratories for research on radioactivity and the use of radium in the treatment of cancer. In 1921 President Harding, on behalf of the women of the United States, presented her with a gram of radium in recognition of her services to science.

Mme. Curie died in 1934 of leukemia, a form of blood cancer, which may have been caused by overexposure to radioactive substances. This occurred shortly before an event that would no doubt have made her very proud. In 1935 the Curies' daughter, Irene Joliot-Curie, and her husband, Frederic Joliot, were awarded the Nobel prize in chemistry.

Figure 1
Marie and Pierre Curie. (Courtesy W.F. Meggers Collection, AIP Niels Bohr Library.)

cles (Fig. 31.6). Because they are deflected more, beta particles must be less massive than alpha particles. Gamma rays are not deflected at all, so they must be uncharged. Actually, a gamma ray or "particle" is a photon, or quantum of energy.

Radioactivity was discovered accidentally by the French physicist Henri Becquerel in 1896. The circumstances were not unlike those surrounding Roentgen's discovery of X-rays. Becquerel noticed that a photographic plate in a light-tight wrapper that he had left in a drawer

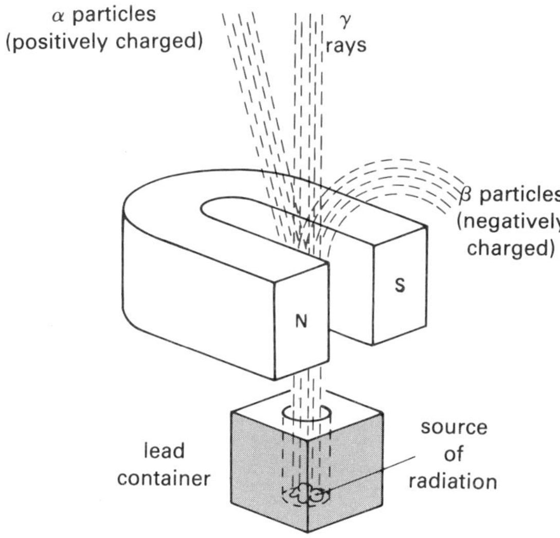

Figure 31.6

Radioactive decay particles. Passing radiations through a magnetic field shows there are positively charged alpha particles, negatively charged electrons, and electrically neutral gamma rays.

with a uranium compound showed signs of exposure when developed. Evidently, some type of radiation, or "rays," from the uranium was able to penetrate the wrapper and expose the film.

A couple of years later, the husband-and-wife team of Pierre and Marie Curie announced the discovery of two new radioactive elements, radium and polonium, which they had isolated from uranium pitchblende ore (see Special Feature 31.1). They had painstakingly isolated 10 mg of radium and a smaller amount of polonium from 8 tons of ore! The Curies and Becquerel shared the 1903 Nobel prize in physics for their work in radioactivity. Madame Curie, as she is commonly known, also received the Nobel prize in chemistry in 1911 for her contributions in chemically isolating radioactive materials.

Let's take a closer look at the ABC's of radioactivity: alpha (α), beta (β), and gamma (γ) decays and their respective particles.

Alpha Decay

An alpha particle is made up of two protons and two neutrons. Hence, it is the same as a helium nucleus ($^4_2$He, sometimes written $^4_2\alpha$). When a nucleus undergoes alpha decay, an energetic alpha particle is emitted and the nucleus is *transmuted* into the nucleus of another element, since two protons are lost. An example of a nucleus that alpha-decays is

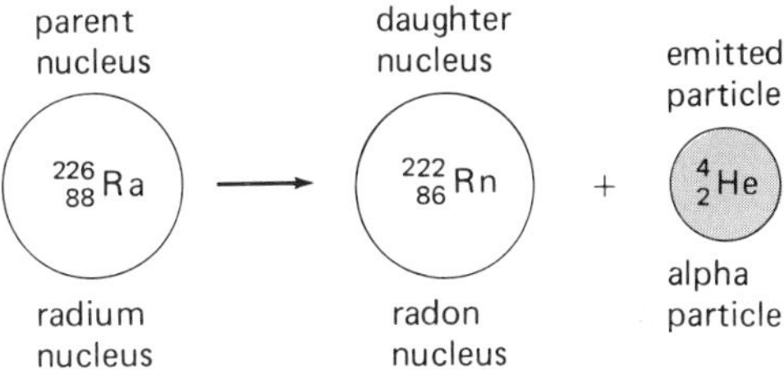

or, simply as a nuclear reaction equation (similar to a chemical reaction equation),

$$^{226}_{88}\text{Ra} \rightarrow {}^{222}_{86}\text{Rn} + {}^4_2\text{He}$$

The original nucleus in a decay process is commonly called the parent nucleus (Ra-226 in this case), and the resulting nucleus is called the daughter nucleus (Rn-222).

Notice that in this reaction, as **in all nuclear reactions,**

(a) **the total number of nucleons is conserved and**

(b) **the total charge is conserved.**

Condition (**a**) means that the superscript mass numbers on each side of the equation are equal ($226 = 222 + 4$). Condition (**b**) requires the subscript proton numbers on each side of the equation to be equal ($88 = 86 + 2$).

Beta Decay

A beta particle is simply an electron. An example of beta decay is

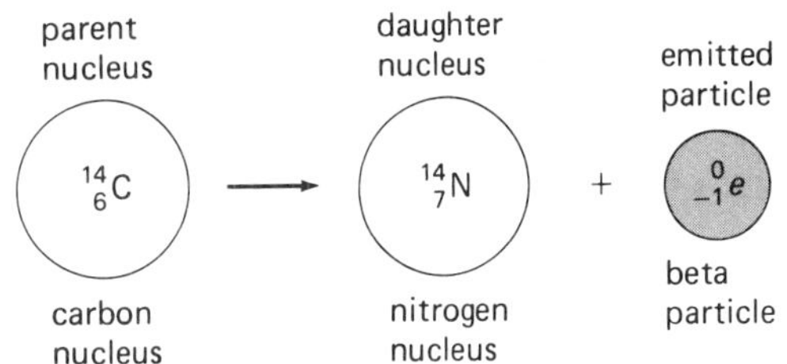

or

$$^{14}_{6}\text{C}_8 \rightarrow {}^{14}_{7}\text{N}_7 + {}^{0}_{-1}e$$

Notice that the -1 subscript of the electron (its *charge* number) allows for the conservation of total charge ($6 = 7 - 1$).

But where does an electron come from within a nucleus? Looking at the proton and neutron numbers of the beta decay reaction, we see that the daughter nucleus has one less neutron and one more proton than the parent nucleus. This indicates that a neutron must have been converted into a proton and an electron in the beta decay process; i.e.,

$$^{1}_{0}n \rightarrow {}^{1}_{1}p + {}^{0}_{-1}e$$

A nuclear force distinct from the strong interaction is associated with beta decay. This is the **weak nuclear force,** or **weak interaction.** All the particles we will study interact through the weak interaction, while some also interact electromagnetically and still fewer also act through the strong interaction. For example, two nuclear protons interact by all four fundamental forces—strong nuclear, electromagnetic, weak nuclear, and gravitational. The relative strengths of the fundamental interactions are given in Table 31.2.

As in classical physics, we generally consider only the important forces and ignore the others or consider them negligible in a particular situation; for example, the strong nuclear and electromagnetic interactions dominate between nucleons. In the case of beta decay involving an electron, there is no strong nuclear force between the nucleons and an electron, so the weak nuclear interaction becomes important. The weak interaction is complex and is not well understood.

Gamma Decay

A gamma "particle" is a quantum of electromagnetic energy. The nucleus does not change in gamma decay; for example:

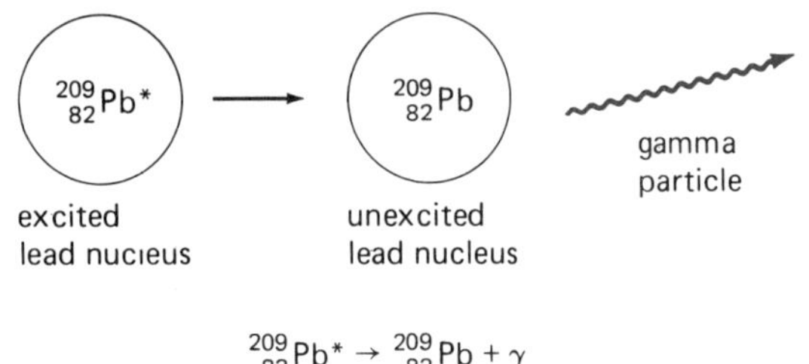

$$^{209}_{82}\text{Pb}^* \rightarrow {}^{209}_{82}\text{Pb} + \gamma$$

The asterisk indicates that the nucleus is in an excited state. It decays to a lower energy state with the emission of a gamma particle, similar to the emission of a photon in the transition of an orbital electron. Gamma rays are made of energetic gamma particles, which have frequencies in the upper X-ray range and higher.

Table 31.2
Relative Strengths of Fundamental Interactions

Interaction	Relative Strength	Range
Strong nuclear	1	Short range, about 10^{-14} m
Electromagnetic	10^{-2}	Infinite ($1/r^2$)
Weak nuclear	10^{-13}	Extremely short range
Gravitational	10^{-40}	Infinite ($1/r^2$)

Decay Rate and Half-Life

Radioactive isotopes decay at different rates. The decay rate of a radioactive sample is characterized by its **half-life ($t_{1/2}$).**

> **One half-life is the time it takes for one half of the nuclei of a sample of a radioactive isotope to decay.**

For example, suppose a sample initially ($t = 0$) contains N_o nuclei. After one half-life, $N_o/2$ parent nuclei would be present; after two half-lives, $N_o/4$ parent nuclei [$(N_o/2)/2 = N_o/4$]; after three half-lives, $N_o/8$ parent nuclei [$(N_o/4)/2 = N_o/8$]; and so on, or $N_o/(2)^n$, where n is the number of half-lives. This process is plotted in Figure 31.7.

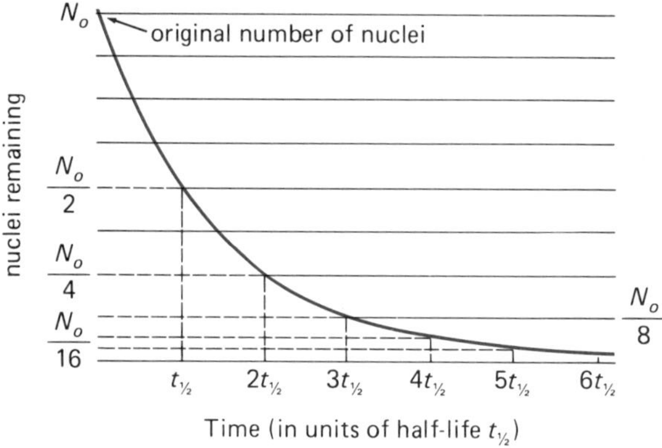

Figure 31.7
Radioactive decay. The graph shows the decay of a radioactive sample as a function of time. In one half-life, one half of the parent nuclei in the sample decay.

Table 31.3
The Half-Lives of Some Radioactive Isotopes

Nuclide		Decay Mode	Half-Life
Beryllium-8	($^{8}_{4}Be$)	α	1×10^{-16} second
Polonium-213	($^{213}_{84}Po$)	α	4×10^{-6} second
Carbon-16	($^{16}_{6}C$)	β	0.75 second
Aluminum-28	($^{28}_{13}Al$)	β	2.24 minutes
Magnesium-28	($^{28}_{12}Mg$)	β	21 hours
Gold-191	($^{191}_{79}(Au)$	β	40 hours
Iodine-131	($^{131}_{53}I$)	β	8 days
Cobalt-60	($^{60}_{27}Co$)	β	5.3 years
Strontium-90	($^{90}_{38}Sr$)	β	28 years
Radium-226	($^{226}_{88}Ra$)	α	1600 years
Carbon-14	($^{14}_{6}C$)	β	5730 years
Uranium-238	($^{238}_{92}U$)	α	4.5×10^{9} years
Rubidium-87	($^{87}_{37}Rb$)	β	4.7×10^{10} years

The resulting curve is an exponential decay curve, similar to the curve of the discharge rate of a capacitor (Chapter 24). The half-lives of some radioactive isotopes are given in Table 31.3. Notice the range of the half-life periods.

The activity of a radioactive sample is the number of nuclear decays per unit time. The common unit of radioactivity is the **curie (Ci)**, and

$$1 \text{ Ci} = 3.7 \times 10^{10} \text{ decays/s}$$

The curie unit was originally defined as the activity of one gram of radium-226.

The SI unit of radioactivity is the **becquerel (Bq)**, which is defined as the number of decays per second, i.e.,

$$1 \text{ Bq} = 1 \text{ decay/s}$$

It follows that

$$1 \text{ Ci} = 3.7 \times 10^{10} \text{ Bq}$$

Although much more convenient, the becquerel unit has not gained widespread use.

The activity of a radioactive sample decreases exponentially with time, as illustrated in Figure 31.7. Hence, after one half-life, the activity is decreased by one half.

EXAMPLE 31.1 A sample of strontium-90 has an activity of 100 μCi (or 3.7 MBq). How long will it take for the activity of the sample to decrease to 25 μCi?

Solution:

$$\begin{array}{lll} \text{Initial activity} & 100\ \mu\text{Ci} & \}\ t_{1/2} \\ \text{At 1 half-life} & 50\ \mu\text{Ci} & \\ \text{At 2 half-lives} & 25\ \mu\text{Ci} & \}\ t_{1/2} \end{array}$$

Hence, the activity will be 25 μCi at two half-lives. From Table 31.3, $t_{1/2} = 28$ years for strontium-90, and

$$2t_{1/2} = 2(28 \text{ y}) = 56 \text{ y}$$

The unvarying decay rates of radioactive isotopes form the basis of radioactive dating. This is discussed in Special Feature 31.2.

31.3 Radiation Detectors and Applications

Since we cannot generally sense the particles of nuclear radiations when they are emitted from decaying nuclei, they must "do" something before we know they are there. For example, the detection of nuclear radiations by photographic films has been mentioned. Nuclear radiations and X-rays expose various film materials, and persons working with radiation often wear "film badges" as a safety measure. The amount of exposure is indicated by the degree of darkening of the film after it has been developed. This is fine, but it is desirable to detect radiations immediately, as well as get a count of the emissions or decays.

Energetic alpha, beta, and gamma rays from nuclear decay processes have the ability to ionize neutral atoms. As a result, they are commonly referred to as ionizing radiations. For example, when an energetic charged alpha or beta particle passes through a gas, it interacts with gas molecules and rips some of the loosely bound outer electrons from the gas atoms. Although some of the freed electrons might recombine or attach themselves to other atoms or molecules, most remain free for a while.

This ionizing property of nuclear radiations forms the basis of a commonly used detection instrument, the **Geiger counter.** Your instructor probably has one of these detectors and will

SPECIAL FEATURE 31.2

Radioactive Dating

Because of their fixed decay rates, radioactive isotopes can be used as nuclear "clocks." For example, suppose you had half a gram of ^{90}Sr. Then you know that 28 years (one half-life) have "ticked off" on the clock since there was 1 g of ^{90}Sr. Similarly, if you had $\frac{1}{4}$ g, you would know that it was $2t_{1/2}$ (two half-lives), or $2 \times 28 = 56$ y, past the 1-g mark on your nuclear "clock." It is this radioactive "ticking" and counting backward that forms the basis of radioactive dating, which allows scientists to determine how old objects are.

An important radioactive-dating process uses the radioisotope carbon-14. Carbon-14 is a natural part of the environment (Fig. 1). It is formed in the upper atmosphere as a result of cosmic ray interactions. (Cosmic rays are particles from outer space, mostly protons.) The carbon-14 enters the biosphere, and all living things—plants and animals—obtain a certain amount of carbon-14. A radioactive isotope behaves chemically just like a stable isotope, and part of the carbon in your body is radioactive carbon-14. The activity of carbon-14 in a living organism remains fairly constant because of life processes.

In each gram of carbon in a body, there is enough carbon-14 to generate about 16 beta emissions per minute (per gram of carbon). Recall that carbon-14 beta decays. However, when the organism dies and the carbon-14 is not replenished by the intake in life processes, the activity decreases.

Suppose an old bone found in an archeological dig has a carbon-14 activity of four beta emissions per minute per gram of carbon. When the bone was a part of a living organism, there were 16 beta emissions per minute. So, counting backward in time ($4 \rightarrow 8 \rightarrow 16$), we know that the carbon-14 in the bone has gone through two half-lives. Carbon-14 has a half-life ($t_{1/2}$) of 5730 years (Table 31.3), and $2t_{1/2} = 2(5730) = 11{,}460$ years. Hence, the bone is about 11,500 years old.

Carbon-14 dating can be used to date objects up to about 50,000 years old. Other radioactive isotopes with longer half-lives can be used to date older objects. For example, ^{48}K and ^{87}Rb are useful in determining the age of geological rock formations.

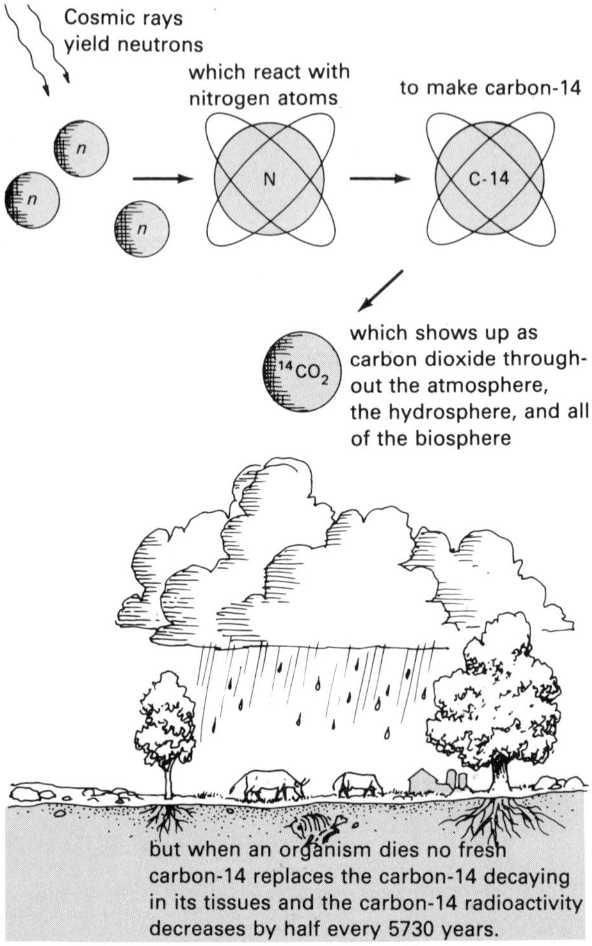

Figure 1
Carbon-14 in the environment. Created in the upper atmosphere, C-14 shows up in the biosphere, where it becomes a part of all living things.

demonstrate it for you. It was developed chiefly by Hans Geiger, a student and colleague of Lord Rutherford.

The basic principle of the Geiger counter is illustrated in Figure 31.8. A voltage is applied across the Geiger tube, which contains argon gas at a low pressure (about 5 percent of normal atmospheric pressure). When an ionizing particle enters the tube through a thin window and "strikes" one or more argon atoms, electrons are ejected. These electrons are accelerated by the applied voltage and produce further ioniza-

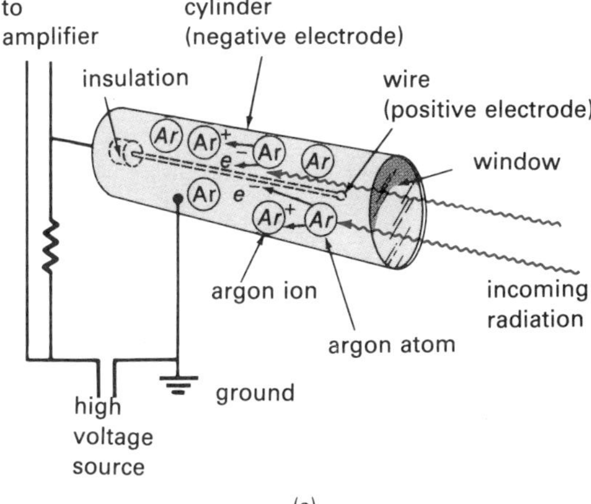

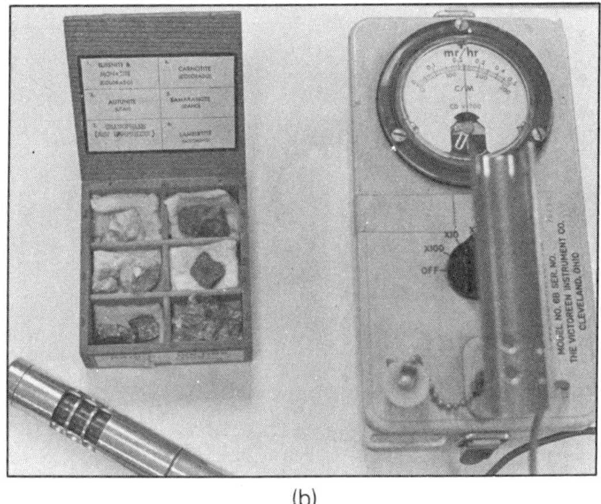

Figure 31.8
Geiger counter. (a) Schematic diagram. Radiation ionizes the gas molecules in the tube, giving rise to an electrical pulse, which is counted. (b) A portable, battery-powered Geiger counter. The tube is lying near natural samples of radioactive ore, and the counter indicates a radiation count.

tions. As a result, an "avalanche" current is set up momentarily in the tube. This current pulse is amplified and counted by the counter circuitry. Hence, radiation particles are detected and counted by counting the current pulses.

One of the main disadvantages of the Geiger counter is its "dead time," referring to the time it takes for the tube to recover, or clear itself, between counts. If a lot of particles enter the Geiger tube at a rapid rate, the tube will not have time to recover, and some particles will not be counted.

Another method of counting that gives a shorter recovery time is used in the scintillation counter. Certain phosphor materials have the ability to convert the energy of a particle into visible light. For example, if an alpha particle falls on a screen coated with zinc sulfide, it produces a tiny flash of light that is bright enough to be seen in the dark with the aid of a magnifying lens.

This was the detection device of Rutherford's scattering experiment, in which he used alpha particle "projectiles" from a natural radioactive source to probe the nucleus. In his experiment, the counting was done visually, which was slow and tiring. It has been said that whenever Rutherford himself did the actual counting, he would "damn" vigorously for a few minutes and then have one of his assistants take over. Overall, more than a million scintillations, or flashes, were counted in the experiment.

This laborious task was eliminated with the development and application of the photomultiplier tube (Fig. 31.9). Photons from the scintillator strike a photoelectric material, and photoelectrons are emitted. The photocurrent is

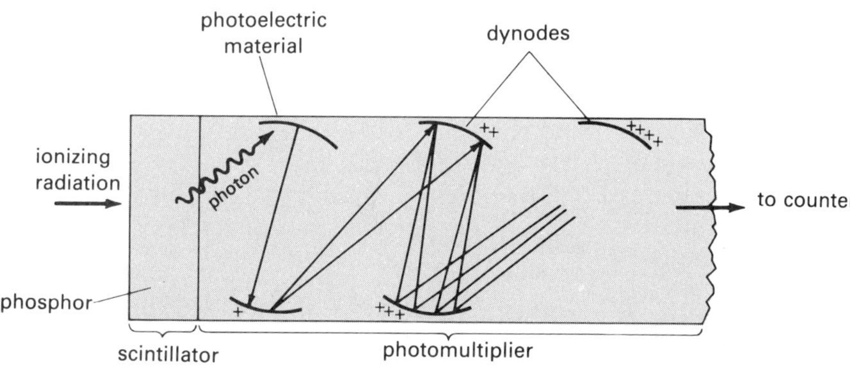

Figure 31.9
Photomultiplier tube. The photocurrent is amplified or multiplied at successive dynodes. See text for description.

amplified in a photomultiplier tube, which consists of a series of electrodes, called "dynodes," at successively higher potentials. The accelerated photoelectrons cause secondary electrons to be emitted when they strike a dynode, and so the current is multiplied. In this manner, weak scintillations are converted into sizable electrical pulses that can be counted electronically. Also, since the energy of the incident particle and the photon is proportionately amplified in the photomultiplier tube, the resulting output pulse gives a measure of the energy of the incident particle.

With the development of semiconducting materials, "solid state" detectors are now available that can measure the energy of radiation very precisely and are capable of very high counting rates (very small dead times)—i.e., they are very "fast" detectors. Charged particles passing through a semiconductor produce electron-hole pairs (solid state "ionization"). With an applied voltage, these electron-hole pairs give rise to electrical pulses that can be amplified and counted.

Since the discovery of radioactivity less than 100 years ago, nuclear decay has been used in a variety of applications. Let's consider a few medical and industrial applications here, and the application to the production of nuclear energy in a following section.

Ionizing radiation can be harmful to living tissue. Ionization can disrupt cell processes and cause genetic damage. The latter may give rise to abnormal cell growth, or cancer. Consequently, special safety precautions must be taken, and a distinctive symbol has been devised to warn of the presence of radioactive materials (Fig. 31.10). Since nuclear radiations can disrupt cell processes, radioactive sources, such as cobalt-60, are used in cancer treatment to kill cancerous cells.

Figure 31.10
Radiation symbol. This symbol is used to indicate the presence of radioactive sources or general radiation areas.

The fact that radioactive isotopes behave chemically just like nonradioactive isotopes of the same element makes them useful as radioactive tracers. For example, an atom with a carbon-14 nucleus will react to form a molecule of carbon dioxide (CO_2) in the same way a nonradioactive atom of carbon-12 would. Radioactively "tagged" molecules can be traced in various processes by detecting the nuclear radiations from these molecules. For example, the location of an underground pipe or a clog can be determined by using a liquid radioactive tracer, and research in the wearing of machine parts can be carried out by tracing processes (Fig. 31.11).

Similarly, in medicine radioactive tracers allow the monitoring of body functions. For example, the iodine uptake of the thyroid gland can be measured by using radioactive iodine-131. The minimum daily requirement of iodine is 150 micrograms. If a patient is given some radioactive iodine, the thyroid gland doesn't know the difference, and a scan of the gland shows how the radioactive iodine is distributed and concentrated in the gland (Fig. 31.11).*

The distribution is altered if there is some abnormality in the gland function. Various other internal parts of the body can be "observed" using radioactive tracers. A detector is used to scan the area in question. The information is stored in a computer and later put together as a scan "picture."

The penetration of radiation depends on the density of a particular material and is inversely proportional to the thickness of the material—the greater the thickness, the less penetration. This follows a given relationship, so radiation absorption can be used in thickness measure-

* Iodine-131 beta-decays into xenon-131, which emits gamma rays. It is the gamma radiation that is monitored in a thyroid scan.

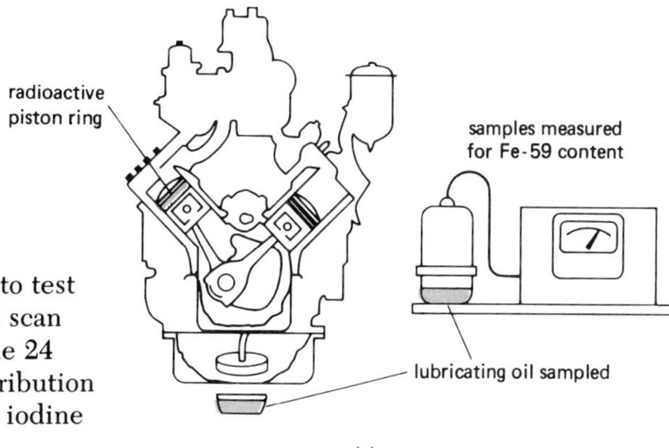

Figure 31.11
Radioactive tracers. (a) Tracers are used to test mechanical wear in engines. (b) Thyroid scan showing distribution of radioactive iodine 24 hours after being administered. The distribution is normal, and the gland is "picking up" iodine normally. [(b) Courtesy of L. Greenberg.]

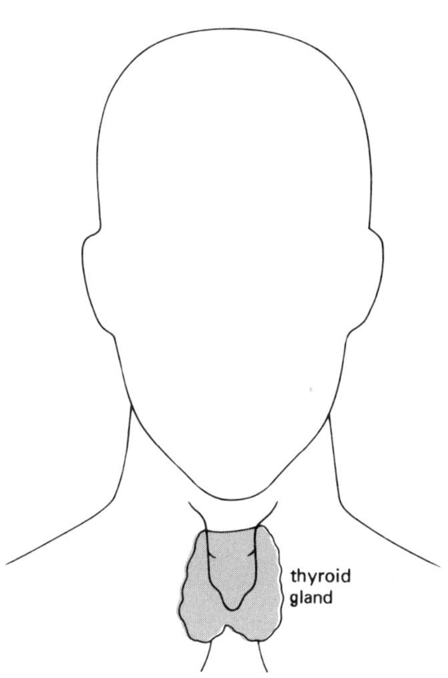

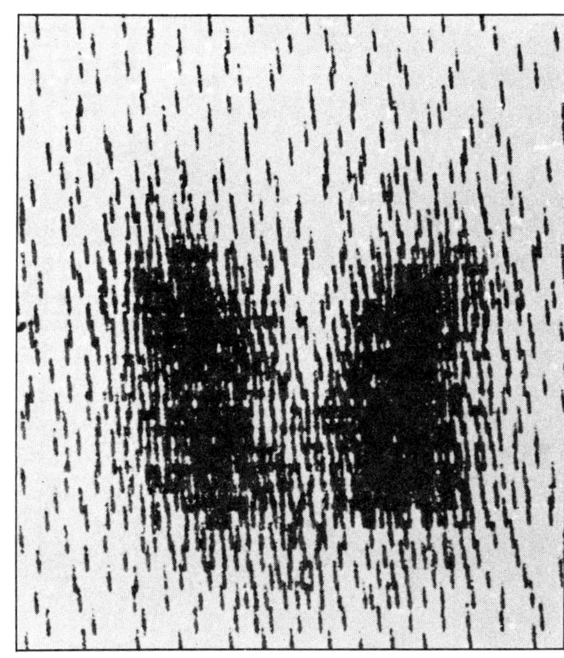

(b)

ment and monitoring—for example, in metal sheet and foil manufacturing (Fig. 31.12). Another common application is discussed in Special Feature 31.3.

31.4 Nuclear Reactions

Thus far, we have considered only nuclear decay reactions, in which a radioactive nucleus changes into a nucleus of another element by the emission of a particle. However, it is possible to initiate a nuclear reaction by adding a particle to a nucleus. This requires the bombardment of nuclei with energetic particles.

Lord Rutherford produced the first induced nuclear reaction in 1919 by bombarding nitrogen-14 with alpha particles from a natural radioactive source. He found that some of the nitrogen had been converted to oxygen by the reaction

$$\underset{\text{(nitrogen)}}{{}^{14}_{7}\text{N}} + \underset{\text{(alpha particle)}}{{}^{4}_{2}\text{He}} \rightarrow \underset{\text{(oxygen)}}{{}^{17}_{8}\text{O}} + \underset{\text{(hydrogen)}}{{}^{1}_{1}\text{H}}$$

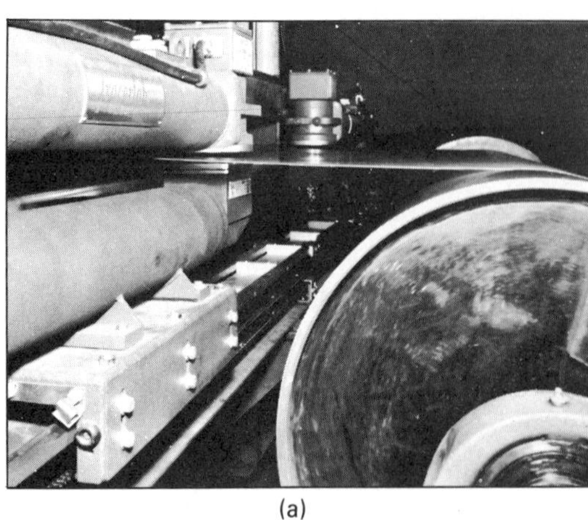

(a)

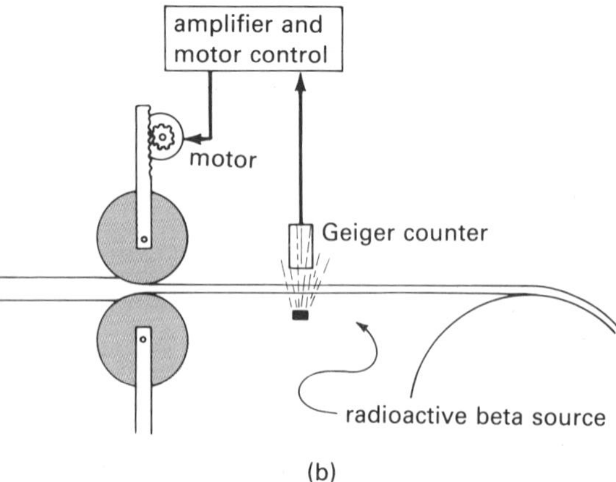

(b)

Figure 31.12
Radioactive thickness measurement. The penetration of radiation depends on the density, and hence the thickness, of a material. As a result, radioactive particles can be used in thickness monitoring and control. (Courtesy G. Holton, F. Rutherford, and F. Watson.)

By using such a bombardment process, we can make many artificial isotopes. This allows the production of artificial radioactive isotopes that can be used in industrial and medical applications. The oxygen-17 in the previous reaction is not radioactive. But in the reaction

$$^{27}_{13}\text{Al} + {}^{1}_{0}n \rightarrow {}^{27}_{12}\text{Mg} + {}^{1}_{1}\text{H}$$

(aluminum) (neutron) (magnesium) (hydrogen)

the magnesium-27 is radioactive, and it beta-decays with a half-life of 9.5 min.

The energetic particles used to initiate nuclear reactions are produced in various types of particle accelerators. Basically, charged particles, e.g., protons, are accelerated through electric potentials to produce beams of charged particles. Neutron beams are made by having charged beams react with a material that produces neutrons.

SPECIAL FEATURE 31.3

The Smoke Alarm

Smoke alarms are installed in many homes. Early warning of smoke is important for personal and property protection. Smoke may be detected by several means. A photoelectric type of smoke detector is described in Chapter 30. However, more common is the ionization type of smoke detector, which uses radioactivity (Fig. 1). A weak radioactive source ionizes the air, setting up a small current in a detector circuit. If smoke is present, the ions become attached to the smoke particles. The slower movement of the heavier, charged smoke particles causes a reduction in the detector current. This is electronically sensed and causes an alarm to sound.

Such smoke detectors are quite sensitive. Smoke vapors from cooking (from burnt foods) often set them off.

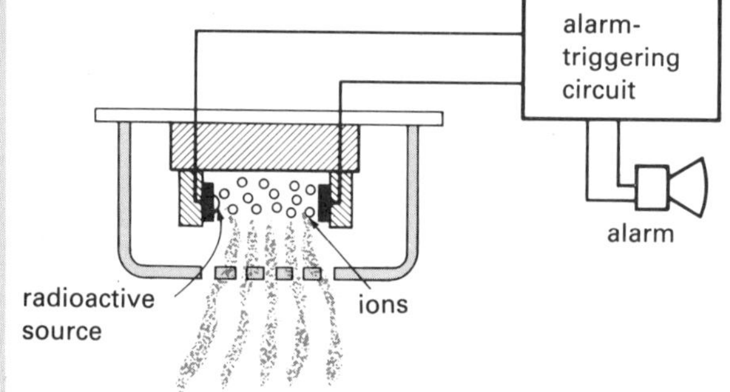

Figure 1
Smoke detection. In an ionization-type smoke detector, a weak radioactive source ionizes the air and sets up a small current. Smoke affects the current and causes an alarm to sound.

An interesting and practical aspect of nuclear reactions is mass-energy conversion. If we analyze the masses of the particles in Lord Rutherford's reaction, we find that the product particles have a total mass that is greater than the total mass of the initial particles*:

	Before Reaction	After Reaction
	$^{14}_{7}$N: 14.00307 u	$^{17}_{8}$O: 16.99913 u
	$^{4}_{2}$He: 4.00260 u	$^{1}_{1}$H: 1.00783 u
Total mass	18.00567 u	18.00696 u

and there is a **mass difference** or **defect** of

$$\Delta m = 18.00696 - 18.00567 = 0.00129 \text{ u}$$

Where does this excess mass come from? The answer is that some of the energy of the incident alpha particle is converted to (rest) mass. According to a theory by Einstein, mass is a form of energy, and the mass-energy conversion is related by the famous equation

$$E = mc^2 \qquad \textbf{(Eq. 31.1)}$$

where c is the speed of light. If 1 u (1.66×10^{-27} kg) of mass could be completely converted to energy, the amount of energy released would be

$$E = mc^2 = (1.66 \times 10^{-27} \text{ kg})(3.00 \times 10^8 \text{ m/s})^2$$
$$= 1.49 \times 10^{-10} \text{ J}$$

or, in more convenient nuclear units,

$$E = (1 \text{ u})c^2 = 931 \text{ MeV}$$

Hence, **1 u of mass is equivalent to 931 MeV of energy.** In nuclear physics, the principle of the conservation of energy is extended to the conservation of mass-energy.

In Lord Rutherford's reaction, the energy equivalence of the additional mass is

$$0.00129 \text{ u } (931 \text{ MeV/u}) = 1.20 \text{ MeV}$$

In this case, energy was converted to mass in the reaction. This is called an **endoergic reaction.**

> In an endoergic nuclear reaction, energy is converted to mass.

* The masses are actually those of neutral atoms (nuclei with electrons). The mass of the electrons subtracts out when the mass difference is figured.

There is also an **exoergic reaction:**

> In an exoergic nuclear reaction, mass is converted to energy.

This process is particularly important because the reaction produces energy. Nuclear energy comes from exoergic reactions. For example, when the total masses of the following reaction are analyzed, it is found that the total mass of the products is less than the total mass of the reactants:

$$^{235}_{92}\text{U} + ^{1}_{0}n \rightarrow ^{140}_{54}\text{Xe} + ^{94}_{38}\text{Sr} + 2^{1}_{0}n \text{ (+ energy)}$$

More total mass — Less total mass

Here, mass is converted into energy. The energy release of this exoergic reaction is on the order of 200 MeV. This is a relatively small amount of energy. But millions of such reactions in nuclear reactors produce large amounts of energy.

31.5 Nuclear Energy

Fission and Chain Reactions

Nuclear reactors in operation today use nuclear fission reactions, such as the preceding uranium-235 reaction.

> In a fission reaction, a heavy nucleus splits into two intermediate nuclei with the emission of two or more neutrons.

The fission process is where the phrase "splitting the atom" originated. U-235 has several fission decay modes, depending on the neutron energy. One mode was shown previously. Another induced fission reaction, with the emission of three neutrons, is

$$^{235}_{92}\text{U} + ^{1}_{0}n \rightarrow ^{141}_{56}\text{Ba} + ^{92}_{36}\text{Kr} + 3^{1}_{0}n \text{ (+ energy)}$$

Some heavy nuclei may undergo fission naturally, but energy production depends on induced fission and a **chain reaction** (Fig. 31.13). If a ^{235}U nucleus is struck by a neutron, ^{236}U is formed, which quickly decays. (The ^{236}U "compound" nucleus is usually omitted in the nuclear reaction equation.) The neutrons released in the fission process induce other fission reactions, and so on, so that a growing "chain" reac-

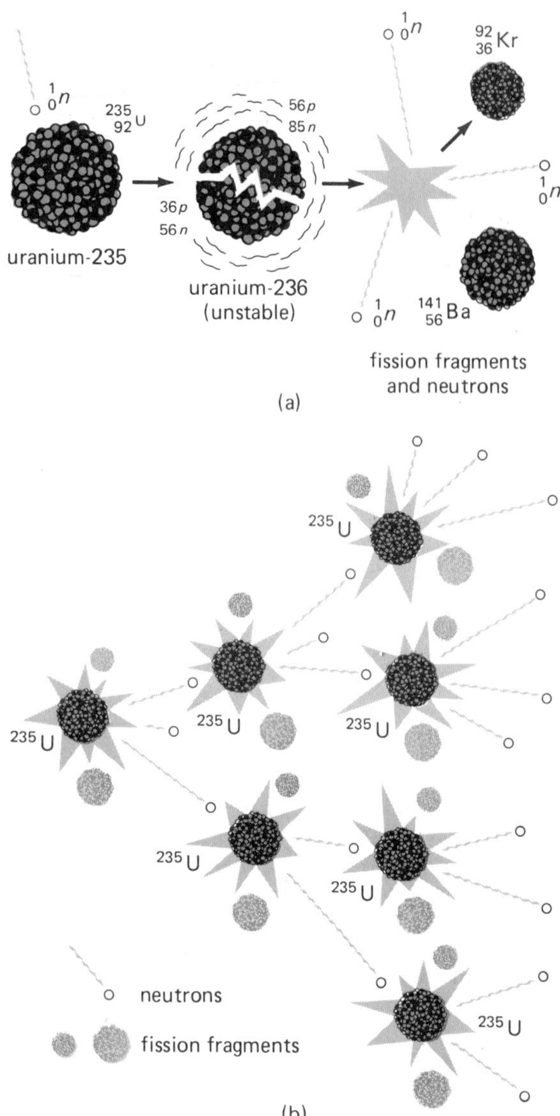

Figure 31.13
Induced fission and a chain reaction. (a) A neutron striking a U-235 nucleus induces fission, which results in two intermediate nuclei and the emission of two or more neutrons. (b) A chain reaction. Neutrons from fissioning nuclei induce fission in other nuclei. The reaction chain grows, with the release of energy with each fission.

tion takes place. Energy is released with each fission reaction. The chain reaction takes place instantaneously, and there is a tremendous release of energy from the millions of fissioning nuclei in a sample.

For a chain reaction to occur, there must be a sufficient amount of the fissionable material present. Otherwise, neutrons would escape from the sample before reacting with a nucleus, and the chain reaction would not proceed. The chain would be "broken," so to speak. Scientists call the amount of fissionable material necessary to sustain a chain reaction a **critical mass.**

Natural uranium is composed of 99.3 percent U-238 and only 0.7 percent of the fissionable U-235. To have more fissionable U-235 nuclei present in a sample, the U-235 is concentrated, or "enriched." Weapons-grade material may be enriched to 99 percent or more, whereas the enriched uranium used in nuclear reactors for the production of electricity is only 3 to 5 percent U-235.

That is why a nuclear reactor cannot explode like a bomb. In a bomb, a critical mass of highly enriched material must be held together for a short time to get an explosive release of energy (Fig. 31.14). Segments of the fissionable material in an atomic bomb, or "nuclear device" (modern terminology), are separated before explosion so there is not a critical mass for the chain reaction. A chemical explosive is used to bring the segments together, and the explosive force on interlocking configurations holds them together long enough for a large fraction of the material to undergo fission. The result is an explosive release of energy.

A bomb is an example of *uncontrolled* fission, which we hope will never again be used for purposes of war. A nuclear reactor is an example of *controlled* fission, in which we control the growth of the chain reaction and the release of energy.

Fission Reactors

The basic design of a fission nuclear reactor is shown in Figure 31.15. Fuel pellets in tubes form the fuel rods in the reactor core. The chain reaction and energy output are controlled by means of boron or cadmium **control rods.** These materials absorb neutrons. Hence, by insertion of the control rods between the fuel rod assemblies, the number of available neutrons for inducing fission and the chain reaction can be controlled.

The control rods can be adjusted so that energy is released at a relatively steady rate. This requires that, on the average, one neutron from each fission event initiate only one other event.

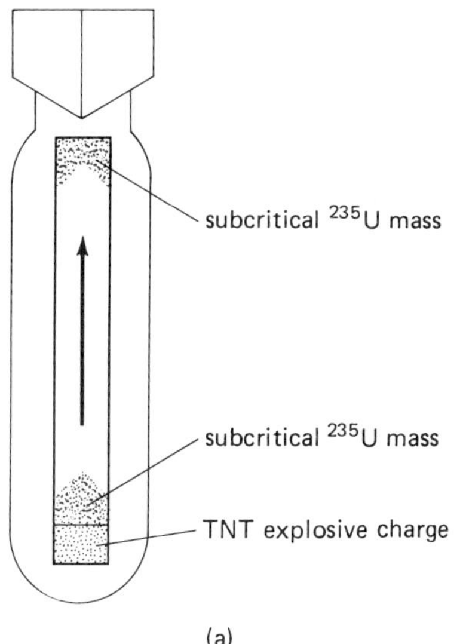

(a)

(b)

(c)

Figure 31.14
Uncontrolled nuclear fission. (a) In an atomic bomb, subcritical masses of fissionable material are brought together by a small explosive charge. (b) An atomic bomb of the "Little Boy" type that was detonated over Hiroshima, Japan, during World War II. The bomb was 28 inches in diameter and 120 inches long. It had a yield equivalent of approximately 20,000 tons of TNT. (Courtesy Los Alamos Scientific Laboratory.) (c) An underwater detonation. Notice the ships in the foreground. (Courtesy Joint Task Force One.)

If more energy is needed, the rods are withdrawn farther. When fully inserted, the control rods can curtail the chain reaction and shut the reactor down. A nuclear reactor can run for about four years before having to be "refueled."

A reactor is basically just an energy source with the energy removed by a coolant. Water flowing through the fuel assemblies and the reactor vessel acts not only as a coolant–heat transfer agent, but also as a **moderator.** The U-235 nuclei react best with "slow" neutrons (kinetic energies of only fractions of an electron volt). The "fast" neutrons emitted from the fission reactions (energies up to 20 MeV with an

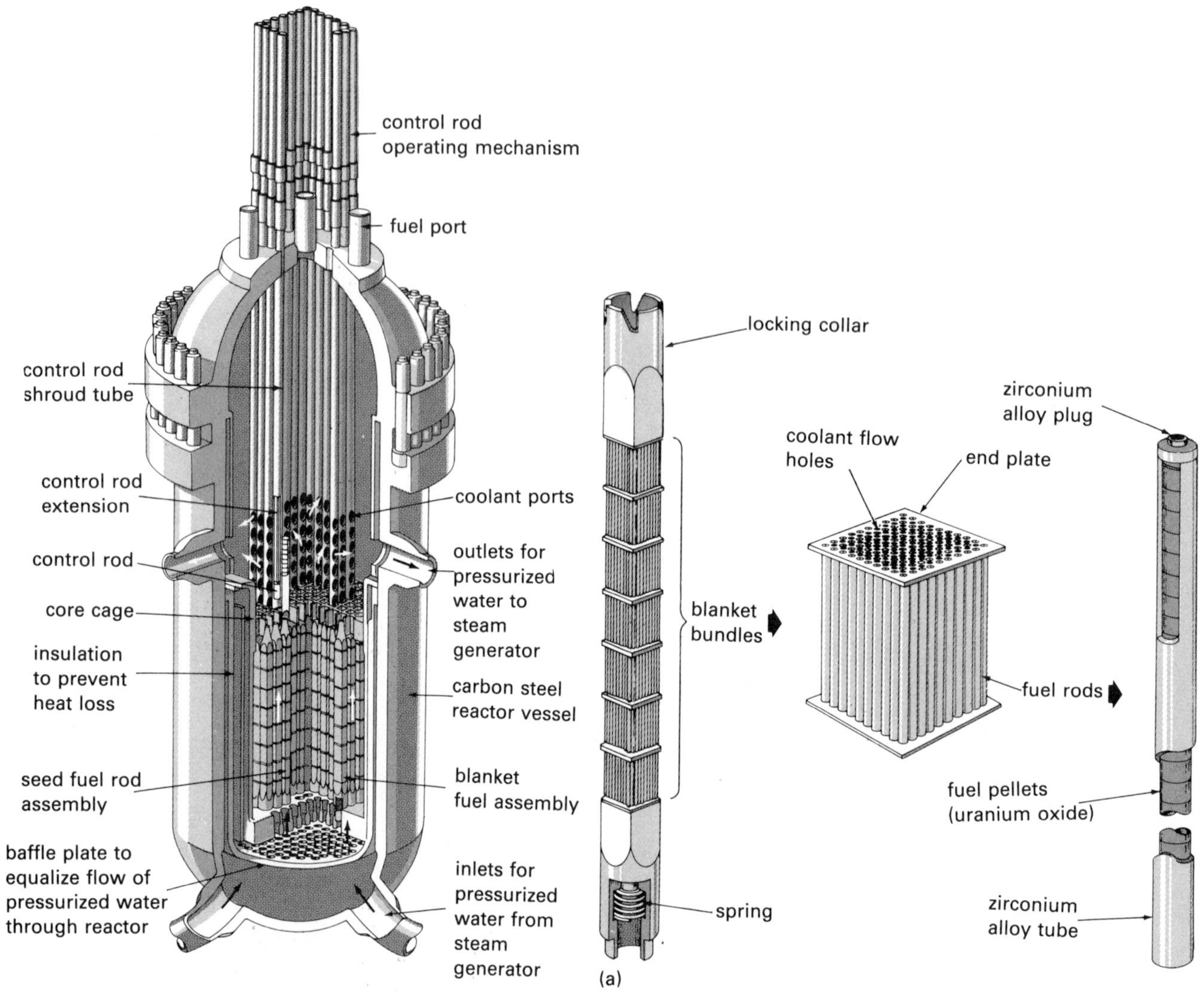

Figure 31.15

Nuclear fission reactor. (a) The basic elements of a reactor. (Courtesy U.S. Department of Energy.)

average of about 2 MeV) are slowed down, or "moderated," by energy losses through collisions.

Hydrogen atoms in water are very effective in slowing down neutrons because the mass of the hydrogen nucleus is about the same as that of a neutron. Recall from Chapter 10 that when particles of nearly equal masses collide, there is a large transfer of energy. It takes only about 20 collisions on the average to slow fast neutrons down to energies less than 1 eV.

Other materials, such as beryllium and graphite (carbon), have been used as moderators. Because these have heavier nuclei, not as much energy is transferred per collision. About 120 collisions with carbon atoms are needed to slow fast neutrons down to an optimal "slow" speed. Although it is not the best moderator, carbon in the form of graphite permits a chain reaction to occur in natural (unenriched) uranium fuel arranged in a large mass of graphite.

Heavy water (D_2O, where D stands for deuterium) can also be used as a moderator in a reactor that uses natural uranium as fuel. The hydrogen nuclei of regular water have a high tendency to capture neutrons when colliding

(b)

Figure 31.15 *(Continued)*
Nuclear fission reactor. (b) A fuel-rod assembly and a partially disassembled reactor. (Courtesy U.S. Department of Energy.)

with them.* As a result, it is impossible to achieve a chain reaction with natural uranium and ordinary water. Ordinary water can be used only with enriched uranium. Deuterium nuclei, on the other hand, do not absorb neutrons very readily, and so heavy water can be used as a moderator in reactors with natural uranium fuel. Most of the reactors in Canada are of this type, called heavy-water reactors, whereas most of the reactors in the United States are "light"-water reactors. The first nuclear reactor for the purpose of electrical generation went into operation in 1957 at Shippingsport, Pennsylvania.

One of the major concerns in reactor safety is that a reactor may get out of control, or in a "runaway condition," because of a malfunction of the control rods or a **loss-of-coolant accident** (LOCA). With an uncontrolled release of energy, the coolant might not be able to handle the excess heat generated, or if there were a loss of coolant, heat could build up in the core. In either case (or both), the reactor core might get so hot that the fuel rods would melt and the fused mass would burn through the reactor vessel and the containment floor.

* ${}^{1}_{0}n + {}^{1}_{1}\text{H} \rightarrow {}^{2}_{1}\text{H} + \gamma$.

Such a *meltdown* would cause leakage of dangerous radioactive materials into the environment. (This is the so-called China Syndrome, in which the fissioning mass might melt its way through the Earth to China. Calculations show that this cannot really happen. Also, as a geographic point, China is not on the opposite side of the United States through the Earth's center.) The introductory photo of this chapter shows the Three Mile Island nuclear facility, where a LOCA happened and where a partial meltdown occurred in 1979.

In more recent memory is the meltdown that occurred at Chernobyl in the Soviet Union in 1986. The reactor was a graphite type with some 1660 fuel assemblies encased in 1700 tons of graphite blocks. (Such reactors are not used commercially in the United States.) The accident resulted from human error. During a reactor shutdown, a power surge built up, which shattered the fuel into fragments and flash boiled the water coolant. The steam pressure blew a hole in the reactor top, exposing the red-hot core to air. This mixed with hydrogen, produced by steam reacting with zirconium metal of the fuel assembly casings. A violent *chemical* explosion was triggered that spewed radioactive material into the air and set the graphite on fire. Air rising from the burning graphite carried radioactive material into the atmosphere and it was carried over Europe by prevailing winds.

The extent of the damage from the Chernobyl accident is not known. Over 30 people died shortly after from radiation exposure, and it is estimated that thousands of fatal cancer cases will develop.

As pointed out earlier, a nuclear reactor cannot explode like a nuclear bomb. The uranium enrichment is too low for the chain reaction to proceed fast enough for the required energy release in a short time. Moreover, to explode, a system must be prevented from flying apart before this time elapses. Should the uncontrolled pressure buildup in a reactor core cause it to burst apart, the resulting separation of the fuel rods would give neutrons more opportunity to escape from the system, and the chain reaction would stop.

The Breeder Reactor

We have about a 50-year known domestic reserve of uranium. As with oil and gas, our domestic supply of uranium will run out in the not-too-distant future. However, it is possible to extend the supply of fissionable material through breeder reactors. A breeder reactor converts nonfissionable material, such as U-238, into fissionable material, and the reactor produces, or "breeds," more fuel than it consumes.

This is not a case of getting something for nothing. Rather, it is a process of converting a nonfuel into a fuel. For example, in an ordinary nuclear reactor there is a great deal of U-238 that "goes along for the ride"; the 3 percent or so of U-235 is the fissionable fuel that does the job. A U-238 nucleus can absorb a fast-moving neutron and undergo a change to plutonium-239 (Pu-239) by the reaction

$$\underset{\text{Fertile material}}{^{238}\text{U}} + {}^{1}n \rightarrow \underset{\text{Unstable products}}{^{239}\text{U} \rightarrow {}^{239}\text{Np}} \rightarrow \underset{\text{Fissionable material}}{^{239}\text{Pu}}$$

Pu-239 is a fissionable material that can do the same job as U-235. The original material is called a *fertile* material. Other fertile materials that can be used in a breeding reaction are Th-232 and Pu-240. These breed fissionable U-233 and Pu-241 isotopes, respectively.

In a U-238 breeder reactor, the moderation (slowing down) of fast neutrons is minimized, and with the presence of an appreciable number of fast neutrons, Pu-239 is bred from U-238 while energy is obtained from U-235 fission. France and several other countries have breeder reactors, but only experimental breeder reactors have been operational in the United States because of limited funding. One of the reasons for the curtailment of the development of the breeder reactor is discussed in the Chapter Supplement.

Fusion and Fusion Reactors

Fusion reactions are another class of nuclear reactions that release energy. In these exoergic reactions, light nuclei are "fused" together to form heavier nuclei with the release of energy. An example is hydrogen fusion, illustrated in Figure 31.16.

Fusion is the source of energy of stars, which include our Sun. In the Sun, the fusion process results in the production of a helium nucleus from four protons (hydrogen nuclei). The proc-

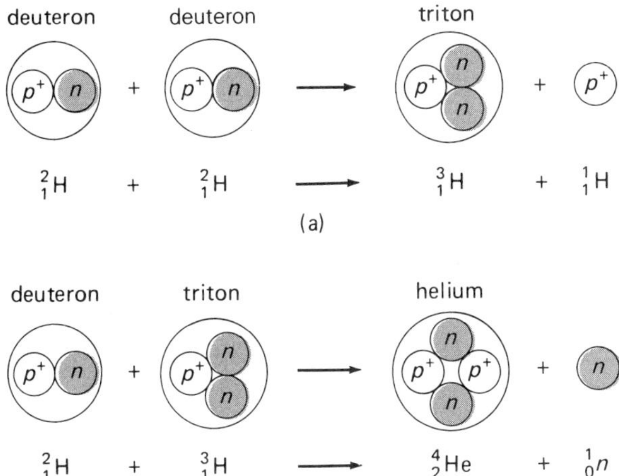

Figure 31.16
Fusion reactions. (a) D-D reaction. (b) D-T reaction.

ess does not take place in a single reaction, but goes through different sets of reactions with the net result of

$$4{}^1_1\text{H} \rightarrow {}^4_2\text{He} + 2({}_{+1}^{\;\;0}e)$$

(The symbol ${}_{+1}^{\;\;0}e$ represents a positron, which is the antiparticle of the electron.)

Many scientists look upon fusion as an ultimate and ideal source of energy for several reasons. Among these are the availability of raw materials and advantages in waste disposal. "Heavy hydrogen," or deuterium (${}^2_1$H), is present in all water (heavy-water molecules). For about every 6500 atoms of ordinary (light) hydrogen in water, there is one atom of deuterium. This may not seem like much, but taking into account the vast amounts of water in the oceans and other surface waters, it is estimated that there is enough deuterium to supply the world's energy needs for more than a million years.

The other heavy isotope of hydrogen, tritium (${}^3_1$H), is very rare in nature. It is radioactive, with a half-life of about 12 years. As "nuclear waste" of the first (D-D) reaction in Figure 31.16, it would be a relatively short-term disposal problem, as compared with fission wastes (see the Chapter Supplement). Notice that the tritium formed by deuterium fusion could immediately react by the second (D-T) reaction with available deuterium. For use as a fusion "fuel," tritium is produced by the nuclear reaction of neutron capture by lithium (Li). As a raw material, lithium is widely distributed over the Earth, being estimated to be more abundant than tin. Should this source of lithium not be adequate, it could be extracted from the oceans, which contain large amounts (but in small concentrations).

The energy releases from the D-D and D-T reactions are about 4.0 MeV and 17.6 MeV, respectively. This is significantly less than the 200-MeV energy release from a typical fission reaction. But keep in mind that a given mass of hydrogen isotopes has many, many more nuclei than an equivalent mass of a "heavy," fissionable isotope.

In the case of fusion, there is no critical mass or size, since there is no chain reaction to maintain. However, there are problems in practical fusion energy production. One of the major ones is in obtaining a self-sustaining reaction. The repulsive electric force between positively charged nuclei opposes their coming together and fusing. For hydrogen (one nuclear proton) this force is not excessive, which is why hydrogen is the fusion candidate. Fusion reactions can be produced in particle accelerators. For example, deuterons (deuterium nuclei) can be accelerated and "slammed" into a solid deuterium target to produce D-D fusion reactions. This shows fusion to be possible, but much more energy is spent in accelerating the deuteron than is produced by the small number of fusion reactions that occur.

To get an appreciable energy output, one might use a confined gas of hydrogen isotopes that are to undergo fusion. If the temperature of the gas is raised, the kinetic energy of the molecules is increased, and hence it is only a matter of attaining a sufficiently high temperature for fusion to occur. In the confined space of the gas, the moving nuclei collide repeatedly until fusion reactions take place. However, it is found that this requires temperatures on the order of 100 million degrees (Fahrenheit, Celsius, or kelvins, take your choice).

At such high temperatures, almost all of the hydrogen-isotope atoms are stripped of their electrons. Such a gas, consisting almost entirely of positively charged ions and free, negatively charged electrons, is called a **plasma.** Plasmas have a number of special properties that have caused them to be referred to as a "fourth phase of matter," a term used in 1879 by William

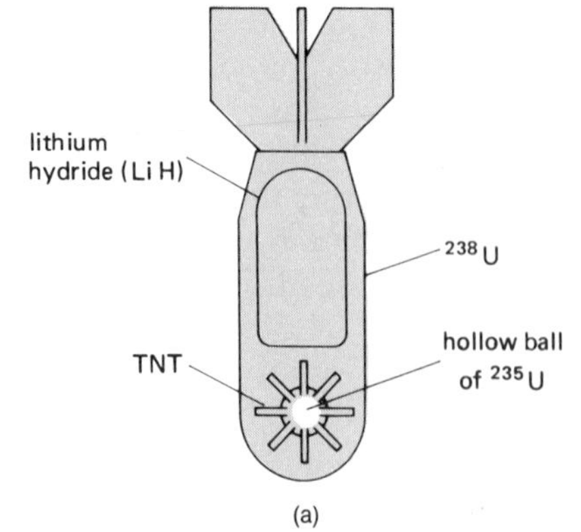

Figure 31.17
Uncontrolled fusion. (a) In a hydrogen bomb, a small fission bomb is used to "trigger" the fusion reaction. (b) A thermonuclear explosion in the Pacific in 1954.

Crookes, an English chemist, who generated plasmas in gas-discharge, or Crookes, tubes. Some of these properties can be used to advantage in potential fusion-reactor techniques, but others create problems.

Large amounts of fusion energy have been released, but this occurred in uncontrolled fusion in the form of the hydrogen (H) bomb. In the H-bomb, a nuclear fission bomb is used to "trigger" the fusion reaction, i.e., to supply the energy to get it started (Fig. 31.17). Because of the high temperatures required to bring about fusion reactions in plasmas, they are called thermonuclear reactions, and fusion weapons are called thermonuclear weapons (or "devices").

The controlled release of fusion energy might be accomplished by adding fuel in small amounts to a fusion reactor, similar to the way we control the amount of fuel in conventional energy (heating) sources, such as a gas furnace. However, fusion reactors are not yet practical. There are major problems with confinement—confinement of an ultra-hot plasma and confinement of sufficient energy in the plasma to initiate and sustain fusion.

No material could be used to "hold" a plasma. Tungsten, the material with the highest melting point, melts at around 3370°C. Also, heat would be readily transferred from the confinement region. Considering all of the various factors, it is calculated that to achieve a net energy production from D-T fusion would require a reactor operating temperature of at least 100 million degrees. Even higher temperatures would be required for D-D fusion.

Two types of confinement, magnetic and inertial, are now under development.

Magnetic Confinement Since the plasma is a gas of charged particles, it can be controlled and manipulated with electric and magnetic fields. In magnetic confinement, magnetic fields are used to hold the plasma in a confined space (a magnetic "bottle" or "ring"). Electric fields pro-

Figure 31.18
Magnetic confinement. An ORMAK fusion research device. The magnetic system used is known as a tokamak (a Russian word for a toroidal magnetic chamber) and is circularly shaped much like a donut. (Courtesy U.S. Department of Energy.)

duce electric currents that raise the temperature of the plasma (Fig. 31.18).

The energy break-even point (getting as much out as is put in) depends on two things: (1) the temperature of the plasma and (2) its density and confinement time. Temperatures of 100 million kelvins have been achieved in tokamaks, and in other experiments the density-confinement time requirement seems to be satisfied. The trick scientists must now perform is to satisfy these requirements at the same time, which may prove more difficult than satisfying them separately.

Inertial Confinement In this technique, it is hoped that laser, electron, or ion beams can be used to initiate fusion. It calls for hydrogen fuel pellets to be dropped into a reactor chamber (Fig. 31.19). Energetic beam pulses would make a pellet "implode," causing compression and high temperatures. Fusion could occur if the pellet stayed together for a sufficient time. This would depend on the inertia of the pellet (hence the name *inertial confinement*). At the present time, lasers and particle beams are not powerful enough to induce fusion by this means, but research has begun.

If you think about it, the technological problems of fusion reactors are enormous, but so are the benefits. Practical energy production from fusion is not expected until well into the 21st century.

A more recent approach to fusion attempts to fuse nuclei without high temperature requirements. This is termed *cold fusion.* One method involves *muons*—negatively charged particles about 200 times more massive than electrons.* Muons, which are produced in particle accelerators, take the place of electrons in muonic molecules of deuterium and tritium in what is called muon-catalyzed fusion.

In 1989 there was an announcement that another type of cold fusion had been achieved in the laboratory. This process involved the electrolysis of heavy water (D_2O) using platinum-palladium electrodes. It is still uncertain whether or not fusion was actually achieved. The process remains under investigation, but there is considerable doubt.

* There is a large number of "elementary" particles that result from nuclear reactions.

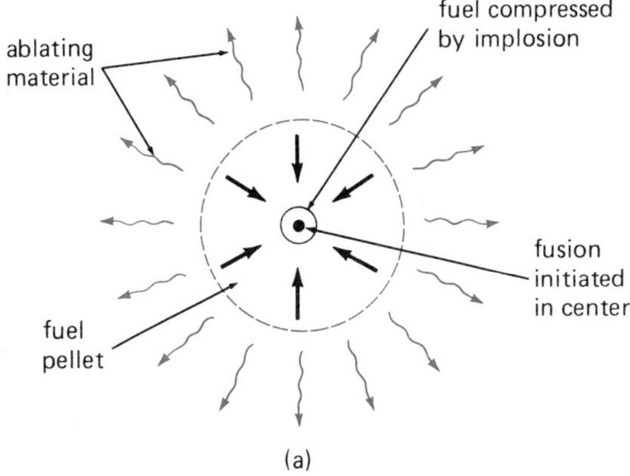

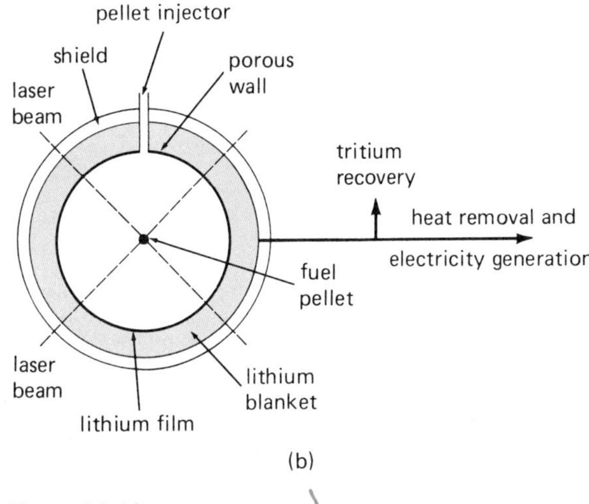

Figure 31.19

Inertial confinement and implosion. (a) The compression of a fuel pellet is enhanced by the use of an outer shell of material—for example, a plastic—that is ablated or vaporized. This drives the inner shell inward and compresses the fusion fuel. (b) A design of a laser reactor cavity. The laser beams compress and heat the fuel pellet. Neutrons released from the fusion are captured in the lithium blanket, creating tritium, which can be recovered and used as fuel.

Important Terms

nucleus the core of an atom, made up generally of protons and neutrons

nucleons a collective term for nuclear protons and neutrons

isotopes nuclei of the same element (same number of protons) with different numbers of neutrons

nuclide a particular nuclear species, or isotope

mass number (*A*) the number of nucleons (protons + neutrons) in a nucleus

proton (atomic) number (*Z*) the number of protons in a nucleus

neutron number (*N*) the number of nucleons in a nucleus

atomic mass unit (u) $\frac{1}{12}$ the mass of a carbon-12 atom where a carbon-12 atom has a mass of 12.0000 u

nuclear force (strong interaction) the attractive fundamental force acting between any pair of nucleons

radioactivity the spontaneous decay of unstable nuclei with the emission of energetic particles

alpha particle a particle consisting of two protons and two neutrons, or a helium nucleus (^4_2He) emitted in radioactive alpha decay

beta particle an electron ($_{-1}^{0}e$) emitted in radioactive beta decay

weak nuclear force (weak interaction) a fundamental interaction occurring between most particles and particularly important in beta decay where this is no strong interaction between nucleons and the beta particle

gamma particle a quantum of electromagnetic energy emitted in radioactive gamma decay

half-life the time it takes for one half of a sample of a radioactive isotope to decay (or for the activity to decrease by one half)

Geiger counter a common radiation detection device based on ionization by particles

endoergic reaction a nuclear reaction in which energy is converted to (rest) mass

exoergic reaction a nuclear reaction in which (rest) mass is converted into energy

fission an exoergic reaction in which a heavy nucleus "splits" (fissions) into two intermediate nuclei with the emission of two or more neutrons

chain reaction a growing series of induced fission reactions due to emitted neutrons from previous fission reactions

critical mass the amount of mass or concentration of fissionable material needed to sustain a chain reaction

control rods rods of neutron-absorbing materials used to control the energy output of a chain reaction

moderator a substance used in a nuclear reactor to moderate (slow) neutrons by collision energy transfer

breeder reactor a reactor that produces more fissionable material than it consumes

fusion an exoergic reaction in which two light nuclei are "fused" (reacted) together to form a heavier nucleus

plasma a gas of positively charged ions and free negatively charged electrons

Important Formulas

mass-energy conversion: $E = (\Delta m)c^2$

in all nuclear reactions: (a) The total number of nucleons is conserved. (b) The total charge is conserved.

conversion factors:
$1 \text{ u} = 1.66 \times 10^{-27}$ kg
$1 \text{ Ci} = 3.7 \times 10^{10}$ decays/s
$= 3.7 \times 10^{10}$ Bq
(1 Bq = 1 decay/s)
1 u = 931 MeV
(equivalent to)

Questions

The Nucleus and the Nuclear Force

1. What are (a) a nucleon and (b) a nuclide?
2. The nucleus contains what percentage of the atomic mass?
3. (a) What determines whether an atom is of a particular element? (b) How is the mass number related to the proton and neutron numbers?
4. What is an atomic mass unit? What are the approximate masses, in atomic mass units, of (a) a proton, (b) a neutron, and (c) an electron?
5. Discuss how the gravitational and electrical forces might contribute to the stability or instability of an atom, taking into account nucleons and orbital electrons.
6. Given only the proton numbers or neutron numbers or mass numbers of several nuclei, would you *always* be able to distinguish between different nuclides? Explain.
7. Explain the role of neutrons in nuclides in terms of nuclear stability.
8. Considering the following nuclides, show that the neutron number generally exceeds the proton number for nuclei with proton numbers greater than 20: lithium-6, bromine-80, silicon-28, titanium-48, fluorine-19, and platinum-179.
9. Tin ($_{50}Sn$) has 23 isotopes. Write the complete nuclear symbol (with neutron number) for isotopes of tin with mass numbers from 114 to 120. (The complete family of isotopes has mass numbers from 108 to 130.)

Radioactivity

10. What is radioactivity, and who discovered it?
11. Speculate on why the Curies named one of the new elements they discovered in 1898 "radium." How about polonium? (*Hint:* it was named after Madame Curie's native country.)

12. (a) Explain how the three decay radiations can be distinguished by using a magnetic field. (b) Could an electric field be used instead? Explain.
13. Is the neutron number always conserved in radioactive decay? How about the mass number and proton number?
14. How can an electron be emitted from a nucleus in beta decay when it is thought that only protons and neutrons exist in the nucleus?
15. Do radioactive isotopes decay linearly with time? Explain.
16. How is the curie unit defined? How many becquerels are there in 0.50 Ci?
17. Explain the principle of carbon-14 radioactive dating.

Radiation Detectors and Applications

18. How does a Geiger counter detect radiation, and what is meant by "dead time"?
19. Discuss the advantages and disadvantages of the following radiation detectors: (a) film badge, (b) Geiger counter, (c) scintillation counter, (d) solid state detector.
20. Give some practical uses of radioactive isotopes in industry and medicine.

Nuclear Reactions

21. In trying to initiate nuclear reactions, would there be any advantage to using protons or hydrogen nuclei instead of alpha particles? Explain. How was the first artificial nuclear reaction initiated?
22. Explain what is meant by the conservation of mass-energy.
23. Distinguish between exoergic and endoergic reactions.
24. Is natural radioactivity an endoergic or exoergic reaction? Explain.

Nuclear Energy

25. What is fission? Distinguish between spontaneous fission and induced fission.
26. Why is it impossible to have a sustained fission chain reaction in a very small piece of fissionable material?
27. Why can't a nuclear reactor blow up or explode like a bomb?
28. Could the moderator (or lack of it) be used to control the chain reaction in a reactor? Explain.
29. Is it necessary always to use enriched uranium in a reactor? Explain.
30. What is meant by "meltdown"? How is this related to the China Syndrome?
31. Explain how a breeder reactor can produce more fuel than it consumes, and write a nuclear reaction used in a breeder reactor.
32. Why is an A-bomb needed to start an H-bomb? What is a thermonuclear reaction?
33. The energy release from each fusion reaction is less than 20 MeV, whereas about 200 MeV of energy is released from a fission reaction. Even so, discuss the advantages that fusion reactors would have over fission reactors.
34. Why would a LOCA be less serious or less likely in a fusion reactor than in a fission reactor?
35. Why are nuclear wastes and nuclear proliferation considered to be problems? What is the United States currently doing about these problems?

Problems

Levels of difficulty are indicated by asterisks for your convenience.

31.1 The Nucleus and the Nuclear Force

1. Write the complete nuclear notations for (a) argon-40, (b) chromium-60, (c) gold-197, and (d) uranium-239.
2. Write the complete nuclear notations for the isotopes antimony-120, -121, -122, -123, and -124.
3. What are the neutron numbers of (a) ^{132}Te, (b) ^{178}Hf, (c) ^{23}Na, and (d) ^{196}Hg?
4. A proton is how many times more massive than an electron?

*5. Considering the masses of the nucleons and electrons of nitrogen-14 individually, what percentage of the mass is in the nucleus?

*6. (a) What is the mass of the carbon-12 atom in kilograms? (b) What is the weight of a carbon-12 atom in pounds? (Recall that 1 kg of mass is equivalent to a weight of 2.2 lb.)

*7. (a) Compare the magnitudes of the gravitational force and the electrical force between two protons in the nucleus. (*Hint:* use a ratio.) (b) What does this tell you about the magnitude of the nuclear force?

*8. Considering the following elements, show that the neutron number generally exceeds the proton number for elements of $Z > 20$: lithium-6, bromine-80, silicon-28, titanium-48; fluorine-19, and platinum-179.

31.2 Radioactivity

*9. Complete the following decay reactions:

(a) $^{64}Cu \rightarrow {}^{64}Zn$ + ____

(b) $^{190}Pt \rightarrow {}^{186}Os$ + ____

(c) $^{123}Te^{*} \rightarrow {}^{123}Te$ + ____

(d) $^{239}Pu \rightarrow {}^{235}U$ + ____

(e) $^{32}P \rightarrow {}^{32}S$ + ____

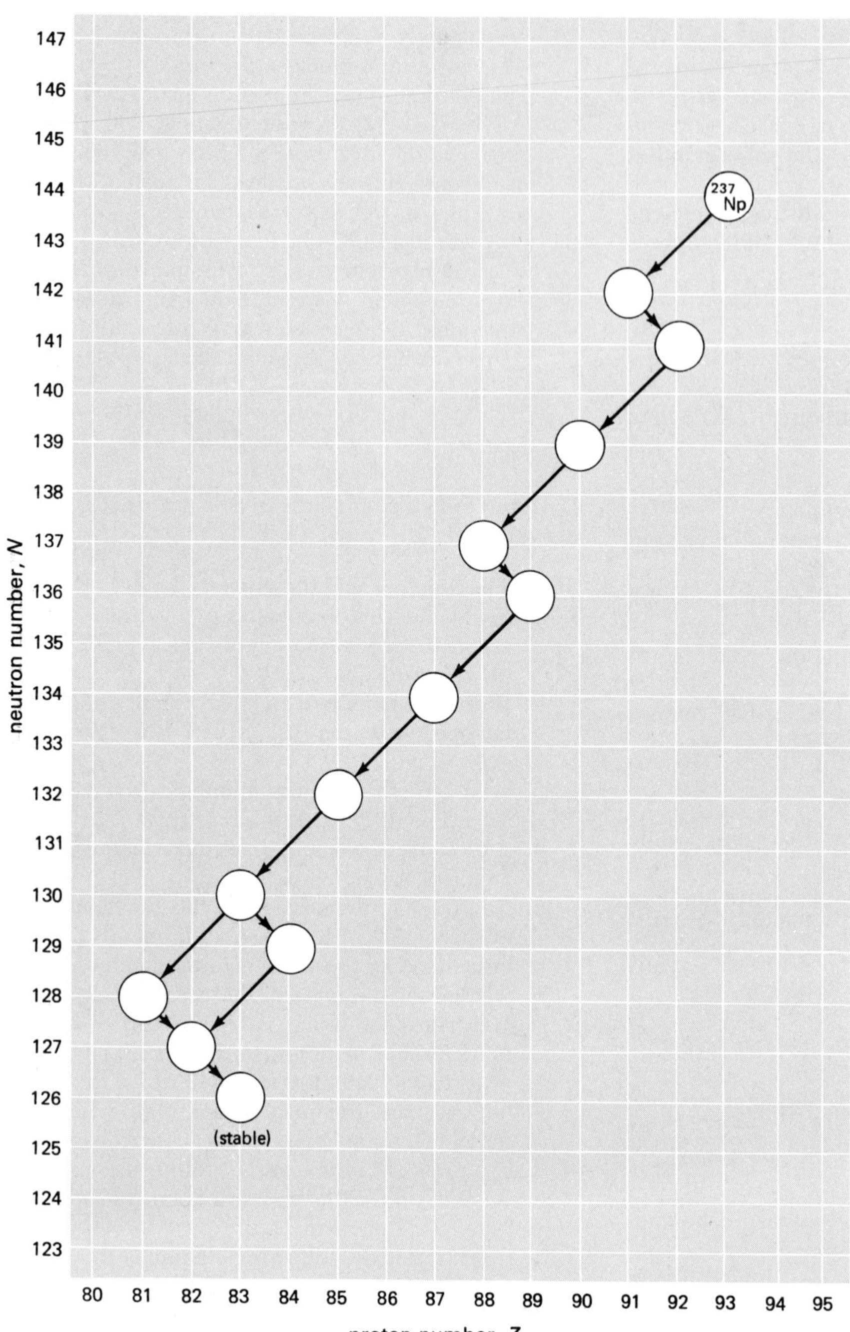

Figure 31.20
See Problem 10.

*10. Complete the radioactive decay series shown in Figure 31.20.

*11. Write the nuclear equations for (a) the alpha decay of $^{226}_{88}Ra$, (b) the beta decay of $^{60}_{27}Co$, and (c) the gamma decay of $^{210}_{84}Po$.

*12. Protactinium-233 ($^{233}_{91}Pa$) undergoes beta decay. (a) What is the daughter nucleus in this process? Write the equation for it. (b) The daughter nucleus then undergoes alpha decay. What is the "granddaughter" nucleus?

*13. Radon-222 ($^{222}_{86}Rn$), a radioactive noble gas, undergoes alpha decay. (a) What is the daughter nucleus of this process? Write the equation for the decay. (b) The daughter nucleus undergoes both beta decay and alpha decay. What is the "granddaughter" nucleus in each case? Write the equations.

*14. Write the nuclear decay equations that show how gamma rays are used to monitor the thyroid gland in iodine-131 uptake. (*Note:* iodine-131 first beta-decays.)

*15. A "mad" scientist wants to sell you a large quantity of gold-194 that he made by artificial nuclear reactions for only one-fourth the current price of gold. Should you buy it? (Justify your answer.)

*16. A sample of cobalt-60 has an activity of 120 μCi. How long will it take for the activity of the sample to decrease to 15 μCi?

*17. (a) Given 10 milligrams of iodine-131, how much of it would remain after 24 days? (b) Iodine-131 beta-decays. What is the daughter nucleus?

*18. A petrified wood carving is found in an archeological dig. It is found that there are 2 beta emissions/g of C/min from a sample of the carving. Approximately how old is the carving?

**19. If a large animal died today, in what year would there be 4 beta emissions/g of C/min from its bones?

**20. A sodium-24 source has an activity of 3.7×10^9 decays/s. (a) What is its activity in curies and becquerels? (b) What would be its activity in 5 days in these units? ($t_{1/2} = 15$ hr).

31.4 Nuclear Reactions

21. What is the equivalent energy of 10 u of mass?

22. Write the induced fission reaction for uranium-235 "splitting" into rubidium-94 and cesium-139.

*23. Complete the following nuclear reactions:

(a) $^{6}_{3}Li + ^{1}_{0}n \rightarrow$ ____ $+ ^{4}_{2}He$

(b) $^{45}_{21}Sc + ^{4}_{2}He \rightarrow$ ____ $+ ^{1}_{1}H$

(c) $^{27}_{13}Al + ^{1}_{0}n \rightarrow ^{28}_{13}Al +$ ____

(d) $^{2}_{1}H + ^{3}_{1}H \rightarrow$ ____ $+ ^{1}_{0}n$

(e) ____ $+ ^{4}_{2}He \rightarrow ^{35}_{17}Cl + ^{1}_{1}H$

(f) $^{235}_{92}U + ^{1}_{0}n \rightarrow ^{141}_{54}Xe +$ ____ $+ 3^{1}_{0}n$

*24. Complete the following nuclear reactions and calculate the mass defects of each:

(a) $^{235}_{92}U + ^{1}_{0}n \rightarrow$ ____ $+ ^{92}_{36}Kr + 3^{1}_{0}n + 200$ MeV

(b) $^{2}_{1}H + ^{2}_{1}H \rightarrow ^{3}_{2}He +$ ____ $+ 3.26$ MeV

**25. Considering Lord Rutherford's reaction,

$$\underset{(14.00307)}{^{14}_{7}N} + \underset{(4.00260)}{^{4}_{2}He} \rightarrow \underset{(16.99913)}{^{17}_{8}O} + \underset{(1.00783)}{^{1}_{1}H}$$

where the numbers under the symbols are the masses in atomic mass units (u), is the reaction endoergic or exoergic? How much energy is released or put into the reaction?

**26. Are the following reactions endoergic or exoergic? If exoergic, how much energy is released?

(a) $$\underset{(199.96828)}{^{200}_{80}Hg} + \underset{(1.00783)}{^{1}_{1}H} \rightarrow \underset{(196.96654)}{^{197}_{79}Au} + \underset{(4.00260)\text{ u}}{^{4}_{2}He}$$

(b) $$\underset{(7.01600)}{^{7}_{3}Li} + \underset{(1.00783)}{^{1}_{1}H} \rightarrow \underset{(4.00260)}{^{4}_{2}He} + \underset{(4.00260)\text{ u}}{^{4}_{2}He}$$

**27. Determine whether the following reactions are endoergic or exoergic. If exoergic, how much energy is released?

(a) $$\underset{(2.01410)}{^{2}_{1}H} + \underset{(2.01410)}{^{2}_{1}H} \rightarrow \underset{(3.01603)}{^{3}_{2}He} + \underset{(1.00867)\text{ u}}{^{1}_{0}n}$$

(b) $$\underset{(3.01603)}{^{3}_{2}He} + \underset{(1.00867)}{^{1}_{0}n} \rightarrow \underset{(2.01410)}{^{2}_{1}H} + \underset{(2.01410)\text{ u}}{^{2}_{1}H}$$

**28. The beryllium-9 nucleus has a mass of 9.01218 u. Is the sum of the masses of the individual nucleons the same? If not, explain the difference. (*Hint:* work is required to separate the nucleus into individual nucleons. This is called the binding energy of the nucleus. Use the mass of a hydrogen atom to account for the electrons.)

Chapter Supplement

Nuclear Wastes and Proliferation

The building and planning of nuclear reactors were greatly curtailed after the Three Mile Island incident. This was not only because of public concern for safety, but also because it was realized that there was not the demand for electricity that had been projected, and new reactors were not needed. People were conserving energy. Another factor was falling fossil-fuel costs. Overall, it was cheaper to build and operate a fossil-fuel plant than a nuclear one, so economics played a role.

Even so, there are about 110 nuclear reactors (some military) operating in the United States (Fig. 31S.1). (About 430 nuclear reactor generating facilities are operating in other countries. The figure shows the percentages of electrical generation from nuclear energy in some countries.) They are creating large amounts of nuclear wastes—the "ashes" from the fission process. A large variety of fission products are formed in nuclear reactors, most of which are radioactive. Some of these products have very

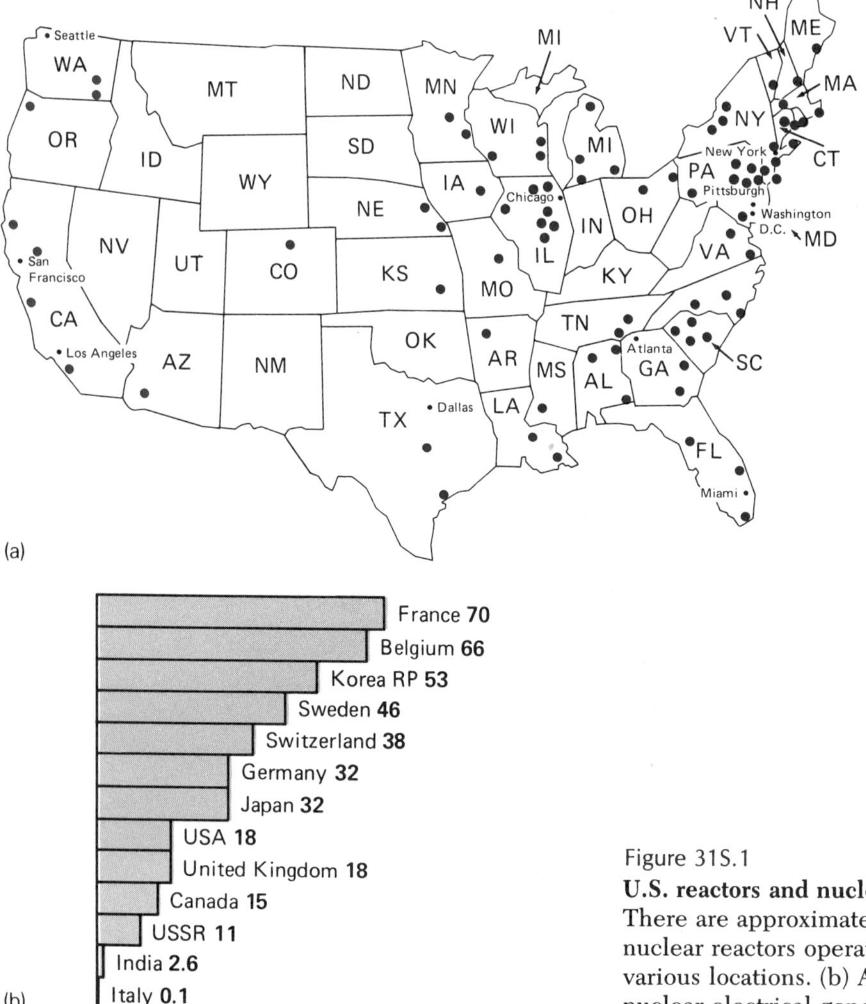

Figure 31S.1
U.S. reactors and nuclear energy production. (a) There are approximately 110 electrical generating nuclear reactors operating in the United States at various locations. (b) A comparative production of nuclear electrical generation of some countries.

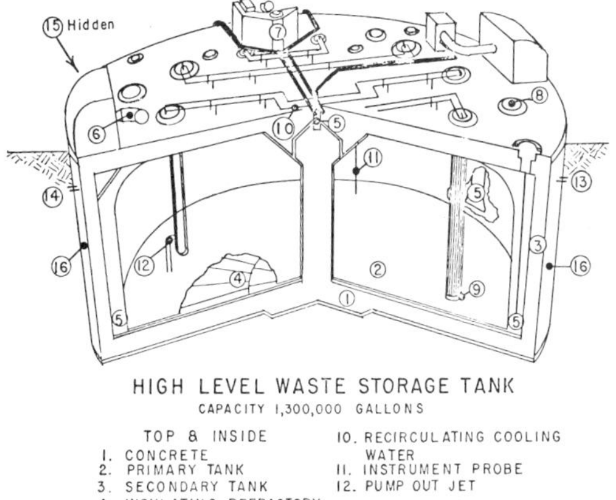

(a)

(b)

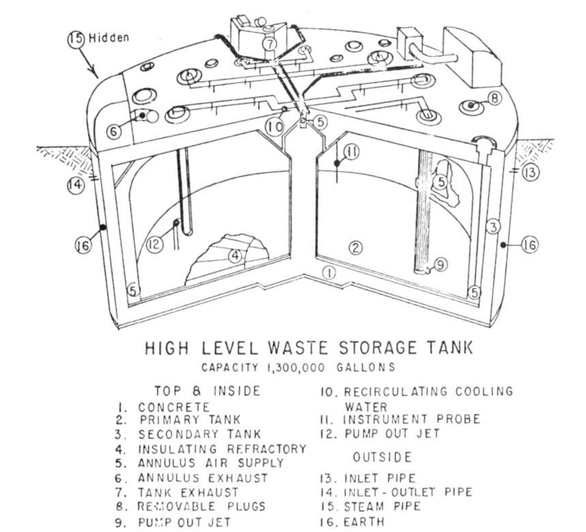

Figure 31S.2
Radioactive wastes. (a) Nuclear wastes are stored in double-walled tanks that are covered with earth. (b) Embedding radioactive waste in glass is a possible method of storage.

long half-lives, up to tens of thousands of years or more. This means that these "high-level" wastes will be around for hundreds of years, and the concentrations make them highly dangerous.

There are still useful fissionable materials in the spent fuel assemblies, and it would be good to recover them. Currently, however, nuclear wastes are being stored in double-walled storage tanks that are covered with earth (Fig. 31S.2). Plans call for the "ultimate" disposal of these wastes to be in underground storage in stable geologic repositories such as salt or rock formations. A bill passed by Congress calls for these repositories to be available by the year 1996.

"Permanent" underground burial requires the development of long-lasting containers or the conversion of wastes into solid, insoluble form—for example, embedded in glass (Fig. 31S.2). Stainless-steel containers are currently being considered. Appropriate geologic formations must be found where the wastes could cause no contamination of underground water should they leak.

In addition to the radioactive fragments of the fission process, the transuranic element plutonium-239 is formed in the normal operation of a reactor. Some U-238 is converted to Pu-239 through "fast" neutron capture and two beta decays. This is one of the reactions promoted in a breeder reactor, discussed previously. Pu-239 is a fissionable material and does the same job as U-235. In the reactor operation, the freshly produced Pu-239 also undergoes fission and produces energy. As a result, it is possible to extend the normal "lifetime" of the reactor fuel elements, and they do not have to be replaced as often.

However, it is possible to remove the fuel elements before the Pu-239 is used up and to separate it through difficult and complicated procedures. The Pu-239 can then be processed into nuclear weapons. (The atomic bomb that destroyed the city of Nagasaki, Japan, used Pu-239 instead of U-235, as had been used in the bomb dropped on Hiroshima. The Pu-239 had been made in an experimental reactor.)

A nuclear power plant reactor can generate up to 200 to 300 kg to Pu-239 a year. A great deal of this is used up in the fission process, but only 5 to 10 kg of Pu-239 is needed to make a nuclear bomb capable of destroying a medium-size city. Thus, a nuclear reactor can be used to produce electricity and/or to produce Pu-239, which could be used as material for weapons. Fissionable materials can also be made in research reactors. Such capability gives rise to what is called the problem of nuclear proliferation.

The potential for nuclear proliferation beyond the developed world became clear in 1974 when India exploded a "nuclear device" as a test for "peaceful applications," according to that government. The explosion showed that countries other than the five "nuclear powers" (those who had tested and demonstrated nuclear weapons) could produce fissionable material and use it in nuclear weapons if they so desired.

In the early days of the "nuclear age," only three countries had the bomb. The United States, Great Britain, and the Soviet Union test-exploded several bombs in the atmosphere after World War II. The large amounts of radioactive fission products that come to Earth from atmospheric explosions, called radioactive fallout, can be carried around the world by winds; they eventually settle out or are carried to the Earth by rain and snow. As you can imagine, radioactive fallout can be dangerous. The greatest source of radioactive fallout in the absence of atmospheric nuclear explosions is coal-fired plants, from which radioactive materials in the coal are vented to the atmosphere in combustion gases and particles.

As a result of public and scientific concern over the environmental effects of nuclear explosions, the three nuclear powers and many other countries signed the Nuclear Test Ban Treaty in 1963, by which they agreed to a moratorium on further tests in the atmosphere. Underground testing was still permitted. The most notable exceptions to the signing of the treaty were France and China. Both countries have since exploded bombs, now more politely called "nuclear devices," in the atmosphere and have joined the known nuclear weapons club.

Midtemperature Solar Systems Test Facility (MSSTF) at Sandia National Laboratories. (Courtesy Sandia Laboratories)

Chapter 32

Solar Energy Technology

The term *energy crisis* is common these days. It generally refers to the depletion of the natural fuel resources needed to meet our energy demands. We are becoming increasingly dependent on imported oil and natural gas, and the world output of these resources is expected to decline in the not-too-distant future. Nuclear resources are also limited, as we learned in the last chapter. Coal is our greatest domestic energy resource, with known reserves that should last for 300 to 400 years. However, pollution problems of varying degrees are associated with all these fuels, such as sulfur in coal, nuclear wastes, and combustion products.

There is one natural energy resource without pollution: solar energy. The abundant energy from the Sun is becoming increasingly important. Many practical applications now exist, and as solar energy technology develops, there will be many more.

32.1 Solar Radiation

The Sun's energy comes from the nuclear fusion, or "burning," of hydrogen deep within its core, where the temperature is on the order of millions of degrees. This energy is transferred outward, giving the Sun a surface temperature of about 6000 K.

The Sun radiates its energy into space as would any other hot body, with a total power output of about 3.8×10^{20} MW (megawatts). The majority of this radiation is in the visible and near infrared regions of the spectrum (Fig. 32.1 and Table 32.1).

The Earth intercepts only a small fraction of this radiation, about one two-billionth. Still, this is about 1.7×10^{11} MW, which is many thou-

Table 32.1
Spectral Distribution of Solar Radiation (at Top of Earth's Atmosphere)

Ultraviolet to gamma	7%
Visible	47%
Near infrared	39%
Infrared to radio	7%

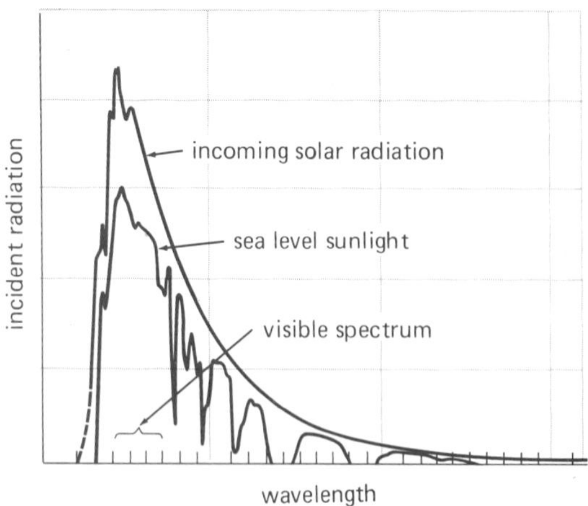

Figure 32.1
Solar intensity versus wavelength. Part of the solar radiation incident on the Earth's atmosphere is absorbed by atmospheric gases. The majority of the sunlight reaching the Earth's surface (sea level) is in the visible and near-infrared regions.

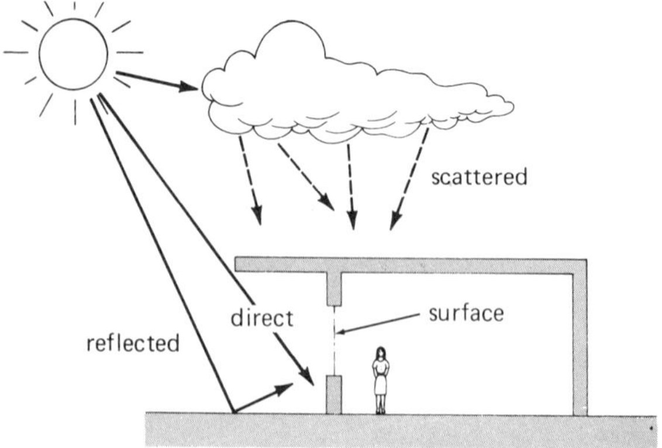

Figure 32.2
Methods by which solar radiation reaches collection surfaces. Radiation may strike the surface directly or be reflected or scattered to the surface.

sand times the electrical generation capacity of the United States. The average intensity of the solar radiation reaching the top of the Earth's atmosphere is about 1.35 kW/m^2 (429 Btu/h-ft^2). About one third of this *in*coming *sol*ar radi*ation*, called **insolation,** is reflected back into space.

As radiation passes through the atmosphere, certain wavelengths are absorbed and scattered by the atmospheric gases. Because of these processes, only about 50 percent of the insolation reaches the Earth's surface at sea level.

The insolation reaching the Earth heats it, and the Earth reradiates energy in the lower-frequency infrared region. A portion of this terrestrial (Earth) infrared radiation is absorbed by atmospheric gases, chiefly water vapor and carbon dioxide, which helps keep the Earth warm. This is the so-called **greenhouse effect.** See Special Feature 32.1.

In solar applications, **radiation reaches the collection surfaces in three ways: as direct radiation, as scattered (sky) radiation, and as reflected radiation** (Fig. 32.2). It must be kept in mind that solar radiation at the Earth's surface is *variable* and *intermittent*. For example, the intensity of the direct radiation received varies during the day and from season to season, depending on the angle of incidence (Fig. 32.3). The incident radiation may be intermittently interrupted by clouds. On an overcast day, the small amount of radiation received is almost totally scattered radiation. And, of course, there is nighttime, when no insolation is received by that portion of the Earth opposite the Sun.

Taking all these factors into account, the average solar energy received per area in the central United States is approximately 5.0 kWh/m^2 (1.6×10^3 Btu/ft^2) per day. That is, the average

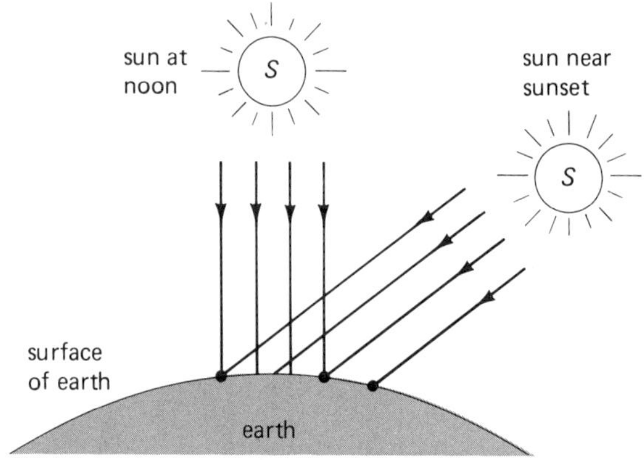

Figure 32.3
Intensity and angle of incidence. When the Sun is high in the sky, the solar radiation is distributed over a smaller area and the intensity (energy/area/time) is greater than when the Sun is low in the sky. When the angle of incidence decreases, e.g., at sunset, radiation is distributed over a larger area and the intensity is less. A similar seasonal effect occurs.

SPECIAL FEATURE 32.1

The Greenhouse Effect

If the Earth continually receives energy from the Sun, you may wonder why the Earth's average temperature does not increase. To maintain a relatively constant long-term average temperature, the Earth must lose energy, which on the average is equal to the amount it receives. This is accomplished through the reradiation of energy back into space. Atmospheric effects on this reradiation are important in preventing large daily temperature variations. On the moon, which has no atmosphere, the daily temperature ranges from about 100°C (212°F, the boiling point of water) on the day (Sun) side to −173°C (−280°F) on the dark side.

Incoming solar radiation warms the atmosphere and the surface of the Earth, and the warm Earth reradiates energy in the form of infrared radiation. The gases of the atmosphere, in particular water vapor and carbon dioxide (CO_2), are "selective absorbers." That is, they allow the visible incoming sunlight to pass through, but they absorb (trap) certain infrared radiations. This atmospheric absorption helps to retain the Earth's energy so that we don't have daily temperature fluctuations as on the moon. Clouds (water vapor) also assist in maintaining the Earth's warmth. In the absence of cloud coverage, nights are "*cold* and clear."

Hence, the atmospheric gases have a "thermostatic" effect in maintaining the Earth's daily temperature variations. We call this process the **greenhouse effect** because glass has absorption properties similar to those of the atmospheric gases. As used in a greenhouse (Fig. 1), glass allows the visible sunlight to pass through, then blocks or absorbs the infrared radiation.

Actually, in a greenhouse the warmth is primarily due to the prevention of the escape of warm air heated by the ground within the glass enclosure. The temperature of a greenhouse in the summer is controlled by painting the glass panels white, which reflects the sunlight, and opening panels to allow the hot air to escape. The interior of a closed greenhouse is quite warm, even on a cold day. You have probably experienced the greenhouse effect in an automobile on a cold, sunny day.

In recent years, concern has been expressed over the emission of CO_2 and other gases that may cause increased absorption and give rise to global warming.

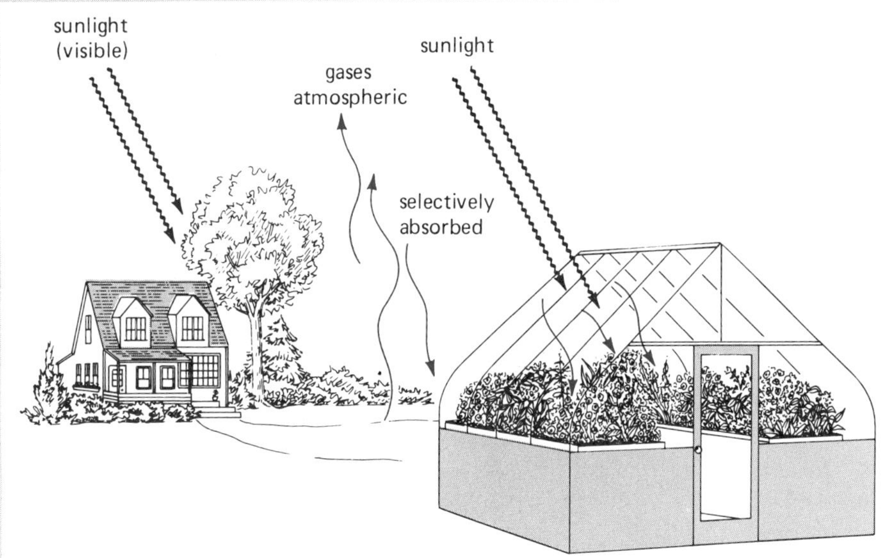

Figure 1
The greenhouse effect. The gases of the atmosphere, particularly water vapor and carbon dioxide, are selective absorbers of radiation with absorption properties similar to those of glass as used in a greenhouse. Visible light is transmitted, and infrared radiation is absorbed.

Table 32.2
Energy Density Conversion Factors

1 kWh/m^2 = 317 Btu/ft^2 = 86 cal/cm^2
1 Btu/ft^2 = 3.2×10^{-3} kWh/m^2 = 0.27 cal/cm^2
1 cal/cm^2 = 3.7 Btu/ft^2 = 0.012 kWh/m^2
(1 kWh = 3.4×10^3 Btu = 857 cal)

solar intensity (energy/area/time, or power/area) is

$$5.0 \text{ kWh/m}^2\text{/day} = 0.21 \text{ kW/m}^2$$

$$1.6 \times 10^3 \text{ Btu/ft}^2\text{-day} = 67 \text{ Btu/h-ft}^2$$

Helpful energy surface density conversion factors are given in Table 32.2.

EXAMPLE 32.1 A 100 ft × 50 ft building in the central United States has a flat roof. What is the average solar energy received by the roof in a month?

Solution: In one month (30 days) with an average solar intensity of 1.6×10^3 Btu/ft^2/day, the energy received per area is

$$\frac{E}{A} = (1.6 \times 10^3 \text{ Btu/ft}^2\text{/day})(30 \text{ days})$$
$$= 4.8 \times 10^4 \text{ Btu/ft}^2$$

Then, with ℓ = 100 ft and w = 50 ft,

$$A = \ell \times w = (100 \text{ ft})(50 \text{ ft}) = 5.0 \times 10^3 \text{ ft}^2$$

and

$$E_t = \left(\frac{E}{A}\right) A = (4.8 \times 10^4 \text{ Btu/ft}^2)(5.0 \times 10^3 \text{ ft}^2)$$
$$= 2.4 \times 10^8 \text{ Btu}$$

or

$$2.4 \times 10^8 \text{ Btu} \left(\frac{1 \text{ kWh}}{3.4 \times 10^3 \text{ Btu}}\right) = 7.0 \times 10^4 \text{ kWh} = 70{,}000 \text{ kWh}$$

Note: an average home without electric heat uses about 1200 to 1500 kWh of electrical energy per month.

32.2 Electricity from Solar Energy

Solar Cells

Solar cells allow the direct conversion of solar radiation to electricity. These cells are solid state junction diodes. Silicon solar cells are the most common. In effect, the principle of the solar cell is the reverse of that of the LED (light-emitting diode; see Chapter 26).

When sunlight strikes the surface of a solar cell, some of the photons are energetic enough to free electrons from the n-type semiconductor (Fig. 32.4). This creates electron-hole pairs. Some of the electron-hole pairs immediately recombine. But if the electron-hole pairs are near the p-n junction, the junction potential barrier (due to diffusion) causes the "charge" pairs to separate. The negative electrons move to the n-type side of the cell and the "positive" holes move to the p side.

The separation of the charges gives rise to a "terminal" voltage. When connected to an external circuit, a current flows as long as sunlight illuminates the cell. Potentials of about 0.5 V are developed by solar cells. Hence, for practical applications solar cells are connected in series arrays. Such arrays are used to supply electricity for satellites, and this use was a major reason for the research and development of solar cells (Fig. 32.5).

The n-type layer of a solar cell is very thin, so light penetrates into the crystal to create electron-hole pairs in the vicinity of the junction. The current flow in the circuit depends on the intensity of the incident light, as in the case of the photocell. The wavelength (or frequency) of the photons is critical.

EXAMPLE 32.2 A minimum energy of about 1.7×10^{-19} J is needed to free an electron in a silicon solar cell. What is the minimum frequency of a photon that will produce an electron-hole pair?

Solution: It is given that $E = 1.7 \times 10^{-19}$ J. Then $E = hf$, and

$$f = \frac{E}{h} = \frac{1.7 \times 10^{-19} \text{ J}}{6.63 \times 10^{-34} \text{ J-s}} = 2.6 \times 10^{14} \text{ Hz}$$

This corresponds to a wavelength of ($\lambda f = c$)

$$\lambda = \frac{c}{f} = \frac{3.0 \times 10^{10} \text{ cm/s}}{2.6 \times 10^{14} \text{ Hz}} = 1.2 \times 10^{-4} \text{ cm}$$

Notice that this is in the infrared region. Hence, infrared radiation with longer wavelengths (longer wavelengths, lower frequencies) will not free electrons.

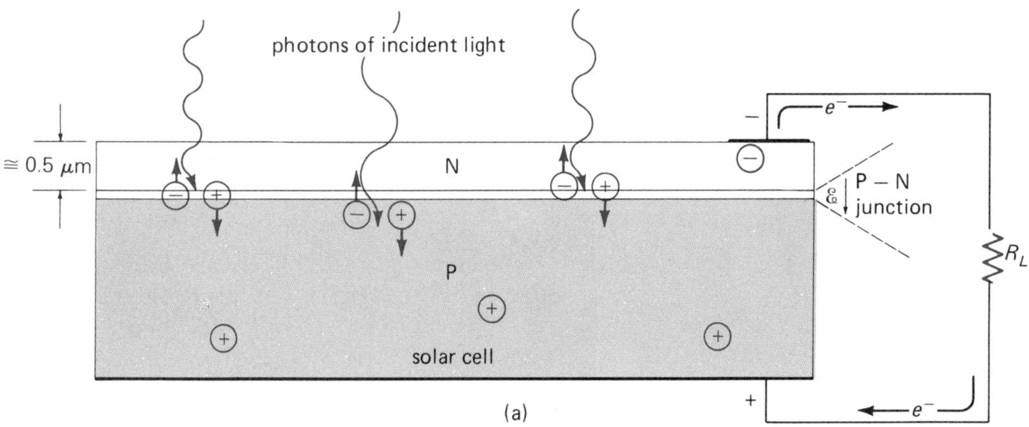

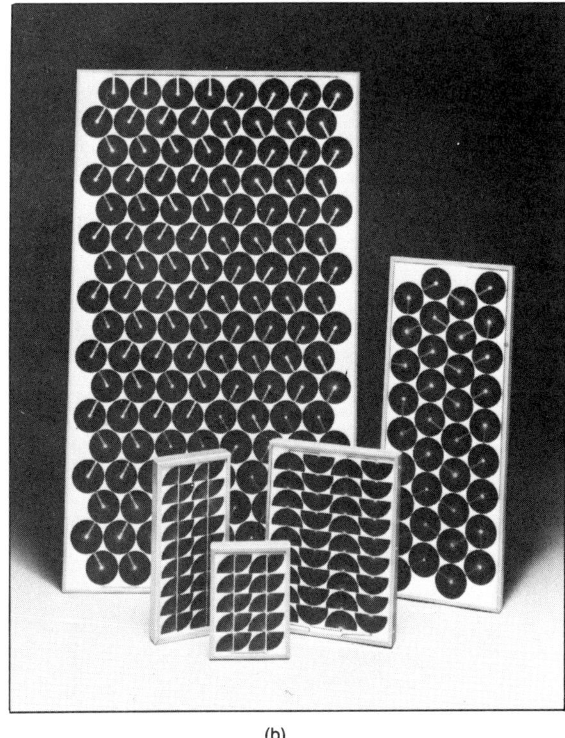

Figure 32.4
The solar cell. (a) When light strikes the surface of a solar cell, photons free electrons and create electron-hole pairs, some of which separate and give rise to a voltage and current in the circuit. (b) A variety of solar cell modules. (Courtesy Applied Solar Energy Corporation.)

There is a limiting efficiency of a solar cell. As shown in the previous example, radiation with wavelengths longer than 1.2×10^{-4} cm will not cause silicon solar cell conduction. Hence, about one fourth of the solar radiation will not activate solar cells (see Fig. 32.1).

Some of the solar radiation incident on the cell is reflected. Solar cell efficiencies of about 18 percent have been achieved using relatively pure silicon crystals, which minimize electron-hole recombination in the crystal. In general, commercially available silicon solar panels rarely exceed an efficiency of 12 percent.

However, efficiencies greater than 25 percent have been reported with the use of a special technique. The high efficiency was achieved by concentrating sunlight on highly efficient cells with lenses. A large area of lenses and a relatively small area of solar cells reduce the need for a large area of expensive cells.

Another technique being researched is the cascade cell. In this approach, two cells are stacked on top of each other with each cell having a different photon energy range, thereby increasing the overall efficiency of the cell combination.

(a)

(b)

Figure 32.5
Satellite solar panels. (a) Solar panels on the Skylab space station photographed against a black sky background. The picture was taken from the command module on a "fly-around" inspection prior to docking. (Courtesy NASA.) (b) Artist's conception of the Power Extension Package (PEP), a large solar-cell array designed to provide in-orbit electric power for the Space Shuttle. Carried inside the Shuttle, the PEP would be unfolded and placed in space when orbit is achieved. (Courtesy Applied Solar Energy Corp.)

A 31 percent sunlight-to-electricity conversion efficiency using a two-layer cell has been reported. The cell has multiple photosensitive layers, each optimized for different wavelengths of light. The upper layer (gallium-arsenide) is sensitive to wavelengths from the ultraviolet through the visible portions of the spectrum, and converts about 27 percent of the light striking it. The unabsorbed light passes through an underlying layer of silicon, which is more sensitive to light, into the near infrared. It is hoped that other combinations will produce even higher efficiencies.

EXAMPLE 32.3 Assuming a 15 percent efficiency of solar cells, what would be the required area of a solar cell array to produce 50 kWh of energy for the daily electrical requirements of a home? (Take the average solar energy density received per day to be 5.0 kWh/m^2.)

Solution: With 15 percent efficiency, the amount of solar energy converted to electrical energy per area in one day is

$$\frac{E_o}{A} = \frac{E}{A}(\text{efficiency}) = (5.0 \text{ kWh/m}^2)(0.15)$$
$$= 0.75 \text{ kWh/m}^2$$

where E_o is the electrical energy output. Then, with $E_o = 50$ kWh,

$$A = \frac{E_o}{0.75 \text{ kWh/m}^2} = \frac{50 \text{ kWh}}{0.75 \text{ kWh/m}^2} = 67 \text{ m}^2$$

which is 67 m^2 (10.8 ft^2/m^2) = 724 ft^2.

As this example shows, solar cells are generally limited to small-scale applications owing to both size and economy. Solar cells are now commonly used to power calculators. Other applications include electrical supplies for devices in remote areas where electricity is not

Figure 32.6
Electricity production. An electrical generating system of arrays of high-efficiency solar collection cells such as this will provide electricity for remote villages in tropical countries. Focusing lenses concentrate the solar radiation, providing 33 times the amount of energy that would normally strike a cell without a lens. Each array produces 10 kW of electrical power. (Courtesy Applied Solar Energy Corp.)

Figure 32.7
Distributed collector system. In this test facility, a parabolic trough collector collects and concentrates the solar radiation on a fluid transfer system. (Courtesy Sandia Laboratories.)

ordinarily available. In some cases, large arrays are practical (Fig. 32.6).

Since the amount of current produced by a solar cell depends on the intensity of light, one possibility, as noted previously, is the use of lenses or mirrors to collect and focus sunlight on a solar cell array. This would reduce the number of solar cells needed for a given output. However, a problem arises since the efficiency of solar cells decreases with increasing temperature. By circulating water to cool the cells and remove the heat energy, which could be used to provide heat and hot water to a home, the efficiency can be increased.

Thermal-Electric Conversion

Solar thermal-electric conversion systems collect solar radiation, converting it first to thermal energy and then to electrical energy. The latter part of the operation is carried out by conventional steam-turbine generators. Hence, such systems must collect enough solar energy to heat a working fluid to a temperature of several hundred degrees ($\simeq$400°C $\simeq$ 750°F).

There are two types of thermal-electric conversion systems under consideration and on which research is being done: distributed collector systems and central receiver systems. In the **distributed collector system,** a large number of solar collectors collect and concentrate the solar energy (Fig. 32.7). The mirrors may be turned toward the Sun by a computer-driven control system. The thermal energy is then transported by a fluid through pipes to an electrical generating facility. Parabolic solar reflectors or converging lenses are used to focus the solar radiation on pipes carrying molten sodium or potassium salts or fluid synthetic oils, which act as high-temperature heat-transfer agents.

Such distribution systems can be used advantageously in desert areas, such as the southwestern United States, which receive large amounts of solar energy. Because of the large land areas needed for practical electrical production, such facilities are called solar farms. There are currently several solar electrical generating farms in the Mojave Desert in California. A typical plant covers 380 acres and produces 80 MW of electricity. These plants supply about 2 percent of southern California's electrical needs, but that amount is more than 90 percent of the world's utility-related solar electricity.

The cost of solar-produced electricity now runs 8 to 9 cents per kilowatt-hour. This is cheaper than electricity from new nuclear plants. In the next few years, it is planned that solar energy will be used to supply electricity to more than 800,000 people and businesses in sunny California.

In a **central receiver system,** a large number of tracking mirrors, called heliostats, concentrate the solar energy on a single receiver atop a tall "solar energy" tower (Fig. 32.8). Here again, a large mirror array or many small instal-

(a)

(b)

Figure 32.8
Solar towers. (a) A French solar tower near Odeillo in southern France, which was built for testing materials under extremely high temperatures. An array of 63 mirrors (heliostats), each measuring 6.0 × 7.5 m, reflects sunlight, which is focused on an aperture in the tower at the center. Temperatures of 7000°F can be produced—enough to melt any known metal. (Courtesy DOE.) (b) A solar tower at Sandia Laboratories, NM, concentrates sunlight from 1775 mirror facets on a steel target mounted on the "power tower," causing molten steel to drip from the target. (Courtesy Sandia Laboratories.)

lations are needed to produce commercial amounts of electric power. Then the electric power must be transmitted over long distances to populated consumer areas. Material costs and efficiency may be limiting factors in solar thermal-electric conversion systems in their competition with conventional fossil-fuel electrical generation.

32.3 Solar Heating and Cooling

Solar energy has been used for some years to provide space heating in buildings and to heat water for homes, particularly in Florida and the southwestern United States. As energy costs rise, more attention is being given not only to solar heating but also to solar cooling. Two basic types of systems are used: passive and active.

Passive Solar Systems

In passive systems, the building itself is designed as a solar collector. The heat energy is then distributed by natural means (conduction, convection, or radiation), without the use of any mechanical or electrical equipment.

There are many passive home designs. In some instances, there are large windows facing south. (No windows face north. Why?) The southern-exposure windows generally have overhangs to protect them from solar radiation in the summer (Fig. 32.9).

In other instances, thermal storage walls are built inside the south-facing windows (Fig. 32.10). The wall blocks and stores the solar energy transmitted through the glass windows. The wall is painted a dark color to improve absorption. Another method uses containers of water as a wall to improve energy storage capacity. Recall from Chapter 16 that water has a relatively large specific heat.

Vents placed at the top and bottom of the wall allow the convectional circulation of air through the house. This circulation can be used for either heat distribution or cooling. Another solar home design uses a rooftop pond system. A shallow pond of water on the roof

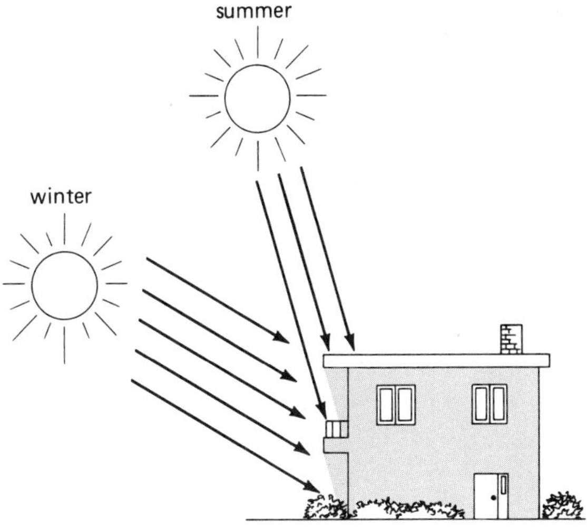

Figure 32.9
Passive solar heating. In a typical passive solar heating design, there are large windows facing south. These southern-exposure windows generally have overhangs to protect them from summer solar radiation.

provides thermal storage for thermal heating or cooling.

Natural convection cycles of fluids are an important aspect of passive systems. Convection loops are used in solar water heaters. In

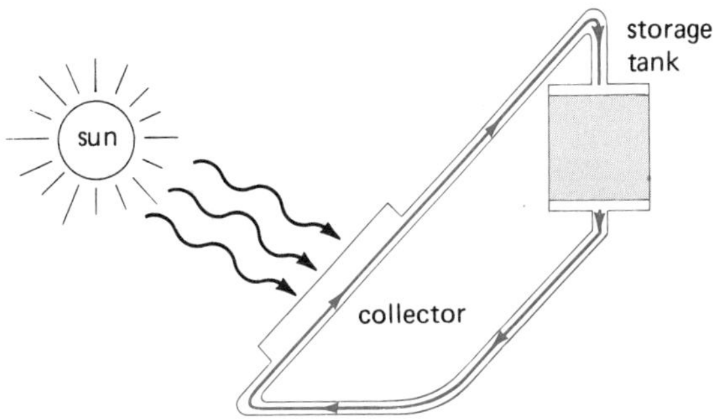

Figure 32.11
Thermosiphoning hot water system. Warm water heated in the collector rises to the storage tank through the convection loop. The warm water stored in the tank can be used for space heating or in the home hot water system.

so-called thermosiphoning hot water systems, warm water rises through a collector to a storage tank (Fig. 32.11). The warm water in the tank may be used for space heating or in the home hot-water system. The water is heated electrically if the passive system does not achieve the required water temperature. Even so, the solar preheating saves conventional energy.

The convectional thermosiphoning action is also used in space heaters. Heated air is circu-

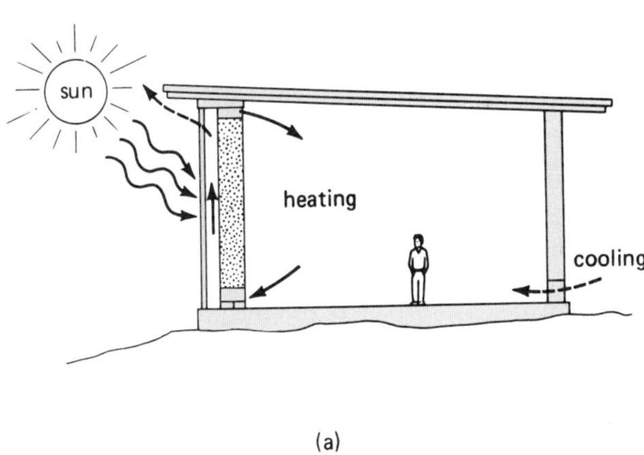

(a)

(b)

Figure 32.10
Thermal storage. (a) A thermal storage wall made of blocks or containing water inside the southern-facing window can be used to store solar energy for heating and cooling. The solid arrows in the room show the air flow for heating and the dashed arrows show the air flow for cooling. (b) A solar-heated home with a water-filled storage wall. The array at the right is a solar water heater. The windmill in the background pumps water.

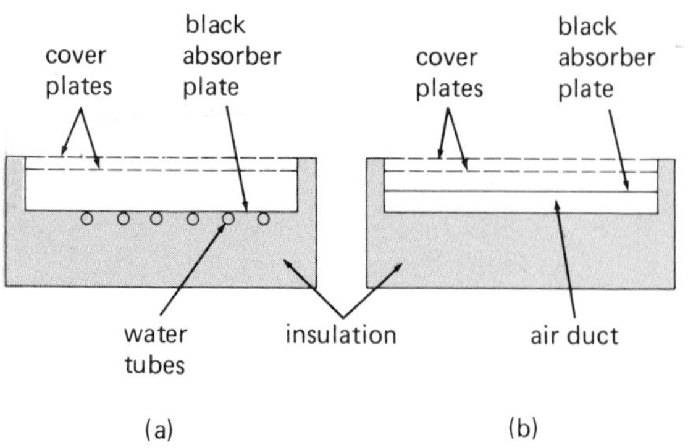

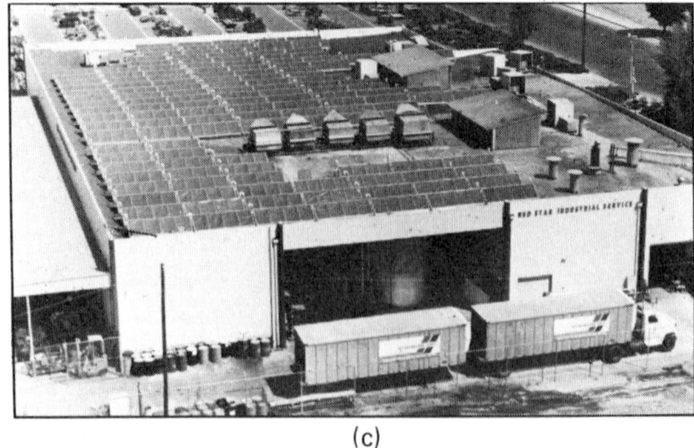
(c)

Figure 32.12
Flat plate collectors. (a) Liquid transfer system. Water flowing through the tubes in contact with the absorber plate transfers heat energy to a storage reservoir. (b) Air transfer system. Air can also be used as a transfer agent. (c) A laundry in California uses 6500 ft^2 of solar energy collectors for one of the nation's largest applications of solar energy for industrial water heating.
(Courtesy DOE, photo by Aratex.)

lated through a bin of rocks for storage. The energy stored in the rock bin can be used for space heating of the home during the night.

Active Solar Systems

Active systems make use of special collectors, storage tanks, pumps, and fans to collect and distribute solar energy. There are two general types of solar collectors. A flat-plate collector has a collecting area approximately as large as the overall collector, whereas a concentrating collector uses mirrors or lenses to focus the solar radiation onto a smaller absorber area. (See the discussion on solar towers in the previous section.)

Typical **flat-plate collectors** are shown in Figure 32.12. The sunlight passes through a transparent cover plate (or glazing) made of glass or some plastic material. The cover plates transmit radiations in the visible and near infrared regions but are relatively opaque to far infrared radiation. The opacity of the plates to far infrared radiation prevents reradiation heat loss. The plates also prevent convectional heat losses due to wind. Antireflecting coatings can reduce reflection losses. The absorber plate is usually made of aluminum that has been painted black. Both water and air are used as heat-transfer agents.

As in passive home systems, there are many designs for active home systems. Figure 32.13 illustrates a typical layout of an active home heating system. The collected energy is transferred to a storage tank. Storage tanks may contain water, rocks (in the case of air heat transfer), or some chemical salt. The type of storage material depends on the particular application and economy. Water is inexpensive, but the water pipes could freeze and burst on a cold night in some climates. Rocks are also cheap and eliminate the problem of frozen pipes. Chemical salts are more expensive but generally have greater heat storage capabilities.

An auxiliary heat supply, usually electrical, is built into the system. This is needed for cold weather when the solar system cannot supply sufficient heat for the home. A series of cloudy or rainy days and the lack of sunlight may reduce the stored heat of the system to a point where it must be supplemented by the auxiliary supply.

Cooling with (solar) heat may sound rather odd. Such cooling (i.e., air conditioning) requires heat to be removed from a low-temperature reservoir (inside the building) and rejected to a high-temperature reservoir (outside the building). Recall from Chapter 18 that this is also the function of a heat pump, or refrigerator, which requires work input. In the case of

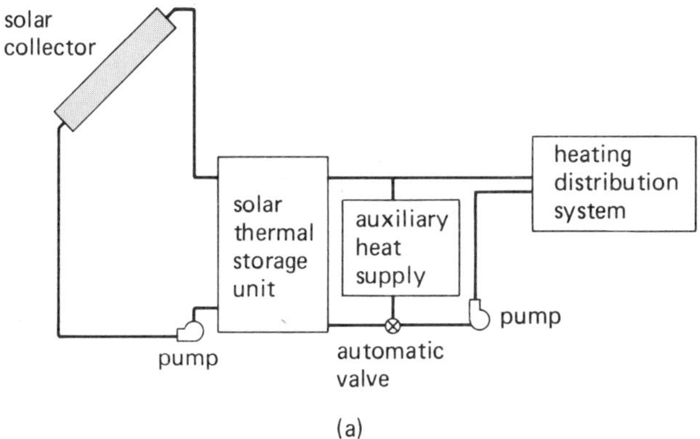

(a)

(b)

Figure 32.13
Active home-heating system. (a) A schematic diagram of a system. Solar energy from the collector is stored in a thermal storage unit until needed and is then distributed. An auxiliary heat supply supplements the heating on cold and cloudy days. (b) The collectors of a solar-heated house.

solar cooling systems, the solar heat energy is used to supply the work to the refrigeration system.

Several cycles or systems can be used in solar cooling. We will look at one of these, absorption cooling, which is described for a natural-gas refrigerator with an ammonia-water absorber in Special Feature 32.2.

In solar air-conditioning systems, a lithium bromide–water solution is used as an absorber because the generator temperature that is required to separate the lithium bromide from the water is less than that for ammonia-water solutions, and low refrigerator temperatures are not necessary. The low volatility of lithium bromide makes the rectification step unnecessary. Such units are now in operation.

In either ammonia or lithium bromide systems, conventional heat sources can be used to supplement or take over the generator heating when the solar heating is inadequate.

The important point to remember in all cases of solar heating and cooling is that, while solar energy may not be a total substitute, it can supplement our other energy supplies. This is economical and conserves our natural resources.

SPECIAL FEATURE 32.2

Cooling from Heat: The Absorption Refrigerator

Solar cooling systems may operate on several different cycles. One of these is the absorption cycle, which is the principle of the natural-gas refrigerator that was once quite common. This refrigerator had a cooling compartment and a freezer at the top, yet if you looked at the bottom you could see a gas flame—a case of cooling from heat. The point to keep in mind is that heat energy, like all forms of energy, is capable of doing work—the essential ingredient in pumping heat out (Chapter 18). The trick is how to do it.

The basic absorption refrigeration cycle illustrated in Figure 1 uses ammonia as the refrigerant and an ammonia-water solution as the absorber. To see how it works, let's start at the freezer (1) in the figure. Here, liquid ammonia evaporates (boils). The evaporation takes place in an atmosphere of hydrogen to speed up the process. The resulting mixture is heavier than either of the two gases alone.

The heavy ammonia-hydrogen mixture then sinks to the absorber (2). Here, water absorbs the ammonia out of the gas mixture and the light hydrogen gas is free to return to the freezer. The heat absorbed in the freezer is liberated in the absorption process.

The ammonia-water solution then drains to the generator (3). Heat from a gas flame (or solar heat source) causes the liquid to boil up through a tube, much as in a percolator coffeepot. Ammonia boils at a lower temperature (−2°C) than water, so the bubbles lifting the liquid to the rectifier (4) are practically pure ammonia.

From the rectifier, the water drains back to the absorber, and the light ammonia gas rises to the condenser (5). Here the ammonia gas is cooled and condensed back to liquid ammonia, which drains back to the freezer (1), and the cycle begins again.

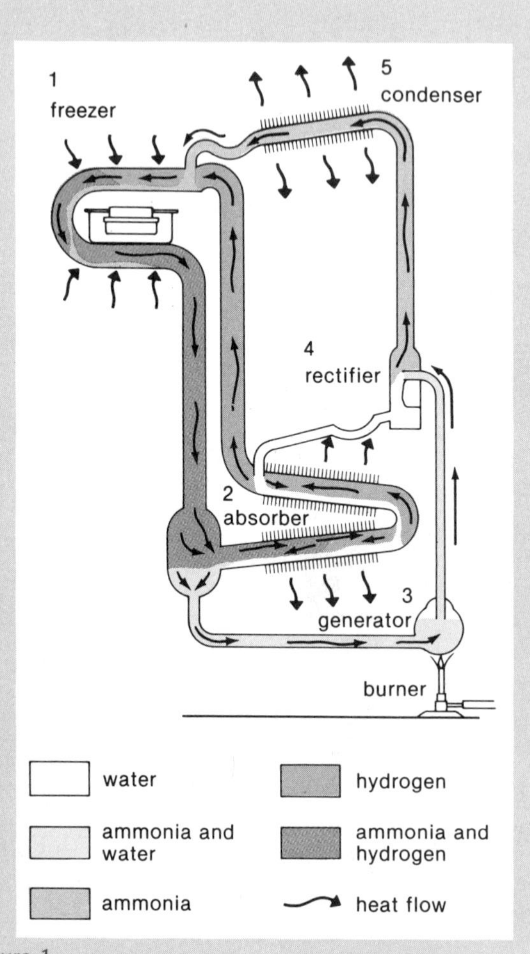

Figure 1
Diagram of an absorption refrigeration system. See text for description. (Courtesy Frigidaire Co., Division of White Consolidated Industries.)

Important Terms

insolation incoming solar radiation

greenhouse effect the selective absorption of terrestrial (infrared) radiation by gases in the atmosphere, chiefly carbon dioxide and water vapor

solar cells solid state diodes that convert light into electrical energy

thermal-electric conversion conversion of solar radiation first to thermal energy and then to electrical energy

distributed collector system a system in which a large number of collectors individually collect the solar energy, which is concentrated by a heat-transfer agent

central receiver system a system in which a large number of collectors (tracking mirrors) concentrate the solar energy on a single collector (tower)

passive system a system in which the building itself is designed as a collector and heat is distributed by natural means

active system a system in which special collectors, storage tanks, and mechanical and electrical equipment (pumps and fans) are used to transport and distribute the heat energy

Important Formulas

average intensity of solar radiation at the top of the atmosphere: $1.35\ \text{kW/m}^2 = 429\ \text{Btu/h-ft}^2$

average solar intensity in (central) U.S. considering all factors: $5.0\ \text{kWh/m}^2\text{/day} = 0.21\ \text{kW/m}^2$
$1.6 \times 10^3\ \text{Btu/ft}^2\text{/day} = 67\ \text{Btu/h-ft}^2$

intensity: $I = \dfrac{(E/A)}{t} = \dfrac{E}{At} = \dfrac{P}{A}$

Questions

Solar Radiation

1. What is the United States' greatest energy resource?
2. What is the source of the Sun's energy?
3. The Earth intercepts what fraction of the Sun's radiation, and what is the spectral distribution of this radiation?
4. Why does only about 50 percent of the insolation reach the Earth's surface?
5. The albedo is the fraction of insolation a body reflects back into space. The Earth's albedo is 0.33 (or 33 percent), as compared to the moon's albedo of 0.07 (or 7 percent). (a) How much insolation is reflected back into space by the Earth? (b) Why does the moon have such a low albedo?
6. In 1815 the volcano Tamboro, located in the East Indies, had the most violent volcanic eruption in recorded history. It blew as much as 35 cubic miles of particulate debris into the atmosphere. The next year was uncommonly cold. Why? (Do you think that a similar situation was predicted when Mount St. Helens erupted in 1980?)
7. Explain why the inside of a car can be quite warm on a cold, sunny day.
8. Is it correct to say that energy from fossil fuels is really solar energy? Explain.

Electricity from Solar Energy

9. In solar applications, is all of the radiation that is received direct sunlight? Explain.
10. Why must solar heating systems generally be supplemented by conventional heating systems?
11. Explain the principle of a solar cell. Why is this the reverse of an LED?
12. Does the operation of a solar cell depend on the frequency of light? Explain.
13. How efficient are solar cells, and is there a limiting efficiency?
14. Does the efficiency of solar cells depend on temperature? Explain.
15. Distinguish solar cell systems from thermal-electric systems.
16. Describe the principles of distributed collector systems and central receiver systems.
17. Solar collectors using fluid ducts are covered with glass or plastic to make use of the greenhouse effect. Explain how this improves the efficiency of a solar collector.

Solar Heating and Cooling

18. Distinguish between passive and active solar systems.
19. Why do southern-exposure windows in passive home designs generally have overhangs?
20. Explain the purpose of, and materials that might be used in, thermal storage walls.
21. What is thermosiphoning, and how is it applied?
22. Explain the operation of flat-plate collectors.
23. Why are the covers of flat-plate collectors purposely selected to be opaque to the far infrared radiation?
24. List some materials used in heat storage units and the advantage of each. Is the specific heat capacity a consideration?
25. How is solar heat energy used in solar cooling systems?
26. Describe the functions of the (a) absorber, (b) generator, (c) rectifier, and (d) condenser in an absorption refrigeration system.
27. What is the purpose of hydrogen in an ammonia-water absorption system?
28. Is heat added to and rejected from an ammonia-water absorption system only one time during a cycle? Explain.
29. Is ammonia water used in solar air-conditioning systems? Explain.

Problems

Levels of difficulty are indicated by asterisks for your convenience.

32.1 Solar Radiation

1. What is the temperature of the Sun's surface in degrees (a) Celsius and (b) Fahrenheit?

*2. The area of the state of Kansas is about 82,000 mi^2. Approximately how many joules of solar energy, on the average, does Kansas receive each day?

*3. A collector plate array in the central United States has an area of 20 m^2. How much solar energy, in kilocalories, is collected on an average day?

*4. An industrial plant in the Southwest has a flat room measuring 300 ft by 200 ft. If the average daily energy surface density in that area is 1800 Btu/ft^2, how much solar energy is received by the roof in a week?

*5. A business in the Midwest uses 1000 kWh of electrical energy a day. What area, in (a) square meters and (b) square feet, would receive an equivalent amount of solar energy on an average day? (c) What would be the lengths of the sides of a square collector in this case?

*6. The area of a sunbather's back is 3000 cm^2. How much energy, in calories, would the bather receive on the back by remaining in the sunlight for 30 minutes? (Assume 50 percent of the insolation reaches the Earth's surface.)

*7. Using the average intensity at the top of the atmosphere, how much solar energy reaches the Earth's surface in one day? (Consider the sky to be clear.)

**8. (a) How much energy does the Sun radiate in one day? (b) Using Einstein's mass-energy equation (Chapter 31), compute how much of the Sun's mass is converted to energy each day.

**9. Using the Sun's output, show explicitly that the Earth intercepts about one two-billionth of this amount. (*Hint:* consider the energy density on a sphere at the Earth's distance from the Sun.)

32.2 Electricity from Solar Energy

10. A minimum energy of 2.0×10^{-19} J is needed to free an electron in a solar cell. What wavelengths of light will not activate the cell?

11. The maximum wavelength that will activate a solar cell is 110 nm. What is the minimum activation energy of the cell in electron volts?

*12. A solar cell array has an area of 300 cm^2. If the cell efficiency is 12 percent, what is the electrical output on an average day in the central United States?

*13. A single solar cell with an area of 4.0 cm^2 has an electrical energy output of 2.8×10^{-4} kWh. If the solar energy density is 4.0 kWh/m^2, what is the efficiency of the cell?

*14. Solar cells with an efficiency of 15 percent are used in an array that is to have a daily output of 1.0 kWh. What is the required area of the array if the solar energy density per day is 6.5 kWh/m^2?

32.3 Solar Heating and Cooling

*15. Design what you think would be an ideal solar home, and point out some of its superior features.

*16. Design a solar-heated classroom building for your campus. Consider both passive and active systems.

Appendix 1: Conversion Factors

Mass

	g	kg	slug
1 gram = (g)	1	10^{-3}	6.852×10^{-5}
1 kilogram = (kg)	1000	1	6.852×10^{-2}
1 slug =	1.459×10^{4}	14.59	1

1 (avoirdupois) pound = 16 (avoirdupois) ounces = 7000 grains = 454 grams
1 (troy or apothecaries') pound = 12 (troy) oz = 0.8229 (avd) lb
1 hundredweight (cwt) = 112 lb
1 short ton = 2000 lb
1 long ton = 2240 lb = 20 cwt
1 metric ton (tonne) = 1000 kg = 2205 lb
1 kg = 2.2 lb
1 u = 1.66×10^{-27} kg = 931 MeV

Length

	cm	m	km	in.	ft	mi
1 centimeter = (cm)	1	10^{-2}	10^{-5}	0.3937	3.281×10^{-2}	6.214×10^{-6}
1 meter = (m)	100	1	10^{-3}	39.37	3.281	6.214×10^{-4}
1 kilometer = (km)	10^{5}	1000	1	3.937×10^{4}	3281	0.6214
1 inch = (in.)	2.54	0.0254	2.54×10^{-5}	1	0.0833	1.58×10^{-5}
1 foot = (ft)	30.48	0.3048	3.048×10^{-4}	12	1	1.89×10^{-4}
1 statute mile = (mi)	1.609×10^{5}	1609	1.609	6.336×10^{4}	5280	1

1 nautical mile = 6076 ft = 1.151 (statute) mi = 1852 m
1 angstrom = 10^{-10} m = 10^{-8} cm
1 yard = 3 ft
1 mil = 10^{-3} in.
1 rod = 16.5 ft
1 fathom = 6 ft

Area

	m^2	cm^2	ft^2	in^2
1 sq. meter = (m^2)	1	10^4	10.76	1550
1 sq. centimeter = (cm^2)	10^{-4}	1	1.076×10^{-3}	0.1550
1 sq. foot = (ft^2)	9.29×10^{-2}	929	1	144
1 sq. inch = (in^2)	6.45×10^{-4}	6.45	6.944×10^{-3}	1

1 sq. mi (mi^2) = 27,878,400 ft^2 = 640 acres
1 acre = 43,560 ft^2
1 hectare = 1 sq. hectometer $(100\ m)^2 = 10^4\ m^2$ = 2.47 acres

Volume

	m^3	cm^3	ft^3	in^3
1 cubic meter = (m^3)	1	10^6	35.31	6.102×10^4
1 cubic cm = (cm^3)	10^{-6}	1	3.531×10^{-5}	6.102×10^{-2}
1 liter = (L)	10^{-3}	10^3	3.531×10^{-2}	61.02
1 cubic foot = (ft^3)	2.832×10^{-2}	2.832×10^4	1	1728
1 cubic inch = (in^3)	1.639×10^{-5}	16.39	5.787×10^{-4}	1

1 U.S. fluid gallon = 4 quarts = 8 pints = 128 fluid ounces = 231 in^3 = 0.134 ft^3
1 British Imperial gallon = 1.2 U.S. gallons = 277.42 in^3

Time

	s	min	h	d	y
1 second = (s)	1	1.667×10^{-2}	2.778×10^{-4}	1.157×10^{-5}	3.169×10^{-8}
1 minute = (min)	60	1	1.667×10^{-2}	6.944×10^{-4}	1.901×10^{-6}
1 hour = (h)	3600	60	1	4.167×10^{-2}	1.141×10^{-4}
1 day = (d)	86,400	1440	24	1	2.738×10^{-3}
1 year = (y)	3.156×10^7	5.259×10^5	8.766×10^3	365.2	1

Density

	g/cm³	kg/m³	slug/ft³	lb/in³	lb/ft³
1 gram per cm³ = (g/cm³)	1	1000	1.940	3.613×10^{-2}	62.43
1 kg per m³ = (kg/m³)	0.001	1	1.940×10^{-3}	3.613×10^{-5}	6.243×10^{-2}
1 slug per ft³ = (slug/ft³)	0.5154	515.4	1	1.862×10^{-2}	32.17
1 pound per in³ = (lb/in³)	27.68	2.768×10^{4}	53.71	1	1728
1 pound per ft³ = (lb/ft³)	1.602×10^{-2}	16.02	3.108×10^{-2}	5.787×10^{-4}	1

Speed

	m/s	km/h	ft/s	mi/h
1 meter per second = (m/s)	1	3.6	3.281	2.237
1 km per hour = (km/h)	0.2778	1	0.9113	0.6214
1 foot per second = (ft/s)	0.3048	1.097	1	0.6818
1 mile per hour = (mi/h)	0.4470	1.609	1.467	1

1 knot = 1 nautical mile/hour

Force

	lb	N	dyne
1 pound = (lb)	1	4.448	4.448×10^{5}
1 newton = (N)	0.2248	1	10^{5}
1 dyne = (dyn)	2.248×10^{-6}	10^{-5}	1

Pressure

	lb/in² (psi)	N/m² (Pa)	atm	mm Hg (torr)	in. Hg
1 pound per in² = (lb/in² or psi)	1	6.895×10^{3}	6.805×10^{-2}	51.71	2.0358
1 newton per m² or pascal = (N/m² or Pa)	1.450×10^{-4}	1	9.869×10^{-6}	7.501×10^{-3}	2.953×10^{-3}
1 atmosphere = (atm)	14.7	1.013×10^{5}	1	760	29.92
1 mm of mercury or torr = (mm Hg or T)	0.01934	1333	1.316×10^{-3}	1	0.03937
1 in. of mercury = (in. Hg)	0.4912	338.64	0.03342	25.4	1

Energy

	J	ft-lb	cal	Btu	kWh
1 joule = (J)	1	0.7376	0.2390	4.481×10^{-4}	2.778×10^{-7}
1 foot-pound = (ft-lb)	1.356	1	0.3240	1.285×10^{-3}	3.776×10^{-7}
1 calorie = (cal)	4.186	3.086	1	3.968×10^{-3}	1.163×10^{-6}
1 British thermal unit = (Btu)	1055	778	252	1	2.930×10^{-4}
1 kilowatt-hour = (kWh)	3.6×10^{6}	2.655×10^{6}	8.60×10^{5}	3413	1

Power

	W	ft-lb/s	hp	Btu/h
1 watt = (W)	1	0.7376	1.341×10^{-3}	3.413
1 ft-lb per second = (ft-lb/s)	1.356	1	1.818×10^{-3}	4.628
1 horsepower = (hp)	745.7	550	1	2545
1 Btu per hour = (Btu/h)	0.2930	0.2161	3.929×10^{-4}	1

1 ton of refrigeration = 12,000 Btu/h

Appendix 2: Physical Constants

Name	Symbol	Value
Acceleration due to gravity	g	9.8 m/s^2; 32 ft/s^2
Universal gravitational constant	G	6.67×10^{-11} $N\text{-}m^2/kg^2$
Standard atmospheric pressure	p_{atm}	1.013×10^5 N/m^2, 14.7 lb/in^2; 30 in. Hg, 760 mm Hg (torr)
Boltzmann's constant	k	1.38×10^{-23} J/K
Mechanical equivalent of heat	—	4.186 J/cal; 778 ft-lb/Btu
Speed of light	c	3.00×10^8 m/s^2
Electronic charge	e	1.60×10^{-19} C
Coulomb's law constant	k	9×10^9 $N\text{-}m^2/C^2$
Permittivity of free space	ε_o	8.85×10^{-12} $C^2/N\text{-}m^2$
Permeability of free space	μ_o	$4\pi \times 10^{-7}$ Wb/A-m
Planck's constant	h	6.63×10^{-34} J-s
Electron mass	m_e	9.11×10^{-31} kg
Proton mass	m_p	1.673×10^{-27} kg
Neutron mass	m_n	1.675×10^{-27} kg
Astronomical data:		
Mass of Earth		6.0×10^{24} kg
Mass of moon		7.4×10^{22} kg $= \frac{1}{81}$ mass of Earth
Mass of Sun		2.0×10^{30} kg
Approximate radius of Earth (average)		4000 mi; 6440 km (3957 mi); (6368 km)
Radius of moon		1.74×10^3 km; 1081 mi
Average distance of Earth from Sun		1.5×10^8 km; 93×10^6 mi
Average distance from moon to Earth		3.8×10^5 km; 2.4×10^5 mi

Appendix 3: Trigonometric Tables and Formulas

Angle					Angle				
Degree	**Radian**	**Sine**	**Cosine**	**Tan-gent**	**Degree**	**Radian**	**Sine**	**Cosine**	**Tan-gent**
0°	.000	0.000	1.000	0.000					
1°	.017	.018	1.000	.018	31°	.541	.515	.857	.601
2°	.035	.035	0.999	.035	32°	.559	.530	.848	.625
3°	.052	.052	.999	.052	33°	.576	.545	.839	.649
4°	.070	.070	.998	.070	34°	.593	.559	.829	.675
5°	.087	.087	.996	.088	35°	.611	.574	.819	.700
6°	.105	.105	.995	.105	36°	.628	.588	.809	.727
7°	.122	.122	.993	.123	37°	.646	.602	.799	.754
8°	.140	.139	.990	.141	38°	.663	.616	.788	.781
9°	.157	.156	.988	.158	39°	.681	.629	.777	.810
10°	.175	.174	.985	.176	40°	.698	.643	.766	.839
11°	.192	.191	.982	.194	41°	.716	.658	.755	.869
12°	.209	.208	.978	.213	42°	.733	.669	.743	.900
13°	.227	.225	.974	.231	43°	.751	.682	.731	.933
14°	.244	.242	.970	.249	44°	.768	.695	.719	.966
15°	.262	.259	.966	.268	45°	.785	.707	.707	1.000
16°	.279	.276	.961	.287	46°	0.803	0.719	0.695	1.036
17°	.297	.292	.956	.306	47°	.820	.731	.682	1.072
18°	.314	.309	.951	.325	48°	.838	.743	.669	1.111
19°	.332	.326	.946	.344	49°	.855	.755	.656	1.150
20°	.349	.342	.940	.364	50°	.873	.766	.643	1.192
21°	.367	.358	.934	.384	51°	.890	.777	.629	1.235
22°	.384	.375	.927	.404	52°	.908	.788	.616	1.280
23°	.401	.391	.921	.425	53°	.925	.799	.602	1.327
24°	.419	.407	.914	.445	54°	.942	.809	.588	1.376
25°	.436	.423	.906	.466	55°	.960	.819	.574	1.428
26°	.454	.438	.899	.488	56°	.977	.829	.559	1.483
27°	.471	.474	.891	.510	57°	.995	.839	.545	1.540
28°	.489	.480	.883	.532	58°	1.012	.848	.530	1.600
29°	.506	.485	.875	.554	59°	1.030	.857	.515	1.664
30°	.524	.500	.866	.577	60°	1.047	.866	.500	1.732

Angle					Angle				
Degree	**Radian**	**Sine**	**Cosine**	**Tan-gent**	**Degree**	**Radian**	**Sine**	**Cosine**	**Tan-gent**
61°	1.065	.875	.485	1.804	76°	1.326	.970	.242	4.011
62°	1.082	.833	.470	1.881	77°	1.344	.974	.225	4.331
63°	1.100	.891	.454	1.963	78°	1.361	.978	.208	4.705
64°	1.117	.899	.438	2.050	79°	1.379	.982	.191	5.145
65°	1.134	.906	.423	2.145	80°	1.396	.985	.174	5.671
66°	1.152	.914	.407	2.246	81°	1.414	.988	.156	6.314
67°	1.169	.921	.391	2.356	82°	1.431	.990	.139	7.115
68°	1.187	.927	.375	2.475	83°	1.449	.993	.122	8.144
69°	1.204	.934	.358	2.605	84°	1.466	.995	.105	9.514
70°	1.222	.940	.342	2.747	85°	1.484	.996	.087	11.43
71°	1.239	.946	.326	2.904	86°	1.501	.998	.070	14.30
72°	1.257	.951	.309	3.078	87°	1.518	.999	.052	19.08
73°	1.274	.956	.292	3.271	88°	1.536	.999	.035	28.64
74°	1.292	.961	.276	3.487	89°	1.553	1.000	.018	57.29
75°	1.309	.966	.259	3.732	90°	1.571	1.000	.000	∞

1. Trigonometric Functions
   $\sin \theta = y/r$
   $\cos \theta = x/r$
   $\tan \theta = y/x$
2. Reduction Formulas
   $\sin \theta = +\cos (\theta - 90°) = -\sin (\theta - 180°) = -\cos (\theta - 270°)$
   $\cos \theta = -\sin (\theta - 90°) = -\cos (\theta - 180°) = +\sin (\theta - 270°)$
   $\tan \theta = +\tan (\theta - 180°)$
3. Fundamental Identities
   $\tan \theta = \sin \theta / \cos \theta$ $\quad \sin^2\theta + \cos^2\theta = 1$
   $\sin 2\theta = 2 \sin \theta \cos \theta$ $\quad \sin^2\theta = \frac{1}{2}(1 - \cos 2\theta)$
   $\cos 2\theta = \cos^2\theta - \sin^2\theta$ $\quad \cos^2\theta = \frac{1}{2}(1 + \cos 2\theta)$
   $= 2 \cos^2\theta - 1$
   $= 1 - 2 \sin^2\theta$
   $\sin (a \pm b) = \sin a \cos b \pm \cos a \sin b$
   $\cos (a \pm b) = \cos a \cos b \mp \sin a \sin b$

Appendix 4: Listing of Elements

Element	Symbol	Atomic No.	Atomic Weight*
Actinium	Ac	89	227
Aluminum	Al	13	26.9815
Americium	Am	95	[243]
Antimony	Sb	51	121.75
Argon	Ar	18	39.948
Arsenic	As	33	74.9216
Astatine	At	85	[210]
Barium	Ba	56	137.34
Berkelium	Bk	97	[249]
Beryllium	Be	4	9.0122
Bismuth	Bi	83	208.980
Boron	B	5	10.811
Bromine	Br	35	79.909
Cadmium	Cd	48	112.40
Calcium	Ca	20	40.08
Californium	Cf	98	[251]
Carbon	C	6	12.01115
Cerium	Ce	58	140.12
Cesium	Cs	55	132.905
Chlorine	Cl	17	35.453
Chromium	Cr	24	51.996
Cobalt	Co	27	58.9332
Copper	Cu	29	63.54
Curium	Cm	96	[247]
Dysprosium	Dy	66	162.50
Einsteinium	Es	99	[254]
Erbium	Er	68	167.26
Europium	Eu	63	151.96
Fermium	Fm	100	[253]
Fluorine	F	9	18.9984
Francium	Fr	87	[223]
Gadolinium	Gd	64	157.25
Gallium	Ga	31	69.72
Germanium	Ge	32	72.59
Gold	Au	79	196.967
Hafnium	Hf	72	178.49
Helium	He	2	4.0026
Holmium	Ho	67	164.930
Hydrogen	H	1	1.00797
Indium	In	49	114.82
Iodine	I	53	126.9044
Iridium	Ir	77	192.2
Iron	Fe	26	55.847
Krypton	Kr	36	83.80
Lanthanum	La	57	138.91
Lawrencium	Lw	103	[257]
Lead	Pb	82	207.19
Lithium	Li	3	6.939
Lutetium	Lu	71	174.97
Magnesium	Mg	12	24.312
Manganese	Mn	25	54.9380
Mendelevium	Md	101	[256]
Mercury	Hg	80	200.59
Molybdenum	Mo	42	95.94
Neodymium	Nd	60	144.24
Neon	Ne	10	20.183
Neptunium	Np	93	[237]
Nickel	Ni	28	58.71
Niobium	Nb	41	92.906
Nitrogen	N	7	14.0067
Nobelium	No	102	[253]
Osmium	Os	76	190.2
Oxygen	O	8	15.9994
Palladium	Pd	46	106.4
Phosphorus	P	15	30.9738
Platinum	Pt	78	195.09
Plutonium	Pu	94	[242]
Polonium	Po	84	210
Potassium	K	19	39.102
Praseodymium	Pr	59	140.907
Promethium	Pm	61	[145]
Protactinium	Pa	91	231
Radium	Ra	88	226.05
Radon	Rn	86	222
Rhenium	Re	75	186.2
Rhodium	Rh	45	102.905
Rubidium	Rb	37	85.47
Ruthenium	Ru	44	101.07
Samarium	Sm	62	150.35
Scandium	Sc	21	44.956
Selenium	Se	34	78.96
Silicon	Si	14	28.086
Silver	Ag	47	107.870
Sodium	Na	11	22.9898
Strontium	Sr	38	87.62
Sulfur	S	16	32.064
Tantalum	Ta	73	180.948
Technetium	Tc	43	[99]
Tellurium	Te	52	127.60
Terbium	Tb	65	158.924
Thallium	Tl	81	204.37
Thorium	Th	90	232.038
Thulium	Tm	69	168.934
Tin	Sn	50	118.69
Titanium	Ti	22	47.90
Tungsten	W	74	183.85
Uranium	U	92	238.03
Vanadium	V	23	50.942
Xenon	Xe	54	131.30
Ytterbium	Yb	70	173.04
Yttrium	Y	39	88.905
Zinc	Zn	30	65.37
Zirconium	Zr	40	91.22

* Atomic weights are based on carbon-12. A value given in brackets denotes the mass number of the longest-lived or best-known isotope.

Answers to Even-Numbered Problems

Answers may vary slightly due to multiple calculations and rounding off to significant figures. In some instances, more figures are reported for clarity.

Chapter 1

2. $w_m = 30$ lb
4. (a) $A = 0.13$ m^2, (b) $V = 0.0042$ m^3
6. (a) $V = 94$ in^3, (b) $A = 98$ in^2
8. $\rho = 0.47$ g/cm^3
10. $\rho_e/\rho_m = 1.5 \times 10^6$
12. $D = 32$ lb/ft^3
14. $V = 5.0$ cm^3
16. $h = 3.6$ ft
18. (a) 2.4 in., (b) 52.5 ft, (c) 6.30 in., (d) 12 mi
20. (a) 30 m pole, (b) 38 L hat, (c) 2.54 cm and 1.61 km, (d) 909 kg
22. 50-m dash longer by 4.3 m
24. Metric ton greater by 200 lb (or 91 kg)
26. 1.0×10^6 cm^3
28. (150 lb person 5 ft 6 in. tall) $m = 68$ kg (4.7 slug) and $h = 1.68$ m
30. 10.6 gal
32. 11.1 yd^2
34. 153 in^3
36. $A = 120$ ft^2
38. (100 m) $\bar{v} = 6.7$ m/s, (100 yd) $\bar{v} = 6.1$ m/s 100-m runner averages 0.6 m/s more.

Chapter 2

2. (a) 0.021, (b) 10,000, (c) 0.0314, (d) 67,300, (e) 0.0000167, (f) 785.6
4. (a) 1.2×10^6, (b) 7.0×10^1, (c) 1.6×10^{-5} (d) 4.3×10^3
6. (a) 9.0×10^9, (b) 4.5, (c) 5.0×10^6 (d) 4.0×10^{-6}, (e) 3.0×10^{-2}
8. (a) 2.0×10^4, (b) 1.0×10^5, (c) 3.0×10^{-3} (d) 0.50×10^3
10. (a) $\pi/2$ rad, (b) $\pi/12$ rad, (c) $2\pi/3$ rad, (d) $3\pi/2$ rad, (e) $5\pi/3$ rad, (f) 3π rad
12. (a) π m, (b) $\pi/3$ m, (c) 2π m, (d) $\pi/2$ m
14. (a) $r = 5.0$ m, (b) sin = 0.60, cos = 0.80, tan = 0.75, (c) sin = 0.80, cos = 0.60, tan = 1.33, (d) θ = arcsin = 37° = 0.65 rad, ϕ = arcsin = 53° = 0.92 rad
16. (a) $x = 6.5$ m, (b) sin = 0.36, cos = 0.93, tan = 0.38, (c) θ = arcsin = 21° = 0.37 rad, ϕ = 69° = 1.2 rad
18. $x = 6.0$ m, $r = 8.5$ m
20. $x = 3.73$ m, $r = 8.83$ m
22. $x = 6.89$ m, $r = 5.79$ m
24. (a) $F_x = -7.50$ N, $F_y = 13.0$ N, (b) $F_x = -0.40$ N, $F_y = -6.00$ N, (c) $F_x = 3.9$ N, $F_y = -1.6$ N, (d) $F_x = 0$, $F_y = -9.0$ N
26. (a) 0.30 (30%), (b) $\theta = 17°$
28. (a) $y = 17$ m, (b) 5 stories
30. $A = 78$ cm, $C = 90$ cm
32. (a) $B = 12$ m, (b) $\theta = 53°$
34. $A = 1.6$ m, $B = 1.2$ m
36. 12.8 m **x** + 16.3 m **y**, or 20.7 m at 52°
38. **F** = 4.2 N **x** + 5.3 N **y**, or 6.8 N at 52°
40. 6.8 N, 52°
42. $d = 5$ blocks, $\theta = 37°$, south of west
44. **F** = −3.5 N **x** + 7.5 N **y**, or $F = 8.5$ N, 65° (4th quadrant)

Chapter Supplement

2. $a = 17.7$ in., $\beta = 15°$, $\gamma = 115°$

Chapter 3

2. $\mathbf{F}_3$ = −4 N **x** + 7 N **y**
4. $\Sigma\mathbf{F} = 0$, in equilibrium
6. $\mathbf{F}_4$ = 4.2 N **x** + 4.2 N **y**
8. $T = 262$ N
10. N = 70.7 N, $f = 70.7$ N
12. N = 66.4 N
14. $w_1 = 5.8$ N
16. $w_1 = 20$ N
18. $T_1 = 89.5$ lb, $T_2 = 73.4$ lb
20. $R = 3.47$ ton, $T = 2.00$ ton
22. $\tau = 7.5$ lb-ft
24. $\tau = 52$ m-N
26. (a) 23 in. position, (b) $R = 1.5$ lb
28. $\Sigma\tau = o$
30. $R = 8.5$ ton (B), 16.5 ton (A)
32. $T = 861$ lb
34. $T = 1.2 \times 10^3$ lb
36. $T_1 = 122$ lb, $T_2 = 158$ lb
38. Four books on top of bottom book, unstable equilibrium

Chapter 4

2. $\bar{v} = 10$ mi/h
4. $\bar{v} = 15$ ft/s, toward other goal line
6. (a) the car, (b) $\bar{v}_a = 100$ km/h, $\bar{v}_b = 90$ km/h
8. $t = 0.042$ h = 150 s
10. (a) $\bar{v} = 0.50$ m/s, (b) $v = 0.50$ m/s in +X direction
12. (a) $x = 0.50$ m, (b) $x = 10$ m, (c) $x = 30$ m
14. $t \approx 2.0 \times 10^8$ y = 200 million years
16. (a) $\bar{v} = 45$ mi/h, (b) $v = 22$ ft/s = 15 mi/h, $\theta = 0°$, east
18. $\bar{a} = 5.0$ m/s^2, along road
20. $v_f = 60$ m/s, in original direction

22. (a) $\bar{v} = 1.0$ m/s, (b) $x = 4.0$ m
24. (a) $= -1.3$ m/s^2, opposite motion, (b) $v_f = 0$, $v_o = 4.0$ m/s
26. (a) $a = 13.9$ m/s^2, (b) $x = 174$ m, (c) $v_f = 55.6$ m/s
28. $x = 194$ ft
30. (a) $a = -35.2$ ft/s^2, (b) $x = 440$ ft
32. (a) $v = 49$ m/s, (b) $y = 123$ m
34. $v_f = 62$ ft/s
36. $y = 30.6$ m or 100 ft
38. $y = 53.8$ ft
40. (a) $x = 0.52$ ft, (b) $v_{x2} = 2.3$ ft/s

Chapter 5

2. $v = 25$ m/s, $\theta - 29°$
4. $y = v_y t = v_y(x/v_x)$, $m = (v_y/v_x)$, $b = 0$ or $b = y_o$
6. $x = 1.2$ m
8. (a) $t_u = 1.0$ s, (b) $t = 2.0$ s, $v = -9.8$ m/s
10. (a) $v = -26$ m/s, (b) $t = 1.8$ s
12. (a) $y = 32$ m, (b) $y = 22$ m
14. $x = 6.9$ ft
16. (a) $v_x = 15$ m/s, (b) $y = 44$ m
18. $R = 210$ ft
20. (a) $y_{max} = 112$ ft, (b) $R = 450$ ft
22. $\theta = 43°$
24. $R = 221$ m
26. (a) $x = 246$ ft, (b) $x = 324$ ft
28. $a = 0.54$ m/s^2
30. $d = 600$ ft
32. $a = 2.3 \times 10^{-3}$ m/s^2
34. $a = 7.8$ m/s^2

Chapter 6

2. $a = 64$ ft/s^2
4. $a_1 = 8a_1$
6. $v_f = 8.0$ m/s
8. $x = 72$ m
10. $F = 24$ N
12. $F = 2200$ N
14. $F = 440$ lb, would break rope
16. $t = 1.0$ s
18. $m_1 = 3.0$ kg (slightly greater)
20. (a) $a = 6.0$ m/s^2, (b) $T = 6.0$ N
22. $a = 13$ m/s^2
24. $a = 2.4$ m/s^2
26. $m = 0.49$ kg
28. (a) $m_2 = 0.40$ kg, (b) $y = 0.025$ m
30. (a) $v_f = 18$ m/s, (b) $v_f = -2.0$ m/s, (c) case a: 26 m, case b: total distance traveled, 6.8 m
32. $a = 3.0$ m/s^2
34. $x = 2.05$ m
36. $g = 7.9$ m/s^2, friction and inertia errors.
38. $F = -5.4$ N, not possible.
40. (a) $F = 4.7 \times 10^{27}$ N, (b) 4.7×10^{27} N, equal and opposite

Chapter Supplement

2. $\theta_2 = 19.9°$

Chapter 7

2. $F = 4.7 \times 10^{-4}$ N
4. $m_2 = 1.2 \times 10^{11}$ kg
6. $x = 0.20$ m
8. $F_{se} = 3.6 \times 10^{22}$ N
10. $g(h) = 0.39$ m/s^2
12. $g = 9.8$ m/s^2
14. $w' = 123$ lb
16. $w'/w = 0.024$ (2.4%)
18. $h = 4.7 \times 10^6$ m
20. Saturn much larger. $R_s = (9.7)R_e$
22. $M_s = 2.0 \times 10^{30}$ kg
24. (a) $R = 152$ lb, (b) $R = 88$ lb
26. $a = 9.8$ m/s^2, upward
28. (a) $f_s = 74$ N, (b) $F > f_s = 74$ N
30. $\mu_s = 0.90$
32. $a = 2.7$ m/s^2
34. (a) $\mu_s = 0.50$, (b) $a = 0.65$ m/s^2
36. $\theta = 33°$
38. $\mu_k = 0.29$
40. $t = 1.5$ s
42. $v = 76$ ft/s
44. $\mu_s mg \geq mv^2/r$, or car slides outward.
46. (a) $v = (gr/\mu_s)^{1/2}$, (b) yes, (c) different μ, (d) $v = 8.4$ m/s

Chapter 8

2. $W = 78$ J
4. $W = 2.8 \times 10^3$ ft-lb
6. (a) $W = 4.5 \times 10^3$ J, (b) $W = 3.6 \times 10^3$ J
8. $W = 2.1 \times 10^2$ J
10. $W = 2.4$ J
12. $W = 68$ ft-lb
14. $KE = 0.90$ J
16. $W = 2.2 \times 10^4$ ft-lb
18. $W = 2.9 \times 10^5$ J
20. $PE = 5.4$ ft-lb
22. $E = 1.8 \times 10^{17}$ J
24. $d_2 = 4d_1$
26. $v = 13$ m/s
28. $h' = 66.7$ m, or by inspection 2/3 way down.
30. (a) $E = 0.13$ J, (b) $v = 1.1$ m/s, (c) no
32. (a) $\bar{F} = 3.75 \times 10^4$ N, (b) $\bar{F} = 2.73 \times 10^4$ N
34. (a) $W = 7.2 \times 10^5$ ft-lb, (b) $\bar{P} = 57$ ft-lb/s
36. $P = 0.11$ hp
38. $P = 60$ W
40. $-E = 1650$ ft-lb (each second)
42. $-E = 1.6 \times 10^6$ J
44. $P = 4.7$ hp

Chapter 9

2. $v_f = 1.1$ m/s
4. $v_b = 4.5 \times 10^4$ m/s
6. (a) Imp $= -1.5$ N-s, (b) $\bar{F} = -15$ N
8. (a) $\Delta p = 1.0$ N-s, (b) $v_f = 40$ m/s
10. $p_x = 0$, $p_y = (2)^{1/2}p$
12. (a) $F = 8.0 \times 10^2$ N, (b) $\Delta p = 4.0$ N-s
14. $v_c = 3.8$ m/s
16. Car comes to rest.

18. $t = 800$ s
20. Equal masses, exchange of momentum.
22. $v_{1f} = -13$ m/s, $v_{2f} = 1.3$ m/s
24. $v_f = 4.7$ m/s
26. $K_i \neq K_f$, not elastic.
28. Similar derivation as Eq. 9.12 with $v_{2o} \neq 0$.
30. (a) $v_f = 15$ km/h, (b) not elastic
32. Derivation. Conservations of momentum and energy.
34. (a) $P_{xf} = 0$, (b) $P_{yf} = P_{yo}$

Chapter 10

2. (a) 1080°, (b) 36°, (c) 7.5°
4. (a) $\theta_{sh} = 10\pi = 31.4$ rad, $\theta_{hh} = \pi/6 = 0.52$ rad (b) $v = \pi/3 = 1.0$ cm/s
6. $\overline{w} = \pi/30 = 0.105$ rad/s
8. $\overline{\alpha} = -2.7\pi = -8.5$ rad/s^2
10. $w = 3.1$ rad/s
12. $\alpha = 0.20$ rad/s^2
14. (a) $a_c = 2.5 \times 10^{-2}$ m/s^2, toward center (b) $a_t = 0.17$ m/s^2, $a_c = 0.23$ m/s^2
16. (a) $r_\perp = 1.3$ ft, (b) $r = 2.2$ ft
18. $I = 25$ kg-m^2
20. $\tau = 1.3 \times 10^3$ m-N
22. (a) $k = R/(2)^{1/2}$, (b) $k = (2/5)^{1/2}R$
24. (a) Pivoted at end. (b) $I_e/I_c = 4$
26. $I = I_1 + I_2$
28. $W = 6.3 \times 10^2$ ft-lb
30. $KE = 7.7 \times 10^2$ J
32. (a) $W = -9.6 \times 10^3$ J, (b) $\tau = -1.5 \times 10^2$ m-N
34. $v = 6.3$ m/s
36. $P = 120$ ft-lb/s
38. $\tau = 4.5$ lb-ft
40. (a) Rotating objects have rotational *KE*, and sliding objects do not, so sliding speed greater. Show mathematically. (b) $v_s > v_h$
42. $L = 0.38$ kg-m^2/s
44. $L = 7.1 \times 10^{33}$ kg-m^2/s
46. $w = w_e/4$
48. $w_f = 46$ rad/s

Chapter 11

2. $F_i = 15$ N
4. IMA = 2.7
6. $F_i = 90$ N
8. AMA = 2.1
10. (a) IMA = 0.43, (b) $F_o = 22$ lb, (c) first class
12. (a) $F_i = 118$ N, (b) third class
14. (a) IMA = 4.0, (b) $F_o = 12$ lb
16. (a) Eff = 0.40, (b) $F_i = 15$ lb
18. IMA = 4
20. (a) IMA = 6.0, (b) $F_o = 91$ N
22. (a) AMA = 4.0, (b) Eff = 0.69
24. (a) IMA = 8.0, (b) $\theta = 7.2°$
26. $t = 0.125$ in.
28. Total IMA = 25
30. Sketch. Five support strands.
32. (a) IMA = 40, (b) $F_o = 2800$ lb
34. $F_i = 0.041$ N
36. 125 strokes
38. $w_o = 250$ rpm
40. (a) Twist belt one turn. (b) Use chain.
42. (a) Total IMA = 0.53, (b) $s = 1440$ in.
44. $w_o = 57.5$ rpm

Chapter 12

2. Stress = 2.3×10^3 lb/in^2
4. $d = 0.10$ in.
6. Strain = -0.0017
8. (a) $F = 2.0 \times 10^2$ N, (b) $F = 3.8 \times 10^2$ N
10. $d = 0.13$ in.
12. (a) $\Delta L = 0.024$ ft = 0.29 in., (b) $\Delta L = 0.038$ ft = 0.46 in.
14. $F_n = 92$ N
16. $\Delta L/L_o = 1.0 \times 10^{-2} = 1.0\%$
18. $L_o = 49.99$ cm
20. $\Delta L/L_o = 6.0 \times 10^{-5}$
22. (a) $\phi = 3.7 \times 10^{-4}$ rad, (b) $d = 1.1 \times 10^{-3}$ in.
24. $d = 1.0 \times 10^{-7}$ m
26. $\tau = 65$ m-N
28. $s = L\phi$ and $s = r\theta$, so $\phi = (r/L)\theta$
30. $F_t = 5.5 \times 10^4$ N
32. $p = 3.6 \times 10^5$ N/m^2
34. $\Delta V = -0.092$ in^3
36. $\Delta V/V_o = 0.11$

Chapter 13

2. $v = 14$ m/s
4. $F = 4.5$ lb
6. $v_2/v_1 = 2$, greater speed in light cord.
8. $F_2 = 4F_1$
10. (a) $T = 10$ min, (b) $f = 1.7 \times 10^{-3}$ Hz, relatively slow
12. (a) $T = 1.3$ s, (b) $f = 0.77$ Hz
14. (a) $T = 1.4$ s, (b) $f = 0.71$ Hz
16. $T_2/T_1 = 0.87$
18. (a) $y = 10$ cm, (b) $y = 0$ cm
20. (a) $f = 1.6$ Hz, (b) $y = 7.0$ cm
22. (a) $y = -(10 \text{ cm}) \sin 10\pi t$, (b) $y = -(10 \text{ cm}) \cos 10\pi t$, (c) $y = (7.1 \text{ cm}) [\sin 10\pi t + \cos 10\pi t]$
24. (a) $y = (8.0 \text{ cm}) \sin \pi t$, (b) $y = (8.0 \text{ cm}) \cos \pi t$
26. $v_{max} = 3.8$ m/s
28. (a) $E = 1.5$ J, (b) $v = 1.1$ m/s
30. (a) $T = 0.14$ s, (b) $f = 7.1$ Hz
32. $\lambda = 0.25$ m
34. $v = 0.75$ m/s, (b) no, wave speed dependent on medium.
36. $v = 0.40$ m/s
38. (a) $f_1 = 4.5$ Hz, (b) $f_3 = 13.5$ Hz
40. (a) $f = 0.64$ Hz, (b) Only one resonance f.
42. $f_n = nv/4L$, $n = 1, 3, 5, 7, \ldots$

Chapter 14

2. $f = 170$ Hz, yes.
4. $v = 3.5 \times 10^3$ m/s

6. $v_{gl} = 4.6 \times 10^3$ m/s, aluminum
8. $t = 0.20$ s
10. $v_w/v_{Al} = 1.2$
12. $t = 6.9 \times 10^{-3}$ s
14. (a) $T_C = 16.7$°C, (b) $T_F = 9.1$°F
16. $t = 2.4$ s
18. $\lambda = 2.4 \times 10^{-5}$ m
20. $I = 2.0 \times 10^{-8}$ W/m^2
22. $I_2 = 1.8 \times 10^{-7}$ W/m^2
24. (a) $IL = 70$ dB, (b) $IL = 110$ dB, (c) $IL = 140$ dB
26. $I_2/I_1 = 10^5$
28. $r_2/r_1 = 10$
30. IL_1, 8 h max; IL_2, 2 h max.
32. (a) $r_2 = 300$ m, (b) $r_2 = 3.0 \times 10^5$ m (IL_2 = 0 dB)
34. $f_1 = 141$ Hz, $f_3 = 423$ Hz
36. $f_b = 4.0$ Hz
38. (a) $f' = 1079$ Hz, (b) $f' = 932$ Hz
40. (a) Mach 1.2, (b) yes
42. $f' = 1174$ Hz
44. $f = 256$ Hz

Chapter 15

2. $V = 3.6 \times 10^{-3}$ ft^3
4. $V_{br} = 3.3$ cm^3
6. sp.gr. = 4.8
8. $w = 2.0 \times 10^2$ lb
10. (a) $F = 6.2 \times 10^{-5}$ N, (b) $h_b = 0.73$ cm
12. (a) $F = 2.0 \times 10^4$ N, (b) 4.5×10^3 lb
14. $p_{abs} = 1.26 \times 10^5$ N/m^2
16. $h = 0.40$ m = 40 cm
18. $h = 749$ mm Hg = 29.6 in. Hg
20. $p_g = 2.0 \times 10^4$ N/m^2 (or 150 torr)
22. (a) $p_g = -1.33 \times 10^4$ N/m^2, (b) $p_{abs} = 8.8 \times 10^4$ N/m^2
24. (a) p depends only on height, so all pressures on bases are the same. (b) Part of weight force balanced by reaction force of container surface on water.
26. $h = 1.6$ cm
28. $w' = 9.0$ lb
30. $V = 2.5 \times 10^4$ ft^3
32. (a) $V_w = 7.9$ m^3 = V_{cube}, (b) $L = 1.99$ m
34. $t = 2047$ s = 34 min
36. $d_2 = 3.2$ cm
38. (a) $v = 14$ m/s, (b) $R = 2.8 \times 10^{-4}$ m^3/s
40. (a) $v_2 = 153$ ft/min = 2.5 ft/s, (b) $p_2 = 318$ lb/ft^2
42. $n_B/n_A = 1.5$
44. Use Bernoulli's and pressure-depth equations to find v_1, then $R = A_1 v_1$.
46. $R = 6.1 \times 10^{-5}$ m^3/s

Chapter 16

2. (a) 140°F, (b) −13°F, (c) 39.2°F, (d) 302°F
4. (a) 560°R, (b) 518°R, (c) 537°R, (d) 573°R
6. $E = 18$ cal
8. $E_{lost} = 0.19$ kcal
10. $\overline{KE} = 6.1 \times 10^{-21}$ J
12. $E_2 = 3E_1$
14. $T_2 = T_1$, no effect
16. (a) Celsius increase, (b) $E_{2C}/E_{2F} = 1.07$
18. $T_2 = 313$°C
20. $V_2 = 0.30$ m^3
22. $\Delta Q = 1500$ cal = 1.5 kcal
24. Al, greater specific heat, absorbs more heat, 1.6 Btu
26. $T_f = 74.8$°F
28. $T_f = 69.6$°F
30. $m_{Pb}/m_{Br} = 6.1$
32. $T_f = 45.2$°C
34. $\Delta Q = 4000$ cal = 4.0 kcal
36. $\Delta Q = 270$ kcal
38. $\Sigma Q_i = 8302$ Btu
40. $\Sigma Q_i = 2597.7$ Btu
42. $T_f = 3.3$°C

Chapter 17

2. $\Delta Q/\Delta t = 2.3$ cal/s
4. $I_{2y} = 16 I_1$
6. $\Delta Q/\Delta t = 806$ Btu/h
8. (a) $\Delta Q/\Delta t = 0.063$ kcal/s = 63 cal/s, (b) m = 0.19 kg
10. (a) Greatest: lead, 4.2×10^{-3} ft^2-h-°F/Btu-in. Least: silver, 3.3×10^{-4} ft^2-h-°F/Btu-in. (b) $(R\text{-value})_{Pb} = 4.2 \times 10^{-3}$, $(R\text{-value})_{Ag} = 3.3 \times 10^{-4}$
12. (a) $L_f = 12.6$ in., (b) $L_g = 8.1$ in.
14. $\Delta L = 5.7$ in.
16. (a) foam: 13.3, (b) air: 23.5, (c) vac: → ∞
18. $I_2 = (1.3)I_1$
20. $T_C = 27$°C
22. (a) Greater A, greater heat flow.
24. $\Delta T = 265$°F
26. $\Delta T_p = (2.7)\Delta T_g$
28. $\Delta L = 0.15$ ft = 1.8 in.
30. $\Delta V/V_o = 0.012 = 1.2\%$
32. (a) $V = 1.0168$ L, (b) $V = 0.9966$ L
34. (a) $A = 4.926$ in^2, (b) $L = 2.219$ in.
36. $\Delta Q = 1.15 \times 10^5$ Btu
38. $\Delta Q_T = 121.5$ Btu = 30.6 kcal
40. (a) $\Delta m = 0.050$ kg, (b) $\Delta m = 0.005$ kg

Chapter 18

2. (a) horizontal line, (b) vertical line, (c) sloping straight line.
4. (a) $W = 260$ J, (b) $W = -200$ J
6. $\Delta U = 10$ Btu
8. $h = 0.22$ ft = 2.6 in.
10. $T_2 = 599$ K
12. $\Delta S = -0.15$ kcal/K = −630 J/K
14. $\Delta S = 28.5$ cal/K = 119 J/K
16. $\Delta S = 9.2$ J/K
18. (a) vertical line, adiabatic process (b) Area = $T\Delta S = \Delta Q$ (+ added, − removed)
20. (a) $Q_c = 390$ Btu, (b) $W = 210$ Btu

22. (a) $W = 3.9 \times 10^2$ Btu, (b) $Q_c = 1.6 \times 10^3$ Btu
24. (a) $Q_{in} = 6.0 \times 10^4$ J, (b) $\varepsilon_{th} = 0.27 = 27\%$
26. $\varepsilon_c = 0.35 = 35\%$
28. $T_h = (1.5)T_c$
30. $\varepsilon_{c1} = 0.35 = 35\%$
32. (a) Yes. (b) Because of its boiling point, work done by gas expansion.
34. (a) $T_c = 300$ K $= 27°$C, (b) $\varepsilon_c = 0.33 = 33\%$, (c) $W = 420$ J
36. (a) cop = 1.5, (b) $W = 1.2$ Btu
38. $(\text{cop})_c = 6.8$
40. $w = 2.1 \times 10^3$ lb = 1.05 ton
42. hp = 6.3 hp

Chapter 19

2. $n = 6.0 \times 10^{13}$
4. $F = 0.80$ N, toward other charge
6. $r = 1.5 \times 10^{-14}$ m
8. (a) No, (b) Yes, $x = 0.20$ m
10. (a) $F = 0.65$ N, $\theta = 45°$ (2nd quad, away from q_4) (b) $F = 1.05$ N, $\theta = 45°$ (2nd quad, toward q_1)
12. $E = 7.5 \times 10^4$ N/C, away from charge
14. (0, 0) due to symmetry.
16. (a) $E = 4.5 \times 10^6$ N/C, $-y$ direction (b) Same as in Problem 9(c), $F = 0$ at $y = 0.50$ m
18. $E = 5.6 \times 10^{-11}$ N/C
20. Zero, by symmetry.
22. (a) $U_1 = -0.60$ J, (b) $W = +0.15$ J
24. $U_T = 0.11$ J
26. $U = 0.27$ J
28. (a) $V_1 = +2.4 \times 10^5$ J/C, (b) $|V| = 8.0 \times 10^4$ J/C
30. $W = 1.6 \times 10^{-16}$ J
32. (a) $a = 6.7 \times 10^9$ m/s², (b) $K = 30$ eV

Chapter 20

2. $C = 5.0 \times 10^{-7}$ F
4. (a) $E = 7.5 \times 10^3$ N/C, (b) $C = 3.3 \times 10^9$ F, $V = 15$ V
6. $d = 5.3 \times 10^{-4}$ m
8. $N = 29$
10. $U = 1.4 \times 10^{-9}$ J
12. (a) $U_2 = (16)U_1$, (b) $U_2 = (0.89)U_1$
14. $V = 32$ V
16. (a) $U = 1.4 \times 10^{-6}$ J, (b) $u_e = 1.8 \times 10^{-3}$ J/m³, (c) $F = 2.8 \times 10^{-3}$ N
18. $K = 2.6$ (polystyrene)
20. (a) $C = 1.3 \times 10^{-8}$ F, (b) $U = 9.2 \times 10^{-7}$ J
22. (a) $C_o = 5.9 \times 10^{-12}$ F, (b) $C = 1.4 \times 10^{-11}$ F, (c) $C = 2.4 \times 10^{-11}$ F
24. (a) $C_s = 1.3\ \mu$F, (b) $C_p = 6.0\ \mu$F
26. 2 in parallel, 15 μF, 25 μF, 30 μF; 3 in parallel, 35 μF
28. (a) $C_T = 0.55\ \mu$F, (b) $Q_T = 6.6\ \mu\text{C} = 6.6 \times 10^{-6}$ C
30. $C_2 = 2.0\ \mu$F
32. Series-parallel combinations: 4.3 μF, 7.1 μF, 8.6 μF, 12 μF, 14 μF, 23 μF.
34. $R = 2.4 \times 10^4\ \Omega$
36. $C_1 = 2.0 \times 10^{-6}$ F, $C_2 = 4.0 \times 10^{-6}$ F
38. (a) $Q = 28\ \mu\text{C} = 2.8 \times 10^{-5}$ C, (b) $6.2\ \mu\text{C} = 6.2 \times 10^{-6}$ C, (c) $14\ \mu\text{C} = 1.4 \times 10^{-5}$ C

Chapter 21

2. $n = 3.1 \times 10^{17}$
4. $n = 1.9 \times 10^{19}$
6. $R = 24\ \Omega$
8. $I_2 = 750$ mA = 0.75 A
10. $R = 3.0\ \Omega$
12. $d = 9.4 \times 10^{-4}$ m
14. $A_{Cu} = 2.5 \times 10^5$ CM
16. $A = 1$ CM $= \pi d^2/4 = \pi(\text{mil})^2/4 = \pi/4$ sq. mil
18. $A = 2.9 \times 10^4$ CM
20. $L = 667$ ft
22. $R = 4.2 \times 10^{-5}\ \Omega$
24. $R = 230\ \Omega$
26. (a) $\alpha_{Cu} = 2.2 \times 10^{-3}$ °F^{-1}, (b) $\alpha_W = 2.5 \times 10^{-3}$ °F^{-1}
28. $\Delta I = 8.0$ A
30. (a) $E = 4.2 \times 10^5$ J, (b) $I = 10.3$ A
32. $P/P_o = 1.16$
34. $E = 0.50$ kWh
36. Greater wattage requirement, hence greater current requirement for same voltage ($P = IV$).
38. \$ = 0.64
40. $L = 0.181$ km = 181 m
42. (a) $R = 2.12\ \Omega$, (b) $w = 5.15 \times 10^3$ lb

Chapter Supplement

2. life = 30 h

Chapter 22

2. (a) $I = 0.33$ A; (b) $V_1 = 1.0$ V, $V_2 = 5.0$ V
4. (a) $V_1 = 12$ V; (b) $I_1 = 3.0$ A, $I_2 = 2.0$ A
6. (a) $R_T = 7.2\ \Omega$, (b) $I_1 = 1.7$ A, $I_2 = 0.80$ A, $I_3 = 0.90$ A, $I_4 = 0.27$ A, $I_5 = 0.53$ A
8. (a) $I_1 = 0.37$ A, $I_2 = 0.19$ A, $I_3 = 0.56$ A; (b) $V_1 = V_2 = 1.6$ V, $V_3 = 1.1$ V, $V_4 = 3.4$ V; (c) $P = 3.4$ W
10. $R_T = 23.3\ \Omega$
12. $I_1 = 3.1$ A, $I_2 = 1.5$ A, $I_3 = 0.15$ A
14. (a) $I_1 = 3.0$ A, $I_2 = 1.0$ A, $I_3 = 2.0$ A; (b) $P_1 = 18$ W, $P_2 = 2.0$ W, $P_3 = 4.0$ W, $P_{v1} = 30$ W, $P_{v2} = -2.0$ W (recharging)
16. $I_{R6} = 0.73$ A
18. $R_x = 800\ \Omega$
20. $V_{AB} = 4.0$ V, $V_{BC} = 8.0$ V
22. (a) $\varepsilon_x = 3.0$ V, (b) $\varepsilon_x = \varepsilon_s$ and $L_x = L_s$

Chapter 23

2. $B = 2\pi \times 10^{-5}$ T
4. (a) $B = 1.0 \times 10^{-5}$ T, out of page, (b) $B = 1.0 \times 10^{-5}$ T, into page
6. A: $B = 1.25 \times 10^{-5}$ T, up; B: $B = 0$; C: $B - 1.25 \times 10^{-5}$ T, down
8. (a) $B - 16\pi \times 10^{-6}$ T $= 5.0 \times 10^{-5}$ T, (b) $B = 0$
10. $B = \pi \times 10^{-3}$ T

12. (a) $n = 2.0 \times 10^3$ turns/m, (b) $N = 300$
14. (a) $B_m = 32\pi = 1.0 \times 10^2$ T, (b) $B = 1.3 \times 10^{-2}$ T
16. (a) Zero, (b) $F = 2.4 \times 10^{-19}$ N
18. $B = 5.0 \times 10^5$ T, upward
20. $F/L = 5.0 \times 10^{-3}$ N/m, down
22. (a) $\tau = 1.0 \times 10^{-6}$ m-N, (b) $\tau = 0$
24. (a) $F_2/L = 1.7 \times 10^{-5}$ N/m, away from other wire;
(b) $F_1/L = 1.7 \times 10^{-5}$ N/m, away;
(c) $F_1/L = 1.7 \times 10^{-5}$ N/m, toward
26. $I_c = 0.20$ A
28. $R_m = 19990 = 19.99$ kΩ
30. (a) Open: $I_c = 0$; shorted: $I_c = V/R_c$; $R_x = R_c$, $I_c = V/R_c$; (b) no

Chapter 24

2. (a) $\Phi = 1.6 \times 10^{-3}$ Wb, (b) $\Phi = 9.0 \times 10^{-4}$ Wb
4. (a) $\varepsilon = -25$ V, (b) $I = 100$ A
6. (a) $\varepsilon = -4.0$ V, (b) $I = 25$ A, (c) clockwise
8. $N = 100$ turns
10. $\varepsilon = -0.50$ V
12. (a) $I = 4.0$ A, (b) $I_s = 6.0$ A
14. (a) $R = 60\ \Omega$, (b) $\varepsilon_b = 100$ V, (c) $P = 1.7 \times 10^2$ W
16. ε_{max} (a) 156 V, 170 V; (b) 311 V, 339 V
18. (a) $\varepsilon = 0$, (b) $I_{eff} = 1.8 \times 10^{-2}$ A
20. (a) $V_s = 360$ V, (b) $P_{out} = 720$ W, (c) $P_{in} = 720$ W
22. (a) $N_s = 6545$ turns, (b) $I_s = 0.81$ A
24. (a) 2 : 100 or 1 : 50, (b) $I_s = 50$ A
26. (a) $N_p/N_s = 10$, (b) $I_p = 0.050$ A
28. $V_s = 920$ V, $I_s = 1.1$ A

Chapter 25

2. $f = 88$ Hz
4. (a) X_C decreases and I increases, (b) $I_2/I_1 = 2.0$
6. $L = 0.080$ H
8. (a) $X_L = 94\ \Omega$, (b) $I = 1.3$ A
10. $L = 0.15$ H
12. (a) $Z = 226\ \Omega$, (b) $I = 0.53$ A
14. (a) $X_C > X_L$, capacitive; (b) $Z = 224\ \Omega$, $\phi = 48°$
16. $f_r = 159$ Hz
18. (a) f_r not dependent on R, (b) $L = 0.51 \times 10^{-3}$ H
20. $P = 56$ W
22. $P = 43$ W
24. $I = 0.23$ A
26. (a) $P = 0.13$ W, (b) $f_r = 59$ Hz, (c) $C = 1.1 \times 10^{-5}$ F
28. (a) percent loss = 0.011%, (b) 68%
30. (a) $\cos \phi = 0.60$, (b) $C = 1.3 \times 10^{-5}$ F
32. $I = 22$ mA, muscular freeze

Chapter 26

2. Diode is connected between ground and a negative voltage, so diode conducts during negative half cycle only.
4. Same as text figure with tubes replaced by solid state diodes.
6. The variation of pressure on carbon granules varies the resistance, producing pulsating dc current of the same form as the sound waves. The output is through the coupling transformer.
8. Since only the tops of the peaks would be detected, the output is a relatively constant voltage with time with the appropriate RC.
10. The iron diaphragm is normally attracted to the magnet. Ac signals through the coils either add to or subtract from the magnetic field, causing the diaphragm to move in and out. These sympathetic oscillations transfer energy to the air to produce sound.

Chapter 27

2. $f_1 = 7.5 \times 10^{14}$ Hz, $f_2 = 4.3 \times 10^{14}$ Hz
4. (a) $\lambda_m = 3.8 \times 10^{-7}$ m, (b) $\lambda_m/\lambda_v = 0.77$
6. (a) $y = A \cos 2\pi x/\lambda$, (b) $y = -A \sin 2\pi x/\lambda$ (c) $y = -A \cos 2\pi x/\lambda$, (d) $y = A \sin 2\pi x/\lambda$
8. $y = (0.25 \text{ cm}) \cos [(4\pi x) \times 10^{-6}]$
10. (a) $\theta = 4\pi$ rad, (b) constructive
12. $\lambda = 6.25 \times 10^{-7}$ m (red)
14. $\Delta y = 0.36 \times 10^{-3}$ m
16. $\Delta y = L\lambda/d$
18. (a) $t = 1.25 \times 10^{-7}$ m, (b) $t = 1.75 \times 10^{-7}$ m
20. $\lambda = 2.6 \times 10^{-7}$ m
22. (a) $\theta_1 = 37°$, (b) θ_2 not possible
24. (a) 2 times, (b) 4 times, (c) no change
26. (550 nm) $F_e = 1.0$ lm, (600 nm) 0.60 lm
28. $F_2 = 5.0$ lm
30. $E = 48$ hlx
32. radio: $E = 125$ fc, TV: $E = 64$ fc

Chapter 28

2. $\theta = 15°$
4. $n - 2.0$
6. flint glass, $c_m = 5.25 \times 10^7$ m/s
8. $n_2 = 1.46$ (carbon tetracholoride)
10. (a) $\theta_c = 24.4°$, (b) $\theta_c = 47.3°$
12. (a) $\theta_c = 62.5°$, (b) $\theta_c = 41.8°$
14. (a) $\theta_3 = 45°$, (b) $x = 4.6$ cm
16. $L = 33$ in.
18. $h_i = 12$ cm, real, inverted
20. $d_i = -6.67$ cm (virtual), $M = 0.33$ (upright)
22. (a) $d_o = 75$ cm, (b) $M = -0.67$
24. (a) $M = 0.90$, 10% reduction, (b) $y_i = 64.8$ in.
26. $\theta_{12} = 10°$
28. Diagrams: (a) real, inverted, $M = 1$; (b) real, inverted, larger; (c) virtual, upright, larger.
30. (a) $d_i = 60$ cm (real), $M = -2$ (inverted); (b) $y_i = 20$ cm

32. $d_i = -16.7$ cm (virtual), $M = 0.67$ (upright)
34. $f = 5.7$ cm
36. Express similar triangles, vertically shaded and horizontally shaded, and equate.
38. (a) from lens 2, $d_i = 10$ cm, (b) inverted, real, and reduced, $M = 2/3$.

Chapter Supplement

2. (a) $f = 0.80$ m, (b) $P = +1.25$ D
4. If the lenses are turned around, the surface signs are reversed, but so are the signs of the R's, so the equation has the same result.

Chapter 29

2. $f = 40$ mm
4. (a) $M = 5$x, (b) Probably not, 3x maximum.
6. $M_t = 200$x
8. $M_t = 150$x
10. $M = 20$x
12. $f_e = 1.6$ cm

Chapter 30

2. $f = 5.0 \times 10^3$ Hz
4. $f_o = 6.0 \times 10^{14}$ Hz
6. (a) $\phi = 3.0 \times 10^{-19}$ J, (b) $f_o = 4.5 \times 10^{14}$ Hz
8. $\phi = 0.53$ eV
10. $E < \phi$, no photoemission
12. $n = 3.0 \times 10^{18}$ photons
14. $nh/2\pi$ = J-s = kg/m^2/s = mvr = L
16. (a) $E_3 = -1.51$ eV, (b) $-E_B = 1.51$ eV
18. (a) $\Delta E = 13.06$ eV, (b) $-E_5 = 0.544$ eV
20. (a) $\lambda = 2627$ nm, (b) No
22. $n_i = 2$
24. $E = 3.03 \times 10^{-19}$ J = 1.89 eV, only absorbed if atom is initially in $n = 2$ and would be excited to $n = 3$.
26. $f_o = 3.6 \times 10^{-19}$ Hz
28. $V = 20 \times 10^3$ V = 20 kV

Chapter 31

2. $^{120}_{51}Sb_{69}$, $^{121}Sb_{70}$, $^{122}Sb_{71}$, $^{123}Sb_{72}$, $^{124}Sb_{73}$
4. $m_p/m_e = 1836$
6. (a) 1.99×10^{-26} kg, (b) 4.38×10^{-26} lb
8. $^{6}_{3}Li_3$ ($Z = N$), $^{80}_{35}Br_{45}$ ($N > Z$), $^{28}_{14}Si_{14}$ ($Z = N$), $^{48}_{22}Ti_{26}$ ($N > Z$), $^{19}_{9}F_{10}$ ($N \approx Z$), $^{179}_{78}Pt_{101}$ ($N > Z$)
10. $^{237}_{93}Np$, ^{233}Pa, ^{233}U, ^{229}Th, ^{225}Ra, ^{225}Ac, ^{221}Fr, ^{217}At, ^{213}Bi, ^{213}Po, ^{209}Tl, ^{209}Pb, ^{209}Bi
12. (a) $^{233}_{91}Pu \rightarrow {}^{233}_{92}U + {}^{0}_{-1}e$, (b) $^{229}_{90}Th$
14. ^{131}I beta decays to excited ^{131}Xe, which emits gammas.
16. 15.9 years
18. 17,190 years
20. (a) 0.10 Ci = 3.7×10^9 Bq, (b) 5.6×10^4 Bq
22. $^{235}_{92}U + {}^{1}_{0}n \rightarrow {}^{94}_{37}Rb + {}^{139}_{55}Cs + 3\,{}^{1}_{0}n$
24. (a) 0.2148 u, (b) $^{1}_{0}n$, 0.00350 u
26. (a) Δm = 6.49 MeV, exoergic; (b) 17.3 MeV, exoergic
28. Δm = 582 MeV, work needed to separate nucleons.

Chapter 32

2. $E/t = 3.8 \times 10^{18}$ J/day
4. $E = 7.6 \times 10^8$ Btu
6. $E = 8.6 \times 10^4$ cal
8. (a) 3.3×10^{31} J, (b) 3.7×10^{14} kg
10. $\lambda > 9.9 \times 10^{-7}$ m
12. $E = 0.018$ kWh
14. $A = 1.03$ m^2

Index

(fn, footnote; t, table; f, figure; p, problem)

N

O

P

Q

R

S

Physical Constants

Name	Symbol	Value
Acceleration due to gravity	g	9.8 m/s^2; 32 ft/s^2
Universal gravitational constant	G	6.67×10^{-11} $N\text{-}m^2/kg^2$
Standard atmospheric pressure	P_{atm}	1.013×10^5 N/m^2, 14.7 lb/in^2; 30 in. Hg, 760 mm Hg (torr)
Boltzmann's constant	k	1.38×10^{-23} J/K
Mechanical equivalent of heat	—	4.186 J/cal; 778 ft-lb/Btu
Speed of light	c	3.00×10^8 m/s^2
Electronic charge	e	1.60×10^{-19} C
Coulomb's law constant	k	9×10^9 $N\text{-}m^2/C^2$
Permittivity of free space	ε_0	8.85×10^{-12} $C^2/N\text{-}m^2$
Permeability of free space	μ_0	$4\pi \times 10^{-7}$ Wb/A-m
Planck's constant	h	6.6×10^{-34} J-s 6.6×10^{-27} erg-s
Electron mass	m_e	9.11×10^{-31} kg
Proton mass	m_p	1.673×10^{-27} kg
Neutron mass	m_n	1.675×10^{-27} kg
Astronomical data:		
Mass of Earth		6.0×10^{24} kg
Mass of moon		7.4×10^{22} kg $= \frac{1}{81}$ mass of Earth
Mass of Sun		2.0×10^{30} kg
Approximate radius of Earth (average)		4000 mi; 6440 km $= 6.4 \times 10^3$ m (3957 mi); (6368 km)
Average distance of Earth from Sun		93×10^6 mi; 1.5×10^8 km
Average distance of moon from Earth		2.4×10^5 mi; 3.8×10^5 km

Trigonometric Relationships

$$\sin\theta = \frac{y}{r} \quad \left(\frac{\text{side opposite}}{\text{hypotenuse}}\right)$$

$$\cos\theta = \frac{x}{r} \quad \left(\frac{\text{side adjacent}}{\text{hypotenuse}}\right)$$

$$\tan\theta = \frac{\sin\theta}{\cos\theta} = \frac{y}{x} \quad \left(\frac{\text{side opposite}}{\text{side adjacent}}\right)$$

$\theta°$ (rad)	$\sin\theta$	$\cos\theta$	$\tan\theta$
0° (0)	0	1	0
30° ($\pi/6$)	0.500	0.866	0.577
45° ($\pi/4$)	0.707	0.707	1.00
60° ($\pi/3$)	0.866	0.500	1.73
90° ($\pi/2$)	1	0	$\rightarrow \infty$

See Trigonometric Tables for other angles.